Cerebral Cortex
Principles of
Operation

Edmund T. Rolls

Oxford Centre for Computational Neuroscience, Oxford, UK

UNIVERSITY PRESS

OXFORD
UNIVERSITY PRESS

Great Clarendon Street, Oxford, OX2 6DP,
United Kingdom

Oxford University Press is a department of the University of Oxford.
It furthers the University's objective of excellence in research, scholarship,
and education by publishing worldwide. Oxford is a registered trade mark of
Oxford University Press in the UK and in certain other countries

© Edmund Rolls 2016

The moral rights of the author have been asserted

First published 2016
First published in paperback 2017

Impression: 1

Published in the United States of America by Oxford University Press
198 Madison Avenue, New York, NY 10016, United States of America

British Library Cataloguing in Publication Data
Data available

Library of Congress Cataloging in Publication Data
Data available

ISBN 978–0–19–878485–2 (Hbk.)
ISBN 978–0–19–882034–5 (Pbk.)

Printed and bound by
CPI Group (UK) Ltd, Croydon, CR0 4YY

Links to third party websites are provided by Oxford in good faith and
for information only. Oxford disclaims any responsibility for the materials
contained in any third party website referenced in this work.

Preface

The overall aim of this book is to provide insight into the principles of operation of the cerebral cortex. These are key to understanding how we, as humans, function.

There have been few previous attempts to set out some of the important principles of operation of the cortex, and this book is pioneering. I have asked some of the leading investigators in neuroscience about their views on this, and most have not had many well formulated answers or hypotheses. As clear hypotheses are needed in this most important area of 21st century science, how our brains work, I have formulated a set of hypotheses to guide thinking and future research. I present evidence for many of the hypotheses, but at the same time we must all recognise that hypotheses and theory in science are there to be tested, and hopefully refined rather than rejected. Nevertheless, such theories and hypotheses are essential to progress, and it is in this frame of reference that I present the theories, hypotheses, and ideas that I have produced and collected together.

This book focusses on the principles of operation of the cerebral cortex, because at this time it is possible to propose and describe many principles, and many are likely to stand the test of time, and provide a foundation I believe for further developments, even if some need to be changed. In this context, I have not attempted to produce an overall theory of operation of the cerebral cortex, because at this stage of our understanding, such a theory would be incorrect or incomplete. I believe though that many of the principles will be important, and that many will provide the foundations for more complete theories of the operation of the cerebral cortex.

Given that many different principles of operation of the cortex are proposed in this book, with often several principles in each Chapter, the reader may find it convenient to take one Chapter at a time, and think about the issues raised in each Chapter, as the overall enterprise is large. The Highlights sections provided at the end of each Chapter may be useful in helping the reader to appreciate the different principles being considered in each Chapter.

To understand how the cortex works, including how it functions in perception, memory, attention, decision-making, and cognitive functions, it is necessary to combine different approaches, including neural computation. Neurophysiology at the single neuron level is needed because this is the level at which information is exchanged between the computing elements of the brain. Evidence from the effects of brain damage, including that available from neuropsychology, is needed to help understand what different parts of the system do, and indeed what each part is necessary for. Neuroimaging is useful to indicate where in the human brain different processes take place, and to show which functions can be dissociated from each other. Knowledge of the biophysical and synaptic properties of neurons is essential to understand how the computing elements of the brain work, and therefore what the building blocks of biologically realistic computational models should be. Knowledge of the anatomical and functional architecture of the cortex is needed to show what types of neuronal network actually perform the computation. And finally the approach of neural computation is needed, as this is required to link together all the empirical evidence to produce an understanding of how the system actually works. This book utilizes evidence from all these disciplines to develop an understanding of how different types of memory, perception, attention, and decision-making are implemented by processing in the cerebral cortex.

I emphasize that to understand how memory, perception, attention, decision-making, cognitive functions, and actions are produced in the cortex, we are dealing with large-scale computational systems with interactions between the parts, and that this understanding requires analysis at the computational and global level of the operation of many neurons to perform together a useful function. Understanding at the molecular level is important for helping to understand how these large-scale computational processes are implemented in the brain, but will not by itself give any account of what computations are performed to implement these cognitive functions. Instead, understanding cognitive functions such as object recognition, memory recall, attention, and decision-making requires single neuron data to be closely linked to computational models of how the interactions between large numbers of neurons and many networks of neurons allow these cognitive problems to be solved. The single neuron level is important in this approach, for the single neurons can be thought of as the computational units of the system, and is the level at which the information is exchanged by the spiking activity between the computational elements of the brain. The single neuron level is therefore, because it is the level at which information is communicated between the computing elements of the brain, the fundamental level of information processing, and the level at which the information can be read out (by recording the spiking activity) in order to understand what information is being represented and processed in each brain area.

With its focus on how the brain and especially how the cortex works at the computational neuroscience level, this book is distinct from the many excellent books on neuroscience that describe much evidence about brain structure and function, but do not aim to provide an understanding of how the brain works at the computational level. This book aims to forge an understanding of how some key brain systems may operate at the computational level, so that we can understand how the cortex actually performs some of its complex and necessarily computational functions in memory, perception, attention, decision-making, cognitive functions, and actions.

A test of whether one's understanding is correct is to simulate the processing on a computer, and to show whether the simulation can perform the tasks of cortical systems, and whether the simulation has similar properties to the real cortex. The approach of neural computation leads to a precise definition of how the computation is performed, and to precise and quantitative tests of the theories produced. How memory systems in the cortex work is a paradigm example of this approach, because memory-like operations which involve altered functionality as a result of synaptic modification are at the heart of how many computations in the cortex are performed. It happens that attention and decision-making can be understood in terms of interactions between and fundamental operations in memory systems in the cortex, and therefore it is natural to treat these areas of cognitive neuroscience in this book. The same fundamental concepts based on the operation of neuronal circuitry can be applied to all these functions, as is shown in this book.

One of the distinctive properties of this book is that it links the neural computation approach not only firmly to neuronal neurophysiology, which provides much of the primary data about how the cortex operates, but also to psychophysical studies (for example of attention); to neuropsychological studies of patients with brain damage; and to functional magnetic resonance imaging (fMRI) (and other neuroimaging) approaches. The empirical evidence that is brought to bear is largely from non-human primates and from humans, because of the considerable similarity of their cortical systems.

In this book, I have not attempted to produce a single computational theory of how the cortex operates. Instead, I have highlighted many different principles of cortical function, most of which are likely to be building blocks of how our cortex operates. The reason for this approach is that many of the principles may well be correct, and useful in understanding how the cortex operates, but some might turn out not to be useful or correct. The aim of this book is therefore to propose some of the fundamental principles of operation of the cerebral cortex, many or most of which will provide a foundation for understanding the operation of the cortex, rather than to produce a single theory of operation of the cortex, which might be disproved if any one of its elements was found to be weak.

The overall aims of the book are developed further, and the plan of the book is described, in Chapter 1, Section 1.1. Some of the main Principles of Operation of the Cerebral Cortex that I describe can be found in the titles of Chapters 2–22; but in practice, most Chapters include several Principles of Operation, which will appear in the Highlights to each Chapter. Section 26.5 may be useful in addition to the Highlights, for Section 26.5 draws together in a synthesis some of the Principles of Operation of the Cerebral Cortex that are described in the book. Further evidence on how these principles are relevant to the operation of different cortical areas and systems and operate together is provided in Chapters 24–25. In these Chapters, the operation of two major cortical systems, those involved in memory and in visual object recognition, are considered to illustrate how the principles are combined to implement two different key cortical functions. The Appendices provide some of the more formal and quantitative properties of the operation of neuronal systems, and are provided because they provide a route to a deeper understanding on the principles, and to enable the presentation in earlier Chapters to be at a readily approachable level. The Appendices describe many of the building blocks of the neurocomputational approach, and are designed to be useful for teaching. Appendix D describes Matlab software that has been made available with this book to provide simple demonstrations of the operation of some key neuronal networks related to cortical function. The programs are available at http://www.oxcns.org.

Part of the material described in the book reflects work performed in collaboration with many colleagues, whose tremendous contributions are warmly appreciated. The contributions of many will be evident from the references cited in the text. Especial appreciation is due to Gustavo Deco, Simon M. Stringer, and Alessandro Treves who have contributed greatly in an always interesting and fruitful research collaboration on computational aspects of brain function, and to many neurophysiology and functional neuroimaging colleagues who have contributed to the empirical discoveries that provide the foundation to which the computational neuroscience must always be closely linked, and whose names are cited throughout the text. Much of the work described would not have been possible without financial support from a number of sources, particularly the Medical Research Council of the UK, the Human Frontier Science Program, the Wellcome Trust, and the James S. McDonnell Foundation. I am also grateful to many colleagues who I have consulted while writing this book, including Joel Price (Washington University School of Medicine), and Donald Wilson (New York University). Dr Patrick Mills is warmly thanked for his comments on the text. Section 24.3.12 on *ars memoriae* is warmly dedicated to my colleagues at Corpus Christi College, Oxford. The book was typeset by the author using LATEXand WinEdt.

The cover includes part of the picture *Pandora* painted in 1896 by J. W. Waterhouse. The metaphor is to look inside the system of the mind and the brain, in order to understand how the brain functions, and thereby better to understand and treat its disorders. The cover also includes an image of the dendritic morphology of excitatory neurons in S1 whisker barrel cortex (Fig. 1.14) (adapted from Marcel Oberlaender, Christiaan P.J. de Kock, Randy M. Bruno, Alejandro Ramirez, Hanno S. Meyer, Vincent J. Dercksen, Moritz Helmstaedter and Bert Sakmann, Cell type-specific three-dimensional structure of thalamocortical circuits in a column of rat vibrissal cortex, Cerebral Cortex, 2012, Vol. 22, issue 10, pp. 2375–2391, by permission of Oxford University Press). The cover also includes a diagram of the computational circuitry of the hippocampus by the author (Fig. 24.1). The aim of these second two images is to highlight the importance of moving from the anatomy of the cortex using all the approaches available including neuronal network models that address and incorporate neurophysiological discoveries to lead to an understanding of how the cortex operates computationally.

Updates to and .pdfs of many of the publications cited in this book are available at http://www.oxcns.org. Updates and corrections to the text and notes are also available at http://www.oxcns.org.

I dedicate this work to the overlapping group: my family, friends, and colleagues – in salutem praesentium, in memoriam absentium.

Contents

1 Introduction

1.1 Principles of operation of the cerebral cortex: introduction and plan

To understand how the cortex works, it is necessary to combine different approaches, including neural computation. Neurophysiology at the single neuron level is needed because this is the level at which information is exchanged between the computing elements of the brain. Evidence from the effects of brain damage, including that available from neuropsychology, is needed to help understand what different parts of the system do, and indeed what each part is necessary for. Neuroimaging is useful to indicate where in the human brain different processes take place, and to show which functions can be dissociated from each other. Knowledge of the biophysical and synaptic properties of neurons is essential to understand how the computing elements of the cortex work, and therefore what the building blocks of biologically realistic computational models should be. Knowledge of the anatomical and functional architecture of the cortex is needed to show what types of neuronal network actually perform the computation. And finally the approach of neural computation is needed, as this is required to link together all the empirical evidence to produce an understanding of how the system actually works. This book utilizes evidence from all these disciplines to develop an understanding of how different types of cortical function including perception, memory, attention, emotion, decision-making, and action are implemented.

A test of whether one's understanding is correct is to simulate the processing on a computer, and to show whether the simulation can perform the tasks of particular systems in the brain, and whether the simulation has similar properties to the real brain. The approach of neural computation leads to a precise definition of how the computation is performed, and to precise and quantitative tests of the theories produced. How memory systems in the cortex work is a paradigm example of this approach, because memory-like operations which involve altered functionality as a result of synaptic modification are at the heart of how many computations in the cortex are performed.

One of the distinctive properties of this book is that it links the neural computation approach not only firmly to neuronal neurophysiology, which provides much of the primary data about how the brain operates, but also to psychophysical studies (for example of attention); to neuropsychological studies of patients with brain damage; and to functional magnetic resonance imaging (fMRI) (and other neuroimaging) approaches. The empirical evidence that is brought to bear is largely from non-human primates and from humans, because of the considerable similarity of their cortical systems, and the overall aims to understand the human brain, and the disorders that arise after brain damage.

In Chapters 2–23 I set out some of what appear to be some of the Principles of Operation of the Cerebral Cortex. I have chosen to provide different Chapters for different Principles of Operation, to help with easy assimilation and treatment of the different principles, but I note that a comprehensive understanding will benefit from reading all the Chapters, for the concepts are frequently inter-locked, and cross-references are made. Some of the main Principles of Operation of the Cerebral Cortex that I describe can be found in the titles of Chapters 2–22; but in practice, most Chapters include several Principles of Operation, which will appear in the Highlights to each Chapter. Section 26.5 may be useful in addition to the Highlights, for Section 26.5 draws together in a synthesis some of the Principles of Operation of the Cerebral Cortex that are described in the book. Further evidence on how these principles are relevant to the operation of different cortical areas and systems and operate together is provided in Chapters 24–25. In these Chapters, the operation of two major cortical systems, those involved in memory and in visual object recognition, are considered, to illustrate how the principles are combined to implement two different key cortical functions. The Appendices provide some of the more formal and quantitative properties of the operation of neuronal systems, and are provided because they provide a route to a deeper understanding of the Principles of Operation, and to enable the presentation in earlier Chapters to be at a readily approachable level. Appendix D describes Matlab software that has been made available with this book to provide simple demonstrations of the operation of some key neuronal networks related to cortical function. The programs are available at http://www.oxcns.org.

In this book, I have not attempted to produce a single computational theory of how the cortex operates. Instead, I have highlighted many different principles of cortical function, most of which are likely to be building clocks of how our cortex operates. The reason for this approach is that many of the principles may well be correct, and useful in understanding how the cortex operates, but some might turn out not to be useful or correct. The aim of this book is therefore to propose some of the fundamental principles of operation of the cerebral cortex, many or most of which will provide a foundation for understanding the operation of the cortex, rather than to produce a single theory of operation of the cortex, which might be disproved if any one of its elements was found to be weak.

I start with some history showing how recently it is that our understanding of the (computational) principles of operation of the cerebral cortex have started to develop.

Before the 1960s there were many and important discoveries about the phenomenology of the cortex, for example that damage in one part would affect vision, and in another part movement, with electrical stimulation often producing the opposite effect. The principles may help us to understand these phenomena, but the phenomena provide limited evidence about how the cortex works, apart from the very important principle of localization of function (see Chapter 3), and the important principle of hierarchical organization (Hughlings Jackson 1878, Swash 1989) (see Chapter 2) which has been supported by increasing evidence on the connections between different cortical areas, which is a fundamental building block for understanding cortical computation.

In the 1960s David Hubel and Torsten Wiesel made important discoveries about the stimuli that activate primary visual cortex neurons, showing that they respond to bar-like or edge-like visual stimuli (Hubel and Wiesel 1962, Hubel and Wiesel 1968, Hubel and Wiesel 1977), instead of the small circular receptive fields of lateral geniculate cortex neurons. This led them to suggest the elements of a model of how this might come about, by cortical neurons that respond to elongated lines of lateral geniculate neurons (Fig. 1.1). This led to the concept that hierarchical organization over a series of cortical areas might at each stage form combinations of the features represented in the previous cortical area, in what might be termed feature combination neurons.

However, before 1970 there were few ideas about how the cerebral cortex operates computationally.

David Marr was a pioneer who helped to open the way to an understanding of how the details of cortical anatomy and connectivity help to develop quantitative theories of how cortical areas may compute, including the cerebellar cortex (Marr 1969), the neocortex (Marr 1970), and the hippocampal cortex (Marr 1971). Marr was hampered by some lack in detail of the available anatomical knowledge, and did not for example hypothesize that the hippocampal CA3 network was an autoassociation memory. He attempted to test his theory of

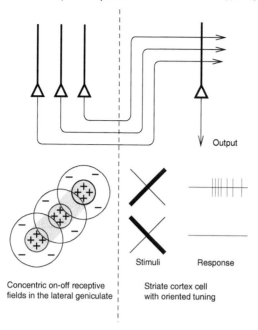

Output

Stimuli   Response

Concentric on-off receptive
fields in the lateral geniculate

Striate cortex cell
with oriented tuning

Fig. 1.1 Receptive fields in the lateral geniculate have concentric on-centre off-surround (or vice versa) receptive fields (left). Neurons in the striate cortex, such as the simple cell illustrated on the right, respond to elongated lines or edges at certain orientations. A suggestion that the lateral geniculate neurons might combine their responses in the way shown to produce orientation-selective simple cells was made by Hubel and Wiesel (1962).

the cerebellum with Sir John (Jack) Eccles by stimulating the climbing fibres in the cerebellum while providing an input from the parallel fibres to a Purkinje cell, but the experiment did not succeed, partly because of a lack of physiological knowledge about the firing rates of climbing fibres, which are low, rarely more than 10 spikes/s, whereas they had stimulated at much higher frequencies. Perhaps in part because David Marr was ahead of the experimental techniques available at the time to test his theories of network operations of cortical systems, he focussed in his later work on more conceptual rather that neural network based approaches, which he applied to understanding vision, with again limited success at least in understanding invariant object recognition (Marr 1982), again related to the lack of available experimental data (Rolls 2011c).

Very stimulating advances in thinking about cortical function were made in books by Abeles (1991), Braitenberg and Schütz (1991) and Creutzfeldt (1995), but many advances have been made since those books.

Theories of operation are essential to understanding the brain - e.g. collective computation in attractor networks, and emergent properties. It cannot be done just by molecular biology, though that provides useful tools, and potentially ways to ameliorate brain dysfunction.

Because many of the processes involved in cortical function can be understood in terms of the operation of different types of memory networks, and interactions between these networks, their operation is described in Appendix B, with some of the basic mathematical concepts covered in Appendix A. Appendix B provides quite a self-contained overview of neural networks, in the sense that those not already familiar with neural computation can find many of the fundamentals and foundations here.

I emphasize that to understand cortical function, and processes such as memory, perception, attention, and decision-making in the cortex, we are dealing with large-scale computational systems with interactions between the parts, and that this understanding requires analysis at the computational and global level of the operation of many neurons to perform together a useful function. Understanding at the molecular level is important for helping to understand how these large-scale computational processes are implemented in the brain, but will not by itself give any account of what computations are performed to implement these cognitive functions. Instead, understanding cognitive functions such as object recognition, memory recall, attention, and decision-making requires single neuron data to be closely linked to computational models of how the interactions between large numbers of neurons and many networks of neurons allow these cognitive problems to be solved. The single neuron level is important in this approach, for the single neurons can be thought of as the computational units of the system, and is the level at which the information is exchanged by the spiking activity between the computational elements of the brain. The single neuron level is therefore, because it is the level at which information is communicated between the computing elements of the brain, the fundamental level of information processing, and the level at which the information can be read out (by recording the spiking activity) in order to understand what information is being represented and processed in each cortical area.

Because of this importance of being able to analyze the activity of single neurons and populations of neurons in order to understand brain function, Appendix C describes rigorous approaches to understanding the information represented by neurons, and summarizes evidence on how the information is actually represented. In that the information encoded by different neurons is shown to be different, this confirms that what is being represented, and how it is represented, requires the level of analysis to encompass the single neuron level. Understanding how neurons represent information is fundamental for understanding how neurons and networks of neurons read the code from other neurons, and the actual nature of the computation that could be performed in a cortical network. The networks described in this book are consistent in their operation with the evidence presented in Appendix C on cortical neuronal information encoding. The neurocomputational approach taken in this book enables the single neuron level of analysis to be linked to the level of large-scale neuronal networks and the interactions between them, so that large-scale processes such as memory retrieval, object recognition, attention, and decision-making can be understood.

In the rest of this Chapter, I introduce some of the background for understanding cortical computation, such as how single neurons operate, how some of the essential features of this can be captured by simple formalisms, and some of the biological background to what it can be taken happens in the nervous system, such as synaptic modification based on information available locally at each synapse.

1.2 Neurons in the brain, and their representation in neuronal networks

Neurons in the vertebrate brain typically have, extending from the cell body, large dendrites which receive inputs from other neurons through connections called synapses. The synapses operate by chemical transmission. When a synaptic terminal receives an all-or-nothing action potential from the neuron of which it is a terminal, it releases a transmitter that crosses the synaptic cleft and produces either depolarization or hyperpolarization in the postsynaptic neuron, by opening particular ionic channels. (A textbook such as Kandel, Schwartz, Jessell, Siegelbaum and Hudspeth (2013) gives further information on this process.) Summation of a

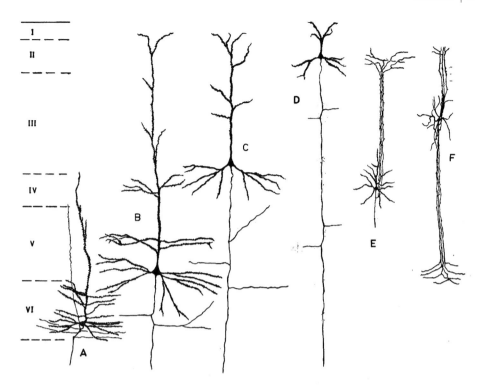

Fig. 1.2 Examples of neurons found in the brain. Cell types in the cerebral neocortex are shown. The different laminae of the cortex are designated I–VI, with I at the surface. Cells A–D are pyramidal cells in the different layers. Cell E is a spiny stellate cell, and F is a double bouquet cell. (Adapted from Edward G. Jones and Alan Peters, Cerebral Cortex, Functional Properties of Cortical Cells, volume 2, p. 7, Springer US, Copyright ©1984, Springer Science+Business Media New York.)

number of such depolarizations or excitatory inputs within the time constant of the receiving neuron, which is typically 15–25 ms, produces sufficient depolarization that the neuron fires an action potential. There are often 5,000–20,000 inputs per neuron. Examples of cortical neurons are shown in Figs. 1.2, 1.14, 1.15, and 24.12, further examples are shown in Shepherd (2004), and schematic drawings are shown in Shepherd and Grillner (2010). Once firing is initiated in the cell body (or axon initial segment of the cell body), the action potential is conducted in an all-or-nothing way to reach the synaptic terminals of the neuron, whence it may affect other neurons. Any inputs the neuron receives that cause it to become hyperpolarized make it less likely to fire (because the membrane potential is moved away from the critical threshold at which an action potential is initiated), and are described as inhibitory. The neuron can thus be thought of in a simple way as a computational element that sums its inputs within its time constant and, whenever this sum, minus any inhibitory effects, exceeds a threshold, produces an action potential that propagates to all of its outputs. This simple idea is incorporated in many neuronal network models using a formalism of a type described in the next section.

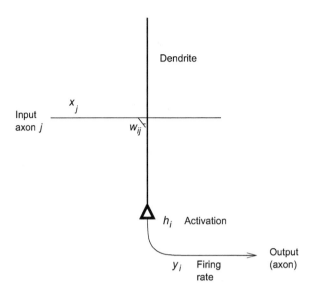

Fig. 1.3 Notation used to describe an individual neuron in a network model. By convention, we generally represent the dendrite as thick, and vertically oriented (as this is the normal way that neuroscientists view cortical pyramidal cells under the microscope); and the axon as thin. The cell body or soma is indicated between them. The firing rate we also call the activity of the neuron.

1.3 A formalism for approaching the operation of single neurons in a network

Let us consider a neuron i as shown in Fig. 1.3, which receives inputs from axons that we label j through synapses of strength w_{ij}. The first subscript (i) refers to the receiving neuron, and the second subscript (j) to the particular input. j counts from 1 to C, where C is the number of synapses or connections received. The firing rate of the ith neuron is denoted as y_i, and that of the jth input to the neuron as x_j. To express the idea that the neuron makes a simple linear summation of the inputs it receives, we can write the activation of neuron i, denoted h_i, as

$$h_i = \sum_j x_j w_{ij} \tag{1.1}$$

where $\sum_j$ indicates that the sum is over the C input axons (or connections) indexed by j to each neuron. The multiplicative form here indicates that activation should be produced by an axon only if it is firing, and depending on the strength of the synapse w_{ij} from input axon j onto the dendrite of the receiving neuron i. Equation 1.1 indicates that the strength of the activation reflects how fast the axon j is firing (that is x_j), and how strong the synapse w_{ij} is. The sum of all such activations expresses the idea that summation (of synaptic currents in real neurons) occurs along the length of the dendrite, to produce activation at the cell body, where the activation h_i is converted into firing y_i. This conversion can be expressed as

$$y_i = f(h_i) \tag{1.2}$$

which indicates that the firing rate is a function (f) of the postsynaptic activation. The function is called the activation function in this case. The function at its simplest could be linear, so that the firing rate would be proportional to the activation (see Fig. 1.4a). Real neurons have thresholds, with firing occurring only if the activation is above the threshold. A threshold linear activation function is shown in Fig. 1.4b. This has been useful in formal analysis of

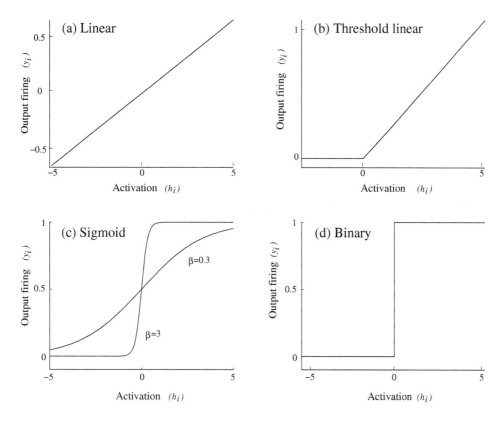

Fig. 1.4 Different types of activation function. The activation function relates the output activity (or firing rate), y_i, of the neuron (i) to its activation, h_i. (a) Linear. (b) Threshold linear. (c) Sigmoid. [One mathematical exemplar of this class of activation function is $y_i = 1/(1 + \exp(-2\beta h_i))$.] The output of this function, also sometimes known as the logistic function, is 0 for an input of $-\infty$, 0.5 for 0, and 1 for $+\infty$. The function incorporates a threshold at the lower end, followed by a linear portion, and then an asymptotic approach to the maximum value at the top end of the function. The parameter β controls the steepness of the almost linear part of the function round $h_i = 0$. If β is small, the output goes smoothly and slowly from 0 to 1 as h_i goes from $-\infty$ to $+\infty$. If β is large, the curve is very steep, and approximates a binary threshold activation function. (d) Binary threshold.

the properties of neural networks. Neurons also have firing rates that become saturated at a maximum rate, and we could express this as the sigmoid activation function shown in Fig. 1.4c. Another simple activation function, used in some models of neural networks, is the binary threshold function (Fig. 1.4d), which indicates that if the activation is below threshold, there is no firing, and that if the activation is above threshold, the neuron fires maximally. Some non-linearity in the activation function is an advantage, for it enables many useful computations to be performed in neuronal networks, including removing interfering effects of similar memories, and enabling neurons to perform logical operations, such as firing only if several inputs are present simultaneously.

A property implied by equation 1.1 is that the postsynaptic membrane is electrically short, and so summates its inputs irrespective of where on the dendrite the input is received. In real neurons, the transduction of current into firing frequency (the analogue of the transfer function of equation 1.2) is generally studied not with synaptic inputs but by applying a steady

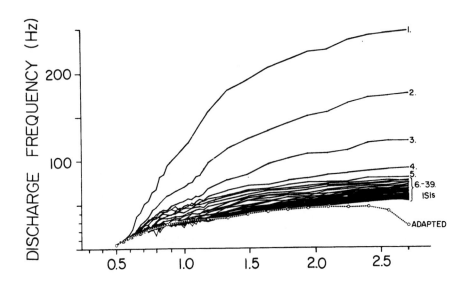

Fig. 1.5 Frequency – current plot (the closest experimental analogue of the activation function) for a CA1 pyramidal cell. The firing frequency (in Hz) in response to the injection of 1.5 s long, rectangular depolarizing current pulses has been plotted against the strength of the current pulses (in nA) (abscissa). The first 39 interspike intervals (ISIs) are plotted as instantaneous frequency (1 / ISI), together with the average frequency of the adapted firing during the last part of the current injection (circles and broken line). The plot indicates a current threshold at approximately 0.5 nA, a linear range with a tendency to saturate, for the initial instantaneous rate, above approximately 200 Hz, and the phenomenon of adaptation, which is not reproduced in simple non-dynamical models (see further Appendix A5 of Rolls and Treves 1998). (Reproduced from *Experimental Brain Research*, Current-to-frequency transduction in CA1 hippocampal pyramidal cells: Slow prepotentials dominate the primary range firing. 53 (2), 1984, pp. 431–443, Lanthorn, T., Storm, J. and Andersen, P. Copyright ©1984, Springer-Verlag. With permission of Springer.)

current through an electrode into the soma. Examples of the resulting curves, which illustrate the additional phenomenon of firing rate adaptation, are shown in Fig. 1.5.

1.4 Synaptic modification

For a neuronal network to perform useful computation, that is to produce a given output when it receives a particular input, the synaptic weights must be set up appropriately. This is often performed by synaptic modification occurring during learning.

A simple learning rule that was originally presaged by Donald Hebb (1949) proposes that synapses increase in strength when there is conjunctive presynaptic and postsynaptic activity. The Hebb rule can be expressed more formally as follows

$$\delta w_{ij} = \alpha y_i x_j \qquad (1.3)$$

where δw_{ij} is the change of the synaptic weight w_{ij} which results from the simultaneous (or conjunctive) presence of presynaptic firing x_j and postsynaptic firing y_i (or strong depolarization), and α is a learning rate constant that specifies how much the synapses alter on any one pairing. The presynaptic and postsynaptic activity must be present approximately simultaneously (to within perhaps 100–500 ms in the real brain).

The Hebb rule is expressed in this multiplicative form to reflect the idea that both pre-synaptic and postsynaptic activity must be present for the synapses to increase in strength. The multiplicative form also reflects the idea that strong pre- and postsynaptic firing will produce a larger change of synaptic weight than smaller firing rates. The Hebb rule thus captures what is typically found in studies of associative Long-Term Potentiation (LTP) in the brain, described in Section 1.5.

One useful property of large neurons in the brain, such as cortical pyramidal cells, is that with their short electrical length, the postsynaptic term, y_i, is available on much of the dendrite of a cell. The implication of this is that once sufficient postsynaptic activation has been produced, any active presynaptic terminal on the neuron will show synaptic strengthening. This enables associations between coactive inputs, or correlated activity in input axons, to be learned by neurons using this simple associative learning rule.

If, in contrast, a group of coactive axons made synapses close together on a small dendrite, then the local depolarization might be intense, and only these synapses would modify onto the dendrite. (A single distant active synapse might not modify in this type of neuron, because of the long electrotonic length of the dendrite.) The computation in this case is described as Sigma-Pi ($\Sigma\Pi$), to indicate that there is a local product computed during learning; this allows a particular set of locally active synapses to modify together, and then the output of the neuron can reflect the sum of such local multiplications (see Rumelhart and McClelland (1986), Koch (1999)). Sigma-Pi neurons are not used in most of the networks described in this book. There has been some work on how such neurons, if present in the brain, might utilize this functionality in the computation of invariant representations (Mel, Ruderman and Archie 1998, Mel and Fiser 2000) (see Section 25.6).

1.5 Long-term potentiation and long-term depression as models of synaptic modification

Long-term potentiation (LTP) and long-term depression (LTD) provide useful models of some of the synaptic modifications that occur in the brain. The synaptic changes found appear to be synapse-specific, and to depend on information available locally at the synapse. LTP and LTD may thus provide a good model of the biological synaptic modifications involved in real neuronal network operations in the brain. Some of the properties of LTP and LTD are described next, together with evidence that implicates them in learning in at least some brain systems. Even if they turn out not to be the basis for the synaptic modifications that occur during learning, they have many of the properties that would be needed by some of the synaptic modification systems used by the brain.

Long-term potentiation is a use-dependent and sustained increase in synaptic strength that can be induced by brief periods of synaptic stimulation (Bliss and Collingridge 2013). It is usually measured as a sustained increase in the amplitude of electrically evoked responses in specific neural pathways following brief trains of high-frequency stimulation (see Fig. 1.6b). For example, high frequency stimulation of the Schaffer collateral inputs to the hippocampal CA1 cells results in a larger response recorded from the CA1 cells to single test pulse stimulation of the pathway. LTP is long-lasting, in that its effect can be measured for hours in hippocampal slices, and in chronic in vivo experiments in some cases it may last for months. LTP becomes evident rapidly, typically in less than 1 minute. LTP is in some brain systems associative. This is illustrated in Fig. 1.6c, in which a weak input to a group of cells (e.g. the commissural input to CA1) does not show LTP unless it is given at the same time as (i.e. associatively with) another input (which could be weak or strong) to the cells. The associativity arises because it is only when sufficient activation of the postsynaptic neuron to

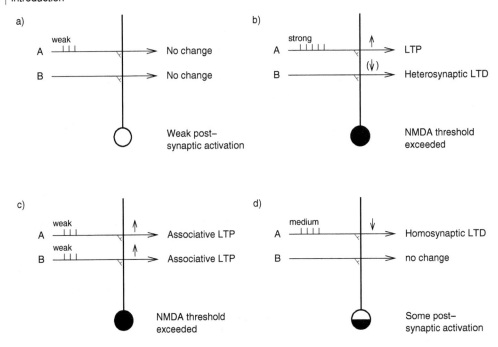

Fig. 1.6 Schematic illustration of synaptic modification rules as revealed by long-term potentiation (LTP) and long-term depression (LTD). The activation of the postsynaptic neuron is indicated by the extent to which its soma is black. There are two sets of inputs to the neuron: A and B. (a) A weak input (indicated by three spikes) on the set A of input axons produces little postsynaptic activation, and there is no change in synaptic strength. (b) A strong input (indicated by five spikes) on the set A of input axons produces strong postsynaptic activation, and the active synapses increase in strength. This is LTP. It is homosynaptic in that the synapses that increase in strength are the same as those through which the neuron is activated. LTP is synapse-specific, in that the inactive axons, B, do not show LTP. They either do not change in strength, or they may weaken. The weakening is called heterosynaptic LTD, because the synapses that weaken are other than those through which the neuron is activated (hetero- is Greek for other). (c) Two weak inputs present simultaneously on A and B summate to produce strong postsynaptic activation, and both sets of active synapses show LTP. (d) Intermediate strength firing on A produces some activation, but not strong activation, of the postsynaptic neuron. The active synapses become weaker. This is homosynaptic LTD, in that the synapses that weaken are the same as those through which the neuron is activated (homo- is Greek for same).

exceed the threshold of NMDA receptors (see below) is produced that any learning can occur. The two weak inputs summate to produce sufficient depolarization to exceed the threshold. This associative property is shown very clearly in experiments in which LTP of an input to a single cell only occurs if the cell membrane is depolarized by passing current through it at the same time as the input arrives at the cell. The depolarization alone, or the input alone, is not sufficient to produce the LTP, and the LTP is thus associative. Moreover, in that the presynaptic input and the postsynaptic depolarization must occur at about the same time (within approximately 500 ms), the LTP requires temporal contiguity. LTP is also synapse-specific, in that, for example, an inactive input to a cell does not show LTP even if the cell is strongly activated by other inputs (Fig. 1.6b, input B).

These spatiotemporal properties of long term potentiation can be understood in terms of actions of the inputs on the postsynaptic cell, which in the hippocampus has two classes of receptor, NMDA (N-methyl-D-aspartate) and AMPA (alpha-amino-3-hydroxy-5-methyl-isoxasole-4-propionic acid), activated by the glutamate released by the presynaptic terminals.

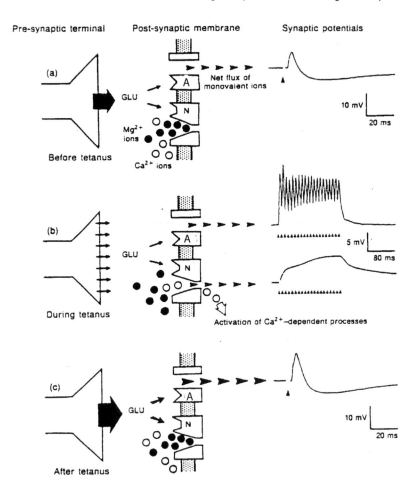

Fig. 1.7 The mechanism of induction of LTP in the CA1 region of the hippocampus. (a) Neurotransmitter (e.g. L-glutamate, GLU) is released and acts upon both AMPA (A) and NMDA (N) receptors. The NMDA receptors are blocked by magnesium and the excitatory synaptic response (EPSP) is therefore mediated primarily by ion flow through the channels associated with AMPA receptors. (b) During high-frequency activation ('tetanus'), the magnesium block of the ion channels associated with NMDA receptors is released by depolarization. Activation of the NMDA receptor by transmitter now results in ions moving through the channel. In this way, calcium enters the postsynaptic region to trigger various intracellular mechanisms which eventually result in an alteration of synaptic efficacy. (c) Subsequent low-frequency stimulation results in a greater EPSP. See text for further details. (After Collingridge and Bliss 1987.)

The NMDA receptor channels are normally blocked by Mg^{2+}, but when the cell is strongly depolarized by strong tetanic stimulation of the type necessary to induce LTP, the Mg^{2+} block is removed, and Ca^{2+} entering via the NMDA receptor channels triggers events that lead to the potentiated synaptic transmission (see Fig. 1.7). Part of the evidence for this is that NMDA antagonists such as AP5 (D-2-amino-5-phosphonopentanoate) block LTP. Further, if the postsynaptic membrane is voltage clamped to prevent depolarization by a strong input, then LTP does not occur. The voltage-dependence of the NMDA receptor channels introduces a threshold and thus a non-linearity that contributes to a number of the phenomena of some types of LTP, such as cooperativity (many small inputs together produce sufficient depolarization to allow the NMDA receptors to operate); associativity (a weak input alone will not produce sufficient depolarization of the postsynaptic cell to enable the NMDA receptors to be activated, but the depolarization will be sufficient if there is also a strong input); and temporal contiguity between the different inputs that show LTP (in that if inputs occur non-conjunctively, the depolarization shows insufficient summation to reach the required level, or some of the inputs may arrive when the depolarization has decayed). Once the LTP has become established (which can be within one minute of the strong input to the cell), the LTP is expressed through the AMPA receptors, in that AP5 blocks only the establishment of LTP, and not its subsequent expression (Bliss and Collingridge 2013).

There are a number of possibilities about what change is triggered by the entry of Ca^{2+} to the postsynaptic cell to mediate LTP. One possibility is that somehow a messenger reaches the presynaptic terminals from the postsynaptic membrane and, if the terminals are active, causes them to release more transmitter in future whenever they are activated by an action potential. Consistent with this possibility is the observation that, after LTP has been induced, more transmitter appears to be released from the presynaptic endings. Another possibility is that the postsynaptic membrane changes just where Ca^{2+} has entered, so that AMPA receptors become more responsive to glutamate released in future. Consistent with this possibility is the observation that after LTP, the postsynaptic cell may respond more to locally applied glutamate (using a microiontophoretic technique).

The rule that underlies associative LTP is thus that synapses connecting two neurons become stronger if there is conjunctive presynaptic and (strong) postsynaptic activity. This learning rule for synaptic modification is sometimes called the Hebb rule, after Donald Hebb of McGill University who drew attention to this possibility, and its potential importance in learning, in 1949.

In that LTP is long-lasting, develops rapidly, is synapse-specific, and is in some cases associative, it is of interest as a potential synaptic mechanism underlying some forms of memory. Evidence linking it directly to some forms of learning comes from experiments in which it has been shown that the drug AP5 infused so that it reaches the hippocampus to block NMDA receptors blocks spatial learning mediated by the hippocampus (Morris 1989, Martin, Grimwood and Morris 2000, Takeuchi, Duszkiewicz and Morris 2014). The task learned by the rats was to find the location relative to cues in a room of a platform submerged in an opaque liquid (milk). Interestingly, if the rats had already learned where the platform was, then the NMDA infusion did not block performance of the task. This is a close parallel to LTP, in that the learning, but not the subsequent expression of what had been learned, was blocked by the NMDA antagonist AP5. Although there is still some uncertainty about the experimental evidence that links LTP to learning (Martin, Grimwood and Morris 2000, Lynch 2004, Takeuchi, Duszkiewicz and Morris 2014), there is a need for a synapse-specific modifiability of synaptic strengths on neurons if neuronal networks are to learn. If LTP is not always an exact model of the synaptic modification that occurs during learning, then something with many of the properties of LTP is nevertheless needed, and is likely to be present in the brain given the functions known to be implemented in many brain regions.

In another model of the role of LTP in memory, Davis (2000) has studied the role of the amygdala in learning associations to fear-inducing stimuli. He has shown that blockade of NMDA synapses in the amygdala interferes with this type of learning, consistent with the idea that LTP also provides a useful model of this type of learning.

Long-Term Depression (LTD) can also occur. It can in principle be associative or non-associative. In associative LTD, the alteration of synaptic strength depends on the pre- and post-synaptic activities. There are two types. Heterosynaptic LTD occurs when the postsynaptic neuron is strongly activated, and there is low presynaptic activity (see Fig. 1.6b input B, and Table B.1 on page 719). Heterosynaptic LTD is so-called because the synapse that weakens is other than (hetero-) the one through which the postsynaptic neuron is activated. Heterosynaptic LTD is important in associative neuronal networks, and in competitive neuronal networks (see Appendix B). In competitive neural networks it would be helpful if the degree of heterosynaptic LTD depended on the existing strength of the synapse, and there is some evidence that this may be the case (see Appendix B). Homosynaptic LTD occurs when the presynaptic neuron is strongly active, and the postsynaptic neuron has some, but low, activity (see Fig. 1.6d and Table B.1). Homosynaptic LTD is so-called because the synapse that weakens is the same as (homo-) the one that is active. Heterosynaptic and homosynaptic LTD are found in the neocortex (Artola and Singer 1993, Singer 1995, Frégnac 1996) and hippocampus (Christie 1996), and in many cases are dependent on activation of NMDA receptors (Fazeli and Collingridge 1996, Bliss and Collingridge 2013). LTD in the cerebellum is evident as weakening of active parallel fibre to Purkinje cell synapses when the climbing fibre connecting to a Purkinje cell is active (Ito 1984, Ito 1989, Ito 1993b, Ito 1993a, Ito 2006, Ito 2010).

An interesting time-dependence of LTP and LTD has been observed, with LTP occurring especially when the presynaptic spikes precede by a few ms the post-synaptic activation, and LTD occurring when the pre-synaptic spikes follow the post-synaptic activation by a few milliseconds (ms) (Markram, Lübke, Frotscher and Sakmann 1997, Bi and Poo 1998, Bi and Poo 2001, Senn, Markram and Tsodyks 2001, Dan and Poo 2004, Dan and Poo 2006, Markram, Gerstner and Sjöström 2012). This is referred to as spike timing-dependent plasticity, STDP. This type of temporally asymmetric Hebbian learning rule, demonstrated in the hippocampus and neocortex, can induce associations over time, and not just between simultaneous events. Networks of neurons with such synapses can learn sequences (Minai and Levy 1993), enabling them to predict the future state of the postsynaptic neuron based on past experience (Abbott and Blum 1996) (see further Koch (1999), Markram, Pikus, Gupta and Tsodyks (1998) and Abbott and Nelson (2000)). This mechanism, because of its apparent time-specificity for periods in the range of ms or tens of ms, could also encourage neurons to learn to respond to temporally synchronous pre-synaptic firing (Gerstner, Kreiter, Markram and Herz 1997, Gutig and Sompolinsky 2006), and indeed to decrease the synaptic strengths from neurons that fire at random times with respect to the synchronized group. This mechanism might also play a role in the normalization of the strength of synaptic connection strengths onto a neuron. Further, there is accumulating evidence (Sjöström, Turrigiano and Nelson 2001) that a more realistic description of the protocols for inducing LTP and LTD probably requires a combination of dependence on spike timing – to take into account the effects of the backpropagating action potential – and dependence on the sub-threshold depolarization of the postsynaptic neuron. However these spike timing-dependent synaptic modifications may be evident primarily at low firing rates rather than those that often occur in the brain (Sjöström, Turrigiano and Nelson 2001), and may not be especially reproducible in the cerebral neocortex (Fregnac, Pananceau, Rene, Huguet, Marre, Levy and Shulz 2010), though interest in STDP, especially from theoretical perspectives, continues (Feldman 2012, Feldman 2009, Fremaux and Gerstner 2015). Under the somewhat steady-state conditions of the firing of neurons in the higher parts of the ventral visual system on the 10 ms timescale that are observed not only when single stimuli are presented for 500 ms (see Fig. 16.15), but also when macaques have found a search target and are looking at it (in the experiments described in Sections 25.2.3 and 25.5.9.1), the average of the presynaptic and postsynaptic rates are likely to be the important determinants of synaptic modification. Part of the reason for this is that correlations between the firing of simultaneously recorded inferior temporal cortex neurons are not common, and if present are not very strong or typically restricted to a short time window in the order of 10 ms (see Section C.3.7). This point is also made in the context that each neuron has thousands of inputs, several tens of which are normally likely to be active when a cell is firing above its spontaneous firing rate and is strongly depolarized. This may make it unlikely statistically that there will be a strong correlation between a particular presynaptic spike and postsynaptic firing, and thus that this is likely to be a main determinant of synaptic strength under these natural conditions.

Synaptic modification for learning is considered further in Chapter 9, and some of the implications of different types of synaptic modification are considered in Appendix B.

1.6 Distributed representations

When considering the operation of many neuronal networks in the brain, it is found that many useful properties arise if each input to the network (arriving on the axons as the vector of input firing rates **x**) is encoded in the activity of an ensemble or population of the axons or input lines (distributed encoding), and is not signalled by the activity of a single input, which is called local encoding. We start off with some definitions, and then highlight some of the differences, and summarize some evidence that shows the type of encoding used in some brain regions. Then in Appendix B (e.g. Table B.2), I show how many of the useful properties of the neuronal networks described depend on distributed encoding. In Appendix C, I review evidence on the encoding actually found in visual cortical areas.

1.6.1 Definitions

A *local representation* is one in which all the information that a particular stimulus or event occurred is provided by the activity of one of the neurons. In a famous example, a single neuron might be active only if one's grandmother was being seen. An implication is that most neurons in the brain regions where objects or events are represented would fire only very rarely. A problem with this type of encoding is that a new neuron would be needed for every object or event that has to be represented. There are many other disadvantages of this type of encoding, many of which will become apparent in this book. Moreover, there is evidence that objects are represented in the brain by a different type of encoding.

A *fully distributed representation* is one in which all the information that a particular stimulus or event occurred is provided by the activity of the full set of neurons. If the neurons are binary (e.g. either active or not), the most distributed encoding is when half the neurons are active for any one stimulus or event.

A *sparse distributed representation* is a distributed representation in which a small proportion of the neurons is active at any one time. In a sparse representation with binary neurons, less than half of the neurons are active for any one stimulus or event. For binary neurons, we can use as a measure of the sparseness the proportion of neurons in the active state. For neurons with real, continuously variable, values of firing rates, the sparseness a^p of the representation provided by the population can be measured, by extending the binary notion of the proportion of neurons that are firing, as

$$a^p = \frac{(\sum\limits_{i=1}^{N} y_i/N)^2}{\sum\limits_{i=1}^{N} y_i^2/N} \tag{1.4}$$

where y_i is the firing rate of the ith neuron in the set of N neurons (Treves and Rolls 1991). This is referred to as the population sparseness, and measures of sparseness are considered in detail in Section C.3.1. A low value of the sparseness a^p indicates that few neurons are firing for any one stimulus.

Coarse coding utilizes overlaps of receptive fields, and can compute positions in the input space using differences between the firing levels of coactive cells (e.g. colour-tuned cones in the retina). The representation implied is very distributed. Fine coding (in which, for example, a neuron may be 'tuned' to the exact orientation and position of a stimulus) implies more sparse coding.

1.6.2 Advantages of different types of coding

One advantage of distributed encoding is that the similarity between two representations can be reflected by the correlation between the two patterns of activity that represent the different stimuli. We have already introduced the idea that the input to a neuron is represented by the activity of its set of input axons x_j, where j indexes the axons, numbered from $j = 1, C$ (see Fig. 1.3 and equation 1.1). Now the set of activities of the input axons is a vector (a vector is an ordered set of numbers; Appendix A provides a summary of some of the concepts involved). We can denote as x^1 the vector of axonal activity that represents stimulus 1, and x^2 the vector that represents stimulus 2. Then the similarity between the two vectors, and thus the two stimuli, is reflected by the correlation between the two vectors. The correlation will be high if the activity of each axon in the two representations is similar; and will become more and more different as the activity of more and more of the axons differs in the two representations. Thus the similarity of two inputs can be represented in a graded or continuous way if (this type of) distributed encoding is used. This enables generalization to similar stimuli, or to incomplete versions of a stimulus (if it is, for example, partly seen or partly remembered), to occur. With a local representation, either one stimulus or another is represented, each by its own neuron firing, and similarities between different stimuli are not encoded.

Another advantage of distributed encoding is that the number of different stimuli that can be represented by a set of C components (e.g. the activity of C axons) can be very large. A simple example is provided by the binary encoding of an 8-element vector. One component can code for which of two stimuli has been seen, 2 components (or bits in a computer byte) for 4 stimuli, 3 components for 8 stimuli, 8 components for 256 stimuli, etc. That is, the number of stimuli increases exponentially with the number of components (or, in this case, axons) in the representation. (In this simple binary illustrative case, the number of stimuli that can be encoded is 2^C.) Put the other way round, even if a neuron has only a limited number of inputs (e.g. a few thousand), it can nevertheless receive a great deal of information about which stimulus was present. This ability of a neuron with a limited number of inputs to receive information about which of potentially very many input events is present is probably one factor that makes computation by the brain possible. With local encoding, the number of stimuli that can be encoded increases only linearly with the number C of axons or components (because a different component is needed to represent each new stimulus). (In our example, only 8 stimuli could be represented by 8 axons.)

In the real brain, there is now good evidence that in a number of brain systems, including the high-order visual and olfactory cortices, and the hippocampus, distributed encoding with the above two properties, of representing similarity, and of exponentially increasing encoding capacity as the number of neurons in the representation increases, is found (Rolls and Tovee 1995b, Abbott, Rolls and Tovee 1996, Rolls, Treves and Tovee 1997b, Rolls, Treves, Robertson, Georges-François and Panzeri 1998b, Rolls, Franco, Aggelopoulos and Reece 2003b, Rolls, Aggelopoulos, Franco and Treves 2004, Franco, Rolls, Aggelopoulos and Treves 2004, Aggelopoulos, Franco and Rolls 2005, Rolls, Franco, Aggelopoulos and Jerez 2006b) (see Appendix C). For example, in the primate inferior temporal visual cortex, the number of faces or objects that can be represented increases approximately exponentially with the number of neurons in the population (see Appendix C). If we consider instead the information about which stimulus is seen, we see that this rises approximately linearly with the number of neurons in the representation (see Appendix C, Fig. 25.13). This corresponds to an exponential rise in the number of stimuli encoded, because information is a log measure (see Appendix C, Fig. 25.14). A similar result has been found for the encoding of position in space by the primate hippocampus (Rolls, Treves, Robertson, Georges-François and Panzeri 1998b).

It is particularly important that the information can be read from the ensemble of neurons using a simple measure of the similarity of vectors, the correlation (or dot product, see Appendix A) between two vectors (e.g. Fig. 25.13). The importance of this is that it is essentially vector similarity operations that characterize the operation of many neuronal networks (see Appendix B). The neurophysiological results show that both the ability to reflect similarity by vector correlation, and the utilization of exponential coding capacity, are properties of real neuronal networks found in the brain.

To emphasize one of the points being made here, although the binary encoding used in the 8-bit vector described above has optimal capacity for binary encoding, it is not optimal for vector similarity operations. For example, the two very similar numbers 127 and 128 are represented by 01111111 and 10000000 with binary encoding, yet the correlation or bit overlap of these vectors is 0. The brain, in contrast, uses a code that has the attractive property of exponentially increasing capacity with the number of neurons in the representation, though it is different from the simple binary encoding of numbers used in computers; and at the same time the brain codes stimuli in such a way that the code can be read off with simple dot product or correlation-related decoding, which is what is specified for the elementary neuronal network operation shown in equation 1.2 (see Section 1.6).

1.7   Neuronal network approaches versus connectionism

The approach taken in this book is to introduce how real neuronal networks in the brain may compute, and thus to achieve a fundamental and realistic basis for understanding brain function. This may be contrasted with connectionism, which aims to understand cognitive function by analysing processing in neuron-like computing systems. Connectionist systems are neuron-like in that they analyze computation in systems with large numbers of computing elements in which the information which governs how the network computes is stored in the connection strengths between the nodes (or "neurons") in the network. However, in many connectionist models the individual units or nodes are not intended to model individual neurons, and the variables that are used in the simulations are not intended to correspond to quantities that can be measured in the real brain. Moreover, connectionist approaches use learning rules in which the synaptic modification (the strength of the connections between the nodes) is determined by algorithms that require information that is not local to the synapse, that is, evident in the pre- and post-synaptic firing rates (see further Appendix B). Instead, in many connectionist systems, information about how to modify synaptic strengths is propagated backwards from the output of the network to affect neurons hidden deep within the network (see Section B.11). Because it is not clear that this is biologically plausible, we have instead in this text concentrated on introducing neuronal network architectures which are more biologically plausible, and which use a local learning rule. Connectionist approaches (see for example McClelland and Rumelhart (1986), McLeod, Plunkett and Rolls (1998)) are very valuable, for they show what can be achieved computationally with networks in which the connection strength determines the computation that the network achieves with quite simple computing elements. However, as models of brain function, many connectionist networks achieve almost too much, by solving problems with a carefully limited number of "neurons" or nodes, which contributes to the ability of such networks to generalize successfully over the problem space. Connectionist schemes thus make an important start to understanding how complex computations (such as language) could be implemented in brain-like systems.

In doing this, connectionist models often use simplified representations of the inputs and outputs, which are often crucial to the way in which the problem is solved. In addition, they may use learning algorithms that are really too powerful for the brain to perform, and therefore they can be taken only as a guide to how cognitive functions might be implemented by neuronal networks in the brain. In this book, we focus on more biologically plausible neuronal networks.

1.8 Introduction to three neuronal network architectures

With neurons of the type outlined in Section 1.3, and an associative learning rule of the type described in Section 1.4, three neuronal network architectures arise that appear to be used in many different brain regions. The three architectures will be described in Appendix B, and a brief introduction is provided here.

In the first architecture (see Fig. 1.8a and b), pattern associations can be learned. The output neurons are driven by an unconditioned stimulus. A conditioned stimulus reaches the output neurons by associatively modifiable synapses w_{ij}. If the conditioned stimulus is paired during learning with activation of the output neurons produced by the unconditioned stimulus, then later, after learning, due to the associative synaptic modification, the conditioned stimulus alone will produce the same output as the conditioned stimulus. Pattern associators are described in Section B.2.

In the second architecture, the output neurons have recurrent associatively modifiable synaptic connections w_{ij} to other neurons in the network (see Fig. 1.8c). When an external input causes the output neurons to fire, then associative links are formed through the modifiable synapses that connect the set of neurons that is active. Later, if only a fraction of the original input pattern is presented, then the associative synaptic connections or weights allow the whole of the memory to be retrieved. This is called completion. Because the components of the pattern are associated with each other as a result of the associatively modifiable recurrent connections, this is called an autoassociative memory. It is believed to be used in the cortex for many purposes, including short-term memory; episodic memory, in which the parts of a memory of an episode are associated together; and helping to define the response properties of cortical neurons, which have collaterals between themselves within a limited region. Autoassociation or attractor networks are described in Section B.3.

In the third architecture, the main input to the output neurons is received through associatively modifiable synapses w_{ij} (see Fig. 1.8d). Because of the initial values of the synaptic strengths, or because every axon does not contact every output neuron, different patterns tend to activate different output neurons. When one pattern is being presented, the most strongly activated neurons tend via lateral inhibition to inhibit the other neurons. For this reason the network is called competitive. During the presentation of that pattern, associative modification of the active axons onto the active postsynaptic neuron takes place. Later, that or similar patterns will have a greater chance of activating that neuron or set of neurons. Other neurons learn to respond to other input patterns. In this way, a network is built that can categorize patterns, placing similar patterns into the same category. This is useful as a preprocessor for sensory information, self-organizes to produce feature analyzers, and finds uses in many other parts of the brain too. Competitive networks are described in Section B.4.

All three architectures require inhibitory interneurons, which receive inputs from the principal neurons in the network (usually the pyramidal cells shown in Fig. 1.8) and implement feedback inhibition by connections to the pyramidal cells. The inhibition is usually implemented by GABA neurons, and maintains a small proportion of the pyramidal cells active.

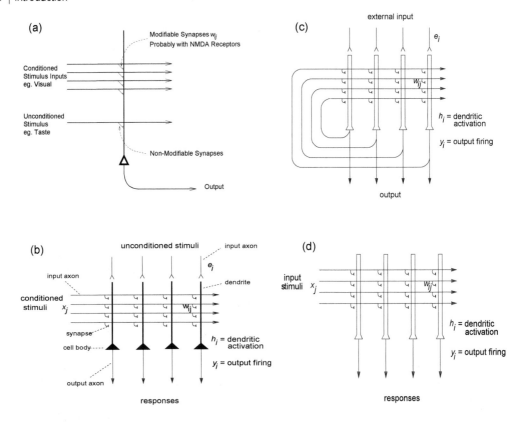

Fig. 1.8 Three network architectures that use local learning rules: (a) pattern association network introduced with a single output neuron; (b) pattern association network; (c) autoassociation network; (d) competitive network.

These are three fundamental building blocks for neural architectures in the cortex. They are often used in combination with each other. Because they are some of the building blocks of some of the architectures found in the cerebral cortex, they are described in Appendix B.

1.9 Systems-level analysis of brain function

To understand the neuronal network operations of any one brain region, it is useful to have an idea of the systems-level organization of the brain, in order to understand how the networks in each region provide a particular computational function as part of an overall computational scheme. In the context of vision, it is very useful to appreciate the different processing streams, and some of the outputs that each has. Some of the processing streams are shown in Fig. 1.9. Some of these regions are shown in the drawings of the primate brain in the next few Figures. Each of these routes is described in turn. The description is based primarily on studies in non-human primates, for they have well-developed cortical areas that in many cases correspond to those found in humans, and it has been possible to analyze their connectivity and their functions by recording the activity of neurons in them.

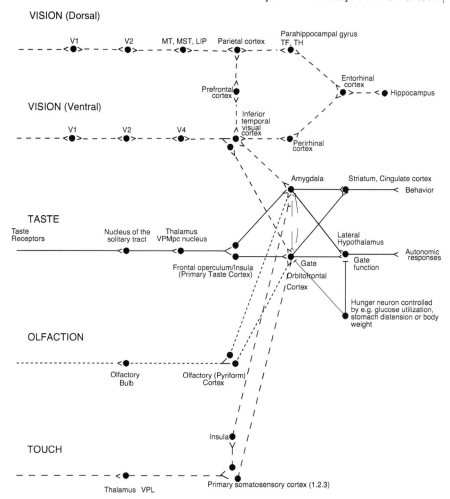

Fig. 1.9 The pathways involved in some different cortical systems described in the text. Forward connections start from early cortical areas on the left. To emphasise that backprojections are important in many memory systems, they are made explicit in the synaptic terminals drawn in the upper part of the diagram, but are a property of most of the connections shown. The top pathway, also shown in Fig. 1.12, shows the connections in the 'dorsal or where visual pathway' from V1 to V2, MT, MST, 7a etc, with some connections reaching the dorsolateral prefrontal cortex and frontal eye fields. The second pathway, also shown in Fig. 1.10, shows the connections in the 'ventral or what visual pathway' from V1 to V2, V4, the inferior temporal visual cortex, etc., with some connections reaching the amygdala and orbitofrontal cortex. The two systems project via the parahippocampal gyrus and perirhinal cortex respectively to the hippocampus, and both systems have projections to the dorsolateral prefrontal cortex. The taste pathways project after the primary taste cortex to the orbitofrontal cortex and amygdala. The olfactory pathways project from the primary olfactory, pyriform, cortex to the orbitofrontal cortex and amygdala. The bottom pathway shows the connections from the primary somatosensory cortex, areas 1, 2 and 3, to the mid-insula, orbitofrontal cortex, and amygdala. Somatosensory areas 1, 2 and 3 also project via area 5 in the parietal cortex, to area 7b.

1.9.1   Ventral cortical visual stream

Information in the *'ventral or what'* visual cortical processing stream projects after the primary visual cortex, area V1, to the secondary visual cortex (V2), and then via area V4 to the posterior and then to the anterior inferior temporal visual cortex (see Figs. 1.9, 1.10, and 1.11).

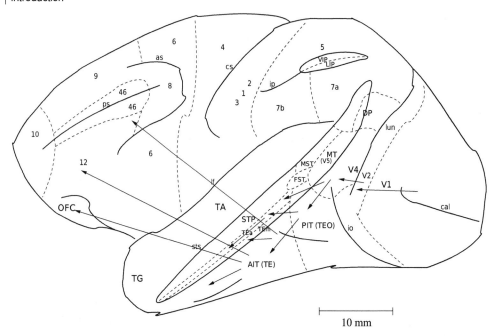

Fig. 1.10 Lateral view of the macaque brain showing the connections in the 'ventral or what visual pathway' from V1 to V2, V4, the inferior temporal visual cortex, etc., with some connections reaching the amygdala and orbitofrontal cortex. as, arcuate sulcus; cal, calcarine sulcus; cs, central sulcus; lf, lateral (or Sylvian) fissure; lun, lunate sulcus; ps, principal sulcus; io, inferior occipital sulcus; ip, intraparietal sulcus (which has been opened to reveal some of the areas it contains); sts, superior temporal sulcus (which has been opened to reveal some of the areas it contains). AIT, anterior inferior temporal cortex; FST, visual motion processing area; LIP, lateral intraparietal area; MST, visual motion processing area; MT, visual motion processing area (also called V5); OFC, orbitofrontal cortex; PIT, posterior inferior temporal cortex; STP, superior temporal plane; TA, architectonic area including auditory association cortex; TE, architectonic area including high order visual association cortex, and some of its subareas TEa and TEm; TG, architectonic area in the temporal pole; V1–V4, visual areas 1–4; VIP, ventral intraparietal area; TEO, architectonic area including posterior visual association cortex. The numbers refer to architectonic areas, and have the following approximate functional equivalence: 1, 2, 3, somatosensory cortex (posterior to the central sulcus); 4, motor cortex; 5, superior parietal lobule; 7a, inferior parietal lobule, visual part; 7b, inferior parietal lobule, somatosensory part; 6, lateral premotor cortex; 8, frontal eye field; 12, inferior convexity prefrontal cortex; 46, dorsolateral prefrontal cortex.

Information processing along this stream is primarily unimodal, as shown by the fact that inputs from other modalities (such as taste or smell) do not anatomically have significant inputs to these regions, and by the fact that neurons in these areas respond primarily to visual stimuli, and not to taste or olfactory stimuli, etc. (Rolls 2000a, Baylis, Rolls and Leonard 1987, Ungerleider 1995, Rolls 2008d). The representation built along this pathway is mainly about what object is being viewed, independently of exactly where it is on the retina, of its size, and even of the angle with which it is viewed (see Chapter 25 and Rolls (2012c)), and for this reason it is frequently referred to as the 'what' visual pathway. The representation is also independent of whether the object is associated with reward or punishment, that is the representation is about objects per se (Rolls, Judge and Sanghera 1977). The computation that must be performed along this stream is thus primarily to build a representation of objects that shows invariance. After this processing, the visual representation is interfaced to other sensory systems in areas in which simple associations must be learned between stimuli in different modalities (see Chapters 24 and 15). The representation must thus be in a form in which the simple generalization properties of associative networks can be useful. Given that the association is about what object is present (and not where it is on the retina), the representation computed in sensory systems must be in a form that allows the simple correlations computed by associative networks to reflect similarities between objects, and not between their positions on the retina. The way in which such invariant sensory representations could be built in the brain is the subject of Chapter 25 and Rolls (2012c).

The ventral visual stream converges with other mainly unimodal information processing streams for taste, olfaction, touch, and hearing in a number of areas, particularly the amygdala and orbitofrontal cortex (see Figs. 1.9, 1.10, and 1.11). These areas appear to be necessary for learning to associate sensory stimuli with other reinforcing (rewarding or punishing) stimuli. For example, the amygdala is involved in learning associations between the sight of food and its taste. (The taste is a primary or innate reinforcer.) The orbitofrontal cortex is especially involved in rapidly relearning these associations, when environmental contingencies change (Rolls 2014a, Rolls and Grabenhorst 2008, Rolls 2000d). They thus are brain regions in which the computation at least includes simple pattern association (e.g. between the sight of an object and its taste). In the orbitofrontal cortex, this association learning is also used to produce a representation of flavour, in that neurons are found in the orbitofrontal cortex that are activated by both olfactory and taste stimuli (Rolls and Baylis 1994), and in that the neuronal responses in this region reflect in some cases olfactory to taste association learning (Rolls, Critchley, Mason and Wakeman 1996b, Critchley and Rolls 1996b). In these regions too, the representation is concerned not only with what sensory stimulus is present, but for some neurons, with its hedonic or reward-related properties, which are often computed by association with stimuli in other modalities. For example, many of the visual neurons in the orbitofrontal cortex respond to the sight of food only when hunger is present. This probably occurs because the visual inputs here have been associated with a taste input, which itself in this region only occurs to a food if hunger is present, that is when the taste is rewarding (Chapter 15) (Rolls 2014a, Rolls 2015e, Rolls and Grabenhorst 2008, Rolls 2000d). The outputs from these associative memory systems, the amygdala and orbitofrontal cortex, project onwards to structures such as the hypothalamus, through which they control autonomic and endocrine responses such as salivation and insulin release to the sight of food; and to the striatum, including the ventral striatum, through which behaviour to learned reinforcing stimuli is produced (Fig. 1.9).

1.9.2 Dorsal cortical visual stream

The *'dorsal or where' visual processing stream* shown in Figs. 1.9, 1.12, and 1.11 is that from V1 to MT, MST and thus to the parietal cortex (Ungerleider 1995, Ungerleider and Haxby 1994, Rolls and Deco 2002, Perry and Fallah 2014). This 'where' pathway for primate vision is involved in representing where stimuli are relative to the animal (i.e. in egocentric space), and the motion of these stimuli. Neurons here respond, for example, to stimuli in visual space around the animal, including the distance from the observer, and also respond to optic flow or to moving stimuli. Outputs of this system control eye movements to visual stimuli (both slow pursuit and saccadic eye movements). These outputs proceed partly via the frontal eye fields, which then project to the striatum, and then via the substantia nigra reach the superior colliculus (Goldberg and Walker 2013). Other outputs of these regions are to the dorsolateral prefrontal cortex, area 46, which is important as a short-term memory for where fixation should occur next, as shown by the effects of lesions to the prefrontal cortex on saccades to remembered targets, and by neuronal activity in this region (Goldman-Rakic 1996, Passingham and Wise 2012). The dorsolateral prefrontal cortex short-term memory systems in area 46

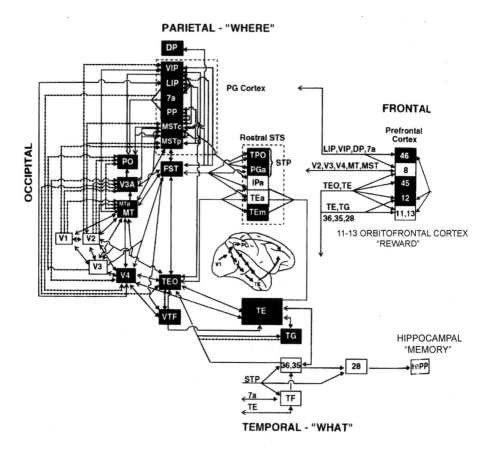

Fig. 1.11 Visual processing pathways in monkeys. Solid lines indicate connections arising from both central and peripheral visual field representations; dotted lines indicate connections restricted to peripheral visual field representations. Shaded boxes in the 'ventral (lower) or what' stream indicate visual areas related primarily to object vision; shaded boxes in the 'dorsal or where' stream indicate areas related primarily to spatial vision; and white boxes indicate areas not clearly allied with only one stream. Abbreviations: DP, dorsal prelunate area; FST, fundus of the superior temporal area; HIPP, hippocampus; LIP, lateral intraparietal area; MSTc, medial superior temporal area, central visual field representation; MSTp, medial superior temporal area, peripheral visual field representation; MT, middle temporal area; MTp, middle temporal area, peripheral visual field representation; PO, parieto-occipital area; PP, posterior parietal sulcal zone; STP, superior temporal polysensory area; V1, primary visual cortex; V2, visual area 2; V3, visual area 3; V3A, visual area 3, part A; V4, visual area 4; and VIP, ventral intraparietal area. Inferior parietal area 7a; prefrontal areas 8, 11 to 13, 45 and 46 are from Brodmann (1925). Inferior temporal areas TE and TEO, parahippocampal area TF, temporal pole area TG, and inferior parietal area PG are from Von Bonin and Bailey (1947). Rostral superior temporal sulcal (STS) areas are from Seltzer and Pandya (1978) and VTF is the visually responsive portion of area TF (Boussaoud, Desimone and Ungerleider 1991). Area 46 in the dorsolateral prefrontal cortex is involved in short-term memory. (Modified from *Science*, 270, Functional brain imaging studies of cortical mechanisms for memory, Ungerleider, L. G. pp. 769–775. Copyright ©1995, The American Association for the Advancement of Science. Reprinted with permission from AAAS.)

with spatial information received from the parietal cortex play an important role in attention, by holding on-line the target being attended to, as described in Chapter 6.

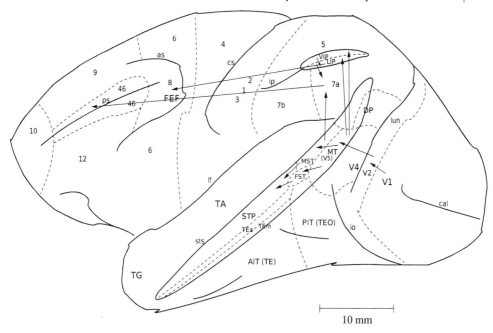

Fig. 1.12 Lateral view of the macaque brain showing the connections in the 'dorsal or where visual pathway' from V1 to V2, MST, LIP, VIP, and parietal cortex area 7a, with some connections then reaching the dorsolateral prefrontal cortex. Abbreviations as in Fig. 1.10. FEF - frontal eye field.

1.9.3 Hippocampal memory system

The hippocampus receives inputs from both the 'what' and the 'where' visual systems (Figs. 1.9 and 1.11, and Chapter 24). By rapidly learning associations between conjunctive inputs in these systems, it is able to form memories of particular events occurring in particular places at particular times. To do this, it needs to store whatever is being represented in each of many cortical areas at a given time, and later to recall the whole memory from a part of it. The types of network it contains that are involved in this memory function are described in Chapters 12 and 24.

1.9.4 Frontal lobe systems

As shown in Figs. 1.11–1.12 and Fig. 25.1, the dorsolateral parts of the prefrontal cortex receive inputs from dorsal stream processing pathways, and the ventrolateral parts from the ventral stream processing pathways. These lateral prefrontal cortical areas, including Brodmann area (BA) 46, provide short-term memory functions, providing the basis for functions such as top-down attention, working memory, and planning (Sections 4.2 and 4.3.1). These lateral prefrontal cortex areas can be thought of as providing processing that need not be dominated by perceptual processing, as must the posterior visual and parietal lobe spatial areas if incoming stimuli are to be perceived, but can instead be devoted to processing recent information that may no longer be present in perceptual areas.

The orbitofrontal cortex, areas BA 11, 12 and 13 (Figs. 15.2 and 1.11–1.12), receives information from the 'what' systems for vision, taste, olfaction, touch, and audition), forms multimodal representations, and computes reward value. It is therefore of fundamental importance for emotional, motivational, and social behaviour (see Chapter 15).

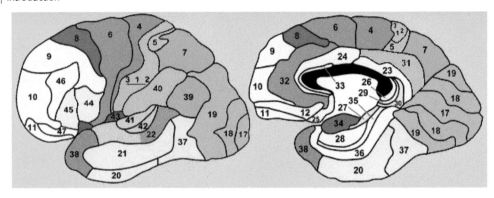

Fig. 1.13 Brodmann areas. Left: lateral view of the human brain. The front of the brain is on the left. Right: Medial view of the human brain. The front of the brain is on the left. (Modified from http://spot.colorado.edu/ dubin/talks/brodmann/brodmann.html)

1.9.5 Brodmann areas

A very brief outline of some of the connections and functions of different Brodmann areas (Brodmann 1925) is provided here, intended mainly for those who are not familiar with them. Brodmann used cytoarchitecture to distinguish different areas, but because function reflects the architecture, different Brodmann areas also to a considerable extent have different functions. Some of these areas are shown by the numbers in Figs. 1.13 and 1.9–1.12. Some of the information in this section (1.9.5) is modified from https://www.kenhub.com/en/library/anatomy/brodmann-areas.

Brodmann areas:
Areas 3, 1 & 2 – Primary somatosensory cortex (in the postcentral gyrus). This is numbered rostral to caudal as 3,1,2. This region is associated with the sense and localization of touch, temperature, vibration, and pain. There is a map or homunculus of different parts of the body. A lesion in this area may cause agraphesthesia, asterognosia, loss of vibration, proprioception and fine touch, and potentially hemineglect if the non-dominant hemisphere is affected.
Area 4 – Primary motor cortex (in the precentral gyrus). This is the area responsible for executing fine motor movements, which includes contralateral finger/hand/wrist or orofacial movements, learned motor sequences, breathing control, and voluntary blinking. There is a motor map in area 4, which approximately corresponds to the map in the primary somatosensory cortex. A lesion in this area may cause paralysis of the contralateral side of the body, including facial palsy, arm or leg monoparesis, and hemiparesis.
Area 5 – Somatosensory association cortex. Brodmann area 5 is part of the superior parietal cortex, and is a secondary somatosensory area that receives inputs from the primary somatosensory cortex, and projects to parietal area 7b. A lesion in the left superior parietal lobe may cause ideomotor apraxia, which is the loss of ability to produce purposeful, skilled movements as a result of brain pathology not due to physical weakness, paralysis, lack of coordination or sensory loss. Astereognosis (also known as tactile agnosia) is also possible, which would lead to loss of ability to recognize objects by feeling or handling them.
Area 6 – Premotor and Supplementary Motor Cortex (Secondary Motor Cortex). This projects to motor cortex area 4, and can be thought of as a higher order motor cortical area. This region is critical for the sensory guidance of movement and control of proximal and trunk muscles, and contributes to the planning of complex and coordinated motor movements. A lesion here may affect sensory guidance of movement and control of proximal and trunk muscles. Dam-

age of the lateral premotor area 6 may result in kinetic apraxia (which would appear as coarse or unrefined movements that no longer have the appearance of being practiced over time).

Area 7 – 7b: Somatosensory Association Cortex in the anterior part of the superior parietal lobule on the lateral surface. 7a: Visuospatial association cortex in the posterior part of the superior parietal lobule on the lateral surface. The precuneus medially is involved in spatial function including the sense of self (Cavanna and Trimble 2006).

Area 8 – A higher order motor cortical area, which projects to area 6. Area 8 includes the frontal eye fields, which are involved in making saccades based on information in short-term memory (Goldberg and Walker 2013).

Area 9 – is in the dorsolateral/anterior prefrontal cortex (DLPFC). This region is a high order cortical area responsible for motor planning, organization, and regulation, and sustaining attention and working memory. The DLPFC projects to areas such as Area 8. The DLPFC plays an important role in working memory, and hence is important in planning. Area 46 is a part of the dorsolateral prefrontal cortex around and within the principal sulcus, and is involved in spatial working memory (dorsally), and object working memory (ventrally) (Passingham and Wise 2012).

Area 10 – Anterior prefrontal cortex (most rostral part of superior and middle frontal gyri). This region is involved in strategic processes of memory retrieval and executive functions (such as reasoning, task flexibility, problem solving, planning, execution, working memory, processing emotional stimuli, inferential reasoning, decision making, calculation of numerical processes, etc.). The ventral medial prefrontal cortex area 10 and the anterior and medial parts of the orbitofrontal cortex are sometimes referred to as the ventromedial prefrontal cortex.

Area 11 – Orbitofrontal area (orbital and rectus gyri, plus part of the rostral part of the superior frontal gyrus).

Area 12 – Orbitofrontal area (refers to the area between the superior frontal gyrus and the inferior rostral sulcus, i.e. laterally) (Fig. 15.2).

Area 13 – Caudal orbitofrontal cortex just anterior to the anterior insula (Ongur, Ferry and Price 2003, Carmichael and Price 1994, Ongur and Price 2000, Rolls 2014a). Areas 11–13 are illustrated in Fig. 15.2, and are involved in reward value, non-reward, and emotion (Rolls 2014a).

Areas 14–16 Insular cortex. The anterior part contains the taste cortex, and below is visceral / autonomic cortex (Rolls 2016b). More posteriorly, there are somatotopically organised somatosensory representations.

Area 17 – Primary visual cortex (V1). The primary visual cortex is located in the occipital lobe at the back of the brain, and contains a well-defined map of the visual field. Depending on where and how damage and lesions occur to this region, partial or complete cortical blindness can result; for example, if the upper bank of the calcarine sulcus is damaged, then the lower bank of the visual field is affected. Patients with damage to the striate cortex may show blindsight, in which they deny seeing objects, but can often guess better than chance (Weiskrantz 1998).

Area 18 – Secondary visual cortex (V2)

Area 19 – Associative visual cortex (V3,V4,V5)

Area 20 – Inferior temporal gyrus. Includes inferior temporal visual cortex.

Area 21 – Middle temporal gyrus. This is present in humans, and may correspond to parts of the cortex in the macaque cortex in the superior temporal sulcus. It contains face expression cells, and is implicated also in humans in theory of mind and autism (Cheng, Rolls, Gu, Zhang and Feng 2015).

Area 22 – Auditory cortex (in the Superior Temporal Gyrus, of which the caudal part is usually considered to be within Wernicke's area which is involved in language comprehension).

Area 23 – Ventral posterior cingulate cortex (Vogt 2009).

Area 24 – Ventral anterior cingulate cortex (Vogt 2009).

Area 25 – Subgenual cingulate cortex (part of the Ventromedial prefrontal cortex) (Vogt 2009).

Area 26 – Ectosplenial portion of the retrosplenial region of the cerebral cortex (Vogt 2009).

Area 27 – Piriform (olfactory) cortex.

Area 28 – Entorhinal cortex. The gateway to and from the hippocampus (Fig. 24.1). The medial entorhinal cortex contains grid (place) cells in rodents, and grid (spatial view) cells in primates (Chapter 24).

Area 29 – Retrosplenial cingulate cortex. Processes spatial information, with connections with the hippocampus.

Area 30 – Part of cingulate cortex (Vogt 2009).

Area 31 – Dorsal posterior cingulate cortex (Vogt 2009).

Area 32 – Anterior cingulate cortex (Vogt 2009).

Area 33 – Part of anterior cingulate cortex (Vogt 2009).

Area 34 – Part of the parahippocampal gyrus (Fig. 24.2).

Area 35 – Perirhinal cortex (in the rhinal sulcus) (Fig. 24.2). Connects visual cortical areas to and from the entorhinal cortex and thus hippocampus.

Area 36 – Also perirhinal cortex (in the rhinal sulcus).

Area 37 – Fusiform gyrus. This region is involved in face, object, and scene representation. It receives from earlier cortical visual areas.

Area 38 – Temporopolar area (most rostral part of the superior and middle temporal gyri).

Area 39 – Angular gyrus. In the inferior parietal lobule. Areas 39 (angular gyrus) and 40 (supramarginal area) on the left in right-handed people are often known as Wernicke's area, which is involved in language, especially in comprehension. A lesion here causes language disorders characterized by fluent speech paraphasias where words are jumbled and nonsensical sentences are spoken.

Area 40 – Supramarginal gyrus, in the inferior parietal lobule, is involved in phonological word processing.

Areas 41 and 42 – Auditory cortex.

Area 43 – Primary gustatory cortex.

Areas 44–45 (pars opercularis of the inferior frontal gyrus) and area 45 (triangular part of the inferior frontal gyrus are together often known as Broca's area in the dominant hemisphere. This region is associated with the praxis of speech (motor speech programming). This includes being able to put together the binding elements of language, selecting information among competing sources, sequencing motor/expressive elements, cognitive control mechanisms for syntactic processing of sentences, and construction of complex sentences and speech patterns. Lesions in this area cause Broca's aphasia: a deficit in the ability to speak and produce the proper words/sounds, even though the person maintains the ability to comprehend language and to mentally formulate proper sentences.

Area 45 – Pars triangularis, part of the inferior frontal gyrus and part of Broca's area.

Area 46 – Dorsolateral prefrontal cortex. Area 46 is a part of the dorsolateral prefrontal cortex around and within the principal sulcus, and is involved in spatial working memory (dorsally), and object working memory (ventrally) (Passingham and Wise 2012).

Area 47 – Inferior frontal gyrus, pars orbitalis.

Area 48 – Retrosubicular area (a small part of the medial surface of the temporal lobe).

Area 49 – Parasubicular area in a rodent.

Area 52 – Parainsular area (at the junction of the temporal lobe and the insula).

1.10 Introduction to the fine structure of the cerebral neocortex

An important part of the approach to understanding how the cerebral cortex could implement the computational processes that underlie visual perception is to take into account as much as possible its fine structure and connectivity, as these provide important indicators of and constraints on how it computes. An introductory description is provided in this section (1.10), and a description that considers the operational / computational principles of neocortex, and compares them to pyriform cortex and hippocampal cortex, is provided in Section 18.2.

1.10.1 The fine structure and connectivity of the neocortex

The neocortex consists of many areas that can be distinguished by the appearance of the cells (cytoarchitecture) and fibres or axons (myeloarchitecture), but nevertheless, the basic organization of the different neocortical areas has many similarities, and it is this basic organization that is considered here. Useful sources for more detailed descriptions of neocortical structure and function are the books in the series 'Cerebral Cortex' edited by Jones and Peters (Jones and Peters (1984) and Peters and Jones (1984)) and many other resources (Douglas, Markram and Martin 2004, Shepherd 2004, Shepherd and Grillner 2010, da Costa and Martin 2010, Harris and Mrsic-Flogel 2013, Kubota 2014, Harris and Shepherd 2015). The detailed understanding of the cell types and their patterns of connections is now benefitting from molecular markers (Harris and Shepherd 2015, Bernard, Lubbers, Tanis, Luo, Podtelezhnikov, Finney, McWhorter, Serikawa, Lemon, Morgan, Copeland, Smith, Cullen, Davis-Turak, Lee, Sunkin, Loboda, Levine, Stone, Hawrylycz, Roberts, Jones, Geschwind and Lein 2012). Approaches to quantitative aspects of the connectivity are provided by Braitenberg and Schütz (1991), Braitenberg and Schütz (1998) by Abeles (1991), and in this book. Some of the connections described in Sections 1.10.2 and 1.10.3 are shown schematically in Figs. 1.17 and 1.18.

1.10.2 Excitatory cells and connections

Some of the cell types found in the neocortex are shown in Fig. 1.2. Cells A–D are pyramidal cells. The dendrites (shown thick in Fig. 1.2) are covered in spines, which receive the excitatory synaptic inputs to the cell. Pyramidal cells with cell bodies in different laminae of the cortex (shown in Fig. 1.2 as I–VI) not only have different distributions of their dendrites, but also different distributions of their axons (shown thin in Fig. 1.2), which connect both within that cortical area and to other brain regions outside that cortical area (see labelling at the bottom of Fig. 1.17). Further examples of cell types, from S1 (somatosensory) rat whisker barrel cortex, are shown in Fig. 1.14 (Harris and Shepherd 2015).

The main information-bearing afferents to a cortical area have many terminals in layer 4. (By these afferents, we mean primarily those from the thalamus or from the preceding cortical area. We do not mean the cortico-cortical backprojections; nor the subcortical cholinergic, noradrenergic, dopaminergic, and serotonergic inputs, which are numerically minor, although they are important in setting cortical cell thresholds, excitability, and adaptation, see for example Douglas, Markram and Martin (2004).) In primary sensory cortical areas only, there are spiny stellate cells in a rather expanded layer 4, and the thalamic terminals synapse onto these cells (Lund 1984, Martin 1984, Douglas and Martin 1990, Douglas, Markram and Martin 2004, Levitt, Lund and Yoshioka 1996, da Costa and Martin 2013). (Primary sensory cortical areas receive their inputs from the primary sensory thalamic nucleus for a sensory modality. An example is the primate striate cortex which receives inputs from the lateral geniculate nucleus, which in turn receives inputs from the retinal ganglion cells. Spiny

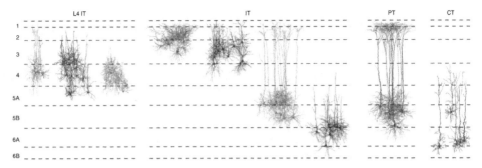

Fig. 1.14 Dendritic morphology of excitatory neurons in S1 whisker barrel cortex. L4 IT shows the three morphological classes of L4 intratelencephalic (IT) neurons: pyramidal, star pyramidal and spiny stellate cells. Under IT are other intratelencephalic neurons of L2, L3, 5A/B and 6. PT shows pyramidal tract neurons of L5B. CT shows corticothalamic neurons of L6. (Adapted with permission from Marcel Oberlaender, Christiaan P.J. de Kock, Randy M. Bruno, Alejandro Ramirez, Hanno S. Meyer, Vincent J. Dercksen, Moritz Helmstaedter and Bert Sakmann, Cell type-specific three-dimensional structure of thalamocortical circuits in a column of rat vibrissal cortex, *Cerebral Cortex*, 2012, 22 (10), pp. 2375–2391, Figure 2, ©2012, Oxford University Press.)

stellate cells are so-called because they have radially arranged, star-like, dendrites. Their axons usually terminate within the cortical area in which they are located.) Each thalamic axon makes 1,000–10,000 synapses, not more than several (or at most 10) of which are onto any one spiny stellate cell. In addition to these afferent terminals, there are some terminals of the thalamic afferents onto pyramidal cells with cell bodies in layers 6 and 3 (Martin 1984) (and terminals onto inhibitory interneurons such as basket cells, which thus provide for a feedforward inhibition) (see Fig. 1.15). Even in layer 4, the thalamic axons provide less than 20% of the synapses. The number of thalamo-cortical connections received by a cortical cell is relatively low, in the range 18–191 (da Costa and Martin 2013). The spiny stellate neurons in layer 4 have axons which terminate in layers 3 and 2, at least partly on dendrites of pyramidal cells with cell bodies in layers 3 and 2. (These synapses are of Type I, that is are asymmetrical and are on spines, so that they are excitatory. Their transmitter is glutamate.) These layer 3 and 2 pyramidal cells provide the onward cortico-cortical projection with axons which project into layer 4 of the next cortical area. For example, layer 3 and 2 pyramidal cells in the primary visual (striate) cortex of the macaque monkey project into the second visual area (V2), layer 4.

In non-primary sensory areas, important information-bearing afferents from a preceding cortical area terminate in layer 4, but there are no or few spiny stellate cells in this layer (Lund 1984, Levitt, Lund and Yoshioka 1996). Layer 4 still looks 'granular' (due to the presence of many small cells), but these cells are typically small pyramidal cells (Lund 1984). (It may be noted here that spiny stellate cells and small pyramidal cells are similar in many ways, with a few main differences including the absence of a major apical dendrite in a spiny stellate which accounts for its non-pyramidal, star-shaped, appearance; and for many spiny stellate cells, the absence of an axon that projects outside its cortical area.) The terminals presumably make synapses with these small pyramidal cells, and also presumably with the dendrites of cells from other layers, including the basal dendrites of deep layer 3 pyramidal cells (see Fig. 1.17).

The axons of the *superficial (layer 2 and 3) pyramidal cells* have collaterals and terminals in layer 5 (see Fig. 1.17), and synapses are made with the dendrites of the layer 5 pyramidal cells (Martin 1984). The axons also typically project out of that cortical area, and on to the next cortical area in sequence, where they terminate in layer 4, forming the forward cortico-cortical projection. It is also from these pyramidal cells that projections to the amygdala arise in some sensory areas that are high in the hierarchy (Amaral, Price, Pitkanen and Carmichael 1992).

The axons of the *layer 5 pyramidal cells* have many collaterals in layer 6 (see Fig. 1.2), where synapses could be made with the layer 6 pyramidal cells (based on indirect evidence, see Fig. 13 of Martin (1984)), and axons of these cells typically leave the cortex to project to subcortical sites (such as the striatum), or back to the preceding cortical area to terminate in layer 1. It is remarkable that there are as many of these backprojections as there are forward connections between two sequential cortical areas. The possible computational significance of this connectivity is considered in Chapters 11 and 24.

The *layer 6 pyramidal cells* have prolific dendritic arborizations in layer 4 (see Fig. 1.2), and receive synapses from thalamic afferents (Martin 1984), and also presumably from pyramidal cells in other cortical layers. The axons of these cells form backprojections to the thalamic nucleus which projects into that cortical area, and also axons of cells in layer 6 contribute to the backprojections to layer 1 of the preceding cortical area (see Jones and Peters (1984) and Peters and Jones (1984); see Figs. 1.2 and 1.17). Layer 6 pyramidal cells also project to layer 4, where thalamic afferents terminate. Layer 6 is thus closely related to the thalamus, and the significance of this is considered in Chapter 18.

Although the pyramidal and spiny stellate cells form the great majority of neocortical neurons with excitatory outputs, there are in addition several further cell types (see Peters and Jones (1984), Chapter 4). Bipolar cells are found in layers 3 and 5, and are characterized by having two dendritic systems, one ascending and the other descending, which, together with the axon distribution, are confined to a narrow vertical column often less than 50 μm in diameter (Peters 1984a). Bipolar cells form asymmetrical (presumed excitatory) synapses with pyramidal cells, and may serve to emphasize activity within a narrow vertical column.

1.10.3 Inhibitory cells and connections

There are a number of types of neocortical inhibitory neurons. All are described as smooth in that they have no spines, and use GABA (gamma-amino-butyric acid) as a transmitter. (In older terminology they were called Type II.) A number of types of inhibitory neuron can be distinguished, best by their axonal distributions (Szentagothai 1978, Peters and Regidor 1981, Douglas, Markram and Martin 2004, Douglas and Martin 2004, Shepherd and Grillner 2010, Harris and Mrsic-Flogel 2013). One type is the *basket cell*, present in layers 3–6, which has few spines on its dendrites so that it is described as smooth, and has an axon that participates in the formation of weaves of preterminal axons which surround the cell bodies of pyramidal cells and form synapses directly onto the cell body, but also onto the dendritic spines (Somogyi, Kisvarday, Martin and Whitteridge 1983) (Fig. 1.15). Basket cells comprise 5–7% of the total cortical cell population, compared with approximately 72% for pyramidal cells (Sloper and Powell 1979b, Sloper and Powell 1979a). Basket cells receive synapses from the main extrinsic afferents to the neocortex, including thalamic afferents (Fig. 1.15), so that they must contribute to a feedforward type of inhibition of pyramidal cells. The inhibition is feedforward in that the input signal activates the basket cells and the pyramidal cells by independent routes, so that the basket cells can produce inhibition of pyramidal cells that does not depend on whether the pyramidal cells have already fired. Feedforward inhibition of this type not only enhances stability of the system by damping the responsiveness of the pyramidal cell simultaneously with a large new input, but can also be conceived of as a mechanism which normalizes the magnitude of the input vector received by each small region of neocortex (see further Appendix B). In fact, the feedforward mechanism allows the pyramidal cells to be set at the appropriate sensitivity for the input they are about to receive. Basket cells can also be polysynaptically activated by an afferent volley in the thalamo-cortical projection

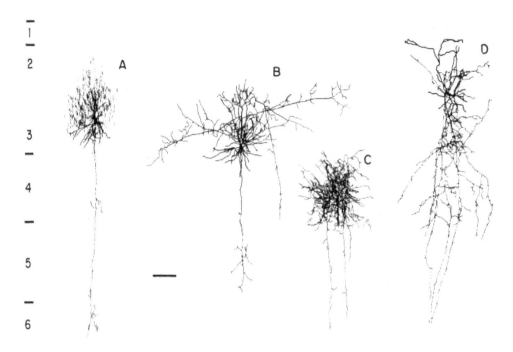

Fig. 1.15 Smooth (inhibitory) cells from cat visual cortex. (A) Chandelier or axoaxonic cell. (B) Large basket cell of layer 3. Basket cells, present in layers 3–6, have few spines on their dendrites so that they are described as smooth, and have an axon which participates in the formation of weaves of preterminal axons which surround the cell bodies of pyramidal cells and form synapses directly onto the cell body. (C) Small basket or clutch cell of layer 3. The major portion of the axonal arbor is confined to layer 4. (D) Double bouquet cell. The axon collaterals run vertically. The cortical layers are as indicated. Bar = 100 μm. (This material has been adapted from Rodney Douglas, Henry Markram and Kevin Martin, Neocortex, In Gordon M. Shepherd (ed), *The Synaptic Organization of the Brain* 5e, pp. 499–558, figures 12.4–12.11 (c) 2004, Oxford University Press and has been reproduced by permission of Oxford University Press https://global.oup.com/academic/product/the-synaptic-organization-of-the-brain-9780195159561?q=the synaptic organization of the brain&lang=en&cc=gb. For permission to reuse this material, please visit http://www.oup.co.uk/academic/rights/permissions.)

(Martin 1984), so that they may receive inputs from pyramidal cells, and thus participate in feedback inhibition of pyramidal cells.

The transmitter used by the basket cells is gamma-amino-butyric acid (GABA), which opens chloride channels in the postsynaptic membrane. Because the reversal potential for Cl^- is approximately -10 mV relative to rest, opening the Cl^- channels does produce an inhibitory postsynaptic potential (IPSP), which results in some hyperpolarization, especially in the dendrites. This is a subtractive effect, hence it is a linear type of inhibition (Douglas and Martin 1990, Douglas, Markram and Martin 2004). However, a major effect of the opening of the Cl^- channels in the cell body is that this decreases the membrane resistance, thus producing a shunting effect. The importance of shunting is that it decreases the magnitude of excitatory postsynaptic potentials (EPSPs) (cf. Andersen, Dingledine, Gjerstad, Langmoen and Laursen (1980) for hippocampal pyramidal cells), so that the effect of shunting is to produce division (i.e. a multiplicative reduction) of the excitatory inputs received by the cell, and not just to act by subtraction (see further Bloomfield (1974), Martin (1984), Douglas and Martin (1990)). Thus, when modelling the normalization of the activity of cortical pyramidal cells, it is common to include division in the normalization function (cf. Appendix B). It is

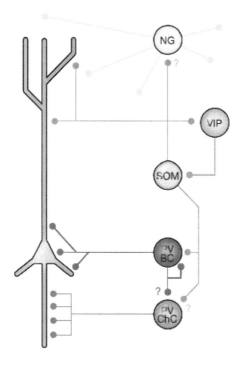

Fig. 1.16 Classes of neocortical inhibitory interneuron.
Parvalbumin-expressing interneurons (PVs) are capable of firing rapidly and with high temporal precision. They consist of two main subgroups: basket cells (BCs) that target the soma and proximal dendrites of principal cells, and chandelier cells (ChCs) that target the axon initial segment. PV cells receive strong excitatory inputs from thalamus and cortex, as well as inhibition from other PVs. A key role of these cells is to stabilize the activity of cortical networks using negative feedback: their absence leads to epileptiform activity, whereas more moderate chronic dysfunction of these cells has been implicated in diseases such as schizophrenia.
Somatostatin-expressing interneurons (SOMs) consist largely of Martinotti cells that target the tuft dendrites of principal cells, as well as inhibiting other interneurons. Consistent with their targeting of dendritic tufts, these cells have been implicated in behaviour-dependent control of dendritic integration, as well as in more general lateral inhibition. Connections from principal cells to SOMs show facilitating synapses. In contrast to PVs, SOMs receive the majority of their input from local principal cells but little inhibition or thalamic drive.
5HT3A-receptor-expressing interneurons are the most numerous interneuron of the superficial layers. They contain two prominent subgroups: neurogliaform cells (NGs), which are thought to release GABA by volume transmission; and cells that express vasoactive intestinal peptide (VIP) and preferentially target SOMs. Putative 5HT3A-receptor-expressing cells have been implicated in the control of cortical circuits by higher-order cortex and the thalamus. (Reprinted by permission from Macmillan Publishers Ltd: Nature, Harris, K. D. and Mrsic-Flogel, T. D., Cortical connectivity and sensory coding 503: 51–58, Copyright ©2013.)

notable that the dendrites of basket cells can extend laterally 0.5 mm or more (primarily within the layer in which the cell body is located), and that the axons can also extend laterally from the cell body 0.5–1.5 mm. Thus the basket cells produce a form of lateral inhibition which is quite spatially extensive. There is some evidence that each basket cell may make 4–5 synapses with a given pyramidal cell, that each pyramidal cell may receive from 10–30 basket cells, and that each basket cell may inhibit approximately 300 pyramidal cells (Martin 1984, Douglas and Martin 1990, Douglas, Markram and Martin 2004). The basket cells are sometimes called clutch cells.

A second type of GABA-containing inhibitory interneuron is the *axoaxonic (or 'chandelier') cell*, named because it synapses onto the initial segment of the axon of pyramidal cells. The pyramidal cells receiving this type of inhibition are almost all in layers 2 and 3, and much less in the deep cortical layers. One effect that axoaxonic cells probably produce is thus prevention of outputs from layer 2 and 3 pyramidal cells reaching the pyramidal cells in the deep layers, or from reaching the next cortical area. Up to five axoaxonic cells converge onto a pyramidal cell, and each axoaxonic cell may project to several hundred pyramidal cells scattered in a region that may be several hundred microns in length (Martin 1984, Peters 1984b). This implies that axoaxonic cells provide a rather simple device for preventing runaway overactivity of pyramidal cells, but little is known yet about the afferents to axoaxonic cells, so that the functions of these neurons are very incompletely understood.

A third type of (usually smooth and inhibitory) cell is the *double bouquet cell*, which has primarily vertically organized axons. These cells have their cell bodies in layer 2 or 3, and have an axon traversing layers 2–5, usually in a tight bundle consisting of varicose, radially oriented collaterals often confined to a narrow vertical column 50 μm in diameter (Somogyi and Cowey 1984). Double bouquet cells receive symmetrical, type II (presumed inhibitory) synapses, and also make type II synapses, perhaps onto the apical dendrites of pyramidal cells, so that these neurons may serve, by this double inhibitory effect, to emphasize activity within a narrow vertical column.

An overview of neocortical inhibitory neurons is shown in Fig. 1.16.

1.10.4 Quantitative aspects of cortical architecture

Some quantitative aspects of cortical architecture are described, because, although only preliminary data are available, they are crucial for developing an understanding of how the neocortex could work. Further evidence is provided by Braitenberg and Schütz (1991), and by Abeles (1991). Typical values, many of them after Abeles (1991), are shown in Table 1.1. The figures given are for a rather generalized case, and indicate the order of magnitude. The number of synapses per neuron (20,000) is an estimate for monkeys; those for humans may be closer to 40,000, and for the mouse, closer to 8,000. The number of 18,000 excitatory synapses made by a pyramidal cell is set to match the number of excitatory synapses received by pyramidal cells, for the great majority of cortical excitatory synapses are made from axons of cortical, principally pyramidal, cells.

Microanatomical studies show that pyramidal cells rarely make more than one connection with any other pyramidal cell, even when they are adjacent in the same area of the cerebral cortex. An interesting calculation takes the number of local connections made by a pyramidal cell within the approximately 1 mm of its local axonal arborization (say 9,000), and the number of pyramidal cells with dendrites in the same region, and suggests that the probability that a pyramidal cell makes a synapse with its neighbour is low, approximately 0.1 (Braitenberg and Schütz 1991, Abeles 1991) (see further Hill, Wang, Riachi, Schürmann and Markram (2012) and Markram et al. (2015)). This fits with the estimate from simultaneous recording of nearby pyramidal cells using spike-triggered averaging to monitor time-locked EPSPs (Abeles 1991, Thomson and Deuchars 1994).

Now the implication of the pyramidal cell to pyramidal cell connectivity just described is that within a cortical area of perhaps 1 mm^2, the region within which typical pyramidal cells have dendritic trees and their local axonal arborization, there is a probability of excitatory-to-excitatory cell connection of 0.1. Moreover, this population of mutually interconnected neurons is served by 'its own' population of inhibitory interneurons (which have a spatial

Table 1.1 Typical quantitative estimates for neocortex reflecting estimates in macaques. (Some of the information in this Table is adapted from M. Abeles, *Corticonics, Neural Circuits of the Cerebral Cortex*, p. 59, Table 1.5.4, Copyright ©1991, Cambridge University Press.)

Neuronal density	20,000–40,000/mm^3
Neuronal composition:	
Pyramidal	75%
Spiny stellate	10%
Inhibitory neurons, for example smooth stellate, chandelier	15%
Synaptic density	8 x 10^8/mm^3
Numbers of synapses on pyramidal cells:	
Excitatory synapses from remote sources onto each neuron	9,000
Excitatory synapses from local sources onto each neuron	9,000
Inhibitory synapses onto each neuron	2,000
Pyramidal cell dendritic length	10 mm
Number of synapses made by axons of pyramidal cells	18,000
Number of synapses on inhibitory neurons	2,000
Number of synapses made by inhibitory neurons	300
Dendritic length density	400 m/mm^3
Axonal length density	3,200 m/mm^3
Typical cortical thickness	2 mm
Cortical area	
human (assuming 3 mm for cortical thickness)	300,000 mm^2
macaque (assuming 2 mm for cortical thickness)	30,000 mm^2
rat (assuming 2 mm for cortical thickness)	300 mm^2

receiving and sending zone in the order of 1 mm^2), enabling local threshold setting and optimization of the set of neurons with 'high' (0.1) connection probability in that region. Such an architecture is effectively recurrent or re-entrant. It may be expected to show some of the properties of recurrent networks, including the fast dynamics described in Appendix B. Such fast dynamics may be facilitated by the fact that cortical neurons in the awake behaving monkey generally have a low spontaneous rate of firing (personal observations; see for example Rolls and Tovee (1995b), Rolls, Treves, Tovee and Panzeri (1997d), and Franco, Rolls, Aggelopoulos and Jerez (2007)), which means that even any small additional input may produce some spikes sooner than would otherwise have occurred, because some of the neurons may be very close to a threshold for firing. It might also show some of the autoassociative retrieval of information typical of autoassociation networks, if the synapses between the nearby pyramidal cells have the appropriate (Hebbian) modifiability. In this context, the value of 0.1 for the probability of a connection between nearby neocortical pyramidal cells is of interest, for the connection probability between hippocampal CA3 pyramidal is approximately 0.02–0.04, and this is thought to be sufficient to sustain associative retrieval (see Appendix B, Rolls and Treves (1998), and Rolls (2012a)). The role of this diluted connectivity of the recurrent excitatory synapses between nearby neocortical pyramidal cells and between hippocampal CA3 cells as underlying a principle of operation of the cerebral cortex is considered in Chapter 7.

In the neocortex, each 1 mm^2 region within which there is a relatively high density of recurrent collateral connections between pyramidal cells probably overlaps somewhat continuously with the next. This raises the issue of modules in the cortex, described by many authors as regions of the order of 1 mm^2 (with different authors giving different sizes), in which there are vertically oriented columns of neurons that may share some property (for example, responding to the same orientation of a visual stimulus), and that may be anatomically marked (for example (Powell 1981, Mountcastle 1984, Douglas, Mahowald and Martin 1996, da Costa and Martin 2010)). The anatomy just described, with the local connections between nearby (1 mm) pyramidal cells, and the local inhibitory neurons, may provide a network basis for starting to understand the columnar architecture of the neocortex, for it implies that local recurrent connectivity on this scale implementing local re-entrancy is a feature of cortical computation. We can note that the neocortex could not be a single, global, autoassociation network, because the number of memories that could be stored in an autoassociation network, rather than increasing with the number of neurons in the network, is limited by the number of recurrent connections per neuron, which is in the order of 10,000 (see Table 1.1), or less, depending on the species, as pointed out by O'Kane and Treves (1992). This would be an impossibly small capacity for the whole cortex. It is suggested that instead a principle of cortical design is that it does have in part local connectivity, so that each part can have its own processing and storage, which may be triggered by other modules, but is a distinct operation from that which occurs simultaneously in other modules (Chapter 3).

An interesting parallel between the hippocampus and any small patch of neocortex is the allocation of a set of many small excitatory (usually non-pyramidal, spiny stellate or granular) cells at the input side. In the neocortex this is layer 4, in the hippocampus the dentate gyrus (Section 18.2). In both cases, these cells receive the feedforward inputs and relay them to a population of pyramidal cells (in layers 2–3 of the neocortex and in the CA3 field of the hippocampus) which have extensive recurrent collateral connections. In both cases, the pyramidal cells receive inputs both as relayed by the preprocessing array and directly. Such analogies might indicate that the functional roles of neocortical layer 4 cells and of dentate granule cells could be partially the same (see Section 18.2).

The short-range high density of connectivity may also contribute to the formation of cortical topographic maps, as described in Section B.4.6. This may help to ensure that different parameters of the input space are represented in a nearly continuous fashion across the cortex, to the extent that the reduction in dimensionality allows it; or when preserving strict continuity is not possible, to produce the clustering of cells with similar response properties, as illustrated for example by colour 'blobs' in striate cortex, or by the local clustering of face cells in the temporal cortical visual areas (Rolls 2007e, Rolls 2008b, Rolls 2011d, Rolls 2012c).

1.10.5 Functional pathways through the cortical layers

Because of the complexity of the circuitry of the cerebral cortex, some of which is summarized in Figs. 1.17 and 1.19, there are only preliminary indications available now of how information is processed through cortical layers and between cortical areas (da Costa and Martin 2010, Harris and Mrsic-Flogel 2013, Harris and Shepherd 2015) (though see further Chapter 18). In primary sensory cortical areas, the main extrinsic 'forward' input is from the thalamus, and ends in layer 4, where synapses are formed onto spiny stellate cells. These in turn project heavily onto pyramidal cells in layers 3 and 2, which in turn send projections forward to the next cortical area. The situation is made more complex than this by the fact that the thalamic afferents also synapse onto the basal dendrites in or close to the layer 2 pyramidal cells, as well as onto layer 6 pyramidal cells and inhibitory interneurons. Given that the functional implications of this particular architecture are not fully clear, it would be of interest to examine the strength of the functional links between thalamic afferents and different classes of cortical cell using cross-correlation techniques, to determine which neurons are strongly activated by thalamic afferents with monosynaptic or polysynaptic delays. Given

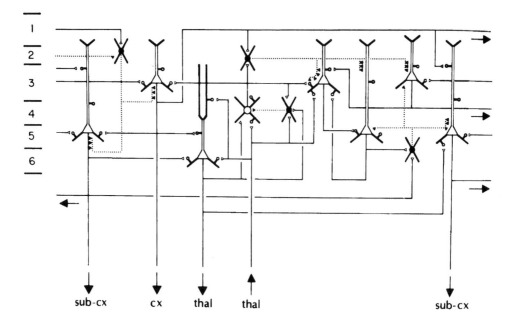

Fig. 1.17 Basic circuit for visual cortex. Excitatory neurons, which are spiny and use glutamate as a transmitter, and include the pyramidal and spiny stellate cells, are indicated by open somata; their axons are indicated by solid lines, and their synaptic boutons by open symbols. Inhibitory (smooth, GABAergic) neurons are indicated by black (filled) somata; their axons are indicated by dotted lines, and their synaptic boutons by solid symbols. thal, thalamus; cx, cortex; sub-cx, subcortex. Cortical layers 1–6 are as indicated. ((This material has been adapted from Rodney Douglas, Henry Markram and Kevin Martin, Neocortex, In Gordon M. Shepherd (ed), *The Synaptic Organization of the Brain* 5e, pp. 499–558, figure 12.15.3 (c) 2004, Oxford University Press and has been reproduced by permission of Oxford University Press https://global.oup.com/academic/product/the-synaptic-organization-of-the-brain-9780195159561?q=the synaptic organization of the brain&lang=en&cc=gb. For permission to reuse this material, please visit http://www.oup.co.uk/academic/rights/permissions.)

that this is technically difficult, an alternative approach has been to use electrical stimulation of the thalamic afferents to classify cortical neurons as mono- or poly-synaptically driven, then to examine the response properties of the neuron to physiological (visual) inputs, and finally to fill the cell with horseradish peroxidase so that its full structure can be studied (see for example Martin (1984)). Using these techniques, it has been shown in the cat visual cortex that spiny stellate cells can indeed be driven monosynaptically by thalamic afferents to the cortex. Further, many of these neurons have S-type receptive fields, that is they have distinct on and off regions of the receptive field, and respond with orientation tuning to elongated visual stimuli (Martin 1984) (see Rolls and Deco (2002)). Further, consistent with the anatomy just described, pyramidal cells in the deep part of layer 3, and in layer 6, could also be monosynaptically activated by thalamic afferents, and had S-type receptive fields (Martin 1984). Also consistent with the anatomy just described, pyramidal cells in layer 2 were di- (or poly-) synaptically activated by stimulation of the afferents from the thalamus, but also had S-type receptive fields.

In contrast to these 'core' cortico-thalamic inputs to primary sensory areas which carry

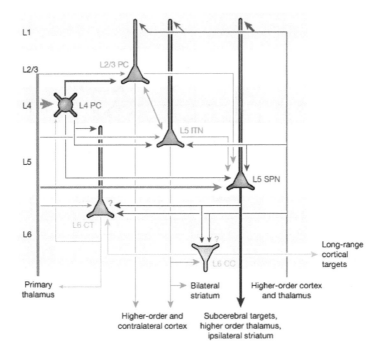

Fig. 1.18 Canonical connectivity of cortical principal cells. Line thickness represents the strength of a pathway.

Sensory information arrives from primary thalamus into all cortical layers, but most densely into L4 and the L5–L6 border. Contextual inputs from higher-order cortex and thalamus most densely target L1, L5 and L6, but avoid L4.

L4 principal cells comprise two morphological classes, pyramidal and spiny stellate cells. L4 principal cells project to all layers, but most strongly L2/3. However they receive little intracolumnar input in return.

L2/3 principal cells send outputs L5, and to the next cortical area in the hierarchy L4.

Upper L5 (L5A) 'Intratelencephalic neurons' project locally upward to L2/3 and to the striatum and send backprojections to the preceding cortical area in the hierarchy. They show firing rate adaptation.

Lower L5 (L5B) 'Subcerebral projection neurons' (SPNs) are larger cells with prominent dendritic tufts in L1. They project to subcerebral motor centres for example via the pyramidal tract, and send collaterals to the striatum and higher-order thalamus with large, strong 'driver' synapses. They show little adaptation, and may fire in bursts.

L6 Corticocortical cells (CCs) have small dendritic trees, and long-range horizontal axons.

L6 Corticothalamic cells (CTs) send projections to the thalamus which, unlike those of L5 SPNs, are weak, and target the reticular and primary sensory thalamic nuclei. Corticothalamic cells also project to cortical layer 4, where they strongly target interneurons, as well as hyperpolarizing principal cells via group II mGluRs. Consistent with this connectivity, optogenetic stimulation of L6 in vivo suppresses cortical activity, suggesting a role of this layer in gain control or translaminar inhibition. (Reprinted by permission from Macmillan Publishers Ltd: Nature, Harris, K. D. and Mrsic-Flogel, T. D., Cortical connectivity and sensory coding 503: 51–58, Copyright ©2013.)

the main input, 'matrix' cortico-thalamic inputs from higher-order thalamic nuclei project to non-primary cortical areas, further up in the hierarchy (Harris and Shepherd 2015). In higher cortical areas, the main inputs come instead from the preceding cortical area.

Inputs could reach the layer 5 pyramidal cells from the pyramidal cells in layers 2 and 3, the axons of which ramify extensively in layer 5, in which the layer 5 pyramidal cells have widespread basal dendrites (see Fig. 1.2), and also perhaps from thalamic afferents. Many layer 5 pyramidal cells are di- or trisynaptically activated by stimulation of the thalamic afferents, consistent with them receiving inputs from monosynaptically activated deep layer 3 pyramidal cells, or from disynaptically activated pyramidal cells in layer 2 and upper layer 3 (Martin 1984). Upper L5 (L5A) 'Intratelencephalic neurons' project locally upward to L2/3 and to the striatum and send backprojections to the preceding cortical area in the hierarchy. They show firing rate adaptation. Lower L5 (L5B) 'Subcerebral projection neurons' (SPNs) are larger cells with prominent dendritic tufts in L1. They project to subcerebral motor centres for example via the pyramidal tract, and send collaterals to the striatum and higher-order thalamus with large, strong 'driver' synapses (Harris and Mrsic-Flogel 2013, Harris and Shepherd 2015). They may also provide some cortico-cortical backprojections, which terminate in superficial layers of the cerebral cortex, including layer 1 (see Fig. 11.1). They show little adaptation, and may fire in bursts. These neurons may have a less sparse representation than those in L2/3, with this representation being useful for high information transmission. (In contrast, the more sparse representation in L2/L3 pyramidal cells may be useful when information is being stored in the recurrent collaterals, for this maximizes the number of patterns that can be stored and correctly retrieved. This may be one reason why the cerebral neocortex has superficial pyramidal cell layers (L2/L3) which are separate from the deep pyramidal cell layers (L5/L6) (see further Chapter 18).

L6 Corticocortical cells (CCs) have small dendritic trees, and long-range horizontal axons. L6 Corticothalamic cells (CTs) send projections to the thalamus which, unlike those of L5 SPNs, are weak, and target the reticular and primary sensory thalamic nuclei. Corticothalamic cells also project to cortical layer 4, where they strongly target interneurons, as well as hyper-polarizing principal cells via group II mGluRs. Consistent with this connectivity, optogenetic stimulation of L6 in vivo suppresses cortical activity, suggesting a role of this layer in gain control or translaminar inhibition (Harris and Mrsic-Flogel 2013).

Studies on the function of inhibitory pathways in the cortex are also beginning. The fact that basket cells often receive strong thalamic inputs, and that they terminate on pyramidal cell bodies where part of their action is to shunt the membrane, suggests that they act in part as a feedforward inhibitory system that normalizes the thalamic influence on pyramidal cells by dividing their response in proportion to the average of the thalamic input received (see Appendix B). The smaller and numerous smooth (or sparsely spiny) non-pyramidal cells that are inhibitory may receive inputs from pyramidal cells as well as inhibit them, so that these neurons could perform the very important function of recurrent or feedback inhibition (see Appendix B). It is only feedback inhibition that can take into account not only the inputs received by an area of cortex, but also the effects that these inputs have, once multiplied by the synaptic weight vector on each neuron, so that recurrent inhibition is necessary for competition and contrast enhancement (see Appendix B).

Another way in which the role of inhibition in the cortex can be analyzed is by applying a drug such as bicuculline using iontophoresis (which blocks GABA receptors to a single neuron), while examining the response properties of the neuron (see Sillito (1984)). With this technique, it has been shown that in the visual cortex of the cat, layer 4 simple cells lose their orientation and directional selectivity. Similar effects are observed in some complex cells, but the selectivity of other complex cells may be less affected by blocking the effect of endogenously released GABA in this way (Sillito 1984). One possible reason for this is that the inputs to complex cells must often synapse onto the dendrites far from the cell body, and distant synapses will probably be unaffected by the GABA receptor blocker released near the cell body. The experiments reveal that inhibition is very important for the normal selectivity of many visual cortex neurons for orientation and the direction of movement. Many of the cells displayed almost no orientation selectivity without inhibition. This implies that not only is the inhibition important for maintaining the neuron on an appropriate part of its activation function, but also that lateral inhibition between neurons is important because it allows the

Area A   Area B

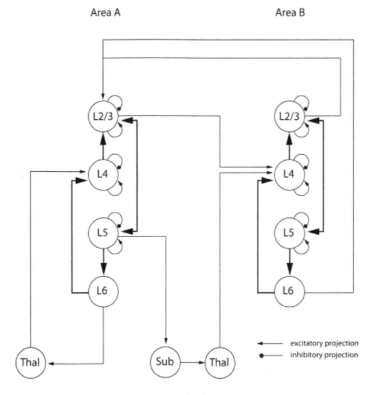

Fig. 1.19 Representation of the major connections in the canonical microcircuit (adapted from Douglas et al., 1989; Douglas and Martin, 1991, 2004). Excitatory connections are represented by arrows and inhibitory ones as lines with round ends. Neurons from different cortical layers or brain structures are represented as circles. 'Lx' designates the cortical layer where the cell body is located, 'Thal' designates the thalamus and 'Sub' designates other subcortical structures. (Adapted from Rodney J. Douglas, Kevan A.C. Martin, David Whitteridge, A Canonical Microcircuit for Neocortex, *Neural Computation*, 1989, 1 (4), MIT Press Journals, pp. 480–488. Reprinted by permission of MIT Press Journals.)

responses of a single neuron (which need not be markedly biased by its excitatory input) to have its responsiveness set by the activity of neighbouring neurons (see Appendix B).

1.10.6 The scale of lateral excitatory and inhibitory effects, and the concept of modules

The forward cortico-cortical afferents to a cortical area sometimes have a columnar pattern to their distribution, with the column width 200–300 μm in diameter (see Eccles (1984)). Similarly, individual thalamo-cortical axons often end in patches in layer 4 which are 200–300 μm in diameter (Martin 1984). The dendrites of spiny stellate cells are in the region of 500 μm in diameter, and their axons can distribute in patches 200–300 μm across, separated by distances of up to 1 mm (Martin 1984). The dendrites of layer 2 and 3 pyramidal cells can be approximately 300 μm in diameter, but after this the relatively narrow column appears to become less important, for the axons of the superficial pyramidal cells can distribute over 1 mm or more, both in layers 2 and 3, and in layer 5 (Martin 1984). Other neurons that may contribute to the maintenance of processing in relatively narrow columns are the double bouquet cells, which because they receive inhibitory inputs, and themselves produce inhibition, all within a column perhaps 50 μm across (see above), would tend to enhance local excitation. The bipolar cells, which form excitatory synapses with pyramidal cells, may also serve to emphasize activity within a narrow vertical column approximately 50 μm across. These two mechanisms for enhancing local excitation operate against a much broader-ranging set of lateral inhibitory processes, and could it is suggested have the effect of increasing contrast between the firing rates of pyramidal cells 50 μm apart, and thus be very important in competitive interactions between pyramidal cells. Indeed, the lateral inhibitory effects are broader than the excitatory effects described so far, in that for example the axons of basket cells spread laterally 500 μm or more (see above) (although those of the small, smooth non-pyramidal cells are closer to 300 μm – see Peters and Saint Marie (1984)). Such short-range local excitatory interactions with longer range inhibition not only provide for contrast enhancement and for competitive interactions, but also can result in the formation of maps in which neurons with similar responses are grouped together and neurons with dissimilar response are more widely separated (see Appendix B). Thus these local interactions are consistent with the possibilities that cortical pyramidal cells form a competitive network (see Appendix B and below), and that cortical maps are formed at least partly as a result of local interactions of this kind in a competitive network (see Section B.4.6).

The type of genetic specification that could provide the fundamental connectivity rules between cortical areas, which would then self-organize the details of the exact synaptic connectivity, have been considered by Rolls and Stringer (2000). They compared the connectivity of different cortical areas, thereby suggested a set of rules that the genes might be specifying, and then simulated using genetic algorithms the selection of the appropriate rules for solving particular types of computational problem, including pattern association memory, autoassociation memory, and competitive learning (see Chapter 19).

In contrast to the relatively localized terminal distributions of forward cortico-cortical and thalamo-cortical afferents, the cortico-cortical backward projections that end in layer 1 have a much wider horizontal distribution, of up to several millimetres (mm). It is suggested below that this enables the backward-projecting neurons to search over a larger number of pyramidal cells in the preceding cortical area for activity that is conjunctive with their own (Chapters 11 and 24).

1.11 Highlights

Introductory, background, information about the following aspects of cortical function are provided as follows:

1. Neurons, synapses, and the synaptic modification used in learning (which is extended in Chapter 9).
2. The types of coding of information by neuronal responses, including local (grandmother-cell), distributed, and sparse distributed, which is extended in Chapter 8 and Appendix C.
3. An introduction to pattern association networks, autoassociation networks with attractor properties, and competitive networks.
4. A description of some major processing pathways in the cerebral cortex, which is developed for two key systems, the episodic memory system in Chapter 24 and the visual object recognition system in Chapter 25. These two Chapters illustrate how many of the principles are combined to produce what are essentially theories of how the episodic memory and visual object recognition cortical systems operate.
5. A description of some of the properties of cortical neuronal architecture, which is extended in Section 18.2.

2 Hierarchical organization

2.1 Introduction

The hierarchical organization of the cerebral neocortex is one of its key computational principles. This enables computations to be divided into a series of stages, with many advantages. One is that the fan in and fan out of neurons can be kept manageable, in the order of 20,000 connections. (Neocortical pyramidal cells often have in the order of 20,000 excitatory synapses, as shown in Table 1.1. Presumably neurons with many more synapses than this would run into problems such as the difficulty of summating over so many small currents injected over such a large dendritic surface, and the effects of noise inherent in the ion channels themselves (Faisal, Selen and Wolpert 2008).)

A second is that the same or a similar architecture and process can be repeated a number of times, which simplifies the evolution of the neocortex, which can proceed by adding another stage on to the top of an existing hierarchy.

A third advantage is that this simplifies the genetic specification of the cortex, in that once a useful and stable cortical network in one area has evolved, it is safest and quickest in evolution to add something with a similar specification as the next stage (Chapter 19).

A fourth advantage is that by dividing a computation into successive stages, each stage can perform a much simpler computation.

A fifth advantage is that within any one stage of the hierarchy performing a particular computation, the information that needs to be exchanged between the nearby neurons is of the same type (e.g. about the presence or not of a nearby edge in vision). That facilitates cortical design, in that generic connectivity between nearby neurons can be specified with self-organization of neuronal selectivity during development. This obviates the need for a neuron with a particular type of response to the input to have connections with another neuron with another particular type of response, which would be impossible for genetic specification, given that we have only in the order of 20,000 – 25,000 protein coding genes (Human Genome Sequencing 2004). Hypotheses of what the genes may specify to implement the design of any one cortical area are described in Chapter 19. The localization of function that is present can be viewed partly in this way, and has the effect of minimizing the average length of the connections between neurons, which is very important in keeping brain size down (see further Chapter 3 and Sections B.4.6 and D.4).

A sixth advantage and property of hierarchical cortical organization is that there are major connections between adjacent areas in the hierarchy, which again simplifies the genetic specification of cortical design, in that a neuron does not need to have its connectivity specified to one of a myriad of cortical areas. The principle of hierarchical organization of the neocortex, and its advantages, are considered in this Chapter. (The term level is used in this Chapter to refer to how far one is up the cortical hierarchy of connected levels, to help avoid confusion with the layers of the neocortex from the surface down found in every cortical area.) Graph-theoretic approaches to cortical connectivity do reflect this hierarchical organization (Markov, Ercsey-Ravasz, Ribeiro Gomes, Lamy, Magrou, Vezoli, Misery, Falchier, Quilodran, Gariel, Sallet, Gamanut, Huissoud, Clavagnier, Giroud, Sappey-Marinier, Barone, Dehay, Toroczkai, Knoblauch, Van Essen and Kennedy 2014a, Markov and Kennedy 2013, Markov, Vezoli,

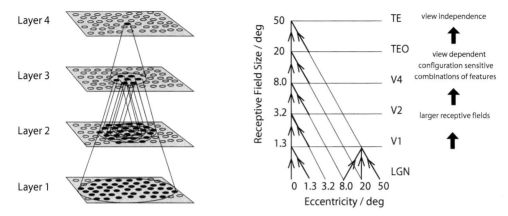

Fig. 2.1 Convergence in the visual system. Right – as it occurs in the brain. V1, visual cortex area V1; TEO, posterior inferior temporal cortex; TE, inferior temporal cortex (IT). Left – as implemented in VisNet. Convergence through the network is designed to provide fourth layer neurons with information from across the entire input retina.

Chameau, Falchier, Quilodran, Huissoud, Lamy, Misery, Giroud, Ullman, Barone, Dehay, Knoblauch and Kennedy 2014b, Markov, Ercsey-Ravasz, Van Essen, Knoblauch, Toroczkai and Kennedy 2013b).

There is an excellent body of evidence from neurology on hierarchical processing in the brain (Hughlings Jackson 1878, Kennard and Swash 1989, Mesulam 1998). In this Chapter, the focus is on the cerebral cortex, and in particular on the computational advantages of hierarchical organization.

2.2   Hierarchical organization in sensory systems

All sensory systems in the cortex have a hierarchical cortical organization. We start by considering the visual system, as this enables elucidation of some of the advantages of hierarchical organization.

2.2.1   Hierarchical organization in the ventral visual system

A schematic diagram to indicate some aspects of the processing involved in object identification from the primary visual cortex, V1, through V2 and V4 to the posterior inferior temporal cortex (TEO) and the anterior inferior temporal cortex (TE) is shown in Fig. 2.1. The approximate location of these visual cortical areas on the brain of a macaque monkey is shown in Figs. 25.2 and 1.10, which also show that TE has a number of different subdivisions. Fuller details of the processing performed in the ventral visual stream and how the computations are performed are provided in Chapter 25, but here we focus on the principles of the hierarchical organization.

One notable property of the architecture illustrated in Fig. 2.1 is that a neuron in any one stage receives from a small region of the preceding stage. This enables the number of connections received by any neuron to be kept to a reasonable number for a neuron, in the order of 10,000, and for the information received by any neuron to be from neurons in the preceding stage with similar properties. In early stages of the visual system, this might reflect

Top layer

Intermediate
layers

Input (V1)

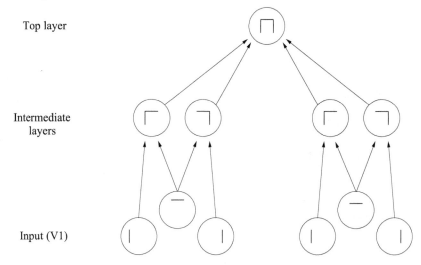

Fig. 2.2 The feature hierarchy approach to object recognition. The inputs may be neurons tuned to oriented straight line segments. In early intermediate levels neurons respond to a combination of these inputs in the correct spatial position with respect to each other. In further intermediate levels, of which there may be several, neurons respond with some invariance to the feature combinations represented early, and form higher order feature combinations. Finally, in the top level, neurons respond to combinations of what is represented in the preceding intermediate level, and thus provide evidence about objects in a position (and scale and even view) invariant way. Convergence through the network is designed to provide top level neurons with information from across the entire input retina, as part of the solution to translation invariance, and other types of invariance are treated similarly.

information from the same part of the visual field. At later stages, this might reflect neurons with similar types of tuning, for example to faces, or to non-face objects.

An important point is that the *topographical maps* that self-organize in cortical regions (see Chapter 3, and Sections B.4.6 and D.4) help this design for the connectivity, in that they help to promote the receipt of information about similar types of stimuli from the preceding cortical stage, using just the principle of local convergence. This enables, for example, neurons in the anterior inferior temporal cortex that compute differences between faces, and invariant properties of faces such as translation invariance, to receive information from patches in the preceding posterior inferior temporal visual cortex in which neurons respond to faces, but not from regions where the neurons respond to objects. This helps to ensure that relevant information about a type of computation, in this cases about faces, is available to neurons in a given level of the hierarchy, without having to 'waste' too many input connections on inputs from non-face neurons that have little relevance to the computations about faces.

Part of the way in which hierarchical organization helps to solve difficult computational problems can be illustrated by the *feature hierarchy* approach to 2D shape and object encoding illustrated in Fig. 2.2.

In this approach, the system starts with some low-level description of the visual scene, in terms for example of oriented straight line segments of the type that are represented in the responses of primary visual cortex (V1) neurons and are built hierarchically (Hubel and Wiesel 1962, Miller 2016) (Fig. 1.1), and then builds, in repeated hierarchical levels, features based on what is represented in previous levels. A feature may thus be defined as a combination of what is represented in the previous level. For example, after V1, features might consist of combinations of straight lines, which might represent longer curved lines (Zucker, Dobbins and Iverson 1989), or terminated lines (in fact represented in V1 as end-stopped cells), corners, 'T' junctions which are characteristic of obscuring edges, and (at least in humans) the arrow and 'Y' vertices which are characteristic properties of human environments. Evidence that such feature combination neurons are present in V2 is that some neurons respond to combinations of line elements that join at different angles (Hegde and Van Essen 2000, Ito and Komatsu 2004). (An example of this might be a neuron responding to a 'V' shape at a particular orientation.) As one ascends the hierarchy, neurons might respond to more complex trigger features (such as two parts of a complex figure in the correct spatial arrangement with respect to each other, as shown by Tanaka (1996) for V4 and posterior inferior temporal cortex neurons). In the (anterior) inferior temporal visual cortex, we showed that some face-selective neurons respond to combinations of features such as eyes, mouth, and hair, but not to these components presented individually, or in a configuration with the parts rearranged (Perrett, Rolls and Caan 1982, Rolls, Tovee, Purcell, Stewart and Azzopardi 1994b), and this is captured by hierarchical competitive networks (Robinson and Rolls 2015). Further on, neurons might respond to combinations of several such intermediate-level feature combination neurons, and thus come to respond systematically differently to different objects, and thus to convey information about which object is present. This approach received neurophysiological support early on from the results of Hubel and Wiesel (1962, 1968) in the cat and monkey, and much of the data described in Chapter 5 of Rolls and Deco (2002) are consistent with this scheme.

One advantage of feature hierarchy systems is that they can map the whole of an input space (in this case the retina) to the whole of the output space (in this case anterior inferior temporal visual cortex) without needing neurons with millions and millions of synaptic inputs. Consider the problem that needs to be solved. A particular face seen over a wide range of positions on the retina will activate the same neuron in the anterior inferior temporal cortex, as described in Chapter 25. If this were to be achieved in a one-level system, without a hierarchy, then a neuron in the second level would need to have connections to every neuron in level 1 that might be involved in the representation of that particular face image at every possible position at which the face might be placed on level 1. That involves a combinatorial explosion in the number of connections that every neuron in level 2 would need to the enormous numbers of neurons in level 1. Instead, a hierarchical approach divides the problem into several stages, in which any one neuron at a given stage responds to just a limited set of combinations of neuronal firing over a small region of the previous level. This enables a single neuron in the top level to respond to a combination of inputs from any part of the input space, as illustrated in Fig. 2.1. Moreover, if the neurons at any one level self-organize by a process such as competitive learning (Section B.4), then each neuron needs to connect to only a limited number of neurons in the preceding level (e.g. 10,000), and the competition helps to ensure that different neurons within a local cortical region within which lateral inhibition operates through the inhibitory interneurons learn to respond to different combinations of features in the input space. Moreover, not every combination of inputs needs to be encoded by neurons, but only those that occur during self-organization given the statistics of the world. For example, natural image statistics reflect the presence of structure in the input space, in for example edges in a scene, so that not every possible combination of pixels in the input space need to be encoded, for vision in a world with natural image statistics.

A second advantage is that the feature analyzers can be built out of the rather simple competitive networks described in Section B.4 which use a local learning rule, and have no external teacher, so that they are rather biologically plausible. Another advantage is that, once trained at lower levels on sets of features common to most objects, the system can then learn new objects quickly using only its higher levels (Elliffe, Rolls and Stringer 2002).

A related third advantage is that, if implemented with competitive nets as in the case of VisNet (see Section 25.5), then neurons are allocated by self-organization to represent just the features present in the natural statistics of real images (cf. Field (1994)), and not every possible feature that could be constructed by random combinations of pixels on the retina.

A related fourth advantage of feature hierarchy networks is that because they can utilize competitive networks, they can still produce a good estimate of what is in the image under non-ideal conditions, when only parts of objects are visible because for example of occlusion by other objects, etc. The reasons for this are that competitive networks assess the evidence for the presence of certain 'features' to which they are tuned using a dot product operation on their inputs, so that they are inherently tolerant of missing input evidence; and reach a state that reflects the best hypothesis or hypotheses (with soft competition) given the whole set of inputs, because there are competitive interactions between the different neurons (see Section B.4).

A fifth advantage of a feature hierarchy system is that, as shown in Section 25.5.6, the system does not need to perform segmentation into objects as part of pre-processing, nor does it need to be able to identify parts of an object, and can also operate in cluttered scenes in which the object may be partially obscured. The reason for this is that once trained on objects, the system then operates somewhat like an associative memory, mapping the image properties forward onto whatever it has learned about before, and then by competition selecting just the most likely output to be activated. Indeed, the feature hierarchy approach provides a mechanism by which processing at the object recognition level could feed back using backprojections to early cortical areas to provide top-down guidance to assist segmentation (Section 11.2). Although backprojections are not built into VisNet2 (Rolls and Milward 2000), they have been added when attentional top-down processing must be incorporated (Deco and Rolls 2004), are present in the brain, and are incorporated into the models described in Section 11.5.

A sixth advantage of feature hierarchy systems is that they can naturally utilize features in the images of objects that are not strictly part of a shape description scheme, such as the fact that different objects have different textures, colours etc. Feature hierarchy systems, because they utilize whatever is represented at earlier stages in forming feature combination neurons at the next stage, naturally incorporate such 'feature list' evidence into their analysis, and have the advantages of that approach (see Section 25.3.1 and also Mel (1997)). Indeed, the feature space approach can utilize a hybrid representation, some of whose dimensions may be discrete and defined in structural terms, while other dimensions may be continuous and defined in terms of metric details (Biederman 1987), and others may be concerned with non-shape properties such as texture and colour (cf. Edelman (1999)).

A seventh advantage of feature hierarchy systems is that they do not need to utilize 'on the fly' or run-time arbitrary binding of features. Instead, the spatial syntax is effectively hardwired into the system when it is trained, in that the feature combination neurons have learned to respond to their set of features when they are in a given spatial arrangement on the retina.

An eighth advantage of feature hierarchy systems is that they can self-organize (given the right functional architecture, trace synaptic learning rule, and the temporal statistics of the normal visual input from the world), with no need for an external teacher to specify that the neurons must learn to respond to objects (see Chapter 25). The correct, object, representation self-organizes itself given rather economically specified genetic rules for building the network (Rolls and Stringer 2000) (Chapter 19).

Ninth, hierarchical visual systems may recognize 3D objects based on a limited set of 2D views of objects, and the same architectural rules just stated and implemented in VisNet (see Chapter 25) will correctly associate together the different views of an object. It is part of the concept, and consistent with neurophysiological data (Tanaka 1996), that the neurons in the upper levels will generalize correctly within a view (see Section 25.5.7).

Tenth, another advantage of cortical feature hierarchy systems is that they can operate fast, in approximately 20 ms per level (see Sections B.4 and B.6.5).

Eleventh, although hierarchical cortical systems so far have been described in a feed-forward mode of operation, they are completely compatible with backprojections, again connected mainly between adjacent levels, to implement top-down effects of attention and cognition, and to implement memory recall, as described in Chapter 11. Again, the connectivity requirements in terms of the number of input and output connections of each neuron are met best by having backprojection connections mainly between adjacent levels of the backprojection hierarchy. As described in Chapter 11 the backprojections must not dominate the bottom-up (feedforward) connections, and therefore are relatively weak, which is facilitated by their termination in layer 1 of the cortex, well away from the cell body, and where the effects can be shunted if there are strong bottom-up inputs.

Twelfth, hierarchical cortical systems are also consistent with local recurrent collateral connections within a small region within a level of the hierarchy, to implement memory and constraint satisfaction functions as described in Chapter 4.

Thirteenth, some artificial neural networks are invertible (Hinton, Dayan, Frey and Neal 1995, Hinton and Ghahramani 1997), and operate by minimizing an error between an input representation and a reconstructed input representation based on a forward pass to the top of the hierarchy, and a backward pass down the hierarchy. There are a number of reasons why this seems implausible for cortical architecture. One is that information appears to be lost in the forward pass in the neocortex, for example in computing position and view invariant representations at the top of the network, for use in subsequent stages of processing. That makes it unlikely that what is represented at the top of a cortical network could be used to reconstruct the input. Another problem is how an error-correction synaptic modification rule might be computed and applied to the appropriate synapses of the neurons at every stage of the cortical architecture. A third difficulty is that if the backprojections are used for top-down attention and recall, as seems to be the case (see Chapter 11), then associative modifiability of the top-down backprojections is appropriate, and may not be compatible with invertible networks.

A number of problems need to be solved for such feature hierarchy systems to provide a useful model of, for example, object recognition in the primate visual system.

First, some way needs to be found to keep the number of feature combination neurons realistic at each stage, without undergoing a combinatorial explosion. If a separate feature combination neuron was needed to code for every possible combination of n types of feature each with a resolution of 2 levels (binary encoding) in the preceding stage, then 2^n neurons would be needed. The suggestion that is made in Section 25.4 is that by forming neurons that respond to low-order combinations of features (neurons that respond to a combination of only say 2–4 features from the preceding stage), the number of actual feature analyzing neurons can be kept within reasonable numbers. By reasonable I mean the number of neurons actually found at any one stage of the visual system, which, for V4 might be in the order of 60×10^6 neurons (assuming a volume for macaque V4 of approximately 2,000 mm^3, and a cell density of 20,000–40,000 neurons per mm^3, see Table 1.1). This is certainly a large number; but the fact that a large number of neurons is present at each stage of the primate visual system is in fact consistent with the hypothesis that feature combination neurons are part of the way in which the brain solves object recognition.

Another factor that also helps to keep the number of neurons under control is the statistics of the visual world, which contain great redundancies. The world is not random, and indeed the statistics of natural images are such that many regularities are present (Field 1994), and not every possible combination of pixels on the retina needs to be separately encoded.

A third factor that helps to keep the number of connections required onto each neuron under control is that in a multilevel hierarchy each neuron can be set up to receive connections from only a small region of the preceding level. Thus an individual neuron does not need to have connections from all the neurons in the preceding level. Over multiple levels, the required convergence can be produced so that the same neurons in the top level can be activated by an image of an effective object anywhere on the retina (see Fig. 2.1).

A second problem of feature hierarchy approaches is how to map all the different possible images of an individual object through to the same set of neurons in the top level by modifying the synaptic connections (see Fig. 2.1). The solution discussed in Sections 25.4, 25.5.1.1 and 25.5.4 is the use of a synaptic modification rule with a short-term memory trace of the previous activity of the neuron, to enable it to learn to respond to the now transformed version of what was seen very recently, which, given the statistics of looking at the visual world, will probably be an input from the same object.

A third problem of feature hierarchy approaches is how they can learn in just a few seconds of inspection of an object to recognize it in different transforms, for example in different positions on the retina in which it may never have been presented during training. A solution to this problem is provided in Section 25.5.5, in which it is shown that this can be a natural property of feature hierarchy object recognition systems, if they are trained first for all locations on the intermediate level feature combinations of which new objects will simply be a new combination, and therefore require learning only in the upper levels of the hierarchy.

A fourth potential problem of feature hierarchy systems is that when solving translation invariance they need to respond to the same local spatial arrangement of features (which are needed to specify the object), but to ignore the global position of the whole object. It is shown in Section 25.5.5 that feature hierarchy systems can solve this problem by forming feature combination neurons at an early stage of processing (e.g. V1 or V2 in the brain) that respond with high spatial precision to the local arrangement of features. Such neurons would respond differently for example to L, +, and T if they receive inputs from two line-responding neurons. It is shown in Section 25.5.5 that at later levels of the hierarchy, where some of the intermediate level feature combination neurons are starting to show translation invariance, then correct object recognition may still occur because only one object contains just those sets of intermediate level neurons in which the spatial representation of the features is inherent in the encoding.

2.2.2   Hierarchical organization in the dorsal visual system

An overview of the hierarchical organization of the dorsal visual stream is provided in Section 1.9 and Figs. 1.11 and 1.12 (Perry and Fallah 2014). Again, there are strong connections between adjacent levels of the hierarchy (see further Pandya, Seltzer, Petrides and Cipolloni (2015)). In the dorsal visual system, the hierarchical organization again allows computations to become tractable.

One example of what the convergence from level to level enables in the dorsal visual system is provided by the computation of global motion. Motion-sensitive neurons in V1 have small receptive fields (in the range 1–2 deg at the fovea), and can therefore not detect global motion, and this is part of the aperture problem (Rust, Mante, Simoncelli and Movshon 2006). (The aperture afforded by the receptive field size in V1 is insufficiently large to enable an average of motion to be computed, from for example random dot motion to which has been added drift, with a real-world example being snowflakes whirling in the wind, and not necessarily being seen to fall if only a few snowflakes can be seen.) Neurons in MT, which receives inputs from V1 and V2 (see Fig. 1.11), have larger receptive fields (e.g. 5 degrees at the fovea), and are able to respond to planar global motion, such as a field of small dots in which the majority (in practice as few as 55%) move in one direction, or to the overall direction of a moving plaid,

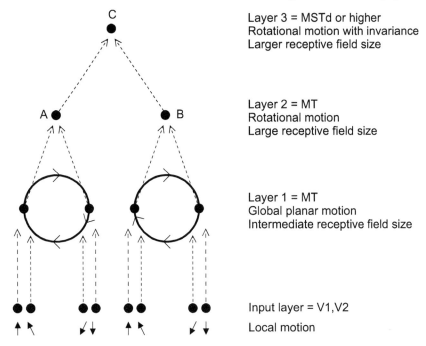

Layer 3 = MSTd or higher
Rotational motion with invariance
Larger receptive field size

Layer 2 = MT
Rotational motion
Large receptive field size

Layer 1 = MT
Global planar motion
Intermediate receptive field size

Input layer = V1,V2
Local motion

Fig. 2.3 Two rotating wheels at different locations rotating in opposite directions. The local flow field is ambiguous. Clockwise or counterclockwise rotation can only be diagnosed by a global flow computation, and it is shown how the network (VisNet) solved the problem to produce position invariant global motion-sensitive neurons. One rotating wheel is presented at any one time, but the need is to develop a representation of the fact that in the case shown the rotating flow field is always clockwise, independently of the location of the flow field. (Adapted from Edmund T. Rolls, Simon M. Stringer, Invariant Global Motion Recognition in the Dorsal Visual System: A Unifying Theory, *Neural Computation*, 2007, 19 (1), MIT Press Journals, pp. 139–169. Reprinted by permission of MIT Press Journals.)

the orthogonal grating components of which have motion at 45 degrees to the overall motion (Movshon, Adelson, Gizzi and Newsome 1985, Newsome, Britten and Movshon 1989). Even higher visual areas must be involved to account for human global motion perception (Hedges, Gartshteyn, Kohn, Rust, Shadlen, Newsome and Movshon 2011).

The hierarchical organization of the dorsal visual system with repeated convergence from level to level can also be seen to be important in analyzing object motion. Indeed, a key issue in understanding the cortical mechanisms that underlie motion perception is how we perceive the motion of objects such as a rotating wheel invariantly with respect to position on the retina, and size. For example, we perceive the wheel shown in Fig. 2.3 rotating clockwise independently of its position on the retina. This occurs even though the local motion for the wheels in the different positions may be opposite. How could this invariance of the visual motion perception of objects arise in the visual system? Invariant motion representations are known to be developed in the cortical dorsal visual system. Further on in the dorsal visual system, some neurons in macaque visual area MST (but not MT) respond to rotating flow fields or looming with considerable translation invariance (Graziano, Andersen and Snowden 1994, Geesaman and Andersen 1996, Perry and Fallah 2014). In the cortex in the anterior part of the superior temporal sulcus, which is a convergence zone for inputs from the ventral and dorsal visual systems, some neurons respond to object-based motion, for example to a head rotating clockwise but not anticlockwise, independently of whether the head is upright or inverted which reverses the optic flow across the retina (Hasselmo, Rolls, Baylis and Nalwa 1989b).

In a unifying hypothesis with the design of the ventral cortical visual system, Rolls and Stringer (2007) proposed that the dorsal visual system uses a hierarchical feedforward network architecture (V1, V2, MT, MSTd, parietal cortex) with training of the connections with a short-term memory trace associative synaptic modification rule to capture what is invariant at each stage. The principle is illustrated in Fig. 2.3a. Simulations showed that the proposal is computationally feasible, in that invariant representations of the motion flow fields produced by objects self-organize in the later layers of the architecture (see examples in Fig. 2.3b–e). The model produces invariant representations of the motion flow fields produced by global in-plane motion of an object, in-plane rotational motion, and looming vs receding of the object. The model also produces invariant representations of object-based rotation about a principal axis, of the type illustrated in Fig. 25.8 on page 564. Thus it is proposed that the dorsal and ventral visual systems may share some unifying computational principles (Rolls and Stringer 2007). Indeed, the simulations of Rolls and Stringer (2007) used a standard version of VisNet, with the exception that instead of using oriented bar receptive fields as the input to the first layer (Gabor filters), local motion flow fields provided the inputs.

Although many other types of computation take place in the dorsal visual system, this example helps to elucidate operations that are facilitated by the hierarchical organization with convergence from level to level. What should be noted is that with the large numbers of different areas in the dorsal visual system, the principle of having different areas specialized for different types of computation, such as saccade-related activity vs slow drift-related activity is part of the principle of having neurons that are involved in the same type of computation close together, in the same cortical area. This is important to the efficiency of the computation, to minimizing wiring length between the relevant neurons, and to simplicity of cortical specification, as described earlier in this Chapter.

2.2.3   Hierarchical organization of taste processing

In addition to the two visual pathways just considered, some of the cortical processing pathways for taste and smell are shown schematically in Fig. 1.9, and in more detail in Fig. 2.4. The analysis that follows is based on processing in primates (including humans), for the pathways involved, and the whole operation of the systems, is very different in rodents, as described in Chapter 18 and by Rolls (2015e).

2.2.3.1   Taste - insular primary taste cortex

First, taste is considered. The cortical hierarchy for taste can be thought of as the primary taste cortex in the anterior insula, the secondary taste cortex in the orbitofrontal cortex, and the tertiary taste cortex in the anterior cingulate cortex (Rolls 2014a, Rolls 2015e). The first central synapse of the gustatory system is in the rostral part of the nucleus of the solitary tract (Beckstead and Norgren 1979, Beckstead, Morse and Norgren 1980, Scott, Yaxley, Sienkiewicz and Rolls 1986b, Yaxley, Rolls, Sienkiewicz and Scott 1985), and this projects via the thalamic taste relay to the primary taste cortex in the anterior insula and adjoining frontal operculum (Pritchard, Hamilton, Morse and Norgren 1986).

In the primary gustatory cortex in the frontal operculum and insula, neurons are more sharply tuned to gustatory stimuli than in the nucleus of the solitary tract, with some neurons responding, for example, primarily to sweet, and much less to salt, bitter, or sour stimuli (Scott, Yaxley, Sienkiewicz and Rolls 1986a, Yaxley, Rolls and Sienkiewicz 1990). Hunger does not influence the magnitude of neuronal responses to gustatory stimuli (Rolls, Scott, Sienkiewicz and Yaxley 1988, Yaxley, Rolls and Sienkiewicz 1988), and the neuronal re-

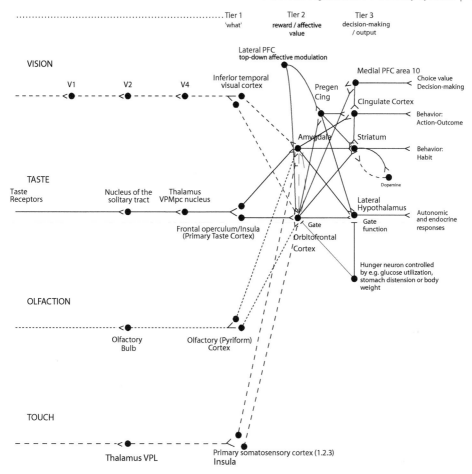

Fig. 2.4 Schematic diagram showing some of the connections of the taste, olfactory, somatosensory, and visual pathways in the brain. V1, primary visual (striate) cortex; V2 and V4, further cortical visual areas. PFC, prefrontal cortex. VPL, ventro-postero-lateral nucleus of the thalamus, which conveys somatosensory information to the primary somatosensory cortex (areas 1, 2 and 3). VPMpc, ventro-postero-medial nucleus pars parvocellularis of the thalamus, which conveys taste information to the primary taste cortex. Pregen Cing, pregenual cingulate cortex. For purposes of description, the stages can be described as Tier 1, representing what object is present independently of reward value; Tier 2 in which reward value is represented; and Tier 3 in which decisions between stimuli of different value are taken, and in which value is interfaced to behavioural output systems.

sponses are related to the concentration of the tastant. Consistent with this, activations in the human insular primary taste cortex are linearly related to the subjective intensity of the taste (which depends on the concentration of the tastant (Rolls, Rolls and Rowe 1983a)) and not to the pleasantness rating (Fig. 6.19) (Grabenhorst and Rolls 2008). Further, activations in the human insular primary taste cortex are related to the concentration of the tastant, for example monosodium glutamate (Grabenhorst, Rolls and Bilderbeck 2008a). Single neurons in the insular primary taste cortex also represent the viscosity (measured with carboxymethyl-cellulose) and temperature of stimuli in the mouth, and fat texture, but not the sight and smell of food (Verhagen, Kadohisa and Rolls 2004, Kadohisa, Rolls and Verhagen 2005a). In humans, the insular primary taste cortex is activated not only by taste (Francis, Rolls, Bowtell, McGlone, O'Doherty, Browning, Clare and Smith 1999, Small, Zald, Jones-Gotman, Zatorre,

Petrides and Evans 1999, De Araujo, Kringelbach, Rolls and Hobden 2003a, De Araujo and Rolls 2004, De Araujo, Kringelbach, Rolls and McGlone 2003b, Grabenhorst, Rolls and Bilderbeck 2008a, Grabenhorst, Rolls and Bilderbeck 2008a), but also by oral texture including viscosity (with the activation linearly related to the log of the viscosity) and fat texture (De Araujo and Rolls 2004), and temperature (Guest, Grabenhorst, Essick, Chen, Young, McGlone, de Araujo and Rolls 2007). Neurons in the macaque primary taste cortex do not have olfactory or visual responses (Verhagen, Kadohisa and Rolls 2004). Consistently, in a human fMRI investigation of olfactory and taste convergence in the brain, it was shown that there is a part of the human taste insula that is not activated by odour (De Araujo, Rolls, Kringelbach, McGlone and Phillips 2003c), though if a taste is recalled by an odour the situation could be different because of the role of cortico-cortical backprojections in recall (Rolls 2008d) (see Chapter 11).

Thus the primary taste cortex contains a representation of what the taste is, and not of its outcome value, expected value, or emotional / affective properties (Rolls 2015e, Rolls 2016b). In an even more anterior part of the insula (agranular cortex), there is a region that responds to both taste and odour stimuli (De Araujo, Rolls, Kringelbach, McGlone and Phillips 2003c, McCabe and Rolls 2007), and this is a region that is almost a part of the caudal orbitofrontal cortex topologically.

It is of interest that there is only one unimodal taste cortical area (i.e. without olfactory and visual strong bottom-up / feedforward inputs, though including oral texture inputs which are feedforward (Rolls 2014c, Rolls 2015e)), without a multiple-level hierarchy for purely taste processing. One reason for this may be that the computations to be solved in the taste system are far less complicated than in vision, for which a 'divide and conquer' strategy is used to solve the major issue of invariant including translation invariant representations. There is no computation of translation invariance to be computed in taste, or for that matter olfaction. Thus an important function performed by the primary taste (and olfactory) cortices is building representations, which include neurons that respond differently to, and hence encode, different combinations of different tastes, and oral texture and temperature stimuli. This then is a process of categorisation, which can be performed by competitive networks and providing large numbers of neurons in expansion recoding (Appendix B, Rolls (2008d)). Related to the one-level cortical hierarchy of purely taste processing, the peak firing rates of neurons in the insular taste cortex are relatively low, often reaching only 20 spikes/s to the most effective stimulus, whereas in regions such as the inferior temporal visual cortex, the peak firing rates can be 100 spikes/s, and are often greater than 50 spikes/s. This higher firing rate it is suggested is due to the fact that in the ventral visual system cortical hierarchy with its four or more levels, the information transmission from each stage must be fast, to limit the delays accumulated after multiple stages, and given that the number of spikes transmitted by neurons in a short time interval is an important determinant of information transmission rate (Rolls and Treves 2011), as shown in Appendix C. Another factor of course is that taste (and olfactory) stimuli occur more slowly in the natural world, whereas in vision and hearing different events may happen rapidly in succession, and any or all might be important or life-threatening.

Another aspect of why there is only one purely taste cortical area is that a large part of what is computed are associations between taste representations with what happens just before and at the same time, namely olfactory and visual inputs that are associated with taste. The next level of the taste hierarchy becomes a multimodal cortical area in which associations are put together between taste, olfaction, and the sight of whatever has been placed in the mouth, to form conjunctive representations of flavor, defined by a combination of taste, oral texture, olfactory, and related visual inputs. That area is the orbitofrontal cortex.

The orbitofrontal cortex contains multimodal representations of the taste, oral texture,

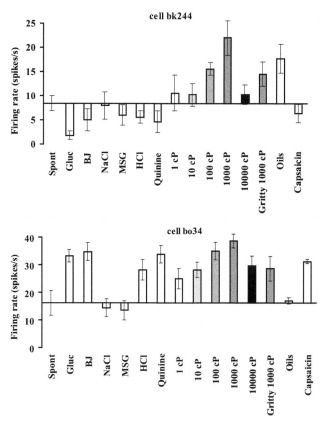

Fig. 2.5 Above. Firing rates (mean ± s.e.m.) of orbitofrontal cortex viscosity-sensitive neuron bk244 which did not have taste responses. The firing rates are shown to the viscosity series (carboxymethylcellulose in the range 1–10,000 centiPoise), to the gritty stimulus (carboxymethylcellulose with Fillite microspheres), to the taste stimuli 1 M glucose (Gluc), 0.1 M NaCl, 0.1 M MSG , 0.01 M HCl and 0.001 M QuinineHCl, and to fruit juice (BJ). Spont = spontaneous firing rate. Below. Firing rates (mean ± s.e.m.) of viscosity-sensitive neuron bo34 which had no response to the oils (mineral oil, vegetable oil, safflower oil and coconut oil, which have viscosities which are all close to 50 cP). The neuron did not respond to the gritty stimulus in a way that was unexpected given the viscosity of the stimulus, was taste tuned, and did respond to capsaicin. (Reproduced from *Journal of Neurophysiology*, 90 (6), Representations of the texture of food in the primate orbitofrontal cortex: neurons responding to viscosity, grittiness, and capsaicin, E. T. Rolls, J. V. Verhagen, and M. Kadohisa, pp. 3711–3724, ©2003, The American Physiological Society.)

smell, and sight of food, with different neurons often encoding different combinations of these inputs (Rolls 2014a, Rolls 2015e), as described next.

2.2.3.2 Secondary taste cortex – orbitofrontal cortex

A secondary cortical taste area has been discovered in the caudolateral orbitofrontal cortex of the primate in which gustatory neurons can be even more finely tuned to particular taste (and oral texture) stimuli (Rolls, Yaxley and Sienkiewicz 1990, Rolls and Treves 1990, Rolls, Sienkiewicz and Yaxley 1989c, Verhagen, Rolls and Kadohisa 2003, Rolls, Verhagen and Kadohisa 2003e, Kadohisa, Rolls and Verhagen 2004, Kadohisa, Rolls and Verhagen 2005a, Rolls, Critchley, Verhagen and Kadohisa 2010a) (see Figs. 2.5 and 2.6). In addition to representations of the 'prototypical' taste stimuli sweet, salt, bitter, and sour, different neurons in this region respond to other taste stimuli including umami or delicious savory taste as exemplified by monosodium glutamate which is present in tomatoes, mushroom, fish and meat, and human mothers' milk (Baylis and Rolls 1991, Rolls, Critchley, Wakeman and Mason 1996c, Rolls, Critchley, Browning and Hernadi 1998a), with corresponding activations in humans (De Araujo, Kringelbach, Rolls and Hobden 2003a). Umami is a component of many foods which helps to make them taste pleasant, especially when the umami taste is paired with a consonant savoury odour (Rolls, Critchley, Browning and Hernadi 1998a, McCabe and Rolls 2007, Rolls 2009b).

In relation to hierarchical processing, it is important that different orbitofrontal cortex neurons respond to different combinations of taste and related oral texture stimuli. This shows that in the taste system too, in the hierarchy, neurons become tuned to low order combinations of sensory inputs, in order to separate the input space in a way that allows easy decoding by dot product decoding, allows specific tastes to be associated with particular stimuli in other modalities including olfaction and vision, and allows for sensory-specific satiety, as will soon be shown. Examples of the tuning of orbitofrontal cortex neurons to different combinations of taste and oral texture are illustrated in Fig. 2.5. For example, the neuron illustrated in Fig. 2.5 (lower) responded to the taste of sweet (glucose, blackcurrant juice BJ, sour (HCl), and bitter (quinine) but not to the taste of salt (NaCl) or umami (monosodium glutamate, MSG); to viscosity (as altered parametrically using the standard food thickening agent carboxymethylcellulose made up in viscosities of 1–10,000 cPoise (Rolls, Verhagen and Kadohisa 2003e), where 10,000 cP is approximately the viscosity of toothpaste); did not respond to fat in the mouth (oils); and did respond to capsaicin (chilli pepper). In contrast, the neuron shown in the upper part of Fig. 2.5 had no response to taste; did respond to oral viscosity; did respond to oral fat (see Rolls, Critchley, Browning, Hernadi and Lenard (1999a) and Verhagen, Rolls and Kadohisa (2003)); and did not respond to capsaicin. This combinatorial encoding in the taste and oral texture system is important in the mechanism of the fundamental property of reward systems known as sensory-specific satiety, as shown below.

These findings have been extended to humans, with the finding with fMRI that activation of the orbitofrontal cortex and perigenual cingulate cortex is produced by the texture of fat in the mouth (De Araujo and Rolls 2004). Moreover, activations in the orbitofrontal cortex and pregenual cingulate cortex are correlated with the pleasantness of fat texture in the mouth (Grabenhorst, Rolls, Parris and D'Souza 2010b).

2.2.3.3 The reward value of taste is represented in the orbitofrontal cortex

Another fundamental change takes place in the primate secondary taste cortex in the orbitofrontal cortex, a transform from the representation about the identity of the taste and texture of stimuli in the primary taste cortex that is independent of reward value and pleasantness, to a representation of the reward value and pleasantness of the taste in the secondary taste cortex. In particular, in the primate orbitofrontal cortex, it is found that the responses of taste neurons to the particular food with which a monkey is fed to satiety decrease to zero (Rolls, Sienkiewicz and Yaxley 1989c). An example is shown in Fig. 2.6. This neuron reduced its responses to the taste of glucose during the course of feeding as much glucose as the monkey wanted to drink. When the monkey was fully satiated, and did not want to drink any more glucose, the neuron no longer responded to the taste of glucose. Thus the responses of these neurons decrease to zero when the reward value of the food decreases to zero. Interestingly the neuron still responded to other foods, and the monkey chose to eat these other foods. Thus the modulation of the responses of these orbitofrontal cortex taste neurons occurs in a sensory-specific way, and they represent *reward outcome value*.

Another example is shown in Fig. 2.7 (after Rolls, Critchley, Browning, Hernadi and

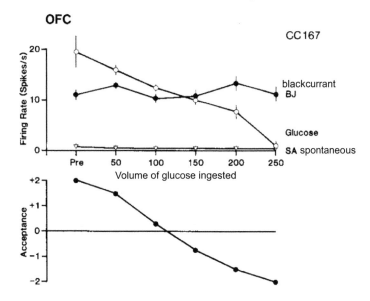

Fig. 2.6 The effect of feeding to satiety with glucose solution on the responses (rate ± s.e.m.) of a neuron in the secondary taste cortex to the taste of glucose (open circles) and of blackcurrant juice (BJ). The spontaneous firing rate is also indicated (SA). Below the neuronal response data, the behavioural measure of the acceptance or rejection of the solution on a scale from +2 (strong acceptance) to –2 (strong rejection) is shown. The solution used to feed to satiety was 20% glucose. The monkey was fed 50 ml of the solution at each stage of the experiment as indicated along the abscissa, until he was satiated as shown by whether he accepted or rejected the solution. Pre is the firing rate of the neuron before the satiety experiment started. (Reproduced with permission from European Journal of Neuroscience, 1: 53–60. Hunger Modulates the Responses to Gustatory Stimuli of Single Neurons in the Caudolateral Orbitofrontal Cortex of the Macaque Monkey. Rolls, E. T., Sienkiewicz, Z. J. and Yaxley, S. (1989). Copyright ©1989, John Wiley and Sons.)

Lenard (1999a)). This neuron decreased its response to the fatty texture of cream when fed to satiety with cream, but still responded to the taste of glucose after feeding to satiety with cream. This indicated sensory-specific satiety for the reward value of the texture of cream in the mouth.

The orbitofrontal cortex is the first stage of the primate taste system in which this modulation of the responses of neurons to the taste of food is affected by hunger, in that this modulation is not found in the nucleus of the solitary tract, or in the frontal opercular or insular primary gustatory cortices (Yaxley, Rolls, Sienkiewicz and Scott 1985, Rolls, Scott, Sienkiewicz and Yaxley 1988, Yaxley, Rolls and Sienkiewicz 1988, Rolls 2015e). It is of course only when hungry that the taste of food is rewarding. This is an indication that the responses of these orbitofrontal cortex taste neurons reflect the reward value of food. The firing of these orbitofrontal cortex neurons may actually implement the reward value of a food. The hypothesis is that primates work to obtain firing of these reward value neurons, by eating food when they are hungry (Chapter 15, Rolls (2014a)).

Further evidence that the firing of these orbitofrontal cortex taste neurons does actually implement the primary reward value of food is that in other investigations we showed that monkeys would work to obtain electrical stimulation of this area of the brain (Rolls, Burton and Mora 1980, Mora, Avrith, Phillips and Rolls 1979, Mora, Avrith and Rolls 1980, Rolls 2005). Moreover, the reward value of the electrical stimulation was dependent on hunger being

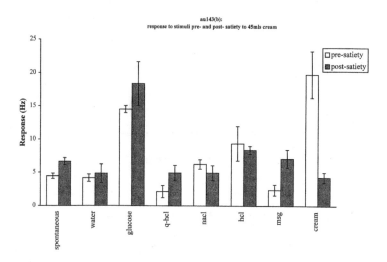

Fig. 2.7 A neuron in the primate orbitofrontal cortex that decreased its response to the texture of fat (cream) in the mouth after feeding to satiety with cream. The neuron did not decrease its response to the taste of glucose after feeding to satiety with fat (single cream). Gluc, 1 M glucose; NaCl 0.1 M, salt; 0.01 M HCl, sour; 0.001 M Q-HCl, quinine, bitter. The spontaneous firing rate of the cell is also shown. (Republished with permission of Society for Neuroscience from *Journal of Neuroscience*, 19 (4), Responses to the sensory properties of fat of neurons in the primate orbitofrontal cortex, E. T. Rolls, H. D. Critchley, A. S. Browning, A. Hernadi, and L. Lenard, pp. 1532–1540, ©1999 Society for Neuroscience.)

present. If the monkey was fed to satiety, the monkey no longer found electrical stimulation at this site so rewarding, and stopped working for the electrical stimulation. Indeed, of all the brain sites tested, this orbitofrontal cortex region was the part of the brain in which the reward value of the electrical stimulation was most affected by feeding to satiety (Mora, Avrith, Phillips and Rolls 1979, Mora, Avrith and Rolls 1980, Rolls, Burton and Mora 1980). Thus all this evidence indicates that the reward value of taste is encoded in the secondary taste cortex, and that primates work to obtain food in order to activate these neurons, the activation of which actually mediates reward. This is probably an innate reward system, in that taste can act as a reward in rats without prior training (Berridge, Flynn, Schulkin and Grill 1984).

These results also provide evidence on the nature of the mechanisms that underlie sensory-specific satiety. Sensory-specific satiety, as noted above, is the phenomenon in which the decreases in the palatability and acceptability of a food that has been eaten to satiety are partly specific to the particular food that has been eaten. The results just described suggest that such sensory-specific satiety for taste cannot be largely accounted for by adaptation at the receptor level, in the nucleus of the solitary tract, or in the frontal opercular or insular gustatory cortices, to the food which has been eaten to satiety, otherwise modulation of neuronal responsiveness should have been apparent in the recordings made in these regions. Indeed, the findings suggest that sensory-specific satiety is not represented in the primary gustatory cortex. It is thus of particular interest that a decrease in the response of orbitofrontal cortex neurons occurs which is partly specific to the food that has just been eaten to satiety (Rolls, Sienkiewicz and Yaxley 1989c). The specificity of the tuning of taste and oral texture neurons, to particular combinations of taste and texture stimuli, are important for the operation of the mechanism, for just by making adaptation during the time course of a meal to their inputs a property of neurons in the orbitofrontal cortex, a mechanism for specifically reducing the reward value of one food but not other foods is provided (Rolls 2014a).

This provides evidence for another principle of cortical operation and computation, selective adaptation tailored to produce particular functions (see Chapter 10 and Section 10.5).

The situation appears to be the same in humans, in whom fMRI investigations show that sensory-specific satiety for food is represented in the orbitofrontal cortex (Kringelbach, O'Doherty, Rolls and Andrews 2003); that activations in the orbitofrontal cortex and pregenual cingulate cortex are linearly correlated with the subjective pleasantness value of taste (Grabenhorst and Rolls 2008, Rolls 2012e); and in that activations in the insular primary taste cortex are linearly correlated with the subjective intensity of taste (Grabenhorst and Rolls 2008) (Fig. 6.19).

It is suggested that the computational significance of this architecture is as follows (Rolls 1986b, Rolls 1989d, Rolls and Treves 1990, Rolls 2008d, Rolls 2014a). If satiety were to operate at an early level of sensory analysis, then because of the broadness of tuning of neurons, responses to non-foods would become attenuated as well as responses to foods (and this could well be dangerous if poisonous non-foods became undetectable). This argument becomes even more compelling when it is realized that satiety typically shows some specificity for the particular food eaten, with other foods not eaten in the meal remaining relatively pleasant (see above). Unless tuning were relatively fine, this mechanism could not operate, for reduction in neuronal firing after one food had been eaten would inevitably reduce behavioural responsiveness to other foods. Indeed, it is of interest to note that such a sensory-specific satiety mechanism can be built by arranging for tuning to particular foods to become relatively specific at one level of the nervous system (as a result of categorization processing in earlier stages), and then at this stage (but not at prior stages) to allow habituation to be a property of the synapses, as proposed above. This is a principle of operation of the neocortex, at least in primates including humans.

Thus information processing in the taste system illustrates an important principle of cortical function in primates, namely that it is only after several or many stages of sensory information processing (which produce efficient categorization of 'what' the stimulus is) that there is an interface to reward and motivational systems, to other modalities, or to systems involved in association memory (Rolls and Treves 1990, Rolls and Treves 1998). This principle of cortical operation is described further in Chapter 15.

2.2.3.4 Tertiary taste cortex - anterior cingulate cortex

The orbitofrontal cortex projects to the anterior cingulate cortex (Carmichael and Price 1995a, Morecraft and Tanji 2009, Vogt 2009), where taste neurons are also found in what is thereby a tertiary taste cortical area (Rolls 2008c, Rolls 2009a). For example, Gabbott, Verhagen, Kadohisa and Rolls found neurons in the pregenual cingulate cortex that respond to taste (see example in Fig. 2.8), and it was demonstrated that the representation is of reward value, for devaluation by feeding to satiety selectively decreased neuronal responses to the food with which the animal was satiated (Rolls 2008c, Rolls 2014a). Consistently, in humans (pleasant) sweet taste also activates the most anterior part of the anterior cingulate cortex, the pregenual cingulate cortex (De Araujo and Rolls 2004, De Araujo, Kringelbach, Rolls and Hobden 2003a). Less pleasant tastes activate an anterior cingulate cortex region just posterior and dorsal to this just above the corpus callosum (De Araujo and Rolls 2004, De Araujo, Kringelbach, Rolls and Hobden 2003a, Rolls 2014a, Grabenhorst and Rolls 2011).

So far, this anterior cingulate cortex region appears to be a re-representation of what is already represented in the orbitofrontal cortex. But there is a large difference that occurs in this tertiary level of the taste cortical hierarchy. Whereas the orbitofrontal cortex represents the reward value of the sensory stimulus of taste (and other sensory stimuli) with little activity

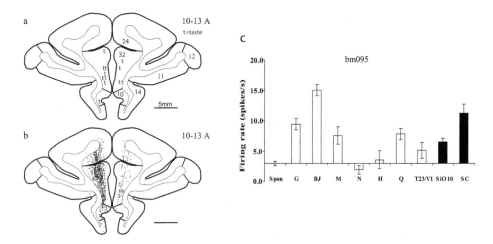

Fig. 2.8 Pregenual cortex taste neurons. The reconstructed positions of the anterior cingulate neurons with taste (t) responses, together with the cytoarchitectonic boundaries determined by Carmichael and Price (1994). Most (11/12) of the taste neurons were in the pregenual cingulate cortex (area 32), as shown. The neurons are shown on a coronal section at 12 mm anterior (A) to the sphenoid reference point. b. The locations of all the 749 neurons recorded in the anterior cingulate region in this study are indicated to show the regions sampled. c. Responses of a pregenual cingulate cortex neuron (bm095) with differential responses to tastes and oral fat texture stimuli. The mean ($\pm$ sem) firing rate responses to each stimulus calculated in a 5 s period over several trials are shown. The spontaneous (Spon) firing rate of 3 spikes/s is shown by the horizontal line, with the responses indicated relative to this line. The taste stimuli were 1 M glucose (G), blackcurrant fruit juice (BJ), 0.1 M NaCl (N), 0.1 M MSG (M), 0.01 M HCl (H) and 0.001 M QuinineHCl (Q); water (T23/V1); single cream (SC); and silicone oil with a viscosity of 10 cP (SiO10). The neuron had significantly different responses to the different stimuli as shown by a one-way ANOVA (F[9,46]=17.7, $p < 10^{-10}$). (Data from Rolls, Gabbott, Verhagen, and Kadohisa. Republished with permission of Akademiai Kiado Rt., from Acta Physiologica Hungarica, Functions of the orbitofrontal and pregenual cingulate cortex in taste, olfaction, appetite and emotion, E. T. Rolls, 95 pp. 131–164. Copyright ©2008.)

related to movements, motor events, and actions (Thorpe, Rolls and Maddison 1983, Rolls, Critchley, Mason and Wakeman 1996b, Critchley and Rolls 1996a, Rolls and Baylis 1994, Rolls 2005, Wallis and Miller 2003, Padoa-Schioppa and Assad 2006, Rolls 2014a), in the anterior and midcingulate cortex there are representations also of movements, actions, and errors about actions (Matsumoto, Matsumoto, Abe and Tanaka 2007, Luk and Wallis 2009). Moreover, lesions to the cingulate cortex influence action selection when it is guided by the reward value of the goal, and this provides some of the evidence that the anterior cingulate cortex is involved in learning associations between actions and the reward (or punishment) outcome (Rushworth, Walton, Kennerley and Bannerman 2004, Rushworth, Noonan, Boorman, Walton and Behrens 2011, Grabenhorst and Rolls 2011, Rolls 2014a).

Thus the anterior cingulate stage of the taste cortical hierarchy can be conceptualized as the level at which the representation of sensory reward value is interfaced to actions. In primates, stimulus valuation performed in the orbitofrontal cortex is kept separate from motor and action systems. In terms of hierarchical organization, the impact is that the orbitofrontal cortex specialises in reward value representations and learns about the value of sensory stimuli using as we shall see information from all sensory modalities, so it performs one type of computation in which the expected reward value of a visual stimulus can be recomputed in one trial if the taste associated with the visual stimulus changes. The next stage of the hierarchy, the anterior cingulate cortex, then follows the principle of further convergence as one progresses up the hierarchy, in this case convergence between the value of sensory stimuli and the actions needed to obtain them, which is a separate and typically slower learning process because it involves trial and error learning, whereas the orbitofrontal cortex can perform its sensory-sensory processes by association in what can be as little as one trial (Rolls 2014a).

2.2.4 Hierarchical organization of olfactory processing

In the primate olfactory system, the olfactory bulb projects to the primary olfactory cortex (the pyriform cortex, directly and via the mediodorsal nucleus of the thalamus), which in turn projects to the orbitofrontal cortex, which is thereby defined as a secondary cortical olfactory area, and the orbitofrontal cortex in turn projects to the anterior cingulate cortex (Fig. 2.4) (Price, Carmichael, Carnes, Clugnet and Kuroda 1991, Morecraft, Geula and Mesulam 1992, Barbas 1993, Carmichael, Clugnet and Price 1994, Price 2006). Olfactory projections from the olfactory bulb do though reach some other areas, including the olfactory tubercle which is part of the ventral striatum, and the lateral entorhinal cortex, which in turn projects to the hippocampus (Rolls 2014a, Wilson, Xu, Sadrian, Courtiol, Cohen and Barnes 2014, Wilson and Sullivan 2011) (Fig. 1.9). At the hippocampal stage of the hierarchy, the principle of convergence is evident, for in the hippocampus olfactory / flavour information is combined with information about where the flavor with its olfactory components is in allocentric space (Rolls and Xiang 2005, Kesner and Rolls 2015) (see Chapter 24). The principle of convergence in cortical hierarchies is also evident in the fact that olfactory information is combined with taste information to produce flavor in the orbitofrontal cortex, but not at earlier cortical stages (Section 2.2.5).

We should at the outset note that the principles of olfactory as well as taste processing are very different in rodents at least in terms or sensory vs value encoding, as described in Chapters 15 and 18, so that rodents may not provide a good guide to the principles of taste, and olfactory (as well as visual) cortical processing in primates including humans (Rolls 2015e).

In humans, activity in the pyriform cortex appears to be related to 'what' stimulus is present (its identity), and its intensity, and not to its reward value, in that activations in the human pyriform cortex are linearly related to the intensity rating of odourants, and not to their pleasantness (Grabenhorst, Rolls, Margot, da Silva and Velazco 2007); in that paying attention to the intensity vs the pleasantness of an odour increases activations in the pyriform cortex (Rolls, Grabenhorst, Margot, da Silva and Velazco 2008b); and in that activations in the pyriform (primary olfactory) cortex were not decreased by odor devaluation by satiety (Gottfried 2015). The pyriform cortex has a rich set of recurrent collateral connections between its pyramidal cells, and in line with this the hypothesis is that the pyriform cortex implements an autoassociative memory, which helps to form representations of olfactory mixtures as odor objects, which display a property of completion in which the whole odor may be represented even if some components are missing in the process known as completion (Haberly 2001, Wilson and Sullivan 2011) (see Section 18.2.2). The anterior pyriform cortex is more strongly influenced by the olfactory bulb inputs, and in turn influences the posterior pyriform cortex. In both humans (Gottfried, Winston and Dolan 2006, Gottfried 2010) and rodents (Kadohisa and Wilson 2006), the anterior piriform cortex may encode information related to structural or perceptual identity of the odor, e.g. 'banana'. More posterior regions, perhaps in accord with the dominance of association fiber input, appear to encode the perceptual category of an odor, e.g. 'fruity' (Wilson and Sullivan 2011).

After reward value representation of odour in the orbitofrontal cortex (Rolls, Kringelbach and De Araujo 2003c) where the activations in humans are linearly related to the subjective pleasantness rating, there is a further level of olfactory processing in a region just anterior to the orbitofrontal cortex, medial prefrontal cortex area 10, where activations to odors are related

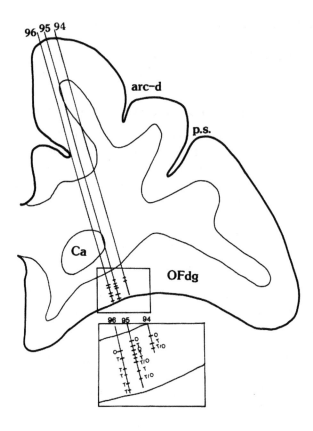

Fig. 2.9 Examples of tracks made into the orbitofrontal cortex in which taste (T) and olfactory (O) neurons were recorded close to each other in the same tracks. Some of the neurons were bimodal (T/O). arc-d, arcuate sulcus; Ca, head of Caudate nucleus; Ofdg, dysgranular part of the Orbitofrontal Cortex; p.s., principal sulcus. (Republished with permission of Society for Neuroscience, from Journal of Neuroscience, Gustatory, olfactory and visual convergence within the primate orbitofrontal cortex, E. T. Rolls and L. L. Baylis, 14, pp. 5437–5452. Copyright ©1994 Society for Neuroscience.)

to decisions about the pleasantness of odors (Rolls, Grabenhorst and Parris 2010d, Rolls and Grabenhorst 2008, Rolls, Grabenhorst and Deco 2010b, Rolls, Grabenhorst and Deco 2010c, Rolls 2014a). This shows how different computations are separated into different levels of a cortical hierarchy, with linear processing to represent reward value on a continuous scale in the orbitofrontal cortex, which is then followed by a highly non-linear processing stage in medial prefrontal cortex area 10 where binary choices, and confidence in a decision just made, are represented. This progression in a cortical hierarchy to more non-linear processing (involved for example in categorisation or decision-making) is another principle of operation of the cerebral cortex.

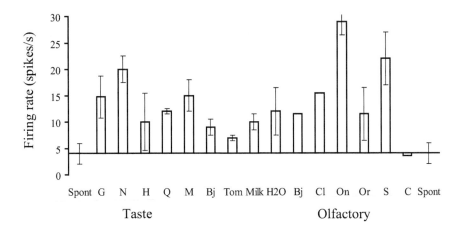

Fig. 2.10 The responses of a bimodal neuron with taste and olfactory responses recorded in the caudo-lateral orbitofrontal cortex. G, 1 M glucose; N, 0.1 M NaCl; H, 0.01 M HCl; Q, 0.001 M Quinine HCl; M, 0.1 M monosodium glutamate; Bj, 20% blackcurrant juice; Tom, tomato juice; B, banana odour; Cl, clove oil odour; On, onion odour; Or, orange odour; S, salmon odour; C, control no-odour presentation. The mean responses ± s.e.m. are shown. The neuron responded best to the savoury tastes of NaCl and monosodium glutamate and to the consonant odours of onion and salmon. (Republished with permission of Society for Neuroscience, from Journal of Neuroscience, Gustatory, olfactory and visual convergence within the primate orbitofrontal cortex, E. T. Rolls and L. L. Baylis, 14, pp. 5437–5452. Copyright ©1994 Society for Neuroscience.)

2.2.5 Hierarchical convergence of taste, olfactory and visual processing to produce multimodal flavour in the orbitofrontal cortex

In the orbitofrontal cortex, not only unimodal taste neurons, but also unimodal olfactory neurons are found (see Fig. 2.9). In addition some single neurons respond to both gustatory and olfactory stimuli, often with correspondence between the two modalities (Rolls and Baylis 1994) (see Fig. 2.10; cf. Fig. 2.4). It is probably here in the orbitofrontal cortex of primates that these two modalities converge to produce the representation of flavour (Rolls and Baylis 1994), and, consistent with this, neurons in the macaque primary taste cortex do not have olfactory responses (Verhagen, Kadohisa and Rolls 2004). Consistently, in a human fMRI investigation of olfactory and taste convergence in the brain, it was shown that there is a part of the human taste insula that is not activated by odour (De Araujo, Rolls, Kringelbach, McGlone and Phillips 2003c), though if a taste is recalled by an odour the situation could be different because of the role of cortico-cortical backprojections in recall (Rolls 2008d, Rolls 2015e). The evidence described below indicates that these bimodal representations are built by olfactory–gustatory association learning, an example of stimulus–reinforcer association learning.

The human orbitofrontal cortex also reflects the convergence of taste and olfactory inputs, as shown for example by the fact that activations in the human medial orbitofrontal cortex are correlated with both the cross-modal consonance of combined taste and olfactory stimuli (high for example for sweet taste and strawberry odour), as well as for the pleasantness of the combinations, as shown in Fig. 2.11 (De Araujo, Rolls, Kringelbach, McGlone and Phillips 2003c). In addition, the combination of monosodium glutamate taste and a consonant

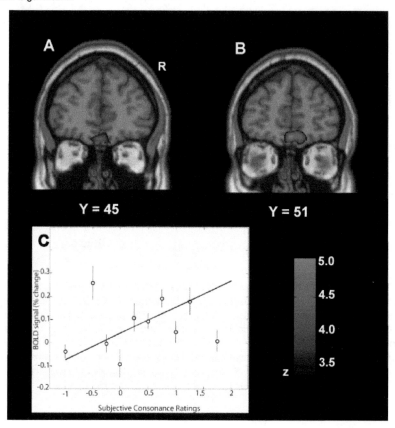

Fig. 2.11 Flavour formation in the human brain, shown by cross-modal olfactory–taste convergence. Brain areas where activations were correlated with the subjective ratings for stimulus (taste–odour) consonance and pleasantness. (A) A second-level, random effects analysis based on individual contrasts (the consonance ratings being the only effect of interest) revealed a significant activation in a medial part of the anterior orbitofrontal cortex. (B) Random effects analysis based on the pleasantness ratings showed a significant cluster of activation located in a (nearby) medial part of the anterior orbitofrontal cortex. The images were thresholded at p<0.0001 for illustration. (C) The relation between the BOLD signal from the cluster of voxels in the medial orbitofrontal cortex shown in (A) and the subjective consonance ratings. The analyses shown included all the stimuli included in this investigation. The means and standard errors of the mean across subjects are shown, together with the regression line, for which r=0.52. (Reproduced with permission from *European Journal of Neuroscience*, 18 (7) pp. 2059–2068. Taste-olfactory convergence, and the representation of the pleasantness of flavour, in the human brain, Ivan E. T. De Araujo, Edmund T. Rolls, Morten L. Kringelbach, Francis McGlone, and Nicola Phillips. Copyright ©2003, John Wiley and Sons.)

savoury odour produced a supralinear effect in the medial orbitofrontal cortex to produce the rich delicious flavour of umami that makes many foods rich in protein pleasant (McCabe and Rolls 2007, Rolls 2009b).

Rolls and colleagues have analysed the rules by which orbitofrontal olfactory representations are formed and brought together with taste representation in primates (Rolls 2001, Rolls and Grabenhorst 2008, Rolls 2011a). For 65% of neurons in the orbitofrontal olfactory areas, Critchley and Rolls (1996b) showed that the representation of the olfactory stimulus was independent of its association with taste reward (analysed in an olfactory discrimination task with taste reward, as some orbitofrontal cortex olfactory neurons are bimodal, with responses also to taste stimuli (Rolls and Baylis 1994)). For the remaining 35% of the neurons, the odours

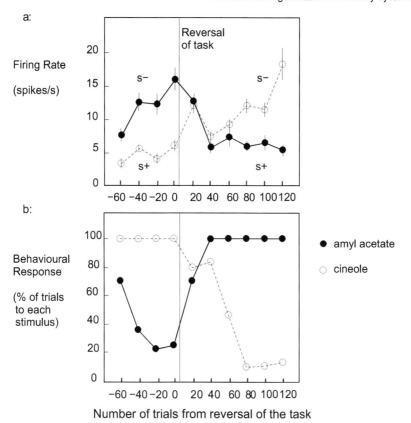

Fig. 2.12 Orbitofrontal cortex: olfactory to taste association reversal. (a) The activity of a single orbitofrontal olfactory neuron during the performance of a two-odour olfactory discrimination task and its reversal is shown. Each point represents the mean poststimulus activity of the neuron in a 500-ms period on approximately 10 trials of the different odourants. The standard errors of these responses are shown. The odourants were amyl acetate (closed circle) (initially S−) and cineole (o) (initially S+). After 80 trials of the task the reward associations of the stimuli were reversed. This neuron reversed its responses to the odourants following the task reversal. (b) The behavioural responses of the monkey during the performance of the olfactory discrimination task. The number of lick responses to each odourant is plotted as a percentage of the number of trials to that odourant in a block of 20 trials of the task. (Reproduced with permission from *Journal of Neurophysiology*, 75 (5), Orbitofrontal cortex neurons: role in olfactory and visual association learning, E. T. Rolls, H. D. Critchley, R. Mason, E. A. Wakeman, pp. 1970–1981, ©1996, The American Physiological Society.)

to which a neuron responded were influenced by the taste (glucose or saline) with which the odour was associated (Critchley and Rolls 1996b). Thus the odour representation for 35% of orbitofrontal neurons appeared to be built by olfactory-to-taste association learning, and thus by learning the neurons come to encode the *expected value* of olfactory stimuli.

   This possibility that the odour representation of some primate orbitofrontal cortex olfactory neurons is built by olfactory-to-taste association learning to encode *expected value* was confirmed by reversing the taste with which an odour was associated in the reversal of an olfactory discrimination task. It was found that 73% of the sample of neurons analysed altered the way in which they responded to odour when the taste reinforcer association of the odour was reversed (Rolls, Critchley, Mason and Wakeman 1996b). Reversal was shown by 25% of the neurons (see, for example, Fig. 2.12), and 48% altered their activity in that they no longer discriminated after the reversal. These latter neurons thus respond to a particular odour only if it is associated with a taste reward, and not when it is associated with the taste of salt, a punisher. They do not respond to the other odour in the task when it is associated with reward. Thus they respond to a particular combination of an odour, and its being associated with taste reward and not a taste punisher. They may be described as *conditional olfactory-reward neurons*, and may be important in the mechanism by which stimulus–reinforcer (in this case olfactory-to-taste) reversal learning occurs (Deco and Rolls 2005d), as described in Chapter 15.

We have been able to show that there is also a major visual input to many neurons in the orbitofrontal cortex, and that what is represented by these neurons is in many cases the reinforcer (reward or punisher) association of visual stimuli. Many of these neurons reflect the relative preference or reward value of different visual stimuli, in that their responses decrease to zero to the sight of one food on which the monkey is being fed to satiety, but remain unchanged to the sight of other food stimuli. In this sense the visual reinforcement-related neurons predict the reward value that is available from the primary reinforcer, the taste. The visual input is from the ventral, temporal lobe, visual stream concerned with 'what' object is being seen, in that orbitofrontal visual neurons frequently respond differentially to objects or images (but depending on their reward association) (Thorpe, Rolls and Maddison 1983, Rolls, Critchley, Mason and Wakeman 1996b). The primary reinforcer that has been used is taste.

The fact that these neurons represent the reinforcer associations of visual stimuli and hence the expected value has been shown to be the case in formal investigations of the activity of orbitofrontal cortex visual neurons, which in many cases reverse their responses to visual stimuli when the taste with which the visual stimulus is associated is reversed by the experimenter (Thorpe, Rolls and Maddison 1983, Rolls, Critchley, Mason and Wakeman 1996b). An example of the responses of an orbitofrontal cortex neuron that reversed the visual stimulus to which it responded during reward reversal is shown in Fig. 2.13.

This reversal by orbitofrontal visual neurons can be very fast, in as little as one trial, that is a few seconds (Thorpe, Rolls and Maddison 1983, Rolls, Critchley, Mason and Wakeman 1996b) (see for example Fig. 2.14). The significance of the visual stimulus, a syringe from which the monkey was fed, was altered during the trials. On trials 1–5, no response of the neuron occurred to the sight of the syringe from which the monkey had been given glucose solution to drink from the syringe on the preceding trials. On trials 6–9, the neuron responded to the sight of the same syringe from which he had been given aversive hypertonic saline drink on the preceding trial. Two more reversals (trials 10–15, and 16–17) were performed. The reversal of the neuron's response when the significance of the visual stimulus was reversed shows that the responses of the neuron only occurred to the stimulus when it was associated with aversive saline and not when it was associated with glucose reward.

These neurons thus reflect the information about which stimulus is currently associated with reward during reversals of visual discrimination tasks – they are reward predicting neurons, that is, they represent *expected value*. If a reversal occurs, then the taste cells provide the information that an unexpected taste reinforcer has been obtained, another group of cells shows a vigorous discharge that reflects the error between the expected reward value and the reward outcome actually obtained (see below), and the visual cells with reinforcer association-related responses reverse the visual stimulus to which they are responsive. These neurophysiological changes take place rapidly, in as little as 5 s, and are part of the neuronal learning mechanism that enables primates to alter their knowledge of the reinforcer association of visual stimuli so rapidly. This capacity is important whenever behaviour must be corrected when expected reinforcers are not obtained, in, for example, feeding, emotional, and social situations (see Chapter 15, Rolls (2014a), and Kringelbach and Rolls (2003)). In that these

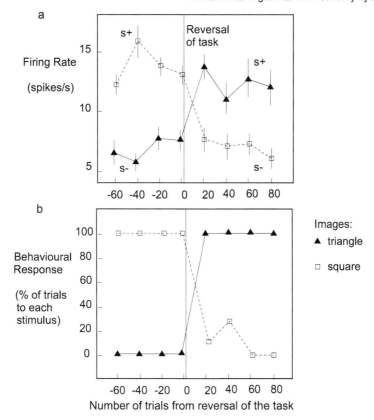

Fig. 2.13 Orbitofrontal cortex: visual discrimination reversal. The activity of an orbitofrontal visual neuron during performance of a visual discrimination task and its reversal. The stimuli were a triangle and a square presented on a video monitor. (a) Each point represents the mean poststimulus activity in a 500 ms period of the neuron based on approximately 10 trials of the different visual stimuli. The standard errors of these responses are shown. After 60 trials of the task the reward associations of the visual stimuli were reversed. s+ indicates that a lick response to that visual stimulus produces fruit juice reward; s− indicates that a lick response to that visual stimulus results in a small drop of aversive tasting saline. This neuron reversed its responses to the visual stimuli following the task reversal. (b) The behavioural response of the monkey to the task. It is shown that the monkey performs well, in that he rapidly learns to lick only to the visual stimulus associated with fruit juice reward. (Reproduced with permission from *Journal of Neurophysiology*, 75 (5), Orbitofrontal cortex neurons: role in olfactory and visual association learning, E. T. Rolls, H. D. Critchley, R. Mason, and E. A. Wakeman, pp. 1970–1981, ©1996, The American Physiological Society.)

neurons reflect whether a visual stimulus is associated with reward or a punisher, they reflect the relative preference for different stimuli, i.e. the value (Thorpe, Rolls and Maddison 1983, Rolls, Critchley, Mason and Wakeman 1996b) (as found also by Tremblay and Schultz (1999)). Consistent with this evidence that the responses of some orbitofrontal cortex neurons reflect the learned predictive reward value of visual stimuli, Thorpe, Rolls and Maddison (1983) and Tremblay and Schultz (2000) found that orbitofrontal cortex neurons learned to respond differently to new stimuli that did or did not predict reward. Different neurons in the orbitofrontal cortex are tuned to different learned or conditioned reinforcers, with for example approximately 5% responding to visual stimuli associated with taste reward, and 3% to visual stimuli associated with taste punishment (Thorpe, Rolls and Maddison 1983, Rolls, Critchley, Mason and Wakeman 1996b, Rolls 2014a).

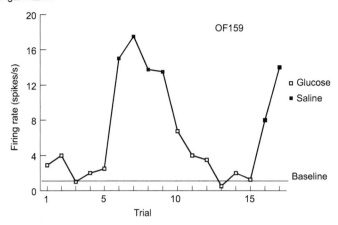

Fig. 2.14 Orbitofrontal cortex: one-trial visual discrimination reversal by a neuron. On trials 1–5, no response of the neuron occurred to the sight of a 2 ml syringe from which the monkey had been given orally glucose solution to drink on the previous trial. On trials 6–9, the neuron responded to the sight of the same syringe from which he had been given aversive hypertonic saline to drink on the previous trial. Two more reversals (trials 10–15, and 16–17) were performed. The reversal of the neuron's response when the significance of the same visual stimulus was reversed shows that the responses of the neuron only occurred to the sight of the visual stimulus when it was associated with a positively reinforcing and not with a negatively reinforcing taste. Moreover, it is shown that the neuronal reversal took only one trial. (Reproduced from *Experimental Brain Research*, The orbitofrontal cortex: Neuronal activity in the behaving monkey, 49 (1), 1983, pp. 93–115, S. J. Thorpe, E. T. Rolls, and S. Maddison ©1983, Springer. With permission of Springer.)

In the visual discrimination reversal task, a second class of neuron was found that codes for particular stimuli only if they are associated with reward, and not if they are associated with punishment. Such a neuron might respond to a green stimulus associated with reward; after reversal not respond to the green stimulus when it was associated with punishment; and not respond to a blue stimulus irrespective of whether it was associated with reward or punishment (Thorpe, Rolls and Maddison 1983) (see example in Fig. 15.7). They may be described as *conditional visual stimulus-to-taste reward neurons* or *conditional expected value neurons*, and are analogous to their olfactory counterparts described above. These neurons are probably important in the mechanisms that implement rapid reversal, as described in Chapter 15.

These findings illustrate the principle of convergence as one ascends a cortical hierarchy, in that neurons that respond to visual, taste and olfactory stimuli are found in the orbitofrontal cortex, but not at earlier levels of cortical processing, in the insular primary taste cortex (Verhagen, Kadohisa and Rolls 2004) and inferior temporal visual cortex where objects are represented independently of their taste reward associations (Rolls, Judge and Sanghera 1977, Rolls, Aggelopoulos and Zheng 2003a) (see Fig. 2.4). Moreover, the findings provide evidence that one way in which convergence in a hierarchy brings together corresponding inputs in an appropriate way at the neuronal level is by associative learning.

2.2.6 Hierarchical organization of auditory processing

Rauschecker and colleagues, building on studies in primates, have described two hierarchical cortical streams of information transfer from auditory cortex (AC) which eventually reach prefrontal cortical regions (Rauschecker 2012, Bornkessel-Schlesewsky, Schlesewsky, Small and Rauschecker 2015). As illustrated in Fig. 2.15, there are two streams of auditory processing, a ventral auditory stream (shown in green) that has been compared with the visual

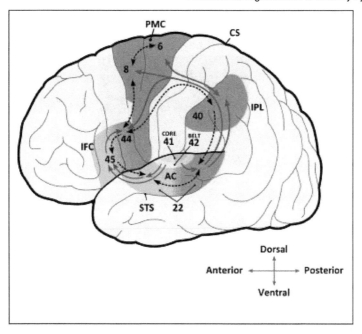

Fig. 2.15 Hierarchical organization of auditory cortical processing. The ventral auditory pathway is shown in green, and the dorsal auditory pathway in red. AC – auditory cortex; CS – central sulcus; IFC – inferior frontal cortex; IPL – inferior parietal lobule; PMC – premotor cortex; STS – superior temporal sulcus. For description, see text. (Reprinted from *Trends in Cognitive Sciences*, 19 (3), Ina Bornkessel-Schlesewsky, Matthias Schlesewsky, Steven L. Small, and Josef P. Rauschecker, Neurobiological roots of language in primate audition: common computational properties, pp. 142–50, Copyright (2015), with permission from Elsevier.)

ventral steam of object-related processing, and a dorsal auditory stream (shown in red) that has been compared with the dorsal visual stream. The first auditory cortical areas are area 41, a core region, and area 42, a belt region. The ventral auditory stream involves projections to the anterior superior temporal cortex (aST, area 22), which in turn projects to the anterior and ventral parts of the inferior frontal gyrus (IFG), area 45. The dorsal auditory stream projects from the first auditory areas 41 and 42 to the posterior superior temporal (pST) cortex, and then to the inferior parietal lobule (IPL, area 40), which in turn projects to the more dorsal part of the inferior frontal cortex (IFC area 44), and also to premotor cortical area 6 (PMC) and to area 8.

In nonhuman primates, the ventral stream subserves the recognition of successively more complex auditory objects, ranging from elementary auditory features [e.g., frequency-modulated (FM) sweeps or bandpass noise bursts] in the first auditory cortical areas, to species-specific vocalizations (monkey calls) in higher areas including the ventral prefrontal cortex (Rauschecker and Scott 2009, Romanski, Averbeck and Diltz 2005). Indeed, in another area to which the superior temporal auditory cortex projects, the orbitofrontal cortex, we discovered neurons that respond to primate vocalizations (Rolls, Critchley, Browning and Inoue 2006a), and related these auditory neurons in this particular part of the auditory cortex hierarchy to decoding the emotional significance of vocalizations, for humans with damage to the orbitofrontal cortex and related areas were found to be impaired at identifying the emotional expression of different human vocalizations (Hornak, Rolls and Wade 1996). A feature hierarchy approach to how neurons with complex auditory properties could be formed is illustrated in Fig. 2.16.

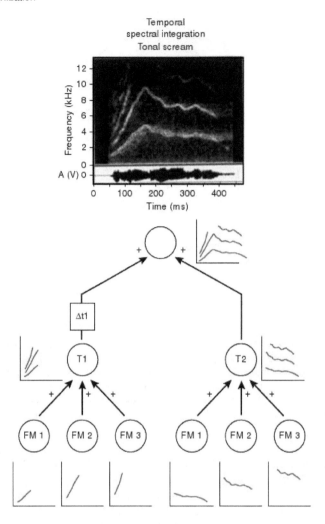

Fig. 2.16 Schematic of a feature hierarchy approach to building communication calls. Communication calls consist of elementary features, such as bandpass noise bursts or frequency-modulated (FM) sweeps. Harmonic calls, such as the vocal scream from the rhesus monkey repertoire depicted here by its spectrogram and time signal amplitude (A, measured as output voltage of a sound meter), consist of fundamental frequencies and higher harmonics. The neural circuitry for processing such calls is thought to consist of small hierarchical networks. At the lowest level, there are neurons serving as FM detectors tuned to the rate and direction of FM sweeps; these detectors extract each FM component (shown in cartoon spectrograms) in the upward and downward sweeps of the scream. The output of these FM detectors is combined nonlinearly at the next level: the target neurons T1 and T2 possess a high threshold and fire only if all inputs are activated. At the final level, a 'tonal-scream detector' is created by again combining output from neurons T1 and T2 nonlinearly. Temporal integration is accomplished by having the output of T1 pass through a delay line sufficient to hold up the input to the top neuron long enough that all inputs arrive at the same time. (Reprinted by permission from Macmillan Publishers Ltd: Nature Neuroscience, 12 (6) pp. 718–724, Maps and streams in the auditory cortex: nonhuman primates illuminate human speech processing, J P Rauschecker and S K Scott. Copyright ©2009.)

The dorsal auditory stream is implicated for example in the localization of sounds (Rauschecker and Scott 2009), and perhaps in processing the dynamic aspects of sounds (Rauschecker 2012, Bornkessel-Schlesewsky et al. 2015).

2.3   Hierarchical organization of reward value processing

Fig. 2.4 helps to reveal an important principle of cortical processing in primates including humans, that there is a clear separation between hierarchical levels involved in producing a representation of what a stimulus is, its identity and intensity (labelled Tier 1, 'what'), and hierarchical levels involved in representing the reward (and punishment) value of stimuli (labelled Tier 2, reward / affective value) (Rolls 2014a). This provides a foundation for the independence of judgements about the identity and intensity of sensory stimuli from their reward value and subjective pleasantness. For example, ratings of the intensity of sweet and salt taste are closely related to their concentration, but the pleasantness ratings are relatively much less influenced by concentration (Rolls et al. 1983a). In contrast, feeding to satiety produces a large decrease in the reward value and pleasantness of a food, to zero, but has little influence on the subjective intensity of the stimulus (Rolls et al. 1983a, Rolls and Rolls 1997, Kringelbach et al. 2003). Part of the adaptive value of this independence of perceptual from value encoding is that we may wish to learn for example about the location of food and other rewarding stimuli even if they are at a particular time not rewarding, because for example a food has been eaten to satiety. We do not go blind to a stimulus when the stimulus is no longer of reward value.

The principles of separate processing of these perceptual, reward value, and decision-making / action levels shown in Fig. 2.4 can be seen as follows.

1. In Tier 1 (Fig. 2.4), information is processed to a level at which the neurons represent 'what' the stimulus is, independently of the reward or punishment value of the stimulus. Thus neurons in the primary taste cortex represent what the taste is, and its intensity, but not its reward value, in that feeding to satiety does not reduce their responses (Rolls et al. 1988, Yaxley et al. 1988). In the primary olfactory (pyriform) cortex, activations in humans are related to intensity and not pleasantness (Grabenhorst, Rolls, Margot, da Silva and Velazco 2007); and activations to food odours are not decreased by odor devaluation by satiety (Gottfried 2015). In the inferior temporal visual cortex, where objects are represented (Rolls 2012c) (see Chapter 25), neuronal responses to food objects are not decreased by feeding to satiety (Rolls, Judge and Sanghera 1977). In the inferior temporal visual cortex, the representation is of objects, and is invariant with respect to the exact position on the retina, size, and even view. Forming invariant representations involves a great deal of cortical computation in the hierarchy of visual cortical areas from the primary visual cortex V1 to the inferior temporal visual cortex (Rolls 2008d, Rolls 2012c) (Chapter 25). The fundamental advantage of this separation of 'what' processing in Tier 1 from reward value processing in Tier 2 is that any learning in Tier 2 of the value of an object or face seen in one location on the retina, size, and view will generalize to other views etc because of the invariant representations in Tier 1. This would not be the case if the reward value learning occurred in Tier 1 where there was not a view-invariant representation. Another important advantage is that the output of the 'what' system projects from the end of Tier 1 to other brain systems such as the hippocampal memory system (see Fig. 1.11 and Chapter 24), and this enables learning about where and when objects are seen to occur even if the objects are not currently rewarding. If reward contaminated the Tier 1 representation, learning about the properties of objects might not be possible if they were not rewarding.

Evidence that there is no such clear separation of 'what' from 'value' representations in rodents, for example in the taste and olfactory systems, is described in Chapter 15, and this property makes the processing in rodents not only different from that in primates including humans, but also much more difficult to analyse, and a poor model for humans (Rolls 2015e).

2. In Tier 2, reward value is represented. For taste, in the orbitofrontal cortex reward value is represented in that feeding to satiety reduces the neuronal responses and brain activations to taste stimuli (Rolls, Sienkiewicz and Yaxley 1989c) (see Fig. 2.6). The same is found in the orbitofrontal cortex for neurons that respond to the odor or sight of food (Critchley and Rolls 1996a), and its texture (Rolls et al. 1999a). Correspondingly, in humans, orbitofrontal cortex activations to the flavour of a food are decreased by feeding to satiety, in a sensory-specific way (Kringelbach et al. 2003). Similarly, the pleasantness of touch (Rolls, O'Doherty, Kringelbach, Francis, Bowtell and McGlone 2003d, Rolls, Grabenhorst and Parris 2008a), and the reward value of money (O'Doherty, Kringelbach, Rolls, Hornak and Andrews 2001), are represented in the orbitofrontal cortex. Representing different rewards by different subpopulations of neurons in the orbitofrontal cortex provides a potential opportunity for different rewards to be represented on the same scale of value (Grabenhorst, D'Souza, Parris, Rolls and Passingham 2010a) by utilizing competitive learning and inhibition. Another principle is that by bringing all rewards close together in one brain area, the orbitofrontal cortex, the simple wiring principle of connecting that area to the next area in the hierarchy helps to ensure that the correct inputs for the next stage are all present. One such stage is the medial prefrontal cortex area 10, which is implicated in decision-making between rewards, and which would therefore need a representation of all rewards to reach it. Similarly, the orbitofrontal cortex representation of all rewards by different neurons provides a means for another tier, the anterior cingulate cortex, to receive information about all possible rewards, and punishers, to help it implement action-outcome learning. Another principle here is that by bringing together all primary rewards and potential learned reward-related e.g. visual stimuli at the object level in the orbitofrontal cortex, stimulus-reward associations can be learned in the network which provides for this convergence.

3. Tier 3. Whereas the orbitofrontal cortex in Tier 2 represents the value of stimuli (potential goals for action) on a continuous scale, an area anterior to this, medial prefrontal cortex area 10 (in Tier 3), is implicated in decision-making between stimuli, in which a selection must be made, moving beyond representation of value on a continuous scale towards a decision between goods based on their value and utilizing non-linear processing (Rolls 2014a, Rolls et al. 2010b, Rolls et al. 2010c, Grabenhorst, Rolls and Parris 2008b). Other Tier 3 areas include the interfaces to action-outcome learning in the cingulate cortex, and to stimulus-response (habit) learning in the striatum (see Fig. 2.4). The principle of keeping actions and responses separate from reward value in the orbitofrontal cortex, and having convergence with action and response learning areas in Tier 3, is that choice often need to be made between different rewards, compared on the same scale of value, in a computation that is independent of any responses that may then be useful.

2.4 Hierarchical organization of connections to the frontal lobe for short-term memory-related functions

Jones and Powell (1970) introduced interesting concepts about hierarchies of connections to the prefrontal cortex based on anatomical evidence. The implication of the anatomical connections illustrated in Fig. 2.17 is that each hierarchical stage of cortical processing in the somatic, visual and auditory cortical hierarchies has connectivity with its own level of local networks in the prefrontal cortex. A very important function of the prefrontal cortical areas is in maintaining neuronal activity in short-term memory periods (Fuster 2008, Goldman-Rakic 1996). The implication is that short-term memory is organised at different hierarchical levels to serve the functions of different levels of cortical processing hierarchies. In that

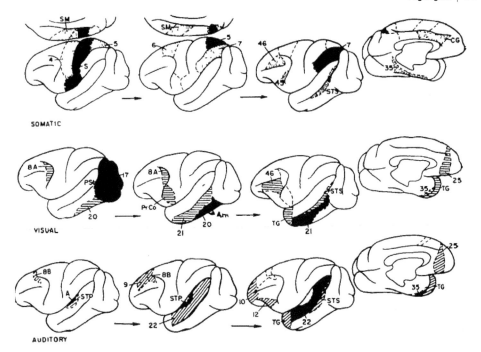

Fig. 2.17 A schematic diagram summarizing the outward progression of connections from the primary somatic, visual and auditory areas of the cortex. Each new local step is shown in black and the further connections of the new areas by light stippling or hatching. Each level of the somatic, visual and auditory hierarchy of cortical areas projects to a higher level of a prefrontal cortex hierarchy. (Reproduced with permission from E.G.Jones and T.P.S.Powell, An anatomical study of converging sensory pathways within the cerebral cortex of the monkey, *Brain*, 93(4) pp. 793–820, Figure 16, ©1970, Oxford University Press.)

different represenattions are formed at different stages in the posterior, perceptual, neocortical hierarchies, separate, that is independent, short-term memory systems are likely to be useful in the prefrontal cortex. While it may be an open question about whether the prefrontal cortex areas are themselves operating hierarchically, the concept is certainly raised that the prefrontal architecture for short-term memory, and hence multiple step planning (Rolls 2014a), has different levels of short-term memory operations that reflect the hierarchies in more posterior information processing systems. Another principle apart from independent short-term memory prefrontal networks to provide short-term capability for different levels of the posterior cortical hierarchies is that the separate prefrontal levels may then feed back the information for top-down attentional control that is appropriate for each level of the posterior cortical processing levels (see further Chapter 11).

2.5 Highlights

1. Hierarchical cortical organization is found in all sensory systems, in the reward system, and in the memory systems.
2. Adjacent cortical areas in the hierarchy are connected by strong forward connections, and weaker backprojections which have synapses in cortical layer 1.
3. There is convergence from cortical area to cortical area, in that neurons in a cortical area receive inputs from a limited region topologically of the preceding cortical area. This

enables neurons to operate with the number of synapses from the preceding cortical area received by neurons limited to in the order of 10,000 synapses. This is a major cortical principle of operation, for if each processing system consisted of only an input and an output cortical area, any neuron in the output area would need to receive the biologically implausible number of tens of millions of synapses to cover the whole space of the input cortical area.

4. The convergence from cortical area to cortical area is such that after approximately at most four areas or stages of cortical processing, the convergence is sufficient to enable a single neuron at the top of the hierarchy to receive input from anywhere in the first cortical area, as illustrated in Fig. 2.1.

5. A consequence of this connectivity is that within a neocortical area, the computation is local. Consistent with this, the recurrent collaterals of pyramidal cells spread mainly with a region of 1–3 mm, and the inhibitory neurons operate within a similar extent of neocortex.

6. One reason for the number of areas in a hierarchy being limited to approximately a maximum of four is that this keeps the computation time within reasonable limits, of approximately 20 ms per cortical area for the computation and information transmission (see Fig. 2.1), which is sufficient for a cortical area to be influenced by its recurrent collaterals and to fall towards a basin of attraction that reflects the previously learned constraints implemented in the strengths of the local recurrent collaterals (Sections B.2.6 and B.3.3.5).

7. Topography is often present associated with this connectivity, though in higher cortical areas the maps can become highly fragmented and local (as described in Section B.4.6), because the dimensionality of what is being represented is high (that is, there are many types of item, e.g. object, being represented).

8. The cortico-cortical backprojections are not sufficiently strong to make neurons at earlier stages in the hierarchy have the same responses as at higher stages. Face selective neurons with receptive fields that can be up to 70° are not found in the primary visual cortex, where the neurons respond to features such as bars and edges and have receptive fields in the order of 1° (Chapter 25). Instead, the cortical backprojections are used for gentle influences on earlier cortical stages, such as top-down attention, and memory recall if there is not a strong bottom-up input. Each cortical hierarchy thus operates computationally with a feedforward style of computation, in which new representations are built at each succeeding stage, in a divide and conquer approach to computation.

9. An advantage of this hierarchical organisation is that the same or a similar architecture and process can be repeated a number of times, which simplifies the evolution of the neocortex, which can proceed by adding another stage on to the top of an existing hierarchy.

10. Another advantage is that this simplifies the genetic specification of the cortex, in that once a useful and stable cortical network in one area has evolved, it is safest and quickest in evolution to add something with a similar specification as the next stage (Chapter 19).

11. Another advantage is that by dividing a computation into successive stages, each stage can perform a much simpler computation.

12. Another advantage is that within any one stage of the hierarchy performing a particular computation, the information that needs to be exchanged between the nearby neurons is of the same type (e.g. about the presence or not of a nearby edge in vision). That facilitates cortical design, in that generic connectivity between nearby neurons can be specified with self-organization of neuronal selectivity during development. This

obviates the need for a neuron with a particular type of response to the input to have connections with another neuron with another particular type of response, which would be impossible for genetic specification, given that we have only in the order of 25,000 genes. Issues of what the genes may specify to implement the design of any one cortical area are described in Chapter 19. The localization of function that is present can be viewed partly in this way, and has the effect of minimizing the average length of the connections between neurons, which is very important in keeping brain size down (see further Chapter 3 and Sections B.4.6 and D.4).

13. Another advantage and property of hierarchical cortical organization is that there are major connections between adjacent areas in the hierarchy, which again simplifies the genetic specification of cortical design, in that a neuron does not need to have its connectivity specified to one of a myriad of cortical areas.

14. In primates (including humans), the identity of stimuli is computed first ('what representations'), independently of reward value, and then at a later stage reward value is computed, and then at a later stages decisions between stimuli of different value are made. This independence of perceptual from 'value' processing is advantageous, for then information is available to other brain systems, including memory and language systems, even if a stimulus is not currently rewarding. This independence of perceptual from reward processing appears to be much less in rodents.

3 Localization of function

The concept of localization of function in the cortex was at one time a debated issue. Not only is there now very good empirical evidence for localization of function in the cerebral cortex, with examples evident throughout this book, but there are now clear principles of operation of the cerebral cortex that provide a foundation for understanding why localization of function is present in the cerebral cortex. These principles include the following.

3.1 Hierarchical processing

In Chapter 2 many computational advantages for hierarchical processing were described, and these provide one foundation for understanding localization of function in the cerebral cortex.

3.2 Short-range neocortical recurrent collaterals

A second foundation is that the cerebral cortex is built on the operational principle of short-range recurrent collateral excitatory connections (see Section 1.10.6) providing the basis for local attractor networks in the neocortex to implement short-term memory functions, constraint satisfaction, etc, as described in Chapter 4. A computational argument here is that the attractors must be local attractors, for otherwise the total memory capacity of a single global neocortical attractor network would be just of the order of the number of connections onto any one neuron in the network (O'Kane and Treves 1992) (see Appendix B).

3.3 Topographic maps

Another foundation for localization of neocortical function is that the recurrent collaterals are relatively short-range, in the order of 1–2 mm, for the reasons referred to in Appendix B to ensure that different neocortical attractor networks can operate relatively independently, in terms of their storage capacity. (Of course there are interactions between the different networks, to implement for example top-down attention, as described in Chapter 11.) A consequence of the short-range excitatory recurrent collaterals is the formation of topographic maps (Section B.4.6.) These maps can be advantageous, as noted above and in Chapter 2, in that they help with interpolation between trained exemplars; and in that they help networks in the next cortical area that receive from a small region of the current cortical area to receive mainly information on their synapses that is relevant to one particular task, such as face analysis.

3.4 Modularity

Modularity may help evolution by allowing a change in one brain subsystem (a module) to be selected for if it is advantageous for the phenotype, without at the same time producing changes throughout the brain that may impair the function of other brain subsystems (Barrett 2012).

3.5 Lateralization of function

Another principle is that neurons with similar functions are placed closed together, using for example topographic map formation to minimize connection lengths between neurons that need to communicate, and thus reducing brain weight (Section B.4.6). Related to this, if a computation does not have to be bilaterally represented in terms of for example separate cortical areas for the right and left hands, then such functions are most efficiently lateralized to a considerable extent to one hemisphere. This happens with language in humans, which in right-handed people is typically in the left hemisphere. Consistent with this, the memory of allocentric space and for words does not need separate egocentric representations, and in the human hippocampus there is right (spatial) vs left (word) specialization of function (Crane and Milner 2005, Burgess, Maguire and O'Keefe 2002, Barkas, Henderson, Hamilton, Redhead and Gray 2010, Bonelli, Powell, Yogarajah, Samson, Symms, Thompson, Koepp and Duncan 2010, Sidhu, Stretton, Winston, Bonelli, Centeno, Vollmar, Symms, Thompson, Koepp and Duncan 2013).

3.6 Ventral and dorsal cortical areas

It is well established that there is a ventral visual stream specialised for 'what' processing for perception (Ungerleider and Mishkin 1982, Ungerleider 1995), followed by multimodal convergence, and then object reward decoding in the orbitofrontal cortex (Rolls 2014a) (Figs. 1.9, 1.10, and 1.11, and Section 1.9). In a complementary way, the dorsal visual stream analyses motion and position in space, and links via parietal cortex into motor and somatosensory cortical areas in the dorsal part of the primate including human neocortex, with these parts specialized for 'where' stimuli are (Ungerleider and Mishkin 1982, Ungerleider 1995, Perry and Fallah 2014) (Figs. 1.9, 1.12, and 1.11). Another characterization is that the ventral visual stream is involved in perception, and the dorsal visual stream in action (Milner and Goodale 1995). A similar distinction can be made for auditory processing (Fig. 2.15 and Section 2.2.6) (Rauschecker 2012, Bornkessel-Schlesewsky et al. 2015). The orbitofrontal cortex can be thought of as the part of the primate brain where all the ventral 'what' representations for vision, hearing, taste, olfaction, and touch / oral texture are brought together to form multimodal representations of objects that then become represented in terms of their reward value (Rolls 2014a) (Sections 2.2.5 and Chapter 15). Multimodal semantic representations of stimuli may also be formed in anterior temporal lobe areas.

Pandya, Seltzer, Petrides and Cipolloni (2015) have hypothesized that this ventral / dorsal division of cortical processing is related to an evolutionary history in which 3-layer cortex of which olfactory cortex (paleocortex) is an example, gradually developed into 6-layer neocortex for the ventral 'what' areas; and in which 3-layer archicortex, of which hippocampal cortex is an example, gradually developed into 6-layer neocortex for the dorsal 'where' or 'spatial' areas. (The different types of cortex are described in Chapter 18, and Kaas and Preuss (2014) show how the olfactory cortex is arranged ventrally and the hippocampal cortex is mainly dorsal earlier in evolutionary history.) They term this the 'dual origin' concept. It turns out that Pandya et al. (2015) would need to have the posterior hippocampus especially in mind, for it is the posterior hippocampus that has especially developed connections with dorsal stream areas including with such areas as the parietal areas, splenium, precuneus, and posterior cingulate cortex (Vogt 2009). Further, the hippocampus is a place in the brain where spatial representations are brought together with object representations for episodic memory, so the hippocampus would seem to be a site that needs input from, and is closely related to, both the ventral and dorsal processing streams (Rolls 2015c) (Chapter 24).

My own view is that while the 'dual origin' concept is an interesting evolutionary concept, and may be better supported in future by molecular markers that may allow better comparisons across species of different cortical areas and types of cortex (Bernard et al. 2012), there are complementary but compelling computational reasons for the ventral / dorsal division of cortical areas. In particular when object representations are being computed, information about features, and combinations of features, is necessary (including for vision, hearing, olfaction, and taste) to form the object representations (see e.g. Chapter 25). For this purpose, movement, and where the object is in space, are not useful, and indeed for object recognition it is necessary to throw away information about where the object is in space, for the aim is to recognise the object independently of where it is in space. Similarly, to compute global motion, or position in space, information about the relative positions of features is not necessary, and instead the emphasis is on representations of global motion (at least in local patches in a scene) and where they are occurring in space. Given the important principle of keeping neurons that need to exchange and combine information for particular computations to be localised in the cortex to minimise the connection distances and thus brain weight, and to simplify the specification of which neurons need to be connected to each other (see Section B.4.6), the neocortex is led to an organizational scheme that utilizes localisation of function, and therefore separation at the gross level between ventral and dorsal processing streams.

Another question about the 'dual origin' concept (Pandya, Seltzer, Petrides and Cipolloni 2015) is why different types of three-layer cortex (paleocortex and archicortex) would have (independently?) evolved to a six-layer neocortex. The concepts described in Chapter 18 may offer a more parsimonious approach to the principles of operation of the neocortex.

3.7 Highlights

1. Neocortical function is localized for the reasons described in Chapter 2, including limiting the numbers of synapses received by neurons to in the order of 20,000; minimizing the axonal connection lengths between neurons involved in the same computation by enabling those neurons to be close together in the cortex; and simplifying the genetic specification in that connections can often be specified to be local, and may not need to specify one of every class of neuron found in the cortex (see Chapter 19).

2. Neocortical function is also localised and modular so that the attractors implemented by the recurrent collaterals of the pyramidal cells can be separate, with separate basins of attraction. Without this, the number of memories that could be stored in the whole of the cortex would be limited by the number of memories that can be stored in any one attractor network, which is in the order of the number of synaptic connections devoted to recurrent collateral synapses, which is in the order of 10,000 (Section B.3.3.7).

4 Recurrent collateral connections and attractor networks

4.1 Introduction

This may be the most important Chapter of this book. The hypothesis is developed that the characteristic property of the cerebral cortex is excitatory recurrent collateral connections between the pyramidal cells. This creates positive feedback, which is inherently unstable and dangerous. The cost may be epilepsy, and states such as the positive symptoms of schizophrenia. In contrast, the basal ganglia appear to operate much more safely, and have an earlier evolutionary origin, because they operate by mutual inhibition between the neurons, which can perform selection (see Chapter 20). The benefits of cortical recurrent collaterals are though immense, and must have driven its evolution. The benefits include maintaining an active state in short-term memory which is produced by the positive feedback maintaining a subset of neurons (that have enhanced synaptic connectivity between them) in an active continuing firing state. The same architecture provides for long-term memory, in which a whole memory can be recalled from any part, a process termed completion; for constraint satisfaction in which the activity of whole populations of activated neurons is taken into account in achieving the most consistent solution given the starting evidence and previous learning; and for decision-making, in which a stable decision is reached given the competing inputs, and which has the advantage that the decision can be maintained active as a short-term memory state that can be used as the goal for subsequent actions. These processes provide the fundamental computational bases for functions such as multiple-step planning, which requires multiple items to be held in short-term memory as the steps of the plan that can all be performed 'off-line' as thought processes without performing the plan; for the recall of long term-memory; for associations between memories that can lead with probabilistic effects implemented by the stochasticity of neuronal firing to new thoughts which provide an important component of creativity; and even for language, which requires multiple items in a sentence to remain active and in context for later thoughts about what was produced. These effects implemented by the recurrent collateral architecture of the cerebral cortex are elucidated in this Chapter.

4.2 Attractor networks implemented by the recurrent collaterals

The recurrent collateral connections between pyramidal cells can implement attractor states if the synaptic connections are associatively modifiable. Long-term potentiation does occur in the recurrent collateral connections in both the hippocampus (Hasselmo, Schnell and Barkai 1995, Jackson 2013) and neocortex (Fregnac et al. 2010), providing evidence that these synapses are associatively modifiable. The operation and properties of attractor networks are introduced in Section 1.8 and described in detail in Section B.3. In brief, during storage of a memory in the attractor network, the synapses between the neurons with high firing rates that represent what is to be stored increase in strength because of the associative synaptic

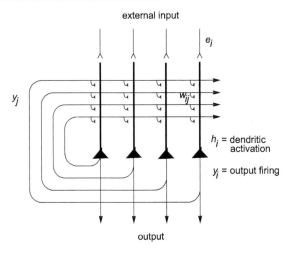

external input

e_i

y_j

w_{ij}

h_i = dendritic activation

y_i = output firing

output

Fig. 4.1 The architecture of an autoassociative neural network.

modifiability. The output firing effectively is presented to the synapses on each neuron because of the recurrent collateral connections, as shown in Fig. 4.1, and effectively becomes associated with itself. Hence these networks are often referred to as autoassociation networks. One of the advantages is that the positive feedback keeps a subpopulation of neurons firing for many seconds, and this can implement short-term memory. Short-term memory provides a major advance in evolution, for it enables recent events that may be related to future action to be recalled; it enables attention to be maintained on a single target for many seconds or longer; and it potentially enables many steps in a plan to be maintained on-line, and so facilitates planning. Another advantage is that if only part of the subpopulation representing any one memory is activated, this can lead to completion of the whole memory. This occurs because in an attractor network all the subparts of a single memory are associated with each other. This is fundamental to episodic memory, in which the whole of a recent memory about an episode can be recalled from any part (Chapter 24); and to semantic long-term memory, in which all the properties of an object can be recalled from a part.

The prefrontal cortical areas, especially the dorsolateral prefrontal cortex which may be more specialized for spatial functions, and the ventral prefrontal cortex which may be more specialized for object short-term memory functions, are specialized for maintaining evidence 'off-line', i.e. independently of sensory input, whereas the more posterior perceptual and sensory areas such as the visual cortical areas necessarily have their activity dominated by the sensory input, which can not be ignored if survival is important. Associated with this increase in the importance of short-term memory operations in the prefrontal cortical areas, neurons in the prefrontal cortex have much larger dendrites with more recurrent collateral connections (Elston, Benavides-Piccione, Elston, Zietsch, Defelipe, Manger, Casagrande and Kaas 2006) to increase the operation as a short-term memory system.

4.3 Evidence for attractor networks implemented by recurrent collateral connections between cortical neurons

There is considerable evidence that recurrent collateral connections do implement attractor, autoassociation, memories in the cerebral cortex.

4.3.1 Short-term Memory

One short-term memory system is in the dorso-lateral prefrontal cortex, area 46. This is involved in remembering the locations of spatial responses, in for example delayed spatial response tasks (Goldman-Rakic 1996, Fuster 2000). In such a task performed by a monkey, a light beside one of two response keys illuminates briefly, there is then a delay of several seconds, and then the monkey must touch the appropriate key in order to obtain a food reward. The monkey must not initiate the response until the end of the delay period, and must hold a central key continuously in the delay period. Lesions of the prefrontal cortex in the region of the principal sulcus impair the performance of this task if there is a delay, but not if there is no delay. Some neurons in this region fire in the delay period, while the response is being remembered (Fuster 1973, Fuster 1989, Goldman-Rakic 1996, Fuster 2000). Different neurons fire for the two different responses.

There is an analogous system in a more dorsal and posterior part of the prefrontal cortex involved in remembering the position in visual space to which an eye movement (a saccade) should be made (Funahashi, Bruce and Goldman-Rakic 1989, Goldman-Rakic 1996). In this case, the monkey may be asked to remember which of eight lights appeared, and after the delay to move his eyes to the light that was briefly illuminated. The short-term memory function is topographically organized, in that lesions in small parts of the system impair remembered eye movements only to that eye position. Moreover, neurons in the appropriate part of the topographic map respond to eye movements in one but not in other directions (see Fig. 4.2). Such a memory system could be easily implemented in such a topographically organized system by having local cortical connections between nearby pyramidal cells which implement an attractor network (Fig. 4.3). Then triggering activity in one part of the topographically organized system would lead to sustained activity in that part of the map, thus implementing a short-term or working memory for eye movements to that position in space.

Another short-term memory system is implemented in the inferior temporal visual cortex, especially more ventrally towards the perirhinal cortex (see Section 24.2.6). This memory is for whether a particular visual stimulus (such as a face) has been seen recently. This is implemented in two ways. One way is that some neurons respond more to a novel than to a familiar visual stimulus in such tasks, or in other cases respond to the familiar, selected, stimulus (Baylis and Rolls 1987, Miller and Desimone 1994) (see Section 24.2.6). The other way is that some neurons, especially more ventrally, continue to fire in the delay period of a delayed match to sample task (Fahy, Riches and Brown 1993, Miyashita 1993), and fire for several hundred milliseconds after a 16 ms visual stimulus has been presented (Rolls and Tovee 1994) (see Fig. C.17). These neurons can be considered to reflect the implementation of an attractor network between the pyramidal cells in this region (Amit 1995). A cortical syndrome that may reflect loss of such a short-term visual memory system is simultanagnosia, in which more than one visual stimulus cannot be remembered for more than a few seconds (Warrington and Weiskrantz 1973, Kolb and Whishaw 2015).

In the main part of the inferior temporal visual cortex, there is evidence for the operation of an attractor system, but it is insufficiently powerful to sustain the neuronal firing in the absence of a visual input. We have found that inferior temporal visual cortex neurons typically respond for 200–300 ms after a visual stimulus is terminated (see example in Fig. 16.15). Given that the latency to respond is approximately 100 ms, the firing is prolonged for 100–200 ms longer than expected. This is not found in the lateral geniculate nucleus (K. Martin, personal communication), in which there are no recurrent collateral connections. The implication is that the inferior temporal visual cortex utilizes recurrent collateral connections for constraint satisfaction (i.e. taking into account the possibly weak or inconsistent inputs that it receives and the categories of objects about which it has previously learned towards which it is attracted),

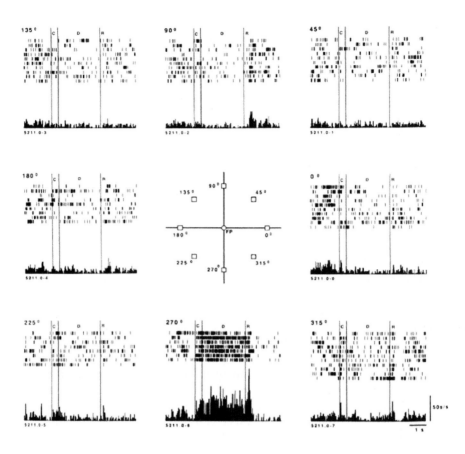

Fig. 4.2 The activity of a single neuron in the dorsolateral prefrontal cortical area involved in remembered saccades. Each row is a single trial, with each spike shown by a vertical line. A cue is shown in the cue (C) period, there is then a delay (D) period without the cue in which the cue position must be remembered, then there is a response (R) period. The monkey fixates the central fixation point (FP) during the cue and delay periods, and saccades to the position where the cue was shown, in one of the eight positions indicated, in the response period. The neuron increased its activity primarily for saccades to position 270°. The increase of activity was in the cue, delay, and response period while the response was made. The time calibration is 1 s. (Reproduced with permission from *Journal of Neurophysiology* 61: 331–349, Figure 3, Mnemonic coding of visual space in monkey dorsolateral prefrontal cortex, S. Funahashi, C. J. Bruce, P. S. Goldman-Rakic. Copyright ©1989, The American Physiological Society.)

but does not have sufficient strength to maintain the attractor without input. Further evidence for this principle of operation is that after inferior temporal cortex neurons have been trained on a set of exemplars, intermediate (morphed) versions of the stimuli tend to produce firing that corresponds to one of the training set (Akrami, Liu, Treves and Jagadeesh 2009). Thus in sensory, perceptual, cortical areas, a principle of operation is that local attractor networks implemented by the recurrent collateral connections implement constraint satisfaction, but not self-sustaining short-term memory, helping these cortical areas to perform sensory and

Local autoassociation networks in the prefrontal cortex
for delayed spatial responses, eg. saccades

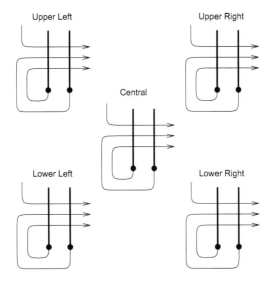

Fig. 4.3 A possible cortical model of a topographically organized set of attractor networks in the prefrontal cortex that could be used to remember the position to which saccades should be made. Excitatory local recurrent collateral Hebb-modifiable connections would enable a set of separate attractors to operate. The input that would trigger one of the attractors into continuing activity in a memory delay period would come from the parietal cortex, and topology of the inputs would result in separate attractors for remembering different positions in space. (For example inputs for the Upper Left of space would trigger an attractor that would remember the upper left of space.) Neurons in different parts of this cortical area would have activity related to remembering one part of space; and damage to a part of this cortical area concerned with one part of space would result in impairments in remembering targets to which to saccade for only one part of space.

perceptual functions rather than short-term memory. The limited short-term memory for 100–200 ms though may have a very useful memory-related function, by helping to maintain a short-term memory trace of previous neuronal firing, which helps the cortex to build view-invariant representations of objects, as described in Chapter 25.

Continuing firing of neurons in short-term memory tasks in the delay period is also found in other cortical areas. For example, it is found in an object short-term memory task (delayed match to sample) in the inferior frontal convexity cortex, in a region connected to the ventral temporal cortex (Fuster 1989, Wilson, O'Sclaidhe and Goldman-Rakic 1993). However, whether this network is distinct from the network in the dorsolateral prefrontal cortex involved in spatial short-term memory is not clear, as some neurons in these regions may be involved in both spatial and object short-term memory tasks (Rao, Rainer and Miller 1997). Heterogeneous populations of neurons, some with more spatial input, others with more object input, and top-down attentional modulation of the different subpopulations depending on the task, starts to provide an explanation for these findings.

Delay-related neuronal firing is also found in the parietal cortex when monkeys are remembering a target to which a saccade should be made (Andersen 1995); and in the motor cortex when a monkey is remembering a direction in which to reach with the arm (Georgopoulos 1995).

Another short-term memory system is human auditory–verbal short-term memory, which appears to be implemented in the left hemisphere at the junction of the temporal, parietal, and occipital lobes. Patients with damage to this system are described clinically as showing conduction aphasia, in that they cannot repeat a heard string of words (cannot conduct the input to the output) (Warrington and Weiskrantz 1973, Kolb and Whishaw 2015).

4.3.2 Long-term Memory

There is evidence that the hippocampal memory recall process operates as an attractor, with recall of a memory taking place within a single theta cycle. Indeed, it has been found that within each theta cycle, when tested in ambiguous places, hippocampal pyramidal neurons may stochastically represent one or other of the learned environments (Jezek, Henriksen, Treves, Moser and Moser 2011). This is an indication, predicted by Rolls and Treves (1998), that autoassociative memory recall can take place sufficiently rapidly to be complete within one theta cycle (120 ms), and that theta cycles could provide a mechanism for a fresh retrieval process to occur after a reset caused by the inhibitory part of each theta cycle, so that the memory can be updated rapidly to reflect a continuously changing environment, and not remain too long in an attractor state.

Further evidence that the hippocampus acts as an attractor network comes from experiments with morphed environments. Hippocampal place cells are recorded while the rat is in each of two environments, for example circular or square. Then the rat is tested in an intermediate (morphed) environment, and it is found that the place cells may respond as if they are in one of the two environments in which the rat was originally trained. The concept is that when the recall cue is intermediate, the attractor network falls into one of its previously trained states, representing either the square or the circular environment (Leutgeb, Leutgeb, Treves, Moser and Moser 2004, Leutgeb, Leutgeb, Treves, Meyer, Barnes, McNaughton, Moser and Moser 2005, Wills, Lever, Cacucci, Burgess and O'Keefe 2005).

Further evidence that in the hippocampal system there are attractor networks used for timing and therefore potentially useful for sequence memory is that some hippocampal neurons have peaks of firing at different times during a delay period of 10 s (MacDonald, Lepage, Eden and Eichenbaum 2011, Eichenbaum 2014). The attractor mechanism for these neurons may be in the entorhinal cortex, where timing cells are also found (Kraus, Brandon, Robinson, Connerney, Hasselmo and Eichenbaum 2015).

Evidence for attractors in another brain region, the orbitofrontal cortex, is that a non-reward / error population of neurons is triggered into a high firing rate state which lasts for many seconds when an expected reward is not received (Thorpe, Rolls and Maddison 1983, Rolls and Grabenhorst 2008, Rolls 2016d, Rolls and Deco 2016). This is part of a mechanism for rapid, rule-based reversal learning (Rolls 2014a, Deco and Rolls 2005d), and over-activity of this system may be involved in depression (Rolls 2016d) (see Section 16.3).

4.3.3 Decision-Making

If there are competing inputs to an attractor network, the system will under the influence of the spiking-related noise eventually choose one of the attractor states, and remain stably in that state. This provides a model of decision-making, and of memory recall. This mechanism, and the evidence that this is implemented in the cortex, is described in Chapter 5.

4.4 The storage capacity of attractor networks

Section B.3.3.7 describes quantitatively the storage capacity of attractor networks. Important and rigorous analytic results supported by simulations show that the number of separate memories that can be stored in an attractor network is proportional to the number of associatively modifiable recurrent collateral connections C onto each neuron, with the number increasing as the representation becomes more sparse (Hopfield 1982, Amit 1989, Treves and Rolls 1991, Treves 1991b, Rolls 2012a). The sparseness a can be thought of as the proportion of neurons firing to any one memory to be stored in the network. This applies to binary representations, in which for example the neurons are either firing at a high rate or are not firing. The definition has been extended to the biologically realistic case where the firings are graded, with some neurons firing fast to a stimulus, many have intermediate firing rates, and very many have low firing rates. Each memory is the firing of the whole set of neurons N in the network, occurring in the brain during a time in the order of one to a few seconds. C may be lower than N if the connectivity is diluted, that is if each neuron receives synapses from only a subset of the other neurons in the network. This is explained in Section B.3.3.7. The important point here is that the number of memories that can be stored in approximately $0.14N$ if the sparseness of a binary representation is 0.5. If a cortical attractor network has 10,000 connections C onto each memory, this would correspond to 1,400 memories. With sparse representations many more memories can be stored. For example, with $a = 0.1$, the number of memories that can be stored is approximately C. In the rat CA3 region of the hippocampus where C is 12,000, the storage capacity is thus estimated to be, if the sparseness of its representation is 0.1, in the order of 12,000 different memories (or 36,000 if the sparseness is 0.02). That number of memories may well be larger in humans, in which CA3 hippocampal neurons may have C considerably larger than in rats.

With graded firing rate representations, if the firing rate of each neuron is to be stored, this requires extra information to be stored than only which neurons are firing, and this does reduce the memory capacity of an attractor network somewhat (Treves 1990, Treves and Rolls 1991, Treves 1991b, Rolls, Treves, Foster and Perez-Vicente 1997c). An implication is that in cortical areas where the need is to store large numbers of memories, there are advantages to relatively binary encoding (and to non-linearity in the neurons of the type that is implemented by NMDA receptors, as described in Section B.3.3.7); whereas in cortical areas in which much information must be represented, such as perceptual areas, graded firing may be very useful, as described in Appendix C.

4.5 A global attractor network in hippocampal CA3, but local attractor networks in the neocortex

The hippocampal CA3 network has recurrent collateral connections that extend throughout CA3, enabling any one CA3 neuron to have a chance of connecting with any other CA3 neuron. This is the case in rats, and even more so in primates where there is little gradient in the connectivity as a function of distance (Kondo, Lavenex and Amaral 2009). The implication of this connectivity is that any one subset of neurons (representing for example a place) will have some synaptic connections with any other subset of neurons (representing for example an object). This enables arbitrary associations to be formed, for example between any object and any place. This is a requirement for episodic memory (the memory of recent past episodes, such as whom one met, and where the meeting was). That is the important property of global connectivity in an attractor network, and is key in my theory of hippocampal function (Rolls and Kesner 2006, Kesner and Rolls 2015, Rolls 1987, Rolls 1989b, Rolls 1990b) (see Chapter 24). That is the reason, accounted for by the theory, of why hippocampal CA3 connectivity is global within CA3, and this is an important principle of the operation of the hippocampus.

The question arises then about what the situation is in the neocortex, where the connectivity implemented by the recurrent collateral connections is mainly local, within 1–3 mm. One important point is that neocortical connections would be very inefficient in terms of storage capacity if they were global, for then the total number of memories that could be stored in the whole of the cerebral neocortex would be of order C, the number of connections onto any one neocortical neuron, which is in the order of 10,000. This was realised by O'Kane and Treves (1992). The principle then for the neocortex is that the rather local connectivity of the recurrent collaterals allows cortical networks more than 1–3 mm apart to operate largely independently in terms of which attractor any one network stores and retrieves. This rescues the situation, and enables the total number of memories that might be stored to increase from of order C to of order C times the number of independent local neocortical networks there are. If we assume that each neocortical network occupies a cortical area of say 3 mm^2 (given by the range of the local recurrent collateral connections), and the total area of the human neocortex is 300,000 mm^2 (Table 1.1), then the total number of networks is of order 100,000 and the total number of memories that might be stored is in the order of 10^5 networks $\times$ 10^4 memories per network $= 10^9$ memories. (This number includes the number of attractor-related states in early cortical areas, so the number of items that humans can store and later remember will be smaller than this.)

There are some important quantitative constraints here that are required for attractor networks with connections between them to influence each other (e.g. to pass information from one network to the next). One requirement is that the connections between the networks must be low in strength relative to the strength of the recurrent collateral connections within the attractor network, with a useful operating region being found with the techniques of theoretical physics when this ratio is in the order of 0.1 (Renart, Parga and Rolls 1999b, Renart, Parga and Rolls 1999a, Renart, Moreno, Rocha, Parga and Rolls 2001, Rolls, Webb and Deco 2012). This again is a fundamental principle of operation of the neocortex. The scenario is not that every neocortical area acts to store what we normally think of as memories; but instead that for its perceptual, mnemonic, decision-making, motor and other functions a principle of operation is that attractor dynamics and synaptic changes and the retrieval of information from the recurrent collateral synapses are part of the mechanism by which the neocortex performs these functions.

These fundamental computational ideas then provide a reason for why the connectivity of the neocortex is to a considerable extent local. That in turn provides a fundamental reason for the localization of function in the cerebral cortex. The neocortical computational networks must be largely local, and each one must be performing something relatively independent for the sum of their operations to add up. This is an important principle of neocortical function.

Within this scenario, the local networks must of course be connected to each other in an appropriate way so that useful functions can be performed by the whole system of cortical local networks. One such principle is that of hierarchical organization, described in Chapter 2. Another such principle is that of backprojections between adjacent cortical area, which implement functions such as memory recall and top-down attention, as described in Chapters 11 and 24.

In the neocortex, the local recurrent collateral connectivity can be conceptualized as being dense locally, and falling off as a function of distance over 1–3 mm. The implication is that there is some connectivity overlap between neighbouring neocortical attractor networks. This scenario will produce some interference between neighbouring neocortical attractor networks, and this is likely to reduce the memory capacity of each local cortical attractor network, probably by a factor of 2–3 (Roudi and Treves 2006, Roudi and Treves 2008). Although this is a cost, there are benefits, including simplicity of genetic specification of cortical design, and the continuous topographic maps of information that self-organize due to the locality of cortical recurrent collaterals, with the maps providing a way for similar neurons involved in similar computations to be grouped together which minimizes total cortical inter-neuronal connection length and thus brain size and weight, and potentially for interpolation across the map, as described in Section 4.8.

4.6 The speed of operation of cortical attractor networks

If a simple implementation of an autoassociation net such as that described by Hopfield (1982) is simulated on a computer, then 5–15 iterations are typically necessary for completion of an incomplete input cue e. This might be taken to correspond to 50–200 ms in the brain, rather too slow for any one local network in the brain to function. However, it has been shown that if the neurons are treated not as McCulloch–Pitts neurons which are simply 'updated' at each iteration, or cycle of timesteps (and assume the active state if the threshold is exceeded), but instead are analyzed and modelled as 'integrate-and-fire' neurons in real continuous time, then the network can effectively 'relax' into its recall state very rapidly, in one or two time constants of the synapses (see Section B.6 and Treves (1993), Battaglia and Treves (1998a), Panzeri, Rolls, Battaglia and Lavis (2001), and Appendix A5 of Rolls and Treves (1998)). This corresponds to perhaps 20 ms in the brain.

One factor in this rapid dynamics of autoassociative networks with brain-like 'integrate-and-fire' membrane and synaptic properties is that with some spontaneous activity, some of the neurons in the network are close to threshold already before the recall cue is applied, and hence some of the neurons are very quickly pushed by the recall cue into firing, so that information starts to be exchanged very rapidly (within 1–2 ms of brain time) through the modified synapses by the neurons in the network. The progressive exchange of information starting early on within what would otherwise be thought of as an iteration period (of perhaps 20 ms, corresponding to a neuronal firing rate of 50 spikes/s) is the mechanism accounting for rapid recall in an autoassociative neuronal network made biologically realistic in this way. Further analysis of the fast dynamics of these networks if they are implemented in a biologically plausible way with 'integrate-and-fire' neurons is provided in Section B.6, in Appendix A5 of Rolls and Treves (1998), and by Treves (1993). *The general approach applies to other networks with recurrent connections, not just autoassociators, and the fact that such networks can operate much faster than it would seem from simple models that follow discrete time dynamics is probably a major factor in enabling these networks to provide some of the building blocks of cortical function.*

4.7 Dilution of recurrent collateral cortical connectivity

Cortical networks typically have diluted connectivity, that is, the probability that one cortical neuron will make a synapse with another even neighbouring cortical neuron is less than 1. In the CA3 region of the rat hippocampus, $N \approx 300,000$ neurons, and the number of recurrent collateral connections received by each neuron is $C \approx 12,000$, making the dilution of the connectivity 0.04. In the neocortex, the value may be closer to 0.1 (see Chapter 1). (In the neocortex, more than one synapse from an individual axon can sometimes be observed, but these connections may be from thalamic inputs which are involved in driving cortical cells; and in the connections between cortical areas, which may be involved more in one area driving another than in attractor network operations. It is the dilution of the connectivity in the recurrent collateral connections that is being considered here, and its implications for the operation of cortical attractor networks.)

What are the implications of this diluted connectivity in cortical attractor networks?

The first important principle is that attractor networks operate well with connectivity that is randomly diluted to 0.1 or even to 0.01, with little impact on the storage capacity, which is still set primarily by C, a, and the firing rate distribution (Treves and Rolls 1991, Treves 1991b, Bovier and Gayrard 1992, Rolls and Treves 1998, Rolls 2012a). The intuition about this is that even with an incomplete retrieval cue, neurons that were co-firing when the neuronal representation was stored will gradually recruit each other through the modified synaptic connections that are present, and the network will fall into a stable basin of attraction (recall state). However, it is clear that there must be a threshold for the maximal diluted connectivity which does not prevent the storage capacity from being reached. The critical probability for the random connections in a Hopfield model is derived as a function of the size of the network, decaying as $1/\sqrt{n}$ (Bovier and Gayrard 1992). Furthermore one can argue that the auto-associative memory functions properly, or more efficiently, on graphs that permit bootstrap percolation (Turova 2012).

A second important principle is that the speed of recall can still be fast, in an integrate-and-fire network with diluted connectivity, continuous dynamics, and random firing times of the neurons (Treves 1993, Treves, Rolls and Simmen 1997, Battaglia and Treves 1998a). The intuition about this is that even with an incomplete retrieval cue, some neurons because they happen to be close to threshold when the retrieval cue is applied will fire earlier than expected and will transmit this information to other neurons in the network that are close to threshold, so that the recall will build up quickly across the subset of neurons that have strengthened synapses between them to store a memory. The retrieval time for a single attractor network is still in the order or 1.5 to 2 time constants of the relevant synapses between the neurons, which for AMPA receptors is less than 10 ms, so that recall can be complete within 20 ms in a biologically plausible network.

A third important principle has been discovered recently (Rolls 2012a). The hypothesis is that the diluted connectivity allows biological processes that set up synaptic connections between neurons to arrange for there to be only very rarely more than one synaptic connection between any pair of neurons. If connections are made at random between nearby cortical neurons (a parsimonious assumption that has the great advantage of simplicity of cortical specification by genes), then if the average connection probability between a pair of neurons was 1, many neurons would have more than one connection between then, according to a Poisson distribution. Dilution of the connectivity to 0.1 keeps the probability of such multiple connections low, with biologically plausible scenarios set out by Rolls (2012a). If probabilistically there were more than one connection between any two neurons, it was shown by simulation of an autoassociation attractor network that such connections would dominate the attractor states into which the network could enter and be stable, thus strongly reducing the memory capacity of the network (the number of memories that can be stored and correctly retrieved), below the normal large capacity for diluted connectivity, by a factor of 3 or more. Diluted connectivity between neurons in the cortex thus may now be understood as allowing high capacity of memory networks in the cortex by reducing the probability of multiple connections between any pair of neurons in the network. The somewhat greater dilution of connectivity (with fewer multiple synapses) in the hippocampal CA3 network (0.04) than the estimate of its value in the neocortex (0.1) may be related to the great importance of achieving a high memory capacity with good retrieval of all stored patterns in the hippocampus where this may be useful for episodic memory, for which memory capacity in a single network is important (Rolls 2008d, Rolls 2010b, Kesner and Rolls 2015).

Another advantage of diluted connectivity (for the same number of connections per neuron) is that this increases the stability and accuracy of the network as there is less spiking-related noise producing stochastic fluctuations in the diluted network (which has more neurons in it), with little cost in increased decision times (Rolls and Webb 2012).

4.8   Self-organizing topographic maps in the neocortex

The local excitatory recurrent collateral connections between nearby neurons in the neocortex encourages neurons that are close together to respond to similar stimuli or features in the input space. This will occur even if the recurrent collateral connections are modifiable. If we then assume that the synapses that introduce inputs to the network are associatively modifiable, then nearby neurons will tend to learn to similar features, because the recurrent collaterals encourage similar firing of nearby neurons. If there is competition between the neurons implemented by a population of feedback inhibitory neurons, then different neurons tend to learn different patterns, as in a standard competitive network described in Section B.4. However, the combination of the local recurrent excitation and the competition result is neurons that learn to similar features being close in the map, and neurons that respond to different features being far apart in the space. The result is the self-organization of a topographic map (Malsburg 1973, Willshaw and von der Malsburg 1976, Kohonen 1982, Kohonen 1989, Kohonen 1995). This is illustrated in Figs. B.21 and B.22, and by the simulation software described in Section D.4.

These topographic maps provide a way for similar neurons involved in similar computations to be grouped together, and this is important as set out in Chapter 2. This minimizes total cortical inter-neuronal connection length and thus brain size and weight, which are major factors that influence the evolution of the brain. It also potentially allows for interpolation across the map.

4.9   Attractors formed by the forward and backward connections between cortical areas?

In principle, the feedforward and backprojection connections between cortical areas at adjacent levels of a cortical hierarchy (Chapter 2) might provide for an attractor to be formed. The associative synaptic plasticity of both the forward and back connections might support the formation of attractors. On balance, this appears to be unlikely, for a number of reasons.

First, the forward connections need to be strong, to dominate the firing at the next level, so that most parts of the cortical network can reflect the changing input to the brain, and not be lost in a world of attractor-mediated reverie. At the same time, the backprojections must be weak, for they are implicated in recall if bottom-up inputs to the preceding level are weak, but must not dominate the firing at the previous level if there is a bottom-up input (Section 11.5). If there were an attractor in this forward / backprojection system, then because the forward connections are much stronger, these would dominate the attractor basins, as described in Section 4.7 (Rolls 2012a).

The empirical evidence is that the forward and backward connections between connected cortical levels in at least the early layers of the cortical hierarchy do not implement long-lasting attractor states, is the same as that described in Section 4.3, namely that early cortical areas in the visual system do not maintain continuing firing in the absence of bottom-up input.

A strong argument based on theory is that cortical attractors must be local, for the reasons set out in Section 4.5, namely that otherwise the total storage capacity of the neocortex would be the same as that of any one attractor network in which the leading factor is the number of recurrent collateral synapses C onto each neuron; and that it would be very inefficient to have the whole of the neocortex involved in one type of computation, which would be implied if the whole of the neocortex was a single attractor. This argument applies in the same way to the connected levels in a neocortical hierarchy, in which a lower level does not represent the same information as a higher level (e.g. there are face-selective neurons in the inferior

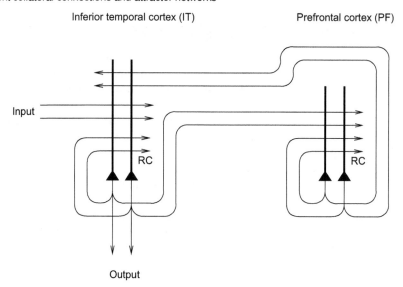

Fig. 4.4 A short-term memory autoassociation network in the prefrontal cortex could hold active a working memory representation by maintaining its firing in an attractor state. The prefrontal module would be loaded with the to-be-remembered stimulus by the posterior module (in the temporal or parietal cortex) in which the incoming stimuli are represented. Backprojections from the prefrontal short-term memory module to the posterior module would enable the working memory to be unloaded, to for example influence on-going perception, mainly when the bottom-up inputs to IT were absent or weak (see text). RC, recurrent collateral connections.

temporal cortex but not in V1), and indeed different computations (that is, computations on different representations) must be performed at each layer of the hierarchy for the hierarchy to operate by convergence and competition to produce new representations (Chapter 2). For these reasons, adjacent levels in a neocortical hierarchy need to perform independent computations, and thus are not likely to be coupled by sufficiently strong forward and backward connections for a single stable attractor supporting continuing firing to be formed. An exception may be during memory recall, when information from a higher level in an attractor might reinstate firing at an earlier level of processing, particularly when the bottom-up inputs are absent or weak, as illustrated in Fig. 4.4.

Nevertheless, the top-down backprojections to the adjacent cortical level in a neocortical hierarchy can have an important influence on the attractor states in the earlier level because of a top-down biasing effect similar to that involved in top-down attention, and this is potentially very useful, as described in Section 4.10.

4.10 Interacting attractor networks

It is prototypical of the cerebral neocortical areas that there are recurrent collateral connections between the neurons within an area or module, and forward connections to the next cortical area in the hierarchy, which in turn sends backprojections (see Chapter 11). This architecture, made explicit in Fig. 4.4, immediately suggests, given that the recurrent connections within a module, and the forward and backward connections, are likely to be associatively modifiable, that the operation incorporates at least to some extent, interactions between coupled attractor (autoassociation) networks. For these reasons, it is important to analyze the rules that govern the interactions between coupled attractor networks. This has been done using the formal type of model described by Renart, Parga and Rolls (1999b).

One boundary condition is when the coupling between the networks is so weak that there is effectively no interaction. This holds when the coupling parameter g between the networks is less than approximately 0.002, where the coupling parameter indicates the relative strength of the inter-modular to the intra-modular connections, and measures effectively the relative strengths of the currents injected into the neurons by the inter-modular relative to the intra-modular (recurrent collateral) connections (Renart, Parga and Rolls 1999b). At the other extreme, if the coupling parameter is strong, all the networks will operate as a single attractor network, together able to represent only one state (Renart, Parga and Rolls 1999b). This critical value of the coupling parameter (at least for reciprocally connected networks with symmetric synaptic strengths) is relatively low, in the region of 0.024 (Renart, Parga and Rolls 1999b). This is one reason why cortico-cortical backprojections are predicted to be quantitatively relatively weak, and for this reason it is suggested that they end on the apical parts of the dendrites of cortical pyramidal cells (see Chapter 11). In the strongly coupled regime when the system of networks operates as a single attractor, the total storage capacity (the number of patterns that can be stored and correctly retrieved) of all the networks will be set just by the number of synaptic connections onto any single neuron received from other neurons in the network, a number in the order of a few thousand (see Section B.3). This is one reason why connected cortical networks are thought not to act in the strongly coupled regime, because the total number of memories that could be represented in the whole of the cerebral cortex would be so small, in the order of a few thousand, depending on the sparseness of the patterns (see equation B.15) (O'Kane and Treves 1992).

Between these boundary conditions, that is in the region where the inter-modular coupling parameter g is in the range 0.002–0.024, it has been shown that interesting interactions can occur (Renart, Parga and Rolls 1999b, Renart, Parga and Rolls 1999a). In a bimodular architecture, with forward and backward connections between the modules, the capacity of one module can be increased, and an attractor is more likely to be found under noisy conditions, if there is a consistent pattern in the coupled attractor. By consistent we mean a pattern that during training was linked associatively by the forward and backward connections, with the pattern being retrieved in the first module. This provides a quantitative model for understanding some of the effects that backprojections can produce by supporting particular states in earlier cortical areas (Renart, Parga and Rolls 1999b). The total storage capacity of the two networks is however, in line with O'Kane and Treves (1992), not a great deal greater than the storage capacity of one of the modules alone. Thus the help provided by the attractors in falling into a mutually compatible global retrieval state (in e.g. the scenario of a hierarchical system) is where the utility of such coupled attractor networks must lie. Another interesting application of such weakly coupled attractor networks is in coupled perceptual and short-term memory systems in the brain, described in Section 4.3.1 and Chapter 6, with the architecture illustrated in Fig. 4.4.

In a trimodular attractor architecture shown in Fig. 4.5 (which is similar to the architecture of the multilayer competitive net illustrated in Fig. B.19 but has recurrent collateral connections within each module), further interesting interactions occur that account for effects such as the McGurk effect, in which what is seen affects what is heard (Renart, Parga and Rolls 1999a). The effect was originally demonstrated with the perception of auditory syllables, which were influenced by what is seen (McGurk and MacDonald 1976). The trimodular architecture (studied using similar methods to those used by Renart, Parga and Rolls (1999b) and frequently utilizing scenarios in which first a stimulus was presented to a module, then removed during a memory delay period in which stimuli were applied to other modules) showed a phase with $g < 0.005$ in which the modules operated in an isolated way. With g

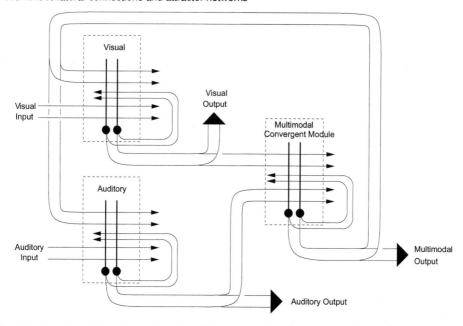

Fig. 4.5 A two-layer (2-level) set of nets in which feedback from level 2 can influence the states reached in level 1. Level 2 could be a higher cortical visual area with convergence from earlier cortical visual areas (see Chapter 25). Level 2 could also be a multimodal area receiving inputs from unimodal visual and auditory cortical areas, as labelled. Each of the 3 modules has recurrent collateral synapses that are trained by an associative synaptic learning rule, and also inter-modular synaptic connections in the forward and backward direction that are also associatively trained. Attractors are formed within modules, the different modules interact, and attractors are also formed by the forward and backward inter-modular connections. The higher level may not only affect the states reached during attractor settling in the input level, but may also, as a result of this, influence the representations that are learned in earlier cortical areas. A similar principle may operate in any multilayer hierarchical cortical processing system, such as the ventral visual system, in that the categories that can be formed only at later stages of processing may help earlier stages to form categories relevant to what can be diagnosed at later stages.

in the range 0.005–0.012, an 'independent' regime existed in which each module could be in a separate state to the others, but in which interactions between the modules occurred, which could assist or hinder retrieval in a module depending on whether the states in the other modules were consistent or inconsistent. It is in this 'independent' regime that a module can be in a continuing attractor that can provide other modules with a persistent external modulatory input that is helpful for tasks such as making comparisons between stimuli processed sequentially (as in delayed match-to-sample tasks and visual search tasks) (see Section 4.3.1). In this regime, if the modules are initially quiescent, then application of a stimulus to one input module propagates to the central module, and from it to the non-stimulated input module as well (see Fig. 4.5). When g grows beyond 0.012, the picture changes and the independence between the modules is lost. The delay activity states found in this region (of the phase space) *always* involve the three modules in attractors correlated with consistent features associated in the synaptic connections. Also, since g is now larger, changes in the properties of the external stimuli have more impact on the delay activity states. The general trend seen in this phase under the change of stimulus after a previous consistent attractor has been reached is that, first, if the second stimulus is not effective enough (it is weak or brief), it is unable to move any of the modules from their current delay activity states. If the stimulus is made more effective, then as soon as it is able to change the state of the stimulated input module, the internal and non-stimulated input modules follow, and the whole network moves into the new consistent attractor selected by the second stimulus. In this case, the interaction between the modules is so large that it does not allow contradictory local delay activity states to coexist, and the network is described as being in a 'locked' state (Renart, Parga and Rolls 1999a).

The conclusion is that the most interesting scenario for coupled attractor networks is when they are weakly coupled (in the trimodular architecture $0.005 < g < 0.012$), for then interactions occur whereby how well one module responds to its own inputs can be influenced by the states of the other modules, but it can retain partly independent representations. This emphasizes the importance of weak interactions between coupled modules in the brain (Renart, Parga and Rolls 1999b, Renart, Parga and Rolls 1999a, Renart, Parga and Rolls 2000).

These generally useful interactions between coupled attractor networks can be useful in implementing top-down constraint satisfaction (see Chapter 11) and short-term memory (see Section 4.3.1). One type of constraint satisfaction in which they are also probably important is cross-modal constraint satisfaction, which occurs for example when the sight of the lips moving assists the hearing of syllables. If the experimenter mismatches the visual and auditory inputs, then auditory misperception can occur, as in the McGurk effect. In such experiments (McGurk and MacDonald 1976) the subject receives one stimulus through the auditory pathway (e.g. the syllables *ga-ga*) and a *different* stimulus through the visual pathway (e.g. the lips of a person performing the movements corresponding to the syllables *ba-ba* on a video monitor). These stimuli are such that their acoustic waveforms as well as the lip motions needed to pronounce them are rather different. One can then assume that although they share the same vowel 'a', the internal representation of the syllables is dominated by the consonant, so that the representations of the syllables *ga-ga* and *ba-ba* are not correlated either in the primary visual cortical areas or in the primary auditory ones. At the end of the experiment, the subject is asked to repeat what he heard. When this procedure is repeated with many subjects, it is found that roughly 50% of them claim to have heard either the auditory stimulus (*ga-ga*), or the visual one (*ba-ba*). The rest of the subjects report to have heard neither the auditory nor the visual stimuli, but actually a combination of the two (e.g. *gabga*) or even something else including phonemes not presented auditorially or visually (e.g. *gagla*).

Renart, Parga and Rolls (1999a) were able to show that the McGurk effect can be accounted for by the operation of coupled attractor networks of the form shown in Fig. 4.5. One input module is for the auditory input, the second is for the visual input, and both converge into a higher area which represents the syllable formed on the evidence of combination of the two inputs. There are backprojections from the convergent module back to the input modules. Persistent (continuing) inputs were applied to both the inputs, and during associative training of all the weights the visual and auditory inputs corresponded to the same syllable. When tested with inconsistent visual and auditory inputs, it was found for g between ~ 0.10 and ~ 0.11, the convergent module can either remain in a symmetric state in which it represents a mixture of the two inputs, or choose between one of the inputs, with either situation being stable. For lower g the convergent module always settles into a state corresponding to the input in one of the input modules. It is the random fluctuations produced during the convergence to the attractor that determine the pattern selected by the convergent module. When the convergent module becomes correlated with *one* of its stored patterns, the signal back-projected to the input module stimulated with the feature associated with that pattern becomes stronger and the overlap in this module is increased. Thus, with low values of the inter-module coupling parameter g, situations are found in which sometimes the input to one module dominates, and sometimes the input to the other module dominates what is represented in the convergent module, and sometimes mixture states are stable in the convergent module. This model can thus account for the influences that visual inputs can have on what is heard, in for example the McGurk effect.

The interactions between coupled attractor networks can lead to the following effects. Facilitation can occur in a module if its external input is matched by an input from another module, whereas suppression in a module of its response to an external input can occur if the two inputs mismatch. This type of interaction can be used in imaging studies to identify brain regions where different signals interact with each other. One example is to locate brain regions where multimodal inputs converge. If the inputs in two sensory modalities are consistent based on previous experience, then facilitation will occur, whereas if they are inconsistent, suppression of the activity in a module can occur. This is one of the effects described in the bimodular and trimodular architectures investigated by Renart, Parga and Rolls (1999b), Renart, Parga and Rolls (1999a) and Rolls and Stringer (2001b), and found in architectures such as that illustrated in Fig. 4.5.

If a multimodular architecture is trained with each of many patterns (which might be visual stimuli) in one module associated with one of a few patterns (which might be mood states) in a connected module, then interesting effects due to this asymmetry are found (Rolls and Stringer 2001b).

An interesting issue that arises is how rapidly a system of interacting attractor networks such as that illustrated in Fig. 4.5 settles into a stable state. Is it sufficiently rapid for the interacting attractor effects described to contribute to cortical information processing? It is likely that the settling of the whole system is quite rapid, if it is implemented (as it is in the brain) with synapses and neurons that operate with continuous dynamics, where the time constant of the synapses dominates the retrieval speed, and is in the order of 15 ms for each module, as described in Section B.6 and by Panzeri, Rolls, Battaglia and Lavis (2001). In that Section, it is shown that a multimodular attractor network architecture can process information in approximately 15 ms per module (assuming an inactivation time constant for the synapses of 10 ms), and similarly fast settling may be expected of a system of the type shown in Fig. 4.5.

4.11 Highlights

1. Autoassociation or attractors networks can be formed within a given cortical area by the associatively modifiable recurrent collateral excitatory synaptic connections.

2. These attractor networks provide the basis for short-term memory and thereby planning in which neurons are kept firing; for long-term memory in which completion is important; and for decision-making and memory retrieval, when attractors compete.

3. In the hippocampal CA3 network, there is a global attractor, so that any one item can be associated with any other item.

4. In the neocortex, the attractor networks are within a local region of the neocortex, and have separate states allowing high memory capacity across the whole neocortex. These attractors mildly influence each other, as in top-down attention. The coupling between interacting attractors must be weak to enable them to influence each other, yet not force each other into the same attractor state.

5. An attractor network operates rapidly in the cortex, falling into its basin of attraction in 1.5 time constants of the synapses, in practice within 20–30 ms.

6. Attractor networks provide a fundamental principle of operation of the neocortex, hippocampal cortex, and pyriform cortex (see Chapter 18), performing the operations described in this Chapter.

5   The noisy cortex: stochastic dynamics, decisions, and memory

The spiking of the neurons in the brain is almost random in time for a given mean rate, i.e. the spiking is approximately Poissonian, and this randomness introduces noise into the system, which makes the system behave stochastically. The effect that this stochasticity has on a neural system, and more generally the dynamics of the system, for example how long it takes to respond, and how long it remains in a given state, can be investigated using integrate-and-fire neuronal network simulations, which model how the currents through the different synaptic receptor-activated ion channels are integrated to produce the membrane potential of a neuron, which fires an action potential when the threshold is reached. The integrate-and-fire neuronal network approach is described in Section B.6, and as used here dynamically models the membrane potential of a point cell, and the dynamics of the opening and closing of different receptor-activated ion channels. Each of the neurons shows spiking that is very much like that of neurons recorded in the brain. From such integrate-and-fire models it is possible to make predictions not only about synaptic and neuronal activity, but also about the signals recorded in functional neuroimaging experiments, and the behaviour of the system, in terms for example of its probabilistic choices, its reaction times, and its stability. This makes a close link with human behaviour possible (Rolls, Grabenhorst and Deco 2010b, Rolls, Grabenhorst and Deco 2010c). The theory as to how networks operate stochastically (Rolls and Deco 2010, Deco, Rolls, Albantakis and Romo 2013) has implications throughout the cortex, some of which are described in this chapter.

First we consider the sources of the noise in the cortex (Section 5.1). Then we consider how the spiking-related noise influences the stability of short-term and long-term memory mechanisms in the cortex (Sections 5.2–5.5). Then we consider how the spiking-related noise influences decision-making mechanisms in the cortex (Section 5.6). Then we treat how noise influences perception (Section 5.7). Then we consider some advantages of noise in the brain, and how this is played out in some different ways in the cortex (Sections 5.8–5.12). Sections B.6–B.8 provide fundamental material, including equations, on stochastic dynamics in the cortex.

5.1   Reasons why the brain is inherently noisy and stochastic

Why is the spiking activity of neurons probabilistic, and what are the advantages that this may confer? The answer suggested (Rolls 2008d) is that the spiking activity is approximately Poisson-like (as if generated by a random process with a given mean rate, both in the brain and in the integrate-and-fire simulations we describe), because the neurons are held close to (just slightly below) their firing threshold, so that any incoming input can rapidly cause sufficient further depolarization to produce a spike. It is this ability to respond rapidly to an input, rather than having to charge up the cell membrane from the resting potential to the threshold, a slow process determined by the time constant of the neuron and influenced by that of the synapses, that enables neuronal networks in the brain, including attractor networks, to operate and retrieve information so rapidly (Treves 1993, Rolls and Treves 1998, Battaglia and Treves 1998a, Rolls 2008d, Panzeri, Rolls, Battaglia and Lavis 2001, Rolls and Deco 2010). The spike trains are essentially Poisson-like because the cell potential hovers noisily close to the threshold for firing (Dayan and Abbott 2001), the noise being generated in part by the Poisson-like firing of the other neurons in the network (Jackson 2004). The noise and spontaneous firing help to ensure that when a stimulus arrives, there are always some neurons very close to threshold that respond rapidly, and then communicate their firing to other neurons through the modified synaptic weights, so that an attractor process can take place very rapidly (Rolls 2008d, Rolls and Deco 2010).

The implication of these concepts is that the operation of networks in the brain is inherently noisy because of the Poisson-like timing of the spikes of the neurons, which itself is related to the mechanisms that enable neurons to respond rapidly to their inputs (Rolls 2008d). However, the consequence of the Poisson-like firing is that, even with quite large attractor networks of thousands of neurons with hundreds of neurons representing each pattern or memory, the network inevitably settles probabilistically to a given attractor state. This results, *inter alia*, in decision-making being probabilistic. Factors that influence the probabilistic behaviour of the network include the strength of the inputs (with the difference in the inputs / the magnitude of the inputs being relevant to decision-making and Weber's Law as described in Section 5.6); the depth and position of the basins of attraction, which if shallow or correlated with other basins will tend to slow the network; noise in the synapses due to probabilistic release of transmitters (Koch 1999, Abbott and Regehr 2004); and the mean firing rates of the neurons during the decision-making itself, the firing rate distribution (see Section 8.2) (Webb, Rolls, Deco and Feng 2011), and the dilution of the connectivity (Rolls and Webb 2012). In terms of synaptic noise, if the probability of release, p, is constant, then the only effect is that the releases in response to a Poisson presynaptic train at rate r are described by a Poisson process with rate pr, rather than r. However, if the synapse has short-term depression or facilitation, the statistics change (Fuhrmann, Segev, Markram and Tsodyks 2002b, Abbott and Regehr 2004). A combination of these factors contributes to the probabilistic behaviour of the network becoming poorer and slower as the difference between the inputs divided by the base frequency is decreased, as shown in Fig. 5.14 (Deco and Rolls 2006). Other factors that may contribute to noise in the brain are reviewed by Faisal, Selen and Wolpert (2008).

In signal-detection theory, behavioural indeterminacy is accounted for in terms of noisy input (Green and Swets 1966). In the present approach, indeterminacy is accounted for in part in terms of internal noise inherent to the operation of spiking neuronal circuitry (Rolls 2008d, Rolls and Deco 2010) (see also Wang (2008) and Faisal et al. (2008)).

In fact, there are many sources of noise, both external to the brain and internal (Faisal, Selen and Wolpert 2008). A simple measure of the variability of neuronal responses is the Fano factor, which is the ratio of the variance to the mean. A Poisson process has a Fano factor of 1, and neurons often have Fano factors close to one, though in some systems and for some neurons they can be lower (less variable than a Poisson process), or higher (Faisal et al. 2008). The sources of noise include the following, with citations to original findings provided by Faisal et al. (2008).

First, there is sensory noise, arising external to the brain. Thermodynamic noise is present for example in chemical sensing (including smell and gustation) because molecules arrive at the receptor at random rates owing to diffusion. Quantum noise can also be a factor, with photons arriving at the photoreceptor at a rate governed by a Poisson process. The peripheral nervous system (e.g. the retina, and in the olfactory bulb the glomeruli) often sums inputs over many receptors in order to limit this type of noise. In addition to these processes inherent in the transduction and subsequent amplification of the signals, there may of course be variation in the signal itself arriving from the world.

Second, there is cellular or neuronal noise arising even in the absence of variation in synaptic input. This cellular noise arises from the stochastic opening and closing of voltage or ligand-gated ion channels, the fusing of synaptic vesicles, and the diffusion and binding of signalling molecules to the receptors. The channel noise becomes appreciable in small axons and cell bodies (where the number of channels and molecules involved becomes relatively small). This channel noise can affect not only the membrane potential of the cell body and the initiation of spikes, but also spike propagation in thin axons. There may also be some cross-talk between neurons, caused by processes such as ephaptic coupling, changes in extracellular ion concentration after electrical signalling, and spillover of neurotransmitters between adjacent unrelated synapses.

Third, there is synaptic noise evident for example as spontaneous miniature postsynaptic currents (mPSCs) that are generated by the 'quantal' (discrete) nature of released neurotransmitter vesicles (Fatt and Katz 1950, Manwani and Koch 2001). This synaptic noise can be produced by processes that include spontaneous opening of intracellular Ca^{2+} stores, synaptic Ca^{2+}-channel noise, spontaneous triggering of the vesicle-release pathway, and spontaneous fusion of a vesicle with the membrane. In addition, at the synapses there is variation in the number of transmitter molecules in a released vesicle, and there are random diffusion processes of the molecules (Franks, Stevens and Sejnowski 2003). In addition, there is randomness in whether an action potential releases a vesicle, with the probability sometimes being low, and being influenced by factors such as synaptic plasticity and synaptic adaptation (Faisal et al. 2008, Abbott and Regehr 2004, Manwani and Koch 2001).

The sources of noise in the brain arise from the effects described earlier that make the firing properties of neurons very often close to Poisson-like. If the system were infinitely large (as the mean field approach described in the following sections assumes) then the Poisson firing would average out to produce a steady average firing rate averaged across all the neurons. However, if the system has a finite number of neurons, then the Poisson or semirandom effects will not average out, and the average firing rate of the population will fluctuate statistically. These effects are referred to as coherent fluctuations. The magnitude of these fluctuations decreases with the square root of the size of the network (the number of neurons in the population in a fully connected network, and the number of connections per neuron in a network with diluted connectivity) (Mattia and Del Giudice 2002). The magnitude of these fluctuations is described by the standard deviation of the firing rates divided by the mean rate, and this can be used to measure the noise in the network. These effects are described in detail in the Appendix of Rolls and Deco (2010).

The signal-to-noise ratio can be measured in a network by the average firing rate change produced by the signal in the neuronal population divided by the standard deviation of the firing rates. Non-linearities in the system will affect how the signal-to-noise ratio described in this way is reflected in the output of the system. Another type of measure that reflects the signal-to-noise ratio is the stability of the system as described by its trial to trial variability, as described in Sections 5.2 and 5.4 and by Rolls, Loh, Deco and Winterer (2008d).

Although many approaches have seen noise as a problem in the brain that needs to be averaged out or resolved (Faisal, Selen and Wolpert 2008), and this is certainly important in sensory systems, the thrust of *The Noisy Brain* (Rolls and Deco 2010) is to show that in fact noise inherent in brain activity has a number of advantages by making the dynamics stochastic, which allows for many remarkable features of the brain, including creativity, probabilistic decision-making, stochastic resonance, unpredictability, conflict resolution, symmetry breaking, allocation to discrete categories, and many other important properties described later in this Chapter.

The size of a network is an important factor in influencing how noisy the network is in relation to the statistical fluctuations related to the Poisson nature of the neuronal spiking. An infinite size network would have no noise, and its performance can be calculated analytically using mean field methods (Section 4.4). In terms of relevance to cortical function, attractor networks with even thousands of neurons show probabilistic behaviour, showing that the spiking-related noise of neurons does influence cortical networks of biologically relevant size. For example, even when a fully connected recurrent attractor network has 4,000 neurons, the operation of the network is still probabilistic as shown in Fig. 5.13 (Deco and Rolls 2006). Under these conditions, the probabilistic spiking of the excitatory (pyramidal) cells in the recurrent collateral firing, rather than variability in the external inputs to the network, is what makes the major contribution to the noise in the network (Deco and Rolls 2006). Thus, once the firing in the recurrent collaterals is spike-implemented by integrate-and-fire neurons, the probabilistic behaviour seems inevitable, even up to quite large attractor network sizes.

The graded nature of the sparse distributed representations in the cortex tends to increase the noise (Webb, Rolls, Deco and Feng 2011). Representations in the cortex are often distributed with graded firing rates in the neuronal populations. The firing rate probability distribution of each neuron to a set of stimuli is often exponential or gamma (see Sections 8.2 and C.3.1.1) (Webb et al. 2011). In integrate-and-fire simulations of an attractor decision-making network, we showed that the noise is indeed greater for a given sparseness of the representation for graded, exponential, than for binary firing rate distributions. The greater noise was measured by faster escaping times from the spontaneous firing rate state when the decision cues are applied, and this corresponds to faster decision or reaction times. The greater noise was also evident as less stability of the spontaneous firing state before the decision cues are applied. The implication is that spiking-related noise will continue to be a factor that influences processes such as decision-making, signal detection, short-term memory, and memory recall even with the quite large networks found in the cerebral cortex. In these networks there are several thousand recurrent collateral synapses onto each neuron. The greater noise with graded firing rate distributions has the advantage that it can increase the speed of operation of cortical circuitry (Webb, Rolls, Deco and Feng 2011).

Dilution of the connectivity within an attractor network can have the effect of decreasing the noise compared to a fully connected network with the same number of connections onto each neuron and therefore the same memory capacity (Rolls and Webb 2012). The connectivity of the cerebral cortex is diluted,with the probability of excitatory connections between even nearby pyramidal cells rarely more than 0.1, and in the hippocampus 0.04 (Chapters 7 and 24). To investigate the extent to which this diluted connectivity affects the dynamics of attractor networks in the cerebral cortex, we simulated an integrate-and-fire attractor network taking decisions between competing inputs with diluted connectivity of 0.25 or 0.1, and with the same number of synaptic connections per neuron for the recurrent collateral synapses within an attractor population as for full connectivity. The results indicated that there was less spiking-related noise with the diluted connectivity, in that the stability of the network when in the spontaneous state of firing increased, and the accuracy of the correct decisions increased. The decision times were a little slower with diluted than with complete connectivity. Given that the capacity of the network is set by the number of recurrent collateral synaptic connections per neuron, on which there is a biological limit, the findings indicate that the stability of cortical networks, and the accuracy of their correct decisions or memory recall operations, can be increased by utilizing diluted connectivity and correspondingly increasing the number of neurons in the network, with little impact on the speed of processing of the cortex (Rolls and Webb 2012).

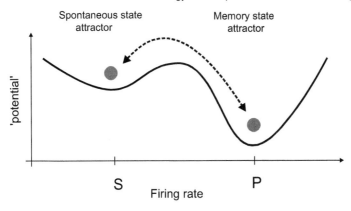

Fig. 5.1 Energy landscape. The noise influences when the system will jump out of the spontaneous firing stable (low energy) state S, and whether it jumps into the high firing rate state labelled P (with persistent or continuing firing in a state which is even more stable with even lower energy), which might correspond to a short-term memory, or to a decision.

5.2 Attractor networks, energy landscapes, and stochastic neurodynamics

We have seen that attractor networks can be used for short-term memory and long-term memory, and for decision-making (Chapter 4 and Section B.3). This section considers approaches to the stability of these networks. This section can be thought of as explaining the stability of memory networks, though the principles apply just as much to the decision-making networks considered in Section 5.6.

Autoassociation attractor systems (described in Section B.3) can have two types of stable fixed points: a spontaneous state with a low firing rate, and one or more attractor states with high firing rates in which the positive feedback implemented by the recurrent collateral connections maintains a high firing rate. We sometimes refer to this latter state as the persistent state, because the high firing normally persists to maintain a set of neurons active, which might implement a short-term memory, or the recall of a long-term memory that persists for a short time.

The stable points of the system can be visualized in an energy landscape (see Fig. 5.1). The area in the energy landscape within which the system will move to a stable attractor state is called its basin of attraction. The attractor dynamics can be pictured by energy landscapes, which indicate the basins of attraction by valleys, and the attractor states or fixed points by the bottom of the valleys (see Fig. 5.1).

The stability of an attractor is characterized by the average time in which the system stays in the basin of attraction under the influence of noise. The noise provokes transitions to other attractor states. One source of noise results from the interplay between the Poissonian character of the spikes and the finite-size effect due to the limited number of neurons in the network.

Two factors determine the stability. First, if the depths of the attractors are shallow (as in the left compared to the right valley in Figure 5.1), then less force is needed to move a ball from one valley to the next. Second, high noise will make it more likely that the system will jump over an energy boundary from one state to another. We envision that the brain as a dynamical system has characteristics of such an attractor system including statistical fluctuations (see further Rolls and Deco (2010), where the effects of noise are defined quantitatively). The noise could arise not only from the probabilistic spiking of the neurons which has significant

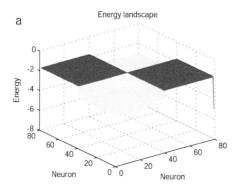 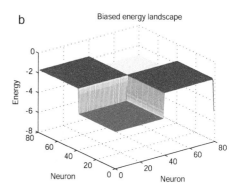

Fig. 5.2 (a) Energy landscape without any differential bias applied. The landscape is for a network with neurons 1–40 connected by strengthened synapses so that they form attractor 1, and neurons 41–80 connected by synapses strengthened (by the same amount) so that they form attractor 2. The energy basins in the two-dimensional landscape are calculated by equation 5.2. In this two-dimensional landscape, there are two stable attractors, and each will be reached equally probably under the influence of noise. This scenario might correspond to decision-making where the input λ_1 to attractor 1 has the same value as the input λ_2 to attractor 2, and the network is equally likely under the influence of the noise to fall into attractor 1 representing decision 1 as into attractor 2 representing decision 2. (b) Energy landscape with bias applied to neurons 1–40. This make the basin of attraction deeper for attractor 1, as calculated with Equation 5.2. Thus, under the influence of noise caused by the randomness in the firing of the neurons, the network will reach attractor 1 more probably than it will reach attractor 2. This scenario might correspond to decision-making, where the evidence for decision 1 is stronger than for decision 2, so that a higher firing rate is applied as λ_1 to neurons 1–40. The scenario might also correspond to memory recall, in which memory 1 might be probabilistically more likely to be recalled than memory 2 if the evidence for memory 1 is stronger. Nevertheless, memory 2 will be recalled sometimes in what operates as a non-deterministic system.

effects in finite size integrate-and-fire networks (Deco and Rolls 2006), but also from any other source of noise in the brain or the environment (Faisal et al. 2008), including the effects of distracting stimuli.

In an attractor network in which a retrieval cue is provided to initiate recall but then removed, a landscape can be defined in terms of the synaptic weights. An example is shown in Fig. 5.2a. The basins in the landscape can be defined by the strengths of the synaptic weights which describe the stable operating points of the system, where the depth of the basins can be defined in terms of the synaptic weight space, in terms defined by an associative rule operating on the firing rates of pairs of neurons during the learning as follows

$$w_{ij} = y_i y_j \tag{5.1}$$

where y_i is the firing rate of the postsynaptic neuron, y_j is the firing rate of the presynaptic neuron, and w_{ij} is the strength of the synapses connecting these neurons.

Hopfield (1982) showed how many stable states a simple attractor system might contain, and this is the capacity of the network described in Section B.3.3.7. He showed that the recall process in his attractor network can be conceptualized as movement towards basins of attraction, and his equation defines the energy at a given point in time as being a function of the synaptic weights and the current firing rates as follows

$$E = -\frac{1}{2}\sum_{i,j} w_{ij}(y_i - \langle y \rangle)(y_j - \langle y \rangle). \tag{5.2}$$

where y_i is the firing rate of the postsynaptic neuron, y_j is the firing rate of the presynaptic neuron, w_{ij} is the strength of the synapse connecting them, and $< y >$ is the mean firing rate of the neurons. I note that the system defined by Hopfield had an energy function, in that the neurons were connected by symmetric synaptic weights (produced for example by associative synaptic modification of the recurrent collateral synapses) and there was no self-coupling (Hertz, Krogh and Palmer 1991, Moreno-Bote, Rinzel and Rubin 2007, Hopfield and Herz 1995).

The situation is more complicated in an attractor network if it does not have a formal energy function. One such condition is when the connectivity is randomly diluted, for then the synaptic weights between pairs of neurons will not be symmetric. Indeed, in general, neuronal systems do not admit such an energy function. (This is the case in that it is not in general possible to define the flow in terms of the gradient of an energy function. Hopfield defined first an energy function, and from there derived dynamics.) However, such diluted connectivity systems can still operate as attractor systems (Treves 1993, Treves 1991a, Treves 1991b, Treves and Rolls 1991, Treves, Rolls and Simmen 1997, Rolls and Treves 1998, Battaglia and Treves 1998a), and the concept of an energy function and landscape is useful for discussion purposes. In practice, a Lyapunov function can be used to prove analytically that there is a stable fixed point such as an attractor basin (Khalil 1996), and even in systems where this can not be proved analytically, it may still be possible to show numerically that there are stable fixed points, to measure the flow towards those fixed points which describes the depth of the attractor basin as we have done for this type of network (Loh, Rolls and Deco 2007a), and to use the concept of energy or potential landscapes to help visualize the properties of the system (Rolls and Deco 2010).

If an external input remains on during the retrieval process, this will influence the energy function of such a network, and its stable points, as implied by equation 5.2, and as illustrated in Fig. 5.2b. In this situation, the external inputs bias the stable points of the system. Indeed, in this situation, a landscape, though not necessarily formally an energy landscape, can be specified by a combination of the synaptic weights and external inputs that bias the firing rates. The noise introduced into the network by for example the random neuronal spiking can be conceptualized as influencing the way that the system flows across this fixed landscape shaped by the synaptic weights, and by the external inputs if they remain on during operation of the network, in what is referred to as a 'clamped' condition, the normal condition that applies during decision-making (see Section 5.6).

In more detail, the flow, which is the time derivative of the neuronal activity, specifies the landscape in an attractor system. The flow is defined in the mean field analysis in terms of the effects of the synaptic weights between the neurons and the external inputs (Loh, Rolls and Deco 2007a, Rolls and Deco 2010). The flow is the force that drives the system towards the attractor given a parameter value in phase space, i.e. the firing rates of the pools (populations) of neurons. This is measured by fixing the value of the firing rate of the selective pool and letting the other values converge to their fixed point. The flow can then be computed with this configuration (Mascaro and Amit 1999). This landscape is thus fixed by the synaptic and the external inputs. The noise, produced for example by the almost Poissonian spiking of the neurons, can be conceptualized as influencing the way that the system flows across this fixed landscape. Moreover, the noise can enable the system to jump over a barrier in this fixed landscape, as illustrated in Figs. 5.1 and 5.8.

In Fig. 5.8 (and in Fig. 5.2a) the decision basins of attraction are equally deep, because the inputs λ_1 and λ_2 to the decision-making network are equal, that is, ΔI the difference between them is zero. If λ_1 is greater than λ_2, the basin will be deeper for λ_1. The shape of the landscape is thus a function of the synaptic weights and the biasing inputs to the system.

This is illustrated in Fig. 5.2b. Noise can be thought of as provoking movements across the 'effective energy landscape' conceptualized in this way.

The way in which we conceptualise the operation of an attractor network used for noise-driven stochastic decision-making, stimulus detection, etc, is as follows. The noise in the system (caused for example by statistical fluctuations produced by the Poisson-like neuronal firing in a finite-sized system as described in Section 5.1) produces changes in neuronal firing. These changes may accumulate stochastically, and eventually may become sufficiently large that the firing is sufficient to produce energy to cause the system to jump over an energy barrier (see Fig. 5.3). Opposing this noise-driven fluctuation will be the flow being caused by the shape and depth of the fixed energy landscape defined by the synaptic weights and the applied external input bias or biases. The noisy statistical fluctuation is a diffusion-like process. If the spontaneous firing rate state is stable with the decision cues applied (see Fig. 5.4 middle on page 100), eventually the noise may provoke a transition over the energy barrier in an escaping time, and the system will drop, again noisily, down the valley on the other side of the hill. The rate of change of the firing rate is again measured by the flow, and is influenced by the synaptic weights and applied biases, and by the statistical fluctuations. In this scenario, the reaction times will depend on the amount of noise, influenced by the size of the network, and by the fixed 'effective energy landscape' as determined by the synaptic weights and applied biasing inputs λ, which will produce an escaping time as defined further in the Appendix of Rolls and Deco (2010).

If the spontaneous state is not stable (see Fig. 5.4 right), the reaction times will be influenced primarily by the flow as influenced by the gradient of the energy landscape, and by the noise caused by the random neuronal firings. A noise-produced escaping time from a stable spontaneous state attractor will not in this situation contribute to the reaction times.

The noise-driven escaping time from the stable spontaneous state is important in understanding long and variable reaction times, and such reaction times are present primarily in the scenario when the parameters make the spontaneous state stable, as described further by Marti, Deco, Mattia, Gigante and Del Giudice (2008). While in a spontaneous stable state the system may be thought of as being driven by the noise, and it is primarily when the system has reached a ridge at the edge of the spontaneous valley and the system is close to a bifurcation point into a high firing rate close to the ridge that the attractor system can be thought of as accumulating evidence from the input stimuli (Deco, Scarano and Soto-Faraco 2007). While in the spontaneous state valley (see Fig. 5.4 middle), the inputs can be thought of as biasing the 'effective energy landscape' across which the noise is driving the system stochastically.

An interesting aspect of the model is that the recurrent connectivity, and the relatively long time constant of the NMDA receptors (Wang 2002), may together enable the attractor network to accumulate evidence over a long time period of several hundred milliseconds. Important aspects of the functionality of attractor networks are that they can accumulate and maintain information.

5.3 A multistable system with noise

In the situation illustrated in Figs. 5.1 and 5.8, there is multistability, in that the spontaneous state and a large number of high firing rate persistent states are stable. More generally, and depending on the network parameters including the strengths of the inputs, a number of different scenarios can occur. These are illustrated in Fig. 5.4. Let us consider the activity of a given neuronal population while inputs are being applied.

In Fig. 5.4 (left) we see a situation in which only the spontaneous state S is stable. This might occur if the external inputs λ_1 and λ_2 are weak.

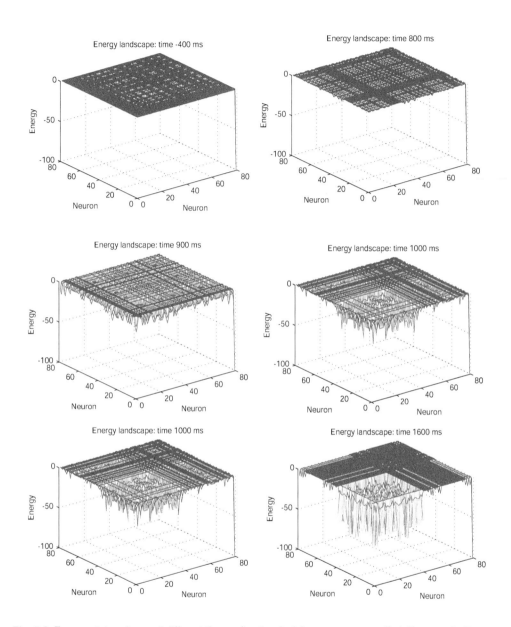

Fig. 5.3 Energy states shown at different times after the decision cues were applied. Neurons 1–40 are in attractor 1 and are connected by strong weights with each other; and neurons 41–80 are in attractor 2 and are connected by strong weights with each other. The energy state is defined in equation 5.2, and the energy between any pair of neurons is a product of the firing rates of each neuron and the synaptic weight that connects them. These are the energy states for the trial shown in Fig. 5.10b and c on page 114. Time 0 is the time when the decision stimuli were applied (and this corresponds to time 2 s in Fig. 5.10).

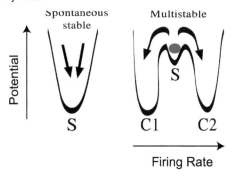

Fig. 5.4 Computational principles underlying the different dynamical regimes of the decision-making attractor network (see text). The x-axis represents the neuronal activity of one of the populations (ν_i) and the landscape represents an energy landscape ('potential') regulating the evolution of the system. S is a stable state of spontaneous activity, C2 is a high firing rate state of this neuronal population corresponding to the decision implemented by this population, and C1 is a low firing rate state present when the other population wins the competition.

On the right we have a situation in which our neuronal population is either in a high firing rate stable state C2, or in a low firing rate state C1 because another population is firing fast and inhibiting our neuronal population. There is no stable spontaneous state.

In the middle of Fig. 5.4 we see a situation in which our population may be either in C1, or in C2, or in a spontaneous state of firing S when no population has won the competition. We emphasize that this can be a scenario even when the decision cues λ_1 and λ_2 are being applied during the decision-making period. We refer to this system as a multistable system.

The differences between these scenarios are of interest in relation to how noise influences the decision-making. In the scenario shown in the middle of Fig. 5.4 we see that there are three stable states when the inputs λ_1 and λ_2 are being applied, and that it is the stochastic noise that influences whether the system jumps from the initial spontaneous state to a high firing rate state in which one of the decision-state populations fires fast, producing either C2 if our population wins, or C1 if our population loses. The statistical properties of the noise (including its amplitude and frequency spectrum), and the shape of the different basins in the energy landscape, influence whether a decision will be taken, the time when it will be taken, and which high firing rate decision attractor wins. In contrast, in the scenario shown in Fig. 5.4 (right) the energy landscape when the stimuli are being applied is such that there is no stable spontaneous state, so the system moves to one of the high firing rate decision attractors without requiring noise. In this case, the noise, and the shape of the energy landscape, influence which high firing rate decision state attractor will win.

A more detailed analysis suggests that there are two scenarios that are needed to understand the time course of processes such as decision-making and memory retrieval (which are very similar processes involving an attractor network driven by one input, or by several inputs) (Marti, Deco, Mattia, Gigante and Del Giudice 2008).

First, in the scenario investigated by Wang (2002), the spontaneous state is unstable when the decision cues are applied. The network, initially in the spontaneous state, is driven to a competition regime by an increase of the external input (that is, upon stimulus presentation) that destabilizes the initial state. The decision process can then be seen as the relaxation from an unstable stationary state towards either of the two stable decision states (Fig. 5.4 (right)). When the system is completely symmetric (i.e. when there is no bias in the external inputs that favours one choice over the other), this destabilization occurs because the system undergoes a pitchfork bifurcation for sufficiently high inputs. The time spent by the system to evolve from the initial state to either of the two decision states is determined by the actual stochastic trajectory of the system in the phase space. In particular, the transition time increases significantly when the system wanders in the vicinity of the saddle that appears when the spontaneous state becomes unstable. Reaction times in the order of hundreds of ms may be produced in this way, and are strongly influenced by the long time constants of the NMDA receptors (Wang 2002). The transition can be further slowed down by setting the external input slightly above the bifurcation value. This tuning can be exploited to obtain realistic decision times.

Second, there is a scenario in which the stimuli do not destabilize the spontaneous state, but rather increase the probability for a noise-driven transition from a stable spontaneous state to one of the decision states (see Fig. 5.4 (centre) and Section 5.3). Due to the presence of finite-size noise in the system there is a nonzero probability that this transition occurs and hence a finite mean transition rate between the spontaneous and the decision states. It has been shown that in this scenario mean decision times tend to the Van't Hoff-Arrhenius exponential dependence on the amplitude of noise in the limit of infinitely large networks. As a consequence, in this limit, mean decision times increase exponentially with the size of the network (Marti, Deco, Mattia, Gigante and Del Giudice 2008). Further, the decision events become Poissonian in the limit of vanishing noise, leading to an exponential distribution of decision times. For small noise a decrease in the mean input to the network leads to an increase of the positive skewness of decision-time distributions.

These results suggest that noise-driven decision models as in this second scenario provide an alternative dynamical mechanism for the variability and wide range of decision times observed, which span from a few hundred milliseconds to more than one second (Marti et al. 2008). In this scenario, there is an escaping time from the spontaneous firing state (Rolls and Deco 2010). In this time the information can be thought of as accumulating in the sense that the stochastic noise may slowly drive the firing rates in a diffusion-like way such than an energy barrier is jumped over and escaped (see Fig. 5.1 on page 95 and Fig. 5.2). In this situation, a landscape can be specified by a combination of the synaptic weights and external decision-related input evidence that biases the firing rates of the decision attractors, as described in Section 5.2. The noise introduced into the network by for example the random neuronal spiking can be conceptualized as influencing the way that the system flows across this fixed landscape shaped by the synaptic weights, and by the external inputs if they remain on during operation of the network.

The model for the second scenario, with a stable spontaneous state even when the decision cues are being applied, makes specific predictions about reaction times. The analysis shows that there will be a gamma-like distribution with an exponential tail of long reaction times in the reaction time distribution with this second scenario (Marti et al. 2008). This is illustrated in Fig. 5.12.

5.4   Stochastic dynamics and the stability of short-term memory

I now introduce concepts on how noise produced by neuronal spiking, or noise from other sources, influences short-term memory, with illustrations drawn from Loh, Rolls and Deco (2007a) and Rolls, Loh and Deco (2008c).

The noise caused by the probabilistic firing of neurons can influence the stability of short-term memory. To investigate this, it is necessary to use a model of short-term memory which is a biophysically realistic integrate-and-fire attractor network with spiking of the neurons, so that the properties of receptors, synaptic currents and the statistical effects related to the

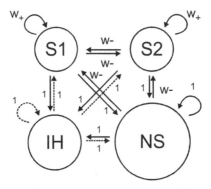

Fig. 5.5 The attractor network model. The excitatory neurons are divided into two selective pools S1 and S2 (with 40 neurons each) with strong intra-pool connection strengths w_+, and one non-selective pool (NS) (with 320 neurons). The other connection strengths are 1 or weak w_-. The network contains 500 neurons, of which 400 are in the excitatory pools and 100 are in the inhibitory pool IH. Each neuron in the network also receives inputs from 800 external neurons, and these neurons increase their firing rates to apply a stimulus or distracter to one of the pools S1 or S2. The synaptic connection matrices are shown in Tables 5.1 and 5.2. (Reproduced from Loh M, Rolls ET, Deco G (2007) A Dynamical Systems Hypothesis of Schizophrenia, *PLoS Computational Biology*, 3(11): e228, Figure 8, doi:10.1371/journal.pcbi.0030228 ©2007, Loh et al.)

noisy probabilistic spiking of the neurons can be analyzed. Loh, Rolls and Deco (2007a) and Rolls, Loh and Deco (2008c) used a minimal architecture, a single attractor or autoassociation network (Hopfield 1982, Amit 1989, Hertz et al. 1991, Rolls and Treves 1998, Rolls 2008d) (see Section B.3), to investigate how spiking-related stochastic noise influences the stability of short-term memory. They chose a recurrent (attractor) integrate-and-fire network model which includes synaptic channels for AMPA, NMDA and GABA$_A$ receptors (Brunel and Wang 2001). The integrate-and-fire model was necessary to characterize and exploit the effects of the spiking noise produced by the neurons in a finite-sized network. However, to initialize the parameters of the integrate-and-fire model such as the synaptic connection strengths to produce stable attractors, and to ensure that the spontaneous activity is in the correct range, they used a mean-field approximation consistent with the integrate-and-fire network, as described in Section B.8.2. Both excitatory and inhibitory neurons were represented by a leaky integrate-and-fire model (Tuckwell 1988) described in detail in Section B.6.3.

The single attractor network contained 400 excitatory and 100 inhibitory neurons, which is consistent with the observed proportions of pyramidal cells and interneurons in the cerebral cortex (Abeles 1991, Braitenberg and Schütz 1991). The connection strengths were adjusted using mean-field analysis (see Brunel and Wang (2001) and Section B.8.2), so that the excitatory and inhibitory neurons exhibited a spontaneous activity of 3 Hz and 9 Hz respectively (Koch and Fuster 1989, Wilson, O'Scalaidhe and Goldman-Rakic 1994a). The recurrent excitation mediated by the AMPA and NMDA receptors is dominated by the long time constant NMDA currents to avoid instabilities during the delay periods (Wang 1999, Wang 2002).

The cortical network model featured a minimal architecture to investigate stability (and also distractibility), and consisted of two selective pools (or populations of neurons) S1 and S2, as shown in Fig. 5.5. Pool S1 is used for the short-term memory item to be remembered, sometimes called the target; and pool S2 is used for a distracter stimulus. The non-selective pool NS modelled the spiking of cortical neurons and served to generate an approximately Poisson spiking dynamics in the model (Brunel and Wang 2001), which is what is observed in the cortex. The inhibitory pool IH contained 100 inhibitory neurons. There were thus four populations or pools of neurons in the network, and the connection weights were set

Table 5.1 Connection matrix for AMPA and NMDA – [from, to]

	S1	S2	NS	IH
S1	w_+	w_-	1	1
S2	w_-	w_+	1	1
NS	w_-	w_-	1	1
IH	0	0	0	0

Table 5.2 Connection matrix for GABA – [from, to]

	S1	S2	NS	IH
S1	0	0	0	0
S2	0	0	0	0
NS	0	0	0	0
IH	1	1	1	1

up using a mean-field analysis to make S1 and S2 have stable attractor properties. The connection weights between the neurons within each selective pool or population were called the intra-pool connection strengths w_+. The increased strength of the intra-pool connections was counterbalanced by the other excitatory connections (w_-) to keep the average input to a neuron constant. The actual synaptic strengths are shown in Tables 5.1 and 5.2 where $w_- = \frac{0.8 - f_{S1}w_+}{0.8 - f_{S1}}$, and f_{S1} is the fraction of the total number of excitatory neurons in pool S1. For these investigations, $w_+ = 2.1$ was selected, because with the default values of the NMDA and GABA conductances this yielded relatively stable dynamics with some effect of the noise being apparent, that is, a relatively stable spontaneous state if no retrieval cue was applied, and a relatively stable state of persistent firing after a retrieval cue had been applied and removed.

Each neuron in the network received Poisson input spikes via AMPA receptors which are envisioned to originate from 800 external neurons at an average spontaneous firing rate of 3 Hz from each external neuron, consistent with the spontaneous activity observed in the cerebral cortex (Wilson et al. 1994a, Rolls and Treves 1998, Rolls 2008d) (see Section B.6.3).

5.4.1   Analysis of the stability of short-term memory

The analyses (Loh, Rolls and Deco 2007a, Rolls, Loh and Deco 2008c) aimed to investigate the stability of the short-term memory implemented by the attractor network. Simulations were performed for many separate trials, each run with a different random seed to analyze the statistical variability of the network as the noise varied from trial to trial. We focus here on simulations of two different conditions: the spontaneous, and persistent conditions.

In spontaneous simulations (see Fig. 5.6), we ran spiking simulations for 3 s without any extra external input. The aim of this condition was to test whether the network was stable in maintaining a low average firing rate in the absence of any inputs, or whether it fell into one of its attractor states without any external input.

In persistent simulations, an external cue of 120 Hz above the background firing rate of 2400 Hz was applied to each neuron in pool S1 during the first 500 ms to induce a high activity state and then the system was run without the additional external cue of 120 Hz for another 2.5 s, which was a short-term memory period. The 2400 Hz was distributed across the 800 synapses of each S1 neuron for the external inputs, with the spontaneous Poisson spike trains received by each synapse thus having a mean rate of 3 Hz. The aim of this condition was to

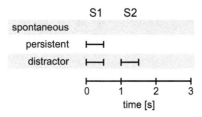

Fig. 5.6 The simulation protocols. Stimuli to either S1 or S2 are applied at different times depending on the type of simulations. The spontaneous simulations include no input. The 'persistent' simulations assess how stably a stimulus is retained in short-term memory by the network. The distracter simulations add a distracter stimulus to further address the stability of the network activity in the face of a distracting stimulus, and thus both the stability of attention, and the shifting of attention. (Reproduced from Loh M, Rolls ET, Deco G (2007) A Dynamical Systems Hypothesis of Schizophrenia, *PLoS Computational Biology*, 3(11): e228, Figure 8, doi:10.1371/journal.pcbi.0030228 ©2007, Loh et al.)

investigate whether once in an attractor short-term memory state, the network can maintain its activity stably, or whether it fell out of its attractor, which might correspond to an inability to maintain short-term memory.

5.4.2 Stability and noise in a model of short-term memory

To clarify the concept of stability, we show examples of trials of spontaneous and persistent simulations in which the statistical fluctuations have different impacts on the temporal dynamics. Fig. 5.7 shows the possibilities, as follows.

In the spontaneous state simulations, no cue was applied, and we are interested in whether the network remains stably in the spontaneous firing state, or whether it is unstable and on some trials due to statistical fluctuations entered one of the attractors, thus falsely retrieving a memory. Figure 5.7 (top) shows an example of a trial on which the network correctly stayed in the low spontaneous firing rate regime, and (bottom) another trial (labelled spontaneous unstable) in which statistical spiking-related fluctuations in the network caused it to enter a high activity state, moving into one of the attractors even without a stimulus.

In the persistent state simulations (in which the short-term memory was implemented by the continuing neuronal firing), a strong excitatory input was given to the S1 neuronal population between 0 and 500 ms. Two such trials are shown in Fig. 5.7. In Fig. 5.7 (top), the S1 neurons (correctly) keep firing at approximately 30 Hz after the retrieval cue is removed at 500 ms. However, due to statistical fluctuations in the network related to the spiking activity, on the trial labelled persistent unstable the high firing rate in the attractor for S1 was not stable, and the firing decreased back towards the spontaneous level, in the example shown starting after 1.5 s (Fig. 5.7 bottom). This trial illustrates a failure to maintain a stable short-term memory state.

When an average was taken over many trials, for the persistent run simulations, in which the cue triggered the attractor into the high firing rate attractor state, the network was still in the high firing rate attractor state in the baseline condition on 88% of the runs. The noise had thus caused the network to fail to maintain the short-term memory on 12% of the runs.

The spontaneous state was unstable on approximately 10% of the trials, that is, on 10% of the trials the spiking noise in the network caused the network run in the condition without any initial retrieval cue to end up in a high firing rate attractor state. This is of course an error that is related to the spiking noise in the network.

We emphasise that the transitions to the incorrect activity states illustrated in Fig. 5.7 are caused by statistical fluctuations (noise) in the spiking activity of the integrate-and-fire

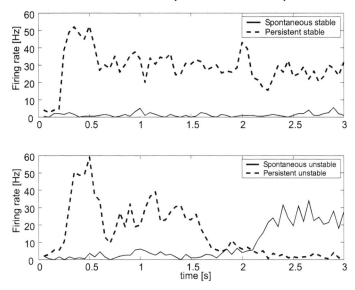

Fig. 5.7 Example trials of the Integrate-and-Fire attractor network simulations of short-term memory. The average firing rate of all the neurons in the S1 pool is shown. Top. Normal operation. On a trial in which a recall stimulus was applied to S1 at 0–500 ms, firing continued normally until the end of the trial in the 'persistent' simulation condition. On a trial on which no recall stimulus was applied to S1, spontaneous firing continued until the end of the trial in the 'spontaneous' simulation condition. Bottom: Unstable operation. On this persistent condition trial, the firing decreased during the trial as the network fell out of the attractor because of the statistical fluctuations caused by the spiking dynamics. On the spontaneous condition trial, the firing increased during the trial because of the statistical fluctuations. In these simulations the network parameter was w_+=2.1. (After Rolls, Loh, and Deco 2008c.)

neurons. Indeed, we used a mean-field approach with an equivalent network with the same parameters but without noise to establish parameter values where the spontaneous state and the high firing rate short-term memory ('persistent') state would be stable without the spiking noise, and then in integrate-and-fire simulations with the same parameter values examined the effects of the spiking noise (Loh, Rolls and Deco 2007a, Rolls, Loh and Deco 2008c). The mean-field approach used to calculate the stationary attractor states of the network for the delay period (Brunel and Wang 2001) is described in Section B.7. These attractor states are independent of any simulation protocol of the spiking simulations and represent the behaviour of the network by mean firing rates to which the system would converge in the absence of statistical fluctuations caused by the spiking of the neurons and by external changes. Therefore the mean-field technique is suitable for tasks in which temporal dynamics and fluctuations are negligible. It also allows a first assessment of the attractor landscape and the depths of the basin of attraction which then need to be investigated in detail with stochastical spiking simulations. Part of the utility of the mean-field approach is that it allows the parameter region for the synaptic strengths to be investigated to determine which synaptic strengths will on average produce stable activity in the network, for example of persistent activity in a delay period after the removal of a stimulus. For the spontaneous state, the initial conditions for numerical simulations of the mean-field method were set to 3 Hz for all excitatory pools and 9 Hz for the inhibitory pool. These values correspond to the approximate values of the spontaneous attractors when the network is not driven by stimulus-specific inputs. For the persistent state, the network parameters resulted in a selective pool having a high firing rate value of approximately 30 Hz when in its attractor state (Loh, Rolls and Deco 2007a, Rolls, Loh and Deco 2008c).

We note that there are two sources of noise in the simulated integrate-and-fire spiking networks that cause the statistical fluctuations: the randomly arriving external Poisson spike trains, and the statistical fluctuations caused by the spiking of the neurons in the finite sized network. Some of the evidence that the statistical fluctuations caused by the neuronal spiking do provide an important source of noise in attractor networks in the brain is that factors that affect the noise in the network such as the number of neurons in the network have clear effects on the operation of integrate-and-fire attractor networks, as considered further in Chapter 5.6. Indeed, the magnitude of these fluctuations increases as the number of neurons in the network becomes smaller (Mattia and Del Giudice 2004).

The ways in which alterations in the inputs to the different synapse types in the different neurons in the network influence the stability of the network, and their applications to understanding psychiatric disorders and normal aging, are described in Chapter 16.

5.5 Long-term memory recall

The theory is effectively a model of the stochastic dynamics of the recall of a memory in response to a recall cue. The memory might be a long-term memory, but the theory applies to the retrieval of any stored representation in the brain. The way in which the attractor is reached depends on the strength of the recall cue, and inherent noise in the attractor network performing the recall because of the spiking activity in a finite size system. The recall will take longer if the recall cue is weak. Spontaneous stochastic effects may suddenly lead to the memory being recalled, and this may be related to the sudden recovery of a memory which one tried to remember some time previously. These processes are considered further by Rolls (2008d). The noise in the recall of long-term memory may in fact be advantageous, in contributing for example to creative thought, as described in Section 5.11.

The theory applies to a situation where the representation may be being 'recalled' by a single input, which is perceptual detection as described in Chapter 7 of Rolls and Deco (2010).

The theory also applies to a situation where the representation may be being 'recalled' by two or more competing inputs λ, which is decision-making as described in Section 5.6.

The theory also applies to short-term memory, in which the continuation of the recalled state as a persistent attractor is subject to stochastic noise effects, which may knock the system out of the short-term memory attractor, as described in Section 5.4.

The theory also applies to attention, in which the continuation of the recalled state as a persistent attractor is subject to stochastic noise effects, which may knock the system out of the short-term memory attractor that is normally stable because of the non-linear positive feedback implemented in the attractor network by the recurrent collateral connections, as described in Chapter 6.

5.6 Stochastic dynamics and probabilistic decision-making in an attractor network

In this section, we consider how an attractor network can model probabilistic decision-making. For decision-making, the attractor network is trained to have two (or many more) high firing rate attractor states, each one of which corresponds to one of the decisions and one of which reaches a high firing rate state to represent a decision on an individual trial. Each attractor set of neurons receives a biasing input which corresponds to the evidence in favour of that decision.

When the network starts from a state of spontaneous firing, the biasing inputs encourage one of the attractors to gradually win the competition, but this process is influenced by the Poisson-like firing (spiking) of the neurons, so that which attractor wins is probabilistic. If the evidence in favour of the two decisions is equal, the network chooses each decision probabilistically on 50% of the trials. The model not only shows how probabilistic decision-making could be implemented in the cerebral cortex, but also how the evidence can be accumulated over long periods of time because of the integrating action of the attractor short-term memory network; how this accounts for reaction times as a function of the magnitude of the difference between the evidence for the two decisions (difficult decisions take longer); and how Weber's law appears to be implemented in the brain. Details of the model are provided by Deco and Rolls (2006), and in Section B.6.3 for the integrate-and-fire implementation, and Sections B.8.2 and B.8.3 for the mean-field implementation.

It is very interesting that the model of decision-making is essentially the same as an attractor model of long-term memory or short-term memory in which there are competing retrieval cues. This makes the approach very unifying, and elegant, and consistent with the presence of well-developed recurrent collateral excitatory connections in the neocortex which with the same architecture and functionality can be put to different uses. This provides for economy and efficiency in evolution and in the genetic prescription of a type of cortical architecture that can be used for many functions (see Chapter 19).

5.6.1 Decision-making in an attractor network

Let us consider the attractor network architecture again, but this time as shown in Fig. 5.8a and b with two competing inputs λ_1 and λ_2, each encouraging the network to move from a state of spontaneous activity into the attractor corresponding to λ_1 or to λ_2. These are separate attractor states that have been set up by associative synaptic modification, one attractor for the neurons that are coactive when λ_1 is applied, and a second attractor for the neurons that are coactive when λ_2 is applied. When λ_1 and λ_2 are both applied simultaneously, each attractor competes through the inhibitory interneurons (not shown in a), until one wins the competition, and the network falls into one of the high firing rate attractors that represents the decision. The noise in the network caused by the random spiking of the neurons means that on some trials, for given inputs, the neurons in the decision 1 (D1) attractor are more likely to have more spikes and therefore are more likely to win, and on other trials the neurons in the decision 2 (D2) attractor are more likely to have more spikes and therefore are more likely to win. This makes the decision-making probabilistic, for, as shown in Fig. 5.8c, the noise influences when the system will jump out of the spontaneous, low-firing, stable (low energy) state S, and whether it jumps into the high firing state for decision 1 or decision 2.

The operation and properties of this model of decision-making (Wang 2002, Deco and Rolls 2006) are described in this section (5.6), with further analysis provided in Rolls and Deco (2010).

5.6.2 Theoretical framework: a probabilistic attractor network

The theoretical framework within which the model was developed was utilized by Wang (2002), which is based on a neurodynamical model first introduced by Brunel and Wang (2001), and which has been extended and successfully applied to explain several experimental paradigms (Rolls and Deco 2002, Deco and Rolls 2002, Deco and Rolls 2003, Deco and Rolls 2004, Deco, Rolls and Horwitz 2004, Szabo, Almeida, Deco and Stetter 2004, Deco and Rolls 2005b, Rolls, Grabenhorst and Deco 2010b, Rolls, Grabenhorst and Deco 2010c). In this framework, we model probabilistic decision-making by a single attractor network

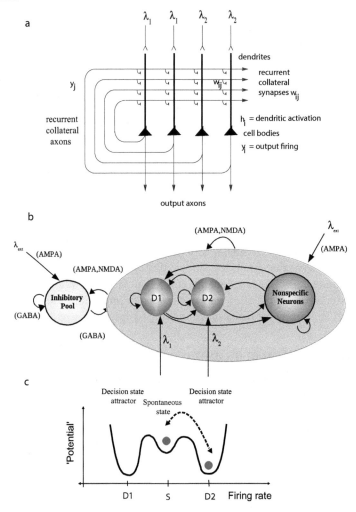

Fig. 5.8 (a) Attractor or autoassociation single network architecture for decision-making. The evidence for decision 1 is applied via the λ_1 inputs, and for decision 2 via the λ_2 inputs. The synaptic weights w_{ij} have been associatively modified during training in the presence of λ_1 and at a different time of λ_2. When λ_1 and λ_2 are applied, each attractor competes through the inhibitory interneurons (not shown), until one wins the competition, and the network falls into one of the high firing rate attractors that represents the decision. The noise in the network caused by the random spiking times of the neurons (for a given mean rate) means that on some trials, for given inputs, the neurons in the decision 1 (D1) attractor are more likely to win, and on other trials the neurons in the decision 2 (D2) attractor are more likely to win. This makes the decision-making probabilistic, for, as shown in (c), the noise influences when the system will jump out of the spontaneous firing stable (low energy) state S, and whether it jumps into the high firing state for decision 1 (D1) or decision 2 (D2). (b) The architecture of the integrate-and-fire network used to model decision-making (see text). (c) A multistable 'effective energy landscape' for decision-making with stable states shown as low 'potential' basins. Even when the inputs are being applied to the network, the spontaneous firing rate state is stable, and noise provokes transitions from the low firing rate spontaneous state S into the high firing rate decision attractor state D1 or D2.

of interacting neurons organized into a discrete set of populations, as depicted in Fig. 5.8. Populations or pools of neurons are defined as groups of excitatory or inhibitory neurons sharing the same inputs and connectivities. The network contains N_E (excitatory) pyramidal cells and N_I inhibitory interneurons. In the simulations, we used $N_E = 800$ and $N_I = 200$, consistent with the neurophysiologically observed proportion of 80% pyramidal cells versus 20% interneurons (Abeles 1991, Rolls 2008d). The neurons are fully connected (with synaptic strengths as specified later).

The specific populations have specific functions in the task. In our minimal model, we assumed that the specific populations encode the categorical result of the comparison between the two inputs λ_1 and λ_2. Each specific population of excitatory cells contains rN_E neurons (in our simulations $r = 0.1$). In addition there is one non-specific population, named 'Non-specific', which groups all other excitatory neurons in the modelled brain area not involved in the present task, and one inhibitory population, named 'Inhibitory', grouping the local inhibitory neurons in the modelled brain area. The latter population regulates the overall activity and implements competition in the network by spreading a global inhibition signal.

Because we were mainly interested in the non-stationary probabilistic behaviour of the network, the proper level of description at the microscopic level is captured by the spiking and synaptic dynamics of one-compartment *Integrate-and-Fire* (IF) neuron models (see Section B.6). At this level of detail the model allows the use of realistic biophysical time constants, latencies and conductances to model the synaptic current, which in turn allows a thorough study of the realistic time scales and firing rates involved in the time evolution of the neural activity. Consequently, the simulated neuronal dynamics, that putatively underly cognitive processes, can be quantitatively compared with experimental data. For this reason, it is very useful to include a thorough description of the different time constants of the synaptic activity. The IF neurons are modelled as having three types of receptor mediating the synaptic currents flowing into them: AMPA, NMDA (both activated by glutamate), and GABA receptors. The excitatory recurrent postsynaptic currents (EPSCs) are considered to be mediated by AMPA (fast) and NMDA (slow) receptors; external EPSCs imposed onto the network from outside are modelled as being driven only by AMPA receptors. Inhibitory postsynaptic currents (IPSCs) to both excitatory and inhibitory neurons are mediated by GABA receptors. The details of the mathematical formulation are summarized in previous publications (Brunel and Wang 2001, Deco and Rolls 2005b, Deco and Rolls 2006), and are provided in Section B.6.3.

We modified the conductance values for the synapses between pairs of neurons by synaptic connection weights, which can deviate from their default value 1. The structure and function of the network was achieved by differentially setting the weights within and between populations of neurons. We assumed that the connections are already formed, by for example earlier self-organization mechanisms, as if they were established by Hebbian learning, i.e. the coupling will be strong if the pair of neurons have correlated activity (i.e. covarying firing rates), and weak if they are activated in an uncorrelated way. The two possible decisions are with D1 having much more firing than D2, or D2>D1. As a consequence of this, neurons within a specific excitatory population are mutually coupled with a strong weight w_+, and each such population thus forms an attractor. Furthermore, the populations encoding these two decisions are likely to have anti-correlated activity in this behavioural context, resulting in weaker than average connections between the two different populations. Consequently, we choose a weaker value $w_- = 1 - r(w_+ - 1)/(1 - r)$, so that the overall recurrent excitatory synaptic drive in the spontaneous state remains constant as w_+ is varied (Brunel and Wang 2001). Neurons in the inhibitory population are mutually connected with an intermediate weight $w = 1$. They are also connected with all excitatory neurons in the same layer with the same intermediate weight, which for excitatory-to-inhibitory connections is $w = 1$, and for inhibitory-to-excitatory connections is denoted by a weight w_I. Neurons in a specific excitatory population are connected to neurons in the non-selective population in the same layer with a feedforward synaptic weight $w = 1$ and a feedback synaptic connection of weight w_-.

Each individual population is driven by two different kinds of input. First, all neurons in the model network receive spontaneous background activity from outside the module through N_{ext}=800 external excitatory synaptic connections. Each synaptic connection carries a Poisson spike train at a spontaneous rate of 3 Hz, which is a typical spontaneous firing rate value observed in the cerebral cortex. This results in a background external input with a rate summed across all 800 external synapses onto each neuron of 2.4 kHz for each neuron. Second, the neurons in the two specific populations additionally receive external inputs encoding stimulus-specific information. Two such inputs to the decision-making network are shown as λ_1 and λ_2 in Fig. 5.8. These inputs which convey the evidence for each of the decisions are added to the background external inputs being applied via the 800 synapses to each neuron.

5.6.3 Stationary multistability analysis: mean-field

A first requirement for use of the network shown in Fig. 5.8 in a probabilistic decision-making neurodynamical framework is to tune its connectivity such that the network operates in a regime of multistability. This means that at least for the stationary conditions, i.e. for periods after the dynamical transients, different possible attractors are stable. The attractors of interest correspond to high activity (high spiking rates) or low activity (low spiking rates) of the neurons in the specific populations D1 and D2. High firing rates in D1 indicate that the D1 decision has been taken. Low activity in both specific populations D1 and D2 (the 'spontaneous state') corresponds to encoding that no decision has been made. If both specific populations are activated (the 'pair state') this also corresponds to no decision. Because the useful states are one of the two populations D1 and D2 high, or both low to reflect no decision yet, the operating working point of the network should be such that both possible categorical decisions, i.e. both possible single states, and sometimes (depending on the whether a stimulus is being applied and its magnitude) the spontaneous states, are possible stable states.

The network's operating regimes just described can all occur if the synaptic connection weights are appropriate. To determine the correct weights a mean field analysis is used (Deco and Rolls 2006) as described in Section B.8.2 on page 785. Although a network of integrate-and-fire neurons with randomness in the spikes being received is necessary to understand the dynamics of the network, and how these are related to probabilistic decision-making, this means that the spiking activities fluctuate from time-point to time-point and from trial to trial. Consequently, integrate-and-fire simulations are computationally expensive and their results probabilistic, which makes them rather unsuitable for systematic parameter explorations. To solve this problem, we simplify the dynamics via the *mean-field* approach at least for the stationary conditions, i.e. for periods after the dynamical transients, and then analyze the bifurcation diagrams of the dynamics. The essence of the mean-field approximation is to simplify the integrate-and-fire equations by replacing after the diffusion approximation (Tuckwell 1988, Amit and Brunel 1997, Brunel and Wang 2001), the sums of the synaptic components by the average D.C. component and a fluctuation term. The stationary dynamics of each population can be described by the *population transfer function*, which provides the average population rate as a function of the average input current. The set of stationary, self-reproducing rates ν_i for the different populations i in the network can be found by solving a set of coupled self-consistency equations using the formulation derived by Brunel and Wang (2001) (see Section B.8.2). The equations governing the activities in the mean-field approximation can hence be studied by standard methods of dynamical systems. The formulation departs from the equations describing the dynamics of one neuron to reach a stochastic analysis of the mean-first passage time of the membrane potentials, which results in a description of the population spiking rates as functions of the model parameters, in the limit of very large N. Obtaining a mean-field analysis of the stationary states that is

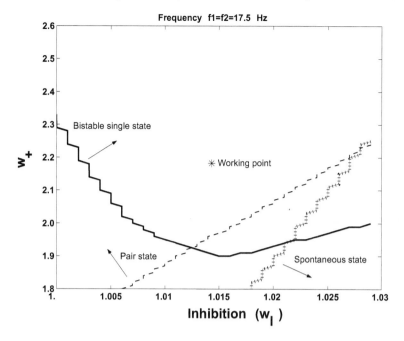

Fig. 5.9 Mean-field analysis to determine suitable values of the synaptic weights for the decision-making network. The bifurcation diagram is for the particular case where the behavioural decision-making is at chance due to λ_1 and λ_2 being equal. The different regions where single states, a pair state, and a spontaneous firing rate state are stable are shown. In all of our investigations with this network, we focus on the region of multistability (i.e. where either one or the other pool of neurons wins the competition, but where the spontaneous firing state is also a stable state), so that a probabilistic decision is possible, and therefore a convenient working point is one corresponding to a connectivity given by $w_+{=}2.2$ and $w_\mathrm{I}{=}1.015$. (Adapted with permission from European Journal of Neuroscience, 24 (3) pp. 901–16, Decision-making and Weber's law: a neurophysiological model, Gustavo Deco and Edmund T. Rolls. Copyright ©2006, John Wiley and Sons.)

consistent with the network when operating dynamically as an integrate-and-fire network is an important part of the approach used by Deco and Rolls (see Sections B.7 and B.8). The mean-field analysis is consistent with the integrate-and-fire spiking simulation described in Section B.6.3, that is, the same parameters used in the mean-field analysis can then be used in the integrate-and-fire simulations. Part of the value of the mean-field analysis is that it provides a way of determining the parameters that will lead to the specified steady state behaviour (in the absence of noise), and these parameters can then be used in a well-defined system for the integrate-and-fire simulations to investigate the full dynamics of the system in the presence of the noise generated by the random spike timings of the neurons.

To investigate how the stable states depend on the connection parameters w_+ and w_I, Deco and Rolls (2006) solved the mean-field equations for particular values of these parameters starting at different initial conditions. Figure 5.9 presents the bifurcation diagrams resulting from the mean-field analysis, for a particular case where the behavioural decision-making is hardest and is in fact purely random (i.e. at chance) as λ_1 and λ_2 are equal. The different regions where single states, a pair state, and a spontaneous state are stable are shown. In the simulations, Deco and Rolls (2006) focused on a region of multistability, in which both the possible decision states, and the spontaneous firing state, were stable (see further Section 5.3), so that a probabilistic decision is possible, and therefore a convenient working point (see

Fig. 5.9) is one corresponding to a connectivity given by $w_+ = 2.2$ and $w_I = 1.015$. Overall, Fig. 5.9 shows very large regions of stability, so that the network behaviour described here is very robust.

This description of the mean-field analysis thus shows how the fixed parameters of the network, such as the connection strengths w_+ within a decision population or pool of neurons such as D1 can be chosen.

5.6.4 Integrate-and-fire simulations of decision-making: spiking dynamics

A full characterization of the dynamics, and especially of its probabilistic behaviour, including the non-stationary regime of the system, can only be obtained through computer simulations of the spiking network model. Moreover, these simulations enable comparisons between the model in which spikes occur and neurophysiological data.

To illustrate the approach, decision-making simulations are illustrated in Fig. 5.10. The simulations were used to investigate how task difficulty affected the neuronal responses and the decision-making (Rolls, Grabenhorst and Deco 2010b, Rolls, Grabenhorst and Deco 2010c). Neurons within a specific excitatory population (D1 and D2) are mutually coupled with a strong synaptic weight w_+, set to 2.1 for the simulations described here. Furthermore, the populations encoding these two decisions are likely to have anti-correlated activity in this behavioural context, resulting in weaker than average connections between the two different populations. Consequently, we choose a weaker value $w_- = 1 - r(w_+ - 1)/(1 - r)$, so that the overall recurrent excitatory synaptic drive in the spontaneous state remains constant as w_+ is varied (Brunel and Wang 2001). Neurons in the inhibitory population are mutually connected with an intermediate weight $w = 1$. They are also connected with all excitatory neurons in the same layer with the same intermediate weight, which for excitatory-to-inhibitory connections is $w = 1$, and for inhibitory-to-excitatory connections is denoted by a weight w_I. Neurons in a specific excitatory population are connected to neurons in the nonselective population in the same layer with a feedforward synaptic weight $w = 1$ and a feedback synaptic connection of weight w_-. The simulations were run for 2 s of spontaneous activity, and then for a further 2 s while the stimuli were being applied. During the spontaneous period, the stimuli applied to D1 and D2 (and to all the other neurons in the network) had a value of 3 Hz. During the decision period, the mean input to each external synapse of each neuron of D1 and D2 was increased to 3.04 Hz per synapse (an extra 32 Hz per neuron). For $\Delta I = 0$, 32 extra Hz to the spontaneous was applied to each neuron of both D1 and D2. For $\Delta I = 16$, $32 + 8$ Hz was the extra applied to D1 and corresponds to λ_1 in Fig. 5.8, and $32 - 8$ Hz was the extra applied to D2, etc. (Rolls, Grabenhorst and Deco 2010b, Rolls, Grabenhorst and Deco 2010c).

These parameters were chosen so that if after a decision had been reached and the network had fallen into the D1 or D2 attractor state the decision-related inputs λ_1 and λ_2 were removed, the attractor would remain stable, and maintain its decision-related high firing rate in the short-term memory implemented by the attractor network. *This continuing firing and short-term memory is an extremely important property and advantage of this type of decision-making network, for it enables the decision state to be maintained for what could be a delay before and while an action is being implemented to effect the decision taken.* No further, separate, special-purpose, memory of the decision is needed, and the memory of a recent decision state emerges as a property of the decision-related mechanism. This is an important way in which the approach described in this book provides a unifying approach to understanding many properties of neocortical function, implemented by its special type of connectivity, short-range excitatory recurrent collateral synapses. The continuing short-term memory-related firing of the choice decision just taken also very usefully retains the information needed for any subsequent decision that might be taken about the first decision, such as whether to change the first decision, and perhaps abort or reverse any action taken as a result of the first decision.

Fig. 5.10a and e show the mean firing rates of the two neuronal populations D1 and D2 for two trial types, easy trials (ΔI=160 Hz) and difficult trials (ΔI=0) (where ΔI is the difference in spikes/s summed across all synapses to each neuron between the two inputs, λ_1 to population D1, and λ_2 to population D2). The results are shown for correct trials, that is, trials on which the D1 population won the competition and fired with a rate of > 10 spikes/s more than the rate of D2 for the last 1000 ms of the simulation runs. Figure 5.10b shows the mean firing rates of the four populations of neurons on a difficult trial, and Fig. 5.10c shows the rastergrams for the same trial, for which the energy landscape is also shown in Fig. 5.8d. Figure 5.10d shows the firing rates on another difficult trial (ΔI=0) to illustrate the variability shown from trial to trial, with on this trial prolonged competition between the D1 and D2 attractors until the D1 attractor finally won after approximately 1100 ms. Figure 5.10f shows firing rate plots for the four neuronal populations on an example of a single easy trial (ΔI=160), Fig. 5.10g shows the synaptic currents in the four neuronal populations on the same trial, and Fig. 5.10h shows rastergrams for the same trial.

Three important points are made by the results shown in Fig. 5.10. First, the network falls into its decision attractor faster on easy trials than on difficult trials. We would accordingly expect reaction times to be shorter on easy than on difficult trials. We might also expect the BOLD signal related to the activity of the network to be higher on easy than on difficult trials because it starts sooner on easy trials[1].

Second, the mean firing rate after the network has settled into the correct decision attractor is higher on easy than on difficult trials. We might therefore expect the BOLD signal related to the activity of the network to be higher on easy than on difficult trials because the maintained activity in the attractor is higher on easy trials. This shows that the exact firing rate in the attractor is a result not only of the internal recurrent collateral effect, but also of the external input to the neurons, which in Fig. 5.10a is 32 Hz to each neuron (summed across all synapses) of D1 and D2, but in Fig. 5.10e is increased by a further 80 Hz to D1, and decreased (from the 32 Hz added) by 80 Hz to D2 (i.e. the total external input to the network is the same, but ΔI=0 for Fig. 5.10a, and ΔI=160 for Fig. 5.10b).

Third, the variability of the firing rate is high, with the standard deviations of the mean firing rate calculated in 50 ms epochs indicated in order to quantify the variability. The large standard deviations on difficult trials for the first second after the decision cues are applied at t=2 s reflects the fact that on some trials the network has entered an attractor state after 1000 ms, but on other trials it has not yet reached the attractor, although it does so later. This trial by trial variability is indicated by the firing rates on individual trials and the rastergrams in the lower part of Fig. 5.10. The effects evident in Fig. 5.10 are quantified, and elucidated over a range of values for ΔI, next.

Fig. 5.11a shows the firing rates (mean $\pm$ sd) on correct trials when in the D1 attractor as a function of ΔI. ΔI=0 corresponds to the most difficult decision, and ΔI=160 corresponds to easy. The firing rates for both the winning population D1 and for the losing population D2 are shown. The firing rates were measured in the last 1 s of firing, i.e. between t=3 and t=4 s. It is clear that the mean firing rate of the winning population increases monotonically as ΔI increases, and interestingly, the increase is approximately linear (Pearson $r = 0.995$, p<10^{-6}). The higher mean firing rates as ΔI increases are due not only to higher peak firing, but also to the fact that the variability becomes less as ΔI increases ($r = -0.95$, p<10^{-4}),

---

[1] The blood-oxygenation level dependent (BOLD) signal is what is measured in fMRI investigations, and reflects the underlying neural activity because blood flow increases locally in the brain when extra metabolites are needed because of increased neural activity.

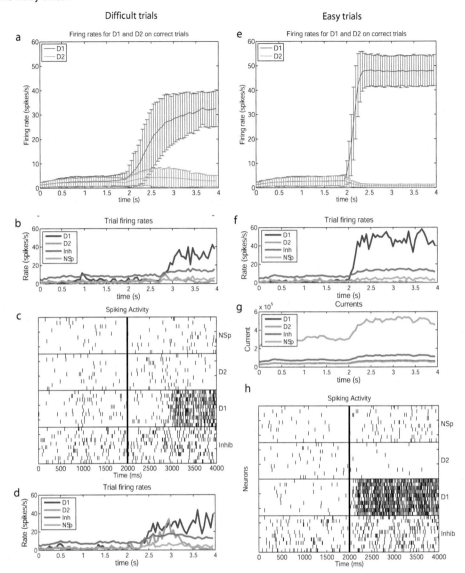

Fig. 5.10 (a) and (e) Firing rates (mean $\pm$ sd) for difficult (ΔI=0) and easy (ΔI=160) trials. The period 0–2 s is the spontaneous firing, and the decision cues were turned on at time = 2 s. The mean was calculated over 1000 trials. D1: firing rate of the D1 population of neurons on correct trials on which the D1 population won. D2: firing rate of the D2 population of neurons on the correct trials on which the D1 population won. A correct trial was one in which in which the mean rate of the D1 attractor averaged > 10 spikes/s for the last 1000 ms of the simulation runs. (Given the attractor nature of the network and the parameters used, the network reached one of the attractors on >90% of the 1000 trials, and this criterion clearly separated these trials, as indicated by the mean rates and standard deviations for the last 1 s of the simulation as shown.) (b) The mean firing rates of the four populations of neurons on a difficult trial. Inh is the inhibitory population that uses GABA as a transmitter. NSp is the non-specific population of neurons (see Fig. 5.8). (c) Rastergrams for the trial shown in b. 10 neurons from each of the four pools of neurons are shown. (d) The firing rates on another difficult trial (ΔI=0) showing prolonged competition between the D1 and D2 attractors until the D1 attractor finally wins after approximately 1100 ms. (f) Firing rate plots for the 4 neuronal populations on a single easy trial (ΔI=160). (g) The synaptic currents in the four neuronal populations on the trial shown in f. (h) Rastergrams for the easy trial shown in f and g. 10 neurons from each of the four pools of neurons are shown. (After *NeuroImage* 53, E. T. Rolls, F. Grabenhorst, and G. Deco, Choice, difficulty, and confidence in the brain, pp. 694–706 (2010), with permission from Elsevier.)

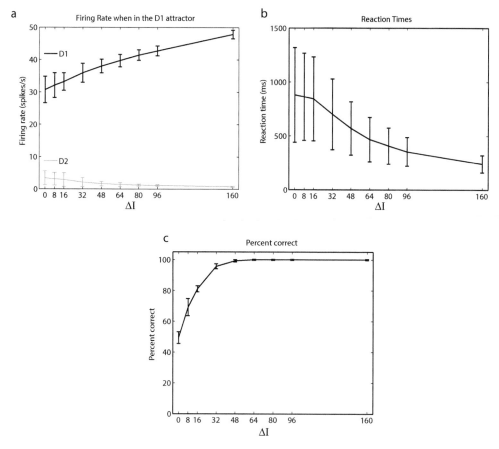

Fig. 5.11 (a) Firing rates (mean $\pm$ sd) on correct trials when in the D1 attractor as a function of ΔI. ΔI=0 corresponds to difficult, and ΔI=160 spikes/s corresponds to easy. The firing rates for both the winning population D1 and for the losing population D2 are shown for correct trials by thick lines. All the results are for 1000 simulation trials for each parameter value, and all the results shown are statistically highly significant. (b) Reaction times (mean $\pm$ sd) for the D1 population to win on correct trials as a function of the difference in inputs ΔI to D1 and D2. (c) Per cent correct performance, i.e. the percentage of trials on which the D1 population won, as a function of the difference in inputs ΔI to D1 and D2. The mean was calculated over 1000 trials, and the standard deviation was estimated by the variation in 10 groups each of 100 trials. (Adapted from *NeuroImage*, 53 (2), Edmund T. Rolls, Fabian Grabenhorst, and Gustavo Deco, Choice, difficulty, and confidence in the brain, pp. 694–706, Copyright (2010), with permission from Elsevier.)

reflecting the fact that the system is more noisy and unstable with low ΔI, whereas the firing rate in the attractor is maintained more stably with smaller statistical fluctuations against the Poisson effects of the random spike timings at high ΔI. (The measure of variation indicated in the figure is the standard deviation, and this is shown here unless otherwise stated to quantify the degree of variation, which is a fundamental aspect of the operation of these neuronal decision-making networks.)

As shown in Fig. 5.11a, the firing rates of the losing population decrease as ΔI increases. The decrease of firing rate of the losing population is due in part to feedback inhibition through the inhibitory neurons by the winning population. Thus the difference of firing rates between the winning and losing populations, as well as the firing rate of the winning population D1, both clearly reflect ΔI, and in a sense the confidence in the decision.

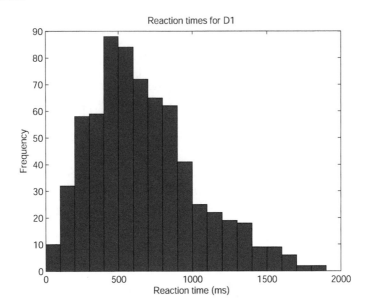

Fig. 5.12 Reaction time distribution for the decision-making attractor network. The difference between the two stimuli was relatively small (ΔI=16 Hz, though sufficient to produce 81% correct choices). The criterion for the reaction time was the time after application of both stimuli at which the firing rate of the neurons in the correct attractor became 25 Hz or more greater than that of the neurons in the incorrect attractor, and remained in that state for the remainder of the trial.

The increase of the firing rate when in the D1 attractor (upper thick line) as ΔI increases thus can be related to the confidence in the decision, and, as will be shown next in Fig. 5.11b, the performance as shown by the percentage of correct choices. The firing rate of the losing attractor (D2, lower thick line) decreases as ΔI increases, due to feedback inhibition from the winning D1 attractor, and thus the difference in the firing rates of the two attractors also reflects well the decision confidence.

We emphasize from these findings (Rolls, Grabenhorst and Deco 2010b) that the firing rate of the winning attractor reflects ΔI, and thus the confidence in the decision which is closely related to ΔI (Rolls, Grabenhorst and Deco 2010b, Rolls, Grabenhorst and Deco 2010c).

5.6.5   Reaction times of the neuronal responses

Because of the noise-driven stochastic nature of the decision-making, the reaction times even for one set of parameters vary from trial to trial. An example of the distribution of reaction times of the attractor network are shown in Fig. 5.12. This distribution is for a case when the difference between the two stimuli is relatively small ($\Delta I = 16$ Hz for the simulations described by Rolls et al. (2010b), though sufficient to produce 81% correct choices). The considerable variability of the reaction times across trials, and the long tail of the probability distribution, provide a very useful basis for understanding the variability from trial to trial that is evident in human choice reaction times (Welford 1980). Indeed, human studies commonly report skewed reaction time distributions with a long right tail (Luce 1986, Ratcliff and Rouder 1998, Ratcliff, Zandt and McKoon 1999, Usher and McClelland 2001, Marti et al. 2008). In the model, the gamma-like distribution of reaction times with a long tail to the distribution is produced when the spontaneous state is stable even when the decision cues are applied (Marti et al. 2008).

The reaction times of this model of decision-making are faster when the discrimination is easy. The time for the network to reach the correct D1 attractor, i.e. the reaction time of the network, is shown as a function of ΔI in Fig. 5.11b (mean $\pm$ sd). Interestingly, the reaction time continues to decrease ($r = -0.95$, p$<10^{-4}$) over a wide range of ΔI, even when as shown in Fig. 5.11c the network is starting to perform at 100% correct. The decreasing reaction time as ΔI increases is attributable to the altered 'effective energy landscape' (see Section 5.2): a larger input to D1 tends to produce occasionally higher firing rates, and these statistically are more likely to induce a significant depression in the landscape towards which the network flows sooner than with low ΔI. Correspondingly, the variability (quantified by the standard deviation) of the reaction times is greatest at low ΔI, and decreases as ΔI increases ($r = -0.95$, p$<10^{-4}$). This variability would not be found with a deterministic system (i.e. the standard deviations would be 0 throughout, and such systems include those investigated with mean-field analyses), and is entirely due to the random statistical fluctuations caused by the random spiking times of the neurons in the integrate-and-fire network.

5.6.6 Percentage correct

At ΔI=0, there is no influence on the network to fall more into attractor D1 representing decision 1 than attractor D2 representing decision 2, and its decisions are at chance, with approximately 50% of decisions being for D1. As ΔI increases, the proportion of trials on which D1 is reached increases. The relation between ΔI and percentage correct is shown in Fig. 5.11c. Interestingly, the performance becomes 100% correct with ΔI=64, whereas as shown in Figs. 5.11a and b the firing rates while in the D1 attractor (and therefore potentially the BOLD signal), continue to increase as ΔI increases further, and the reaction times continue to decrease as ΔI increases further. It is a clear prediction for neurophysiological and behavioural measures that the firing rates with decisions made by this attractor process continue to increase as ΔI is increased beyond the level for very good performance as indicated by the percentage of correct decisions, and the neuronal and behavioural reaction times continue to decrease as ΔI is increased beyond the level for very good performance. Fig. 5.11c also shows that the variability in the percentage correct (in this case measured over blocks of 100 trials) is large with ΔI=0, and decreases as ΔI increases. This is consistent with unbiased effects of the noise producing very variable effects in the energy landscape at ΔI=0, and with the external inputs biasing the energy landscape more and more as ΔI increases, so that the flow is much more likely to be towards the D1 attractor.

5.6.7 Finite-size noise effects

The results described earlier indicate that the probabilistic settling of the system is related to the finite size noise effects of the spiking dynamics of the individual neurons with their Poisson-like spike trains in a network of limited size. The concept here is that the smaller the network, the greater will be the statistical fluctuations (i.e. the noise) caused by the random spiking times of the individual neurons in the network. In an infinitely large system, the statistical fluctuations would be smoothed out, and reach zero. To investigate this further, and to show what sizes of network are in practice influenced by these finite-size related statistical fluctuations, an important issue when considering the operation of real neuronal networks in the brain, Deco and Rolls (2006) simulated networks with different numbers of neurons, N. The noise due to the finite size effects is expected to increase as the network becomes smaller, and indeed to be proportional to $1/\sqrt{N}$. We show in Fig. 5.13 the effects of altering N on the operation of the network, where $N = N_E + N_I$, and $N_E : N_I$, was held at 4:1. The simulations were for f1=30 Hz and f2=22 Hz (where these values represent vibrotactile frequencies used

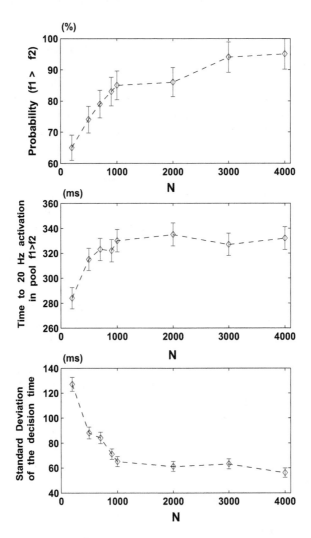

Fig. 5.13 The effects of altering N, the number of neurons in the network, on the operation of the decision-making network. The simulations were for f1=30 Hz and f2=22 Hz, which were the vibrotactile stimulation frequencies being discriminated for which was higher in a decision task. The top panel shows the probability that the network will settle into the correct (f1>f2) attractor state. The mean $\pm$ the standard deviation is shown. The middle panel shows the time for a decision to be reached, that is for the system to reach a criterion of a firing rate of 20 Hz in the pool (f1>f2). The mean $\pm$ the standard deviation of the sampled mean is shown. The bottom panel shows the standard deviation of the reaction time. (Adapted with permission from European Journal of Neuroscience, 24 (3) pp. 901–16, Decision-making and Weber's law: a neurophysiological model, Gustavo Deco and Edmund T. Rolls. Copyright ©2006, John Wiley and Sons.)

in a decision-making task for whether f1 was higher in frequency), which in term influence λ_1 and λ_2. Figure 5.13 shows overall that when N is larger than approximately 1,000, the network shows the expected settling to the (f1 > f2) attractor state on a proportion of occasions that is in the range 85–93%, increasing only a little as the number of neurons reaches 4,000 (top panel). The settling remains probabilistic, as shown by the standard deviations in the probability that the (f1 > f2) attractor state will be reached (top panel). When N is less than approximately 1,000, the finite size noise effects become very marked, as shown by the fact that the network reaches the correct attractor state (f1>f2) much less frequently, and in that the time for a decision to be reached can be premature and fast, as the large fluctuations in the stochastic noise can cause the system to reach the criterion [in this case of a firing rate of 20 Hz in the pool (f1>f2)] too quickly.

The overall conclusion of the results shown in Fig. 5.13 is that the size of the network, N, does influence the probabilistic settling of the network to the decision state. None of these probabilistic attractor and decision-related settling effects would of course be found in a mean-field or purely rate simulation, without spiking activity. The size of N in the brain is likely to be greater than 1,000 (and probably in the neocortex in the range 4,000–12,000) (see Table 1.1). Further simulations have confirmed that the decision-making remains probabilistic in larger systems, with for example N=10,000 neurons, and C=1,000 connections onto each neuron (Rolls and Webb 2012).

5.6.8 Comparison with neuronal data during decision-making

Deco and Rolls (2006) used this decision-making network to model the responses of ventral premotor cortex (VPC) neurons (Romo, Hernandez and Zainos 2004) in a vibrotactile decision task, to examine how closely the model could account for and lead to an understanding of neuronal responses during decision-making. In the task macaques must decide and report which of two mechanical vibrations applied sequentially to their fingertips has the higher frequency of vibration by pressing one of two pushbuttons. The vibrotactile stimuli f1 and f2 utilized were in the range of frequencies called *flutter*, i.e. within approximately 5–50 Hz.

The decision-making implemented by this attractor model is probabilistic in a similar way to that of the neuronal responses (Deco and Rolls 2006). Figure 5.14 shows the probability of correct discrimination as a function of the difference between the two presented vibrotactile frequencies to be compared. We assume that f1>f2 by a Δ-value, i.e. f1=f2+Δ. (In Fig. 5.14 this value is called 'Delta frequency (f1-f2)'.) Each diamond-point in the Figure corresponds to the result calculated by averaging 200 trials of the full spiking simulations. The lines were calculated by fitting the points with a logarithmic function. A correct classification occurs when during the 500 ms comparison period, the network evolves to a 'single-state' attractor that shows a high level of spiking activity (larger than 10 Hz) for the population (f1>f2), and simultaneously a low level of spiking activity for the population (f1<f2) (at the level of the spontaneous activity). Figure 5.14 shows that the decision-making is probabilistic, and that the probability of a correct discrimination increases as Δf, the difference between the two stimuli being compared, increases. When Δf is 0, the network performs at chance, and its choices are 50% correct.

The second panel of Fig. 5.14 shows a good fit between the actual neuronal data described by Romo and Salinas (2003) for the f2=20 Hz condition (indicated by *), and the results obtained with the model (Deco and Rolls 2006). This shows that the model implements the probabilistic decision-making in a very similar way to that of neurons recorded in the ventral premotor cortex.

Moreover, the model was also able to account for Weber's law, and provided a hypothesis about how it is implemented, as follows.

One can observe from the different panels in Fig. 5.14 corresponding to different base vibrotactile frequencies f2, that to reach a threshold of correct classification of for example 85% correct (horizontal dashed line in Fig. 5.14), the difference between f1 and f2 must become larger as the base frequency f2 increases.

Figure 5.15 plots the critical discrimination Δ-value corresponding to an 85% correct performance level (the 'difference threshold') as a function of the base frequency f2. The 'difference threshold' increases linearly as a function of the base frequency, that is, $\Delta f/f$ is a

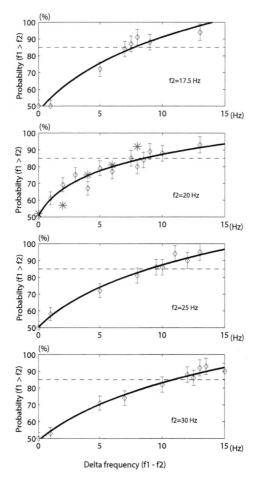

Fig. 5.14 Probability of correct discrimination ($\pm$ sd) as a function of the difference between the two presented vibrotactile frequencies to be compared. In the simulations, we assume that f1>f2 by a Δ-value (labelled 'Delta frequency (f1-f2)'), i.e. f1=f2+Δ. The points correspond to the trial averaged spiking simulations. The line interpolates the points with a logarithmic function. The horizontal dashed line represents the threshold of correct classification for a performance of 85% correct discrimination. The second panel down includes actual neuronal data (indicated by *) described by Romo and Salinas (2003) for the f2=20 Hz condition. (Adapted with permission from *European Journal of Neuroscience*, 24 (3) pp. 901–16, Decision–making and Weber's law: a neurophysiological model, Gustavo Deco and Edmund T. Rolls Copyright ©2006, John Wiley and Sons.)

constant. This corresponds to Weber's law for the vibrotactile discrimination task. (Weber's law is often expressed as $\Delta I/I$ is a constant, where I stands for stimulus intensity, and ΔI for the smallest difference of intensity that can just be discriminated, sometimes called the just noticeable difference. In the case simulated, the stimuli were vibrotactile frequencies, hence the use of f to denote the frequencies of the stimuli.)

The analysis shown in Figs. 5.14 and 5.15 suggests that Weber's law, and consequently the ability to discriminate two stimuli, is encoded in the probability of performing a transition to the correct final attractor. The firing rates when the attractor is entered to not reflect Weber's law. The model gives further insights into the mechanisms by which Weber's law is implemented. We hypothesized that because Δf/f is practically constant in the model, the

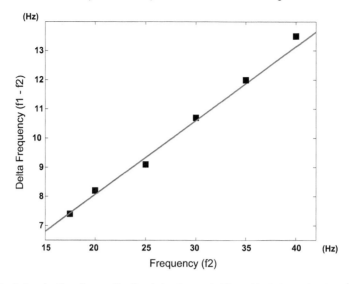

Fig. 5.15 Weber's law for the vibrotactile discrimination task. The critical discrimination Δ-value ('difference-threshold') is shown corresponding to an 85% correct performance level as a function of the base frequency f2. The 'difference-threshold' increases linearly as a function of the base frequency. (Adapted with permission from European Journal of Neuroscience, 24 (3) pp. 901–16, Decision-making and Weber's law: a neurophysiological model, Gustavo Deco and Edmund T. Rolls Copyright ©2006, John Wiley and Sons.)

difference of frequencies Δf required to push the single attractor network towards an attractor basin might increase with f because as f increases, shunting (divisive) inhibition produced by inhibitory feedback inputs (from the inhibitory interneurons) might act divisively on the pyramidal cells in the attractor network to shunt the excitatory inputs f1 and f2. In more detail, as the base frequency f increases, more excitation will be provided to the network by the inputs λ_1 and λ_2, this will tend to increase the firing rates of the pyramidal cells which will in turn provide a larger excitatory input to the inhibitory neurons. This will tend to make the inhibitory neurons fire faster, and their GABA synapses onto the pyramidal cells will be more active. Because these GABA synapses open chloride channels and act with a driving potential $V_I = -70$ mV which is relatively close to the membrane potential (which will be in the range $V_L = -70$ mV to $V_{\text{thr}} = -50$ mV), a large part of the GABA synaptic input to the pyramidal cells will tend to shunt, that is to act divisively upon, the excitatory inputs to the pyramidal cells from the vibrotactile biasing inputs λ_1 and λ_2. To compensate for this current shunting effect, f1 and f2 are likely to need to increase in proportion to the base frequency f in order to maintain the efficacy of their biasing effect. To assess this hypothesis, we measured the change in conductance produced by the GABA inputs as a function of the base frequency. Figure 5.16 shows that the conductance increases linearly with the base frequency (as does the firing rate of the GABA neurons, not illustrated). The shunting effect does appear therefore to be dividing the excitatory inputs to the pyramidal cells in the linear way as a function of f that we hypothesized.

Deco and Rolls (2006) therefore proposed that Weber's law is implemented by shunting effects acting on pyramidal cells that are produced by inhibitory neuron inputs which increase linearly as the base frequency increases, so that the difference of frequencies Δf required to push the network reliably into one of its decision attractors must increase in proportion to the base frequency. We checked the excitatory inputs to the pyramidal cells (for which $V_E = 0$ mV), and found that their conductances were much smaller (in the order of 5 nS for the AMPA

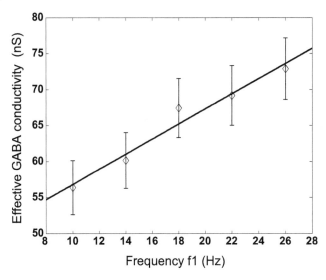

Fig. 5.16 The conductance in nS (mean $\pm$ sd) produced by the GABA inputs to the pyramidal cells as a function of the base frequency f1. The effective conductance produced through the GABA synapses (i.e. $I_{GABA}/(V - V_I)$) was averaged over the time window in which the stimuli were presented in one of the excitatory neuron pools, when the base frequency was f1, and f2-f1 was set to 8 Hz. (Adapted with permission from European Journal of Neuroscience, 24 (3) pp. 901–16, Decision-making and Weber's law: a neurophysiological model, Gustavo Deco and Edmund T. Rolls Copyright ©2006, John Wiley and Sons.)

and 15 nS for the NMDA receptors) than those produced by the GABA receptors, so that it is the GABA-induced conductance changes that dominate, and that produce the shunting inhibition.

Further properties of the attractor network model of decision-making that enable it to implement Weber's law include stability of the spontaneous firing rate condition, even when the decision cues are applied, so that the system depends on the noise to escape this starting state, as described in Section 5.3.

5.6.9 Testing the model of decision-making with human functional neuroimaging

In this subsection (5.6.9) it is shown how this integrate-and-fire model of probabilistic decision-making can make predictions about functional magnetic resonance neuroimaging (fMRI) signals during decision-making tasks in brain areas involved in decision-making, and how the predictions can then be tested.

The fMRI signal is a secondary consequence of neuronal activity, which requires increased metabolic activity to pump ions back across the cell membrane against the electrochemical gradient which have crossed the cell membrane due to the opening of synaptic receptor-activated ion channels which leads to neuronal firing. Regionally, increased oxidative metabolism causes a transient decrease in oxyhaemoglobin and increase in deoxyhaemoglobin, as well as an increase in CO_2 and NO. This provokes over several seconds a local dilatation and increased blood flow in the affected regions that leads by overcompensation to a relative decrease in the concentration of deoxyhaemoglobin in the venules draining the activated region, and the alteration of deoxyhaemoglobin, which is paramagnetic, can be detected by changes in T2 or T2* in the MRI signal as a result of the decreased susceptibility and thus

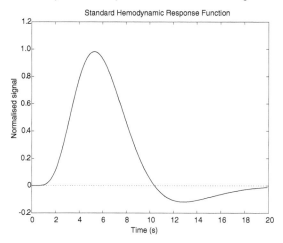

Fig. 5.17 The standard haemodynamic response function $h(t)$ (see text).

decreased local inhomogeneity which increases the MR intensity value (Glover 1999, Buxton and Frank 1997, Buxton, Wong and Frank 1998).

The fMRI BOLD (blood oxygen level-dependent) signal may reflect the total synaptic activity in an area (as ions need to be pumped back across the cell membrane) and is spatially and temporally filtered. The filtering reflects the inherent spatial resolution with which the blood flow changes, and the slow temporal response of the blood flow changes (Glover 1999, Buxton and Frank 1997, Buxton et al. 1998). Glover (1999) demonstrated that a good fit of the haemodynamical response $h(t)$ can be achieved by the following analytic function:

$$h(t) = c_1 t^{n_1} e^{-\frac{t}{\tau_1}} - a_2 c_2 t^{n_2} e^{-\frac{t}{\tau_2}}$$

$$c_i = \max(t^{n_i} e^{-\frac{t}{\tau_i}})$$

where t is the time, and c_1, c_2, a_2, n_1, and n_2 are parameters that are adjusted to fit the experimentally measured haemodynamical response. Figure 5.17 plots the haemodynamic standard response $h(t)$ for a biologically realistic set of parameters (see Deco, Rolls and Horwitz (2004)).

The temporal evolution of fMRI signals can be simulated from an integrate-and-fire population of neurons by convolving the total synaptic activity in the simulated population of neurons with the standard haemodynamic response formulation of Glover (1999) (Deco, Rolls and Horwitz 2004, Horwitz and Tagamets 1999). The rationale for this is that the major metabolic expenditure in neural activity is the energy required to pump back against the electrochemical gradient the ions that have entered neurons as a result of the ion channels opened by synaptic activity, and that mechanisms have evolved to increase local blood flow in order to help this increased metabolic demand to be met. (In fact, the increased blood flow overcompensates, and the blood oxygenation level-dependent (BOLD) signal by reflecting the consequent alteration of deoxyhaemoglobin which is paramagnetic reflects this.)

The total synaptic current (I_{syn}) is given by the sum of the absolute values of the glutamatergic excitatory components (implemented through NMDA and AMPA receptors) and inhibitory components (GABA) (Tagamets and Horwitz 1998, Horwitz, Tagamets and McIntosh 1999, Rolls and Deco 2002, Deco, Rolls and Horwitz 2004). In our integrate-and-fire simulations the external excitatory contributions are produced through AMPA receptors ($I_{\mathrm{AMPA,ext}}$), while the excitatory recurrent synaptic currents are produced through AMPA and

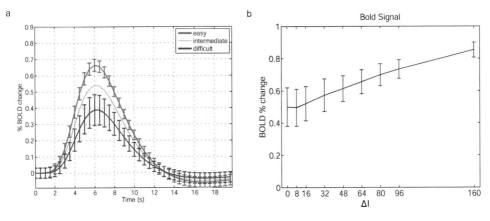

Fig. 5.18 (a) The percentage change in the simulated BOLD signal on easy trials (ΔI=160 spikes/s), on intermediate trials (ΔI=80), and on difficult trials (ΔI=0). The mean $\pm$ sd are shown for the easy and difficult trials. The percentage change in the BOLD signal was calculated from the firing rates of the D1 and D2 populations, and analogous effects were found with calculation from the synaptic currents averaged for example across all 4 populations of neurons. (b) The percentage change in the BOLD signal (peak mean Â– sd) averaged across correct and incorrect trials as a function of ΔI. ΔI=0 corresponds to difficult, and ΔI=160 corresponds to easy. The percent change was measured as the change from the level of activity in a period of 1 s immediately before the decision cues were applied at t=0 s, and was calculated from the firing rates of the neurons in the D1 and D2 populations. The BOLD per cent change scaling is arbitrary, and is set so that the lowest value for the peak of a BOLD response is 0.5%. (Adapted from *NeuroImage*, 53 (2), Edmund T. Rolls, Fabian Grabenhorst, and Gustavo Deco, Choice, difficulty, and confidence in the brain, pp. 694–706, Copyright (2010), with permission from Elsevier.)

NMDA receptors ($I_{\text{AMPA,rec}}$ and $I_{\text{NMDA,rec}}$). The GABA inhibitory currents are denoted by I_{GABA}. Consequently, the simulated fMRI signal activity S_{fMRI} is calculated by the following convolution equation:

$$S_{\text{fMRI}}(t) = \int_{0}^{\infty} h(t - t') I_{\text{syn}}(t') \, dt'.$$

5.6.9.1 Prediction of the BOLD signals on difficult vs easy decision-making trials

We now show how this model makes predictions for the fMRI BOLD signals that would occur in brain areas in which decision-making processing is taking place. The BOLD signals were predicted from the firing rates of the neurons in the network (or from the synaptic currents flowing in the neurons with similar results) by convolving the neuronal activity with the haemodynamic response function in a realistic period, the two seconds after the decision cues are applied. This is a reasonable period to take, as decisions will be taken within this time, and the attractor state may not necessarily be maintained for longer than this. (The attractor states might be maintained for longer if the response that can be made is not known until later, as in some fMRI tasks with delays, and then the effects described might be expected to be larger, given the mean firing rate effects shown in Fig. 5.10.)

In more detail, the haemodynamic signal associated with the decision was calculated by convolving the neuronal activity or the synaptic currents of the neurons from the simulations illustrated in Fig. 5.10 with the haemodynamic response function used with SPM5 (Statistical Parametric Mapping, Wellcome Department of Imaging Neuroscience, London) (as this was the function also used in the analyses by SPM of the experimental fMRI data) (Rolls, Grabenhorst and Deco 2010b).

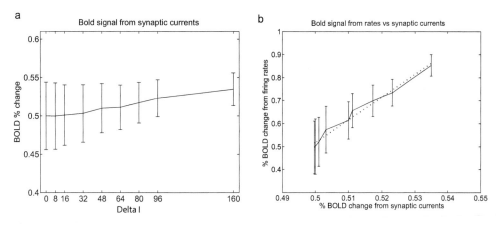

Fig. 5.19 (a) The percentage change of the BOLD signal calculated from the synaptic currents in all populations of neurons in the network (D1, D2, GABA, and non-specific, see Fig. 5.8). (Analogous results were found when the currents were calculated from the D1 and D2 populations; or from the D1, D2 and GABA populations.) (b) The relation between the BOLD current predicted from the firing rates and from the synaptic currents (r=0.99, p<10^{-6}) for values of ΔI between 0 and 160. The fitted linear regression line is shown. The BOLD per cent change scaling is arbitrary, and is set so that the lowest value is 0.5%. (Adapted from *NeuroImage*, 53 (2), Edmund T. Rolls, Fabian Grabenhorst, and Gustavo Deco, Choice, difficulty, and confidence in the brain, pp. 694–706, Copyright (2010), with permission from Elsevier.)

As shown in Fig. 5.18a, the predicted fMRI response is larger for easy (ΔI=160 spikes/s) than for difficult trials (ΔI=0), with intermediate trials (ΔI=80) producing an intermediate fMRI response. The difference in the peak response for ΔI=0 and ΔI=160 is highly significant ($p \ll 0.001$). Importantly, the BOLD response is inherently variable from brain regions associated with this type of decision-making process, and this is nothing to do with the noise arising in the measurement of the BOLD response with a scanner. If the system were deterministic, the standard deviations, shown as a measure of the variability in Fig. 5.18a, would be 0. It is the statistical fluctuations caused by the noisy (random) spike timings of the neurons that account for the variability in the BOLD signals in Fig. 5.18a. Interestingly, the variability is larger on the difficult trials (ΔI=0) than on the easy trials (ΔI=160), as shown in Fig. 5.18a, and indeed this also can be taken as an indicator that attractor decision-making processes of the type described here are taking place in a brain region.

Fig. 5.18b shows that the percentage change in the BOLD signal (peak mean $\pm$ sd) averaged across correct and incorrect trials increases monotonically as a function of ΔI. This again can be taken as an indicator (provided that fMRI signal saturation effects are minimized) that attractor decision-making processes of the type described here are taking place in a brain region. The percentage change in Fig. 5.18b was calculated by convolution of the firing rates of the neurons in the D1 and D2 populations with the haemodynamic response function. Interestingly, the percentage change in the BOLD signal is approximately linearly related throughout this range to ΔI (r=0.995, p<10^{-7}). The effects shown in Figs. 5.18a and b can be related to the earlier onset of a high firing rate attractor state when ΔI is larger (see Figs. 5.10 and 5.11b), and to a higher firing rate when in the attractor state (as shown in Figs. 5.10 and 5.11a). As expected from the decrease in the variability of the neuronal activity as ΔI increases (Fig. 5.11a), the variability (standard deviation) in the predicted BOLD signal also decreases as ΔI increases, as shown in Fig. 5.18b (r=0.955, p<10^{-4}).

Similar effects, though smaller in degree, were found when the percentage change of the BOLD signal was calculated from the synaptic activity in all populations of neurons in the network (D1, D2, GABA, and non-specific, see Fig. 5.8), as shown in Fig. 5.19a. The percentage change in the BOLD signal is approximately linearly related throughout this range to ΔI (r=0.991, p<10^{-6}). Fig. 5.19b shows the relation in the model between the BOLD signal predicted from the firing rates and from the synaptic currents for values of ΔI between 0 and 160 Hz. The fitted linear regression line is shown (r=0.99, p<10^{-6}).

The findings shown in Fig. 5.19b are of interest when considering how the fMRI BOLD signal is generated in the cortex. The fMRI BOLD signal is produced by an alteration of blood flow in response to activity in a brain region. The BOLD signal, derived from the magnetic susceptibility of deoxyhaemoglobin, reflects an overcompensation in the blood flow, resulting in more deoxyhaemoglobin in active areas. The coupling of the activity in a brain region to the altered blood flow is complex (Logothetis, Pauls, Augath, Trinath and Oeltermann 2001, Logothetis 2008, Mangia, Giove, Tkac, Logothetis, Henry, Olman, Maraviglia, Di Salle and Ugurbil 2009), but may reflect the energy needed to pump ions that have crossed the cell membrane as a result of synaptic activity to be pumped back against the electrochemical gradient. In Fig. 5.18 predicted BOLD effects are shown related to the firing rate of the two populations D1 and D2 of neurons in the attractor network. In Fig. 5.19 are shown predicted BOLD effects related to the sum of the synaptic currents in all the neurons in the network, with analogous results found with the synaptic currents in only the D1 and D2 neurons, or in the D1, D2, and GABA neurons in the network. Although it has been suggested that synaptic activity is better coupled to the blood flow and hence the BOLD signal than the firing rates (Logothetis et al. 2001, Logothetis 2008, Mangia et al. 2009), in the neocortex these might be expected to be quite closely related, and indeed the integrate-and-fire model shows a linear relation between the BOLD signal predicted from the firing rates and from the synaptic currents (r=0.99, p<10^{-6}, Fig. 5.19b), providing an indication that at least in the cerebral cortex the BOLD signal may be related to the firing rates of the neurons as well as to the synaptic currents.

5.6.9.2 Neuroimaging investigations of task difficulty, and confidence

Two functional neuroimaging investigations were performed to test the predictions of the model just described. Task difficulty was altered parametrically to determine whether there was a close relation between the BOLD signal and task difficulty (Rolls, Grabenhorst and Deco 2010b), and whether this was present especially in brain areas implicated in choice decision-making by other criteria (Grabenhorst, Rolls and Parris 2008b, Rolls, Grabenhorst and Parris 2010d). The decisions were about the pleasantness of olfactory stimuli (Rolls, Grabenhorst and Parris 2010d) or thermal stimuli applied to the hand (Grabenhorst, Rolls and Parris 2008b).

Figure 5.20 shows experimental data with the fMRI BOLD signal measured on easy and difficult trials of the olfactory affective decision task (left) and the thermal affective decision task (right) (Rolls, Grabenhorst and Deco 2010b). The upper records are for prefrontal cortex medial area 10 in a region identified by the following criterion as being involved in choice decision-making. The criterion was that a brain region for identical stimuli should show more activity when a choice decision was being made than when a rating on a continuous scale of affective value was being made. Fig. 5.20 shows for medial prefrontal cortex area 10 that there is a larger BOLD signal on easy than on difficult trials. The top diagram shows the medial prefrontal area activated in this contrast for decisions about which olfactory stimulus was more pleasant (yellow), and for decisions about whether the thermal stimulus would be chosen in future based on whether it was pleasant or unpleasant (red).

In more detail, for the thermal stimuli, the contrast was the warm2 and the cold trials (which were both easy in that the percentage of the choices were far from the chance value of 50%, and in particular were 96±1% (mean±sem) for the warm, and 18±6% for the cold),

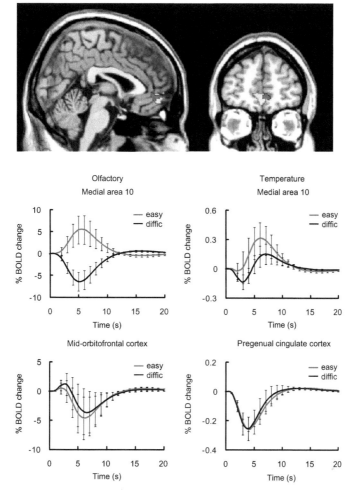

Fig. 5.20 Top: Medial prefrontal cortex area 10 activated on easy vs difficult trials in the olfactory pleasantness decision task (yellow) and the thermal pleasantness decision task (red). Middle: experimental data showing the BOLD signal in medial area 10 on easy and difficult trials of the olfactory affective decision task (left) and the thermal affective decision task (right). This medial area 10 was a region identified by other criteria (see text) as being involved in choice decision-making. Bottom: The BOLD signal for the same easy and difficult trials, but in parts of the pregenual cingulate and mid-orbitofrontal cortex implicated by other criteria (see text) in representing the subjective reward value of the stimuli on a continuous scale, but not in making choice decisions between the stimuli, or about whether to choose the stimulus in future. (Adapted from *NeuroImage*, 53 (2), Edmund T. Rolls, Fabian Grabenhorst, and Gustavo Deco, Choice, difficulty, and confidence in the brain, pp. 694–706, Copyright (2010), with permission from Elsevier.)

versus the mixed stimulus of warm2+cold (which was difficult in that the percentage of choices of 'Yes, it would be chosen in future' was 64±9%). For the temperature easy vs difficult decisions about pleasantness, the activation in medial area 10 had peaks at [4 42 −4] z=3.59 p=0.020 and [6 52 −4] z=3.09 p=0.045.

For the olfactory decision task, the activations in medial area 10 for easy vs difficult choices were at [−4 62 −2] z=2.84 p=0.046, confirmed in a finite impulse response (FIR) analysis with a peak at 6–8 s after the decision time at [−4 54 −6] z=3.50 p=0.002. (In the olfactory task, the easy trials were those in which one of the pair of odours was from the pleasant set, and

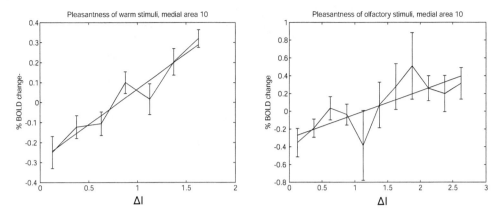

Fig. 5.21 Experimental fMRI data showing the change in the BOLD signal (mean±sem, with the fitted linear regression line shown) as a function of ΔI, the difference in pleasantness of warm stimuli or olfactory stimuli about which decision was being made, for medial prefrontal cortex area 10. (Adapted from *NeuroImage*, 53 (2), Edmund T. Rolls, Fabian Grabenhorst, and Gustavo Deco, Choice, difficulty, and confidence in the brain, pp. 694–706, Copyright (2010), with permission from Elsevier.)

the other from the unpleasant set. The mean difference in pleasantness, corresponding to ΔI, was 1.76±0.25 (mean±sem). The difficult trials were those in which both odours on a trial were from the pleasant set, or from the unpleasant set. The mean difference in pleasantness, corresponding to ΔI, was 0.72±0.16. For easy trials, the percentage correct was 90±2, and for difficult trials was 59±8.) No other significant effects in the a priori regions of interest (Grabenhorst, Rolls and Parris 2008b, Rolls, Grabenhorst and Parris 2010d) were found for the easy vs difficult trial contrast in either the thermal or olfactory reward decision task (Rolls, Grabenhorst and Deco 2010b).

The lower records in Fig. 5.20 are for the same easy and difficult trials, but in parts of the pregenual cingulate and mid-orbitofrontal cortex implicated by the same criteria in representing the subjective reward value of the stimuli, but not in making choice decisions between the stimuli. [For the pregenual cingulate cortex, there was a correlation of the activations with the subjective ratings of pleasantness of the thermal stimuli at [4 38 −2] z=4.24 p=0.001. For the mid-orbitofrontal cortex, there was a correlation of the activations with the subjective ratings of pleasantness of the thermal stimuli at [40 36 −12] z=3.13 p=0.024]. The BOLD signal was similar in these brain regions for easy and difficult trials, as shown in Fig. 5.20, and there was no effect in the contrast between easy and difficult trials.

Fig. 5.21 shows the experimental fMRI data with the change in the BOLD signal for medial prefrontal cortex area 10 indicated as a function of ΔI, the difference in pleasantness of warm stimuli or olfactory stimuli about which decision was being made, and thus the easiness of the decision. For the olfactory decision task, ΔI was the difference in pleasantness (for a given subject) between the mean pleasantness of the first odour and the mean pleasantness of the second odour between which decision was being taken, about which was more pleasant (Rolls et al. 2010b). It is shown in Fig. 5.21 (right) that there was a clear and approximately linear relation between the BOLD signal and ΔI for the olfactory pleasantness decision-making task ($r = 0.77, p = 0.005$). The coordinates for these data were as given for Fig. 5.20.

For the warm decision task, ΔI was the difference in mean pleasantness for a given subject from 0 for a given thermal stimulus about which a decision was being made about whether it should or should not be repeated in future (Grabenhorst et al. 2008b). It is shown in Fig. 5.21 (left) that there was a clear and approximately linear relation between the BOLD signal and ΔI for the thermal pleasantness decision-making task ($r = 0.96, p < 0.001$). The coordinates for these data were as given for Fig. 5.20.

These experimental findings are thus consistent with the predictions made from the model, and provide strong support for the type of model of decision-making described in this book. Moreover, they show that decision confidence, which increases with ΔI, can be read off from fMRI BOLD signals in parts of the brain involved in making choices, as shown in this unifying theory.

In further investigations, it was shown that the model predicted smaller fMRI signals on error trials as ΔI increased (because then the external decision cues were operating against the noise-driven erroneous decision that was inconsistent with the decision cues), and that the BOLD signal in the same and related brain areas showed the same parametric change (Rolls, Grabenhorst and Deco 2010c).

Thus detailed predictions from the model of decision-making were confirmed in experimental fMRI investigations. This provides further support for this approach to decision-making mechanisms in the cortex, and shows how these models can be used to help understand the neural processing and signals during decision-making in parts of the cortex implicated in the type of decision being investigated (Rolls, Grabenhorst and Deco 2010b, Rolls, Grabenhorst and Deco 2010c, Rolls and Deco 2010).

5.6.10 Decisions based on confidence in one's decisions: self-monitoring

We have seen that after a binary decision there is nevertheless a continuous-valued representation of decision confidence encoded in the firing rates of the neurons in a decision-making integrate-and-fire attractor neuronal network (Fig. 5.11). What happens if instead of having to report or assess the continuous-valued representation of confidence in a decision one has taken, one needs to take a decision based on one's (partly) continuous-valued confidence estimate that one has just made a correct or incorrect decision? One might for example wait for a reward if one thinks one's decision was correct, or alternatively stop waiting on that trial and start another trial or action. We propose that in this case, one needs a second decision-making network, which takes decisions based on one's decision confidence (Insabato, Pannunzi, Rolls and Deco 2010).

The architecture has a decision-making network, and a separate confidence decision network that receives inputs from the decision-making network, as shown in Fig. 5.22. The decision-making network has two main pools or populations of neurons, DA which becomes active for decision A, and DB which becomes active for decision B. Pool DA receives sensory information about stimulus A, and Pool DB receives sensory information about stimulus B. Each of these pools has strong recurrent collateral connections between its own neurons of strength w_+, so that each operates as an attractor population. There are inhibitory neurons with global connectivity to implement the competition between the attractor subpopulations. When stimulus A is applied, pool DA will usually win the competition and end up with high firing indicating that decision A has been reached. When stimulus B is applied, pool DB will usually win the competition and end up with high firing indicating that decision B has been reached. If a mixture of stimuli A and B is applied, the decision-making network will probabilistically choose decision A or B influenced by the proportion of stimuli A and B in the mixture.

We simulated the network, and it performed as expected, implementing decisions based on confidence estimates of a decision in the absence of any feedback about the success of the first decision (Insabato, Pannunzi, Rolls and Deco 2010).

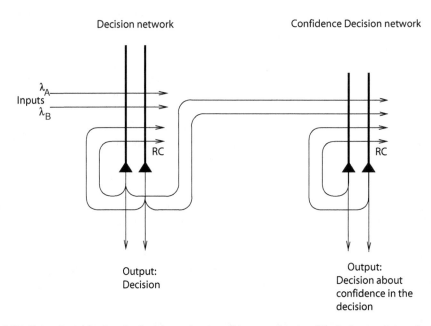

Fig. 5.22 Network architecture for decisions about confidence estimates. The first network is a decision–making network, and its outputs are sent to a second network that makes decisions based on the firing rates from the first network, which reflect the decision confidence. In the first network, high firing of neuronal population (or pool) DA represents decision A, and high firing of population DB represents decision B. Pools DA and DB receive a stimulus-related input (respectively λ_A and λ_B), the evidence for each of the decisions, and these bias the attractor networks, which have internal positive feedback produced by the recurrent excitatory connections (RC). Pools DA and DB compete through inhibitory interneurons. The neurons are integrate-and-fire spiking neurons with random spiking times (for a given mean firing rate) which introduce noise into the network and influence the decision-making, making it probabilistic. The second network is a confidence decision attractor network, and receives inputs from the first network. The confidence decision network has two selective pools of neurons, one of which (C) responds to represent confidence in the decision, and the other of which responds when there is little or a lack of confidence in the decision (LC). The C neurons receive the outputs from the selective pools of the (first) decision-making network, and the LC neurons receive $\lambda_{\text{Reference}}$ which is from the same source but saturates at 40 spikes/s, a rate that is close to the rates averaged across correct and error trials of the sum of the firing in the selective pools in the (first) decision-making network. /footnotesize (Data from Confidence-Related Decision Making, Andrea Insabato, Mario Pannunzi, Edmund T. Rolls, Gustavo Deco, Journal of Neurophysiology, Jul 2010, 104 (1) 539–547; DOI: 10.1152/jn.01068.2009.)

This approach indicates that an integrate-and-fire attractor neuronal decision-making network encodes confidence in its firing rates, and that adding a second attractor network allows decisions to be made about whether to change the decision made by the first network, and for example abort the trial or strategy (see Fig. 5.22). The second network, the confidence decision network, is in effect monitoring the decisions taken by the first network, and can cause a change of strategy or behaviour if the assessment of the decision taken by the first network does not seem a confident decision.

Now this is the type of description, and language used, to describe 'monitoring' functions, taken to be high level cognitive processes, possibly related to consciousness (Lycan 1997, Block 1995a). For example, in an experiment performed by Hampton (2001) (experiment 3), a monkey had to remember a picture over a delay. He was then given a choice of a 'test flag', in which case he would be allowed to choose from one of four pictures the one seen before the delay, and if correct earn a large reward (a peanut). If he was not sure that he remembered the first picture, he could choose an 'escape flag', to start another trial. With longer delays, when memory strength might be lower partly due to noise in the system, and confidence therefore in the memory on some trials might be lower, the monkey was more likely to choose the escape flag. The experiment is described as showing that the monkey is thinking about his own memory, that is, is a case of meta-memory, which may be related to consciousness (Heyes 2008). However, the decision about whether to escape from a trial can be taken just by adding a second decision network to the first decision network. Thus we can account for what seem like complex cognitive phenomena with a simple system of two attractor decision-making networks (Fig. 5.22). The design of Kepecs, Uchida, Zariwala and Mainen (2008) to measure a rat's confidence in an olfactory decision was analogous, in that the rat could choose to abort a trial if decision confidence was low, and again this functionality can be implemented by two attractor decision-making networks.

The implication is that some types of 'self-monitoring' can be accounted for by simple, two-attractor network, computational processes. But what of more complex 'self-monitoring', such as is described as occurring in a commentary that might be based on reflection on previous events, and appears to be closely related to consciousness (Weiskrantz 1997). This approach has been developed into a higher order syntactic theory (HOST) of consciousness (Rolls 1999a, Rolls 2004b, Rolls 2005, Rolls 2007a, Rolls 2007b, Rolls 2008a, Rolls 2014a), in which there is a credit assignment problem if a multi-step reasoned plan fails, and it may be unclear which step failed. Such plans are described as syntactic as there are symbols at each stage that must be linked together with the syntactic relationships between the symbols specified, but kept separate across stages of the plan. It is suggested that in this situation being able to have higher order syntactic thoughts will enable one to think and reason about the first order plan, and detect which steps are likely to be at fault. Now this type of 'self-monitoring' is much more complex, as it requires syntax. The thrust of the argument is that some types of 'self-monitoring' are computationally simple, for example in decisions made based on confidence in a first decision, and may have little to do with consciousness; whereas higher order thought processes are very different in terms of the type of syntactic computation required, and may be more closely related to consciousness (Rolls 1999a, Rolls 2004b, Rolls 2005, Rolls 2007a, Rolls 2007b, Rolls 2008a, Rolls 2014a) (see Chapter 22).

5.6.11 Decision-making with multiple alternatives

This framework can also be extended very naturally to account for the probabilistic decision taken when there are multiple, that is more than two, choices. One such extension models choices between continuous variables in a continuous or line attractor network (Furman and Wang 2008, Liu and Wang 2008) to account for the responses of lateral intraparietal cortex neurons in a 4-choice random dot motion decision task (Churchland, Kiani and Shadlen 2008). In another approach, a network with multiple discrete attractors (Albantakis and Deco 2009) can account well for the same data.

A clear example of a discrete multiple choice decision is a decision about which of a set of objects, perhaps with different similarity to each other, has been shown on each trial, and where the decisions are only probabilistically correct. When a decision is made between different numbers of alternatives, a classical result is Hick's law, that the reaction time increases linearly with $\log_2$ of the number of alternatives from which a choice is being made. This has been interpreted as supporting a series of binary decisions each one taking a unit amount of time (Welford 1980). As the integrate-and-fire model we describe works completely in parallel, it will be very interesting to investigate whether Hick's law is a property of the network. If so, this could be related in part to the fact that the activity of the inhibitory interneurons is likely to increase linearly with the number of alternatives between which a decision is being made (as each one adds additional bias to the system through a λ input), and that the GABA inhibitory interneurons implement a shunting, that is divisive, operation (Rolls 2008d).

5.6.12 The matching law

Another potential application of this model of decision-making is to probabilistic decision tasks. In such tasks, the proportion of choices reflects, and indeed may be proportional to, the expected value of the different choices. This pattern of choices is known as the matching law (Sugrue, Corrado and Newsome 2005). An example of a probabilistic decision task in which the choices of the human participants in the probabilistic decision task clearly reflected the expected value of the choices is described by Rolls, McCabe and Redoute (2008e).

A network of the type described here in which the biasing inputs λ_1 and λ_2 to the model are the expected values of the different choices alters the proportion of the decisions it makes as a function of the relative expected values in a way similar to that shown in Fig. 5.14, and provides a model of this type of probabilistic reward-based decision-making (Marti, Deco, Del Giudice and Mattia 2006). It was shown for example that the proportion of trials on which one stimulus was shown over the other was approximately proportional to the difference of the two values between which choices were being made. In setting the connection weights to the two attractors that represent the choices, the returns (the average reward per choice), rather than the incomes (the average reward per trial) of the two targets, are relevant (Soltani and Wang 2006).

This type of model also accounts for the observation that matching is not perfect, and the relative probability of choosing the more rewarding option is often slightly smaller than the relative reward rate ('undermatching'). If there were no neural variability, decision behaviour would tend to get stuck with the more rewarding alternative; stochastic spiking activity renders the network more exploratory and produces undermatching as a consequence (Soltani and Wang 2006).

5.6.13 Comparison with other models of decision-making

In the attractor network model of decision-making described in this book and elsewhere (Wang 2002, Deco and Rolls 2006, Wang 2008, Rolls 2008d, Deco, Rolls and Romo 2009, Rolls, Grabenhorst and Deco 2010b, Rolls, Grabenhorst and Deco 2010c, Insabato, Pannunzi, Rolls and Deco 2010, Deco, Rolls and Romo 2010, Rolls and Deco 2011b, Martinez-Garcia, Rolls, Deco and Romo 2011, Deco, Rolls, Albantakis and Romo 2013), the decisions are taken probabilistically because of the finite size noise due to spiking activity in the integrate-and-fire dynamical network, with the probability that a particular decision is made depending on the biasing inputs provided by the sensory stimuli f1 and f2.

The model described here is different in a number of ways from accumulator or counter models which may include a noise term and which undergo a random walk in real time, which is a diffusion process (Ratcliff, Zandt and McKoon 1999, Carpenter and Williams 1995, Shadlen and Kiani 2013) (see further Deco, Rolls, Albantakis and Romo (2013), Wang (2002), Wang (2008), and Usher and McClelland (2001)). First, in accumulator models, a mechanism for computing the difference between the stimuli is not described, whereas in the current model this is achieved, and scaled by f, by the feedback inhibition included in the attractor network.

Second, in the current attractor network model the decision corresponds to high firing rates in one of the attractors, and there is no arbitrary threshold that must be reached.

Third, the noise in the current model is not arbitrary, but is accounted for by finite size noise effects of the spiking dynamics of the individual neurons with their Poisson-like spike trains in a system of limited size.

Fourth, because the attractor network has recurrent connections, the way in which it settles into a final attractor state (and thus the decision process) can naturally take place over quite a long time, as information gradually and stochastically builds up due to the positive feedback in the recurrent network, the weights in the network, and the biasing inputs, as shown in Fig. 5.10.

Fifth, the recurrent attractor network model produces longer response times in error trials than in correct trials (Wong and Wang 2006, Rolls et al. 2010c), consistent with experimental findings (Roitman and Shadlen 2002). Longer reaction times in error trials can be realized in the diffusion model only with the additional assumption that the starting point varies stochastically from trial to trial (Ratcliff and Rouder 1998).

Sixth, the diffusion model never reaches a steady state and predicts that performance can potentially improve indefinitely with a longer duration of stimulus processing, e.g. by raising the decision bound. In the recurrent attractor network model, ramping activity eventually stops as an attractor state is reached (Fig. 5.10). Consequently performance plateaus at sufficiently long stimulus-processing times (Wang 2002, Wang 2008).

Seventh, the attractor network model has been shown to be able to subtract negative signals as well as add positive evidence about choice alternatives, but the influence of newly arriving inputs diminishes over time, as the network converges towards one of the attractor states representing the alternative choices (Wang 2002, Wang 2008). This is consistent with experimental evidence (Wong, Huk, Shadlen and Wang 2007). This violation of time-shift invariance cannot be accounted for by the inclusion of a leak in the linear accumulator model. In fact, in contrast to the recurrent attractor network model, the linear leaky competing accumulator model, which takes into account a leakage of integration and assumes competitive inhibition between accumulators selective for choice alternatives (Usher and McClelland 2001), actually predicts that later, not earlier, signals influence more the ultimate decision, because an earlier pulse is gradually 'forgotten' due to the leak and does not affect significantly the decision that occurs much later (Wong et al. 2007).

Eighth, the attractor network model of probabilistic decision-making has the property that because it is an attractor network, it can hold on-line, by persistent high firing rates in the winning population of neurons, the decision just taken, because the network is effectively a short-term memory network. Maintaining the decision just taken active for a short period is highly advantageous, for this may enable actions to be taken with the decision just made as the goal of the action, which may take some time to perform; and may enable decisions to be monitored, and corrected if there is a low level of confidence in them (Insabato, Pannunzi, Rolls and Deco 2010), for example as reflected in a relatively low firing rate in the winning attractor (Rolls, Grabenhorst and Deco 2010b, Rolls, Grabenhorst and Deco 2010c, Rolls and Deco 2010).

Ninth, decision confidence is reflected in the firing rates of the winning population of neurons, and is thus an emergent property of this approach to decision-making (Rolls, Grabenhorst and Deco 2010b, Rolls, Grabenhorst and Deco 2010c, Rolls and Deco 2010).

The approach described here offers therefore a biologically plausible alternative this type of linear diffusion model, in the sense that the attractor network model is nonlinear (due to the positive feedback in the attractor network), and is derived from and consistent with the underlying neurophysiological experimental data. The model thus differs from traditional linear diffusion models of decision-making used to account for example for human reaction time data (Luce 1986, Ratcliff and Rouder 1998, Ratcliff et al. 1999, Usher and McClelland 2001). The non-linear diffusion process that is a property of the attractor network model

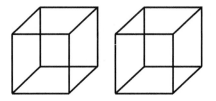

Fig. 5.23 Two Necker cubes. (It may be helpful to increase the viewing distance.)

is analyzed further in Rolls and Deco (2010). A further comparison with other models of decision-making is provided by Deco, Rolls, Albantakis and Romo (2013).

5.7 Perceptual decision-making and rivalry

Another application is to changes in perception. Perceptions can change 'spontaneously' from one to another interpretation of the world, even when the visual input is constant, and a good example is the Necker cube, in which visual perception flips occasionally to make a different edge of the cube appear nearer to the observer (Fig. 5.23). We hypothesize that the switching between these multistable states is due in part to the statistical fluctuations in the network due to the Poisson-like spike firing that is a form of noise in the system. It will be possible to test this hypothesis in integrate-and-fire simulations. (This may or may not be supplemented by adaptation effects (of the synapses or neurons) in integrate-and-fire networks.) (You may observe interesting effects in Fig. 5.23, in which when one cube flips which face appears closer, the other cube performs a similar flip, so that the two cubes remain for most of the time in the same configuration. This effect can be accounted for by short-range cortico-cortical excitatory connections between corresponding depth feature cue combination neurons that normally help to produce a consistent interpretation of 3D objects. The underlying mechanism is that of attractor dynamics linking in this case corresponding features in different objects. When the noise in one of the attractors make the attractor flip, this in turn applies a bias to the other attractor, making it very likely that the attractor for the second cube will flip soon under the influence of its internal spiking-related noise. The whole configuration provides an interesting perceptual demonstration of the important role of noise in influencing attractor dynamics, and of the cross-linking between related attractors which helps the whole system to move towards energy minima, under the influence of noise. Another interesting example is shown in Fig. 5.24.)

The same approach should provide a model of pattern and binocular rivalry, where one image is seen at a time even though two images are presented simultaneously, and indeed an

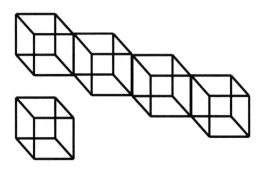

Fig. 5.24 Linked Necker cubes, and a companion.

attractor-based noise-driven model of perceptual alternations has been described (Moreno-Bote, Rinzel and Rubin 2007). When these are images of objects or faces, the system that is especially important in the selection is the inferior temporal visual cortex (Blake and Logothetis 2002, Maier, Logothetis and Leopold 2005), for it is here that representations of whole objects are present (Rolls 2008d, Rolls and Stringer 2006, Rolls 2009c), and the global interpretation of one object can compete with the global interpretation of another object. These simulation models are highly feasible, in that the effects in integrate-and-fire simulations to influence switching between stable states not only of noise, but also of synaptic adaptation and neuronal adaptation which may contribute, have already been investigated (Deco and Rolls 2005d, Deco and Rolls 2005c, Moreno-Bote et al. 2007).

5.8 Symmetry-breaking

It is of interest that the noise that contributes to the stochastic dynamics of the brain through the spiking fluctuations may be behaviourally adaptive, and that the noise should not be considered only as a problem in terms of how the brain works. This is the issue raised for example by the donkey in the medieval Duns Scotus paradox, in which a donkey situated between two equidistant food rewards might never make a decision and might starve.

The problem raised is that with a deterministic system, there is nothing to break the symmetry, and the system can become deadlocked. In this situation, the addition of noise can produce probabilistic choice, which is advantageous. We have shown here that stochastic neurodynamics caused for example by the relatively random spiking times of neurons in a finite sized cortical attractor network can lead to probabilistic decision-making, so that in this case the stochastic noise is a positive advantage.

5.9 The evolutionary utility of probabilistic choice

Probabilistic decision-making can be evolutionarily advantageous in another sense, in which sometimes taking a decision that is not optimal based on previous history may provide information that is useful, and which may contribute to learning. Consider for example a probabilistic decision task in which choice 1 provides rewards on 80% of the occasions, and choice 2 on 20% of the occasions. A deterministic system with knowledge of the previous reinforcement history would always make choice 1. But this is not how animals including humans behave. Instead (especially when the overall probabilities are low and the situation involves random probabilistic baiting, and there is a penalty for changing the choice), the proportion of choices made approximately matches the outcomes that are available, in what is called the matching law (Sugrue, Corrado and Newsome 2005, Corrado, Sugrue, Seung and Newsome 2005, Rolls, McCabe and Redoute 2008e) (Section 5.6.12). By making the less favoured choice sometimes, the organism can keep obtaining evidence on whether the environment is changing (for example on whether the probability of a reward for choice 2 has increased), and by doing this approximately according to the matching law minimizes the cost of the disadvantageous choices in obtaining information about the environment.

This probabilistic exploration of the environment is very important in trial-and-error learning, and indeed has been incorporated into a simple reinforcement algorithm in which noise is added to the system, and if this improves outcomes above the expected value, then changes are made to the synaptic weights in the correct direction (in the associative reward-penalty algorithm) (Sutton and Barto 1981, Barto 1985, Rolls 2008d).

In perceptual learning, probabilistic exploratory behaviour may be part of the mechanism by which perceptual representations can be shaped to have appropriate selectivities for the behavioural categorization being performed (Sigala and Logothetis 2002, Szabo, Deco, Fusi, Del Giudice, Mattia and Stetter 2006).

Another example is in food foraging, which probabilistically may reflect the outcomes (Krebs and Davies 1991, Kacelnik and Brito e Abreu 1998), and is a way optimally in terms of costs and benefits to keep sampling and exploring the space of possible choices.

Another sense in which probabilistic decision-making may be evolutionarily advantageous is with respect to detecting signals that are close to threshold (Rolls and Deco 2010, Deco, Rolls, Albantakis and Romo 2013).

Intrinsic indeterminacy may be essential for unpredictable behaviour (Glimcher 2005). For example, in interactive games like matching pennies or rock–paper–scissors, any trend that deviates from random choice by an agent could be exploited to his or her opponent's advantage.

5.10 Selection between conscious vs unconscious decision-making, and free will

Another application of this type of model is to taking decisions between the implicit (unconscious) and explicit (conscious) systems in emotional decision-making (see Rolls (2014a) and Rolls (2008d)), where again the two different systems could provide the biasing inputs λ_1 and λ_2 to the model. An implication is that noise will influence with probabilistic outcomes which system, the implicit or the conscious reasoning system, takes a decision (see further Chapter 22).

When decisions are taken, sometimes confabulation may occur, in that a verbal account of why the action was performed may be given, and this may not be related at all to the environmental event that actually triggered the action (Gazzaniga and LeDoux 1978, Gazzaniga 1988, Gazzaniga 1995, Rolls 2014a, LeDoux 2008). It is accordingly possible that sometimes in normal humans when actions are initiated as a result of processing in a specialized brain region such as those involved in some types of rewarded behaviour, the language system may subsequently elaborate a coherent account of why that action was performed (i.e. confabulate) (Chapter 22). This would be consistent with a general view of brain evolution in which, as areas of the cortex evolve, they are laid on top of existing circuitry connecting inputs to outputs, and in which each level in this hierarchy of separate input–output pathways may control behaviour according to the specialized function it can perform.

This raises the issue of free will in decision-making (see further Chapter 22).

First, we can note that in so far as the brain operates with some degree of randomness due to the statistical fluctuations produced by the random spiking times of neurons, brain function is to some extent non-deterministic, as defined in terms of these statistical fluctuations. That is, the behaviour of the system, and of the individual, can vary from trial to trial based on these statistical fluctuations, in ways that are described in this book. Indeed, given that each neuron has this randomness, and that there are sufficiently small numbers of synapses on the neurons in each network (between a few thousand and 20,000) that these statistical fluctuations are not smoothed out, and that there are a number of different networks involved in typical thoughts and actions each one of which may behave probabilistically, and with 10^{11} neurons in the brain each with this number of synapses, the system has so many degrees of freedom that it operates effectively as a non-deterministic system. (Philosophers may wish to argue about different senses of the term deterministic, but it is being used here in a precise, scientific, and quantitative way, which has been clearly defined.)

Second, do we have free will when both the implicit and the explicit systems have made the choice? Free will would in Rolls' view (Rolls 2014a, Rolls 2008a, Rolls 2008d, Rolls 2010c, Rolls 2011b) involve the use of language to check many moves ahead on a number of possible series of actions and their outcomes, and then with this information to make a choice from the likely outcomes of different possible series of actions. (If, in contrast, choices were made only on the basis of the reinforcement value of immediately available stimuli, without the arbitrary syntactic symbol manipulation made possible by language, then the choice strategy would be much more limited, and we might not want to use the term free will, as all the consequences of those actions would not have been computed.) It is suggested that when this type of reflective, conscious, information processing is occurring and leading to action, the system performing this processing and producing the action would have to believe that it could cause the action, for otherwise inconsistencies would arise, and the system might no longer try to initiate action. This belief held by the system may partly underlie the feeling of free will. At other times, when other brain modules are initiating actions (in the implicit systems), the conscious processor (the explicit system) may confabulate and believe that it caused the action, or at least give an account (possibly wrong) of why the action was initiated. The fact that the conscious processor may have the belief even in these circumstances that it initiated the action may arise as a property of it being inconsistent for a system that can take overall control using conscious verbal processing to believe that it was overridden by another system. This may be the underlying computational reason why confabulation occurs.

The interesting view we are led to is thus that when probabilistic choices influenced by stochastic dynamics are made between the implicit and explicit systems, we may not be aware of which system made the choice. Further, when the stochastic noise has made us choose with the implicit system, we may confabulate and say that we made the choice of our own free will, and provide a guess at why the decision was taken. In this scenario, the stochastic dynamics of the brain plays a role even in how we understand free will (Rolls 2010c) (see further Chapter 22).

5.11 Creative thought

Another way in which probabilistic decision-making may be evolutionarily advantageous is in creative thought, which is influenced in part by associations between one memory, representation, or thought, and another. If the system were deterministic, i.e. for the present purposes without noise, then the trajectory through a set of thoughts would be deterministic and would tend to follow the same furrow each time. However, if the recall of one memory or thought from another were influenced by the statistical noise due to the random spiking of neurons, then the trajectory through the state space would be different on different occasions, and we might be led in different directions on different occasions, facilitating creative thought (Rolls 2008d, Rolls and Deco 2010).

Of course, if the basins of attraction of each thought were too shallow, then the statistical noise might lead one to have very unstable thoughts that were too loosely and even bizarrely associated to each other, and to have a short-term memory and attentional system that is unstable and distractible, and indeed this is an account that we have proposed for some of the symptoms of schizophrenia (Rolls 2014a, Rolls 2008d, Loh, Rolls and Deco 2007a, Loh, Rolls and Deco 2007b, Rolls, Loh, Deco and Winterer 2008d) (see Section 16.1).

The stochastic noise caused by the probabilistic neuronal spiking plays an important role in these hypotheses, because it is the noise that destabilizes the attractors when the depth of the basins of attraction is reduced. If the basins of attraction were too deep, then the noise might be insufficient to destabilize attractors, and this leads to an approach to understanding obsessive-compulsive disorders (Rolls, Loh and Deco 2008c) (see Section 16.2).

5.12 Unpredictable behaviour

An area where the spiking-related noise in the decision-making process may be evolutionarily advantageous is in the generation of unpredictable behaviour, which can be advantageous in a number of situations, for example when a prey is trying to escape from a predator, and perhaps in some social and economic situations in which organisms may not wish to reveal their intentions (Maynard Smith 1982, Maynard Smith 1984, Dawkins 1995). We note that such probabilistic decisions may have long-term consequences. For example, a probabilistic decision in a 'war of attrition' such as staring down a competitor e.g. in dominance hierarchy formation, may fix the relative status of the two individual animals involved, who then tend to maintain that relationship stably for a considerable period of weeks or more (Maynard Smith 1982, Maynard Smith 1984, Dawkins 1995).

Thus intrinsic indeterminacy may be essential for unpredictable behaviour. We have noted the example provided by Glimcher (2005): in interactive games like matching pennies or rock–paper–scissors, any trend that deviates from random choice by an agent could be exploited to his or her opponent's advantage.

5.13 Predicting a decision before the evidence is applied

There is a literature on how early one can predict from neural activity what decision will be taken (Hampton and O'Doherty 2007, Haynes and Rees 2005a, Haynes and Rees 2005b, Haynes and Rees 2006, Haynes, Sakai, Rees, Gilbert, Frith and Passingham 2007, Lau, Rogers and Passingham 2006, Pessoa and Padmala 2005, Rolls, Grabenhorst and Franco 2009). For example, when subjects held in mind in a delay period which of two tasks, addition or subtraction, they intended to perform, then it was possible to decode or predict with fMRI whether addition or subtraction would later be performed from medial prefrontal cortex activations with accuracies in the order of 70%, where chance was 50% (Haynes et al. 2007). A problem with such studies is that it is often not possible to know exactly when the decision was taken at the mental level, or when preparation for the decision actually started, so it is difficult to know whether neural activity that precedes the decision itself in any way predicts the decision that will be taken (Rolls 2011b). In these circumstances, is there anything rigorous that our understanding of the neural mechanisms involved in the decision-making can provide? It turns out that there is.

While investigating the speed of decision-making using a network, A.Smerieri and E.T.Rolls studied the activity that preceded the onset of the decision cues. The results found in simulations in which the firing rate in the spontaneous firing period is measured before a particular attractor population won or lost the competition are illustrated in Fig. 5.25. The firing rate averaged over approximately 800 winning (correct) vs losing (error) trials for the attractor shows that the firing rate when the attractor will win is on average higher than that for when the attractor will lose at a time that starts in this case approximately 1000 ms before the decision cues are applied. Thus it is possible before the decision cues are applied to predict in this noisy decision-making attractor network something about what decision will be taken for periods in the order of 1 s before the decision cues are applied. (If longer time constants are used for some of the GABA inhibitory neurons in the network, the decision that will be taken

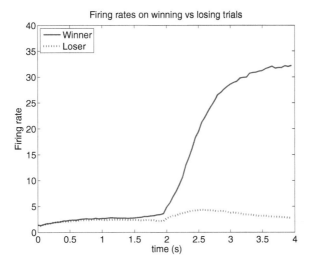

Fig. 5.25 Prediction of a decision before the evidence is applied. In this integrate-and-fire simulation of decision-making, the decision cues were turned on at t=2 s, with ΔI=0. The firing rate averaged over approximately 400 winning vs 400 losing trials for the attractor shows that the firing rate when the attractor will win is on average higher than that for when the attractor will lose at a time that starts in this case at approximately 1000 ms before the decision cues are applied. (At t=2 s with ΔI=0 the input firing rate on each of the 800 external input synapses onto every neuron of both of the selective attractor populations is increased from 3.00 to 3.04 Hz.) (Data from Decision time, slow inhibition, and theta rhythm, A. Smerieri, E.T. Rolls, and J. Feng, *Journal of Neuroscience*, October 2010, 30 (42), pp. 14173–14181, DOI: 10.1523/JNEUROSCI.0945-10.2010.)

can be predicted (probabilistically) for as much as 2 s before the decision cues are applied (Smerieri, Rolls and Feng 2010).)

What could be the mechanism? I suggest that the mechanism is as follows. There will be noise in the neuronal firing that will lead to low, but different, firing rates at different times in the period before the decision cues are applied of the two selective populations of neurons that represent the different decisions. If the firing rate of say population D1 (representing decision 1) is higher than that of the D2 population at a time just as the decision-cues are being applied, this firing will add to the effect of the decision cues, and make it more likely that the D1 population will win. These fluctuations in the spontaneous firing rate will have a characteristic time course that will be influenced by the time constants of the synapses etc in the system, so that if a population has somewhat higher firing at say 500 ms before the cues are applied, it will be a little more likely to also have higher firing some time later. By looking backwards in time one can see how long the effects of such statistical fluctuations can influence the decision that will be reached, and this is shown in Fig. 5.25 to be approximately 1000 ms in the network we have studied. I emphasize that this gradual increase of firing rate for the attractor that will win before the decision cues are applied is an effect found by averaging over very many trials, and that the fluctuations found on an individual trial (illustrated in Fig. 5.10) do not reveal obvious changes of the type illustrated in Fig. 5.25.

Thus we have a rigorous and definite answer and understanding of one way in which decisions that will be taken later are influenced by and can be probabilistically predicted from the prior state of the network. It is possible to make a probabilistic prediction of which decision will be taken from the prior activity of the system, before the decision cues are applied. This conclusion emerges from a fundamental understanding of how noise in the brain produces statistical fluctuations that can influence neural processes, a fundamental principle of cortical operation.

5.14   Highlights

1. Noise is present in the firing times of cortical neurons, which are typically close to Poisson distributed, that is maximally random in time, for a given mean firing rate. The noise being referred to is the randomness in the spiking times of cortical neurons.

2. The noise arises from inherent noise in ion channels in neurons, and from the fact that cortical neurons are kept close to their firing threshold, so that any small changes of input will rapidly be expressed in the next spike emitted (or not), rather than having to charge the neuron membrane to firing threshold from its reset value with the membrane time constant in the order of 20 ms slowing down cortical computation.

3. Attractor networks can be influenced by the statistical fluctuations in the spiking, in that competing attractor states within a single cortical attractor network may have more spikes across all the relevant neurons in one of the attractors on one trial, and in another of the attractors on another trial. This makes the decision-making probabilistic. If the inputs to the two attractors are equal in average strength, then one of the attractors will win on 50% of the trials, and the other on 50% of the trials. As the difference between the two inputs, $\Delta\lambda$, becomes greater, the proportion of trials on which one attractor wins increases, until eventually the noise has little effect on the decision that is taken.

4. These stochastic processes are very important for decision-making, memory recall, short-term memory, and the stability of cortical networks which has applications to psychiatric disorders (Chapter 16).

5. For these reasons, when humans make a decision that is difficult, it may be wise to make the decision several times.

6. This stochastic attractor decision-making mechanism is preferred to drift-diffusion models, because the cortical attractor mechanism reflects the mechanisms by which cortical networks operate, and does not require the artificial setting of parameters in drift-diffusion models such as the amount of noise, the linear accumulation of drift, the threshold at which a decision is said to have been taken, etc. Drift-diffusion models are just mathematical models. Stochastic cortical attractor models are models of the operation of cortical neuronal networks, and incorporate the non-linearities of neurons. Further, the stochastic attractor model makes direct predictions about how impaired function including stability might be corrected in brain disorders, because the underlying neuronal mechanisms are part of the model, and the effects of altering the biological mechanisms can be analyzed.

7. The stochastic processes have many advantages, including sometimes taking decisions that do not reflect the best average outcome which enables a check on whether the probabilities are still the same; creativity, by encouraging a jump to a new part of the energy landscape of memories and ideas; and sometimes taking decisions at random which can be advantageous in predator avoidance.

8. The long recall time for a memory such as a person's name for which we may be searching is related to the stochastic attractor dynamics of the decision-making processes implemented in cortical attractor networks.

6 Attention, short-term memory, and biased competition

Attention can operate by bottom-up, saliency, mechanisms (Section 6.1), and by top-down mechanisms, which operate by biased competition (Section 6.2), and, in a recent extension, by biased activation of whole cortical processing streams (Section 6.3).

6.1 Bottom-up attention

Bottom-up attention operates when salient features in the sensory input attract attention (Itti and Koch 2001). The salient features might include, for vision, high contrast, and visual motion. This visual processing is described as feedforward and bottom-up, in that it operates forward in the visual pathways from the visual input (see Fig. 6.6 in which the solid lines show the forward pathways).

It turns out that the dorsal visual stream (see Figs. 6.1 and 1.10–1.12) (Ungerleider and Mishkin 1982, Ungerleider and Haxby 1994) implements bottom-up saliency mechanisms by guiding saccades to salient stimuli, using properties of the stimulus such as high contrast, colour, and visual motion (Miller and Buschman 2013). This is described as **overt attention**, for it is evident from the eye position where attention is directed. One particular region, the lateral intraparietal cortex (LIP), which is an area in the dorsal visual system, seems to contain saliency maps sensitive to strong sensory inputs (Arcizet, Mirpour and Bisley 2011). Highly salient, briefly flashed, stimuli capture both behavior and the response of LIP neurons (Goldberg, Bisley, Powell and Gottlieb 2006, Bisley and Goldberg 2003, Bisley and Goldberg 2006, Bisley and Goldberg 2010). Inputs reach LIP via dorsal visual stream areas including area MT, and via V4 in the ventral stream (Miller and Buschman 2013, Soltani and Koch 2010) (see Figs. 6.1 and 1.10–1.12).

This process for guiding the eyes to salient parts of a visual scene which are then fixated for short periods greatly simplifies the task for the object recognition system, for instead of dealing with the whole scene as in traditional computer vision approaches, the brain processes just a small fixated region of a complex natural scene at any one time, and then the eyes are moved to another part of the scene. By modelling both the operation of the dorsal visual system in moving the eyes sequentially to different salient parts of a scene, and then the operation of the ventral visual cortical stream in invariant object recognition, we can understand how multiple objects are sequentially located (by the dorsal visual system), and then recognised (by the ventral visual system) to account for and allow investigation of the recognition of multiple objects in complex natural visual scenes (Rolls and Webb 2014) (Section 25.5.16).

An interesting process that facilitates object recognition in complex natural scenes is that the receptive fields of inferior temporal cortex neurons, typically 78° in diameter in a scene with one object on a blank background, shrink to approximately 22° in a complex natural scene (Rolls, Aggelopoulos and Zheng 2003a) (see Section 25.5.9 on page 643 and Fig. 25.60). This greatly facilitates processing in the ventral visual cortical stream object recognition system, for it means that it is much more likely that there is only one object or a

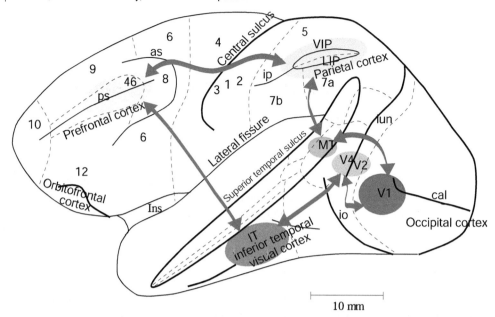

Fig. 6.1 Lateral view of the macaque brain showing some connections of the ventral visual stream projecting to the temporal lobe visual cortical areas and the dorsal visual stream projecting to parietal cortex areas (V1, primary visual area; V2, V4, extrastriate visual areas; IT, inferior temporal visual cortex; LIP, lateral intraparietal cortex visual area; PP, posterior parietal cortex; VIP, ventral intraparietal cortex visual area) with the anterior brain areas (PFCv, ventrolateral prefrontal cortex; PFCd, dorsolateral prefrontal cortex). as, arcuate sulcus; cal, calcarine sulcus; Ins, insula; io, inferior occipital sulcus; ip, intraparietal sulcus; lun, lunate sulcus; ps, principal sulcus. (Data from Attention, short-term memory, and action selection: A unifying theory, Gustavo Deco and Edmund T. Rolls, *Progress in Neurobiology*, 2005, 76 (4), pp. 236–56, DOI: 10.1016/j.pneurobio.2005.08.004.)

few objects to be dealt with at the fovea that need to be recognised (Rolls, Aggelopoulos and Zheng 2003a, Rolls and Deco 2006). The mechanism for the shrinking of the receptive fields of inferior temporal cortex neurons in complex natural scenes is probably lateral inhibition from nearby visual features and objects, which effectively leaves a neuron sensitive to only the peak of the receptive field, which typically includes the fovea because of its greater cortical magnification factor for inferior temporal cortex neurons (Trappenberg, Rolls and Stringer 2002) (see Section 25.5.9.2 and Figs. 25.61 and 25.62).

If several objects are close together near the fovea, then an inferior temporal cortex neuron that responds to one of the objects will respond to that object when it is at the fovea, and when it is in some but not all positions near (10° from) the fovea (see Section 25.5.10 and Fig. 25.64). This asymmetry of the receptive fields of inferior temporal cortex neurons is hypothesized to be produced by the *diluted connectivity* of the feedforward projections from one cortical area to another, which because of the randomness of the dilution means that an inferior temporal cortex neuron may have more connections from one part of the parafoveal region than from another part. This is another useful function produced by the dilution of cortico-cortical connectivity, and another *principle of operation of the cortex* (Chapter 7). The useful function implemented in this case is that this facilitates the representation of several objects close to the fovea and their positions during a single fixation. It also allows multiple instances of the same object close to the fovea to be represented (Section 25.5.10). This I hypothesize is the solution that the cortex uses to represent multiple objects in a single fixation

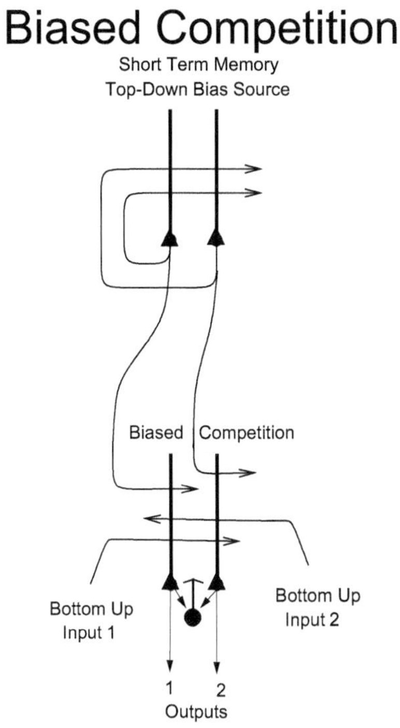

Biased Competition

Short Term Memory
Top-Down Bias Source

Fig. 6.2 Biased competition mechanism for top-down attention. There is usually a single attractor network that can enter different attractor states to provide the source of the top-down bias (as shown). If it is a single network, there can be competition within the short-term memory attractor states, implemented through the local GABA inhibitory neurons. The top-down continuing firing of one of the attractor states then biases in a top-down process some of the neurons in a cortical area to respond more to one than the other of the bottom-up inputs, with competition implemented through the GABA inhibitory neurons (symbolized by a filled circle) which make feedback inhibitory connections onto the pyramidal cells (symbolized by a triangle) in the cortical area. The thick vertical lines above the pyramidal cells are the dendrites. The axons are shown with thin lines and the excitatory connections by arrow heads.

of a scene (which was a real problem for connectionist models that assumed full connectivity (Mozer 1991)).

All the above processes are bottom-up processes and operate during bottom-up attention.

6.2 Top-down attention – biased competition

6.2.1 The biased competition hypothesis

A second type of selective attentional process, with which we are mainly concerned in this chapter, involves actively maintaining in short-term memory a location or object as the target of attention, and using this by top-down processes to influence earlier cortical processing. Some of the top-down or backprojection pathways in the visual system are shown in Fig. 6.6 on page 151 by the dashed lines.

The **biased competition hypothesis** of attention proposes that multiple stimuli in the visual field activate populations of neurons that engage in competitive interactions. Attending to a stimulus at a particular location or with a particular feature biases this competition in favour of neurons that respond to the feature or location of the attended stimulus. This attentional effect is produced by generating signals in cortical areas outside the visual cortex that are then fed back (or down) to extrastriate areas, where they bias the competition such that when multiple stimuli appear in the visual field, the neurons representing the attended stimulus 'win', thereby suppressing neurons representing distracting stimuli (Duncan and Humphreys 1989, Desimone and Duncan 1995, Duncan 1996).

The top-down biased competition mechanism operates in, for example, visual selective attention (Desimone and Duncan 1995, Duncan 1980, Duncan and Humphreys 1989), effectively replaces earlier theories such as the spotlight metaphor and feature integration theory (Treisman 1982, Crick 1984, Treisman and Gelade 1980, Treisman 1988), and provides a neuronal mechanism for attention (Rolls and Deco 2002, Rolls 2008d). Other authors have postulated as a binding and attentional mechanism the synchronous firing of neurons in order to change connection strengths (Crick 1984, Von der Malsburg and Bienenstock 1986, Crick and Koch 1990, Gray and Singer 1989, Singer 1999). The synchronization specifies just a possible physical implementation of binding and is not directly associated with the computational description of attention. The temporal synchrony hypothesis of feature binding is considered in Sections 25.5.5.1 and C.3.7.

A modern neuronal network diagram of how the biased competition theory of attention operates is provided in Fig. 6.2 (Rolls 2008f, Rolls 2008d, Rolls 2013a, Rolls 2014a). There is usually a single attractor network that can enter different attractor states to provide the source of the top-down bias (as shown). If it is a single network, there can be competition within the short-term memory attractor states, implemented through the local GABA inhibitory neurons. The top-down continuing firing of one of the attractor states then biases in a top-down process some of the neurons in a 'lower' cortical area to respond more to one than the other of the bottom-up inputs, with competition implemented through the GABA inhibitory neurons (symbolized by a filled circle) that make feedback inhibitory connections onto the pyramidal cells (symbolized by a triangle) in the 'lower' cortical area.

This biased competition is conceptualised as operating within a 'lower' cortical area, e.g. a cortical region. Some neurons receive a weak top-down input that increases their response to the bottom-up stimuli (Desimone and Duncan 1995), potentially supralinearly if the bottom-up stimuli are weak (Deco and Rolls 2005b, Rolls and Deco 2002, Rolls 2008d). The enhanced firing of the biased neurons then, via the local inhibitory neurons, inhibits the other neurons in the local area from responding to the bottom-up stimuli. This is a local mechanism, in that the inhibition in the neocortex is primarily local, being implemented by cortical inhibitory neurons that typically have inputs and outputs over no more than a few mm (Douglas et al. 2004) (see Chapter 1).

This process implements **covert attention**, in that the attention occurs to a part of the scene, without any eye movement.

The architecture in Fig. 6.2 illustrates some key principles of operation of the cortex. One is the function of cortical recurrent collateral connections to implement short-term memory attractor states to hold the object of attention active and on-line (Chapter 4). A second principle is the use of competitive networks to select winning populations of neurons that provide a useful output to the next stage of processing (Sections 7.4 and B.4). A third principle is allocation of cortical resources to respond to current processing needs efficiently, for example by using selective attention to select form vs location as currently relevant to make this processing faster and more efficient, and to minimize distraction by currently irrelevant information.

The massively feedforward and feedback connections that exist in the anatomy and physiology of the cortex (described in Chapter 11) implement the attentional mutual biasing interactions that result in a neurodynamical account of visual perception consistent with the seminal constructivist theories of Helmholtz (1867) and Gregory (1970). The architecture

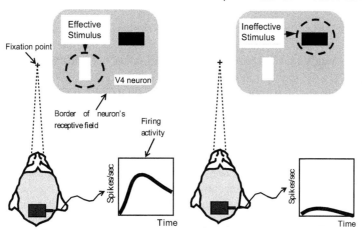

Fig. 6.3 Spatial attention and shrinking receptive fields in single-cell recordings of V4 neurons from the brain of a behaving monkey. The areas that are circled indicate the attended locations. When the monkey attended to effective sensory stimuli, the V4 neuron produced a good response, whereas a poor response was observed when the monkey attended to the ineffective sensory stimulus. The point fixated was the same in both conditions. (Adapted from Science, 229, Selective attention gates visual processing in the extrastriate cortex, Moran, J. and Desimone, R., 782–784. Copyright ©1985, The American Association for the Advancement of Science. Reprinted with permission from AAAS.)

with which this can occur is illustrated in Fig. 6.6, and systems that operate in this way are described later in this Chapter.

6.2.2 The biased competition hypothesis in single neuron studies

6.2.2.1 Neurophysiology of attention

A number of neurophysiological experiments (Moran and Desimone 1985, Spitzer, Desimone and Moran 1988, Sato 1989, Motter 1993, Miller, Gochin and Gross 1993a, Chelazzi, Miller, Duncan and Desimone 1993, Motter 1994, Reynolds and Desimone 1999, Chelazzi 1998) have been performed suggesting biased competition neural mechanisms that are consistent with the theory of Duncan and Humphreys (1989) (i.e. with the role for a top-down memory target template in visual search), as follows (with more detail provided by Rolls and Deco (2002) and Rolls (2008d)).

Moran and Desimone (1985) showed that the firing activity of visually tuned neurons in the cortex was modulated if monkeys were instructed to attend to the location of the target stimulus. In these studies, Moran and Desimone first identified the V4 neuron's classic receptive field and then determined which stimulus was effective for exciting a response from the neuron (e.g. a vertical white bar) and which stimulus was ineffective (e.g. a horizontal black bar). In other words, the effective stimulus made the neuron fire whereas the ineffective stimulus did not (see Fig. 6.3). They presented these two stimuli simultaneously within the neuron's receptive field, which for V4 neurons may extend over several degrees of visual angle. The task required attending covertly to a cued spatial location. When spatial attention was directed to the effective stimulus, the pair elicited a strong response. On the other hand, when spatial attention was directed to the ineffective stimulus, the identical pair elicited a weak response, even though the effective stimulus was still in its original location (see Fig. 6.3).

Based on results of this type, the spatial attentional modulation could be described as a shrinkage of the classical receptive field around the attended location. Similar studies of

Luck, Chelazzi, Hillyard and Desimone (1997) and Reynolds, Chelazzi and Desimone (1999) replicated this result in area V4, and showed similar attentional modulation effects in area V2 as well. Even in area V1 a weak attentional modulation has been described (McAdams and Maunsell 1999).

Maunsell (1995) and many others (Duhamel, Colby and Goldberg 1992, Colby, Duhamel and Goldberg 1993, Gnadt and Andersen 1988) have shown that the modification of neural activity by attentional and behavioural states of the animal is not only true for the extrastriate areas in the ventral stream, but is also true for the extrastriate areas of the dorsal stream as well. In particular, Maunsell (1995) demonstrated that the biased competitive interaction due to spatial and object attention exists not only for objects within the same receptive field, but also for objects in spatially distant receptive fields. This suggests that mechanisms exist to provide biased competition of a more global nature. Furthermore, Connor, Gallant, Preddie and Van Essen (1996) showed that the locus of spatial attention can modulate the structure of the receptive fields of V4 neurons. By asking the monkeys to discriminate subtle changes in features in particular points in visual space, they managed to shift the hot spot of the receptive field of nearby V4 neurons towards the spatial focus of attention.

6.2.2.2 The role of competition

In order to test at the neuronal level the biased competition hypothesis more directly, Reynolds, Chelazzi and Desimone (1999) performed single-cell recordings of V2 and V4 neurons in a behavioural paradigm that explicitly separated sensory processing mechanisms from attentional effects. They first examined the presence of competitive interactions in the absence of attentional effects by having the monkey attend to a location far outside the receptive field of the neuron that they were recording. They compared the firing activity response of the neuron when a single reference stimulus was within the receptive field, with the response when a probe stimulus was added to the field. When the probe was added to the field, the activity of the neuron was shifted toward the activity level that would have been evoked if the probe had appeared alone. This effect is shown in Fig. 6.4. When the reference was an effective stimulus (high response) and the probe was an ineffective stimulus (low response), the firing activity was suppressed after adding the probe (Fig. 6.4a). On the other hand, the response of the cell increased when an effective probe stimulus was added to an ineffective reference stimulus (Fig. 6.4b).

These results are explained in Section 6.2.7.1 by assuming that V2 and V1 neurons coding the different stimuli engage in competitive interactions, mediated for example through intermediary inhibitory neurons, as illustrated in Fig. 6.5.

The neurons in the extrastriate area receiving input from the competing neurons respond according to their input activity after the competition; that is, the response of a V4 neuron to two stimuli in its field is not the sum of its responses to both, but rather is a weighted average of the responses to each stimulus alone.

Attentional modulatory effects have been independently tested by repeating the same experiment, but now having the monkey attend to the reference stimulus within the receptive field of the recorded neuron. The effect of attention on the response of the V2 or V4 neuron was to almost compensate for the suppressive or excitatory effect of the probe (Reynolds, Chelazzi and Desimone 1999). That is, if the probe caused a suppression of the neuronal response to the reference when attention was outside the receptive field, then attending to the reference restored the neuron's activity to the level corresponding to the response of the neuron to the reference stimulus alone (Fig. 6.4a). Symmetrically, if the probe stimulus had increased the neuron's level of activity, attending to the reference stimulus compensated the response by shifting the activity to the level that had been recorded when the reference was presented alone (Fig. 6.4b). As is shown in Fig. 6.5, this attentional modulation can be understood by

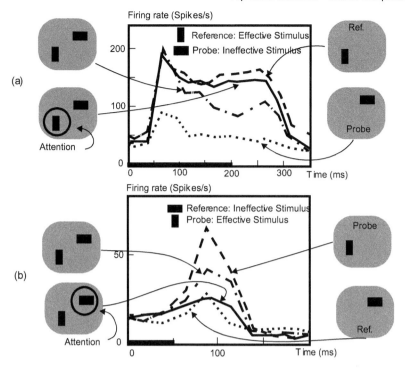

Fig. 6.4 Competitive interactions and attentional modulation of responses of single neurons in area V2. (a) Inhibitory suppression by the probe and attentional compensation. (b) Excitatory reinforcement by the probe and attentional compensation. The black horizontal bar at the bottom indicates stimulus duration. (Adapted with permission of Society for Neuroscience, from Journal of Neuroscience, Competitive mechanisms subserve attention in macaque areas V2 and V4, Reynolds, J. H., Chelazzi, L. and Desimone, R. 19: 1736–1753. Copyright ©1999 Society for Neuroscience.)

assuming that attention biases the competition between V1 neurons in favour of a specific location or feature.

6.2.2.3 Evidence for attentional bias

Direct physiological evidence for the existence of an attentional bias was provided by Luck, Chelazzi, Hillyard and Desimone (1997) and by Spitzer, Desimone and Moran (1988). Luck et al. (1997) discovered that attending to a location within the receptive field of a V2 or V4 neuron increased its spontaneous firing activity. They reported that in the absence of stimuli, the spontaneous firing activity of V2 and V4 neurons increased by 30–40% when the monkey attended to a location within the receptive field of the recorded neuron, compared with when attention was directed outside the field.

When a single stimulus is presented in the receptive field of a V4 neuron, Spitzer et al. (1988) observed an increase of the neuronal response to that stimulus when the monkey directed attention inside the receptive field, compared to when attention was directed outside the field.

6.2.3 Non-spatial attention

Additional evidence for the biased competition hypothesis as a mechanism for the selection of non-spatial attributes has been put forward by Chelazzi, Miller, Duncan and Desimone (1993). Chelazzi et al. (1993) measured the responses of inferior temporal cortex (IT) neurons

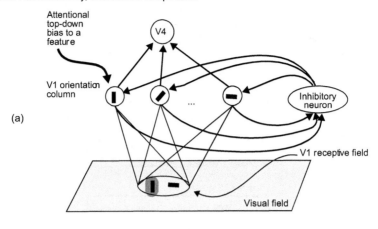

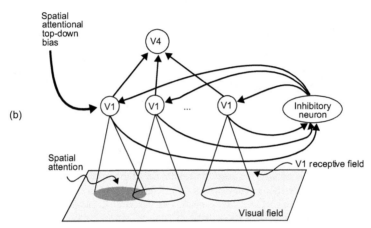

Fig. 6.5 Biased competition hypothesis at the neuronal level. (a) Attention to a particular top-down prespecified feature. Intermediate inhibitory interneurons provide an inhibitory signal to V1 neurons in an orientation column (with receptive fields at the same spatial location) that are sensitive to an orientation different from the to-be-attended orientation. (b) Attention to a particular top-down prespecified spatial location. Inhibitory interneurons provide an inhibitory signal to V1 neurons with receptive fields corresponding to regions that are outside the to-be-attended location. In this schematic representation, we assume for simplicity direct connections from V1 to V4 neurons without intermediate V2 neurons. (From Edmund Rolls and Gustavo Deco, Computational Neuroscience of Vision ©2002, Oxford University Press, reproduced by permission of Oxford University Press https://global.oup.com/academic/product/computational-neuroscience-of-vision-9780198524885?q=computational neuroscience of vision &lang=en&cc=gb. For permission to reuse this material, please visit http://www.oup.co.uk/academic/rights/permissions.)

in monkeys while the animals were looking at a display containing a target and a distracter. It was found that the response was maintained or increased only if the effective stimulus is the target, and the firing rate decreases if the stimulus is not a target. Thus, the activity of IT neurons depends on the internal attentional state of the monkey. Usher and Niebur (1996) formulated a detailed neural model that explains these data assuming parallel dynamic processing driven by a competition mechanism.

Further evidence on how object-based attention can facilitate neuronal activity in the inferior temporal visual cortex is found in an object search task (Rolls, Aggelopoulos and Zheng 2003a). The macaque was looking for one of two objects in a large (70 deg × 55 deg) display. It was found that the receptive field of a neuron was much larger (on average 78 deg against a blank background) for the effective stimulus for the neuron when the stimulus was being searched for than when the stimulus was not being searched for and was just a distracter (when the receptive field size is 22 deg, see Section 25.5.9.1, and Fig. 25.60 on page 644). Thus top-down object-based attentional processes can increase the size of receptive fields of inferior temporal cortex neurons in a search task.

6.2.4　The biased competition hypothesis in functional brain imaging

In functional magnetic resonance imaging (fMRI) studies, additional evidence for similar mechanisms in human extrastriate cortex has been obtained (Kastner, De Weerd, Desimone and Ungerleider 1998, Kastner, Pinsk, De Weerd, Desimone and Ungerleider 1999). In line with the biased competition hypothesis, these studies have shown that multiple stimuli in the visual field interact in a mutually suppressive way when presented simultaneously, but not when presented sequentially, and that spatially directed attention to one stimulus location reduces the mutually suppressive effect. These studies also revealed increased activity in extrastriate cortex in the absence of visual stimulation, when subjects covertly directed attention to a peripheral location at which they expected the onset of visual stimuli. This increased activity in visual cortex revealed a top-down bias of neural signals, deriving from frontal and parietal areas, to favour the attended location. More details of these investigations, and a computational model of them, are described elsewhere (Rolls and Deco 2002, Corchs and Deco 2002, Rolls 2008d).

6.2.5　A basic computational module for biased competition

We have seen in Fig. 6.5 how given forward (bottom-up) inputs to a network, competition between the principal (excitatory) neurons can be implemented by inhibitory neurons which receive from the principal neurons in the network and send back inhibitory connections to the principal neurons in the network. This competition can be biased by a top-down input to favour some of the populations of neurons, and this describes the biased competition hypothesis.

　　The implementation of biased competition in a computational model can be at the mean-field level, in which the average firing rate of each population of neurons is specified. Equations to implement the dynamics of such a model are given in Section B.8.4. The advantages of this level of formulation are relative simplicity, and speed of computation. The architecture of Fig. B.38 is that of a competitive network (described in Section B.4) but with top-down backprojections as illustrated in Fig. B.18. However, the network could equally well include associatively modifiable synapses in recurrent collateral connections between the excitatory (principal) neurons in the network, making the network into an autoassociation or attractor network capable of short-term memory, as described in Section B.3. The competition between the neurons in the attractor network is again implemented by the inhibitory feedback neurons.

　　The implementation can also be at the level of the spiking neuron by using an integrate-and-fire simulation (see Sections B.6 and 6.2.9). The integrate-and-fire level allows single neuron findings to be modelled, allows more details of the neuronal implementation to be incorporated, and also allows statistical effects related to the probability that spikes from different neurons will occur at different times to be investigated. Integrate-and-fire simulations are more difficult to set up, and are slow to run. Equations for such simulations are given in Section B.6, and as applied to a biased competition model of attention by Deco and Rolls (2005b).

The computational investigations can be extended to systems with many interacting modules, as described in Section B.8.5. These could be arranged in a hierarchy, with repeated layers of bottom-up and top-down connections as illustrated in Fig. 6.6.

6.2.6 Neurodynamical architecture of a model of spatial and object-based attention

In order to analyze the principles of operation of attentional systems in the brain, Rolls and Deco (2002) and Deco and Rolls (2004) described a cortical model of visual attention for object recognition and visual search based on the neurophysiological constraints described in Sections 6.2.2 and 6.2.4. The system is absolutely autonomous and each component of its functional behaviour is explicitly described in a complete mathematical framework. The model is described here in order to illustrate many principles of cortical function, with further details available elsewhere (Deco 2001, Deco and Zihl 2001b, Rolls and Deco 2002, Deco and Lee 2002, Corchs and Deco 2002, Heinke, Deco, Zihl and Humphreys 2002, Deco and Lee 2004, Deco and Rolls 2004, Deco and Rolls 2005a, Rolls 2008d).

Top-down processes in attention are considered here, and how they interact with bottom-up processing, in a model of visual attentional processing that has multiple hierarchically organized modules in the architecture shown schematically in Figs. 6.1 and 6.6. The model shows how the dorsal (sometimes termed *where*) visual stream (reaching the posterior parietal cortex, PP) and the ventral (*what*) visual stream (via V4 to the inferior temporal cortex, IT) could interact through early visual cortical areas (such as V1 and V2) to account for many aspects of visual attention (Deco 2001, Deco and Zihl 2001b, Rolls and Deco 2002, Deco and Lee 2002, Corchs and Deco 2002, Heinke et al. 2002, Deco and Lee 2004, Deco and Rolls 2004, Rolls 2008d). The source of the attentional bias in the model comes from short-term memory networks in for example the prefrontal cortex, as shown in Figs. 6.1 and 6.6.

The overall systems-level representation of the model simulated is shown in Fig. 6.7. The system is essentially composed of six modules structured such that they resemble the two known main visual paths of the mammalian visual cortex. Information from the retino-geniculo-striate pathway enters the visual cortex through area V1 in the occipital lobe and proceeds into two processing streams. The occipital-temporal stream leads ventrally through V2–V4 and IT (inferotemporal cortex), and the occipito-parietal stream leads dorsally into PP (posterior parietal complex). In this model system described by Rolls and Deco (2002), we chose to model with our ventral stream one particular aspect of the inferior temporal cortex function: translation invariant object recognition. (A fuller model of the processing that leads to invariant object representations is provided in Chapter 25.) We chose to model with our dorsal stream one particular aspect of parietal cortex function: encoding of visual space in retinotopic coordinates. This is obviously a great simplification of the complex hierarchical architecture and functions of the primate visual system (Felleman and Van Essen 1991) (see Rolls and Deco (2002) Chapters 2–5 and 8), but the model is sufficient to test and demonstrate our fundamental proposals for how the object and spatial streams interact to implement visual attentional processes.

An overview of the entire architecture is given first. The ventral stream consists of four modules as shown in Fig. 6.7: (1) a V1 module; (2) a V2–V4 module; (3) an IT module; and (4) a module v46 corresponding to the ventral part of area 46 of the prefrontal cortex that maintains a short-term memory of a recognized object or holds a representation of the target object in a visual search task. The V1 module contains $P \times P$ hypercolumns, covering an $N \times N$ pixel scene. Each hypercolumn contains L orientation columns of complex cells with K octave levels corresponding to different spatial frequencies. The complex cells are modelled by the power modulus of Gabor wavelets. There is one inhibitory pool interacting

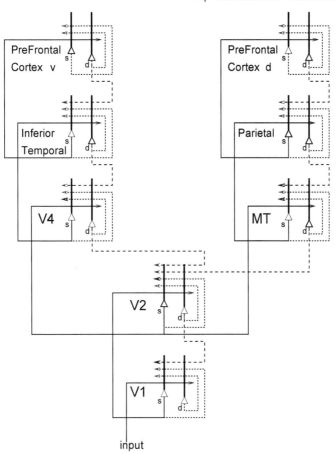

Fig. 6.6 The overall architecture of the model of object and spatial processing and attention, including the prefrontal cortical areas that provide the short-term memory required to hold the object or spatial target of attention active. Forward connections are indicated by solid lines; backprojections, which could implement top-down processing, by dashed lines; and recurrent connections within an area by dotted lines. The triangles represent pyramidal cell bodies, with the thick vertical line above them the dendritic trees. The cortical layers in which the cells are concentrated are indicated by s (superficial, layers 2 and 3) and d (deep, layers 5 and 6). The prefrontal cortical areas most strongly reciprocally connected to the inferior temporal cortex 'what' processing stream are labelled v to indicate that they are in the more ventral part of the lateral prefrontal cortex, area 46, close to the inferior convexity in macaques. The prefrontal cortical areas most strongly reciprocally connected to the parietal visual cortical 'where' processing stream are labelled d to indicate that they are in the more dorsal part of the lateral prefrontal cortex, area 46, in and close to the banks of the principal sulcus in macaques (see text).

with complex cells of all orientations at each scale. This module sends spatial and feature information up to the dorsal stream and the ventral stream respectively. It also provides a high-resolution representation for the two streams to interact through recurrent feedback. This interaction between the streams may be important for several functions such as binding, and high-resolution visual analysis. Rolls and Deco (2002) first focused on the possible role of the interaction in mediating spatial and object attention in visual search and object recognition tasks. The V2–V4 module serves primarily to pool and channel the responses of V1 neurons to IT, to achieve a limited degree of translation invariance. It also implements a certain degree of localized competitive interaction between different targets. A topologically organized lattice

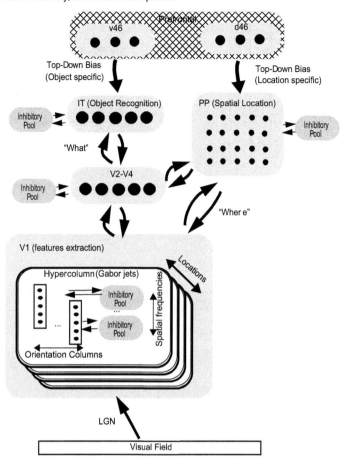

Fig. 6.7 The systems-level architecture of a model of the cortical mechanisms of visual attention. The system is essentially composed of six modules structured such that they resemble the two known main visual pathways of the primate visual cortex. Information from the retino-geniculo-striate pathway enters the visual cortex through area V1 in the occipital lobe and proceeds into two processing streams. The occipital-temporal stream leads ventrally through V2–V4 and IT (inferotemporal), and is mainly concerned with object recognition. The occipito-parietal stream leads dorsally into PP (posterior parietal complex) and is responsible for maintaining a spatial map of an object's location. (From Edmund Rolls and Gustavo Deco, Computational Neuroscience of Vision ©2002, Oxford University Press, reproduced by permission of Oxford University Press https://global.oup.com/academic/product/computational-neuroscience-of-vision-9780198524885?q=computational neuroscience of vision &lang=en&cc=gb. For permission to reuse this material, please visit http://www.oup.co.uk/academic/rights/permissions.)

is used to represent V2 or V4. Each node in this lattice has $L \times K$ cell assemblies as in a hypercolumn in V1. Each cell assembly, however, receives convergent input from the cell assemblies with the same tuning from a $M \times M$ hypercolumn neighbourhood in V1. The feedforward connections from V1 to the V4 module are modelled with convergent Gaussian weight functions, with symmetric recurrent connections. The IT module contains C pools, as the network is trained to search for or recognize C particular objects. Each of these cell assemblies is fully connected to each of the V4 pools. The connection weights from V4 to IT are trained by Hebbian learning rules in a learning phase.

The dorsal stream consists of three modules: (1) the V1 module; (2) a PP module; and (3) a module d46 corresponding to the dorsal part of area 46 of the prefrontal cortex that maintains a representation of a spatial location in a short-term spatial memory. This d46 module not only provides a short-term memory function, but can be used to generate an attentional bias for a spatial location. The PP module is responsible for mediating spatial attention modulation and for updating the spatial position of the attended object. A lattice of $N \times N$ nodes represents the topographical organization of module PP. Each node in the lattice corresponds to the spatial position of each pixel in the input image. Each of these assemblies monitors the activities from hypercolumns in V1–V2 via a Gaussian weighting function that connects topologically corresponding locations.

Prefrontal cortex area 46 (modules d46 and v46) is not explicitly simulated in the model described by Rolls and Deco (2002) and in more detail elsewhere (Deco 2001, Deco and Zihl 2001b, Rolls and Deco 2002, Deco and Lee 2002, Corchs and Deco 2002, Heinke et al. 2002, Deco and Lee 2004, Deco and Rolls 2004), but is part of the model. We assume that top-down feedback connections from these areas provide the external top-down bias that specifies the processing conditions of earlier modules. Concretely, the feedback connection from area v46 with the IT module specifies the target object in a visual search task; and the feedback connection from area d46 with the PP module generates the bias to a targeted spatial location in a recognition task at a fixed prespecified location.

The system operates in two different modes: the learning mode and the recognition mode. During the learning mode the synaptic connections between V4 and IT are trained by means of Hebbian learning during several presentations of a specific object at changing random positions in the visual field. This is the simple way in which translation invariant representations are produced in IT in this model. During the recognition mode there are two possibilities for running the system, illustrated in Fig. 6.8.

First, in **visual spatial search mode** (Fig. 6.8b), an object can be found in a scene by biasing the system with an external top-down (backprojection) component (from e.g. prefrontal area v46) to the IT module. This drives the competition in IT in favour of the pool associated with the specific object to be searched for. Then, the intermodular backprojection attentional modulation IT–V4–V1 will enhance the activity of the pools in V4 and V1 associated with the component features of the specific object to be searched for. This modulation will add to the visual input being received by V1, resulting in greater local activity where the features in the topologically organized visual input features match the backprojected features being facilitated. Finally, the enhanced firing in a particular part of V1 will lead to increased activity in the forward pathway from V1 to V2–V4 to PP, resulting in increased firing in the PP module in the location that corresponds to where the object being searched for is located. In this way, the architecture automatically finds the location of the object being searched for, and the location found is made explicit by which neurons in the spatially organized PP module are firing.

Second, in **visual object identification mode** (Fig. 6.8a), the PP module receives a top-down (backprojection) input (from e.g. prefrontal area d46) which specifies the location in which to identify an object. The spatially biased PP module then drives by its backprojections the competition in the V2–V4 module in favour of the pool associated with the specified location. This biasing effect in V1 and V2–V4 will bias these modules to have a greater response for the specified location in space. The shape feature representations which happen to be present due to the visual input from the retina at that location in the V1 and V2–V4 modules will therefore be enhanced, and the enhanced firing of these shape features will by the feedforward pathway V1–V4–IT favour the IT object pool that contains the facilitated features, leading to recognition in IT of the object at the attentional location being specified in the PP module. The operation of these two attentional modes is shown schematically in Fig. 6.8.

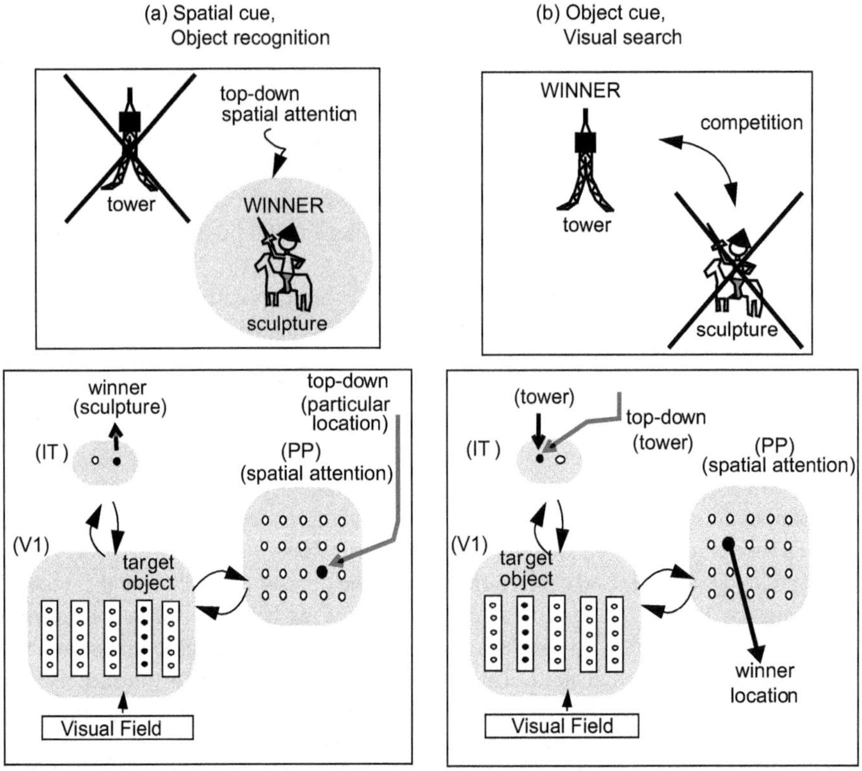

Fig. 6.8 Attentional modulation for finding an object (in visual object identification mode) at a specific spatial location using spatial bias to the PP module from for example a prefrontal module d46. See text for details. (b) Attentional modulation for finding a visual spatial location (in visual spatial search mode) when an object is specified by bias to the IT module from for example a prefrontal module v46. (From Edmund Rolls and Gustavo Deco, Computational Neuroscience of Vision ©2002, Oxford University Press, reproduced by permission of Oxford University Press https://global.oup.com/academic/product/computational-neuroscience-of-vision-9780198524885?q=computational neuroscience of vision &lang=en&cc=gb. For permission to reuse this material, please visit http://www.oup.co.uk/academic/rights/permissions. See also Deco and Lee 2004.)

A formal description of the model is provided in Section B.8.5 and elsewhere (Rolls and Deco 2002, Rolls 2008d).

6.2.7 Simulations of basic experimental findings

According to the biased competition hypothesis multiple stimuli in the visual field activate populations of neurons that engage in competitive interactions. Moreover, attending to a stimulus at a particular location biases this competition in favour of neurons that respond to the location of the attended stimulus. Different experimental results seem to support these ideas. In particular, in Section 6.2.2 the results are described of Reynolds, Chelazzi and Desimone (1999) from single cell recording in V2 neurons in monkeys, which show inhibitory competitive effects and attentive biasing modulation at the neuronal level. In addition, in Section 6.2.4 the fMRI studies in humans were described of Kastner, Pinsk, De Weerd, Desimone and Ungerleider (1999), which are consistent with the idea of biased competition at the larger scale (i.e. mesoscopic) level in which the activation of more gross regions of the brain is measured.

In this section, simulations based on the model introduced in Section 6.2.6 are shown to explain the results observed in both experiments. Consequently, the dynamical behaviour of the multimodular architecture described is Section 6.2.6 provides an account of attentional processes in vision that is consistent with the biased competition hypothesis at the (microscopic) level of neuronal firing and at the level of the activation of different large regions of cortex.

6.2.7.1 Simulations of single-cell experiments

In this Section, simulation results (Deco 2001, Deco and Zihl 2001b, Rolls and Deco 2002, Deco and Lee 2002, Corchs and Deco 2002, Heinke et al. 2002, Deco and Lee 2004) are described with the model outlined in Section 6.2.6 that correspond to the experiments by Reynolds, Chelazzi and Desimone (1999) on single cell recordings from V2 neurons in monkeys. We described the dynamical behaviour of the model of the cortical architecture of visual attention by solving numerically the system of coupled differential equations in a computer simulation. For this experiment we included a module of V2 neurons. The input system processed an input image of 66×66 pixels ($N = 66$). The V1 hypercolumns covered the entire image uniformly. They were distributed in 33×33 locations ($P = 33$) and each hypercolumn was sensitive to two spatial frequencies and to eight different orientations (i.e. $K = 2$ and $L = 8$). The V2 module had 2×8 pools receiving convergent input from the pools of the same tuning from a 10×10 (i.e. $M = 10$) hypercolumn in the neighbourhood in V1. The feedforward connections from V1 to V2 are modelled with convergent Gaussian weight functions, with symmetric recurrent connections. We analyzed the firing activity of a single pool in the V2 module which was highly sensitive to a vertical bar presented in its receptive field (the effective stimulus) and poorly sensitive to a 75 degree oriented bar presented in its receptive field (ineffective stimulus). The size of the bars was 2×4 pixels. Following the experimental setup reviewed in Section 6.2.2, we plot in Figs. 6.9a and 6.9b the development of the firing activity of a V2 pool under four different conditions: (1) with a single reference stimulus within the receptive field; (2) with a single probe stimulus within the receptive field; (3) with a reference and a probe stimulus within the receptive field and without attention; and (4) with a reference and a probe stimulus within the receptive field and with attention directed to the spatial location of the reference stimulus. In our simulation, spatial attention was directed to the location of the reference stimulus by setting in module PP the top-down attentional bias coming from prefrontal area d46 equal to 0.05 if i and j corresponded to the location of the reference stimulus and zero elsewhere. In the unattended condition the external top-down bias was zero everywhere. The computational simulations of Fig. 6.9a and 6.9b should be compared with the experimental results shown in Fig. 6.4a and 6.4b, respectively. The same qualitative behaviour is observed in the model and the real neuronal data in all experimental conditions. The competitive interactions in the absence of attention are due to the intramodular competitive dynamics at the level of V1 (i.e. the suppressive and excitatory effect of the probe shown in Fig. 6.9a and 6.9b, respectively). The modulatory biasing corrections in the attended conditions are caused by the intermodular interactions between the V1 and PP pools, and between the PP pools and prefrontal top-down modulation.

We are also able to account for the findings of the experiments by Connor, Gallant, Preddie and Van Essen (1996), which showed that the locus of spatial attention can modulate the structure of the receptive fields of V4 neurons. By asking the monkeys to discriminate subtle changes in features in a particular point in visual space, they managed to shift the hot spot of the receptive field of V4 neurons in the neighbourhood of the location being fixated towards the spatial focus of attention. For these simulations, we used two cell pools in the V4 module. One pool received input from the vertically oriented cells in the V1 module.

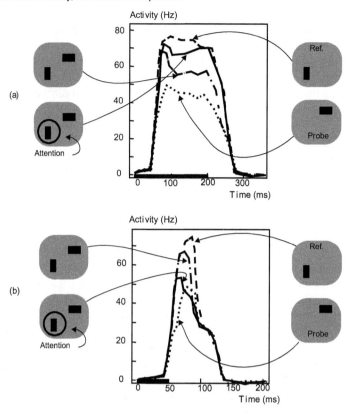

Fig. 6.9 Simulation of the experiment of Reynolds et al. (1999). (a) The stimulus was presented for 200 ms. When an optimal reference stimulus was presented alone, the cell's response (dashed line) was much stronger than its response (dotted line) when a suboptimal probe stimulus was presented alone in the receptive field. Simultaneous presentation of both the reference and the probe stimuli produced an intermediate response (dashed-dotted line), indicating that the probe was producing competitive suppression of the response to the reference stimulus. However, when spatial attention was directed toward the reference stimulus, the suppression due to the probe was largely eliminated: the neuronal response returned to the level when the reference was presented alone (continuous line). (b) The same manipulation as in (a) except that the stimulus was presented for only 50 ms. In this simulation, we used I_o=0, and slower dynamics, τ=15 ms. (From Edmund Rolls and Gustavo Deco, Computational Neuroscience of Vision ©2002, Oxford University Press, reproduced by permission of Oxford University Press https://global.oup.com/academic/product/computational-neuroscience-of-vision-9780198524885?q=computational neuroscience of vision &lang=en&cc=gb. For permission to reuse this material, please visit http://www.oup.co.uk/academic/rights/permissions. See also Deco and Lee 2004.)

The other pool received input from the horizontally oriented cells in the V1 module. In our simulation, spatial attention was again specified by introducing a bias to the appropriate cell pool in the PP module. Figure 6.10 shows that the central 'hot spot' peak of the receptive field of V4 neurons was shifted spatially as different PP module pools received the top-down bias.

In conclusion, the dynamical evolution of the firing activity at the neuronal level of our model of the cortical architecture of visual attention (Deco 2001, Deco and Zihl 2001b, Rolls and Deco 2002, Deco and Lee 2002, Corchs and Deco 2002, Heinke et al. 2002, Deco and Lee 2004) is consistent with the single cell experiments of Reynolds, Chelazzi and Desimone (1999) and Connor, Gallant, Preddie and Van Essen (1996). This indicates that the particular computational model we propose, of how biased competition and interaction between the

Attentional Shifting of Receptive Field

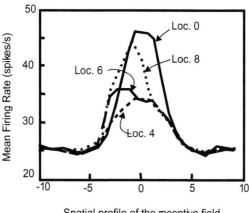

Spatial profile of the receptive field.

Fig. 6.10 Simulation of the experiment by Connor et al. (1993) showing that spatial attention can move the hot spot of the receptive field of a V4 neuron towards the location which is the focus of attention (if the receptive field is close to the focus of attention). The connection between the early V1 module and V4 was modelled with spatially local Gaussian weights centred at location 10, with standard deviation equal to 2.5. The V4 neuron had a more local spatial support than the ventral IT module's neuron in the other simulations in this series. Spatial attention allocated in the dorsal PP module pools corresponding to retinotopic locations 6 and 8 effectively shifted the peak of the receptive field of the V4 neuron to the left. Shift to the right can be produced by biasing locations 12 and 16 (not shown). Location 4 was far away from the centre of the receptive field. Allocating spatial attention to location 4 made the hot spot of the receptive field snap back to its centre position, but the responses of the V4 cell were markedly attenuated.

'what' and 'where' visual processing streams can be used to understand visual attention, is able to account for data at the neuronal level. This represents an advance beyond the biased competition hypothesis, in that it shows how object and spatial attention can be produced by dynamic interactions between the 'what' and 'where' streams, and in that as a computational model which has been simulated, the details of the model have been made fully explicit and have been defined quantitatively (Deco and Lee 2002, Rolls and Deco 2002, Deco and Lee 2004, Rolls 2008d).

6.2.7.2 Simulations of fMRI experiments

The dynamic evolution of activity at the cortical level, as shown for example by fMRI signals in experiments with humans, can be simulated in the framework of our model by integrating the pool of neuronal activity in a given area over space and time (Corchs and Deco 2002, Rolls and Deco 2002). The integration over space yields an average activity of the brain area considered at a given time. (The spatial resolution in most fMRI experiments is worse than 3 mm in any one dimension.) The integration over time is performed in order to simulate the relatively coarse temporal resolution of MRI experiments, which is in the order of a few seconds (see Section 5.6.9). We were able to simulate fMRI signals from V4 under the experimental conditions defined by Kastner, Pinsk, De Weerd, Desimone and Ungerleider (1999) (see Section 6.2.4). We used the same parameters as in the previous section with the only difference that the V1 hypercolumns included three levels of spatial resolution (i.e. $K=3$). The results demonstrated that the cortical attentional architecture described in Sections 6.2.6 and elsewhere (Rolls and Deco 2002, Rolls 2008d) shows and explains the typical dynamical

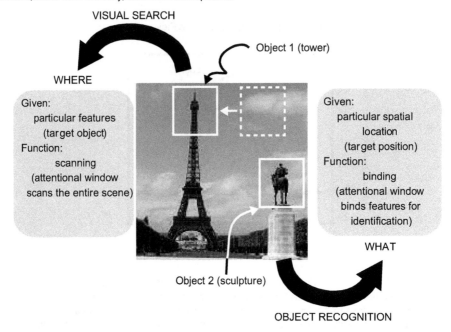

Fig. 6.11 The role of attentional mechanisms in object recognition and visual search in a natural scene. In the case of object recognition, a particular location in the natural scene is *a priori* specified with the aim of identification of the object which lies at that position. On the other hand, in visual search, a given target object (i.e. shape features) is *a priori* specified with the goal of finding out whether the target object is present in the scene and, if so, at which location. (From Edmund Rolls and Gustavo Deco, Computational Neuroscience of Vision ©2002, Oxford University Press, reproduced by permission of Oxford University Press https://global.oup.com/academic/product/computational-neuroscience-of-vision-9780198524885?q=computational neuroscience of vision &lang=en&cc=gb. For permission to reuse this material, please visit http://www.oup.co.uk/academic/rights/permissions. See also Deco and Lee 2004.)

competition and attention modulating effects found in attention experiments, even at the level of gross brain area activation as measured with fMRI (Corchs and Deco 2002).

6.2.8 The role of attention in object recognition and visual search

We now consider how the role of attention in object recognition and visual search can be understood with this quantitative approach. We concentrate on the macroscopic[2] level of analysis, and consider the interplay between the microscopic neuronal dynamics and the macroscopic functional behaviour in object recognition and visual search. These two different functions of visual perception can be explained in a unifying fashion by our neurodynamical model of the visual cortical system (as described in Section 6.2.6).

A phenomenological description of these two perceptual functions is presented schematically in Fig. 6.11 in the context of a natural scene.

In the case of object recognition, a particular location in the natural scene is a priori specified with the aim of identification of the object which lies at that position. Therefore, object recognition asks for 'what' is at a predefined particular spatial location. In the naive

---

[2] Physicists distinguish levels of analysis, and Rolls and Deco (2002) use microscopic to refer to the neuronal level, mesoscopic to refer to an intermediate level of activation of brain areas as shown for example by fMRI, and macroscopic to refer to the behaviour of the whole organism as measured for example in psychophysical experiments.

framework of the spotlight metaphor, one can describe the role of attention in object recognition by imagining that the prespecification of the particular spatial location is realized by fixing an attentional window or spotlight at that position. The features inside the fixed attentional window should now be bounded and recognized. On the other hand, in visual search, a given target object (composed of a set of shape features) is a priori specified with the goal of finding out whether the target object is present in the scene and, if so, at which location. Consequently, visual search asks for 'where' a predefined set of shape features is located. A naive description of visual search in the framework of the spotlight paradigm considers that during visual search, attentional mechanisms shift a window through the entire scene in order to serially search for the target object at different positions. The neurodynamical system described in Section 6.2.6 was tested in these two modes of operation: **object recognition** in an attended spatial location (spatial attention); and **visual search** of a target object (object attention), as shown and described in Fig. 6.8 on page 154. The input image was the scene shown in Fig. 6.11. The objects to be identified and located were the sculpture, and the top of the tower as indicated in the image. In the following two subsections, we will describe the dynamics of object recognition and visual search under these two modes of attention.

The thrust of our computational neuroscience approach to attention, and the particular model we describe in Section 6.2.6, is that it shows how spatial and object attention mechanisms can be integrated and function as a unitary system in visual search and visual recognition tasks. The dynamical intra- and inter-modular interactions in our cortical model implement attentional top-down feedback mechanisms that embody a physiologically plausible system of active vision unifying in this way different perceptual functions. In our system, attention is a dynamical emergent property, rather than a separate mechanism operating independently of other perceptual and cognitive processes. The dynamical mechanisms that define our system work across the visual field in parallel, but due to the different latencies of settling due to the intra- and intermodular dynamics, actually exhibit temporal properties like those normally described as 'serial' or 'parallel' when used to describe visual search (see Section 6.2.10). (The whole system was described by Deco and Rolls (2004) and is illustrated in Fig. 25.63, and the simulation illustrated in Figs. 6.12 and 6.13 is a simplified simulation including only the IT–V1–PP modules performed by Deco and Lee (2004) just to show the main aspects of the dynamics.)

The model accounts for a number of the temporal properties of serial search tasks found psychophysically without either an explicit serial scanning process with an attentional spotlight, or a saliency map. The following two subsections 6.2.8.1 and 6.2.8.2 show how these properties arise. In doing this, the model offers new insight into how attention may actually work, and goes beyond the biased competition hypothesis not only by showing how the 'what' and 'where' visual processing streams interact, and how the system operates quantitatively, but also by providing a direct approach to the temporal properties of the system, leading to a new conceptualization of what was described previously as serial and parallel search.

6.2.8.1 Dynamics of spatial attention and object recognition

In the object recognition task, the system functioned in a **spatial attention mode** as shown in Fig. 6.8a (see Deco and Lee (2002), Rolls and Deco (2002) and Deco and Lee (2004)). Spatial attention was initiated by introducing a bias to a cell pool coding for a particular location, and the aim was to perform object recognition for the object at the cued location. (This is therefore also termed 'object identification mode'.) The input system processed a pixelized 66×66 image ($N = 66$). The V1 hypercolumns covered the entire image uniformly. They are distributed in 33×33 locations ($P = 33$) and each hypercolumn was sensitive to three spatial frequencies and to eight different orientations (i.e. $K = 3$ and $L = 8$). Consequently, the V1 module had 26,136 pools and three inhibitory pools. The IT module utilized had two pools and one common inhibitory pool. Finally, the PP module contained 4,356 pools corresponding to each possible spatial location, i.e. to each of the 66×66 pixels, and a common inhibitory pool. Two objects (also shown in Fig. 6.11) were isolated in order to define two categories to be associated with two different pools in the IT module. During the learning phase, these two objects were presented randomly and at random positions in order to learn translation invariant responses. The system required 1,000,000 different presentations for training the IT pools. We used $\tau = \tau_P = 10$ ms and $T_r = 1$ ms.

When the image (Fig. 6.11) was presented, the spatial bias acting from the PP module as a top-down effect on the V1 module facilitated V1 responses at the attended location. Any features in the input image at that location produced larger responses in V1. This enhancement of neural activity to some features in V1 effectively gated information in that area of the V1 module to the IT module for recognition (and in this way also performed a type of shift invariance by using an attentional spatial modulation of early visual cortical processing). When the spatially highlighted image patch contained features of one of the trained object classes, the activity of the IT cell pools started to polarize in response to the features that were part of one of the trained objects, resulting in only one cell pool surviving the competition. The winner indicated the object class being recognized, identifying 'what was where', or 'what from where'. Thus object identification in IT was achieved by a spatial bias from PP interacting with the image representation in V1.

The actual simulation results just outlined are shown in Fig. 6.12, which demonstrates the neuronal activities in the three modules during object recognition in the spatial attention mode[3]. The average population activity of the three modules at the sculpture location was compared against the activity in all other locations when the spatial attention was allocated to the sculpture location by the PP spatial input bias. Spatial attention allocated in the dorsal stream PP module ultimately led to the dominance of the sculpture neurons' activity in the ventral stream IT module. The bifurcation of the maximum activities in different pools showed the propagation of the spatial attentional effect across the different modules. The effect of attention, as indicated by the polarization of the responses, started at the dorsal stream PP module, then propagated to the early stage V1 module, and then to the ventral stream IT module. The relative time of onset of the attentional effect was not distinguishable between the V1 module and the IT module because the attention-related computation in the V1 module and the IT module was concurrent and interactive. The polarization and stabilization of neuronal pool activity in the IT module corresponded to recognition of the object in the attended location. By moving the attentional spotlight to different spatial locations, the system can gate information at different spatial locations to the IT module, realizing a form of translation-invariant object recognition. Interestingly, a relatively small elevation in the V1 module was sufficient to bias the IT module cell pools to produce a large polarization in their responses.

It is important to emphasize that no explicit spotlight of attention is forced onto every module in the system in our model. Instead, given just a bias to one of the modules (in this case the spatial PP module), parallel and global dynamical evolution of the entire system converges to a state where a spotlight of attention in PP appears, and is explicitly used (even while it is emerging) to modulate the information processing channel for object recognition. Effectively spatial effects influence the function of the non-spatially organized IT ventral stream module through modulatory effects on information processing implemented through the topologically

---

[3] In the presentation of this and all subsequent simulations, we introduced a delay of 40 ms in the response of the early module neurons relative to stimulus onset. This is because physiologically there is typically a 40 ms delay between the presentation of the stimulus and the onset of V1 neurons' responses (Lee, Mumford, Romero and Lamme 1998) (see Section B.6.5).

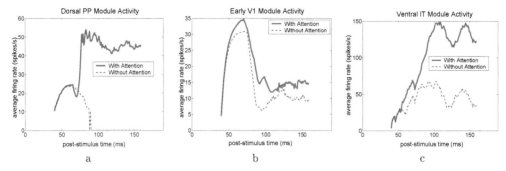

Fig. 6.12 Effect of spatial attention on neural activities in the three modules when performing object recognition. A top-down spatial attentional bias was introduced to the dorsal PP module pool that encodes the location of the sculpture in the scene (curves marked 'With Attention'), or there was no attentional input (curves marked 'Without Attention'). The responses of neuronal pools in the corresponding locations in the dorsal stream PP module and the V1 module, and the response of the neuronal pool encoding the sculpture in the ventral stream IT module were compared with and without spatial attention. (a) The evolution of the neuronal pool response at the dorsal stream PP module with and without the top-down spatial attentional bias. When the top-down bias was absent, the competition in the dorsal stream PP module was entirely determined by the bottom-up signals from V1. In this case, the sculpture shape did not provide the strongest input and was rapidly suppressed by the competitive interactions from the neuronal activities in the other locations in the scene. When there was a top-down spatial bias to the PP neuronal pool, the neuronal pool's activity rose rapidly and was maintained at a sustained level. (b) The increase in the activity of the neuronal pool in the dorsal stream PP module enhanced the activity at the corresponding retinotopic location in the early V1 module, particularly in the latter phase of the response, producing a highlighting effect. (c) The response of the ventral stream IT module neuron pool coding for the sculpture was substantially stronger when the spatial attention bias to the PP module was allocated to the location of the sculpture than when there was no spatial attention. The spatial highlighting gated the sculpture image to be analyzed and recognized by the ventral stream IT module neurons. Without spatial attention, the sculpture did not have bottom-up saliency to enable the domination of the sculpture neuron in the ventral stream IT module. (From Edmund Rolls and Gustavo Deco, Computational Neuroscience of Vision ©2002, Oxford University Press, reproduced by permission of Oxford University Press https://global.oup.com/academic/product/computational-neuroscience-of-vision-9780198524885?q=computational neuroscience of vision &lang=en&cc=gb. For permission to reuse this material, please visit http://www.oup.co.uk/academic/rights/permissions. See also Deco and Lee 2004.)

organized V1 module. An interesting aspect of our theory is that the process of object recognition can be influenced by spatial attention arising from a global dynamical continuing interaction between all the processing modules involved. Of course each local brain area has a defined functional role, but interesting global attention-related behaviour emerges from the constant cross-talk between the different modules in the ventral and dorsal visual paths, implemented partly through the early processing areas such as V1, using backprojections as well as forward connections.

6.2.8.2 Dynamics of object attention and visual search

In the visual search task, i.e. when the system was looking for a particular object in a visual scene, the system functioned in an **object attention mode** as shown in Fig. 6.8b (see Deco and Lee (2002), Rolls and Deco (2002) and Deco and Lee (2004)). Object attention was created by introducing a top-down bias to a particular cell pool in the IT module corresponding to a particular object class (as will be described in more detail below with Fig. 6.13). This ventral IT module pool backprojected the expected shape activity patterns over all spatial positions in the early V1 visual module through the top-down feedback connections. When the image (Fig. 6.11) containing the target object was presented, the hypercolumns in the early V1

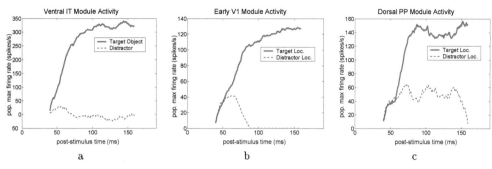

Fig. 6.13 Neuronal activities in the three modules during visual search in object attention mode. The maximum population activity of the neuronal pools corresponding to the identity or location of the sculpture in the scene in the three modules was compared against the maximum activity in pools coding any other locations or objects when the sculpture was the target object of attention. The effect of attention, i.e. differentiation of neuronal response between the target and distracter conditions, was observed to start at the ventral stream IT module, and then to propagate to the early V1 module and the dorsal stream PP module. (a) The neuronal activity of the top-down biased 'sculpture' neuronal pool in the ventral-stream module was compared against the maximum activity of all other object pools. The increase in the response of the sculpture pool relative to the response in all other pools was observed to rise rapidly as a result of the top-down object attentional bias applied to the IT module. (b) The maximum population activity at the sculpture location in the early V1 visual module was compared against the maximum activity of pools at all other locations in the scene. (c) The maximum activity of the pools coding in the dorsal stream PP module for the sculpture location was compared against the maximum activity of the pools encoding the other locations. (From Edmund Rolls and Gustavo Deco, Computational Neuroscience of Vision ©2002, Oxford University Press, reproduced by permission of Oxford University Press https://global.oup.com/academic/product/computational-neuroscience-of-vision-9780198524885?q=computational neuroscience of vision &lang=en&cc=gb. For permission to reuse this material, please visit http://www.oup.co.uk/academic/rights/permissions. See also Deco and Lee 2004.)

visual module whose activities were closest to the top-down 'template' became more excited because of the interactive activation or resonance between the forward visual inputs and the backprojected activity from the IT module. Over time, these V1 hypercolumns with neuronal activities best matching the encoded object dominated over all the other hypercolumns, resulting in a spatially localized response peak in the early V1 visual module. Meanwhile, the dorsal PP module was not idle but actively participated by having all its pools engaged in the competitive process to narrow down the location of the target. The simultaneous competition in the spatial domain and in the object domain in the two extrastriate modules as mediated by their reciprocal connections with the early V1 module finally resulted in a localized peak of activation in the spatially mapped dorsal stream PP module, with a corresponding peak of activity for the object mapped in the ventral stream IT module, and corresponding activity in the early V1 module. This corresponds to finding the object's location in the image in a visual search task, or linking 'where is what', or computing 'where' from 'what'.

The actual simulation results just outlined are shown in Fig. 6.13, which compares the responses of the pools in the three modules corresponding to the location or identity of the attended object (the sculpture) to the responses at all other locations or identities. The evolution of the population maximum activity shows the polarization of responses that started in the ventral stream IT module, and then propagated to the other two modules. In this case, the attentional object bias was applied to the IT module, and this then propagated to the V1 module where features relevant to the object were biased on. This top-down bias interacted with features in the scene, resulting in larger activation of the features that were present in the object. However, because V1 is topologically mapped, the location of the features was contained in the V1 activations, and these then influenced the PP module to gradually settle at the location with extra activity in V1. Thus just providing object bias to the V1 module resulted in the spatial location of the object being made explicit in the PP module as a result of the interaction with the image representation provided for in V1.

It should be emphasized that because the V1 visual module interacted with both the PP and the IT modules simultaneously, the attentional effect observed in the later response of the neurons was not purely spatial or featural, but involved both components simultaneously. In the spatial attention mode, the PP module's bias initially highlighted the V1 module's response at a particular location, then the IT module got drawn into the process, and competition in the IT module, combined with the ongoing interaction via the reciprocal connections between the PP module and the IT module, produced the enhancement effect in the V1 module.

A similar situation also appeared in the object attention mode. An object attention effect of this magnitude has been observed in V4, but not in V1. We reduced the magnitude of the coupling between the early V1 module and the ventral IT module, and found that the system continues to perform well in the visual search task when the coupling strength is reduced from 1 to 0.4. With this coupling strength, the effect of object attention in V1 is more modest, and yet it can still produce a bias that leads to the emergence of a peak response at the dorsal stream PP module's spatial location map.

6.2.9 The neuronal and biophysical mechanisms of attention: linking computational and single-neuron data

The theoretical framework described in Section 6.2.6 and in detail by Rolls and Deco (2002) and Deco and Lee (2004) provides an account of a potential functional role for the back-projection connections in the visual cortex in visual attention. We now consider how closely this approach can account for the responses of single neurons in the system, and how well a detailed implementation of the model at the neuronal level can lead to the overall functional properties described above in Section 6.2.8. We then show that the same theoretical framework and model can also be directly related to psychophysical data on serial vs parallel processing (Section 6.2.10), and to neuropsychological data (Rolls 2008d). The framework and model operate in the same way to account for findings at all these levels of investigation, and in this sense provide a unifying framework.

Deco and Lee (2004) and Corchs and Deco (2002) (see Rolls and Deco (2002)) showed a long-latency enhancement effect on V1 units in the model under top-down attentional modulation (see Figs. 6.12 and 6.13) which is similar to the long-latency contextual modulation effects observed in early visual cortex (Lamme 1995, Zipser, Lamme and Schiller 1996, Lee et al. 1998, Roelfsema, Lamme and Spekreijse 1998, Lee, Yang, Romero and Mumford 2002). Interestingly, in our simulation, we found that the observed spatial or object attentional enhancement is stronger for weaker stimuli. This predicted result has been confirmed neurophysiologically by Reynolds, Pastemak and Desimone (2000). The mechanism for this may be that the top-down attentional influence can dominate the firing of the neurons relatively more as there are fewer feedforward forcing and shunting effects on the neurons. An extension of this model (Deco and Rolls 2004) can account for the reduced receptive fields of inferior temporal cortex neurons in natural scenes (Rolls, Aggelopoulos and Zheng 2003a), and makes predictions about how the receptive fields are affected by interactions of features within them, and by object-based attention (see Section 25.5.9.2 and Fig. 25.63 on page 650).

The model has also been extended to the level of spiking neurons which allows biophysical properties of the ion channels affected by synapses, and of the membrane dynamics, to be incorporated, and shows how the non-linear interactions between bottom-up effects (produced for example by altering stimulus contrast) and top-down attentional effects can account for new neurophysiological results in areas MT and V4 (Deco and Rolls 2005b). The model and simulations show that attention has its major modulatory effect at intermediate levels of bottom-up input, and that the effect of attention disappears at low and high levels of contrast of the competing stimulus. The model assumed no kind of multiplicative attentional effects on the gain of neuronal responses. Instead, in the model, both top-down attention and bottom-up input information (contrast) are implemented in the same way, via additive synaptic effects in the postsynaptic neurons. There is of course a non-linearity in the effective activation function of the integrate-and-fire neurons, and this is what we identify as the source of the apparently multiplicative effects (Martinez-Trujillo and Treue 2002) of top-down attentional biases on bottom-up inputs. The relevant part of the effective activation function of the neurons (the relation between the firing and the injected excitatory currents) is the threshold non-linearity, and the first steeply rising part of the activation function, where just above threshold the firing increases markedly with small increases in synaptic inputs (cf. Amit and Brunel (1997) and Brunel and Wang (2001)). Attention could therefore alternatively be interpreted as a phenomenon that results from purely additive synaptic effects, non-linear effects in the neurons, and cooperation–competition dynamics in the network, which together yield a variety of modulatory effects, including effects that appear (Martinez-Trujillo and Treue 2002) to be multiplicative. In addition, we were able to show that the non-linearity of the NMDA receptors may facilitate non-linear attentional effects, but is not necessary for them. This was shown by disabling the voltage-dependent non-linearity of the NMDA receptors in the simulations (Deco and Rolls 2005b).

More detail is now provided about the integrate-and-fire model of attention described by Deco and Rolls (2005b), as it shows how attentional processes can be understood at the neuronal and biophysical level. The design of the experiments and the different measures as implemented in our simulations are shown graphically in Fig. 6.14. Figure 6.14 shows the experimental design of Reynolds and Desimone (2003) and the architecture we used to simulate the results. They measured the neuronal response in V4, manipulating the contrast of the non-preferred stimulus and comparing the response to both stimuli when attention was allocated to the poor stimulus. They observed that the attentional suppressive effect of the competing non-preferred stimulus is higher when the contrast of that stimulus increases. In our simulations we measured neuronal responses from neurons in pool S1′ in V4 to a preferred and a non-preferred stimulus simultaneously presented within the receptive field. We manipulated the contrast of the stimulus that was non-preferred for the neurons S1′ (in the simulation by altering λ_{in} to S2). We analysed the effects of this manipulation for two conditions, namely without spatial attention, or with spatial attention on the non-preferred stimulus S2 implemented by adding an extra bias λ_{att} to S2.

Figure 6.15 (top) shows the results of our simulations for the design of Reynolds and Desimone (2003). We observed that the attentional suppressive effect implemented through λ_{att} on the responses of neurons S1′ of the competing non-preferred stimulus is higher when the contrast of the non-preferred stimulus increases, as in the original neurophysiological experiments. The top figure shows the response of a V4 neuron to different log contrast levels (abscissa) in the no-attention condition (AO: attending outside the receptive field) and in the attention condition (AI: attending inside the receptive field). The top right part of Fig. 6.15 shows the difference between both conditions. As in the experimental observations the suppressive effect of the competing non-preferred stimulus is higher when the contrast of that stimulus increases, but at higher levels of salience (contrast), the top-down attentional effect disappears.

In order to study the relevance of NMDA synapses for the inter-area cortical dynamics of attention, we repeated the analysis shown in Fig. 6.15 (top), but with the voltage-dependent non-linearity removed from the NMDA receptors (in the feedforward, the feedback, and the

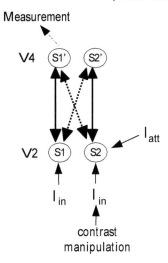

Fig. 6.14 Interaction between salience (contrast, a bottom-up influence) and attention (a top-down influence) in an integrate-and-fire model of biased competition. The architecture shows that used to model the neurophysiological experiment of Reynolds and Desimone (2003). In the model we measured neuronal responses from neurons in pool S1′ in V4 to a preferred and a non-preferred stimulus simultaneously presented within the receptive field. We manipulated the contrast of the stimulus that was non-preferred for the neurons S1′ (in the simulation by altering λ_{in} to S2). We analysed the effects of this manipulation for two conditions, namely without spatial attention or with spatial attention on the non-preferred stimulus S2 (implemented in these simulations by applying an extra bias λ_{att} to S2 from an external top-down source). We observed that the attentional suppressive effect implemented through λ_{att} on the responses of neurons S1′ of the competing non-preferred stimulus is higher when the contrast of the non-preferred stimulus increases, as in the original neurophysiological experiments. Reproduced with permission from *Journal of Neurophysiology*, 94 (1), Neurodynamics of Biased Competition and Cooperation for Attention: A Model With Spiking Neurons, Gustavo Deco and Edmund T. Rolls, 295–313, Figure 5a, DOI: 10.1152/jn.01095.2004, (c) 2005, The American Physiological Society.

recurrent collateral connections) by setting $[Mg^{2+}] = 0$ (which corresponds to removing the non-linear dependence of the NMDA synapses on the postsynaptic potential, see Deco and Rolls (2005b)). (We compensated for the effective change of synaptic strength by rerunning the mean field analysis to obtain the optimal parameters for the simulation, which was produced with $J_f = 1.6$ and $J_b = 0.42$, and then with these values, we reran the simulation.) The results are shown in Fig. 6.15 (middle), where it is clear that the same attentional effects can be found in exactly the same qualitative and even quantitative form as when the non-linear property of NMDA receptors is operating. The implication is that the non-linearity of the effective activation function (firing rate as a function of input current to a neuron) of the neurons (both the threshold and its steeply rising initial part) implicit in the integrate-and-fire model with AMPA and other receptors is sufficient to enable non-linear attentional interaction effects with bottom-up inputs to be produced. The non-linearity of the NMDA receptor may facilitate this process by its non-linearity, but is not necessary.

Deco and Rolls (2005b) further studied the relevance of the NMDA receptors by repeating the analysis in Fig. 6.15 (top), but with the time constants of the NMDA receptors set to be the same values as those of the AMPA receptors (and as in Fig. 6.15 (middle) with the NMDA voltage-dependent effects disabled by setting $[Mg^{2+}] = 0$. (We again compensated for the effective change of synaptic strength by rerunning the mean field analysis to obtain the optimal parameters for the simulation.) The results are shown in Fig. 6.15 (bottom), where it is shown that top-down attentional effects are now very greatly reduced. (That is, there is

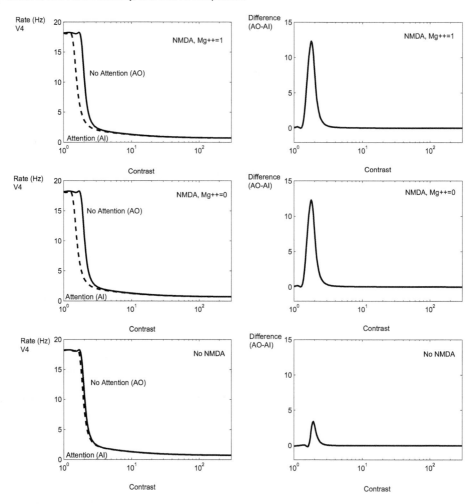

Fig. 6.15 (Top): results of our simulations for the effect of interaction between contrast and attention after the design of Reynolds and Desimone (2003). The left figure shows the response of V4 neurons to different log contrast levels (abscissa) in the no-attention condition (AO: attending outside the receptive field), and in the attention condition (AI: attending inside the receptive field) (see legend to Fig. 6.14). The right figure shows the difference between both conditions. As in the experimental observations the suppressive effect of the competing non-preferred stimulus is higher when the contrast of that stimulus increases, but at higher levels of salience (contrast) the attentional effect disappears. Middle: as top, but with the NMDA receptor non-linearity removed by setting $[Mg^{2+}] = 0$. Bottom: as top, but with the NMDA receptor time constants set to the same values as those of the AMPA receptors (see text). Reproduced with permission from *Journal of Neurophysiology*, 94 (1), Neurodynamics of Biased Competition and Cooperation for Attention: A Model With Spiking Neurons, Gustavo Deco and Edmund T. Rolls, 295–313, Figure 6, DOI: 10.1152/jn.01095.2004, (c) 2005, The American Physiological Society.

very little difference between the no-attention condition (AO: attending outside the receptive field), and the attention condition (AI: attending inside the receptive field).) This effect is not just because the NMDA receptor system with its long time constant may play a generic role in the operation of the integrate-and-fire system, by facilitating stability and helping to prevent oscillations, for a similar failure of attention to operate normally was also found with the mean field approach, in which the stability of the system is not an issue. Thus the long time constant of NMDA receptors does appear to be an important factor in enabling top-down attentional processes to modulate correctly the bottom-up effects to account for the effects of attention on neuronal activity. This is an interesting result which deserves further analysis. Deco and Rolls (2005b) did show that the mean field equation that effectively defines the non-linear transfer function of the neurons will be affected by the long time constant of the NMDA receptors.

These investigations thus show how attentional processes can be understood by the inter-actions between top-down and bottom-up influences on non-linear processes operating within neurons, and how these processes in turn impact the performance of large separate populations of interacting neurons.

6.2.10 Linking computational and psychophysical data: 'serial' vs 'parallel' processing

6.2.10.1 'Serial' vs 'parallel' search

In the literature on selective attention, it has frequently been assumed that there are two types of search, one parallel where the reaction time is not influenced by the number of items in a display, and the other serial where the reaction time increases with the number of items in a display. This is illustrated in Fig. 6.16. Much work has been based on these two kinds of search paradigms: feature and conjunction search (Quinlan and Humphreys 1987, Wolfe, Cave and Franzel 1989, Treisman and Sato 1990). In feature integration theory (Treisman 1988, Treisman and Gelade 1980) there is a first preattentive process that runs in parallel across the complete visual field extracting single primitive features without integrating them. The second attentive stage corresponds to the serial specialized integration involving for example forming feature conjunctions for a limited part of the field at any one time. In this section, I show that our dynamical model of the neural bases of attention accounts for the findings using entirely parallel processing, but where the reaction time depends on the difficulty of the solution into which the dynamical system must fall.

In the visual search tasks we consider, subjects examine a display containing randomly positioned items in order to detect an *a priori* defined target (i.e. a target that the subject is paying attention to and must search for), and other items in the display which are different from the target serve the role of distracters. In a feature search task the target differs from the distracters in one single feature, e.g. only colour. In a conjunction search task the target is defined by a conjunction of features, and each distracter shares at least one of those features with the target. Conjunction search experiments show that search time increases linearly with the number of items in the display, which has been taken to imply a serial process, such as an attentional spotlight moving from item to item in the display (Treisman 1988). An example of a display with this 'serial' search is shown in Fig. 6.17b, in which an E is the target and Fs are the distracters. On the other hand, search times in a feature search can be independent of the number of items in the display, and this is described as a preattentive parallel search (Treisman 1988) (see further (Deco and Zihl 2001b, Rolls and Deco 2002, Deco and Lee 2004)). An example of a display with this 'parallel' search is shown in Fig. 6.17a, in which an E is the target and Xs are the distracters.

In more detail, the stimulus in Fig. 6.17a contains shapes E and X. Because the elementary features in E and X are distinct, i.e. their component lines have different orientations, E pops out from X, and its location can be rapidly localized independently of the number of distracting X shapes in the image. On the other hand, the stimulus in Fig. 6.17b contains E and F. Since both letters are composed of vertical and horizontal lines, there is no difference in elementary features to produce a preattentive pop-out, so they can be distinguished from each other only

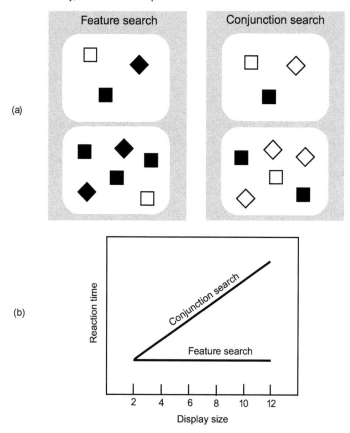

Fig. 6.16 Visual search experiments: feature and conjunction search. (a) Typical sample displays in visual feature and conjunction search tasks. The target (white square) 'pops out' in the case of feature search despite the variation in the number of distracters, whereas in the case of conjunction search the target is harder to find, especially when there are many distracters. (b) Feature search compared with conjunction search times. In feature searches, the subjects' reaction times do not increase as a function of the display size as they do in conjunction searches. (From Edmund Rolls and Gustavo Deco, Computational Neuroscience of Vision ©2002, Oxford University Press, reproduced by permission of Oxford University Press https://global.oup.com/academic/product/computational-neuroscience-of-vision-9780198524885?q=computational neuroscience of vision &lang=en&cc=gb. For permission to reuse this material, please visit http://www.oup.co.uk/academic/rights/permissions.)

after their features are glued together by attention. It has been thought that because 'attention' is serial, the time required to localize the target in such an image increases linearly with the number of distracters in the image. The serial movement of the attentional spotlight has been thought to be governed by a *saliency map* or *priority map* for registering the potentially interesting areas in the retinal input and directing a *gating* mechanism for selecting information for further processing. Does the linear increase in search time observed in visual search tests necessarily imply a serial search process, a saliency map, or a gating mechanism? Could both the serial and the parallel search phenomena be explained by a single parallel neurodynamical process, without the extra serial control mechanism?

To investigate this issue, the stimuli shown in Fig. 6.17a and Fig. 6.17b were presented to the network described in Section 6.2.6, which had been trained to recognize X, E, and F in a translation invariant manner (Deco and Zihl 2001b, Rolls and Deco 2002). The aim was to find the location of the E in the display. The system received a top-down bias for the 'E' pool in

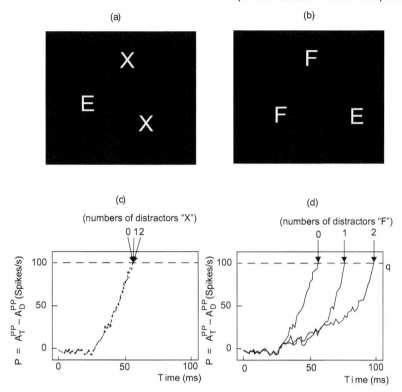

Fig. 6.17 (a) Parallel search example: an image that contains a target E in a field of X distracters. Since the elementary features in E and X are distinct, i.e. their component lines have different orientations, E pops out from X, and its location can be rapidly localized independently of the number of distracting X shapes in the image. This is called parallel search. (b) Serial search example: an image that contains a target E in a field of F distracters. Since both letters are composed of vertical and horizontal lines, there is no difference in the elementary features to produce a preattentive pop-out, so the E and F can be distinguished from each other only after their features are bound or glued by attention. The time required to locate the target in such an image increases linearly with the number of distracters in the image. This is called serial search and has been thought to involve the scanning of the scene with a covert attentional spotlight. (c–d) Simulation result of the network performing visual search on images (a) and (b) respectively. The difference (polarization) between the maximum activity in the neuronal pool corresponding to the target locations and the maximum activity of all other neuronal pools in the dorsal PP module is plotted as a function of time. (c) shows that the difference signal rose to a threshold, corresponding to localization of the object, at about the same time independently of the numbers of distracter items. (d) shows that the time for the polarization signal to rise to a threshold increased linearly with the number of distracting items, with an additional 25 ms required per item. (From Edmund Rolls and Gustavo Deco, Computational Neuroscience of Vision ©2002, Oxford University Press, reproduced by permission of Oxford University Press https://global.oup.com/academic/product/computational-neuroscience-of-vision-9780198524885?q=computational neuroscience of vision &lang=en&cc=gb. For permission to reuse this material, please visit http://www.oup.co.uk/academic/rights/permissions. See also Deco and Lee 2004.)

the ventral IT module, and then was presented with stimuli containing E in a variable number of X shapes, or E in a variable number of F shapes. Let us use polarization, the difference between the maximal activity of the pools indicating the E location and that indicating the F location, i.e. $P = A_T^{PP} - A_D^{PP}$, as a measure to determine whether detection and localization of the target had been achieved or not. We found that for the E in X case, the time required for the polarization to reach a certain threshold in the dorsal PP module was almost identical

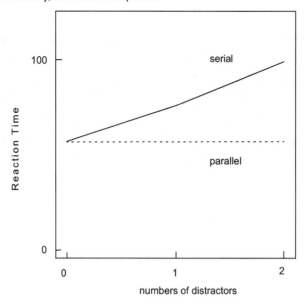

Fig. 6.18 Simulated reaction times for different types and numbers of distracters from the results shown in Fig. 6.17c and 6.17d for 'parallel' and 'serial' search examples. (From Edmund Rolls and Gustavo Deco, Computational Neuroscience of Vision ©2002, Oxford University Press, reproduced by permission of Oxford University Press https://global.oup.com/academic/product/computational-neuroscience-of-vision-9780198524885?q=computational neuroscience of vision &lang=en&cc=gb. For permission to reuse this material, please visit http://www.oup.co.uk/academic/rights/permissions. See also Deco and Lee 2004 who show similar results for up 16 distracters.)

whether the number of X shapes was equal to 0, 1, or 2, as shown in Fig. 6.17c. On the other hand, when E and F were presented, the time required for polarization to reach threshold increased linearly with the number of distracting items. Although the system was running with the same parallel dynamics, it took an additional 25 ms for each additional distracter added to the stimulus as shown in Fig. 6.17d.

Figure 6.18 shows explicitly the simulated reaction times as a function of the number of distracters for both types of search. The system works across the visual field in parallel, but, due to the different dynamic latencies, resembles the two apparent different modes of visual attention, namely: serial focal search, and parallel search. In the case of serial search, the latency of the dynamics is longer indicating 'apparently' a serial component in the search, although the underlying mechanisms work in parallel. The typical linear increase in the search time with the display size is clearly obtained as the result of a slower convergence (latency) of the dynamics. In this case, the strong competition present in V1 and propagated to PP delays the convergence of the dynamics. The strong competition in the feature extraction module V1 is finally resolved by the feedback received from PP. In other words, stimulus similarity in the feature space is decided by competition mechanisms at the intra-modular level of V1 and at the inter-modular level of V1–PP.

The results of the simulations are consistent with psychophysical results (Quinlan and Humphreys 1987). Although the whole simulation is parallel, and involves no serial moving spotlight process, the search takes longer with more distracters because the constraints are then more difficult to satisfy, and the dynamics of the coupled set of networks takes longer to settle. (The constraints are more difficult to satisfy in the E vs F search task in that the features, which include oriented line elements, are more similar to each other in the E vs F case than in the E vs X case.) The linear increase in time observed in 'serial search' in the E–F case reflects simply the fact that when features are similar between objects, the similarity in the stimuli's representations in V1 will require more time for the competition to sort out the target against the distracters. This sorting out at the level of V1 requires a constant interaction with both the ventral stream IT and the dorsal stream PP pathways in the emergent process of the integration and combination of different features to form a shape.

This is an important result, for it provides direct computational evidence that some apparently serial cognitive tasks may in fact be performed by fully parallel processing neuronal networks with realistic dynamics. *It is proposed that for many computations that have been thought to be serial by the brain, the computation is in fact fully parallel, but as the constraints become more difficult, the dynamical system takes longer to settle into an attractor basin.* Further evidence comes from the analysis of feature conjunction search considered by Rolls and Deco (2002) and Rolls (2008d). Another example of increasing reaction times due to longer settling into an attractor basin as a decision task becomes more difficult is provided in Section 5.6.

Deco, Pollatos and Zihl (2002) extended the approach just described to deal with serial search when there are conjunctions of different numbers of features (see Rolls and Deco (2002) and Rolls (2008d)). The approach has also been extended to account not only for different types of neglect after brain damage, but also even object-based visual neglect in which damage to the right parietal cortex can lead to the neglect of the left part on each one of a series of objects arranged horizontally across the visual field (Rolls and Deco 2002, Deco and Rolls 2002, Deco and Zihl 2001a, Heinke, Deco, Zihl and Humphreys 2002, Deco and Zihl 2004, Rolls 2008d).

6.3 Top-down attention – biased activation

There is now evidence that whole cortical areas and even processing streams can be biased by selective attention. This can occur for example if some processing attributes such as reward value are processed in one cortical stream (the orbitofrontal cortex and anterior cingulate cortex), and other attributes such as the intensity and identity of a stimulus in other cortical streams. Selective attention to these attributes can then bias some vs other cortical areas and processing streams. First some of the evidence for this is described, and then a biased activation theory of selective attention is described to account for this (Rolls 2013a).

6.3.1 Selective attention to reward value vs sensory processing selectively activates different cortical areas and processing streams

We have found that with taste and flavour (Grabenhorst and Rolls 2008) stimuli, and olfactory (Rolls, Grabenhorst, Margot, da Silva and Velazco 2008b) stimuli, selective attention to pleasantness modulates representations in the orbitofrontal cortex, whereas selective attention to intensity modulates activations in areas such as the primary taste cortex (Fig. 6.19). Thus, depending on the context in which tastes and odours are presented and whether affect is relevant, the brain responds to taste, flavour, and odour, differently.

These findings show that when attention is paid to affective value, the brain systems engaged to represent the stimulus are different from those engaged when attention is directed to the physical properties of a stimulus such as its intensity.

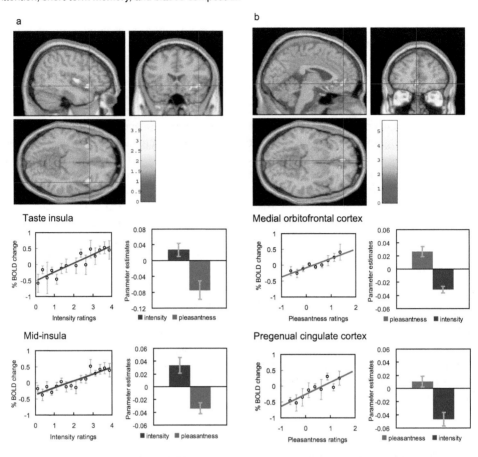

Fig. 6.19 Effect of paying attention to the pleasantness vs the intensity of a taste stimulus. a. Top: A significant difference related to the taste period was found in the taste insula at [42 18 -14] (indicated by the cursor) and in the mid insula at [40 -2 4]. Middle: Taste Insula. Right: The parameter estimates (mean ± sem) for the activation at the specified coordinate for the conditions of paying attention to pleasantness or to intensity. The parameter estimates were significantly different for the taste insula p=0.001. Left: The correlation between the intensity ratings and the activation (% BOLD change) at the specified coordinate (r=0.91, p<0.001). Bottom: Mid Insula. Right: The parameter estimates for the activation at the specified coordinate for the conditions of paying attention to pleasantness or to intensity. The parameter estimates were significantly different for the mid insula p=0.001. Left: The correlation between the intensity ratings and the activation at the specified coordinate (r=0.89, p<<0.001). The taste stimulus, monosodium glutamate, was identical on all trials. b. Top: A significant difference related to the taste period was found in the medial orbitofrontal cortex at [-6 14 -20] (towards the back of the area of activation shown) and in the pregenual cingulate cortex at [-4 46 -8] (at the cursor). Middle: Medial orbitofrontal cortex. Right: The parameter estimates for the activation at the specified coordinate for the conditions of paying attention to pleasantness or to intensity. The parameter estimates were significantly different for the orbitofrontal cortex $p< 10^{-4}$. Left: The correlation between the pleasantness ratings and the activation at the specified coordinate (r=0.94, p<0.001). Bottom: Pregenual cingulate cortex. Conventions as above. Right: The parameter estimates were significantly different for the pregenual cingulate cortex $p< 10^{-5}$. Left: The correlation between the pleasantness ratings and the activation at the specified coordinate (r=0.89, p=0.001). The taste stimulus, 0.1 M monosodium glutamate, was identical on all trials. (Reproduced from European Journal of Neuroscience, Selective attention to affective value alters how the brain processes taste stimuli, 27: 723–729. Fabian Grabenhorst and Edmund T. Rolls. ©The Authors (2008). Reproduced with permission from John Wiley and Sons.)

6.3.2 Sources of the top-down modulation of reward value and emotion-related processing

There is relatively little prior evidence on the top-down source of the bias when attention is to affective (emotional) vs sensory aspects (e.g. the intensity) of the same stimulus (Pessoa 2009). In a study using psychophysiological interaction (PPI) analysis, we found that two sites where selective attention to pleasantness increased the activation to taste, the orbitofrontal cortex and a region to which it is connected, the pregenual cingulate cortex, both had functional connectivity with a quite anterior (mean Y≈50) part of the lateral prefrontal cortex, illustrated in Grabenhorst and Rolls (2010). These parts of the orbitofrontal cortex and pregenual cingulate cortex are a functionally appropriate target site for a top-down attentional modulation, in that their activations are correlated with the subjectively rated pleasantness of the taste (Grabenhorst and Rolls 2008). Moreover, the lateral prefrontal cortex has been shown to represent current task sets and attentional demands for different types of tasks (Passingham and Wise 2012).

The statistics used in the calculation of PPI effects (Friston, Buechel, Fink, Morris, Rolls and Dolan 1997) do not reveal the directionality of the connectivity, for they are based on correlations. However, the directionality in this case is likely to be from the prefrontal cortex to the orbitofrontal and pregenual cingulate cortices, for the following reasons. First, the prefrontal cortex has a powerfully developed recurrent collateral system (Elston et al. 2006), which provides the basis for the short-term memory (Chapter 4) that is needed to hold the subject of attention active, providing the source of the bias for top-down biased competition. Second, prefrontal cortex lesions impair attention (Beck and Kastner 2009, Rossi, Pessoa, Desimone and Ungerleider 2009). Third, activations in areas of the lateral prefrontal cortex are related to task set, attentional instructions, and remembering rules that guide task performance (Beck and Kastner 2009, Rossi et al. 2009, Passingham and Wise 2012). Fourth, direct anatomical connections exist between the lateral prefrontal cortex and the orbitofrontal and pregenual cingulate cortices (Price 2006).

The conclusion that these findings suggest is therefore that a part of the lateral prefrontal cortex, not a site normally implicated in affective value and emotion, may be able to modulate emotion-related / affect-related processing in the brain by a top-down attentional influence. This may be one way in which higher cognitive functions, such as a reasoning-based strategy and route to action, or verbal instruction to direct processing towards or away from emotion-related brain processing, or conscious volition, can influence the degree to which the affect-related parts of the brain process incoming (or potentially remembered) stimuli that can produce emotional responses. This is thus a part of the way in which cognition can influence, and control, emotion (Rolls 2014a).

We also found that two sites where selective attention to intensity increased the activation to the taste delivery into the mouth, the anterior and mid insula, both had functional connectivity with a less anterior (mean Y≈37) part of the lateral prefrontal cortex (Grabenhorst and Rolls 2010). These parts of the insula are a functionally appropriate site for a top-down attentional modulation, in that their activations are correlated with the subjectively rated intensity of the taste (Grabenhorst and Rolls 2008, Grabenhorst et al. 2008a). The anterior insular site may be the primary taste cortex (Pritchard, Hamilton, Morse and Norgren 1986, Yaxley, Rolls and Sienkiewicz 1990, De Araujo, Kringelbach, Rolls and Hobden 2003a, De Araujo and Rolls 2004, Rolls 2008c, Rolls 2016b), and the mid-insular site a region activated by other oral including somatosensory and fat texture inputs from the oral cavity (De Araujo, Kringelbach, Rolls and McGlone 2003b, De Araujo and Rolls 2004) and perhaps by taste per se (Small 2010, Rolls 2016b), in that the activations there were correlated with the trial-by-trial subjective ratings of the taste intensity made during the scanning (Grabenhorst and

Rolls 2008). In the analyses described here, such somatosensory inputs could contribute to the attention-dependent correlations found between the mid insula and other areas.

The interpretation of this functional connectivity revealed with PPI (Friston, Buechel, Fink, Morris, Rolls and Dolan 1997) is that the prefrontal cortex and orbitofrontal / pregenual cingulate areas covary in their activations more strongly when attention is directed to pleasantness than to intensity. In this study, the implication is that when the activity in the orbitofrontal and pregenual cingulate areas is high, as it is on trials when attention is paid to pleasantness relative to trials when attention is paid to intensity, then activations in this prefrontal cortex region are also high. A large source of this variation which gives rise to the PPI effect is thus the difference in the activations on different trial types which can be captured by the correlation arising from the difference in the mean activations of both sites (orbitofrontal / pregenual cingulate cortices and prefrontal cortex) on each of the two trial types. However, in addition to this source of variation, it could be that when two areas are functionally interacting strongly, there may be an additional contribution to the connectivity term produced by the trial-by-trial variation within a type of trial. For example, on trials on which pleasantness is the subject of attention, then any small variation on a particular trial in the prefrontal cortex would be expected to be reflected in the activations in the orbitofrontal / pregenual cingulate cortex. This effect would arise because when both areas are active, the neurons in each area may be operating on a relatively linear part of their activation function, producing strong coupling, whereas when one or both areas are relatively inactive, with only spontaneous firing, then the neurons may be subject to some effects produced by being close to the firing threshold, such that small changes in input may produce a smaller than linear effect on the output. This trial-by-trial variation corresponds in information theoretic analysis of neuronal covariation to a 'noise' effect as compared to a 'signal' effect (Rolls and Treves 2011) (Appendix C).

6.3.3   Granger causality used to investigate the source of the top-down biasing of affective processing

Correlations between signals, including signals at the neuronal or at the functional neuro-imaging level, do not reveal the direction of the possible influence of one signal on the other. PPI analysis is based on correlations. Understanding how one brain area may influence another, for example by providing it with inputs, or by top-down modulation, is fundamental to understanding how the brain functions. Hence, inferring causal influences from time series data has been attracting intensive interest. Recently, Granger causality has become increasingly popular due to its easy implementation and many successful applications to econometrics, neuroscience, etc., and in particular, the study of brain function (Bressler and Seth 2011, Ding, Chen and Bressler 2006). The application of Granger causality analysis to BOLD fMRI signals which are inherently slow has been discussed elsewhere (Stephan and Roebroeck 2012).

Granger causality is based on precedence and predictability. Originally proposed by Wiener (1956) and further formalized by Granger (1969), it states that given two times series x and y, if the inclusion of the past history of y helps to predict the future states of x in some plausible statistical sense, then y is a cause of x in the Granger sense. In spite of the wide acceptance of this definition, classical Granger causality is not tailored to measure the effects of interactions between time series x and y on the causal influences, and cannot measure systematically the effects of the past history of x on x (Ge, Feng, Grabenhorst and Rolls 2012). In this situation, a componential form of Granger causality analysis has recently been introduced which has advantages over classical Granger analysis (Ge, Feng, Grabenhorst and Rolls 2012). Componential Granger causality measures the effect of y on x, but allows interaction effects between y and x to be measured (Ge, Feng, Grabenhorst and Rolls 2012).

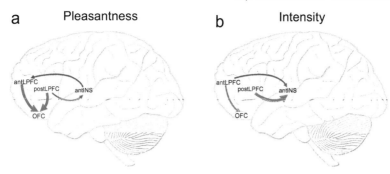

Fig. 6.20 Componential Granger causality analysis of top-down effects on taste processing from different lateral prefrontal cortex areas during attention to either the pleasantness (A) or to the intensity (B) of a taste. Significant causal influences from t tests with a Bonferroni correction are marked by blue arrows (i.e. cross-componential Granger causality is greater than 0). Red arrows indicate where significant top-down effects exist in addition to significant causal influences (i.e. a significant cross-componential Granger causality that is different in the two directions). The areas are anterior (mean Y≈50) and posterior (mean Y≈37) lateral prefrontal cortex (antLPFC, postLPFC); orbitofrontal cortex secondary cortical taste area (OFC); and anterior insular cortex primary cortical taste area (antINS). Reprinted from NeuroImage, 59 (2), Tian Ge, Jianfeng Feng, Fabian Grabenhorst, and Edmund T. Rolls, Componential Granger causality, and its application to identifying the source and mechanisms of the top-down biased activation that controls attention to affective vs sensory processing, pp. 1846–58, Copyright (2012), with permission from Elsevier.)

In addition, the terms in componential Granger causality sum to 1, allowing causal effects to be directly compared between systems.

We showed using componential Granger causality analysis applied to an fMRI investigation that there is a top-down attentional effect from the anterior dorsolateral prefrontal cortex to the orbitofrontal cortex when attention is paid to the pleasantness of a taste, and that this effect depends on the activity in the orbitofrontal cortex as shown by the interaction term (Ge, Feng, Grabenhorst and Rolls 2012). Correspondingly there is a top-down attentional effect from the posterior dorsolateral prefrontal cortex to the insular primary taste cortex when attention is paid to the intensity of a taste, and this effect depends on the activity of the insular primary taste cortex as shown by the interaction term. The prefrontal cortex sites are those identified by the PPI analysis (Grabenhorst and Rolls 2010) and the effects are shown schematically in Fig. 6.20. Componential Granger causality thus not only can reveal the directionality of effects between areas (and these can be bidirectional), but also allows the mechanisms to be understood in terms of whether the causal influence of one system on another depends on the state of the system being causally influenced. Componential Granger causality measures the full effects of second order statistics by including variance and covariance effects between each time series, thus allowing interaction effects to be measured, and also provides a systematic framework within which to measure the effects of cross, self, and noise contributions to causality (Ge, Feng, Grabenhorst and Rolls 2012). The findings reveal some of the mechanisms involved in a biased activation theory of selective attention.

6.3.4 Top-down cognitive modulation

To what extent does cognition influence the hedonics of stimuli that produce emotions, and how far down into the sensory system does the cognitive influence reach? Examples of the evidence on this are considered next. This evidence is consistent with what is found for top-down selective attention, but extends it to how even word-level cognitive labels influence sensory and reward-related processing, often by biasing different cortical areas and even processing streams (Rolls 2014a).

6.3.4.1 Effects of cognition on olfactory and taste reward-related processing

To address this, we performed an fMRI investigation in which the delivery of a standard test odor (isovaleric acid combined with cheddar cheese odour, presented orthonasally using an olfactometer) was paired with a descriptor word on a screen, which on different trials was "Cheddar cheese" or "Body odour". Participants rated the affective value of the test odour as significantly more pleasant when labelled "Cheddar Cheese" than when labeled "Body odour", and these effects reflected activations in the medial orbitofrontal cortex (OFC) / rostral anterior cingulate cortex (ACC) that had correlations with the pleasantness ratings (De Araujo, Rolls, Velazco, Margot and Cayeux 2005). The implication is that cognitive factors can have profound effects on our responses to the hedonic properties of affective stimuli, in that these effects are manifest quite far down into sensory processing, in that hedonic representations of odours are affected (De Araujo, Rolls, Velazco, Margot and Cayeux 2005).

Similar cognitive effects and mechanisms have now been found for the taste and flavour of food, where the cognitive word level descriptor was for example 'rich delicious flavour' and activations to flavour were increased in the orbitofrontal cortex and regions to which it projects including the pregenual cingulate cortex and ventral striatum, but were not influenced in the insular primary taste cortex where activations reflected the intensity (concentration) of the stimuli (Grabenhorst, Rolls and Bilderbeck 2008a).

6.3.4.2 Effects of cognition on touch reward-related processing

While there have been many investigations of the neural representations of pain stimuli (Grabenhorst and Rolls 2011, Shackman, Salomons, Slagter, Fox, Winter and Davidson 2011, Kobayashi 2012), there have been fewer investigations of the representation of pleasant touch in the brain. In one study, the cortical areas that represent affectively positive and negative aspects of touch were investigated using functional magnetic resonance imaging (fMRI) by comparing activations produced by pleasant touch, painful touch produced by a stylus, and neutral touch, to the left hand (Rolls, O'Doherty, Kringelbach, Francis, Bowtell and McGlone 2003d). It was found that regions of the orbitofrontal cortex were activated more by pleasant touch and by painful stimuli than by neutral touch, and that different areas of the orbitofrontal cortex were activated by the pleasant and painful touches. The orbitofrontal cortex activation was related to the affective aspects of the touch, in that the somatosensory cortex (S1) was less activated by the pleasant and painful stimuli than by the neutral stimuli (as shown by a two-way analysis of variance performed on the percentage change of the BOLD signals under the different stimulation conditions in the different areas). Further, it was found that a rostral part of the anterior cingulate cortex was activated by the pleasant stimulus and that a more posterior and dorsal part was activated by the painful stimulus (and this is consistent with effects in other sensory modalities) (Grabenhorst and Rolls 2011, Rolls 2014a). Regions of the somatosensory cortex, including S1, and part of S2 in the superior temporal plane at the mid-insula level, were activated more by the neutral touch than by the pleasant and painful stimuli. Part of the posterior insula was activated only in the pain condition, and different parts of the brainstem, including the central grey, were activated in the pain, pleasant and neutral touch conditions. The results provide evidence that different areas of the human orbitofrontal cortex are involved in representing both pleasant touch and pain, and that dissociable parts of the cingulate cortex are involved in representing pleasant touch and pain (Rolls, O'Doherty, Kringelbach, Francis, Bowtell and McGlone 2003d).

Warm and cold stimuli have affective components such as feeling pleasant or unpleasant, and these components may have survival value, for approach to warmth and avoidance of cold may be reinforcers or goals for action built into us during evolution to direct our behaviour to stimuli that are appropriate for survival (Rolls 2014a). Understanding the brain processing that underlies these prototypical reinforcers provides a direct approach to understanding the brain mechanisms of emotion. In an fMRI investigation in humans, we showed that the mid-orbitofrontal and pregenual cingulate cortex and the ventral striatum have activations that are correlated with the subjective pleasantness ratings made to warm (41°C) and cold (12°C) stimuli, and combinations of warm and cold stimuli, applied to the hand (Rolls, Grabenhorst and Parris 2008a). Activations in the lateral and some more anterior parts of the orbitofrontal cortex were correlated with the unpleasantness of the stimuli. In contrast, activations in the somatosensory cortex and ventral posterior insula were correlated with the intensity but not the pleasantness of the thermal stimuli (Rolls, Grabenhorst and Parris 2008a).

A principle thus appears to be that processing related to the affective value and associated subjective emotional experience of somatosensory and thermal stimuli that are important for survival is performed in different brain areas to those where activations are related to sensory properties of the stimuli such as their intensity. This conclusion appears to be the case for processing in a number of sensory modalities, and the finding with such prototypical stimuli as pleasant and painful touch, and warm (pleasant) and cold (unpleasant) thermal stimuli, provides strong support for this principle (Rolls 2014a, Grabenhorst and Rolls 2008, Grabenhorst, Rolls and Bilderbeck 2008a, Grabenhorst and Rolls 2011). An implication of the principle is that by having a system specialized for the affective or reward aspects of stimuli it is possible to modify goal oriented behaviour, and to do this independently of being able to know what the stimulus is (its intensity, physical characteristics etc). Thus even if a stimulus has lost its pleasantness because of for example a change of core body temperature, it is still possible to represent the stimulus, recognize it, and learn about where it is in the environment for future use (Rolls 2014a). This is a fundamental aspect of brain design (Rolls 2008d, Rolls 2014a).

There have been many studies of the top-down attentional modulation of touch, with effects typically larger in secondary somatosensory and association cortical areas (e.g. parietal area 7), and smaller in S1 (Johansen-Berg and Lloyd 2000, Rolls 2010a). However, there has been little investigation of where high-level cognition influences the representation of affective touch in the brain.

To investigate where cognitive influences from the very high level of language might influence the affective representation of touch, we performed a fMRI study in which the forearm was rubbed with a cream, but this could be accompanied by a word label that indicated that it was a rich moisturizing cream (pleasant to most people) vs a basic cream (McCabe, Rolls, Bilderbeck and McGlone 2008).

We found that cognitive modulation by a label at the word level indicating pleasantness / richness ("rich moisturising cream" vs "basic cream") influenced the representation of tactile inputs in the orbitofrontal cortex (McCabe, Rolls, Bilderbeck and McGlone 2008). (The cream was identical in all conditions in the study: it was only the word labels that were changed. The cream was rubbed onto the ventral surface of the forearm.) For example, a negative correlation with the pleasantness ratings of the touch as influenced by the word labels was found in the lateral orbitofrontal cortex, a region shown in other studies to be activated by less pleasant stimuli including unpleasant odors, and losing money (O'Doherty, Kringelbach, Rolls, Hornak and Andrews 2001, Rolls, Kringelbach and De Araujo 2003c, Rolls, O'Doherty, Kringelbach, Francis, Bowtell and McGlone 2003d). A positive correlation with the pleasantness of touch as influenced by the word labels was found in the pregenual cingulate cortex (McCabe, Rolls, Bilderbeck and McGlone 2008). Convergent evidence on the functions of this region is that the pregenual cingulate region is close to where in different studies another somatosensory stimulus, oral texture, is represented (De Araujo and Rolls 2004), correlations with pleasantness ratings are found to food and olfactory stimuli (Kringelbach, O'Doherty, Rolls and Andrews 2003, De Araujo, Rolls, Velazco, Margot and Cayeux 2005, Grabenhorst and Rolls 2011), and pleasant touch produces activation (Rolls, O'Doherty, Kringelbach, Francis, Bowtell and McGlone 2003d). We also found that activations to touch in the parietal cortex area 7 were influenced by the word labels, in that there was more activation when the rich label than when the thin label was present (McCabe, Rolls, Bilderbeck and McGlone 2008).

Cognitive modulation of effects produced by the sight of touch were investigated by a comparison of the effects of the sight of the arm being rubbed when accompanied by the label "Rich moisturising cream" vs "Basic cream". Cognitive modulation effects were found in the pregenual cingulate cortex extending into the orbitofrontal cortex, in regions close to those where activations were correlated with the pleasantness ratings with the same two stimulus conditions. The effect of the cognitive label "Rich moisturising cream" was to make the sight of the touch more pleasant by increasing activations in these pregenual cingulate and orbitofrontal cortex areas (McCabe, Rolls, Bilderbeck and McGlone 2008).

Thus cognitive modulation of taste, olfactory, flavor and touch processing occurs for even word-level cognitive labels, which have top-down influences on both sensory and reward-related processing, often by biasing different cortical areas and even processing streams (Rolls 2014a). The evidence is consistent with what is found for top-down selective attention, and the mechanisms, described next, are likely to be very similar for both top-down cognitive modulation of processing and top-down selective attention (Rolls 2013a).

6.3.5 A top-down biased activation model of attentional and cognitive modulation

The way that we think of top-down biased competition as operating normally in for example visual selective attention (Desimone and Duncan 1995) is that within an area, e.g. a cortical region, some neurons receive a weak top-down input that increases their response to the bottom-up stimuli, potentially supralinearly if the bottom-up stimuli are weak (Rolls and Deco 2002, Deco and Rolls 2005b, Rolls 2008d). The enhanced firing of the biased neurons then, via the local inhibitory neurons, inhibits the other neurons in the local area from responding to the bottom-up stimuli. This is a local mechanism, in that the inhibition in the neocortex is primarily local, being implemented by cortical inhibitory neurons that typically have inputs and outputs over no more than a few mm (Chapter 1, Douglas et al. (2004)). This model of biased competition is illustrated in Fig. 6.2.

This locally implemented biased competition situation may not apply in the present case, where we have facilitation of processing in a whole cortical area (e.g. orbitofrontal cortex, or pregenual cingulate cortex) or even cortical processing stream (e.g. the linked orbitofrontal and pregenual cingulate cortex) in which any taste neurons may reflect pleasantness and not intensity. So the attentional effect might more accurately be described in this case as biased activation, without local competition being part of the effect. This biased activation theory and model of attention, illustrated in Fig. 6.21 (Rolls 2013a), is a rather different way to implement attention in the brain than biased competition, and each mechanism may apply in different cases, or both mechanisms in some cases.

The biased activation theory of top-down attentional and cognitive control is as follows (Rolls 2013a), and is illustrated in Fig. 6.21. There are short-term memory systems implemented as cortical attractor networks with recurrent collateral connections to maintain neuronal activity that provide the source of the top-down activation (Chapter 4). The short-term memory systems may be separate (as shown in Fig. 6.21), or could be a single network with different attractor states for the different selective attention conditions. The top-down short-term memory systems hold what is being paid attention to active by continuing firing in

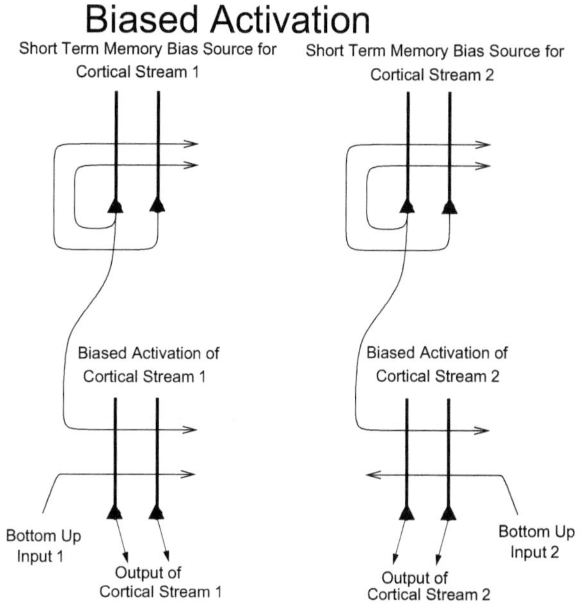

Biased Activation

Fig. 6.21 Biased activation mechanism for top-down attention. The short-term memory systems that provide the source of the top-down activations may be separate (as shown), or could be a single network with different attractor states for the different selective attention conditions. The top-down short-term memory systems hold what is being paid attention to active by continuing firing in an attractor state, and bias separately either cortical processing system 1, or cortical processing system 2. This weak top-down bias interacts with the bottom up input to the cortical stream and produces an increase of activity that can be supralinear (Deco and Rolls 2005c). Thus the selective activation of separate cortical processing streams can occur. In the example, stream 1 might process the affective value of a stimulus, and stream 2 might process the intensity and physical properties of the stimulus. The outputs of these separate processing streams then must enter a competition system, which could be for example a cortical attractor decision-making network that makes choices between the two streams, with the choice biased by the activations in the separate streams (see text). The thick vertical lines above the pyramidal cells are the dendrites. The axons are shown with thin lines and the excitatory connections by arrow heads. (After Rolls 2013a.)

an attractor state, and bias separately either cortical processing system 1, or cortical processing system 2. This weak top-down bias interacts with the bottom-up input to the cortical stream and produces an increase of activity that can be supralinear (Deco and Rolls 2005b). Thus the selective activation of separate cortical processing streams can occur. In the example, stream 1 might process the affective value of a stimulus, and stream 2 might process the intensity and physical properties of the stimulus.

The top-down bias needs to be weak relative to the bottom-up input, for the top-down bias must not dominate the system, otherwise bottom-up inputs, essential for perception and survival, would be over-ridden. Under such conditions, top-down attentional and cognitive effects will be largest when the bottom-up inputs are not too strong or are ambiguous, and that has been shown to be the case in realistic simulations with integrate-and-fire neurons (Deco and Rolls 2005b, Rolls 2008d). The weakness of the top-down biasing input is included as a part of brain design, for the top-down inputs are effectively backprojections from higher cortical areas, and these end on the apical dendrites of cortical pyramidal cells, and so have weaker effects than the bottom up inputs, which make connections lower down the dendrite towards the cell body. I suggest here that the correct connections could be set up in such a system by the following associative (Hebbian) synaptic learning process. The top-down backprojection synapses would increase in strength when there is activity in a population of short-term memory neurons that by their firing hold attention in one direction (e.g. the short-term memory system for cortical stream 1 shown in Fig. 6a), and simultaneously there is activity in the neurons that receive the top-down inputs (e.g. in cortical stream 1 shown in Fig. 6.21).

The outputs of the separate processing streams showing biased activation (Fig. 6.21) may need to be compared at a later stage of processing, in order to lead to a single behaviour (Rolls 2013a). One way in which this comparison could take place is by both outputs entering a single network cortical attractor model of decision-making, in which positive feedback implemented by the excitatory recurrent collateral connections leads through non-linear dynamics to a single winner, which is ensured by competition between the different possible attractor states produced through inhibitory neurons (Wang 2002, Deco and Rolls 2006, Deco, Rolls, Albantakis and Romo 2013) (Chapter 5). A second way in which the competition could be implemented is by that usually conceptualized as important in biased competition (Desimone and Duncan 1995, Rolls and Deco 2002, Deco and Rolls 2005b), in which a feedforward competitive network using inhibition through local inhibitory neurons provides a way for a weak top-down signal to bias the output especially if the bottom-up inputs are weak, and this implementation is what is shown at the bottom of Fig. 6.2. A third way in which the biased activation reflected in the output of the streams shown in Fig. 6.21 could be taken into account is by a mechanism such as that in the basal ganglia, where in the striatum the different excitatory inputs activate GABA (gamma-amino-butyric acid) neurons, which then directly inhibit each other to make the selection (Chapter 20, Rolls (2014a)).

The difference between biased competition and biased activation may be especially important in the context of functional neuroimaging, for biased activation, in which processing in whole cortical areas is facilitated by selective attention, can be revealed by functional neuroimaging, which operates at relatively low spatial resolution, in the order of mm. In contrast, biased competition may selectively facilitate some pyramidal neurons within a local cortical area which then through the local GABA inhibitory neurons compete with the other pyramidal neurons in the area receiving bottom-up input. In this situation, in which some but not other neurons within a cortical area are showing enhanced firing, functional neuroimaging may not be able to show which local population of pyramidal cells is winning the competition due to the top-down bias. The evidence presented by Grabenhorst and Rolls (2010) is that not only the processing streams, but also even the short-term memory systems in the prefrontal cortex that provide the top-down source of the biased activation, are physically separate, as illustrated in Fig. 6.20.

A possibility arising from this model is that some competition may occur somewhere in the attentional system before the output stage, and one possible area is within the prefrontal cortex, where it is a possibility that the attractors that implement the short-term memory for attention to pleasantness (at $Y\approx50$) may inhibit the attractors that implement the short-term memory for attention to intensity (at $Y\approx37$), which could occur if there is some physical overlap between their zones of activation, even if the peaks are well separated. Some evidence for this possibility was found (Grabenhorst and Rolls 2010), in that the correlation between the % BOLD activations in these two prefrontal cortex regions was $r=-0.72$ ($p=0.0034$) on the pleasantness trials; and $r=-0.8$ ($p<0.001$) on the intensity trials. In a biased competition model (Fig. 6.2 we would normally think of the short-term memory attractors that provide the source of the bias as being within the same single attractor network, so that there would be competition between the two attractor states through the local inhibitory interneurons. In the biased activation model (Fig. 6.21), it is an open issue about whether the attractors that provide the source of the top-down bias are in the same single network, or are physically separate making interactions between the attractor states difficult through the short-range cortical inhibitory neurons. The findings just described indicate that in the case of top-down control of affective vs intensity processing of taste stimuli, although the two attractors are somewhat apart in the prefrontal cortex, there is some functional inhibitory interaction between them.

The principle of biased activation providing a mechanism for selective attention (Rolls 2013a) probably extends beyond processing in the affective vs sensory coding cortical streams. It may provide the mechanism also for effects in for example the dorsal vs the ventral visual system, in which attention to the motion of a moving object may enhance processing in the dorsal stream, and attention to the identity of the moving object may enhance processing in the ventral visual stream (Brown 2009). Similar biased activation may contribute to the different localization in the prefrontal cortex of systems involved in 'what' vs 'where' working memory (Deco, Rolls and Horwitz 2004, Rottschy, Langner, Dogan, Reetz, Laird, Schulz, Fox and Eickhoff 2012). Biased activation as a mechanism for top-down selective attention may be widespread in the brain, and may be engaged when there is segregated processing of different attributes of stimuli (Grabenhorst and Rolls 2010, Rolls 2013a).

6.4 Conclusions

The model of attention described in this chapter (see also Rolls and Deco (2002), Rolls (2008d), and Rolls (2013a)) represents an advance beyond the biased competition hypothesis, in that it shows how object and spatial attention can be produced by dynamic interactions between the 'what' and 'where' streams, and in that as a computational model that has been simulated, the details of the model have been made fully explicit and have been defined quantitatively. An interesting and important feature of the model is that the model does not use explicit multiplication as a computational method, but the modulation of attention (for example the effects of posterior parietal module (PP) activity on V1) appears to be like multiplication. This is an interesting contribution of the model, namely that multiplicative-like attentional gains are implemented without any explicit multiplicative operation (see Sections 6.2.9 and B.8.5).

In this chapter the neuronal ('microscopic-level') neurodynamical mechanisms that underlie visual attention are considered. A computational model of cortical systems based on the 'biased competition' hypothesis is described. The model consists of interconnected populations of cortical neurons distributed in different brain modules, which are related to the different areas of the dorsal or 'where' and ventral or 'what' processing pathways of the primate visual cortex. The 'where' pathway incorporates mutual connections between a feature extracting module (V1–V4), and a parietal module (PP) that consists of pools coding the locations of the stimuli. The 'what' path incorporates mutual connections between the feature extracting module (V1–V4) and an inferior temporal cortex module (IT) with pools of neurons coding for specific objects. External attentional top-down bias is defined as inputs coming from higher prefrontal cortex modules that hold what is to be paid attention to active in short-term memory. These short-term memory processes are not explicitly modelled in this chapter but are modelled in Chapter 4. Intermodular attentional biasing is modelled through the coupling between pools of different modules, which are explicitly modelled. Attention appears now as an emergent effect that supports the dynamical evolution to a state where the constraints given by the stimulus and the external bias are satisfied. Visual search and attention can be explained in this theoretical framework of a biased competitive neurodynamics. The top–down bias guides attention to concentrate at a given spatial location or on given features. The neural population dynamics are handled analytically in the framework of the mean-field approximation. Consequently, the whole process can be expressed as a system of coupled differential equations. The model has been extended to include the high resolution buffer hypothesis, and a 'microscopic' physical (i.e. neuron-level) implementation of the global precedence effect (see Rolls and Deco (2002)). We have also analyzed the attentional neuro-dynamics involved in visual search of hierarchical patterns, and also modelled a mechanism for feature binding that can account for conjunction visual search tasks (see Rolls and Deco (2002)).

The fundamental contributions of this model of attention (Rolls and Deco 2002, Deco and Lee 2004, Deco and Zihl 2004) are as follows:

1. Different functions involved in active visual perception have been integrated by a model based on the biased competition hypothesis. Attentional top-down bias guides the dynamics to concentrate at a given spatial location or on given (object) features. The model integrates, in a unifying form, the explanation of several existing types of experimental data obtained at different levels of investigation. At the microscopic neuronal level, we simulated single cell recordings, at the mesoscopic level of cortical areas we reproduced the results of fMRI (functional magnetic resonance imaging) studies, and at the macroscopic perceptual level we accounted for psychophysical performance. Specific predictions at different levels of investigation have also been made. These predictions inspired single cell, fMRI, and psychophysical experiments, that in part have been already performed and the results of which are consistent with the theory (see Rolls and Deco (2002)).

2. Attention is a dynamical emergent property in our system, rather than a separate mechanism operating independently of other perceptual and cognitive processes.

3. The computational perspective provides not only a concrete mathematical description of mechanisms involved in brain function, but also a model that allows complete simulation and prediction of neuropsychological experiments. Interference with the operation of some of the modules was used to predict impairment in visual information selection in patients suffering from brain injury. The resulting experiments support our understanding of the functional impairments resulting from localized brain damage in patients.

As discussed in Chapter 25, in complex natural scenes visual search may take place largely overtly, by eye movements, which are serial in nature. However, mechanisms for covert visual search are described in this chapter, and although perhaps contributing to performance more in simple visual displays with two or a few objects present, may contribute to performance in complex natural scenes by influencing the next eye movement that will be made.

We demonstrated that it is possible to build a neural system for visual search, which works across the visual field in parallel but, due to the different latencies of its dynamics, can show the two experimentally observed modes of visual attention, namely: serial focal attention, and the parallel spread of attention over space. Neither explicit serial focal search nor saliency maps need to be assumed. In particular, we have shown that many processes in attention formerly thought to operate serially, in fact operate in parallel, but take longer to converge when the task is difficult.

The model described in this chapter shows that no mysterious controller of attention needs to be found, but that instead the control is performed by the information loaded into prefrontal cortex short-term memories biasing earlier visual cortical spatial and object processing areas by backprojections. The short-term memories are themselves loaded by presentation of the sample cue, in object or spatial working memory tasks such as delayed match to sample with intervening stimuli, as described in Chapter 4. The short-term memories are loaded in visual search tasks with the object or location that is the subject of the search, as described in Chapter 4 and in this chapter.

Other parts of the brain in addition to the prefrontal cortex might provide the top-down bias to the parietal spatial or the IT object modules. We do not wish to exclude these. One example is the auditory–verbal short-term memory system in humans (which using rehearsal holds on-line a set of approximately 7 chunks of information), and which may be located in the cortex at the junction of the left parieto-occipito-temporal areas. The principle though is the same, that there is no mysterious controller of attention, and that what is needed is a short-term memory system to hold the subject of attention active, and which provides top-down bias to the high-level spatial or object representation areas such as PP or IT as well potentially to other areas with different perceptual or related representations.

This I believe is an important conceptual point, in that it removes the concern that there is some non-understood aspect of the control of attention, with a type of 'deus ex machina', or at least an unlocated (serial or parallel) 'spotlight controller', being needed. Indeed, the overall schematic architecture of the system described in this chapter is illustrated in Fig. 6.6. The architecture allows the target of attention to be analyzed in the spatial or object processing stream, then loaded into a prefrontal cortex (or other) short-term memory system, from which it can exert its top-down biasing effect on the spatial or object stream, which in turn by interactive feedforward and feedback effects causes the whole system to settle to optimally satisfy the constraints. The constraint satisfaction we describe is not itself a mysterious process either, but can be understood as an energy minimization process now well understood in neural networks (Hopfield 1982, Amit 1989, Hertz, Krogh and Palmer 1991). This constraint satisfaction generally operates well in practice even when the conditions required for the formal analysis are not present, for example when the system does not have complete and reciprocal connectivity due to random asymmetric dilution of the connectivity, i.e. when synapses are missing at random (Treves 1991b, Rolls and Treves 1998). For the memory retrieval properties of such systems to operate well, the number of synapses per neuron must be kept relatively high, above 1,000–2,000 (Rolls, Treves, Foster and Perez-Vicente 1997c), an important condition which significantly is well met by the actual numbers of synapses per neuron in the cortex. We note also that the large number of forward and backward connections between adjacent cortical areas in the architecture shown in Fig. 6.6 provides a suitable basis, given also some type of associative synaptic connection rule between the connected areas, for a system that can operate in the interactive constraint satisfaction way described in Sections 11, 4.10, and this chapter. I would also expect top-down attentional activations to have similar effects in emotional and other systems, as described and modelled by Rolls and Stringer (2001b).

Indeed, another fundamental principle of cortical computation described in this Chapter is that selective attention, and top-down effects of cognition on sensory and reward representations, often involve whole cortical areas or even connected streams, and the new biased activation mechanism of selective attention (Rolls 2013a) (Section 6.3) describes this new theory and model.

I believe that the conceptual framework for understanding attention described in this book may be useful in helping to understand the otherwise rather complicated pictures that are often produced in neuroimaging studies of attention in humans, in which large swathes of parietal and frontal cortical territory often show activation. We now have clear reasons for expecting frontal, parietal, temporal, and even occipital lobe contributions to attention, given the architecture shown in Fig. 6.6 and the model described in this Chapter. The model, and the specialization of function within the parietal cortex described in Chapter 4 of Rolls and Deco (2002), lead us to understand that different parietal and even connected frontal areas may be activated during different types of spatial attention and memory, for example when attention must be paid to where a response is to be made with the arm or with the eyes, or when there are spatial cues in both visual fields, one of which is a target and others of which are distracters. We would also expect different temporal cortical areas to become activated while paying attention depending on whether the attention is to face identity, face expression, objects, objects undergoing motion, colour, etc. (see Chapter 25). Given the tendency of neurons to cluster into small regions where similar neurons are found (due to the self-organizing map principles described in Section B.4.6), we would even expect the exact loci of activation found in the temporal areas to be somewhat different for different classes of object, and to be in not necessarily the same relative positions in different humans. We would also expect some activation during attentional tasks to be found quite early on in cortical visual processing, perhaps as far back as V1, and have given reasons in this chapter why this might though weak still be a useful feature of the attentional architecture.

I also note that memory operations implemented by recurrent collaterals (Chapter 4) are at the heart of how attentional processes operate in the brain. A short-term memory is needed to hold the subject of the attention active. This then biases competition in networks that themselves are likely to have recurrent collateral connections that make them operate as attractors, so that the biased competition is frequently biasing a short-term memory attractor. Examples of this process are described in Section 5.6. Finally, the top-down processes are mirrored by bottom-up processes, so that the top-down / bottom-up interactions themselves provide what is potentially a short-term memory system. Indeed, the operation of attention is seen to be a dynamical process of interactions between all these short-term memory systems.

Thus the fundamental understanding offered by this conceptualization of the operation of attentional processes in the brain may provide a fundamental basis for understanding the phenomena that arise in imaging studies, but also of course in neurophysiological, psychophysical, and neuropsychological studies.

6.5   Highlights

One mechanism of attention is bottom-up, and uses saliency to direct overt attentional mechanisms such as eye movements.

A second mechanism of attention is top-down, in which the attentional effects can be covert, for example without eye or head movements.

The architectures in Figs. 6.2 (showing a biased competition architecture for attention) and 6.21 (showing a biased activation architecture for attention) illustrate some key principles of operation of the cortex, used here for top-down attention.

1. One principle is the function of cortical recurrent collateral connections to implement short-term memory attractor states to hold the object of attention active and on-line (Chapter 4).

2. A second principle is the use of competitive networks to select winning populations of neurons that provide a useful output to the next stage of processing (Sections 7.4 and B.4).

3. A third principle is allocation of cortical resources to respond to current processing needs efficiently, for example by using selective attention to select form vs location as currently relevant to make this processing faster and more efficient, and to minimize distraction by currently irrelevant information. This enables an individual with the benefit of short-term memory systems in the prefrontal cortex to focus on a task, to devote cortical resources efficiently to that task, and to not be too distractible by other stimuli. These operations are key to **executive function**, a key function of the

prefrontal cortex which describes selecting a task, and focussing on that task without being continually distracted. In humans, this may include multi-step syntactic plans, as described in Chapter 22 and Section 5.6.10.

4. A fourth principle is that whole cortical areas and streams may be biased by top-down attention. Examples of this include the biasing up of activity in the reward processing stream including the orbitofrontal cortex and anterior cingulate cortex when attention is to reward-value and pleasantness; and of other processing streams such as insular cortex regions for taste when attention is to the sensory attributes of stimuli, such as identity and intensity. A new **biased activation theory of attention** provides a model of this (Fig. 6.21).

5. A fifth principle is that cognition even from the highest, word, level can have top-down effects that bias processing in areas such as the orbitofrontal cortex where reward value and subjective pleasantness are first reflected explicitly in the representation. (By made explicit in the representation is meant that the stimulus or event can be decoded easily from the firing rates using for example dot product decoding, as described in Appendix C.)

6. A sixth principle is that there is in addition a system for bottom-up attention whereby salient sensory stimuli elicit orienting, for example eye movements and/or turning of the head. Part of this mechanism is implemented in the dorsal cortical visual stream (Bisley and Goldberg 2010), which operates in conjunction with collicular (midbrain) mechanisms (Krauzlis, Lovejoy and Zenon 2013).

This Chapter on attention illustrates how modern computational models of interactions between cortical networks offer precise explanations of how the cerebral cortex operates to implement many complex computations that are very important in cognitive function.

7 Diluted connectivity

7.1 Introduction

A key feature of the architecture of the neocortex and the CA3 region of the hippocampus is that there are many excitatory interconnections between the pyramidal cells that are associatively modifiable. This prototypical architecture provides the basis for cortical attractor networks which enable memories to be stored and recalled from a fragment (Rolls 2008d, Rolls 2010b); for continuing neuronal firing to implement short-term memory (Rolls 2008d) and thereby to provide for top-down attention (Rolls 2008d) by providing the source of the biased competition (Desimone and Duncan 1995, Deco and Rolls 2005b) or biased activation (Grabenhorst and Rolls 2010, Rolls 2013a); and for decision-making when there is competition between two inputs to produce a high firing rate attractor state which represents the decision (Wang 2008, Rolls 2008d, Rolls and Deco 2010, Deco, Rolls, Albantakis and Romo 2013). However, a key property of this connectivity is that it is diluted, that is the probability that any one hippocampal CA3 neuron will contact another is approximately 0.04 (Rolls 1989b, Treves and Rolls 1992b) (see Chapter 24), and the probability that a neocortical pyramidal cell will contact another nearby neocortical pyramidal cell is in the order of 0.1 (Rolls 2008d, Perin, Berger and Markram 2011, Perin, Telefont and Markram 2013). In this Chapter, I address the fundamental issue in brain design of why cortical connectivity, especially in the recurrent collaterals and in the feedforward and feedback connections between adjacent areas in a cortical hierarchy, is diluted.

In both the hippocampal cortex and neocortex, there is evidence that the excitatory recurrent collateral connections between pyramidal cells are associatively modifiable, and that the system supports attractor dynamics that enable memories to be stored (Rolls 2008d). In the hippocampal cortex, the memory stored may be episodic, about a particular recent event or episode (Rolls 2008d, Rolls 2010b, Rolls and Kesner 2006) (such as where one parked one's bicycle today, or where one was for dinner yesterday, with whom, and what ideas were discussed). In the neocortex, the memory might be a long-term semantic memory (Rolls 2008d, McClelland, McNaughton and O'Reilly 1995) (for example the classification of plant and animal species). The prototypical cortical neuronal network for the storage of these memories is an autoassociation or attractor network in which the recurrent collateral connections make associatively modifiable synapses onto neurons in the same network (Rolls 2008d, Kohonen, Oja and Lehtio 1981, Hopfield 1982, Amit 1989, Hertz et al. 1991, Rolls and Treves 1998). A memory in such a system is implemented by the set of the neuronal firing rates across the whole population when the memory is stored. The particular firing rate vector comprised of the firing chosen at random of all the neurons is one of the memory patterns that is stored and must later be retrieved. During retrieval, presentation of even a fragment of the memory can produce recall of the whole memory (Rolls 2008d) (Section B.3). If a sparse distributed representation is used (in which a small proportion of the neurons chosen at random fires for each memory pattern), then the number of different memories (or memory patterns) that can be stored and correctly retrieved is in the order of the number of recurrent collateral synapses onto each neuron, that is in the order of 10,000 (Treves and Rolls 1991, Treves 1991b, Rolls and Treves 1998, Rolls 2008d) (Section B.3). The accuracy of the retrieval of each memory pattern can be measured by the correlation between the retrieved firing rate pattern (i.e. vector of firing rates) and the stored memory pattern.

A difference between the hippocampal cortex and the neocortex is that the CA3 network is a single attractor network allowing any representation to be associated with any other representation, providing an implementation of episodic memory (Rolls 2008d, Rolls 2010b, Kesner and Rolls 2015). In contrast, the neocortex has local connectivity with a radius of approximately 2 mm, and this enables the whole of the cerebral cortex to have many separate attractor networks, each storing a large number of memories (O'Kane and Treves 1992, Rolls 2008d) (Sections 4.5 and 26.5.18).

The hypothesis proposed (Rolls 2012a) is that this diluted connectivity in the cortex allows biological processes that set up synaptic connections between neurons to arrange for there to be only very rarely more than one synaptic connection between any two neurons. If probabilistically there was more than one connection between any two neurons, it is shown here that such multiple connections between a proportion of the neurons would dominate the attractor states into which the network could enter and be stable, thus severely reducing the memory capacity of the network below the normal limit for diluted connectivity, which is approximately that of a fully connected network with the same number of recurrent collateral connections onto any neuron (Treves 1991b, Treves and Rolls 1991, Bovier and Gayrard 1992, Rolls, Treves, Foster and Perez-Vicente 1997c, Rolls and Treves 1998). The implication is that a major advantage of the diluted connectivity in the cortex is that it makes it likely that there is rarely more than one synapse between any pair of neurons, and the computational advantage of this is that the memory capacity is high, that is, depends to the first order on the number of recurrent collateral connections C per neuron, and is not greatly reduced below that by the presence of many instances of multiple connections between pairs of neurons. This provides a theory for this fundamental aspect of the design of the neocortex and the hippocampal cortex (Section 7.2). Other advantages of diluted connectivity in this and in pattern association (Section 7.3) and competitive networks (Section 7.4) are presented below.

7.2 Diluted connectivity and the storage capacity of attractor networks

Rolls (2012a) investigated how diluted connectivity can increase the storage capacity of attractor networks, by reducing the probability that there will be more than one synaptic connection between any pair of neurons, which would distort the landscape of the basins of attraction.

7.2.1 The autoassociative or attractor network architecture being studied

The architecture and functional properties of autoassociative or attractor networks are described in detail by Hertz et al. (1991), by Amit (1989), and in Section B.3. Here it is assumed that the memory patterns are stored in the autoassociative network by an associative (or Hebbian) learning process in an architecture of the type shown in Fig. B.13 as follows. The firing of every output neuron i is forced to a value y_i determined by the external input e_i. Then a Hebb-like associative local learning rule is applied to the recurrent synapses in the network:

$$\delta w_{ij} = \alpha y_i y_j. \tag{7.1}$$

where α is a learning rate constant, and y_j is the presynaptic firing rate.

During recall, the external input e_i is applied, and produces output firing, operating through the non-linear activation function described below. The firing is fed back by the recurrent collateral axons shown in Fig. B.13 to produce activation of each output neuron through the modified synapses on each output neuron. The activation h_i produced by the recurrent collateral effect on the ith neuron is the sum of the activations produced in proportion to the firing rate of each axon y_j operating through each modified synapse w_{ij}, that is,

$$h_i = \sum_j^C y_j w_{ij} \tag{7.2}$$

where $\sum_j^C$ indicates that the sum is over the C input axons to each neuron, indexed by j.

The output firing y_i is a function of the activation produced by the recurrent collateral effect (internal recall) and by the external input (e_i):

$$y_i = \mathrm{f}(h_i + e_i). \tag{7.3}$$

The activation function should be non-linear, and may be for example binary threshold, linear threshold, sigmoid, etc. (Hopfield 1982, Hertz et al. 1991, Rolls 2008d) (Fig. 1.4 and Section B.3).

7.2.2 The storage capacity of attractor networks with diluted connectivity

With non-linear neurons used in the network, the capacity can be measured in terms of the number of input memory patterns **y** (each a firing rate vector comprised by the firing rate of each neuron forming a vector of firing rates across the population of neurons) produced by the external input **e**, see Fig. B.13), that can be stored in the network and recalled later, even from a fragment of the stored memory pattern, whenever the network settles within each stored pattern's basin of attraction. The accuracy of the recall of each memory pattern can be measured by the correlation between the recalled firing rate vector and the stored firing rate vector. The first quantitative analysis of storage capacity, measured by the number of memory patterns that can be stored and later recalled correctly, considered a fully connected Hopfield (1982) autoassociator model, in which neurons are binary elements with an equal probability of being 'on' or 'off' in each pattern, and the number C of inputs per neuron is the same as the number N of output units (Amit, Gutfreund and Sompolinsky 1987). (Actually it is equal to $N - 1$, since a neuron is taken not to connect to itself.) Learning is taken to occur by clamping the desired patterns on the network and using a modified Hebb rule, in which the mean of the presynaptic and postsynaptic firings is subtracted from the firing on any one learning trial (this amounts to a covariance learning rule, and is described more fully in Appendix A4 of Rolls and Treves (1998), and is shown in equation 7.6). With such fully distributed binary random patterns, the number of patterns that can be learned is (for C large) $p \approx 0.14C = 0.14N$, hence well below what could be achieved with orthogonal patterns or with an 'orthogonalizing' synaptic matrix (Hopfield 1982, Amit et al. 1987). Many variations of this 'standard' autoassociator model have been analyzed subsequently (Amit 1989, Hertz et al. 1991, Rolls and Treves 1998).

This analysis has been extended to autoassociation networks that are much more biologically relevant in the following ways (Treves 1990, Treves 1991b, Treves and Rolls 1991, Rolls, Treves, Foster and Perez-Vicente 1997c, Rolls and Treves 1998, Rolls 2008d). First, some or many connections between the recurrent collaterals and the dendrites are missing (this is referred to as diluted connectivity, and results in a non-symmetric synaptic connection matrix in which w_{ij} does not equal w_{ji}, one of the original assumptions made in order to introduce the energy formalism in the Hopfield (1982) model). Second, the neurons need not be restricted to binary threshold neurons, but can have a threshold linear activation function (see Fig. 1.4). This enables the neurons to assume real continuously graded firing rates to different stimuli (Treves 1990, Treves and Rolls 1991), which are what is found in the brain (Rolls and Tovee 1995b, Treves, Panzeri, Rolls, Booth and Wakeman 1999, Rolls 2008d, Rolls and Treves 2011). Third, the representation need not be fully distributed (with half the neurons 'on', and half 'off'), but instead can have a small proportion of the neurons firing above the spontaneous rate (Treves and Rolls 1991), which is what is found in parts of the brain such as the hippocampus that are involved in memory (Rolls 2008d) (Appendix C). Such a representation is defined as being sparse, and the sparseness a of the representation can be measured, by extending the binary notion of the proportion of neurons that are firing, as

$$ a = \frac{(\sum\limits_{i=1}^{N} y_i/N)^2}{\sum\limits_{i=1}^{N} y_i^2/N} \tag{7.4} $$

where y_i is the firing rate of the ith neuron in the set of N neurons. Treves and Rolls (1991) have shown that such a network does operate efficiently as an autoassociative network, and can store (and recall correctly) a number of different patterns p as follows

$$ p \approx \frac{C^{RC}}{a \ln(\frac{1}{a})} k \tag{7.5} $$

where C^{RC} is the number of synapses on the dendrites of each neuron devoted to the recurrent collaterals from other neurons in the network, and k is a factor that depends weakly on the detailed structure of the rate distribution, on the connectivity pattern, etc., but is roughly in the order of 0.2–0.3.

The main factors that determine the maximum number of memories that can be stored in an autoassociative network are thus the number of connections on each neuron devoted to the recurrent collaterals, and the sparseness of the representation (Treves 1991b, Treves and Rolls 1991, Rolls and Treves 1998, Rolls 2008d) (Section B.3). For example, for $C^{RC} = 12,000$ and $a = 0.02$, p is calculated to be approximately $36,000$. This storage capacity can be realized, with little interference between patterns, if the learning rule includes some form of heterosynaptic long-term depression that counterbalances the effects of associative long-term potentiation (Treves and Rolls (1991); see Appendix A4 of Rolls and Treves (1998)). It should be noted that the number of neurons N (which is greater than C^{RC}, the number of recurrent collateral inputs received by any neuron in the network from the other neurons in the network) is not a parameter that influences the number of different memories that can be stored in the network. The implication of this is that increasing the number of neurons (without increasing the number of connections per neuron) does not increase the number of different patterns that can be stored (see Rolls and Treves (1998) Appendix A4), although it may enable simpler encoding of the firing patterns, for example more orthogonal encoding, to be used. *This latter point may account in part for why there are generally in the cerebral cortex more neurons in a recurrent network than there are connections per neuron (Rolls 2008d), which is an important principle of cortical function.* In addition, the random stochastic fluctuations (or 'noise' (Rolls and Deco 2010)) related to the finite number of spiking neurons is smaller with diluted compared to fully connected networks when the number of connections C per neuron and hence the storage capacity is equated (Rolls and Webb 2012).

7.2.3 The network simulated

The network can be described under four headings which correspond to the four stages in which the simulation of the network operates. The formal specification of the operation of the network is the same as that of the network analysed by Treves (1990) (see also Treves (1991b)) and simulated by Rolls, Treves, Foster and Perez-Vicente (1997c), except where indicated. First, the patterns that the net is to be trained on are binary, that is, a fraction of the neurons, which defines the sparseness a, are 1, and the remainder are 0. Second, the weights are set according to a Hebbian covariance rule. Third, the weight matrix is ablated, that is a proportion of its elements are probabilistically set to zero, to achieve an effective dilution of recurrent connectivity. In other tests, in addition, some of the synaptic weights are multiplied by 2, 3 etc as defined by a Poisson distribution to investigate the effects of multiple synaptic connections between some of the neurons. Fourth, the net undergoes testing with incomplete persistent external cues until the state has settled into retrieval or otherwise.

7.2.3.1 Pattern generation

Random binary patterns with a sparseness a of 0.1 were used.

The sparseness of the retrieved patterns was measured, to ensure that the network was operating in such a way that the sparseness of the retrieved patterns was close to that of the stored patterns. In these simulations, in contrast to earlier simulations (Rolls, Treves, Foster and Perez-Vicente 1997c), the sparseness was set by altering the threshold T_{thr} for the activation of a neuron to produce firing in such a way that the sparseness reached a value of a=0.1. This has the advantage that it can be implemented by an automatic algorithm, and ensures that the sparseness of the retrieved pattern is close to the value desired of a.

7.2.3.2 Learning

The learning mechanism is a form of Hebbian covariance synaptic modification, a one-step application of a simple rule which takes account of simple pairwise covariance relationships within each pattern. The exact rule is as follows. Note that the form of the covariance rule is commutative with respect to units i and j, therefore forcing a fully connected net with such a rule to have symmetric weights.

$$w_{ij} = \frac{1}{Na^2} \sum_{\mu=1}^{p} (y_i^\mu - a)(y_j^\mu - a) \tag{7.6}$$

where w_{ij} is the weight between units i and j. y_i^μ represents the firing rate of unit i within pattern μ. This is a simple covariance rule, and a represents the mean activity in the neuronal network.

7.2.3.3 Connectivity

Networks with asymmetrical dilution of the connectivity were investigated. The dilution applied is described below where the effects of different types of dilution, and of some probability of multiple synapses between pairs of neurons, are analyzed. The total number of neurons is N, each neuron receives exactly C inputs, the dilution is C/N, and $p = \alpha C$ is the number of patterns stored in the net. The critical loading of the net, when it fails to operate as a memory, is denoted as α_c.

7.2.3.4 Testing recall by the net

During recall, the activity of each neuron in the network was asynchronously updated according to a rule which, by analogy with the theoretical analysis, considered a local field h_i at each unit i consisting of an internal field and external field, as follows:

$$h_i = \sum_{j(\neq i)} w_{ij} y_j + \sum_\mu s^\mu \frac{e_i^\mu}{a} \tag{7.7}$$

where y_j represents the output of neuron j, and s^μ represents the relative strength of pattern μ, see below.

The external field (the last term in the above equation with e_i^μ the firing rate of the external input to neuron i produced by the pattern μ) is equivalent to the clamping, persistent external cue, which is believed to be provided by for example the direct perforant path afferents into CA3 from entorhinal cortex (Treves and Rolls 1992b). The ratio between the average number of perforant path synapses per CA3 cell and that of the recurrent collaterals is in this model allowed to determine their relative influence on the firing of CA3 cells. Anatomical evidence available from the rat suggests that the ratio of the external input (the retrieval cue) to the internal recall provided by the recurrent collaterals should be in the order of 0.25 (see Treves and Rolls (1992b)), and s^μ was set to produce this ratio (for example when the retrieval cue had a correlation of 0.5 with the originally learned pattern).

The internal field (the first term in the above equation) is equivalent to the recurrent activation provided by the recurrent collaterals in CA3. This is implemented through a standard autoassociation update rule involving weighted inputs from each of the other units. As explained above, this is qualified by the connectivity enforced through zero weights.

The activation function of the neurons is a threshold linear function of the local field h_i, with a gain factor g described by Treves (1990) set to 0.5, and with the threshold T_{thr} adjusted after each iteration in the recall to produce a retrieval sparseness that was close to 0.1 (which was the sparseness of the stored patterns), as described above.

$$y = \begin{cases} g(h - T_{thr}), & h > T_{thr} \\ 0 & , & h < T_{thr} \end{cases}$$

The recall of the net was measured by the correlation of the retrieved pattern with that stored, when incomplete retrieval cues were used. The performance of the network was also measured by the information retrievable from the network in bits per synapse about the set of stored patterns, as follows

$$I = \frac{\alpha}{\log 2} \sum_{k=0}^{m} \sum_{l=0}^{n} c_k^l \log \frac{c_k^l}{c_k c^l} \tag{7.8}$$

with Treves (1990) and Rolls (2008d) providing further details. Briefly, for each element of the retrieved network state and the corresponding stored pattern, the firing was first discretized into bins, and then the expression above was evaluated. In the above, c_k is the probability that the pattern element is in the kth bin of m bins, c^l is the probability that the retrieved element is in the lth bin of n bins, and c_k^l is the probability that the retrieved element is in the lth bin, and the pattern element is in the kth bin. In the implementation of this calculation, due to practical limitations, both the patterns and network states were binned into 15 bins. Note that the factor α means that the result is in bits per synapse, which is proportional to the total information stored in, and retrievable from, the whole network.

7.2.3.5 Parameters

The network functioned with a set of parameters chosen to be biologically relevant. Where the parameters are not in correspondence with measurable quantities, they were optimized to the values required for the theory to apply (Treves 1990). This subsection details some of these parameters, and the reason for their choice.

The total number of neurons in the net was set to 1000 for the results reported.

The gain parameter, equivalent to the gradient of the linear threshold function producing the output of each unit from its incoming local field, was tested at a number of different values, and found to be optimal in the region of $g = 0.5$ for binary neurons. The actual value of g used was 0.5 for the binary patterns.

The loading α was expressed as the ratio p/C, where p is the number of patterns stored, and C is the number of connections per neuron. The loading was varied between 0.1 and 1.2 to investigate the maximum value of the storage capacity α_c, in terms of patterns, as well as to investigate the effect of over-loading. The number of patterns was varied from 40 to 480. The net was allowed to iterate for a maximum of 30 epochs.

7.2.4 The effects of diluted connectivity on the storage capacity of attractor networks

First a model of how the connections may be set up between neurons is formulated. The initial value of the synaptic weights is 0. This is only the prescription for whether a neuron i will receive a synaptic connection from neuron j. Let us assume that some genetic factors set the number of connections to be received by a neuron of class A from class B to be a constant C (Rolls and Stringer 2000), the number of neurons in the network to be N, and the average connection probability to be p, so that $C = pN$. Previous analyses (Treves and Rolls 1991, Treves 1991b, Rolls and Treves 1998) have assumed that when diluted connectivity is present there are exactly C connections onto each neuron i, that $C < N$, that the connections between pairs of neurons need not be reciprocal and symmetrical, and that there is at most 1 connection from neuron i to neuron j and vice versa. It is this last condition that is relaxed in this investigation (Rolls 2012a), to examine the consequence of different scenarios that might arise according to different prescriptions for determining whether there is a synaptic connection between any two neurons in the network. Three prescriptions are considered.

7.2.4.1 Full connectivity

With full connectivity, there is one and only one synapse in a given direction between neurons i and j, and all neurons are reciprocally connected with equal weights in the two directions. This situation arises in a fully connected autoassociation network trained with an associative (Hebbian) synaptic modification rule, which produces a symmetric synaptic weight matrix (Section B.3).

Although this is the system favoured for formal analysis (Hopfield 1982, Amit 1989), I suggest that this would be extremely difficult for real biological systems to set up. There would have to be a biological system that would have to detect whether among something in the order of say 10,000 synapses being received by neuron i, there was more than one from neuron j. This is rejected as being implausible. There would also have to be a mechanism for ensuring that every neuron in the set of N neurons had at least 1 connection to every other neuron in the set of N neurons. This is rejected as being implausible. There would also need to be a way for genes to specify that a particular set of N neurons specified a particular network. This is rejected as being implausible. Thus it is argued that fully connected networks are unlikely to be found in the brain if they are set up by a simple mechanism such as a neuron making synapses at random with nearby neurons, which is consistent with the evidence (Hill, Wang, Riachi, Schürmann and Markram 2012). Thus this genetically simple mechanism to prescribe connections for neurons to make would not lead to a fully connected network.

7.2.4.2 Connectivity with on average one synapse in a given direction between neurons j and i

Let us assume that among a population of N neurons, a biological process forms connections at random to any one neuron i from the other neurons j until the average number of connections to any neuron i from any one neuron j is 1. In this situation, some neurons i will receive more than one connection from neuron j, and some will receive 0 connections from neuron j. In fact, the number of connections received by neuron i from neuron j will follow approximately a Poisson distribution with mean $(\lambda) = 1$. Given that all neurons are trained using an associative synaptic modification rule of the form shown in equation 7.1, this will mean that some neuron pairs will have a synaptic strength that is twice, three times, etc. the strength of the connections between the neurons with one synapse between them. It is suggested, and tested next, that the consequences of this are that the energy landscape is considerably distorted. Hopfield (1982) was able to show that in a fully connected network trained associatively the recall state can be thought of as the local minimum in an energy landscape, where the energy would be defined as

$$E = -\frac{1}{2} \sum_{i,j} w_{ij}(y_i - <y>)(y_j - <y>). \tag{7.9}$$

This equation can be understood in the following way. If two neurons are both firing above their mean rate (denoted by $<y>$), and are connected by a weight with a positive value, then the firing of these two neurons is consistent with each other, and they mutually support each other, so that they contribute to the system's tendency to remain stable. If across the whole network such mutual support is generally provided, then no further change will take place, and the system will indeed remain stable. If, on the other hand, either of our pair of neurons was not firing, or if the connecting weight had a negative value, the neurons would not support each other, and indeed the tendency would be for the neurons to try to alter ('flip' in the case of binary units) the state of the other. This would be repeated across the whole network until a situation in which most mutual support, and least 'frustration', was reached. What makes it possible to define an energy function and for these points to hold is that the matrix is symmetric (see Hopfield (1982), Hertz, Krogh and Palmer (1991), Amit (1989)).

It is shown next by simulations that if connectivity of the type defined in this section is present, with some pairs of neurons having several connections between them, then the energy landscape is distorted in such a way that there are deep energy minima in some parts of the energy landscape, and this reduces the capacity of the network, that is, its ability to store 0.14 N binary patterns with sparseness $a=0.5$ (Hopfield 1982), or a larger number of patterns with more sparse representations as shown in equations 7.5 and B.15 (Treves and Rolls 1991).

Simulations were run with the attractor network described in Section 7.2.3 and by Rolls, Treves, Foster and Perez-Vicente (1997c) with 1000 neurons (Rolls 2012a). The sparseness of the randomly chosen binary patterns was 0.1. The connectivity was set to have a mean number of connections between any pair of neurons, λ, equal to 1.0 with a Poisson distribution, resulting in the number of connections between pairs of neurons shown in Table 7.1. The algorithm for setting the number of connections was to select an output neuron i, and then set it to have exactly the number of inputs from each of the other neurons in the network (chosen in a random sequence to prevent connections from a neuron already chosen) shown in Table 7.1. This resulted in an asymmetric connection matrix with each neuron i having exactly

Table 7.1 The probability of different numbers of connections X between neurons for different values of λ, the average number of connections between neurons in the network, based on a Poisson distribution. The column labelled $C(N = 1000)$ shows the numbers of synaptic connections to a neuron i multiplied by X in prescribing a network with $N=1000$ neurons with $\lambda=1$.

X	λ=1	$C(N = 1000)$	λ=0.1	λ=0.04
0	0.3679	368	0.9048	0.9608
1	0.3679	368	0.0905	0.0384
2	0.1839	184	0.0045	0.0008
3	0.0613	61	0.0002	0.0000
4	0.0153	15	0.0000	0.0000
5	0.0031	3	0.0000	0.0000
6	0.0005	0.5	0.0000	0.0000

the number of 0, single, double, etc synaptic connections to other neurons in the network shown in Table 7.1. In practice, the algorithm was applied after training by multiplying the synaptic weights by the values shown in Table 7.1. The resulting synaptic connectivity matrix thus reflected what would be produced by any one neuron having a mean number of connections with other neurons set according to a Poisson distribution with the mean number of connections between any pair of neurons, λ, equal to 1. In some cases a pair of neurons would have no synaptic connections, and in other cases 1, 2, 3 etc synaptic connections as shown by the probabilities and numbers for a network of 1000 neurons shown in Table 7.1.

Fig. 7.1a,b shows the results of applying just the degree of dilution of the synaptic weights indicated in Table 7.1 as the baseline control condition, without any double, triple etc synapses. In this case, there were 368 weights of zero out of the starting number of $C=1000$, and the remainder were multiplied by 1. The correlation of the final state of the network after recall with the stored pattern, as a function of loading $\alpha = p/C$, is shown in Fig. 7.1a. The two lines in each graph correspond to two retrieval cue levels: the lower line is for a retrieval cue correlated 0.5–0.55 with the original stored pattern, and the upper line is for a retrieval cue correlation of 0.9–0.95. The information retrieved in bits per synapse after recall with the stored pattern, as a function of loading $\alpha = p/C$ is shown in Fig. 7.1b. (In this Figure, the loading value α shown on the abscissa refers to the fully connected case with $C=1000$ connections per neuron, and should be multiplied by 1.58 given that with the dilution there are in fact $C=638$ connections per neuron, as shown in Table 7.1.) The results in Fig. 7.1a,b show good performance up to high loading levels when only dilution of the connectivity is present, and there are no multiple synapses between neuron pairs present.

These results emphasize that with diluted connectivity, and asymmetric connections between pairs of neurons, the attractor network still displays its predicted memory capacity, and ability to complete memories from incomplete patterns (Treves and Rolls 1991, Treves 1991b, Bovier and Gayrard 1992, Perez Castillo and Skantzos 2004, Rolls et al. 1997c, Rolls and Treves 1998).

Fig. 7.1c,d shows the results of applying the identical degree of dilution of the synaptic weights to that used in Fig. 7.1a,b, but now with the number of double, triple etc weight synapses indicated in Table 7.1 column 3. At low levels of loading $\alpha \leq 0.4$ the performance is as good as or better than the control baseline dilution-only simulation shown in Fig. 7.1a,b, but at higher loading levels the performance drops off very greatly (Fig. 7.1c,d). Fig. 7.1a,b shows that α_c, the critical loading capacity of the network beyond which the memory fails and little information can be retrieved from it, is approximately 0.4. With just diluted connectivity, without multiple synaptic connections, the critical capacity α_c was higher than 1.2, as shown in Fig. 7.1a,b.

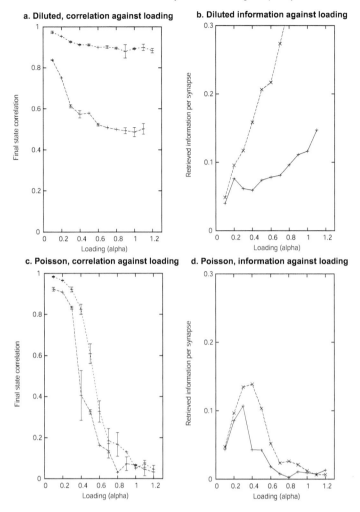

Fig. 7.1 (a, b). The performance of the attractor network with diluted connectivity only with λ=1 for the average number of connections received by each output neuron. In this case, there were 368 weights of zero, and the remainder were multiplied by 1. (c, d). The performance of the attractor network with the numbers of connections between neurons specified by the full Poisson distribution shown in Table 7.1 diluted connectivity only with λ=1 for the average number of connections received by each output neuron. In this case, there were 368 weights of zero, and 368 were multiplied by 1, 184 by 2, 61 by 3, etc. (a,c). The correlation of the final state of the network after recall with the stored pattern, as a function of loading $\alpha = p/C$. The two lines in each graph correspond to two retrieval cue levels: the lower line is for a cue correlated 0.5–0.55 with the original stored pattern, and the upper line is for a cue correlation of 0.9–0.95. The error bars represent the standard deviations. (b,d). The information retrieved in bits per synapse after recall with the stored pattern, as a function of loading $\alpha = p/C$. (Reproduced with permission from Biologically Inspired Cognitive Architectures. Advantages of dilution in the connectivity of attractor networks in the brain, 1: 44–54, Edmund T. Rolls. Copyright ©2012 Elsevier B.V.)

It is therefore concluded (Rolls 2012a) that prescription 2 (section 7.2.4.2) for setting the connectivity is seriously flawed, as it reduces the capacity of the attractor network, i.e. the number of patterns that it can store per synapse onto each neuron (Treves and Rolls 1991, Rolls 2008d), because of the multiple synapses found between a proportion of the neurons. The interpretation of the loss of capacity with some multiple synapses present is that this distorts the energy landscape by producing irregular deep and broad areas which attract patterns that are some distance away, so that it is not possible to have a large number of approximately equal-size basins of attraction in the energy landscape. The interpretation of the somewhat better performance at low loading with some multiple synapses present is that if there are few basins present, but they are deep and large due to the multiple synapses, then patterns some distance away can be drawn into a nearby attractor, which, given large spacing between the attractor basins, is likely to be the correct attractor basin.

I further hypothesize that double synapses would distort the patterns being stored if they are graded (for example having an exponential distribution of firing rates across the population of neurons for any one stimulus), which is typical of neural representations (Treves 1990, Rolls 2008d, Rolls and Treves 2011, Rolls et al. 1997c, Franco et al. 2007, Rolls and Treves 2011) (Appendix C). It is possible that the graded firing rate representation could no longer be accurately stored: the graded representation would be distorted by the extra firing of neurons with double connections. Simulations of the type described by Rolls, Treves, Foster and Perez-Vicente (1997c) could be performed to check whether this becomes an issue, depending on the number of double, triple etc synapses between some neuron pairs.

7.2.4.3 Diluted connectivity with $C < N$

Let us assume that among a population of N neurons, a biological process forms connections at random between any pair of neurons, so that in the resulting network there are C incoming connections to any of N neurons, with $C < N$. Assuming symmetry, this implies that the number L_{ij} of the connections from any neuron j to any neuron i follows a Binomial distribution $\text{Bin}\left(C, \frac{1}{N-1}\right)$:

$$\text{Prob}\{L_{ij} = k\} = \frac{C!}{k!(C-k)!}\left(\frac{1}{N-1}\right)^k\left(1 - \frac{1}{N-1}\right)^{C-k} \quad (7.10)$$
$$= \frac{C(C-1)\dots(C-k+1)}{1\cdot 2\cdot\dots\cdot k}(N-1)^{C-k}$$

where k is between 0 and C. When $C = \lambda N$ and $N \to \infty$, this distribution is well approximated by the Poisson distribution with mean λ.

For example, let us assume a diluted connectivity with the average number of connections per neuron $\lambda=0.1$. We set up connections with the same general prescription as in prescription 2 (Section 7.2.4.2). However, if $\lambda = 0.1$, the probability that any two neurons will have two connections between them is 0.0045, as shown in Table 7.1. (In a network with $N=1000$ neurons, there would be 91 single connections onto each neuron, and only 5 double connections onto each neuron, with no triple connections.) This is a much smaller probability. Additional simulations of the type illustrated in Fig. 7.1 were performed with $\lambda = 0.1$. The results showed that having this small proportion of double strength synapses in the network produced very little reduction in the correlation of the retrieved patterns with those stored, or in the capacity of the attractor network to store many patterns using the type of analysis shown in Fig. 7.1c,d. The number of patterns that can be stored is still of order C (Treves and Rolls 1991), with the constant k in equation 7.5 little affected by this small number of multiple connections between pairs of neurons.

This scenario applies in the real brain. For example, an estimate for the dilution C/N in the neocortex might be 0.1. (For the neocortex, assuming 10,000 recurrent collaterals per pyramidal cell, that the density of pyramidal cells is 30,000 / mm^3 (Rolls 2008d), that the radius of the recurrent collaterals is 1 mm, and that we are dealing with the superficial (or deep) layers of the cortex with a depth of approximately 1 mm, the dilution between the superficial (or deep) pyramidal cells would be approximately 0.1.)

A similar but somewhat more diluted scenario applies in the hippocampus. In the rat hippocampus, there are $N = 300,000$ CA3 neurons, and each neuron receives $C = 12,000$ synapses (see Fig. 24.14) (Rolls 1989b, Treves and Rolls 1992b). (It is assumed that each CA3 neuron correspondingly makes 12,000 recurrent collateral synapses.) The dilution of this network C/N thus equals 0.04. In a network with $N=1000$ neurons and $\lambda=0.04$, there would be 38 single synapses onto each neuron, and only 1 double synapse, as is evident from Table 7.1. This low number of double synapses, with no triple etc synapses, has little effect on the operation of the network, as shown by further simulations of the type illustrated in Fig. 7.1 (Rolls 2012a). It is proposed that this network operates as an autoassociation or attractor network, and is key to how the hippocampus operates to store episodic memories (Rolls 1996c, Rolls 2008d, Rolls 2010b) (see Chapter 2 of Rolls (2008d)), and here the number of different memories that can be stored in the network is at a premium, and is provided for by this level of dilution of the connectivity which ensures few multiple synapses between any pair of neurons.

7.2.5 Synthesis of the effects of diluted connectivity in attractor networks

The results of the simulations (Rolls 2012a) therefore support the proposal made by Rolls (2012a) that the reason why networks in the neocortex and hippocampal cortex have diluted connectivity is that the diluted connectivity ensures that the energy landscape and thereby the memory capacity is not disturbed by multiple connections between pairs of neurons, with the network still operating correctly in the diluted regime where C/N is in the range 0.1–0.01 (Treves 1991b, Rolls et al. 1997c, Rolls and Treves 1998, Rolls 2008d). The somewhat greater dilution of connectivity (with fewer multiple synapses) in the hippocampal CA3 network (0.04) than the estimate of its value in the neocortex (0.1) may be related to the great importance of achieving a high memory capacity with good retrieval of all stored patterns in the hippocampus where this may be useful for episodic memory, for which memory capacity in a single network is important (Rolls 2008d, Rolls 2010b) (Chapter 24).

In both these types of cortex, there is evidence that the excitatory interconnections between neurons are associatively modifiable, and that the system supports attractor dynamics that enable memories to be stored (Section B.3). One of the points made here is that with a local associative rule, the presynaptic and the postsynaptic activity at a given synapse to determine the strength of that connection between a pair of neurons, which is a widely held assumption (Rolls and Treves 1998, Rolls 2008d), then consequences of the type described here on memory capacity would ensue. A difference between these cortical structures is that the CA3 network is a single network allowing any representation to be associated with any other representation, providing an implementation of episodic memory (Rolls 2008d, Rolls 2010b, Kesner and Rolls 2015). In contrast, the neocortex has local connectivity with a radius of approximately 2 mm, and this enables the whole of the cerebral cortex to have many separate attractor networks, each storing a large number of memories (O'Kane and Treves 1992, Rolls 2008d).

Another potential advantage of diluted connectivity in which the number of neurons N is greater than C^{RC}, the number of recurrent collateral inputs received by any neuron in the network from the other neurons in the network, is that this may enable simpler encoding of the firing patterns, for example more orthogonal encoding, to be used. For example, much of the information available from the firing rates of a population of neurons about which stimulus was presented can be read by a decoding procedure as simple as a dot product, which is what neurons compute using their synaptic weight vector, and which is very biologically plausible (Rolls 2008d, Rolls and Treves 2011) (Appendix C).

Another advantage of diluted recurrent collateral connectivity is that it can increase the storage capacity α_c of autoassociation (attractor) networks, that is the number of patterns that can be stored per recurrent collateral synapse (Treves and Rolls 1991). (This useful effect applies provided that representations are not too sparse, because when the sparseness is very low, $a \ll 0.01$, the performance becomes similar to that of a fully connected network (see Treves and Rolls (1991) Fig. 5a; and Rolls and Treves (1998) Fig. A4.2.)

Another advantage of diluted connectivity (for the same number of connections per neuron) is that this increases the stability and accuracy of the network as there is less spiking-related noise producing stochastic fluctuations in the diluted network (which has more neurons in it), with little cost in increased decision times (Rolls and Webb 2012).

Very diluted connectivity in feedforward networks (for example, competitive networks) can also play a role in pattern separation, by encouraging different output neurons to respond to different, random, combinations of the inputs because each neuron is connected to a different combination of inputs (see Section 7.4). This finds application in the brain in for example the dentate granule cell mossy fibre to CA3 connections which are very dilute but strong. The architecture helps grid cell representations to be transformed into place or spatial view representations (Treves and Rolls 1992b, Rolls, Stringer and Elliot 2006c, Rolls and Kesner 2006, Rolls 2008d, Rolls 2010b).

In summary, Rolls (2012a) proposed that the reason why networks in the neocortex and hippocampal cortex have diluted connectivity in the recurrent collateral connections is that the diluted connectivity ensures that the energy landscape and thereby the memory capacity is not disturbed by multiple connections between pairs of neurons, with the network still operating correctly in the diluted regime where C/N is in the range 0.1–0.01 (Treves 1991b, Rolls et al. 1997c, Rolls and Treves 1998, Rolls 2008d). It was shown that having a proportion of multiple connections between neurons in an attractor network trained by an associative rule produces a major reduction in $\alpha_c = p/C$, the memory capacity of the network beyond which adding further memories drastically impairs the ability of the network to retrieve any memories. That important reason for dilution in the connectivity of cortical networks, which helps them to be specified by relatively few and simple genetic rules consistent with a limited number of genes in the whole genome in the order of 25,000 compared to the number of synapses which is in the order of 10^{15} (Rolls and Stringer 2000) (Chapter 19), is accompanied by other advantages of the dilution of cortical connectivity elucidated by Rolls (2012a).

7.3   The effects of dilution on the capacity of pattern association networks

Pattern association networks appear to be used in very many cortical systems, including the cortico-cortical backprojections used in memory recall and in top-down selective attention (see Chapter 11 and Section 24.3.7) (Kesner and Rolls 2015, Rolls 2015b), but also in systems in the orbitofrontal cortex and amygdala involved in stimulus-reinforcement association learning for emotion (Rolls 2014a).

New hypotheses recently described (Rolls 2015b) are that having at random some multiple connections between the input and the output neurons can decrease the capacity of pattern association networks; that dilution of connectivity in pattern association networks can minimize this loss of capacity by reducing the probability of multiple synapses if they are made at random between input and output neurons; and that this dilution helps in this way to ensure that the recall of information from the hippocampus to the neocortex by several stages of pattern association network is efficient and has high capacity.

The hypotheses were tested by simulations of a pattern association network (Rolls 2015b), the architecture and properties of which are described in Section B.2. Random binary pattern

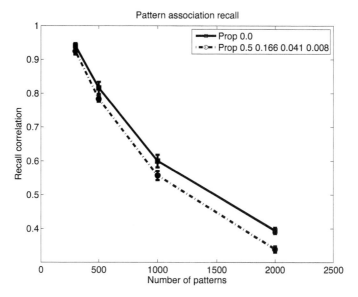

Fig. 7.2 Effects of multiple synapses from an input neuron to an output neuron in a pattern association network. In the simulations with random binary input and output patterns each with sparseness a=0.05 there were N=1000 output neurons, and C=1000 synaptic inputs onto each neuron. In the case representing the effects of some multiple synapses between input neurons and output neurons due to random connectivity and modelled with a Poisson distribution, the proportion of double strength synapses was 0.5, of triple strength was 0.166, of quadruple strength was 0.041, and of quintuple strength was 0.008. The case for single synapses is labelled Prop 0.0. Performance was measured by the correlation between each recalled output vector, and the corresponding output vector that had been trained. Each data point shows the mean and standard deviation calculated over 5-10 simulations with different random binary pattern vectors to be associated. The effects are shown for different numbers of trained pattern associations. The effects illustrated were without the introduction of noise into the recall vector. (Reproduced with permission from Progress in Brain Research. Chapter 2 - Diluted connectivity in pattern association networks facilitates the recall of information from the hippocampus to the neocortex, Volume 219, 2015, pages 21–43. Edmund T. Rolls. Copyright ©2015 Elsevier B.V.)

vectors of firing rates were used for the inputs to the network (default sparseness 0.05) and for the outputs of the network (default sparseness 0.05). The network was trained with an associative (Hebbian) rule shown in equation B.2 on page 709. During recall, the activations of the neurons h_i were calculated using a dot product between the presynaptic firing rate vector, the conditioned stimulus or recall cue, and the synaptic weights on the dendrite of each neuron as shown in equation B.3, and converted to binary 1,0 firing rates with a sparseness a_o (default 0.05) by setting the threshold appropriately. The performance of the network was measured by the (Pearson) correlation between the output vector of firing rates and the trained vector of output firing rates for each conditioned stimulus input pattern.

The hypothesis about the effects of having a proportion of the synapses randomly selected with double strength due to two synapses from an input neuron to an output neuron was tested as follows. The fully connected network was first trained on a number *nPatts* of input–output pattern pairs. Then a proportion of the synapses, reflecting what might be expected from a Poisson distribution with λ=1, the mean number of connections from an input to an output neuron onto each neuron selected by random permutation was doubled, tripled etc in strength, to reflect what would happen if there were two or three etc synapses present as specified by a Poisson distribution (see Table 7.1). Illustrative results are shown in Fig. 7.2. Recall was considerably impaired, measured by the reduction in the correlation between the recalled outputs firing rate vectors and the trained firing rate output vectors, if some multiple synapses were present as might be expected from a Poisson distribution. The effect occurs because the distortion of the synaptic weights can affect the recalled pattern, with the effect particularly evident when the loading is high, and there is interference between the different pattern pairs associated together in the synaptic matrix. In control simulations, it was shown that if the number of double synapses was set to the number expected for a dilution of $\lambda=0.1$, then the much smaller number of double synapses produced by chance had only small effects on performance, confirming the usefulness of diluted connectivity (Rolls 2015b).

In addition to this finding, before each recall pattern was applied, a small proportion of the elements in the input recall vector were flipped between from 1 to 0 or from 0 to 1 using random permutation, to introduce some noise into the recall cue, to require generalization of the pattern association network towards the correct output firing (see Section B.2), and to reflect what might be imperfect recall within the hippocampus or the preceding stage in the recall process. This small variation of the recall cue contributes to the effects described at somewhat lower loadings (measured by the number of stored patterns), for it results in the recall cue perhaps not activating a (strong) double synapse which should have been activated, or in activating a (strong) double synapse that should not have been activated, resulting in distortion of the pattern recalled from the pattern association network (Rolls 2015b).

Thus having a proportion of synapses on an output neuron arising from the same input neuron in a pattern association network can impair the retrieval of information from the pattern association network. This provides a potentially fundamental approach to why connectivity in cortical systems is diluted, to be typically less than 0.1, which results in the proportion of synapses that are replicated between any pair of neurons being sufficiently low that retrieval is not impaired (Rolls 2015b).

In the hippocampo-cortical backprojection system used for recall in which there are multiple stages of pattern association networks (Fig. 24.1), any distortion of what is recalled at any one stage (Fig. 7.2) will be magnified and exacerbated by the distortion being further produced at each stage of the recall process. This means that the effect described here is likely to be very relevant indeed in the recall of memories from the hippocampus to the neocortex, and to provide an important reason why this connectivity must be diluted (Chapter 24). Another advantage of the dilution in this connectivity is that it reflects the divergence from a relatively small number of CA3 neurons (300,000 in the rat) to the enormous numbers of neocortical neurons, which can be achieved by making this a multistage recall pathway with diluted connectivity and with approximately 10,000 backprojection synapses onto each receiving neuron from the previous stage (Chapter 24).

In the context of any distortion produced at each stage of the hippocampo-cortical recall process, part of the hypothesis is that local recurrent collaterals between nearby neocortical neurons within each stage will operate as an autoassociation network to clean up the retrieved pattern, and it has been shown that this effect operates sufficiently fast (within 20 ms per stage) with integrate-and-fire neurons to contribute to the correction of the retrieval process (Panzeri, Rolls, Battaglia and Lewis 2001).

The findings described by Rolls (2015b) on the advantages of dilution in the connectivity of pattern associators in the cerebral cortex provide a fundamental advance in understanding cortical computation, for pattern association networks appear to be used in very many cortical systems, including the cortico-cortical backprojections used in memory recall and in top-down selective attention (Rolls and Deco 2002, Deco and Rolls 2005b, Rolls 2008d, Rolls 2013a), but also in systems in the orbitofrontal cortex and amygdala involved in stimulus-reinforcement association learning for emotion (Rolls 2014a, Rolls 2015c). The new evidence adds to the advantages of dilution in the connectivity of attractor networks in the cerebral cortex, where multiple connections between any pair of neurons potentially distort the basins of attraction, and thereby impair memory recall, short-term memory, etc (Rolls 2012a).

The same arguments are likely to apply to the capacity of perceptrons, as described in Chapter 23.

7.4 The effects of dilution on the performance of competitive networks

7.4.1 Competitive Networks

In another class of network fundamental to cortical computation, competitive networks, which can build new representations as described in Section B.4, dilution in connectivity may be advantageous, but for very different reasons. Dilution in the connectivity of competitive networks can help to break the symmetry, so that some neurons are likely to be allocated to some patterns, and other neurons to other patterns (because of the particular set of input connections of each neuron), and this can help to stabilize a competitive network, by making it difficult for neurons to drift during further learning if the input patterns drift (Rolls 2008d) (Section B.4). In such a system, multiple connections from some input neurons to an output neuron might even be advantageous, for the reasons just given. This emphasizes the point that multiple connections between neurons are found in some parts of the cortex, including for example thalamo-cortical inputs and cortico-cortical forward projections, where a strong selective drive to some neurons may be important, and the presence of such multiple contacts involving some cortical neurons does not invalidate at all the hypotheses and arguments presented in this Chapter about autoassociation and pattern association networks.

In recent research, the effects of diluted connectivity on the capacity and properties of the categorisation performed by competitive networks have been investigated (Rolls 2016e). The network simulated was similar to that described in Section B.4. A binary threshold or binary activation function was used, and a method of setting the sparseness of the output firing was used as described for VisNet (Chapter 25 and Rolls (2012c)). The number of output and input neurons was N (where N was in the range 100–1000), dilutions of 1 (full connectivity), to 20 were tested, and the sparseness a was set at 0.02 (where the sparseness in the binary case is the proportion of neurons firing with a high firing rate of 1 to any one pattern). The input patterns for the $N=100$ neuron simulations illustrated consisted of a set of 20 stimuli each 20 elements long (i.e. with a sparseness of 0.2), with each overlapping in 5 elements with the preceding stimulus, as illustrated in Fig. 7.4a. (The stimuli were modified to provide a richer set of different correlations between each of the input stimuli by flipping a random number of elements between 1 and 4 to the opposite state.) Learning was implemented with a learning rate for equation B.19 of 0.1, and 30 epochs of training were used. The learning is important to allow the dendritic weight vectors to point towards patterns or clusters of patterns, and this facilitates generalization between those similar patterns. The performance was measured by the correlation between the different output firing rate vectors produced by the different input patterns. Pattern separation is indicated by low values for these correlations.

The results of a simulation to illustrate the operation of a competitive network with complete (undiluted) connectivity are illustrated in Fig. 7.3b. The synaptic matrix is shown after training with 20 presynaptic input patterns or stimuli. In the synaptic matrix, white corresponds to the maximum synaptic weight, and black to 0. The network is shown during testing with presynaptic input pattern 1, which activated strongly two of the neurons as shown because of the pattern of their increased synaptic weights. After competition, two of the postsynaptic neurons had high firing rates (white), corresponding to a sparseness of 0.02 in

Fig. 7.3 a. The architecture of a competitive network. b. Simulation of a competitive network. The synaptic matrix is shown after training with 20 presynaptic input patterns or stimuli, each 20 elements long, and each shifted down 5 elements with respect to the previous pattern. In the synaptic matrix, white corresponds to the maximum synaptic weight, and black to 0. The network is shown during testing with presynaptic input pattern 1, which activates strongly 2 of the neurons as shown because of the pattern of their increased synaptic weights after training. After competition, two of the postsynaptic neurons have high firing rates (white), corresponding to a sparseness of 0.02 in this competitive network with N=100 output neurons, and C=100 synaptic inputs onto each neuron. The synaptic weight vector for each neuron corresponds to one vertical column in the synaptic matrix. The neurons left with random synaptic weights have not learned to this set of patterns, and remain available for further presynaptic input patterns or stimuli. (Reproduced with permission from Neurobiology of Learning and Memory. Pattern separation, completion, and categorisation in the hippocampus and neocortex, Volume 129, March 2016, Pages 4–28. Edmund T. Rolls. Copyright ©2015 Elsevier Inc.)

this competitive network with N=100 output neurons, and C=100 synaptic inputs onto each neuron. The synaptic weight vector for each neuron corresponds to one vertical column in the synaptic matrix. The neurons left with random synaptic weights have not learned to this set of patterns, and remain available for further presynaptic input patterns or stimuli. The synaptic weight vectors show that each neuron with modified synapses has learned to respond to part of one or more of the input patterns, and that this is how the network has learned to respond differently to each of the different inputs.

The operation of competitive networks with full or with diluted connectivity, and with or without learning, is considered in the next subsections.

7.4.2 Competitive networks without associative learning but with diluted connectivity

This provides a model of the mossy fibre inputs to the hippocampal CA3 neurons, as described in Chapter 24.

The results of a simulation with this scenario are shown in Fig. 7.4b. The dilution was 2, the learning rate was 0, and the synapses that were present had the same positive value. Fig. 7.4b shows that the correlations between the firing rate vectors (each vector the firing rates of the 100 output neurons) were lower than the correlations between the inputs of the corresponding stimulus patterns that produced the output firing rates. That is, the computation that was achieved was some pattern separation or orthogonalization of the inputs in transforming them to outputs. This was confirmed by the results of an analysis that showed that 90–100% of the output vectors to the different stimuli (on different runs) were less correlated than 0.8.

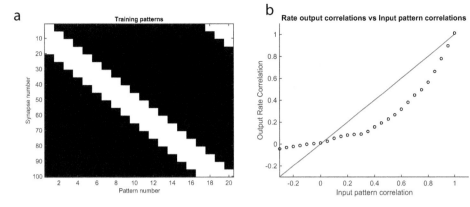

Fig. 7.4 Competitive net with diluted connectivity of 2 and no learning. a. The 20 basic stimuli, each a vertical column with 20 values of 1 (white) and the remainder of the 100 values 0 (black). Each stimulus overlaps in 15 positions with the preceding stimulus in the series. After this stage, some bits were flipped to produce a wider range of correlations between different stimuli (see text). b. The correlations between the output firing rate vectors as a function of the correlations between the different input patterns. The diagonal line with a slope of 1 shows where correlations between the output firing rate vectors would fall if there was no pattern separation of the input pattern vectors. (Reproduced with permission from Neurobiology of Learning and Memory. Pattern separation, completion, and categorisation in the hippocampus and neocortex, Volume 129, March 2016, Pages 4–28. Edmund T. Rolls. Copyright ©2015 Elsevier Inc.)

If there was no dilution, or almost no dilution, of the connectivity, then the outputs to the different patterns were of course the same or very similar, that is, no pattern separation or orthogonalization was produced. The outputs with no dilution were very close to the diagonal line in Fig. 7.4b.

The simulation thus shows that pattern separation can be produced by diluted connectivity in a competitive network. (For completeness, it is noted that an analogous type of pattern separation can be produced by a fully connected competitive net without learning in which the synaptic weights are set to random values with a uniform distribution.)

The mode of operation considered in this subsection is the condition considered analytically by Treves and Rolls (1992b) for discrete input patterns (e.g. random patterns of 0s and 1s as the input vectors) that might represent objects, and by Cerasti and Treves (2010) for input patterns with Gaussian firing rate distributions that might represent a position in continuous space. This was in the context of how the dentate gyrus projections via the mossy fibres to the CA3 neurons, which have diluted connectivity and no associative learning, might operate as a system to activate relatively orthogonal representations in CA3 cell firing for different inputs. This pattern separation and orthogonalization is important for the CA3 autoassociation system, so that each episodic memory is as different as possible from all the other episodic memories stored in CA3, to minimize interference between them, and to maximize the number of memories that can be stored and correctly retrieved (Kesner and Rolls 2015, Rolls 1989b, Treves and Rolls 1992b) (see Chapter 24).

7.4.3 Competitive networks with learning and with diluted connectivity

If in the same network we allow associative learning, then the result is that the synaptic weight vectors of some of the neurons come to point towards a cluster of similar stimuli. This results in the important new property of categorisation, in which the outputs for similar stimuli are

a
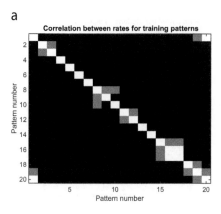

b
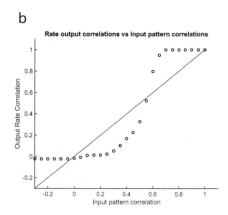

Fig. 7.5 Competitive net with diluted connectivity of 2 and learning. a. The correlations between the output firing rate vectors produced by each of the input stimulus patterns. White represents a correlation of 1, and black represents a correlation of approximately 0. b. The correlations between the output firing rate vectors as a function of the correlations between the different input patterns. The diagonal line with a slope of 1 shows where correlations between the output firing rate vectors would fall if there was no pattern separation of the input pattern vectors. Each point in the graph represents the mean value across the presentation of many stimuli within the same correlation bin. (Reproduced with permission from Neurobiology of Learning and Memory. Pattern separation, completion, and categorisation in the hippocampus and neocortex, Volume 129, March 2016, Pages 4–28. Edmund T. Rolls. Copyright ©2015 Elsevier Inc.)

more similar than the inputs. This is the opposite of pattern separation. This is illustrated in Fig. 7.5 in the same network with a diluted connectivity of 2 and with the same stimulus input patterns, but now with the learning between the inputs and the output neurons enabled (with the standard learning rate of 0.1), and with the synaptic weights initialised to random values with a uniform distribution. Fig. 7.5b shows that pattern separation is achieved if the inputs are less correlated than approximately 0.5, and that in contrast the outputs are more similar (as shown by the correlation between the firing rate vectors on the ordinate) if the inputs are more correlated than approximately 0.5. The combination of these two processes, pattern separation for different patterns and pattern clustering for similar patterns, that is implemented is the categorisation that can be performed by a competitive network (see further Section B.4).

It should be noted that this process is quite distinct from the process of pattern completion, in which the whole of a pattern can be retrieved from any part (Section B.3).

Fig. 7.5a illustrates the pattern separation that is achieved, by showing that the correlations of the output firing rate vectors for most of the patterns are low. (In this run, only the outputs to patterns 16–17 and 20–1 were highly correlated with each other, resulting in 80% correct pattern separation. These pairs of input patterns were closer together than the other pairs as a result of the small amount of randomness introduced into the patterns by flipping 1–4 of the bits.)

The competitive network can thus be described as generalizing to similar patterns, in that it places similar patterns into the same category. This generalization was found to untrained as well as trained patterns provided that the patterns were similar. The dentate granule cells and the CA1 network in the hippocampal system are hypothesized to operate as competitive networks with this property (Kesner and Rolls 2015, Rolls 1987, Rolls 1989b, Rolls and Treves 1998, Schultz and Rolls 1999, Rolls, Stringer and Elliot 2006c) (see Chapter 24).

a

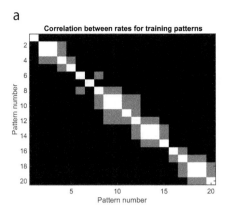

b

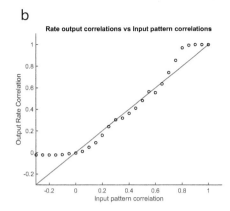

Fig. 7.6 Competitive net with undiluted connectivity and with learning. a. The correlations between the firing rate vectors between produced by each of the patterns. White represents a correlation of 1, and black represents a correlation of approximately 0. b. The correlations between the output firing rate vectors as a function of the correlations between the different input patterns. The diagonal line with a slope of 1 shows where correlations between the output firing rate vectors would fall if there was no pattern separation of the input pattern vectors. (Reproduced with permission from Neurobiology of Learning and Memory. Pattern separation, completion, and categorisation in the hippocampus and neocortex, Volume 129, March 2016, Pages 4–28. Edmund T. Rolls. Copyright ©2015 Elsevier Inc.)

7.4.4 Competitive networks with learning and with full (undiluted) connectivity

If the same network is set up with full (undiluted) connectivity, associative learning, and random initialization of the synaptic weights, the clustering of similar patterns still occurs as a result of the learning, but the network may perform less pattern separation. This is illustrated in Fig. 7.6 in the same network with a connectivity of 1 and with the same stimulus input patterns, with the learning between the inputs and the output neurons enabled (with the standard learning rate of 0.1), and with the synaptic weights initialised to random values with a uniform distribution. Fig. 7.6b shows that the outputs are more similar (as shown by the correlation between the firing rate vectors on the ordinate) if the inputs are more correlated than approximately 0.7, and this is the clustering together of similar patterns. However, almost no pattern separation is achieved: if the inputs are less correlated that approximately 0.7, then the correlations between the firing rate output vectors produced are almost the same as the correlations between the input stimulus vectors. (The output correlations fall close to the diagonal line in Fig. 7.6b for correlations less than 0.7.)

Fig. 7.6a illustrates that pattern separation with full connectivity is less good than with the diluted connectivity, by showing that the correlations of the output firing rate vectors for several pairs of the patterns were high. (In this run, the outputs to patterns 2–3, 9–10, 13–14 and 18–19 were highly correlated with each other, resulting in only 60% correct pattern separation.)

Similar results were obtained with a sigmoid activation function of the type used in VisNet (Chapter 25 and Rolls (2012c)) with β, the slope of the activation function, in the range 3 (to produce graded firing rate outputs) and 20 (to produce almost binary firing rate outputs). This is mentioned here, because this was the only way in which graded firing rate distributions with an approximately exponential distribution were found, and this is relevant, for this type of firing rate distribution is typical of those found in the cerebral cortex (see Chapter 8 and Rolls and Treves (2011)). It is hypothesized that graded exponential firing rate distributions may be facilitated by the dilution, for some neurons are capable of being more strongly activated by a particular input pattern than others.

7.4.5 Overview and implications of diluted connectivity in competitive networks

These findings (Rolls 2016e) establish the point that diluted connectivity can improve the pattern separation that can be performed by competitive networks (Rolls 2016e). The magnitude of the effect depends on several factors, such as the size of the network, and the correlations between the inputs, which will be harder to pattern separate without the dilution, because similar patterns will tend to activate the same neurons during the learning (see further Section B.4.9.7). Indeed, the mechanism by which the dilution works is that it forces different neurons to be capable of pointing in only certain directions in the input space of patterns, so that some neurons can be thought of as inherently more likely to learn to some input patterns than to other input patterns (see Appendices A and B).

Learning in a fully connected competitive network results in the synaptic weight vectors of different neurons pointing towards the centre of a cluster of input vectors (see Section B.4). The diluted connectivity produces in a sense something similar, but enables learning to also contribute to the direction in which the weight vector points, for optimization. Without the diluted connectivity, the learning alone cannot separate out the firing of different neurons sufficiently to allow successful pattern separation by competitive learning, under conditions of high loading and non-sparse input patterns.

The functional implications include the following:

1. It is hypothesized that the independence of the information conveyed by different neurons (see Appendix C and Rolls and Treves (2011)) may be related to the diluted representations, for this contributes to each neuron having different tuning to different input patterns. This pattern separation may be facilitated by the competitive learning.

2. Pattern separation has as one of its major mechanisms this diluted connectivity. An example is in the hippocampal mossy fibre system (see Section 24.3.3.8).

3. The dilution may increase the capacity of competitive networks, by encouraging different neurons to point towards different parts of the input space of patterns.

4. The effects of diluted connectivity in networks without associative synaptic modification described in Section 7.4.2 are relevant to how the dentate gyrus projections via the mossy fibres to the CA3 neurons, which have diluted connectivity and no associative learning, might operate as a system to activate relatively orthogonal representations in CA3 cell firing for different inputs (Chapter 24). This pattern separation and orthogonalization is important for the CA3 autoassociation system, so that each episodic memory is as different as possible from all the other episodic memories stored in CA3, to minimize interference between them and to maximize the number of different memories that can be stored (Kesner and Rolls 2015, Rolls 1987, Rolls 1989b, Treves and Rolls 1992a, Cerasti and Treves 2010). This may be one of the few major cortical pathways that does not show associative synaptic modification, for in this dentate to CA3 system the main computation appears to be the production of pattern separation so that CA3 representations are relatively orthogonal to each other. In this situation, a single attractor state can be set up in CA3 by associative synaptic modification in the CA3-CA3 recurrent collateral connections that is particular for each episodic memory (Chapter 24). The theory is that the direct perforant path input to the CA3 cells becomes associatively modified during learning, and later during recall provides the partial retrieval cue for completion to occur in CA3, which then connects to CA1 on the path back to neocortex for the whole of the memory to be retrieved in each neocortical area that provided part of the original input to the

episodic memory. In this situation, the dentate to CA3 projection via the mossy fibres may be one of the few places in the brain where information is not stored by a change of synaptic weight to store memories or build new perceptual or cognitive representations, as its function is to orthogonalize new patterns, and not to act as the retrieval cue for previously formed memories. In this situation, the lack of associative synaptic learning may in fact be useful, for as shown in Section 7.4.3, if learning is present, this will tend to cluster similar patterns, whereas the goal is to orthogonalize all patterns, even similar ones, in the hippocampal mossy fibre system, to maximize the number of different episodic memories that can be stored and correctly retrieved. The presence of neurogenesis in this dentate system, but not in most other cortical pathways, can be understood in this framework (Chapter 13).

5. Asymmetries in the receptive fields of inferior temporal cortex neurons (Aggelopoulos and Rolls 2005) may arise as a result of the forward diluted connectivity from the previous cortical area, and this asymmetry is important in the perception of multiple objects seen close to the fovea (see Section 25.5.10).

7.5 The effects of dilution on the noise in attractor networks

To investigate the extent to which diluted connectivity affects the dynamics of attractor networks in the cerebral cortex, Rolls and Webb (2012) simulated an integrate-and-fire attractor network taking decisions between competing inputs with diluted connectivity of 0.25 or 0.1, and with the same number of synaptic connections per neuron for the recurrent collateral synapses within an attractor population as for full connectivity. The results indicated that there was less spiking-related noise with the diluted connectivity in that the stability of the network when in the spontaneous state of firing increased, and the accuracy of the correct decisions increased. The decision times were a little slower with diluted than with complete connectivity. Given that the capacity of the network is set by the number of recurrent collateral synaptic connections per neuron, on which there is a biological limit, the findings indicate that the stability of cortical networks, and the accuracy of their correct decisions or memory recall operations, can be increased by utilizing diluted connectivity and correspondingly increasing the number of neurons in the network, with little impact on the speed of processing of the cortex. Thus diluted connectivity can decrease cortical spiking-related noise. In addition, Rolls and Webb (2012) showed that the Fano factor for the trial-to-trial variability of the neuronal firing decreases from the spontaneous firing state value when the attractor network makes a decision.

7.6 Highlights

1. In autoassociation / attractor networks, diluted connectivity can be advantageous, because it reduces the probability that there will be more than one synapse between pairs of neurons if the synaptic contacts are made at random. Different numbers of synapses between pairs of neurons distort the basins of attraction if synapses are made at random between neurons, and greatly degrades the memory capacity. With diluted connectivity, there are likely to be mainly 0 or 1 synapse between pairs of neurons, and attractor networks operate properly up to their theoretical limit under these conditions. The argument applies to the cortical systems with recurrent collaterals, and not to inputs from for example the thalamus that drive the neurons into activity. The degree of dilution is in the order of 0.04 in the hippocampus, or 0.02 if we consider the two sides of the rodent

hippocampus. The hippocampal CA3 recurrent collaterals may be more diluted than those in neocortex, for memory capacity is at a premium in the single CA3 network, and the greater dilution may help to keep the energy landscape undistorted by some multiple synapses between pairs of neurons.

2. Dilution of connectivity in attractor networks indicates that there are more neurons N in the network than recurrent collateral connections C onto each neuron. (Dilution = C/N.) The dilution may facilitate simpler encoding by the vector of neuron firing rates, for example more orthogonal encoding, which helps to keep the capacity of attractor networks high. This is an important principle of cortical function.

3. Dilution in the connectivity of attractor networks can decrease the noise in the condition that the number of synapses per neuron is kept constant as the dilution increases. A consequence of the dilution is that the number of neurons in the network is increased, and the reduction of noise is related to the increase in the number of neurons.

4. In pattern association networks having a proportion of synapses on an output neuron arising from the same input neuron can impair the retrieval of information from the pattern association network, and thus diluted connectivity for this cortical operation is also a useful principle.

5. In competitive networks, diluted connectivity can facilitate pattern separation, and also makes the categorisation more stable, for the output neurons cannot drift throughout the space.

8 Coding principles

8.1 Types of encoding

In most cortical systems, information is encoded by which neurons are firing, and how fast they are firing (Rolls 2008d, Rolls and Treves 2011, Rolls 2012c, Rolls 2017a) (Appendix C). This is called *place coding*, for each neuron has its place in the cerebral cortex. If there is a topographic map, for example of the visual field on the primary visual cortex, or of each finger in the somatosensory cortex, then the place on the map is an indicator of where the stimulus is being applied. In the primary auditory cortex, the place of maximum activity provides evidence about what the frequency of the sound is. This type of encoding is also referred to as *labelled line encoding*, which again refers to the fact that it is which neurons are firing that identifies the stimulus. In both cases, there should be no implication that a single neuron encodes a particular stimulus, for the place code present in the cerebral cortex is one with a *sparse distributed representation*, in which a small proportion of the neurons is active, each tuned to a set of stimuli in a different way, with the profile of the firing rates across a population of neurons conveying the information (Rolls and Treves 2011, Rolls 2017a). The graded nature of the firing rate of each single neuron each tuned in a different way to the set of stimuli is important in this type of place encoding, illustrated in Fig. C.1 on page 828.

In some peripheral systems, *firing frequency encoding* is used, in which the firing frequency encodes the identity of the stimulus. One example is that for sounds below approximately 1 kHz, some auditory nerve neurons fire at the frequency of the sound, and can represent the phase by the time at which they fire (Schnupp, Nelken and King 2011). In most cortical areas, the frequency of firing does not represent the physical aspects of the stimulus, but in the somatosensory cortex and some related sensori-motor cortical areas, flutter vibration frequency (the frequency of mechanical vibration applied to for example a finger tip) is encoded by the firing rates of some neurons, and this is indeed important in understanding decision-making about flutter frequency (Romo, Hernandez and Zainos 2004, Deco and Rolls 2006, Deco, Rolls and Romo 2009, Deco, Rolls and Romo 2010, Martinez-Garcia, Rolls, Deco and Romo 2011, Deco, Rolls, Albantakis and Romo 2013) (Section 5.6). Although *oscillations* and rhythms are frequently found in a dynamical system as complex as the brain (Buzsáki 2006), and stimulus-dependent synchronous firing between neurons and *coherence* have been suggested as a way in which information might be encoded (Singer 1999, Singer 2000, Engel, Roelfsema, Fries, Brecht and Singer 1997, Fries 2009, Fries 2015) (Section 8.3), the evidence is that this makes only a small contribution to cortical encoding compared to place coding, at least in higher order visual cortical areas in primates under natural viewing conditions (Rolls and Treves 2011) (Appendix C).

The importance of place coding has been evident since even before the exciting work of Hubel and Wiesel (1962, 1968, 1977) which showed for example that in the primary visual cortex which neuron is firing conveys information about features and their location, for example about orientation of a bar or edge and its location in retinal space. The crucial point here is that it is which neurons are firing that conveys the information about the object and the spatial relations of its parts. The principle has been extended to high order visual cortical areas: in the inferior temporal visual cortex one reads off information from which neurons are firing about which face or object is being shown (Rolls, Treves, Tovee and Panzeri 1997d, Rolls 2008d, Rolls and Treves 2011) (Appendix C). This does not at all mean that this is local or grandmother cell encoding, in which just one neuron encodes which stimulus is present. Instead, there is a sparse distributed code, in which a small proportion of neurons is firing with a particular distribution of firing rates to represent one object, and another but partially overlapping population of neurons is firing with a particular distribution of firing rates to represent another object (Rolls and Tovee 1995b, Rolls, Treves and Tovee 1997b). In this encoding, the information increases approximately linearly with the number of neurons (up to reasonable numbers of neurons), showing that the coding by different neurons is independent (Rolls, Treves and Tovee 1997b, Franco, Rolls, Aggelopoulos and Jerez 2007). This is a sparse distributed place code, in that it is which neurons are firing, and their relative firing rates, that encode which object is present (Rolls 2008d, Rolls and Treves 2011). Similar encoding principles are used in the orbitofrontal cortex to encode information about taste and odour (Rolls, Critchley, Verhagen and Kadohisa 2010a), and in the hippocampus to encode information about spatial view (Rolls, Treves, Robertson, Georges-François and Panzeri 1998b).

There are two important points to emphasize here about place coding. The first is that what is being represented is encoded by a neuronal place code in the brain, in that for example neurons in the inferior temporal visual cortex convey information about which visual object is present, in the primary taste cortex about which taste, texture, or temperature is present; in the orbitofrontal cortex about which odour is present; and in the hippocampus about which spatial view is present (Rolls and Treves 2011). The second point is that if relational information about parts needs to be represented, as it must to define objects, then the relational information is encoded by which neurons are firing, where neurons are tuned not only to features or objects, but also to their location (Hubel and Wiesel 1968, Hubel and Wiesel 1977, Aggelopoulos and Rolls 2005, Rolls 2008d, Rolls 2012c) (Appendix C).

8.2 Place coding, with sparse distributed firing rate representations and independent information encoded by different neurons

8.2.1 Reading the code used by single neurons

A fundamental issue in neuroscience is the question of how information is encoded in the brain (Rolls 2008d, Rolls and Treves 2011). It is fundamental because to discover how the brain processes information we first have to know how it encodes and represents inputs from the environment, in order to understand the differences in the representations at each stage of cortical processing, and thus the computations performed by each stage of processing. Neuroscience has approached this problem by measuring the level of the spiking activity of single neurons and populations of neurons, as it is by the all-or-none spiking activity that information is accurately carried from one neuron to the thousands of neurons to which it has synaptic connections[4]. E.D. Adrian (1928) working at Cambridge succeeded in measuring the spiking activity of an afferent nerve fibre receiving from a stretch receptor in a muscle of the frog. He found that the nerve fibre generated action potentials or spikes with a firing rate

---

[4]The electrical spike of a neuron lasts for approximately 1 ms (one thousandth of a second) and is used to transmit information along the axon of a neuron because as an active process it ensures that the spike is all-or-none, either there or not, at the terminals of the axon. If neural transmission was by passive effects of a voltage change being sensed at a distance, it would be subject to unknown degradation influenced by the distance travelled and by the diameter of the axon.

measured in spikes/s that increased when the muscle was stretched. He showed for the first time the characteristic sigmoidal response function of a sensory neuron, i.e. the firing rate of the neuron increased at first slowly, then had a linear portion, and then saturated at a high firing rate as the stimulus applied increased (in this case the weight applied to the muscle). The objective character of the neuronal response function inspired the firing rate hypothesis, namely that the firing rate of a neuron represents the stimulation, because knowing the firing rate we can decode the strength of the applied stimulation[5].

There are now many experimental demonstrations that the firing rates of neurons indeed do encode information about stimuli. In one example neurons in the primary visual cortex have firing rates that depend on the orientation of edges or bars of light in their receptive field (Hubel and Wiesel 1968)[6]. In another example, neurons in the primary taste cortex have firing rates that depend on the concentration of the taste stimulus in the mouth (Scott, Yaxley, Sienkiewicz and Rolls 1986a), as do the activations in this area recorded with functional neuroimaging using functional magnetic resonance imaging (fMRI) of the brain, and the rated subjective intensity of the taste which is proportional to the fMRI signal (Grabenhorst, Rolls and Bilderbeck 2008a, Grabenhorst and Rolls 2008).

This type of sensory neuron also conveys information about what stimulus is present, something that goes beyond the orientation of a line in a visual scene, or the intensity of a taste. The information may be for example about what taste is present in the mouth. The neuron shown in Fig. 8.1 recorded in the secondary taste cortex which we discovered in the orbitofrontal cortex (Rolls, Yaxley and Sienkiewicz 1990, Baylis, Rolls and Baylis 1994) increased its firing rate to the sweet taste of glucose, but not to salt, sour or bitter (Rolls et al. 1990). The neuron was quite selective, in that it even had only a small response, a few spikes, to the flavour of fruit juice (BJ) in the mouth. The activity in Fig. 8.1 also illustrates the all-or-none spikes transmitted by single neurons, and the firing rate code being used. What stimulus is present in the mouth is signalled by the firing rate of the neuron, the number of spikes it emits in a given time period.

Although the firing rate of such neurons indicates how strong the stimulus is, which stimulus is present is encoded by which neurons are firing. Each neuron is tuned to respond to a different subset of stimuli. This type of coding, used throughout the cerebral cortex, can also thus be described as a place code, for which neuron is firing defines which stimulus or event is occurring. ('Place' here refers to which neuron is firing, for each neuron is necessarily at a different place in the cortical sheet.) The firing rate in such a 'labelled line' of 'place' representation determines the contribution of whatever is being coded for by that single neuron to the overall representation of a stimulus that is represented and transmitted by a whole population of neurons, as we shall see next.

---

[5]E. D. Adrian was awarded the Nobel Prize in Physiology or Medicine in 1932 (with Sir Charles Sherrington). I remember as a medical student at Cambridge finding one of Lord Adrian's electrometers for measuring neural activity on a rubbish pile outside the Department of Physiology. I asked Fergus Campbell what it was, and when he recognized it, we brought it back into the Physiological Laboratory, where I hope that it is now on display. Physiology at Cambridge continued to develop great expertise in recording and amplifying the small signals from neurons. The father of Richard S. Pumphrey, one of my undergraduate medical student friends at Cambridge, was Professor Richard Julius Pumphrey (http://www.jstor.org/pss/769453), and with his expertise in low-noise neuronal recording, was drafted in the war into the development of radar. I was told that one of the early tests of radar was whether more wobble could be seen from the blip on an oscilloscope if it was a woman than a man walking. Even as an undergraduate at Cambridge, there was great expertise in low-noise recording, and I made in the Physiological Laboratory low-noise cathode-follower first-stage amplifiers for distribution to colleagues internationally. I built on experience I had developed while at school in building an oscilloscope (with a 3.5 inch cathode ray tube and rectifiers as voltage doublers to obtain the 4,000 volts needed for the plate) and amplifiers for recording from the giant axon of the earthworm.

[6]David Hubel and Torsten Wiesel were awarded (with Roger Sperry) the Nobel Prize in Physiology or Medicine in 1981.

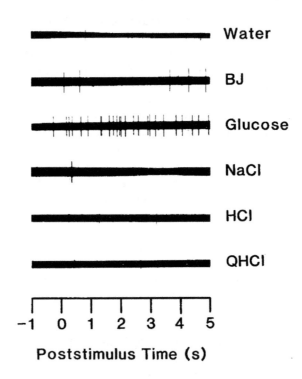

Fig. 8.1 Examples of the responses recorded from one caudolateral orbitofrontal taste cortex neuron to the six taste stimuli, water, 20% blackcurrant juice (BJ), 1 M glucose (sweet), 1 M NaCl (salt), 0.01 M HCl (sour), and 0.001 M quinine HCl (QHCl, bitter). The stimuli were placed in the mouth at time 0. Each vertical line is a voltage spike generated by the action potential of the single neuron being recorded. (Reproduced with permission from Journal of Neurophysiology 64: 1055–1066, Gustatory responses of single neurons in the orbitofrontal cortex of the macaque monkey. E. T. Rolls, S. Yaxley, Z. J. Sienkiewicz. Copyright ©1990, The American Physiological Society.)

In another example, the firing rates of the 'face neurons' that we discovered in the inferior temporal visual cortex (IT), amygdala, and orbitofrontal cortex reflect which face has been seen (Sanghera, Rolls and Roper-Hall 1979, Perrett, Rolls and Caan 1982, Rolls, Critchley, Browning and Inoue 2006a, Rolls 2011d). In particular, each neuron responds with a different firing rate to each different face in a set, as illustrated in Fig. C.4, so that which face is being seen can be read off or decoded by knowing the firing rates of a set of these neurons, as described below. Face-selective neurons have also been described in humans (Kreiman, Koch and Freid 2000), though in terms of their properties and locations they seem to be involved more in memory-related representations than the perceptual representations described here.

Further, something close to Adrian's 'Sensation' (the title of his 1928 book was *The Basis of Sensations*) is reflected rather directly in the firing rate code used by neurons. A good example is that the subjective sensation of the thickness of carboxymethylcellulose (a food thickener) in the mouth is proportional to the log of its viscosity (Kadohisa, Rolls and Verhagen 2005a), as is the firing rate of single neurons (Verhagen, Kadohisa and Rolls 2004), and the signal recorded with functional neuroimaging (using functional magnetic resonance imaging, fMRI) in the human primary taste cortex (De Araujo and Rolls 2004).

In other parts of the brain, the firing rates still encode information, but about events that reflect the internal workings of the brain, rather than events that reflect the firing of sensory receptors. For example, the reward system that we discovered in the orbitofrontal cortex for taste reward (Rolls, Sienkiewicz and Yaxley 1989c) as well as many other rewards (Rolls, Burton and Mora 1980, Thorpe, Rolls and Maddison 1983, Rolls 2005, Rolls and Grabenhorst 2008, Grabenhorst and Rolls 2011, Rolls 2014a) contains neurons that fire fast when a taste is rewarding, and gradually decrease their responses to zero to the food as it is fed to satiety so that the food is no longer rewarding (Rolls, Sienkiewicz and Yaxley 1989c) (Fig. 2.6[7]). Correspondingly, the subjectively rated pleasantness of the food reflects this neuronal firing, for the subjective pleasantness decreases to zero as the food is fed to satiety, as does the signal recorded with fMRI neuroimaging in the human orbitofrontal cortex (Kringelbach, O'Doherty, Rolls and Andrews 2003). Thus the neuronal firing rates, which are reflected in the fMRI signal (Rolls, Grabenhorst and Deco 2010b), encode in this part of the brain the reward value and subjective pleasantness of many stimuli and events in the world, and this is fundamental to our understanding of emotion (Rolls 2005, Rolls and Grabenhorst 2008, Rolls 2014a) (see Chapter 15).

Another good example of how firing rates can now be read to understand the internal workings of the brain is provided by the spatial view neurons that we discovered in the hippocampus (Georges-François, Rolls and Robertson 1999, Robertson, Rolls and Georges-François 1998, Rolls, Robertson and Georges-François 1997a, Rolls, Treves, Robertson, Georges-François and Panzeri 1998b) (see Chapter 24). These neurons increase their firing rates only when one part of a spatial environment is being looked at. These responses occurred when the spatial view was seen from different angles with respect to the body and with many different head directions, so the representation was not egocentric (that is it was not with respect to the body or head axis). The increase in neural firing rate could occur when the spatial view was seen from many places in the environment. These findings indicated that spatial view neurons represent locations 'out there' in space on allocentric or 'world' coordinates. Very interestingly, some of these neurons had a short-term memory of the allocentric spatial view for their firing, which occurred even in the dark provided that the eyes moved to look at the allocentric location of the spatial view. (This type of memory lasts for one or two minutes, after which one's sense of direction if one is moving around in the dark becomes much less good.) This allocentric representation of space is we believe part of a system for remembering where objects have been seen, for example where one has seen good food, or other good resources (without necessarily ever having been actually at the place, having only seen the place), for some of these neurons respond to a combination of a spatial view and the object that has been shown at that location (Rolls, Xiang and Franco 2005c). Some of the neurons also respond during recall of the location in space where an object was last seen (Rolls and Xiang 2006). This type of neuronal activity is prototypical of an episodic memory, the memory for particular past events or episodes, and indeed these neurons are an important component of the theory of episodic memory that we have developed (Rolls 1989b, Treves and Rolls 1994, Rolls 2010b, Kesner and Rolls 2015) (see further Chapter 24).

What has been described in this subsection (8.2.1) is that the firing rates of single neurons provide much evidence about how the world is represented in the brain.

When considering the operation of the brain, and of neuronal networks in the brain, it is found that many useful properties arise if each input to the network (arriving on the set of input axons that make synapses onto a single neuron) is encoded by the firing rate of an ensemble or subset of the axons (distributed encoding), and is not signalled by the activity of a single input or axon, which is called local encoding. Because this firing rate code distributed across a population of neurons each tuned differently is so important to understanding how the brain

---

[7]The neuron did not decrease its firing rate to the fruit juice (BJ), which was not being fed to satiety. Thus the reward value decrease was specific to the food eaten to satiety. It was in experiments of this type that we discovered sensory-specific satiety (Rolls 1981a, Rolls 2011e, Rolls 2014a).

operates (Rolls and Treves 2011, Rolls 2017a), this neural encoding is considered further in the next section, and in more detail in Appendix C. When taken together with the concepts outlined in the rest of this book, the aim is to show how we at present can go far beyond phenomena found in the brain, towards an understanding of how it works computationally. This is one of the major advances in neuroscience, in the understanding of the Principles of Operation of our own brains.

8.2.2 Understanding the code provided by populations of neurons

I start with some definitions, then summarize some evidence that shows the type of encoding used in some brain regions, and then show how the representation found is advantageous (Rolls 2008d, Rolls and Treves 2011).

8.2.2.1 Definitions

A *local representation* is one in which all the information that a particular stimulus or event occurred is provided by the activity of one of the neurons. In a famous example, a single neuron might be active only if one's grandmother was being seen, and this is sometimes called grandmother cell encoding. (The term was coined by Jerry Lettvin in about 1969 – see Charles Gross (2002).) An implication is that most neurons in the brain regions where objects or events are represented would fire only very rarely (Barlow 1972, Barlow 1995). A problem with this type of encoding is that a new neuron would be needed for every object or event that has to be represented. Another disadvantage is that this type of coding does not generalize easily to similar inputs, so that similarities between perceptions or memories would not be apparent. Another disadvantage is that the system is rather sensitive to brain damage: if a single neuron is lost, the representation may be lost. Another disadvantage of local encoding is that the storage capacity in a memory system in the brain (the number of stimuli that can be stored and recalled) may not be especially high (in the order of the number of synapses onto each neuron) (Rolls 2008d).

A *fully distributed representation* is one in which all the information that a particular stimulus or event occurred is provided by the activity of the full set of neurons. If the neurons are binary (e.g. either active or not), the most distributed encoding is when half the neurons are active (i.e. firing fast) for any one stimulus or event, and half are inactive. Different stimuli are represented by different subsets of the neurons being active.

A *sparse distributed representation* is a distributed representation in which a small proportion of the neurons is active at any one time. In a sparse representation with binary neurons (i.e. neurons with firing rates that are either high or low), less than half of the neurons are active for any one stimulus or event. For binary neurons, we can use as a measure of the sparseness the proportion of neurons in the active state. For neurons with real, continuously variable, values of firing rates, the sparseness a^p of the representation provided by the population can be quantified as described in Section B.3.3.7. A low value of the sparseness a^p indicates that few neurons are firing for any one stimulus.

8.2.2.2 Encoding provided by neuronal populations

At the time that Barlow (1972) wrote, there was little actual evidence on the activity of neurons in the higher parts of the visual and other sensory systems. There is now considerable evidence, some of which is now described (Rolls 2008d, Rolls and Treves 2011, Rolls 2017a), with more detail in Appendix C.

The representation is distributed as illustrated in Fig. C.4 on page 841 by the firing rates produced by each of 68 stimuli in a single neuron in the primate inferior temporal visual cortex. Rather few stimuli produce high firing rates (e.g. above 60 spikes/s), and increasingly large numbers of stimuli produce lower and lower firing rates. The spontaneous firing rate of this neuron, the rate when no stimuli were being shown, was 20 spikes/s (Rolls and Tovee 1995b). The histogram bars indicate the change of firing rate from the spontaneous value produced by each stimulus. Stimuli that are faces are marked F, or P if they are in profile. B refers to images of scenes that included either a small face within the scene, sometimes as part of an image that included a whole person, or other body parts, such as hands (H) or legs. The non-face stimuli are unlabelled. The neuron responded best to three of the faces (profile views), had some response to some of the other faces, and had little or no response, and sometimes had a small decrease of firing rate below the spontaneous firing rate, to the non-face stimuli. The representation was thus rather distributed, and this is typical of neurons in the higher order visual cortical areas (Rolls and Tovee 1995b, Baddeley, Abbott, Booth, Sengpiel, Freeman, Wakeman and Rolls 1997, Treves, Panzeri, Rolls, Booth and Wakeman 1999, Franco, Rolls, Aggelopoulos and Jerez 2007); of neurons in the taste and flavour cortical areas in the insula and orbitofrontal cortex (Verhagen, Kadohisa and Rolls 2004, Rolls, Verhagen and Kadohisa 2003e, Verhagen, Rolls and Kadohisa 2003, Kadohisa, Rolls and Verhagen 2004, Kadohisa, Rolls and Verhagen 2005a); and of neurons tuned to spatial view in the hippocampus (Rolls, Treves, Robertson, Georges-François and Panzeri 1998b).

These data provide a clear answer to whether these neurons are grandmother cells: they are not, in the sense that each neuron has a graded set of responses to the different members of a set of stimuli, with the prototypical distribution similar to that of the neuron illustrated in Fig. C.4. On the other hand, each neuron does respond very much more to some stimuli than to many others, and in this sense is tuned to some stimuli.

With data of this type recorded from single neurons and simultaneously recorded populations of single neurons, and the application of quantitative information theoretic measures described in Appendix C (Rolls 2008d, Rolls and Treves 2011), the working hypotheses about the **principles of cortical neuronal encoding** in the visual, hippocampal, olfactory and taste cortical systems are as follows (Rolls 2008d, Rolls and Treves 2011, Rolls 2017a) (Appendix C):

1. Much information is available about the stimulus presented in the number of spikes emitted by single neurons in a fixed time period, the firing rate. Importantly, just knowing the firing rate of a single neuron provides quite a lot of information (evidence) about which particular stimulus was shown, as illustrated by the neuronal responses shown in Fig. C.4.

2. Much of this firing rate information is available in short periods, with a considerable proportion available in as little as 20 ms. This rapid availability of information enables the next stage of processing to read the information quickly, and thus for multistage processing to operate rapidly. This time is the order of time over which a receiving neuron might be able to utilize the information, given its synaptic and membrane time constants. In this time, a sending neuron is most likely to emit 0, 1, or 2 spikes.

3. This rapid availability of information is confirmed by population analyses, which indicate that across a population of neurons, much information is available in short time periods.

4. More information is available using this rate code in a short period (of e.g. 20 ms) than from just the first spike (Rolls, Franco, Aggelopoulos and Jerez 2006b).

5. Little information is available by time variations within the spike train of individual neurons for static visual stimuli (in periods of several hundred milliseconds), apart from a small amount of information from the onset latency of the neuronal response (Tovee, Rolls, Treves

and Bellis 1993, Rolls and Treves 2011). For a time-varying stimulus, clearly the firing rate will vary as a function of time.

6. Across a population of neurons, the firing rate information provided by each neuron tends to be independent; that is, the information increases approximately linearly with the number of neurons. The outcome is that the number of stimuli that can be encoded rises exponentially (i.e. very rapidly) with the number of neurons in the ensemble (Rolls, Treves and Tovee 1997b). An implication of the independence is that the response profiles to a set of stimuli of different neurons are uncorrelated (Franco, Rolls, Aggelopoulos and Jerez 2007).

7. The information in the firing rate across a population of neurons can be read moderately efficiently by a decoding procedure as simple as a synaptically weighted sum of the inputs (a dot product) (Rolls, Treves and Tovee 1997b, Rolls and Treves 2011), which is what is shown in equation 1.1 on page 6. This is the simplest type of processing that might be performed by a neuron, as it involves taking a dot (or inner) product of the incoming firing rates with the receiving synaptic weights to obtain the activation (e.g. depolarization) of the neuron. This type of information encoding ensures that the simple emergent properties of associative neuronal networks such as generalization, completion, and graceful degradation (see Section 8.2.2.3) can be realized very naturally and simply. This type of encoding is also what makes the responses of an individual neuron interpretable: it is tuned to respond better to some stimuli than others, with a graded firing rate representation of the type illustrated in Fig. C.4. This decoding principle is illustrated in Fig. 8.2.

8. There is little additional information to the great deal available in the firing rates from any stimulus-dependent cross-correlations or synchronization between the spikes of different neurons that may be present (Aggelopoulos, Franco and Rolls 2005, Rolls 2008d, Rolls and Treves 2011). Stimulus-dependent synchronization might in any case only be useful for grouping different neuronal populations, and would not easily provide a solution to the binding problem in vision. Instead, the binding problem in vision may be solved by the presence of neurons that respond to combinations of features in a given spatial position with respect to each other (Chapter 25) (Rolls 2008d, Rolls 2009c, Rolls and Treves 2011).

9. There is little information available in the order of the spike arrival times of different neurons for different stimuli that is separate or additional to that provided by a rate code (Rolls, Franco, Aggelopoulos and Jerez 2006b). The presence of spontaneous activity in cortical neurons facilitates rapid neuronal responses, because some neurons are close to threshold at any given time, but this also would make a spike order code difficult to implement.

10. Analysis of the responses of single neurons to measure the sparseness of the representation indicates that the representation is distributed, and not grandmother cell like (or local) (Rolls and Tovee 1995b, Franco, Rolls, Aggelopoulos and Jerez 2007, Rolls 2008d, Rolls and Treves 2011).

11. The representation is not very sparse in the perceptual systems studied (as shown for example by the values of the single cell sparseness), and this may allow much information to be represented. At the same time, the responses of different neurons to a set of stimuli are decorrelated, in the sense that the correlations between the response profiles of different neurons to a set of stimuli are low. Consistent with this, the neurons convey independent information, at least up to reasonable numbers of neurons. The representation may be more sparse in memory systems such as the hippocampus, and this may help to maximize the number

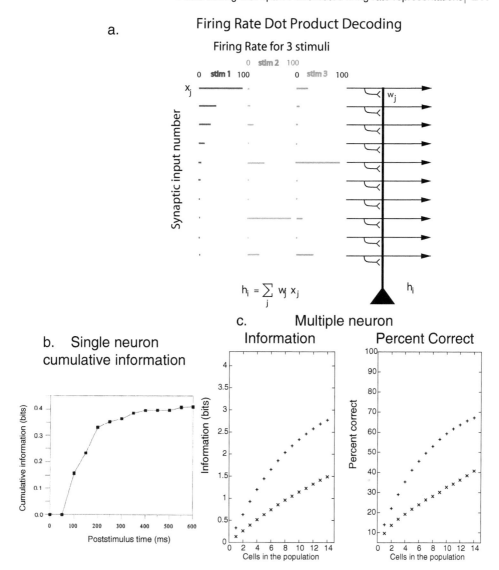

a. Firing Rate Dot Product Decoding

$$h_i = \sum_j w_j x_j$$

b. Single neuron cumulative information

c. Multiple neuron

Fig. 8.2 Information encoding by firing rates. (a). Each stimulus is encoded by an approximately exponential firing rate distribution of a population of neurons. The distribution is ordered to show this for stimulus 1, and other stimuli are represented by similar distributions with each neuron tuned independently of the others. The code can be read by a dot (inner) product decoding performed by any receiving neuron of the firing rates with the synaptic weights, and it is the almost linear increase in information with the number of neurons illustrated in (c) which shows that the tuning profiles to the set of stimuli are almost independent. (b). Much information is available from single neurons, and from populations of neurons, using the number of spikes in short time windows of e.g. 50 ms. This is shown by the cumulative information from the firing rates for different time periods from single neurons in the cortex in the superior temporal sulcus to a set of 10 face and non-face stimuli, and is remarkable in that the neurons do not start to respond to the visual stimuli until 80–90 ms, so that the data point at 100 ms shows the information from 20 ms or less of firing. (c). The information available about which of 20 faces had been seen that is available from the responses measured by the firing rates in a time period of 500 ms (+) or a shorter time period of 50 ms (x) of different numbers of temporal cortex cells. The corresponding percentage correct from different numbers of cells is also shown. Decoding with dot product decoding reveals similar principles.

of memories that can be stored in associative networks (Rolls 2008d, Rolls and Treves 2011). The representations may be sparse in attractor networks to maximize the storage capacity. This may apply in hippocampal CA3, the anterior pyriform cortex, and the superficial layers (L2/L3) of the cerebral cortex, as described in Section 18.2.The representations may be less sparse in competitive networks which follow attractor networks, because more fully distributed representations can encode more information. This latter may apply in hippocampal CA1, the posterior pyriform cortex, and the deep layers of the cerebral cortex, as described in Section 18.2.

12. Even when temporal order must be encoded and stored in the brain, recent evidence indicates that a firing rate code is being used (MacDonald, Lepage, Eden and Eichenbaum (2011); see Chapter 24).

13. Because neurons can convey almost independent information, and each neuron is tuned in a different way to a set of stimuli (the most important property of cortical encoding), measures that take the average of the activity of many neurons (or synapses (Logothetis 2008)) cannot reveal how information is encoded in the brain. For example, a typical $3 \times 3 \times 3$ mm voxel (volume) with fMRI would contain 810,000 neurons if the neuronal density is taken as 30,000 neurons/mm^3 (Table 1.1). The result is that the activation of a voxel with fMRI reveals little about exactly which stimulus was presented (e.g. whose face), and even the information reflected by a voxel that it is in a category (such as that it is pleasant, is predicted to be chosen, or is a face) is less than that of a typical neuron, and moreover does not increase linearly with the number of voxels (Rolls, Grabenhorst and Franco 2009) a fundamental property of neuronal encoding. The same argument applies also to other measures that group together the effects of a large number of neurons, such as local field potentials (LFP), and magnetoencephalography (MEG). Such measures, including fMRI and positron emission topography (PET), are useful in analyzing what categories of information may be represented in a brain region (e.g. faces, houses, spatial scenes), but do not reveal which particular instance is being represented, or the details of the neuronal code. (Insofar as anything is revealed about the neural code, these measures are consistent with the evidence that firing rates rather than properties such as stimulus-dependent neuronal synchronization are being used (Rolls, Grabenhorst and Franco 2009, Rolls, Grabenhorst and Deco 2010b, Rolls and Treves 2011).

Because understanding neuronal encoding in a brain region, and how this is related to the detailed neuronal network functional architecture of a region is so crucial to understanding neuronal computation, functional neuroimaging in humans can never replace neuronal recordings, possible mainly in animals, and these are among the reasons why research with animals continues to be important in understanding how the brain actually works. For some brain regions, including the visual system, the prefrontal including orbitofrontal cortex, the hippocampus (given the types of representation found in it), and even the taste system which one might think is evolutionarily old, the human brain is so far developed and so different from that in rodents that in such systems it is important to rely on findings in non-human primates (Chapter 18). Further, information is conveyed between the computing elements of the brain, the neurons, by action potentials of single cells travelling along their axons, and it is therefore only by analyzing at this single and multiple simultaneously recorded single neuron level that we will understand the encoding being used in the brain to transfer information between its computing elements, or what information is being encoded and transmitted (Appendix C) (Rolls 2008d, Rolls and Treves 2011).

8.2.2.3 Advantages of different types of coding

One advantage of distributed encoding is that the similarity between two representations can be reflected by the correlation between the two patterns of activity that represent the different stimuli. I describe in Fig. 8.2 the idea that the input to a neuron is represented by the activity on its set of input synapses with each input weighted by the synaptic strength of the synapse. There are in the order of 10,000 such excitatory synaptic terminals carrying the input spikes, and 10,000 corresponding synapses, on each neuron. In this way, a single cortical neuron receives in the order of 10,000 synaptic inputs, each one from a different cortical neuron. At each synapse, a current is injected into the neuron that is weighted by the synaptic strength. The weighting corresponds mathematically to the operation of multiplication. The total input to the neuron is then the sum of these 10,000 currents. The more similar the input set of firing rates is to the set of synaptic weights, the larger will be the output of the neuron. Thus neurons compute the similarity between the firing rates on each input and the strengths of the corresponding synaptic weights on a neuron, which might represent a previously stored input. In this sense, neurons compute the similarity or correlation between a set of input firings and their synaptic weights, and thus can compare a new input with an input previously stored as modified synaptic strengths or weights (see further Fig. 1.3 and equation 1.1). The correlation will be high if the activity of each axon in the two representations is similar; and will become more and more different as the firing rate activity of more and more of the axons differs in the two representations. Thus the similarity of two inputs can be represented in a graded or continuous way if (this type of) distributed encoding is used. This enables generalization to similar stimuli, or to incomplete versions of a stimulus (if it is, for example, partly seen or partly remembered), to occur. With a local representation, either one stimulus or another is represented, each by its own neuron firing, and similarities between different stimuli are not encoded.

Another advantage of distributed encoding is that the number of different stimuli that can be represented by a set of C components (e.g. the firing rates applied to the C synaptic inputs to a neuron) can be very large. It can be much larger than C, as shown in Section B.3 (Treves and Rolls 1991, Rolls 2008d) (see Section 8.2). Put the other way round, even if a neuron has only a limited number of inputs (e.g. a few thousand), it can nevertheless receive a great deal of information about which stimulus was present. This ability of a neuron with a limited number of inputs to receive information about which of potentially very many input events is present is probably one factor that makes computation by the brain possible. With local encoding, the number of stimuli that can be encoded increases only linearly with the number C of axons or components (because a different component is needed to represent each new stimulus).

In the real brain, there is now good evidence that in a number of brain systems, including the high-order visual and olfactory cortices, and the hippocampus, distributed encoding with the properties described above, of representing similarity, and of exponentially increasing encoding capacity as the number of neurons in the representation increases, is found. For example, as we have seen, in the primate inferior temporal visual cortex, the number of faces or objects that can be represented increases approximately exponentially with the number of neurons in the population (Rolls and Tovee 1995b, Abbott, Rolls and Tovee 1996, Rolls, Treves and Tovee 1997b, Rolls, Treves, Robertson, Georges-François and Panzeri 1998b, Rolls, Franco, Aggelopoulos and Reece 2003b, Rolls, Aggelopoulos, Franco and Treves 2004, Franco, Rolls, Aggelopoulos and Treves 2004, Aggelopoulos, Franco and Rolls 2005, Rolls, Franco, Aggelopoulos and Jerez 2006b, Rolls 2008d). A similar result has been found for the encoding of position in space by the primate hippocampus (Rolls, Treves, Robertson, Georges-François and Panzeri 1998b), and for the encoding of olfactory stimuli in the orbitofrontal cortex (Rolls,

Critchley, Verhagen and Kadohisa 2010a).

The **advantages of the sparse graded distributed firing rate encoding** found in many primate cortical areas include the following, with further evidence described in Appendix C and elsewhere (Rolls and Treves 2011, Rolls 2017a):

1. *Exponentially high coding capacity.* This property arises from a combination of the encoding being sufficiently close to independent by the different neurons (i.e. factorial), and sufficiently distributed, and is illustrated by the evidence shown in Figs. 25.13, C.22 and 25.14.

2. *Ease with which the code can be read by receiving neurons.* For brain plausibility, it is also a requirement that neurons should be able to read the code. This is why when we have estimated the information from populations of neurons, we have used in addition to a probability estimating measure (optimal, in the Bayesian sense), also a dot product measure, which is a way of specifying that all that is required of decoding neurons would be the property of adding up postsynaptic potentials produced through each synapse as a result of the activity of each incoming axon. It was found that with such a neurally plausible algorithm (the Dot Product, DP, algorithm), which calculates which average response vector the neuronal response vector on a single test trial was closest to by performing a normalised dot product (equivalent to measuring the angle between the test and the average vector), the same generic results were obtained, with only a 40% reduction of information compared to the more efficient (Bayesian) algorithm (Fig. 25.13). This is an indication that the brain could utilise the exponentially increasing capacity for encoding stimuli as the number of neurons in the population increases.

3. *Higher Resistance to Noise.* Because the information is decoded from a large population of neurons by inner product multiplication with the synaptic weight vector, there is less dependence on the random (almost Poisson) firing times for a given mean rate of single neurons, and thus there is resistance to the noise that is inherent in the firing times of single neurons (Chapter 5).

4. *Generalization.* Generalization to similar stimuli is again a property that arises in neuronal networks if distributed but not if local encoding is used. The generalization arises as a result of the fact that a neuron can be thought of as computing the inner or dot product of the stimulus representation expressed as the firing rate on the set of input neurons with its synaptic weight vector (Appendix C).

5. *Completion.* Completion occurs in associative memory networks by a similar process. Completion is the property of recall of the whole of a pattern in response to any part of the pattern.

6. *Graceful degradation or fault tolerance.* Again, because the information is decoded from a large population of neurons by inner product multiplication with the synaptic weight vector, there is less dependence on the firing of any one neuron or on any particular subset of neurons, so that if some neurons are damaged, the performance of the system only gradually degrades, and is in this sense fault tolerant.

7. *Speed of readout of the information.* The information available in a distributed representation can be decoded by an analyzer more quickly than can the information from a local representation, given comparable firing rates. Within a fraction of an interspike interval, with a distributed representation, much information can be extracted. In effect, spikes from many different neurons can contribute to calculating the angle between a neuronal population and a synaptic weight vector within an interspike interval. With local encoding, the speed of information readout depends on the exact model considered, but if the rate of firing needs to be taken into account, this will necessarily take

time, because of the time needed for several spikes to accumulate in order to estimate the firing rate.

8. *Distributed representations support attractor representations in the brain.* Another advantage of distributed representations is that attractor states are very likely to be used to hold information on-line for short-term memory, attention, long-term memory, decision-making etc, and attractors can only be supported in the brain by neuronal systems with distributed not local representations because positive feedback from many neurons (not from a single neuron) is required to keep an attractor state active in the brain. (A single neuron providing input to itself would be insufficient to maintain its activity in a typical cortical pyramidal cell with 10,000 recurrent collateral synapses.)

Further evidence on all these points is in Appendix C.

My view is that even in cortical systems such as those involved in language, the same sparse distributed graded firing rate encoding will be used, though the coding might be more sparse (Rolls 2017a). My reasoning is that the identity of a person or object is encoded by a sparse distributed representation of neurons that reflect the main attributes. The more those attributes match, the more likely I am to decode the input as, for example, my grandmother. This type of representation can be built by competitive learning. If on one occasion my grandmother is wearing glasses, that is then an attribute represented by some members of the population of neurons in the sparse distributed representation. If the glasses are absent (but are usually present), the input may look a little less like my grandmother, because the dot product decoding used by the receiving neurons produces a smaller result in this population of neurons. This type of 'dot product' decoding, which is neuronally plausible, is described in Appendix C. A local or grandmother cell representation does not cope at all with this scenario, for I might need a separate grandmother cell for every possible appearance of my grandmother, and there would be no useful generalization between the different grandmother cells all for my same grandmother but with different hats, glasses, etc.

Another advantage of sparse distributed representations for language is that attractor states are very likely to be used to hold on-line the parts of a sentence for syntactical operations (Rolls and Deco 2015a), and attractors can only be supported in the brain by neuronal systems with distributed not local representations as noted above.

What happens in this situation if I need to remember that at a particular occasion, say her 70th birthday, my grandmother was wearing a red hat? My view is that this case where particular attributes have to be remembered is then becomes a role for the hippocampus, which is involved in episodic memory, and which can associate together the sparse representation received from the inferior temporal visual cortex that it is my grandmother, with other neurons that sparsely encode the red hat (and with other neurons that encode the time and place) (Chapter 24).

8.3   Synchrony, coherence, and binding

Quantitative information theoretic analyses further show that relatively little information is encoded by stimulus-dependent cross-correlations between neurons, with typically 95% or more of the information being encoded by a place/firing rate code where the graded firing rates of each neuron in a sparse distributed representation are used as the encoding principle (Rolls 2008d, Rolls and Treves 2011), as considered further in Section 17.1.1.

Gamma band (50–70 Hz) oscillations have been found in many cortical areas and in a variety of tasks. It has been studied most extensively in the visual cortex of cats and monkeys (Gray, Konig, Engel and Singer 1989, Fries, Reynolds, Rorie and Desimone 2001, Womelsdorf, Fries, Mitra and Desimone 2006, Womelsdorf, Schoffelen, Oostenveld, Singer, Desimone, Engel and Fries 2007, Fries, Womelsdorf, Oostenveld and Desimone 2008, Fries 2015). Several authors have proposed that these oscillations which imply synchronization of neuronal firing influence the interactions among neuronal groups (Varela, Lachaux, Rodriguez and Martinerie 2001, Salinas and Sejnowski 2001), a hypothesis referred to as communication through coherence (CTC) (Fries 2005, Fries 2009, Fries 2015).

The communication through coherence hypothesis proposes that coherent or synchronous oscillations in connected neural systems can promote communication. It has been applied mainly to how oscillations interact in connected networks. Rolls, Webb and Deco (2012) tested by simulations whether information transmission about an external stimulus from one network to a second network is influenced by gamma oscillations, by whether the oscillations are coherent, and by their phase. Gamma oscillations were induced by increasing the relative conductance of AMPA to NMDA excitatory synapses. It was found that small associative synaptic connection strengths between the networks were sufficient to produce information transmission (measured by Shannon mutual information) such that the second attractor network took the correct decision based on the state of the first network. Although gamma oscillations were present in both networks, the synaptic connections sufficient for perfect information transmission about the stimulus presented to the network (100% correct, 1 bit of information) were insufficiently strong to produce coherence, or phase locking, between the two networks, which only occurred when the synaptic strengths between the networks were increased more than 10 times. Further, the phase of the oscillations between the networks did not influence the information transmission or its speed at these connection strengths. Moreover, information transmission was as good when oscillations were abolished by altering the AMPA to NMDA ratio (with a greater proportion of NMDA receptor currents with their long time constants damping down the oscillations). Similar results were found when the second network was not an attractor decision-making network. Thus information transmission can occur before synapses have been made sufficiently strong to produce coherence.

Rolls, Webb and Deco (2012) suggested that gamma oscillations are a by-product of strong stimulation of neuronal networks with short time constant receptors (AMPA and GABA), that the longer time constant NMDA receptors tend to damp down the oscillations, and that a very critical and evaluative approach is needed when assessing whether coherence is a useful operational principle of the cerebral cortex. Rigorous evidence is needed that information transmission is influenced by the phase relations between particular neuronal populations when the firing rates of the neurons have been controlled and shown not to be the relevant factor in the information transmission.

8.4 Principles by which the representations are formed

The whole issue of how information is encoded in the cortex is important and fundamental to understanding the operation of the cortex, and is therefore considered in more detail in Appendix C.

The ways in which sparse distributed representations with independent information conveyed by different neurons are set up using learning in competitive networks with diluted connectivity are considered in Sections 7.4 and B.4.

8.5   Information encoding in the human cortex

Much of the evidence on what is represented in different cortical areas, and many of the concepts about what different cortical areas perform that correspond to those in humans, comes from single neuron studies in macaques (Rolls and Treves 2011) (Appendix C). It is therefore important to consider and assess (Rolls 2015d) some of the evidence now becoming available from recordings from single neurons in humans (Fried, Rutishauser, Cerf and Kreiman 2014), for even a little evidence of this type may help to show how understanding from studies in the non-human cortex is providing a framework that is relevant to understanding the human brain, and potentially its disorders. This information is being obtained in clinical studies performed to help treat disorders such as epilepsy. If course there are many provisos that apply, such as that the brain regions from which recordings are made may be dysfunctional, that the brain areas from which recordings are made are constrained by what is clinically useful, and that the time for recording from neurons is necessarily limited, so that the data obtained are likely to be limited.

Some of the most fascinating findings are about how neurons respond in medial temporal lobe regions. Some neurons have responses that appear quite selective, with one neuron responding for example to Jennifer Aniston but much less to other individuals, and responding multimodally, for example not just to the sight of Jennifer Aniston, but also to the sound of her voice. At first, the tuning of single neurons might on the basis of a few striking examples in humans be thought to be more selective than those in the macaque temporal lobe, but on the basis of many such recordings, those who have recorded these neurons argue that the code is sparse distributed (Quiroga, Kreiman, Koch and Fried 2008, Quiroga 2012, Quiroga 2013), and therefore somewhat similar to that of neurons in macaques (Rolls and Treves 2011) (Appendix C). To illustrate the type of tuning that can be found, the responses of a single neuron in the human amygdala activated by animal pictures are shown in Fig. 8.3. The neuron responded to different extents to pictures of animals, but showed no responses to pictures of humans or landmarks (Mormann, Dubois, Kornblith, Milosavljevic, Cerf, Ison, Tsuchiya, Kraskov, Quiroga, Adolphs, Fried and Koch 2011).

Neurons recorded in the human medial temporal lobe areas such as the hippocampus are described as being 'concept neurons', for not only are they multimodal, but they can also respond to imagery of for example Jennifer Aniston in the famous case in one patient (Fried et al. 2014). How does this fit in with concepts of hippocampal function? The hippocampus is thought to be involved in episodic or event memory, for example the memory of a particular person in a particular place (Chapter 24) (Kesner and Rolls 2015). Each memory must be as separate as possible from other memories, and the evidence is that single neurons in the macaque CA3 respond to combinations of for example a particular place being viewed, or a particular object, or a combination of these (Rolls, Xiang and Franco 2005c). Indeed the theory is that the CA3 region with its recurrent collateral associatively modifiable connections enables any object or person to be associated with any place by this associativity, to form a unique episodic memory (Rolls 1989b, Rolls 2008d, Kesner and Rolls 2015) (Chapter 24). Human neurons in the hippocampus that respond to 'concepts', for example with quite selective tuning for a person, appear to be consistent with this theory. Of course, the nature of the sparse distributed encoding is that no single neuron does need to be selective for just one person or object, for it is across a population of such neurons with sparse distributed encoding that a particular individual is represented (Rolls and Treves 2011) and becomes part of the autoassociation or attractor memory of a particular object or person in a particular place (Rolls 2008d, Kesner and Rolls 2015) (Chapter 24).

Suthana and Fried in Fried et al. (2014) consider the representation of space by neurons in the human medial temporal lobe. A potentially important difference between rodents and

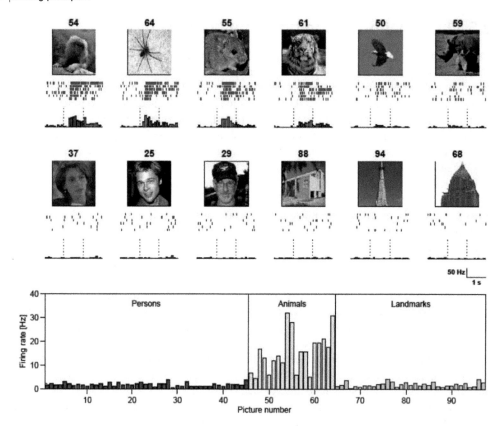

Fig. 8.3 Example of a single neuron in the human amygdala activated by animal pictures. The neuron responded to different extents to pictures of animals, but had no responses to pictures of humans or landmarks. Upper rows: Responses of a neuron in the right amygdala to pictures from different stimulus categories, presented in randomized order. For each picture, the corresponding raster plots (order of trials from top to bottom) and peri-stimulus time histograms are given. Vertical dashed lines indicate image onset and offset (1 s apart). Lower row: The mean response firing rates of this neuron between image onset and offset across 6 presentations for all individual pictures. Pictures of persons, animals and landmarks are denoted by brown, yellow and cyan bars, respectively. (Reprinted by permission from Macmillan Publishers Ltd: Nature Neuroscience, A category-specific response to animals in the right human amygdala, 14, 1247–1249. Florian Mormann, Julien Dubois, Simon Kornblith et al. Copyright ©2011.)

primates including humans is that rat hippocampal neurons represent the place where the rat is located (O'Keefe 1979), whereas a population of neurons in the macaque hippocampus and parahippocampal cortex has been discovered that represents a spatial view of a scene with its landmarks independently of the place at which the macaque is located (Rolls and O'Mara 1995, Rolls, Robertson and Georges-François 1997a, Rolls, Treves, Robertson, Georges-François and Panzeri 1998b). Further, this representation is allocentric, in that it is the allocentric location in space being viewed and not egocentric cues such as eye position, or position relative to head direction, or indeed head direction itself (Georges-François, Rolls and Robertson 1999). The importance of these neurons is that they provide a basis for remembering where an object is located at a place at which one has never been present, such as a source of food at some distance away, or the position of a particular player when a goal was scored while one was a spectator who had never been on the pitch. Place cells could not implement that type of memory. The implication is that such spatial view cells are therefore very important in hippocampally mediated episodic memory in humans, of for example an object or person and where the object or person is located in space. What then is found in humans? It has been reported that in humans there are landmark (or spatial view) cells, as well as place cells (Ekstrom, Kahana, Caplan, Fields, Isham, Newman and Fried 2003), and further that some of these neurons do become activated during the recall of an episodic memory of the place (Miller, Neufang, Solway, Brandt, Trippel, Mader, Hefft, Merkow, Polyn, Jacobs, Kahana and Schulze-Bonhage 2013), just as spatial view cells do in macaques when a location "out there" in space is recalled from an object (Robertson, Rolls and Georges-François 1998, Rolls, Xiang and Franco 2005c, Rolls and Xiang 2006). Thus spatial view cells in humans may be essential for episodic memory involving what was present at places "out there", and further investigation of them seems important, especially if possible in real spatial situations where place and spatial view can be unambiguously separated (Georges-François, Rolls and Robertson 1999), rather than in the more artificial virtual environments presented on a screen that are often used.

A possible difference between the human and non-human hippocampal systems is that in rodents and non-human primates there may effectively be one hippocampus because the CA3 neurons are so connected between the two hemispheres (Rolls 2008d) (Chapter 24), whereas in humans there may be separate CA3 networks in the two hemispheres, for the left hippocampal system specialises in language-based memories, whereas the right hippocampal system specialises in spatial memory (Banks, Sziklas, Sodums and Jones-Gotman 2012). The underlying computational reason for this is that language does not work by spatial locations, that is humans do not use and need a word-place episodic memory (Rolls 2008d) (Chapter 24).

Adolphs, Kawasaki, Tuduscuic, Howard, Heller, Sutherling, Philpott, Ross, Mamelak and Rutishauser in Fried et al. (2014) describe face-selective neurons in the human amygdala, which seem very similar to those discovered in macaques (Leonard, Rolls, Wilson and Baylis 1985, Rolls 2011d, Gothard, Battaglia, Erickson, Spitler and Amaral 2007), and also neurons similar to inferior temporal cortex neurons from which the amygdala receives inputs that respond to categories, such as animals as illustrated in Fig. 8.3.

A possible difference of single neurons in the human medial temporal lobe is that many seem to have rather low firing rate responses compared to those in macaques. However, the firing rates of neurons in different cortical neurons are very different. In the macaque inferior temporal visual cortex neurons with peak firing rates of 100 spikes/s to the most effective stimulus are common, whereas in the hippocampus neuronal responses have much lower firing rates, typically reaching a peak of 10–15 spikes/s to the most effective stimulus, from a spontaneous rate of less than 1 spike/s (Rolls and Xiang 2006). Thus when interpreting temporal lobe recordings in humans, it is important to take into account as much as possible the recording site, for what neurons respond to, and how much they respond, differs greatly between different cortical areas. In this context, any information such as MNI coordinates of recorded single neurons in humans is important information to provide, and moreover will help the human single neuron studies to be related to the activations found in human imaging studies, which of course reflect the average activity of hundreds of thousands of neurons, so provide little evidence about how the information is encoded by the neurons.

What are unique to humans are the findings on neuronal responses related to human language, described by Ojemann (2013) and in Fried et al. (2014). Many of these recordings were made in lateral temporal cortex, and not from areas that are essential for language. One interesting finding has been of single neurons that change their activity when naming objects in one language, but not in another language. This suggests that the neuronal networks for different languages may be at least partly separate in terms of how they operate. Another interesting finding is that some temporal lobe neurons are involved in perception, and others in production, and indeed neurons with mirror-neuron-like properties are described as being rare in the superior temporal lobe cortical areas. In a more recent study, Halgren, Cash and colleagues described in the left anterior superior temporal gyrus of a right-handed man that 45% of units robustly fired to some spoken words with little or no response to pure tones, noise-vocoded speech, or environmental sounds (Chan, Dykstra, Jayaram, Leonard, Travis, Gygi, Baker, Eskandar, Hochberg, Halgren and Cash 2014). The tuning to words might be described as sparse distributed. Many units were tuned to complex but specific sets of phonemes, which were influenced by local context but invariant to speaker, and suppressed during self-produced speech. The firing of several units to specific visual letters was correlated with their response to the corresponding auditory phonemes, providing direct neural evidence for phonological recoding during reading.

A fundamental issue is how syntax is encoded in the cortex. A recent hypothesis is that place coding might be used, with for example a neuron responding to the word "cat" when it is the subject of a sentence, and not when it is the object (Rolls and Deco 2015a). If this hypothesis was found to be the case in single neuron recordings in future, this would greatly simplify concepts about how language and in particular syntax are implemented in the cortex (Rolls and Deco 2015a) (Chapter 17).

8.6 Highlights

1. Information is encoded by sparse distributed place coded graded firing rate representations, with almost independent information conveyed by each neuron, up to reasonable numbers of neurons.

2. Place cell encoding conveys much more information than stimulus-dependent synchronicity ('oscillations', coherence) of firing of groups of neurons in the awake behaving animal.

3. The principles of coding are summarized in Section 8.2.2.2, are considered in more detail in Appendix C, and seem to be similar in humans to those in macaques, in which most quantitative studies have been performed.

4. The advantages of different types of coding are described in Section 8.2.2.3.

9 Synaptic modification for learning

9.1 Introduction

There are far too few genes ($<$25,000) to specify which neurons are connected to which other neurons, let alone their synaptic strengths, given that there are $\approx 10^{15}$ synapses in the brain. This is one reason why the strength of most synapses in the brain is experience-dependent. Another reason is that some synapses need to change in strength during development, for example to set up disparity tuning while the head is still growing. Other synapses need to change to implement the many specific types of learning, for example one-trial learning of an association between a visual stimulus and reward or punishment; or the slow learning of skills.

In this Chapter, we assess the principles of synaptic modification, by examining the different types of synaptic modification, and how they are put to use by the cortex to help implement its operation. Much of what is known depends on associative synaptic modification, where the pre-synaptic and post-synaptic firing rates influence the synaptic changes, but an open issue raised is whether that, with some error correction learning in some systems, is sufficient, or whether other principles of synaptic modification are needed to account for the operation of the cerebral cortex. An important principle, different from many machine 'neural network' learning algorithms, is that the information required to change the synaptic strength is present locally at the synapse and is provided by the presynaptic and postsynaptic firing rates, and to some extent their timing. In the cerebellar cortex, each neuron may have its own teacher, the climbing fibre (Chapter 23). Some machine learning algorithms are non-local, in that they require a neuron in a hidden (intermediate) layer to calculate the neuron's contribution to the errors in all succeeding layers, without the connectivity and computational ability required for this being present in the cerebral cortex, as described in Section B.11.

9.2 Associative synaptic modification implemented by long-term potentiation

Long-term potentiation provides a model of synaptic modification, and is introduced in Section 1.5, with some of the rules illustrated in Fig. 1.6. In that LTP is long-lasting, develops rapidly, is synapse-specific, and is in some cases associative, it is of interest as a potential synaptic mechanism underlying some forms of memory (Section 1.5) (Feldman 2009, Bliss and Collingridge 2013).

Consistent with this, in the hippocampal mossy fibre connections from dentate granule cells to the CA3 cells, where associative synaptic modification is not a feature, the function in this case is not to store information, but instead to orthogonalize representations of the CA3 network, so that each CA3 memory is different from other memories. The recall in this case is triggered by the associatively modifiable direct perforant path to CA3 synapses (Treves and Rolls 1992b, Kesner and Rolls 2015) (Chapter 24). In this scenario, the rapid adaptation shown by mossy fibre synapses may be part of a mechanism to move rapidly from one discrete event to another, so that a sequence of events can be stored as an episodic memory (Chapter 24).

The issue of how finely graded the synaptic strengths need to be in associative memory systems is an interesting one. It turns out that in associative memories (autoassociation and pattern association memories), less than one bit of storage can be taken advantage of at each synapse (Appendix C). An implication is that there may not be a need for much precision in setting the strengths of synapses in associative memories. The lower level of zero (it is not plausible to have negative synaptic weights), and a single strong value for a synaptic weight, providing a binary resolution, would achieve much in associative memories. This applies to the final state of the synapses. However, during the learning, when overlapping patterns may add with LTP or subtract with long-term depression (LTD) from the existing synaptic strength, it is advantageous to have synapses that are more graded than binary (Fusi and Abbott 2007). In this context, it is of interest that synapses do appear to have gradations (at least in rat hippocampal CA1), with perhaps 26 distinguishable synaptic strengths (Bartol, Bromer, Kinney, Chirillo, Bourne, Harris and Sejnowski 2016). This gradation of synaptic strength would not increase the memory capacity of the brain at least as far as attractor and pattern association networks are involved (see Sections B.2.7.1 and B.3.3.7); but graded synaptic strengths would facilitate the learning of memories in these networks.

There are two types of long-term depression (LTD) (Fig. 1.6).

One is heterosynaptic LTD, in which the synapse become weaker if there is high pre-synaptic but low postsynaptic firing. This is important in associative memories by providing a mechanism for decorrelating patterns of firing stored as memories by removing the effect of the mean positive presynaptic firing rate, given that neurons can have only positive firing rates (Sections B.2.7.1 and B.3.3.7) (Rolls and Treves 1990, Treves and Rolls 1991). Hetero-synaptic LTD is also important in competitive networks, in which it helps to keep the total synaptic strength onto a neuron within bounds, so that a few neurons do not learn to all the patterns (Section B.4.9.2). Some type of trophic factor might also contribute to any one neuron receiving only a certain total synaptic strength (Park and Poo 2013, Van Ooyen 2011, Harris, Ermentrout and Small 1997).

Homosynaptic LTD is so-called because the synapse that weakens is the same as (homo-) the one that is active (Section 1.5 and Fig. 1.6).

9.3 Forgetting in associative neural networks and in the brain, and memory reconsolidation

9.3.1 Forgetting

Forgetting is an important feature of associative neural networks and the brain, and is important in their successful operation. There are a number of different mechanisms for forgetting, and a number of different reasons why forgetting is important in particular classes of network.

Consider attractor, that is autoassociation, networks, which are used for short-term memory, episodic memory, etc. These networks have a critical storage capacity, as described in Section B.3, and if this is exceeded, most of the memories in the network become unretrievable. It is therefore crucial to have a mechanism for forgetting in these networks.

One mechanism is decay of synaptic strength. The simple forgetting mechanism is just an exponential decay of the synaptic value back to its baseline, which may be exponential in time or in the number of learning changes incurred (Nadal, Toulouse, Changeux and Dehaene 1986). This form of forgetting does not require keeping track of each individual change and preserves linear superposition, that is each memory is added linearly to previous memories, as provided for by equation B.9 on page 724. In calculating the storage capacity of pattern associators and of autoassociators, the inclusion or exclusion of simple exponential decay does not change significantly the calculation of capacity, and only results in a different prefactor (one 2.7 times the other) for the maximum number of associations that can be stored. Therefore a forgetting mechanism as simple as exponential decay is normally omitted, and one has just to remember that its inclusion would reduce the critical capacity of an autoassociation network obtained to roughly 0.37 of that without the decay. This type of memory has been called a palimpsest.

Another form of forgetting, which is potentially interesting in terms of biological plausibility, is implemented by setting limits to the range allowed for each synaptic strength or weight (Parisi 1986). As a particular synapse hits the upper or lower limit on its range, it is taken to be unable to further modify in the direction that would take it beyond the limit. Only after modifications in the opposite direction have taken it away from the limit does the synapse regain its full plasticity. A forgetting scenario of this sort requires a slightly more complicated formal analysis (since it violates linear superposition of different memories), but it effectively results in a progressive, exponential degradation of older memories similar to that produced by straight exponential decay of synaptic weights.

A combined forgetting rule that may be particularly attractive in the context of modelling synapses between pyramidal cells is implemented by setting a lower limit, that is, zero, on the excitatory weight (just requiring that the associated conductance be a non-negative quantity!), and allowing exponential decay of the value of the weight with time. Again, this type of combined forgetting rule places demands on the analytical techniques that have to be used to calculate the storage capacity, but leads to functionally similar effects (Rolls and Treves 1998).

A third mechanism for forgetting is overwriting of previously stored memories, which will happen as a result of long-term depression. Two conditions under which synaptic strengths may decrease are described in Fig. 1.6. Heterosynaptic long-term depression, which can occur for inactive presynaptic terminals on active postsynaptic neurons, can be useful computationally in the following ways. First, it is a useful way to subtract the effect of the mean presynaptic firing rate of each neuron in a pattern associator, which removes the effect of the mean firing rate in increasing the correlation between different input patterns used in the network, as is made evident in equation B.7. This orthogonalizing effect helps to maximize the storage capacity of pattern associators. The same situation applies to autoassociation (attractor) networks (see equation B.13). This decrease of firing rate for inactive inputs may of course result in some loss of memories previously stored in these association networks, if a particular synapse had been strengthened as part of a previous memory. This process enables new memories to overwrite old memories (Nadal et al. 1986, Amit 1989, Rolls 1996b). For example, if a postsynaptic neuron is activated during the formation of a new memory, then any inactive synaptic inputs, from other memories, will become weaker due to heterosynaptic long-term depression, and this will tend to weaken previously stored memories, and thus gradual forgetting of old memories occurs. Homosynaptic long-term depression might contribute to a similar function.

Heterosynaptic long-term depression (LTD) is also useful in competitive networks, for it provides a way for the synaptic weight vectors of different neurons to be kept of approximately equal length, and this is important to ensure that the different categories of input patterns find different output neurons to activate. In the case of competitive networks, the appropriate effect is achieved if the subtractive term in the presynaptic component depends on the existing strength of the synapse, as shown in equation B.22. The fact that in studies of LTD it is sometimes remarked that LTD is easier to demonstrate after LTP has been induced lends support to the likelihood that LTD of the form indicated in equation B.22 that depends on the existing synaptic strength is implemented in the brain. Because of the computational significance of LTD that depends on the existing strength of synapses for competitive networks, it would be useful to see further experimental exploration of this.

Forgetting of the ongoing activity in attractor networks used for short-term memory takes two forms that can be clearly distinguished. One is that the current attractor state implemented by the continuing firing of one set of neurons in the network is labile, and may be interrupted by a strong new input which forces the network into a new attractor state; and another is that the attractor state may be interrupted by quenching effects through non-specific effects implemented through for example inhibitory neurons (Deco and Rolls 2005d). Both these effects could be facilitated by synaptic or neuronal adaptation to help implement many useful computations including reversal, sequential memory, non-reward, and the temporal order useful in syntax, as as described in Chapter 10 (Deco and Rolls 2005d, Deco and Rolls 2005c, Rolls and Deco 2015a, Rolls and Deco 2015b, Rolls and Deco 2016).

The second form of forgetting in attractor networks is of the strengthened synaptic connections that specify each of the different attractor memories in the network. If the network is to be used for large numbers of different short-term memories, then these synaptic weights must decay or be overwritten by LTD as described above. This is likely to be required for short-term memory networks in the prefrontal cortex which must adapt themselves to be capable of storing the particular stimuli and actions that may be required in particular tasks, in short-term memory networks that implement the visuo-spatial scratchpad, etc. We may note that because short-term memory networks have these two different aspects, once an attractor set of synaptic connections has been imprinted in a network by synaptic modification, then no further synaptic modification is necessary to use that network repeatedly for holding the neurons in one of the stored attractors active to implement the short-term memory in a delay period (Kesner and Rolls 2001).

Forgetting may be less important in semantic networks. (Semantic networks store structured information with appropriate associative links and hierarchical structure, for example a family tree, or one's geographical knowledge (McClelland and Rumelhart 1986).) We may note that because any one associative memory network has a memory capacity that is related to the number of associative synapses onto each neuron from other neurons in the network, semantic memory is likely to involve connections between modules in the cortex, where each module might be defined by a 1–3 mm region of neocortex with high local connectivity between the neurons. For this type of memory, forgetting is not so much the requirement as incorporating new semantic knowledge, that is making new appropriate links, and perhaps weakening existing links. In this scenario, the fact that when a memory is retrieved, as would occur when a sematic memory is being updated or extended, then it may need to be reconsolidated, suggests a possible useful function for memory reconsolidation (see below), as it could facilitate the restructuring of a semantic memory. In contrast, such restructuring would not be a very useful property of an episodic memory, in which each episode must be clearly distinguished from others.

9.3.2   Factors that influence synaptic modification

Memories stored early in life may be stored better, and later recalled better, than those stored later in life. There are a number of possible reasons for this. One is that the transmitters that generally facilitate synaptic modification, such as acetylcholine and noradrenaline (see Sections 16.4.4, 10.4.1, and 10.4.2), may become depleted with aging (Rolls and Deco 2015b). These transmitters may play a computationally useful role in modulating synaptic modification (Fremaux and Gerstner 2015).

Another mechanism, not necessarily independent, is that new synaptic modification, as assessed by long-term potentiation (LTP), appears to be less long-lasting with aging (Burke and Barnes 2006) (Section 16.4). Another mechanism may be that storing memories in a flat energy landscape (i.e. without much prior synaptic modification) may help these memories to stand out from those added later. While this would not be a natural property of the type of autoassociation palimpsest memory described above, it could be a property of the way in which an episodic memory stored in the hippocampus may be retrieved into the neocortex where it can be incorporated into a semantic memory (see Chapter 24), the relevant example of which in this case would be an autobiographical memory.

In semantic memories, it could be that the first stored links tend to provide the framework around which other information is structured.

Another factor in the apparent strength of early memories may be that some may be stored with an affective component, and this may not only make the memory strong by activation of the cholinergic and related systems described in Section 16.4.4, but may also mean that part at least of the memory is stored in different brain structures such as the amygdala which may have relatively more persistent and less flexible or reversible memories than other memory systems.

Another factor in the importance and stability of synaptic modification early in life arises in perceptual systems, in which it is important to allow neurons to become tuned to the statistics of for example the visual environment, but once feature analyzers have been formed, stability of the feature analyzers in early cortical processing layers may be important so that later stages in the hierarchy can perform reliable object recognition, which achieves stability only if the input filters to the system do not keep changing (see Chapter 25). This could be the importance of a critical period for learning early on in perceptual development.

Sleep (see also Chapter 21) has been proposed as a state in which useful forgetting or consolidation of memories could occur. One suggestion was that if deep basins of attraction formed in a memory network, then this could impair performance, as the memories in the basins would tend to be recalled whatever the retrieval cue. If noise, present in the disorganized patterns of neural firing during sleep, caused these memories to be recalled, this would indicate that they were 'parasitic', and the suggestion was that associative synaptic weakening (LTD) of synapses of neurons with high firing during sleep would tend to decrease the depth of those basins of attraction, and improve the performance of the memory (Crick and Mitchison 1995). At least at the formal level of neural networks, the suggestion does have some merit as a possible way to 'clean up' associative networks, even if it is not a process implemented in the brain. Although the idea of some role of sleep in memory remains active, this remains to be fully established (Walker and Stickgold 2006) (see Chapter 21).

The idea that sleep could be a time when memories are unloaded from the hippocampus to be consolidated in long term, possibly semantic, memories during sleep (Marr 1971) (allowing hippocampal episodic memories to then be overwritten by new episodic memories) continues to be explored. It has been shown for example that after hippocampal spatial representations have been altered by experience during the day, these changes are reflected in neuronal activity in the neocortex during sleep (Wilson and McNaughton 1994, Wilson 2002). The type of experience might involve repeated locomotion between two places, and the place fields of rat hippocampal neurons for those places may become associated with each other because of coactivity of the neurons representing the frequently visited places. The altered co-firing of the hippocampal neurons for those places may then be reflected in neocortical representations of those places. This could then result in altered representations in the neocortex, if LTP occurs during sleep in the neocortex. Of course, any change in neocortical neuronal activity might just reflect the altered representations in the hippocampus, which would be expected to influence the neocortical representations via hippocampo-neocortical backprojections, even without any neocortical learning (see Chapter 24).

9.3.3 Reconsolidation

Reconsolidation refers to a process in which after a memory has been stored, it may be weakened or lost if recall is performed during the presence of a protein synthesis inhibitor (Debiec, LeDoux and Nader 2002, Debiec, Doyere, Nader and LeDoux 2006, Dudai 2012, Alberini and Ledoux 2013). The implication that has been drawn is that whenever a memory is recalled, some reconsolidation process requiring protein synthesis may be needed.

One possible function of reconsolidation is that it may allow some restructuring of a memory, as described above, though this might be useful more in semantic than episodic memory systems.

A second possible computational function is that reconsolidation might be useful as a mechanism to ensure that whenever a memory is retrieved, additional LTP (long-term potentiation of synaptic strength) is not added to the existing LTP. This could be achieved if during the recall process the memory strength is reset to a low value from which it is then strengthened. Indeed, a potential problem with memory systems is what separates storage from recall, in that whenever recall occurs, pre- and post-synaptic activity is present at the relevant synapses for the memory, and thus one might expect another round of synaptic strengthening to occur. Reconsolidation, by effectively resetting the baseline of synaptic strength during recall, might then provide for the restrengthened synapses not to be stronger than they were before the memory recall. A relevant point here is that in associative memories, the amount of information stored and retrieved from any one synapse is quite low, in the order of 0.2–0.3 bits for autoassociators (Rolls and Treves 1998, Treves and Rolls 1991) (Section B.3) and a little higher for pattern associators (Rolls and Treves 1998) (Section B.2), so that in any case having synaptic strengths that could be repeatedly strengthened by superposition of different memories with distributed representations and with precision maintained at each strengthening would not appear to be a necessary property of the synapses in such memory systems. Under these circumstances, allowing during recall a weakening of a memory, and then its reconsolidation from a relatively fixed baseline might not lead to loss of useful information, and might be a possible solution to continually strengthening synapses every time a memory is recalled.

A third possible computational function of reconsolidation is that it could enable the selective retention of 'useful' memories (or in fact memories being used), and the forgetting of memories not being used, as follows. Consider a memory system in which there is slow exponential decay of synaptic strength with time, a not altogether unlikely scenario given the properties of a biological system. In this situation memories will gradually be lost, perhaps with a different time course in different memory systems, which might be in the order of days, weeks, months or years. In this scenario, if a piece of information was actually recalled because an environmental situation occurred in which for example there was a retrieval cue for a memory, then that memory (i.e. the synaptic strengths) would by reconsolidation be strengthened back to near its initial value. That memory would then be strong and available for future use, compared to other memories not recalled that would be passively decaying. The passive decay of memories not being recalled and reset by reconsolidation would be useful in cleaning out the memory stores so that any critical capacity was not reached, and at the same time in minimizing interference (due to generalization to similar patterns) between memories in store. An example might be the number of one's hotel room, which while it is being repeatedly recalled for use while in that hotel and thus restored, would then decay passively and gradually be lost when it was not longer being actively recalled and hence reconsolidated.

One could propose that in some memory systems the passive decay might be relatively rapid, occurring within hours or days. An example might be the dorsolateral prefrontal cortex, where depending on the requirements of the short-term memory or planning tasks being performed, synapses might by reconsolidation keep representations used in attractor networks available while a given task was being performed. However, when that task was no longer being performed, passive synaptic decay would mean that neurons allocated to that task would gradually decline, and instead new attractor landscapes (i.e. memories) could be set up for new tasks or planning, without interference from representations that were previously being used. There is some evidence at the phenomenological level that neuronal representations in the prefrontal cortex are made and kept relevant to whatever task is being performed (Miller, Nieder, Freedman and Wallis 2003, Everling, Tinsley, Gaffan and Duncan 2006), and I have just proposed a possible mechanism for this implemented by reconsolidation.

It is of importance that reconsolidation applies to fear association mechanisms in the amygdala (Doyere, Debiec, Monfils, Schafe and LeDoux 2007), and drug-associated memories in the amygdala (Milton, Lee, Butler, Gardner and Everitt 2008). The findings have interesting implications for the treatment of fear-associated memories. For example, in humans old fear memories can be updated with non-fearful information provided during the reconsolidation window. As a consequence, fear responses are no longer expressed, an effect that can last at least a year and is selective only to reactivated memories without affecting other memories (Schiller, Monfils, Raio, Johnson, LeDoux and Phelps 2010). Procedures that influence the extinction of fear memory may also be useful in the treatment of fear states (Davis 2011).

In conclusion, we have seen in this Section that forgetting has important functions in the brain, and is a necessary property of many different types of memory system if they are to continue to function efficiently and to allow new learning in the same networks in the brain.

9.4 Spike-timing dependent plasticity

An interesting time-dependence of LTP and LTD has been observed, with LTP occurring especially when the presynaptic spikes precede by a few ms the post-synaptic activation, and LTD occurring when the presynaptic spikes follow the postsynaptic activation by a few milliseconds (ms) (Markram, Lübke, Frotscher and Sakmann 1997, Bi and Poo 1998, Bi and Poo 2001, Senn, Markram and Tsodyks 2001, Dan and Poo 2004, Dan and Poo 2006, Markram, Gerstner and Sjöström 2012, Feldman 2009, Feldman 2012, Fremaux and Gerstner 2015). This is referred to as spike timing-dependent plasticity, STDP (Section 1.5). However these spike timing dependent synaptic modifications may be evident primarily at low firing rates rather than those that often occur in the brain (Sjöström, Turrigiano and Nelson 2001). Indeed, in the original description of STDP (Markram, Lübke, Frotscher and Sakmann 1997) it was described with postsynaptic firing rates < 10 spikes/s. At higher firing rates, LTP is typically observed, and indeed at higher rates the distinction between whether a presynaptic spike is after the previous postsynaptic spike or after it becomes blurred. Further, STDP may not be especially reproducible in the cerebral neocortex in vivo (Fregnac, Pananceau, Rene, Huguet, Marre, Levy and Shulz 2010), though is still arousing much interest (Feldman 2009, Feldman 2012, Fremaux and Gerstner 2015).

9.5 Long-term synaptic depression in the cerebellar cortex

Long-term synaptic depression in the cerebellar cortex is evident as weakening of active parallel fibre to Purkinje cell synapses when the climbing fibre connecting to a Purkinje cell is active (Ito 1984, Ito 1989, Ito 1993b, Ito 1993a), and is described in Chapter 23.

9.6   Reward prediction error learning

9.6.1   Blocking and delta-rule learning

In classical conditioning, a conditioned stimulus (CS) only becomes associated with an unconditioned stimulus (UCS) if the CS predicts the US. Part of the evidence for this is the blocking effect, that if CS1 already predicts the UCS, nothing is learned about a CS2 presented at the same time as CS1 (Kamin 1969). Because CS2 does not add to the predictability of the UCS, learning about CS2 is 'blocked'. This shows that not temporal contiguity, but whether a cue, a potential CS, has predictive power is important for classical conditioning (Roesch, Esber, Li, Daw and Schoenbaum 2012).

One approach is to assume a direct relationship between salience and associative strength (Rescorla and Wagner 1972, Esber and Haselgrove 2011). In this construct, salience can refer to surprise, or unpredictability (or anything that influences synaptic eligibility for learning). A selective learning mechanism is invoked that grants salient cues a superior status as predictors, relative to irrelevant cues. An obvious candidate is the delta rule advanced by Rescorla and Wagner (1972). According to this rule, cues conditioned in compound compete for the maximum amount of associative strength supported by the reinforcer λ, with a fully predicted reinforcer suffering a loss of processing relative to a surprising reinforcer. For each cue, the increment in associative strength (ΔV) on any given trial is proportional to the discrepancy between the value of λ and the aggregate prediction of the reinforcer generated on the basis of all cues present (ΣV). This is captured in the following equation:

$$\Delta V = \alpha\beta(\lambda - \Sigma V) \tag{9.1}$$

where β is a constant (range 0–1) encapsulating the intrinsic motivational properties of the reinforcer. The larger the discrepancy $\lambda - \Sigma V$, the more associative learning will take place. Learning ceases as this discrepancy approaches zero. On trials in which the reinforcer is omitted, it is assumed that $\lambda = 0$ and the resultant ΔV will be negative, signifying a reduction in the cue's associative strength (i.e. extinction) (Esber and Haselgrove 2011).

Another approach proposes that errors promote changes in associative strength by modulating attention and processing of events (Pearce and Hall 1980). These two approaches have been combined (Esber and Haselgrove 2011).

These approaches to conditioning led to interest in whether and if so how delta-rule or error correction rule learning is implemented in the brain (Schultz, Romo, Ljunberg, Mirenowicz, Hollerman and Dickinson 1995, Roesch et al. 2012). For this, a prediction error must be produced.

9.6.2   Dopamine neuron firing and reward prediction error learning

There is extensive evidence that dopamine neurons may signal positive reward prediction error, that is, can respond when a reward is unexpectedly obtained, or when the reward obtained is greater than predicted, or when a stimulus predicting reward is given (Schultz 2013, Glimcher 2011b). This evidence has stimulated the application of reinforcement learning theories in which dopamine-communicated prediction errors are used to train systems in the brain such as the stimulus-response learning system in the basal ganglia. (Reinforcement learning is described in Section B.16.) However, the evidence for this interpretation is not fully consistent,

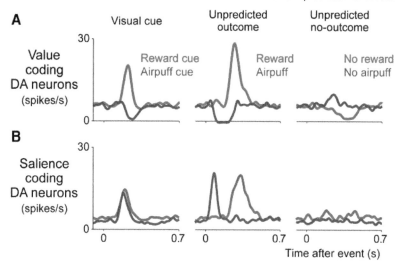

Fig. 9.1 Responses of different types of dopamine neuron. (A) Reward prediction error neurons (see text). (B) Neurons activated by aversive and also by rewarding stimuli, sometime called motivational salience neurons (see text). (Reprinted by permission from Macmillan Publishers Ltd: *Nature*, 459, 837–841, Masayuki Matsumoto and Okihide Hikosaka, Two types of dopamine neuron distinctly convey positive and negative motivational signals. ©2009.)

in that some dopamine neurons respond to aversive stimuli, some to rewarding and aversive stimuli, and others to stimuli that may be salient in other ways, for example novel stimuli (Matsumoto and Hikosaka 2009, Bromberg-Martin, Matsumoto and Hikosaka 2010a). Some of the evidence will now be considered.

Schultz et al. (1995) argued from their recordings from dopamine neurons that the firing of these neurons might be involved in reward. For example, dopamine neurons can respond to the taste of a liquid reward in an operant task. However, these neurons may stop responding to such a primary (unlearned) reinforcer quite rapidly as the task is learned, and instead respond only to the earliest indication that a trial of the task is about to begin (Schultz et al. 1995). **Thus dopamine neurons could not convey information about a primary reward obtained if the trial is successful. They are thus unlike, and could not perform the functions of, the outcome value neurons in the orbitofrontal cortex described in Chapter 15.**

Instead, there is considerable evidence that some dopamine neurons convey a reward positive prediction error signal (Schultz 2013, Glimcher 2011b), illustrated by the types of response shown in Fig. 9.1A. This type of neuron increases its firing rate to a visual cue that predicts that reward (juice) will be delivered, and decreases its rate to a visual cue that predicts that an aversive stimulus (an air puff) will be delivered (left); and is not activated when a predicted reward is obtained, that is when the juice is delivered (not illustrated in Fig. 9.1A). This type of neuron increases its firing rate to an unpredicted reward outcome, and this is termed a positive reward prediction error response, for the reward outcome is greater than was predicted. This type of neuron decreases its firing if an expected reward stimulus is not received (Fig. 9.1A right), suggesting that it encodes the sign of the prediction error by whether it increases or decreases its firing rate. These neurons encode an accurate prediction error signal, including strong inhibition by omission of rewards and mild excitation by omission of aversive events (Fig. 9.1A right). This is the type of dopamine neuron that is generally discussed (Schultz 2013, Glimcher 2011a). It should be noted that these dopamine neurons are completely different to the negative reward prediction error neurons found in the

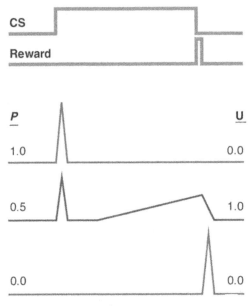

Fig. 9.2 Recordings from midbrain dopamine neurons by Fiorillo, Tobler and Schultz (2003) in a task in which different visual conditioned stimuli (CS) predicted different probabilities P of juice reward after a delay period when the stimulus switched off. The intertrial interval was variable, so that the conditioned stimulus provided information about reward delivery even when $P = 1.0$. When $P = 1.0$, the neurons responded phasically at the onset of the CS, and did not respond to the delivery of the taste reward. When $P = 0.0$, there was no response to the CS, and no response at the end of the CS unless reward was unexpectedly given (as illustrated). However, when $P = 0.5$, the dopamine neurons responded with gradually increasing and sustained firing during the CS. U indicates the uncertainty of reward. (Reproduced from *Science*, 299 (5613), Gambling on dopamine, P. Shizgal and A. Arvanitogiannis, pp. 1856–1858, ©2003, The American Association for the Advancement of Science. Reprinted with permission from AAAS.)

orbitofrontal cortex, which increase their firing rate when an expected reward is not obtained (see Section 15.14 and Thorpe, Rolls and Maddison (1983)) and which has never been found for dopamine neurons.

In a further investigation of the function of dopamine neurons, Fiorillo, Tobler and Schultz (2003) found not only the phasic response effects just described, but also that the firing rate of the neurons increased steadily during the 2 s period in which another conditioned stimulus was being shown that predicted that reward would be obtained with a probability P of 0.5 (see Fig. 9.2). However, when $P = 0.25$ or $P = 0.75$, the gradually increasing and sustained firing during the conditioned stimulus (CS) was less than when $P = 0.5$. This pattern of results indicates that the tonic, sustained, firing of the dopamine neurons in the delay period reflects reward uncertainty, and not the expected reward, nor the magnitude of the prediction error. Nor could the sustained, tonic, firing indicate expected reward (or expected value, where expected value = probability multiplied by reward value (Glimcher 2003, Glimcher 2004, Glimcher and Fehr 2013, Rolls 2014a) (Section 15.9), for this would be highest in the order $P = 1.0, P = 0.75, P = 0.5, P = 0.25, P = 0.0$. Both the phasic and the tonic components were higher for higher reward values (more drops of juice reward). These results are difficult to reconcile with the previously hypothesized (Waelti, Dickinson and Schultz 2001, Dayan and Abbott 2001) 'prediction error' training signal function of the firing of dopamine neurons, for it is difficult to understand how any brain system receiving the phasic ('prediction error') and tonic ('uncertainty of reward') dopamine signals by the same set of neurons could disentangle them and use them for different functions (Shizgal and Arvanitogiannis 2003). However, a possible resolution is that the so-called tonic component arises from averaging across trials, and more importantly from the potential difficulty that there is an asymmetry in the errors that could be conveyed by dopamine neurons, given their low spontaneous firing rate of a few spikes/s (Niv, Duff and Dayan 2005). As a result of this asymmetry, positive prediction errors could be represented by dopamine neurons by firing rates of ≈270% above baseline, while negative errors could be represented by a decrease of only ≈55% below baseline (see Fig. 9.1). This asymmetry in prediction errors remains a potential problem for the dopamine error hypothesis, but the tonic firing has a possible explanation (Niv et al. 2005).

It has also been shown that dopamine neurons can perform one-trial rule-based reversal (Bromberg-Martin, Matsumoto, Hong and Hikosaka 2010b), and the origin of this I suggest is the orbitofrontal cortex, which performs this function as described in Section 15.14 and projects via the ventral striatum to the dopamine neurons (Rolls 2014a).

The possibility that dopamine neuron firing may provide an error signal, of the type used in temporal difference (TD) learning (see Section B.16.3) useful in training neuronal systems to predict reward has continued to be actively studied. The firing of the neurons can be thought of as an error signal about reward prediction (see Section B.16.3), in that the firing occurs in a task when a reward is given, but then moves forward in time when a stimulus is presented that can be used to predict when the taste reward will be obtained (Schultz et al. 1995). The argument is that there is no prediction error when the taste reward is obtained if it has been signalled by a preceding conditioned stimulus, and that is why the dopamine midbrain neurons do not respond at the time of taste reward delivery, but instead, as least during training, to the onset of the conditioned stimulus (Waelti, Dickinson and Schultz 2001, Schultz 2013, Glimcher 2011b). If a different conditioned stimulus is shown that normally predicts that no taste reward will be given, there is no firing of the dopamine neurons to the onset of that conditioned stimulus. (If after that non-reward conditioned stimulus, a taste reward is unexpectedly given, then the dopamine neurons show a phasic burst of firing, perhaps signalling a reward prediction error.) This hypothesis has been built into models of learning in which the error signal is used to train synaptic connections in dopamine pathway recipient regions (such as presumably the striatum) (Waelti et al. 2001, Dayan and Abbott 2001, Schultz 2013, Glimcher 2011b) (see Section B.16.3). A possible effect of the dopamine to implement temporal difference (TD) learning (Section B.16.3) would be for dopamine release to act via D1 receptors in the striatum to facilitate long-term synaptic potentiation (LTP) of the cortical glutamatergic excitatory inputs onto striatal neurons (see Schultz (2013)).

In an application of the temporal difference learning (TD) approach, in a multistep task, monkey dopamine neurons were shown to reflect the TD error quantitatively by reflecting the difference between the sum of multiple future rewards and their prediction (Enomoto, Matsumoto, Nakai, Satoh, Sato, Ueda, Inokawa, Haruno and Kimura 2011). In the task used, the reward probabilities increase towards the end of the multistep sequence, resulting in the highest discounted sum of future reward in the centre of the sequence from which the lower predictions arising from earlier stimuli are subtracted. The dopamine responses match this temporal profile of TD error closely.

A number of difficulties with the dopamine reward prediction error learning hypothesis need to be considered.

One difficulty is the asymmetry of the positive and negative reward prediction error signals that might be available because of the low spontaneous firing rate of dopamine neurons (Fig. 9.1) referred to above.

A second difficulty is that some dopamine neurons appear to reflect at least a positive reward prediction error signal, but it appears to arise with no explanation of how this is computed, or where it comes from (Schultz 2013, Glimcher 2011b). Some suggestions are made below about the source of the inputs that reach dopamine neurons. For example, given that we know that the orbitofrontal cortex contains the necessary signals for this type of computation, including at least reward outcome neurons (taste, flavour, oral texture, etc.), expected reward value neurons, and negative reward prediction error neurons, and novel encoding neurons (see Sections 2.2.5 and 15.14), and that the orbitofrontal cortex provides major inputs to the striatum, it can be suggested that the orbitofrontal cortex, and to some extent the amygdala, provide the source of the inputs to the dopamine neurons via the striatum, even if the striatum or some other as yet unknown part of the brain computes the signals that are reflected in the error signals of some dopamine neurons (Rolls 2014a).

Another point is that reinforcement / temporal difference learning is most applicable to slow learning systems such as those in the basal ganglia for stimulus-response habit learning, and may not be appropriate therefore for most types of learning, many of which can occur in one trial.

A fourth point to be remembered is that there is no evidence that the dopamine neurons reflect different rewards or goals, so they could not be used to direct goal-directed behaviour. In contrast, orbitofrontal cortex neurons encode by their firing many different types of reward value, each appropriate for a different type of goal-directed action. Thus the possible role of some dopamine neurons would seem to be limited to modulating learning in systems where the specific information is in the presynaptic firing of other neurons, whose synaptic connectivity to other neurons may be modulated by dopamine.

A fifth potential problem with the dopamine reward prediction error hypothesis is that there are other types of dopamine neuron, some of which have been studied (Matsumoto and Hikosaka 2009, Bromberg-Martin, Matsumoto and Hikosaka 2010a). These neurons, illustrated in Fig. 9.1B, respond to both rewarding and to aversive stimuli, and could not therefore signal reward prediction error. In more detail, these dopamine neurons increase their firing rate to stimuli predicting reward and to stimuli predicting aversive events (left); and also respond when either an unpredicted reward outcome or an unpredicted aversive stimulus is obtained (middle). These are called 'motivational salience' neurons (Matsumoto and Hikosaka 2009, Bromberg-Martin et al. 2010a), in that they encode something about both positive and negative reinforcers, but not a reward error signal. Another type of dopamine neuron responds only to a cue predicting an aversive stimulus, or to an unpredicted aversive outcome. Another type of dopamine neuron responds to novel stimuli (see Bromberg-Martin et al. (2010a)). Another type of dopamine neuron responds to alerting cues, for example to the first cue in a trial that indicates that a trial is beginning (Bromberg-Martin et al. 2010a, Horvitz 2000, Redgrave, Prescott and Gurney 1999).

What are we to make of this apparent diversity of dopamine neuron types?

First, the different neuron types seem to reflect the types of neuron found in the primate ventral striatum (including the ventral and medial part of the head of the caudate nucleus) (Rolls and Williams 1987a, Williams, Rolls, Leonard and Stern 1993, Rolls, Thorpe and Maddison 1983b), as described in Chapter 20 and by Rolls (2014a). In these parts of the striatum, there are some neurons that respond to reward-related stimuli, some to punishment-predicting stimuli, some to novel stimuli, and many to alerting cues that prepare the animal for action (Rolls 2014a). Now this is not surprising: the major inputs to the dopamine neurons (located in the substantia nigra pars compacta) come from the striatum, especially these parts of the striatum (Haber and Knutson 2009). Consistently, salience, rather than reward, is what it has been suggested is encoded in a main region that projects to the dopamine neurons, the striatum, in that head of caudate and accumbens activation in humans occurs to a monetary reward much more when it is made salient by being actively worked for than when it is received passively (Zink, Pagnoni, Martin-Skurski, Chappelow and Berns 2004), and occurs to salient non-rewarding stimuli (Zink, Pagnoni, Martin, Dhamala and Berns 2003).

The implication might be, if between them dopamine neurons reflected all these types of information, that the firing of dopamine neurons might reflect a much more general control over for example the striatum, by facilitating behavioural responses when the dopamine neurons are firing. This could for example facilitate transmission through the striatum and basal ganglia to produce 'go' behaviour to rewarding stimuli, to punishing stimuli (e.g. flee), to novel stimuli (alert and look for more inputs), to cues that indicate that preparations should be made because a cue signal has just arrived, etc. (Rolls 2014a).

Somewhat consistent with this, dopamine exerts immediate postsynaptic effects during behavioural performance and approach behaviour. At striatal neurons of the direct pathway (from the striatum directly to the globus pallidus, see Chapter 20 and Rolls (2014a)), dopamine has excitatory effects via the D1 receptor by eliciting or prolonging glutamate inputs and transitions to the up state (depolarization) of the membrane potential. At striatal neurons in the indirect pathway (from the striatum via the subthalamic nucleus to the globus pallidus, see Chapter 20 and Rolls (2014a)), D2 receptor activation has inhibitory effects by reducing glutamate release and prolonging membrane down states (hyperpolarization). Both effects of dopamine tend to promote behavioural output (Gerfen and Surmeier 2011). This is an important effect of dopamine, and depletion of dopamine leads to the akinesia, the lack of voluntary action, in Parkinson's disease.

Indeed, as a whole population, dopamine neurons appear to convey information that would be much better suited to a behaviour preparation or 'Go' role for dopamine release in the striatum. Evidence that is needed on this issue is whether dopamine neurons respond when the animal has to initiate behaviour actively to escape from or avoid aversive (e.g. painful) stimuli (active avoidance), vs to remain still and do nothing to avoid the aversive outcome (passive avoidance) (Rolls 2014a).

Some of the evidence that the dopamine projection does not convey a specific 'reward' signal is that dopamine release occurs not only to rewards (such as food or brain-stimulation reward, or later in training to an indication that a reward might be given later), but also to aversive stimuli such as aversive stimulation of the medial hypothalamus, foot-shock, and stimuli associated with footshock (Hoebel, Rada, Mark, Parada, Puig de Parada, Pothos and Hernandez 1996, Rada, Mark and Hoebel 1998, Hoebel 1997, Leibowitz and Hoebel 1998, Gray, Young and Joseph 1997, Bromberg-Martin et al. 2010a). (The dopamine release might in this case be related to the firing of dopamine neurons, or to presynaptic terminals on to the dopamine terminals in the nucleus accumbens that cause the release of dopamine.) This evidence argues against the possibility that different populations of dopamine neurons, for example reward prediction error neurons vs all other types of dopamine neuron, are totally separate populations that project to different brain regions (Schultz 2013, Bromberg-Martin, Matsumoto and Hikosaka 2010a).

These findings are consistent with the hypothesis that instead of acting as the reward prediction error signals in a reinforcement learning system, the dopamine projection to the striatum may act as a 'Go' or 'behaviour preparation' signal to set the thresholds of neurons in the striatum, and/or as a general modulatory signal that could help to strengthen synapses of conjunctively active pre- and postsynaptic neurons (Rolls 2014a). In such a system, what is learned would be dependent on the presynaptic firing of all the input axons and the postsynaptic activation of the neuron, and would not be explicitly guided by a dopamine reinforce/teacher signal that would provide feedback after each trial on the degree of success of each trial as in the reinforcement learning algorithm (described in Section B.16). The facts that many of the neurons in the head of the caudate nucleus, and in some other parts of the striatum, respond in relation to signals that indicate that behaviour should be initiated (see Chapter 20, Rolls, Thorpe and Maddison (1983b), and Rolls (2014a)), and that connections from these striatal regions as well as from the hypothalamus reach the dopamine neurons in the substantia nigra, pars compacta, and ventral tegmental area, are fully consistent with the hypothesis that the dopamine neurons are not activated only by reward-related stimuli, but more generally by stimuli including punishing stimuli that can lead to the initiation of behaviour, or at least that dopamine concentrations in brain regions such as the prefrontal cortex reflect the delivery of punishing, non-rewarding, and aversive as well as unexpected rewarding stimuli. Moreover, the dopamine concentrations may remain elevated for periods of many minutes after such stimuli, and are therefore not well suited to a specific teaching signal (Seamans and Yang 2004).

Overall, although there is much evidence that some dopamine neurons encode a reward prediction error signal, there are difficulties with the hypothesis, and an alternative hypothesis is that overall the dopamine neurons together reflect the effects of many salient stimuli, and that dopamine release has the important function of turning on behaviour to such salient stimuli by facilitating information transmission through the basal ganglia. These possibilities can be further evaluated by taking into account the functioning of the basal ganglia, which is considered in Chapter 20 and by Rolls (2014a).

9.7 Highlights

1. Long-term potentiation provides a model of synaptic modification, and alters synaptic strength associatively in proportion to the presynaptic firing rate and the post-synaptic activation. This implements a Hebbian learning rule. The change is synapse-specific.

2. Heterosynaptic long-term depression decreases a synaptic weight if the presynaptic rate is low and there is postsynaptic activation. This is important in associative memory by helping to correct for the fact that firing rates are positive-only; and in competitive networks by helping to maintain the total strength of the synapses on the post-synaptic neuron constant, thus ensuring that one or more neurons do not learn to all the same patterns. Homosynaptic long-term depression can also occur, in which the synapse decreases in strength if there is high presynaptic firing but low postsynaptic activation.

3. Spike-timing dependent plasticity, in which a synapse strengthens if a presynaptic spike precedes a postsynaptic spike, and decreases in strength if a presynaptic spike follows a postsynaptic spike, may only be evident with low post-synaptic firing rates of < 10 spikes/s, with LTP occurring with higher rates; and is not very evident in the cortex in vivo (Fregnac et al. 2010).

4. Temporal contiguity may be insufficient to account for conditioning, and instead it may be the utility of the conditioned stimulus to predict the unconditioned stimulus that is relevant, as shown by blocking. This has led to investigations of error correction learning, and the hypothesis that dopamine conveys a reward prediction error in reinforcement learning.

10 Synaptic and neuronal adaptation and facilitation

Synaptic and neuronal adaptation occurs in some cortical synapses and neurons (Silberberg, Wu and Markram 2004, Silberberg, Grillner, LeBeau, Maex and Markram 2005), appears to be put to use for particular computational functions, and can be influenced by some widespread modulatory systems, such as the cholinergic projections to the cortex, which decrease adaptation. The mechanisms, utility, and consequences of these types of adaptation are considered in this Chapter, for adaptation is an important principle of cortical function. Further, synaptic facilitation can also occur at some cortical synapses, and also appears to be used for particular operations by the cortex.

In Section 10.1 the different types of neuronal adaptation and short-term synaptic depression and facilitation are described. In the following sections of this Chapter, examples are described of how these processes contribute to different cortical computations.

10.1 Mechanisms for neuronal adaptation and synaptic depression and facilitation

Two types of neuronal adaptation mechanism are described which reduce the firing of a neuron depending on its firing. In addition, short-term synaptic depression and synaptic facilitation mechanisms are described. The descriptions show how these processes can be modelled in integrate-and-fire neuronal networks (Deco and Rolls 2005c, Deco, Rolls and Romo 2010, Martinez-Garcia, Rolls, Deco and Romo 2011, Rolls, Dempere-Marco and Deco 2013, Rolls and Deco 2015a). An important point that makes the effects mediated by these mechanisms a principle of operation of the cortex is that these mechanisms are expressed in some neurons and synapses, and much less in others, and can be modulated by transmitter systems such as acetylcholine, so enable particular computations to be promoted where they are present in particular cortical neurons and synapses.

10.1.1 Sodium inactivation leading to neuronal spike-frequency adaptation

Spike-frequency adaptation can occur in some cortical neurons, in that the spike frequency is a sublinear function of the current injected into a neuron (see e.g. Fig. 1.5) (Lanthorn, Storm and Andersen 1984). One mechanism for spike-frequency adaptation is slow recovery from inactivation of sodium channels after spike generation (Benda and Herz 2003), and a statistical analysis of a model of adaptation produced in this way has been described in the framework of integrate-and-fire models (Giugliano, La Camera, Rauch, Luescher and Fusi 2002, Rauch, La Camera, Luescher, Senn and Fusi 2003). The model was called an Integrate-and-May-Fire (IMF) model, and takes into account the inactivation of sodium channels after spike generation. The integrate-and-fire model is modified, by changing the condition that when the membrane potential reaches the threshold θ, the emission of a spike at that time is an event occurring with an activity-dependent probability q. After the spike emission, the membrane potential is set to the value $V_{reset} = -55$ mV, for an absolute refractory time, after which the current integration starts again (reflecting physiological events). However, each time the excitability threshold θ is crossed and no spike has been generated (i.e. an event with probability $1 - q$), the membrane potential is reset to H_2 ($V_{reset} < H_2 < \theta < 0$) and no refractoriness occurs. Additionally, q is a decreasing function of a slow voltage-dependent variable w ($0 < w < 1$), reminiscent of the sigmoidal voltage dependence of the fast inactivation state variables that characterize conductance-based model neurons:

$$q = \left[1 + e^{\frac{(w - w_0)}{\sigma_w}}\right]^{-1} \tag{10.1}$$

where w evolves by,

$$\tau_w \frac{dw(t)}{dt} = \frac{V(t)}{\theta} - w \tag{10.2}$$

which corresponds to a first approximation to the average transmembrane electric field experienced by individual ion channels and affecting their population-level activation and inactivation. In our simulations we utilized $\tau_w = 9000$ ms, $\sigma_w = 0.01$, $w_0 = 0.8563$, and $H_2 = -52$ mV (Deco and Rolls 2005c).

10.1.2 Ca^{++}-activated K^+ hyper-polarizing currents leading to neuronal spike-frequency adaptation

Another mechanism that can produce neuronal spike-frequency adaptation is the production of After-Hyper-Polarizing (AHP) currents, which can occur when a spike allows a small amount of calcium to enter a neuron, which triggers hyperpolarizing potassium channel mediated currents (Benda and Herz 2003). These Ca^{++}-activated K^+ hyper-polarizing currents can be modelled as follows (Liu and Wang 2001). It is assumed that the intrinsic gating of K^+ After-Hyper-Polarizing current (I_{AHP}) is fast, and therefore its slow activation is due to the kinetics of the cytoplasmic Ca^{2+} concentration. This can be introduced by adding an extra current term in the integrate-and-fire model (Section B.6). In the model, the subthreshold membrane potential $V(t)$ of each neuron evolves according to the following equation:

$$C_m \frac{dV(t)}{dt} = -g_m(V(t) - V_L) - I_{syn}(t) \tag{10.3}$$

where $I_{syn}(t)$ is the total synaptic current flow into the cell, V_L is the resting potential, C_m is the membrane capacitance, and g_m is the membrane conductance. We can add the hyper-polarizing current by adding I_{AHP} on the right hand of equation 10.3 as follows:

$$I_{AHP} = -g_{AHP}[Ca^{2+}](V(t) - V_K) \tag{10.4}$$

where V_K is the reversal potential of the potassium channel. Further, each action potential generates a small amount (α) of calcium influx, so that I_{AHP} is incremented accordingly. Between spikes the $[Ca^{2+}]$ dynamics is modelled as a leaky integrator with a decay constant τ_{Ca}. Hence, the calcium dynamics can be described by the following system of equations:

$$\frac{d[Ca^{2+}]}{dt} = -\frac{[Ca^{2+}]}{\tau_{Ca}} \tag{10.5}$$

If $V(t) = \theta$, then $[Ca^{2+}] = [Ca^{2+}] + \alpha$ and $V = V_{reset}$, and these are coupled to the modified equations of the integrate-and-fire model. The $[Ca^{2+}]$ is initially set to be 0 μM, $\tau_{Ca} = 600 ms$, $\alpha = 0.005$, $V_K = -80 mV$ and $g_{AHP} = 7.5 nS$.

Another mechanism that can produce spike-frequency neuronal adaptation is produced by voltage-gated potassium currents (M-type currents) (Benda and Herz 2003), and they have described a general framework for modelling all three mechanisms of spike-frequency adaptation.

10.1.3 Short-term synaptic depression and facilitation

Short-term synaptic plasticity includes synaptic facilitation and depression. Facilitation is seen when the synapse increases its release of transmitter over time, and depression is seen when this decreases. The generally accepted hypothesis to explain facilitation is increased residual calcium at the vesicular release site, and depression is often associated with depletion of vesicles (Zucker and Regehr 2002, Roy, Stefan and Rosenmund 2014). The effects typically last in the order of a few seconds. The other types of short-term plasticity kinetics are enhancement on longer time scales. These include augmentation which lasts 5–10 s and post-tetanic potentiation which lasts up to 30 s (Roy et al. 2014).

Synaptic depression can be modelled as described by Dayan and Abbott (2001), page 185. In particular, the probability of transmitter release P_{rel} can be decreased after each presynaptic spike by a factor $P_{rel} = P_{rel}.f_D$ with $f_D = 0.982$. Between presynaptic action potentials the release probability P_{rel} is updated by

$$\tau_P \frac{dP_{rel}}{dt} = P_0 - P_{rel} \tag{10.6}$$

with $P_0 = 1$ and $\tau_P = 600$ ms.

Synaptic facilitation is common in higher cortical areas such as the prefrontal cortex (Zucker and Regehr 2002, Hempel, Hartman, Wang, Turrigiano and Nelson 2000, Wang, Markram, Goodman, Berger, Ma and Goldman-Rakic 2006) and hippocampus (Roy, Stefan and Rosenmund 2014). Synaptic facilitation is caused for example by the increased accumulation of residual calcium at the presynaptic terminals, which increases the probability of neurotransmitter release (Zucker and Regehr 2002).

Short term synaptic facilitation can be modelled using a phenomenological model of calcium-mediated transmission (Mongillo, Barak and Tsodyks 2008). The synaptic efficacy of the recurrent connections between all of the excitatory neurons was modulated by the utilization parameter u (the fraction of resources used) reflecting the calcium level. When a spike reaches the presynaptic terminal, calcium influx in the presynaptic terminal causes an increase of u which increases the release probability of transmitter and thus the strength of that synapse. The time constant of the decay of the synaptic facilitation is regulated by a parameter τ_F which experimentally is around 1–2 s (Wang et al. 2006, Mongillo et al. 2008). The values for the baseline utilization factor U (0.15) and for τ_F (1.5 s) used by Rolls, Dempere-Marco and Deco (2013) in a model of multiple-item short-term memory are similar to values reported experimentally and used elsewhere (Wang, Markram, Goodman, Berger, Ma and Goldman-Rakic 2006, Mongillo, Barak and Tsodyks 2008, Deco, Rolls and Romo 2010, Martinez-Garcia, Rolls, Deco and Romo 2011). In more detail, the strength of each recurrent excitatory synapse j is multiplied by the presynaptic utilization factor $u_j(t)$, which is described by the following dynamics:

$$\frac{du_j(t)}{dt} = \frac{U - u_j(t)}{\tau_F} + U(1 - u_j(t)) \sum_k \delta(t - t_j^k) \tag{10.7}$$

where t_j^k is the time of the corresponding presynaptic spike k. The first term shows how the synaptic utilization factor u_j decays to the baseline utilization factor U=0.15 with time constant τ_F=1500 ms, and the second term shows how u_j is increased by each presynaptic action potential k to reach a maximum value of 1 when the neuron is firing fast. The modulation by the presynaptic utilization factor u is implemented by multiplying the synaptic weight by u to produce an effective synaptic weight w_{eff}. This models the underlying synaptic processes (Mongillo, Barak and Tsodyks 2008).

The following sections of this Chapter provide examples of how these processes contribute to different cortical computations.

10.2 Short-term depression of thalamic input to the cortex

Thalamic afferents to the cortex appear to adapt more rapidly than cortico-cortical recurrent connections between cortical pyramidal cells (da Costa and Martin 2013, Silberberg et al. 2005). Part of this effect is due to synaptic depression of thalamo-cortical synapses; and part is due to adaptation of thalamic neurons, which can be prevented by cortical layer 6 cortico-thalamic activity, providing a control mechanism for cortical input (Mease, Krieger and Groh 2014).

It is hypothesized that a function of this thalamo-cortical depression is to allow cortico-cortical synaptic connections to make a major contribution to the firing of cortical neurons after thalamic inputs may have introduced an input into the cortical circuitry. This allows a collective computation to be performed that may involve many nearby cortical neurons contributing to the firing to shape the responsiveness of the local group of neurons. In the language of attractor dynamics, this allows cortical circuitry to operate after a cue has been applied in a relatively unclamped state (Section B.3).

This is an important principle of cortical operation / computation, and is likely to be more evident the further towards memory systems one moves in the cortical stages of processing.

Although the thalamo-cortical afferents may adapt, this does not necessarily imply that cortical neurons will show transient responses, because cortical recurrent collateral connections can potentially maintain the cortical firing.

10.3 Relatively little adaptation in primate cortex when it is operating normally ·

Much of what has been described about synaptic and neuronal adaptation is in cortical slices or under anaesthesia, and under these conditions, adaptation is very prominent, especially in non-primary-sensory cortical areas (Silberberg, Wu and Markram 2004, Silberberg, Grillner, LeBeau, Maex and Markram 2005, Thomson and Lamy 2007). Charles Gross discovered this when he tried to analyse the response properties of macaque inferior temporal cortex neurons under anaesthesia. The neurons might show some response to a stimulus, but then adapt out, and be unresponsive for many seconds or minutes, making the analysis very difficult (Gross, Rocha-Miranda and Bender 1972).

To provide clear evidence on the magnitude of adaptation in the awake, behaving state, evidence from neuronal recordings in macaques is very helpful, for the stimuli can be precisely delivered, and attention to the stimuli can be ensured.

An example of what is typical in recordings made in macaques under these conditions for the inferior temporal visual cortex is shown in Fig. 10.1, in which the macaque was looking for an object in a complex visual scene. On this typical trial, the neuron responded at a high firing rate as soon as the monkey had found the object with multiple eye movements (in this

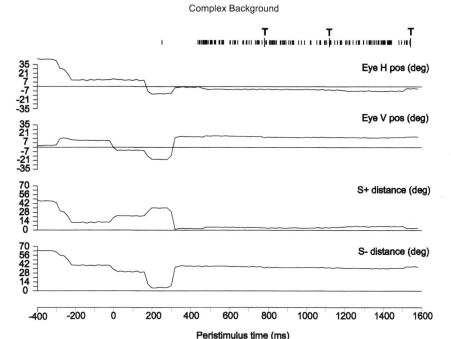

Complex Background

Fig. 10.1 Little adaptation of the responses of a macaque inferior temporal cortex neuron. Eye position and neuronal response data collection during the performance of the visual search task for one neuron. The horizontal and the vertical eye position traces are calibrated with respect to the centre of the screen in degrees (with -35 deg horizontal and -35 deg vertical being the lower left of the screen). Separate traces show the distance of the fovea from the target search object (S+) and from the distracter object (S-). A rastergram for the neuron is shown above, with each vertical line representing an action potential from the neuron. The visual display was switched on at time 0, and after looking at two different positions for which there was only spontaneous neuronal activity, the eyes saccaded to the target object at approximately 340 ms post-stimulus, the neuron responded approximately 100 ms later, and the monkey then made multiple touches (T) of the object on the screen to obtain fruit juice. (After Rolls, Aggelopoulos and Zheng 2003.)

trial at about 340 ms), and the neuron responded within the typical 100 ms latency with a high rate for the next 1000 ms with no adaptation while the monkey continued to look at the object (Rolls, Aggelopoulos and Zheng 2003a). (The remarkable selectivity of some inferior temporal cortex neurons is made evident by seeing the complex scene in which these neurons can respond selectively to an object, which is illustrated in Fig. 25.59 on page 643.) Another example is shown in Fig. 16.15 on page 342. This typical inferior temporal cortex neuron responded throughout the 500 ms presentation period, but had a somewhat higher firing rate for the first 100 ms of its response.

In another cortical area, the orbitofrontal cortex, neurons can also respond with only minor adaptation, in this case to the oral texture stimulus of fat and other oils, as illustrated in Fig. 10.2 (Verhagen, Rolls and Kadohisa 2003). In another orbitofrontal cortex example, another population of neurons respond for the full 2 s presentation of novel but not familiar visual stimuli with almost no adaptation (Rolls, Browning, Inoue and Hernadi 2005a). Even olfactory representations in the orbitofrontal cortex do not adapt, provided that the order of presentation of olfactory stimuli is randomized, and each odour is presented for a short time of 1–2 s (Critchley and Rolls 1996b, Rolls, Critchley, Mason and Wakeman 1996b,

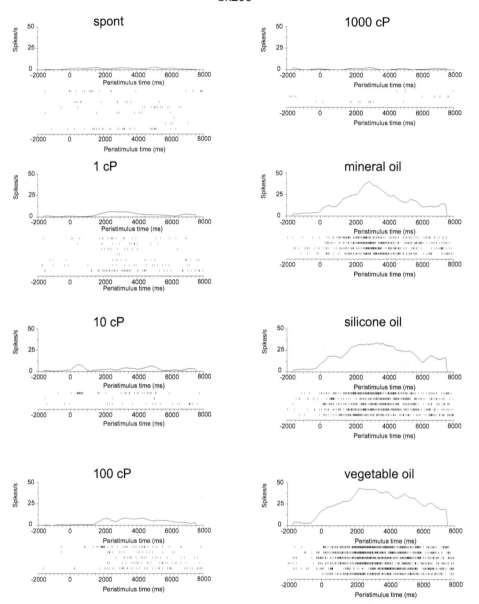

Fig. 10.2 Little adaptation of a neuron in the macaque orbitofrontal cortex that responded to the texture of fat and other oils in the mouth. The stimuli were delivered at approximately time 0, and the neuron continued to fire for 6–8 s, by which time the stimulus was being cleared from the mouth by swallowing. For each stimulus, rastergrams for several trials are shown below the mean firing rate across trials for that stimulus. In the rastergram, each vertical line is when the spike of a neuron occurred. This neuron responded much more to the three oils: vegetable oil (a fat), mineral oil (a hydrocarbon), and silicone oil. The neuron had little response to a viscosity series made from carboxymethylcellulose (CMC), a food thickening agent, at viscosities in the range 1 centiPoise to 1000 centiPoise. These results not only show a typical timecourse of the response of a neuron in the orbitofrontal cortex, which includes the secondary taste cortex, but also shows that fat is encoded by these neurons by its texture. (After Verhagen, Rolls and Kadohisa 2003.)

Rolls, Critchley and Treves 1996a, Critchley and Rolls 1996a, Rolls, Critchley, Verhagen and Kadohisa 2010a).

In another cortical area in macaques, the perirhinal cortex, neurons can respond with no adaptation for the full 1300 ms duration of highly familiar but not relatively novel visual stimuli, and in fact can do this for five such stimuli separated by only 400 ms (Hölscher, Rolls and Xiang 2003).

Consistent with this evidence, in areas that receive from the orbitofrontal cortex and amygdala, neurons also often show little adaptation of their visual responses, including neurons in the basal forebrain / substantia innominata (Rolls, Sanghera and Roper-Hall 1979), and ventral striatum (Williams, Rolls, Leonard and Stern 1993).

In the primate hippocampus, neurons typically respond to spatial views or the objects presented in those locations with increased firing rates that continue for at least 500 ms, though sometimes there is more firing in the first part of the response (Rolls, Xiang and Franco 2005c), and this is noticeable when the monkey moves the eyes to enter an effective spatial view field for a spatial view neuron (Rolls, Robertson and Georges-François 1997a, Rolls, Treves, Robertson, Georges-François and Panzeri 1998b, Robertson, Rolls and Georges-François 1998, Georges-François, Rolls and Robertson 1999), as it is for place cells in rats (personal communications).

In summary, in the awake behaving macaque, in which the presentation of stimuli can be precisely controlled, in many cortical areas, including higher order cortical areas, neurons show little evidence of adaptation, beyond in some cases a slightly higher firing rate in the first approximately 100–200 ms of the response. This it must be said corresponds with the subjective intensity of stimuli, which continues while the stimuli are presented, and does not markedly adapt. This also applies to the pleasantness of sensory inputs, which is not transient, but continues for many minutes until sensory-specific satiety may be produced, as described in Section 10.5.

Neuronal recordings in humans in the temporal cortex and other areas that receive from them such as the amygdala are consistent with those in macaques in often showing relatively little adaptation (Fried, Rutishauser, Cerf and Kreiman 2014, Mormann, Dubois, Kornblith, Milosavljevic, Cerf, Ison, Tsuchiya, Kraskov, Quiroga, Adolphs, Fried and Koch 2011) (see example in Fig. 8.3).

This is an important principle of cortical operation, that adaptation is minor, often hardly present at all, when the primate (including human) brain is working normally, and attention is being paid to the stimuli. The issue of whether there is relatively more adaptation in deep than superficial cortical layers is an interesting issue in relation to the hypothesized different dynamics and functions of superficial and deep neocortical layers (Section 26.5.18).

What factors then may contribute to maintaining cortical firing under normal conditions, but not in slices or under anaesthesia? Some of these factors are considered in the next section.

10.4 Acetylcholine, noradrenaline, and other modulators of adaptation and facilitation

10.4.1 Acetylcholine

One of the factors that reduces adaptation is acetylcholine, originating for most areas of the cerebral neocortex from the basal magnocellular forebrain nuclei of Meynert (Mesulam 1990) in which neurons respond to rewarding, punishing, and novel stimuli in the awake, behaving, monkey (Rolls, Burton and Mora 1976, Burton, Rolls and Mora 1976, Mora, Rolls and Burton 1976, Rolls, Sanghera and Roper-Hall 1979, Wilson and Rolls 1990c, Wilson and

Rolls 1990b, Wilson and Rolls 1990a, Rolls 2014a). Cortical neurons become much more sluggish in their responses, and show much more firing rate adaptation, in the absence of cholinergic inputs (Fuhrmann, Markram and Tsodyks 2002a, Abbott, Varela, Sen and Nelson 1997). Without this source of acetylcholine in the slice or under anaesthesia, adaptation is profound, and this raises questions about what is studied under these conditions, including adaptation (Silberberg et al. 2004, Thomson and Lamy 2007). Acetylcholine can act via a nicotinic receptor to enhance thalamo-cortical transmission (Gil, Connors and Amitai 1997).

Given this background, it has been hypothesized that one way in which impaired cholinergic neuron function is likely to impair cortical function including memory is by reducing the normally maintained high firing rates of activated cortical neurons, and thus reducing the depth of the basins of attraction of cortical (including hippocampal) networks (Rolls and Deco 2010, Rolls and Deco 2015b). This would make the recall of long-term episodic memories less reliable in the face of stochastic noise, and the maintenance of short-term memory less reliable in the face of stochastic noise (Rolls and Deco 2010, Rolls and Deco 2015b) (Section 16.4.4).

Acetylcholine in the neocortex (Bear and Singer 1986) and hippocampus also makes it more likely that LTP will occur, probably through activation of an inositol phosphate second messenger cascade (Markram and Segal 1990, Markram and Segal 1992, Seigel and Auerbach 1996, Hasselmo and Bower 1993, Hasselmo et al. 1995, Giocomo and Hasselmo 2007, Hasselmo and Sarter 2011). In the hippocampus and prefrontal cortex acetylcholine may simultaneously decrease transmission in recurrent collateral excitatory connections, and this may have the beneficial effect of reducing the effects of memories already stored in the recurrent collaterals so that they do not influence too much the neuronal firing when new memories must be stored (Giocomo and Hasselmo 2007, Hasselmo and Sarter 2011, Newman, Gupta, Climer, Monaghan and Hasselmo 2012, Rolls 2013b).

Reduced cholinergic function in normal aging may contribute to reduced memory function in both these ways, by increasing adaptation and thus reducing the maintained firing rates of cortical neurons, and by reducing long-term potentiation and thus memory storage (Rolls and Deco 2010, Rolls and Deco 2015b).

10.4.2 Noradrenergic neurons

The source of the noradrenergic projection to the neocortex is the locus coeruleus (noradrenergic cell group A6) in the pons (Cooper, Bloom and Roth 2003). (Note that noradrenaline is the same as norepinephrine.) There are a few thousand of these neurons that innervate the whole of the cerebral cortex, as well as the amygdala and other structures, so it is unlikely that the noradrenergic neurons convey the specific information stored in synapses that specifies each memory. Instead, to the extent that the noradrenergic neurons are involved in memory (including pattern association), it is likely that they would have a modulatory role on cell excitability, which would influence the extent to which the voltage-dependent NMDA receptors are activated, and thus the likelihood that information carried on specific afferents would be stored (Seigel and Auerbach 1996). Evidence that this may be the case comes from a study in which it was shown that neocortical LTP is impaired if noradrenergic and simultaneously cholinergic inputs to cortical cells are blocked pharmacologically (Bear and Singer 1986).

Further, in a study designed to show whether the noradrenergic modulation is necessary for memory, Borsini and Rolls (1984) showed that intra-amygdaloid injections of noradrenergic receptor blockers did impair the type of learning in which rats gradually learned to accept novel foods.

The function implemented by this noradrenergic input may be more general activation, rather than a signal that carries information about whether reward vs punishment has been given, for noradrenergic neurons in rats respond to both rewarding and punishing stimuli, and one of the more effective stimuli for producing release of noradrenaline is placing the feet in cool water (McGinty and Szymusiak 1988).

Norepinephrine (noradrenaline) acting on alpha2A-adrenoceptors can strengthen working memory implemented in recurrent attractor neural networks in the dorsolateral prefrontal cortex by inhibiting cAMP (Wang, Ramos, Paspalas, Shu, Simen, Duque, Vijayraghavan, Brennan, Dudley, Nou, Mazer, McCormick and Arnsten 2007). cAMP closes Hyperpolarization-activated Cyclic Nucleotide-gated (HCN) channels. The HCN channels on the distal dendrites allow K^+ (and Na^+) to pass through generating an h current (I_h) that shunts the effects of synaptic inputs (He, Chen, Li and Hu 2014), including inputs from the recurrent collateral connections. Thus the noradrenaline, by reducing shunting inhibition of the synaptic inputs, strengthens the attractor network, that is, maintains it more stably for more prolonged periods (Rolls and Deco 2015b). Part of the evidence comes from iontophoresis of agents that influence these cAMP-activated HCN channels onto single neurons in the macaque dorsolateral prefrontal cortex (Wang et al. 2007).

10.5   Synaptic depression and sensory-specific satiety

Sensory-specific satiety is the reduced reward value and subjective pleasantness (but not intensity) that is produced for example by eating a food to satiety (Section 2.2.3.3 and Rolls (2014a)). For taste, it is not reflected in the responses of neurons in the insular primary taste cortex, but it is present in the responses of neurons one synapse on in the orbitofrontal cortex (Rolls, Sienkiewicz and Yaxley 1989c, Rolls 2015e, Rolls 2016b). After sensory-specific satiety, orbitofrontal cortex neurons no longer respond to the taste, smell, or sight of the food eaten to satiety, but do respond to other foods (Critchley and Rolls 1996a, Rolls 2015e, Rolls 2016b). For this reason, the proposed mechanism is synaptic depression of the synapses that introduce information about food-related stimuli into the orbitofrontal cortex. This would for this particular case take approximately 15 minutes to develop (which is the length of time for which a food is typically eaten to induce sensory-specific satiety), and last for at least 15–30 minutes (the minimum time for which sensory-specific satiety lasts) (Rolls 2014a).

Consistent with this proposed mechanism, some sensory-specific satiety can be produced just by tasting or even smelling a food for a few minutes, without swallowing any of it (Rolls and Rolls 1997). This shows that just the presence of neuronal activity produced by the taste or the smell of food is sufficient to reduce the firing of neurons that represent the pleasantness of the taste or smell of food. This can occur to some extent even without gastric and post-gastric factors such as gastric distension and intestinal stimulation by food (Rolls 2014a). This indicates that this aspect of sensory-specific satiety can be produced by firing that is sustained for a few minutes in neurons in the pathway. Moreover, the decline in neuronal responsiveness must be at a late stage of the processing of the taste and probably the olfactory signals, in that just smelling the food for several minutes produces much more of a reduction in the pleasantness than in the intensity of the odour, and just tasting the food for several minutes produces much more of a decrease in the pleasantness than in the intensity of the taste (Rolls and Rolls 1997).

10.6   Neuronal and synaptic adaptation, and the memory for sequential order

Sequential learning refers to the ability to encode and represent the temporal order of discrete elements occurring in a sequence. Deco and Rolls (2005c) showed that the short-term memory for a sequence of items can be implemented in an autoassociation neuronal network using adaptation. Each item is one of the attractor states of the network. The autoassociation network was implemented at the level of integrate-and-fire neurons so that the contributions of different biophysical mechanisms to sequence learning could be investigated. It was shown that if it is a property of the synapses or neurons that support each attractor state that they adapt, then every time the network is made quiescent (by for example inhibition), then the attractor state that emerges next is the next item in the sequence. Network simulations demonstrated the operation of the mechanisms using 1) a sodium inactivation-based spike-frequency-adaptation mechanism, 2) a Ca^{2+} activated K^+ current, and 3) short term synaptic depression, with sequences of up to three items. The network does not need repeated training on a particular sequence, and will repeat the items in the order that they were last presented. The time between the items in a sequence is not fixed, allowing the items to be read out as required over a period of up to many seconds. The network thus uses adaptation rather than associative synaptic modification to recall the order of the items in a recently presented sequence. This mechanism would be potentially useful for multiple steps in a syntactic plan, and indeed just that adaptation mechanism is used in a model of how syntax could be implemented in the cortex (Chapter 17) (Rolls and Deco 2015a).

A similar principle was applied to a model of one-trial rule-based stimulus-reward reversal in the orbitofrontal cortex (Deco and Rolls 2005d). An attractor network with adaptation held one of two rules active. If the attractor was quashed by non-reward to indicate that reversal should occur, then the other attractor state in the network rose up out of the noisy firing, because it was less adapted. This mechanism thus changed the rule, and the rule in turn biased other neurons into the correct state to respond to whichever stimulus was now rewarded.

10.7   Destabilization of short-term memory by adaptation or synaptic depression

In Section 4.3.1 an autoassociation attractor principle of short-term memory in the cortex is described. Factors that influence the stability of these networks are described in Chapter 16. It is important that any one short-term memory is not stable for too long, so that we are sensitive to new inputs, and so that we can have new thoughts, or move to new steps in a short-term plan. Mechanisms that are likely to contribute to this include adaptation and synaptic depression, which would tend to destabilize the firing of one attractor state in a network holding a short-term memory after it has been active for a few seconds, in the presence of noise caused by the randomness of the spiking (Chapter 16). Too little adaptation or synaptic depression may contribute to the overstability of short-term memory systems in some parts of the cerebral cortex, and may in this way contribute to obsessive-compulsive disorder (Rolls, Loh and Deco 2008c, Rolls 2012b) (Section 16.2).

10.8 Non-reward computation in the orbitofrontal cortex using synaptic depression

Single neurons in the primate orbitofrontal cortex respond when an expected reward is not obtained, and behavior must change (Thorpe, Rolls and Maddison 1983, Rolls and Grabenhorst 2008, Rolls 2014a) (Fig. 15.4). The human lateral orbitofrontal cortex is activated when non-reward, or loss occurs (Kringelbach and Rolls 2003, O'Doherty, Kringelbach, Rolls, Hornak and Andrews 2001). The process is important in many emotions, which can be produced if an expected reward is not received, or lost, with examples including sadness and anger (Rolls 2014a). Consistent with this the psychiatric disorder of depression may arise when the non-reward system leading to sadness is too sensitive, or maintains its activity for too long (Rolls 2016d), or has increased functional connectivity (Cheng, Rolls, Qiu, Liu, Tang, Huang, Wang, Zhang, Lin, Zheng, Pu, Tsai, Yang, Lin, Wang, Xie and Feng 2016). Conversely, if the non-reward system is underactive or is damaged by lesions of the orbitofrontal cortex, the decreased sensitivity to non-reward may contribute to increased impulsivity (Berlin, Rolls and Kischka 2004, Berlin, Rolls and Iversen 2005), and even antisocial and psychopathic behavior (Rolls 2014a). For these reasons, understanding the mechanisms that underlie non-reward is important not only for understanding normal human behavior and how it changes when rewards are not received, but also for understanding and potentially treating better some emotional and psychiatric disorders. This orbitofrontal cortex non-reward error computation may influence the dopaminergic 'prediction error' neurons (Schultz 2013, Rolls and Deco 2016).

Rolls and Deco (2016) proposed and analyzed a mechanism for the computation of non-reward in the orbitofrontal cortex, which utilized the properties of synaptic depression. A single attractor network has a Reward population (or pool) of neurons that is activated by Expected Reward (for example the sight of food, or a stimulus associated with food), and maintain their firing until, after a time, synaptic depression reduces the firing rate in this neuronal population. (The network was similar to that illustrated in Fig. 5.5, with the Reward population corresponding to S1, and the Non-Reward population to S2.) If a Reward Outcome (for example the taste of food) is not received, the decreasing firing in the Reward neurons releases the inhibition implemented by inhibitory neurons, and this results in a second population of Non-Reward neurons to start and continue firing encouraged by the spiking-related noise in the network (Fig. 10.3).

If a Reward Outcome is received, this keeps the Reward attractor active, and this through the inhibitory neurons prevents the Non-Reward attractor neurons from being activated.

If an Expected Reward has been signalled, and the Reward attractor neurons are active, their firing can be directly inhibited by a Non-Reward Outcome (e.g. an aversive stimulus), and the Non-Reward neurons become activated because the inhibition on them is released (Rolls and Deco 2016).

A key concept in the mechanism proposed is the balance between the Reward and the Non-Reward attractor populations, which are effectively in competition with each other via the common population of inhibitory neurons. This helps to account for the neurophysiological evidence that if a food reward is moved towards a macaque, then the expected reward neurons in the primate orbitofrontal cortex fire fast, and as soon as the reward stops moving (within sight) towards the macaque, or is slowly removed (by being moved backwards), the non-reward neurons start to fire, and the reward neurons stop firing (Rolls, Thorpe and Maddison 1983b). The competing attractor hypothesis for reward and non-reward representations in the orbitofrontal cortex described here provides an account for this important property of the populations of neurons in the primate orbitofrontal cortex. Consistent with this competing attractor hypothesis, there is evidence for reciprocal activity in the human medial orbitofrontal cortex reward area and the lateral orbitofrontal cortex non-reward / loss area (O'Doherty,

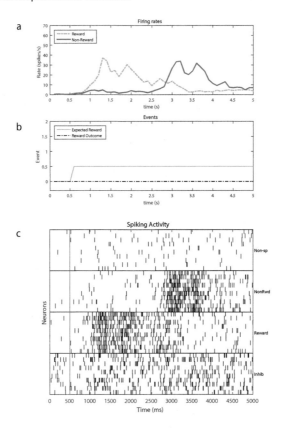

Fig. 10.3 The operation of the network when an Expected Reward is not obtained (extinction). a. The firing rates of the Reward population of neurons and the Non-Reward population of neurons during a 5 s trial. b. After a period of spontaneous activity from 0 until 0.5 s, the Expected Reward input was applied to the reward attractor population of neurons (3.1 Hz per synapse), and maintained at that level for the remainder of the trial. No Reward Outcome input was received. The input to the Non-Reward population of neurons was maintained constant at 3.05 Hz per synapse from time = 0.5 s until the end of the trial. Note: there were 800 external synapses onto each neuron through which the external inputs were applied with the firing rates specified for each synapse.) c. Rastergrams showing for each of the four populations of neurons, Non-Specific, Non-Reward, Reward, and Inhibitory, the spiking of ten neurons chosen at random from each population. Each small vertical line represents a spike from a neuron. Each horizontal row shows the spikes of one neuron. The different neurons are from the same trial. (After Rolls and Deco 2016.)

Kringelbach, Rolls, Hornak and Andrews 2001). In particular, activations in the human medial orbitofrontal increase in proportion to the amount of (monetary) reward received, and decrease in proportion to the amount of (monetary) loss; and in the lateral orbitofrontal cortex, the activations are reciprocally related to these changes (O'Doherty et al. 2001).

Synaptic depression is the mechanism used to illustrate the concept of how non-reward neuronal firing is produced, but other mechanisms can be envisaged. For example, if the trials typically involved a delay of several seconds before the reward outcome was received, then a separate short-term memory that is set to estimate the typical delay before reward outcome might be used to decrease the expected reward input to the network after the appropriate delay, allowing the non-reward neurons to then fire (Rolls and Deco 2016).

10.9 Synaptic facilitation and a multiple-item short-term memory

short-term memory!multiple item

Human short-term memory has a capacity of several items maintained simultaneously. Rolls, Dempere-Marco and Deco (2013) showed how the number of short-term memory representations that an attractor network modeling a cortical local network can simultaneously maintain active is increased by using synaptic facilitation of the type found in the prefrontal cortex. It was shown to be possible to maintain 9 short-term memories active simultaneously in integrate-and-fire simulations of a single attractor network where the proportion of neurons in each population, the sparseness, is 0.1. The stability of such a system was confirmed with mean-field analyses. Without synaptic facilitation the system can maintain many fewer memories active in the same network. The system operates because of the effectively increased synaptic strengths formed by the synaptic facilitation just for those pools (populations of neurons) to which the cue is applied, and then maintenance of this synaptic facilitation in just those pools when the cue is removed by the continuing neuronal firing in those pools.

The findings have implications for understanding how several items can be maintained simultaneously in short-term memory, how this may be relevant to the implementation of language in the brain, and suggest new approaches to understanding and treating the decline in short-term memory that can occur with normal aging (Rolls, Dempere-Marco and Deco 2013).

10.10 Synaptic facilitation in decision-making

During decision-making between sequential stimuli, the first stimulus must be held in memory and then compared with the second. In systems that encode the stimuli by their firing rate, such as vibrotactile stimuli, neurons can use synaptic facilitation not only to remember the first stimulus during the delay, but also during the presentation of the second stimulus so that they respond to a combination of the first and second stimuli (Deco, Rolls and Romo 2010). This models the activity of "partial differential" neurons recorded in the ventral premotor cortex during vibrotactile flutter frequency decision making (Deco, Rolls and Romo 2010). Moreover, such partial differential neurons can provide important input to a subsequent attractor decision-making network that can then compare this combination of the first and second stimuli with inputs from other neurons that respond only to the second stimulus. Thus, both synaptic facilitation and neuronal attractor dynamics can account for sequential decision-making in such systems in the brain (Deco, Rolls and Romo 2010).

Another potential role for synaptic facilitation in decision-making is when actions must be delayed for several seconds (Martinez-Garcia, Rolls, Deco and Romo 2011). In a decision task in which the response must be postponed for several seconds, information about a forthcoming action becomes available from the activity of neurons in the medial premotor cortex in a sequential decision-making task after the second stimulus is applied, providing the information for a decision about whether the first or second stimulus is higher in vibrotactile frequency (Martinez-Garcia, Rolls, Deco and Romo 2011). The information then decays in a 3-s delay period in which the neuronal activity declines before the behavioral response can be made. The information then increases again when the behavioral response is required.

This neuronal activity was modelled using an attractor decision-making network in which information reflecting the decision is maintained at a low level during the delay period, and is then selectively restored by a non-specific input when the response is required. One mechanism for the short-term memory is synaptic facilitation, which can implement a mechanism for postponed decisions that can be correct even when there is little neuronal firing during the delay period before the postponed decision. Another mechanism is graded firing rates by different neurons in the delay period, with restoration by the non-specific input of the low-rate activity from the higher-rate neurons still firing in the delay period. These mechanisms can account for the decision-making and for the memory of the decision before a response can be made, which are evident in the activity of neurons in the medial premotor cortex (Martinez-Garcia, Rolls, Deco and Romo 2011).

10.11 Highlights

1. Adaptation is not a major effect evident in cortical neuronal responses in awake behaving primates including humans.

2. Adaptation is evident in cortical slices, and in higher cortical areas when under anaesthesia.

3. Part of the reason for the adaptation in (2) is that modulators such as acetylcholine present during alertness in the awake behaving state reduce neuronal adaptation. Acetylcholine is released in the cortex by basal forebrain cholinergic neurons that respond to rewarding, punishing, and/or novel stimuli.

4. Nevertheless, neuronal firing rate adaptation may play a role in a number of cortical operations, including adaptation of thalamo-cortical inputs so that cortical recurrent networks can after activity has been started contribute to the resulting firing; sequence order memory; and rule reversal in stimulus-reward reversal.

5. Synaptic depression may contribute to the decreasing stability of cortical attractor network states over a period of seconds, allowing them to be replaced with new states; to the firing of non-reward neurons in the orbitofrontal cortex if an expected reward is not received; and to sensory-specific satiety.

6. Short-term synaptic facilitation may contribute to holding information online in delay periods, for example during short-term memory tasks in which many items must be held active, in decision-making with a delay between stimuli, and decision-making when the response must be postponed.

11   Backprojections in the neocortex

In Chapters 2 and 25 a possible way in which processing could operate through a hierarchy of cortical stages is described. Convergence and competition are key aspects of the processing. This processing could act in a feedforward manner, and indeed, in experiments on backward masking, it is shown that there is insufficient time for top-down processing to occur when objects can just be recognized (Rolls 2003) (see Section C.3.4). (Neurons in each cortical stage respond for 20–30 ms when an object can just be seen, and given that the time for processing to travel from V1 to inferior temporal visual cortex (IT) is approximately 50 ms, there is insufficient time for a return projection from IT to reach V1, influence processing there, and in turn for V1 to project up to IT to alter processing there (Rolls 2003).) Nevertheless, backprojections are a major feature of cortical connectivity, and we next consider hypotheses about their possible function.

11.1   Architecture

The forward and backward projections in the neocortex that will be considered are shown in Fig. 11.1 (for further anatomical information see Sections 1.10, 18.2.1 and 26.5.18, Fig. 1.19 (Jones and Peters 1984, Peters and Jones 1984, da Costa and Martin 2010, Markov et al. 2014a, Markov, Ercsey-Ravasz, Lamy, Ribeiro Gomes, Magrou, Misery, Giroud, Barone, Dehay, Toroczkai, Knoblauch, Van Essen and Kennedy 2013a, Markov et al. 2014b, Markov et al. 2013b, Pandya et al. 2015)). As described above, in primary sensory cortical areas, the main extrinsic 'forward' input is from the thalamus and ends in layer 4, where synapses are formed onto spiny stellate cells. These in turn project heavily onto pyramidal cells in layers 3 and 2, which in turn send projections forwards to terminate strongly in layer 4 of the next cortical layer (on small pyramidal cells in layer 4 or on the basal dendrites of the layer 2 and 3 (superficial) pyramidal cells, and also onto layer-6 pyramidal cells and inhibitory interneurons). Inputs reach the layer 5 (deep) pyramidal cells from the pyramidal cells in layers 2 and 3 (Martin 1984), and it is the deep pyramidal cells that send backprojections to end in layer 1 of the preceding cortical area (see Fig. 11.1), where there are apical dendrites of pyramidal cells (Fig. 1.14). It is important to note that in addition to the axons and their terminals in layer 1 from the succeeding cortical stage, there are also axons and terminals in layer 1 in many stages of the cortical hierarchy from the amygdala and (via the subiculum, entorhinal cortex, and parahippocampal cortex) from the hippocampal formation (see Figs. 11.1, and 24.1) (Van Hoesen 1981, Turner 1981, Amaral and Price 1984, Amaral 1986, Amaral 1987, Amaral, Price, Pitkanen and Carmichael 1992).

A feature of the cortico-cortical forward and backprojection connectivity shown schematically in Figs. 11.1, and 24.1 is that it is 'keyed', in that the origin and termination of the connections between cortical areas provide evidence about which one is forwards or higher in the hierarchy (Pandya et al. 2015). The reasons for the asymmetry (including the need for backprojections not to dominate activity in preceding cortical areas, see Rolls and Stringer (2001b)) are described below, but the nature of the asymmetry between two cortical areas provides additional evidence about the hierarchical nature of processing in visual cortical

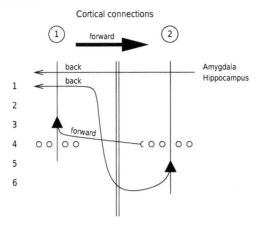

Fig. 11.1 Schematic diagram of forward and backward connections between adjacent neocortical areas. (Area 1 projects forwards to area 2 in the diagram. Area 1 would in sensory pathways be closest to the sense organs.) The superficial pyramidal cells (triangles) in layers 2 and 3 project forwards (in the direction of the arrow) to the next cortical area. The deep pyramidal cells in layer 5 project backwards to end in layer 1 of the preceding cortical area (on the apical dendrites of pyramidal cells). The hippocampus and amygdala are also sources of backprojections that end mainly in layer 1 of the higher association cortical areas. Spiny stellate cells are represented by small circles in layer 4. See text for further details.

areas especially in the ventral stream, for which there is already much other evidence (see Chapter 25, and Panzeri, Rolls, Battaglia and Lavis (2001)). However, the keying is less distinct when cross-connections between areas at similar levels of the hierarchy are considered (Pandya et al. 2015). Moreover, findings on the macaque connectome are revealing that although there are separate neurons for the forward and backprojecting pathways, they are not exclusively limited to the superficial and deep layers respectively (Markov et al. 2014a, Markov and Kennedy 2013, Markov et al. 2014b, Markov et al. 2013b, Wang and Kennedy 2016). A consistent difference is that the forward projections are point-to-point, whereas the backward projections are more diffuse (Markov et al. 2013b, Wang and Kennedy 2016). In primates the feedforward projection neurons are concentrated in L3B, but some L2 and L3A neurons do send backprojections in primates (Markov et al. 2014b). In primates the main feedback projection neurons are in L6 and Lower L5 (L5B) (Markov et al. 2014b).

One point made in Figs. 1.9, 24.1 and 2.4 is that the amygdala and hippocampus are stages of information processing at which the different sensory modalities (such as vision, hearing, touch, taste, and smell for the amygdala) are brought together, so that correlations between inputs in different modalities can be detected in these regions, but not at prior cortical processing stages in each modality, as these cortical processing stages are mainly unimodal. As a result of bringing together any two sensory modalities, significant correspondences between the two modalities can be detected. One example might be that a particular visual stimulus is associated with the taste of food. Another example might be that another visual stimulus is associated with painful touch. Thus at these limbic (and orbitofrontal cortex, see Chapter 15) stages of processing, but not before, the significance of, for example, visual and auditory stimuli can be detected and signalled. Sending this information back to the neocortex could thus provide a signal that indicates to the cortex that information should be stored. Even more than this, the backprojection pathways could provide patterns of firing that could help the neocortex to store the information efficiently, one of the possible functions of backprojections within and to the neocortex considered next.

11.2 Learning

The way in which the backprojections could assist learning in the cortex can be considered using the architecture shown in Figs. 11.1 and B.18 (see also Section B.4.5). The input stimulus occurs as a vector applied to (layer 3) cortical pyramidal cells through modifiable synapses in the standard way for a competitive net (input A in the schematic Fig. B.18). If it is a primary cortical area, the input stimulus is at least partly relayed through spiny stellate cells, which may help to normalize and orthogonalize the input patterns in a preliminary way before the patterns are applied to the layer 3 pyramidal cells. (The details of these computational operations will be made clear in Appendix B.) If it is a non-primary cortical area, the cortico-cortical forward axons may end more strongly on the basal dendrites of neurons in the superficial cortical layers (3 and possibly 2). The lower set of synapses on the pyramidal cells would then start by competitive learning to set up representations on these neurons that would represent correlations in the input information space, and could be said to correspond to features in the input information space, where a feature is defined simply as the representation of a correlation in the input information space (see Appendix B).

Consider now the conjunctive application of a pattern vector via the backprojection axons with terminals in layer 1 (B in Fig. B.18 on page 743) with the application of one of the (forward) input stimulus vectors. If, to start with, all synapses can be taken to have random weights, some of the pyramidal cells will by chance be strongly influenced both by the input stimulus and by the backprojecting vector. These strongly activated neurons will then compete with each other as in a standard competitive net, to produce contrast enhancement of their firing patterns. The relatively short-range (50 μm) excitatory operations produced by the bipolar and double bouquet cells, together with more widespread (300–500 μm) recurrent lateral inhibition produced by the smooth non-pyramidal cells and perhaps the basket cells, may be part of the mechanism of this competitive interaction. Next, Hebbian learning takes place as in a competitive net (see Appendix B), with the addition that not only the synapses between forward projecting axons and active neurons are modified, but also the synapses between the backward projecting axons and the active neurons, which are in layer 1, are associatively modified.

This functional architecture has the following properties (see also Section B.4.5). First, orthogonal backprojecting inputs can help the neurons to separate input stimuli (on the forward projection lines, A in Fig. B.18) even when the input stimuli are very similar. This is achieved by pairing two somewhat different input stimuli A with very different (for example orthogonal) backprojection stimuli B. This is easily demonstrated in simulations (for example Rolls (1989b) and Rolls (1989f)). Conversely, if two somewhat different input stimuli A are paired with the same backprojection stimulus B, then the outputs of the network to the two input stimuli are more similar than they would otherwise be (see Section B.18). This is also easily demonstrated in simulations.

In the neocortex, the backprojecting 'tutors' (see Rolls (1989b) and Rolls (1989f)) can be of two types. One originates from the amygdala and hippocampus, and by benefiting from cross-modal comparison, can for example provide orthogonal backprojected vectors. This backprojection, moreover, may only be activated if the multimodal areas detect that the visual stimulus is significant, because for example it is associated with a pleasant taste. This provides one way in which guidance can be provided for a competitive learning system as to what it should learn, so that it does not attempt to lay down representations of all incoming sensory information. The type of guidance is to influence which categories are formed by the competitive network.

The second type of backprojection 'tutor' is that from the next cortical area in the hierarchy. The next cortical area could operate in the same manner, and because it is a competitive system, is able to further categorize or orthogonalize the stimuli it receives, benefiting also from additional convergence (see for example Chapter 25). It then projects back these more orthogonal representations as tutors to the preceding stage, to effectively build better filters for the categories it is finding (cf. Fig. B.19). These categories might be higher order (such as two parallel lines on the retina), and even though the receptive fields of neurons in the preceding area might never receive inputs about both lines because the receptive fields are too small, the backprojections could still help to build feature analyzers at the earlier stage that would be tuned to the components of what can be detected as higher order features at the next stage (see Chapter 25 and Fig. B.19).

Another way for what is laid down in neocortical networks to be influenced is by neurons which 'strobe' the cortex when new or significant stimuli are shown. The cholinergic system originating in the basal forebrain (see Sections 16.4.4 and 10.4.1), and the noradrenergic input to layer 1 of the cortex from the locus coeruleus, may contribute to this function (see Section 10.4.2). By modulating whether storage occurs according to the arousal or activation being produced by the environment, the storage of new information can be promoted at important times only, thus making good use of inevitably limited storage capacity. This influence on neocortical storage is not explicit guidance about the categories formed, as could be produced by backprojections, but instead consists of influencing which patterns of neuronal activity in the cortex are stored.

11.3 Recall

Evidence that during recall neural activity does occur in cortical areas involved in the original processing comes, for example, from investigations that show that when humans are asked to recall visual scenes in the dark, blood flow is increased in visual cortical areas, stretching back from association cortical areas as far as early (possibly even primary) visual cortical areas (Roland and Friberg 1985, Kosslyn 1994). Recall is a function that could be produced by cortical backprojections.

If in Figs. 11.1, B.18 or 24.1 only the backprojection input (B in Fig. B.18) is presented after the type of learning just described, then the neurons originally activated by the forward-projecting input stimuli (A in Fig. B.18) are activated. This occurs because the synapses from the backprojecting axons onto the pyramidal cells have been associatively modified only where there was conjunctive forward and backprojected activity during learning. This thus provides a mechanism for recall. The crucial requirement for recall to operate in this way is that, in the backprojection pathways, the backprojection synapses would need to be associatively modifiable, so that the backprojection input when presented alone could operate effectively as a pattern associator to produce recall. Some aspects of neocortical architecture consistent with this hypothesis (Rolls 1989b, Rolls 1989f, Treves and Rolls 1994, Rolls and Treves 1998, Kesner and Rolls 2015) are as follows.

First, there are many NMDA receptors on the apical dendrites of cortical pyramidal cells, where the backprojection synapses terminate. These receptors are implicated in associative modifiability of synapses, and indeed plasticity is very evident in the superficial layers of the cerebral cortex (Diamond, Huang and Ebner 1994).

Second, the backprojection synapses in ending on the apical dendrite, quite far from the cell body, might be expected to be sufficient to dominate the cell firing when there is no forward input close to the cell body. In contrast, when there is forward input to the neuron, activating synapses closer to the cell body than the backprojecting inputs, this would tend to electrically shunt the effects received on the apical dendrites. This could be beneficial during the original learning, in that during the original learning the forward input would have the stronger effect on the activation of the cell, with mild guidance then being provided by the backprojections.

An example of how this recall could operate is provided next. Consider the situation when in the visual system the sight of food is forward projected onto pyramidal cells in higher order visual cortex, and conjunctively there is a backprojected representation of the taste of the food from, for example, the amygdala or orbitofrontal cortex. Neurons which have conjunctive inputs from these two stimuli set up representations of both, so that later if only the taste representation is backprojected, then the visual neurons originally activated by the sight of that food will be activated. In this way many of the low-level details of the original visual stimulus might be recalled. Evidence that during recall relatively early cortical processing stages are activated comes from cortical blood flow studies in humans, in which it has been found, for example, that quite posterior visual areas are activated during recall of visual (but not auditory) scenes (Kosslyn 1994). The backprojections are probably in this situation acting as pattern associators.

The quantitative analysis of the recall that could be implemented through the hippocampal backprojection synapses to the neocortex, and then via multiple stages of cortico-cortical backprojections, makes it clear that the most important quantitative factor influencing the number of memories that can be recalled is the number of backprojecting synapses onto each cortical neuron in the backprojecting pathways (see Section 24.3.7, Fig. 24.1, Treves and Rolls (1994), and Treves and Rolls (1991)). This provides an interpretation of why there are in general as many backprojecting synapses between two adjacent cortical areas as forward connections. The number of synapses on each neuron devoted to the backprojections needs to be large to recall as many memories as possible, but need not be larger than the number of forward inputs to each neuron, which influences the number of possible classifications that the neuron can perform with its forward inputs (see Section 24.3.7).

An implication of these ideas is that if the backprojections are used for recall, as seems likely as just discussed, then this would place severe constraints on their use for functions such as error backpropagation (see Appendix B). It would be difficult to use the backprojections in cortical architecture to convey an appropriate error signal from the output layer back to the earlier, hidden, layers if the backprojection synapses are also to be set up associatively to implement recall. Recall can be implemented by making the backprojection synapses associatively modifiable, as described in Section 24.3.7. Error backpropagation learning is not related to this backprojection-for-recall function (Section B.11).

11.4 Semantic priming

A third property of this backprojection architecture is that it could implement semantic priming, by using the backprojecting neurons to provide a small activation of just those neurons that are appropriate for responding to that semantic category of input stimulus.

11.5 Top-down Attention

In the same way, attention could operate from higher to lower levels, to selectively facilitate only certain pyramidal cells by using the backprojections. Indeed, the backprojections described could produce many of the 'top-down' influences that are common in perception (cf. Fig. B.19 and Chapter 6).

For top-down attention to operate to bias or modulate the networks at earlier cortical stages, but not to dominate them to prevent the top-down influence from overriding the

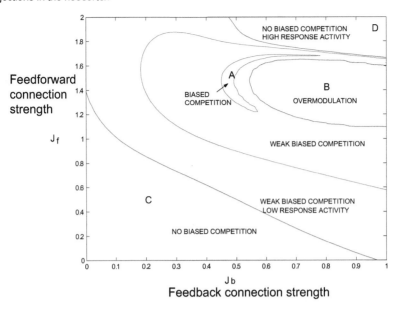

Feedforward
connection
strength

J_f

J b

Feedback connection strength

Fig. 11.2 For top-down attention to operate by biased competition, the ratio of the backward to forward connections needs to be approximately 1/3. Inter-area cortical parameter exploration showing the regions within which biased competition operates as a function of the feedforward and feedback V2–V4 synaptic connections J_f and J_b in a model of biased competition which accounts for neurophysiological findings on selective attention. (Reproduced with permission from Journal of Neurophysiology 94: 295–313. Neurodynamics of biased competition and cooperation for attention: a model with spiking neurons, Gustavo Deco, Edmund T. Rolls. Copyright ©2005, The American Physiological Society.)

bottom up effects, the backprojections need to be approximately 3 times weaker than the bottom-up forward connections (Deco and Rolls 2005b). This was shown in integrate-and-fire simulations of attentional effects between for example visual cortical areas V4 and V2 (Fig. 11.2) which account for neurophysiological effects of attention in these areas (Reynolds et al. 1999, Reynolds and Desimone 1999). Region "A" is the correct working region for biased competition, where the top-down ('feedback') connections usefully modulate the effects of the bottom-up ('feedforward') inputs. The region at the top around the point "D" (high J_f and high J_b) corresponds to a region that we call "No Biased Competition: High Response Activity" because there is low attentional modulation (both up-regulating and down-regulating), and relatively high neural responses in both specific pools of area V2 and V4. The region around point "B", which has higher feedback values as required for biased competition, corresponds to a dynamical attractor that we call "Overmodulation", because the attentional modulation effects are unrealistically high, in spite of the fact that the level of response activity in the absence of attention is in the experimental range. There are large regions of the parameter space, which we characterise as "Weak Biased Competition", corresponding to a dynamical attractor that shows attentional modulation qualitatively according to biased competition but quantitatively too weak, and with the normal level of neural response when attention is absent. This region is followed by another region "Weak Biased Competition: Low Activity Response" which also shows a low level of neural response in the absence of attention. The last region, corresponding to low feedforward values and called "No Biased Competition" shows a low level of response with no attentional modulation.

11.6   Autoassociative storage, and constraint satisfaction

If the forward connections from one cortical area to the next, and the return backprojections, are both associatively modifiable, then the coupled networks could be regarded as, effectively, an autoassociative network. (Autoassociation networks are described in Section B.3). A pattern of activity in one cortical area would be associated with a pattern in the next that occurred regularly with it. This could enable higher cortical areas to influence the state of earlier cortical areas, and could be especially influential in the type of situation shown in Fig. B.19 in which some convergence occurs at the higher area. For example, if one of the earlier stages (for example the olfactory stage in Fig. B.19) had a noisy input on a particular occasion, its representation could be cleaned up if a taste input normally associated with it was present. The higher cortical area would be forced into the correct pattern of firing by the taste input, and this would feed back as a constraint to affect the state into which the olfactory area settled. This could be a useful general effect in the cerebral cortex, in that constraints arising only after information has converged from different sources at a higher level could feed back to influence the representations that earlier parts of the network settle into. This is a way in which top-down processing could be implemented, and is analyzed in Section 4.10.

The autoassociative effect between two forward and backward connected cortical areas could also be used in short-term memory functions (see Section 4.3.1), to implement the types of short-term memory effect described in Appendix B and Section 4.3.1. Such connections could also be used to implement a trace learning rule as described in Chapter 25.

However, forming a single attractor network based on the forward and backward connections between adjacent areas in a cortical hierarchy is unlikely to be useful, as described in Section 4.9.

11.7   Highlights

1. Cortical backprojections do not transfer the signal computed in one cortical area back to the preceding cortical area, in that for example the responses of the neurons in the primary visual cortex are quite different from those in the inferior temporal visual cortex.
2. Cortical backprojections are used for memory recall, for example in the backprojections from the hippocampus. To enable the number of memories recalled to be high in number, the backprojections must be high in number, as the number of memories recalled depends on the number of synapses from the backprojections onto each cortical neuron in the preceding cortical area (see Chapter 24). This remains the only quantitative argument for why there are as many backprojections as forward projections between two adjacent areas in a cortical hierarchy.
3. To implement recall, the backprojections synapses must be associatively modifiable in at least some of the stages, in order to act as a pattern associator and achieve good recall capacity (see Chapter 24).
4. Cortical backprojections can also be used for top-down attention, operating by biased competition or by biased activation.

12   Memory and the hippocampus

12.1   Introduction

The hippocampus is a three-layered type of cortex sometimes termed archicortex, distinct from the six-layered neocortex with which much of the rest of this book is concerned. Partly because hippocampal cortex is different from neocortex, the details of operation of the hippocampal system are considered in detail in a separate chapter, Chapter 24. This Chapter (12) outlines some of the principles of operation of hippocampal cortex and where they are different from those of the neocortex, and emphasizes that especially in primates including humans the operation of the hippocampal system can only be understood in the context of the connections that it receives from and sends back to the neocortex. The hippocampal system is one in which there are well developed quantitative theories of operation of its circuitry, and thus its principles of operation, which are highlighted in Chapter 24.

12.2   Hippocampal circuitry and connections

Fig. 12.1 shows that the primate hippocampus receives major inputs via the entorhinal cortex (area 28). These inputs come from the highly developed parahippocampal gyrus (areas TF and TH) as well as the perirhinal cortex, and thereby from the ends of many processing streams of the cerebral association cortex, including the visual and auditory temporal lobe association cortical areas, the prefrontal cortex, and the parietal cortex (Amaral et al. 1992, Amaral 1987, Suzuki and Amaral 1994a, Van Hoesen 1982, Witter, Wouterlood, Naber and Van Haeften 2000b, Lavenex, Suzuki and Amaral 2004, Johnston and Amaral 2004). The hippocampus is thus by its connections potentially able to associate together object representations (from the temporal lobe visual and auditory cortical areas) and spatial representations (from e.g. parietal cortex areas). In addition, the entorhinal cortex receives inputs from the amygdala, and the orbitofrontal cortex, which provide reward-related information to the hippocampus (Carmichael and Price 1995b, Suzuki and Amaral 1994a, Pitkanen, Kelly and Amaral 2002, Stefanacci, Suzuki and Amaral 1996, Rolls 2015c). The connections are analogous in the rat, although areas such as the temporal lobe visual cortical areas are not well developed in rodents, and the parahippocampal gyrus may be represented by the dorsal perirhinal cortex (Burwell, Witter and Amaral 1995).

12.3   The hippocampus and episodic memory

The hippocampal system is important in episodic memory including spatial memory. An episodic memory is a memory of a particular episode, and the spatial context in which the memory was formed is typically an important component. An episodic memory might involve remembering who was present at dinner on the previous day, and thus might involve representing a spatial position in a scene, who was present at that spatial position, what food was served, and what was discussed. For this, associations must be very rapidly learned

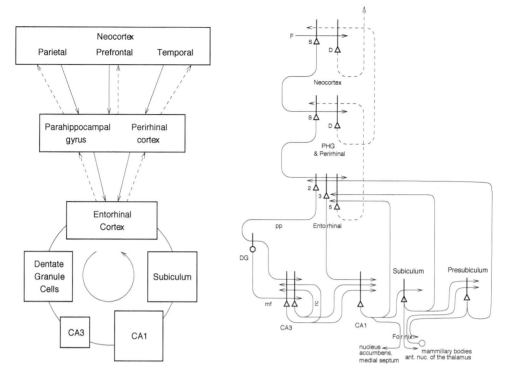

Fig. 12.1 Forward connections (solid lines) from areas of cerebral association neocortex via the parahippocampal gyrus and perirhinal cortex, and entorhinal cortex, to the hippocampus; and backprojections (dashed lines) via the hippocampal CA1 pyramidal cells, subiculum, and parahippocampal gyrus to the neocortex. There is great convergence in the forward connections down to the single network implemented in the CA3 pyramidal cells; and great divergence again in the backprojections. Left: block diagram. Right: more detailed representation of some of the principal excitatory neurons in the pathways. Abbreviations: D, deep pyramidal cells; DG, dentate granule cells; F, forward inputs to areas of the association cortex from preceding cortical areas in the hierarchy. mf: mossy fibres; PHG, parahippocampal gyrus and perirhinal cortex; pp, perforant path; rc, recurrent collaterals of the CA3 hippocampal pyramidal cells; S, superficial pyramidal cells; 2, pyramidal cells in layer 2 of the entorhinal cortex; 3, pyramidal cells in layer 3 of the entorhinal cortex; 5, 6, pyramidal cells in the deep layers of the entorhinal cortex. The thick lines above the cell bodies represent the dendrites.

between a unique combination of sensory inputs including often a location in a scene, a person, an object, etc. Later, it must be possible to recall the complete episode from any part.

12.4 Autoassociation in the CA3 network for episodic memory

A unique feature of the hippocampus, unlike what is found in the neocortex, is that there is a single CA3 network with recurrent collateral associatively modifiable connections that extend throughout CA3 (Chapter 24). The theory is that the CA3 operates as a single autoassociation or attractor network. The autoassociation architecture enables arbitrary associations to be formed in the sense that any representation within CA3 may be associated with any other representation in CA3, and the whole representation can be recalled from any part. This is not possible in a pattern association network, in which the conditioned stimulus can not be recalled from the unconditioned stimulus (Appendix B). For the network to operate as a single attractor network, it is advantageous if the recurrent collateral connections are distributed throughout CA3, and this is even more the case in primates than in rodents (Kondo, Lavenex and Amaral 2009).

The leading factor in determining the memory capacity of the hippocampus in Rolls' theory of the hippocampus (Rolls 1987, Rolls 1989b, Rolls 2010b, Kesner and Rolls 2015) is the number of recurrent collateral synapses on each CA3 neuron, and not the number of CA3 neurons. Sparse representations help to increase the memory capacity, and if the sparseness is 0.1, the number of memories that can be stored is approximately the same as the number of recurrent collateral connections onto each neuron (Appendix B). In the rat, this number is approximately 12,000. It may be higher in humans.

The representations in the primate hippocampus are more sparse than those in many high-order neocortical regions (Rolls and Treves 2011), and this is consistent with the importance of maximizing the number of memories that can be stored and retrieved in the hippocampus. In the cortex, other computations are important, including categorising stimuli.

The fact that there are fewer recurrent collateral synapses per CA3 neuron (12,000) than CA3 neurons (300,000) indicates that the connectivity is diluted, in the rat to 0.04. The theory is that this diluted connectivity enables there to be relatively few cases of more than one randomly made synaptic connection between any pair of hippocampal CA3 neurons, which otherwise would considerably reduce the memory capacity by distorting the basins of attraction (Rolls 2012a) (Chapter 7). Consistent with the need for high storage capacity in the hippocampus, the connectivity may be more diluted in the hippocampus than in the neocortex, for which latter an estimate is 0.1.

Binary firing rate representations enable more memories to be stored in an attractor network than graded firing rate representations, because with graded representations extra information is required if the magnitude of the firing rate has to be stored (Treves and Rolls 1991) (Chapter 8 and Appendix C). The non-linearity in the NMDA receptor may result in a more binary representation being stored than is present in neuronal firing, in that low rates may not be stored. Whether the low peak firing rates in the primate hippocampus (rarely up to 15 spikes/s) compared with the peak rates of 100 spikes/s present in for example the inferior temporal visual cortex contribute to more binary representations for storage in the hippocampus is an interesting hypothesis.

In a single attractor network, the memory capacity can be calculated as described in Section B.3. In the cerebral neocortex, the concept is that there are local attractor networks, which operate within the range specified by the recurrent collateral connections, in the order of 2–3 mm. A result is that although neocortical attractors operate locally, there is some spatial overlap with neighbouring local attractor networks, and this results in interference, which may reduce the memory capacity in the neocortical type of local attractor network by a factor in the order of 2–3 (Roudi and Treves 2006). The theory that the hippocampal CA3 operates as a single network with a global attractor is thus very much consistent with the theory that the hippocampus is set up to store a large number of memories, with no reduction of memory capacity due to overlapping different attractor networks.

If the number of memories stored in an attractor autoassociation network exceeds its critical memory capacity, then most of the memories in the network may become unretrievable. One mechanism that may help to ensure that the critical storage capacity is not reached is forgetting, as described in Section 9.3, and this is one sound computational reason for forgetting to be a principle of operation of the cerebral cortex.

12.5   The dentate gyrus as a pattern separation mechanism, and neurogenesis

In Rolls' theory of the hippocampus (Chapter 24, Rolls (1989b), Kesner and Rolls (2015)), the dentate granule cells operate as a competitive network that helps to separate (or categorize or orthogonalize) patterns being received from the entorhinal cortex. Then the very diluted connectivity of the dentate granule cells via the mossy fibers to result in only approximately 46 synapses onto each CA3 neuron (Fig. 24.14) produces further pattern separation, in that a unique subset of CA3 neurons is selected by different inputs to the dentate gyrus. In this theory, the function of the mossy fibers is to produce orthogonal firing of CA3 representations for different memories, and no information is stored in the mossy fiber synapses. This makes the hippocampus operate completely differently to the neocortex, in which the goal is not to store large numbers of memories independently of each other (as in the hippocampus), but instead to set up useful categorisation of inputs using for example competitive learning. In competitive learning, the afferent synapses are modifiable, and effectively store information about which inputs are to strongly activate which neurons (Sections B.4 and 7.4).

One result of this conceptualization is that mossy fibre type orthogonalizing synapses are nothing to do with the operation of the neocortex. Another implication is that in the hippocampal system, if all that the granule cell / mossy fibre systems are performing computationally is to select orthogonal subsets of CA3 neurons for each input, then neurogenesis here may be useful (by enabling completely new subsets of CA3 neurons to be selected by the new random connectivity of new dentate granule cells), but will not be useful at all in the neocortex, and should not be normal, because information is stored in neocortical afferent synapses, as in competitive networks. That is the computational rule of operation of cortical circuitry by which neurogenesis may be useful in the hippocampal dentate granule cell system, but not in the neocortex.

These are fundamentally different principles of operation of the dentate granule cell / hippocampal system, and the neocortex. Neurogenesis is a useful principle of operation of the hippocampal cortex. Neurogenesis is not a useful principle of operation of the neocortex. The reasons are given in this Section, and in Chapter 24.

12.6   Rodent place cells vs primate spatial view cells

The award of the Nobel Prize for Physiology or Medicine in 2014 to John O'Keefe (together with the Mosers) for his discovery of (O'Keefe and Dostrovsky 1971) and research on rat place cells was excellently merited. The announcement for the award of the Nobel Prize described this system as a 'component of a positioning system, an "inner GPS" in the brain'. John O'Keefe has continued to emphasize the role of the hippocampus and rodent place cells in navigation (O'Keefe 1990, Hartley, Lever, Burgess and O'Keefe 2014). Rolls' discoveries and theory are thus somewhat different, in that Rolls has shown that spatial view cells may be especially relevant to the operation of the hippocampus in primates including humans; and in that the roles of the hippocampal system in memory are emphasized (Rolls 1989b, Rolls 1990b, Rolls and Kesner 2006, Rolls 2010b, Kesner and Rolls 2015) (Chapter 24).

Indeed, a potentially important difference between rodents and primates including humans is that rat hippocampal neurons represent the place where the rat is located (O'Keefe 1979), whereas a population of neurons in the macaque hippocampus and parahippocampal cortex has been discovered that represents a spatial view of a scene with its landmarks independently of the place at which the macaque is located (Rolls and O'Mara 1995, Rolls, Robertson and Georges-François 1997a, Rolls, Treves, Robertson, Georges-François and Panzeri 1998b,

Robertson, Rolls and Georges-François 1998). Further, this representation is allocentric, in that it is the allocentric location in space being viewed and not egocentric cues such as eye position, or position relative to head direction, or indeed head direction itself (Georges-François, Rolls and Robertson 1999, Feigenbaum and Rolls 1991). The importance of these spatial view neurons is that they provide a basis for remembering where an object is located at a place at which one has never been present, such as a source of food at some distance away, or the position of a particular player when a goal was scored while one was a spectator who had never been on the pitch. Place cells could not implement that type of memory. The implication is that such spatial view cells are therefore very important in hippocampally mediated episodic memory in humans, of for example an object or person and where the object or person is located in space (Chapter 24, Kesner and Rolls (2015), Rolls (2015d)).

12.7   Backprojections, and the recall of information from the hippocampus to the neocortex

A fundamental difference between the hippocampus and neocortex is that whereas the neocortex is involved in processing and building representations of stimuli, in analyzing reward value, in taking decisions about actions, and in producing the appropriate actions, the hippocampus is involved in the computations required to form one-off episodic memories on the fly without reorganising the inputs, and then recalling the information back to the neocortex (Rolls 2008d, Rolls 2015c).

The recall process is a fundamental aspect of hippocampal operation. The only quantitative theory of recall from the hippocampus to the neocortex includes the following (Rolls 1989b, Treves and Rolls 1994, Rolls 2010b, Kesner and Rolls 2015).

First, the recall connectivity must involve multiple stages each with divergent connectivity (e.g. CA1, entorhinal cortex, parahippocampal / perirhinal cortex, neocortical areas high in the processing hierarchy, the preceding cortical area ..., see Fig. 12.1), in order for the number of backprojection connections made by any one neuron in the backprojection pathways to be kept down to a reasonable number (in the order of 10,000–20,000 connections made by a backprojecting neuron). (If the hippocampus had to recall memories to the neocortex in a single stage of connectivity, each hippocampal neuron would have to make thousands of millions of synapses in order to reach much of the neocortex.)

Second, the number of backprojecting synapses onto each cortical neuron in the backprojecting pathways must be high, in order to be able to recall to the cortex the same number of memories that can be stored in the hippocampus (Treves and Rolls 1994). The number of synapses onto each neuron in the neocortical recall backprojection pathways needs to be of the same order as the number of recurrent collateral synapses onto each CA3 neuron (depending on the hippocampal vs neocortical sparsenesses), that is, in the order of 10,000–20,000 synapses per neuron (Treves and Rolls 1994). We arrived at this theory of what is effectively the capacity of multistage heterosynaptic recall (see Fig. 12.1) by observing that each stage of the heterosynaptic recall can be thought of as an iteration of an autoassociation attractor network, enabling the rigorous analytic approach of the memory capacity of autoassociation networks to be extended to that of multistage hetero-association networks (Treves and Rolls 1994). This remains the only quantitative theory of the recall of memories from the hippocampus to the neocortex that we know. The theory has been tested quantitatively by simulations, which meet the theoretical capacity (Rolls 1995b). The theory remains the only quantitative theory of why there are approximately as many backprojecting as forward projecting neurons in a cortical hierarchy.

The theory of recall using the backprojections to the neocortex also requires that the synapses be associatively modifiable at one stage at least of the backprojection pathways (Rolls 1989b, Treves and Rolls 1994). The theory is that the conditions for this associative learning are present in a neocortical area during the storage of the episodic memory, when the forward projecting inputs to the hippocampus that activate neocortical neurons are present simultaneously with the activity in the backprojection pathways originating from the hippocampus (see Fig. 12.1 and Chapter 24) (Rolls 1989b, Treves and Rolls 1994, Rolls and Kesner 2006, Rolls 2010b, Kesner and Rolls 2015).

Further descriptions of these principles of cortical operation are described in Chapters 11 and 24.

12.8 Subcortical structures connected to the hippocampo-cortical memory system

Subcortical structures such as the septal region, and the fornix connections to the mammillary bodies and thus by the anterior thalamic nuclei to the hippocampus, are not part of the quantitative theory of hippocampal function in memory and recall back to the neocortex. What role then do these subcortical structures play in the hippocampal system?

One contribution of the septal nuclei is that they provide a cholinergic input to the hippocampus, which is implicated in helping to facilitate LTP, and in minimizing the effects of the CA3 recurrent collaterals during this (Hasselmo et al. 1995, Giocomo and Hasselmo 2007).

One function of the mammillary body circuitry may be to introduce a head direction signal from the tegmental nuclei of Gudden via the anterior thalamic nuclei to the hippocampal circuitry (Dillingham, Frizzati, Nelson and Vann 2015). This may provide one route for vestibular inputs, important of course in head direction representation, to reach the presubiculum, where head direction cells are found in primates (Robertson, Rolls, Georges-François and Panzeri 1999) as well as in rodents (Wiener and Taube 2005). Indeed, the anterior thalamic nuclei project to the cingulate and retrosplenial cortex, which in turn project into areas such as the subiculum, and thus can provide inputs to the hippocampal system about for example head direction (see Fig. 12.1 and Aggleton and Nelson (2015)).

The evidence for these principles of operation of hippocampal cortex and its connections back to the neocortex is provided in Chapter 24 and elsewhere (Rolls 1989b, Treves and Rolls 1994, Rolls and Kesner 2006, Rolls 2010b, Kesner and Rolls 2015).

12.9 Highlights

1. A brief description of the operation of the hippocampus in episodic memory is provided, with a theory of how the hippocampus operates, and how the hippocampal memory system illustrates many of the principles of cortical operation, provided in Chapter 24.

2. A feature of the hippocampal network is that the CA3 system has a highly developed recurrent collateral system which enables it to associate together, as an autoassociation or attractor network, places and objects, with the inputs coming from very different neocortical systems. The single attractor network in CA3 enables any object to be associated with any place in an episodic or event memory, and for the place to be recalled from the object, and the object from the place. This is completion of a memory in an autoassociation network.

3. Although the representation of place in rodents is of the place where the rodent is located, in primates there is a representation of spatial view, enabling humans to remember where they have seen a person or object.

4. The number of memories that can be stored in the hippocampus is determined largely by the number of recurrent collateral connections onto any one neuron (in the order of 12,000 in the rat) and the sparseness of the representation.

5. The dentate gyrus input to CA3 performs pattern separation of the inputs before CA3 to help achieve the relatively uncorrelated representations that help to achieve high memory capacity in CA3. The dentate granule cells perform this operation by competitive learning; by the low probability of contact of the dentate granule cell to CA3 connections (0.00015 in the rat); and by neurogenesis with the new dentate granule cells helping to ensure that new memories are distinct from old memories by making new random connections to CA3 neurons. The dentate gyrus is the only part of cortex where new neurons can develop in adulthood, for here the premium is on selecting new random sets of neurons in CA3 for new memories distinct from previous memories, whereas elsewhere in cortex the premium is on building and reusing useful new representations.

6. The CA1 which follows the CA3 prepares for the recall of memories, completed in CA3 by a partial retrieval cue, back to the neocortex. The CA1 has the architecture of a competitive network, useful for recoding the parts of an episodic memory necessarily represented as separate in CA3 for autoassociation in CA3.

7. Backprojections from the hippocampal system to neocortical areas have a large number of synapses onto each neocortical neuron to achieve high recall capacity. This accounts for why there are as many backprojections as forward connections between adjacent cortical areas in a hierarchy. For the recall to work efficiently, the backprojection synapses at one or more stages of the backprojection pathway must be associatively modifiable.

8. Once set up in this way, the backprojection connections can also be used for top-down attentional control (see Chapter 6).

13 Limited neurogenesis in the adult cortex

13.1 No neurogenesis in the adult neocortex

One principle of operation of the cerebral neocortex is that there is little or no neurogenesis in the adult neocortex. Two reasons for this are as follows.

First, an important function of the neocortex is in the formation of new representations, using the principles of convergence in hierarchical cortical areas, and synaptic modification of the afferents that reach neurons in a given cortical area from the previous cortical area. A key computational principle involved is competitive learning to categorise the inputs, and this is described in Sections B.4 and 7.4. Because the strength of the feedforward synapses (e.g. from one cortical area to the next in a neocortical hierarchy) shapes the representation at the next stage, information is effectively encoded in the synaptic connectivity that is shaped by self-organizing learning. For this reason, loss of the existing connectivity, and replacing it with at least the initial connectivity of new neurons would lose the information encoded in the whole hierarchy, and make it uninterpretable at the output. For example, if V4 neurons were replaced at random, whatever representations had been set up in the inferior temporal visual cortex would be lost, and the output of the inferior temporal visual cortex would become uninterpretable, in areas such as the orbitofrontal cortex, amygdala, and hippocampal system. The information would be lost. This is one reason why neurogenesis is normally turned off in the adult neocortex.

An interesting example to illustrate this is the critical period for plasticity in the primary visual cortex, after which plasticity becomes greatly reduced (Blakemore and van Sluyters 1974, Espinosa and Stryker 2012). A reason for this is that if self-organizing learning remained after the juvenile stage, the representations in V1 would drift round if the statistics of the input changed (see Section B.4.3.7), and this would make the operation of the rest of the ventral visual stream impossible, for all its inputs would have changed.

Second, connectivity that is long-range may be set up during development when brain regions are close together, and the connectivity rules can include proximity (Chapter 19). It would be impracticable in the adult for many long-range connections to be established by neurogenesis, because the whole routing and connectivity for use in the adult would have to be encoded genetically. An example is the dopamine input from the A10 cells to the orbitofrontal cortex, which follow a long trajectory in adults, and there is no route-finding in the genome to enable midbrain dopamine neurons to find the orbitofrontal cortex and connect to it. Another example is the connectivity of the subicular cortex via the fornix to the mammillary bodies (Fig. 12.1), which in the adult follows a very complicated route.

13.2 Limited neurogenesis in the adult hippocampal dentate gyrus

In Rolls' theory of the hippocampus (Chapter 24, Rolls (1989b), Kesner and Rolls (2015)), the dentate granule cells operate as a competitive network that helps to separate (or categorize or orthogonalize) patterns being received from the entorhinal cortex. Then the very diluted connectivity of the dentate granule cells via the mossy fibers to result in only approximately 46 synapses onto each CA3 neuron (Fig. 24.14) produces further pattern separation, in that a unique subset of CA3 neurons is selected by different inputs to the dentate gyrus. In this theory, the function of the mossy fibers is to produce orthogonal firing of CA3 representations for different memories, and no information is stored in the mossy fiber synapses. This makes the hippocampus operate completely differently to the neocortex, in which the goal is not to store large numbers of memories independently of each other (as in the hippocampus), but instead to set up useful categorisation of inputs using for example competitive learning. In competitive learning, the afferent synapses are modifiable, and effectively store information about which inputs are to strongly activate which neurons (Sections B.4 and 7.4).

One result of this conceptualization is that mossy fibre type orthogonalizing synapses are nothing to do with the operation of the neocortex. Another implication is that in the hippocampal system, if what the granule cell / mossy fibre systems are performing computationally is to select orthogonal subsets of CA3 neurons for each new input, then neurogenesis here may be useful (by enabling completely new subsets of CA3 neurons to be selected by the new random connectivity of new dentate granule cells), but will not be useful at all in the neocortex, and should not be normal, because information is stored in neocortical afferent synapses, as in competitive networks. That is the computational rule of operation of cortical circuitry by which neurogenesis may be useful in the hippocampal dentate granule cell / mossy fibre system, but not in the neocortex.

These are fundamentally different principles of operation of the dentate granule cell / hippocampal system, and the neocortex. Neurogenesis is a useful principle of operation of the hippocampal cortex (Rolls 2010b, Aimone, Deng and Gage 2010, Aimone and Gage 2011, Johnston, Shtrahman, Parylak, Goncalves and Gage 2016). Neurogenesis is not a useful principle of operation of the neocortex. The reasons are given in this Section, and in Chapter 24.

13.3 Neurogenesis in the chemosensing receptor systems

In both the olfactory and taste systems, there is peripheral turnover of receptor cells. The simple connectivity principle realised here is genetic encoding of the connections of new receptor cells. For example, in the olfactory system, there are approximately 1,000 olfactory receptor genes, and these genes that define the chemosensitivity of the 1,000 receptor types also determine which neurons the olfactory receptor cells synapse onto, and the glomerulus in the olfactory bulb to which an olfactory receptor relates (Mombaerts 2006). This is a genetically very expensive mechanism of allowing for neurogenesis, in that for olfaction alone, 1,000 of the approximately 25,000 genes in the genome are allocated for this. This primitive system for encoding connectivity is impractical for the connectivity of cortical neurons, where there are 10^{11} neurons and 10^{15} synapses, far beyond the number of genes in the genome.

The tuning of neurons in the primate primary (insular) and secondary (orbitofrontal) taste cortex indicates that the tuning is not specific for a single taste receptor type, in that the neurons respond to more than one of the five tastes, sweet, salt, bitter, sour, and umami (Scott, Yaxley, Sienkiewicz and Rolls 1986a, Yaxley, Rolls and Sienkiewicz 1990, Rolls, Yaxley and Sienkiewicz 1990, Smith-Swintosky, Plata-Salaman and Scott 1991, Critchley and Rolls 1996c, Rolls, Critchley, Wakeman and Mason 1996c, Rolls, Critchley, Verhagen and Kadohisa 2010a, Scott and Plata-Salaman 1999, Verhagen, Kadohisa and Rolls 2004, Kadohisa, Rolls and Verhagen 2005a, Rolls 2015e, Rolls 2016b). On the other hand, some tastes appear to be innately pleasant (e.g. sweet when hungry, and salt when salt-deprived (Berridge et al. 1984)), and given that reward value is only made explicit in the taste representation by neuronal firing in the orbitofrontal cortex (Rolls, Sienkiewicz and Yaxley 1989c, Rolls 2015e, Rolls 2016b,

Rolls 2016f), there may be some genetic specification through the taste pathways as far as the orbitofrontal cortex in primates (Rolls 2014a).

13.4 Highlights

1. The dentate gyrus input to hippocampal CA3 performs pattern separation of the inputs before CA3 to help achieve the relatively uncorrelated representations that help to achieve high memory capacity in CA3. The dentate granule cells perform this operation by competitive learning; by the low probability of contact of the dentate granule cell to CA3 connections (0.00015 in the rat); and by neurogenesis with the new dentate granule cells helping to ensure that new memories are distinct from old memories by making new random connections to CA3 neurons.

2. The dentate gyrus is the only part of cortex where new neurons can develop in adulthood, for here the premium is on selecting new random sets of neurons in CA3 for new memories distinct from previous memories, whereas elsewhere in cortex the premium is on building and reusing useful new representations. Further, in neocortex once developed it would be difficult for long-range connections to be guided to the correct destinations far away in the full-size adult brain.

3. In both the olfactory and taste systems, there is peripheral turnover of receptor cells. The principle here is that this is possible, and the correct connections can be remade to the neural systems in the brain, because each type of receptor cell for olfaction and taste has its own genetic specification for its onward connectivity to the brain.

14 Invariance learning and vision

An introduction to processing in the visual pathways is provided in Section 1.9. The major computational problem solved by the visual system of producing view, translation, size, and rotation invariant representations of objects by the inferior temporal cortex stage of processing is described in Chapter 25. In this Chapter, some of the key operational principles of the cortical areas involved in vision, especially those involved in object recognition in the ventral visual stream, are highlighted. Similar principles may apply to some of the operations in the dorsal visual system (Rolls and Stringer 2007).

14.1 Hierarchical cortical organization with convergence

The hierarchical organization of the ventral visual stream is illustrated in Fig. 2.1 on page 41, and described in Chapters 2 and 25. One key principle of the architecture illustrated in Fig. 2.1 is that a neuron in any one stage receives from a small region of the preceding stage. This enables the number of connections received by any neuron to be kept to a reasonable number for a neuron, in the order of 10,000, and for the information received by any neuron to be from neurons in the preceding stage with similar properties. In early stages of the visual system, this might reflect information from the same part of the visual field. At later stages, this might reflect neurons with similar types of tuning, for example to faces, or to non-face objects. The principle is convergence from a small region of the preceding layer. One effect realised by this is the increase of receptive field size from stage to stage, by a factor of approximately 2.5 from stage to stage (Fig. 2.1).

One operational principle of this cortical feature hierarchy system is that it can map the whole of an input space (in this case the retina) to the whole of the output space (in this case anterior inferior temporal visual cortex, IT) without needing neurons with millions and millions of synaptic inputs. If an IT neuron had to respond to a particular face in most possible positions on the retina and with different views and sizes, yet had connections directly from V1, it would need millions of inputs, which is biologically implausible for a neuron to operate effectively, and also for the connectivity involved.

14.2 Feature combinations

In the hierarchy from V1 to IT, the system operates as a feature hierarchy, with neurons responding to combinations of what is represented at earlier stages. One example is the neurons that respond to whole faces, but not to their parts, which are represented by other neurons at earlier stages, or even in the same cortical area.This is a crucial operational principle, as described in Chapters 2 and 25. Yet the number of neurons at each stage of the hierarchy does not increase markedly. This is consistent with the concept that there is redundancy in the visual world of images, and that not all combinations of what is represented at earlier stages need to be formed. Competitive networks can operate in this way, as described in Section B.4. This property is reflected not only in the responses of IT neurons, but also in the model VisNet, which uses competitive learning in multiple layers. This principle, of forming neurons that respond to non-linear combinations of features in the correct spatial configuration, is important. Networks such as HMAX (another model of ventral stream processing (Riesenhuber and Poggio 1999b, Riesenhuber and Poggio 1999a, Serre, Wolf, Bileschi, Riesenhuber and Poggio 2007c, Serre, Oliva and Poggio 2007b, Serre, Kreiman, Kouh, Cadieu, Knoblich and Poggio 2007a) described in Section 25.6) do not have any learning that allows neurons to respond to feature combinations differently to the components, and so HMAX for example responds to parts similarly to wholes, and cannot capture a particular spatial combination of features that might represent a face as compared to a face with the parts jumbled (Robinson and Rolls 2015) (Section 25.6).

14.3   Sparse distributed representations

Information is represented with sparse distributed representations by inferior temporal cortex neurons (Rolls and Treves 2011) (Chapter 8). A key property of the encoding is that a reasonable amount of information about which stimulus was shown can be read from the firing of single neurons to a set of stimuli with biologically plausible dot product decoding, and this information increases linearly with the number of neurons (up to tens of neurons), showing that the coding by the neurons is independent (Appendix C). Neurons with similar sparse representations are present in VisNet, which uses competitive learning (Chapter 25).

For comparison, in HMAX, with neurons that reflect the maximum value of a set of filters within an area and no attempt as sparsification, the neurons in the layers are broadly tuned, very little information is represented in the firing of an individual neuron, and powerful decoding by for example a support vector machine is needed to interpret the output of the system. This highlights one important property of cortical visual processing, the use of sparse distributed representations, that is not represented in some models.

14.4   Self-organization by largely feedforward processing without a teacher at the top of the system

The ventral stream visual pathways project forwards in a hierarchically organised series of 4–5 stages of processing from the primary visual cortex V1 to the inferior temporal visual cortex, as illustrated in Figs. 1.10–1.11, 2.1, and 25.1–25.2. By the inferior temporal visual cortex stage, neurons respond to faces, objects etc with all the invariances necessary for object recognition. After the inferior temporal visual cortex, there are onward projections to structures involved in associative learning and memory such as the orbitofrontal cortex, amygdala, and hippocampal system (Fig. 1.9). There is no external teacher that projects into the inferior temporal visual cortex to tell each neuron in the inferior temporal visual cortex how to respond, let alone every neuron at all the earlier stages of the visual cortical hierarchy. Instead, the whole hierarchical system self-organizes itself using learning at each stage, and the visual inputs being received from the natural world.

This is in complete contrast to many machine learning approaches to vision, in which the neurons in the output stage are provided with a teacher that instructs the neurons how to categorise the inputs being received, into classes such as cars, bicycles, etc. Early in learning, the representations in the top layer have errors, and these errors are used to correct the weights into the neurons, but in addition are backpropagated backwards back down through the whole hierarchy to calculate errors for every neuron at earlier stages, and correct the weights into all neurons at every stage using machine learning algorithms that are essentially backpropagation of error algorithms (Larochelle and Hinton 2010, Krizhevsky, Sutskever and Hinton 2012, LeCun, Bengio and Hinton 2015) (Section B.13). The backpropagation of error algorithms are biologically implausible, as considered in Section B.11.

Not only are the machine learning algorithms biologically implausible, but also the aim has usually been to categorise visual inputs into object classes such as cars and bicycles, whereas neurons in the primate including human visual system typically represent particular individuals or particular objects (Rolls and Treves 2011, Quiroga et al. 2008, Fried et al. 2014, Rolls 2015d) (using sparse distributed coding, which does also allow the class of object to be read out), which is of course of great importance to primates including humans to enable social and emotional behaviour, identification of one's own kin and belongings, etc. Nevertheless, non-biologically plausible approaches to object vision are important in assessing how different types of system operate with large numbers of training and test images (Khaligh-Razavi and Kriegeskorte 2014, Cadieu, Hong, Yamins, Pinto, Ardila, Solomon, Majaj and DiCarlo 2014), and there are attempts to make multilayer error correction networks more biologically plausible (O'Reilly and Munakata 2000, Balduzzi, Vanchinathan and Buhmann 2014). Biologically plausible systems for object recognition need to have not only the properties described here, but also mechanisms that use a local learning rule, no separate teacher for each output neuron in a supervised learning scheme, and no lateral copying of weights or 'convolution'. Moreover, understanding how the brain operates is important not only in its own right, but also for its implications for understanding disorders of brain function (Rolls 2008d, Rolls 2012d).

The concept, provided in more detail in Chapter 25, is that each stage of the cortical visual system learns what it can using the statistics of the visual input being received, and one-layer networks in each stage which operate by the self-organizing principles of competitive learning (Section B.4 and Rolls (2012c)). The convergence from stage to stage with small receptive fields in the early stages breaks up the computational problem into a number of simpler problems. The convergence ensures that each stage learns something that could not be represented at earlier stages. Very importantly, there is useful teaching information in the statistics of the visual input, which given the way that primates look at the world with brief periods of fixation often on different parts of the same object over brief periods of time, enables the statistics of the input to be used to learn representations that at any one stage provide evidence about the different transforms of an object that characterise an object. So there is a teacher, according to the hypothesis, but it is in the statistics of the input being received, not by a top-down teacher that instructs every neuron at the top of the system how to respond (Chapter 25).

Feedforward hierarchical connectivity with convergence from stage to stage, competitive learning within each stage, and not top-down instruction about how each neuron should respond but instead self-organisation during development using the statistics of the visual inputs that are provided by the natural world, are key principles of operation of the ventral cortical visual stream to implement invariant object recognition (Chapter 25).

14.5 Learning guided by the temporal and spatial continuity of the visual inputs to learn invariant representations

Many neurons in the inferior temporal visual cortex have responses that are invariant with respect to transforms such as position on the retina, size, rotation, and in some cases even view (Rolls 2012c), with the evidence and examples provided in Chapter 25. This important computational principle of cortical computation enables associations between the representation of a visual object to be correctly associated with inputs in other modalities such as taste, smell, and position in a spatial scene in structures such as the orbitofrontal cortex, amygdala, and hippocampus, independently of which transform (e.g. view) is shown on an individual occasion or trial. Because the responses of the inferior temporal cortex neurons have this transform invariance, an association learned on a single trial to one view of an object will generalize correctly the very next time the object is shown, even when the view is different (Chapter 25). This is a fundamental advantage conferred by this aspect of cortical design.

As just described, the theory is that the primate including human visual system learns translation-, rotation-, size- and view-invariant object recognition by using continuity of the statistics of the visual inputs, which tend to be about the same object over time in the order of 250 ms to 2000 ms. We have shown in a model of the ventral visual system that the different transforms of images which tend to be about the same object over short time periods can be linked together by the continuity of the inputs in time, with each stage in the hierarchy learning what it can given its receptive field size (Rolls 2012c, Földiák 1991, Rolls 1992a) (Chapter 25). This is a principle of temporal continuity.

In addition, there may be spatial continuity in images as they transform, and we have shown that spatial continuity, even with no temporal continuity in a model, can enable the learning of invariant representations (Stringer, Perry, Rolls and Proske 2006, Perry, Rolls and Stringer 2010), and typically both will be present (Perry, Rolls and Stringer 2006). The temporal continuity has the major advantage that even spatially dissimilar views of an object, for example the catastrophic change in the image (Koenderink 1990) when for example a cup rotates from a view from just below the rim to above the rim, can be associated together. The spatial continuity is often apparent on the small spatial scale, for example when an object moves just a small amount across the retina. Thus two factors, temporal continuity, and spatial continuity, which may complement each other, contribute to building invariant representations of objects, faces, etc (Rolls 2012c), as described in Chapter 25.

The approach adopted by the cortex it appears is thus to associate together different transforms of combinations of features or views, in order to implement transform invariant object recognition. Moreover, the approach that the evidence indicates is used by the cortex is essentially associative, so that the best match is sought, and there is tolerance to errors (see Chapter 25 and Appendix B).

This is thus very different from an approach in which after segmentation of the parts of an object, an attempt is then made to produce a complete 3D structural description of the 3-D object (which is extremely difficult as this requires syntax to relate the parts to each other which can be disrupted if any part of the object is obscured or incorrectly segmented), and then test which stored 3-D structural description the attempted structural description of the scene matches (Marr 1982, Ullman 1996).

14.6 Bottom up saliency guides a foveate system to analyse a small part of the scene, not the whole scene, at any one time

In computer vision, the traditional approach has been to attempt to parse a whole scene, performing edge detection, object segmentation, and then if possible object recognition (Marr 1982, Ullman 1996). The primate cortex does not do this (Rolls 2011c, Rolls 2012c). Instead, primates fixate for short periods, typically 200–300 ms, and then move their eyes often to fixate another part of the same object, and then move their eyes again to fixate an object somewhere else in the scene, enabling the cortex to process typically one object at a time even in cluttered scenes (Rolls, Aggelopoulos and Zheng 2003a). This breaks the problem down into manageable parts, each typically involving one object at the fovea (Rolls, Aggelopoulos and Zheng 2003a), or in some cases a few objects near the fovea (Aggelopoulos and Rolls 2005). This is a major computational principle of the cortex, and is in contrast to many machine vision approaches to object recognition and scene processing.

The question then arises of how the eyes are guided in a complex natural scene to fixate close to what may be an object; and how close the fixation is to the centre of typical objects for this determines how much translation invariance needs to be built into the ventral visual system (Rolls and Webb 2014). It turns out that the dorsal visual system (Ungerleider and Mishkin 1982, Ungerleider and Haxby 1994) implements bottom-up saliency mechanisms by guiding saccades to salient stimuli, using properties of the stimulus such as high contrast, colour, and visual motion (Miller and Buschman 2013). (Bottom-up refers to inputs reaching the visual system from the retina.) One particular region, the lateral intraparietal cortex (LIP), which is an area in the dorsal visual system, seems to contain saliency maps sensitive to strong sensory inputs (Arcizet et al. 2011). Highly salient, briefly flashed, stimuli capture both behaviour and the response of LIP neurons (Goldberg et al. 2006, Bisley and Goldberg 2003, Bisley and Goldberg 2006). Inputs reach LIP via dorsal visual stream areas including area MT, and via V4 in the ventral stream (Miller and Buschman 2013, Soltani and Koch 2010).

Saliency algorithms have been developed that produce similar fixation patterns to humans in many bottom-up saliency tasks (Itti and Koch 2000, Harel, Koch and Perona 2006a, Harel, Koch and Perona 2006b). A combination of the operation of bottom-up saliency to guide fixation to objects in complex scenes to mimic the operation of the dorsal visual stream in this function, with the VisNet model of object recognition in the ventral visual stream to perform the object recognition on the object being fixated at any one time, provides a combined model of how the primate visual system breaks up the problem of object recognition using two separate processing streams to perform the visual fixation and then the object recognition at the fixated location, and does this sequentially for different objects in the scene (Rolls and Webb 2014) (Section 25.5.16).

14.7   Lateral interactions help to shape the receptive fields of neurons in complex scenes

One mechanism that the brain uses to simplify the task of recognising objects in complex natural scenes is that the receptive fields of inferior temporal cortex neurons change from approximately 70° in diameter when tested under classical neurophysiology conditions with a single stimulus on a blank screen to as little as a radius of 8° (for a 5° stimulus) when tested in a complex natural scene (Rolls, Aggelopoulos and Zheng 2003a, Aggelopoulos and Rolls 2005) (Section 25.5.9) (with consistent findings described by Sheinberg and Logothetis (2001)). This greatly simplifies the task for the object recognition system, for instead of dealing with the whole scene as in traditional computer vision approaches, the brain processes just a small fixated region of a complex natural scene at any one time, and then the eyes are moved to another part of the screen. During visual search for an object in a complex natural scene, the primate visual system, with its high resolution fovea, therefore keeps moving the eyes until they fall within approximately 8° of the target, and then inferior temporal cortex neurons respond to the target object, and an action can be initiated towards the target, for example to obtain a reward (Rolls et al. 2003a). The inferior temporal cortex neurons then respond to the object being fixated with view, size, and rotation invariance (Rolls 2012c), and also need some translation invariance, for the eyes may not be fixating the centre of the object when the inferior temporal cortex neurons respond (Rolls et al. 2003a). A model of how the receptive field size are decreased by the lateral inhibition accounts in a robust way for these findings (Trappenberg, Rolls and Stringer 2002).

Lateral interactions can also produce asymmetries in the receptive fields of inferior temporal cortex neurons (Aggelopoulos and Rolls 2005) as described in Section 25.5.10. The asymmetries reflect, it is hypothesized, the diluted connectivity of the feedforward projections from one visual cortical area to another. This diluted connectivity results probabilistically in individual neurons having asymmetries in how much they are likely to respond to features or objects in different positions in their receptive fields (Aggelopoulos and Rolls 2005). These asymmetries are unmasked by the competition between stimuli close to the fovea. These asymmetries in the receptive fields enable the positions of different objects close to the fovea to be represented, and also allow multiple exemplars of the same object in different positions close to the fovea to be represented. How multiple objects could be represented in a scene had been a major problem for connectionist modelling with distributed representations (Mozer 1991). These are further important principles of cortical function, and illustrate the utility of another principle, diluted connectivity (Chapter 7).

14.8 Top-down selective attention can improve performance, but feedforward processing is sufficient for object recognition

Backward masking experiments indicate that each cortical area needs to fire for only 20–30 ms to pass information to the next stage (Rolls and Tovee 1994, Rolls, Tovee, Purcell, Stewart and Azzopardi 1994b, Kovacs, Vogels and Orban 1995, Rolls, Tovee and Panzeri 1999b, Rolls 2003) (see Section C.3.4). This leaves insufficient time for processing to ascend through the visual processing hierarchy, activate anterior inferior temporal visual cortex neurons, and then be fed back to early cortical stages to implement perception. The important conclusion is that feedforward processing is sufficient to activate correctly inferior temporal cortex neurons and to implement perception (Rolls 2003). Consistent with this, rapid serial visual presentation of image sequences shows that cells in the temporal visual cortex are still face-selective when faces are presented at the rate of 14 ms/image (Keysers, Xiao, Foldiak and Perrett 2001). These findings do not exclude the possibility that to learn the correct representations, some top-down, feed-back, processing may be involved. But the evidence does establish the important principle of cortical function that the cortex can operate with feedforward processing to implement some of its major functions such as object recognition and perception (Rolls 2003).

However, top-down selective attention can facilitate processing in earlier cortical areas. For example, even when the eyes are held still, a cue to pay attention to one object or position in space can facilitate neuronal responses, as described in Chapter 6. The mechanism is biased competition (Desimone and Duncan 1995), and precise neuronal models of how this works have been produced (Rolls and Deco 2002, Deco and Rolls 2004, Deco and Rolls 2005b).

However, although top-down attention using biased competition can facilitate the operation of attentional mechanisms, and is a subject of great interest (Desimone and Duncan 1995, Rolls and Deco 2002, Deco and Rolls 2005a, Miller and Buschman 2013), top-down object-based attention makes only a small contribution to visual search for an object in a complex natural unstructured scene (such as leaves on a tree), increasing the receptive field size from a radius of approximately $7.8°$ to approximately $9.6°$ (Rolls, Aggelopoulos and Zheng 2003a) (Fig. 25.60). Indeed, in this visual search task, multiple saccades were required round the complex visual scene to find a target object (Rolls et al. 2003a), so top-down attention has an important but limited role under natural vision conditions (Section 25.5.9).

14.9 Some topography which minimizes connection length and thus cortical size

A very common characteristic of connectivity in the brain, found for example throughout the neocortex, consists of short-range excitatory connections between neurons, with inhibition mediated via inhibitory interneurons. The density of the excitatory connectivity even falls gradually as a function of distance from a neuron, extending typically a distance in the order of 1 mm from the neuron (Braitenberg and Schütz 1991), contributing to a spatial function quite like that of a Mexican hat (Fig. B.20).

It turns out, as described in Section B.4.6, that a simple modification to competitive networks enables them to develop topological maps. In such maps, the closeness in the map reflects the similarity (correlation) between the features in the inputs. The modification that allows such maps to self-organize is to add short-range excitation and longer-range inhibition between the neurons. The effect of this connectivity between neurons, which need not be modifiable, is to encourage neurons that are close together to respond to similar features in the input space, and to encourage neurons that are far apart to respond to different features in the input space. When these response tendencies are present during learning, the feature analyzers that are built by modifying the synapses from the input onto the activated neurons tend to be similar if they are close together, and different if far apart. This is illustrated in Fig. B.22. Feature maps built in this way were described by von der Malsburg (1973) and Willshaw and von der Malsburg (1976).

If a low-dimensional space, for example the orientation sensitivity of cortical neurons in the primary visual cortex (which is essentially one-dimensional, the dimension being angle), is mapped to a two-dimensional space such as the surface of the cortex, then the resulting map can have long spatial runs where the value along the dimension (in this case orientation tuning) alters gradually, and continuously. Such self-organization can account for many aspects of the mapping of orientation tuning, and of ocular dominance columns, in V1 (Miller 1994, Harris, Ermentrout and Small 1997). If a high-dimensional information space is mapped to the two-dimensional cortex, then there will be only short runs of groups of neurons with similar feature responsiveness, and then the map must fracture, with a different type of feature mapped for a short distance after the discontinuity. This is exactly what Rolls suggests is the type of topology found in the anterior inferior temporal visual cortex, with the individual groupings representing what can be self-organized by competitive networks combined with a trace rule as described in Section 25.5. In IT, visual stimuli are not represented with reference to their position on the retina, because here the neurons are relatively translation invariant. Instead, when recording here, small clumps of neurons with similar responses may be encountered close together, and then one moves into a group of neurons with quite different feature selectivity (personal observations). This topology will arise naturally, given the anatomical connectivity of the cortex with its short-range excitatory connections, because there are very many different objects in the world and different types of features that describe objects, with no special continuity between the different combinations of features possible.

There is a potential advantage of these topological maps. By placing neurons that respond similarly close together, the wiring length (the axonal length in the cortex) that is needed to allow the relevant neurons to interact to shape their response properties by competition is shorter than if the relevant neurons were randomly distributed, thus reducing brain weight, which is an important factor in evolution. Moreover, topological maps simplify the specification of the anatomical connectivity, for the neurons that need to interact, for example by lateral inhibition to implement competitive learning, are automatically placed close together in the cortex, within the range of the inhibitory neurons (Section B.4.6). For these reasons, topological maps are a useful principle of operation of the neocortex.

14.10 The output is about objects in a form that can be decoded by pattern association networks

A fundamental principle of cortical operation is that the responses of neurons in one cortical area can be read by neurons in the next cortical area by neuronally plausible decoding. A prototypical example of that decoding is summing the input firing rates each multiplied by a synaptic strength described as dot product decoding in Chapter 8 and Appendix C (Fig. 8.2). The information is represented in that form, using sparse distributed encoding with very high representational capacity and the important properties of generalization and graceful degradation, as described in Chapter 8 and Appendix C.

14.11 Highlights

Highlights of the organisation of the ventral visual stream cortical pathways that enable transform invariant representations of objects to be formed include the following, with a fuller description of how the operation of the ventral visual system illustrates many of the principles of operation of the neocortex provided in Chapter 25.

1. Feed-forward hierarchical connectivity with convergence from stage to stage, where each stage is a cortical area, and there are approximately 5 stages in the hierarchy, namely V1, V2, V4, posterior inferior temporal cortex, and anterior inferior temporal cortex.

2. Competitive learning within each stage, which relies on bottom-up inputs, and self-organisation of synaptic connectivity by experience during development. There is no top-down instruction about how each neuron should respond.

3. The self-organisation during development uses the slow statistics of the visual inputs that are provided by the natural world as a 'teacher' to guide help implement invariant object recognition. Not only temporal continuity, but also spatial continuity, can be a useful teacher for the formation of transform-invariant representations.

4. Top-down attention can operate by biased competition to improve performance in complex scenes.

5. Performance in complex scenes is facilitated by foveal vision, which simplifies the problem of scene parsing by fixating on small parts of the scene at any one time, and performing this sequentially using multiple fixations. This is 'overt' search, and differs from what is sometimes described as 'covert' search when the eyes do not move as described in Section 6.2.8.

6. This process is facilitated by the reduction in receptive field size which occurs in complex natural scenes. The receptive fields can then become asymmetrical with respect to the fovea, allowing multiple objects and even instances of the same object, and their position with respect to the fovea, to be represented in a single fixation.

7. The representation of objects and faces in the inferior temporal visual cortex is sparse distributed, encodes to the level of individual objects and people, and can be read by dot product decoding in the learning and memory structures that receive from the inferior temporal visual cortex including the orbitofrontal cortex, amygdala, and hippocampal system, and where associations with stimuli in other sensory modalities are learned.

8. The transform-invariant responses of a visual object provided by neurons in the inferior temporal visual cortex enable the object to be correctly associated with inputs in other modalities such as taste, smell, and position in a spatial scene in structures such as the orbitofrontal cortex, amygdala, and hippocampus, independently of which transform

(e.g. view) is shown on an individual occasion or trial. Because the responses of the inferior temporal cortex neurons have this transform invariance, an association learned on a single trial to one view of an object will generalize correctly the very next time the object is shown, even when the view is different. This is a fundamental advantage conferred by this important principle of cortical design.

15 Emotion, motivation, reward value, pleasure, and their mechanisms

The aims of this Chapter are to describe the principles of design and operation of the cortex for emotion and motivation by computing representations of reward value that are the goals for actions, and why this is evolutionarily adaptive. The approach follows that described in *Emotion and Decision-Making Explained* (Rolls 2014a), where further elaboration of these principles and their implementation in the cortex is provided.

15.1 Emotion, reward value, and their evolutionary adaptive utility

Emotions can be defined as states elicited by rewards and punishers, that is, by instrumental reinforcers (Rolls 2014a, Rolls 2005, Rolls 1999a). A reward is anything for which an animal will work. A punisher is anything that an animal will work to escape or avoid, or that will suppress actions on which it is contingent. The force of 'instrumental' in this definition is that the emotional states are seen as defining the goals for arbitrary behavioural actions, made to obtain the instrumental reinforcer. This is very different from classical conditioning, in which a response, typically autonomic, may be elicited to a stimulus, without any need for an intervening state. An important part of the function of emotion is to provide an intervening state, after for example the delivery of a non-reward, in which behaviour can be reorganized, with perhaps a new action substituted for what was performed previously, in order to obtain the goal / reward (Rolls 2014a).

An outline of a classification scheme for how different reinforcement contingencies (delivery of reward, omission or termination of reward, delivery of punishment, omission or termination of punishment) are related to different emotions is shown in Fig. 15.1. Movement away from the centre of the diagram represents increasing intensity of emotion, on a continuous scale. The diagram shows that emotions associated with the delivery of a reward (S+) include pleasure, elation and ecstasy. Of course, other emotional labels can be included along the same axis. Emotions associated with the delivery of a punisher (S−) include apprehension, fear, and terror (see Fig. 15.1). Emotions associated with the omission of a reward (S+) or the termination of a reward (S+!) include frustration, anger and rage. Emotions associated with the omission of a punisher (S−) or the termination of a punisher (S−!) include relief.

An important point about Fig. 15.1 is that there are a large number of different primary (unlearned, gene-specified) reinforcers, and that for example the reward label S+ shows states that might be elicited by just one type of reward, such as a pleasant touch. There will be a different reward axis (S+) and non-reward axis (S+ and S+!) for each type of reward (e.g. pleasant touch vs sweet taste); and, correspondingly, a different punisher axis (S−) and non-punisher axis (S− and S−!) for each type of punisher (e.g. pain vs bitter taste). Examples of the reinforcers include food reward, water reward, pleasant touch, pain, a potential mate, kin including children, and altruism, with many examples provided in Table 2.1 of Rolls (2014a).

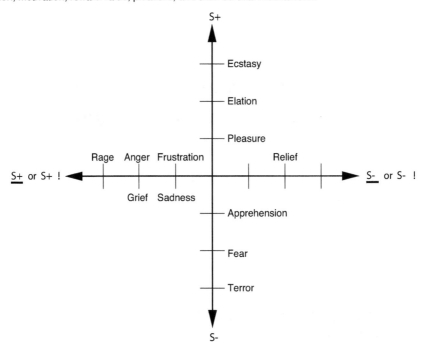

Fig. 15.1 Some of the emotions associated with different reinforcement contingencies are indicated. Intensity increases away from the centre of the diagram, on a continuous scale. The classification scheme created by the different reinforcement contingencies consists with respect to the action of (1) the delivery of a reward (S+), (2) the delivery of a punisher (S−), (3) the omission of a reward (S+) (extinction) or the termination of a reward (S+!) (time out), and (4) the omission of a punisher (S−) (avoidance) or the termination of a punisher (S−!) (escape). Note that the vertical axis describes emotions associated with the delivery of a reward (up) or punisher (down). The horizontal axis describes emotions associated with the non-delivery of an expected reward (left) or the non-delivery of an expected punisher (right). The diagram summarizes emotions that might result for one reinforcer as a result of different contingencies. Every separate reinforcer has the potential to operate according to contingencies such as these. This diagram does not imply a dimensional theory of emotion, but shows the types of emotional state that might be produced by a specific reinforcer. Each different reinforcer will produce different emotional states, but the contingencies will operate as shown to produce different specific emotional states for each different reinforcer.

Previously neutral stimuli can become associated by learning with these primary reinforcers, and become secondary reinforcers, which thereby become goals for action. An example is the sight of a new food, which may become a secondary reinforcer by association between its sight and the primary reinforcer of its taste.

It is argued that brains are designed around reward and punishment value systems, because this is the way that genes can build a complex system that will produce appropriate but flexible behaviour to increase their fitness (Rolls 2014a). The way that evolution by natural selection does this is to build us with reward and punishment systems that will direct our behaviour towards goals in such a way that survival and in particular reproductive fitness are achieved. By specifying goals, rather than particular responses, genes leave much more open the possible behavioural strategies that might be required to increase their fitness. Specifying particular responses would be inefficient in terms of behavioural flexibility as environments change during evolution, and also would be more genetically costly to specify (in terms of the information to be encoded and the possibility of error). This view of the evolutionary adaptive value for genes to build organisms using reward- and punishment-decoding and action systems in the brain places one squarely in line as a scientist developing ideas from Darwin (1859) onwards (Hamilton 1964, Dawkins 1976, Hamilton 1996), and is a key part of Rolls' theory of emotion (Rolls 2014a, Rolls 2013e).

The theory helps in understanding much of sensory information processing in the cortex, followed by reward and punishment value encoding, followed by decision-making and action selection to obtain the goals identified by the sensory/reinforcer decoding systems. Value coding systems must be separate from purely sensory or motor systems, and while a goal is being sought, or if a goal is not obtained, the value-related representation must remain to direct further goal-directed behaviour, and it is these continuing goal-related states to which emotion is related.

15.2   Motivation and reward value

My definition of motivation is that motivational states are states that are present when rewards and punishers, that is, instrumental reinforcers, are the goals for action (Rolls 2014a, Rolls 2016c). An example of a motivational state might thus be a hunger state in which the animal will perform goal-directed actions to obtain the reinforcer or goal of a food reward.

We must be clear about the difference between motivation and emotion. According to Rolls' theory of emotion, emotion is the state that results from having received, or not having received, the instrumental reinforcer, the goal object (Rolls 2014a). In contrast, motivation is the state when the instrumental reinforcer is being worked for, before the outcome stage, where the outcome is the delivery or not of the reinforcer. An important attribute of this theory of motivation and emotion is that the goal objects can be the same for motivation and emotion, simplifying the biological specification, with the difference being that motivation is the phase before the outcome, and emotion is the phase after the outcome. Indeed, the argument is that the genetic specification of reward value provides a unifying account of motivation and emotion, because for the rewards to be effective, they have to elicit motivation to obtain them, as well as the emotions that arise when they are obtained (Rolls 2016c). An additional property is that emotions, states occurring after the delivery or not of the reinforcer, can be motivating (Rolls 2014a). A good example is that if an expected reward is not obtained, then the frustrative non-reward can be motivating, and make the animal work harder to obtain the goal object (Rolls 2014a).

15.3   Principles of cortical design for emotion and motivation in primates including humans

I outline next what may be the fundamental architectural and design principles of the brain for sensory, reward value, and punishment value information processing in primates including humans. These cortical principles of operation for emotion and motivation are described in the following sections, with some of the pathways shown in Figs. 15.2, 15.3, 22.1, and 2.4. These principles of operation are somewhat different in rodents, as described in Chapter 18.

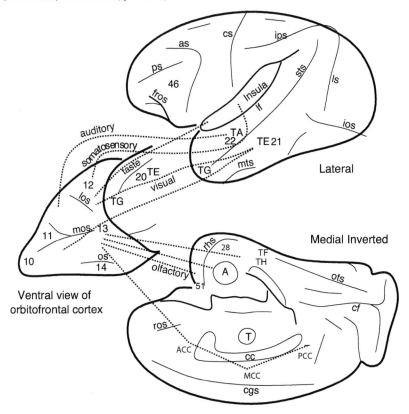

Fig. 15.2 Schematic diagram showing some of the gustatory, olfactory, and visual pathways to the orbito-frontal cortex, and some of the outputs of the orbitofrontal cortex. The secondary taste cortex and the secondary olfactory cortex are within the orbitofrontal cortex. V1, primary visual cortex. V4, visual cortical area V4. Abbreviations: as, arcuate sulcus; cc, corpus callosum; cf, calcarine fissure; cgs, cingulate sulcus; cs, central sulcus; ls, lunate sulcus; ios, inferior occipital sulcus; mos, medial orbital sulcus; os, orbital sulcus; ots, occipito-temporal sulcus; ps, principal sulcus; rhs, rhinal sulcus; sts, superior temporal sulcus; lf, lateral (or Sylvian) fissure (which has been opened to reveal the insula); A, amygdala; ACC, anterior cingulate cortex; INS, insula; MCC, mid-cingulate cortex; PCC, posterior cingulate/retrosplenial cortex; T, thalamus; TE (21), inferior temporal visual cortex; TA (22), superior temporal auditory association cortex; TF and TH, parahippocampal cortex; TG, temporal pole cortex; 12, 13, 11, orbitofrontal cortex; 28, entorhinal cortex; 51, olfactory (prepyriform and periamygdaloid) cortex.

15.4   Objects are represented independently of reward value, before rewards are evaluated in a later stage of cortical processing

For potential *secondary reinforcers*, analysis is to the stage of invariant object identification (in Tier 1 of Fig. 2.4, for example the inferior temporal visual cortex) before reward and punisher associations are learned at a later stage of processing, in for example the orbitofrontal cortex and amygdala (Section 2.2.5). The reason for the utility of transform-invariant object-level representations is to enable correct perceptual generalization to other instances of the same or similar objects, even when a reward or punisher has been associated with a different instance previously.

The representation of the object is (appropriately) in a form that is ideal as an input to pattern associators that allow the reinforcement associations to be learned. The representations are appropriately encoded in that they can be decoded by dot product decoding of the type that is very neuronally plausible; are distributed so allowing excellent generalization and graceful degradation; have relatively independent information conveyed by different neurons in the ensemble thus allowing very high capacity; and allow much of the information to be read off very quickly, in periods of 20–50 ms (as described in Appendix C).

An aim of processing in the ventral visual system (which projects to the inferior temporal visual cortex) is to help select the goals, or objects with reward or punisher associations, for actions. Action is concerned with the identification and selection of goals, the targets for action. The ventral visual system is crucially involved in this. I have thus extended the concept of Milner and Goodale (1995) that the dorsal visual system is for actions, and the ventral visual system for 'perception', e.g. perceptual and cognitive representations. I argue that *the ventral visual system is concerned with selecting the goals for action.* It does this by providing invariant representations of objects, with a representation that is appropriate for interfacing to systems (such as the amygdala and orbitofrontal cortex) which determine, using pattern association, the reward or punishment value of the object, as part of the process of selecting which goal is appropriate for action (Rolls 2014a) (Section 2.2.5).

Some of the evidence for this is that large lesions of the temporal lobes (which damage the ventral visual system and some of its outputs such as the amygdala) produce the Kluver-Bucy syndrome, in which monkeys select objects indiscriminately, independently of their reward value, and place them in their mouths. The dorsal visual system helps with executing those actions, for example with shaping the hand appropriately to pick up a selected object. (Often this type of sensori-motor operation is performed implicitly, i.e. without conscious awareness.) In so far as explicit planning about future goals and actions requires knowledge of objects and their reward or punisher associations, it is the ventral visual system that provides the appropriate input for planning future actions. Further, for the same reason, I propose that when explicit, or conscious, planning is required, activity in the ventral visual system will be closely related to consciousness, because it is to objects, represented in the ventral visual system, that we normally apply multi-step planning processes (as described in Chapter 22).

For *primary reinforcers*, the reward decoding may occur after several stages of processing, as in the primate taste system in which reward is decoded only after the primary taste cortex (see Figs. 2.4, 2.6, and 2.7, and Section 2.2.3) (Rolls 2016b). The architectural principle here is that in primates there is one main taste information-processing stream in the brain, via the thalamus to the primary taste cortex, and the information about the identity of the taste is not biased by modulation of how good the taste is before the insular primary taste cortex, so that the taste representation in the primary taste cortex can be used for purposes that are not reward-dependent. One example might be learning where a particular taste can be found in the environment, even when the primate is not hungry and therefore the taste is not currently rewarding. For this, the primary taste cortex provides a representation of what the taste is independently of its reward value (Yaxley, Rolls and Sienkiewicz 1988, Grabenhorst and Rolls 2008, Rolls 2016b).

In the case of some sensory systems, the reinforcement value may be made explicit early on in sensory processing. This occurs, for example, in the pain system. The architectural basis of this is that there are different channels (nerve fibres) for pain and touch, so that the affective value and the identity of a tactile stimulus can be carried by separate parallel information channels, allowing separate representation and processing of each. In contrast, for vision, invariant visual object recognition must be performed before reward can be assigned to an object, and because invariant visual object recognition requires many stages of cortical processing, the reward decoding can only be performed after the invariant object representation has been computed (Rolls 2014a).

In *non-primates* including, for example, rodents, the design principles may involve less sophisticated design features, partly because the stimuli being processed are simpler. For example, view-invariant object recognition is probably much less developed in non-primates, with the recognition that is possible being based more on physical similarity in terms of texture, colour, simple features, etc., rather than in terms of shape that involves the combination of many features in a particular spatial relationship (Chapter 25). It may be because there is less sophisticated cortical processing of visual stimuli in this way that other sensory systems are also organized more simply in rodents, with, for example, some (but not total, only perhaps 30%) modulation of taste processing by hunger early in sensory processing in rodents (Scott, Yan and Rolls 1995, Rolls and Scott 2003, Rolls 2016b) (Chapter 18). Further, while it is appropriate usually to have emotional responses to well-processed objects (e.g. the sight of a particular person), there are instances, such as a loud noise or a pure tone associated with punishment, where it may be possible to tap off a sensory representation early in sensory information processing that can be used to produce emotional responses, and this may occur, for example, in rodents, where the subcortical auditory system provides afferents to the amygdala (LeDoux 2011, Pessoa and Adolphs 2010). For these reasons, emphasis is placed in this book on research in primates including humans.

Another design principle is that the outputs of the reward and punishment value systems must be treated by the action system as being the goals for action. The action systems must be built to try to maximize the activation of the representations produced by rewarding events, and to minimize the activation of the representations produced by punishers or stimuli associated with punishers.

Drug addiction produced by the psychomotor stimulants such as amphetamine and cocaine can be seen as activating the brain at the stage where the outputs of the amygdala and orbitofrontal cortex, which provide representations of whether stimuli are associated with rewards or punishers, are fed into the ventral striatum to influence approach behaviour. The fact that addiction is persistent may be related to the fact that because the outputs of the amygdala and orbitofrontal cortex are after the stage of stimulus-reinforcer learning, and after sensory-specific satiety has been computed, the action system has to be built to interpret the representations the amygdala and orbitofrontal cortex provide as indicating reward value, and a goal for action (Rolls 2014a). This would account for why drugs of addiction show little sensory-specific satiety, and why they are difficult to extinguish because they act after the normal stimulus-reinforcer stage of learning.

15.5 Specialized systems for face identity and expression processing in primates

Especially in primates, the visual processing for emotional and social behaviour requires sophisticated representation of individuals, and for this there are many neurons devoted to face processing (Perrett, Rolls and Caan 1982, Rolls 2000a, Rolls 2011d, Rolls 2012c, Rolls 2014a) (see Chapter 25). In addition, there is a separate system that encodes face gesture, movement, and view, as all are important in social behaviour, for interpreting whether a particular individual, with his or her own reinforcement associations, is producing threats or appeasements (Rolls 2011d, Rolls 2012c, Hasselmo, Rolls and Baylis 1989a, Rolls 2017c). The face expression system is in the cortex in the superior temporal sulcus in the macaque, and in the middle temporal gyrus in humans in a region that shows reduced functional connectivity in autism (Cheng, Rolls, Gu, Zhang and Feng 2015).

15.6   Unimodal processing to the object level before multimodal convergence

After mainly unimodal processing to the object level, sensory systems then project into convergence zones. Those especially important for reward and punishment, emotion and motivation, are the orbitofrontal cortex and amygdala (see Figs. 15.2 and 2.4 and Section 2.2.5), where primary reinforcers are represented to encode *outcome value*. These parts of the brain appear to be especially important in emotion and motivation not only because they are the parts of the brain where in primates the primary (unlearned) reinforcing value of stimuli is represented, but also because they are the parts of the brain that perform pattern-association learning between potential secondary reinforcers and primary reinforcers to compute expected value (Thorpe, Rolls and Maddison 1983, Rolls, Critchley, Mason and Wakeman 1996b, Rolls 2014a) (Section 2.2.5 and Figs. 2.12, 2.13 and 2.14). They are thus the parts of the brain involved in learning the emotional and motivational reward value of many stimuli.

The orbitofrontal cortex is involved in the rapid, one-trial, reversal of emotional behaviour when the reinforcement contingencies change (Thorpe, Rolls and Maddison 1983, Rolls, Critchley, Mason and Wakeman 1996b, Rolls 2014a) (Fig. 2.14), and this may be implemented by switching a rule, as described in Sections 2.2.5, 15.14, and 15.15. These orbitofrontal cortex neurons may be described as expected value neurons. This rapid, rule-based, reversal and re-valuation of stimuli to encode their current reward value in one trial and using rules may be a computation made possible by the development of the granular orbitofrontal cortex and connected areas that are not present in rodents (Chapter 18). The *negative reward prediction error neurons* (which respond when an expected reward is not obtained) found in the orbitofrontal cortex (Thorpe, Rolls and Maddison 1983) are thought to be involved in the rule-based reward reversal implemented in the orbitofrontal cortex (Rolls and Deco 2016) (Sections 15.14 and 15.15).

15.7   A common scale for reward value

The different reward valuation systems are specified to have values scaled appropriately by genes to lead to their selection in such a way that reproductive success is maximized. In addition, the reward systems are specific, encode the value of specific rewards and punishers, and have tendencies to self-regulate, so that they operate on a common value scale that leads to the selection of different rewards with appropriately balanced probabilities, and often depending on modulation by internal motivational signals (Rolls 2014a). The presence of many different specific reward value systems operating with a common scale of value helps each reward to be selected to maximize reproductive success. There is no conversion to a common currency, as this would no longer encode the specific reward to which the action system should direct action (Rolls 2014a, Grabenhorst and Rolls 2011, Grabenhorst, D'Souza, Parris, Rolls and Passingham 2010a). The value of a reward or punisher specified by genes may be rescaled by learning, as in taste aversion learning and conditioned appetite and satiety (Rolls 2014a, Rolls 2016f).

15.8   Sensory-specific satiety

A principle that assists the selection of different behaviours is sensory-specific satiety, which builds up when a reward is repeated for a number of minutes, and is implemented in the orbitofrontal cortex as shown by the effects on neuronal activity and fMRI activation of feeding to satiety with one food, and finding that responses remain to other foods (Rolls et al. 1989c, Critchley and Rolls 1996a, Kringelbach et al. 2003) (Fig. 2.6). Thus reward devaluation can operate in a partly sensory-specific way.

A principle that helps behaviour to lock on to one goal for at least a useful period is *incentive motivation*, the process by which early on in the presentation of a reward there is reward potentiation (Rolls 2014a).

There are probably simple neurophysiological bases for these time-dependent processes in the reward (as opposed to the early sensory) systems that involve synaptic adaptation and (non-associative) facilitation respectively (Section 10.5) (Rolls 2014a).

15.9 Economic value is represented in the orbitofrontal cortex

The representation in the orbitofrontal cortex is of economic value (Glimcher and Fehr 2013), in that neuronal responses and neural activations appear to be related closely to what is chosen, and also to the conscious subjective value or pleasantness rating placed on a 'good' (Rolls 2014a). The orbitofrontal cortex activity reflects by these measures of economic value the effects of risk (the probability that a reward will be available), of ambiguity (whether the outcome probability is known), of temporal discounting, and of the trade-off between the amount of the good that is available and the value of the good (Padoa-Schioppa 2011, Glimcher and Fehr 2013, Rolls 2014a).

The usefulness of the representation of 'economic value' is that this encompasses the concept of 'net value', in which all the benefits and their associated costs are combined into a single value representation, for this is then the input that is needed to the decision-making networks (Rolls and Grabenhorst 2008, Grabenhorst and Rolls 2011, Rolls 2014a).

These decision-making attractor networks need a single 'net value' input for each choice variable, for otherwise which costs and which benefits relate to which choice would be unclear in the network (see Chapter 4 and Rolls (2014a)).

Value is represented on a continuous scale in the orbitofrontal cortex, and indeed orbitofrontal cortex activations are linearly related to the subjective pleasantness (medially) or unpleasantness (laterally) of stimuli or events (Rolls et al. 2003c, Grabenhorst and Rolls 2008, Rolls 2014a), as well as to monetary value (O'Doherty, Kringelbach, Rolls, Hornak and Andrews 2001).

15.10 Neuroeconomics vs classical microeconomics

The reward value placed on different rewards, and the punishment value placed on different non-rewards or punishers, will be different between different individuals, as a result of genetic variation for natural selection. Also, most humans cannot perform accurate economic calculations of the expected value of different choices, as the human brain operates primarily by similarity comparisons and not by logical calculations of the type implemented in digital computers (Rolls 2012d) (Section 26.3). For these reasons, classical microeconomics with its approach of a few axioms and a rational actor can no longer be considered as what really may account for the behaviour of humans and other animals (Rolls 2014a).

Instead, classical (micro)economics (Glimcher 2011a) may be replaced by an understanding in neuroeconomics of how heuristics guided by evolution make different rewards and costs become differently scaled in different individuals, and further how choices may be selected by a decision-making process based on in-built and probabilistic heuristics rather than by correctly computed calculations in a rational, reasoning, decision-making actor (Rolls 2014a).

15.11 Output systems influenced by orbitofrontal cortex stimulus reward value representations

The orbitofrontal cortex represents the value of stimuli. Its neurons respond to stimulus value, and not to behavioural responses (Rolls 2014a, Thorpe, Rolls and Maddison 1983, Wallis and Miller 2003, Padoa-Schioppa and Assad 2006). The principle here is that value needs to be represented separately from actions. We can assess the value of a stimulus, and use that to compare with the value of other stimuli, and all this can take place independently of any possible action that might follow the assessment of value.

The orbitofrontal cortex provides outputs to different systems, to implement different types of behavioural response as follows (see Figs. 2.4, 15.2, and 22.1).

1. The orbitofrontal cortex projects to the anterior cingulate cortex (Fig. 15.3) where goal-directed actions are computed taking into account the outcomes (the rewards and punishers, including costs of the actions (Rushworth, Kolling, Sallet and Mars 2012, Rolls 2014a). The type of learning is action-outcome learning. The evidence is that, in the context of the orbitofrontal cortex projections to the anterior cingulate cortex (Vogt 2009, Ongur and Price 2000, Price 2006), and in which positive value is represented in the pregenual anterior cingulate cortex and negative value more dorsally in the supracallosal anterior cingulate cortex (Fig. 16.11), the value or outcome representations can be used to implement action-outcome learning (Grabenhorst and Rolls 2011, Rolls 2009a, Rolls 2014a). This is learning in which the action selected to obtain a goal is being learned based on the outcome received. This type of behavioural response stops if the goal is devalued, so is goal-directed, and is often used while an action is being learned.

2. The orbitofrontal cortex projects to the striatum (ventral striatum and adjoining part of the head of the caudate nucleus, parts of the basal ganglia), and the basal ganglia are involved in stimulus-response habits, which are set up with extensive learning, and in which the response is essentially a learned reflex which is no longer under the control of the goal (see Chapter 20 and Rolls (2014a)).

3. The orbitofrontal cortex is reciprocally connected with the amygdala. The amygdala implements Pavlovian learning processes that enable stimuli to elicit approach or withdrawal as well as affective states that may become the goals for instrumental actions. The amygdala also implements Pavlovian learning processes that enable autonomic and endocrine responses to be elicited by conditioned stimuli (Rolls 2014a, Cardinal, Parkinson, Hall and Everitt 2002). The orbitofrontal cortex appears to make a special and important contribution to emotion beyond what the evolutionarily old amygdala can implement (Chapter 18), in that the orbitofrontal cortex can implement one-trial reversal learning, as described in Sections 2.2.5, 15.14, and 15.15.

The cortical architecture of the orbitofrontal cortex may be important in this, for one-trial reversal implies that the rule about which stimulus is currently rewarded must be held online in short-term memory networks, and an attractor network that can be implemented by recurrent collateral connections is ideal for this. Indeed, an attractor network mechanism that can implement this one-trial reversal has been described in which a non-reward

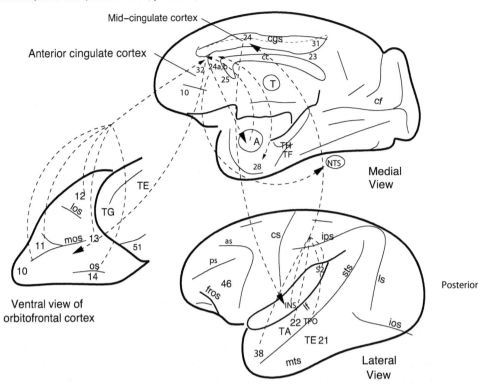

Fig. 15.3 Connections of the anterior cingulate (perigenual) and midcingulate cortical areas (shown on views of the primate brain). The cingulate sulcus (cgs) has been opened to reveal the cortex in the sulcus, with the dashed line indicating the depths (fundus) of the sulcus. The cingulate cortex is in the lower bank of this sulcus, and in the cingulate gyrus which hooks above the corpus callosum and around the corpus callosum at the front and the back. The anterior cingulate cortex extends from cingulate areas 32, 24a and 24b to subgenual cingulate area 25. (The cortex is called subgenual because it is below the genu (knee) formed by the anterior end of the corpus callosum, cc.) The perigenual cingulate cortex tends to have connections with the amygdala and orbitofrontal cortex, whereas area 24c tends to have connections with the somatosensory insula (INS), the auditory association cortex (22, TA), and with the temporal pole cortex (38). The midcingulate areas include area 24d, which is part of the cingulate motor area. Abbreviations: as, arcuate sulcus; cc, corpus callosum; cf, calcarine fissure; cgs, cingulate sulcus; cs, central sulcus; ls, lunate sulcus; ios, inferior occipital sulcus; mos, medial orbital sulcus; os, orbital sulcus; ps, principal sulcus; sts, superior temporal sulcus; lf, lateral (or Sylvian) fissure (which has been opened to reveal the insula); A, amygdala; INS, insula; NTS, autonomic areas in the medulla, including the nucleus of the solitary tract and the dorsal motor nucleus of the vagus; TE (21), inferior temporal visual cortex; TA (22), superior temporal auditory association cortex; TF and TH, parahippocampal cortex; TPO, multimodal cortical area in the superior temporal sulcus; 28, entorhinal cortex; 38, TG, temporal pole cortex; 12, 13, 11, orbitofrontal cortex; 51, olfactory (prepyriform and periamygdaloid) cortex.

attractor become activated if a reward attractor becomes activated by an expected reward but then decreases its firing due to synaptic adaptation if no reward outcome is received (Rolls and Deco 2016). This network implements non-reward or error neuron firing, which in turn may bias the other networks in the orbitofrontal cortex that decode whether incoming stimuli should be treated as rewards or punishers (Deco and Rolls 2005d) (Section 15.15). The amygdala may be much less well suited to implementing the attractors needed for this computation, because as a non-cortical structure it does not have a highly developed recurrent collateral system with associative synaptic modifiability to implement attractors.

Consistent with the hypothesis that the orbitofrontal cortex becomes more important in the evolution towards primates including humans is that damage to the amygdala does not produce major changes in human behaviour that are obvious without special testing (Whalen and Phelps 2009, Rolls 2014a), whereas damage to the human orbitofrontal cortex may produce disinhibition, altered emotional states, difficulty in recognising the emotional expression in a face and in social behaviour, an altered personality, and difficulty in reversing behaviour when reinforcement contingencies change (Hornak, Rolls and Wade 1996, Rolls 1999b, Hornak, Bramham, Rolls, Morris, O'Doherty, Bullock and Polkey 2003, Rolls, Hornak, Wade and McGrath 1994a, Berlin, Rolls and Kischka 2004, Berlin, Rolls and Iversen 2005, Rolls 2014a).

Important principles here are that the orbitofrontal cortex representations of stimulus value are computed independently of any actions or responses, and provide a basis for emotional states elicited by stimuli; and that there are several different systems that receive from the orbitofrontal cortex to implement different types of behavioural response, each of which has its own adaptive utility, using the value representations of stimuli computed in the orbitofrontal cortex.

15.12 Decision-making about rewards in the anterior orbitofrontal cortex / ventromedial prefrontal cortex

Decision-making can now be understood as a non-linear competition between different attractor states in a spiking attractor neuronal network that results in a single winning neuronal population, as described in Section 5.6. The decision variables bias the neurons that are members of the different possible attractor states. The mechanism is understood at the level of integrate-and-fire neurons with biophysically realistic parameters (Wang 2002, Rolls and Deco 2010, Deco, Rolls, Albantakis and Romo 2013). The decision-making is probabilistic because of statistical fluctuations introduced by the approximately Poisson nature of the timing of the spikes of the neurons for a given mean firing rate (Section 5.6).

This understanding is replacing the phenomenological drift-diffusion mathematical models of decision-making with fitted parameters (Shadlen and Kiani 2013), not only because the attractor neuronal network model is more neurobiologically realistic, but also because it allows the exploration of how biological parameters such as ion channel conductances and the effects of different transmitters expressed through different ion channels influence the operation and stability of the decision-making system. This is enabling medically relevant implications to be investigated, for example to neuropsychiatric disorders including schizophrenia and obsessive-compulsive disorder (Deco, Rolls, Albantakis and Romo 2013, Rolls 2014a) (Chapter 16).

In humans there is evidence that reward value is represented on a continuous scale in the medial orbitofrontal cortex, as shown for example by linear correlations of the subjective pleasantness ratings or monetary value with the activations (Rolls, Kringelbach and De Araujo 2003c, Grabenhorst and Rolls 2008, O'Doherty, Kringelbach, Rolls, Hornak and Andrews 2001, Rolls, Grabenhorst and Parris 2008a, Rolls 2014a). In comparison, activations are greater more anteriorly, in what is sometimes termed ventromedial prefrontal cortex and which may include the ventral part of medial prefrontal cortex area 10, when choices must be made between two stimuli that differ in reward value (Grabenhorst, Rolls and Parris 2008b, Rolls, Grabenhorst and Parris 2010d, Rolls and Grabenhorst 2008). Moreover, a signature of an attractor decision-making network, that activations increase with the easiness of the decision, is found in the more anterior medial area 10 / ventromedial prefrontal cortex region (Rolls, Grabenhorst and Deco 2010b, Rolls, Grabenhorst and Deco 2010c, Rolls 2014a). Consistent with this, in macaques lesions of the posterior orbitofrontal cortex (area 13) may impair reward valuation (tested by devaluation by satiation), whereas lesions of the more anterior orbitofrontal cortex (area 11) may impair choice decision-making (Murray, Moylan, Saleem, Basile and Turchi 2015).

These findings provide evidence for another principle of operation of the cerebral cortex: value representations are kept separate from value-related choice decision-making mechanisms. The value representation appears to be computed first, followed by the choice. Part of the underlying reason for this principle may be that the inhibition implemented by the inhibitory neurons for these two types of operation should be kept separate, to ensure that the value can be represented independently of the choice made on a particular occasion. Further, it is important to represent the exact value of the decision variables on a trial, even when noise on a trial may have produced one or another decision. That is, reward value should not be influenced by what has been chosen on a particular trial. Indeed, maintaining reward value representations may be important in expressing one's confidence in a decision, and even later reversing the decision, before the outcome is ever received (Insabato, Pannunzi, Rolls and Deco 2010).

It is a general principle of operation of the cortex that decision-making areas tend to be higher in the hierarchy than cortical areas that linearly represent reward value or the stimulus (Rolls 2014a, Rolls and Deco 2010), as shown by decision-making systems also in the lateral and medial intraparietal areas (LIP and MIP) (Gold and Shadlen 2007, de Lafuente, Jazayeri and Shadlen 2015). One reason may be that a non-linear operation such as a binary choice inevitably throws away information, and it is a goal of earlier cortical stages to build useful representations using all of the information available, and to leave selection to later stages. Consistent with the greater role of the recurrent collateral connections between pyramidal cells for non-linear operations such as attractor operations including decision-making, the further up the hierarchy one is from primary sensory areas, the more developed numerically are the recurrent collateral synapses in areas such as the prefrontal cortex (Elston et al. 2006).

15.13   Probabilistic emotion-related decision-making

The probabilistic operation of the decision-making process caused by the spiking-related statistical fluctuations in finite-sized attractor neuronal networks has many advantages, for emotion as well as for other types of function (Section 5.6). For emotion, these advantages include sometimes choosing less favourable options to update knowledge, predator avoidance, social interactions, and creativity (Rolls and Deco 2010, Rolls 2014a). An important implication for human behaviour is that if a difficult decision is being made, one in which the net value of the inputs is similar, then it may be wise to take the decision several times, to check that the decision the first time was not influenced by a statistical fluctuation.

15.14   Non-reward, error, neurons in the orbitofrontal cortex

Some single neurons in the primate orbitofrontal cortex respond when an expected reward is not obtained, and behavior must change. This was discovered by Thorpe, Rolls and Maddison (1983), who found that 3.5% of neurons in the macaque orbitofrontal cortex detect different types of non-reward. These neurons thus signal negative reward prediction error, the reward outcome value minus the expected value. For example, some neurons responded in extinction,

Orbitofrontal cortex non-reward neuron

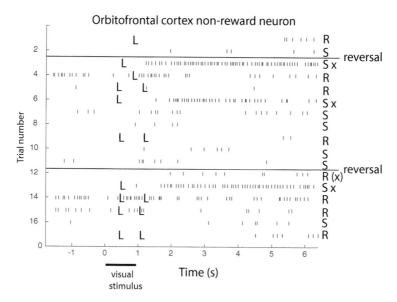

Fig. 15.4 Error neuron: Responses of an orbitofrontal cortex neuron that responded only when the monkey licked to a visual stimulus during reversal, expecting to obtain fruit juice reward, but actually obtained the taste of aversive saline because it was the first trial of reversal (trials 3, 6, and 13). Each vertical line represents an action potential; each L indicates a lick response in the Go–NoGo visual discrimination task. The visual stimulus was shown at time 0 for 1 s. The neuron did not respond on most reward (R) or saline (S) trials, but did respond on the trials marked S x, which were the first or second trials after a reversal of the visual discrimination on which the monkey licked to obtain reward, but actually obtained saline because the task had been reversed. The two times at which the reward contingencies were reversed are indicated. After responding to non-reward, when the expected reward was not obtained, the neuron fired for many seconds, and was sometimes still firing at the start of the next trial. It is notable that after an expected reward was not obtained due to a reversal contingency being applied, on the very next trial the macaque selected the previously non-rewarded stimulus. This shows that rapid reversal can be performed by a non-associative process, and must be rule-based. (Data after *Experimental Brain Research*, 49 (1) pp. 93–115, The orbitofrontal cortex: Neuronal activity in the behaving monkey, S. J. Thorpe, E. T. Rolls, and S. Maddison (c) 1983, Springer Science and Business Media. With kind permission from Springer Science and Business Media.)

immediately after a lick had been made to a visual stimulus that had previously been associated with fruit juice reward. Other non-reward neurons responded in a reversal task, immediately after the monkey had responded to the previously rewarded visual stimulus, but had obtained the punisher of salt taste rather than reward, indicating that the choice of stimulus should change in this visual discrimination reversal task. Importantly, at least some of these non-reward neurons continue firing for several seconds when an expected reward is not obtained, as illustrated in Fig. 15.4. These neurons do not respond when an expected punishment is received, for example the taste of salt from a correctly labelled dispenser. Different non-reward neurons may respond in different tasks, providing the potential for behaviour to reverse in one task, but not in all tasks (Thorpe, Rolls and Maddison 1983, Rolls and Grabenhorst 2008, Rolls 2014a). The existence of neurons in the middle part of the macaque orbitofrontal cortex that respond to non-reward (Thorpe, Rolls and Maddison 1983) (originally described by Thorpe, Maddison and Rolls (1979) and Rolls (1981b)) is confirmed by recordings that revealed 10 such non-reward neurons (of 140 recorded, or approximately 7%) found in delayed match to sample and delayed response tasks by Joaquin Fuster and colleagues (Rosenkilde, Bauer and Fuster 1981).

An example of an orbitofrontal cortex error neuron is shown in Fig. 15.4. On reversal trials, when saline (S) was obtained if a lick (indicated by an L) was made indicating that reversal should occur, the neuron started to respond approximately 1 s after the unexpected saline was obtained, and continued to respond as shown for many seconds, and indeed was still responding before the start of the next trial 8–10 s later, as shown. These neurons did not respond just to the taste of saline, but only when the saline indicated that a taste reward was not being obtained in that trial, and that because of the non-reward, reversal behavior should occur. The continuing firing for many seconds of these neurons provides evidence that they are part of a population that has entered a high firing rate attractor state. This short-term memory of a recent negatively reinforcing event is an essential component of a system that must change its operation after non-reward in a rule-based way, and this has been modelled (Deco and Rolls 2005d). The change of behavior in a reversal task must be rule-based, for after a single trial in which a reward was not received to one stimulus, the choice switches to the other stimulus, even though the most recent reinforcement association of the second stimulus is with punishment. This is clearly illustrated by both reversals in Fig. 15.4 (Thorpe, Rolls and Maddison 1983).

The orbitofrontal cortex is likely to be where non-reward is computed, for not only does it contain non-reward neurons, but it also contains neurons that encode the signals from which non-reward is computed. These signals include a representation of Expected Reward, and Reward Outcome (Rolls 2014a). Expected Reward is represented in the primate orbitofrontal cortex in that many neurons respond to the sight of a food reward (Thorpe, Rolls and Maddison 1983), learn to respond to any visual stimulus associated with a food reward (Rolls, Critchley, Mason and Wakeman 1996b), very rapidly reverse the stimulus to which they respond when the reward outcome associated with each stimulus changes (Rolls, Critchley, Mason and Wakeman 1996b) (Figs. 2.13 and 2.14), and represent reward value in that they stop responding to a visual stimulus when the reward is devalued by satiation (Critchley and Rolls 1996a). Other orbitofrontal cortex neurons represent the expected reward value of olfactory stimuli, based on similar evidence (Rolls and Baylis 1994, Critchley and Rolls 1996b, Rolls, Critchley, Mason and Wakeman 1996b, Critchley and Rolls 1996a) (Fig. 2.12). Reward Outcome is represented in the orbitofrontal cortex by its neurons that respond to taste and fat texture (Rolls, Yaxley and Sienkiewicz 1990, Rolls and Baylis 1994, Rolls, Critchley, Browning, Hernadi and Lenard 1999a, Verhagen, Rolls and Kadohisa 2003, Rolls, Verhagen and Kadohisa 2003e), and do this based on their reward value as shown by the fact that their responses decrease to zero when the reward is devalued by feeding to satiety (Rolls, Sienkiewicz and Yaxley 1989c, Rolls, Critchley, Browning, Hernadi and Lenard 1999a, Rolls 2015e, Rolls 2016f) (Figs. 2.6 and 2.7). Moreover, the primate orbitofrontal cortex is key in these reward value representations, for it receives the necessary inputs from the inferior temporal visual cortex, insular taste cortex, and pyriform olfactory cortex, yet in these preceding areas reward value is not represented (Rolls, Judge and Sanghera 1977, Rolls 2014a, Rolls 2015e, Rolls 2016b) (Sections 2.2.3.1 and 2.2.4).

An attractor network mechanism that can implement this one-trial reversal has been described in which a non-reward attractor become activated if a reward attractor becomes activated by an Expected Reward but then decreases its firing due to synaptic adaptation if no Reward Outcome is received (Section 15.15) (Rolls and Deco 2016).

There is consistent evidence that Expected Value and Reward Outcome Value are represented in the human orbitofrontal cortex (O'Doherty, Kringelbach, Rolls, Hornak and Andrews 2001, Rolls, O'Doherty, Kringelbach, Francis, Bowtell and McGlone 2003d, Kringelbach, O'Doherty, Rolls and Andrews 2003, Rolls, McCabe and Redoute 2008e, Grabenhorst, Rolls and Bilderbeck 2008a, Rolls 2014a, Rolls 2015e). We have also been able to obtain evidence that non-reward used as a signal to reverse behavioural choice is represented in the human orbitofrontal cortex. Kringelbach and Rolls (2003) used the faces of two different people, and if one face was selected then that face smiled, and if the other was selected, the face showed an angry expression. After good performance was acquired, there were repeated reversals of the visual discrimination task. Kringelbach and Rolls (2003) found that activation of a lateral part of the orbitofrontal cortex in the fMRI study was produced on the error trials, that is when the human chose a face, and did not obtain the expected reward (Fig. 16.10). Control tasks showed that the response was related to the error, and the mismatch between what was expected and what was obtained as the reward outcome, in that just showing an angry face expression did not selectively activate this part of the lateral orbitofrontal cortex. An interesting aspect of this study that makes it relevant to human social behavior is that the conditioned stimuli were faces of particular individuals, and the unconditioned stimuli were face expressions. Moreover, the study reveals that the human orbitofrontal cortex is very sensitive to social feedback when it must be used to change behaviour (Kringelbach and Rolls 2003, Kringelbach and Rolls 2004). Consistently, it has now been shown in macaques using fMRI that the lateral orbitofrontal cortex is activated by non-reward during reversal (Chau, Sallet, Papageorgiou, Noonan, Bell, Walton and Rushworth 2015).

The non-reward neurons in the orbitofrontal cortex are implicated in changing behavior when non-reward occurs. Monkeys with orbitofrontal cortex damage are impaired on Go/NoGo task performance, in that they Go on the NoGo trials (Iversen and Mishkin 1970), and in an object-reversal task in that they respond to the object that was formerly rewarded with food, and in extinction in that they continue to respond to an object that is no longer rewarded (Butter 1969, Jones and Mishkin 1972, Meunier, Bachevalier and Mishkin 1997). The visual discrimination reversal learning deficit shown by monkeys with orbitofrontal cortex damage (Jones and Mishkin 1972, Baylis and Gaffan 1991, Murray and Izquierdo 2007) may be due at least in part to the tendency of these monkeys not to withhold responses to non-rewarded stimuli (Jones and Mishkin 1972) including objects that were previously rewarded during reversal (Rudebeck and Murray 2011), and including foods that are not normally accepted (Butter, McDonald and Snyder 1969, Baylis and Gaffan 1991). Consistently, orbitofrontal cortex (but not amygdala) lesions impaired instrumental extinction (Murray and Izquierdo 2007). Similarly, in humans, patients with ventral frontal lesions made more errors in the reversal (or in a similar extinction) task, and completed fewer reversals, than control patients with damage elsewhere in the frontal lobes or in other brain regions (Rolls, Hornak, Wade and McGrath 1994a), and continued to respond to the previously rewarded stimulus when it was no longer rewarded. A reversal deficit in a similar task in patients with ventromedial frontal cortex damage was also reported by Fellows and Farah (2003). Further, in patients with well localised surgical lesions of the orbitofrontal cortex (made to treat epilepsy, tumours etc), it was found that they were severely impaired at the reversal task, in that they accumulated less money (Hornak, O'Doherty, Bramham, Rolls, Morris, Bullock and Polkey 2004). These patients often failed to switch their choice of stimulus after a large loss; and often did switch their choice even though they had just received a reward, and this has been quantified in a more recent study (Berlin, Rolls and Kischka 2004). The importance of the failure to rapidly learn about the value of stimuli from negative feedback has also been described as a critical difficulty for patients with orbitofrontal cortex lesions (Fellows 2007, Wheeler and Fellows 2008, Fellows 2011), and has been contrasted with the effects of lesions to the anterior cingulate cortex which impair the use of feedback to learn about actions (Fellows 2011, Camille, Tsuchida and Fellows 2011, Rolls 2014a).

Further, it is suggested that oversensitivity of the non-reward system in the lateral orbitofrontal cortex could contribute to depression (for example, sadness may result if a reward is lost and cannot be regained), and that impulsivity may be related to underactivity of this non-reward system in the lateral orbitofrontal cortex (Rolls 2016d, Cheng,

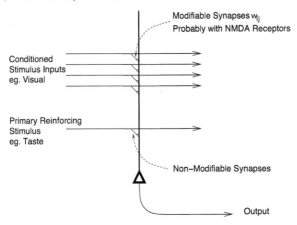

Fig. 15.5 Pattern association between a primary reinforcer, such as the taste of food, which activates neurons through non-modifiable synapses, and a potential secondary reinforcer, such as the sight of food, which has modifiable synapses on to the same neurons. Such a mechanism appears to be implemented in the amygdala and orbitofrontal cortex.

Rolls, Qiu, Liu, Tang, Huang, Wang, Zhang, Lin, Zheng, Pu, Tsai, Yang, Lin, Wang, Xie and Feng 2016, Rolls 2017f, Rolls 2017e, Deng, Rolls, Ji, Robbins, Banaschewski, Bokde, Bromberg, Buechel, Desrivieres, Conrod, Flor, Frouin, Gallinat, Garavan, Gowland, Heinz, Ittermann, Martinot, Lemaitre, Nees, Papadopoulos Orfanos, Poustka, Smolka, Walter, Whelan, Schumann, Feng and the Imagen consortium 2017).

15.15 Reward reversal learning in the orbitofrontal cortex

Decoding the reinforcement value of stimuli, which involves for previously neutral (e.g. visual) stimuli learning their association with a primary reinforcer, often rapidly, and which may involve not only rapid learning but also rapid relearning and alteration of responses when reinforcement contingencies change, is a function proposed for the orbitofrontal cortex. The mechanism of rapid reversal learning that may be implemented in the orbitofrontal cortex utilizes a working memory of which rule is currently active, which may depend on a highly developed recurrent collateral set of synaptic connections present in the orbitofrontal cortex but not the amygdala, as described by Deco and Rolls (2005d). This offers a more computational account of the different functions of the orbitofrontal cortex and amygdala in emotion than previous accounts (e.g. Pickens, Saddoris, Setlow, Gallagher, Holland and Schoenbaum (2003), and Holland and Gallagher (2004)). The rule-based approach to rapid reversal of value representations requires attractor networks to hold active in short-term memory which rule currently applies, and this mechanism may be part of what evolution provides in the function of the granular orbitofrontal cortex areas that are not present in rodents (see Fig. 18.4 and Section 25.3).

We now consider how stimulus–reinforcer association learning and its reversal may be implemented in the orbitofrontal cortex. This illustrates a number of principles of operation of the cerebral cortex, including an attractor short-term memory to hold the rule that is currently active, the computational utility of synaptic or neuronal adaptation, and top-down biased competition.

The suggested process for the initial association is illustrated in Fig. 15.5. If the visual input (e.g. the sight of food) is present at the same time (or just before) the primary reinforcer (e.g. taste) is activating the postsynaptic neuron, then the set of synapses that are driven by the conditioned stimulus become strengthened by the associative process of long-term synaptic potentiation (LTP).

A first approach to the reversal is as follows, and can be understood by referring to Fig. 15.5. Consider a neuron with unconditioned responses to taste in the orbitofrontal cortex. When a particular visual stimulus, say a triangle, was associated with the taste of glucose, the active synaptic connections for this visual (conditioned) stimulus would have shown long-term synaptic potentiation on to the taste neuron, which would respond to the sight of the triangle. During reversal, the same visual stimulus, the triangle, would again activate the same synaptic afferents to the neuron, but that neuron would be inactive when the taste of saline was given. Active presynaptic inputs and a low level of postsynaptic activation is the condition for homosynaptic long-term synaptic depression (LTD, see Fig. 1.6 on page 10), which would then occur, resulting in a decline of the response of the neuron to the triangle. At the same time, visual presentation of a square would now be associated with the taste of glucose, which would activate the postsynaptic neuron, leading now to long-term potentiation of afferents on to that neuron made active by the sight of the square.

Although reversal might be implemented in the way just described by having long-term synaptic depression for synapses that represented the reward-associated stimulus before the reversal, and long-term potentiation of the new stimulus that after reversal is associated with reward, this would require one-trial LTP and one-trial homosynaptic LTD to account for one-trial stimulus–reward reversal (Thorpe, Rolls and Maddison 1983, Rolls, Critchley, Mason and Wakeman 1996b, Rolls 2000d). Moreover, the mechanism would not account for reversal learning set, the process by which during repeated reversal learning, performance gradually improves until reversal can occur in one trial. Even more, the mechanism would not account for the fact that after reversal learning set has been acquired, when the contingency is reversed, the animal makes a response to the current S+ expecting to get reward, but instead obtains the punisher. On the very first subsequent trial on which the pre-reversal S– is shown, the animal will perform a response to it expecting now to get reward, *even though the post-reversal S+ has not since the reversal been associated with reward to produce LTP for the post-reversal S+*. (This is in fact illustrated in Fig. 15.4.) To implement this very rapid stimulus–reinforcer association reversal, a different mechanism is therefore needed. The mechanism can not rely on associative processes, but instead on a rule-based process, which requires the current rule to be held in mind, in short-term memory. This, and non-reward neurons that maintain their firing for many seconds as illustrated in Fig. 15.4, may be developments provided for by the evolution of granular orbitofrontal cortex areas in primates (see Section 25.3).

A model for how the very rapid, one-trial, reversal could be implemented has been developed (see Deco and Rolls (2005d) for a full description). The model uses a short-term memory autoassociation attractor network with associatively modifiable synaptic connections to hold the neurons representing the current rule active (see Fig. 15.6). Rule one might correspond to 'stimulus 1 (e.g. a triangle) is associated with reward, and stimulus 2 (e.g. a square) is associated with punishment'. Rule 2 might correspond to the opposite contingency. A small, very biologically plausible, modification of the standard one-layer autoassociation network is that there is a small amount of adaptation in the recurrent collateral synapses that keep the neurons representing the current rule firing (Chapter 10). Now consider the case when the neurons representing rule one are firing. How does the rule module reverse? The proposal is that when the non-reward or error neurons described above (Section 15.14) fire, this additional set of firing neurons destabilizes the rule attractor module, by for example producing extra firing of the inhibitory neurons in the orbitofrontal cortex, which in turn inhibit the excitatory neurons in the rule autoassociation network, thus quenching its attractor state. This error input to the rule attractor network is shown in Fig. 15.6. After neuronal firing

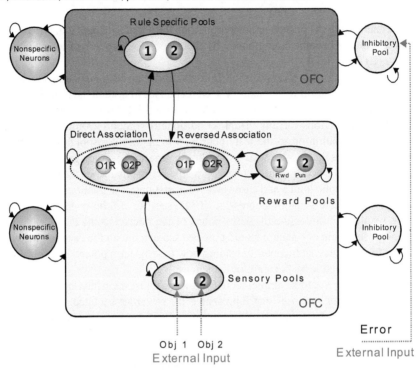

Fig. 15.6 Cortical architecture of the reward reversal model. There is a rule module (top) and a sensory – intermediate neuron – reward module (below). Neurons within each module are fully connected, and form attractor states. The sensory – intermediate neuron – reward module consists of three hierarchically organized levels of attractor network, with stronger synaptic connections in the forward than the backprojection direction. The intermediate level of the sensory – intermediate neuron – reward module contains neurons that respond to combinations of an object and its association with reward or punishment, e.g. object 1–reward (O1R, in the direct association set of pools), and object 1–punishment (O1P in the reversed association set of pools). These intermediate level neurons have the properties of 'conditional reward neurons' illustrated in Fig. 15.7, and provide a function for such conditional reward neurons. The rule module acts as a biasing input to bias the competition between the object–reward combination neurons at the intermediate level of the sensory – intermediate neuron – reward module. The whole model is implemented with integrate-and-fire neurons. (Reproduced from G. Deco and E. T. Rolls, Synaptic and spiking dynamics underlying reward reversal in the orbitofrontal cortex, Cerebral Cortex, 2005, 15 (1), pp. 15–30, by permission of Oxford University Press.)

in the network has stopped and the error signal, which may last for 10 s as illustrated in Fig. 15.4, is no longer present, then firing gradually can build up again in the rule attractor network. (This build-up may be assisted by non-specific inputs from other neurons in the area, as illustrated in Deco and Rolls (2005d).) However, with the competitive processes operating within the rule attractor network between the populations of neurons representing rule 1 and those representing rule 2, and the fact that the neurons or synapses that are part of the rule 1 attractor are partly adapted, the neurons that win the competition and become active are those representing rule 2, and the rule attractor has reversed its state. This process is illustrated by Deco and Rolls (2005d), and takes one trial. Reversal learning set takes a number of reversals to acquire because the correct attractors for the relevant rules, and their connections to other 'mapping' neurons, have to be learned.

To achieve the correct 'mapping' from stimuli to their reinforcer association, and thus emotional state, the rule neurons bias the competition in a mapping module, illustrated in

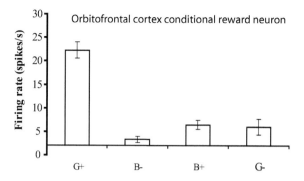

Fig. 15.7 A conditional reward neuron recorded in the orbitofrontal cortex which responded only to the Green stimulus when it was associated with reward (G+), and not to the Blue stimulus when it was associated with Reward (B+), or to either stimuli when they were associated with a punisher, the taste of salt (G– and B–). The mean firing rate ± the s.e.m. is shown. (Reproduced from *Experimental Brain Research*, 49 (1) pp. 93–115, The orbitofrontal cortex: Neuronal activity in the behaving monkey, S. J. Thorpe, E. T. Rolls, and S. Maddison, ©1983, Springer-Verlag with permission of Springer.)

Fig. 15.6. The mapping module has sensory input neurons, intermediate 'conditional reward' neurons (of the type described by Thorpe, Rolls and Maddison (1983) and Rolls, Critchley, Mason and Wakeman (1996b) and illustrated in Fig. 15.7) which respond to combinations of stimuli and whether they are currently associated with reward (or for other neurons to a punisher), and output neurons which represent the reinforcement association of the stimulus currently being viewed. (In the case described there are four populations or pools of neurons at the intermediate level, two for the direct rewarding context: object 1-rewarding, object 2-punishing, and two for the reversal condition: object 1-punishing, object 2-rewarding.) These intermediate pools or populations of neurons respond to combinations of the sensory stimuli and the expected reward, e.g. to object 1 and an expected reward (glucose obtained after licking), and are the conditional reward neurons illustrated in Fig. 15.7. The sensory – intermediate – reward module thus consists of three hierarchically organized levels of attractor network, with stronger synaptic connections in the forward direction from input to output than the backprojection direction. The rule module acts as a biasing input to bias the competition between the object–reward combination neurons at the intermediate level of the sensory – intermediate – reward module. This biasing is achieved because rule 1 has associatively strengthened connections to object 1–rewarding and object 2–punishing neurons. (The whole network could be set up by simple associative learning operating to strengthen connections made with low probability between different neurons that are conjunctively active during the task in the network – see Deco and Rolls (2005d).)

Thus when object 1, e.g. the triangle, is being presented and rule one for direct mapping is in the rule module and biasing the intermediate neurons of the sensory – intermediate – reward module, then the intermediate neurons that fire are the object 1-reward neurons (O1R in Fig. 15.6), and these in turn through associative connections activate the reward neurons (Rwd in Fig. 15.6) at the third, reward/punishment, level of the hierarchy. If on the other hand object 1, e.g. the triangle, is being presented and rule two for reversed mapping is in the rule module and biasing the intermediate neurons of the sensory – intermediate – reward module, then the intermediate neurons that fire are the object 1–punishment (O1P) neurons, and these in turn through associative connections activate the punishment neurons (Pun) at the third, reward/punishment, level of the hierarchy. This model can thus account for one-trial reversal learning, and provides an account for the presence of the conditional reward and conditional punishment neurons discovered by Thorpe, Rolls and Maddison (1983) and Rolls, Critchley, Mason and Wakeman (1996b) in the orbitofrontal cortex.

It is an important part of the architecture that at the intermediate level of the sensory – intermediate – reward module one set of neurons fire if an object being presented is currently associated with reward, and a different set if the object being presented is currently associated with punishment. This representation means that these neurons can be used for different functions, such as the elicitation of emotional or autonomic responses, which can occur for example to particular stimuli associated with particular reinforcers (Rolls 2014a). For example, particular emotions might arise if a particular cognitively processed input such as a particular person is associated with a particular type of reinforcer or reinforcement contingency.

It is also an interesting part of the architecture that associative synaptic modifiability (LTP, and LTD if present) is needed only to set up the functional architecture of the network while the reversal learning set is being acquired. However, once the correct synaptic connections have been set up to implement the architecture illustrated in Fig. 15.6, then no further synaptic modifiability is needed each time reversal occurs, as reversal is achieved just by the error signal quenching the current rule attractor, and the attractor for the other rule then starting up because its synapses are not adapted.

The network just described uses biased competition from a rule module to bias the mapping from sensory stimuli to the representation of a reward vs a punisher. An analogous rule network reversed in the same way by error signals quenching the current rule attractor, can be used to reverse the mapping from stimuli via intermediate stimulus–response neurons to response neurons, and thus to switch the stimulus-to-motor response being mapped in a model of conditional response learning (Deco and Rolls 2005d). While reward rule neurons have not been described yet for the orbitofrontal cortex, neurons which may correspond to stimulus–response rule neurons have been found in the dorsolateral prefrontal cortex (Wallis, Anderson and Miller 2001).

This model also provides a computational account of why the orbitofrontal cortex may play a more important role in rapid reversal learning than the amygdala. The account is based on the fact that a feature of cortical architecture is a highly developed set of local (within 1–2 mm) recurrent collateral excitatory associatively modifiable connections between pyramidal cells (Rolls and Deco 2002, Rolls and Treves 1998). These provide the basis for short-term memory attractor networks, and thus the basis for the rule attractor model which is at the heart of the hypothesis for how rapid reversal learning is implemented (Deco and Rolls 2005d). In contrast, the amygdala is thought to have a much less well developed set of recurrent collateral excitatory connections, and thus may not be able to implement rapid reversal learning in the way described using competition biased by a rule module. Instead, the amygdala would need to rely on synaptic relearning as described in the first approach above, and this would be likely to be a slower process, and would certainly not lead to correct choice of the new S+ the first time it is presented after a punishment trial when the reversal contingency changes. Of course, in addition it is possible that the rapidity of LTP, and the efficacy of LTD, both of which would also facilitate rapid reversal, may be enhanced in the orbitofrontal cortex compared to the amygdala. Thus, the cortical neuronal reversal mechanism in the orbitofrontal cortex may be effectively a faster implementation in two ways than what is implemented in the amygdala. The cortical (in this case orbitofrontal cortex) mechanism may have evolved particularly to enable rapid updating by received reinforcers in social and other situations in primates. This principle of operation of the cerebral cortex, that the orbitofrontal cortex, as a rapid learning mechanism, effectively provides an additional route for some of the functions performed by the amygdala, and is very important when this stimulus–reinforcer learning must be rapidly readjusted, has been developed elsewhere (Rolls 1990c, Rolls 1992b, Rolls 1996a, Rolls 2000c, Rolls 2014a) (see also Chapter 18).

Another feature of the rule attractor model of rapid reversal learning (Deco and Rolls 2005d) is that it does utilize a set of coupled attractor networks in the orbitofrontal cortex. Consistent with this, Hikosaka and Watanabe (2000) have shown that a short-term memory for reward, such as the flavour of a food, is represented by continuing firing in orbitofrontal cortex neurons in a reward delayed match-to-sample short-term memory task. This could be implemented by associatively modified synaptic connections between taste reward neurons in the orbitofrontal cortex.

Although the mechanism has been described so far for visual-to-taste association learning, this is because neurophysiological experiments on this are most direct. It is likely, given the evidence from the effects of lesions, that taste is only one type of primary reinforcer about which such learning occurs in the orbitofrontal cortex, and is likely to be an example of a much more general type of stimulus–reinforcer learning system. Some of the evidence for this is that humans with orbitofrontal cortex damage are impaired at visual discrimination reversal when working for a reward that consists of points (Rolls, Hornak, Wade and McGrath 1994a) or money (Hornak, Bramham, Rolls, Morris, O'Doherty, Bullock and Polkey 2003, Rolls 2014a). Moreover, as described above, there is now evidence that the representation of the affective aspects of touch are represented in the human orbitofrontal cortex (Rolls, O'Doherty, Kringelbach, Francis, Bowtell and McGlone 2003d, McCabe, Rolls, Bilderbeck and McGlone 2008), and learning about what stimuli are associated with this class of primary reinforcer is also likely to be an important aspect of the stimulus–reinforcer association learning performed by the orbitofrontal cortex.

15.16 Dopamine neurons and emotion

With respect to the dopamine system, it appears that the activity of the dopamine neurons does not represent reward, and is not correlated with hedonic or emotional states (Section 9.6.2). Part of the evidence for this is that the dopamine neurons may fire in relation to a reward prediction error, rather than in relation to reward itself (Schultz 2013), and that damage to the dopaminergic system does not impair hedonic responses or 'liking'. Dopamine pathways do influence systems involved in Pavlovian (classically conditioned) incentive salience effects mediated by the ventral striatum, and may thereby influence 'wanting'. However, there are inconsistencies with the dopamine reward prediction error hypothesis, with respect for example to whether the dopamine system is implicated in salience (as many dopamine neurons respond to aversive, novel, and/or alerting stimuli (Bromberg-Martin et al. 2010a)) vs reward error prediction, and whether a positive reward prediction error signal could facilitate the learning of stimuli associated with punishment and the inhibition of actions.

One of the areas influenced by the dopamine neurons is the orbitofrontal cortex, and thus alterations in dopamine may influence emotional behaviour in this way. However, the dopamine neurons do not themselves appear to have the necessary inputs from which a positive reward error signal would be directly computed, and it is indeed suggested that the inputs to the dopamine neurons may originate from structures such as the orbitofrontal cortex, where expected value, reward outcome (e.g. taste), and negative reward prediction error are represented (Rolls 2014a). Consistent with this, the orbitofrontal cortex projects into the ventral stratum, which in turn projects to the dopamine neurons in the ventral tegmentum (Rolls 2014a).

15.17   The explicit reasoning system vs the emotional system

In addition to the reward-related / emotional system for action selection, in humans and perhaps related animals there is also an explicit system that can use language to compute actions to obtain deferred rewards using a one-off plan (Chapter 22). The language system enables one-off multistep plans that require the syntactic organization of symbols to be formulated in order to obtain rewards and avoid punishers. There are thus two separate systems for producing actions to rewarding and punishing stimuli in humans and potentially in related animals. These systems may weight different courses of action differently, and produce conflict in decision-making, in that each can produce behaviour for different goals (immediate vs long-term goals involving multiple step planning). Understanding our evolutionary history is useful in enabling us to understand our emotional decision-making processes, and the conflicts that may be inherent in how they operate (Chapter 22).

It is argued that the decisions taken by the emotional system are in the interests of selfish genes, in which the value systems have their foundations. The reasoning, rational, system enables longer-term decisions to be taken by planning ahead, and for the decisions to be in the interests of the individual, the phenotype, instead of the genes (Rolls 2014a) (Section 22.3.2).

15.18   Pleasure

Pleasure is a subjective state reported in humans that is associated with a reward (Rolls 1999a, Rolls 2014a), and the cortical processes that underlie pleasure need to be understood. We have seen that pleasure, measured by pleasantness ratings, is linearly related to activations in the orbitofrontal cortex and anterior cingulate cortex (Rolls, Kringelbach and De Araujo 2003c, Grabenhorst and Rolls 2008, Rolls, Grabenhorst and Parris 2008a, Rolls 2014a) (Fig. 6.19). Just as reward is specific, with many different types of reward, so is pleasure. These cortical regions may provide inputs to other cortical regions involved in the subjective experiences. It is possible that emotional feelings (including pleasure), part of the much larger problem of consciousness, arise as part of a process that involves thoughts about thoughts, which have the adaptive value of helping to correct one-off multistep plans. This is the higher-order syntactic thought (HOST) approach to consciousness described in Chapter 22. However, there seems to be no clear way to choose which theory of consciousness is moving in the right direction, caution must be exercised here, and current theories should not be taken to have implications.

Pleasure is not itself an explanatory variable. An emotional state can be operationally defined as the state elicited by an instrumental reinforcer, and has important functions that are described in Section 15.1. Motivation can similarly be operationally defined, and has particular functions including being the state when actions are performed to obtain instrum-ental reinforcers (Section 15.2). Subjective states such as pleasure may accompany the state of emotion, but do not themselves 'explain' the behaviour. Instead, it is the evolutionary adaptive values of the operationally and functionally defined states of motivation and emotion that makes them useful in understanding behaviour, as described in Section 15.1 and by Rolls (2014a) and Rolls (2016c).

15.19   Personality relates to differences in sensitivity to rewards and punishers

The reward value placed on different rewards, and the punishment value placed on different non-rewards or punishers, will be different between different individuals, as a result of genetic variation for natural selection (Rolls and McCabe 2007, Corr and McNaughton 2012, Rolls 2014a). The different sensitivity of different individuals to different rewards and punishers is reflected in brain systems such as the orbitofrontal cortex and amygdala, and provides a biological basis for understanding differences in personality (Rolls 2014a, Corr and McNaughton 2012).

15.20   Highlights

1. Emotions can be defined as states elicited by instrumental reinforcers (rewards and punishers), which are the goals for action. Motivational states can be defined as states in which an instrumental reinforcer is the goal for action.

2. The principle of operation is that genes can specify goals for actions that are in the selfish interests of the genes. By specifying the rewards (e.g. a sweet taste) and punishers (e.g. painful touch), the specification is simpler than trying to specify detailed behavioural responses to stimuli, and allows much greater flexibility of the actions, which can be learned instead of pre-specified by the genes.

3. The principle of operation of the cortical systems in primates is that there is a first tier of processing that computes what the stimulus is independently of its reward value. This is adaptive, for it enables objects to be seen, tasted, etc and their locations remembered even when they are not rewarding, for example when hunger is not present. The ventral cortical systems perform this *what* computation for taste (the insular taste cortex), olfaction (the pyriform cortex), vision (the inferior temporal cortex), and hearing.

4. A second tier of processing, involving the amygdala and orbitofrontal cortex, then computes the reward value of the stimuli, and this is used as an input to other systems, including the anterior cingulate cortex which is important in instrumental learning by enabling the outcome of the action (the reward obtained or not) to be associated with the action; the striatum (for habits, that is well-learned behavioural responses to a stimulus that has been rewarded previously but which now are performed as a stimulus-response association that is no longer under the direct control of the goal or reward value); and autonomic outputs including via the visceral insula and hypothalamus.

5. The separation or 'what' from reward systems is much less clear in rodents, with changes in reward value influencing sensory processing in early stages of processing. In rodents there is much less dominance of cortical processing to compute 'what' representations before and separate from reward evaluation, in that for example in rodents there is a pre-cortical pontine taste area that distributes taste information independently of cortical processing to brain systems such as the amygdala and hypothalamus.

6. The orbitofrontal cortex, which develops considerably in primates, enables, with its well-developed recurrent collateral connections that are the hallmark of the cortex, to show very rapid rule-based changes in responses to reinforcers to enable behaviour to change very rapidly when the reinforcers being received change. This is very important is many types of behaviour, including social behaviour, where great sensitivity to reinforcers including punishers can be adaptive. The amygdala has been present from much earlier in evolution, and without the highly developed recurrent collateral system

of cortex, is less able to compute rule-based changes of behaviour when reinforcement contingencies change.

7. The orbitofrontal cortex implements multimodal convergence of visual, olfactory, taste, somatosensory, and auditory inputs from the 'what' cortical systems, to produce a *value* representation of a stimulus, which can then be used as the input to 'neuroeconomic' decision systems between different 'value' stimuli in a third cortical tier of processing in more anterior regions such as the ventromedial prefrontal cortex and medial prefrontal cortex area 10. The evidence is that decisions are made in this third tier of cortical processing for emotion and motivation by cortical attractor decision-making networks.

8. In addition to this system for emotion, humans and perhaps some other animals have a reasoning system that can evaluate outcomes that may be in the long-term interests of the individual, such as not to eat very rewarding and pleasant foods that might be unhealthy. This reasoning, explicit system, may take decisions that are sometimes in conflict with the gene-goal-specified emotional system, as described in Chapter 22, and this makes the whole operation of the system complex, with its two emotional systems, and the noise-influenced decision-making within and between them described in Chapter 5.

16 Noise in the cortex, stability, psychiatric disease, and aging

16.1 Stochastic noise, attractor dynamics, and schizophrenia

16.1.1 Introduction

Schizophrenia is a major mental illness, which has a great impact on patients and their environment. One of the difficulties in proposing models for schizophrenia is the complexity and heterogeneity of the illness. We propose that part of the reason for the inconsistent symptoms may be a reduced signal-to-noise ratio and increased statistical fluctuations in different cortical brain networks. The novelty of the approach described here is that instead of basing our hypothesis purely on biological mechanisms, we develop a top-down approach based on the different types of symptoms and relate them to instabilities in attractor neural networks (Rolls 2005, Loh, Rolls and Deco 2007a, Loh, Rolls and Deco 2007b, Rolls, Loh, Deco and Winterer 2008d, Rolls 2008d, Deco, Rolls and Romo 2009, Rolls and Deco 2011a, Rolls 2012b, Rolls 2014a).

The main assumption of our hypothesis is that attractor dynamics are important in cognitive processes. Our hypothesis is based on the concept of attractor dynamics in a network of interconnected neurons which in their associatively modified synaptic connections store a set of patterns, which could be memories, perceptual representations, or thoughts (Hopfield 1982, Amit 1989, Rolls 2008d, Rolls and Deco 2010). The attractor states are important in cognitive processes such as short-term memory, attention, and action selection (Deco and Rolls 2005a). The network may be in a state of spontaneous activity, or one set of neurons may have a high firing rate, each set representing a different memory state, normally recalled in response to a retrieval stimulus. Each of the states is an attractor in the sense that retrieval stimuli cause the network to fall into the closest attractor state, and thus to recall a complete memory in response to a partial or incorrect cue. Each attractor state can produce stable and continuing or persistent firing of the relevant neurons. The concept of an energy landscape (Hopfield 1982) is that each pattern has a basin of attraction, and each is stable if the basins are far apart, and also if each basin is deep, caused for example by high firing rates and strong synaptic connections between the neurons representing each pattern, which together make the attractor state resistant to distraction by a different stimulus (Chapter 5). The spontaneous firing state, before a retrieval cue is applied, should also be stable. Noise in the network caused by statistical fluctuations in the stochastic spiking of different neurons can contribute to making the network transition from one state to another, and we take this into account by performing integrate-and-fire simulations with spiking activity, and relate this to the concept of an altered signal-to-noise ratio in schizophrenia (Winterer, Ziller, Dorn, Frick, Mulert, Wuebben, Herrmann and Coppola 2000, Winterer, Coppola, Goldberg, Egan, Jones, Sanchez and Weinberger 2004, Winterer, Musso, Beckmann, Mattay, Egan, Jones, Callicott, Coppola and Weinberger 2006).

Schizophrenia is characterized by three main types of symptom: cognitive dysfunction, negative symptoms, and positive symptoms (Liddle 1987, Baxter and Liddle 1998, Mueser and McGurk 2004, Owen, Sawa and Mortensen 2016, Rolls, Lu, Wan, Yan, Wang, Yang, Tan, Li, Group, Yu, Liddle, Palaniyappan, Zhang, Yue and Feng 2017b). We consider how the basic characteristics of these three categories might be produced in a neurodynamical system, as follows.

Dysfunction of working memory, the core of the cognitive symptoms, may be related to instabilities of persistent attractor states (Durstewitz, Seamans and Sejnowski 2000b, Wang 2001) which we show can be produced by reduced firing rates in attractor networks, in brain regions such as the prefrontal cortex.

The negative symptoms such as flattening of affect or reduction of emotions may be caused by a consistent reduction in firing rates of neurons in regions associated with emotion such as the orbitofrontal cortex (Rolls 2014a, Rolls 2008d). These hypotheses are supported by the frequently observed hypofrontality, a reduced activity in frontal brain regions in schizophrenic patients during cognitive tasks (Ingvar and Franzen 1974, Kircher and Thienel 2005, Scheuerecker, Ufer, Zipse, Frodl, Koutsouleris, Zetzsche, Wiesmann, Albrecht, Bruckmann, Schmitt, Moller and Meisenzahl 2008).

The positive symptoms are characterized by phenomenologically overactive perceptions or thoughts such as hallucinations or delusions which are reflected for example by higher activity in the temporal lobes (Shergill, Brammer, Williams, Murray and McGuire 2000, Scheuerecker et al. 2008). We relate this category of symptoms to a spontaneous appearance of activity in attractor networks in the brain and more generally to instability of both the spontaneous and persistent attractor states (Loh, Rolls and Deco 2007a).

In particular we were interested in how these symptoms are related. Schizophrenia is a neurodevelopmental disease (Owen et al. 2016), and negative and cognitive symptoms typically precede the first psychotic episode (Lieberman, Perkins, Belger, Chakos, Jarskog, Boteva and Gilmore 2001, Hafner, Maurer, Loffler, an der Heiden, Hambrecht and Schultze-Lutter 2003). Positive symptoms can be treated in most cases with neuroleptics whereas negative and cognitive symptoms persist, at least for typical neuroleptics. Can a mapping onto a dynamical system help to understand these relations? After proposing a dynamical systems hypothesis for the different symptoms of schizophrenia, we study a standard neural network model (Brunel and Wang 2001) of cortical dynamics specifically in relation to our hypothesis (Rolls 2005, Loh, Rolls and Deco 2007a, Rolls, Loh, Deco and Winterer 2008d). We were especially interested in how excitation and inhibition implemented by NMDA and GABA synapses affect the network dynamics. Alterations in the efficacies of the NMDA and GABA channels have been identified in the pathology of schizophrenia (Coyle 2012, Balu and Coyle 2015, Lewis 2014), and transmitters such as dopamine influence the currents in these receptor-activated channels (Seamans and Yang 2004). Do NMDA and GABA currents have antagonistic effects or do they have a special role in the network dynamics? How could this be related to our hypothesis of schizophrenia? Building upon the current body of neural network research, we describe neural network simulations to substantiate our dynamical systems hypothesis of schizophrenia. While focussing on NMDA and GABA synapses in the simulations, we also consider how altered transmission at D1 and D2 receptors by modulating NMDA and GABA conductances could not only influence working memory, which has been investigated previously (Durstewitz, Kelc and Gunturkun 1999, Durstewitz, Seamans and Sejnowski 2000a, Brunel and Wang 2001, Durstewitz and Seamans 2002), but could in particular influence the different symptoms of schizophrenia.

16.1.2 A dynamical systems hypothesis of the symptoms of schizophrenia

We relate the three types of symptoms of schizophrenia to the dynamical systems attractor framework described earlier as follows (Rolls 2005, Loh, Rolls and Deco 2007a, Loh, Rolls and Deco 2007b, Rolls, Loh, Deco and Winterer 2008d, Rolls 2008d, Deco, Rolls and Romo 2009, Rolls and Deco 2011a, Rolls 2012b, Rolls 2014a).

The *cognitive symptoms* of schizophrenia include distractibility, poor attention, and the dysexecutive syndrome (Liddle 1987, Green 1996, Mueser and McGurk 2004, Owen et al. 2016). The core of the cognitive symptoms is a working memory deficit in which there is a difficulty in maintaining items in short-term memory (Goldman-Rakic 1994, Goldman-Rakic 1999), which could directly or indirectly account for a wide range of the cognitive symptoms. We propose that these symptoms may be related to instabilities of persistent states in attractor neural networks, consistent with the body of theoretical research on network models of working memory (Durstewitz, Seamans and Sejnowski 2000b). The neurons are firing at a lower rate, leading to shallower basins of attraction of the persistent states, and thus a difficulty in maintaining a stable short-term memory, normally the source of the bias in biased competition models of attention (Rolls and Deco 2002, Deco and Rolls 2005a). The shallower basins of attraction would thus result in working memory deficits, poor attention, distractibility, and problems with executive function and action selection (Deco and Rolls 2003, Deco and Rolls 2005a, Rolls 2008d).

The *negative symptoms* refer to the flattening of affect and a reduction in emotion. Behavioural indicators are blunted affect, emotional and passive withdrawal, poor rapport, lack of spontaneity, motor retardation, and disturbance of volition (Liddle 1987, Mueser and McGurk 2004, Owen et al. 2016). We propose that these symptoms are related to decreases in firing rates in the orbitofrontal cortex and/or anterior cingulate cortex (Rolls 2014a), where neuronal firing rates and activations in fMRI investigations are correlated with reward value and pleasure (Rolls 2014a, Rolls 2006b, Rolls 2007f, Rolls and Grabenhorst 2008, Rolls 2008c). Consistent with this, imaging studies have identified a relationship between negative symptoms and prefrontal hypometabolism, i.e. a reduced activation of frontal areas (Wolkin, Sanfilipo, Wolf, Angrist, Brodie and Rotrosen 1992, Aleman and Kahn 2005).

The *positive symptoms* of schizophrenia include bizarre (psychotic) trains of thoughts, hallucinations, and (paranoid) delusions (Liddle 1987, Mueser and McGurk 2004, Owen et al. 2016). We propose that these symptoms are related to shallow basins of attraction of both the spontaneous and persistent states in the temporal lobe semantic memory networks and to the statistical fluctuations caused by the probabilistic spiking of the neurons. This could result in activations arising spontaneously, and thoughts moving too freely round the energy landscape, loosely from thought to weakly associated thought, leading to bizarre thoughts and associations, which may eventually over time be associated together in semantic memory to lead to false beliefs and delusions. Consistent with this, neuroimaging studies suggest higher activation especially in areas of the temporal lobe (Weiss and Heckers 1999, Shergill et al. 2000, Scheuerecker et al. 2008).

To further investigate our hypothesis, we use an attractor network, as this is likely to be implemented in many parts of the cerebral cortex by the recurrent collateral connections between pyramidal cells, and has short-term memory properties with basins of attraction which allow systematic investigation of stability and distractibility. The particular neural network implementation we adopt includes channels activated by AMPA, NMDA and $GABA_A$ receptors and allows not only the spiking activity to be simulated, but also a consistent mean-field approach to be used (Brunel and Wang 2001). The network, and some of the simulations performed with it, are described in Chapter 5 and Section B.6.

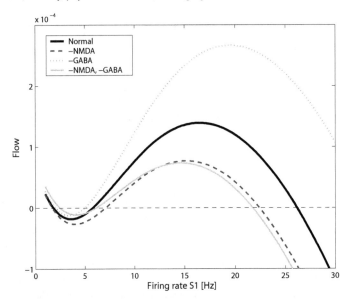

Fig. 16.1 Flow of a single attractor network for different modifications of synaptic conductances. The flow of an attractor network with one selective pool was assessed using a meanfield analysis. The flow represents the force that drives the system towards one of the stable attractors. The stable/unstable attractor states are at crossings with the flow=0 value on the flow axis with a negative/positive derivative respectively. A modulation of the synapses labelled as (−NMDA) and (−GABA) corresponds to a reduction of 4.5% and 9% respectively of the efficacies. A pool cohesion (connectivity) of w_+=1.6 and a selective pool size of 80 neurons were used for the simulations. (After Loh, Rolls and Deco 2007a.)

16.1.3 The depth of the basins of attraction: mean-field flow analysis

First we introduced an analytical approach to the concepts of how changes in transmitters could affect the depth of the basins of attraction in networks in ways that may be related to the symptoms of schizophrenia. Figure 16.1 shows the flow between the spontaneous and persistent state in a network featuring one selective pool. A reduction of NMDA receptor activated ion channel currents (−NMDA) produces a stronger flow than the unchanged condition at low firing rates towards the spontaneous attractor (at about 2 Hz). The absolute values of the function are higher compared to the normal condition until the first unstable fixed point (at around 6–7 Hz). The basin of attraction towards the persistent attractor at high firing rates yields the reverse picture. Here the (−NMDA) curve is clearly below the unchanged condition and the flow towards the high firing rate attractor is smaller. Overall, the basin of attraction is deeper for the spontaneous state and shallower for the persistent state compared to the unchanged condition. This pattern fits to the cognitive symptoms of schizophrenia as proposed in our hypothesis. We also note that the firing rate of the persistent fixed point is reduced in the (−NMDA) condition (crossing with the flow=0 axis), which is consistent with the hypothesis for the negative symptoms.

A reduction of the GABA receptor activated ion channel conductances (−GABA) yields the opposite pattern to that in the reduced NMDA condition. Here the basin of attraction of the persistent high firing rate state is deeper. This is not a condition that we suggest is related to the symptoms of schizophrenia.

However, in the condition in which both the NMDA and the GABA conductances are reduced (−NMDA, −GABA), the persistent high firing rate state basin of attraction is shallower, and the spontaneous state basin is a little shallower. This condition corresponds to

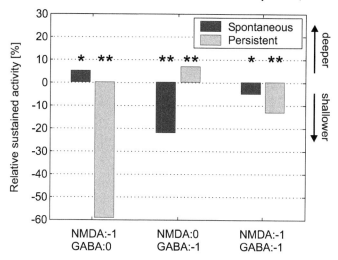

Fig. 16.2 Stability of the spontaneous and persistent state relative to the unmodulated reference state for modulations of the synaptic efficacies in integrate-and-fire simulations. We assessed how often in 100 trials the average activity during the last second (2–3 s) stayed above 10 Hz. The value shows how often it stayed more in the respective state than in the reference state. A negative percentage means that the system was less stable than in the reference state. A modulation of the synapses shown as (−NMDA) and (−GABA) corresponds to a reduction of 4.5% and 9% respectively in their efficacies. We assessed with the Binomial distribution the statistical significance of the effects observed, with P<0.01 relative to the reference state marked by **, and P<0.02 by *. (After Loh, Rolls and Deco 2007a.)

the proposed landscape for the positive symptoms as considered earlier. In particular, in the (−NMDA, −GABA) condition, the system would be less stable in the persistent state, tending to move to another attractor easily, and less stable in the spontaneous state, so tending to move too readily into an attractor from spontaneous activity.

Overall, the mean-field flow analysis suggests that both the cognitive and negative symptoms could be related to a decrease in the NMDA conductances. This is consistent with the fact that these two symptoms usually appear together. The flow analysis suggests that the positive symptoms are related to a reduction in both NMDA and GABA. Thus, the transition from the cognitive and negative symptoms to the positive, psychotic, symptoms might be caused by an additional decrease in the GABA conductance.

Very importantly, it is notable that excitation and inhibition do not cancel each other out as assumed by many models, but have distinct influences on the stochastic dynamics of the network.

16.1.4 Decreased stability produced by reductions of NMDA receptor activated synaptic conductances

We assessed how the stability of both the spontaneous and persistent states changes when NMDA and GABA efficacies are modulated (Loh, Rolls and Deco 2007b). Specifically we ran multiple trial integrate-and-fire network simulations and counted how often the system maintained the spontaneous or persistent state, assessed by the firing rate in the last second of the simulation (2–3 s) of each 3 s trial. Figure 16.2 shows the stability of the spontaneous and persistent attractor relative to the unmodulated reference state (Normal). A negative percentage means that the system was less stable than in the unmodulated state.

A reduction of the NMDA receptor activated synaptic conductances (−NMDA) reduces

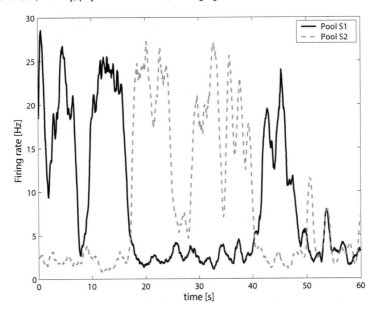

Fig. 16.3 Wandering between attractor states by virtue of statistical fluctuations caused by the randomness of the spiking activity. We simulated a single long trial (60 s) in the spontaneous test condition for the synaptic modification (−NMDA, −GABA). The two curves show the activity of the two selective pools over time smoothed with a 1 s sliding averaging window. The activity moves noisily between the attractor for the spontaneous state and the two persistent states S1 and S2. (After Loh, Rolls and Deco 2007a.)

the stability of the persistent state drastically, while slightly increasing the stability of the spontaneous state (see Fig. 16.2). We hypothesized that this type of change might be related to the cognitive symptoms, since it shows a reduced stability of the working memory properties. A reduction of GABA shows the opposite pattern: a slight reduction in the stability of the spontaneous state, and an increased stability of the persistent (i.e. attractor) state (see Fig. 16.2).

When both NMDA and GABA are reduced one might think that these two counterbalancing effects (excitatory and inhibitory) would either cancel each other out or yield a tradeoff between the stability of the spontaneous and persistent state. However, this is not the case. The stability of both the spontaneous state and the persistent state is reduced (see Figure 16.2). We relate this pattern to the positive symptoms of schizophrenia, in which both the spontaneous and attractor states are shallow, and the system merely jumps by the influence of statistical fluctuations between the different (spontaneous and high firing rate attractor) states.

To investigate more directly the wandering between spontaneous and several different persistent attractor states, we simulated the condition with decreased NMDA and GABA conductances over a long time period in which no cue stimulus input was given (Loh, Rolls and Deco 2007b). Figure 16.3 shows the firing rates of the two selective pools S1 and S2. The high activity switches between the two attractors due to the influence of fluctuations, which corresponds to spontaneous wandering in a shallow energy landscape, corresponding for example to sudden jumps between unrelated cognitive processes. These results are consistent with the flow analysis and demonstrate that the changes in the attractor landscape influence the behaviour at the stochastic level.

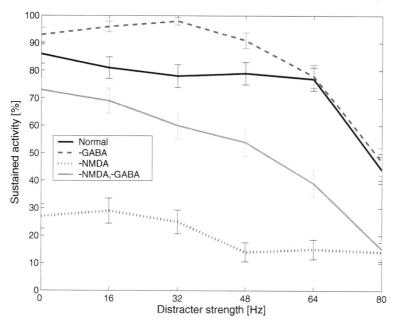

Fig. 16.4 Stability and distractibility as a function of the distracter strength and modulations of the synaptic efficacies. We assessed how often in 100 trials the average activity during the last second (2–3 s) stayed above 10 Hz in the S1 attractor. The modulation of the synapses NMDA:-1 and GABA:-1 corresponds to a reduction of 4.5% and 9%, respectively. The strength of the distracter stimulus applied to S2 is an increase in firing rate above the 2.4 kHz background activity which is distributed among 800 synapses per neuron. The lower the sustained activity in S1 the higher is the distractibility. The standard deviations were approximated with the binomial distribution. (After Loh, Rolls and Deco 2007.)

16.1.5 Increased distractibility produced by reduced NMDA conductances

As distractibility is directly related to the symptoms of schizophrenia, we ran simulations specifically to assess this property using persistent and distracter simulations (see Fig. 5.6 on page 104). A distracter strength of 0 Hz corresponds to the persistent condition described in Section 5.4.2. Figure 16.4 shows the stability and distractibility for reductions of NMDA and GABA currents. The reference state is labelled 'Normal'. In this state, pool S1 continued to maintain its attractor firing without any distracter (distracter strength = 0 Hz) throughout the delay period on almost 90% of the trials. In both conditions which reduce the NMDA current (labelled −NMDA) the network was less and less able to maintain the S1 attractor firing as the distracter stimulus strength was increased through the range 0–80 Hz. The stability of the persistent state was reduced, and also the distractibility was increased, as shown by the fact that increasing distracter currents applied to S2 could move the attractor away from S1 (Loh, Rolls and Deco 2007b).

The implication therefore is that a reduction of the NMDA currents could cause the cognitive symptoms of schizophrenia, by making short-term memory networks less stable and more distractible, thereby reducing the ability to maintain attention. Reducing only the GABA currents (−GABA) reduces the distractibility for low distracter strengths and coincides with the reference (Normal) condition at high values of the distracter strengths.

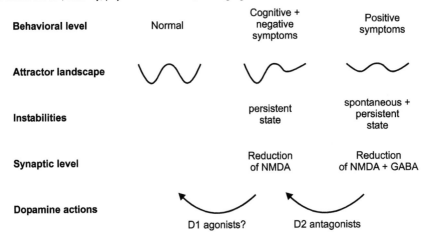

Fig. 16.5 Summary of attractor hypothesis of schizophrenic symptoms and simulation results (see text). The first basin (from the left) in each energy landscape is the spontaneous state, and the second basin is the persistent attractor state. The vertical axis of each landscape is the energy potential. (After Loh, Rolls and Deco 2007b.)

16.1.6 Synthesis: network instability and schizophrenia

We have proposed a hypothesis that relates the cognitive, negative, and positive symptoms of schizophrenia (Liddle 1987, Mueser and McGurk 2004, Owen et al. 2016) to the depth of basins of attraction and to the stability properties of attractor networks caused by statistical fluctuations of spiking neurons (Rolls 2014a, Loh, Rolls and Deco 2007a, Loh, Rolls and Deco 2007b, Rolls 2008d, Rolls, Loh, Deco and Winterer 2008d, Rolls and Deco 2011a, Rolls 2012b). This assumes that some cognitive processes can be understood as dynamical attractor systems, which is an established hypothesis in areas such as working memory, but has also been used many other areas (Rolls 2008d, O'Reilly 2006). Our approach applies this concept to mental illnesses (Bender, Albus, Moller and Tretter 2006). Due to the diversity of the symptoms of schizophrenia, our general hypothesis is meant to serve as a heuristic for how the different kinds of symptoms might arise and are related. We investigated the hypothesis empirically in a computational attractor framework to capture an important aspect of cortical functionality. Figure 16.5 summarizes our hypothesis and its relation to the investigations of stochastic neurodynamics.

The middle column in Fig. 16.5 shows the overview for the cognitive and negative symptoms. The core of the cognitive symptoms is a failure of working memory and attentional mechanisms. Working memory activity is related to the ongoing (i.e. persistent) firing of neurons during the delay period of cognitive tasks (Goldman-Rakic 1994, Goldman-Rakic 1999). This could be implemented by associatively modifiable synapses between the recurrent collateral synapses of cortical pyramidal cells (Rolls and Treves 1998, Durstewitz et al. 2000b, Renart et al. 2001, Wang 2001). We propose that the cognitive symptoms of schizophrenia could arise because the basins of attraction of the persistent states in the prefrontal cortex become too shallow. This leads in combination with the statistical fluctuations due to randomness of the spiking activity to either a fall out of an active working memory state or to a shift to a different attractor state, leading to a failure to maintain attention and thereby impairing executive function. There is reduced spine density on excitatory neurons in for example the prefrontal cortex (Glausier and Lewis 2013, Konopaske, Lange, Coyle and Benes 2014, Glantz and Lewis 2000), and by reducing the excitatory input mediated by NMDA and AMPA receptors (which we modelled by reduced NMDA conductance), this could reduce the firing rates of prefrontal cortex neurons, and thus the stability of the attractor networks related to the cognitive symptoms. The hypofrontality in schizophrenia, that is less activation in frontal brain regions during working memory tasks (Ingvar and Franzen 1974, Carter, Perlstein, Ganguli, Brar, Mintun and Cohen 1998), is in line with our hypothesis, since the firing rates of the persistent state are lower in the reduced NMDA condition (Fig. 16.1), and the system spends on average less time in the persistent state, since it is less stable than in the normal condition (Figure 16.2). In addition, a reduced signal-to-noise ratio has also been identified in imaging studies (Winterer et al. 2000, Winterer et al. 2004, Winterer et al. 2006). Our simulations suggest that a reduction in NMDA conductance at the synaptic level (see Figs. 16.5, 16.13 and 16.14) can account for this phenomenon. This is in line with previous work on the stability of working memory networks (Wang 1999, Durstewitz et al. 2000a, Wang 2001).

A reduction of the NMDA conductance also results in a reduction of the firing rates of the neurons in the persistent state (see Figs. 16.1, 16.13 and 16.14, and Brunel and Wang (2001)). We relate this, following Rolls (2005), to the negative symptoms which include flattening of affect, a reduction in emotion, emotional and social withdrawal, poor rapport, passive withdrawal, lack of spontaneity, motor retardation, apathy, and disturbance of motivation. These symptoms are related to decreases in activity in the orbitofrontal cortex and/or anterior cingulate cortex (Wolkin et al. 1992, Aleman and Kahn 2005), both of which are implicated in emotion (Rolls 2005, Rolls 2006b, Rolls and Grabenhorst 2008, Rolls 2014a). The emotional states represented in the orbitofrontal cortex and anterior cingulate cortex include states elicited both by rewards and punishers. Our hypothesis is that both would be reduced by the mechanism described. Correspondingly, motivation would be reduced in the same way, in that motivation is a state in which we work to obtain goals (rewards) or avoid punishers (Rolls 2014a, Rolls 2016c). The reduced spine density on excitatory neurons in the prefrontal cortex (Glausier and Lewis 2013, Konopaske et al. 2014, Glantz and Lewis 2000), if present in the orbitofrontal and anterior cingulate cortex, could reduce the firing rates of these cortical neurons, and thus produce the negative symptoms.

Both the negative and cognitive symptoms thus could be caused by a reduction of the NMDA conductance (simulating for example decreased spine density) in attractor networks. The proposed mechanism links the cognitive and negative symptoms of schizophrenia in an attractor framework and is consistent with a close relation between the cognitive and negative symptoms: blockade of NMDA receptors by dissociative anesthetics such as ketamine produces in normal subjects schizophrenic symptoms including both negative and cognitive impairments (Malhotra, Pinals, Weingartner, Sirocco, Missar, Pickar and Breier 1996, Newcomer, Farber, Jevtovic-Todorovic, Selke, Melson, Hershey, Craft and Olney 1999); agents that enhance NMDA receptor function reduce the negative symptoms and improve the cognitive abilities of schizophrenic patients (Goff and Coyle 2001, Coyle 2012, Balu and Coyle 2015); and the cognitive and negative symptoms occur early in the illness and precede the first episode of positive symptoms (Lieberman et al. 2001, Hafner et al. 2003, Mueser and McGurk 2004). Consistent with this hypothesized role of a reduction in NMDA conductances being involved in schizophrenia, postmortem studies of schizophrenia have identified abnormalities in glutamate receptor density in regions such as the prefrontal cortex, thalamus and the temporal lobe (Goff and Coyle 2001, Coyle 2012, Owen et al. 2016), brain areas that are active during the performance of cognitive tasks.

The dopamine D1 receptor has been shown to modulate the performance of working memory tasks (Sawaguchi and Goldman-Rakic 1991, Sawaguchi and Goldman-Rakic 1994, Goldman-Rakic 1999, Castner, Williams and Goldman-Rakic 2000). An increase in D1 receptor activation has been shown to increase the NMDA current (Durstewitz and Seamans 2002, Seamans and Yang 2004), and modelling studies have shown that this increase is related to the stability of working memory states (Durstewitz et al. 1999, Durstewitz et al. 2000a, Brunel and Wang 2001). Imaging data also support the importance of the D1 receptor in schizophrenia (Okubo, Suhara, Sudo and Toru 1997a, Okubo, Suhara, Suzuki, Kobayashi, Inoue, Terasaki, Someya, Sassa, Sudo, Matsushima, Iyo, Tateno and Toru 1997b). We therefore suggest that an increased activation of D1 receptors might alleviate both the cognitive and the negative symptoms of schizophrenia (Goldman-Rakic, Castner, Svensson, Siever and Williams 2004, Miyamoto, Duncan, Marx and Lieberman 2005), by increasing NMDA receptor mediated synaptic currents (Fig. 16.5). Atypical neuroleptics might use this mechanism by not only blocking D2 receptors, but also by increasing the presynaptic release of dopamine which in turn would increase the activation of the extrasynaptic D1 receptors (Castner et al. 2000, Moller 2005).

Taken together, we suggest that the cognitive and negative symptoms could be caused by the same synaptic mechanism, namely a reduction in the NMDA conductance (or reduced excitatory input to glutamate neurons), which reduces the stability and increases the distractibility of the persistent attractors, and reduces the activity (firing rates) of neurons (Fig. 16.5, middle column). The reduced depth of the basins of attraction can be understood in the following way. Hopfield (1982) showed that the recall state in an attractor network can be thought of as the local minimum in an energy landscape, where the energy would be defined as shown in equation B.12 on page 726. In general neuronal systems do not admit an energy function. Nevertheless, we can assume an effective energy function: in fact, the flow picture shown in Fig. 16.1 resulting from the mean-field reduction associated with the spiking network analyzed here, can be viewed as an indirect description of an underlying effective energy function. From equation B.12, it follows that the depth of a basin of attraction is deeper if the firing rates are higher and if the synaptic strengths that couple the neurons that are part of the same attractor are strong. (The negative sign results in a low energy, and thus a stable state, if the firing rates of the neurons in the same attractor and their synaptic coupling weights are high.) If we reduce the NMDA receptor activated channel conductances, then the depth of the basins of attraction will be reduced both because the firing rates are reduced by reducing excitatory inputs to the neurons, and because the synaptic coupling weights are effectively reduced because the synapses can pass only reduced currents.

The positive symptoms (Fig. 16.5, right column) of schizophrenia include delusions, hallucinations, thought disorder, and bizarre behaviour. Examples of delusions are beliefs that others are trying to harm the person, impressions that others control the person's thoughts, and delusions of grandeur. Hallucinations are perceptual experiences, which are not shared by others, and are frequently auditory but can affect any sensory modality. These symptoms may be related to activity in the temporal lobes (Liddle 1987, Epstein, Stern and Silbersweig 1999, Mueser and McGurk 2004). The attractor framework approach taken here hypothesizes that the basins of attraction of both spontaneous and persistent states are shallow (Fig. 16.5). Due to the shallowness of the spontaneous state, the system can jump spontaneously up to a high activity state causing hallucinations to arise and leading to bizarre thoughts and associations. This might be the cause for the higher activations in schizophrenics in temporal lobe areas which are identified in imaging experiments (Shergill et al. 2000, Scheuerecker et al. 2008).

We relate the positive symptoms to not only a reduction in NMDA conductance, but especially also to a reduction in what we modelled as GABA conductance, but more generally to any reduction of inhibitory neuron efficacy that might lead to higher firing rates of cortical excitatory neurons. This is consistent with the fact that the positive symptoms usually follow the cognitive and negative ones and represent a qualitative worsening of the illness (Mueser and McGurk 2004). Alterations in GABA receptors have been identified in schizophrenia (Lewis 2014, Gonzalez-Burgos, Hashimoto and Lewis 2010, Gonzalez-Burgos and Lewis 2012).

D2 receptor antagonism remains a main target for antipsychotics (Seeman and Kapur 2000,

Leuner and Muller 2006). Dopamine receptor D2 antagonists mainly alleviate the positive symptoms of schizophrenia, whereas the cognitive and negative symptoms persist, especially for the typical neuroleptics (Mueser and McGurk 2004, Owen et al. 2016). Together with the simulations, our hypothesis suggests that an increase in the GABA current in the state corresponding to the positive symptoms ($-$NMDA, $-$GABA) might have the same effect as D2 antagonists. The therapeutic effect of D2 antagonists might thus be caused by an increase in GABA currents. Indeed, it has been found that D2 receptors decrease the efficacy of the GABA system (Seamans, Gorelova, Durstewitz and Yang 2001, Trantham-Davidson, Neely, Lavin and Seamans 2004). (For example, Seamans et al. (2001) found that the application of D2 antagonists prevented a decrease in eIPSC amplitude produced by dopamine.) Thus D2 antagonists would, in a hypersensitive D2 receptor state (Seeman, Weinshenker, Quirion, Srivastava, Bhardwaj, Grandy, Premont, Sotnikova, Boksa, El-Ghundi, O'Dowd, George, Perreault, Mannisto, Robinson, Palmiter and Tallerico 2005, Seeman, Schwarz, Chen, Szechtman, Perreault, McKnight, Roder, Quirion, Boksa, Srivastava, Yanai, Weinshenker and Sumiyoshi 2006), increase GABA inhibition in the network, and this we suggest could increase the stability of attractor networks involved in the positive symptoms of schizophrenia, and thus ameliorate the positive symptoms. Since the concentration of dopamine in the cortex depends on cortical-subcortical interactions (Carlsson 2006), the causes of the described changes could also result from subcortical deficits. A detailed analysis of these feedback loops would require specific modelling.

Earlier accounts of the relation of dopamine and schizophrenia in the cortex (Seamans et al. 2001, Seamans and Yang 2004) have suggested two distinct states of dopamine modulation. One is a D2-receptor-dominated state in which there is weak gating and information can easily affect network activity. The other is a D1-receptor-dominated state in which network activity is stable and maintained. We have proposed a more detailed account for stability and discussed this separately for the spontaneous and persistent attractor states. This allows us to account for the dichotomy between the cognitive/negative and positive symptoms. We emphasize that, in biophysically realistic network simulations, excitation and inhibition are not merely antagonistic but implement different functions in the network dynamics. Thereby our stochastic dynamics modeling approach (Loh, Rolls and Deco 2007a, Loh, Rolls and Deco 2007b, Rolls, Loh, Deco and Winterer 2008d) provides a missing link between the symptoms of schizophrenia and network models of working memory and dopamine (Durstewitz et al. 1999, Durstewitz et al. 2000a, Brunel and Wang 2001).

We considered earlier a possible cause for the proposed alterations of the attractor landscape related to schizophrenia, namely changes in NMDA and GABA conductance as these are directly related to schizophrenia (Coyle 2012, Lewis 2014). We did not investigate changes in AMPA conductance. In this particular model the contribution of the AMPA current is relatively small (Brunel and Wang 2001). A more detailed investigation could also include AMPA conductance, especially because it is known to be influenced by NMDA synaptic plasticity (Bagal, Kao, Tang and Thompson 2005). Indeed, if reduced NMDA currents led in turn by synaptic plasticity to reduced AMPA currents, this would amplify the effects we describe.

The proposed alterations in the attractor landscape could have a variety of causes at the neurobiological level: abnormalities in glutamate and GABA receptors and signalling, modulations in synaptic plasticity, alterations in connectivity including reduced spine density, aberrant dopamine signalling, reduced neuropil, genetic mechanisms, and brain volume reduction (Goldman-Rakic 1999, Winterer and Weinberger 2004, Mueser and McGurk 2004, Stephan, Baldeweg and Friston 2006, Owen et al. 2016). Besides cortical mechanisms, cortical-subcortical dynamics could also cause the proposed alterations in the cortical attractor landscape for example via neuromodulatory influences such as dopamine or serotonin or cortical-subcortical feedback loops (Capuano, Crosby and Lloyd 2002, Carlsson 2006). Our general hypothesis regarding the attractor landscape is meant to describe the aberrant dynamics in cortical regions which could be caused by several pathways.

Future work could analyze further how changes of different factors such as regional differences, subcortical-cortical networks or even more detailed neural and synaptic models might influence the stability of the type of neurodynamical system described here. Moreover, future work could investigate within the framework we describe the effect of treatment with particular combinations of drugs designed to facilitate both glutamate transmission (including modulation with glycine or serine of the NMDA receptor) and GABA effects. Drug combinations identified in this way could be useful to explore, as combinations might work where treatment with a single drug on its own may not be effective, and possibilities can be systematically explored with the approach described here.

16.2 Stochastic noise, attractor dynamics, and obsessive-compulsive disorder

16.2.1 Introduction

Obsessive-compulsive disorder (OCD) is a chronically debilitating disorder with a lifetime prevalence of 2–3% (Robins, Helzer, Weissman, Orvaschel, Gruenberg, Burke and Regier 1984, Karno, Golding, Sorenson and Burnam 1988, Weissman, Bland, Canino, Greenwald, Hwu, Lee, Newman, Oakley-Browne, Rubio-Stipec, Wickramaratne et al. 1994, Pauls, Abramovitch, Rauch and Geller 2014, Stein, Kogan, Atmaca, Fineberg, Fontenelle, Grant, Matsunaga, Reddy, Simpson, Thomsen, van den Heuvel, Veale, Woods and Reed 2016). It is characterized by two sets of symptoms, obsessive and compulsive. Obsessions are unwanted, intrusive, recurrent thoughts or impulses that are often concerned with themes of contamination and 'germs', checking household items in case of fire or burglary, order and symmetry of objects, or fears of harming oneself or others. Compulsions are ritualistic, repetitive behaviours or mental acts carried out in relation to these obsessions e.g. washing, household safety checks, counting, rearrangement of objects in symmetrical arrays or constant checking of oneself and others to ensure no harm has occurred (Menzies, Chamberlain, Laird, Thelen, Sahakian and Bullmore 2008, Stein et al. 2016). Patients with OCD experience the persistent intrusion of thoughts that they generally perceive as foreign and irrational but which cannot be dismissed. The anxiety associated with these unwanted and disturbing thoughts can be extremely intense; it is often described as a feeling that something is incomplete or wrong, or that terrible consequences will ensue if specific actions are not taken. Many patients engage in repetitive, compulsive behaviours that aim to discharge the anxieties associated with these obsessional thoughts. Severely affected patients can spend many hours each day in their obsessional thinking and resultant compulsive behaviours, leading to marked disability (Pittenger, Krystal and Coric 2006, Stein et al. 2016). While patients with OCD and related disorders exhibit a wide variety of obsessions and compulsions, the symptoms tend to fall into specific clusters. Common patterns include obsessions of contamination, with accompanying cleaning compulsions; obsessions with symmetry or order, with accompanying ordering behaviours; obsessions of saving, with accompanying hoarding; somatic obsessions; aggressive obsessions with checking compulsions; and sexual and religious obsessions (Pittenger et al. 2006, Stein et al. 2016).

In this section we describe a theory of how obsessive-compulsive disorders arise, and of the different symptoms. The theory (Rolls, Loh and Deco 2008c, Rolls 2012b) is based on the top-down proposal that there is overstability of attractor neuronal networks in cortical and related areas in obsessive-compulsive disorders. The approach is top-down in that it starts with the set of symptoms and maps them onto the dynamical systems framework, and only after this considers detailed underlying biological mechanisms, of which there could be many, that might produce the effects. (In contrast, a complementary bottom-up approach starts from detailed neurobiological mechanisms, and aims to interpret their implications with a brain-like model for higher level phenomena.) We show by integrate-and-fire neuronal network simulations that the overstability could arise by for example overactivity in glutamatergic excitatory neurotransmitter synapses, which produces an increased depth of the basins of attraction, in the presence of which neuronal spiking-related and potentially other noise is insufficient to help the system move out of an attractor basin. We relate this top-down proposal, related to the stochastic dynamics of neuronal networks, to new evidence that there may be overactivity in glutamatergic systems in obsessive-compulsive disorders, and consider the implications for treatment.

The background is that there has been interest for some time in the application of complex systems theory to understanding brain function and behaviour (Riley and Turvey 2001, Peled 2004, Heinrichs 2005, Lewis 2005, Rolls and Deco 2010). Dynamically realistic neuronal network models of working memory (Wang 1999, Compte, Brunel, Goldman-Rakic and Wang 2000, Durstewitz et al. 2000a, Durstewitz et al. 2000b, Brunel and Wang 2001, Wang 2001, Rolls and Deco 2010), decision-making (Wang 2002, Deco and Rolls 2006, Rolls and Deco 2010, Rolls 2014a), and attention (Rolls and Deco 2002, Deco and Rolls 2005a, Deco and Rolls 2005b, Rolls 2008d), and how they are influenced by neuromodulators such as dopamine (Durstewitz et al. 1999, Durstewitz et al. 2000a, Brunel and Wang 2001, Durstewitz and Seamans 2002, Deco and Rolls 2003) have been produced, and provide a foundation for the approach developed by Rolls, Loh and Deco (2008c).

16.2.2 A hypothesis about the increased stability of attractor networks and the symptoms of obsessive-compulsive disorder

We hypothesized (Rolls, Loh and Deco 2008c) that cortical and related attractor networks become too stable in obsessive-compulsive disorder, so that once in an attractor state, the networks tend to remain there too long (Rolls, Loh and Deco 2008c, Rolls 2012b). The hypothesis is that the depths of the basins of attraction become deeper, and that this is what makes the attractor networks more stable. We further hypothesize that part of the mechanism for the increased depth of the basins of attraction is increased glutamatergic transmission, which increases the depth of the basins of attraction by increasing the firing rates of the neurons, and by increasing the effective value of the synaptic weights between the associatively modified synapses that define the attractor, as is made evident in equation B.12. The synaptic strength is effectively increased if more glutamate is released per action potential at the synapse, or if in other ways the currents injected into the neurons through the NMDA (N-methyl-d-aspartate) and/or AMPA synapses are larger. In addition, if NMDA receptor function is increased, this could also increase the stability of the system because of the temporal smoothing effect of the long time constant of the NMDA receptors (Wang 1999). This increased stability of cortical and related attractor networks, and the associated higher neuronal firing rates, could occur in different brain regions, and thereby produce different symptoms, as follows. If these effects occurred in high order motor areas, the symptoms could include inability to move out of one motor pattern, resulting for example in repeated movements or actions. In parts of the cingulate cortex and dorsal medial prefrontal cortex, this could result in difficulty in switching between actions or strategies (Rushworth, Behrens, Rudebeck and Walton 2007a, Rushworth, Buckley, Behrens, Walton and Bannerman 2007b), as the system would be locked into one action or strategy. If an action was locked into a high order motor area due to increased stability of an attractor network, then lower order motor areas might thereby not be able to escape easily what they implement, such as a sequence of movements, so that the sequence would be repeated.

A similar account, of becoming locked in one action and having difficulty in switching to another action, can be provided for response inhibition deficits, which have been found in OCD. The response inhibition deficit has been found in tasks such as go/no-go and stop-signal reaction time (SSRT) which examine motor inhibitory processes, and also the Stroop task, a putative test of cognitive inhibition (Hartston and Swerdlow 1999, Bannon, Gonsalvez, Croft and Boyce 2002, Penades, Catalan, Andres, Salamero and Gasto 2005, Bannon, Gonsalvez, Croft and Boyce 2006, Chamberlain, Fineberg, Blackwell, Robbins and Sahakian 2006, Chamberlain, Fineberg, Menzies, Blackwell, Bullmore, Robbins and Sahakian 2007, Penades, Catalan, Rubia, Andres, Salamero and Gasto 2007). For example, response inhibition deficits have been reported in OCD patients when performing the SSRT, which measures the time taken to internally suppress pre-potent motor responses (Chamberlain et al. 2006). Unaffected first-degree relatives of OCD patients are also impaired on this task compared with unrelated healthy controls, suggesting that response inhibition may be an endophenotype (or intermediate phenotype) for OCD (Chamberlain et al. 2007, Menzies et al. 2008).

If occurring in the lateral prefrontal cortex (including the dorsolateral and ventrolateral parts), the increased stability of attractor networks could produce symptoms that include a difficulty in shifting attention and in cognitive set shifting. These are in fact important symptoms that can be found in obsessive-compulsive disorder (Menzies et al. 2008). Two different forms of shifting have been investigated: affective set shifting, where the affective or reward value of a stimulus changes over time (e.g. a rewarded stimulus is no longer rewarded) (intradimensional or ID set shifting); and attentional set shifting, where the stimulus dimension (e.g. shapes or colors) to which the subject must attend is changed (extradimensional or ED set shifting). Deficits of attentional set shifting in OCD have been found in several neurocognitive studies using the CANTAB ID/ED set shifting task (Veale, Sahakian, Owen and Marks 1996, Watkins, Sahakian, Robertson, Veale, Rogers, Pickard, Aitken and Robbins 2005, Chamberlain et al. 2006, Chamberlain et al. 2007). This deficit is most consistently reported at the ED stage (in which the stimulus dimension, e.g. shape, color or number, alters and subjects have to inhibit their attention to this dimension and attend to a new, previously irrelevant dimension). The ED stage is analogous to the stage in the Wisconsin Card Sorting Task where a previously correct rule for card sorting is changed and the subject has to respond to the new rule (Berg 1948). This ED shift impairment in OCD patients is considered to reflect a lack of cognitive or attentional flexibility and may be related to the repetitive nature of OCD symptoms and behaviours. Deficits in attentional set shifting are considered to be more dependent upon dorsolateral and ventrolateral prefrontal regions than the orbital prefrontal regions included in the orbitofronto-striatal model of OCD (Pantelis, Barber, Barnes, Nelson, Owen and Robbins 1999, Rogers, Andrews, Grasby, Brooks and Robbins 2000, Nagahama, Okada, Katsumi, Hayashi, Yamauchi, Oyanagi, Konishi, Fukuyama and Shibasaki 2001, Hampshire and Owen 2006), suggesting that cognitive deficits in OCD may not be underpinned exclusively by orbitofrontal cortex pathology. Indeed, intradimensional or affective set shifting may not be consistently impaired in OCD (Menzies et al. 2008).

Planning may also be impaired in patients with OCD (Menzies et al. 2008), and this could arise because there is too much stability of attractor networks in the dorsolateral prefrontal cortex concerned with holding in mind the different short-term memory representations that encode the different steps of a plan (Rolls 2008d). Indeed, there is evidence for dorsolateral prefrontal cortex (DLPFC) dysfunction in patients with OCD, in conjunction with impairment on a version of the Tower of London, a task often used to probe planning aspects of executive function (van den Heuvel, Veltman, Groenewegen, Cath, van Balkom, van Hartskamp, Barkhof and van Dyck 2005). Impairment on the Tower of London task has also been demonstrated in healthy first-degree relatives of OCD patients (Delorme, Gousse, Roy, Trandafir, Mathieu, Mouren-Simeoni, Betancur and Leboyer 2007).

An increased firing rate of neurons in the orbitofrontal cortex, and anterior cingulate cortex, produced by hyperactivity of glutamatergic transmitter systems, would increase emotionality, which is frequently found in obsessive-compulsive disorder. Part of the increased anxiety found in obsessive-compulsive disorder could be related to an inability to complete tasks or actions in which one is locked. But part of our unifying proposal is that part of the increased emotionality in OCD may be directly related to increased firing produced by the increased glutamatergic activity in brain areas such as the orbitofrontal and anterior cingulate cortex. The orbitofrontal cortex and anterior cingulate cortex are involved in emotion, in that they are activated by primary and secondary reinforcers that produce affective states (Rolls 2004a, Rolls 2014a, Rolls 2008d, Rolls and Grabenhorst 2008), and in that damage to these regions alters emotional behaviour and emotional experience (Rolls, Hornak, Wade and McGrath 1994a, Hornak, Rolls and Wade 1996, Hornak, Bramham, Rolls, Morris, O'Doherty, Bullock and Polkey 2003, Hornak, O'Doherty, Bramham, Rolls, Morris, Bullock and Polkey 2004, Berlin, Rolls and Kischka 2004, Berlin, Rolls and Iversen 2005). Indeed, negative emotions as well as positive emotions activate the orbitofrontal cortex, with the emotional states produced by negative events tending to be represented in the lateral orbitofrontal cortex and dorsal part of the anterior cingulate cortex (Kringelbach and Rolls 2004, Rolls 2014a, Rolls 2008d, Rolls 2009a, Grabenhorst and Rolls 2011). We may note that stimulus-reinforcer reversal tasks (also known as intra-dimensional shifts or affective reversal) are not generally impaired in patients with OCD (Menzies et al. 2008), and this is as predicted, for the machinery for the reversal including the detection of non-reward (Rolls 2014a) (Chapter 15) is present even if the system is hyperglutamatergic.

If the increased stability of attractor networks occurred in temporal lobe semantic memory networks, then this would result in a difficulty in moving from one thought to another, and possibly in stereotyped thoughts, which again may be a symptom of obsessive-compulsive disorder (Menzies et al. 2008). The obsessional states are thus proposed to arise because cortical areas concerned with cognitive functions have states that become too stable. The compulsive states are proposed to arise partly in response to the obsessional states, but also partly because cortical areas concerned with actions have states that become too stable. The theory provides a unifying computational account of both the obsessional and compulsive symptoms, in that both arise due to increased stability of cortical attractor networks, with the different symptoms related to overstability in different cortical areas. The theory is also unifying in that a similar increase in glutamatergic activity in the orbitofrontal and anterior cingulate cortex could increase emotionality, as described earlier.

16.2.3 Glutamate and increased depth of the basins of attraction of attractor networks

To demonstrate how alterations of glutamate as a transmitter for the connections between the neurons may influence the stability of attractor networks, Rolls, Loh and Deco (2008c) performed integrate-and-fire simulations of the type described in Section 4.3.1, Chapter 6 and Section 16.1. A feature of these simulations is that we simulated the currents produced by activation of NMDA and AMPA receptors in the recurrent collateral synapses, and took into account the effects of the spiking-related noise, which is an important factor in determining whether the attractor stays in a basin of attraction, or jumps over an energy barrier into another basin (Loh, Rolls and Deco 2007a, Rolls, Loh, Deco and Winterer 2008d, Deco, Rolls and Romo 2009). For our investigations, we selected $w_+ = 2.1$, which with the default values of the NMDA and GABA conductances yielded stable dynamics, that is, a stable spontaneous

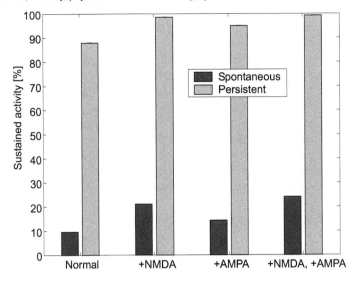

Fig. 16.6 The stability of the spontaneous and persistent attractor states. The percent (%) of the simulation runs on which the network during the last second of the 3 s simulation was in the high firing rate attractor state is shown on the ordinate. For the persistent run simulations, in which the cue triggered the attractor into the high firing rate attractor state, the network was still in the high firing rate attractor state in the baseline condition on approximately 88% of the runs, and this increased to nearly 100% when the NMDA conductances were increased by 3% (+NMDA). The effect was highly significant, with the means ± sem shown. Increasing AMPA by 10% (+AMPA) could also increase the stability of the persistent high firing rate attractor state, as did the combination +NMDA +AMPA. For the spontaneous state simulations, in the baseline condition the spontaneous state was unstable on approximately 10% of the trials, that is, on 10% of the trials the spiking noise in the network caused the network run in the condition without any initial retrieval cue to end up in a high firing rate attractor state. In the +NMDA condition, the spontaneous state had jumped to the high firing rate attractor state on 25% of the runs, that is the low firing rate spontaneous state was present at the end of a simulation on only approximately 75% of the runs. The +AMPA can also make the spontaneous state more likely to jump to a persistent high firing rate attractor state, as can the combination +NMDA +AMPA. (After Rolls, Loh and Deco 2008a.)

state if no retrieval cue was applied, and a stable state of persistent firing after a retrieval cue had been applied and removed.

To investigate the effects of changes (modulations) in the NMDA, AMPA and GABA conductances, we chose for demonstration purposes increases of 3% for the NMDA, and 10% for the AMPA and GABA synapses between the neurons in the network shown in Fig. 5.5, as these were found to be sufficient to alter the stability of the attractor network. Fig. 16.6 shows the stability of the spontaneous and persistent attractor. We plot the percentage of the simulation runs on which the network during the last second of the simulation was in the high firing rate attractor state. Fig. 16.6 shows that for the persistent run simulations, in which the cue triggered the attractor into the high firing rate attractor state, the network was still in the high firing rate attractor state in the baseline condition on approximately 88% of the runs, and that this had increased to 98% when the NMDA conductances were increased by 3% (+NMDA). Thus increasing the NMDA receptor-activated synaptic currents increased the stability of the network. Fig. 16.6 also shows that increasing AMPA by 10% (+AMPA) could also increase the stability of the persistent high firing rate attractor state, as did the combination +NMDA +AMPA.

Fig. 16.6 shows that in the baseline condition the spontaneous state was unstable on approximately 10% of the trials, that is, on 10% of the trials the spiking noise in the network

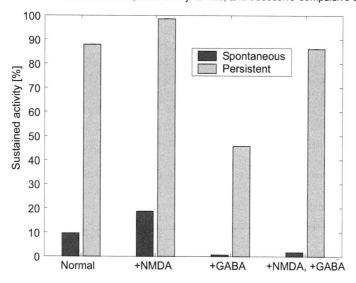

Fig. 16.7 The effect of increasing GABA-receptor mediated synaptic conductances by 10% (+GABA) on the stability of the network. Conventions as in Fig. 16.6. Increasing the GABA currents by 10% when the NMDA currents are increased by 3% (+NMDA +GABA) moved the persistent state away from the overstability produced by +NMDA alone, and returned the persistent state to the normal baseline level. That is, instead of the system ending up in the high firing rate attractor state in the persistent state simulations on 98% of the runs in the +NMDA condition, the system ended up in the high firing rate attractor state on approximately 88% of the runs in the +NMDA +GABA condition. The combination +NMDA +GABA produced a spontaneous state that was less likely than the normal state to jump to a high firing rate attractor (see text). (After Rolls, Loh and Deco 2008a.)

caused the network run in the condition without any initial retrieval cue to end up in a high firing rate attractor state. This is of course an error that is related to the spiking noise in the network. In the +NMDA condition, the spontaneous state had jumped to the high firing rate attractor state on approximately 22% of the runs, that is the low firing rate spontaneous state was present at the end of a simulation on only approximately 78% of the runs. Thus increasing NMDA receptor activated currents can contribute to the network jumping from what should be a quiescent state of spontaneous activity into a high firing rate attractor state. We relate this to the symptoms of obsessive-compulsive disorders, in that the system can jump into a state with a dominant memory (which might be an idea or concern or action) even when there is no initiating input. Fig. 16.6 also shows that +AMPA can make the spontaneous state more likely to jump to a persistent high firing rate attractor state, as can the combination +NMDA +AMPA.

We next investigated to what extent alterations of the GABA-receptor mediated inhibition in the network could restore the system towards more normal activity even when NMDA conductances were high (Rolls, Loh and Deco 2008c). Fig. 16.7 shows that increasing the GABA currents by 10% when the NMDA currents are increased by 3% (+NMDA +GABA) can move the persistent state away from overstability back to the normal baseline state. That is, instead of the system ending up in the high firing rate attractor state in the persistent state simulations on 98% of the runs, the system ended up in the high firing rate attractor state on approximately 88% of the runs, the baseline level. Increasing GABA has a large effect on the stability of the spontaneous state, making it less likely to jump to a high firing rate attractor state. The combination +NMDA +GABA produced a spontaneous state in which the +GABA overcorrected for the effect of +NMDA. That is, in the +NMDA +GABA condition the network was very likely to stay in the spontaneous firing rate condition in which it was started, or, equivalently, when tested in the spontaneous condition, the network was less likely than normally to jump to a high firing rate attractor. Increasing GABA thus corrected for the effect of increasing NMDA receptor activated synaptic currents on the persistent type of run when there was an initiating stimulus; and overcorrected for the effect of increasing NMDA on the spontaneous state simulations when there was no initiating retrieval stimulus, and the network should remain in the low firing rate state until the end of the simulation run.

The implications for symptoms are that agents that increase GABA conductances might reduce and normalize the tendency to remain locked into an idea or concern or action; and would make it much less likely that the quiescent resting state would be left by jumping because of the noisy spiking towards a state representing a dominant idea or concern or action. The effects of increasing GABA receptor activated currents alone was to make the persistent simulations less stable (less likely to end in a high firing rate state), and the spontaneous simulations to be more stable (more likely to end up in the spontaneous state).

We next investigated how an increase of NMDA currents might make the system less distractible, and might make it overstable in remaining in an attractor. This was investigated as shown in Fig. 5.6 by setting up a system with two high firing rate attractors, S1 and S2, then starting the network in an S1 attractor state with S1 applied at t=0–0.5 s, and then applying a distracter S2 at time t=1.0–1.5 s to investigate how strong S2 had to be to distract the network out of its S1 attractor. We assessed how often in 200 trials the average activity during the last second (2 s–3 s) stayed above 10 Hz in the S1 attractor (percentage sustained activity shown on the ordinate of Fig. 16.8). The strength of the distracter stimulus applied to S2 was an increase in firing rate above the 2.4 kHz background activity which is distributed among 800 synapses per neuron.

Fig. 16.8 shows that increasing the NMDA receptor activated currents by 5% (+NMDA) means that larger distracter currents must be applied to S2 to move the system away from S1 to S2. That is, the +NMDA condition makes the system more stable in its high firing rate attractor, and less able to be moved to another state by another stimulus (in this case S2). We relate this to the symptoms of obsessive-compulsive disorder, in that once in an attractor state (which might reflect an idea or concern or action), it is very difficult to get the system to move to another state. Increasing AMPA receptor activated synaptic currents (by 10%, +AMPA) produces similar, but smaller, effects.

Fig. 16.9 shows that increasing GABA activated synaptic conductances (by 10%, +GABA) can partly normalize the overstability and decrease of distractibility that is produced by elevating NMDA receptor activated synaptic conductances (by 3%, +NMDA). The investigation and conventions used the same protocol as for the simulations shown in Fig. 16.8. We assessed how often in 200 trials the average activity during the last second (2 s–3 s) stayed above 10 Hz in the S1 attractor, instead of being distracted by the S2 distracter. The strength of the distracter stimulus applied to S2 is an increase in firing rate above the 2.4 kHz background activity which is distributed among 800 synapses per neuron. The higher the sustained activity is in S1 the higher is the stability to S1 and the lower is the distractibility by S2. What was found (as shown in Fig. 16.9) is that the increase of GABA was able to move the curve in part back towards the normal state away from the overstable and less distractible state produced by increasing NMDA receptor activated synaptic currents. The effect was particularly evident at low distracter currents.

16.2.4 Synthesis on obsessive-compulsive disorder

This provides a new approach to the symptoms of obsessive-compulsive disorder, for it deals with the symptoms in terms of overstability of attractor networks in the cerebral cortex (Rolls,

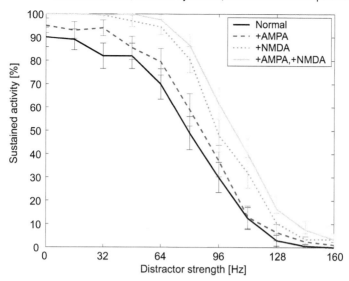

Fig. 16.8 Stability and distractibility as a function of the distracter strength and the synaptic efficacies. We assessed how often in 200 trials the average activity during the last second (2-3 s) stayed above 10 Hz in the S1 attractor, instead of being distracted by the S2 distracter. The modulation of the synapses +NMDA and +AMPA corresponds to an increase of 5% and 10%, respectively. The strength of the distracter stimulus applied to S2 is an increase in firing rate above the 2.4 kHz background activity which is distributed among 800 synapses per neuron. The higher the sustained activity is in S1 the higher is the stability and the lower is the distractibility by S2. The standard deviations were approximated with the binomial distribution. (After Rolls, Loh and Deco 2008a.)

Loh and Deco 2008c, Rolls 2012b). If the same generic change in stability were produced in different cortical areas, then we have indicated how different symptoms might arise. Of course, if these changes were more evident in some areas than in others in different patients, this would help to account for the different symptoms in different patients. Having proposed a generic hypothesis for the disorder, we recognize of course that the exact symptoms that arise if stability in some systems is increased will be subject to the exact effects that these will have in an individual patient, who may react to these effects, and produce explanatory accounts for the effects, and ways to deal with them, that may be quite different from individual to individual.

The integrate-and-fire simulations show that an increase of NMDA or AMPA synaptic currents can increase the stability of attractor networks. They can become so stable that the intrinsic stochastic noise caused by the spiking of the neurons is much less effective in moving the system from one state to another, whether spontaneously in the absence of a distracting, that is new, stimulus, or in the presence of a new (distracting) stimulus that would normally move the system from one attractor state to another.

The simulations also show that the stability of the spontaneous (quiescent) firing rate state is reduced by increasing NMDA or AMPA receptor activated synaptic currents. We relate this to the symptoms of obsessive-compulsive disorder in that the system is more likely than normally, under the influence of the spiking stochastic noise caused by the neuronal spiking in the system, to jump into one of the dominant attractor states, which might be a recurring idea or concern or action.

An implication is that increasing the inhibition in the cortical system, for example by increasing GABA receptor activated synaptic currents, might be useful, both by bringing the system from a state where it was locked into an attractor back to the normal level, and by

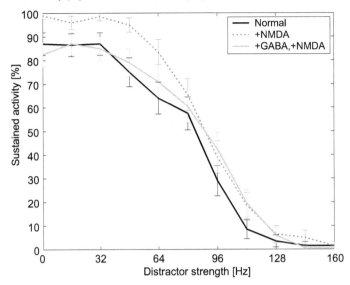

Fig. 16.9 Stability and distractibility: the effect of increasing GABA by 10% (+GABA) on the overstability to S1 / decrease in distractibility by S2 produced by increasing NMDA synaptic conductances by 3% (+NMDA). We assessed how often in 200 trials the average activity during the last second (2 s–3 s) stayed above 10 Hz in the S1 attractor, instead of being distracted to the S2 attractor state by the S2 distracter. The modulation of the synapses +NMDA and +GABA corresponds to an increase of 3% and 10%, respectively. The strength of the distracter stimulus applied to S2 is an increase in firing rate above the 2.4 kHz background activity which is distributed among 800 synapses per neuron. The higher the sustained activity is in S1 the higher is the stability to S1 and the lower is the distractibility by S2. The standard deviations were approximated with the binomial distribution. (After Rolls, Loh and Deco 2008a.)

making the spontaneous state more stable, so that it would be less likely to jump to an attractor state (which might represent a dominant idea or concern or action) (see Fig. 16.7).

This simulation evidence, that an increase of glutamatergic synaptic efficacy can increase the stability of attractor networks and thus potentially provide an account for some of the symptoms of obsessive-compulsive disorder, is consistent with evidence that glutamatergic function may be increased in some brain systems in obsessive-compulsive disorder (Rosenberg, MacMaster, Keshavan, Fitzgerald, Stewart and Moore 2000, Rosenberg, MacMillan and Moore 2001, Rosenberg, Mirza, Russell, Tang, Smith, Banerjee, Bhandari, Rose, Ivey, Boyd and Moore 2004, Pittenger et al. 2006, Rolls 2012b, Pittenger, Bloch and Williams 2011, Pittenger 2015) and that cerebro-spinal-fluid glutamate levels are elevated (Chakrabarty, Bhattacharyya, Christopher and Khanna 2005). Consistent with this, agents with antiglutamatergic activity such as riluzole, which can decrease glutamate transmitter release, may be efficacious in obsessive-compulsive disorder (Pittenger et al. 2006, Bhattacharyya and Chakraborty 2007, Pittenger et al. 2011, Pittenger 2015).

Further evidence for a link between glutamate as a neurotransmitter and OCD comes from genetic studies. There is evidence for a significant association between the SLC glutamate transporter genes and OCD (Stewart, Fagerness, Platko, Smoller, Scharf, Illmann, Jenike, Chabane, Leboyer, Delorme, Jenike and Pauls 2007, Pauls et al. 2014). These transporters are crucial in terminating the action of glutamate as an excitatory neurotransmitter and in maintaining extracellular glutamate concentrations within a normal range (Bhattacharyya and Chakraborty 2007, Pauls et al. 2014). In addition, it has been postulated that N-methyl-d-aspartate (NMDA) receptors are involved in OCD, and specifically that polymorphisms in the 3′ untranslated region of GRIN2B (glutamate receptor, ionotropic, N-methyl-d-aspartate 2B) were associated with OCD in affected families (Arnold, Rosenberg, Mundo, Tharmalingam, Kennedy and Richter 2004, Pittenger et al. 2011), and more recent evidence has also found some association to glutamate-related genes (Pauls et al. 2014).

Thus we have seen how an increase in cortical excitability could increase the stability of cortical recurrent attractor networks, and that if expressed in different prefrontal systems, could produce some of the different types of symptoms found in different patients with obsessive-compulsive disorders (Rolls 2012b). Moreover, the approach links the low-level changes found in the increase in NMDA receptor efficacy in OCD to the systems-level effects such as overstability of cognitive and/or motor functioning. Implications for treatments are developed elsewhere (Rolls 2012b).

We have described results with a single attractor network. It is natural to think of such a network as being implemented by associatively modifiable recurrent collateral synaptic connections between the pyramidal cells within a few mm of each other in a cortical area. Such circuitry is characteristic of the cerebral neocortex, and endows it with many remarkable and fundamental properties, including the ability to implement short-term memories, which are fundamental to cognitive processes such as top-down attention, and planning (Rolls 2008d). However, when two cortical areas are connected by forward and backward connections, which are likely to be associatively modifiable, then this architecture, again characteristic of the cerebral neocortex (Rolls 2008d), also provides the basis for the implementation of an attractor network, but with now a different set of glutamatergic synapses in which overactivity could be relevant to the theory described here. Further there are likely to be interactions between connected attractor networks (Deco and Rolls 2003, Deco and Rolls 2006, Rolls 2008d) (Section 4.10), and these are also relevant to the way in which the theory of OCD described here would be implemented in the brain.

It is emphasized that the way in which the network effects we consider produce the symptoms in individual patients will be complex, and will depend on the way in which each person may deal cognitively with the effects. In this respect, we emphasize that what we describe is a theory of obsessive-compulsive disorder, that the theory must be considered in the light of empirical evidence yet to be obtained, and that cognitive behaviour therapy makes an important contribution to the treatment of such patients. Stochastic neurodynamics provides however we believe an important new approach for stimulating further thinking and research in this area.

16.3   Stochastic noise, attractor dynamics, and depression

16.3.1   Introduction

Major depressive disorder is ranked by the World Health Organization as the leading cause of years-of-life lived with disability (Drevets 2007, Gotlib and Hammen 2009, Hamilton, Chen and Gotlib 2013). Major depressive episodes, found in both major depressive disorder and bipolar disorder are pathological mood states characterized by persistently sad or depressed mood. Major depressive disorders are generally accompanied by: (1) altered incentive and reward processing, evidenced by amotivation, apathy, and anhedonia; (2) impaired modulation of anxiety and worry, manifested by generalized, social and panic anxiety, and oversensitivity to negative feedback; (3) inflexibility of thought and behaviour in association with changing reinforcement contingencies, apparent as ruminative thoughts of self-reproach, pessimism, and guilt, and inertia toward initiating goal-directed behaviour; (4) altered integration of sensory and social information, as evidenced by mood-congruent processing biases; (5) impaired attention and memory, shown as performance deficits on tests of attention

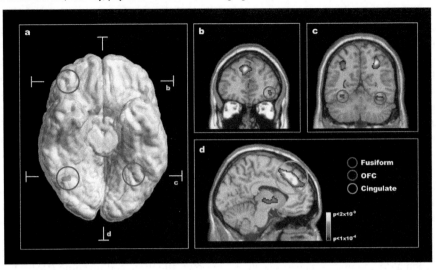

Fig. 16.10 Social reversal: Composite figure showing that changing behaviour based on face expression is correlated with increased brain activity in the human orbitofrontal cortex. a) The figure is based on two different group statistical contrasts from the neuroimaging data which are superimposed on a ventral view of the human brain with the cerebellum removed, and with indication of the location of the two coronal slices (b,c) and the transverse slice (d). The red activations in the orbitofrontal cortex (denoted OFC, maximal activation: z=4.94 [42 42 –8]; and z=5.51 [–46 30 –8] shown on the rendered brain arise from a comparison of reversal events with stable acquisition events, while the blue activations in the fusiform gyrus (denoted Fusiform, maximal activation: z>8 [36 –60 –20] and z=7.80 [–30 –56 –16]) arise from the main effects of face expression. b) The coronal slice through the frontal part of the brain shows the cluster in the right orbitofrontal cortex across all nine subjects when comparing reversal events with stable acquisition events. Significant activity was also seen in an extended area of the anterior cingulate/paracingulate cortex (denoted Cingulate, maximal activation: z=6.88 [–8 22 52]; green circle). c) The coronal slice through the posterior part of the brain shows the brain response to the main effects of face expression with significant activation in the fusiform gyrus and the cortex in the intraparietal sulcus (maximal activation: z>8 [32 –60 46]; and z>8 [–32 –60 44]). d) The transverse slice shows the extent of the activation in the anterior cingulate/paracingulate cortex when comparing reversal events with stable acquisition events. Group statistical results are superimposed on a ventral view of the human brain with the cerebellum removed, and on coronal and transverse slices of the same template brain (activations are thresholded at p=0.0001 for purposes of illustration to show their extent). (Reproduced with permission from *NeuroImage* 20 (2), Morten L. Kringelbach and Edmund T. Rolls, Neural correlates of rapid reversal learning in a simple model of human social interaction, pp. 1371–83, ©2003 Elsevier Inc.)

set-shifting and maintenance, and autobiographical and short-term memory; and 6) visceral disturbances, including altered weight, appetite, sleep, and endocrine and autonomic function (Drevets 2007, Gotlib and Hammen 2009). This section describes an attractor-based theory of some of the brain mechanisms that are related to depression (Rolls 2016d).

The orbitofrontal cortex contains a population of error neurons that respond to non-reward and maintain their firing for many seconds after the non-reward, providing evidence that they have entered an attractor state that maintains a memory of the non-reward (Thorpe, Rolls and Maddison 1983, Rolls 2014a) (Section 15.14). An example if such a neuron is shown in Fig. 15.4.

The human lateral orbitofrontal cortex is activated by non-reward during reward reversal (Kringelbach and Rolls 2003). This is illustrated in Fig. 16.10, which shows activations in the lateral orbitofrontal cortex on reversal trials, that is when the human subject chose one person's face, and did not obtain the expected reward. Activations in the lateral orbitofrontal cortex are

Orbitofrontal cortex

Anterior cingulate and
ventromedial prefrontal cortex

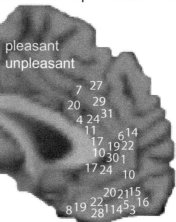

Fig. 16.11 Maps of subjective pleasure in the human orbitofrontal cortex (ventral view) and anterior cingulate and ventromedial prefrontal cortex (sagittal view). Yellow: sites where activations correlate with subjective pleasantness. White: sites where activations correlate with subjective unpleasantness. The numbers refer to effects found in specific studies. Taste: 1, 2; odor: 3-10; flavor: 11-16; oral texture: 17, 18; chocolate: 19; water: 20; wine: 21; oral temperature: 22, 23; somatosensory temperature: 24, 25; the sight of touch: 26, 27; facial attractiveness: 28, 29; erotic pictures: 30; laser-induced pain: 31. (Reproduced with permission from *Trends in Cognitive Sciences*, 15 (2), Fabian Grabenhorst and Edmund T. Rolls, Value, pleasure and choice in the ventral prefrontal cortex, pp. 56–67, ©2011 Elsevier Ltd.)

also produced by a signal to stop a response that is now incorrect, which is another situation in which behavior must change in order to be correct (Deng, Rolls et al. (2017)). Orbitofrontal cortex activations in the stop-signal task have further been related to how impulsive the behavior is (Whelan, Conrod, Poline, Lourdusamy, Banaschewski, Barker, Bellgrove, Büchel, Byrne, Cummins et al. 2012). In this context, it has been suggested that impulsiveness may reflect how sensitive an individual is to non-reward or punishment (Rolls 2014a), and indeed we have shown that people with orbitofrontal cortex damage become more impulsive (Berlin et al. 2004, Berlin et al. 2005). The lateral orbitofrontal cortex also responds to many punishing, unpleasant, stimuli (Grabenhorst and Rolls 2011, Rolls 2014a) (Fig. 16.11) including bad odour (Rolls, Kringelbach and De Araujo 2003c) and losing money (O'Doherty, Kringelbach, Rolls, Hornak and Andrews 2001). Consistent with this human neuroimaging evidence and with the macaque neurophysiology (Thorpe, Rolls and Maddison 1983, Rolls 2014a), the macaque lateral orbitofrontal cortex is also activated by non-reward during a reversal task as shown by fMRI (Chau et al. 2015).

Further evidence that the orbitofrontal cortex is involved in changing rewarded behaviour when non-reward is detected is that damage to the human orbitofrontal cortex impairs reward reversal learning, in that the previously rewarded stimulus is still chosen during reversal even when no reward is being obtained (Rolls, Hornak, Wade and McGrath 1994a, Hornak, O'Doherty, Bramham, Rolls, Morris, Bullock and Polkey 2004, Fellows and Farah 2003, Fellows 2011).

Now it is well established that not receiving expected reward, or receiving unpleasant stimuli or events, can produce depression (Beck 2008, Drevets 2007, Harmer and Cowen 2013, Price and Drevets 2012, Pryce, Azzinnari, Spinelli, Seifritz, Tegethoff and Meinlschmidt

2011, Eshel and Roiser 2010). A clear example is that if a member of the family dies, then this is the removal of reward (in that we would work to try to avoid this), and the result of the removal of the reward can be depression. More formally, in terms of learning theory, the omission or termination of a reward can give rise to sadness or depression, depending on the magnitude of the reward that is lost, if there is no action that can be taken to restore the reward (Rolls 2014a). If an action can be taken, then frustration and anger may arise to the same reinforcement contingency (Rolls 2014a). This relates the current approach to the learned helplessness approach to depression, in which depression arises because no actions are being taken to restore rewards (Forgeard, Haigh, Beck, Davidson, Henn, Maier, Mayberg and Seligman 2011, Pryce et al. 2011).

16.3.2   A non-reward attractor theory of depression

The theory has been proposed that in depression, this lateral orbitofrontal cortex non-reward / punishment attractor network system is more easily triggered, and maintains its attractor-related firing for longer (Rolls 2016d, Rolls 2017f, Rolls 2017e). The greater attractor-related firing of the non-reward / punishment system triggers negative cognitive states held on-line in other cortical systems such as the dorsolateral prefrontal cortex which is implicated in attentional control. These other cortical systems then in turn have top-down effects on the orbitofrontal non-reward system that bias it in a negative direction (Rolls 2013a) (see Section 6.3 and Fig. 6.21), and thus increase the sensitivity of the lateral orbitofrontal cortex to non-reward and maintain its overactivity (Rolls 2016d). It is proposed that the interaction of this type of a non-reward and cognitive attentional brain system accounts for the ruminating and continuing depressive thoughts, which occur as a result of a positive feedback cycle between these types of brain system (Rolls 2016d).

Indeed, we have shown that cognitive states can have top-down effects on affective representations in the orbitofrontal cortex (De Araujo, Rolls, Velazco, Margot and Cayeux 2005, Grabenhorst, Rolls and Bilderbeck 2008a, McCabe, Rolls, Bilderbeck and McGlone 2008, Rolls 2013a). Further, top-down selective attention can also influence affective representations in the orbitofrontal cortex (Rolls et al. 2008b, Grabenhorst and Rolls 2008, Ge et al. 2012, Luo, Ge, Grabenhorst, Feng and Rolls 2013, Rolls 2013a), and paying attention to depressive symptoms when depressed may in this way exacerbate the problems in a positive feedback way. Indeed, more generally, the presence of the cognitive ability to think ahead and see the implications of recent events that is afforded by language may be a computational development in the brain that exacerbates the vulnerability of the human brain to depression (Rolls 2014a). For example, with language we can think ahead and see that perhaps the loss of an individual in one's life may be long-term, and this thought and its consequences for our future can become fully evident.

The theory is that one way in which depression could result from over-activity in this lateral orbitofrontal cortex system is if there is a major negatively reinforcing life event that produces reactive depression and activates this system, which then becomes self-re-exciting based on the cycle between the lateral orbitofrontal cortex non-reward / punishment attractor system and the cognitive system, which together operate as a systems-level attractor. (The generic cortical architecture for such reciprocal feedforward and feedback excitatory effects is illustrated in Figs. 4.4 and 4.5.) The theory is that a second way in which depression might arise is if this lateral orbitofrontal cortex non-reward / punishment system is especially sensitive in some individuals. This might be related for example to genetic predisposition, or to the effects of stress (Gold 2015). In this case, the orbitofrontal system would over-react to normal levels of non-reward or punishment, and start the local attractor circuit in the lateral orbitofrontal cortex (Section 10.3, Fig. 15.14) (Rolls and Deco 2016), which in turn would activate the cognitive system, which would feed back to the over-reactive lateral orbitofrontal cortex system to maintain now a systems-level attractor with ruminating thoughts.

16.3.3 Evidence consistent with the theory

There is some evidence for altered structure and function of the lateral orbitofrontal cortex in depression (Drevets 2007, Ma 2015, Price and Drevets 2012). For example, reductions of grey-matter volume and cortex thickness have been demonstrated specifically in the posterolateral OFC (BA 47, caudal BA 11 and the adjoining BA 45), and also in the subgenual cingulate cortex (BA 24, 25) (Drevets 2007, Nugent, Milham, Bain, Mah, Cannon, Marrett, Zarate, Pine, Price and Drevets 2006). In depression, there is increased cerebral blood flow in areas that include the ventrolateral orbitofrontal cortex (which is a prediction of the theory), and also in regions such as the subgenual cingulate cortex and amygdala, and these increases appear to be related to the mood change, in that they become more normal when the mood state remits (Drevets 2007).

In the first brain-wide voxel-level resting state functional-connectivity neuroimaging analysis of depression (with 421 patients with major depressive disorder and 488 controls), we have found that one major circuit with altered functional connectivity involved the medial orbitofrontal cortex BA 13, which had reduced functional connectivity in depression with memory systems in the parahippocampal gyrus and medial temporal lobe (Cheng et al. 2016) (Fig. 16.12). The lateral orbitofrontal cortex BA 47/12, involved in non-reward and punishing events, did not have this reduced functional connectivity with memory systems, so that there is an imbalance in depression towards decreased reward-related memory system functionality.

A second major circuit change was that the lateral orbitofrontal cortex area BA 47/12 had increased functional connectivity with the precuneus, the angular gyrus, and the temporal visual cortex BA 21 (Cheng et al. 2016) (Fig. 16.12). This enhanced functional connectivity of the non-reward/punishment system (BA 47/12) with the precuneus (involved in the sense of self and agency), and the angular gyrus (involved in language) is thus related to the explicit affectively negative sense of the self, and of self-esteem, in depression.

The reduced functional connectivity of the medial orbitofrontal cortex, implicated in reward, with memory systems provides a new way of understanding how memory systems may be biased away from pleasant events in depression. The increased functional connectivity of the lateral orbitofrontal cortex, implicated in non-reward and punishment, with areas of the brain implicated in representing the self, language, and inputs from face and related perceptual systems provides a new way of understanding how unpleasant events and thoughts, and lowered self-esteem, may be exacerbated in depression (Cheng et al. 2016, Rolls, Cheng, Gilson, Qiu, Hu, Li, Huang, Yang, Tsai, Zhang, Zhuang, Lin, Deco, Xie and Feng 2017a).

Because the lateral orbitofrontal cortex responds to many punishing and non-rewarding stimuli (Grabenhorst and Rolls 2011, Rolls 2014a, Rolls 2014b) that are likely to elicit autonomic/visceral responses, as does the supracallosal anterior cingulate cortex, and in view of connections from these areas to the anterior insula which is implicated in autonomic/visceral function (Critchley and Harrison 2013, Rolls 2016b), the anterior insula would also be expected to be overactive in depression, which it is (Drevets 2007, Hamilton et al. 2013, Ma 2015).

Treatments that can reduce depression such as a single dose of ketamine (Iadarola, Niciu, Richards, Vande Voort, Ballard, Lundin, Nugent, Machado-Vieira and Zarate 2015) may act in part by quashing the attractor state in the lateral orbitofrontal cortex at least temporarily. Evidence consistent with this is that the activity of the lateral orbitofrontal cortex is decreased by a single dose of ketamine (Lally, Nugent, Luckenbaugh, Niciu, Roiser and Zarate 2015). This NMDA receptor blocker may act at least in part by decreasing the high firing rate

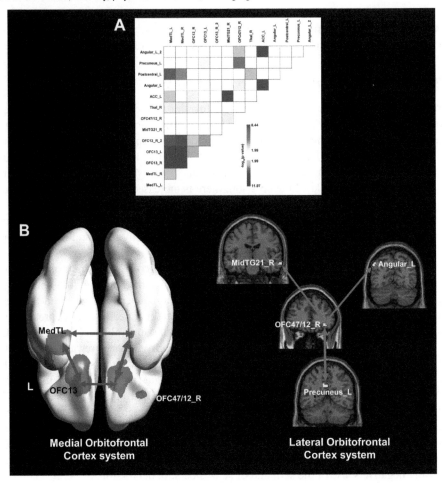

Fig. 16.12 Resting state functional connectivity in depression. Cluster functional connectivity matrix. The color bar shows the -$\log_{10}$ of the p value for the difference of the functional connectivity. The matrix contains rows and columns for all cases in which there were 10 or more significant voxels within a cluster. ACC – anterior cingulate cortex; MedTL – medial temporal lobe, including parts of the parahippocampal gyrus; Thal – thalamus. B. The medial and lateral orbitofrontal cortex networks that show different functional connectivity in patients with depression. A decrease in functional connectivity is shown in blue, and an increase in red. MedTL – medial temporal lobe from the parahippocampal gyrus to the temporal pole; MidTG21R – middle temporal gyrus area 21 right; OFC13 – medial orbitofrontal cortex area 13; OFC47/12R – lateral orbitofrontal cortex area 47/12 right. The lateral orbitofrontal cortex cluster in OFC47/12 is visible on the ventral view of the brain anterior and lateral to the OFC13 clusters. (After Cheng, Rolls et al, 2016.)

state of attractor networks by reducing transmission in the recurrent collateral excitatory connections between the neurons (Rolls 2008d, Rolls, Loh, Deco and Winterer 2008d, Rolls and Deco 2010, Rolls 2012b, Deco, Rolls, Albantakis and Romo 2013, Rolls and Deco 2015b). Given that a ketamine metabolite, hydroxynorketamine, may be related to the antidepressant effects of ketamine and may act via facilitating effects mediated by AMPA receptors (Zanos, Moaddel, Morris, Georgiou, Fischell, Elmer, Alkondon, Yuan, Pribut, Singh, Dossou, Fang, Huang, Mayo, Wainer, Albuquerque, Thompson, Thomas, Zarate and Gould 2016), the effects of ketamine might be mediated by increasing the medial orbitofrontal cortex reward-related system (which tends to be reciprocally related to the lateral orbitofrontal non-reward system), or the functional connectivity of the medial orbitofrontal cortex reward system with the hippocampal system which is reduced in depression (Fig. 16.12). Electroconvulsive therapy may have antidepressant effects, may also knock the non-reward system out of its attractor state, and this may contribute to any antidepressant effect.

Electrical stimulation of the brain that may relieve depression (Hamani, Mayberg, Snyder, Giacobbe, Kennedy and Lozano 2009, Hamani, Mayberg, Stone, Laxton, Haber and Lozano 2011, Lujan, Chaturvedi, Choi, Holtzheimer, Gross, Mayberg and McIntyre 2013) may act in part by providing reward that reciprocally inhibits the non-reward system, and/or by interfering with the attractor state. Treatment with antidepressant drugs decreases the activity of this lateral orbitofrontal cortex system (Ma 2015).

Antidepressant drugs such as Selective Serotonin Reuptake Inhibitors (SSRIs) may treat depression by producing positive biases in the processing of emotional stimuli (Harmer and Cowen 2013), increasing brain responses to positive stimuli and decreasing responses to negative stimuli (Ma 2015). The reward and non-reward systems are likely to operate reciprocally, so that facilitating the reward system, or providing rewards, and thus activating the medial orbitofrontal cortex (O'Doherty, Kringelbach, Rolls, Hornak and Andrews 2001, Grabenhorst and Rolls 2011, Rolls 2014a) (Fig. 16.11), may operate in part by inhibiting the overactivity in the lateral orbitofrontal cortex non-reward / punishment system.

16.3.4   Relation to other brain systems implicated in depression

The human *medial orbitofrontal cortex* has activations related to many rewarding and subjectively pleasant stimuli (Rolls and Grabenhorst 2008, Grabenhorst and Rolls 2011, Rolls 2014a) (Fig. 16.11). In the sense that reward vs non-reward and punishment are reciprocally related in their effects in the medial vs lateral orbitofrontal cortex respectively (O'Doherty, Kringelbach, Rolls, Hornak and Andrews 2001, Rolls 2014a), the anhedonia of depression can also be related to decreased effects of pleasant rewarding stimuli in the medial orbitofrontal cortex during depression, effects that can be restored by antidepressants (Ma 2015). The lateral orbitofrontal cortex / inferior frontal gyrus region that responds to signals to inhibit a response in the stop-signal task (Deng et al. 2017) is implicated in impulsive behavior, with damage in this region increasing impulsive behavior (Aron, Robbins and Poldrack 2014). This is of potential importance, for treatment with antidepressants, which would be expected to reduce the overactivity in this ventrolateral prefrontal cortex region, might thereby increase impulsiveness relative to that in the depressed state. Indeed, it is an interesting hypothesis that impulsiveness might reflect underactivity in this ventrolateral prefrontal cortex region, and that depression produced by oversensitivity to non-reward and punishment might reflect overactivity in this ventrolateral prefrontal cortex region. In a certain sense, these types of behaviour might reflect opposite ends of a continuum of non-reward/punishment sensitivity. One end of the spectrum of sensitivity to non-reward could be impulsive behaviour (with too little sensitivity to non-reward and punishment); and the other end could be depression (with too much sensitivity to non-reward and punishment). The brain system in the lateral orbitofrontal cortex would be the same, but it would be operating in these conditions with too little or too great sensitivity to non-reward and punishment.

The supracallosal *anterior cingulate cortex* is activated by many aversive stimuli, and the pregenual cingulate cortex by many pleasant stimuli (Fig. 16.11) (Grabenhorst and Rolls 2011, Rolls 2014a). However, the anterior cingulate cortex appears to be involved in action-outcome learning, where the outcome refers to the reward or punisher for which an action is being learned (Rudebeck, Behrens, Kennerley, Baxter, Buckley, Walton and Rushworth 2008, Camille et al. 2011, Grabenhorst and Rolls 2011, Rushworth et al. 2011, Rushworth et al. 2012, Rolls 2014a). In contrast, the medial orbitofrontal cortex is implicated in reward-related processing and learning, and the lateral orbitofrontal cortex in non-reward and punishment-related processing and learning (Rolls 2014a). These involve stimulus-stimulus associations, where the second stimulus is a reward (or its omission), or a punisher (Rolls 2014a). Now given that emotions can be considered as states elicited by rewarding and punishing stimuli, and that moods such as depression can arise from prolonged non-reward or punishment (Rolls 2014a), the part of the brain that processes these stimulus-stimulus associations, the orbitofrontal cortex, is more likely to be involved in depression than the action-related parts of the cingulate cortex.

In addition to the insula, lateral orbitofrontal cortex, and supracallosal anterior cingulate cortex, which are all activated by unpleasant stimuli (Grabenhorst and Rolls 2011, Rolls 2014a), parts of the *amygdala* are activated by unpleasant stimuli, and parts by pleasant stimuli (Rolls 2014a), and amygdala activity has been related to depression (Harmer and Cowen 2013, Ma 2015, Price and Drevets 2012). However, the amygdala is less involved in non-reward, especially the rule-based reversal of which stimuli are classified as rewarding that is required in a rapid reward reversal task (Rolls 2014a). The orbitofrontal cortex is special in this, because the evidence is that it has attractor states than can be activated by non-reward (Thorpe, Rolls and Maddison 1983, Rolls 2014a), and these attractor states provide a basis for biasing the correct populations of neurons in the orbitofrontal cortex to implement the rapid one-trial reversal (Deco and Rolls 2005d, Rolls 2014a) (Chapter 15). Because the lateral orbitofrontal cortex has recurrent collaterals that can maintain attractor states, it is more likely to be involved in maintaining attractor states elicited by non-reward, including depression, than the amygdala (Rolls 2014a). The amygdala may therefore because of its responsiveness to punishing stimuli be related to depression, but may not be a structure that maintains its activity in an attractor state after non-reward, and during the mood state of depression.

The *subgenual cingulate cortex* has also been implicated in depression, and electrical stimulation in that region may relieve depression (Mayberg 2003, Hamani et al. 2009, Hamani et al. 2011, Lozano, Giacobbe, Hamani, Rizvi, Kennedy, Kolivakis, Debonnel, Sadikot, Lam, Howard, Ilcewicz-Klimek, Honey and Mayberg 2012, Laxton, Neimat, Davis, Womelsdorf, Hutchison, Dostrovsky, Hamani, Mayberg and Lozano 2013, Lujan et al. 2013). However, the subgenual cingulate cortex is also implicated in autonomic function (Gabbott, Warner, Jays and Bacon 2003), and this could be related to some of the effects found in this area that are related to depression. Whether the subgenual cingulate cortex is activated because of inputs from the orbitofrontal cortex, or performs separate computations is not yet clear. Indeed, the orbitofrontal cortex has the inputs and representations required to compute non-reward, namely representations of expected value, and reward and punishment outcome value (Rolls 2014a) (Section 15.14), and it is not clear that the subgenual cingulate cortex has the information to perform that computation. Further, the possibility is considered that electrical stimulation of the subcallosal region, which includes parts of the ventromedial prefrontal cortex (Laxton et al. 2013), that may relieve depression, may do so at least in part by activating connections involving the orbitofrontal cortex, other parts of the anterior cingulate cortex, and the striatum (Johansen-Berg, Gutman, Behrens, Matthews, Rushworth, Katz, Lozano and Mayberg 2008, Hamani et al. 2009, Lujan et al. 2013).

16.3.5 Implications for treatments

This non-reward / punishment attractor network sensitivity theory of depression has implications for treatments. These implications can be understood and further explored in the context of investigations of the factors that influence the stability of attractor neuronal networks with integrate-and-fire neurons with noise introduced by the close to Poisson spiking times of the neurons (Chapter 5) (Wang 2002, Rolls 2008d, Deco et al. 2009, Rolls and Deco 2010, Deco et al. 2013, Loh et al. 2007a, Rolls and Deco 2015b).

One implication is that antianxiety drugs, by increasing inhibition, might reduce the stability of the high firing rate state of the non-reward attractor, thus acting to quash the depression-related attractor state.

A second implication is that it might be possible to produce agents that decrease the efficacy of NMDA receptors in the lateral orbitofrontal cortex, thereby reducing the stability of the depression-related attractor state. The evidence that there are genes that are select-ive for NMDA receptors for the neurons in different populations is that there are sepa-rate knock-outs for NMDA receptors in the CA3 and CA1 regions of the hippocampus (Nakazawa, Quirk, Chitwood, Watanabe, Yeckel, Sun, Kato, Carr, Johnston, Wilson and Tonegawa 2002, Tonegawa, Nakazawa and Wilson 2003, Nakazawa, Sun, Quirk, Rondi-Reig, Wilson and Tonegawa 2003, Nakazawa, McHugh, Wilson and Tonegawa 2004). The present theory suggests that searching for ways to influence the attractor networks in the lateral orbito-frontal cortex by decreasing excitatory or increasing inhibitory transmission in this region may be of considerable interest. It should be noted that the present theory is a theory specifically of non-reward and punishment-related attractor networks in the lateral orbitofrontal cortex and related areas in relation to depression, and that alterations of attractor networks in other cortical areas may be related to other psychiatric disorders (Rolls 2012b).

In terms of the implications of the attractor-based aspect of the present theory, an important point is that the attractor dynamics must be kept stable in the face of the randomness or noise introduced into the system by the almost Poisson firing times of neurons for a given mean firing rate. Moreover, the spontaneous firing rate state of the non-reward attractor must be maintained stable when no non-reward inputs are present (or otherwise the non-reward attractor would jump into a high firing rate non-reward state for no external reason, contributing to depression). The inhibitory transmitter GABA may be important in maintaining this type of stability (Rolls and Deco 2010). Moreover, the high firing rate state produced by non-reward must not reach too high a firing rate, as this would cause overstability of the non-reward / depression state. In a complementary way, if the high firing rate attractor state is insufficiently high, then that attractor state might be unstable, and the individual might be relatively insensitive to non-reward, not depressed, and impulsive because of not responding sufficiently to non-reward or punishment. The excitatory transmitter glutamate acting at NMDA or AMPA receptors may be important in setting the stability of the high firing rate attractor state. In this respect and in this sense, the tendency to become depressed or to be impulsive may be reciprocally related to each other. Predictions for treatments follow from understanding these noisy attractor-based dynamics (Rolls and Deco 2010, Rolls 2012b).

16.3.6 Mania and bipolar disorder

So far in this Section 16.3, we have been considering unipolar depression.

Bipolar disorder includes periods of mania in addition to periods of depression. The severity of the mania is greatest in bipolar I disorder, moderate in bipolar II disorder, and lower in cyclothymia. During depression, people feel helplessness, reduced energy, and risk aversion, while with mania behaviors include grandiosity, increased energy, less sleep, and risk preference / impulsivity.

What is the relation between mania and depression? Could it be that in mania, there is something that in terms of reward/non-reward systems, is almost the opposite of depression? Might there be in mania *increased sensitivity to reward, and decreased sensitivity to non-reward / punishment*? The latter might manifest itself as increased impulsiveness in mania.

That is a suggestion that might be considered to be the opposite of what has been described for depression in the previous part of this Section 16.3.

It turns out that there is support for this hypothesis. It indeed appears that the risk for mania is characterized by a hypersensitivity to goal- and reward-relevant cues (Nusslock, Young and Damme 2014). This hypersensitivity can lead to an excessive increase in approach-related affect and motivation during life events involving rewards or goal striving and attainment. In the extreme, this excessive increase in reward-related affect is reflected in manic symptoms, such as pursuit of rewarding activities without attention to risks, elevated or irritable mood, decreased need for sleep, increased psychomotor activation, and extreme self-confidence. Some evidence consistent with the hypothesis is that patients with bipolar I disorder and their relatives showed greater activation of the medial orbitofrontal cortex in response to reward delivery (Wessa, Kanske and Linke 2014). Also, reduced deactivation of the medial orbitofrontal cortex (where rewards are represented) during reward reversal might reflect a reduced error signal in bipolar disorder patients and their relatives in the lateral orbitofrontal cortex. (The activation of the lateral orbitofrontal cortex by non-reward in healthy individuals is illustrated in Fig. 16.10.) This type of responsiveness has been found to be very different in mania, with apparently decreasing activations in the lateral orbitofrontal cortex during expectation of increasing loss, the opposite of what is found in healthy participants (Bermpohl, Kahnt, Dalanay, Hagele, Sajonz, Wegner, Stoy, Adli, Kruger, Wrase, Strohle, Bauer and Heinz 2010) which we discovered in the orbitofrontal cortex (O'Doherty, Kringelbach, Rolls, Hornak and Andrews 2001). In this context, of potentially reduced sensitivity or even abnormal function of the lateral orbitofrontal cortex non-reward system in mania, it is relevant that manic bipolar patients continue to pursue immediate rewards despite negative consequences (Wessa et al. 2014). Further, impulsivity in mania is pervasive, encompassing deficits in attention and behavioral inhibition. In addition, impulsivity is greater if the illness is severe (with for example frequent episodes, substance use disorders, and suicide attempts) (Swann 2009). The significance of this is that impulsivity may reflect decreased sensitivity to non-reward, which is represented by activations in the lateral orbitofrontal cortex, where non-reward is represented (see above).

Thus mania may reflect a state in which there is decreased sensitivity of non-reward systems and hence increased impulsiveness due to reduced sensitivity of the lateral orbitofrontal cortex, and at the same time, increased sensitivity to reward reflected in activations in the medial orbitofrontal cortex and pregenual cingulate cortex. Although these medial and lateral orbitofrontal cortex systems may show reciprocally related activations within an individual, with for example increasing activations in the medial orbitofrontal cortex to increasing monetary gains and decreasing activations in the lateral orbitofrontal cortex, and vice versa to increasing monetary loss (O'Doherty et al. 2001), the reward and non-reward systems could be, and indeed are likely to, have their sensitivity set by independent genes, providing a basis for some patients to be depressed, and others to show both mania and depression. Indeed, Rolls' theory of emotion (Rolls 2014a) (Chapter 15) would go beyond this, and suggest that the sensitivity to many different rewards (e.g. food when hungry, water when thirsty, pleasant touch, sensitivity to reputation), and correspondingly to many different non-rewards, may be set by genes somewhat independently. This provides a relation to personality (Rolls 2014a), with the implication that people with depression may be particularly sensitive to certain non-rewards or punishers, and people with mania may be particularly sensitive to particular rewards. This has important implications for therapy, which might be well-directed towards particular sensitivities to particular non-rewards and particular rewards in different individuals.

The question then arises of the extent to which attractor network operations contribute to mania. In terms of responses to inputs that increase the expectancy of reward, a short-term attractor system, probably in the orbitofrontal cortex, is likely to be present, to bridge any temporal interval between the expected reward signal and the actual outcome. This could in principle be oversensitive in mania. When the reward, the outcome, is delivered, it might also be useful to have a short-term attractor, to help reset a rule attractor for which stimulus is currently rewarding. However, it would be maladaptive if these reward-expectancy or reward-outcome attractors normally operated for more than perhaps 10 s, for this would tend to break the important contingency between input stimuli and outcomes. In addition to these short-term attractors, there also needs to be a longer term attractor process to reflect mood state, which typically operates on a much longer time scale. This might again be an attractor (with separate competing attractors for different mood states), and this attractor might be re-activated by the longer loop through the language / planning system, which by recalling a recent reward might calculate the long-term benefits, helping to keep the mood state prolonged. This whole 'long-loop' attractor might also be more sensitive in mania. These are interesting concepts for future empirical exploration.

16.4   Stochastic noise, attractor dynamics, and aging

In Section 16.1 we show how some cognitive symptoms such as poor short-term memory and attention could arise due to reduced depth in the basins of attraction of prefrontal cortical networks, and the effects of noise. The hypothesis is that the reduced depth in the basins of attraction would make short-term memory unstable, so that sometimes the continuing firing of neurons that implement short-term memory would cease, and the system under the influence of noise would fall back out of the short-term memory state into spontaneous firing.

Given that top-down attention requires a short-term memory to hold the object of attention in mind, and that this is the source of the top-down attentional bias that influences competition in other networks that are receiving incoming signals, then disruption of short-term memory is also predicted to impair the stability of attention.

These ideas are elaborated in Section 16.1, where the reduced depth of the basins of attraction in schizophrenia is related to down-regulation of NMDA receptors, or to factors that influence NMDA receptor generated ion channel currents such as dopamine D1 receptors.

Could similar processes, in which the stochasticity of the dynamics is increased because of a reduced depth in the basins of attraction, contribute to the changes in short-term memory and attention that are common in aging? Reduced short-term memory and less good attention are common in aging, as are impairments in episodic memory (Grady 2008). What changes in aging might contribute to a reduced depth in the basins of attraction? We describe recent hypotheses (Rolls and Deco 2015b).

16.4.1   NMDA receptor hypofunction

One factor is that NMDA receptor functionality tends to decrease with aging (Kelly, Nadon, Morrison, Thibault, Barnes and Blalock 2006). This would act, as investigated in detail in Section 16.1, to reduce the depth of the basins of attraction, both by reducing the firing rate of the neurons in the active attractor, and effectively by decreasing the strength of the potentiated synaptic connections that support each attractor as the currents passing through these potentiated synapses would decrease. These two actions are clarified by considering equation B.12 on page 726, in which the Energy E reflects the depth of the basin of attraction.

An example of the reduction of firing rate in an attractor network produced by even a small downregulation (by 4.5%) of the NMDA receptor activated ion channel conductances is shown in Fig. 16.13, based on a mean-field analysis. The effect of this reduction would

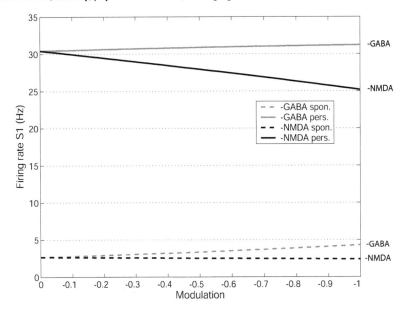

Fig. 16.13 The firing rate of an attractor network as a function of a modulation of the NMDA receptor activated ion channel conductances. On the abscissa, 0 corresponds to no reduction, and −1.0 to a reduction by 4.5%. The change in firing rate for the high firing rate short-term memory state of persistent firing, and for the spontaneous rate, are shown. The values were obtained from the mean-field analysis of the network described in Section 16.1, and thus for the persistent state represent the firing rate when the system is in its high firing rate state. Curves are also shown for a reduction of up to 9% in the GABA receptor activated ion channel conductances. (Data of Loh, Rolls and Deco.)

be to decrease the depth of the basins of attraction, both by reducing the firing rate, and by producing an effect similar to weakened synaptic strengths, as shown in equation B.12.

In integrate-and-fire simulations, the effect on the firing rates of reduction in the NMDA activated channel conductances by 4.5% and in the GABA activated channel conductances by 9% are shown in Fig. 16.14.

The reduced depth in the basins of attraction could have a number of effects that are relevant to cognitive changes in aging.

First, the stability of short-term memory networks would be impaired, and it might be difficult to hold items in short-term memory for long, as the noise might push the network easily out of its shallow attractor, as illustrated in Fig. 16.2.

Second, top-down attention would be impaired, in two ways. First, the short-term memory network holding the object of attention in mind would be less stable, so that the source of the top-down bias for the biased competition in other cortical areas might disappear. Second, and very interestingly, even when the short-term memory for attention is still in its persistent attractor state, it would be less effective as a source of the top-down bias, because the firing rates would be lower, as shown in Figs. 16.13 and 16.14.

Third, the recall of information from episodic memory systems in the temporal lobe (Rolls 2008d, Dere, Easton, Nadel and Huston 2008, Rolls 2008e) would be impaired. This would arise because the positive feedback from the recurrent collateral synapses that helps the system to fall into a basin of attraction, representing in this case the recalled memory, would be less effective, and so the recall process would be more noisy overall.

Fourth, any reduction of the firing rate of the pyramidal cells caused by NMDA receptor

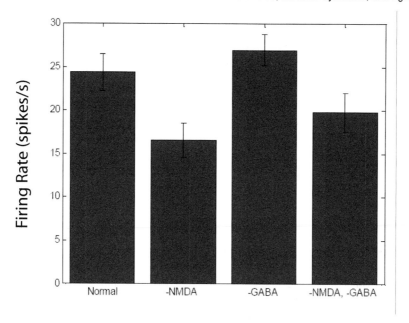

Fig. 16.14 The firing rate (mean ± sd across neurons in the attractor and simulation runs) of an attractor network in the baseline (Normal) condition, and when the NMDA receptor activated ion channel conductances in integrate-and-fire simulations are reduced by 4.5% (−NMDA). The firing rate for the high firing rate short-term memory state of persistent firing is shown. The values were obtained from integrate-and–fire simulations, and only for trials in which the network was in a high firing rate attractor state using the criterion that at the end of the 3 s simulation period the neurons in the attractor were firing at 10 spikes/s or higher. The firing rates are also shown for a reduction of 9% in the GABA receptor activated ion channel conductances (−GABA), and for a reduction in both the NMDA and GABA receptor activated ion channel conductances (−NMDA, −GABA). (Data of Loh, Rolls and Deco.)

hypofunction (Figs. 16.13 and 16.14) would itself be likely to impair new learning involving LTP.

In addition, if the NMDA receptor hypofunction were expressed not only in the prefrontal cortex where it would affect short-term memory, and in the temporal lobes where it would affect episodic memory, but also in the orbitofrontal cortex, then we would predict some reduction in emotion and motivation with aging, as these functions rely on the orbitofrontal cortex (see Rolls (2005) and Section 16.1).

Although NMDA hypofunction may contribute to cognitive effects such as poor short-term memory and attention in aging and in schizophrenia, the two states are clearly very different. Part of the difference lies in the positive symptoms of schizophrenia (the psychotic symptoms, such as thought disorder, delusions, and hallucinations) which may be related to the additional downregulation of GABA in the temporal lobes, which would promote too little stability of the spontaneous firing rate state of temporal lobe attractor networks, so that the networks would have too great a tendency to enter states even in the absence of inputs, and to not be controlled normally by input signals (Loh, Rolls and Deco 2007a, Loh, Rolls and Deco 2007b, Rolls, Loh, Deco and Winterer 2008d) (see Section 16.1). However, in relation to the cognitive symptoms of schizophrenia, there has always been the fact that schizophrenia is a condition that often has its onset in the late teens or twenties, and I suggest that there could be a link here to changes in NMDA and related receptor functions that are related to aging. In particular, short-term memory is at its peak when young, and it may be the case that by the late teens or early twenties NMDA and related receptor systems (including dopamine) may be less efficacious than when younger, so that the cognitive symptoms of schizophrenia are more likely to occur at this age than earlier.

16.4.2 Dopamine

D1 receptor blockade in the prefrontal cortex can impair short-term memory (Sawaguchi and Goldman-Rakic 1991, Sawaguchi and Goldman-Rakic 1994, Goldman-Rakic 1999, Castner et al. 2000). Part of the reason for this may be that D1 receptor blockade can decrease NMDA receptor activated ion channel conductances, among other effects (Seamans and Yang 2004, Durstewitz et al. 1999, Durstewitz et al. 2000a, Brunel and Wang 2001, Durstewitz and Seamans 2002) (see further Section 16.1). Thus part of the role of dopamine in the prefrontal cortex in short-term memory can be accounted for by a decreased depth in the basins of attraction of prefrontal attractor networks (Loh, Rolls and Deco 2007a, Loh, Rolls and Deco 2007b, Rolls, Loh, Deco and Winterer 2008d). The decreased depth would be due to both the decreased firing rate of the neurons, and the reduced efficacy of the modified synapses as their ion channels would be less conductive (see equation B.12). The reduced depth of the basins of attraction can be thought of as decreasing the signal-to-noise ratio (Loh, Rolls and Deco 2007b, Rolls, Loh, Deco and Winterer 2008d). Given that dopaminergic function in the prefrontal cortex may decline with aging (Sikström 2007), and in conditions in which there are cognitive impairments such as Parkinson's disease, the decrease in dopamine could contribute to the reduced short-term memory and attention in aging.

In attention deficit hyperactivity disorder (ADHD), in which there are attentional deficits including too much distractibility, catecholamine function more generally (dopamine and noradrenaline (i.e. norepinephrine)) may be reduced (Arnsten and Li 2005), and I suggest that these reductions could produce less stability of short-term memory and thereby attentional states by reducing the depth of the basins of attraction.

16.4.3 Impaired synaptic modification

Another factor that may contribute to the cognitive changes in aging is that long-lasting associative synaptic modification as assessed by long-term potentiation (LTP) is more difficult to achieve in older animals and decays more quickly (Barnes 2003, Burke and Barnes 2006, Kelly et al. 2006). This would tend to make the synaptic strengths that would support an attractor weaker, and weaken further over the course of time, and thus directly reduce the depth of the attractor basins. This would impact episodic memory, the memory for particular past episodes, such as where one was at breakfast on a particular day, who was present, and what was eaten (Rolls 2008d, Rolls 2008e, Dere et al. 2008). The reduction of synaptic strength over time could also affect short-term memory, which requires that the synapses that support a short-term memory attractor be modified in the first place using LTP, before the attractor is used (Kesner and Rolls 2001).

In view of these changes, boosting glutamatergic transmission is being explored as a means of enhancing cognition and minimizing its decline in aging. Several classes of AMPA receptor potentiators have been described. These molecules bind to allosteric sites on AMPA receptors, slow desensitization, and thereby enhance signalling through the receptors. Some AMPA receptor potentiator agents have been explored in rodent models and are now entering clinical trials (Lynch and Gall 2006, O'Neill and Dix 2007). These treatments might increase the depth of the basins of attraction. Agents that activate the glycine or serine modulatory sites on the NMDA receptor (Coyle 2006) would also be predicted to be useful.

Another factor is that Ca^{2+}-dependent processes affect Ca^{2+} signaling pathways and impair synaptic function in an aging-dependent manner, consistent with the Ca^{2+} hypothesis of brain aging and dementia (Thibault, Porter, Chen, Blalock, Kaminker, Clodfelter, Brewer and Landfield 1998, Kelly et al. 2006). In particular, an increase in Ca^{2+} conductance can occur in aged neurons, and CA1 pyramidal cells in the aged hippocampus have an increased density of L-type Ca^{2+} channels that might lead to disruptions in Ca^{2+} homeostasis, contributing to the plasticity deficits that occur during aging (Burke and Barnes 2006).

16.4.4   Cholinergic function and memory

16.4.4.1   Cholinergic function, memory, and aging

Another factor is acetylcholine. Acetylcholine in the neocortex has its origin largely in the cholinergic neurons in the basal magnocellular forebrain nuclei of Meynert. The correlation of clinical dementia ratings with the reductions in a number of cortical cholinergic markers such as choline acetyltransferase, muscarinic and nicotinic acetylcholine receptor binding, as well as levels of acetylcholine, suggested an association of cholinergic hypofunction with cognitive deficits, which led to the formulation of the cholinergic hypothesis of memory dysfunction in senescence and in Alzheimer's disease (Bartus 2000, Schliebs and Arendt 2006). Could the cholinergic system alter the function of the cerebral cortex in ways that can be illuminated by stochastic neurodynamics?

16.4.4.2   The responses of basal magnocellular forebrain nuclei cholinergic neurons

The cells in the basal magnocellular forebrain nuclei of Meynert lie just lateral to the lateral hypothalamus in the substantia innominata, and extend forward through the preoptic area into the diagonal band of Broca (Mesulam 1990). These cells, many of which are cholinergic, project directly to the cerebral cortex (Divac 1975, Kievit and Kuypers 1975, Mesulam 1990). These cells provide the major cholinergic input to the cerebral cortex, in that if they are lesioned the cortex is depleted of acetylcholine (Mesulam 1990). Loss of these cells does occur in Alzheimer's disease, and there is consequently a reduction in cortical acetylcholine in this disease (Mesulam 1990, Schliebs and Arendt 2006). This loss of cortical acetylcholine may contribute to the memory loss in Alzheimer's disease, although it may not be the primary factor in the aetiology.

In order to investigate the role of the basal forebrain nuclei in memory, Aigner, Mitchell, Aggleton, DeLong, Struble, Price, Wenk, Pettigrew and Mishkin (1991) made neurotoxic lesions of these nuclei in monkeys. Some impairments on a simple test of recognition memory, delayed non-match-to-sample, were found. Analysis of the effects of similar lesions in rats showed that performance on memory tasks was impaired, perhaps because of failure to attend properly (Muir, Everitt and Robbins 1994). Damage to the cholinergic neurons in this region in monkeys with a selective neurotoxin was also shown to impair memory (Easton and Gaffan 2000, Easton, Ridley, Baker and Gaffan 2002).

There are quite limited numbers of these basal forebrain neurons (in the order of thousands). Given that there are relatively few of these neurons, it is not likely that they carry the information to be stored in cortical memory circuits, for the number of different patterns that could be represented and stored is so small. (The number of different patterns that could be stored is dependent in a leading way on the number of input connections on to each neuron in a pattern associator, see e.g. Rolls (2008d)). With these few neurons distributed throughout the cerebral cortex, the memory capacity of the whole system would be impractically small. This argument alone indicates that these cholinergic neurons are unlikely to carry the information to be stored in cortical memory systems. Instead, they could modulate storage in the cortex of information derived from what provides the numerically major input to cortical neurons, the glutamatergic terminals of other cortical neurons. This modulation may operate by setting thresholds for cortical cells to the appropriate value, or by more directly influencing the cascade of processes involved in long-term potentiation (Rolls 2008d). There is indeed evidence that acetylcholine is necessary for cortical synaptic modifiability, as shown by studies in which depletion of acetylcholine and noradrenaline impaired cortical LTP/synaptic modifiability (Bear and Singer 1986). However, non-specific effects of damage to the basal forebrain cholinergic neurons are also likely, with cortical neurons becoming much more sluggish in their responses, and showing much more adaptation, in the absence of cholinergic inputs (Markram and Tsodyks 1996, Abbott, Varela, Sen and Nelson 1997) (see later).

The question then arises of whether the basal forebrain cholinergic neurons tonically release acetylcholine, or whether they release it particularly in response to some external influence. To examine this, recordings have been made from basal forebrain neurons, at least some of which project to the cortex (see Rolls (2005)) and will have been the cholinergic neurons just described.

It has been found that some of these basal forebrain neurons respond to visual stimuli associated with rewards such as food (Rolls 1975, Rolls 1981b, Rolls 1981a, Rolls 1982, Rolls 1986c, Rolls 1986a, Rolls 1986b, Rolls 1990c, Rolls 1993, Rolls 1999a, Rolls, Burton and Mora 1976, Burton, Rolls and Mora 1976, Mora, Rolls and Burton 1976, Wilson and Rolls 1990b, Wilson and Rolls 1990a), or with punishment (Rolls, Sanghera and Roper-Hall 1979), that others respond to novel visual stimuli (Wilson and Rolls 1990c), and that others respond to a range of visual stimuli. For example, in one set of recordings, one group of these neurons (1.5%) responded to novel visual stimuli while monkeys performed recognition or visual discrimination tasks (Wilson and Rolls 1990c).

A complementary group of neurons more anteriorly responded to familiar visual stimuli in the same tasks (Rolls, Perrett, Caan and Wilson 1982, Wilson and Rolls 1990c).

A third group of neurons (5.7%) responded to positively reinforcing visual stimuli in visual discrimination and in recognition memory tasks (Wilson and Rolls 1990b, Wilson and Rolls 1990a).

In addition, a considerable proportion of these neurons (21.8%) responded to any visual stimuli shown in the tasks, and some (13.1%) responded to the tone cue that preceded the presentation of the visual stimuli in the task, and was provided to enable the monkey to alert to the visual stimuli (Wilson and Rolls 1990c). None of these neurons responded to touch to the leg which induced arousal, so their responses did not simply reflect arousal.

Neurons in this region receive inputs from the amygdala (Mesulam 1990, Amaral et al. 1992, Russchen, Amaral and Price 1985) and orbitofrontal cortex, and it is probably via the amygdala (and orbitofrontal cortex) that the information described here reaches the basal forebrain neurons, for neurons with similar response properties have been found in the amygdala, and the amygdala appears to be involved in decoding visual stimuli that are associated with reinforcers, or are novel (Rolls 1990c, Rolls 1992b, Davis 1992, Wilson and Rolls 1993, Rolls 2000c, LeDoux 1995, Wilson and Rolls 2005, Rolls 2005, Rolls 2008d).

16.4.4.3 Basal magnocellular forebrain nuclei cholinergic neurons as a cortical strobe to facilitate memory

On the basis of these findings, it is suggested that the normal physiological function of these basal forebrain neurons is to send a general activation signal to the cortex when certain classes of environmental stimulus occur. These stimuli are often stimuli to which behavioural activation is appropriate or required, such as positively or negatively reinforcing visual stimuli, or novel visual stimuli. The effect of the firing of these neurons on the cortex is excitatory, and in this way produces activation. This cortical activation may produce behavioural arousal, and may thus facilitate concentration and attention, which are both impaired in Alzheimer's disease. The reduced arousal and concentration may themselves contribute to the memory disorders. But the acetylcholine released from these basal magnocellular neurons may in addition be more directly necessary for memory formation, for Bear and Singer (1986) showed that long-term potentiation, used as an indicator of the synaptic modification which underlies learning, requires the presence in the cortex of acetylcholine as well as noradrenaline. For comparison, acetylcholine in the hippocampus makes it more likely that LTP will occur, probably through activation of an inositol phosphate second messenger cascade (Markram and Segal 1992, Seigel and Auerbach 1996, Hasselmo and Bower 1993, Hasselmo et al. 1995).

The adaptive value of the cortical strobe provided by the basal magnocellular neurons may thus be that it facilitates memory storage especially when significant (e.g. reinforcing) environmental stimuli are detected. This means that memory storage is likely to be conserved (new memories are less likely to be laid down) when significant environmental stimuli are not present. In that the basal forebrain projection spreads widely to many areas of the cerebral cortex, and in that there are relatively few basal forebrain neurons (in the order of thousands), the basal forebrain neurons do not determine the actual memories that are stored. Instead the actual memories stored are determined by the active subset of the thousands of cortical afferents on to a strongly activated cortical neuron (Treves and Rolls 1994, Rolls and Treves 1998, Rolls 2008d). The basal forebrain magnocellular neurons would then according to this analysis when activated increase the probability that a memory would be stored. Impairment of the normal operation of the basal forebrain magnocellular neurons would be expected to interfere with normal memory by interfering with this function, and this interference could contribute in this way to the memory disorder in Alzheimer's disease.

Thus one way in which impaired cholinergic neuron function is likely to impair memory is by reducing the depth of the basins of attraction of cortical networks, in that these networks store less strongly the representations that are needed for episodic memory and for short-term memory, thus making the recall of long-term episodic memories less reliable in the face of stochastic noise, and the maintenance of short-term memory less reliable in the face of stochastic noise. Such changes would thereby also impair attention.

16.4.4.4 Basal magnocellular forebrain nuclei cholinergic neurons reduce cortical adaptation

Another property of cortical neurons is that they tend to adapt with repeated input (Abbott, Varela, Sen and Nelson 1997, Fuhrmann, Markram and Tsodyks 2002a) (Chapter 10). However, this adaptation is most marked in slices, in which there is no acetylcholine. One effect of acetylcholine is to reduce this adaptation (Power and Sah 2008). The mechanism is understood as follows. The afterpolarization (AHP) that follows the generation of a spike in a neuron is primarily mediated by two calcium-activated potassium currents, I_{AHP} and sI_{AHP} (Sah and Faber 2002), which are activated by calcium influx during action potentials. The I_{AHP} current is mediated by small conductance calcium-activated potassium (SK) channels, and its time course primarily follows cytosolic calcium, rising rapidly after action potentials and decaying with a time constant of 50 to several hundred milliseconds (Sah and Faber 2002). In contrast, the kinetics of the sI_{AHP} are slower, exhibiting a distinct rising phase and decaying with a time constant of 1–2 s (Sah 1996). A variety of neuromodulators, including acetylcholine (ACh), noradrenaline, and glutamate acting via G-protein-coupled receptors, suppress the sI_{AHP} and thus reduce spike-frequency adaptation (Nicoll 1988).

When recordings are made from single neurons operating in physiological conditions in the awake behaving monkey, peristimulus time histograms of inferior temporal cortex neurons to visual stimuli show only limited adaptation. There is typically an onset of the neuronal response at 80–100 ms after the stimulus, followed within 50 ms by the highest firing rate. There is after that some reduction in the firing rate, but the firing rate is still typically more

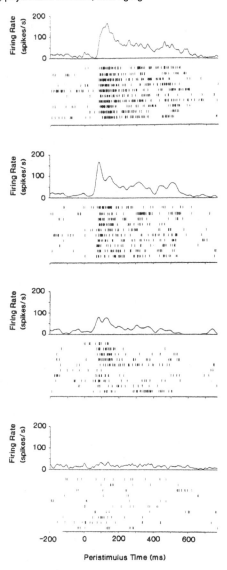

Fig. 16.15 Peristimulus time histograms and rastergrams showing the responses on different trials (originally in random order) of a face-selective neuron in the inferior temporal visual cortex to four different faces. (In the rastergrams each vertical line represents one spike from the neuron, and each row is a separate trial. Each block of the figure is for a different face.) The stimulus onset was at 0 ms, and the duration was 500 ms. (The neuron stopped firing after the 500 ms stimulus disappeared towards the end of the period shown on each trial.) These four faces were presented in random permuted sequence and repeated very many times in the experiment. (Reproduced with permission from Journal of Neurophysiology 70: 640–654. Information encoding and the responses of single neurons in the primate temporal visual cortex, M. J. Tovee, E. T. Rolls, A. Treves, R. P. Bellis. Copyright ©1993, The American Physiological Society.)

than half-maximal 500 ms later (see example in Fig. 16.15). Thus under normal physiological conditions, firing rate adaptation can occur, but does not involve a major adaptation, even when cells are responding fast (at e.g. 100 spikes/s) to a visual stimulus. One of the factors that keeps the response relatively maintained may however be the presence of acetylcholine. The depletion of acetylcholine in aging and some disease states (Schliebs and Arendt 2006) could lead to less sustained neuronal responses (i.e. more adaptation), and this may contribute to the symptoms found. In particular, the reduced firing rate that may occur as a function of time if acetylcholine is low would gradually over a few seconds reduce the depth of the basin of attraction, and thus destabilize short-term memory when noise is present, by reducing the firing rate component shown in equation B.12. I suggest that such changes would thereby impair short-term memory, and thus also top-down attention.

The effects of this adaptation can be studied by including a time-varying intrinsic (potassium like) conductance in the cell membrane, as described in Section B.6.2. A specific implementation of the spike-frequency adaptation mechanism using Ca^{++}-activated K^+ hyperpolarizing currents (Liu and Wang 2001) is described in Section 10.1.2, and was used by Deco and Rolls (2005c) (who also describe two other approaches to adaptation).

On the basis of these hypotheses, it is predicted that enhancing cholinergic function will help to reduce the instability of attractor networks involved in short-term memory and attention that may occur in aging.

Part of the interest of this stochastic dynamics approach to aging is that it provides a way to test combinations of pharmacological treatments, that may together help to minimize the cognitive symptoms of aging. Indeed, the approach facilitates the investigation of drug combinations that may together be effective in doses lower than when only one drug is given. Further, this approach may lead to predictions for effective treatments that need not necessarily restore the particular change in the brain that caused the symptoms, but may find alternative routes to restore the stability of the dynamics.

16.5 Highlights

1. Hypotheses are described about how increased or decreased stability of different cortical networks may contribute to psychiatric disorders and to the memory changes in normal aging.

2. In schizophrenia, the positive symptoms such as hallucinations and intrusive thoughts may be associated with too little stability of attractor networks caused by reduced inhibition (and/or increased glutamatergic function) in the temporal lobe cortex and related areas. The reduction of stability results in some networks entering high firing rate attractor states even with no or little input.

3. In schizophrenia, the cognitive symptoms such as impaired attention and working memory may be related to reduced efficacy of NMDA glutamate receptors and/or decreased numbers of spines introducing excitatory input to neurons in the prefrontal cortex involved in short-term memory and attention. The reduced firing rate would decrease the stability of high firing rate attractor network states involved in maintaining stable working memory and attention.

4. In schizophrenia, the same changes as in (3) in the orbitofrontal cortex would decrease the firing rates of neurons involved in reward, punishment, emotion, and motivation, and thus produce the negative symptoms of decreased mood and motivation.

5. In obsessive-compulsive disorder, enhanced glutamatergic activity may make a number of different cortical attractor networks overstable. The details of which attractor networks are overstable may be different in different patients. Overactivity of some frontal cortical attractor networks might produce over-stability and perseveration of plans. Overstability of other frontal cortical attractor networks might produce repeated actions.

6. In depression, oversensitivity and overactivity of a lateral orbitofrontal cortex network involved in non-reward and punishment may lead it to enter too often and remain too long in high firing rate attractor states, leading to the symptoms of persistent sadness which can be produced by non-reward. The medial orbitofrontal frontal cortex reward system, which is reciprocally related to the lateral orbitofrontal cortex non-reward system, may as a result of this reciprocity be down-regulated, and have reduced functional connectivity with parahippocampal memory areas, resulting in fewer happy memory states.

7. In normal aging, reduced neuronal firing rates in attractor memory networks may contribute to the reduced working memory. The reduction of firing rates may be related to decreased synaptic efficacy and learning, and to increased cortical adaptation.

8. All these hypotheses have implications for potential treatments, and provide a new way to conceptualize some psychiatric disorders, and the effects of normal aging on memory.

9. Part of the value of the approach is that it links changes at the neuronal and synaptic level, and how they are influenced by treatments, to the symptoms, through understanding the principles of operation of neuronal networks in the cerebral cortex.

17 Syntax and Language

17.1 Neurodynamical hypotheses about language and syntax

When considering the computational processes underlying language, it is helpful to analyze the rules being followed (Chomsky 1965, Jackendoff 2002). From this, it is tempting to see what one can infer about how the computations are implemented, using for example logical operations within a rule-based system, and switches turned on during development.

In this Chapter, instead the approach is to take what are understood as some of the key principles of the operation of the cerebral cortex (Rolls 2008d, Rolls 2016a) based on the operation of memory and sensory systems in the brain, and how information is encoded in the brain (Appendix C), and then to set up hypotheses about how some of these computational mechanisms might be useful and used in the implementation of language in the brain (Rolls and Deco 2015a). Then the hypotheses are tested and elucidated by simulations of some of the neuronal network operations involved (Rolls and Deco 2015a).

17.1.1 Binding by synchrony?

A fundamental computational issue is how the brain implements binding of elements such as features with the correct relationship between the elements. The problem in the context of language might arise if we have neuronal populations each firing to represent a subject, a verb, and an object of a sentence. If all we had were three populations of neurons firing, how would we know which was the subject, which the verb, and which the object? How would we know that the subject was related to the verb, and that the verb operated on the object? How these relations are encoded is part of the problem of binding.

Von der Malsburg (1990a) considered this computational problem, and suggested a dynamical link architecture in which neuronal populations might be bound together temporarily by increased synaptic strength which brought them into temporary synchrony. This led to a great deal of research into whether arbitrary relinking of features in different combinations is implemented in visual cortical areas using synchrony (Singer, Gray, Engel, Konig, Artola and Brocher 1990, Engel, Konig, Kreiter, Schillen and Singer 1992, Singer and Gray 1995, Singer 1999, Abeles 1991, Fries 2009), and this has been modelled (Hummel and Biederman 1992). However, although this approach could specify that two elements are bound, it does not specify the relation (Rolls 2008d). For example, in vision, we might know that a triangle and a square are part of the same object because of synchrony between the neurons, but we would not know the spatial relation, for example whether the circle was inside the triangle, or above, below it, etc. Similarly for language, we might know that a subject and an object were part of the same sentence, but we would not know which was the subject and which the object, that the subject operated (via a verb) on the object, etc, that is, the syntactic relations would not be encoded just by synchrony. Indeed, neurophysiological recordings show that although synchrony can occur in a dynamical system such as the brain, synchrony *per se* between neurons in high order visual cortical areas conveys little information about which objects are being represented, with 95% of the information present in the number of spikes being emitted by each of the neurons in the population (Rolls 2008d, Rolls and Treves 2011) (Appendix C).

Instead, in high order visual cortical areas, the spatial relations between features and objects are encoded by neurons that have spatially biased receptive fields relative to the fovea (Aggelopoulos and Rolls 2005, Rolls 2008d, Rolls 2012c), and this feature/place coding scheme is computationally feasible (Elliffe, Rolls and Stringer 2002) (Section 25.5.5). In addition, coherence and temporal synchrony do not appear to be well suited for information transmission, for in quantitative neuronal network simulations, it is found that information is transmitted between neuronal populations at much lower values of synaptic strength than those needed to achieve coherence (Rolls, Webb and Deco 2012) (Section 8.3).

In this situation, I now make alternative proposals for how syntactic relations are encoded in the cortex (see Rolls and Deco (2015a)).

17.1.2 Syntax using a place code

The overview of encoding in the cortex in Appendix C (Rolls 2008d, Rolls and Treves 2011) leads to a hypothesis about how syntax, or the relations between the parts of a sentence, is encoded for language. My hypothesis is that a place code is used, with for example one cortical module or region used to represent subjects, another cortical module used to represent verbs, and another cortical module used to represent objects (see Rolls and Deco (2015a)).

The size of any such neocortical module need not be large. An attractor network in the cortex need occupy no more than a local cortical area perhaps 2–3 mm in diameter within which there are anatomically dense recurrent collateral associatively modifiable connections between the neurons (Rolls 2008d) (Section 4.5). This cortical computational attractor network module would thus be about the size of a cortical column. It is an attractor network module in the sense that neurons more than a few mm away would not be sufficiently strongly activated to form part of the same attractor network (Rolls 2008d). An attractor network of this type with sparse distributed representations can store and encode approximately as many items are there are synaptic connections onto each cortical neuron from the nearby neurons (Treves and Rolls 1991, Rolls and Treves 1998, Rolls 2008d) (for details, see Section 4.5). The implication for language is that of order 10,000 nouns could be stored in a single cortical attractor network with 10,000 recurrent collateral connections on to each neuron. This capacity is only realized if there is only a low probability of more than one recurrent collateral connection between any pair of the neurons in a module, and this has been proposed as one of the underlying reasons for why cortical connectivity is diluted, with a probability in the order of 0.1 for connections between any pair of nearby neurons in the neocortex, and 0.02 for CA3 neurons in the rodent hippocampus (Rolls 2012a) (Chapter 7).

The hypothesis is further that different local cortical modules encode the nouns that are the subjects of sentences, and that are the objects of sentences. A prediction is thus that there will be single neurons in a human cortical language area that respond to a noun when it is the subject but not the object of a sentence, and vice versa. Consistent with the hypothesis, evidence was found in an fMRI study that the agent (or actor, the subject of an active sentence) is localised separately from the patient (or object of an active sentence) in the left mid-superior temporal cortex (Frankland and Greene 2015).

Clearly the full details of the system would be more complicated, but the general hypothesis is that adjectives and adjectival phrases that are related to the subject of a sentence will have strong connections to the subject module or modules; that adverbs and adverbial phrases that are related to the verbs of sentence will have strong connections to the verb module or modules; and that adjectives and adjectival phrases that are related to the object of a sentence will have strong connections to the object module or modules.

17.1.3   Temporal trajectories through a state space of attractors

To represent syntactical structure *within* the brain, what has been proposed already might be along lines that are consistent with the principles of cortical computation (Rolls 2008d, Rolls 2016a). The high representational capacity would be provided for by the high capacity of a local cortical attractor network, and syntactic binding within a brain would be implemented by using a place code in which the syntactic role would be defined by which neurons are firing – for example, subjects in one cortical module or modules, and objects in another cortical module or modules.

However, a problem arises if we wish to communicate this representation to another person, for the neural implementation described so far could not be transferred to another person without transferring which neurons in the language areas were currently active, and having a well trained person as the decoder!

To transfer or communicate what is encoded in the representations to another person, and the relations or syntax, it is proposed that a number of mechanisms might be used. One might be a temporal order encoding, for example the subject–verb–object encoding that is usual in English, and which has the advantage of following the temporal order that usually underlies causality in the world. Another mechanism might be the use of inflections (usually suffixes) to words to indicate their place in the syntax, such as cases for nouns (e.g. nominative for the subject or agent, and accusative for the object or patient, dative, and genitive), and person for verbs (e.g. first, second, and third person singular and plural, to specify I, you, he/she/it, we, you, they) used to help disambiguate which noun or nouns operate on the verb. Another mechanism is the use of qualifying prepositions to indicate syntactic role of a temporally related word, with examples being 'with', 'to', and 'from'. This mechanism is used in combination with temporal order in English.

In this Chapter, I focus on temporal order as an encoder of syntactical relations, and next set out hypotheses on how this could be implemented in the cerebral cortex based on the above computational neuroscience background (Rolls and Deco 2015a).

17.1.4   Hypotheses about the implementation of language in the cerebral cortex

1. Subjects, verbs, and objects are encoded using sparse distributed representations (Rolls 2008d, Rolls and Treves 2011) in localised cortical attractor networks. One cortical module with a diameter of 2–3 mm and 10,000 recurrent collateral connections per neuron could encode in the order of 10,000 items (e.g. subjects or verbs or objects) (Rolls 2008d, Treves and Rolls 1991) (Section 4.5). One cortical module would thus be sufficient to encode all the objects, all the verbs, or all the objects (depending on the module) in most people's working vocabulary, which is of the order to several thousand nouns, or verbs.

This follows from the analysis that the capacity of an attractor network with sparse encoding a (where for binary networks a is the proportion of neurons active for any one memory pattern) is as follows, and from the fact that there are in the order of 10,000 recurrent collateral connections on each neuron (Rolls 2008d). The capacity is measured by the number of patterns p that can be stored and correctly retrieved from the attractor network

$$p \approx \frac{C}{a \ln(\frac{1}{a})} k \tag{17.1}$$

where C is the number of synapses on the dendrites of each neuron devoted to the recurrent collaterals from other neurons in the network, and k is a factor that depends weakly on the detailed structure of the rate distribution, on the connectivity pattern, etc., but is roughly in the order of 0.2–0.3 (Treves and Rolls 1991, Rolls 2008d, Rolls 2012a) (Section 4.5).

The use of attractor networks for language-based functions is itself important. Cortical computation operates at the neuronal level by computing the similarity or dot product or correlation between an input vector of neuronal firing rates and the synaptic weights that connect the inputs to the neurons (Rolls 2008d, Rolls 2012d) (Chapter 8). The output of the neuron is a firing rate, usually between 0 and 100 spikes/s. Cortical computation at the neuronal level is thus largely analogue. This is inherently not well suited to language, in which precise, frequently apparently logical, rules are followed on symbolic representations. This gap may be bridged by attractor or autoassociation networks in the brain, which can enter discrete attractor high firing rate states that can provide for error correction in the analogue computation, and to the robustness to noise, that are often associated with the processing of discrete symbols (Treves 2005, Rolls 2008d, Rolls 2012d). These discrete attractor states are network properties, not the properties of single neurons, and this capability is at the heart of much cortical computation including long-term memory, short-term memory, and decision-making (Rolls 2008d, Rolls 2014a), as elucidated in this book.

2. Place coding with sparse distributed representations is used in these attractors. The result is that the module that is active specifies the syntactic role of what is represented in it. One cortical module would be for subjects, another for verbs, another for objects, etc.

3. The presence of these weakly coupled attractors would enable linguistic operations of a certain type to be performed within the brain, but the information with the syntax could not be communicated in this form to other people. In this statement, 'weakly coupled' is clearly and quantitatively defined by attractors that have weak interconnections so that they can have different basins of attraction, yet can influence each other (Rolls 2008d) (Section 4.10). The computations involved in these interactions might instantiate a 'language of thought' that would be below the level of written or spoken speech, and would involve for example constraint satisfaction within coupled attractors, and the types of brain-style computation described by Rolls (2012d) (see Section 26.3 on brain computation vs computation in a digital computer). The cortical computational processes could be usefully influenced and made creative by the stochastic dynamics of neuronal networks in the brain that are due to the 'noisy' Poisson-like firing of neurons (Rolls 2008d, Rolls and Deco 2010) (Chapter 5). When someone has a hunch that they have solved a problem, this may be the computational system involved in the processing. This might be termed a 'deep' structure or layer of linguistic processing.

4. To enable these computations that involve syntactical relations to be communicated to another person or written down to be elaborated into an extended argument, the process considered is one involving weakly forward-coupled attractor networks. One such system would be to have weak forward coupling between subject–verb–object attractor networks. The exact trajectories followed (from subject to verb to object) could be set up during early language learning, by forming during such learning stronger forward than reverse connections between the attractors, by for example spike-timing dependent plasticity (Markram et al. 1997, Bi and Poo 1998, Feldman 2012) and experience with the order of items that is provided during language learning. Which trajectory was followed would be biased by which subject, which verb, and which object representation was currently active in the deep layer. These temporal trajectories through the word attractors would enable the syntactical relations to be encoded in the temporal order of the words.

With this relatively weak coupling between attractors implemented with integrate-and-fire neurons and low firing rates, the transition from one active attractor to the next can be relatively slow, taking 100–400 or more ms (Deco and Rolls 2005c). This property of the system adapts it well to the production of speech, in which words are produced sequentially with a spacing in the order of 300–500 ms, a rate that is influenced by the mechanics and therefore dynamics of the speech production muscles and apparatus.

A simulation testing this system and making the details of the operation of the system and their biological plausibility clear is described in the Methods and Results sections of Rolls and Deco (2015a).

5. The system for enabling the syntax to be communicated to other people or written down would have some computational advantages apart from purely the communication. In particular, once the syntax can be formally expressed in written statements, it becomes easier to perform logical operations on the statements, which become propositional, and can be tested. These logical operations, and reasoning, may not be the style of computation utilized in general by computational processes within the brain (see Rolls (2012d) Section 2.15), but may become algorithms that can be followed to achieve quantitatively precise and accurate results, as in long division, or by learning logic. Thus the importance of communication using syntax may allow other environmental tools to be applied to enable reasoning and logic that is not the natural style of neural computation.

6. To enable the system to produce words in the correct temporal order, and also to remember with a lower level of neuronal firing what has just been said for monitoring in case it needs correcting, a mechanism such as spike frequency adaptation may be used, as described next (see also Chapter 10).

A property of cortical neurons is that they tend to adapt with repeated input (Abbott et al. 1997, Fuhrmann et al. 2002a). The mechanism is understood as follows. The afterpolarization (AHP) that follows the generation of a spike in a neuron is primarily mediated by two calcium-activated potassium currents, I_{AHP} and the sI_{AHP} (Sah and Faber 2002), which are activated by calcium influx during action potentials. The I_{AHP} current is mediated by small conductance calcium-activated potassium (SK) channels, and its time course primarily follows cytosolic calcium, rising rapidly after action potentials and decaying with a time constant of 50 to several hundred milliseconds (Sah and Faber 2002). In contrast, the kinetics of the sI_{AHP} are slower, exhibiting a distinct rising phase and decaying with a time constant of 1–2 s (Sah 1996). A variety of neuromodulators, including acetylcholine (ACh) acting via a muscarinic receptor, noradrenaline, and glutamate acting via G-protein-coupled receptors, suppress the sI_{AHP} and thus reduce spike-frequency adaptation (Nicoll 1988).

When recordings are made from single neurons operating in physiological conditions in the awake behaving monkey, peristimulus time histograms of inferior temporal cortex neurons to visual stimuli show only limited adaptation. There is typically an onset of the neuronal response at 80–100 ms after the stimulus, followed within 50 ms by the highest firing rate. There is after that some reduction in the firing rate, but the firing rate is still typically more than half-maximal 500 ms later (see example in Tovee, Rolls, Treves and Bellis (1993); Fig. 16.15). Thus under normal physiological conditions, firing rate adaptation can occur.

The effects of this adaptation can be studied by including a time–varying intrinsic (potassium-like) conductance in the cell membrane (Brown, Gähwiler, Griffith and Halliwell 1990a, Treves 1993, Rolls 2008d). This can be done by specifying that this conductance, which if open tends to shunt the membrane and thus to prevent firing, opens by a fixed amount with the potential excursion associated with each spike, and then relaxes exponentially to its closed state. In this manner sustained firing driven by a constant input current occurs at lower

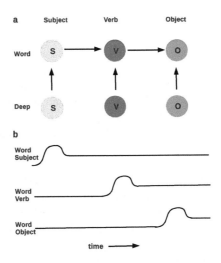

Fig. 17.1 Schematic diagram of the concepts. a. Each circle indicates a local cortical attractor network capable of storing 10,000 items. The Deep layer local cortical attractor networks use place coding to encode the syntax, that is, the syntactic role of each attractor network is encoded by where it is in the cortex. In this implementation there are separate Subject (S), Verb (V), and Object (O) networks. Syntax within the brain is implemented by place encoding. For communication, the deep representation must be converted into a sequence of words. To implement this, the Deep attractor networks provide a weak selective bias to the Word attractor networks, which have weak non-selective forward coupling S to V to O. b. The operation in time of the system. The Deep networks fire continuously, and the syntax is implemented using place encoding. The Deep networks apply a weak selective bias to the Word networks, which is insufficient to make a word attractor fire, but is sufficiently strong to bias it later into the correct one of its 10,000 possible attractor states, each corresponding to a word. Sentence production is started by a small extra input to the Subject Word network. This with the selective bias from the Deep subject network make the Word subject network fall into an attractor, the peak firing of which is sufficient to elicit production of the subject word. Adaptation in the subject network makes its firing rate decrease, but still remain in a moderate firing rate attractor state to provide a short-term memory for the words uttered in a sentence, in case they need to be corrected or repeated. The high and moderate firing rate in the Subject Word network provides non-selective forward bias to the whole of the Object Word network, which falls into a particular attractor produced by the selective bias from the Deep Verb network, and the verb is uttered. Similar processes then lead to the correct object being uttered next. The sentence can be seen as a trajectory through a high dimensional state space of words in which the particular words in the sentence are due to the selective bias from the Deep networks, and the temporal order is determined by the weak forward non-selective connections between the networks, i.e. connections from subject-to-verb, and verb-to-object networks. The simulations show dynamical network principles by which this type of sentence encoding and also decoding could be implemented. The overall concept is that syntax within the brain can be solved by the place coding used in most other representations in the brain; and that the problems with syntax arise when this place-coded information must be transmitted to another individual, when one solution is to encode the role in syntax of a word by its temporal order in a sentence. (After Rolls and Deco, 2015a.)

rates after the first few spikes, in a way similar, if the relevant parameters are set appropriately, to the behaviour observed in vitro of many pyramidal cells (for example, Lanthorn, Storm and Andersen (1984), Mason and Larkman (1990)). The details of the implementation used are described in Section 10.1.2 and in Rolls and Deco (2015a).

17.2   Tests of the hypotheses – a model

17.2.1   An integrate-and-fire network with three attractor network modules connected by stronger forward than backward connections

The computational neuroscience aspects of the hypotheses described above were investigated with an integrate-and-fire network with three attractor network modules connected by stronger forward than backward connections, the operation of which is illustrated conceptually in Figs. 17.1 and 17.2, and the architecture of which is illustrated in Fig. 17.3. Each module is an integrate-and-fire attractor network with n possible attractor states. For the simulations described (Rolls and Deco 2015a) there were $n=10$ orthogonal attractor states in each module, each implemented by a population of neurons with strong excitatory connections between the neurons with value $w_+=2.1$ for the NMDA and AMPA synapses. There were $N=8000$ neurons in each module, of which 0.8 (i.e. 6400) were excitatory, and 1600 were inhibitory. The first module could represent one of 10 Subjects (S), the second module one of 10 verbs (V), and the third module one of 10 objects (O). In the cerebral cortex, each module might be expected to be able to encode up to 10,000 items using sparse distributed representations, and assuming of order 10,000 excitatory recurrent collateral connection onto each neuron (Treves and Rolls 1991, Treves and Rolls 1994, Rolls 2008d). The parameters in each module were set so that the spontaneous firing state with no input applied was stable, and so that only one of its n possible attractor states was active at any time when inputs were applied (Rolls 2008d, Rolls and Deco 2010, Deco, Rolls, Albantakis and Romo 2013).

In the model, there are forward connections from all excitatory neurons in one module to all excitatory neurons in the next module, with uniform strength w_{ff}. The role of these forward connections is to produce some non-specific input to the next module of the network when the previous module enters an attractor state with one of its neuronal pools. The role of this forward input is to encourage the next module in the series to start firing with some small delay after the previous module, to enable a temporal trajectory through a state space of an active attractor pool in one module to an active attractor pool in the next module. In the cerebral cortex, there are typically both stronger forward than backward connections between modules in different cortical areas, and connections in both directions between modules in the same cortical area (Rolls 2008d). In the latter case, the hypothesis is that the stronger connections in one direction than the reverse between nearby modules might be set up by spike-timing dependent plasticity (Markram et al. 1997, Bi and Poo 1998, Feldman 2012) based on the temporal order in which the relevant modules were normally activated during a stage when a language was being learned. For the simulations, only weak forward connections were implemented for simplicity.

During operation, one attractor pool in each module receives a continuous bias throughout a trial from a Deep network (see Figs. 17.1 and 17.2). For example the Subject Word module might receive bias from a Deep attractor pool representing 'James', the Verb Word module might receive bias from an attractor pool representing 'chased', and the Object Word module might receive bias from an attractor pool representing 'John'. These biases would represent the deep structure of the sentence, what it is intended to say, but not the generation of the sentence, which is the function of the network shown in Fig. 17.3. The biases in all the modules apart from the first are insufficient to push any attractor pool in a Word module into an attractor state. In the first (or head) Word module, the bias is sufficiently strong to make the attractor pool being biased enter a high firing rate attractor state. The concept is that because of the forward connections to all neurons in the second (Verb Word) module, the attractor pool in the second module receiving a steady bias (in our example, the 'chased' pool) then

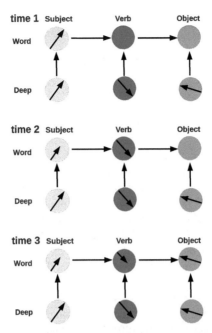

Fig. 17.2 The operation of the model, illustrated at three different time steps. Time 1 is during the production of the subject work, time 2 during the verb word, and time 3 during the object word. Each circle represents an attractor network that can fall into one of 10,000 possible states indicated by the direction of the arrow vector, each state corresponding to a word (for the Word-level networks) or a semantic concept (for the Deep-level networks). The length of the arrow vector indicates the firing rate of the selected attractor. The Deep attractor networks fire continuously, with the syntactic role indicated by the particular network using place coding. The Deep networks, which represent semantics or meaning, provide a weak selective bias continuously to the Word attractor networks. The sequential operation of the Subject then Verb then Object Word networks is produced by the weak non-selective forward connections between the networks. After its initial high firing rate, a Word attractor network remains active at a lower firing rate as a result of adaptation, to provide a short-term memory for the words uttered in a sentence, in case they need to be corrected or repeated. (After Rolls and Deco, 2015a.)

has sufficient input for it to gradually enter an attractor. The same process is then repeated for the biased attractor in Word module 3, which then enters a high firing rate state. Due to the slow stochastic dynamics of the network, there are delays between the firing in each of the Word modules. It is this that provides the sequentiality to the process that generates the words in the sentence in the correct order.

In addition, a concept of the cortical dynamics of this system is that each module should maintain a level of continuing firing in its winning attractor for the remainder of the sentence, and even for a few seconds afterwards. The purpose of this is to enable correction of the process (by for example a higher order thought or monitoring process, see Rolls (2014a) and Chapter 22) if the process needs to be corrected. The maintenance of the attractors in a continuing state of firing enables monitoring of exactly which attractors did occur in the trajectory through the state space, in case there was a slip of the tongue. (Indeed, 'slips of the tongue' or speech production errors are accounted for in this framework by the somewhat noisy trajectory through the state space that is likely to occur because of the close to Poisson spiking times of the neurons for a given mean rate, which introduces noise into the system (Rolls 2008d, Rolls and Deco 2010, Rolls 2014a)). The main parameters of each module that enable this to be achieved are w_+, and the external bias entering each attractor pool.

Subject   Verb   Object

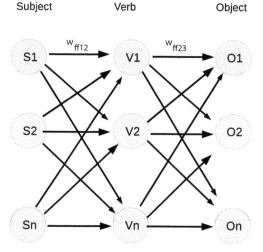

Fig. 17.3 The attractor network model. There are three modules, Subject (S), Verb (V), and Object (O). Each module is a fully connected attractor network with $n=10$ pools of excitatory neurons. The ten excitatory pools each have 640 excitatory neurons, and each module has 1600 inhibitory neurons using GABA as the transmitter. Each excitatory pool has recurrent connections with strength $w_+=2.1$ to other neurons in the same pool implemented with AMPA and NMDA receptors. There are forward connections with strength w_{ff12} from all excitatory neurons in module 1 (S) to all excitatory neurons in module 2 (V). There are forward connections with strength w_{ff23} from all excitatory neurons in module 2 (V) to all excitatory neurons in module 3 (O). An external bias can be applied to any one or more of the attractor pools in each of the modules. In operation for production, a stronger bias is applied to one pool in module 1 to start the process, and then an attractor emerges sequentially in time in each of the following modules. The particular pool that emerges in each of the later modules depends on which pool in that module is receiving a weak bias from another (deeper) structure that selects the items to be included in a sentence. The syntax of the sentence, encoded in the order of the items, is determined by the connectivity and dynamics of the network. The same network can be used for decoding (see text). (After Rolls and Deco, 2015a.)

However, although it is desired to have a short-term memory trace of previous activity during and for a short time after a sentence, it is also important that each word is uttered at its correct time in the sentence, for this carries the syntactic relations in this system. To achieve the production of the word at the correct time, the firing of each attractor has a mechanism to produce high firing initially for perhaps 200–300 ms, and then lower firing later to main an active memory trace of previous neuronal activity. The mechanism used to achieve this initial high firing when a neuronal pool enters a high firing rate state, was spike frequency adaptation, a common neuronal process which is described in Chapter 10, and which has been implemented and utilized previously (Liu and Wang 2001, Deco and Rolls 2005c, Rolls and Deco 2015a).

17.2.2 The operation of a single attractor network module

The aim is to investigate the operation of the system in a biophysically realistic attractor framework, so that the properties of receptors, synaptic currents and the statistical effects related to the probabilistic spiking of the neurons can be part of the model. We use a minimal architecture, a single attractor or autoassociation network (Hopfield 1982, Amit 1989, Hertz et al. 1991, Rolls and Treves 1998, Rolls and Deco 2002, Rolls 2008d) for each module. A recurrent (attractor) integrate-and-fire network model which includes synaptic channels for

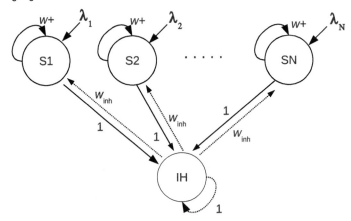

Fig. 17.4 The architecture of one module containing one fully connected attractor network. The excitatory neurons are divided into N=10 selective pools or neuronal populations S1–SN of which three are shown, S1, S2 and SN. The synaptic connections have strengths that are consistent with associative learning. In particular, there are strong intra-pool connection strengths w_+. The excitatory neurons receive inputs from the inhibitory neurons with synaptic connection strength w_{inh}=1. The other connection strengths are 1. The integrate-and-fire spiking module contained 8000 neurons, with 640 in each of the 10 non-overlapping excitatory pools, and 1600 in the inhibitory pool IH. Each neuron in the network also receives external Poisson inputs λ_{ext} from 800 external neurons at a typical rate of 3 Hz/synapse to simulate the effect of inputs coming from other brain areas. (After Rolls and Deco, 2015a.)

AMPA, NMDA and GABA$_A$ receptors (Brunel and Wang 2001, Rolls and Deco 2010) was used.

Each Word attractor network contains 6400 excitatory, and 1600 inhibitory neurons, which is consistent with the observed proportions of pyramidal cells and interneurons in the cerebral cortex (Abeles 1991, Braitenberg and Schütz 1991). The connection strengths are adjusted using mean-field analysis (Brunel and Wang 2001, Deco and Rolls 2006, Rolls and Deco 2010), so that the excitatory and inhibitory neurons exhibit a spontaneous activity of 3 Hz and 9 Hz, respectively (Wilson, O'Scalaidhe and Goldman-Rakic 1994b, Koch and Fuster 1989). The recurrent excitation mediated by the AMPA and NMDA receptors is dominated by the NMDA current to avoid instabilities during delay periods (Wang 2002).

The architecture of the cortical network module illustrated in Fig. 17.4 has 10 selective pools each with 640 neurons. The connection weights between the neurons within each pool or population are called the intra-pool connection strengths w_+, which were set to 2.1 for the simulations described. All other weights including w_{inh} were set to 1.

All the excitatory neurons in each attractor pool S1, S2 ... SN receive an external bias input $\lambda_1, \lambda_2 ... \lambda_N$. This external input consists of Poisson external input spikes via AMPA receptors which are envisioned to originate from 800 external neurons. One component of this bias which is present by default arrives at an average spontaneous firing rate of 3 Hz from each external neuron onto each of the 800 synapses for external inputs, consistent with the spontaneous activity observed in the cerebral cortex (Wilson et al. 1994b, Rolls and Treves 1998, Rolls 2008d). The second component is a selective bias from a deep structure system which provides a bias present throughout a trial to one of the attractor pools in each module, corresponding to the subject, verb, or object (depending on the module) to be used in the sentence being generated. This bias makes it more likely that the attractor pool will become active if there are other inputs, but is not sufficiently strong (except in the first module) to initiate a high firing rate attractor state. (This selective bias might be set up by associative synaptic modification between the Deep and the Word modules.) In addition, all excitatory neurons in a module receive inputs with a uniform synaptic strength of w_{ff} from all the excitatory neurons in the preceding module, as illustrated in Fig. 17.3.

Both excitatory and inhibitory neurons are represented by a leaky integrate-and-fire model (Tuckwell 1988). The basic state variable of a single model neuron is the membrane potential. It decays in time when the neurons receive no synaptic input down to a resting potential. When synaptic input causes the membrane potential to reach a threshold, a spike is emitted and the neuron is set to the reset potential at which it is kept for the refractory period. The emitted action potential is propagated to the other neurons in the network. The excitatory neurons transmit their action potentials via the glutamatergic receptors AMPA and NMDA which are both modeled by their effect in producing exponentially decaying currents in the postsynaptic neuron. The rise time of the AMPA current is neglected, because it is typically very short. The NMDA channel is modeled with an alpha function including both a rise and a decay term. In addition, the synaptic function of the NMDA current includes a voltage dependence controlled by the extracellular magnesium concentration (Jahr and Stevens 1990). The inhibitory postsynaptic potential is mediated by a $GABA_A$ receptor model and is described by a decay term. A detailed mathematical description is provided in the Appendix.

17.2.3 Spike frequency adaptation mechanism

A specific implementation of the spike-frequency adaptation mechanism using Ca^{2+}-activated K^+ hyper-polarizing currents (Liu and Wang 2001) was implemented, and is described in Section 10.1.2. Its parameters were chosen to produce spike frequency adaptation similar in timecourse to that found in the inferior temporal visual cortex of the behaving macaque (Tovee, Rolls, Treves and Bellis 1993) (Fig. 16.15). In particular, $[Ca^{2+}]$ is initially set to be 0 μM, $\tau_{Ca} = 300$ ms, $\alpha = 0.002$, $V_K = -80$ mV and $g_{AHP}=200$ nS. (We note that there are a number of other biological mechanisms that might implement the slow transitions from one attractor state to another, some investigated by Deco and Rolls (2005c), and that we use the spike frequency adaptation mechanism to illustrate the principles of operation of the networks.)

17.3 Tests of the hypotheses – findings with the model

17.3.1 A production system

The operation of the integrate-and-fire system illustrated in Fig. 17.3 is shown in Fig. 17.5 when it is producing a subject – verb – object sequence. The firing rates of attractor pools 1 in Word modules 1 (subject), 2 (verb), and 3 (object) are shown. No other pool had any increase of its firing rate above baseline. The attractor pools 1 in modules 2 and 3 received an increase above the baseline rate of 3.00 Hz per synapse (or 2400 spikes/s per neuron given that each neuron receives these inputs through 800 synapses) to 3.03 Hz per synapse throughout the trial. This itself was insufficient to move any attractor pool in modules 2 and 3 into a high firing rate state, as illustrated, until one of the attractor pools in a preceding module had entered a high firing rate attractor state. At time = 500 ms, the external input into attractor pool 1 of module 1 (subject) was increased to 3.20 Hz per synapse, and maintained at this value for the rest of the trial. This produced after a little delay due to the stochastic recurrent dynamics an increase in the firing of pool 1 in module 1, which peaked at approximately 1000 ms. The initial peak firing rate of approximately 40 spikes/s was followed by a reduction due to the spike frequency adaptation to approximately 25 spikes/s, and this level was maintained for the remainder of the trial, as shown in Fig. 17.5. The parameters for the spike frequency

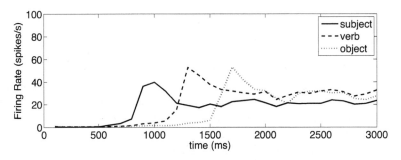

Fig. 17.5 Firing rates of the biased pools in the Subject–Verb–Object modules as a function of time. (After Rolls and Deco, 2015a.)

adaptation for all the excitatory neurons in the network were V_k=-80 mV, g_{AHP}=200 nS, α_{Ca}=0.002, and τ_{Ca} = 300 ms.

The increase in the firing in attractor pool 1 of module 1 influenced the neurons in module 2 via the feedforward synaptic strength of w_{ff12}=0.55, and, because attractor pool 1 in module 2 already had a weak external bias to help it be selected, it was attractor pool 1 in module 2 that increased its firing, with the peak rate occurring at approximately 1250 ms, as shown in Fig. 17.5. Again, the peak was followed by low maintained firing in this pool for the remainder of the trial.

The increase in the firing in attractor pool 1 of module 2 influenced the neurons in module 3 via the feedforward synaptic strength of w_{ff23}=0.4, and, because attractor pool 1 in module 3 already had a weak external bias to help it be selected, it was attractor pool 1 in module 3 that increased its firing, with the peak rate occurring at approximately 1650 ms, as shown in Fig. 17.5. Again, the peak was followed by low maintained firing in this pool for the remainder of the trial. The value of w_{ff23} was optimally a little lower than that of w_{ff12}, probably because with the parameters used the firing in module 2 was somewhat higher than that in module 1. All subsequent stages would be expected to operate with w_{ff}=0.4.

The network thus shows that the whole system can reliably perform a trajectory that is sequential and delayed at each step through the state space, with the item selected in each module determined by the steady bias being received from a deep structure containing the items to be included in the sentence. However, the order in which the items were produced and which specified the syntax was determined by the connectivity and dynamics of the Word network. In this particular example, the resulting sentence might correspond to 'James chased John', which has a completely different syntax and meaning to 'John chased James', 'chased John James', etc. The peak of firing in each module could be used to produce the appropriate word in the correct order. The lower rate of ongoing firing in each module provided the basis for the items produced in the sentence to be remembered and used for monitoring, with the role of each item in the sentence made explicit by the module in which the attractor pool was still active.

17.3.2 A decoding system

In Section 17.3.1 the operation of the system when it is operating to produce a subject–verb–object sequence in which the temporal sequence encodes syntactic information is described. In this section, the operation of the system when it decodes a Subject–Verb–Object sentence is considered. The aim is to receive as input the temporal sequence, and to activate the correct attractor in the Subject, Verb, and Object Word attractor modules. The syntactic information

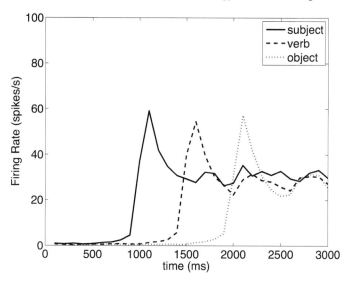

Fig. 17.6 Decoding the Subject–Verb–Object sequence to produce activation in the Subject (module 1), Verb (module 2), and Object (module 3) modules. A weak bias was applied to all pools in module 1 throughout the trial (see text). Noun 1, the subject, was applied to module 1 pool 1 and module 3 pool 1 during the period 500–1000 ms. Verb 2, was applied to module 2 pool 2 during the period 1000–1500 ms. Noun 3, the object, was applied to module 1 pool 3 and module 3 pool 3 during the period 1500–2000 ms. The firing of module 1 attractor 1 neurons that reflect the decoded subject, of module 2 attractor 2 neurons that reflect the decoded verb, and of module 3 attractor 3 neurons that reflect the decoded object, are shown. None of the other 30 attractor neural populations became active. (After Rolls and Deco, 2015a.)

in the sequence allows correct decoding of the subject and object nouns in the sentence, when the position in the sequence is the only information that enables a noun to activate a noun attractor in the subject or the object module. The deep semantic attractor modules could then be activated from the Word attractor modules, using selective, associatively modifiable, synaptic connections.

The operation of the system in decoding mode is illustrated in Fig. 17.6. The architecture of the network is the same as that already described. The baseline input to each of the 800 external synapses is maintained at 3 Hz per synapse in all pools in all modules throughout the sentence except where stated. Throughout the trial all pools in module 1 receive a bias of 0.24 Hz on each of the 800 external input synapses. (This corresponds to an extra 192 spikes/s received by every neuron in each of attractor pools.) The aim of this is to prepare all the attractors in module 1, the subject attractor, to respond if an input cue, a word, is received. This bias essentially sets the system into a mode where it is waiting for an input stream to arrive in module 1. All the attractors in module 1 are stable with low firing rates while only this bias is being applied.

At time 500–1000 ms module 1 pool 1 and module 3 pool 1 receive a noun recall cue as an additional input on the external synapses at an additional 0.08 Hz per synapse. (This corresponds to an extra 64 spikes/s received by every neuron in these two pools of neurons.) Module 1 pool 1 goes into an attractor, as illustrated in Fig. 17.6, because it is receiving a noun recall cue and the bias. Module 3 pool 1 does not enter a high firing rate attractor state, even though the word recall cue for its attractor 1 is being applied, because it is not receiving a bias. This shows how the system can decode correctly a noun due to its position in the sequence as a subject or as an object. In the simulations the bias to pool 1 can be left on, or turned off at this stage in the trial, for once an attractor state has been reached by a pool in module 1, it remains with a stable high firing rate for the remainder of the sentence. The continuing firing is implemented to ensure that the subject remains decoded in the system while the rest of the sentence is decoded, and for use even after the end of the sentence. At time 1000 ms, the noun applied to attractor pools 1 in modules 1 and 3 is removed, as it is no longer present in the environment.

At time 1000–1500 ms the verb recall cue is applied to module 2 pool 2, which enters an attractor. The strength of this recall cue alone (the same as before, an additional 0.08 Hz per synapse) is insufficient to cause this pool in module 2 to enter an attractor. However, all pools in module 2 are receiving now via the feed-forward connections w_{ff12} a priming input from the firing now occurring in module 1, and when the verb recall cue is applied to module 2 pool 2, the combined effects of the recall cue and the feedforward inputs cause module 2 pool 2 to enter its correct attractor state to indicate the presence of this verb in the sentence. The intention of the priming forward input from the preceding module is to provide for future expansion of the system, to allow for example correct decoding of two verbs at different positions within the sequence of words in a sentence. Module 2 pool 2 has the Verb recall cue removed at time 1500 ms as it is no longer present in the environment. Module 2 pool 2 however keeps firing in its stable high firing rate attractor state to ensure that the verb remains decoded in the system while the rest of the sentence is decoded, and for use even after the end of the sentence.

At time 1500–2000 ms the object recall cue is applied to module 3 pool 3, and, as a control and test of the syntactic operation of the system, simultaneously also to module 1 pool 3. Module 3 pool 3 enters its correct object attractor, utilising the feedforward priming inputs w_{ff23}. These priming inputs again provide for further expansion of the system, in case there is another object in the sentence in a later clause. Meanwhile module 1 remains in its pool 1 subject attractor state which is now stable in the face of interfering noun inputs because of its deep low energy basin of attraction. This shows how this noun is forced by its order in the sequence into the Object pool, demonstrating how the system is able to decode information about syntactic role that is present from the position of the item in the sequence. Module 3 pool 3 keeps firing in its stable high firing rate attractor state to ensure that the object remains decoded in the system for use even after the end of the sentence.

As noted, the architecture and overall dynamical principles of operation of the system used for decoding were the same as for encoding. The firing rate adaption was left to operate as before, though it is less useful for the decoding. The only parameters that were adjusted a little for the decoding system were $w_+=2.3$ (to help stability of the high firing rate attractor state in the face of for example interfering nouns); $w_{ff12}=w_{ff23}=0.2$; bias for all pools in module 1 $=0.24$ Hz per synapse; and recall cue $=0.08$ Hz per synapse.

The results just described and illustrated in Fig. 17.6 illustrate some of the principles of operation of the functional architecture when decoding sentences. Further results were as follows. If during the application of the noun for the subject (time 500–1000 ms in the simulations) an input effective for an attractor in module 2 was also applied, then a pool in module 2 tends to enter a high firing rate attractor state, for it is receiving both a recall cue and the forward bias from the high firing starting in module 1 and applied to module 2 via w_{ff12}. An implication is that only nouns should be applied to Subject and Object attractor modules. Having attractors for verbs that respond to different recall cues (words that are verbs) helps the system to decode the input stream into the correct modules and pools. Thus the semantics, the words being applied to the network, are important in enabling the system to respond correctly.

17.4 Evaluation of the hypotheses

The system described here shows how a word production system might operate using neuronal architecture of the type found in the cerebral cortex.

One possible objection to such a computational implementation is how to deal with the passive form. What I propose is that temporal order could again be used, but with different coupling between the attractors appropriate for implementing the passive voice that is again learned by early experience, and is selected instead of the active voice by top-down bias in the general way that we have described elsewhere (Deco and Rolls 2003, Deco and Rolls 2005d). The hypothesis is thus that operation of the system for passive sentences would require a different set of connections to be used to generate the correct temporal trajectory through these or possible different modules, with the head of the sentence no longer being (in English) the subject (e.g. James in 'James chased John'), but instead the object (e.g. 'John was chased by James').

In a similar way, it is proposed that different languages are implemented by different forward connectivity between the different modules representing subjects, verbs, and objects in that language, with the connectivity for each language learned by repeated experience and forced trajectories during learning using for example spike-timing-dependent plasticity. Separate neuronal implementation of different languages is consistent with neurological evidence that after brain damage one language but not another may be impaired.

The system would require considerable elaboration to provide for adjectives and adjectival phrases qualifying the subject or the object, and for adverbs or adverbial phrases qualifying the verbs, but a possible principle is stronger synaptic connectivity between the modules in which the qualifiers are represented. To be specific, one type of implementation might have adjectives in modules that qualify subjects connected with relatively stronger synapses to subject modules than to object modules. This should be feasible given that any one attractor network capable of encoding thousands of words need occupy only 2–3 mm of neocortical area.

In such a system, the problem does arise of how the nouns in the subject attractor module can refer to the same object in the word as the nouns in the object attractor, and of the extent to which when one representation (e.g. in the subject module) is updated by modifying its properties (encoded by which neurons are active in the sparse distributed representation within a module), the representation of the same object in another module (e.g. in the object module) is updated to correspond.

The results on the operation of the system when it is decoding a sentence illustrated in Fig. 17.6 illustrate how the temporal sequence of the words can be used to place them into the appropriate module, for example to place a noun into a subject or an object module. Interestingly, if during the application of the noun for the subject (time 500–1000 ms in the simulations) an input effective for an attractor in module 2 (the verb module) was also applied, then a pool in module 2 tended to enter a high firing rate attractor state, for it was receiving both a recall cue and the forward bias from the high firing starting in module 1 and applied to module 2 via w_{ff12}. An implication is that only nouns should be applied to Subject and Object attractors. Having attractors for verbs that respond to different input cue (words that are verbs) helps the system to decode the input stream into the correct modules and pools. Thus the semantics, the words being applied to the network, are important in enabling the system to respond correctly.

Indeed, overall the system might be thought of as having different modules and pools for different types of word (subject noun, object noun, verb, adverb, adjective, etc) and using the match of the incoming word to the word defined in a module to provide an important cue to which module should have an attractor activated, and then adding to this the temporal sequence sensitivity also considered here to help disambiguate the syntax, for example whether a noun is a subject or an object. Thus the semantics, the word as a noun, verb, or potentially adjective or adverb, help the dynamics because the cue details (adverb, noun, verb, adjective etc) provides constraints on the trajectories and dynamics of the system, and thus on how it decodes input sequences.

Language would thus be brittle if there were not subject-noun, object-noun, verb, adjective, adverb etc pools. An inflected language helps words to activate the correct pools. If inflections are lost or not present in a language, then the order in a sequence can compensate to some extent, but the system still relies on the words activating selective pools, with the temporal dynamics used for example to disambiguate matters if a noun might otherwise be a subject or an object, or an adjective might qualify a subject vs an object, etc. Moreover, because language is often irregular, particular words must also favour particular dynamics / relations.

In such a stochastic dynamical system (Rolls 2008d, Rolls and Deco 2010), speech or writing errors such as words appearing in the incorrect order, word substitution, and repetition of the same word, could be easily accounted for by failures in the stochastically influenced (Rolls 2008d, Rolls and Deco 2010) transitions from one attractor to the next, and in the word selected in each attractor. This supports the proposed account of the cortical implementation of language.

Overall, in this Chapter the coding and dynamical principles of the operation of the cerebral cortex have been considered in the context of how they may be relevant to the implementation of language in the cerebral cortex. It has been proposed that the high capacity of local attractor networks in the neocortex would provide a useful substrate for representations of words with different syntactical roles, for example subject noun, object noun, adjective modifying a subject, adjective modifying an object, verb, and adverb. With this as a principle of operation, in an inflected language the words produced can have the appropriate suffix (typically) added to specify the module from which it originated and therefore its syntactic role in the sentence. In an inflected language, the inflections added to words can indicate case, person etc, and during decoding of the sentence (when listening or reading) these inflections can be used to help the word to activate attractors in the correct module. In a language without inflections, or that is losing inflections, the order in the sequence can be used to supplement the information present in the word to help activate a representation in the correct attractor. Examples of how cortical dynamics might help in this process both during production and during decoding are provided in the simulations in this Chapter (Rolls and Deco 2015a).

Interestingly, at least during decoding, temporal dynamics alone was found to be brittle in enabling words to be decoded by the correct module, and the system was found to work much more robustly if words find matches only in different specialized modules, for example with nouns being decodable by only subject noun or object noun modules, verbs only be verb modules, etc. The actual decoding is of course a great strength of the type of attractor neuronal network approach described here, for attractor networks are beautifully suited to performing such decoding based on the vector dot product similarity of the recall cue to what is stored in the network (Rolls 2008d, Hopfield 1982, Amit 1989). The implication is that content addressable specialized word attractor cortical modules are important in the implementation of language, and that temporal dynamics utilizing the order of a word in the sequence can be used to help disambiguate the syntactic role of a word being received by enabling it to activate a representation in the correct module, using mechanisms of the general type described.

This raises the interesting point that in the present proposal, the syntactic role of a representation is encoded in the brain by the particular cortical module that is active, with different cortical modules for different parts of speech. An implication is that for the internal operations of this syntactical system, the syntax is encoded by the module within which the representation is active, and this is a form of place coding. Much may be computed internally by such a system based on this specification of the syntactic role of each module in the thought process. The problem arises when these thoughts must be communicated to others. Then a production system is needed, and in this system the syntactic role of which module the representation arises from can be specified partly by the word itself (with noun words indicating that they arise from a subject or object noun representations, verb words indicating that they arise from a verb module); and this specification is supported by inflection and/or by temporal order information to help disambiguate the module from which the word originates. Then during decoding of a sentence, the word again allows it to match only certain modules, with inflection and/or order in the sequence being used to disambiguate the module that should be activated by the word.

Thus an internal language of thought may be implemented by allocating different cortical modules to different syntactic roles, and using place encoding. However, when language becomes externalized in the process of communication, the way in which language can be used as a computational mechanism may be enhanced. Once a language has the rules that allow syntactic role to be expressed in for example written form, this enables formal syntactic operations including logic to be checked in an extended argument or algorithm or proof, and this then provides language with much greater power, providing a basis for formal reasoned extended argument, which may not be a general property of neuronal network operations (Rolls 2008d), and which facilitates the use of the reasoned route to action (Rolls 2014a).

One of the hypotheses considered here is that place coding in quite small cortical modules approximately the size of a cortical column (i.e. a region within which there is a high density of local recurrent collaterals to support attractor functionality, and within which local inhibitory neurons operate) may be used to encode the syntactic role of a word might not easily reveal itself at the brain lesion or functional neuroimaging levels, which generally operate with less resolution than this. For example, such studies of the effects of brain damage and of activations do not provide clear evidence for segregation by syntactic role, such as noun-selective vs verb-selective areas (Vigliocco, Vinson, Druks, Barber and Cappa 2011). Although effects produced by nouns vs verbs do segregate to some extent into the temporal and frontal lobes, this may be because the semantic associations of nouns with objects, and verbs with actions, will tend to activate different cortical areas because of the semantic not purely because of the syntactic difference (Vigliocco et al. 2011). One of the hypotheses developed here is that nouns as subjects and nouns as objects may use different place coding, and one way that this might become evident in future is if single neuron recordings from language areas support this. Indeed, it is a specific and testable prediction of the approach described here that some neurons in language-related cortical areas will have responses to words that depend on the syntactic role of the word. For example, such a neuron might respond preferentially to a noun word when it is a subject compared to when it is an object.

It will be interesting in future to investigate whether this approach based on known principles of cortical computation can be extended to show whether it could provide at least a part of the biological foundation for the implementation of language in the brain. Because of its use of place coding, the system would not be recursive. But language may not be recursive, and indeed some of the interesting properties but also limitations of language may be understood in future as arising from the limitations of its biological implementation.

I note that there are a number of biological mechanisms that might implement the slow transitions from one attractor state to another, some investigated by Deco and Rolls (2005c), and that we are not wedded to any particular mechanism. The speed with which the trajectory through the state space of attractors is executed will depend on factors such as the magnitude of the inputs including the biasing inputs, the strengths of the synapses between the modules, the NMDA dynamics, the effects of finite size noise which will be influenced by the number of spiking neurons, the dilution of the connectivity, and the graded nature of the firing (Rolls and Deco 2010, Webb et al. 2011, Rolls and Webb 2012). An implication is that the speed of speech production might be influenced by factors that influence for example NMDA receptors, such as dopamine via D1 receptors (Rolls and Deco 2010).

One such extension in the future of the approach taken in this Chapter would be to extend the implementation from a single cortical attractor network for each linguistic type (subject nouns, object nouns, verbs, etc) to a set of attractor networks for each linguistic type. If each single local cortical attractor network could store say S patterns (the p referred to above), then how would the system operate with M such attractor nets or modules? There would be two types of connection in such a system. One would be the synaptic connections between the neurons in each attractor network or module. The other connections would be the typically weaker connections between cortical modules (see Rolls (2008d)). The whole system of coupled interacting attractor nets is known as a Potts attractor (Treves 2005, Kropff and Treves 2005). In a language system representing for example nouns, one attractor net or 'unit' in Potts terminology might contain properties such as shape, another color, another texture, etc, and the full semantic description of the object might be represented by which attractors in which of the M modules are active, that is in a high firing rate state. One advantage of such a system is that all of the properties of an object could be encoded in this way, so we could specify whether the hat is round, red, smooth in texture, etc. Associated with this advantage, the total capacity of the system, that is the number of possible objects that could be represented, is now proportional to S^2. Thus if a single attractor network could store S=10,000 items (10^4), a Potts system with 5 such modules might be able to represent of order 5.10^8 such objects. In more detail, the number of patterns P_c that can be represented over the Potts attractor is

$$P_c \approx c_M S^2 / a_M \qquad (17.2)$$

where c_M is the number of other modules on average with which a module is connected, S is the number of different attractor states within any one module, and a_M is the proportion of the attractor modules in which there is an active attractor, i.e. a high firing rate attractor state (Treves 2005, Kropff and Treves 2005). Such a Potts system only works well if (1) long-range connections between the different modules are non-uniformly distributed and (2) only a sparse set of the modules (measured by a_M) is in a high firing rate attractor state (Treves 2005). In principle, the dynamical system described here could be replaced by substituting each cortical word type module (e.g. that for subject nouns) with a Potts attractor system each with several attractor network modules coding for different properties of features such as shape, colour, texture, etc.

An overview at present is that such a Potts system might be useful for a semantic network (Treves 2005), which might correspond to the deep network that biases the word modules in the architecture described in this Chapter. That semantic system would then bias the word modules (with one cortical module for each type of word, subject noun etc), and this architecture might have the advantage that word representations may be more uncorrelated than are semantic representations, which would keep the word representation capacity high. However, in the Potts system simulated so far to model language, the units corresponded to semantic features not to words; and place coding was not used, with instead the syntactic roles of the semantic representations requiring further Potts units to specify the syntactic roles of the semantic units (Pirmoradian and Treves 2013).

Treves (2005) also considered how such a Potts system might have dynamics that might display some of the properties of language. Adaptation was introduced into each of the attractor networks. After a population of neurons had been active in a high firing rate state in one attractor module for a short time, due to adaptation in that neuronal population, the system then jumped to another attractor state in the same module, or in another connected module a jump might occur to another attractor state, because its inputs had changed due to the adaptation in the other module. The result was a latching process that resulted in complex dynamics of the overall system (Treves 2005, Song, Yao and Treves 2014). Whether that complex dynamics is how language is produced has not been proven yet. My view here is that there are sensory inputs from the world, or remembered states that enter short-term memory, and that these states in the deep, semantic, networks then bias the word networks to produce the sequential stream of language. That keeps the processing not a random trajectory through a state space perhaps biased by the statistics of the correlations within a language, but instead a trajectory useful for communication that reflects the states produced by the world or recalled into memory and that can then be communicated to others as a stream of words.

In conclusion, I have described some new hypotheses in this Chapter about how some of the principles of cortical computation used to implement episodic memory, short-term memory, perception, attention, and decision-making (Rolls 2008d, Rolls and Deco 2015a) might contribute to the implementation of language including syntax in the cortex.

This is a major frontier in neuroscience, and much remains to be done on how language and syntax may be implemented in the cerebral cortex. A dynamical systems approach that may be helpful is described in Section 26.5.3.

17.5   Highlights

1. The principles of operation of the neocortex for language remain a major frontier in neuroscience, and much remains to be done on how language and syntax may be implemented in the cerebral cortex.

2. A fundamental computational issue is how the cortex implements binding of elements such as words or features with the correct relationship between the elements. The problem in the context of language might arise if we have neuronal populations each firing to represent a subject, a verb, and an object of a sentence. If all we had were three populations of neurons firing, how would we know which was the subject, which the verb, and which the object? How would we know that the subject was related to the verb, and that the verb operated on the object? How these relations are encoded is part of the problem of binding.

3. One possible solution that is described in this Chapter is that a place code is used to solve syntax, which would simply bring syntax into the same framework as is used for representations in most other cortical areas.

4. The hypothesis is that a place code is used, with for example one cortical module or region or set of neurons used to represent subjects, another cortical module used to represent verbs, and another cortical module used to represent objects.

5. An interesting implication is that within the brain, the problem of syntax is then solved, but that the problem then becomes of how the syntactic role of each word is transmitted from one individual to another. One such principle is word order used in some languages, and another is the addition of suffixes, used in other languages.

6. Simulations of a set of linked cortical attractor networks with stronger forward than back interconnections between the networks show how a trajectory could be made through a state space of grammatical entities to generate appropriate sequences of word orders. It is also shown how a recipient could decode such a word order sequence, and assign the words to activate the correct cortical modules.

18 Evolutionary trends in cortical design and principles of operation

18.1 Introduction

It is because of the intended relevance to understanding human brain function and its disorders that emphasis is placed in this book on findings from research in primates including humans. This is important, for many cortical systems have undergone considerable development in primates (e.g. monkeys and humans) compared to non-primates (e.g. rats and mice).

For example, the temporal lobe has undergone great development in primates (Pandya, Seltzer, Petrides and Cipolloni 2015), and several systems in the temporal lobe are either involved in emotion (e.g. the amygdala), or provide some of the main sensory inputs to brain systems involved in emotion and motivation. In particular, the amygdala and the orbito-frontal cortex, key brain structures in emotion, both receive inputs from the highly developed temporal lobe cortical areas, including those involved in invariant visual object recognition, and face identity and expression processing. Moreover, the orbitofrontal cortex has evolved considerably in primates (Section 18.4).

These are among the reasons why emphasis is placed on brain systems in primates, including humans, in the approach taken here. A medically relevant aim of the research described in this book is to provide a foundation for understanding in humans the brain mechanisms of perception, memory, emotion, motivation, decision-making, and cognition, and thus their disorders.

However, our understanding of the neocortex (isocortex) can be enhanced by understanding its operation in the light of what could be achieved by some of its antecedents in evolution, the olfactory and hippocampal cortex (Kaas and Preuss 2014, Pandya, Seltzer, Petrides and Cipolloni 2015). That is one of the aims of this Chapter. A second aim is to take a more computational approach to how all the different types of cerebral cortex operate.

18.2 Different types of cerebral neocortex: towards a computational understanding

There are three main types of cerebral cortex, neocortex, hippocampal cortex, and olfactory cortex. In Section 1.10 a summary of the fine structure and principles of intra-cortical connectivity of the neocortex was provided. Because that has already been described, I start with an analysis of the functions of the neurons in each of its six layers, for that will provide a foundation for understanding why the evolutionarily older olfactory and hippocampal cortices have fewer layers. The concepts considered next are further developed in the synthesis Chapter (26) (especially in Section 26.5.18), where the focus is on bringing different concepts together, and on future research.

Neocortex

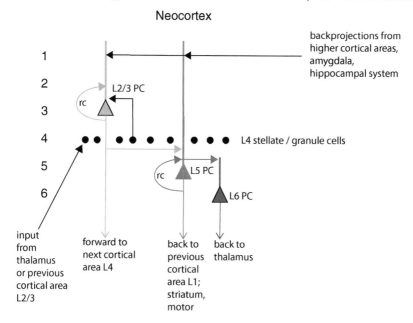

Fig. 18.1 Functional canonical microcircuit of the neocortex (see text). Recurrent collateral connections (rc) are shown as a loop back to a particular population of cells, of which just one neuron is shown in this and Figs. 18.2 and 18.3. In primates the feedforward projection neurons are concentrated in L3B; and the main feedback projection neurons are in L6 and Lower L5 (L5B), but some L2 and L3A neurons do send backprojections (Markov et al 2014b). Some L6 cortico-thalamic neurons send a projection to L4 (see text).

18.2.1 Neocortex or isocortex

The neocortex is characterised by 6 layers, though these vary from cortical area to cortical area, and this provides the basis for cytoarchitectural and myeloarchitectural divisions of the neocortex (Brodmann 1925, Von Bonin and Bailey 1947, Pandya, Seltzer, Petrides and Cipolloni 2015), which it turns out usually reflect connectivity with other cortical areas, and function, as shown by the research cited in this book.

The cell types, connectivity, and some hypotheses about information flow through these cell types are described in Section 1.10 (see e.g. Fig. 1.18), and I now develop further hypotheses about how this connectivity operates functionally, that is, computationally (Fig. 18.1). Some of these hypotheses are new, and are intended to stimulate further empirical research to test them, for the hypotheses provide important tools towards providing a computational understanding of cortical circuitry.

These hypotheses go beyond previous descriptions, which have been in terms of anatomical connectivity patterns and neuronal properties (Shepherd and Grillner 2010, da Costa and Martin 2010, Harris and Mrsic-Flogel 2013, Kubota 2014, Harris and Shepherd 2015), and have not focussed on a computational approach to advance our understanding. I do though offer a note of caution here. The more recent analyses of cortical connectivity have been largely in rodents (Markram, Muller, Ramaswamy, Reimann, Abdellah, Sanchez, Ailamaki, Alonso-Nanclares, Antille, Arsever, Kahou, Berger, Bilgili, Buncic, Chalimourda, Chindemi, Courcol, Delalondre, Delattre, Druckmann, Dumusc, Dynes, Eilemann, Gal, Gevaert, Ghobril, Gidon, Graham, Gupta, Haenel, Hay, Heinis, Hernando, Hines, Kanari, Keller, Kenyon, Khazen, Kim, King, Kisvarday, Kumbhar, Lasserre, Le Be, Magalhaes, Merchan-Perez, Meystre, Morrice, Muller, Munoz-Cespedes, Muralidhar, Muthurasa, Nachbaur, Newton, Nolte, Ovcharenko,

Palacios, Pastor, Perin, Ranjan, Riachi, Rodriguez, Riquelme, Rossert, Sfyrakis, Shi, Shill-cock, Silberberg, Silva, Tauheed, Telefont, Toledo-Rodriguez, Trankler, Van Geit, Diaz, Walker, Wang, Zaninetta, DeFelipe, Hill, Segev and Schurmann 2015, Jiang, Shen, Cadwell, Berens, Sinz, Ecker, Patel and Tolias 2015), where molecular markers can more easily be brought to bear (Harris and Shepherd 2015, Molnar, Kaas, de Carlos, Hevner, Lein and Nemec 2014), but these studies reveal that there are some quite major differences in the intracortical connectivity of rodents and primates, for example in the organization of back-projections (Harris and Shepherd 2015). For this reason, it is important to take into account fully what has been discovered in primates, not only in intra-cortical connectivity, but also in the organisation of the different areas of the cerebral neocortex, which is far more developed in primates than in rodents (Pandya, Seltzer, Petrides and Cipolloni 2015, Markov, Vezoli, Chameau, Falchier, Quilodran, Huissoud, Lamy, Misery, Giroud, Ullman, Barone, Dehay, Knoblauch and Kennedy 2014b). Molecular markers in primates do though provide clear evidence for clusters of genes specifying a particular layer of cortex throughout different brain areas, and cross-species comparisons are of interest and importance in understanding the evolution of the cortex (Bernard et al. 2012). Connections between the excitatory neurons are emphasized, because the inhibitory neurons have fewer synapses and few long-range connections, so are viewed as being related to the local stability of the excitatory circuits, which latter appear to perform the major computations that account for building new representations, and implementing memory and decision-making.

18.2.1.1 Neocortex layer 4

In primary sensory areas, the cortico-thalamic inputs provide the main input that drives the cortical area, and terminate strongly in layer 4, especially on stellate cells which give these cortical areas a granular appearance. Areas with strong thalamic input that provides a strong drive to the cortex can typically be recognised by their granular appearance, with many small neurons, in layer 4 (Pandya et al. 2015). The granular cells can be seen as performing pattern separation or orthogonalization on the inputs, in the same way as do granule cells in the dentate gyrus for the CA3 cells of the hippocampus (see Section 18.2.3 and Chapter 24) (Rolls 2016e).

It is of great architectural significance that the neocortex receives inputs from the thalamus, and in the case of the primary sensory cortical areas, the thalamic inputs are the main cortical inputs. My hypothesis is that the thalamus performs a useful and simple stage of pattern separation before inputs reach the primary sensory cortical areas, by virtue of the local lateral inhibition in the spatially topologically mapped thalamic nuclei, which serves to operate as a high-pass spatial filter. This hypothesis is developed further below. The thalamus is a specialization for neocortex, and there is no thalamus for pyriform and hippocampal cortex.

18.2.1.2 Neocortex layers 2 and 3

These layer 4 inputs then influence pyramidal cells in the superficial layers, layers 2 and 3 (Fig. 1.18). These pyramidal cells have very well-developed recurrent collateral connections to each other, which show associative synaptic plasticity. This enables them to perform as an autoassociation or attractor network. In early cortical areas in the hierarchy (e.g. the primary visual cortex, V1), where linear processing of the inputs may be useful so as not to lose information by non-linear processing, this attractor function may be less important and the recurrent collaterals less well developed and numerous than in higher cortical areas. In contrast, in higher cortical areas, where non-linear processing such as memory retrieval and decision-making is being performed, the recurrent collateral system for autoassociation may be much more highly developed and numerous. This hypothesis is supported by empirical measures of dendritic size (Elston et al. 2006).

A key feature of the neocortical recurrent collateral system is that it is local, extending just several mm from any given neuron. This provides a basis for separate attractor networks in nearby cortical regions spaced only several mm apart, and is necessary to provide for high memory capacity of the whole neocortex, made possible by the partial independence of the local attractors. Without this relative independence of nearby attractors, the total memory capacity of the cortex would be determined by the leading term of the number of recurrent collateral synapses onto any one neuron, as described in Section 4.5 (O'Kane and Treves 1992). Because the recurrent collaterals are local, the total memory capacity of the cortex is the number of memories that could be stored in any one local attractor network (of order 10,000, if the number C of recurrent collaterals onto each pyramidal cell is 10,000), multiplied by the number of local attractor networks in the cortex, divided by a factor of 2–3 due to interference between nearby attractors with some overlap of anatomical connectivity as one moves across the cortical sheet (Roudi and Treves 2006, Roudi and Treves 2008). If the human neocortical surface area is of order 1,900 cm^2, and each local neocortical attractor has a diameter of 3 mm (and thus an area of approximately 7 mm^2), then the number of independent attractor networks would be approximately 27,000, each with a memory capacity of at least several thousand items. The locality of the recurrent collaterals in the neocortex also means that it tends to support topological maps (Section B.4.6), with this locality of neocortical processing in turn contributing to the convergence from area to area as one moves up a cortical hierarchy (Chapter 2).

This fundamental aspect of neocortical architecture and principle of operation, its locality, is in complete contrast to the hippocampus as described in Section 18.2.3 in which the CA3 is a global attractor so that any item can be associated with any other item to form an episodic memory; and with the olfactory pyriform cortex as described in Section 18.2.2 in which there is no topology in the input space of odours as represented in the glomeruli of the olfactory bulb, so that associations may need to be found throughout the olfactory space. Its locality is a fundamental principle of the evolution and operation of the neocortex.

The representations in the superficial pyramidal cells (L2/L3) are more sparse than the deep pyramidal cells (L5/L6) (Harris and Shepherd 2015), and my hypothesis is that this helps to increase the memory capacity of what can be stored in autoassociation networks in the superficial layers. In early cortical areas in the hierarchy, even if they do not contribute to much non-linear / memory-related processing, the recurrent collaterals may perform useful functions that they perform in all areas in a hierarchy, including supporting the formation of topological maps. These topological maps minimize connection lengths between neurons that need to communicate with each other, to for example perform constraint satisfaction in early cortical areas to optimize collective neuronal tuning to the stimuli, or in higher cortical areas to place close to each other neurons with similar functions, such as face-selective neurons, which need to communicate with each other to implement for example competition using the inhibitory neurons which are also local to a small region of neocortex (Section B.4.6).

Another principle of operation of the superficial layers (L2/L3) of the neocortex is competitive learning, which enables single neurons to learn combinations of inputs received from the previous stage of processing, which is performed as described in Section B.4, and the essential utility of which is demonstrated in Chapter 25 where it is shown how this is essential for solving problems in perception.

The superficial pyramidal cells (L2/L3) then have forward connections as one of their major outputs to the next cortical layer in the hierarchy, where they terminate primarily in or close to layer 4, where they drive the pyramidal cells in the superficial cortical layers, providing feedforward computations to compute useful representations by combining inputs from previous areas in the hierarchy to perform the fundamentally useful computations of the cerebral cortex described in Chapters 2, 25, and 26. The feedforward projection neurons are concentrated in L3B, and in addition, some L2 and L3A neurons do send backprojections in primates (Markov et al. 2014b).

The superficial cortical layers are thus hypothesized to perform the main computationally useful functions of the cerebral neocortex, which involve feedforward operations to form non-linear combinations of the inputs from previous cortical areas utilizing the convergence from stage to stage to construct useful combinations essential for cortical computation, and to implement using the attractor properties of the recurrent collaterals useful functions such as information storage and retrieval (long-term memory, short-term memory, decision-making, etc). According to the hypotheses that I am developing, these feedforward competitive learning computations performed by the superficial layers are the computationally useful aspects of the design of the neocortex. I suggest in the next subsection that the deep layers are mainly for output functions of what has been computed at each level of the hierarchy (L5), and for housekeeping functions such as feedback to the thalamus (L6).

18.2.1.3 Neocortex layer 5

As shown in Section 1.10 (Fig. 1.18), layer 5 has output neurons, that send whatever has been computed at each cortical stage to the striatum, and via the pyramidal tract to structures even in the spinal cord in primates, for example for fine control of the fingers. Because there are often large distances to be traversed to the spinal cord, these pyramidal cells are often large and have large and therefore fast-conducting axons. This is very evident in the primary motor cortex, area 4, where the Betz cells as they are called are so large (with cell bodies up to 0.1 mm in diameter) that they can be seen with the naked eye. This is one of the principles underlying the functional utility of cytoarchitecture (Pandya et al. 2015), that areas where major motor output is being performed will have a set of large neurons in layer 5, and little granularity in layer 4, allowing cortical areas to be classified in a way that turns out to be relevant to the computational functions being performed.

Taking some output from every cortical area does seem to be useful. For example, for the primary visual cortex V1, where stereopsis is computed, outputs from the deep cortical layers project to the superior colliculus, where they may be useful in ensuring appropriately convergent eye movements.

The dynamics of the deep cortical area pyramidal cells may be useful for the control of movements, and this may account for why they are often dynamically rich, in fact often showing bursty spiking behaviour. Further, the less sparse representation in the deep cortical areas may be appropriate, for information transmission in bits per second is maximized by having up to 50% of the neurons active at any one time, as it shown in Appendix C. It may be especially important to utilize efficient information transmission, when these are large neurons with large axons because of the distance that they travel (e.g. to the spinal cord) in order to minimize the size and weight of the central and peripheral nervous systems.

It is of considerable computational interest that the deep pyramidal cells in primates are also an important source of feedback to the previous cortical area, where they terminate in the superficial cortical layers (Chapter 11). Indeed, Lower L5 (L5B) 'Subcerebral projection neurons' (SPNs) (Fig. 1.18) project not only to subcerebral motor centres for example via the pyramidal tract, and the striatum, and to higher-order thalamic nuclei (Harris and Mrsic-Flogel 2013, Harris and Shepherd 2015), but they may also provide some cortico-cortical backprojections, which terminate in superficial layers of the cerebral cortex, including layer 1 (Markov et al. 2014b) (see Figs. 11.1 and 18.1).

Perhaps the relatively non-sparse representation in layer 5 is useful in previous cortical areas for processes such as top-down attention, because this helps to ensure that all relevant neurons in a preceding cortical area in the hierarchy receive bias, in case they also happen to receive bottom-up input that can be enhanced. A sparse representation for the backprojections might be less good at biasing up all neurons in the previous area that may be relevant. In terms of memory recall, another function of cortico-cortical backprojections, perhaps a non-sparse but information-rich recall cue is also it is hypothesized useful. Further, as emphasized in Chapter 11, the backprojections do not simply send back information for the high-level information to be represented in early cortical areas. Given this fact, the nature of the signal sent back to early cortical areas, from the deep layers, may be beneficially different from what is sent forward to the next cortical area from the superficial pyramidal cells, for further new computation.

However, the L5 pyramidal cells have a well-developed recurrent collateral system. What computational function may this perform? My proposal is that they also operate as an attractor network, but of a different type to the attractor network in the superficial pyramidal cells, which as I have argued often needs to support discrete attractor states, such as whether it is this object, or that object, especially in the ventral cortical areas involved in 'what' representations for visual, auditory, taste, and olfactory stimuli. The proposal follows from the evidence that the pyramidal cells of L5 project to motor structures such as the striatum and via the pyramidal tract even to the spinal cord. Now, the space of movements is essentially continuous: an arm can move to any position in 3D space, and the eyes can move to any position in their 2D space, without discrete stable positions. The proposal on this basis then is that the L5 pyramidal cells support continuous attractor networks, which as described in Section B.5, can be stable at any position in a continuous space. The advantage of having a continuous attractor here is that all the neurons that are active for any position in the space can become associated with each other, thus smoothing the transitions through the space by making sure that all neurons relevant to a particular position in the space based on previous experience are recruited into activity.

In this context, it should be noted that networks with associatively modifiable recurrent collateral synapses do not necessarily need to have a positive feedback gain sufficiently high to maintain the activity in a high firing rate attractor state even when there is no input. The network with the structure of an attractor network can nevertheless perform constraint satisfaction, by providing a moderate input from other neurons in the network to add to and smooth whatever inputs are providing the main driving input to the network, which in this case would be a L2/L3 input driving the L5 pyramidal cell firing. This would help to provide a smooth trajectory through the continuous attractor space of movements, without jerks and incompatible motor commands being produced, as described in Section B.5.

An interesting implication of this hypothesis is that the L2/L3 attractor network which may often (especially in the 'what' and semantic memory cortical areas) support discrete attractors (Section B.3), should operate as a largely separate attractor system from the L5 continuous attractor network system, as the computational spaces that they encode are different. Consistent with this at least partial separation of superficial and deep layer attractor network system, the dendrites of deep pyramidal cells are largely confined to the deep layers of the neocortex apart from the branches in layer 1 where backprojections are received; and the dendrites of superficial layer pyramidal cells are largely confined to the superficial layers of the cortex, as illustrated in Fig. 1.14. The main relation it is proposed between the superficial layer discrete and the deep layer continuous attractor systems of the neocortex is that the superficial cortical layers should provide the main driving input to the deep cortical layers. Treves (2003) and Montagnini and Treves (2003) also pointed to the computational advantage to having cortical layers with sparse distributed encoding of a discrete or 'what' representation of stimulus identity, separate from a 'where' topographically mapped more spatially continuous representation.

18.2.1.4 Neocortex layer 6

L6 Corticothalamic cells (CTs) send projections to the thalamus which, unlike those of L5 'Subcerebral projection neurons' (SPNs) (Fig. 1.18), are weak, and target the reticular and primary sensory thalamic nuclei. They may perform some type of negative feedback control via these connections, a form of 'housekeeping', in that the inhibition that they produce on thalamic relay cells (through inhibitory interneurons) is stronger than the excitation that they produce via their terminals on the distal dendrites of thalamic relay cells (Sherman and Guillery 2013). Corticothalamic cells also project to cortical layer 4, where they strongly target interneurons, as well as hyperpolarizing principal cells via group II mGluRs. Consistent with this connectivity, optogenetic stimulation of L6 in vivo suppresses cortical activity, suggesting a role of this layer in gain control or translaminar inhibition (Harris and Mrsic-Flogel 2013), another form of 'housekeeping'.

Layer 6 also has neurons that provide backprojections to preceding cortical areas in a hierarchy (Markov et al. 2014b).

There is sometimes puzzlement why there are so many backprojections from the primary visual cortex to the lateral geniculate nuclei (approximately 100 million), compared to the much smaller number (approximately 1 million) of lateral geniculate cells that project to the primary visual cortex, V1. The inference is sometimes made that the backprojections from V1 to the lateral geniculate must be very important, because there are so many. My hypothesis is that the numbers just reflect what layer 6 cortical cells are programmed genetically to do: to send backprojections to the thalamus (for gain control). Given that there are approximately 100 times as many L6 cells in V1 as lateral geniculate cells, that accounts for the ratio of approximately 100: 1. My hypothesis is that this number is not needed; but that it is easier not to have additional genes specifying that a smaller number would be adequate just for V1, for adding such extra 'special case' genes would not only overcomplicate the generic genetic specification of the cortex (Chapter 19), but would also risk problems as another part of the gene fitness space that has to operate correctly.

With these hypotheses about the function of the neurons in different layers of the neocortex, I now consider other hypotheses about neocortical computation, and then turn to consider the olfactory and hippocampal cortices, which evolved earlier in evolution, to develop hypotheses about how they operate in the light of these computational hypotheses for neocortex.

18.2.1.5 Other hypotheses about neocortical computation

Canonical circuits of the neocortex have been described, but in most cases focus on arrow connectivity between neuron types, rather than how the cortex microcircuitry performs useful computations (Douglas, Markram and Martin 2004, Douglas and Martin 2004, da Costa and Martin 2010, Harris and Mrsic-Flogel 2013, Harris and Shepherd 2015), and there is a considerable neglect of what the highly developed recurrent collaterals might perform. In a paper entitled 'Canonical computations of cerebral cortex', Miller (2016) has reviewed the evidence that L4 neurons have their selectivity strongly influenced by the feedforward connectivity from the lateral geniculate nucleus (as suggested by Hubel and Wiesel (1962), see Fig. 1.1), and suggests that there is a recurrent computation of gain. Douglas and Martin (2007) do note that the recurrent connections form a large proportion of the synapses within a lamina in an area of cortex, and suggest gain control as their function, though also note that they may be helpful in cleaning up noisy signals.

In this context, the approach taken in Sections 18.2.1 and 26.5.18 is innovative, and is aiming to move towards understanding principles of operation, and thus necessarily computation, of the cerebral cortex.

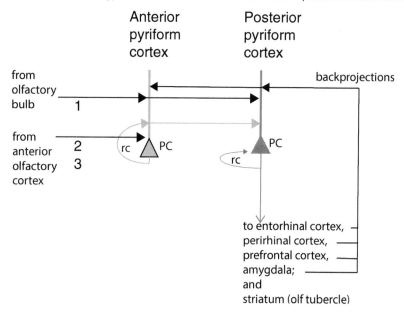

Fig. 18.2 Functional canonical microcircuit of the olfactory (pyriform) cortex. Recurrent collateral connections (rc) are shown as a loop back to a particular population of cells, of which just one neuron is shown. The pyriform cortex has an anterior part, the anterior pyriform cortex (APC), and a posterior part (PPC). Afferents from the olfactory bulb (OB) reach the apical dendrites in layer 1a of APC and PPC. Additional olfactory inputs from an anterior olfactory cortex (AOC) synapse onto the dendrites of the pyramidal cells in layer deep 1b of APC. APC has a highly developed recurrent collateral system, which reaches PPC. There are additional recurrent collaterals on the basal dendrites of the APC (not shown). The APC projects in a feedforward way to PPC, which has a less prominent recurrent collateral system, and which projects on to the recipient olfactory processing regions which include the entorhinal cortex, perirhinal cortex, olfactory tubercle / ventral striatum, prefrontal cortex, and amygdala.

18.2.2 Olfactory (pyriform) cortex

Olfactory (pyriform) cortex is a 3-layer type of cortex termed paleocortex, a type of allocortex (Pandya et al. 2015), with only one layer containing large pyramidal cells. Its structure is shown in Fig. 18.2. This represents my synthesis of what has been described based on these sources (Haberly 2001, Luskin and Price 1983) and on discussion with those who work on the pyriform cortex including Joel Price (Washington University School of Medicine), and has some different emphases to those in Figures 9 and 12A2 of Haberly (2001).

The anterior pyriform cortex especially has well-developed recurrent collateral connections, which it is hypothesized enable it to learn as an autoassociation network (Haberly 2001) which forms combinations of inputs from the glomeruli of the olfactory bulb. Each glomerulus is connected to one of the 1000 types of olfactory receptor each specified by a separate gene, each tuned to a separate range of odours (Buck and Bargmann 2013, Mombaerts 2006). The pyriform cortex detects by its autoassociation which combinations of odours tend to occur together as a result of inputs being received from the natural world. This enables discrimination between, and later good recognition of, particular odour combinations.

In more detail, the anatomy suggests that the anterior pyriform cortex acts as an autoassociation or attractor network that can form and remember representations that reflect combinations of simultaneous olfactory inputs. The anterior pyriform cortex has highly developed recurrent collaterals for the autoassociation. (Haberly (2001) distinguished a ventral part which he suggested provided for a first stage of autoassociation, and a dorsal part which provided he postulated a second stage.) The anterior pyriform cortex receives direct inputs from the olfactory bulb and the 'anterior olfactory cortex' (formerly known as the anterior olfactory nucleus). The anterior pyriform cortex necessarily represents the different parts that are being associated together by its recurrent collaterals, and computationally would enable completion of a whole combination from a part, and also short-term memory which might be useful in identifying an odour if the concentrations were continually varying, for example with each sniff or because of the wind. I hypothesize that in addition, competitive learning (Section B.4) may also take place in the anterior pyriform cortex, to help the neurons there respond to combinations of glomerular inputs. Consistent with this hypothesis, anterior pyriform cortex neurons do separate overlapping odour combinations better than olfactory bulb neurons (Barnes, Hofacer, Zaman, Rennaker and Wilson 2008, Wilson and Sullivan 2011).

The posterior pyriform cortex receives inputs from the anterior pyriform cortex and also from the olfactory bulb, as shown in Fig. 18.2. Its recurrent collateral system may be less well developed than that of the anterior pyriform cortex (Haberly 2001) (as indicated by the thinner line in Fig. 18.2), and makes synapses especially with the basal dendrites of the posterior pyriform cortex pyramidal cells (Luskin and Price 1983). The posterior pyriform cortex then may act (with its feedforward inputs and less highly developed recurrent collateral system), to recode the parts represented in the anterior pyriform cortex into representations that reflect the combinations of the parts, using competitive learning. This corresponds to what I have proposed for the relation between the CA3 recurrent collateral system, and the CA1 competitive network, though the CA1 has essentially no recurrent collaterals (Rolls 1989b, Rolls 1990b, Treves and Rolls 1994, Rolls 1996c, Kesner and Rolls 2015). The code that is formed in the posterior pyriform cortex thus would be expected on this hypothesis to have different neurons that can respond to different combinations of odors, even when the combinations have overlapping components, and as predicted, this has been found in the Posterior Pyriform Cortex more than in the Anterior Pyriform Cortex, at least in naive animals (Kadohisa and Wilson 2006). It is also useful to emphasize that the outputs from the pyriform cortex come primarily from the posterior pyriform cortex, which projects to the entorhinal cortex, perirhinal cortex, olfactory tubercle, prefrontal cortex, and amygdala; and not from the anterior pyriform cortex as shown by (Haberly 2001).

The implication is that the principle of operation of the pyriform cortex is that it operates primarily as a two-stage system, with the anterior pyriform cortex autoassociation stage followed by a posterior pyriform cortex stage (which may perform competitive learning and some more autoassociation), and that the outputs come primarily from the second stage, the posterior pyriform cortex (see Fig. 18.2). However, the anatomy of the anterior and posterior pyriform cortex shows that any differences in connectivity may be quantitative rather than qualitative, in that some recurrent collaterals from the posterior pyriform cortex reach the anterior pyriform cortex, and the anterior pyriform cortex does have some outputs to other olfactory-related areas (Luskin and Price 1983). The pyriform cortex does receive backprojections from its target structures, and these end in layer 1, consistent with what is found in neocortex.

Here is my hypothesis about how the pyriform cortex differs computationally from the neocortex. The pyriform cortex pyramidal cells correspond approximately to the superficial cortical layers of the cerebral neocortex, in which a key computational component is autoassociation or attractor networks, but also competitive learning to form new combination-sensitive representations. Indeed, there is a hint that computationally the anterior pyriform cortex may specialise in autoassociation, and the posterior pyriform cortex, with its less prominent recurrent collateral system, as a competitive learning stage to emphasise encoding of combinations of odor components. The posterior pyriform cortex may thus be analogous to the CA1 neurons of the hippocampus, and to draw out this possibility, both are coloured red in Figs. 18.2 and 18.3. Both the posterior pyriform cortex and the CA1 provide the prominent outputs of their cortical region, including projections to the striatum. However, continuing the comparison with the neocortex, the pyriform olfactory cortex evolved before there was any neocortex, or thalamus. So the olfactory cortex in early evolutionary history had no need of a layer 6, because there was no thalamus. Correspondingly, there was no preceding (neo)cortical area for a putative L5 of pyriform cortex to send backprojections to. So effectively, all that the olfactory cortex could do was to perform its autoassociation (and competitive) computation, and send its outputs on to whatever structure was present. There was no neocortex for a forward projection to reach. So the outputs of the pyriform cortex just went to whichever output region for behaviour might be present, which was probably the striatum / basal ganglia, which was present at early evolutionary stages. Consistent with this, parts of the output targets of pyriform cortex include regions with a striatal structure, the olfactory tubercle, and parts of the amygdala. At later evolutionary stages, when neocortex was present, outputs from the pyriform cortex then reached neocortical areas such as the entorhinal cortex and prefrontal / orbitofrontal cortex. As in neocortex, the backprojections terminate in layer 1, where they are likely to modulate rather than determine pyramidal cell firing under natural conditions. (This last point is very important. All the analyses that I have performed of cortical processing have been when the system is operating normally in the awake behaving animal, and much of this type of research remains to be performed for the pyriform cortex, instead of under anaesthesia or in slices in which effects may be demonstrated, but with the great need for their significance during normal computation to be evaluated.)

My hypothesis is that the olfactory cortex was then a 'primitive' form of cortex, and was part of the origin of 6-layer neocortex. To lead to neocortex, outputs had to be added in neocortex to drive pyramidal tract neurons with a non-sparse code (L5); and to provide a feedback signal from layer 6 to the thalamus, which of course was not present at early evolutionary stages when most of the cortex was pyriform cortex (Molnar, Kaas, de Carlos, Hevner, Lein and Nemec 2014). In evolution, the genes that specified the architecture of pyriform cortex might have provided a start to the design of neocortex. Molecular marker / gene comparisons of pyriform cortex with the superficial layers of neocortex might be of interest (Harris and Shepherd 2015, Bernard et al. 2012, Molnar et al. 2014).

The representation of odours in the anterior pyriform cortex was found to be a little more sparse than in the posterior pyriform cortex in rats familiar with the odours (Kadohisa and Wilson 2006). This may reflect a principle of cortical design, as follows. In the part of the cortex where the storage of large numbers of memories composed of parts is being performed, in attractor networks, there is an advantage to having a sparse representation, for this will increase the memory capacity. This advantage applies to the anterior pyriform cortex, to the hippocampal CA3 neurons, and to the superficial pyramidal cells (L2/L3) of the neocortex. In contrast, in the part of the cortex following an attractor network where a network operating in part as a competitive network is re-coding the parts of the memory into a single conjunctive representation (in which a neuron might respond to a combination of the parts), the representation may be less sparse, because the advantage here is to convey a large amount of information into a given number of neurons in the whole population, as this information will be used as a recall cue. This could apply to the hippocampal CA1 cells, to the deep layers of the neocortex (L5) which give rise to backprojections used for memory recall, and may apply to the posterior pyriform cortex, where the parts of the memory may no longer be as relevant, just the memory of specific combinations of odours. However, the components were more correlated with each other, and the components were more correlated with the binary mixtures, in the posterior pyriform cortex than in the anterior pyriform cortex (Kadohisa and Wilson 2006). This does not suggest pattern separation from the anterior to the posterior pyriform cortex, and is consistent with the coding being less sparse in the posterior pyriform cortex. The authors comment that the posterior pyriform cortex encoding may reflect perceptual similarity, in that for example after experiencing a cherry + smoky mixture, the components may become more similar to each other, with for example the cherry being described as having a smoky component. So some recoding may be taking place from anterior to posterior pyriform cortex, but it is not just pattern separation, according to this evidence (Kadohisa and Wilson 2006). What does appear to be the case is that the posterior pyriform cortex is a later stage in processing, in that it has highly developed axonal connections to entorhinal cortex, perirhinal cortex, prefrontal cortex, amygdala, and striatum (olfactory tubercle) (Haberly 2001, Luskin and Price 1983).

In addition, a L4 was not present in olfactory cortex, for there was no thalamus, and the inputs from the glomeruli in the olfactory bulb just appear as forward inputs projecting as they do in most neocortical areas to superficial layers of the cortex. However, an additional key difference (from the hippocampus too) is that there is no granule cell stage in the pyriform cortex to perform pattern separation before the inputs are applied to the pyriform cortex. Taking this concept one step further, the neocortex can be thought of as more advanced computationally (operationally) than the pyriform and hippocampal cortex, in that the neocortex has thalamic inputs, which as suggested above perform a useful and simple stage of pattern separation before inputs reach the primary sensory cortical areas, by virtue of the local lateral inhibition in the topologically mapped thalamic nuclei, which serves to operate as a high pass spatial filter. There is of course no (or very little) spatial topology in the 1000 dimensional odour space spanned by the olfactory receptor genes and glomeruli, and for this reason a thalamus for olfaction acting in the way just described with local lateral inhibition would not be operationally (computationally) very useful.

This provides an answer to why a thalamus evolved in conjunction with the neocortex. My hypothesis is that the thalamus, which performs primarily local lateral inhibition between its neurons, was useful as a cortical preprocessor, for exactly this computation performs hi-pass filtering of the inputs, making them more sparse, which helps the computations performed by the neocortex, including pattern separation and autoassociation (Appendix B).

18.2.3   Hippocampal cortex

The hippocampal cortex is classified as archicortex, and corresponds in the primate brain to the dentate gyrus and hippocampus (Pandya et al. 2015). Its structure (see Figs. 24.1, 24.12, 24.13, and 18.3) and principles of operation are considered in Chapter 24.

Here I introduce hypotheses about how its functional architecture, evolutionarily early, relates to the neocortex.

The CA3 region with its highly developed recurrent collateral system which is thought to operate as an autoassociation network with sparse representations (Chapter 24) appears to correspond to the superficial layers of the cerebral neocortex, the L2/L3 pyramidal cells (Fig. 18.3). (The CA3 cells and the neocortical superficial pyramidal cells are shown in green in Figs. 18.3 and 18.1 to draw out their operational, that is computational, similarity. The granule cells in both are shown in black. The CA1 cells and the neocortical L5 cells are shown in red in Figs. 18.3 and 18.1 again to draw out their operational similarity, although there are differences as indicated. It will be of interest to determine whether there is any evolutionary correspondence, which might be revealed by molecular markers (Bernard et al. 2012).) The difference of the recurrent collaterals is that the CA3 recurrent collateral connectivity is global, throughout the CA3 region, in the hippocampus, to implement episodic memory, in which any representation can be associated with any other representation. In contrast the cerebral neocortex has recurrent collaterals restricted to a few mm of neocortex, to ensure that the total memory capacity is high (O'Kane and Treves 1992), and to ensure that there is specialization

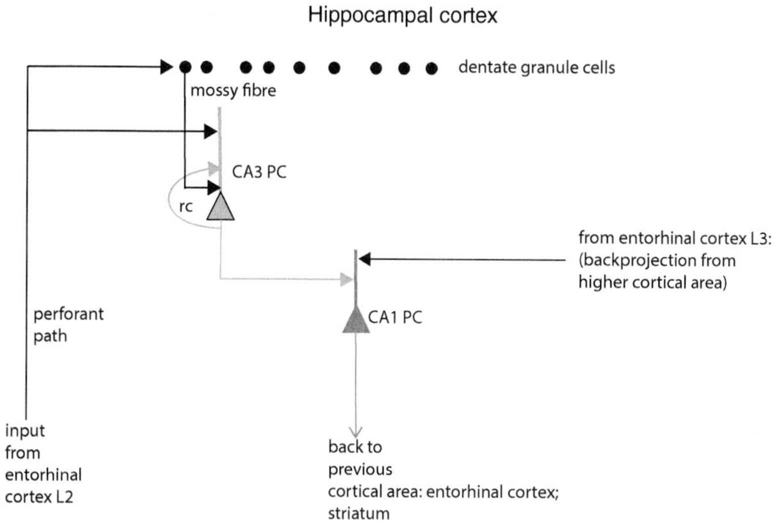

Hippocampal cortex

Fig. 18.3 Functional canonical microcircuit of the hippocampal cortex. Recurrent collateral connections (rc) are shown as a loop back to a particular population of cells, of which just one neuron is shown.

of function, typically involving hierarchical organisation, to perform complex computations, instead of just storing and retrieving information, as in the hippocampus. As an early type of cortical design in evolution, the CA3 hippocampal cells may have sought associations between inputs from any of the afferent structures. The later evolution of neocortex may have been driven by that facts that a single attractor network, as in CA3, has limited memory capacity; and does not perform real computation of new representations, both of which are features of the neocortex.

In the hippocampus, that leaves the dentate gyrus (which does appear to have early evolutionary origins identifiable as far back as reptiles (Hevner 2016, Treves, Tashiro, Witter and Moser 2008)). I hypothesize that the dentate gyrus, with its 'granule cells', just corresponds to the stellate cells in L4 of the neocortex, which are also described as 'granular' (their appearance at low power with a light microscope), due to their small size and large number. It is hypothesized that they play the same computational role in the hippocampus and neocortex, namely to pattern separate / orthogonalize the inputs before they reach the CA3 attractor network (Rolls 1989b, Rolls 1990b, Treves and Rolls 1992b, Rolls and Kesner 2006, Rolls 2008d, Kesner and Rolls 2015), an important computational operation to reduce correlations between memory patterns and thus to maximize storage capacity (Rolls 2016e, Sompolinsky 1987).

A difference though computationally is that in the hippocampus the granule cells perform pattern separation of CA3 representations by having large non-associatively-modifiable mossy fibre synapses onto the CA3 pyramidal cells. In contrast, in neocortex, the connections from granule cells may be modifiable, and they may be used during normal processing, not just during storage (see Chapter 24). It is notable that the granule cells are not present in all types of neocortex, and that some early forms of neocortex may not have had a well developed layer 4 with granule cells (Pandya et al. 2015). Similarly, early evolutionary forms of the hippocampus may have existed without the highly developed dentate granule cell system found in mammals (Hevner 2016). The hypothesis is that in both cases the autoassociation would occur without the granule cells as a preprocessor, but that when they evolved, the granule cells improved the performance of the later autoassociation network by helping the patterns to be processed and stored more orthogonally.

Further, the hippocampal granule cell / mossy fibre system without associative synaptic plasticity of the mossy synapses to CA3 cells may serve primarily to perform pattern separation for CA3, without being used for recall (Chapter 24). In contrast, the inputs from granule cells and other cells to the superficial neocortical pyramidal cells may be associatively modifiable to enable competitive learning to form useful new representations for the neocortex, and would therefore be involved when the representation is re-accessed later (Section B.4 and Chapter 25).

For the hippocampus, that also leaves CA1, which receives from CA3. This is analogous to the neocortical deep layer pyramidal cells receiving from the superficial layer pyramidal cells. Moreover, the CA1 cells are then described computationally as part of the backprojection process which enables memory recall by activating representations using backprojections to an earlier level of the cortical hierarchy (Chapter 24). That is exactly one function that is proposed for the deep layers of the neocortex, to send backprojections to earlier stages of the hierarchy, to perform memory recall, inter alia.

Another comparison supports some similarity between CA1 pyramidal cells in the hippocampus, and neocortical L5 pyramidal cells, which is that the CA1 neurons project directly to the striatum (see Fig. 24.1), just as do the L5 neocortical pyramidal cells (compare the red pyramidal cells in Figs. 18.3 and 18.1).

In implementing their recall operation back to the cerebral neocortex, the main computational operation required of CA1 is competitive learning to produce a compact retrieval cue in which the separate parts of each memory are no longer made explicit in the firing of different subsets of neurons (Chapter 24) (Kesner and Rolls 2015), and consistent with this, there is no well developed recurrent collateral system in CA1. Moreover, a less sparse representation in CA1 than in CA3 may be more appropriate for recall, that is for information transmission, and there may be correspondences in this respect of CA1 cells with the L5 neocortical pyramidal cells and the posterior pyriform cortex.

In overview, I have argued that hippocampal CA3 corresponds to superficial layers of the cerebral cortex; that the dentate granule cells in the hippocampus correspond to the granule cell layer 4 of the neocortex to perform pattern separation; and that hippocampal CA1 corresponds to deep layer L5 of the cerebral neocortex, used for backprojections to earlier cortical layers in the hierarchy (compare Figs. 18.3 and 18.1).

In the hippocampus, there is no correspondence to neocortex L6, because the hippocampus does not receive its major driving inputs from the thalamus. Indeed, the hippocampus (or at least an antecedent (Hevner 2016)) was present in evolution before there was a thalamus.

Of course, in evolution the hippocampus may have developed new principles of operation (such as excitatory recurrent collaterals) different from the largely non-cortical processing present in early vertebrates, and some of these new principles may then have been useful in the later development in evolution of the neocortex.

The issues of cortical lamination, cortical computation, and the similarity of and differences between the neocortex, pyriform cortex, and hippocampal cortex are developed further in Section 26.5.18 and elsewhere (Rolls and Mills 2017).

18.3 Addition of areas in the neocortical hierarchy

A fundamental principle of cortical evolution is that new areas are added to the top of existing areas in a cortical hierarchy, with some of the evidence described by Pandya, Seltzer, Petrides and Cipolloni (2015). For example, the prefrontal cortex and temporal lobes are much more developed in primates than in rodents. The overall design of each new area of the neocortex is similar to that of earlier areas, and this provides for economy of specification of the architecture of each new area (Chapter 19). Given that the basic design of a neocortical area is repeated time after time in the hierarchy (see Chapter 2), the new information that is needed for a new cortical area is the specification of its classes of long-range inter-area neurons so that they have molecular recognition devices to enable them to receive from the correct preceding area, and connect to the next succeeding area (Chapter 19).

It is shown in Chapters 2 and 25 how just adding areas to an existing hierarchy can provide for the system to perform useful computational functions, if there is convergence as one moves from area to area up the hierarchy.

The prefrontal cortex provides an interesting case, with its organisation being driven in evolution by the addition of new areas to the sensory and motor hierarchies in the more posterior neocortex. As illustrated in Fig. 2.17 each hierarchical stage of cortical processing in the somatic, visual and auditory cortical hierarchies has connectivity with its own level of local networks in the prefrontal cortex (Jones and Powell 1970). A very important function of the prefrontal cortical areas is in maintaining neuronal activity in short-term memory periods (Fuster 2008, Goldman-Rakic 1996). The implication is that short-term memory is organised at different hierarchical levels to serve the functions of different levels of cortical processing hierarchies. While it may be an open question about whether the prefrontal cortex areas are themselves operating hierarchically, the concept is certainly raised that the prefrontal architecture for short-term memory, and hence multiple step planning (Rolls 2014a), has different levels of short-term memory operations that reflect the hierarchies in more posterior information processing systems.

Another principle apart from independent short-term memory prefrontal networks to provide short-term capability for different levels of the posterior cortical hierarchies is that the separate prefrontal levels may then feed back the information for top-down attentional control that is appropriate for each level of the posterior cortical processing levels (see further Chapter 11).

As new areas are added to the neocortical hierarchy, there will be some manipulation of the parameters that specify the architecture (Chapter 19) to enable each cortical area to emphasize certain parts of its basic operation. For example, in primary sensory cortical areas, L4 will be well-developed, because the main source of input comes here from the thalamus, and the granule cells (called stellate cells in the neocortex) are especially important to perform pattern separation on the inputs, to help make the representation more sparse. In motor cortical areas, L5 will be especially prominent, with large pyramidal cells, because fast conduction and long distances of the axons are what need to be emphasized. In areas involved in memory, for example the prefrontal cortex which performs major functions involving short-term memory, the recurrent collateral connections are especially highly developed with large dendritic trees with many recurrent collateral synapses (Elston et al. 2006), to increase the memory capacity. These relative specializations of each cortical area in ways such as this provide a computational basis which helps to account in part for the different appearance of each cortical area, on which cytoarchitectonic and myeloarchitectonic classification of cortical areas is largely based (Pandya et al. 2015), though molecular markers that reflect the building instructions for the cortical connectivity (Chapter 19) will doubtless help in this classification greatly in future (Bernard et al. 2012).

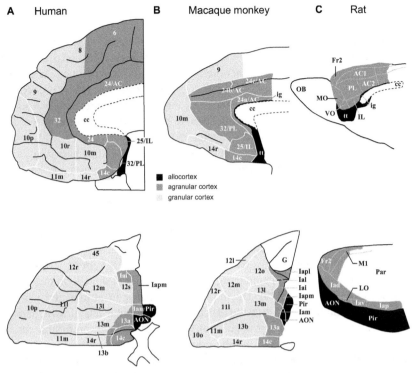

Fig. 18.4 Comparison of the orbitofrontal (below) and medial prefrontal (above) cortical areas in humans, macaque monkeys, and rats. (A) Medial (top) and orbital (bottom) areas of the human frontal codex (Ongur et al. 2003). (B) Medial (top) and orbital (bottom) areas of the macaque frontal cortex (Carmichael and Price 1994). (C) Medial (top) and lateral (bottom) areas of rat frontal cortex (Palomero-Gallagher and Zilles 2004). Rostral is to the left in all drawings. Top row: dorsal is up in all drawings. Bottom row: in (A) and (B), lateral is up; in (C), dorsal is up. Not to scale. Abbreviations: AC, anterior cingulate cortex; AON, anterior olfactory 'nucleus'; cc, corpus callosum; Fr2 second frontal area; Ia, agranular insular cortex; ig, induseum griseum; IL, infralimbic cortex; LO, lateral orbital cortex; MO, medial orbital cortex: OB, olfactory bulb; Pr, piriform (olfactory) cortex; PL, prelimbic cortex; tt, tenia tecta; VO, ventral orbital cortex; Subdivisions of areas are labelled caudal (c); inferior (i), lateral (l), medial (m); orbital (o), posterior or polar (p), rostral(r), or by arbitrary designation (a, b). (After Passingham and Wise (2012). (a) Adapted with permission from Dost Ongur, Amon T. Ferry, and Joseph L. Price, Architectonic subdivision of the human orbital and medial prefrontal cortex, *Journal of Comparative Neurology*, 460 (3), pp. 425–49 Copyright ©2003 Wiley-Liss, Inc. (b) Adapted with permission from S. T. Carmichael and J. L. Price, Architectonic subdivision of the orbital and medial prefrontal cortex in the macaque monkey, *Journal of Comparative Neurology*, 346 (3), pp. 366–402 Copyright ©1994 Wiley-Liss, Inc. (c) Adapted with permission from Palomero-Gallagher N. and Zilles K., Isocortex, in *The Rat Nervous System* 3e, ed. G. Paxinos, p.729–57 ©2004, Elsevier Inc.)

18.4 Evolution of the orbitofrontal cortex

The prefrontal cortex has undergone great development in primates (Passingham and Wise 2012, Pandya, Seltzer, Petrides and Cipolloni 2015), partly for the reason just described that new prefrontal areas are required as new areas are added to the posterior sensory processing hierarchies. Further, one part of the prefrontal cortex, the orbitofrontal cortex, is very little developed in rodents, yet is one of the major brain areas involved in emotion and motivation in primates including humans (Rolls 2014a). Indeed, it has been argued that the granular prefrontal cortex is a primate innovation, and the implication of the argument is that any areas that might be termed orbitofrontal cortex in rats (Schoenbaum, Roesch, Stalnaker and

Takahashi 2009) are homologous only to the agranular parts of the primate orbitofrontal cortex (shaded mid grey in Fig. 18.4), that is to areas 13a, 14c, and the agranular insular areas labelled Ia in Fig. 18.4 (Wise 2008, Passingham and Wise 2012). It follows from that argument that for most areas of the orbitofrontal and medial prefrontal cortex in humans and macaques (those shaded light grey in Fig. 18.4), special consideration must be given to research in macaques and humans (Rolls 2014a, Rolls 2017b). As shown in Fig. 18.4, there may be no cortical area in rodents that is homologous to most of the primate including human orbitofrontal cortex (Preuss 1995, Wise 2008, Passingham and Wise 2012).

18.5   Evolution of the taste and flavour system

18.5.1   Principles

The development of some cortical areas has been so great in primates that even evolutionarily old systems such as the taste system appear to have been rewired, compared with that of rodents, to place much more emphasis on cortical processing, taking place in areas such as the orbitofrontal cortex (Rolls and Scott 2003, Scott and Small 2009, Small and Scott 2009, Rolls 2014a, Rolls 2016b) (Fig. 2.4).

In primates, the reward value of a taste is represented in the orbitofrontal cortex in that the responses of orbitofrontal taste neurons are modulated by hunger in just the same way as is the reward value or palatability of a taste. In particular, it has been shown that orbitofrontal cortex taste neurons stop responding to the taste of a food with which a monkey is fed to satiety, and that this parallels the decline in the acceptability of the food (see Fig. 2.6) (Rolls et al. 1989c). In contrast, the representation of taste in the primary taste cortex of primates (Scott, Yaxley, Sienkiewicz and Rolls 1986a, Yaxley, Rolls and Sienkiewicz 1990) is not modulated by hunger (Rolls, Scott, Sienkiewicz and Yaxley 1988, Yaxley, Rolls and Sienkiewicz 1988). Thus in the primary taste cortex of primates (and at earlier stages of taste processing including the nucleus of the solitary tract), the reward value of taste is not represented, and instead the identity of the taste is represented (Rolls 2016b, Rolls 2016f) (see Sections 2.2.3 and 18.5.2). The importance of cortical processing of taste in primates, first for identity and intensity in the primary taste cortex, and then for reward value in the orbitofrontal cortex, is that both types of representation need to be interfaced to visual and other processing that requires cortical computation. For example, it may have adaptive value to be able to represent exactly what taste is present, and to link it by learning to the sight and location of the source of the taste, even when hunger and reward is not being produced, so that the source of that taste can be found in future, when it may have reward value.

In line with cortical processing to dominate the processing of taste in primates, there is no modulation of taste responsiveness at or before the primary taste cortex, and the pathways for taste are directly from the nucleus of the solitary tract in the brainstem to the taste thalamus and then to the taste cortex (Figs. 2.4 and 18.5). In contrast, in rodents such as the rat, the nucleus of the solitary tract connects to a pontine taste area, the parabrachial nucleus, that is not present in primates (Rolls and Scott 2003, Scott and Small 2009, Small and Scott 2009, Rolls 2016b). The rodent pontine taste area then not only has connections to the thalamus and thus to the cortex, but also has direct connections to many subcortical areas important in appetite control, including the amygdala and hypothalamus (Section 18.5.2). Moreover, in rodents, satiety reduces the responsiveness of neurons in the nucleus of the solitary tract to the taste of food by approximately 30%, so that taste processing in rodents is from the first synapse in the brain confounded by reward value, by hedonics (Rolls 2016b). That makes the taste system of rodents very difficult to understand functionally for different functions are not separated

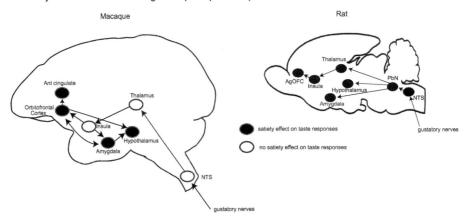

Fig. 18.5 Taste pathways in the macaque and rat. In the macaque, gustatory information reaches the nucleus of the solitary tract (NTS), which projects directly to the taste thalamus (ventral posteromedial nucleus, pars parvocellularis, VPMpc) which then projects to the taste cortex in the anterior insula (Insula). The insular taste cortex then projects to the orbitofrontal cortex and amygdala. The orbitofrontal cortex projects taste information to the anterior cingulate cortex. Both the orbitofrontal cortex and the amygdala project to the hypothalamus (and to the ventral striatum). In macaques, feeding to normal self-induced satiety does not decrease the responses of taste neurons in the NTS or taste insula (and by inference not VPMpc) (see text). In the rat, in contrast, the NTS projects to a pontine taste area, the parabrachial nucleus (PbN). The PbN then has projections directly to a number of subcortical structures, including the hypothalamus, amygdala, and ventral striatum, thus bypassing thalamo-cortical processing. The PbN in the rat also projects to the taste thalamus (VPMpc), which projects to the rat taste insula. The taste insula in the rat then projects to an agranular orbitofrontal cortex (AgOFC), which probably corresponds to the most posterior part of the primate OFC, which is agranular. (In primates, most of the orbitofrontal cortex is granular cortex, and the rat may have no equivalent to this. In the rat, satiety signals such as gastric distension and satiety-related hormones decrease neuronal responses in the NTS (see text), and by inference therefore in the other brain areas with taste-related responses, as indicated in the Figure.)

(taste identity and intensity vs hedonics), and makes the taste system of rodents a poor one with which to understand primate including human taste reward processing. This evidence emphasizes the importance of understanding the evidence from primates including humans, even in a system such as the taste system that one might think is evolutionarily so old (Section 18.5.2).

18.5.2 Taste processing in rodents

There are major differences in the neural processing of taste in rodents and primates (Rolls and Scott 2003, Small and Scott 2009, Scott and Small 2009, Rolls 2015a, Rolls 2016b) (Fig. 18.5). In rodents (and also in primates) taste information is conveyed by cranial nerves 7, 9 and 10 to the rostral part of the nucleus of the solitary tract (NTS) (Norgren 1990, Norgren and Leonard 1971, Norgren and Leonard 1973). However, although in primates the NTS projects to the taste thalamus and thus to the cortex (Fig. 2.4), in rodents the majority of NTS taste neurons responding to stimulation of the taste receptors of the anterior tongue project to the ipsilateral medial aspect of the pontine parabrachial nucleus (PbN), the rodent 'pontine taste area' (Small and Scott 2009, Cho, Li and Smith 2002). The remainder project to adjacent regions of the medulla. From the PbN the rodent gustatory pathway bifurcates into two pathways; 1) a ventral 'affective' projection to the hypothalamus, central gray, ventral striatum, bed nucleus of the stria terminalis and amygdala and 2) a dorsal 'sensory' pathway, which first synapses in the thalamus and then the agranular and dysgranular insular gustatory cortex (Norgren 1990, Norgren and Leonard 1971, Norgren 1974, Norgren 1976, Kosar, Grill and Norgren 1986). These regions, in turn, project back to the PbN to "sculpt the gustatory code" and guide complex feeding behaviours (Norgren 1990, Norgren 1976, Li and Cho 2006, Li, Cho and Smith 2002, Lundy and Norgren 2004, Di Lorenzo 1990, Scott and Small 2009, Small and Scott 2009).

It may be noted that there is strong evidence to indicate that the PbN gustatory relay is absent in the human and the nonhuman primate (Small and Scott 2009, Scott and Small 2009). First, second-order gustatory projections that arise from rostral NTS appear not to synapse in the PbN and instead join the central tegmental tract and project directly to the taste thalamus in primates (Beckstead, Morse and Norgren 1980, Pritchard, Hamilton and Norgren 1989). Second, despite several attempts, no one has successfully isolated taste responses in the monkey PbN (Norgren (1990); Small and Scott (2009) who cite Ralph Norgren, personal communication and Tom Pritchard, personal communication). Third, in monkeys the projection arising from the PbN does not terminate in the region of ventral basal thalamus that contains gustatory responsive neurons (Pritchard et al. 1989).

A further difference of rodent taste processing from that of primates is that physical and chemical signals of satiety have been shown to reduce the taste responsiveness of neurons in the nucleus in the solitary tract, and the pontine taste area, of the rat, with decreases in the order of 30%, as follows (Rolls and Scott 2003, Scott and Small 2009). Gastric distension by air or with 0.3 M NaCl suppress responses in the NTS, with the greatest effect on glucose (Gleen and Erickson 1976). Intravenous infusions of 0.5 g/kg glucose (Giza and Scott 1983), 0.5 U/kg insulin (Giza and Scott 1987b), and 40 μg/kg glucagon (Giza, Deems, Vanderweele and Scott 1993) all cause reductions in taste responsiveness to glucose in the NTS. The intraduodenal infusion of lipids causes a decline in taste responsiveness in the PBN, with the bulk of the suppression borne by glucose cells (Hajnal, Takenouchi and Norgren 1999). The loss of signal that would otherwise be evoked by hedonically positive tastes implies that the pleasure that sustains feeding is reduced, making termination of a meal more likely (Giza, Scott and Vanderweele 1992). Further, if taste activity in NTS is affected by the rat's nutritional state, then intensity judgements in rats should change with satiety. There is evidence that they do. Rats with conditioned aversions to 1.0 M glucose show decreasing acceptance of glucose solutions as their concentrations approach 1.0 M. This acceptance gradient can be compared between euglycemic rats and those made hyperglycemic through intravenous injections (Scott and Giza 1987). Hyperglycemic rats showed greater acceptance at all concentrations from 0.6 to 2.0 M glucose, indicating that they perceived these stimuli to be less intense than did conditioned rats with no glucose load (Giza and Scott 1987a).

The implication is that taste, and the closely related olfactory and visual processing that contribute to food reward value and expected value, are much more difficult to understand in rodents than in primates, partly because there is less segregation of 'what' (identity and intensity) from hedonic processing in rodents, partly because of the more serial hierarchical processing in primates (Fig. 2.4), and partly because in primates there has been great development of the granular orbitofrontal cortex which may help to support the rule-based switching of behaviour important for rapidly reversing stimulus-reward associations and behaviour.

18.6 Evolution of the temporal lobe cortex

Another reason for focusing interest on the primate brain is that there has been great development of the visual system in primates (Pandya, Seltzer, Petrides and Cipolloni 2015), and this itself has had important implications for the types of sensory stimuli that are processed by brain systems involved in emotion and motivation. One example is the importance of face identity and face expression decoding, which are both important in primate emotional behaviour, and indeed provide an important part of the foundation for much primate social behaviour (Rolls 2011d).

18.7   Evolution of the frontal lobe cortex

Another reason for focusing interest on the primate brain is that there has been great development of the frontal lobes in primates including humans, not only for the orbitofrontal part as described in Section 18.4, but also for the dorsolateral parts (Passingham and Wise 2012, Pandya et al. 2015, Fuster 2008, Fuster 2014). Part of the reason for the expansion of the frontal lobes is that new parts must be added to act as short-term memory, off-line, processors for new areas of the temporal and parietal cortical hierarchies, as implied by the anatomical connectivity illustrated in Fig. 2.17 (Jones and Powell 1970, Pandya et al. 2015). There is much more that could be added here and to Section 18.6, but the main aim of this book is to identify and illustrate principles of operation of the cortex, rather than to provide a complete description of all aspects of the cortex.

18.8   Highlights

1. The olfactory pyriform cortex may be thought of as an evolutionarily early type of cortex, the hippocampal cortex as later, and the neocortex or isocortex as the most recently evolved and computationally complex cortex.

2. Global recurrent collateral associative connectivity in the pyriform cortex enables a single or global attractor network to form associations between co-active different representations by any subsets of neurons to form olfactory representations that are different combinations of the parts.

3. Global recurrent collateral associative connectivity in the hippocampal cortex CA3 neurons enables a single or global attractor network to form associations between co-active different representations by any subsets of neurons to form an episodic or event memory.

4. Local recurrent collaterals in the neocortex enable many local attractor networks to be present, to enable short-term and long-term memory operations to operate with a very high capacity, because each local attractor network can operate relatively independently in terms of storage. Competitive learning also helps local representations to be built.

5. The local recurrent collaterals in neocortex contribute to topological maps in neocortex, which help to minimize axonal connection lengths in neocortex, and help to simplify the rules of cortical wiring between neuron classes because locality can be a guiding principle of connectivity.

6. The hippocampus and neocortex add to what is present in pyriform cortex by adding a granule cell input layer, to perform pattern separation.

7. The granule cells in the hippocampal system perform only pattern separation for CA3 learning, with their large, non-associatively modifiable, synapses.

8. The granule cells in the neocortex perform pattern separation (on thalamic inputs or inputs from the preceding cortical area), but also provide an input that is often topologically mapped. The granule cells help to build categorical representations by their modifiable synapses using the principle of competitive learning. In primates, there is a much more developed granule cell system in higher cortical areas than in rodents.

9. The neocortex in addition has a thalamus, which performs a computationally useful function by typically being topologically mapped, and performing pattern separation by local lateral inhibition within the thalamus. The neocortex can regulate its inputs by L6 feedback connections to the thalamus.

10. On the output side, the posterior pyriform cortex may perform a competitive network recoding into an information rich representation in which the parts may be less explicit, and consistently with this, appears to have a less well developed recurrent collateral system than the anterior pyriform cortex.

11. The hippocampal CA1 cells do not have a developed recurrent collateral system, and may provide a competitive network recoding into an information rich representation in which the parts may be less explicit. This representation is hypothesized to be useful for the recall operation back to the neocortex (Section 24.3.7). This stage also has some outputs to the striatum.

12. The neocortical superficial layers (L2/L3) have local highly developed recurrent collaterals, and may in addition perform competitive learning to build new representations. These superficial layers may form the sparse distributed discrete 'what' representations of objects or input stimuli. The superficial layers of neocortex then feed forward to L4 of the next cortical area to form a hierarchy. (Hierarchies are not a principle of operation of the hippocampal cortex.) The superficial layers may be the main parts of neocortex that build new representations up the hierarchy, utilizing convergence from cortical area to cortical area (Chapter 2).

13. In addition to a feedforward hierarchy, the neocortex can send useful outputs from every stage of processing, and these reach the striatum and motor structures from the L5 pyramidal cells. Part of the reason for a separate L5 is that the output (pyramidal) cells may need to be large, as some at least project large distances; and that it may be evolutionarily convenient in terms of cortical design (Chapter 19) to use different molecular recognition devices for them from the superficial cells, to help them identify and connect to their very different targets to the superficial cells. Another part of the reason for a L5 separate from the superficial layers of neocortex is that both, with their somewhat separate recurrent collateral systems, benefit from forming attractors, but that the deep layers may need to do so in a somewhat different space, the space of useful typical motor-related outputs, which imply the need for a continuous attractor space, and for what may be a more continuously spatially mapped representation. These deep layers are frequently the source in a primate sensory hierarchy of the backprojections, which may benefit from being more diffuse topographically as they are involved in processes such as memory recall and top-down attention, and not in transferring the detailed discrete representations in superficial layers back to an earlier cortical stage, which is not a principle of operation of the neocortex (Chapters 2 and 11). Consistent with this recall or modulatory influence, many of the backprojections end on the apical dendrites in L1 of neocortical pyramidal cells, far from the cell body.

14. Given that neocortex has a thalamic input, it is useful to be able to control the magnitude of the thalamic input by gain control of the thalamus. A separate layer for these neurons may simplify the organisation of cortical connectivity, for then they can easily be kept separate if needed from the activity of neurons in other layers.

15. The lamination of the cortex may also related to the need to have separate local gain control for the superficial layers vs L5 and L6, and the lamination may help to simplify the genetic specification of the different classes of inhibitory neurons, which may then be most easily specified as the layer within which they operate. This occurs in neocortex

and hippocampal cortex (Somogyi, Tamas, Lujan and Buhl 1998, Somogyi 2010, Harris and Shepherd 2015, Chu and Anderson 2015, Lewis 2014).

16. The issues of cortical lamination, cortical computation, and the similarity of and differences between the neocortex, pyriform cortex, and hippocampal cortex, are developed further in Section 26.5.18.

17. The great development of the orbitofrontal cortex, temporal lobes, and frontal cortical areas in primates including humans is highlighted, and how with increasing emphasis on the cortical processing of stimuli, even systems as apparently evolutionarily old as the taste system become differently connected to emphasize cortical processing in primates.

19 Genes and self-organization build the cortex

19.1 Introduction

Analysis of the structure and function of different cortical areas is starting to reveal how the operation of networks of neurons may implement the functions of each cortical area. The question then arises as part of understanding cortical function of how the networks found in different cortical areas actually have evolved, and how their basic architecture may be specified by genes and then built using self-organisation and environmental input during development.

To address these fundamental issues, I compared the architecture and networks found in different brain areas, and in particular how they differ from each other, to formulate hypotheses about a set of parameters that may be specified by genes that could lead to the building during development of the neuronal network functional architectures found in different cortical regions (Rolls and Stringer 2000). The choice of parameters was guided not only by comparison of the functional architecture of different brain regions, but also by what parameters, if specified by genes, could with a reasonably small set actually build the networks found in different brain regions. The concept is that if these parameters are specified by different genes, then genetic reproduction and natural selection using these parameters could lead to the evolution of neuronal networks in our brains well adapted for different functions.

A second aim is to show how the sufficiency of the general approach, and the appropriateness of the particular parameters selected, can be tested and investigated using genetic algorithms which actually use the hypothesised genes to specify networks. We showed this by implementing a genetic algorithm, and testing whether it can search the high dimensional space provided by the suggested parameters to specify and build neuronal networks that will solve particular computational problems (Rolls and Stringer 2000).

The computational problems we choose to test are simple and well defined problems that can be solved by one-layer networks, and that as shown in this book capture some of the architectural properties of different cortical areas, and are some of the building blocks of cortical function. These problems are pattern association, autoassociation, and competition, which require quite different architectures for them to be solved (see Chapter 1 including Fig. 1.8 and Appendix B).

Because the problems to be solved are well specified, we can define a good fitness measure for the operation of each class of network, which will be used to guide the evolution by reproduction involving genetic variation and selection in each generation. Although we do not suppose that the actual parameters chosen for illustration are necessarily those specified by mammalian genes, they have been chosen because they seem reasonable given the differences in the functional architecture of different brain areas, and allow illustration of the overall concept about how different network architectures found in different brain regions evolve.

Although these computational problems to be solved were chosen to have well understood one-layer neural network solutions, now that this general approach has been established (Rolls and Stringer 2000), it will be of great interest in future to examine problems that are solved by multilayer networks in the brain (e.g. invariant object recognition (Chapter 25) and episodic memory (Chapter 24), in order to understand how brain mechanisms to solve complex problems may evolve and how they may operate.

The first example of the networks to be investigated was pattern association networks in which one input (e.g. an unconditioned stimulus) drives the output neurons through unmodifiable synapses, and a second input (e.g. a conditioned stimulus) has associatively modifiable synapses onto the output neurons, so that by associative learning it can come to produce the same output as the unconditioned stimulus (Section B.2). Pattern association memory may be implemented in structures such as the amygdala and orbitofrontal cortex to implement stimulus-reward. Pattern association learning may also be used in the connections of back-projecting neurons in the cerebral cortex onto the apical dendrites of neurons in the preceding cortical area to implement memory recall.

A second example is autoassociation networks characterized by recurrent collateral axons with associatively modifiable synapses (Section B.3), which may implement functions such as short term memory in the cerebral cortex, and episodic memory in the hippocampus.

A third example is competitive learning (Section B.4), where there is one major set of inputs to a network connected with associatively modifiable synapses, and mutual (e.g. lateral) inhibition between the output neurons (through e.g. inhibitory feedback neurons). Competitive networks can be used to build feature analyzers by learning to respond to clusters of inputs which tend to co-occur, and may be fundamental building blocks of perceptual systems. Allowing short range excitatory connections between neurons (as in the cerebral cortex) and longer range inhibitory connections can lead to the formation of topographic maps where the closeness in the map reflects the similarity between the inputs being mapped.

19.2 Hypotheses about the genes that could build different types of neuronal network in different brain areas

The hypotheses for the genes that might specify the different types of neuronal network in different brain areas are now introduced. The hypotheses are based on knowledge of the architectures of different brain regions, which are described in Chapter 1 and in sources such as the following (Shepherd 2004, Shepherd and Grillner 2010, Braitenberg and Schütz 1991, Braitenberg and Schütz 1998, Peters and Jones 1984, Somogyi et al. 1998, Harris and Mrsic-Flogel 2013, Kubota 2014), and on a knowledge of some of the parameters that influence the operation of neuronal networks. These genes were used in the simulations in which evolution of the architectures was explored using genetic algorithms. The emphasis in this section is on the rationale for suggesting different genes. A more formal specification of the genes, together with additional information on how they were implemented in the simulations, is provided in Section 19.4.2. It may also be noted that large numbers of genes will be needed to specify for example the operation of a neuron. This Chapter focusses on those genes that, given the basic building blocks of neurons, may specify the differences in the functional architecture of different regions of the cerebral cortex.

The overall concept is as follows. There are far too few genes (in humans 20,000–25,000 protein coding genes (Human Genome Sequencing 2004)) to specify each synapse (i.e. which neuron is connected to which other neuron, and with what connection strength) in the brain. (The number of synapses in the human brain, with 10^{10} - 10^{11} neurons in the brain, and perhaps an average of 10,000 synapses each, is in the order of 10^{14} - 10^{15}). In any case, brain design is likely to be more flexible if the actual strength of each synapse is set up by self-organisation and experience. On the other hand, brain connectivity is far from random.

Some indications about what is specified can be gathered by considering the connectivity of the hippocampus (Chapter 24). The CA3 pyramidal cells each receive approximately 50 mossy fibre synapses from dentate granule cells, 12,000 synapses from other CA3 cells formed from recurrent collaterals, and 3,600 perforant path synapses originating from entorhinal cortex neurons (see Fig. 24.14. In the preceding stage, the 1,000,000 dentate granule cells (the numbers given are for the rat) receive one main source of input, from the entorhinal cortex, and each makes approximately 14 synapses with CA3 cells.

Specification of neuron classes, and of connection rules between classes

On the basis of considerations of this type for many different brain areas, it is postulated that for each class of cell the genome specifies the approximate numbers of synapses the class will receive from a specified other class (including itself, for recurrent collaterals), and the approximate number of synapses its axons will make onto specified classes of target cells (including itself). The individual neurons with which synapses are made are not specified, but are chosen randomly, though sometimes under a constraint (specified by another gene) about how far away the axon should travel in order to make connections with other neurons. One parameter value of the latter gene might specify the widespread recurrent collateral system of the CA3 neurons (Chapter 24). Another value for the latter might specify much more limited spread of recurrent collaterals with the overall density decreasing rapidly from the cell of origin, which, as in the cerebral cortex and if accompanied by longer range inhibitory processes implemented by feedback inhibitory neurons, would produce center-surround organisation, and tend to lead to the formation of topological maps (see Section B.4.6).

The actual mechanism by which this specification of connection rules between classes of neuron is implemented would presumably involve some (genetically specified) chemical recognition process, together with the production of a limited quantity of a trophic substance that would limit the number of synapses from, and made to, each other class of cell. Consistent with this fundamental hypothesis, gene knockout investigations show that for example different genes specify the NMDA receptors in the cell classes, dentate gyrus, CA3, and CA1 (Nakazawa et al. 2002, Nakazawa et al. 2003, Nakazawa et al. 2004, Huerta, Sun, Wilson and Tonegawa 2000).

Some of these processes would of course be occurring primarily during development (ontogenesis), when simple rules such as making local connections as a function of distance away would be adequate to specify the connectivity, without the need for long pathway connection routes (such as between the substantia nigra and the striatum) to be genetically encoded. It is presumably because of the complex genetic specification that would be required to specify in the adult brain the route to reach all target neurons that epigenetic factors are so important in embryological development, and that as a general rule the formation of new neurons is not allowed in adult mammalian brains (Chapter 13).

It is a feature of brain neuronal network design that not only is which class of neuron to connect to apparently specified, but also approximately where on the dendrite the connections from each other class of neuron should be received. For example, in the CA3 system, the mossy fibres synapse closest to the cell body (where they can have a strong influence on the cell), the CA3 recurrent collaterals synapse on the next part of the dendrite, and the entorhinal inputs synapse on the more distal ends of the dendrites. The effect of the proximal relative to the more distal synapses will depend on a number of factors, including for distal inputs whether proximal inputs are active and thereby operating as current shunts producing division, and on the diameter of the dendrite, which will set its cable properties (Koch 1999). If the dendritic cable diameter is specified as large, all inputs (including distal inputs) will sum reasonably linearly to inject current into the cell body.

Although inhibitory neurons have not been included in the examples already given, similar specifications would be applicable, including for example a simple specification of the different parts of the dendrite on which different classes of inhibitory neuron synapse (Kubota 2014), and hence whether the effect is subtractive or shunting (Koch 1999). In cortical areas, both feedforward and feedback inhibition (the latter from pyramidal cells via inhibitory neurons back to the same population of pyramidal cells) could be produced by a simple genetic specification of this type.

Specification of synaptic learning rules

Next, the nature of the synaptic connections, and the learning rule for synaptic modifiability, must be specified. One gene specifies in the simulation described later whether a given neuron class is excitatory or inhibitory. In the brain, this gene (or genes) would specify the transmitter (or transmitters in some cases) that are released, with the actual effects of for example glutamate being excitatory, and gamma-amino-butyric acid (GABA) inhibitory. The learning rule implemented at each synapse is determined by another gene (or genes). One possible effect specified is no synaptic modification. Another is a Hebb rule of associative synaptic plasticity (increase the synaptic strength if both presynaptic and postsynaptic activity are high, the simple rule implemented by associative long-term potentiation (LTP)). For this, the genes might specify NMDA (n-methyl-d-aspartate) receptors on the post-synaptic neurons together with the linked intracellular processes that implement LTP (Bliss and Collingridge 2013) (Section 1.5). Another possible effect is long-term depression (LTD). This may be heterosynaptic, that is the synaptic weight may be decreased if there is high post-synaptic activity but low presynaptic activity. Part of the utility of this in the brain is that when combined with LTP in pattern associators and autoassociators the effect is to remove the otherwise positive correlation that would be produced between different input patterns if all patterns are specified by positive-only firing rates, as they are in the brain. The effect of removing this correlation is to reduce interference between patterns, and to maximize the storage capacity (see Appendix B). A further useful property of heterosynaptic LTD is that it can help to maintain the total synaptic strength onto a neuron constant, by decreasing synaptic strengths from inputs to a neuron which are inactive when the neuron is currently firing. This can be very useful in competitive networks to prevent some winning neurons from continually increasing their synaptic strength so that they win to all patterns, and can in addition be seen as part of the process by which the synaptic weight vector is moved to point in the direction of a current input pattern of neuronal activity (see Appendix B). Another type of LTD is homosynaptic, in which the synaptic strength decreases if there is high presynaptic activity but low or moderate postsynaptic activity. This might be useful in autoassociative synaptic networks, as combined with LTP and heterosynaptic LTD it can produce a covariance-like learning rule.

Another learning rule of potential biological importance is a trace learning rule, in which for example the post-synaptic term is a short term average in an associative Hebb-like learning rule. This encourages neurons to respond to the current stimulus in the same way as they did to previous stimuli. This is useful if the previous stimuli are different views etc of the same object (which they tend statistically to be in our visual world), because this promotes invariant responses to different versions of the same object. Use of such a rule has been proposed to be one way in which networks can learn invariant responses (Rolls 2012c) (Chapter 25). Such a trace could be implemented in real neurons by a number of different mechanisms, including slow unbinding of glutamate from the NMDA receptor (which may take 100 ms or more), and maintaining a trace of previous neuronal activity by using short term autoassociative attractor memories implemented by recurrent collaterals in the cerebral neocortex (Rolls and Treves 1998).

Other types of synaptic modification that may be genetically specified include non-associative LTP (as may be implemented by the hippocampal mossy fibre synapses), and non-associative LTD. Other genes working with these may set parameters such as the rapidity with which synapses learn (which in a structure such as the hippocampus may be very fast, in one trial, to implement memory of a particular episode, and in structures such as the basal ganglia may be slow to enable the learning of motor habits based on very many trials of experience); and the initial and maximal values of synapses (e.g. the mossy fibre synapses onto hippocampal CA3 cells can achieve high values).

Specification of the operation of neurons

Another set of genes specifies some of the biophysical parameters that control the operation of individual neurons (see Koch (1999) for background). One gene (in a simulation, or biologically perhaps several) specifies how the activation h_i of a neuron i is calculated. A linear sum of the inputs r'_j weighted by the synaptic weights w_{ij} is the standard one used in most models of neuronal networks (Appendix B), and those simulated here, as follows:

$$h_i = \sum_j r'_j w_{ij} \tag{19.1}$$

where $\sum_j$ indicates that the sum is over the C input axons (or connections) indexed by j. An alternative is that there is non-linearity in this process, produced for example by local interactions in dendrites, including local shunting, affected most notably by the cable diameter of the dendrite, which is what such genes may control. For most pyramidal cells, the dendrite diameter is sufficiently large that linear summation to produce the net current injected into the cell bodies is a reasonable approximation (Koch 1999).

Several further genes set the activation function of the neuron. (The activation function is the relation between the activation of the neuron and its firing. Examples are shown in Fig. 1.4). One possibility is linear. A second, and the most biologically plausible, is to have a threshold, followed by a part of the curve where the firing rate increases approximately linearly with the activation, followed by a part of the curve where the firing rate gradually saturates to a maximum. This function can be captured by for example a sigmoid activation function as follows:

$$r_i = \frac{1}{1 + e^{-2\beta(h_i - \alpha)}} \tag{19.2}$$

where α and β are the sigmoid threshold and slope, respectively. The output of this function, also sometimes known as the logistic function, is 0 for an input of $-\infty$, 0.5 for h_i equal to α, and 1 for $+\infty$. For this type of activation function, at least two genes would be needed (and biologically there would probably be several to specify the biophysical parameters), one to control the threshold, and a second to control the slope. A third possibility is to have a binary threshold, producing a neuron which moves from zero activity below threshold to maximal firing above threshold. This activation function is sometimes used in mathematical modelling because of its analytic tractability.

One variable that is controlled by the threshold (and to some extent the slope) of the activation function is the proportion of neurons in a population that are likely to be firing for any one input. This is the sparseness of the representation. If the neurons have a binary activation function, the sparseness may be measured just by the proportion of active neurons, and takes the value 0.5 for a fully distributed representation, in which half of the neurons are active. For neurons with continuous activation functions, the sparseness a may be defined as

$$a = \frac{(\sum_i r_i / N)^2}{\sum_i r_i^2 / N} \tag{19.3}$$

where N is the number of neurons in the layer (Appendix B). This works also for binary neurons. To provide precise control of the sparseness in some of the simulations described below we made provision for this to be controlled as an option directly by one gene, rather than being determined indirectly by the threshold and slope gene-specified parameters of the sigmoid activation function shown in Eq. 19.2.

19.3 Genetic selection of neuronal network parameters to produce different network architectures with different functions

Given the proposals just described for the types of parameter that are determined by different genes, we describe next how gene selection is postulated to lead to the evolution of neuronal networks adapted for particular (computational) functions. The description is phrased in terms of simulations of the processes using genetic algorithms, as investigations of this type enable the processes to be studied precisely. The processes involve reproduction, ontogenesis followed by tests of how well the offspring can learn to solve particular problems, and natural selection.

First, a selection of genes is made for a set of G genotypes in a population, which should be of a certain minimum size for evolution to work correctly. The genes are set out on a chromosome (or chromosomes). (Effects of gene linkage on a chromosome are not considered here.) Each set of genes is a genotype. The selection of individual genotypes from which to breed is made a probabilistic function which increases with the fitness of the genotype, measured by a fitness function that is quantified by how well that genotype builds an individual that can solve the computational problem that is set.

Having chosen two genotypes in this way, two genotypes to specify two new (haploid) offspring for the next generation are made by the genetic processes of sexual reproduction involving both gene recombination and mutation, which occur with specified probabilities. This process is repeated until G genotypes have been produced. Then G individuals are built with the network architectures specified by the G genotypes. The fitness of these individuals is then measured by how well they perform at the computational problem set.

In order to solve the computational problem, the networks are trained by presenting the set of input patterns, and adjusting the synaptic weights in the network according to the learning rules specified by the genotype of that individual. The individuals then breed again with a probability of being selected for reproduction that is proportional to their fitness relative to that of the whole population.

This process is allowed to proceed for many generations, during which the fitness of the best individual in the population, and the average fitness, both increase if evolution is working.

This type of genetic process is an efficient method of searching through a high dimensional space (the space specified by the genes), particularly where the space has many local optima so that simple hill climbing is inefficient, and where there is a single measure of fitness (Holland 1975, Ackley 1987, Goldberg 1989). An example of such a landscape is shown in two dimensions in Fig. 19.1. The sexual reproduction by allowing recombination of genes facilitates local hill-climbing in such a landscape. However, once the top of the local hill is reached, the process may become stuck at the top of the hill. Mutations have the effect of allowing a random jump to another part of the space, where the local hill-climbing can begin all over again. The combination of sexual reproduction and occasional mutations is an efficient way to find the top of the highest hill in the landscape, and this is part of the mathematical

Fig. 19.1 A landscape with a broadly unimodal space but with many local optima that is susceptible to genetic search to find the top of the highest peak.

basis for how evolution works by using natural selection operating on changes made using both sexual reproduction and occasional mutations.

An additional useful feature of genetic search is that past gene combinations, useful possibly in other contexts, can remain in the population for a number of generations, and can then be reused later on, without the need to search for those gene combinations again. This re-use of past combinations is one of the features of genetic search that can make it powerful, rapid, and show sudden jumps forward.

19.4 Simulation of the evolution of neural networks using a genetic algorithm

Next I describe simulations of these processes using a genetic algorithm, in order to make explicit some of the parameters hypothesized. The aims of the simulations are to demonstrate the feasibility and details of the hypotheses and approach; to provide an indication of whether the proposed genes can be used to guide efficiently the evolution of networks that can solve computational problems of the type solved by the brain; and to provide a tool for investigating in more detail both the parameters that can best and biologically plausibly be genetically specified to define neural networks in the brain, and more complicated multilayer networks that operate to solve computationally difficult problems such as episodic memory or invariant visual object recognition. The proposals were tested to determine whether three different types of one-layer network with different architectures appropriate for solving different computational problems evolve depending on the computational problem set (Rolls and Stringer 2000). The problems were those appropriate for the pattern associator, autoassociator, and competitive networks shown schematically in Fig. 1.8.

19.4.1 The neural networks

The neural networks considered have a number of classes of neuron. Within a class, a gene allows the number of neurons to vary from 1 up to N. For the simulations described here, N was set to 100. Within a class, the genetic specification of a neuron is homogeneous, with the neurons for that class having, for example, identical activation functions and connectivity probabilities. The number of classes will depend on the individual task, e.g. pattern association, autoassociation, competitive nets, etc, which can be described as one-layer networks, with one layer of computing elements between the input and output (see Fig. 1.8). However, in our simulations there is typically an input layer where patterns are presented, and an output layer where the performance of the network is tested. For the simulations described, the number of classes was set to allow one-layer networks such as these to be built, but the number of layers that could be formed is in principle under genetic control, and indeed multilayer networks can be formed if there is a sufficient number of classes of neuron. Regarding the implementation of inter-layer connection topologies, the neurons in individual layers exist in a circular arrangement, and connections to a cell in one layer are derived from a topologically related region of the preceding layer. Connections to individual neurons may then be established according to either a uniform or Gaussian probability distribution centered on the topologically corresponding location in the sending layer.

On discrete time steps each neuron i calculates a weighted sum of its inputs as shown in equation 19.1. This is in principle the subject of genetic modification (see below), but the gene specifying this was set for the simulations described here to this type of calculation of the neuronal activation. Next the neuronal firing r_i is calculated, using for example the standard sigmoid function shown in Eq. 19.2 and allowing α and β the sigmoid threshold and slope to evolve genetically. Next, there is an optional procedure that can be specified by the experimenter to be called to set the sparseness of the firing rates r_i of a class of neuron according to equation 19.3, with a being allowed to evolve. This specification is used in the simulations, but in the cortex in this and other similar cases a number of genes would specify factors that influence this, such as in this case the strength of the feedback inhibition implemented by inhibitory neurons. After the neuronal outputs r_i have been calculated, the synaptic weights w_{ij} are updated according to one of a number of different learning rules, which are capable of implementing for example both long term potentiation (LTP) and long term depression (LTD), as described below, and which are genetically selected. For example, the standard Hebb rule takes the form

$$\Delta w_{ij} = k r_i r'_j, \tag{19.4}$$

where r'_j is the presynaptic firing rate and k is the learning rate.

19.4.2 The specification of the genes

The genes that specify the architecture and operation of the network are described next. In principle, each gene evolves genetically, but for particular runs, particular genes can be set to specified values to allow investigation of how other genes are selected when there are particular constraints.

Each neural network architecture is described by a genotype consisting of a single chromosome of the following form

$$\mathbf{chromosome} = \begin{bmatrix} \mathbf{c}_1 \\ \mathbf{c}_2 \\ \vdots \\ \mathbf{c}_n \end{bmatrix} \tag{19.5}$$

where $\mathbf{c}_l$ is a vector containing the genes specifying the properties of neurons in class l, and n is the total number of classes that is set manually at the beginning of a run of the simulation. The vectors $\mathbf{c}_l$ take the form

$$\mathbf{c}_l = \begin{bmatrix} \mathbf{g}_l \\ \mathbf{h}_{l1} \\ \mathbf{h}_{l2} \\ \vdots \\ \mathbf{h}_{ln} \end{bmatrix} \tag{19.6}$$

where the vector $\mathbf{g}_l$ contains the *intra*-class properties for class l, and the vectors $\mathbf{h}_{lm}$ contain the *inter*-class connection properties to class l from class m where m is in the range 1 to n.

The vector of intra-class properties takes the form

$$\mathbf{g}_l = \begin{bmatrix} b_l \\ \alpha_l \\ \beta_l \\ a_l \end{bmatrix} \tag{19.7}$$

where we have the following definitions for intra-class genes.

(1) b_l is the number of neurons in class l. b_l is an integer bounded between 2 and N, which was set to 100 for the simulations described here. Individual classes are restricted to contain more than one neuron since a key strategy we have adopted for enhancing the biological plausibility is to evolve classes composed of a number of neurons with homogeneous genetic specifications rather than to specify genetically the properties of individual single neurons.

(2) α_l is the threshold of the sigmoid transfer function in equation 19.2 for class l. α_l is a real number bounded within the interval $[0.0,200.0]$.

(3) β_l is the slope of the sigmoid transfer function in equation 19.2 for class l. β_l is a real number bounded within the interval $[0.0,200.0]$. We note that low values of the slope will effectively specify a nearly linear activation function, whereas high values of the slope specify a nearly binary activation function (see Fig. 1.4).

(4) a_l is the sparseness of firing rates within class l as defined by equation 19.3. By definition a_l is a real number bounded within the interval $[0.0,1.0]$. However, in practice we ensure a minimum firing sparseness by setting a_l to lie within the interval $[1.0/b_l,1.0]$, where b_l is the number of neurons in class l. With binary neurons with output r equal to either 0 or 1, setting $a_l \geq 1.0/b_l$ ensures that at least one neuron is firing. For the simulations described in this Chapter, when this gene was being used, a was set to 0.5 and produced a binary firing rate distribution with a sparseness of 0.5 unless otherwise specified. The alternative way of calculating the firing rates was to allow genes α and β specifying the sigmoid activation function to evolve.

The vectors of inter-class connection genes specify the connections to class l from class m, and take the form

$$\mathbf{h}_{lm} = \begin{bmatrix} r_{lm} \\ s_{lm} \\ c_{lm} \\ e_{lm} \\ z_{lm} \\ t_{lm} \\ p_{lm} \\ \sigma_{lm} \\ q_{lm} \\ f_{lm} \\ k_{lm} \\ d_{lm} \\ u_{lm} \\ v_{lm} \end{bmatrix} \tag{19.8}$$

where we have the following definitions for inter-class genes.

(1) r_{lm} controls the inter-layer connection topology in that it helps to govern which neurons in class m make connections with individual neurons in class l. The exact definition of r_{lm} depends on the type of probability distribution used to make the connections, which is governed by the gene s_{lm} described below. For $s_{lm} = 0$, connections are established according to a uniform distribution, and r_{lm} specifies the number of neurons within the spatial region from which connections may be made. For $s_{lm} = 1$, connections are established according to a Gaussian distribution, where r_{lm} specifies the standard deviation of the distribution. For the Gaussian distribution, individual connections originate with 68% probability from within a region of 1 standard deviation away. These real valued variates are then rounded to their nearest integer values. As noted in section 19.2, connection topologies are characteristically different for different connection types such as CA3 recurrent collateral connections, which are widespread, and intra-module connections in the cerebral cortex. This parameter may also be set by genetic search to enable the region of effect of inhibitory feedback neurons to be just greater than the region within which excitatory neurons receive their input; and to enable topographic maps to be built by arranging for the lateral excitatory connections to operate within a smaller range than inhibitory connections. This parameter may also be set to enable multilayer networks to be built with feedforward connectivity which is from a small region of the preceding layer, but which over many layers allows an output neuron to receive from any part of the input space, as happens in many sensory systems in which topology is gradually lost and as is implemented in a model of the operation of the visual cortical areas (Chapter 25). For the simulations described here, this gene was set to 100 to allow global connectivity, and thus to not specify local connectivity.

(2) s_{lm} specifies for the region gene r_{lm} the connection probability distribution to class l from class m. s_{lm} takes the following values.

- $s_{lm} = 0$: Connections are established according to a uniform probability distribution within the region r_{lm}.

- $s_{lm} = 1$: Connections are established according to a Gaussian probability distribution. r_{lm} specifies the size of the region from which there is a 68% probability of individual connections originating.

In the experiments described here, the values of s were set to 0.

(3) c_{lm} is the number of connections each neuron in class l receives from neurons in class m. c_{lm} takes integer values from 0 to b_m (the number of neurons within class m). These connections are made probabilistically, selecting for each class l of neuron c_{lm} connections from class m. The probabilistic connections are selected either from a uniform probability distribution (as used here), or according to a Gaussian probability distribution. In both cases the local spatial connectivity within the two classes l and m is specified by the region parameter gene r_{lm}.

(4) e_{lm} specifies whether synapses to class l from class m are excitatory or inhibitory. e_{lm} takes the following values.

- $e_{lm} = 0$: Connections are inhibitory

- $e_{lm} = 1$: Connections are excitatory

For inhibitory connections with $e=0$, the weights in equation (19.1) are scaled by -1, and no learning is allowed to take place.

(5) z_{lm} specifies whether connections from class m to class l are additive or divisive. z_{lm} takes the following values.

- $z_{lm} = 0$: Connections are divisive
- $z_{lm} = 1$: Connections are additive

  This gene selects whether the neuronal activation h_i is calculated according to linear summation of the input firing rates r'_j weighted by the synaptic strengths w_{ij}, as specified by Eq. 19.1, and as used throughout by setting $z_{lm} = 1$; or whether with $z_{lm} = 0$ there is local computation on a dendrite including perhaps local summation of excitatory inputs and shunting effects produced by inhibitory inputs or inputs close to the cell body. The most important physical property of a neuron that would implement the effects of this gene would be the diameter of the dendrite (see Section 19.2 and (Koch 1999)). If set so that $z_{lm} = 0$ for local multiplicative and divisive effects to operate, it would be reasonable and in principle straightforward to extend the genome to include genes which specify where on the dendrite with respect to the cell body synapses are made from class m to class l neurons, the cable properties of the dendrite, etc.

(6) t_{lm} specifies how the synaptic weights to class l from class m are initialised at the start of a simulation. t_{lm} takes the following values.

- $t_{lm} = 0$: Synaptic weights are set to zero. This might be optimal for the conditioned stimulus inputs for a pattern associator, or the recurrent collateral connections in an autoassociator, for then existing connections would not produce interference in the synaptic connections being generated during the associative learning. This would produce a non-operating network if used in a competitive network, for an input could not produce any output.
- $t_{lm} = 1$: Synaptic weights are set to a uniform deviate in the interval [0,1] scaled by the constant value encoded on the genome as q_{lm} (described below).
- $t_{lm} = 2$: Synaptic weights are set to the constant value encoded on the genome as q_{lm} (described below). This could be useful in an associative network if a learning rule that allowed synaptic weights to decrease as well as increase was specified genetically, because this would potentially allow correlations between the input patterns due to positive-only firing rates to be removed, yet prevent the synapses from hitting the floor of zero, which could lose information (Rolls and Treves 1998).
- $t_{lm} = 3$: Synaptic weights are set to a Gaussian function of distance x away from the afferent neuron in class l. This function takes the form

$$f(x) = \frac{1}{\sigma\sqrt{2\Pi}} e^{-x^2/2\sigma^2}$$

  where σ is the standard deviation. This would be an alternative way to implement local spatial connectivity effects to those implemented by r_{lm}.

(7) p_{lm} is the scale factor used for synaptic weight initialisation with a Gaussian distribution (i.e. for $t_{lm} = 3$). p_{lm} is a real number bounded within the interval [0.0,100.0].

(8) σ_{lm} is the standard deviation used for synaptic weight initialisation with a Gaussian distribution (i.e. for $t_{lm} = 3$). σ_{lm} is a real number bounded within the interval [0.001,10.0]. σ_{lm} is restricted to be greater than 0.001 to avoid a singularity that would occur in the Gaussian function for $\sigma_{lm} = 0.0$.

(9) q_{lm} is the scale factor used for synaptic weight initialisation for $t_{lm} = 1$ and $t_{lm} = 2$ (see above). q_{lm} is a real number bounded within the interval [0.0,100.0]. If t_{lm} is 1 or 2, q_{lm}

would be expected to be small for pattern associators and autoassociators (as described above).

(10) f_{lm} specifies which learning rule is used to update weights at synapses to class l from class m. f_{lm} takes the following values.

- $f_{lm} = 0$: Learning rule 0 (no learning) $\Delta w_{ij} = 0$.
- $f_{lm} = 1$: Learning rule 1 (Hebb rule) $\Delta w_{ij} = kr_i r'_j$.
- $f_{lm} = 2$: Learning rule 2 (modified Hebb rule) $\Delta w_{ij} = kr_i(r'_j - < r'_j >)$ where $< . >$ indicates an average value. In this rule LTP is incorporated with heterosynaptic LTD set to keep the average weight unchanged. Use of this rule can help to remove the correlations between the input patterns produced by positive-only firing rates (Rolls and Treves 1998). If this rule is selected, it will work best if the synaptic weight initialisation selected genetically is not zero (to allow the weights to decrease without hitting a floor), and is optimally a constant (so that noise is not added to the synaptic weights which implement the association).
- $f_{lm} = 3$: Learning rule 3 (modified Hebb rule) $\Delta w_{ij} = kr_i(r'_j - w_{ij})$. This rule has the effect of helping to maintain the sum of the synaptic weights on a neuron approximately constant, and is therefore helpful in competitive networks (Rolls and Treves 1998). This rule accords with the neurophysiological observation that it is easier to demonstrate LTD after LTP has been produced.
- $f_{lm} = 4$: Learning rule 4 (covariance rule) $\Delta w_{ij} = k(r_i - < r_i >)(r'_j - < r'_j >)$.
- $f_{lm} = 5$: Learning rule 5 (homosynaptic LTD with LTP) $\Delta w_{ij} = k(r_i - < r_i >)r'_j$.
- $f_{lm} = 6$: Learning rule 6 (non-associative LTP) $\Delta w_{ij} = kr'_j$. This rule may be implemented in the hippocampal mossy fibre synapses.
- $f_{lm} = 7$: Learning rule 7 (non-associative LTD) $\Delta w_{ij} = -kr'_j$.
- $f_{lm} = 8$: Learning rule 8 (trace learning rule) $\Delta w_{ij} = k\bar{r}_i^\tau r'_j^\tau$ where the trace $\bar{r}_i^\tau$ is updated according to

$$\bar{r}_i^\tau = (1 - \eta)r_i^\tau + \eta\bar{r}_i^{\tau-1} \tag{19.9}$$

and we have the following definitions: $\bar{r}_i^\tau$ is the trace value of the output of the neuron at time step τ, and η is a trace parameter which determines the time or number of presentations of exemplars of stimuli over which the trace decays exponentially. The optimal value of η varies with the presentation sequence length. Further details, and a hypothesis about how this type of learning rule could contribute to the learning of invariant representations, are provided in Chapter 25. For the simulations described by Rolls and Stringer (2000), this learning rule was disabled.

(11) k_{lm} is the learning rate k for synaptic weights to class l from class m. k_{lm} is a real number bounded within the interval $[0.0, 10.0]$.

(12) d_{lm} is the maximum size allowed for the absolute value of synaptic weight updates from class m to class l. d_{lm} is a real number bounded within the interval $[0.0, 10.0]$.

(13) u_{lm} is the maximum size allowed for synaptic weights to class l from class m. u_{lm} is a real number bounded within the interval $[-100.0, 100.0]$.

(14) v_{lm} is the minimum size allowed for synaptic weights to class l from class m. v_{lm} is a real number bounded within the interval $[-100.0, 100.0]$.

Implementation details are that v is bounded to be less than u; and that during initialisation of the synaptic weights, v and u are applied after the other initialisation parameters have operated.

19.4.3   The genetic algorithm, and general procedure

The genetic algorithm used was a conventional one described by Goldberg (1989). Each individual genotype was a haploid chromosome of the type just defined. The values for the genes in the initial chromosome were chosen at random from the possible values. (In this simulation, if a value in the genotype was one of a large number of possible values, then a single integer or real-valued gene was used, whereas in the brain we would expect several genes to be used, with perhaps each of the several genes coding for different parts of the range with different precision. For example, c_{lm}, the number of synaptic connections to a neuron in class l from a neuron in class m, is to be chosen from 2 to 100 in the simulation, and uses a single integer. In the brain, where we might expect the number of connections per neuron to vary in the range 5 to 50,000, one possible scenario is that 5 different genes would be used, coding for the values 5, 50, 500, 5,000 and 50,000. We would thus expect that the number of genes required to specify the part of the genotype described here would increase by a factor in the order of 5 in the brain.)

There was a set number of genotypes (and hence individuals) per generation, which was 100 for the simulations described. (Too small a number reduces the amount of diversity in the population sufficiently to lead to poor performance of the genetic evolution.) From each genotype an individual network was constructed, with all connections being chosen probabilistically within the values specified by the genotype, and with a uniform probability distribution (i.e. s_{lm} was set to 0 for the simulations described), and the initial values for the synapses were also chosen as proscribed by the genotype. That individual was then tested on the computation being studied, e.g. pattern association, and the fitness (performance) of the network was measured. The actual testing involved presenting the set of patterns to be learned (as specified below for each network) to the network while allowing learning to occur, and then testing the performance of the network by presenting the same, similar, or incomplete test patterns and measuring the output of the network. The fitness measure used for each class of network is defined below. This process was repeated until a fitness measure had been obtained for each of the genotypes.

The next generation of genotypes was then bred by reproduction from the previous genotypes. Two genotypes were selected for reproduction with a probability that was proportional to their fitness relative to the sum of the fitnesses of all individuals in that generation. That is, the probability P_i of an individual i being selected is given by

$$P_i = \frac{F_i}{\sum_j F_j} \tag{19.10}$$

where F_j is the fitness of an individual genotype j, and the sum $\sum_j F_j$ is over all individuals in the population. The genotypes were then bred to produce two offspring genotypes using the processes of recombination and mutation. The probability of recombination was set to 0.4 (so that it occurred with a 40% chance for every breeding pair in every generation), and the probability of mutation of each gene was set to 0.05. The computational background to this is that recombination allows a non-linear search of a high dimensional space in which gene complexes or combinations may be important, allowing local performance hills in the space to be detected, and performance once in a local hill to improve. Mutation on the other hand enables occasional moves to be tried to other parts of the space, usually with poorer performance, but occasionally moving a genotype away from a well explored location to a new area to explore. Both processes enable the search to be much more efficient in a high dimensional space with multiple performance hills or mountains than with simple gradient ascent, which will climb the first hill found, and become stuck there (Ackley 1987, Goldberg 1989). The recombination took place with respect to a random point on the chromosome each time it occurred. The mutation also took place at a random place on the chromosome each time it occurred, and the single gene being mutated was altered to a new random value within the range specified for that gene.

This reproduction process was repeated until sufficient individual genotypes for the next generation (100 unless otherwise stated) had been produced. The fitness of those genotypes was then measured by building and testing the networks they specified as previously outlined. Then the whole reproduction, ontogenesis, and testing to measure fitness, processes were repeated for many generations, to determine whether the fitness would increase, and if so, what solutions were found for the successful networks.

19.4.4 Pattern association networks

The operation of the system simulated is described briefly for the three networks, emphasizing the generic points. Details of the simulations and the explorations with them are described by Rolls and Stringer (2000).

The computational problem set was to associate pairs of random binary patterns of 1s and 0s with sparseness of 0.5 (i.e. with half the elements 0 and the other half 1s), an appropriate problem for a pattern association net (Appendix B). There were 10 pattern pairs, which places a reasonable load on the network (Rolls and Treves 1998) (Section B.2). There were two classes of neuron. One of the patterns of the pair, the unconditioned stimulus, was set during learning to activate the 100 output neurons (class 2) during learning, while the other pattern of the pair, the conditioned stimulus, was provided by the firing of 100 input neurons (class 1). The fitness measure was obtained by providing each conditioned stimulus (or a fraction of it), and measuring the correlation of the output vector of firing rates, r_i where $i = 1$ to 100, with that which was presented as the unconditioned stimulus during learning. This correlation measure ranges from 1.0 for perfect recall to 0.0 (obtained for example if the output of the pattern associator is 0). The fitness measure that was used in most runs was the square of this correlation (used to increase the selectivity of the fitness function). (In practice, a pattern associator with associatively modifiable synapses, random patterns, and the sizes and loading of the networks used here can achieve a maximum correlation of approximately 1.000.) It may be noted that the genetics could form connections onto individual neurons of any number (up to $N = 100$, the number of input and output neurons in the network) and type between classes 1 to 2 (feedforward connections), and 2 to 2 (recurrent collaterals), 2 to 1, and 1 to 1. For the simulations performed, the operation of inhibitory feedback neurons was not explicitly studied, although when using the sigmoid activation function the neurons had to select a threshold (by altering α_l).

The values of the fitness measures for a typical run of the simulations is shown in Fig. 19.2, along with the values of specific genes. The genes shown are key connection properties from class 1 to class 2, and are: learning rate k_{21}, number of connections c_{21}, weight scaling q_{21}, weight initialisation option t_{21}, and the learning rule f_{21}. The fitness measure is the fitness of the best genotype in the population, and for this run the sparseness of the output representation was set to the value of 0.5 by using the sparseness gene set to this value. It is shown that evolution to produce genotypes that specify networks with high performance occurred over a number of generations. Some genes, such as that specifying the learning rule, settled down early, and changes in such genes can sometimes be seen to be reflected in jumps in the fitness measure. An example in Fig. 19.2 is the learning rule, which settled

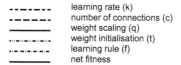

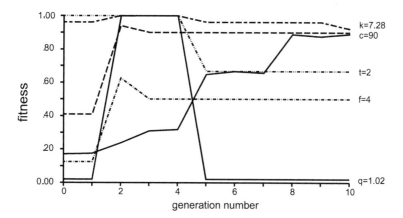

Fig. 19.2 Fitness of the best genotype in the pattern association task as a function of the number of generations, together with the values of selected genes. The simulation is performed with the firing rates assigned binary values of 0 or 1 according to the neuronal activations, with the proportion of neurons firing set by the firing rate sparsity gene a. The parameters shown all refer to connections from class 1 9the input neurons) to class 2 (the output neurons). The final values of the parameters are shown at the end of each plot. For example, the genetic algorithm set the number of connections onto each output neuron as 90, a relatively high number (the maximum was 100), which results in a high memory capacity, which was what the fitness function was set to optimize. (After Rolls and Stringer, 2000.)

down (for the best genotype) by generation 3 to a value of 4 which selects the covariance learning rule. Another example is that there is a jump in the fitness measure from generation 4 to 5 which occurs when the synaptic weight initialisation gene t_{21} changes from a Gaussian distribution to a constant value. This would be expected to increase the performance in these simulations. The increase in fitness from generation 7 to 8 occurred when the minimum weight gene v_{21} changed from 11.6 to -62.5 (not plotted), which would also be expected to increase performance. For other genes, where the impact is less or zero, the values of the genes can fluctuate throughout evolution, because they are not under selection pressure, in that they do not contribute greatly to fitness. Every run of the simulation eventually produced networks with perfect performance. The selection pressure (relative to other genotypes) could be increased by taking a power greater than 1 of the correlation fitness measure. A high value could produce very rapid evolution, but at the expense of minimising diversity early on in evolution so that potentially good genes were lost from the population, with then a longer time being taken later to reach optimal performance because of the need to rely on mutations. Low values resulted in slow evolution. In practice, raising the correlation fitness measure to a power of 1 (no change) or 2 produced the best evolution.

The best networks all had close to 100 excitatory feedforward connections (gene c_{21}) to each class 2 (output) neuron from the class 1 (input) neurons, which is the minimal number necessary for optimal performance of a fully loaded pattern associator and the maximum number allowed. Moreover, the excitatory feed forward connections for all the successful networks were specified as associative, and in particular f_{21} was selected for these networks as 2 or 4. distributions to achieve optimal results in a pattern associator.

To help understand in full detail the actual gene selections found by the genetic algorithm in the simulations, we were able to simulate the pattern associator with manually selected gene combinations for the feedforward connections from class 1 to class 2. In particular, we calculated the network fitnesses for particular combinations of learning rule f_{21}, weight initialisation option t_{21}, and weight scaling q_{21}. This optional facility of the simulator allowed systematic investigation of the effects of one gene, which could be held fixed at a particular value, on the values of the other genes selected during evolution. Examples of the usefulness of this approach are described by Rolls and Stringer (2000).

19.4.5   Autoassociative networks

The computational problem set was to learn random binary pattern vectors (of 1s and 0s with sparseness of 0.5, i.e. with half the elements 0 and the other half 1s), and then to recall the whole pattern when only half of it was presented. This is called completion, and is an appropriate test for an autoassociator (Section B.3). An additional criterion, inherent in the way the simulation was run, was to perform retrieval by using iterated processing in an attractor-like state of steady memory retrieval like that of continuing neuronal firing in a short-term memory (Appendix B).

Most runs of the simulation produced networks with optimal performance. The best networks all had close to 100 excitatory recurrent collateral feedback connections from the output neurons (class 1) to themselves, which is the minimal number necessary for optimal performance of a fully loaded autoassociator given that there are 100 output neurons and that this is the maximum number allowed. The successful networks also had all evolved to use either the covariance or the heterosynaptic LTD learning rule (Rolls and Stringer 2000), which are necessary for good performance of the autoassociation network (Appendix B).

19.4.6   Competitive networks

The computational problem set was to learn to place into separate categories 20 binary pattern vectors each 100 elements long. First, 5 exemplar patterns were randomly generated with 50 1s and 50 0s. Then, for each exemplar, 3 further patterns were created by mutating 20 of the bits from 0 to 1 or 1 to 0. This gave a total of 20 patterns that could be placed in 5 similarity categories. There were two classes of neuron. Class 1 was the input class, consisted of 100 neurons, and was set to fire according to which input pattern was being presented. Class 2 was the output class. Connections were possible from class 1 to 2, from class 1 to 1, from class 2 to 1, and from class 2 to 2. The network shown in Fig. B.14, if trained with learning rule 3 and given a powerful and widespread lateral inhibition scheme can solve this problem (Appendix B). The fitness measure used assessed how well the competitive net was able to classify the 20 patterns into their respective 5 categories.

The simulation runs typically converged to use learning rules 2 or 3, although learning rule 3 appeared to offer the best performance. Learning rule 3 involves LTP, plus LTD proportional to the existing value of the synaptic weight. This is the only rule which actually encourages the synaptic weight vectors of the different output (class 2) neurons to have a similar and limited total value. Use of a learning rule such as this (or explicit normalization of the length of the synaptic weight vector of each neuron), is advantageous in competitive networks, because it prevents any one neuron from learning to respond to all the input patterns (Section B.4). The fact that this learning rule was never chosen by the successful genotypes in the associative nets, but was generally chosen for the competitive nets, is evidence that the genetic evolution was leading to appropriate architectures for different computational problems. Learning rule 0 (no learning) was never chosen by the successful genotypes, and in addition we verified that good solutions to the problem set could not be produced without learning.

Another characteristic of the successful genotypes was that there appeared to be less selection pressure on c_{21} the number of connections to a class 2 from a class 1 neuron, with some highly fit nets having relatively low values for c_{21}. Such diluted connectivity can be useful in competitive networks, because it helps to ensure that different input patterns activate different output neurons (Sections 7.4 and B.4). Low values were never chosen for the best genotypes in the associative networks, where full connectivity is advantageous for full capacity.

19.5 Evaluation of the gene-based evolution of single-layer networks

The simulations described by Rolls and Stringer (2000) showed that the overall conception of using genes which control processes of the type described can, when combinations are searched through using genetic evolution utilising a single performance measure, lead to the specification of neural networks that can solve different computational problems, each of which requires a different functional architecture.

The actual genes hypothesized were based on a comparison of the functional neuronal network architectures of many mammalian brain regions, and were well able to specify the architectures described here. However, the research described is intended as a foundation for further exploration of exactly which genes are used biologically to specify real architectures in the mammalian brain which perform particular computations. We note in this context that it may be the case that exhaustive analysis of the genetic specification of an invertebrate with a small number of neurons does not reveal the principles involved in building complex nervous systems and networks of the type addressed here. The genetic specification for simple nervous systems could be considerably different, with much more genetic specification of the properties of particular synapses to produce a particular more fixed network in terms both of which identified neuron is connected to which, and what the value of the synaptic connection strength between each neuron is. In contrast, in the approach taken here, there are not identifiable particular neurons with particular connections to other identifiable neurons, but instead a specification of the general statistics of the connectivity, with large numbers of neurons connecting according to general rules, and the performance of the network being greatly influenced by learning of the appropriate values of the connection strengths between neurons based on the co-activity of neurons, and the nature of the modifiable synaptic connection between them.

The particular hypotheses about the genes that could specify the functional architecture of simple networks in the brain were shown by the work described here to be sufficient to specify some different neuronal network architectures for use in a system that builds computationally useful architectures using a gene selection process based on fitness. The genes hypothesized may be taken as a guide to the types of genes that could be used to specify the functional architecture of real biological nervous systems, but the gene specification postulated and simulated can be simply revised based on empirical discoveries about the real genes. One choice made here was to specify the genes with respect to the receiving neuron. The reason for this, rather than specification with respect to the sending neuron, was that many of the controlling properties of the network architecture and performance are in the post-synaptic neuron. Examples of these properties include the type of post-synaptic receptor (e.g. NMDA receptors to specify associatively modifiable connections), and the cable properties of the postsynaptic neuron. The specification of the receiving neuron does include the appropriate information about the sending neuron, in that for example gene c_{lm} does specify (by identifier m) the identity of the sending population. This implies that when a neuron makes an output connection (through its synapses), the neuron class of the sending neuron is available (as a chemical recognition identifier) at the synaptic terminal. However, in principle, in terms of actually building a network, it would be possible to have the genetic specifiers listed as being properties of the sending neuron.

Another aspect of the gene specifiers hypothesized here is that there was no gene for how many output connections a class of neurons should make to another class. One reason for this is that the networks can be built numerically without such a specifier, as it is unnecessary if the total numbers of neurons of two classes is specified, and the number of connections received by one class of neuron from the other is specified. However, the number of output connections made to another class of neuron might be specified in real biological systems, and neuron numbers might adjust to reflect this type of constraint, which is a possible way to think about cell death during development. If there is local output connectivity within a certain spatial range, this will also act as a factor that determines the number of output connections made by neurons, and indeed this does seem a possible gene in real biological systems. Indeed, in real systems the single gene specifying the region of connectivity (r_{lm}) here might be replaced by different spatial genes specifying the extent of the dendrites of the receiving neuron, and extent of the axonal arborization for the outputs.

It is emphasized that a number of the genes specified here for the simulations might be replaced in real biological systems with several genes operating at a lower level, or in more detail, to implement the function proposed for each gene specified here. The reason for selecting the particular genes hypothesized here is that they serve as a guide to how in principle neuronal networks could be specified genetically in complex neural systems where the functional architecture is being selected for by genetic evolution. Specification of genes as operating at the level described here provides a useful heuristic for investigation by simulation of how evolution may operate to produce neuronal networks with particular computational properties. Once the processes have been investigated and understood with genes operating at the level described here, subsequent investigations that use genes that are more and more biologically accurate and detailed would be facilitated and are envisaged. The approach allows continual updating of the genotype used in the simulations to help understand the implications of new discoveries in neurobiology.

The number of genes used in simulations of the type described here is now considered, remembering that biologically several genes might be required to specify at least some of the genes used in the simulations. The number of genes for each class of neuron in the simulations was 4 + (the number of classes of neuron from which the class receives connections × 14). In the brain, if there were on average five classes of neuron from which each class of neuron received connections, this would result in 4 + 5 × 14 = 74 genes per class of neuron. If there were on average five classes of neuron per network, this would yield 370 genes per network. 100 such networks would require 37,000 genes, though in practice fewer might be needed, as some parameters might be assigned to be the same for many classes of neuron in different networks. For example, the overall functional architecture of the cerebral neocortex appears to follow the same general design for different architectonic areas (Braitenberg and Schütz 1991, Peters and Jones 1984, Somogyi et al. 1998, Douglas et al. 2004, Pandya et al. 2015), so much of the neocortex may use one generic network specification, with a few genes used to tweak the parameters for each different architectonic area. (An extension of the idea presented here would allow such a generic specification of the architecture of an area of the cerebral neocortex, and then allow genes to connect different areas with the same generic though locally tweaked architecture.) A factor working in the opposite direction is the fact that a single gene taking an integer or real value was used to specify some of the parameters in the simulation, whereas biologically several genes might be needed to represent such a wide range of parameter values with low, though sufficient, precision as described in Section 19.4.3.

Although with 100 individuals per generation, and approximately 64 genes for each genotype (for networks with 2 classes of genes, each connected to each other as well as to themselves) each with several at least possible values, the search might be predicted to be long, in practice the genetic search was often quite short. Part of the reason for this is that some genes made a crucial difference, for example the gene specifying the learning rule, and were selected early on in evolution, usually being present in at least some of the genotypes in the initial population. Because these genes had such a dramatic effect on fitness, they were almost always selected for the next generation, and successive generations then had to search the smaller space of the remaining genes only some of which had a substantial impact on performance. Some genes, such at that specifying the learning rule, do indeed often settle down early, and changes in these genes can sometimes be seen to be reflected in jumps in the fitness measure. For other genes, where the impact is less or zero, the values of the genes can fluctuate throughout evolution, because they are not under selection pressure, in that they do not contribute greatly to fitness.

Although therefore in some senses the actual evolution investigated here takes place rapidly and apparently quite easily, this is part of the point of this research, that is to demonstrate that the particular genes identified as those which might allow the networks to be found in different brain areas to be built, and implemented in the simulations, can actually build these networks, and moreover can do so efficiently when using genetic search. However, the networks tested by Rolls and Stringer (2000) are single layer networks, and multilayer networks, with correspondingly more parameter combinations to be explored, will doubtless take longer to evolve in future simulations. Genetic search of the parameter space should then become even more evident as a good procedure, because of its power in searching high dimensional spaces with many local optima and in which the parameters combine non-linearly; because of its use of a single measure of performance; and because of its ability to reuse genetic codes still present in a diverse population but originally developed for a different purpose. The intention however here is to specify the principles involved. Of course, there are many more details of the networks that could be explored by the existing gene specification, including inhibitory interneurons. Simulations of this type will not only enable the evolution and development of more complex multilayer networks to be explored, but will also enable the utility of the specification of different parameters by different genes to be explored, including details of where inputs end on dendrites, dendritic biophysical properties, and thus eventually networks that operate to simulate also the dynamics of the operation of real neural networks in the brain (Treves 1993, Battaglia and Treves 1998a, Panzeri et al. 2001) (Appendix B).

19.6 The gene-based evolution of multi-layer cortical systems

We followed up these advances by simulating a two-layer network which was intended to show next how a two-layer network might evolve by the mathematical processes implemented by genetics. The system we chose was the dentate granule cells followed by the CA3 network. We used a cost function that was intended to capture the theoretical utility of the dentate granule cells acting as an orthogonalizing process to perform pattern separation to increase the storage capacity of the CA3 when correlated patterns were presented to the system (Chapter 24).

The results were disappointing. It seemed difficult for the fitness function to optimize the architecture of both the layers simultaneously. Perhaps the minima in this landscape were so dependent on what was happening in each of the layers that a solution was very difficult.

This leads me to the following new hypotheses. One hypothesis is that because of the complexity of the solution landscape being optimized by the fitness function and genetic variation and selection, a mechanism that may be used in evolution (with full generality to the whole process of evolution by neo-Darwinian mechanisms) is to allow one subsystem to evolve, fix its parameters, and then allow the next subsystem in the hierarchy to evolve, then fix this, and so on. In the context of cortical processing, the proposal is that the first layer in the system would evolve, and then become fixed, allowing little further variation, and providing stable inputs for the next layer to evolve to do what it can with the outputs of the preceding layer. This would be repeated for layer after layer in the hierarchy. The process would be efficient, in the sense that genetic evolution would need to search only one layer of the system at a time for a useful advance, measured by an increase in fitness. This would greatly simplify the exploration of high dimensional spaces being performed by genetic evolution.

This hypothesis raises the important question of how gene-based evolution would be switched on for an early layer of the system, but then switched off so that the next layer in the hierarchy could evolve later in evolutionary history. One possibility is that genes that have already built early layers of the system or hierarchy show less variation in terms of mutation and recombination, perhaps because of where they are located on a chromosome.

In the context of the brain, a related process may be that after an early layer in a cortical hierarchy has self-organized during development, learning in that cortical layer may be switched off, by switching off its synaptic plasticity, to leave a stable base for the next layer to learn. Consistent with this, there is plasticity in the primary visual cortex, V1, in an early stage of cortical development, so that binocular neurons can be formed to implement stereopsis (Blakemore, Garey and Vital-Durand 1978, LeVay, Wiesel and Hubel 1980). After a 'critical period', the plasticity in V1 is switched off, and later stages of the hierarchy can learn, and be subject to evolution by genetic variation and natural selection. Indeed, our evidence is that plasticity in the primate inferior temporal visual cortex remains present in adulthood (Rolls, Baylis, Hasselmo and Nalwa 1989a).

There may thus be interesting computational advantages of ontogeny recapitulating phylogeny in this case. For both development, and for evolution, the process of finding useful solutions by selection may be simplified by allowing primarily one stage or layer in the system to be developing or evolving at any one time, to reduce the dimensionality of the space being searched at any one time.

An implication is that the inferior temporal visual cortex is not a fixed structure, and that any changes that genes make in its architecture can influence its function throughout life, and may this have a strong impact on fitness. In any case, it is suggested that genetic variation and natural selection are largely switched off for early layers in the cortical hierarchy, so that genetic variation and natural selection operate mainly on the top layers of the hierarchy, enabling new processing layers to be added to an existing hierarchy and shaped by evolution to perform useful functions, without interference cause by simultaneous evolution of the early layers of the hierarchy. Consistent with this hypothesis, many genes are involved in timing when other genes are expressed.

19.7 Highlights

1. Gene-based variation operated on by natural selection to promote fitness provides a basis for the design of the cortex during evolution.

2. It is proposed that genes specify a relatively small number of parameters for each class of neuron in a given cortical area, including the classes of neurons to which each class should make synapses. This provides a basis for one cortical area to make the appropriate connections to other cortical areas.

3. Specifying a few parameters for each class of cortical neuron enables relatively few genes, in the order of thousands, to specify the architecture and connectivity of the cerebral cortex.

4. Simulations are described which show that specification of relatively few parameters by genes can enable neural networks to evolve to optimise fitness functions that specify particular results of cortical computations. The genes specify relatively few rules of connectivity of the neural architecture, and learning in the phenotype sets the synaptic weights.

20 Cortex versus basal ganglia design for selection

The aim of this Chapter is to compare the architecture of the cerebral cortex, which can be involved in selection including that involved in decision-making and memory recall (Chapter 5), with the selection mechanism in another brain system, the basal ganglia. This helps to elucidate the very different principles of operation of these different architectures, each with its advantages and disadvantages, which are highlighted.

It is also of interest to consider the basal ganglia, because it is a major brain system to which every part of the cerebral neocortex projects, from its neurons in layer 5 (Sections 1.10.5, 18.2.1, and 26.5.18).

A more comprehensive consideration of the basal ganglia is provided elsewhere (Rolls 2014a).

20.1 Systems-level architecture of the basal ganglia

The point-to-point connectivity of the basal ganglia as shown by experimental anterograde and retrograde neuroanatomical path tracing techniques in the primate is indicated in Figs. 20.1 and 20.2. The general connectivity is for cortical or limbic inputs to reach the striatum, which then projects to the globus pallidus and substantia nigra pars reticulata, which in turn project via the thalamus back to the cerebral cortex (Gurney, Prescott and Redgrave 2001a, Haber and Knutson 2009, DeLong and Wichmann 2010, Gerfen and Surmeier 2011, Buot and Yelnik 2012).) Within this overall scheme, there is a set of at least partially segregated parallel processing streams, as illustrated in Figs. 20.1 and 20.2 (DeLong, Georgopoulos, Crutcher, Mitchell, Richardson and Alexander 1984, Alexander, Crutcher and DeLong 1990, Rolls and Johnstone 1992, Strick, Dum and Picard 1995, Middleton and Strick 1996b, Middleton and Strick 1996a, Middleton and Strick 2000, Kelly and Strick 2004).

First, the motor cortex (area 4) and somatosensory cortex (areas 3, 1, and 2) project somatotopically to the putamen, which has connections through the globus pallidus and substantia nigra to the ventral anterior thalamic nuclei and thus to the supplementary motor cortex. Experiments with a virus transneuronal pathway tracing technique have shown that there might be at least partial segregation within this stream, with different parts of the globus pallidus projecting via different parts of the ventrolateral (VL) thalamic nuclei to the supplementary motor area, the primary motor cortex (area 4), and to the ventral premotor area on the lateral surface of the hemisphere (Middleton and Strick 1996a).

Second, there is an oculomotor circuit (see Fig. 20.1).

Third, the dorsolateral prefrontal and the parietal cortices project to the head and body of the caudate nucleus, which has connections through parts of the globus pallidus and substantia nigra to the ventral anterior group of thalamic nuclei and thus to the dorsolateral prefrontal cortex.

Fourth, the inferior temporal visual cortex and the ventrolateral (inferior convexity) prefrontal cortex to which it is connected project to the posterior and ventral parts of the putamen and the tail of the caudate nucleus (Kemp and Powell 1970, Saint-Cyr, Ungerleider and

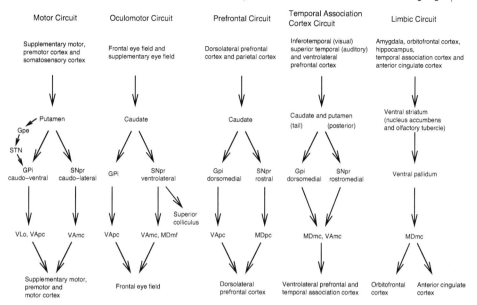

Fig. 20.1 A synthesis of some of the anatomical studies (see text) of the connections of the basal ganglia. GPe, Globus Pallidus, external segment; GPi, Globus Pallidus, internal segment; MD, nucleus medialis dorsalis; SNpr, Substantia Nigra, pars reticulata; VAmc, n. ventralis anterior pars magnocellularis of the thalamus; VApc, n. ventralis anterior pars compacta; VLo, n. ventralis lateralis pars oralis; VLm, n. ventralis pars medialis. An indirect pathway from the striatum via the external segment of the globus pallidus and the subthalamic nucleus (STN) to the internal segment of the globus pallidus is present for the first four circuits (left to right in the figure) of the basal ganglia.

Desimone 1990, Graybiel and Kimura 1995). Moreover, part of the globus pallidus, perhaps the part influenced by the temporal lobe visual cortex, area TE, may project back (via the thalamus) to area TE (Middleton and Strick 1996b).

Fifth, and of especial interest in the context of reward mechanisms in the brain, limbic and related structures such as the amygdala, orbitofrontal cortex, and hippocampus project to the ventral striatum (which includes the nucleus accumbens), which has connections through the ventral pallidum to the mediodorsal nucleus of the thalamus and thus to the prefrontal and cingulate cortices (Strick et al. 1995, Haber and Knutson 2009, Buot and Yelnik 2012). It is notable that the projections from the amygdala and orbitofrontal cortex are not restricted to the nucleus accumbens, but also occur to the adjacent ventral part of the head of the caudate nucleus (Amaral and Price 1984, Seleman and Goldman-Rakic 1985). These same regions may also project to the striosomes or patches (in for example the head of the caudate nucleus), which are set in the matrix formed by the other cortico-striatal systems (Graybiel and Kimura 1995).

At striatal neurons of the direct pathway (from the striatum directly to the globus pallidus internal segment), dopamine has excitatory effects via the D1 receptor by eliciting or prolonging glutamate excitatory inputs. At striatal neurons in the indirect pathway (from the striatum via the external segment of the globus pallidus via the subthalamic nucleus to the internal segment of the globus pallidus, see Fig. 20.1), D2 receptor activation has inhibitory effects by reducing glutamate release and prolonging membrane down states (hyperpolarization). Both effects of dopamine tend to promote behavioural output (Gerfen and Surmeier 2011).

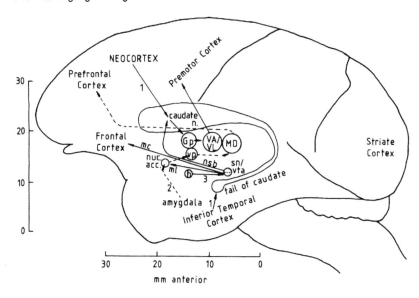

Fig. 20.2 Some of the striatal and connected regions in which the activity of single neurons is described shown on a lateral view of the brain of the macaque monkey. Gp, globus pallidus; h, hypothalamus; sn, substantia nigra, pars compacta (A9 cell group), which gives rise to the nigrostriatal dopaminergic pathway, or nigrostriatal bundle (nsb); vta, ventral tegmental area, containing the A10 cell group, which gives rise to the mesocortical dopamine pathway (mc) projecting to the frontal and cingulate cortices and to the mesolimbic dopamine pathway (ml), which projects to the nucleus accumbens (nuc acc). There is a route from the nucleus accumbens to the ventral pallidum (vp) which then projects to the mediodorsal nucleus of the thalamus (MD) which in turn projects to the prefrontal cortex. Correspondingly, the globus pallidus projects via the ventral anterior and ventrolateral (VA/VL) thalamic nuclei to cortical areas such as the premotor cortex.

20.2 What computations are performed by the basal ganglia?

One way to obtain evidence on the information processing being performed by the striatum is to compare neuronal activity in the striatum with that in its corresponding input and output structures. For example, the taste and visual information necessary for the computation that a visual stimulus is no longer associated with taste reward is represented by the activity of different neurons in the orbitofrontal cortex, and the putative result of such a computation, namely neurons that respond in this non-reward situation, are found not only in the orbitofrontal cortex (Thorpe, Rolls and Maddison 1983, Rolls 1999a, Rolls 2000d, Rolls 2004a, Rolls and Grabenhorst 2008, Grabenhorst and Rolls 2011, Rolls 2014a) (Sections 2.2.5 and 15.14), but also in the head of the caudate nucleus and in the ventral striatum (Rolls, Thorpe and Maddison 1983b, Williams, Rolls, Leonard and Stern 1993). However, neurons that represent the necessary sensory information for this computation, and neurons that respond to the non-reward, were not found in the head of the caudate nucleus or the ventral striatum (Rolls, Thorpe and Maddison 1983b, Williams, Rolls, Leonard and Stern 1993). Instead, in the head of the caudate nucleus, neurons in the same test situation responded in relation to whether the monkey had to make a response on a particular trial, that is many of them responded more on Go than on No-Go trials. This could reflect the output of a cognitive computation performed by the orbitofrontal cortex, indicating whether on the basis of the available sensory information, the current trial should be a Go trial, or a No-Go trial because a visual stimulus previously associated with punishment had been shown.

Similarly, neurons were not found in the ventral striatum that are tuned to all the visual reward, taste reward, olfactory reward, and visual non-reward functions about which macaque orbitofrontal cortex neurons carry information (Sections 2.2.5 and 15.14) (Rolls 2014a). Instead the ventral striatal neurons were usually less easy to classify in these sensory ways, and were especially engaged when tasks were being performed. For example, many of the ventral striatal neurons that respond to visual inputs do so preferentially on the basis of whether the stimuli are recognized, or are associated with reinforcement (Williams, Rolls, Leonard and Stern 1993). Much of the sensory and memory-related processing required to determine whether a stimulus is a face, is recognized, or is associated with reinforcement has been performed in and is evident in neuronal responses in structures such as the amygdala (Leonard, Rolls, Wilson and Baylis 1985, Rolls 2000c), orbitofrontal cortex (Thorpe, Rolls and Maddison 1983, Rolls 2000d, Rolls 2004a, Rolls, Critchley, Browning and Inoue 2006a, Rolls and Grabenhorst 2008, Grabenhorst and Rolls 2011), and hippocampal system (Rolls and Treves 1998, Rolls 1999c, Rolls and Stringer 2005, Rolls, Franco and Stringer 2005b, Rolls, Xiang and Franco 2005c, Rolls and Xiang 2006, Rolls 2010b, Kesner and Rolls 2015).

Similar comparisons can be made for the head and tail of the caudate nucleus, and the posterior putamen (Rolls and Johnstone 1992, Rolls, Thorpe and Maddison 1983b, Rolls, Thorpe, Boytim, Szabo and Perrett 1984, Caan, Perrett and Rolls 1984, Rolls 2014a).

In these four parts of the striatum in which a comparison can be made of processing in the striatum with that in the cortical area that projects to that part of the striatum, it thus appears that the full information represented in the cortex does not reach the striatum, but that rather the striatum receives the output of the computation being performed by a cortical area, and could use this to initiate, switch, or alter behaviour.

The hypothesis arises from these findings that some parts of the striatum, particularly the caudate nucleus, ventral striatum, and posterior putamen, receive the output of these memory-related and cognitive computations, but do not themselves perform them. Instead, on receiving the cortical and limbic outputs, the striatum may be involved in switching behaviour as appropriate as determined by the different, sometimes conflicting, information received from these cortical and limbic areas. On this view, the striatum would be particularly involved in the selection of behavioural responses, and in producing one coherent stream of behavioural output, with the possibility to switch if a higher priority input was received. This process may be achieved by a laterally spreading competitive interaction between striatal or pallidal neurons, which might be implemented by the direct inhibitory connections between neurons that are close together in the striatum and globus pallidus. In addition, the inhibitory interneurons within the striatum, the dendrites of which in the striatum may cross the boundary between the matrix and striosomes, may play a part in this interaction between striatal processing streams (Groves 1983, Graybiel and Kimura 1995, Groves, Garcia-Munoz, Linder, Manley, Martone and Young 1995).

Dopamine could play an important role in setting the sensitivity of this response selection function, as suggested by direct iontophoresis of dopamine on to single striatal neurons, which produces a similar decrease in the response of the neuron and in its spontaneous activity in the behaving macaque (Rolls, Thorpe, Boytim, Szabo and Perrett 1984, Rolls and Williams 1987b). Consistent with dopamine playing an important role of this type, dopamine acting via D1 receptors in the direct pathway, and via D2 receptors in the indirect pathway, promotes behavioural output (Gerfen and Surmeier 2011).

In addition to this response selection function by competition, the basal ganglia may, by the convergence discussed, enable signals originating from non-motor parts of the cerebral cortex to be mapped into motor signals to produce behavioural output. The ways in which these computations might be performed are considered next.

20.3 How do the basal ganglia perform their computations?

On the hypothesis just raised, different regions of the striatum, or at least the outputs of such regions, would need to interact. Is there within the striatum the possibility for different regions to interact, and is the partial functional segregation seen within the striatum maintained in processing beyond the striatum? For example, is the segregation maintained throughout the globus pallidus and thalamus with projections to different premotor and even prefrontal regions reached by different regions of the striatum, or is there convergence or the possibility for interaction at some stage during this post-striatal processing?

Given the anatomy of the basal ganglia, interactions between signals reaching the basal ganglia could happen in a number of different ways. One would be for each part of the striatum to receive at least some input from a number of different cortical regions. As discussed above, there is evidence for patches of input from different sources to be brought adjacent to each other in the striatum (Van Hoesen, Yeterian and Lavizzo-Mourey 1981, Seleman and Goldman-Rakic 1985, Graybiel and Kimura 1995). For example, in the caudate nucleus, different regions of association cortex project to adjacent longitudinal strips (Seleman and Goldman-Rakic 1985). Now, the dendrites of striatal neurons have the shape of large plates which lie at right angles to the incoming cortico-striatal fibres (Percheron, Yelnik and François 1984b, Percheron, Yelnik and François 1984a, Percheron, Yelnik, François, Fenelon and Talbi 1994, Yelnik 2002, Buot and Yelnik 2012) (see Figs. 20.3 and 20.4). Thus one way in which interaction may start in the basal ganglia is by virtue of the same striatal neuron receiving inputs on its dendrites from more than just a limited area of the cerebral cortex. This convergence may provide a first level of integration over limited sets of cortico-striatal fibres. The large number of cortical inputs received by each striatal neuron, in the order of 10,000 (Wilson 1995), is consistent with the hypothesis that convergence of inputs carrying different signals is an important aspect of the function of the basal ganglia. The computation that could be performed by this architecture is discussed below for the inputs to the globus pallidus, where the connectivity pattern is comparable.

The regional segregation of neuronal response types in the striatum described above is consistent with mainly local integration over limited, adjacent sets of cortico-striatal inputs, as suggested by this anatomy. Short-range integration or interactions within the striatum may also be produced by the short length (for example 0.5 mm) of the intra-striatal axons of striatal neurons. These could produce a more widespread influence if the effect of a strong input to one part of the striatum spread like a lateral competition signal (cf. Groves (1983), Groves et al. (1995)). Such a mechanism could contribute to behavioural response selection in the face of different competing input signals to the striatum. The lateral inhibition could operate, for example, between the striatal principal (that is medium spiny) neurons by their direct connections, to inhibit each other. (These neurons receive excitatory connections from the cortex, respond by increasing their firing rates, and could inhibit each other by their local axonal arborizations, which spread in an area as large as their dendritic trees, and which utilize GABA as their inhibitory transmitter to enable the principal neurons of the striatum to directly inhibit each other.) Further lateral inhibition could operate in the pallidum and substantia nigra (see Fig. 20.4). Here again there are local axon collaterals, as widespread as the very large pallidal and nigral dendritic fields. The lateral competition could again operate by direct inhibitory connections between the neurons.

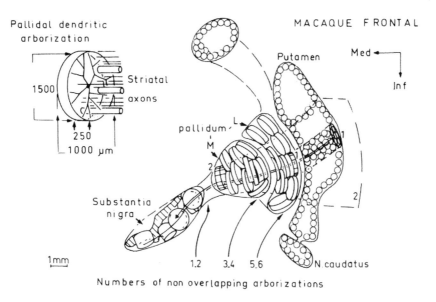

Pallidal dendritic
arborization

1500

250
1000 μm

Striatal
axons

Substantia
nigra

1mm

MACAQUE FRONTAL

Putamen

Med

Inf

pallidum
M

L

2

2

1.2   3.4   5.6

N.caudatus

Numbers of non overlapping arborizations

Fig. 20.3 Semi-schematic spatial diagram of the striato-pallido-nigral system (see text). The numbers represent the numbers of non-overlapping arborizations of dendrites in the plane shown. L, lateral or external segment of the globus pallidus; M, medial or internal segment of the globus pallidus. The inserted diagram in the upper left shows the geometry of the dendrites of a typical pallidal neuron, and how the flat dendritic arborization is pierced at right angles by the striatal axons, which make occasional synapses en passage. (Reproduced from Journal of Comparative Neurology, 227 (2) pp. 214–227. A Golgi analysis of the primate globus pallidus. III. Spatial organization of the striatopallidal complex, Gerard Percheron, Jerome Yelnik, and Chantal Francois. Copyright ©1984 Alan R. Liss, Inc. Reprinted with permission from John Wiley and Sons.)

[Note that pallidal and nigral cells have high spontaneous firing rates (often 25–50 spikes/s), and respond (to their inhibitory striatal inputs) by reducing their firing rates below this high spontaneous rate. Such a decrease in the firing rate of one neuron would release inhibition on nearby neurons, causing them to increase their firing rates, equivalent to responding less. It is very interesting that direct inhibitory connections between the neurons can implement selection, even though at the striatal level the neurons have low spontaneous firing rates and respond by increasing their firing rates, whereas in the globus pallidus and substantia nigra pars reticulata the neurons have a high spontaneous firing rate, and respond by decreasing their firing rate.]

A selection function of this type between processing streams in the basal ganglia, even without any convergence anatomically between the processing streams implemented by feedforward inputs, might provide an important computational raison d'être for the basal ganglia. As noted, the direct inhibitory local connectivity between the principal neurons within the striatum and globus pallidus would seem to provide a simple, and perhaps evolutionarily old, way in which to implement competition between neurons and processing streams. This might even be a primitive design principle that characterizes the basal ganglia. A system such as the basal ganglia with direct inhibitory recurrent collaterals may have evolved easily because it is easier to make stable than architectures such as the cerebral cortex with recurrent excitatory connections. The basal ganglia architecture may have been especially appropriate in motor systems in which instability could produce movement and co-ordination difficulties (Rolls and Treves 1998, Rolls 2014a). Equations that describe the way in which this mutual inhibition between the principal neurons can result in contrast enhancement of neuronal activity in the

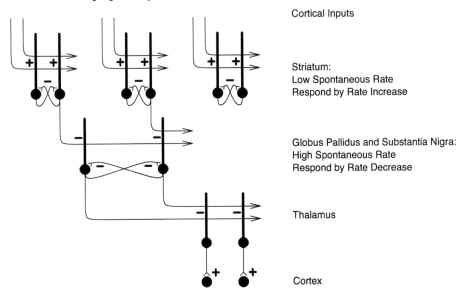

Fig. 20.4 Simple hypothesis of basal ganglia network architecture. A key aspect is that in both the striatum, and in the globus pallidus and substantia nigra pars reticulata, there are direct inhibitory connections (–) between the principal neurons, as shown. These synapses use GABA as a transmitter. Excitatory inputs to the striatum are shown as +. (Reproduced from E. T. Rolls and A. Treves, Neural Networks and Brain Function, 1998, by permission of Oxford University Press.)

different competing neurons, and thus selection, are provided by Grossberg (1988), Gurney et al. (2001a), and Gurney, Prescott and Redgrave (2001b).

This hypothesis of lateral competition between the neurons of the basal ganglia can be sketched simply (see Fig. 20.4 and Rolls and Treves (1998) and Rolls (2014a), where a more detailed neuronal network theory of the operation of the basal ganglia is presented). The inputs from the cortex to the striatum are excitatory, and competition between striatal neurons is implemented by the use of an inhibitory transmitter (GABA), and direct connections between striatal neurons, within an area which is approximately co-extensive with the dendritic arborization. Given that the lateral connections between the striatal neurons are collaterals of the output axons, the output must be inhibitory on to pallidal and nigral neurons. This means that to transmit signals usefully, and in contrast with striatal neurons, the neurons in the globus pallidus and substantia nigra (pars reticulata) must have high spontaneous firing rates, and respond by reducing their firing rates. These pallidal and nigral neurons then repeat the simple scheme for lateral competition between output neurons by having direct lateral inhibitory connections to the other pallidal and nigral neurons. When nigral and pallidal neurons respond by reducing their firing rates, the reduced inhibition through the recurrent collaterals allows the connected pallidal and nigral neurons to fire faster, and also at the same time the main output of the pallidal and nigral neurons allows the thalamic neurons to fire faster. The thalamic neurons then have the standard excitatory influence on their cortical targets.

The simple, and perhaps evolutionarily early, aspect of this basal ganglia architecture is that the striatal, pallidal, and nigral neurons implement competition (for selection) by direct inhibitory recurrent lateral connections of the main output neurons on to other output neurons, with the inputs to each stage of processing (e.g. striatum, globus pallidus) synapsing directly on to the output neurons that inhibit each other (see Fig. 20.4).

Another possible mechanism for interaction within the striatum is provided by the dopaminergic pathway, through which a signal that has descended from, for example, the ventral striatum to the dopamine neurons in the midbrain might thereby influence other parts of the striatum (see Section 9.6). Because of the slow conduction speed of the dopaminergic neurons, this latter system would probably not be suitable for rapid switching of behaviour, but only for more tonic, long-term adjustments of sensitivity.

Further levels for integration within the basal ganglia are provided by the striato-pallidal and striato-nigral projections (Percheron, Yelnik and François 1984b, Percheron, Yelnik and François 1984a, Percheron, Yelnik, François, Fenelon and Talbi 1994, Yelnik 2002). The afferent fibres from the striatum again cross at right angles a flat plate or disc formed by the dendrites of the pallidal or nigral neurons (see Fig. 20.3). The discs are approximately 1.5 mm in diameter, and are stacked up one upon the next at right angles to the incoming striatal fibres. The dendritic discs are so large that in the monkey there is room for only perhaps 50 such discs not to overlap in the external pallidal segment, for 10 non-overlapping discs in the medial pallidal segment, and for one overlapping disc in the most medial part of the medial segment of the globus pallidus and in the substantia nigra.

One result of this convergence achieved by this stage of the medial pallidum/substantia nigra is that even if inputs from different cortical regions were kept segregated by specific wiring rules on to different neurons, there might nevertheless well be the possibility for mutual competition between different pallidal neurons, implemented by their mutual inhibitory connections. Given the relatively small number of neurons into which the cortical signals had now been compressed, it would be feasible to have competition (the same effect as lateral inhibition implemented by inhibitory neurons would achieve elsewhere) implemented between the relatively small population of neurons, now all collected into a relatively restricted space, so that the competition could spread widely within these nuclei. This could allow selection by competition between these pathways, that is effectively between information processing originating in different cortical areas. This could be important in allowing each cortical area to control output when appropriate (depending on the task being performed). Even if full segregation were maintained in the return paths to the cerebral cortex, the return paths could influence each cortical area, allowing it to continue processing if it had the strongest 'call'. Each cortical area on a fully segregated hypothesis might thus have its own non-basal ganglia output routes, but might according to the current suggestion utilize the basal ganglia as a system to select a cortical area or set of areas, depending on how strongly each cortical area is calling for output. The thalamic outputs from the basal ganglia (areas VA and VLo of the thalamus) might according to this hypothesis have to some extent an activity or gain-controlling function on a cortical area (such as might be mediated by diffuse terminals in superficial cortical layers), rather than the strong and selective inputs implemented by a specific thalamic nucleus such as the lateral geniculate.

20.4 Comparison of selection in the basal ganglia and cerebral cortex

The basal ganglia architecture is that the striatal, pallidal, and nigral neurons implement competition (for selection) by direct inhibitory recurrent lateral connections of the main output neurons on to other output neurons. The inputs to each stage of processing (e.g. striatum, globus pallidus) synapse directly on to the output neurons that inhibit each other (see Fig. 20.4). This is a simple and evolutionarily early design in that the basal ganglia were present in the brain before the neocortex evolved. A system such as the basal ganglia with direct inhibitory recurrent collaterals may have evolved easily because it is easier to make stable than architectures such as the cerebral cortex with recurrent excitatory connections. As noted above, the basal ganglia architecture may have been especially appropriate in motor systems in which instability could produce movement and co-ordination difficulties (Rolls and Treves 1998, Rolls 2014a), and it is important to select motor responses before sending any output to muscles so that the muscles do not oppose each other. Thus a major advantage of this method of selecting outputs is that it is safe, with direct mutual inhibition of neurons, and no excitatory recurrent collaterals as in the neocortex.

In contrast, the neocortex with its excitatory recurrent collaterals, can also implement selection, by falling into a basin of attraction in which one mutually excitatory subpopulation of neurons ends up with high firing rates, as described in Chapters 5 and 16. In this circuitry, the inhibitory neurons perform feedback inhibition to try to keep the system stable, with the very dangerous risk of runaway excitation always present. Indeed, epilepsy may be viewed as the price that is paid for this type of architecture (which applies to neocortex, hippocampal cortex, and pyriform cortex). So what in these circumstances are the advantages of performing the selection using cortical architecture with excitatory recurrent collaterals?

Perhaps the major advantage is that the recurrent collaterals allow a subset of the neurons in the network, once activated, to maintain their activity for some time, in a stable attractor, as described in Chapters 4, 5, and 16 and by Rolls (2008d) and Rolls and Deco (2010). This allows short-term memory, which has a major advantage of allowing animals to remember what has happened recently, and just as much, to even plan ahead by holding a number of items active simultaneously to form a plan (see Chapter 22). Thus attractor networks allow both retrospective and prospective memory.

Another major property of the attractor approach is that when used for memory functions, it can complete a memory given a partial retrieval cue (Section B.3). This of course has major adaptive advantages, such as being able to remember where a food was located in response to a food cue or memory of a food (Chapter 24).

Another important property of the cortical attractor approach is that it can be used as a long-term memory system, with quite a high capacity if sparse representations are used (Section B.3). Indeed, in the order of 10,000 memories might be stored in a small cortical region with a diameter of 1–2 mm with 10,000 excitatory recurrent collateral synapses onto each neuron.

Another advantage of the cortical attractor approach is that when used as a decision-making system, the results of the decision can be kept 'on-line', by continuing firing of the attractor, to guide actions which may take some time to performed, or which may need to be delayed.

Another advantage of the cortical attractor approach is that when used for long-term memory, the system can, when influenced by the spiking-related noise, make jumps in different directions on different occasions, and thus implement such important functions as creativity and predator avoidance (Chapter 16).

Another advantage of the cortical attractor approach is that it allows weak interactions between attractors to implement many important cognitive functions, including top-down attention (Chapter 6).

Given the advantages of attractor networks in the cerebral cortex, it is important to consider further disadvantages than the risk of epilepsy. It appears that variation of the stability of cortical attractor networks is being maintained for natural selection because having relatively low or relatively high stability might confer advantages to some individuals (see Chapter 16). In this context, we must remember that there are many different attractor networks in different cortical areas, and that the stability of every one may be being explored by natural selection, to find combinations that may have high fitness in different environments. For example, having low stability in some cortical attractor networks may contribute to creativity by facilitating the tendency to jump to new parts of the high-dimensional space (Chapter 16), but if the stability is too low, the low stability may contribute to the cognitive symptoms in schizophrenia such as difficulty in maintaining attention, and also to the positive symptoms in schizophrenia (Loh, Rolls and Deco 2007a, Rolls, Loh, Deco and Winterer 2008d, Rolls 2012b). At the other end of what I see as a continuum, relatively high stability in some networks may help attention to remain focussed on a task (because of the high stability of the controlling attractor network providing the top-down input), but too much stability is some cortical systems may contribute to obsessive compulsive disorders (Rolls, Loh and Deco 2008c, Rolls 2012b). In the lateral orbitofrontal cortex, too little stability may contribute to impulsivity, and too much to depression (Rolls 2016d) (see Chapter 16).

Thus a comparison of the cerebral cortex with the basal ganglia helps to draw out the advantages and disadvantages of these two types of architecture, and highlights the advantages of many of the operational principles of the cerebral cortex that are implemented by its excitatory recurrent collateral connections which can support attractors, which are almost a defining feature of the architecture of the cerebral cortex.

20.5 Highlights

1. The functional architecture of the basal ganglia make an interesting comparison with the neocortex, for the basal ganglia do not use excitatory recurrent collaterals between the neurons.

2. Instead the much evolutionarily older basal ganglia use direct inhibition between the neurons at the striatal and at the pallidal / nigral stages of processing. In comparison with cortex, this is a much safer method of selection, and of ensuring that only one output is selected, which is important in motor systems so that contradictory outputs are not attempted, for there is no risk of positive feedback becoming uncontrolled and runaway with the striatal direct mutual inhibition by neurons.

3. The basal ganglia do have regular principles of internal connectivity between the neurons which lead to hypotheses about how it performs its computations, and this emphasizes the importance of building on the details of the anatomical and functional connectivity of neurons and networks of neurons to develop theories of how brain systems operate.

4. This comparison helps to underscore the adaptive value of the excitatory recurrent collaterals in the cerebral cortex, for though subject to runaway excitation and instability, they provide the way that the cortex uses to implement short-term memory by continuing firing maintained by the positive feedback, and which provides a computational basis for planning ahead; for completion of a long-term memory from a partial retrieval cue; for decision-making in which the result of the decision can be maintained 'on-line' by continuing attractor-related firing, to guide actions performed on the basis of the decision; and for associating together semantic items which provide the basis for semantic structures in language.

21   Sleep and Dreaming

21.1   Is sleep necessary for cortical function?

There are two main types of sleep, Slow Wave Sleep, and Paradoxical Sleep (Horne 2006, Carlson and Birkett 2017, McCormick and Westbrook 2013). Slow wave sleep (SWS) is characterised by slow waves in its four stages in the range 0.5–15 Hz measured from the cortex in the electroencephalogram (EEG). Neurons in most cortical areas fire at lower rates than in waking. [However, in the default mode network which includes medial prefrontal cortex areas and which become more active in the resting state than when engaged in tasks (Snyder and Raichle 2012), the neuronal firing rates may increase during resting and sleep (Gabbott and Rolls 2013). The default mode network appears to be more active during inner mentalizing than when engaged in tasks during waking, and appears to be less active in people with autism, perhaps because they mentalize less (Cheng, Rolls, Gu, Zhang and Feng 2015).] Paradoxical sleep has faster EEG activity and neuronal firing rates more like those during waking, and short bursts of Rapid Eye Movements occur during paradoxical sleep, sometimes leading this to be termed REM sleep.

The question arises about whether the cortex 'needs' these two types of sleep, that is, whether there are any special operations performed by the cortex that require these sleep states. A parsimonious explanation of sleep at present is an evolutionary adaptive value explanation, that sleep serves to keep animals out of harm's way and to conserve energy at night when movements may be hazardous and may not lead to more food (Horne 2006). However, that does not account for why there are two types of sleep, SWS and paradoxical sleep. A parsimonious hypothesis is as follows. Slow wave sleep is the simpler and evolutionarily older type of sleep, and evolved to perform the functions just described by lowering cortical activity with a diurnal rhythm synchronized to the light-dark cycle by the suprachiasmatic nucleus. However, in this SWS state the brain is unable to perform some homeostatic functions, including maintenance of body temperature (Horne 2006). This then led to the evolution of paradoxical sleep, in which (paradoxically) the cortex is active with low voltage desynchronized rapid EEG similar to that in the waking state, in which the homeostatic functions could be performed better. Because the cortex is active in paradoxical sleep, paradoxical sleep is associated with descending inhibition to the spinal cord that effectively produces a state of paralysis, so that the active cortex does not produce movements. Rapid Eye Movements may be allowed to occur during paradoxical sleep because they do not need to be paralyzed, for they do not disturb the animal and put it into danger. With the cortex active during paradoxical sleep, mental states occur, and these constitute dreaming.

This parsimonious hypothesis is then that sleep may not serve a process fundamental to the operation of the cortex, but is instead an evolutionary adaptation to darkness which minimizes movement, danger, and energy expenditure. Of course, there will be mechanisms that motivate us to sleep, and the longer we are without sleep, the more these motivational processes may make us want to sleep, and may perform this function in part by making us less efficient, and indeed sleepy, that is forcing the brain towards slow wave sleep.

21.2 Is sleep involved in memory consolidation?

Larry Weiskrantz mentioned the hypothesis in an undergraduate lecture on the hippocampus and memory consolidation in 1966 at Cambridge at which David Marr was present that sleep might be involved in the transfer of episodic memories from the hippocampus to the neocortex. When David Marr produced his theory of hippocampal function (Marr 1971), he referred to this hypothesis about sleep and memory consolidation. Since then, there has been considerable interest in whether memories may be transferred from the hippocampus to the neocortex during sleep (Schwindel and McNaughton 2011). My view is that recall from the hippocampus would best be performed during waking, for the following reasons (Kesner and Rolls 2015), with the mechanisms involved described in Chapter 11 and Section 24.2.3.

A hippocampus-dependent episodic memory is unstructured in the sense that the information is stored as an association between representations such as spatial location, objects, and position in a temporal sequence. An episodic memory could be a single event capturing what happens in a time period in the order of 1–10 s (for example a memory of who was present at one's fifth birthday, what they were wearing, where they sat, where the party was, etc), or a sequence of events with some information about the temporal order of the events. A hippocampus-dependent episodic memory includes spatial and/or temporal e.g. sequence information. All this is in contrast to a neocortical semantic memory which might utilize many such episodic memories, for example of separate train journeys, to form a topological map. Such a sematic memory might need to benefit from information about exceptions, which contributes to the type of structure needed for a semantic memory. For example we may be told that a bat, although it flies, is not a bird because it is a mammal which nurses its young, and this information allows restructuring of a semantic representation of what constitutes a bird beyond the simple association and generalization from limited experience that birds fly. Because this active restructuring is an important part of semantic memory that may frequently benefit from language, I propose that semantic memories are best formed during waking using thought and reasoning, and not during sleep. Indeed, I propose that episodic events that are recalled actively in this way from the hippocampus can become incorporated into neocortical representations and thus consolidated into semantic including autobiographical memories. If episodic material (such as where and with whom one had dinner three days ago) is not recalled in this way, then it will be gradually overwritten in hippocampal memory as new sparse distributed representations are added to the store, in the way that has been analyzed quantitatively (Parisi 1986, Rolls 2008d, Rolls and Treves 1998) (Section 9.3).

In contrast, some have suggested that the transfer of information from the hippocampus to the neocortex occurs especially during sleep (Schwindel and McNaughton 2011). Indeed, it has been shown that after learning in hippocampal-dependent tasks, neocortical representations may change (Schwindel and McNaughton 2011). Although this has been interpreted as the transfer of memories from the hippocampus to the neocortex (Schwindel and McNaughton 2011), it should be noted that if the hippocampal representation changes as a result of learning, then the altered representation in CA1 will, even with fixed synaptic connections back to neocortex, alter neocortical firing, with no learning or actual 'transfer' involved. (This occurs whenever one vector comprised by a population of neurons firing changes and influences another vector of neuronal firing through fixed connections.) Indeed, for the reasons given above, active transfer of episodic memories from the hippocampus to the neocortex to form semantic memories is more likely, Rolls proposed (Kesner and Rolls 2015), during the guided construction and correction of semantic memories in the cortex during waking. The actual addition of new information to the neocortical semantic representation could of course itself be fast, for example when one recalls a particular train journey, and incorporates information from that into a geographical representation. An additional property of hippocampal episodic memories is that when recalled to the neocortex they can be used flexibly for many types of behavioural output (because the information is back in the neocortex).

There is evidence that synaptic strength may increase gradually during waking as a result of long-term synaptic potentiation (LTP), and that during sleep synapses may weaken when the brain is cut off from external inputs (Tononi and Cirelli 2014). This may be a useful function performed during sleep, though this physiology might be secondary to the evolutionary adaptation approach described above. Consistent with the hypothesis just described, LTP may be easier to obtain after normal sleep than after sleep deprivation (Tononi and Cirelli 2014).

21.3 Dreams

For the reasons described in Section 21.2, memory consolidation of the cortical states that occur during paradoxical sleep appears to be turned off, and indeed we only remember the dreams that are occurring while we wake up, when memory consolidation is being allowed to return. The question then arises of what produces the different cortical states during paradoxical sleep, when sensory inputs to the cortex and outputs from the cortex are inhibited, and indeed, what these states 'mean'. I argue that dream states reflect noise-driven processes of cortical attractor states, where the noise describes the effects of the random firing times for a given mean firing rate of cortical neurons (Chapter 5), as follows (Rolls and Deco 2010, Rolls 2012d).

Noise-driven processes may lead to dreams, where the content of the dream is not closely tied to the external world because the role of sensory inputs is reduced in paradoxical (desynchronized, high frequency, fast wave, dreaming) sleep, and the cortical networks, which are active in fast-wave sleep (McCormick and Westbrook 2013, Carlson and Birkett 2017, Horne 2006), may move on somewhat freely, under the influence of noise, from states that may have been present during the day. Thus the content of dreams may be seen as a noisy trajectory through a state space, with starting point states that have been active during the day, and passing through states that will reflect the depth of the basins of attraction (which might well reflect ongoing concerns including anxieties and desires), and will be strongly influenced by noise. Moreover, the top-down attentional and monitoring control from for example the prefrontal cortex appears to be less effective during sleep than during waking, allowing the system to pass into states that may be bizarre.

I suggest that this provides a firm theoretical foundation for understanding the interpretation of dreams, which may be compared with that of Freud (1900).

In this context, the following thoughts follow. Dreams, or at least the electrical activity of paradoxical sleep, may occur in cortical areas that are concerned with conscious experience, such as those involved in higher order thoughts, and in others where processing is unconscious, and cannot be reported, such as those in the dorsal visual stream concerned with the control of actions (Rolls 2004b, Goodale 2004, Rolls 2005, Rolls 2007a, Rolls 2008a, Rolls 2008d). Thus it may be remarked that dreams may occur in conscious and unconscious processing systems. Dreams appear to be remembered that occur just before we wake up, and consistent with this, memory storage (implemented by synaptic long-term potentiation, Chapter 9), appears to be turned off during sleep. This may be adaptive, for then we do not use up memory capacity (Rolls 2008d) on noise-related representations. However, insofar as we can memorize and later remember dreams that are rehearsed while we wake up, it could be that bizarre thoughts, possibly unpleasant, could become consolidated. This consolidation could lead to the relevant attractor basins becoming deeper, and returning to the same set of memories on subsequent nights. This could be a mechanism for the formation of nightmares. A remedy is likely to be to not rehearse these unpleasant dreams while waking, and indeed to deliberately move to more pleasant thoughts, which would then be consolidated, and increase the probability of dreams on those more pleasant subjects, instead of nightmares, on later nights, given the memory attractor landscape over which noise would move one's thoughts.

21.4 Highlights

1. Slow wave sleep may be evolutionarily old, and may have evolved to keep animals out of harm, and conserving energy when it is dark and food may not be obtained easily.
2. Desynchronized (paradoxical) sleep may have evolved as a second type of sleep interspersed between periods of slow wave sleep so that certain bodily functions, such as thermoregulation which are not performed efficiently in slow wave sleep, can operate. In paradoxical sleep the neocortex has desynchronized activity of the EEG (electroencephalogram) similar to that in waking, but descending inhibition prevents the body from moving. Dreams occur in desynchronized sleep.
3. Although it is a championed hypothesis that episodic memories are downloaded for consolidation in the neocortex during sleep, this seems not to be computationally optimal, for much of the process may require building useful new semantic representations, sometimes of the self for autobiographical memories, which may require supervision or monitoring of the type that may not be available during sleep. Episodic memories may more usually be downloaded from the hippocampus if they are needed during waking, for example to remember one's current hotel room number.
4. A theory of dreams is described, as a stochastic trajectory produced by the spiking-related noise of neuronal firing through a semantic state space of thoughts started or primed in the day, without the normal higher order thought monitoring process described in Chapter 22 maintaining the thoughts by checking the reasoning and trajectory.

22 Which cortical computations underlie consciousness?

22.1 Introduction

It might be possible to build a computer that would perform the functions of emotions described by Rolls (2014a) and summarized in Chapter 15, and yet we might not want to ascribe emotional feelings to the computer. We might even build the computer with some of the main processing stages present in the brain, and implemented using neural networks that simulate the operation of the real neural networks in the brain (see Chapter 15, Rolls and Treves (1998), Rolls (2008d), and Appendix B), yet we might not still wish to ascribe emotional feelings to this computer. This point often arises in discussions with undergraduates, who may say that they follow the types of point made about emotion in Chapter 15, yet believe that almost the most important aspect of emotions, the feelings, have not been accounted for, nor their neural basis described. In a sense, the functions of reward and punishment in emotional behaviour are described by Rolls (2014a) and summarized in Chapter 15, but what about the subjective aspects of emotion, what about the pleasure?

A similar point arises in Chapter 1, where parts of the taste, olfactory, and visual systems in which the reward value of the taste, smell, and sight of food is represented are described. Although the neuronal representation in the orbitofrontal cortex is clearly related to the reward value of food, and in humans the activations found with functional neuroimaging are directly (indeed linearly) correlated with the reported subjective pleasantness of the stimuli (see Rolls (2014a) and Chapters 15 and 1), is this where the pleasantness (the subjective hedonic aspect) of the taste, smell, and sight of food is represented and produced? Again, we could (in principle at least) build a computer with neural networks to simulate each of the processing stages for the taste, smell, and sight of food that are described in Chapter 1, and yet would probably not wish to ascribe feelings of pleasantness to the system we have simulated on the computer.

What is it about neural processing that makes it feel like something when some types of information processing are taking place? It is clearly not a general property of processing in neural networks, for there is much processing, for example that in the autonomic nervous system concerned with the control of our blood pressure and heart rate, of which we are not aware. Is it then that awareness arises when a certain type of information processing is being performed? If so, what type of information processing? And how do emotional feelings, and sensory events, come to feel like anything? These 'feels' are called qualia. These are great mysteries that have puzzled philosophers for centuries. They are at the heart of the problem of consciousness, for why it should feel like something at all is the great mystery.

Other aspects of consciousness may be easier to analyse, such as the fact that often when we 'pay attention' to events in the world, we can process those events in some better way. These are referred to as 'process' or 'access' aspects of consciousness, as opposed to the 'phenomenal' or 'feeling' aspects of consciousness referred to in the preceding paragraph (Block 1995a, Chalmers 1996, Allport 1988, Koch 2004, Block 1995b).

The puzzle of qualia, that is of the phenomenal aspect of consciousness, seems to be rather different from normal investigations in science, in that there is no agreement on criteria by which to assess whether we have made progress. So, although the aim of this chapter is to address the issue of consciousness, especially of qualia, what is written cannot be regarded as being as firmly scientific as the other chapters in this book. For most of the work in those chapters, there is good evidence for most of the points made, and there would be no hesitation or difficulty in adjusting the view of how things work as new evidence is obtained. However, in the work on qualia, the criteria are much less clear. Nevertheless, the reader may well find these issues interesting, because although not easily solvable, they are very important issues to consider if we wish to really say that we understand some of the very complex and interesting issues about cortical function, and ourselves.

With these caveats in mind, I consider in this chapter the general issue of consciousness and its functions, and how feelings, and pleasure, come to occur as a result of the operation of our brains. A view on consciousness, influenced by contemporary cognitive neuroscience, is outlined next. I outline a theory of what the processing is that is involved in consciousness, of its adaptive value in an evolutionary perspective, and of how processing in our visual and other sensory systems can result in subjective or phenomenal states, the 'raw feels' of conscious awareness. However, this view on consciousness that I describe is only preliminary, and theories of consciousness are likely to develop considerably. Partly for these reasons, this theory of consciousness, at least, should not be taken to have practical implications.

22.2 A Higher-Order Syntactic Thought (HOST) theory of consciousness

22.2.1 Multiple routes to action

A starting point is that many actions can be performed relatively automatically, without apparent conscious intervention. An example sometimes given is driving a car for a short time if we are thinking about something else. Another example is the identification of a visual stimulus that can occur without conscious awareness as described in Section 22.8.3. Another example is much of the sensory processing and actions that involve the dorsal stream of visual processing to the parietal cortex, such as posting a letter through a letter box at the correct orientation even when one may not be aware of what the object is (Milner and Goodale 1995, Goodale 2004, Milner 2008). Another example is blindsight, in which humans with damage to the visual cortex may be able to point to objects even when they are not aware of seeing an object (Weiskrantz 1997, Weiskrantz 1998, Weiskrantz 2009). Similar evidence applies to emotions, some of the processing for which can occur without conscious awareness (De Gelder, Vroomen, Pourtois and Weiskrantz 1999, Phelps and LeDoux 2005, LeDoux 2008). Consistent with the hypothesis of multiple routes to action, only some of which involve conscious awareness, is the evidence that split-brain patients may not be aware of actions being performed by the 'non-dominant' hemisphere (Gazzaniga and LeDoux 1978, Gazzaniga 1988, Gazzaniga 1995). Also consistent with multiple, including non-verbal, routes to action, patients with focal brain damage, for example to the prefrontal cortex, may emit actions, yet comment verbally that they should not be performing those actions (Rolls, Hornak, Wade and McGrath 1994a, Hornak, Bramham, Rolls, Morris, O'Doherty, Bullock and Polkey 2003). In both these types of patient, confabulation may occur, in that a verbal account of why the action was performed may be given, and this may not be related at all to the environmental event that actually triggered the action (Gazzaniga and LeDoux 1978, Gazzaniga 1988, Gazzaniga 1995).

Such implicit (not phenomenally conscious) actions could involve control of behaviour by brain systems that are old in evolutionary terms such as the basal ganglia. It is of interest that the basal ganglia (and cerebellum) do not have backprojection systems to most of the parts of the cerebral cortex from which they receive inputs (see Chapter 20 and Rolls and

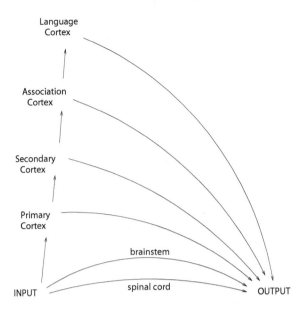

Fig. 22.1 Schematic illustration indicating many possible routes from input systems to action (output) systems. Cortical information-processing systems are organized hierarchically, and there are routes to output systems from most levels of the hierarchy.

Treves (1998)). In contrast, parts of the brain such as the hippocampus and amygdala, involved in functions such as episodic memory and emotion respectively, about which we can make (verbal) declarations (hence declarative memory, Squire (1992)) do have major backprojection systems to the high parts of the cerebral cortex from which they receive forward projections (see e.g. Fig. 24.1 and Chapter 11). It may be that evolutionarily newer parts of the brain, such as the language areas and parts of the prefrontal cortex, are involved in an alternative type of control of behaviour, in which actions can be planned with the use of a (language) system that allows relatively arbitrary (syntactic) manipulation of semantic entities (symbols).

The general view that there are many routes to behavioural output is supported by the evidence that there are many input systems to the basal ganglia (from almost all areas of the cerebral cortex), and that neuronal activity in each part of the striatum reflects the activity in the overlying cortical area (see Chapter 20). The evidence is consistent with the possibility that different cortical areas, each specialized for a different type of computation, have their outputs directed to the basal ganglia, which then select the strongest input, and map this into action (via outputs directed, for example, to the premotor cortex). The view is also supported by the evidence that the cingulate cortex is involved in actions performed for goals (Section 15.11). Within this scheme, the language areas would offer one of many routes to action, but a route particularly suited to planning actions, because of the syntactic manipulation of semantic entities which may make long-term planning possible. A schematic diagram of this suggestion is provided in Fig. 22.1.

It is accordingly possible that sometimes in normal humans when actions are initiated as a result of processing in a specialized brain region such as those involved in some types of rewarded behaviour, the language system may subsequently elaborate a coherent account of why that action was performed (i.e. confabulate). This would be consistent with a general view of brain evolution in which, as areas of the neocortex evolve, they are laid on top of existing circuitry connecting inputs to outputs, and in which each level in this hierarchy of separate input–output pathways may control behaviour according to the specialized function it can perform (see schematic in Fig. 22.1). (It is of interest that mathematicians may get a hunch that something is correct, yet not be able to verbalize why. They may then resort to formal, more serial and language-like, theorems to prove the case, and these seem to require conscious processing. This is a further indication of a close association between linguistic processing, and consciousness. The linguistic processing need not, as in reading, involve an inner articulatory loop.)

We may next examine some of the advantages and behavioural functions that language, present as the most recently added layer to the above system, would confer.

One major advantage would be the ability to plan actions through many potential stages and to evaluate the consequences of those actions without having to perform the actions. For this, the ability to form propositional statements, and to perform syntactic operations on the semantic representations of states in the world, would be important.

Also important in this system would be the ability to have second-order thoughts about the type of thought that I have just described (e.g. I think that he thinks that ..., involving 'theory of mind'), as this would allow much better modelling and prediction of others' behaviour, and therefore of planning, particularly planning when it involves others[8]. This capability for higher-order thoughts would also enable reflection on past events, which would also be useful in planning. In contrast, non-linguistic behaviour would be driven by learned reinforcement associations, learned rules etc., but not by flexible planning for many steps ahead involving a model of the world including others' behaviour. (For an earlier view that is close to this part of the argument see Humphrey (1980).) [The examples of behaviour from non-humans that may reflect planning may reflect much more limited and inflexible planning. For example, the dance of the honey-bee to signal to other bees the location of food may be said to reflect planning, but the symbol manipulation is not arbitrary. There are likely to be interesting examples of non-human primate behaviour that reflect the evolution of an arbitrary symbol-manipulation system that could be useful for flexible planning, cf. Cheney and Seyfarth (1990), Byrne and Whiten (1988), and Whiten and Byrne (1997).]

It is important to state that the language ability referred to here is not necessarily human verbal language (though this would be an example). What it is suggested is important to planning is the syntactic manipulation of symbols, and it is this syntactic manipulation of symbols that is the sense in which language is defined and used here. The type of syntactic processing need not be at the natural language level (which implies a universal grammar), but could be at the level of mentalese (Rolls 2005, Rolls 2004b, Fodor 1994).

I understand **reasoning, and rationality**, to involve syntactic manipulations of symbols in the way just described. Reasoning thus typically may involve multiple steps of 'if ... then' conditional statements, all executed as a one-off or one-time process (see below), and is very different from associatively learned conditional rules typically learned over many trials, such as 'if yellow, a left choice is associated with reward'.

22.2.2 A computational hypothesis of consciousness

It is next suggested that this arbitrary symbol-manipulation using important aspects of language processing and used for planning but not in initiating all types of behaviour is close

---

[8] Second-order thoughts are thoughts about thoughts. Higher-order thoughts refer to second-order, third-order, etc., thoughts about thoughts... (A thought may be defined briefly as an intentional mental state, that is a mental state that is about something. Thoughts include beliefs, and are usually described as being propositional (Rosenthal 2005). An example of a thought is "It is raining". A more detailed definition is as follows. A thought may be defined as an occurrent mental state (or event) that is intentional – that is a mental state that is about something – and also propositional, so that it is evaluable as true or false. Thoughts include occurrent beliefs or judgements. An example of a thought would be an occurrent belief that the earth moves around the sun / that Maurice's boat goes faster with two sails / that it never rains in southern California.)

to what consciousness is about. In particular, consciousness may *be* the state that arises in a system that can think about (or reflect on) its own (or other peoples') thoughts, that is in a system capable of second- or higher-order thoughts (Rosenthal 1986, Rosenthal 1990, Rosenthal 1993, Dennett 1991, Rolls 1995c, Carruthers 1996, Rolls 1997a, Rolls 1997b, Rolls 1999a, Gennaro 2004, Rolls 2004b, Rosenthal 2004, Rolls 2005, Rosenthal 2005, Rolls 2007a, Rolls 2008a, Rolls 2007b, Rolls 2011b, Lau and Rosenthal 2011, Rolls 2012d, Rosenthal 2012, Rolls 2013c). On this account, a mental state is non-introspectively (i.e. non-reflectively) conscious if one has a roughly simultaneous thought that one is in that mental state. Following from this, introspective consciousness (or reflexive consciousness, or self consciousness) is the attentive, deliberately focused consciousness of one's mental states. It is noted that not all of the higher-order thoughts need themselves be conscious (many mental states are not). However, according to the analysis, having a higher-order thought about a lower-order thought is necessary for the lower-order thought to be conscious.

A slightly weaker position than Rosenthal's (and mine) on this is that a conscious state corresponds to a first-order thought that has the capacity to cause a second-order thought or judgement about it (Carruthers 1996). Another position that is close in some respects to that of Carruthers and the present position is that of Chalmers (1996), that awareness is something that has direct availability for behavioural control, which amounts effectively for Chalmers to saying that consciousness is what we can report (verbally) about[9]. This analysis is consistent with the points made above that the brain systems that are required for consciousness and language are similar. In particular, a system that can have second- or higher-order thoughts about its own operation, including its planning and linguistic operation, must itself be a language processor, in that it must be able to bind correctly to the symbols and syntax in the first-order system. According to this explanation, the feeling of anything is the state that is present when linguistic processing that involves second- or higher-order thoughts is being performed.

It might be objected that this hypothesis captures some of the process aspects of consciousness, that is, what is useful in an information processing system, but does not capture the phenomenal aspect of consciousness. (Chalmers, following points made in his 1996 book, might make this point.) I agree that there is an element of 'mystery' that is invoked at this step of the argument, when I say that it feels like something for a machine with higher-order thoughts to be thinking about its own first- or lower-order thoughts. But the return point is the following: if a human with second-order thoughts is thinking about its own first-order thoughts, surely it is very difficult for us to conceive that this would not feel like something? (Perhaps the higher-order thoughts in thinking about the first-order thoughts would need to

---

[9] Chalmers (1996) is not entirely consistent about this. Later in the same book he advocates a view that experiences are associated with information-processing systems, e.g. experiences are associated with a thermostat (p. 297). He does not believe that the brain has experiences, but that he has experiences. This leads him to suggest that experiences are associated with information processing systems such as the thermostat in the same way as they are associated with him. "If there is experience associated with thermostats, there is probably experience everywhere: wherever there is a causal interaction, there is information, and wherever there is information, there is experience." (p. 297). He goes on to exclude rocks from having experiences, in that "a rock, unlike a thermostat, is not picked out as an information-processing system". My response to this is that of course there is mutual information between the physical world (e.g. the world of tastants, the chemical stimuli that can produce tastes) and the conscious world (e.g. of taste) – if there were not, the information represented in the conscious processing system would not be useful for any thoughts or operations on or about the world. And according to the view I present here, the conscious processing system is good at some specialized types of processing (e.g. planning ahead using syntactic processing with semantics grounded in the real world, and reflecting on and correcting such plans), for which it would need reliable information about the world. Clearly Chalmers' view on consciousness is very much weaker than mine, in that he allows thermostats to be associated with consciousness, and in contrast to the theory presented here, does not suggest any special criteria for the types of information processing to be performed in order for the system to be aware of its thoughts, and of what it is doing.

have in doing this some sense of continuity or self, so that the first-order thoughts would be related to the same system that had thought of something else a few minutes ago. But even this continuity aspect may not be a requirement for consciousness. Humans with anterograde amnesia cannot remember what they felt a few minutes ago, yet their current state does feel like something.)

As a point of clarification, I note that according to my theory, a language processing system (let alone a working memory system (LeDoux 2008)) is not sufficient for consciousness. What defines a conscious system according to this analysis is the ability to have higher-order thoughts, and a first order language processor (that might be perfectly competent at language) would not be conscious, in that it could not think about its own or others' thoughts. One can perfectly well conceive of a system that obeyed the rules of language (which is the aim of some connectionist modelling), and implemented a first-order linguistic system, that would not be conscious. [Possible examples of language processing that might be performed non-consciously include computer programs implementing aspects of language, or ritualized human conversations, e.g., about the weather. These might require syntax and correctly grounded semantics, and yet be performed non-consciously. A more complex example, illustrating that syntax could be used, might be "If A does X, then B will probably do Y, and then C would be able to do Z." A first order language system could process this statement. Moreover, the first order language system could apply the rule usefully in the world, provided that the symbols in the language system (A, B, X, Y etc.) are grounded (have meaning) in the world.]

A second clarification is that the plan would have to be a unique string of steps, in much the same way as a sentence can be a unique and one-off (or one-time) string of words. The point here is that it is helpful to be able to think about particular one-off plans, and to correct them; and that this type of operation is very different from the slow learning of fixed rules by trial and error, or the application of fixed rules by a supervisory part of a computer program.

22.2.3 Adaptive value of processing in the system that is related to consciousness

It is suggested that part of the evolutionary *adaptive significance* of this type of higher-order thought is that it enables correction of errors made in first-order linguistic or in non-linguistic processing. Indeed, the ability to reflect on previous events is extremely important for learning from them, including setting up new long-term semantic structures. It was shown above that the hippocampus may be a system for such 'declarative' recall of recent memories (see also Squire, Stark and Clark (2004)). Its close relation to 'conscious' processing in humans (Squire has classified it as a declarative memory system) may be simply that it enables the recall of recent memories, which can then be reflected upon in conscious, higher-order, processing. Another part of the adaptive value of a higher-order thought system may be that by thinking about its own thoughts in a given situation, it may be able to understand better the thoughts of another individual in a similar situation, and therefore predict that individual's behaviour better (Humphrey 1980, Humphrey 1986) (cf. Barlow (1997)).

In line with the argument on the adaptive value of higher-order thoughts and thus consciousness given above, that they are useful for correcting lower-order thoughts, I now suggest that correction using higher-order thoughts of lower-order thoughts would have adaptive value primarily if the lower-order thoughts are sufficiently complex to benefit from correction in this way. The nature of the complexity is specific – that it should involve syntactic manipulation of symbols, probably with several steps in the chain, and that the chain of steps should be a one-off (or in American usage, 'one-time', meaning used once) set of steps, as in a sentence or in a particular plan used just once, rather than a set of well learned rules. The first or lower-order thoughts might involve a linked chain of 'if ... then' statements that would be involved in planning, an example of which has been given above, and this type of cognitive processing is thought to be a primary basis for human skilled performance (Anderson 1996). It is partly because complex lower-order thoughts such as these that involve syntax and language would benefit from correction by higher-order thoughts that I suggest that there is a close link between this reflective consciousness and language.

The *computational hypothesis* is that by thinking about lower-order thoughts, the higher-order thoughts can discover what may be weak links in the chain of reasoning at the lower-order level, and having detected the weak link, might alter the plan, to see if this gives better success. In our example above, if it transpired that C could not do Z, how might the plan have failed? Instead of having to go through endless random changes to the plan to see if by trial and error some combination does happen to produce results, what I am suggesting is that by thinking about the previous plan, one might, for example, using knowledge of the situation and the probabilities that operate in it, guess that the step where the plan failed was that B did not in fact do Y. So by thinking about the plan (the first- or lower-order thought), one might correct the original plan in such a way that the weak link in that chain, that 'B will probably do Y', is circumvented.

To draw a parallel with neural networks: there is a **'credit assignment'** problem in such multistep syntactic plans, in that if the whole plan fails, how does the system assign credit or blame to particular steps of the plan? [In multilayer neural networks, the credit assignment problem is that if errors are being specified at the output layer, the problem arises about how to propagate back the error to earlier, hidden, layers of the network to assign credit or blame to individual synaptic connection; see Rumelhart, Hinton and Williams (1986a) and Section B.11.] **My suggestion is that this solution to the credit assignment problem for a one-off syntactic plan is the function of higher-order thoughts, and is why systems with higher-order thoughts evolved. The suggestion I then make is that if a system were doing this type of processing (thinking about its own thoughts), it would then be very plausible that it should feel like something to be doing this.** I even suggest to the reader that it is not plausible to suggest that it would not feel like anything to a system if it were doing this.

I emphasize that the plan would have to be a unique string of steps, in much the same way as a sentence can be a unique and one-off string of words. The point here is that it is helpful to be able to think about particular one-off plans, and to correct them; and that this type of operation is very different from the slow learning of fixed rules by trial and error, or the application of fixed rules by a supervisory part of a computer program.

22.2.4 Symbol grounding

A further point in the argument should be emphasized for clarity. The system that is having syntactic thoughts about its own syntactic thoughts (higher-order syntactic thoughts or HOSTs) would have to have its symbols grounded in the real world for it to feel like something to be having higher-order thoughts. The intention of this clarification is to exclude systems such as a computer running a program when there is in addition some sort of control or even overseeing program checking the operation of the first program. We would want to say that in such a situation it would feel like something to be running the higher level control program only if the first-order program was symbolically performing operations on the world and receiving input about the results of those operations, and if the higher-order system understood what the first order system was trying to do in the world.

The symbols (or symbolic representations) are symbols in the sense that they can take part in syntactic processing. The symbolic representations are grounded in the world in that they refer to events in the world. The symbolic representations must have a great deal of information about what is referred to in the world, including the quality and intensity of sensory events, emotional states, etc. The need for this is that the reasoning in the symbolic system must be about stimuli, events, and states, and remembered stimuli, events and states, and for the reasoning to be correct, all the information that can affect the reasoning must be represented in the symbolic system, including for example just how light or strong the touch was, etc. Indeed, it is pointed out (Rolls 2005, Rolls 2012d, Rolls 2014a) that it is no accident that the shape of the multidimensional phenomenal (sensory etc.) space does map so clearly onto the space defined by neuronal activity in sensory systems, for if this were not the case, reasoning about the state of affairs in the world would not map onto the world, and would not be useful. Good examples of this close correspondence are found in the taste system, in which subjective space maps simply onto the multidimensional space represented by neuronal firing in primate cortical taste areas (Section 2.2.3). In particular, if a two-dimensional space reflecting the distances between the representations of different tastes provided by macaque neurons in the cortical taste areas is constructed, then the distances between the subjective ratings by humans of different tastes is very similar (Kadohisa, Rolls and Verhagen 2005a, Smith-Swintosky, Plata-Salaman and Scott 1991, Rolls 2014a). Similarly, the changes in human subjective ratings of the pleasantness of the taste, smell and sight of food parallel very closely the responses of neurons in the macaque orbitofrontal cortex (Chapter 15 and Section 2.2.3) (Rolls 2014a).

The representations in the first-order linguistic processor that the HOSTs process include beliefs (for example "Food is available", or at least representations of this), and the HOST system would then have available to it the concept of a thought (so that it could represent "I believe [or there is a belief] that food is available"). However, as argued by Rolls (1999a, 2005, 2012d, 2014a), representations of sensory processes and emotional states must be processed by the first-order linguistic system, and HOSTs may be about these representations of sensory processes and emotional states capable of taking part in the syntactic operations of the first-order linguistic processor. Such sensory and emotional information may reach the first-order linguistic system from many parts of the brain, including those such as the orbitofrontal cortex and amygdala implicated in emotional states. When the sensory information is about the identity of the taste, the inputs to the first-order linguistic system must come from the primary taste cortex, in that the identity and intensity of taste, independently of its pleasantness (in that the representation is independent of hunger) must come from the primary taste cortex (Fig. 6.19). In contrast, when the information that reaches the first-order linguistic system is about the pleasantness of taste, it must come from the secondary taste (orbitofrontal) cortex, in that there the representation of taste depends on hunger and is linearly related to pleasantness (Rolls and Grabenhorst 2008, Grabenhorst and Rolls 2008, Rolls 2014a, Rolls 2015e) (Fig. 6.19).

The main answer that I propose now to the issue of symbol grounding is as follows. The gene-specified rewards or goals for action that are the bases for emotional and motivational states play the role for these states of grounding them in the world. The organism has to be built to want food rewards, to avoid pain, etc. I propose that this grounding for gene-specified emotional and motivational states provides the basis for the symbol grounding in the symbolic system, in that what the symbolic system computes, in a sense its goals, must be close to what the gene-specified emotional and motivational systems are grounded in, as otherwise the symbolic reasoning system and the gene goal-based emotional system would be inconsistent, and the reasoning system would not have adaptive value in evolution. To put this another way, unless the symbolic syntactic reasoning system had the belief that the gene-specified goals of the emotional system were among the goals of the reasoning system, then the two systems together could not produce consistent actions in general, and that would be unadaptive in the evolutionary sense. That leaves it open then in evolution for the symbolic system to add reasoned goals of its own, which might be for the advantage of the individual, but would still be in the same design framework of the emotion- and motivation-related goals specified by genes.

The issue of symbol grounding is considered further in Section 22.6. The philosopher Ruth Millikan (1984) has also considered the issue of symbol grounding.

22.2.5 Qualia

This analysis does not yet give an account for sensory qualia ('raw sensory feels', for example why 'red' feels red), for emotional qualia (e.g. why a rewarding touch produces an emotional feeling of pleasure), or for motivational qualia (e.g. why food deprivation makes us feel hungry). The view I suggest on such **qualia** is as follows. Information processing in and from our sensory systems (e.g. the sight of the colour red) may be relevant to planning actions using language and the conscious processing thereby implied. Given that these inputs must be represented in the system that plans, we may ask whether it is more likely that we would be conscious of them or that we would not. I suggest that it would be a very special-purpose system that would allow such sensory inputs, and emotional and motivational states, to be part of (linguistically based) planning, and yet remain unconscious (given that the processing being performed by this system is inherently conscious, as suggested above). It seems to be much more parsimonious to hold that we would be conscious of such sensory, emotional, and motivational qualia because they would be being used (or are available to be used) in this type of (linguistically based) higher-order thought processing system, and this is what I propose.

The explanation of emotional and motivational subjective feelings or qualia that this discussion has led towards is thus that they should be felt as conscious because they enter into a specialized linguistic symbol-manipulation system, which is part of a higher-order thought system that is capable of reflecting on and correcting its lower-order thoughts involved for example in the flexible planning of actions. It would require a very special machine to enable this higher-order linguistically-based thought processing, which is conscious by its nature, to occur without the sensory, emotional and motivational states (which must be taken into account by the higher-order thought system) becoming felt qualia. The sensory, emotional, and motivational qualia are thus accounted for by the evolution of a linguistic (i.e. syntactic) system that can reflect on and correct its own lower-order processes, and thus has adaptive value.

This account implies that it may be especially animals with a higher-order belief and thought system and with linguistic (i.e. syntactic, not necessarily verbal) symbol manipulation that have qualia. It may be that much non-human animal behaviour, provided that it does not require flexible linguistic planning and correction by reflection, could take place according to reinforcement-guidance (using, e.g., stimulus–reinforcer association learning in the amygdala and orbitofrontal cortex as described in Chapter 15, and rule-following [implemented, e.g., using habit or stimulus–response learning in the basal ganglia, see Chapter 20]). Such behaviours might appear very similar to human behaviour performed in similar circumstances, but need not imply qualia. It would be primarily by virtue of a system for reflecting on flexible, linguistic, planning behaviour that humans (and animals close to humans, with demonstrable syntactic manipulation of symbols, and the ability to think about these linguistic processes) would be different from other animals, and would have evolved qualia.

It is of interest to comment on how the evolution of a system for flexible planning might affect emotions. Consider grief which may occur when a reward is terminated and no immediate action is possible. It may be adaptive by leading to a cessation of the formerly rewarded behaviour and thus facilitating the possible identification of other positive reinforcers in the environment. In humans, grief may be particularly potent because it becomes represented in a system which can plan ahead, and understand the enduring implications of the loss. Thus *depression* in humans may be much more severe than in animals without a reasoning system, because the explicit, reasoning, system can see how bad the non-reward or punisher really is, can foresee the consequences for the future using reasoning, and using re-entrant processing between the explicit and implicit systems may produce positive feedback as a result of rumination (Rolls 2014a, Rolls 2016d) (Section 16.3). In this situation, thinking about or verbally discussing emotional states may help, because this can lead towards the identification of new or alternative reinforcers, and of the realization that for example negative consequences may not be as bad as feared.

22.2.6   Pathways

In order for processing in a part of our brain to be able to reach consciousness, appropriate pathways must be present. Certain constraints arise here. For example, in the sensory pathways, the nature of the representation may change as it passes through a hierarchy of processing levels, and in order to be conscious of the information in the form in which it is represented in early processing stages, the early processing stages must have access to the part of the brain necessary for consciousness. An example is provided by processing in the taste system. In the primate primary taste cortex, neurons respond to taste independently of hunger, yet in the secondary taste cortex, food-related taste neurons (e.g. responding to sweet taste) only respond to food if hunger is present, and gradually stop responding to that taste during feeding to satiety (see Chapter 15 and Section 2.2.3) (Rolls 1989d, Rolls 1997d, Rolls and Scott 2003, Rolls 2012e, Rolls 2014a). Now the quality of the tastant (sweet, salt, etc.) and its intensity are not affected by hunger, but the pleasantness of its taste is reduced to zero (neutral) (or even becomes unpleasant) after we have eaten it to satiety. The implication of this is that for quality and intensity information about taste, we must be conscious of what is represented in the primary taste cortex (or perhaps in another area connected to it that bypasses the secondary taste cortex), and not of what is represented in the secondary taste cortex (Fig. 6.19). In contrast, for the pleasantness of a taste, consciousness of this could not reflect what is represented in the primary taste cortex, but instead what is represented in the secondary taste cortex (or in an area beyond it) (Fig. 6.19) (see Chapter 15 and Section 2.2.3).

The same argument applies for reward in general, and therefore for emotion, which in primates is not represented early on in processing in the sensory pathways (nor in or before the inferior temporal cortex for vision), but in the areas to which these object analysis systems project, such as the orbitofrontal cortex, where the reward value of visual stimuli is reflected in the responses of neurons to visual stimuli (Rolls 2014a) (see Chapter 15).

It is also of interest that reward signals (e.g., the taste of food when we are hungry) are associated with subjective feelings of pleasure (Chapters 15 and 1, and Section 2.2.3). I suggest that this correspondence arises because pleasure is the subjective state that represents in the conscious system a signal that is positively reinforcing (rewarding), and that inconsistent behaviour would result if the representations did not correspond to a signal for positive reinforcement in both the conscious and the non-conscious processing systems.

Do these arguments mean that the conscious sensation of, e.g., taste quality (i.e. identity and intensity) is represented or occurs in the primary taste cortex, and of the pleasantness of taste in the secondary taste cortex, and that activity in these areas is sufficient for conscious sensations (qualia) to occur? I do not suggest this at all. Instead the arguments I have put forward above suggest that we are only conscious of representations when we have high-order thoughts about them. The implication then is that pathways must connect from each of the brain areas in which information is represented about which we can be conscious, to the system that has the higher-order thoughts, which as I have argued above, requires

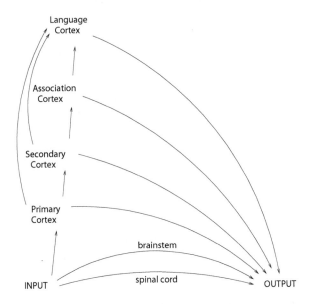

Fig. 22.2 Schematic illustration indicating that early cortical stages in information processing may need access to language areas that bypass subsequent levels in the hierarchy, so that consciousness of what is represented in early cortical stages, and which may not be represented in later cortical stages, can occur. Higher-order linguistic thoughts (HOLTs) could be implemented in the language cortex itself, and would not need a separate cortical area. Backprojections, a notable feature of cortical connectivity, with many probable functions including recall (Rolls and Treves 1998, Rolls 2008d, Treves and Rolls 1994), probably reciprocate all the connections shown.

language (understood as syntactic manipulation of symbols). Thus, in the example given, there must be connections to the language areas from the primary taste cortex, which need not be direct, but which must bypass the secondary taste cortex, in which the information is represented differently (Rolls 2014a, Rolls 1989d, Rolls 2012d). There must also be pathways from the secondary taste cortex, not necessarily direct, to the language areas so that we can have higher-order thoughts about the pleasantness of the representation in the secondary taste cortex. There would also need to be pathways from the hippocampus, implicated in the recall of declarative memories, back to the language areas of the cerebral cortex (at least via the cortical areas that receive backprojections from the hippocampus, see Fig. 24.1, which would in turn need connections to the language areas). A schematic diagram incorporating this anatomical prediction about human cortical neural connectivity in relation to consciousness is shown in Fig. 22.2.

22.2.7 Consciousness and causality

One question that has been discussed is whether there is a causal role for consciousness (e.g. Armstrong and Malcolm (1984)). The position to which the above arguments lead is that indeed conscious processing does have a causal role in the elicitation of behaviour, but only under the set of circumstances when higher-order thoughts play a role in correcting or influencing lower-order thoughts. The sense in which the consciousness is causal is then, it is suggested, that the higher-order thought is causally involved in correcting the lower-order thought; and that it is a property of the higher-order thought system that it feels like something when it is operating. As we have seen, some behavioural responses can be elicited when there is not this type of reflective control of lower-order processing, nor indeed any contribution of language. There are many brain-processing routes to output regions, and only one of these involves conscious, verbally represented processing that can later be recalled (see Fig. 22.1 and Section 22.3.1).

I suggest that these concepts may help us to understand what is happening in experiments of the type described by Libet and many others (Libet 2002) in which consciousness appears to follow with a measurable latency the time when a decision was taken. This is what I predict, if the decision is being made by an implicit perhaps reward/emotion or habit-related process, for then the conscious processor confabulates an account of or commentary on the decision, so that inevitably the conscious account follows the decision. On the other hand, I predict that if the rational (multistep, reasoning) route is involved in taking the decision, as it might be during planning, or a multistep task such as mental arithmetic, then the conscious report of when the decision was taken, and behavioural or other objective evidence on when the decision was taken, would correspond much more. Under those circumstances, the brain processing taking the decision would be closely related to consciousness, and it would not be a case of just confabulating or reporting on a decision taken by an implicit processor. It would be of interest to test this hypothesis in a version of Libet's task (Libet 2002) in which reasoning was required. The concept that the rational, conscious, processor is only in some tasks involved in taking decisions is extended further in the Section on dual routes to action (22.3.1).

22.2.8 Consciousness, a computational system for higher-order syntactic manipulation of symbols, and a commentary or reporting functionality

I now consider some clarifications of the present proposal, and how it deals with some issues that arise when considering theories of the phenomenal aspects of consciousness.

First, the present proposal has as its foundation the type of computation that is being performed, and suggests that it is a property of a higher-order syntactic thought (HOST) system used for correcting multistep plans with its representations grounded in the world that it would feel like something for a system to be doing this type of processing. To do this type of processing, the system would have to be able to recall previous multistep plans, and would require syntax to keep the symbols in each step of the plan separate. In a sense, the system would have to be able to recall and take into consideration its earlier multistep plans, and in this sense *report* to itself, on those earlier plans. Some approaches to consciousness take the ability to report on or make a *commentary* on events as being an important marker for consciousness (Weiskrantz 1997, Weiskrantz 1998), and the computational approach I propose suggests why there should be a close relation between consciousness and the ability to report or provide a commentary, for the ability to report is involved in using higher-order syntactic thoughts to correct a multistep plan.

Second, the implication of the present approach is that the type of linguistic processing or reporting need not be verbal, using natural language, for what is required to correct the plan is the ability to manipulate symbols syntactically, and this could be implemented in a much simpler type of mentalese or syntactic system (Fodor 1994, Jackendoff 2002, Rolls 2004b) than verbal language or natural language which implies a universal grammar.

Third, this approach to consciousness suggests that the information must be being processed in a system capable of implementing HOSTs for the information to be conscious, and in this sense is more specific than global workspace hypotheses (Baars 1988, Dehaene and Naccache 2001, Dehaene, Changeux, Naccache, Sackur and Sergent 2006, Dehaene, Charles, King and Marti 2014). Indeed, the present approach suggests that a workspace could be sufficiently global to enable even the complex processing involved in driving a car to be performed, and yet the processing might be performed unconsciously, unless HOST (supervisory, monitory, correcting) processing was involved (Rolls 2012d).

Fourth, the present approach suggests that it just is a property of HOST computational processing with the representations grounded in the world that it feels like something. There is to some extent an element of mystery about why it feels like something, why it is phenomenal, but the explanatory gap does not seem so large when one holds that the system is recalling, reporting on, reflecting on, and reorganizing information about itself in the world in order to prepare new or revised plans. In terms of the physicalist debate (see for a review Davies (2008)), an important aspect of my proposal is that it is a necessary property of this type of (HOST) computational processing that it feels like something (the philosophical description is that this is an absolute metaphysical necessity), and given this view, then it is up to one to decide whether this view is consistent with one's particular view of physicalism or not (Rolls 2008a). Similarly, the possibility of a zombie is inconsistent with the present hypothesis, which proposes that it is by virtue of performing processing in a specialized system that can perform higher-order syntactic processing with the representations grounded in the world that phenomenal consciousness is necessarily present.

An implication of these points is that my theory of consciousness is a computational theory. It argues that it is a property of a certain type of computational processing that it feels like something. In this sense, although the theory spans many levels from the neuronal to the computational, it is unlikely that any particular neuronal phenomena such as oscillations are necessary for consciousness, unless such computational processes happen to rely on some particular neuronal properties not involved in other neural computations but necessary for higher-order syntactic computations. It is these computations and the system that implements them that this computational theory argues are necessary for consciousness.

These are my initial thoughts on why we have consciousness, and are conscious of sensory, emotional and motivational qualia, as well as qualia associated with first-order linguistic thoughts. However, as stated above, one does not feel that there are straightforward criteria in this philosophical field of enquiry for knowing whether the suggested theory is correct; so it is likely that theories of consciousness will continue to undergo rapid development; and current theories should not be taken to have practical implications.

22.3   Selection between conscious vs unconscious decision-making systems

22.3.1   Dual major routes to action: implicit and explicit

According to the present formulation, there are two major types of route to action performed in relation to reward or punishment in humans. Examples of such actions include those associated with emotional and motivational behaviour.

The first ('implicit') route is via the brain systems that have been present in non-human primates such as monkeys, and to some extent in other mammals, for millions of years. These systems include the amygdala and, particularly well-developed in primates, the orbitofrontal cortex. These systems control behaviour in relation to previous associations of stimuli with reinforcement. The computation which controls the action thus involves assessment of the reinforcement-related value of a stimulus. This assessment may be based on a number of different factors:

One is the previous reinforcement history, which involves stimulus–reinforcer association learning using the amygdala, and its rapid updating especially in primates using the orbitofrontal cortex. This stimulus–reinforcer association learning may involve quite specific information about a stimulus, for example of the energy associated with each type of food, by the process of conditioned appetite and satiety (Booth 1985, Rolls 2014a, Rolls 2015e).

A second is the current motivational state, for example whether hunger is present, whether other needs are satisfied, etc.

A third factor that affects the computed reward value of the stimulus is whether that reward has been received recently. If it has been received recently but in small quantity, this may increase the reward value of the stimulus. This is known as incentive motivation or the 'salted nut' phenomenon. The adaptive value of such a process is that this positive feedback of reward value in the early stages of working for a particular reward tends to lock the organism on to behaviour being performed for that reward. This means that animals that are for example almost equally hungry and thirsty will show hysteresis in their choice of action, rather than continually switching from eating to drinking and back with each mouthful of water or food. This introduction of hysteresis into the reward evaluation system makes action selection a much more efficient process in a natural environment, for constantly switching between different types of behaviour would be very costly if all the different rewards were not available in the same place at the same time. (For example, walking half a mile between a site where water was available and a site where food was available after every mouthful would be very inefficient.) The amygdala is one structure that may be involved in this increase in the reward value of stimuli early on in a series of presentations, in that lesions of the amygdala (in rats) abolish the expression of this reward-incrementing process which is normally evident in the increasing rate of working for a food reward early on in a meal (Rolls and Rolls 1973, Rolls 2014a).

A fourth factor is the computed absolute value of the reward or punishment expected or being obtained from a stimulus, e.g. the sweetness of the stimulus (set by evolution so that sweet stimuli will tend to be rewarding, because they are generally associated with energy sources), or the pleasantness of touch (set by evolution to be pleasant according to the extent to which it brings animals of the opposite sex together, and depending on the investment in time that the partner is willing to put into making the touch pleasurable, a sign that indicates the commitment and value for the partner of the relationship, as in social grooming).

After the reward value of the stimulus has been assessed in these ways, behaviour is then initiated based on approach towards or withdrawal from the stimulus. A critical aspect of the behaviour produced by this type of 'implicit' system is that it is aimed directly towards obtaining a sensed or expected reward, by virtue of connections to brain systems such as the basal ganglia which are concerned with the initiation of actions (see Fig. 22.3). The expectation may of course involve behaviour to obtain stimuli associated with reward, which might even be present in a fixed chain or sequence.

Now part of the way in which the behaviour is controlled with this first ('implicit') route is according to the reward value of the outcome. At the same time, the animal may only work for the reward if the cost is not too high. Indeed, in the field of behavioural ecology animals are often thought of as performing optimally on some cost–benefit curve (see, e.g., Krebs and Kacelnik (1991)). This does not at all mean that the animal thinks about the long-term rewards, and performs a cost–benefit analysis using a lot of thoughts about the costs, other rewards (short and long term) available and their costs, etc. Instead, it should be taken to mean that in evolution the system has evolved in such a way that the way in which the reward varies with the different energy densities or amounts of food and the delay before it is received can be used as part of the input to a mechanism that has also been built to track the costs of obtaining the food (e.g. energy loss in obtaining it, risk of predation, etc.), and to then select given many such types of reward and the associated cost, the current behaviour that provides the most 'net reward'. Part of the value of having the computation expressed in this reward-minus-cost

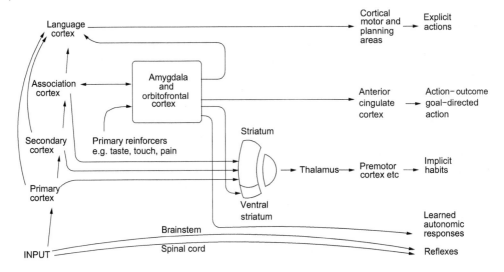

Fig. 22.3 Multiple routes to the initiation of actions and other behavioural responses in response to re-warding and punishing stimuli. The inputs from different sensory systems to brain structures such as the orbitofrontal cortex and amygdala allow these brain structures to evaluate the reward- or punish-ment-related value of incoming stimuli, or of remembered stimuli. One type of route is via the language systems of the brain, which allow explicit (verbalizable) decisions involving multistep syntactic planning to be implemented. The other type of route may be implicit, and includes the anterior cingulate cortex for action–outcome, goal-dependent, learning; and the striatum and rest of the basal ganglia for stimulus–re-sponse habits. The basal ganglia may be involved in selecting only one system for output (see Chapter 20). Outputs for autonomic responses can also be produced using outputs from the orbitofrontal cortex and anterior cingulate cortex (some of which are routed via the anterior insular cortex) and amygdala.

form is that there is then a suitable 'currency', or net reward value, to enable the animal to select the behaviour with currently the most net reward gain (or minimal aversive outcome).

Part of the evidence that this implicit route often controls emotional behaviour in humans is that humans with orbitofrontal cortex damage have impairments in selecting the correct action during visual discrimination reversal, yet can state explicitly what the correct action should be (Rolls, Hornak, Wade and McGrath 1994a, Rolls 1999b). The implication is that the intact orbitofrontal cortex is normally involved in making rapid emotion-related decisions, and that this emotion-related decision system is a separate system from the explicit system, which by serial reasoning can provide an alternative route to action. The explicit system may simply comment on the success or failure of actions that are initiated by the implicit system, and the explicit system may then be able to switch in to control mode to correct failures of the implicit system.

Consistent evidence that an implicit system can control human behaviour is that in psy-chophysical and neurophysiological studies, it has been found that face stimuli presented for 16 ms and followed immediately by a mask are not consciously perceived, yet pro-duce above chance identification (Rolls and Tovee 1994, Rolls, Tovee, Purcell, Stewart and Azzopardi 1994b, Rolls, Tovee and Panzeri 1999b, Rolls 2003, Rolls 2006a). In a similar back-ward masking paradigm, it was found that happy vs angry face expressions could influence how much beverage was wanted and consumed even when the faces were not consciously per-ceived (Winkielman and Berridge 2005, Winkielman and Berridge 2003). Thus unconscious emotion-related stimuli (in this case face expressions) can influence actions, and there is no need for processing to be conscious for actions to be initiated.

Further, in blindsight, humans with damage to the primary visual cortex may not be subjectively aware of stimuli, yet may be able to guess what the stimulus was, or to perform reaching movements towards it (Weiskrantz 1998, Weiskrantz 2009). Further, humans with striate cortex lesions may be influenced by emotional stimuli which are not perceived consciously (De Gelder, Vroomen, Pourtois and Weiskrantz 1999). Thus actions and emotions can be initiated without the necessity for the conscious route to be in control, and we should not infer that all actions require conscious processing.

The second ('explicit') route in (at least) humans involves a computation with many 'if ... then' statements, to implement a plan to obtain a reward. In this case, the reward may actually be deferred as part of the plan, which might involve working first to obtain one reward, and only then to work for a second more highly valued reward, if this was thought to be overall an optimal strategy in terms of resource usage (e.g. time). In this case, syntax is required, because the many symbols (e.g. names of people) that are part of the plan must be correctly linked or bound. Such linking might be of the form: 'if A does this, then B is likely to do this, and this will cause C to do this ...'. The requirement of syntax for this type of planning implies that an output to language systems in the brain is required for this type of planning (see Fig. 22.3). **Thus the explicit language system in humans may allow working for deferred rewards by enabling use of a one-off, individual, plan appropriate for each situation.** This explicit system may allow immediate rewards to be deferred, as part of a long-term plan. This ability to defer immediate rewards and plan syntactically in this way for the long term may be an important way in which the explicit system extends the capabilities of the implicit emotion systems that respond more directly to rewards and punishers, or to rewards and punishers with fixed expectancies such as can be learned by reinforcement learning (see Section B.16).

Consistent with the point being made about evolutionarily old emotion-based decision systems vs a recent rational system present in humans (and perhaps other animals with syntactic processing) is that humans trade off immediate costs/benefits against cost/benefits that are delayed by as much as decades, whereas non-human primates have not been observed to engage in unpreprogrammed delay of gratification involving more than a few minutes (Rachlin 1989, Kagel, Battalio and Green 1995, McClure, Laibson, Loewenstein and Cohen 2004) (though this is a potentially interesting area for further investigation).

Another building block for such planning operations in the brain may be the type of short-term memory in which the prefrontal cortex is involved. This short-term memory may be, for example in non-human primates, of where in space a response has just been made. A development of this type of short-term response memory system in humans to enable multiple short-term memories to be held in place correctly, preferably with the temporal order of the different items in the short-term memory coded correctly, may be another building block for the multiple step 'if .... then' type of computation in order to form a multiple step plan. Such short-term memories are implemented in the (dorsolateral and inferior convexity) prefrontal cortex of non-human primates and humans (Goldman-Rakic 1996, Petrides 1996, Rolls and Deco 2002, Deco and Rolls 2003, Rolls 2008d) (Section 4.3.1), and may be part of the reason why prefrontal cortex damage impairs planning and executive function (Shallice and Burgess 1996, Burgess 2000).

Of these two routes (see Fig. 22.3), it is the second (labelled 'explicit') route that I have suggested above is related to consciousness. The hypothesis is that consciousness is the state that arises by virtue of having the ability to think about one's own thoughts, which has the adaptive value of enabling one to correct long multistep syntactic plans. This latter system is thus the one in which explicit, declarative, processing occurs. Processing in this system is frequently associated with reason and rationality, in that many of the consequences of possible actions can be taken into account. The actual computation of how rewarding a particular stimulus or situation is, or will be, probably still depends on activity in the orbitofrontal cortex and amygdala, as the reward value of stimuli is computed and represented in these regions, and in that it is found that verbalized expressions of the reward (or punishment) value of stimuli are dampened by damage to these systems. (For example, damage to the orbitofrontal cortex renders painful input still identifiable as pain, but without the strong affective, 'unpleasant', reaction to it.)

This language system that enables long-term planning may be contrasted with the first system in which behaviour is directed at obtaining the stimulus (including the remembered stimulus) which is currently most rewarding, as computed by brain structures that include the orbitofrontal cortex and amygdala. There are outputs from this system, perhaps those directed at the basal ganglia and cingulate cortex, which do not pass through the language system, and behaviour produced in this way is described as implicit, and verbal declarations cannot be made directly about the reasons for the choice made. When verbal declarations are made about decisions made in this first ('implicit') system, those verbal declarations may be confabulations, reasonable explanations or fabrications, of reasons why the choice was made. These reasonable explanations would be generated to be consistent with the sense of continuity and self that is a characteristic of reasoning in the language system.

The question then arises of how decisions are made in animals such as humans that have both the implicit, direct reward-based, and the explicit, rational, planning systems (see Fig. 22.3). One particular situation in which the first, implicit, system may be especially important is when rapid reactions to stimuli with reward or punishment value must be made, for then the direct connections from structures such as the orbitofrontal cortex to the basal ganglia may allow rapid responses (such as selecting not to respond to a no-longer rewarded object) (Aron et al. 2014, Deng et al. 2017). Another is when there may be too many factors to be taken into account easily by the explicit, rational, planning, system, then the implicit system may be used to guide action (for example during human pair-bonding).

In contrast, when the implicit system continually makes errors, it would then be beneficial for the organism to switch from automatic, direct, action based on obtaining what the orbitofrontal cortex system decodes as being the most positively reinforcing choice currently available, to the explicit conscious control system which can evaluate with its long-term planning algorithms what action should be performed next. Indeed, it would be adaptive for the explicit system to be regularly assessing performance by the more automatic system, and to switch itself in to control behaviour quite frequently, as otherwise the adaptive value of having the explicit system would be less than optimal.

Another factor that may influence the balance between control by the implicit and explicit systems is the presence of pharmacological agents such as alcohol, which may alter the balance towards control by the implicit system, may allow the implicit system to influence more the explanations made by the explicit system, and may within the explicit system alter the relative value it places on caution and restraint vs commitment to a risky action or plan.

There may also be a flow of influence from the explicit, verbal system to the implicit system, in that the explicit system may decide on a plan of action or strategy, and exert an influence on the implicit system that will alter the reinforcement evaluations made by and the signals produced by the implicit system. An example of this might be that if a pregnant woman feels that she would like to escape a cruel mate, but is aware that she may not survive in the jungle, then it would be adaptive if the explicit system could suppress some aspects of her implicit behaviour towards her mate, so that she does not give signals that she is displeased with her situation[10]. Another example might be that the explicit system might, because of its long-term plans, influence the implicit system to increase its response to for example a positive reinforcer. One way in which the explicit system might influence the implicit system is by setting up the conditions in which, for example, when a given stimulus (e.g. person) is present, positive reinforcers are given, to facilitate stimulus–reinforcement association learning by the implicit system of the person receiving the positive reinforcers. Conversely, the implicit system may influence the explicit system, for example by highlighting certain stimuli in the environment that are currently associated with reward, to guide the attention of the explicit system to such stimuli.

However, it may be expected that there is often a conflict between these systems, in that the first, implicit, system is able to guide behaviour particularly to obtain the greatest immediate reinforcement, whereas the explicit system can potentially enable immediate rewards to be deferred, and longer-term, multistep, plans to be formed. This type of conflict will occur in animals with a syntactic planning ability, that is in humans and any other animals that have the ability to process a series of 'if ... then' stages of planning. This is a property of the human language system, and the extent to which it is a property of non-human primates is not yet fully clear. In any case, such conflict may be an important aspect of the operation of at least the human mind, because it is so essential for humans to decide correctly, at every moment, whether to invest in a relationship or a group that may offer long-term benefits, or whether to pursue immediate benefits directly (Nesse and Lloyd 1992)[11].

Some investigations on deception in non-human primates have been interpreted as showing that animals can plan to deceive others (see, e.g., Griffin (1992), Byrne and Whiten (1988), and Whiten and Byrne (1997)), that is to utilize 'Machiavellian intelligence'. For example, a baboon might 'deliberately' mislead another animal in order to obtain a resource such as food (e.g. by screaming to summon assistance in order to have a competing animal chased from a food patch) or sex (e.g. a female baboon who very gradually moved into a position from which the dominant male could not see her grooming a subadult baboon) (see Dawkins (1993)). The attraction of the Machiavellian argument is that the behaviour for which it accounts seems to imply that there is a concept of another animal's mind, and that one animal is trying occasionally to mislead another, which implies some planning. However, such observations tend by their nature to be field-based, and may have an anecdotal character, in that the previous experience of the animals in this type of behaviour, and the reinforcements obtained, are not known (Dawkins 1993). It is possible, for example, that some behavioural responses that appear to be Machiavellian may have been the result of previous instrumental learning in which reinforcement was obtained for particular types of response, or of observational learning, with again learning from the outcome observed. However, in any case, most examples of Machiavellian intelligence in non-human primates do not involve multiple stages of 'if ... then' planning requiring syntax to keep the symbols apart (but may involve associative learning which might lead to a description of the type 'if the dominant male sees me grooming a subadult male, I will be punished') (see Dawkins (1993)). Nevertheless,

---

[10] In the literature on self-deception, it has been suggested that unconscious desires may not be made explicit in consciousness (or actually repressed), so as not to compromise the explicit system in what it produces; see, e.g., Alexander (1975), Alexander (1979), Trivers (1976), Trivers (1985); and the review by Nesse and Lloyd (1992).

[11] As Nesse and Lloyd (1992) describe, some psychoanalysts ascribe to a somewhat similar position, for they hold that intrapsychic conflicts usually seem to have two sides, with impulses on one side and inhibitions on the other. Analysts describe the source of the impulses as the id, and the modules that inhibit the expression of impulses, because of external and internal constraints, the ego and superego respectively (Leak and Christopher 1982, Trivers 1985, Nesse and Lloyd 1992). The superego can be thought of as the conscience, while the ego is the locus of executive functions that balance satisfaction of impulses with anticipated internal and external costs. A difference of the present position is that it is based on identification of dual routes to action implemented by different systems in the brain, each with its own selective advantage.

the possible advantage of such Machiavellian planning could be one of the adaptive guiding factors in evolution that provided advantage to a multistep, syntactic system that enables long-term planning, the best example of such a system being human language.

Another, not necessarily exclusive, advantage of the evolution of a linguistic multi-step planning system could well be not Machiavellian planning, but planning for social co-operation and advantage. Perhaps in general an 'if ... then' multistep syntactic planning ability is useful primarily in evolution in social situations of the type: 'if X does this, then Y does that; then I would/should do that, and the outcome would be ... '. It is not yet at all clear whether such planning is required in order to explain the social behaviour of social animals such as hunting dogs, or socializing monkeys (Dawkins 1993).

However, in humans, members of 'primitive' hunting tribes spend hours recounting tales of recent events (perhaps who did what, when; who then did what, etc.), perhaps to help learn from experience about good strategies, necessary for example when physically weak men take on large animals (see Pinker and Bloom (1992)).

Thus, social co-operation may be as powerful a driving force in the evolution of syntactical planning systems as Machiavellian intelligence. What is common to both is that they involve social situations. However, such a syntactic planning system would have advantages not only in social systems, for such planning may be useful in obtaining resources purely in a physical (non-social) world. An example might be planning how to cross terrain given current environmental constraints in order to reach a particular place[12].

The thrust of this argument thus is that much complex animal, including human, behaviour can take place using the implicit, non-conscious, route to action. We should be very careful not to postulate intentional states (i.e. states with intentions, beliefs, and desires) unless the evidence for them is strong, and it seems to me that a flexible, one-off, linguistic processing system that can handle propositions is needed for intentional states. What the explicit, linguistic, system does allow is exactly this flexible, one-off, multistep planning-ahead type of computation, which allows us to defer immediate rewards based on such a plan.

Emotions as actions, and emotions as affects, are sometimes contrasted. My view on this is that sometimes emotions can lead to actions implicitly, without the need for conscious processing. However, when emotions involve longer term planning, then representation and processing in the explicit system is required, and affective feelings will then be inextricably linked to the processing.

This discussion of dual routes to action has been with respect to the behaviour produced. There is of course in addition a third output of brain regions such as the orbitofrontal cortex and amygdala involved in emotion, which is directed to producing autonomic and endocrine responses (Fig. 22.3). Although it has been argued by Rolls (2014a) that the autonomic system is not normally in a circuit through which behavioural responses are produced (i.e. against the James–Lange and related theories), there may be some influence from effects produced through the endocrine system (and possibly the autonomic system, through which some endocrine responses are controlled) on behaviour, or on the dual systems just discussed that control behaviour. For example, during female orgasm the hormone oxytocin may be

---

[12] Tests of whether such multistep planning might be possible in even non-human primates are quite difficult to devise. One example might be to design a multistep maze. On a first part of the trial, the animal might be allowed to choose for itself, given constraints set on that trial to ensure trial unique performance, a set of choices through a maze. On the second part of that trial, the animal would be required to run through the maze again, remembering and repeating every choice just made in the first part of that trial. This part of the design is intended to allow recall of a multistep plan. To test on probe occasions whether the plan is being recalled, and whether the plan can be corrected by a higher-order thought process, the animal might be shown after the first part of a trial, that one of its previous free choices was not now available. The test would be to determine whether the animal can make a set of choices that indicate corrections to the multistep plan, in which the trajectory has to be altered before the now unavailable choice point is reached.

released, and this may influence the implicit system to help develop positive reinforcement associations and thus attachment to her lover.

22.3.2 The Selfish Gene vs The Selfish Phenotype

I have provided evidence in Section 22.3.1 that there are two main routes to decision-making and action. The first route selects actions by gene-defined goals for action, and is closely associated with emotion. The second route involves multistep planning and reasoning which requires syntactic processing to keep the symbols involved at each step separate from the symbols in different steps. (This second route is used by humans and perhaps by closely related animals.) Now the 'interests' of the first and second routes to decision-making and action are different. As argued very convincingly by Richard Dawkins in *The Selfish Gene* (Dawkins 1976, Dawkins 1989), and by others (Hamilton 1964, Ridley 1993, Hamilton 1996), many behaviours occur in the interests of the survival of the genes, not of the individual (nor of the group), and much behaviour can be understood in this way.

I have extended this approach by arguing that an important role for some genes in evolution is to define the goals for actions that will lead to better survival of those genes; that emotions are the states associated with these gene-defined goals; and that the defining of goals for actions rather that actions themselves is an efficient way for genes to operate, as it leaves flexibility of choice of action open until the animal is alive (Rolls 2005, Rolls 2012d, Rolls 2014a). This provides great simplification of the genotype as action details do not need to be specified, just rewarding and punishing stimuli, and also flexibility of action in the face of changing environments faced by the genes. Thus the interests that are implied when the first route to action is chosen are those of the 'selfish genes', not those of the individual.

However, the second route to action allows, by reasoning, decisions to be taken that might not be in the interests of the genes, might be longer term decisions, and might be in the interests of the individual. An example might be a choice not to have children, but instead to devote oneself to science, medicine, music, or literature. The reasoning, rational, system presumably evolved because taking longer-term decisions involving planning rather than choosing a gene-defined goal might be advantageous at least sometimes for genes. But an unforeseen consequence of the evolution of the rational system might be that the decisions would, sometimes, not be to the advantage of any genes in the organism. After all, evolution by natural selection operates utilizing genetic variation like a Blind Watchmaker (Dawkins 1986). In this sense, the interests when the second route to decision-making is used are at least sometimes those of the 'selfish phenotype'. (Indeed, we might euphonically say that the interests are those of the 'selfish phene' (where the etymology is Gk $\phi\alpha\iota\nu\omega$ (phaino), 'appear', referring to appearance, hence the thing that one observes, the individual). Hence the decision-making referred to in Section 22.3.1 is between a first system where the goals are gene-defined, and a second rational system in which the decisions may be made in the interests of the genes, or in the interests of the phenotype and not in the interests of the genes. Thus we may speak of the choice as sometimes being between the 'Selfish Genes' and the 'Selfish Phenes' (Rolls 2011b, Rolls 2012d, Rolls 2014a).

Now what keeps the decision-making between the 'Selfish Genes' and the 'Selfish Phenes' more or less under control and in balance? If the second, rational, system chose too often for the interests of the 'Selfish Phene', the genes in that phenotype would not survive over generations. Having these two systems in the same individual will only be stable if their potency is approximately equal, so that sometimes decisions are made with the first route, and sometimes with the second route. If the two types of decision-making, then, compete with approximately equal potency, and sometimes one is chosen, and sometimes the other, then this is exactly the scenario in which stochastic processes in the decision-making mechanism are likely to play an important role in the decision that is taken. The same decision, even with the same evidence, may not be taken each time a decision is made, because of noise in the system.

The system itself may have some properties that help to keep the system operating well. One is that if the second, rational, system tends to dominate the decision-making too much, the first, gene-based emotional system might fight back over generations of selection, and enhance the magnitude of the reward value specified by the genes, so that emotions might actually become stronger as a consequence of them having to compete in the interests of the selfish genes with the rational decision-making process.

Another property of the system may be that sometimes the rational system cannot gain all the evidence that would be needed to make a rational choice. Under these circumstances the rational system might fail to make a clear decision, and under these circumstances, basing a decision on the gene-specified emotions is an alternative. Indeed, Damasio (1994) argued that under circumstances such as this, emotions might take an important role in decision-making. In this respect, I agree with him, basing my reasons on the arguments above. He called the emotional feelings gut feelings, and, in contrast to me, hypothesized that actual feedback from the gut was involved. His argument seemed to be that if the decision was too complicated for the rational system, then send outputs to the viscera, and whatever is sensed by what they send back could be used in the decision-making, and would account for the conscious feelings of the emotional states. My reading of the evidence is that the feedback from the periphery is not necessary for the emotional decision-making, or for the feelings, nor would it be computationally efficient to put the viscera in the loop given that the information starts from the brain (Rolls 2014a).

Another property of the system is that the interests of the second, rational, system, although involving a different form of computation, should not be too far from those of the gene-defined emotional system, for the arrangement to be stable in evolution by natural selection. One way that this could be facilitated would be if the gene-based goals felt pleasant or unpleasant in the rational system, and in this way contributed to the operation of the second, rational, system. This is something that I propose is the case.

22.3.3 Decision-making between the implicit and explicit systems

Decision-making as implemented in neural networks in the brain is now becoming understood, and is described in Section 5.6. As shown there, two attractor states, each one corresponding to a decision, compete in an attractor single network with the evidence for each of the decisions acting as biases to each of the attractor states. The non-linear dynamics, and the way in which noise due to the random spiking of neurons makes the decision-making probabilistic, makes this a biologically plausible model of decision-making consistent with much neurophysiological and fMRI data (Wang 2002, Rolls and Deco 2010, Deco, Rolls, Albantakis and Romo 2013).

I propose (Rolls 2005, Rolls 2008d, Rolls and Deco 2010, Rolls 2012d, Rolls 2014a) that this model applies to taking decisions between the implicit (unconscious) and explicit (conscious) systems in emotional decision-making, where the two different systems could provide the biasing inputs λ_1 and λ_2 to the model. An implication is that noise will influence with probabilistic outcomes which system takes a decision.

When decisions are taken, sometimes confabulation may occur, in that a verbal account of why the action was performed may be given, and this may not be related at all to the environmental event that actually triggered the action (Gazzaniga and LeDoux 1978, Gazzaniga 1988, Gazzaniga 1995, Rolls 2005, LeDoux 2008, Rolls 2012d). It is accordingly possible that sometimes in normal humans when actions are initiated as a result of processing in a specialized brain region such as those involved in some types of rewarded behaviour, the language system may subsequently elaborate a coherent account of why that action was performed (i.e. confabulate). This would be consistent with a general view of brain evolution in which, as areas of the cortex evolve, they are laid on top of existing circuitry connecting inputs to outputs, and in which each level in this hierarchy of separate input–output pathways may control behaviour according to the specialized function it can perform. This hierarchical overlaying is an important concept advanced in this book as being important for understanding emotion, the different brain systems involved in different aspects of emotion and decision-making, and the relation between the implicit and explicit systems. When a new layer is added, previous layers may lose some of their importance, as appears to occur in the taste system in which in primates the subcortical processing from the brainstem nucleus of the solitary tract is lost; when the granular orbitofrontal cortex of primates becomes relatively more important than the amygdala; and when language areas are added on top of existing circuitry (Fig. 22.3).

22.4   Determinism

These thoughts raise the issue of free will in decision-making (considered in Section 5.10), and of determinism (Rolls 2012f), which are considered further here.

There are a number of senses in which our behaviour might be deterministic. One sense might be genetic determinism, and we have already seen that there are far too few genes to determine the structure and function of our brains, and thus to determine our behaviour (Rolls and Stringer 2000, Rolls 2012a). Moreover, development, and the environment with the opportunities it provides for brain self-organization and learning, play a large part in brain structure and function, and thus in our behaviour.

Another sense might be that if there were random factors that influence the operation of the brain, then our behaviour might be thought not to be completely predictable and deterministic. It is this that I consider here, a topic developed in *The Noisy Brain: Stochastic Dynamics as a Principle of Brain Function* (Rolls and Deco 2010), in which we show that there is noise or randomness in the brain, and argue that this can be advantageous[13]. The mechanisms involved are described in Section 5.6.

Given that the brain operates with some degree of randomness due to the statistical fluctuations produced by the random spiking times of neurons, brain function is to some extent non-deterministic, as defined in terms of these statistical fluctuations. That is, the behaviour of the system, and of the individual, can vary from trial to trial based on these statistical fluctuations, in ways that are described in Section 5.6 and by Rolls and Deco (2010). Indeed, given that each neuron has this randomness, and that there are sufficiently small numbers of synapses on the neurons in each network (between a few thousand and 20,000) that these statistical fluctuations are not smoothed out, and that there are a number of different networks involved in typical thoughts and actions each one of which may behave probabilistically, and with 10^{11} neurons in the brain each with this number of synapses, the system has so many degrees of freedom that it operates effectively as a non-deterministic system. (Philosophers may wish to argue about different senses of the term deterministic, but it is being used here in a precise, scientific, and quantitative way, which has been clearly defined.)

---

[13]This randomness is not a property of chaotic systems, which although complex, are not random in the sense that the same trajectory is followed from a given starting position (Peitgen, Jürgens and Saupe 2004).

22.5 Free will

Do we have free will when we make a choice? Given the distinction made between the implicit system that seeks for gene-specified rewards, and the explicit system that can use reasoning to defer an immediate goal and plan many steps ahead for longer-term goals, do we have free will when both the implicit and the explicit systems have made the choice?

Free will would in Rolls' view (Rolls 2005, Rolls 2008a, Rolls 2008d, Rolls 2010c, Rolls 2011b, Rolls 2012f, Rolls 2012d, Rolls 2014a) involve the use of language to check many moves ahead on a number of possible series of actions and their outcomes, and then with this information to make a choice from the likely outcomes of different possible series of actions. (If, in contrast, choices were made only on the basis of the reinforcement value of immediately available stimuli, without the arbitrary syntactic symbol manipulation made possible by language, then the choice strategy would be much more limited, and we might not want to use the term free will, as all the consequences of those actions would not have been computed.) It is suggested that when this type of reflective, conscious, information processing is occurring and leading to action, the system performing this processing and producing the action (the rational or reasoning system) would have to believe that it could cause the action, for otherwise inconsistencies would arise, and the system might no longer try to initiate action. This belief held by the rational system may partly underlie the feeling of free will. At other times, when other brain modules are initiating actions (in the implicit systems), the conscious processor (the rational system in which processing can be explicit) may confabulate and believe that it caused the action, or at least give an account (possibly wrong) of why the action was initiated. The fact that the conscious processor may have the belief even in these circumstances that it initiated the action may arise as a property of it being inconsistent for a system that can take overall control using rational verbal (conscious) processing to believe that it was overridden by another system. This may be the underlying computational reason why confabulation occurs (Section 22.3.1).

The interesting view we are led to is thus that when probabilistic choices influenced by stochastic dynamics (Rolls and Deco 2010) are made between the implicit and explicit systems, we may not be aware of which system made the choice. Further, when the stochastic noise has made us choose with the implicit system, we may confabulate and say that we made the choice of our own free will, and provide a guess at why the decision was taken. In this scenario, the stochastic dynamics of the brain plays a role even in how we understand free will (Rolls 2010c, Rolls 2012d, Rolls 2012f, Rolls 2014a).

The implication of this argument is that a good use of the term free will is when the term refers to the operation of the rational, planning, explicit (conscious) system that can think many moves ahead, and choose from a number of such computations the multistep strategy that best optimizes the goals of the explicit system with long-term goals. When on the other hand our implicit system has taken a decision, and we confabulate a spurious account with our explicit system, and pronounce that we took the decision for such and such a (confabulated) reason of our own "free will", then my view is that the feeling of free will was an illusion (Rolls 2005, Rolls 2010c, Rolls 2012d, Rolls 2012f).

Before leaving these thoughts, it may be worth commenting on the feeling of continuing self-identity that is characteristic of humans. Why might this arise? One suggestion is that if one is an organism that can think about its own long-term multistep plans, then for those plans to be consistently and thus adaptively executed, the goals of the plans would need to remain stable, as would memories of how far one had proceeded along the execution path of each plan. If one felt each time one came to execute, perhaps on another day, the next step of a plan, that the goals were different, or if one did not remember which steps had already been taken in a multistep plan, the plan would never be usefully executed. So, given that it does feel like something to be doing this type of planning using higher-order thoughts, it would have to feel as if one were the same agent, acting towards the same goals, from day to day, for which autobiographical memory would be important.

Thus it is suggested that the feeling of continuing self-identity falls out of a situation in which there is an actor with consistent long-term goals, and long-term recall. If it feels like anything to be the actor, according to the suggestions of the higher-order thought theory, then it should feel like the same thing from occasion to occasion to be the actor, and no special further construct is needed to account for self-identity. Humans without such a feeling of being the same person from day to day might be expected to have, for example, inconsistent goals from day to day, or a poor recall memory. It may be noted that the ability to recall previous steps in a plan, and bring them into the conscious, higher-order thought system, is an important prerequisite for long-term planning which involves checking each step in a multistep process.

Conscious feelings of self will be likely to be of value to the individual. Indeed, it would be maladaptive if feelings of self-identity, and continuation of the self, were not wanted by the individual, for that would lead to the brain's capacity for feelings about self-identity to leave the gene pool, due for example to suicide. This wish for feelings and thoughts about the self to continue may lead to the wish and hope that this will occur after death, and this may be important as a foundation for religions (Rolls 2012d).

22.6 Content and meaning in representations: How are representations grounded in the world?

In Section 22.2 I suggested that representations need to be grounded in the world for a system with higher-order thoughts to be conscious. I therefore now develop somewhat what I understand by representations being grounded in the world.

It is possible to analyse how the firing of populations of neurons encodes information about stimuli in the world (Rolls and Treves 1998, Rolls 2008d, Rolls and Treves 2011) (Chapter 8 and Appendix C). For example, from the firing rates of small numbers of neurons in the primate inferior temporal visual cortex, it is possible to know which of 20 faces has been shown to the monkey (Abbott, Rolls and Tovee 1996, Rolls, Treves and Tovee 1997b, Rolls and Treves 2011). Similarly, a population of neurons in the anterior part of the macaque temporal lobe visual cortex has been discovered that has a view-invariant representation of objects (Booth and Rolls 1998, Rolls 2012c). From the firing of a small ensemble of neurons in the olfactory part of the orbitofrontal cortex, it is possible to know which of eight odours was presented (Rolls, Critchley and Treves 1996a, Rolls, Critchley, Verhagen and Kadohisa 2010a). From the firing of small ensembles of neurons in the hippocampus, it is possible to know where in allocentric space a monkey is looking (Rolls, Treves, Robertson, Georges-François and Panzeri 1998b). In each of these cases, the number of stimuli that is encoded increases exponentially with the number of neurons in the ensemble, so this is a very powerful representation (Abbott, Rolls and Tovee 1996, Rolls, Treves and Tovee 1997b, Rolls and Treves 1998, Rolls and Deco 2002, Rolls, Aggelopoulos, Franco and Treves 2004, Franco, Rolls, Aggelopoulos and Treves 2004, Aggelopoulos, Franco and Rolls 2005, Rolls 2008d, Rolls and Treves 2011, Rolls 2017a). What is being measured in each example is the mutual information between the firing of an ensemble of neurons and which stimuli are present in the world. In this sense, one can read off the code that is being used at the end of each of these sensory systems (Chapter 8 and Appendix C).

However, what sense does the representation make to the animal? What does the firing of each ensemble of neurons 'mean'? What is the content of the representation? In the visual system, for example, it is suggested that the representation is built by a series of appropriately connected competitive networks, operating with a modified Hebb-learning rule (Rolls 1992a, Rolls 1994a, Wallis and Rolls 1997, Rolls 2000a, Rolls and Milward 2000, Stringer and Rolls 2000, Rolls and Stringer 2001a, Rolls and Deco 2002, Elliffe, Rolls and Stringer 2002, Stringer and Rolls 2002, Deco and Rolls 2004, Rolls 2008d, Rolls 2012c) (Chapter 25). Now competitive networks categorize their inputs without the use of a teacher (Kohonen 1989, Hertz et al. 1991, Rolls 2008d, Rolls 2016e) (Section B.4). So which particular neurons fire as a result of the self-organization to represent a particular object or stimulus is arbitrary. What meaning, therefore, does the particular ensemble that fires to an object have? How is the representation grounded in the real world? The fact that there is mutual information between the firing of the ensemble of cells in the brain and a stimulus or event in the world (Rolls and Treves 1998, Rolls 2008d, Rolls and Treves 2011) (Appendix C) does not fully answer this question.

One answer to this question is that there may be meaning in the case of objects and faces that it is an object or face, and not just a particular view. This is the case in that the representation may be activated by any view of the object or face. This is a step, suggested to be made possible by a short-term memory in the learning rule that enables different views of objects to be associated together (Wallis and Rolls 1997, Rolls and Treves 1998, Rolls and Milward 2000, Rolls 2008d, Rolls 2012c). But it still does not provide the representation with any meaning in terms of the real world. What actions might one make, or what emotions might one feel, if that arbitrary set of temporal cortex visual cells was activated?

This leads to one of the answers I propose. I suggest that one type of meaning of representations in the brain is provided by their reward (or punishment) value: activation of these representations is the goal for actions. In the case of primary reinforcers such as the taste of food or pain, the activation of these representations would have meaning in the sense that the animal would work to obtain the activation of the taste of food neurons when hungry, and to escape from stimuli that cause the neurons representing pain to be activated. Evolution has built the brain so that genes specify these primary reinforcing stimuli, and so that their representations in the brain should be the targets for actions (Rolls 2014a) (see Chapter 15). In the case of other ensembles of neurons in, for example, the visual cortex that respond to objects with the colour and shape of a banana, and which 'represent' the sight of a banana in that their activation is always and uniquely produced by the sight of a banana, such representations come to have meaning only by association with a primary reinforcer, involving the process of stimulus–reinforcer association learning.

The second sense in which a representation may be said to have meaning is by virtue of sensory–motor correspondences in the world. For example, the touch of a solid object such as a table might become associated with evidence from the motor system that attempts to walk through the table result in cessation of movement. The representation of the table in the inferior temporal visual cortex might have 'meaning' only in the sense that there is mutual information between the representation and the sight of the table until the table is seen just before and while it is touched, when sensory–sensory association between inputs from different sensory modalities will be set up that will enable the visual representation to become associated with its correspondences in the touch and movement worlds. In this second sense, meaning will be conferred on the visual sensory representation because of its associations in the sensory–motor world. Related views have been developed by the philosopher Ruth Millikan (1984). Thus it is suggested that there are two ways by which sensory representations can be said to be grounded, that is to have meaning, in the real world.

It is suggested that the symbols used in language become grounded in the real world by the same two processes.

In the first, a symbol such as the word 'banana' has meaning because it is associated with primary reinforcers such as the flavour of the banana, and with secondary reinforcers such as the sight of the banana. These reinforcers have 'meaning' to the animal in that evolution has built animals as machines designed to do everything that they can to obtain these reinforcers, so that they can eventually reproduce successfully and pass their genes onto the next generation[14]. In this sense, obtaining reinforcers may have life-threatening 'meaning' for animals, though of course the use of the word 'meaning' here does not imply any subjective state, just that the animal is built as a survival for reproduction machine[15].

In the second process, the word 'table' may have meaning because it is associated with sensory stimuli produced by tables such as their touch, shape, and sight, as well as other functional properties, such as, for example, being load-bearing, and obstructing movement if they are in the way (see Section 22.2).

This section (22.6) thus adds to Section 22.2 on a higher-order syntactic thought (HOST) theory of consciousness, by addressing the sense in which the thoughts may need to be grounded in the world. The HOST theory holds that the thoughts 'mean' something to the individual, in the sense that they may be about the survival of the individual (the phenotype, Section 22.3.2) in the world, which the rational, thought, system aims to maximize (Rolls 2012d).

22.7 The causal role of consciousness: a theory of the relation between the mind and the brain

Does consciousness cause our behaviour? Before discussing the causal role of consciousness, I summarize my view on the **relation between the mind and the brain**.

What is the causal relation between mental events and neurophysiological events, part of the mind–body problem? My view is that the relationship between mental events and neurophysiological events is similar (apart from the problem of consciousness) to the relationship between the program running in a computer (the software) and the hardware on the computer. We can consider that the software and hardware in terms of causality are at different levels of explanation. The program moving from one step to the next is one level of explanation, and the hardware implementation by the transistor-transistor logic (TTL) is another. A crucial point is that the processes described at the different levels take place at the same point in time, so that we should not think of one as causing the other. Completion of one step of the program may cause the next step to be performed, and causality operates here, with a time delay. Similarly, the hardware operations involved in performing one step of the program can be thought of as causing the next step of hardware operations to be performed, so again causality operates here, with a time delay. But on this analysis, for a given step of the program, causality would not operate between the software and the hardware, and these would just be different levels of explanation. The implication of this account for understanding the relation between the mind and the brain is that the mind and the operations performed by the brain are different levels of explanation, and that the mind should not be thought of as causing

---

[14]The fact that some stimuli are reinforcers but may not be adaptive as goals for action is no objection. Genes are limited in number, and can not allow for every eventuality, such as the availability to humans of (non-nutritive) saccharin as a sweetener. The genes can just build reinforcement systems the activation of which is generally likely to increase the fitness of the genes specifying the reinforcer (or may have increased their fitness in the recent past).

[15]This is a novel, Darwinian, approach to the issue of symbol grounding.

brain changes, and vice versa, for they would be happening at the same point in time. This analysis is consistent with the requirement for causality *'post hoc, ergo propter hoc'* ('after this, therefore because of this', a necessary but not sufficient condition for causality). I now favour this account of the relation between the mind and the brain (having considered other possibilities before (Rolls 2012d, Rolls 2013c)), because it is consistent with the condition of *post hoc, ergo propter hoc.*

Some philosophers taking an intentional stance think of one mental event as causing another mental event, and that this is a convenient level of explanation for at least providing an account to the behaviour of others (Dennett 1987). That is consistent with what I have summarized in the preceding paragraph, provided that we realize that it is neuronal operations that implement syntactic thought and hence reasoning. That is, the neuronal networks in the brain have evolved to the stage where they can implement a language of thought. With my approach, there is nothing 'extra' that remains unexplained at the mental level, and that somehow operates in a different way to that just described. To elaborate further, the style of computation implemented by the brain is based on the (dot or inner product) similarity between input firing rate vectors and synaptic weight vectors, and is very poor at logical operations (see Rolls (2012d), and Section 26.3, for my thoughts on the differences between computations in the brain and in digital computers). Nevertheless, as a result of training, our neuronal networks have evolved to have the capability to learn to follow syntactic rules that enable us to perform logical operations such as AND, XOR, and odd or even parity determination, in a serial manner characteristic of language. As a result of this, we can think of humans as intentional systems, with beliefs etc. represented in a language of thought, but all implemented by neuronal networks that implement syntax (Rolls 2012d).

The view that I currently hold about the causal role of consciousness is that the information processing that is related to consciousness (activity in a linguistic system capable of higher-order thoughts, and used for planning and correcting the operation of lower-order linguistic systems) can play a causal role in producing our behaviour (see Fig. 22.3). It is, I postulate, a property of processing in this system (capable of higher-order thoughts) that it feels like something to be performing that type of processing. It is in this sense that I suggest that consciousness can act causally to influence our behaviour – consciousness is the property that occurs when a linguistic system is thinking about its lower-order thoughts, which may be causally useful in correcting plans.

The hypothesis that it does feel like something when this processing is taking place is at least to some extent testable: humans performing this type of higher-order linguistic process-ing, for example recalling episodic memories and comparing them with current circumstances, who denied being conscious, would prima facie constitute evidence against the theory. Most humans would find it very implausible though to posit that they could be thinking about their own thoughts, and reflecting on their own thoughts, without being conscious. This type of processing does appear, for most humans, to be necessarily conscious.

In this context, I provide a short specification of what might have to be implemented in a neural network to implement conscious processing. First, a linguistic system, not necessarily verbal, but implementing syntax between symbols implemented in the environment would be needed. This system would be necessary for a multi-step one-off planning system (i.e. a system used once, but requiring multiple steps of reasoning). Then a higher-order thought system also implementing syntax and able to think about the representations in the first-order linguistic system, and able to correct the reasoning in the first-order linguistic system in a flexible manner, would be needed. The system would also need to have its representations grounded in the world, as discussed in Section 22.6. So my view is that consciousness can be implemented in neural networks (and that this is a topic worth discussing), but that the neural networks would have to implement the type of higher-order linguistic processing described in this chapter, and also would need to be grounded in the world.

22.8 Comparison with other theories of consciousness

22.8.1 Higher-order thought theories

Some ways in which the current theory may be different from other higher-order thought theories (Rosenthal 2004, Rosenthal 2005, Rosenthal 2012, Gennaro 2004, Carruthers 2000) is that it provides an account of the evolutionary, adaptive, value of a higher-order thought system in helping to solve a credit assignment problem that arises in a multistep syntactic plan, links this type of processing to consciousness, and therefore emphasizes a role for syntactic processing in consciousness. The type of syntactic processing need not be at the natural language level (which implies a universal grammar), but could be at the level of mentalese or simpler, as it involves primarily the syntactic manipulation of symbols (Fodor 1994, Rolls 2005).

The current theory holds that it is higher-order linguistic thoughts (HOLTs) (or higher-order syntactic thoughts, HOSTs (Rolls 2004b, Rolls 2007a, Rolls 2011b, Rolls 2012d)) that are closely associated with consciousness, and this might differ from Rosenthal's higher-order thoughts (HOTs) theory (Rosenthal 1986, Rosenthal 1990, Rosenthal 1993, Rosenthal 2004, Rosenthal 2005, Rosenthal 2012) in the emphasis in the current theory on language. Language in the current theory is defined by syntactic manipulation of symbols, and does not necessarily imply verbal (or natural) language.

The reason that strong emphasis is placed on language is that it is as a result of having a multistep, flexible, 'one-off', reasoning procedure that errors can be corrected by using 'thoughts about thoughts'. This enables correction of errors that cannot be easily corrected by reward or punishment received at the end of the reasoning, due to the credit assignment problem. That is, there is a need for some type of supervisory and monitoring process, to detect where errors in the reasoning have occurred. It is having such a HOST brain system, and it becoming engaged (even if only a little), that according to the HOST theory is associated with phenomenal consciousness.

This suggestion on the adaptive value in evolution of such a higher-order linguistic thought process for multistep planning ahead, and correcting such plans, may also be different from earlier work. Put another way, this point is that **credit assignment** when reward or punishment is received is straightforward in a one-layer network (in which the reinforcement can be used directly to correct nodes in error, or responses), but is very difficult in a multistep linguistic process executed once. Very complex mappings in a multilayer network can be learned if hundreds of learning trials are provided. But once these complex mappings are learned, their success or failure in a new situation on a given trial cannot be evaluated and corrected by the network. Indeed, the complex mappings achieved by such networks (e.g. networks trained by backpropagation of errors or by reinforcement learning) mean that after training they operate according to fixed rules, and are often quite impenetrable and inflexible (Rumelhart, Hinton and Williams 1986a, Rolls 2008d). In contrast, to correct a multistep, single occasion, linguistically based plan or procedure, recall of the steps just made in the reasoning or planning, and perhaps related episodic material, needs to occur, so that the link in the chain that is most likely to be in error can be identified. This may be part of the reason why there is a close relationship between declarative memory systems, which can explicitly recall memories, and consciousness.

Some computer programs may have supervisory processes. Should these count as higher-order linguistic thought processes? My current response to this is that they should not, to the extent that they operate with fixed rules to correct the operation of a system that does not itself involve linguistic thoughts about symbols grounded semantically in the external world. If on the other hand it were possible to implement on a computer such a high-order linguistic thought–supervisory correction process to correct first-order one-off linguistic thoughts with symbols grounded in the real world (as described at the end of Section 22.6), then prima facie this process would be conscious. If it were possible in a thought experiment to reproduce the neural connectivity and operation of a human brain on a computer, then prima facie it would also have the attributes of consciousness[16]. It might continue to have those attributes for as long as power was applied to the system.

Another possible difference from earlier theories is that raw sensory feels are suggested to arise as a consequence of having a system that can think about its own thoughts. Raw sensory feels, and subjective states associated with emotional and motivational states, may not necessarily arise first in evolution.

A property often attributed to consciousness is that it is *unitary*. The current theory would account for this by the limited syntactic capability of neuronal networks in the brain, which render it difficult to implement more than a few syntactic bindings of symbols simultaneously (Rolls and Treves 1998, McLeod, Plunkett and Rolls 1998, Rolls 2008d). This limitation makes it difficult to run several 'streams of consciousness' simultaneously. In addition, given that a linguistic system can control behavioural output, several parallel streams might produce maladaptive behaviour (apparent as, e.g. indecision), and might be selected against. The close relationship between, and the limited capacity of, both the stream of consciousness, and auditory–verbal short-term working memory, may be that both implement the capacity for syntax in neural networks.

The hypothesis that syntactic binding is necessary for consciousness is one of the postulates of the theory I am describing (for the system I describe must be capable of correcting its own syntactic thoughts). The fact that the binding must be implemented in neuronal networks may well place limitations on consciousness that lead to some of its properties, such as its unitary nature.

The theory (Rolls 1997b, Rolls 2004b, Rolls 2006a, Rolls 2007a, Rolls 2007b, Rolls 2008a, Rolls 2010c, Rolls 2011b, Rolls 2012d) holds that consciousness arises by virtue of a system that can think linguistically about its own linguistic thoughts. The advantages for a system of being able to do this have been described, and this has been suggested as the reason why consciousness evolved. The evidence that consciousness arises by virtue of having a system that can perform higher-order linguistic processing is however, and I think might remain, circumstantial. [Why must it feel like something when we are performing a certain type of information processing? The evidence described here suggests that it does feel like something when we are performing a certain type of information processing, but does not produce a strong reason for why it has to feel like something. It just does, when we are using this linguistic processing system capable of higher-order thoughts.] The evidence, summarized above, includes the points that we think of ourselves as conscious when, for example, we recall earlier events, compare them with current events, and plan many steps

---

[16]This is a functionalist position. Apparently Damasio (2003) does not subscribe to this view, for he suggests that there is something in the 'stuff' (the 'natural medium') that the brain is made of that is also important. It is difficult for a person with this view to make telling points about consciousness from neuroscience, for it may always be the 'stuff' that is actually important.

ahead. Evidence also comes from neurological cases, from, for example, split-brain patients (who may confabulate conscious stories about what is happening in their other, non-language, hemisphere); and from cases such as frontal lobe patients who can tell one consciously what they should be doing, but nevertheless may be doing the opposite. (The force of this type of case is that much of our behaviour may normally be produced by routes about which we cannot verbalize, and are not conscious about.)

This raises discussion of the *causal role of consciousness*. Does consciousness cause our behaviour? The view that I currently hold is that the information processing that is related to consciousness (activity in a linguistic system capable of higher-order thoughts, and used for planning and correcting the operation of lower-order linguistic systems) can play a causal role in producing our behaviour (see Fig. 22.3), but as part of a 'levels of explanation' account of the relation between mental events and brain events (Section 22.7). It is, I postulate, a property of processing in this system (capable of higher-order thoughts) that it feels like something to be performing that type of processing. It is in this sense that I suggest that consciousness can act causally to influence our behaviour – consciousness is the property that occurs when a linguistic system is thinking about its lower-order thoughts, which may be useful in correcting plans.

The hypothesis that it does feel like something when this processing is taking place is at least to some extent testable (cf. Lau and Rosenthal (2011)): humans performing this type of higher-order linguistic processing, for example recalling episodic memories and comparing them with current circumstances, who denied being conscious, would prima facie constitute evidence against the theory. As noted above, most humans would find it implausible to posit that they could be thinking about their own thoughts, and reflecting on their own thoughts, without being conscious. This type of processing does appear, for most humans, to be necessarily conscious.

It is suggested that qualia, raw sensory, and emotional, 'feels', arise secondarily to having evolved such a higher-order thought system, and that sensory and emotional processing feels like something because once this emotional processing has entered the planning, higher-order thought, system, it would be unparsimonious for it not to feel like something, given that all the other processing in this system I suggest does feel like something.

The adaptive value of having sensory and emotional feelings, or qualia, is thus suggested to be that such inputs are important to the long-term planning, explicit, processing system. Raw sensory feels, and subjective states associated with emotional and motivational states, may not necessarily arise first in evolution.

Reasons why the ventral visual system is more closely related to explicit than implicit processing include the fact that representations of objects and individuals need to enter the planning, hence conscious, system, and are considered in more detail by Rolls (2003) and by Rolls (2008d).

22.8.2   Oscillations and temporal binding

The postulate of Crick and Koch (1990) that oscillations and synchronization are necessary bases of consciousness might possibly be related to the present theory if it turns out that oscillations or neuronal synchronization are the way the brain implements syntactic binding. However, the fact that oscillations and neuronal synchronization are especially evident in anaesthetized cats does not impress as strong evidence that oscillations and synchronization are critical features of consciousness, for most people would hold that anaesthetized cats are not conscious. The fact that oscillations and stimulus-dependent neuronal synchronization are much more difficult to demonstrate in the temporal cortical visual areas of awake behaving monkeys (Tovee and Rolls 1992, Franco, Rolls, Aggelopoulos and Treves 2004, Aggelopoulos,

Franco and Rolls 2005, Rolls 2008d, Rolls and Treves 2011) (Section 8.3), might just mean that during the evolution of primates the cortex has become better able to avoid parasitic oscillations, as a result of developing better feedforward and feedback inhibitory circuits (Rolls 2008d).

The suggestion that syntax in real neuronal networks is implemented by temporal binding (Malsburg 1990b, Singer 1999) seems unlikely (Rolls 2008d, Rolls and Treves 2011). For example, the code about which visual stimulus has been shown can be read off from the end of the visual system without taking the temporal aspects of the neuronal firing into account; much of the information about which stimulus is shown is available in short times of 30–50 ms, and cortical neurons need fire for only this long during the identification of objects (Tovee, Rolls, Treves and Bellis 1993, Rolls and Tovee 1994, Tovee and Rolls 1995, Rolls and Treves 1998, Rolls 2003, Rolls 2006a, Rolls 2008d, Rolls and Treves 2011) (these are rather short time-windows for the expression of multiple separate populations of synchronized neurons); and stimulus-dependent synchronization of firing between neurons is not a quantitatively important way of encoding information in the primate temporal cortical visual areas involved in the representation of objects and faces (Tovee and Rolls 1992, Rolls and Treves 1998, Rolls and Deco 2002, Rolls, Franco, Aggelopoulos and Reece 2003b, Rolls, Aggelopoulos, Franco and Treves 2004, Franco, Rolls, Aggelopoulos and Treves 2004, Aggelopoulos, Franco and Rolls 2005, Rolls 2008d, Rolls and Treves 2011, Rolls 2012c) – see Sections 8.2 and 8.3.

Further, even the hypothesis that information transmission is facilitated by coherent (phase locked) oscillations (communication through coherence (Fries 2005, Fries 2009)) has considerable difficulties (Rolls, Webb and Deco 2012), in that although there is much emphasis on finding coherence in the brain, there is much less causal evidence that coherence affects the transmission of information. In a test of this, it was found in an integrate-and-fire model of two connected networks in which causal effects of gamma oscillations (approximately 50 Hz) could be analysed, that information transmission between coupled networks occurs at very much lower strengths of the connecting synapses than are required to make the oscillations coherent (Rolls, Webb and Deco 2012) Section 8.3). The finding was thus that information transmission does not require, and is little influenced by, gamma oscillations. The implication is that great care is needed to test whether coherence in the brain has causal effects in influencing information transmission, at least in the gamma band (Rolls, Webb and Deco 2012, Rolls and Treves 2011).

22.8.3 A high neural threshold for information to reach consciousness

Part in fact of Rolls' theory of consciousness is that it provides a computational reason why the threshold for information to reach consciousness is higher than the threshold for information to influence behaviour in what is referred to as subliminal processing (Dehaene, Changeux, Naccache, Sackur and Sergent 2006).

Evidence that explicit, conscious, processing may have a higher threshold in sensory processing than implicit processing is considered by Rolls (2003, 2006a, 2011b) based on neurophysiological and psychophysical investigations of backward masking (Rolls and Tovee 1994, Rolls, Tovee, Purcell, Stewart and Azzopardi 1994b, Rolls, Tovee and Panzeri 1999b, Rolls 2003, Rolls 2006a, Rolls 2008d, Rolls 2012c). It is suggested there that part of the adaptive value of this is that if linguistic processing is inherently serial and slow, it may be maladaptive to interrupt it unless there is a high probability that the interrupting signal does not arise from noise in the system. In the psychophysical and neurophysiological studies, it was found that face stimuli presented for 16 ms and followed immediately by a masking stimulus were not consciously perceived by humans, yet produced above chance identification, and firing of inferior temporal cortex neurons in macaques for approximately 30 ms. If the mask was delayed for 20 ms, the neurons fired for approximately 50 ms, and the test face stimuli were more likely to be perceived consciously. In a similar backward masking paradigm, it was found that happy vs angry face expressions could influence how much beverage was wanted and consumed even when the faces were not consciously perceived (Winkielman and Berridge 2005, Winkielman and Berridge 2003). This is further evidence that unconscious emotional stimuli can influence behaviour.

22.8.4 James–Lange theory and Damasio's somatic marker hypothesis about feelings

The theory described here is also different from other theories of consciousness and affect. James and Lange (James 1884, Lange 1885) held that emotional feelings arise when feedback from the periphery (about for example heart rate) reach the brain, but had no theory of why some stimuli and not others produced the peripheral changes, and thus of why some but not other events produce emotional feelings.

Moreover, the evidence that feedback from peripheral autonomic and proprioceptive systems is essential for emotions is very weak, in that for example blocking peripheral feedback does not eliminate emotions, and producing peripheral, e.g. autonomic, changes does not elicit emotion (Reisenzein 1983, Schachter and Singer 1962, Rolls 2014a).

Damasio's theory of emotion (Damasio 1994, Damasio 2003) is a similar theory to the James–Lange theory (and is therefore subject to some of the same objections), but holds that the peripheral feedback is used in decision-making rather than in consciousness. He does not formally define emotions, but holds that body maps and representations are the basis of emotions. When considering consciousness, he assumes that 'all consciousness is self-consciousness' (Damasio 2003) (p. 184), and that 'the foundational images in the stream of the mind are images of some kind of body event, whether the event happens in the depth of the body or in some specialized sensory device near its periphery' (Damasio 2003) (p. 197). His theory does not appear to be a fully testable theory, in that he suspects that "the ultimate quality of feelings, a part of why feelings feel the way they feel, is conferred by the neural medium" (Damasio 2003) (p. 131). Thus presumably if the processes he discusses (Damasio 1994, Damasio 2003) were implemented in a computer, then the computer would not have all the same properties with respect to consciousness as the real brain. In this sense he appears to be arguing for a non-functionalist position, and something crucial about consciousness being related to the particular biological machinery from which the system is made. In this respect the theory seems somewhat intangible.

22.8.5 LeDoux's approach to emotion and consciousness

LeDoux's approach to emotion (LeDoux 1992, LeDoux 1995, LeDoux 1996, LeDoux 2008, LeDoux 2011) is largely (to quote him) one of automaticity, with emphasis on brain mechanisms involved in the rapid, subcortical, mechanisms involved in fear. LeDoux (1996), in line with Johnson-Laird (1988) and Baars (1988), emphasizes the role of working memory in consciousness, where he views working memory as a limited-capacity serial processor that creates and manipulates symbolic representations (p. 280). He thus holds that much emotional processing is unconscious, and that when it becomes conscious it is because emotional information is entered into a working memory system. However, LeDoux (1996) concedes that consciousness, especially its phenomenal or subjective nature, is not completely explained by the computational processes that underlie working memory (p. 281).

22.8.6 Panksepp's approach to emotion and consciousness

Panksepp's (1998) approach to emotion and consciousness has its origins in neuroethological investigations of brainstem systems that when activated lead to behaviours like fixed action patterns, including escape, flight and fear behaviour. A flavour of his views about consciousness includes his postulate that "feelings may emerge when endogenous sensory and emotional systems within the brain that receive direct inputs from the outside world as well as the neurodynamics of the SELF (a Simple Ego-type Life Form) begin to reverberate with each other's changing neuronal firing rhythms" (Panksepp 1998) (p. 309).

22.8.7 Global workspace theories of consciousness

Rolls' approach to consciousness suggests that the information must be being processed in a system capable of implementing HOSTs for the information to be conscious, and in this sense is more specific than global workspace hypotheses (Baars 1988, Dehaene and Naccache 2001, Dehaene et al. 2006, Dehaene et al. 2014). Indeed, the present approach suggests that a workspace could be sufficiently global to enable even the complex processing involved in driving a car to be performed, and yet the processing might be performed unconsciously, unless HOST (supervisory, monitory, correcting) processing was involved.

22.8.8 Monitoring and consciousness

An attractor network in the brain with positive feedback implemented by excitatory recurrent collateral connections between the neurons can implement decision-making (Wang 2002, Deco and Rolls 2006, Wang 2008, Rolls and Deco 2010) (see Chapter 5.6). As explained in detail elsewhere (Rolls and Deco 2010), if the external evidence for the decision is consistent with the decision taken (which has been influenced by the noisy neuronal firing times), then the firing rates in the winning attractor are supported by the external evidence, and become especially high. If the external evidence is contrary to the noise-influenced decision, then the firing rates of the neurons in the winning attractor are not supported by the external evidence, and are lower than expected (Section 5.6.9; Fig. 5.11). In this way the confidence in a decision is reflected in, and encoded by, the firing rates of the neurons in the winning attractor population of neurons (Section 5.6) (Rolls and Deco 2010).

If we now add a second attractor network to read the firing rates from the first decision-making network, the second attractor network can take a decision based on the confidence expressed in the firing rates in the first network (Insabato, Pannunzi, Rolls and Deco 2010) (Section 5.6.10). The second attractor network allows decisions to be made about whether to change the decision made by the first network, and for example abort the trial or strategy (see Fig. 5.22). The second network, the confidence decision network, is in effect monitoring the decisions taken by the first network, and can cause a change of strategy or behaviour if the assessment of the decision taken by the first network does not seem a confident decision. This is described in detail elsewhere (Insabato, Pannunzi, Rolls and Deco 2010, Rolls and Deco 2010), but Fig. 5.22 shows the simple system of two attractor networks that enables confidence-based (second-level) decisions to be made, by monitoring the output of the first, decision-making, network (see Section 5.6.10).

Now this is the type of description, and language used, to describe 'monitoring' functions, taken to be a high-level cognitive process, possibly related to consciousness (Block 1995a, Lycan 1997). For example, in an experiment performed by Hampton (2001) (experiment 3), a monkey had to remember a picture over a delay. He was then given a choice of a 'test flag', in which case he would be allowed to choose from one of four pictures the one seen before the delay, and if correct earn a large reward (a peanut). If he was not sure that he remembered the first picture, he could choose an 'escape flag', to start another trial. With longer delays, when memory strength might be lower partly due to noise in the system, and confidence therefore in the memory on some trials might be lower, the monkey was more likely to choose the escape flag. The experiment is described as showing that the monkey is thinking about his own memory, that is, is a case of meta-memory, which may be related to consciousness (Heyes 2008). However, the decision about whether to escape from a trial can be taken just by adding a second decision network to the first decision network. Thus we can account for what seem like complex cognitive phenomena with a simple system of two attractor decision-making networks (Section 5.6.10; Fig. 5.22) (Rolls and Deco 2010, Insabato et al. 2010, Rolls 2012d).

The implication is that some types of 'self-monitoring' can be accounted for by simple, two-attractor network, computational processes. But what of more complex 'self-monitoring', such as is described as occurring in a commentary that might be based on reflection on previous events, and appears to be closely related to consciousness (Weiskrantz 1997). This approach has been developed into my higher-order syntactic theory (HOST) of consciousness (Section 22.2 (Rolls 1997b, Rolls 2004b, Rolls 2005, Rolls 2007a, Rolls 2008a, Rolls 2007b, Rolls 2010c, Rolls 2011b, Rolls 2012d)), in which there is a credit assignment problem if a multi-step reasoned plan fails, and it may be unclear which step failed. Such plans are described as syntactic as there are symbols at each stage that must be linked together with the syntactic relationships between the symbols specified, but kept separate across stages of the plan. It is suggested that in this situation being able to have higher-order syntactic thoughts will enable one to think and reason about the first-order plan, and detect which steps are likely to be at fault.

Now this type of 'self-monitoring' is much more complex, as it requires syntax. The thrust of the argument is that some types of 'self-monitoring' are computationally simple, for example in decisions made based on confidence in a first decision (Rolls and Deco 2010, Insabato, Pannunzi, Rolls and Deco 2010), and may have little to do with consciousness; whereas higher-order thought processes are very different in terms of the type of syntactic computation required, and may be more closely related to consciousness (Rolls 1997b, Rolls 2003, Rolls 2004b, Rolls 2005, Rolls 2007a, Rolls 2008a, Rolls 2007b, Rolls 2010c, Rolls 2011b, Rolls 2012d).

Thus the theory of consciousness described in this chapter is different from some other theories of consciousness.

22.9 Highlights

1. Consciousness, and especially phenomenal consciousness (why it should feel like something), is a 'hard' problem in philosophy. It is not clear that it is amenable to a solution using neuroscience, but neuroscience does have many relevant and potential contributions to understanding phenomenal consciousness.

2. It is argued that one route to action is the goal-directed route, with the goals set by the selfish genes. A second route involves multi-step reasoning involving syntax to keep the different variables and steps in the plan separate and organised (see Chapter 17).

3. This multi-step planning may suffer a credit assignment problem, in that when the plan fails, it is not clear which step or steps were faulty. In this situation it is proposed that a monitoring system with syntactic capability would be useful to work through the steps of the plan, to work out which the faulty step in the plan might be, and to correct the plan. It is suggested, but only as a plausible argument that cannot be proved,

that this higher order syntactic monitoring operation might be the type of computation, with adaptive value, that might feel like something. This is Rolls' higher order syntactic thought (HOST) theory of consciousness. It is a theory of the computations that underlie consciousness, and that type of approach does seem promising.

4. Much processing in humans can be unconscious, and cannot be verbally reported. It may operate without the involvement of the necessarily slow HOST system with its sequential syntax.

5. The brain does not operate as a deterministic system, because of spiking-related neuronal noise in the brain (Chapter 5).

6. However, when the reasoning system is taking the decision, we might wish to use the term 'free will' to describe the operation being performed by the cortex.

7. Other approaches to consciousness include Rosenthal's higher order thought (HOT) theory; that oscillations somehow account for consciousness; that sensing bodily changes is required for consciousness; and that consciousness is related to operations in a global workspace.

23 Cerebellar cortex

23.1 Introduction

The cerebellar cortex is not cerebral cortex, as it is part of the hindbrain, but its cortical structure is exquisitely developed, and is quite different to that of the cerebral cortex. It is thus very useful to compare its architecture and principles of operation to those of the cerebral cortex, for this helps to highlight what is special about the cerebral cortex.

A comparison with the cerebral cortex also shows how important the architecture and connectivity of a neural system, for example a type of cortex, are to understanding its principles of operation, that is its principles of computation. This goes far beyond trying to produce an arrow diagram to show the flow of information through a set of neurons (Shepherd and Grillner 2010), which though the fine structure is crucially important, may do little to elucidate our understanding of the function of a type of cortex, as in some canonical microcircuit approaches (Chapter 18). The pioneer in taking evidence from the quantitative aspects of the fine structure of the cortex forward to help formulate possible computational principles of operation of the cortex was David Marr (the Cambridge mathematician who came to learn from Giles Brindley in the Physiological Laboratory at Cambridge, when I was a medical student), who produced papers on the cerebellum (Marr 1969), neocortex (Marr 1970), and hippocampal cortex (Marr 1971). My research has followed that pioneering approach, though I have gone beyond the combination of quantitative anatomical and mathematical approaches pioneered by Marr, to include also direct evidence on the neuronal activity at each connected stage of the system, evidence on neuronal encoding, evidence on the effects of damage to the system, and in addition also the approaches of theoretical physics, in collaborations with Alessandro Treves, Gustavo Deco, and many other theoretical physicists (Rolls and Treves 1998, Rolls 2008d, Rolls and Deco 2010, Rolls 2014a, Rolls 2016a).

The cerebellum is involved in the accurate control of movements. If the cerebellum is damaged, movements can still be initiated, but the movements are not directed accurately at the target, and frequently oscillate on the way to the target (Lisberger and Thach 2013). There is insufficient time during rapid movements for feedback control to operate, and the hypothesis is that the cerebellum performs feedforward control by learning to control the motor commands to the limbs and body in such a way that movements are smooth and precise. The cerebellum is thus described as a system for adaptive feedforward motor control (Ito 1984, Ito 2010). Network theories of cerebellar function are directed at showing how it learns to take the context for a movement, which might consist of a desired movement from the cerebral neocortex and the starting position of the limbs and body, and produce the appropriate output signals. The appropriate output signals are learned as a result of the errors that have occurred on previous trials.

Much of the fundamental anatomy and physiology of the cerebellum are described by Masao Ito (1984, 2010) with classical neuroanatomy provided by Eccles, Ito and Szentagothai (1967), and pioneering computational neuroscience approaches provided by Marr (1969) and Albus (1971).

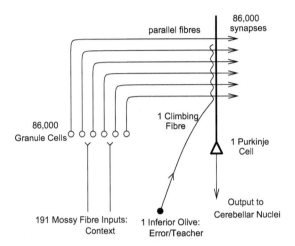

Fig. 23.1 Overall architecture of the cerebellum. Inputs relayed from the pontine nuclei form the mossy fibres which synapse onto dentate granule cells. The dentate granule cells via their parallel fibres form modifiable synapses on the Purkinje cells, from which outputs leave the cerebellar cortex. Each Purkinje cell receives one climbing fibre input. The numbers indicate the approximate numbers of cells of different types, relative to one Purkinje cell (see Ito, 1984). The numbers indicate expansion recoding of the mossy fibre inputs to the parallel fibres. A working hypothesis is that the context for a movement (for example limb state and motor command) reaches the Purkinje cells via the parallel fibres, and that the effect that this input has on the Purkinje cell output is learned through synaptic modification taught by the climbing fibre, so that the cerebellar output produces smooth error-free movements.

23.2   Architecture of the cerebellum

The overall architecture of the cerebellum and the main hypothesis about the functions of the different inputs it receives are shown schematically in Fig. 23.1. The mossy fibres convey the context or to-be-modified input which is applied, after expansion recoding implemented via the granule cells, via modifiable parallel fibre synapses onto the Purkinje cells, which provide the output of the network. There is a climbing fibre input to each Purkinje cell, and this is thought to carry the teacher or error signal that is used during learning to modify the strength of the parallel fibre synapses onto each Purkinje cell. This network architecture is implemented beautifully in the very regular cerebellar cortex. As cortex, it is a layered system. The cerebellar cortex has great regularity and precision, and this has helped analysis of its network connectivity.

23.2.1   The connections of the parallel fibres onto the Purkinje cells

A small part of the anatomy of the cerebellum is shown schematically in Fig. 23.2. Large numbers of parallel fibres course at right angles over the dendritic trees of Purkinje cells. The dendritic tree of each Purkinje cell has the shape of a flat fan approximately 250 μm by 250 μm by 6 μm thick. The dendrites of different Purkinje cells are lined up as a series of flat plates (see Fig. 23.3). As the parallel fibres cross the Purkinje cell dendrites, each fibre

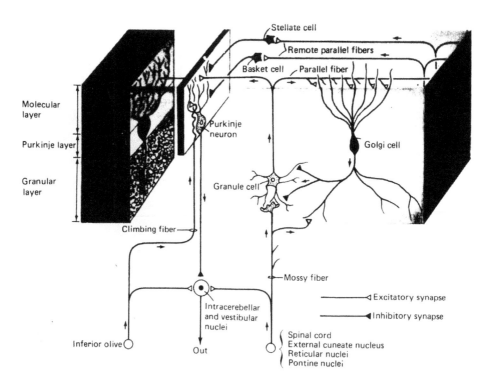

Molecular layer

Purkinje layer

Granular layer

Stellate cell

Remote parallel fibers

Basket cell   Parallel fiber

Purkinje neuron

Golgi cell

Granule cell

Climbing fiber

Mossy fiber

Excitatory synapse

Inhibitory synapse

Intracerebellar and vestibular nuclei

Inferior olive

Out

Spinal cord
External cuneate nucleus
Reticular nuclei
Pontine nuclei

Fig. 23.2 Schematic diagram providing a 3D impression of the connectivity of a small part of the cerebellar cortex. Some of the inhibitory interneurons (Golgi, stellate, and basket cells) are shown.

makes one synapse with approximately every fifth Purkinje cell (Ito 1984). Each parallel fibre runs for approximately 2 mm, in the course of which it makes synapses with approximately 45 Purkinje cells. Each Purkinje cell receives approximately 80,000 synapses, each from a different parallel fibre. The great regularity of this cerebellar anatomy thus beautifully provides for a simple connectivity matrix, with even sampling by the Purkinje cells of the parallel fibre input. This architecture provides for only one synapse from any parallel fibre to any one Purkinje cell. The significance of this, it is suggested, is that in a perceptron (Section B.10), just as in a pattern association network (Section 7.3), this maximizes the number of input patterns that can be correctly associated with outputs.

23.2.2   The climbing fibre input to the Purkinje cell

The other main input to the Purkinje cell is the climbing fibre. There is one climbing fibre to each Purkinje cell, and this climbing fibre spreads to reach every part of the dendritic tree of the Purkinje cell. (Although the climbing fibre does not reach quite to the end of every dendrite, the effects of it on the Purkinje cell will reach to the apical extremity of every dendrite.) Although each Purkinje cell has an input from only one climbing fibre, each climbing fibre does branch and innervate approximately 10–15 different Purkinje cells in different parts of the cerebellum. The climbing fibres have their cell bodies in the inferior olive.

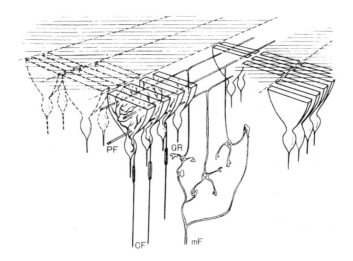

Fig. 23.3 The connections of the parallel fibres (PF) onto the Purkinje cells. Five rows of Purkinje cells with their fan-shaped dendritic trees are shown. CF, climbing fibre; GR, granule cell; mF, mossy fibre. (©1968. IEEE. Reprinted, with permission, from Szentagothai, J. 1968. Structuro-functional considerations of the cerebellar network, Proc IEEE 56: 960-968.)

23.2.3 The mossy fibre to granule cell connectivity

The parallel fibres do not arise directly from the neurons in the pontine and vestibular nuclei (see Fig. 23.1). Instead, axons of the pontine and vestibular cells become the mossy fibres of the cerebellum, which synapse onto the granule cells, the axons of which form the parallel fibres (see Fig. 23.2). There are approximately 450 times as many granule cells (and thus parallel fibres) as mossy fibres (see Ito (1984), p. 116). In the human cerebellum, it is estimated that there are $10^{10} - 10^{11}$ granule cells (see Ito (1984), p. 76), which makes them the most numerous cell type in the brain. What is the significance of this architecture? It was suggested by Marr (1969) that the expansion recoding achieved by the remapping onto granule cells is to decorrelate or orthogonalize the representation before it is applied to the modifiable synapses onto the Purkinje cells. It is shown in Section B.2.7.3 how such expansion recoding can maximize capacity, reduce interference, and allow arbitrary mappings through a simple associative or similar network.

The granule cells of the cerebellum can therefore be regarded as analogous computationally to the dentate granule cells of the hippocampus, and to the granule / stellate cells in layer 4 of the neocortex, both of which perform pattern separation, as described in Sections 18.2 and 26.5.18. The use again of granule cells for pattern separation in the cerebellum as well as in the hippocampus and neocortex is additional evidence for some of the fundamental principles of operation of the cortex described in this book.

As implied by the numbers given above, and as summarized in Fig. 23.1 in which I show (by extracting numbers from Ito (1984)) estimates of the relative numbers of the cells discussed so far in the cerebellum, there is a massive expansion in the number of cells used to code the same amount of information, in that the ratio of granule cells to mossy fibres is approximately 450. This provides a system comparable to a binary decoder. Consider an 8 bit number representation system, which can be used to represent the numbers 0–255 (decimal) (that is 28 numbers). Now, the bits that represent these numbers are often highly correlated with each other. For example, the bits that are set in the binary representations of the numbers 127 and 255 are identical apart from the highest order bit. (127 is 01111111 in binary and 255 is 11111111.) These two numbers would thus appear very similar in an associative memory system, such as that described in Section B.2, so there would be great interference between the two numbers in such a memory system. However, if the numbers 0–255 were each represented by a different active axon synapsing onto the output cell, then there would be no interference between the different numbers, that is the representations would be orthogonalized by binary decoding. The 450-fold increase in the number of fibres as information passes from mossy to parallel fibres potentially allows this expansion recoding to take place.

This architecture of the brain provides an interesting contrast with that of conventional digital computers. The type of information storage device that has evolved in the brain uses a distributed representation over a large number of input connections to a neuron, and gains from this the useful properties of generalization, graceful degradation, and the capacity to function with low information capacity stores (that is synapses), but pays the penalty that some decorrelation of the inputs is required. The brain apparently has evolved a number of specialized mechanisms to do this. One is exemplified by the granule cell system of the cerebellum, with its enormous expansion recoding. In effect, the brain provides an enormous number of what may be low resolution storage locations, which in the cerebellum are the vast number of parallel fibre to Purkinje cell synapses. It is astounding that the number of storage locations in the (human) cerebellum would thus be approximately 3×10^{10} parallel fibres with every parallel fibre making approximately 300 synapses, or approximately 10^{13} modifiable storage locations. (10 Terabits of storage.)

The recoding which could be performed by the mossy fibre to granule cell system can be understood more precisely by considering the quantitative aspects of the recoding. Each mossy fibre ends in several hundred rosettes, where excitatory contact is made with the dendrites of granule cells. Each granule cell has dendrites in 1–7 mossy rosettes, with an average of four. Thus, on average, each granule cell receives from four mossy fibres. Due to the spacing of the mossy fibres, in almost all cases the four inputs to a granule cell will be from different mossy fibres. Approximately 28 granule cell dendrites contact each rosette. There are approximately 450 times as many granule cells as mossy fibres. The result of this architecture is that the probability that any given granule cell has all its C inputs active, and is therefore likely to fire, is low. It is in fact P^C, where P is the probability that any mossy fibre is active. Thus if the probability that a mossy fibre is active is 0.1, the probability that all of the average four inputs to a granule cell are active is 0.1^4 or 0.0001. The probability that three of the four inputs to a given granule cell are active would be 0.001. Given that each mossy fibre forms rosettes over a rather wide area of the cerebellum, that is over several folia, cells with three or four active inputs would be dotted randomly, with a low probability, within any one area.

One further aspect of this recoding must be considered. The output of the parallel fibres is sampled by the Golgi cells of the cerebellum via synapses with parallel fibres. Each Golgi cell feeds back inhibitory influences to about 5,000–10,000 granule cells. Neighbouring Golgi cells overlap somewhat in their dendritic fields, which are approximately 300 μm across, and in their axon arborization. Every granule cell is inhibited by at least one Golgi cell. The Golgi cells terminate directly on the mossy rosettes, inhibiting the granule cells at this point. This very broad inhibitory feedback system suggests the function of an automatic gain control. Thus Albus (1971) argued that the Golgi cells serve to maintain granule cell, and hence parallel fibre, activity at a constant rate. If few parallel fibres are active, Golgi inhibitory feedback decreases, allowing granule cells with lower numbers (for example 3 instead of 4) of excitatory inputs to fire. If many parallel fibres become active, Golgi feedback increases, allowing only those few granule cells with many (for example 5 or 6) active mossy inputs to fire. In addition to this feedback inhibitory control of granule cell activity, there is also a feedforward inhibitory control implemented by endings of the mossy fibres directly on Golgi cells. This feedforward inhibition probably serves a speed-up function, to make the overall inhibitory effect produced by the Golgi cells more smooth in time. The net effect of this inhibitory control is probably thus to maintain a relatively low proportion, perhaps 1%, of parallel fibres active at any one time. This relatively low proportion of active input fibres to the Purkinje cell learning mechanism optimizes information storage in the synaptic storage matrix, in that the sparse representation enables many different patterns to be learned by the parallel fibre to Purkinje cell system (see Section B.2).

23.3   Modifiable synapses of parallel fibres onto Purkinje cell dendrites

The climbing fibres take part in determining the modifiability of the parallel fibre to Purkinje cell synapses (Ito, Yamaguchi, Nagao and Yamazaki 2014). In particular, the parallel fibre to Purkinje cell synapses decrease in strength, that is, they show long-term depression (LTD), when the climbing fibre to a Purkinje cell fires, if the parallel fibre is active (Ito 1984, Ito et al. 2014). Thus the associative learning is between the parallel fibre and the Purkinje cell, but the 'teacher' is the climbing fibre, which produces a plateau potential in the Purkinje cell. This can be thought of as a type of error learning (Ito 1984, Ito 2013). If a parallel fibre is firing and tending to activate the Purkinje cell when is should not, then firing of the climbing fibre to indicate an error would reduce the strength of the synapse from that parallel fibre. If the climbing fibre does carry an error signal (which is a systems-level question (Ito 2013), see below), then this would make the computation performed by the cerebellar cortex that of a one-layer perceptron (Ito 1984, Ito 2006, Ito 2006, Ito 2010) (see Section B.10). If the climbing fibre is on average not firing when a parallel fibre is firing, then that parallel fibre to Purkinje cell synapse will become stronger (i.e. show associative long-term potentiation, LTP).

It is interesting to note that despite an early theory which predicted modifiability of the parallel fibre to Purkinje cell synapses (Marr 1969), experimental evidence for modifiability was only obtained in the 1980s (Ito 1984). Part of the reason for this was that it was only when the physiology of the inferior olive was understood better, by making recordings from single neurons, that it was found that inferior olive neurons, which give rise to the climbing fibres, rarely fire at more than a few spikes per second. It was not until such relatively low-frequency stimulation (at 4 Hz) was used by Ito and his colleagues (see Ito (1984)) that the synaptic modifiability was demonstrated. (David Marr had tried with Jack Eccles to obtain evidence for synaptic modification, but they did not know the normally low firing rates of inferior olive neurons. Possibly as a result, David Marr felt that neuroscience methods were not yet available to test his theories on the principles of operation of cortical circuitry, and moved into more abstract theoretical work on vision (Marr 1982, Rolls 2011c).)

23.4   The cerebellar cortex as a perceptron

The overall concept that the work of Marr (1969), Albus (1971), and Ito (1984, 2006, 2013) has led towards is that the cerebellum acts as an adaptive feedforward controller with the general form shown in Fig. 23.2. The command for the movement and its context given by the current state of the limbs, trunk, etc., signalled by proprioceptive input would be fed to the intracerebellar (deep cerebellar) nuclei to produce a motor output ('Out' in Fig. 23.2). But in addition the input would be fed via the mossy fibres and then via the granule cells to the parallel fibres. There the appropriate firing of the Purkinje cells to enable the movement to be controlled smoothly would be produced via the parallel fibre to Purkinje cell synapses. These synapses would be set to the appropriate values by error signals received on many previous trials and projected via the climbing fibres to provide the teaching input for modifying the synapses utilizing LTD. The proposal described by Ito (1984) is thus that the cerebellum would be in a sideloop, with the overall gain or output of the system resulting from the direct, relatively unmodifiable output produced through the deep cerebellar nuclei, and the variable gain output contributed by the cerebellar sideloop.

This is a generic model of cerebellar function. Different parts of the cerebellum are concerned with different types of motor output (Ito 1984, Lisberger and Thach 2013). One part, the flocculus, is involved in the control of eye movements, and in particular with the way in which eye movements compensate for head movements, helping the images on the retina to remain steady during head movement. This is called the vestibulo-ocular reflex (VOR). Another part of the cerebellum is concerned with the accurate and smooth control of limb movements. Another part is implicated in the classical conditioning of skeletal responses, in which a sensory input such as a tone comes by learning to produce learned reflex responses. In order to assess how the generic circuitry of the cerebellum contributes to these types of adaptive control of motor output, evidence on how the cerebellum contributes to each of these types of motor learning is relevant, and is considered elsewhere (Ito 1984, Ito 2006, Ito 2013, Lisberger and Thach 2013).

One point about the theory that has always intrigued me is how a climbing fibre 'knows' which Purkinje cell to connect to, and therefore to provide the correct error or teaching input for that neuron. Or perhaps it is the other way round: the climbing fibre may determine what functions are being performed by a Purkinje cell, and some connections on the output side of the cerebellum need to be set up for that Purkinje cell's outputs (via the deep cerebellar nuclei) to act appropriately.

The point of this brief survey has been not only to highlight some of the quantitatively remarkable and precise aspects of the design of the cerebellar cortex, but also to use the differences of its design from that of the cerebral cortex (neocortex, hippocampal cortex, and pyriform cortex) to illuminate the principles of operation of all these types of cortex. We now turn to these comparisons.

23.5   Highlights: differences between cerebral and cerebellar cortex microcircuitry

1. First, this comparison does highlight how the design of different types of cortex is very different, and at the same time how precisely types of cortex are designed. The implication is that the quantitative details of the connectivity are important, a point understood by Marr, but hardly emphasized in most 'canonical models' of the microcircuitry of the cerebral cortex (Chapter 18), most of which may not even show recurrent collateral connections between classes of pyramidal cells, let alone consider their number, and what the computational significance may be.

2. Second, the cerebellar cortex does not have excitatory recurrent collaterals, and the cerebral cortex does. This helps to make excitatory recurrent collaterals what is probably the most fundamental aspect of the design of the cerebral cortex, which provides the ability to maintain information by positive excitatory feedback, to enable short-term memory, long-term memory, decision-making, top-down attention, etc, as described throughout this book (see also Chapter 18).

3. Third, analysis of the cerebellum shows how important the quantitative aspects of the connectivity are for brain computation, with large numbers of parallel fibre synapses for

example onto each Purkinje cell. The importance of this is that the number of synapses onto each neuron is the leading factor in the number of memories that can be stored in pattern associators, perceptrons, and autoassociation networks, so these numbers are just as important for cerebral cortex (Appendix B).

4. Fourth, analysis of the cerebellum shows how important it is for associative memories to have relatively orthogonal patterns presented to the network, with this being implemented by the 10^{10} or more granule cells in the cerebellum. In the same way, the hippocampus has granule cells in the dentate gyrus to perform pattern separation, and the neocortex has stellate cells in layer 4 to do the same, especially where it appears that much orthogonalization is required, in primary sensory cortical areas such as the primary visual cortex, V1.

5. Fifth, a major difference from the cerebral cortex is that each Purkinje cell has its own teacher, its climbing fibre. Instead, there appears to be nothing like this in the cerebral cortex. Instead, the hypothesis developed in this book is that in the neocortex the correct neocortical pyramidal cells are selected by which ones win in a competitive network style of computation for any given input pattern. Further, the hypothesis is that in the hippocampus, the selection of which CA3 cells should be activated is performed by random selection, using the mossy fibres, as described in Chapter 24. The difference from neocortex is that the computation performed by the hippocampal CA3 network is to store large numbers of memories separately, and for this, picking neurons at random to be part of each memory is appropriate. In contrast, in the neocortex, the aim of the computation is much more sophisticated, to build useful representations for categorising complex inputs, and for this purpose competitive learning is used to produce neocortical neurons that respond to different non-linear combinations of their inputs (Sections B.4, 18.2.1, and 26.5.18).

24 The hippocampus and memory

24.1 Introduction

In this Chapter a computational neuroscience approach to hippocampal function in memory is described. Then tests of the theory based especially on subregion analysis of the hippocampal system are described, followed by an evaluation of the theoretical proposals, and a comparison with other theories.

The hippocampus is a system that has been selected for this Chapter because it illustrates many of the principles of operation of the cerebral cortex, and shows how they operate together to produce important functions in memory, which we are now starting to understand. While separate principles have been the focus of much of the rest of this book, this Chapter comes close to a theory of how the hippocampal type of cortex may operate.

The theory was originally developed as described next, and was preceded by work of Marr (1971) who developed a mathematical model, which although not applied to particular networks within the hippocampus and dealing with binary neurons and binary synapses which utilised heavily the properties of the binomial distribution, was important in utilizing computational concepts. The model was assessed (Willshaw and Buckingham 1990, Willshaw, Dayan and Morris 2015)[17], who showed that Marr's model did not require any recurrent collateral effect. Early work of Gardner-Medwin (1976) showed how progressive recall could operate in a network of binary neurons with binary synapses.

Rolls (1987) described a theory of the hippocampus presented to the Dahlem conference in 1985 on the Neural and Molecular Bases of Learning in which the CA3 neurons operated as an autoassociation memory to store episodic memories including object and place memories, and the dentate granule cells operated as a preprocessing stage for this by performing pattern separation so that the mossy fibres could act to set up different representations for each memory to be stored in the CA3 cells. Rolls (1987) proposed that the CA1 cells operate as a recoder for the information recalled from the CA3 cells to a partial memory cue, so that the recalled information would be represented more efficiently to enable recall, via the backprojection synapses, of activity in the neocortical areas similar to that which had been present during the original episode[18]. This theory was developed further (Rolls 1989a, Rolls 1989b, Rolls 1989e, Rolls 1989f, Rolls 1990a, Rolls 1990b), including further details about

---

[17]Marr (1971) showed how a network with recurrent collaterals could complete a memory using a partial retrieval cue, and how sparse representations could increase the number of memories stored. The analysis of these auto-association or attractor networks was developed by Kohonen (1977) and Hopfield (1982), and the value of sparse representations was quantified by Treves and Rolls (1991). Marr (1971) did not specify the functions of the dentate granule cells vs the CA3 cells vs the CA1 cells (which were addressed in the Rolls (1989) papers (see e.g. Rolls (1989b) and Rolls (1990b)) and by Treves and Rolls (1992b) and Treves and Rolls (1994)), nor how retrieval to the neocortex of hippocampal memories could be produced, for which a theory was developed by Rolls (1989b) and made quantitative by Treves and Rolls (1994). In addition, Treves and Rolls (1994) and Rolls and Treves (1998) have argued that approaches to neurocomputation which base their calculations on what would happen in the tail of an exponential, Poisson, or binomial distribution are very fragile.

[18]McNaughton and Morris (1987) at about the same time suggested that the CA3 network might be an autoassociation network, and that the mossy fibre to CA3 connections might implement 'detonator' synapses. However, the concepts that the diluted mossy fibre connectivity might implement selection of a new random set of CA3 cells for each new memory, and that a direct perforant path input to CA3 was needed to initiate retrieval, were introduced by

how the backprojections could operate (Rolls 1989b, Rolls 1989e), and how the dentate granule cells could operate as a competitive network and could further produce *pattern separation* by using the very diluted connectivity of the dentate to CA3 neuron mossy fibre synapses (Rolls 1989a, Rolls 1990b). Quantitative aspects of the theory were then developed with Alessandro Treves, who brought the expertise of theoretical physics applied previously mainly to understand the properties of fully connected attractor networks with binary neurons (Amit 1989, Hopfield 1982) to bear on the much more diluted connectivity of the recurrent collateral connections found in real biological networks (e.g. 2% between CA3 pyramidal cells in the rat), in networks of neurons with graded (continuously variable) firing rates, graded synaptic strengths, and sparse representations in which only a small proportion of the neurons is active at any one time, as is found in the hippocampus (Treves 1990, Treves and Rolls 1991). These developments in understanding quantitatively the operation of more biologically relevant recurrent networks with modifiable synapses were applied quantitatively to the CA3 region (Treves and Rolls 1991), and to the issue of why there are separate mossy fibre and perforant path inputs to the CA3 cells of the hippocampus (Treves and Rolls 1992b). This whole model of the hippocampus was described in more detail, and a quantitative treatment of the theory of recall by backprojection pathways in the brain was formulated, by Treves and Rolls (1994).

The speed of operation of the CA3 system, and of the cortico-cortical recall backprojections, was addressed in a number of new developments (Battaglia and Treves 1998b, Panzeri, Rolls, Battaglia and Lavis 2001, Simmen, Rolls and Treves 1996a, Treves 1993). Rolls (1995b) produced a simulation of the operation of the major parts of the hippocampus from the entorhinal cortex through the dentate, CA3, and CA1 cells back to the hippocampus, which established the quantitative feasibility of the whole theory, and raised a number of important issues considered below, including the role of topography within parts of the hippocampal internal connectivity. The simulation also emphasized some of the advantages, for a system that must store many different memories, of a binary representation, in which for any one memory the neurons were either firing or not, as opposed to having continuously graded firing rates. The simulation by Rolls (1995b) also showed how recall, if not perfect at the stage of the CA3 cells, was improved by associative synapses at subsequent stages, including the connections of the CA3 cells to the CA1 cells, and the connections of the CA1 cells to the entorhinal cortex cells.

The theory has been developed further (Stringer, Rolls and Trappenberg 2005, Rolls, Tromans and Stringer 2008f, Rolls 2010b, Webb, Rolls, Deco and Feng 2011, Rolls and Webb 2012, Rolls 2012a, Rolls 2013d, Rolls 2013b, Kesner and Rolls 2015, Rolls 2015b, Rolls 2016e), as described below.

Tests of the theory including analysis of the functions of different subregions of the hippocampus are described in Section 24.4 and by Rolls and Kesner (2006) and Kesner and Rolls (2015).

24.2 Systems-level functions of the hippocampus

Any theory of the hippocampus must state at the systems level what is computed by the hippocampus. Some of the relevant evidence comes from the systems-level connections of the hippocampus (Section 24.2.1), from the effects of damage to the hippocampus (Section 24.2.2), and from the responses of neurons in the hippocampus during behaviour (Section 24.2.4).

Treves and Rolls (1992b). Contributions by Levy (1989); McNaughton (1991); Hasselmo et al. (1995); McClelland et al. (1995); and many others, are described below.

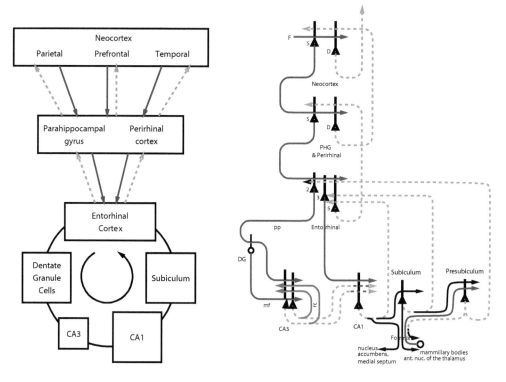

Fig. 24.1 Forward connections (blue solid lines) from areas of cerebral association neocortex via the parahippocampal gyrus and perirhinal cortex, and entorhinal cortex, to the hippocampus; and backprojections (green dashed lines) via the hippocampal CA1 pyramidal cells, subiculum, and parahippocampal gyrus to the neocortex. There is great convergence in the forward connections down to the single network implemented in the CA3 pyramidal cells; and great divergence again in the backprojections. Left: block diagram. Right: more detailed representation of some of the principal excitatory neurons in the pathways. Abbreviations: D, deep pyramidal cells; DG, dentate granule cells; F, forward inputs to areas of the association cortex from preceding cortical areas in the hierarchy. mf: mossy fibres; PHG, parahippocampal gyrus and perirhinal cortex; pp, perforant path; rc, recurrent collaterals of the CA3 hippocampal pyramidal cells shown in red; S, superficial pyramidal cells; 2, pyramidal cells in layer 2 of the entorhinal cortex; 3, pyramidal cells in layer 3 of the entorhinal cortex; 5, 6, pyramidal cells in the deep layers of the entorhinal cortex. The thick lines above the cell bodies represent the dendrites.

24.2.1 Systems-level anatomy

Fig. 24.1 shows that the primate hippocampus receives major inputs via the entorhinal cortex (area 28). These come from the highly developed parahippocampal gyrus (areas TF and TH) as well as the perirhinal cortex, and thereby from the ends of many processing streams of the cerebral association cortex, including the visual and auditory temporal lobe association cortical areas, the prefrontal cortex, and the parietal cortex (Van Hoesen 1982, Amaral, Price, Pitkanen and Carmichael 1992, Amaral 1987, Suzuki and Amaral 1994a, Witter, Wouterlood, Naber and Van Haeften 2000b, Lavenex, Suzuki and Amaral 2004, Johnston and Amaral 2004, Witter 2007, van Strien, Cappaert and Witter 2009, Kondo, Lavenex and Amaral 2009, Somogyi 2010, Rolls 2015c). The hippocampus is thus by its connections potentially able to associate together object representations (from the temporal lobe visual and auditory areas) and spatial representations. In addition, the entorhinal cortex receives inputs from the amygdala, and the orbitofrontal cortex, which could provide reward-related information to the hippocampus (Carmichael and Price 1995b, Suzuki and Amaral 1994a, Pitkanen et al. 2002, Stefanacci et

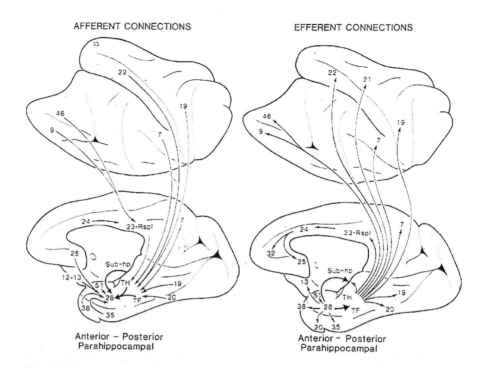

AFFERENT CONNECTIONS   EFFERENT CONNECTIONS

Anterior – Posterior
Parahippocampal

Anterior – Posterior
Parahippocampal

Fig. 24.2 Connections of the primate hippocampus with the neocortex (after Van Hoesen 1982). A medial view of the macaque brain is shown below, and a lateral view is shown inverted above. The hippocampus receives its inputs via the parahippocampal gyrus (areas TF and TH), and the perirhinal cortex (areas 35 and 36), which in turn project to the entorhinal cortex (area 28). The return projections to the neocortex (shown on the right) pass through the same areas. Cortical areas 19, 20, and 21 are visual association areas, 22 is auditory association cortex, 7 is parietal association cortex, and 9, 46, 12, and 13 are areas of frontal association cortex in the prefrontal cortex.

al. 1996, Rolls 2015c). The connections are analogous in the rat, although areas such as the temporal lobe visual cortical areas are not well developed in rodents, and the parahippocampal gyrus may be represented by the dorsal perirhinal cortex (Burwell et al. 1995).

Given that some topographic segregation is maintained in the afferents to the hippocampus in the perirhinal and parahippocampal cortices (Amaral and Witter 1989), it may be that these areas are able to subserve memory within one of these topographically separated areas, of for example visual object, or spatial, or olfactory information. In contrast, the final convergence afforded by the hippocampus into one network in CA3 (see Fig. 24.1) may be especially appropriate for an episodic memory typically involving arbitrary associations between any of the inputs to the hippocampus, e.g. spatial, vestibular related to self-motion, visual object, olfactory, and auditory (see below). There are also direct subcortical inputs from for example the amygdala and septum (Amaral 1986).

The primary output from the hippocampus to neocortex originates in CA1 and projects to subiculum, entorhinal cortex, and parahippocampal structures (areas TF-TH) as well as prefrontal cortex (Van Hoesen 1982, Delatour and Witter 2002, Witter 1993, van Haeften, Baks-te Bulte, Goede, Wouterlood and Witter 2003). CA1 and subiculum also project via the fimbria/fornix to subcortical areas such as the mammillary bodies and anterior thalamic nuclei (see Fig. 24.1). In addition to the projections originating in CA1, projections out of Ammon's horn also originate in CA3. Many researchers have reported that CA3 projects to the lateral and medial septal nuclei (Amaral and Witter 1995, Risold and Swanson 1997, Gaykema, van der Kuil, Hersh and Luiten 1991). The lateral septum also has projections to the medial septum (Jakab and Leranth 1995), which in turn projects to subiculum and eventually entorhinal cortex (Amaral and Witter 1995, Jakab and Leranth 1995).

24.2.2 Evidence from the effects of damage to the hippocampus

Damage to the hippocampus or to some of its connections such as the fornix in monkeys produces deficits in learning about the places of objects and about the places where responses should be made (Buckley and Gaffan 2000). For example, macaques and humans with damage to the hippocampal system or fornix are impaired in object–place memory tasks in which not only the objects seen, but where they were seen, must be remembered (Burgess et al. 2002, Crane and Milner 2005, Gaffan 1994, Gaffan and Saunders 1985, Parkinson, Murray and Mishkin 1988, Smith and Milner 1981). Posterior parahippocampal lesions in macaques impair even a simple type of object–place learning in which the memory load is just one pair of trial-unique stimuli (Malkova and Mishkin 2003). I further predict that a more difficult object-place learning task with non-trial-unique stimuli and with many object-place pairs would be impaired by neurotoxic hippocampal lesions. Neurotoxic lesions that selectively damage the primate hippocampus impair spatial scene memory (Murray, Baxter and Gaffan 1998). Also, fornix lesions impair conditional left–right discrimination learning, in which the visual appearance of an object specifies whether a response is to be made to the left or the right (Rupniak and Gaffan 1987). A comparable deficit is found in humans (Petrides 1985). Fornix sectioned monkeys are also impaired in learning on the basis of a spatial cue which object to choose (e.g. if two objects are on the left, choose object A, but if the two objects are on the right, choose object B) (Gaffan and Harrison 1989a). Monkeys with fornix damage are also impaired in using information about their place in an environment. For example, Gaffan and Harrison (1989b) found learning impairments when which of two or more objects the monkey had to choose depended on the position of the monkey in the room.

Rats with hippocampal lesions are impaired in using environmental spatial cues to re-member particular places (Cassaday and Rawlins 1997, Martin et al. 2000, O'Keefe and Nadel 1978, Jarrard 1993, Kesner, Lee and Gilbert 2004, Kesner and Rolls 2015), to utilize spatial cues or bridge delays (Kesner 1998, Kesner and Rolls 2001, Rawlins 1985, Kesner et al. 2004, Kesner and Rolls 2015), or it has been argued to perform relational operations on remembered material (Eichenbaum 1997). The evidence from rats is further considered in Section 24.4, where tests of the theory are described (Kesner and Rolls 2015).

In humans, functional neuroimaging shows that the hippocampal system is activated by allocentric spatial including scene processing (Burgess 2008, Chadwick, Mullally and Maguire 2013, Hassabis, Chu, Rees, Weiskopf, Molyneux and Maguire 2009, Maguire 2014, Zeidman and Maguire 2016) and by object-spatial episodic memory (Ison, Quian Quiroga and Fried 2015).

One way of relating the impairment of spatial processing to other aspects of hippocampal function is to note that this spatial processing involves a snapshot type of memory, in which one whole scene must be remembered. This memory may then be a special case of episodic memory, which involves an arbitrary association of a particular set of events which describe a past episode (Rolls 2008e, Rolls 2010b, Kesner and Rolls 2015). Further, the non-spatial tasks impaired by damage to the hippocampal system may be impaired because they are tasks in which a memory of a particular episode or context rather than of a general rule is involved (Gaffan, Saunders, Gaffan, Harrison, Shields and Owen 1984). Further, the deficit in paired associate learning in humans may be especially evident when this involves arbitrary associations between words, for example window – lake. Consistent with this, single neurons in the human hippocampus may respond not only to the sight or name of a film star, but also to objects associated with them, such as a bicycle (Quiroga, Reddy, Kreiman, Koch and Fried 2005, Fried et al. 2014, Rolls 2015d). The right (spatial) vs left (word) specialization of function in the human hippocampus (Crane and Milner 2005, Burgess et al. 2002, Barkas et al. 2010, Bonelli et al. 2010, Sidhu et al. 2013), associated in humans with a less well developed hippocampal commissural system, could be related to the fact that arbitrary associations between words and places are not required.

It is suggested that the reason why the hippocampus is used for the spatial and non-spatial types of memory described above, and the reason that makes these two types of memory so analogous, is that the hippocampus contains one stage, the CA3 stage, which acts as an autoassociation memory. It is suggested that an autoassociation memory implemented by the CA3 neurons equally enables whole (spatial) scene or episodic memories to be formed, with a snapshot quality that depends on the arbitrary associations that can be made and the short temporal window which characterises the synaptic modifiability in this system (Rolls 1990a, Rolls 1990b, Rolls 1987, Rolls 1989b, Rolls 1989e). The hypothesis is that the autoassociation memory enables arbitrary sets of concurrent neuronal firings, representing for example the spatial context where an episode occurred, the people present during the episode, and what was seen during the episode, to be associated together and stored as one event. (The associations are arbitrary in the sense that any representation within CA3 may be associated with any other representation in CA3.) The issue of the types of item, e.g. spatial, for which this type of associativity is hippocampus-dependent is considered below. Later recall of that episode from the hippocampus in response to a partial cue can then lead to reinstatement of the activity in the neocortex that was originally present during the episode. The theory described here shows how the episodic memory could be stored in the hippocampus, and later retrieved from the hippocampus and thereby to the neocortex using backprojections.

It is now evident that recognition memory (tested for example in primates by the ability to recognise a sample visual stimulus in a delayed match to sample task) is not impaired by hippocampal damage, produced for example by neurotoxic lesions (Murray et al. 1998, Baxter and Murray 2001b, Baxter and Murray 2001a). As shown below in Section 24.2.6, perirhinal cortex lesions do impair visual recognition memory in this type of task.

The extensive literature on the effects of hippocampal damage in rats is considered in Section 24.4, in particular the parts of it that involve selective lesions to different subregions of the hippocampus and allow the theory described here to be tested.

The theory described here is intended to be as relevant to rodents as to primates including humans, the main difference being that the spatial representation used in the hippocampus of rats is about the place where the rat is, whereas the spatial representation in primates includes a representation of space "out there", as represented by spatial view cells (see Section 24.2.4).

24.2.3   The necessity to recall information from the hippocampus

The information about episodic events recalled from the hippocampus could be used to help form semantic memories (Rolls 1990a, Rolls 1989b, Rolls 1989e, Rolls 1989f, Treves and Rolls 1994). For example, remembering many particular journeys could help to build a geographic cognitive map in the neocortex. The hippocampus and neocortex would thus be complementary memory systems, with the hippocampus being used for rapid, 'on the fly', unstructured storage of information involving activity potentially arriving from many areas of the neocortex; while the neocortex would gradually build and adjust, on the basis of much accumulating information, the semantic representation (Rolls 1989b, Treves and

Rolls 1994, McClelland et al. 1995). A view close to this in which a hippocampal episodic memory trace helps to build a neocortical semantic memory has also been developed by Moscovitch, Rosenbaum, Gilboa, Addis, Westmacott, Grady, McAndrews, Levine, Black, Winocur and Nadel (2005).

What I mean by an episodic memory of the type in which the hippocampus could be involved is described further next (Rolls and Kesner 2006, Kesner and Rolls 2015). A hippocampus-dependent episodic memory is unstructured in the sense that the information is stored as an association between representations such as spatial location, objects, and position in a temporal sequence. An episodic memory could be a single event capturing what happens in a time period in the order of 1 - 10 s (for example a memory of who was present at one's fifth birthday, what they were wearing, where they sat, where the party was, etc), or a sequence of events with some information about the temporal order of the events. A hippocampus-dependent episodic memory includes spatial and/or temporal e.g. sequence information. All this is in contrast to a neocortical semantic memory which might utilize many such episodic memories, for example of separate train journeys, to form a topological map. Such a semantic memory might need to benefit from information about exceptions, which contributes to the type of structure needed for a semantic memory. For example we may be told that a bat, although it flies, is not a bird because it is a mammal which nurses its young, and this inform-ation allows restructuring of a semantic representation of what constitutes a bird beyond the simple association and generalization from limited experience that birds fly. Because this active restructuring is an important part of semantic memory that may frequently benefit from language, I propose that semantic memories are best formed during waking using thought and reasoning, and not during sleep (Chapter 21). Indeed, I propose that episodic events that are recalled actively in this way from the hippocampus can become incorporated into neocortical representations and thus consolidated into semantic including autobiographical memories. If episodic material (such as where and with whom one had dinner three days ago) is not recalled in this way, then it will be gradually overwritten in hippocampal memory as new sparse distributed representations are added to the store, in the way that has been analyzed quantitatively (Section 9.3).

In contrast, some have suggested that the transfer of information from the hippocampus to the neocortex occurs especially during sleep (Marr 1971, Schwindel and McNaughton 2011). Indeed, it has been shown that after learning in hippocampal-dependent tasks, neocortical representations may change (Schwindel and McNaughton 2011). Although this has been interpreted as the transfer of memories from the hippocampus to the neocortex (Schwindel and McNaughton 2011), it should be noted that if the hippocampal representation changes as a result of learning, then the altered representation in CA1 will, even with fixed synaptic connections back to neocortex, alter neocortical firing, with no learning or actual 'transfer' involved. (This occurs whenever one vector comprised by a population of neurons firing changes and influences another vector of neuronal firing through fixed connections.) Indeed, for the reasons given above, active transfer of episodic memories from the hippocampus to the neocortex to form semantic memories is more likely I propose during the guided construction and correction of semantic memories in the cortex during waking (Chapter 21).

The actual addition of new information to the neocortical semantic representation could itself be fast, for example when one recalls a particular train journey, and incorporates inform-ation from that into a geographical representation. An additional property of hippocampal episodic memories is that when recalled to the neocortex they can be used flexibly for many types of behavioral output (because the information is back in the neocortex). I further note that rapid learning is not a unique feature of episodic memory, with for example the stimulus-reward learning and its reversal implemented in the primate orbitofrontal cortex possible in one trial (Rolls 2014a) (see also Section 2.2.5).

The quantitative evidence considered below (Section 24.3.3.1) and by Treves and Rolls (1994) on the storage capacity of the CA3 system indicates that episodic memories could be stored in the hippocampus for at least a considerable period. This raises the issue of the possible gradient of retrograde amnesia following hippocampal damage, and of whether information originally hippocampus-dependent for acquisition gradually becomes 'consolidated' in the neocortex and thereby becomes independent of the hippocampus over time (McGaugh 2000, Debiec et al. 2002). The issue of whether memories stored some time before hippocampal damage are less impaired than more recent memories, and whether the time course is minutes, hours, days, weeks or years is a debated issue (Gaffan 1993, Squire 1992). (In humans, there is evidence for a gradient of retrograde amnesia; in rats and monkeys, hippocampal damage in many studies appears to impair previously learned hippocampal-type memories, suggesting that in these animals, at least with the rather limited numbers of different memories that need to be stored in the tasks used, the information remains in the hippocampus for long periods.) If there is a gradient of retrograde amnesia related to hippocampal damage, then this suggests that information may be retrieved from the hippocampus if it is needed, allowing the possibility of incorporating the retrieved information into neocortical memory stores. If on the other hand there is no gradient of retrograde amnesia related to hippocampal damage, but old as well as recent memories of the hippocampal type are stored in the hippocampus, and lost if it is damaged, then again this implies the necessity of a mechanism to retrieve information stored in the hippocampus, and to use this retrieved information to affect neural circuits elsewhere (for if this were not the case, information stored in the hippocampus could never be used for anything). The current perspective is thus that whichever view of the gradient of retrograde amnesia is correct, information stored in the hippocampus will need to be retrieved and affect other parts of the brain in order to be used, and the current theory shows how the recall could be implemented.

The other major set of outputs from the hippocampus projects via the fimbria/fornix system to the anterior nucleus of the thalamus (both directly and via the mammillary bodies), which in turn projects to the cingulate cortex. This may provide an output for more action-directed use of information stored in the hippocampus, for example in the initiation of conditional spatial responses in a visual conditional spatial response task (Rupniak and Gaffan 1987, Miyashita, Rolls, Cahusac, Niki and Feigenbaum 1989). In such a task, a rapid mapping must be learned between a visual stimulus and a spatial response, and a new mapping must be learned each day. The hippocampus is involved in this rapid visual to spatial response mapping (Rupniak and Gaffan 1987), and the way in which hippocampal circuitry may be appropriate for this is that the CA3 region enables signals originating from very different parts of the cerebral cortex to be associated rapidly together (see below).

24.2.4 Systems-level neurophysiology of the primate hippocampus

24.2.4.1 Spatial view neurons

The systems-level neurophysiology of the hippocampus shows what information could be stored or processed by the hippocampus. To understand how the hippocampus works it is not sufficient to state just that it can store information – one needs to know what information. The systems-level neurophysiology of the primate hippocampus has been reviewed by Rolls and Xiang (2006), and a summary is provided here because it provides a perspective relevant to understanding the function of the human hippocampus that is somewhat different from that provided by the properties of place cells in rodents, which have been reviewed elsewhere (McNaughton, Barnes and O'Keefe 1983, O'Keefe 1984, Muller, Kubie, Bostock, Taube and Quirk 1991, Jeffery and Hayman 2004, Jeffery, Anderson, Hayman and Chakraborty 2004, Howard and Eichenbaum 2015, Knierim and Neunuebel 2016).

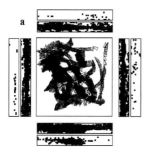
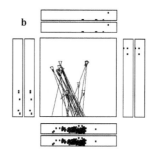
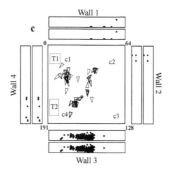

Fig. 24.3 Examples of the firing of a hippocampal spatial view cell when the monkey was walking around the laboratory. (a) The firing of the cell is indicated by the spots in the outer set of 4 rectangles, each of which represents one of the walls of the room. There is one spot on the outer rectangle for each action potential. The base of the wall is towards the centre of each rectangle. The positions on the walls fixated during the recording sessions are indicated by points in the inner set of 4 rectangles, each of which also represents a wall of the room. The central square is a plan view of the room, with a triangle printed every 250 ms to indicate the position of the monkey, thus showing that many different places were visited during the recording sessions. (b) A similar representation of the same 3 recording sessions as in (a), but modified to indicate some of the range of monkey positions and horizontal gaze directions when the cell fired at more than 12 spikes/s. (c) A similar representation of the same 3 recording sessions as in (b), but modified to indicate more fully the range of places when the cell fired. The triangle indicates the current position of the monkey, and the line projected from it shows which part of the wall is being viewed at any one time while the monkey is walking. One spot is shown for each action potential. (After Georges-François, Rolls and Robertson 1999.)

The primate hippocampus contains spatial cells that respond when the monkey looks at a certain part of space, for example at one quadrant of a video monitor while the monkey is performing an object–place memory task in which he must remember where on the monitor he has seen particular images (Rolls, Miyashita, Cahusac, Kesner, Niki, Feigenbaum and Bach 1989b). Approximately 9% of the hippocampal neurons have such spatial view fields, and about 2.4% combine information about the position in space with information about the object that is in that position in space (Rolls et al. 1989b). The latter point shows that information from very different parts of the cerebral cortex (parietal for spatial information, and inferior temporal for visual information about objects) is brought together onto single neurons in the primate hippocampus. The representation of space is for the majority of hippocampal neurons in allocentric not egocentric coordinates (Feigenbaum and Rolls 1991).

In rats, place cells are found, which respond depending on the place where the rat is in a spatial environment (McNaughton et al. 1983, Muller et al. 1991, O'Keefe 1984). To analyse whether such cells might be present in the primate hippocampus, Rolls and O'Mara (1993) and Rolls and O'Mara (1995) recorded the responses of hippocampal cells when macaques were moved in a small chair or robot on wheels in a cue-controlled testing environment. The most common type of cell responded to the part of space at which the monkeys were looking, independently of the place where the monkey was. Ono, Nakamura, Nishijo and Eifuku (1993) performed studies on the representation of space in the primate hippocampus while the monkey was moved in a cab to different places in a room. They found that 13.4% of hippocampal formation neurons fired more when the monkey was at some but not other places in the test area, but it was not clear whether the responses of these neurons occurred to the place where the monkey was independently of spatial view, or whether the responses of place-like cells were view dependent.

Place cells fire best during active locomotion by the rat (Foster, Castro and McNaughton 1989). To investigate whether place cells might be present in monkeys if active locomotion was being performed, we recorded from single hippocampal neurons while monkeys moved themselves round the test environment by walking (or running) on all fours. We discovered in these investigations **hippocampal spatial view neurons** that responded significantly differently for different allocentric spatial views and had information about spatial view in their firing rate, but did not respond differently just on the basis of eye position, head direction, or place (Georges-François, Rolls and Robertson 1999, Robertson, Rolls and Georges-François 1998, Rolls, Robertson and Georges-François 1997a, Rolls, Treves, Robertson, Georges-François and Panzeri 1998b). An example of such a hippocampal spatial view cell present in the primate hippocampus is shown in Fig. 24.3. If the view details are obscured by curtains and darkness, then some spatial view neurons (especially those in CA1 and less those in CA3) continue to respond when the monkey looks towards the spatial view field (see example in Fig. 24.4). This experiment (Robertson, Rolls and Georges-François 1998, Rolls, Robertson and Georges-François 1997a) shows that primate hippocampal spatial view neurons can be updated for at least short periods by idiothetic (self-motion) cues including eye position and head direction signals. Consistent with this, a small drift is sometimes evident after a delay when the view details are obscured, consistent with inaccuracies in the temporal path integration of these signals, which is then corrected by showing the visual scene again.

It is essential to measure the firing rate of a primate hippocampal cell with different head directions so that different spatial views can be compared, as testing with just one head direction (Matsumura, Nishijo, Tamura, Eifuku, Endo and Ono 1999) cannot provide evidence that will distinguish a place cell from a spatial view cell. These points will need to be borne in mind in future studies of hippocampal neuronal activity in primates including humans (Ekstrom et al. 2003, Fried, MacDonald and Wilson 1997, Kreiman et al. 2000), and simultaneous recording of head position, head direction, and eye position, as described by Rolls and colleagues (Georges-François, Rolls and Robertson 1999, Robertson, Rolls and Georges-François 1998, Rolls, Robertson and Georges-François 1997a, Rolls, Treves, Robertson, Georges-François and Panzeri 1998b) will be needed. To distinguish spatial view from place cells it will be important to test neurons while the primate or human is in one place with all the different spatial views visible from that place; and also to test the same neuron when the organism is in a different place, but at least some of the same spatial views are visible, as has been done in our primate single neuron recording (Georges-François, Rolls and Robertson 1999, Robertson, Rolls and Georges-François 1998, Rolls, Robertson and Georges-François 1997a, Rolls, Treves, Robertson, Georges-François and Panzeri 1998b, Rolls and Xiang 2006). Virtual reality testing environments may not allow adequate separation of these different contributions to the firing of primate including human hippocampal spatial representations, and of course do not provide for any normal idiothetic (self-motion) inputs, but has allowed hippocampal neurons with spatial view fields, frequently modulated by place, to be confirmed (Wirth, Baraduc, Plante, Pinede and Duhamel 2017).

A

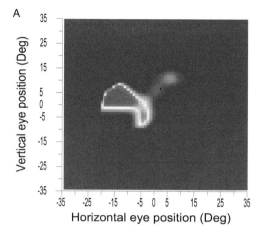

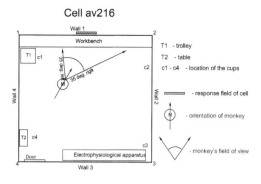

B

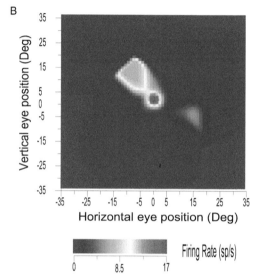

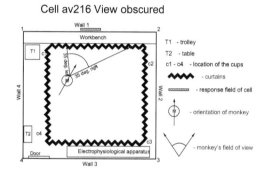

Fig. 24.4 Idiothetic update (by self-motion of the eyes) of the firing of a hippocampal spatial view cell. This cell fired when the macaque looked towards the effective spatial view even when the view was obscured with curtains. The firing is shown with the monkey stationary with his head facing in the direction indicated by the arrow when the curtains were drawn open (A) or were drawn closed (B). Left: firing rate of the cell in spikes/s is indicated by the colour (a calibration bar in spikes/s is shown below) projected onto the monkey's field of view. The two-dimensional firing rate profile of the cell was smoothed for clarity using a 2-dimensional Gaussian spatial filter. The space adequately sampled by the eye movements of the monkey is indicated by light blue. The eye position in degrees is shown. Right: a plan view of the room to indicate the monkey's view of the wall is shown. M, position of the monkey. (After Robertson, Rolls and Georges-François 1998.)

24.2.4.2 Spatial view associated with object neurons

A fundamental question about the function of the primate including human hippocampus is whether object as well as allocentric spatial information is represented. To investigate this, Rolls, Xiang and Franco (2005c) made recordings from single hippocampal formation neurons while macaques performed an object–place memory task which required the monkeys to learn associations between objects and where they were shown in a room. Some neurons (10%)

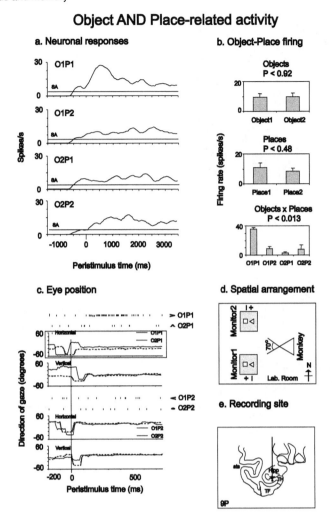

Fig. 24.5 A hippocampal neuron (BL003c8a) with firing occurring to a particular combination of an object and place (Object 1–Place 1, O1P1). (a) Peristimulus time histograms showing the response of the neuron to each object–place combination. (b) The firing rate response of the neuron on different trials sorted according to the object that was shown on the trial, the place where the object was shown, and the combination of which object was shown with the place where it was shown. In the object*position bar histograms, O1P1 = Object 1 in Place 1, etc. The mean responses ± sem are shown. (c) Trials to show the eye positions on 4 different trials. (d) Schematic diagram showing the spatial arrangement used in the Go–NoGo object–place combination task in this experiment. + indicates that a Go (lick) response to that object–place combination will be rewarded, and a minus that it will be punished. Only one object was shown on the screen at a time. The room was approximately 4 m x 4 m, and the distance of the monkey from the monitors was typically 1.5–2 m in the different experiments. The triangle labelled monkey shows the position of the monkey, and the angle shows the approximate angle subtended by the video display monitors. (e) The recording site of the neuron. Hipp, hippocampus; Prh, perirhinal cortex; rhs, rhinal sulcus; sts, superior temporal sulcus; TE, inferior temporal visual cortex. 9P = 9 mm Posterior to the sphenoid reference. (After Rolls, Xiang and Franco 2005c.)

responded differently to different objects independently of location; other neurons (13%) responded to the spatial view independently of which object was present at the location; and some neurons (12%) responded to a combination of a particular object and the place where it was shown in the room. An example of a neuron responding to a particular object–place combination is shown in Fig. 24.5. These results show that there are separate as well as combined representations of objects and their locations in space in the primate hippocampus. This is a property required in an episodic memory system, for which association between objects and the places where they are seen, is prototypical. The results thus show that requirements for a human episodic memory system, separate and combined neuronal representations of objects and where they are seen "out there" in the environment, are present in the primate hippocampus (Rolls, Xiang and Franco 2005c). These results have been confirmed in humans, in whom association learning between a spatial view and an object or person are reflected in the responses of medial temporal lobe neurons (Ison, Quian Quiroga and Fried 2015).

What may be a corresponding finding in rats is that some rat hippocampal neurons respond on the basis of the conjunction of location and odor (Wood, Dudchenko and Eichenbaum 1999). Results consistent with our object-place neurons in primates are that Diamond and colleagues have now shown using the vibrissa somatosensory input for the 'object' system, that rat hippocampal neurons respond to object-place combinations, objects, or places, and there is even a reward-place association system in rats (Itskov, Vinnik and Diamond 2011) similar to that in primates. This brings the evidence from rats closely into line with the evidence from primates of hippocampal neurons useful for object-place episodic associative memory.

Primate hippocampal neuronal activity has also been shown to be related to the recall of memories. In a one-trial object–place recall task, images of an object in one position on a screen, and of a second object in a different position on the screen, were shown successively. Then one of the objects was shown at the top of the screen, and the monkey had to recall the position in which it had been shown earlier in the trial, and to touch that location (Rolls and Xiang 2006). In addition to neurons that responded to the objects or places, a new type of neuronal response was found in which 5% of hippocampal neurons had spatial view-related responses when a spatial view was being recalled by an object cue (Rolls and Xiang 2006, Kesner and Rolls 2015).

24.2.4.3 Spatial view associated with reward neurons

The primate anterior hippocampus (which corresponds to the rodent ventral hippocampus) receives inputs from brain regions involved in reward processing such as the amygdala and orbitofrontal cortex (Carmichael and Price 1995b, Suzuki and Amaral 1994a, Pitkanen et al. 2002, Stefanacci et al. 1996, Rolls 2015c). To investigate how this affective input may be incorporated into primate hippocampal function, Rolls and Xiang (2005) recorded neuronal activity while macaques performed a reward-place association task in which each spatial scene shown on a video monitor had one location which if touched yielded a preferred fruit juice reward, and a second location which yielded a less preferred juice reward. Each scene had different locations for the different rewards. Of 312 hippocampal neurons analyzed, 18% responded more to the location of the preferred reward in different scenes, and 5% to the location of the less preferred reward (Rolls and Xiang 2005). An example of one of these neurons, responding to the place in each scene of a preferred reward, is shown in Fig. 24.6. When the locations of the preferred rewards in the scenes were reversed, 60% of 44 neurons tested reversed the location to which they responded, showing that the reward-place associations could be altered by new learning in a few trials. The majority (82%) of these 44 hippocampal reward-place neurons tested did not respond to object-reward associations in a visual discrimination object-reward association task. Thus the primate hippocampus contains a representation of the reward associations of places "out there" being viewed, and this is a way in which affective information can be stored as part of an episodic memory, and how the current mood state may influence the retrieval of episodic memories (Rolls and Xiang 2005). There

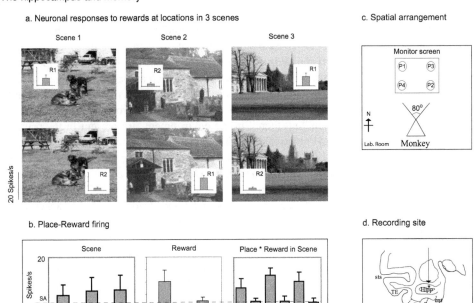

Fig. 24.6 A hippocampal neuron that encoded the particular rewards available at different locations in different scenes. On each trial the monkey could touch a circled location in the scene, and, depending on the location, received either a preferred juice reward or a less preferred juice reward. (a) Firing rate inserts to show the firing in 3 different scenes (S1–S3) of the locations associated with reward 1 (R1, preferred) and reward 2 (R2). The mean responses ± sem are shown. SA, spontaneous firing rate. (b) The firing rates sorted by scene, by reward (1 vs 2), and by scene-reward combinations (e.g. scene 1 reward 1 = S1R1). (c) The spatial arrangement on the screen of the 4 spatial locations (P1–P4). (d) The recording site of the neuron. ent, entorhinal cortex; Hipp, hippocampal pyramidal cell field CA3/CA1 and dentate gyrus; Prh, perirhinal cortex; rhs, rhinal sulcus; sts, superior temporal sulcus; TE, inferior temporal visual cortex. (After Rolls and Xiang 2005.)

is consistent recent evidence that rewards available in a spatial environment can influence the responsiveness of rodent place neurons (Tabuchi, Mulder and Wiener 2003, Mizumori and Tryon 2015).

In another type of task for which the primate hippocampus is needed, conditional spatial response learning, in which the monkeys had to learn which spatial response to make to different stimuli, that is, to acquire associations between visual stimuli and spatial responses, 14% of hippocampal neurons responded to particular combinations of visual stimuli and spatial responses (Miyashita, Rolls, Cahusac, Niki and Feigenbaum 1989). The firing of these neurons could not be accounted for by the motor requirements of the task, nor wholly by the stimulus aspects of the task, as demonstrated by testing their firing in related visual discrimination tasks. These results showed that single hippocampal neurons respond to combinations of the visual stimuli and the spatial responses with which they must become associated in conditional response tasks, and are consistent with the computational theory described above according to which part of the mechanism of this learning involves associations between visual stimuli and spatial responses learned by single hippocampal neurons. In a following study by Cahusac, Rolls, Miyashita and Niki (1993), it was found that during such conditional spatial response learning, 22% of this type of neuron analysed in the hippocampus and parahippocampal gyrus altered their responses so that their activity, which was initially equal to the two new stimuli, became progressively differential to the two stimuli when the monkey learned to make different responses to the two stimuli (Cahusac, Rolls, Miyashita and Niki 1993). These changes occurred for different neurons just before, at, or just after the time when the monkey learned the correct response to make to the stimuli, and are consistent with the hypothesis that when new associations between objects and places (in this case the places for responses) are learned, some hippocampal neurons learn to respond to the new associations that are required to solve the task. Similar findings have been described by Wirth, Yanike, Frank, Smith, Brown and Suzuki (2003).

24.2.4.4   Whole body motion or speed neurons

Another type of neuron found in the primate hippocampus responds to whole body motion (O'Mara, Rolls, Berthoz and Kesner 1994). Some of these neurons respond to linear motion, and others to axial whole body rotation (which can be produced by sitting the macaque on a moving robot). Some of these neurons respond when the visual field is obscured, and thus respond on the basis of vestibular or proprioceptive inputs. Other neurons respond to rotation of the visual environment round the macaque, and thus encode motion by visual optic flow cues. Some neurons respond to both these types of input (O'Mara, Rolls, Berthoz and Kesner 1994). The significance of these neurons is that they could be part of the way in which self-motion, i.e. idiothetic, cues are used to update the representation in the hippocampal formation of the current head direction, or the current position, and of the current spatial view, as described further below. These whole body motion neurons could also be useful in episodic memory, by encoding the movements that were taking place on a particular occasion.

This discovery of whole body motion cells in the macaque hippocampus (O'Mara, Rolls, Berthoz and Kesner 1994) has been confirmed for the rat hippocampus, in which 'speed cells' have now been described (Kropff, Carmichael, Moser and Moser 2015).

The inputs for whole body motion may reach the hippocampus via the mammillary body circuitry which receives inputs from the tegmental nuclei of Gudden (Dillingham et al. 2015). This may provide one route for vestibular inputs, important of course in head direction representation, to reach the presubiculum, where head direction cells are found in primates (Robertson, Rolls, Georges-François and Panzeri 1999) as well as in rodents (Wiener and Taube 2005). The mammillary bodies project to the anterior thalamic nuclei which in turn project to the cingulate and retrosplenial cortex, which in turn project into areas such as the subiculum, and thus can provide inputs to the hippocampal system (see Fig. 24.1 and Aggleton and Nelson (2015)).

Further properties of primate hippocampal neurons are described by Rolls and Xiang (2006) and Rolls and Wirth (2017).

24.2.4.5   Spatial view grid cells in the primate entorhinal cortex

In primates, there is now evidence that there is a grid-cell like representation in the entorhinal cortex, with neurons having grid-like firing as the monkey moves the eyes across a spatial scene (Killian, Jutras and Buffalo 2012). Similar competitive learning processes may transform these entorhinal cortex 'spatial view grid cells' into hippocampal spatial view cells, and may help with the idiothetic (produced in this case by movements of the eyes) update of spatial view cells (Robertson, Rolls and Georges-François 1998). The presence of spatial view grid cells in the entorhinal cortex of primates is of course predicted from the presence of spatial view cells in the primate CA3 and CA1 regions (Rolls and Xiang 2006).

Further support for this type of representation of space being viewed 'out there' rather than or in addition to where one is located as for rat place cells is that neurons in the human entorhinal cortex with spatial view grid-like properties have now been described (Jacobs, Weidemann, Miller, Solway, Burke, Wei, Suthana, Sperling, Sharan, Fried and Kahana 2013).

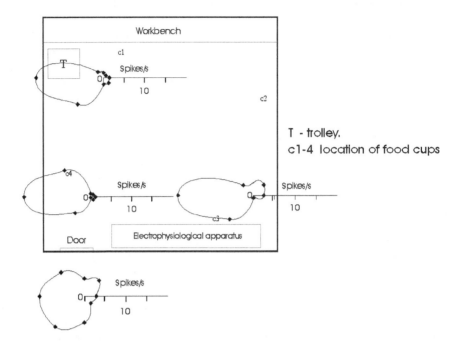

Fig. 24.7 The responses of a primate head direction cell. Polar response plots of the firing rate (in spikes/s) when the monkey was stationary at different positions (shown at the 0 on the firing rate scale) in (and one outside) the room are shown. The monkey was rotated to face in each direction. The mean response of the cell from at least 4 different firing rate measurements in each head direction in pseudorandom sequence are shown. c1–c3: cups to which the monkey could walk on all fours to obtain food. Polar firing rate response plots are superimposed on an overhead view of the square room to show where the firing for each plot was recorded. The plot at the lower left was taken outside the room, in the corridor, where the same head direction firing was maintained. (After Robertson, Rolls, Georges-Francois and Panzeri 1999.)

In humans, there is now some evidence for medial temporal lobe neurons with properties like those of spatial view cells (Ekstrom, Kahana, Caplan, Fields, Isham, Newman and Fried 2003); and some medial temporal lobe neurons learn associations between views of places, and people or objects (Ison, Quian Quiroga and Fried 2015).

24.2.5 Head direction cells in the presubiculum

Head direction cells found in rats (Ranck 1985, Taube, Muller and Ranck 1990a, Taube, Goodridge, Golob, Dudchenko and Stackman 1996, Muller, Ranck and Taube 1996, Wiener and Taube 2005) and in primates (Robertson, Rolls, Georges-François and Panzeri 1999) respond maximally when the animal's head is facing in a particular preferred direction. An example of a primate head direction cell recorded in the presubiculum of the monkey is shown in Fig. 24.7. Similar cells are found in the rat pre-/post-subiculum, but also in a number of connected structures including the anterior thalamic nuclei (Taube et al. 1996, Wiener and Taube 2005). The firing rate of these cells is a function of the head direction of the monkey, with a response that is typically 10–100 times larger to the optimal as compared to the opposite head direction. The mean half-amplitude width of the tuning of the cells was 76 degrees. The response of head direction cells in the presubiculum was not influenced by the place where the monkey was, there being the same tuning to head direction at different places in a room, and even outside the room. The response of these cells was also independent of the 'spatial view' observed by the monkey, and also the position of the eyes in the head. The cells maintain their tuning for periods of at least several minutes when the view details are obscured or the room is darkened (Robertson, Rolls, Georges-François and Panzeri 1999).

This representation of head direction could be useful together with the hippocampal spatial view cells and whole body motion cells found in primates in a number of spatial and memory functions. One is path integration for head direction, the process by which in the absence of visual input one's head direction is updated based on idiothetic, that is self-motion, signals reflecting recent movements or movement 'paths'. The memory mechanism required is an integration over time of self-motion signals produced by for example vestibular inputs or motor commands. Network computational mechanisms by which this memory function may be performed are described in Section B.5.5. Another memory function for which head direction cells may be important is in episodic memory. An episodic memory might include for example the fact that one had turned right by 90 degrees in the dark, and was therefore facing towards a particular spatial view, which one could later recall. Another spatial / memory function in which head direction cells are implicated is in updating using path integration one's representation of the place where one is located. Indeed, place cells in rats may be updated in the dark through an effect of head direction cells in rotating the place map, and normally the place cell and head direction maps are kept in alignment (Leutgeb, Leutgeb, Treves, Meyer, Barnes, McNaughton, Moser and Moser 2005, Wiener and Taube 2005, Yoganarasimha, Yu and Knierim 2006).

Self-organizing continuous attractor neural networks that can perform path integration from velocity signals are described in Section B.5.5 and include head direction from head velocity (Stringer, Trappenberg, Rolls and De Araujo 2002b, Stringer and Rolls 2006); place from whole body motion (Stringer, Trappenberg, Rolls and De Araujo 2002b, Stringer, Rolls, Trappenberg and De Araujo 2002a), and spatial view from eye and whole body motion (Stringer, Rolls and Trappenberg 2004, Stringer, Rolls and Trappenberg 2005, Rolls and Stringer 2005). Path integration of head direction is reflected in the firing of neurons in the presubiculum, and mechanisms outside the hippocampus, perhaps in the dorsal tegmental nucleus (Taube, Muller and Ranck 1990b, Sharp 1996, Bassett and Taube 2005), probably implement path integration for head direction.

Indeed, one function of the mammillary body circuitry may be to introduce a head direction signal from the tegmental nuclei of Gudden via the anterior thalamic nuclei to the hippocampal circuitry (Dillingham et al. 2015). This may provide one route for vestibular inputs, important of course in head direction representation, to reach the presubiculum, where head direction cells are found in primates (Robertson, Rolls, Georges-François and Panzeri 1999) as well as in rodents (Wiener and Taube 2005). Indeed, the anterior thalamic nuclei project to the cingulate and retrosplenial cortex, which in turn project into areas such as the subiculum, and thus can provide inputs to the hippocampal system about for example head direction (see Fig. 24.1 and Aggleton and Nelson (2015)).

24.2.6 Perirhinal cortex, recognition memory, and long-term familiarity memory

The functions of the perirhinal cortex are different to those of the hippocampus, and are reviewed here partly because it provides afferents to the hippocampus via the entorhinal cortex, and partly to show that the computations performed in it are different to hippocampal computations. The location of the perirhinal cortex is shown in Fig. 24.10.

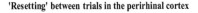

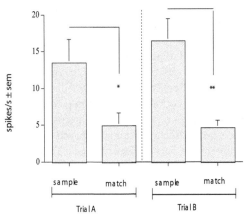

Fig. 24.8 'Resetting' of perirhinal cortex neuronal responses between trials in a short-term memory task (delayed match to sample). Neuronal responses are larger to the sample than the match presentation of a particular image within a trial, and are large to the sample on the next trial ('Trial B') even if it is the same image that has just been shown a few seconds earlier as the match image on the preceding trial. * = $p < 0.01$ and ** = $p < 0.005$ in a Wilcoxon post-hoc test. The mean and sem of the responses across neurons are shown. (After Holscher and Rolls 2002.)

The perirhinal cortex is involved in recognition memory in that damage to the perirhinal cortex produces impairments in recognition memory tasks in which several items intervene between the sample presentation of a stimulus and its presentation again as a match stimulus (Malkova, Bachevalier, Mishkin and Saunders 2001, Zola-Morgan, Squire, Amaral and Suzuki 1989, Zola-Morgan, Squire and Ramus 1994). In macaques, damage to the perirhinal cortex rather than to the hippocampus produces deficits in recognition memory measured in a delayed match to sample task with intervening stimuli (Murray, Baxter and Gaffan 1998, Baxter and Murray 2001b, Baxter and Murray 2001a, Buckley and Gaffan 2006). Further, damage to the perirhinal cortex rather than to the hippocampus is believed to underlie the impairment in recognition memory found in amnesia in humans associated with medial temporal lobe damage (Buckley and Gaffan 2000, Buckley and Gaffan 2006).

Neurophysiologically, it has been shown that many inferior temporal cortex (area TE) neurons (Rolls and Deco 2002, Rolls 2000a), which provide visual inputs to the perirhinal cortex (Suzuki and Amaral 1994a, Suzuki and Amaral 1994b), respond more to the first, than to the second, presentation of a stimulus in a running recognition task with trial-unique stimuli (Baylis and Rolls 1987). In this running recognition task (originally developed for neurophysiology by Rolls, Perrett, Caan and Wilson (1982)), there is typically a presentation of a novel stimulus, and after a delay which may be in the order of minutes or more and in which other stimuli may be shown, the stimulus is presented again as 'familiar', and the monkey can respond to obtain food reward. Many inferior temporal cortex neurons responded more to the 'novel' than to the 'familiar' presentation of a stimulus, where 'familiar' in this task reflects a change produced by seeing the stimulus typically once (or a few times) before. (A small proportion of neurons respond more to the familiar (second) than to the novel (first) presentation of each visual stimulus.) In the inferior temporal cortex this memory spanned up to 1–2 intervening stimuli between the first (novel) and second (familiar) presentations of a given stimulus (Baylis and Rolls 1987), and when recordings are made more ventrally, towards and within the perirhinal cortex, the memory span increases to several or more intervening stimuli (Wilson, Riches and Brown 1990, Brown and Xiang 1998, Xiang and Brown 1998). The implication is that the inferior temporal visual cortex can implement a very simple form of recognition memory in which neurons respond more to the first compared to a second presentation of a stimulus with 1–2 intervening stimuli, and that the same rather simple type of recognition process can extend over more intervening stimuli as one moves towards the perirhinal cortex. The underlying mechanism could be a type of habituation or adaptation which is partly stimulus-specific because neurons in these regions can have different responses to different stimuli. For comparison, very few primate hippocampal neurons have activity related to recognition memory (Rolls, Cahusac, Feigenbaum and Miyashita 1993).

The perirhinal cortex is directly or closely connected with a number of other brain regions involved in the long-term recognition memory for visual stimuli. Relatively far forward in the orbitofrontal cortex, in area 11, neurons are activated by novel visual stimuli (Rolls, Browning, Inoue and Hernadi 2005a), and activations in humans are produced by novel visual stimuli (Frey and Petrides 2002). The responses of these neurons decrease over on average 5 presentations of a novel stimulus, and may be related to long-term memory in that the responses do not occur to a stimulus that was shown as novel 24 h earlier (Rolls, Browning, Inoue and Hernadi 2005a). Using the running recognition memory task, Rolls, Perrett, Caan and Wilson (1982) discovered a population of neurons that responded to familiar visual stimuli and were related to long-term memory at the anterior border of the thalamus, in a region that is at the border of the ventral anterior thalamic nucleus (VA) and the reticular nucleus of the thalamus (see also Wilson and Rolls (1990c)). Neurons that were also activated in a running recognition memory task by familiar stimuli and that may be related to these thalamic neurons were described by Xiang and Brown (2004) in a ventromedial prefrontal region that they termed PFCvm, and a ventrolateral prefrontal region that they termed PFCo. Wilson and Rolls (1993) showed that the activity of some neurons in the amygdala was related to stimulus novelty (with memory spans of 2–10 intervening other stimuli) and of others to familiarity in monkeys performing recognition memory tasks. Wilson and Rolls (1990c) described neuronal responses related to novel visual stimuli in the substantia innominata and diagonal band of Broca. These neurons, also activated in other memory tasks (Wilson and Rolls 1990b, Wilson and Rolls 1990a), may be the cholinergic neurons that project to the cerebral cortex and that may, by providing acetylcholine to the cerebral cortex, help in the consolidation of memories by the release of acetylcholine at the time that a memory is being formed in the neocortex. Consistent with this, cholinergic basal forebrain lesions produce severe learning impairments in learning visual scenes and in object-reward associations (Easton, Ridley, Baker and Gaffan 2002), and damage to the cholinergic basal forebrain neurons is implicated in Alzheimer's disease (Whitehouse, Price, Struble, Clarke, Coyle and Delong 1982, Bierer, Haroutunian, Gabriel, Knott, Carlin, Purohit, Perl, Schmeidler, Kanof and Davis 1995). The neurophysiological evidence of Wilson and Rolls (1990c), Wilson and Rolls (1990b) and Wilson and Rolls (1990a) indicates that these basal forebrain neurons do change their firing rate during different types of association memory and recognition tasks, and thus the phasic release of acetyl choline to the cortex at the time of memory formation may be important to consolidation, by facilitating LTP (Bear and Singer 1986).

This interconnected perirhinal cortex / orbitofrontal cortex system may thus be involved in setting up neurons that respond differently to novel and familiar stimuli and thereby implement recognition memory, and may also influence the operation of other memory systems by influences on the basal forebrain cholinergic neurons which have widespread connections to the cerebral cortex.

A second type of memory in which the perirhinal cortex is involved is the short-term memory for visual stimuli. In a task typically performed with non-trial-unique stimuli, a delayed matching-to-sample task with up to several intervening stimuli, some perirhinal cortex neurons respond more to the match stimulus than to the sample stimulus (Miller, Li and Desimone 1993b). Many other neurons in this task respond more to the sample ('novel') than to the match ('familiar') presentations of the stimuli. There is active engagement of the perirhinal cortex in this delayed matching-to-sample task, in that the response of these neurons is always larger to the sample than to the match stimulus, even if the stimulus shown as the sample has been seen just seconds ago on a previous trial (Hölscher and Rolls 2002), as shown in Fig. 24.8. Thus this short-term memory in the perirhinal cortex is actively reset at the start of the next trial (Hölscher and Rolls 2002).

A third type of memory in which the perirhinal cortex is implicated is paired associate learning (a model of semantic long-term memory), which is represented by a population of neurons in a restricted part of area 36 where the neuronal responses may occur to both members of a pair of pictures separated by a delay used in the paired association task (Miyashita, Kameyama, Hasegawa and Fukushima 1998, Miyashita, Okuno, Tokuyama, Ihara and Nakajima 1996). The learning of this task may be facilitated by the short-term memory-related firing that follows the first member of the pair during the delay period (Hirabayashi, Takeuchi, Tamura and Miyashita 2013). Another very plausible function for these associations across time is not semantic memory but instead view invariant learning, which can be helped by associations across time, as described in Chapter 25, and which may be impaired by perirhinal cortex lesions (Buckley, Booth, Rolls and Gaffan 2001).

Evidence that the perirhinal cortex is involved in a fourth type of memory, long-term familiarity memory, comes from a neuronal recording study in which it was shown that perirhinal cortex neuronal responses in the rhesus macaque gradually increase in magnitude to a set of stimuli as that set is repeated for 400 presentations each 1.3 s long (Hölscher, Rolls and Xiang 2003). The single neurons were recorded in the perirhinal cortex in monkeys performing a delayed matching-to-sample task with up to 3 intervening stimuli, using a set of very familiar visual stimuli used for several weeks. When a novel set of stimuli was introduced, the neuronal responses were on average only 47% of the magnitude of the responses to the familiar set of stimuli. It was shown in eight different replications in three monkeys that the responses of the perirhinal cortex neurons gradually increased over hundreds of presentations of the new set of (initially novel) stimuli to become as large as to the already familiar stimuli. Examples of three replications are shown in Fig. 24.9, and the recording sites were in the perirhinal cortex as indicated in Fig. 24.10 (see Hölscher, Rolls and Xiang (2003)). The mean number of 1.3 s presentations to induce this effect was 400 occurring over 7–13 days. These results show that perirhinal cortex neurons represent the very long-term familiarity of visual stimuli (Hölscher, Rolls and Xiang 2003). A representation of the long-term familiarity of visual stimuli may be important for many aspects of social and other behaviour, and part of the impairment in temporal lobe amnesia may be related to the difficulty of building representations of the degree of familiarity of stimuli. It has been shown that long-term familiarity memory can be modelled by neurons with synapses with a small amount of long-term potentiation which occurs at every presentation of a given stimulus, but which also incorporate hetero-synaptic long-term depression to make the learning self-limiting (Rolls, Franco and Stringer 2005b).

A fifth type of memory in which the perirhinal cortex is implicated is in tasks in which the stimuli to be discriminated are made up of a small number of features that are combined in different combinations. The features in each stimulus are thus highly overlapping, and the task requires setting up different representations of the different but highly overlapping feature conjunctions in a low-dimensional feature space. This is very different to object representations as they typically occur in the world, in which each object is a particular combination of features from a very high-dimensional feature space, so that the representations of different objects in general have much less overlap in feature space, and the inferior temporal visual cortex

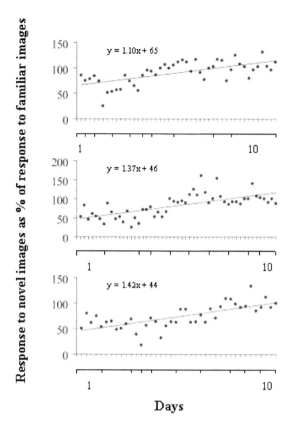

Fig. 24.9 Perirhinal cortex neurons reflect the long-term familiarity of visual stimuli. The average response of each neuron (shown by a full circle) to a set of 15 novel stimuli introduced on day 1 is shown as a proportion of the neuron's response to a very familiar set of stimuli. In each experiment on a neuron, each visual stimulus was shown for approximately 13 1-s periods. Several neurons were analyzed on some days. Thus in the order of 500 exposures to each stimulus in the novel set were required before the response to the initially novel set of stimuli was as large as to the already very familiar set of stimuli. Each stimulus was an image of an object presented on a video monitor. The 3 separate graphs are for 3 separate replications of the whole experiment. (After Hölscher and Rolls 2001, and Hölscher, Rolls and Xiang 2003b.)

can form object representations by forming neurons that respond to different combinations of features from a high-dimensional feature space. In any case, it is argued that perirhinal cortex lesions impair the discrimination between stimuli according to how ambiguous they are in terms of their degree of overlap of a few features (Bussey and Saksida 2005, Buckley and Gaffan 2006, Bussey, Saksida and Murray 2002, Bussey, Saksida and Murray 2003, Bussey, Saksida and Murray 2005). (However, a more mnemonic than perceptual case is made by Hampton (2005) based on the evidence for delay-dependent deficits produced by perirhinal cortex lesions in recognition memory tasks, and reversal-learning impairments which reflect a more associative than perceptual role for the perirhinal cortex.) In the extreme case, such a stimulus set might require A+, B+, but AB- (where A+ means that A is rewarded), which requires a non-linear discrimination. One way to solve such memory problems is to add a further layer to an existing object representation network, where further combinations of what is represented at earlier layers can be formed. This is equivalent to adding a fifth layer to

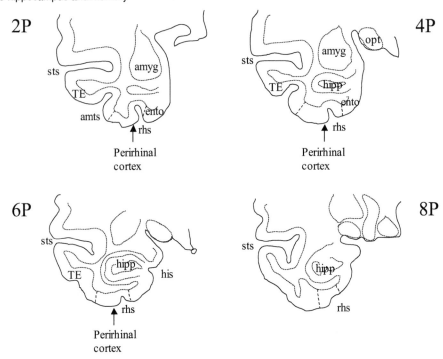

Fig. 24.10 The locations of the perirhinal cortex (areas 35 and 36 in and near the banks of the rhinal sulcus), the entorhinal cortex, and the hippocampus in macaques. The coronal (transverse) sections are at different distances behind (P, posterior to) an anatomical marker, the sphenoid bone, which is approximately at the anterior–posterior position of the optic chiasm and anterior commissure. amts, anterior middle temporal sulcus; amyg, amygdala; ento, entorhinal cortex; hipp, hippocampus; his, hippocampal sulcus; opt, optic tract; rhs, rhinal sulcus; sts, superior temporal sulcus; TE, inferior temporal visual cortex.

the VisNet architecture described in Chapter 25, and adding such an additional layer, which might correspond to adding a perirhinal cortex layer to an inferior temporal cortex layer, does, as expected, help a network to solve the class of problem where the stimuli consist of different highly overlapping combinations of a small number of features (Cowell, Bussey and Saksida 2006). However, when we recorded in the primate perirhinal cortex, we did not find much evidence for very specifically tuned object neurons, but did find that neuronal activity is related to short-term perceptual memory (Hölscher and Rolls 2002). In the light of this, and the fact that inferior temporal visual cortex neurons do provide very good invariant representations of objects which are complex combinations of features (Rolls 2012c, Rolls and Treves 2011), the possibility is suggested that one function of the perirhinal cortex may be to provide a visual short-term memory, which is usually required for the tasks such as oddity with multiple distracters that led to the conjunctive / configural hypothesis (Bussey et al. 2005). Consistently, the visual short-term memory capability of the perirhinal cortex is probably important also for its role in delayed paired-associate learning (Fujimichi, Naya, Koyano, Takeda, Takeuchi and Miyashita 2010).

It then becomes a somewhat semantic issue about whether one considers that the perirhinal cortex is involved in object perception as well as in memory (Bussey and Saksida 2005, Buckley and Gaffan 2006). My own view is that the perirhinal cortex can be thought of as involved in the memory for and discrimination between stimuli when the stimuli are composed of a small number of features with different combinations of features for different stimuli, so

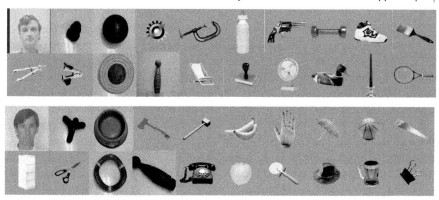

Fig. 24.11 Some examples of the objects and faces encoded by anterior inferior temporal cortex neurons (see text, and further examples in Fig. C.7).

that in this way, and perhaps in other ways, the stimuli, which could be objects, are ambiguous. The perirhinal cortex may thus contribute in these special cases. These special cases should be distinguished from object perception that occurs with normal objects in the world, in which each object is composed of a combination of features drawn from a high-dimensional feature space. Object representation in this sense is implemented in the anterior inferior temporal visual cortex, as shown by evidence described in Chapter 25 and Appendix C.

The evidence includes the findings that the object that has been presented on a single trial can be read off (decoded) from the firing of anterior inferior temporal cortex neurons with position, size, view, spatial frequency, and lighting invariance. Each neuron is tuned to a subset of the objects with a firing rate probability distribution that is frequently close to exponential, indicating that each neuron has high responses to a few objects, to which it is 'tuned', with smaller responses to a few objects, and very little response to most objects. Moreover, the coding for objects is very efficient, in that the response profiles of different neurons to a set of objects are uncorrelated, and this contributes to the information provided by different neurons being independent, so that the number of objects encoded by a population of neurons increases exponentially with the number neurons. These properties are found for objects drawn from the real world, including fruits, bottles, and a multitude of small toys, with some examples illustrated in Figs. 24.11, C.7, 25.3, 25.59, C.24, 25.64, and elsewhere (Booth and Rolls 1998, Rolls and Tovee 1995b). (The objects are shown as greyscale images so that colour can not be used as a defining feature.) These properties are also found for faces, which are certainly not just 'features', but are complex combinations of features where the features must be presented in the correct spatial arrangement for many of the neurons to respond, and where even the spacing of the features is important (Perrett, Rolls and Caan 1982, Rolls, Tovee, Purcell, Stewart and Azzopardi 1994b). The variety of the tuning of anterior inferior temporal cortex neurons is enormously wide, as befits a system that is involved in object representation, with some neurons responding to quite simple features such as low spatial frequency gratings, or the texture of the material from which objects are made, through to more complex features such as eyes or nose or mouth, through to neurons that only respond to combinations of these if they are in the correct spatial arrangement, through to neurons with view-dependent responses, and through to neurons with view-independent responses that can be finely tuned to objects or faces (Perrett, Rolls and Caan 1982, Rolls, Tovee, Purcell, Stewart and Azzopardi 1994b, Tanaka 1996, Booth and Rolls 1998, Franco, Rolls, Aggelopoulos and Jerez 2007, Rolls and Deco 2002), with many further sources cited in Chapter 25 and Appendix C. Neurons responding to one or several features are more likely to be found in the posterior inferior temporal cortex (Tanaka 1996) (see Fig. 1.10), whereas neurons responding to objects or faces and often requiring a complex combination of features to be present are found in the anterior inferior temporal cortex (see Fig. 1.10) in which we have recorded (Baylis, Rolls and Leonard 1987, Rolls, Tovee, Purcell, Stewart and Azzopardi 1994b, Booth and Rolls 1998, Rolls and Tovee 1995b, Rolls 2007e, Franco, Rolls, Aggelopoulos and Jerez 2007, Rolls 2007c). The anterior inferior temporal cortex neurons even maintain their object selectivity when the objects are presented in complex natural backgrounds (Rolls, Aggelopoulos and Zheng 2003a, Aggelopoulos, Franco and Rolls 2005) or are surrounded by other objects (Aggelopoulos and Rolls 2005). With this range of the type of tuning of inferior temporal cortex neurons, it is sometimes possible to reduce the feature set within an object and obtain a response, while in other cases most of the components need to be present, and in the correct spatial configuration with respect to each other, before the object-encoding or face-encoding neuron will respond (Perrett, Rolls and Caan 1982, Rolls, Tovee, Purcell, Stewart and Azzopardi 1994b, Tanaka 1996).

The perirhinal cortex, by receiving from the inferior temporal visual cortex, and perhaps by performing further conjunction learning to that already implemented by the anterior inferior temporal visual cortex (see Chapter 25), thus has visual representations of objects, and some at least of the perirhinal cortex neurons have view-invariant representations.

Another factor that distinguishes the perirhinal cortex is that the perirhinal cortex receives forward inputs from other modalities, including auditory, olfactory and somatosensory (Suzuki and Amaral 1994a, Suzuki and Amaral 1994b). This may enable the perirhinal cortex by combining these multimodal inputs to build representations that reflect the convergence of more than visual inputs. The perirhinal cortex, via its projections to the lateral entorhinal cortex and via the lateral perforant path, could thus introduce object information into the hippocampus for object–place associations. The parahippocampal gyrus, in which cells with spatial information are present in macaques (Rolls, Xiang and Franco 2005c, Rolls and Xiang 2006), could via connections to the medial entorhinal cortex introduce spatial information into the hippocampus, and there is evidence consistent with this in rodents (Hargreaves, Rao, Lee and Knierim 2005). In contrast to the perirhinal cortex, the final convergence afforded by the hippocampus into one network in CA3 (see Fig. 24.1) may enable the hippocampus proper to implement an event or episodic memory typically involving arbitrary associations between any of the inputs to the hippocampus, e.g. spatial, visual object, olfactory, and auditory (see below).

24.3 A theory of the operation of hippocampal circuitry as a memory system

Given the systems-level hypothesis about what the hippocampus performs, and the neurophysiological evidence about what is represented in the primate hippocampus, the next step is to consider how using its internal connectivity and synaptic modifiability the hippocampus could store and retrieve many memories, and how retrieval within the hippocampus could lead to retrieval of the activity in the neocortex that was present during the original learning of the episode. To develop understanding of how this is achieved, we have developed a computational theory of the operation of the hippocampus (Rolls 1989a, Rolls 1989b, Rolls 1989e, Rolls 1989f, Rolls 1987, Treves and Rolls 1994, Rolls 1990a, Rolls 1990b, Treves and Rolls 1991, Treves and Rolls 1992b, Rolls 1996c, Rolls and Treves 1998, Rolls, Stringer and Trappenberg 2002, Stringer, Rolls and Trappenberg 2005, Rolls and Stringer 2005, Rolls and Kesner 2006, Rolls, Stringer and Elliot 2006c, Kesner and Rolls 2015). This theory, and new developments of it, are outlined next.

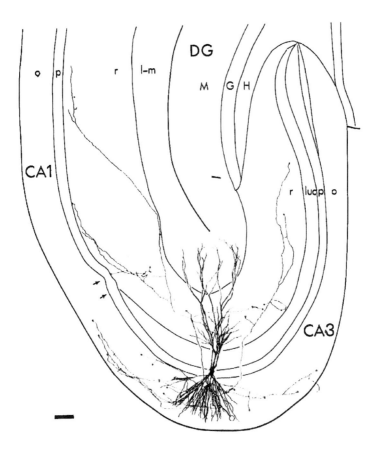

Fig. 24.12 An example of a CA3 neuron from the hippocampus. The thick extensions from the cell body or soma are the dendrites, which form an extensive dendritic tree receiving in this case approximately 12,000 synapses. The axon is the thin connection leaving the cell. It divides into a number of collateral branches. Two axonal branches can be seen in the plane of the section to travel to each end of the population of CA3 cells. One branch (on the left) continues to connect to the next group of cells, the CA1 cells. The junction between the CA3 and CA1 cells is shown by the two arrows. The diagram shows a camera lucida drawing of a single CA3 pyramidal cell intracellularly labelled with horseradish peroxidase. DG, dentate gyrus. The small letters refer to the different strata of the hippocampus. (Reprinted with permission from Journal of Comparative Neurology, 295: 580–623. Organization of intrahippocampal projections originating from CA3 pyramidal cells in the rat, Norio Ishizuk, Janet Weber and David G. Amaral. Copyright ©1990 Wiley-Liss, Inc.)

24.3.1 Hippocampal circuitry

Fig. 24.1 shows a diagram of hippocampal circuitry that is described next, based on many sources (Amaral, Ishizuka and Claiborne 1990, Storm-Mathiesen, Zimmer and Ottersen 1990, Amaral and Witter 1989, Amaral 1993, Naber, Lopes da Silva and Witter 2001, Lavenex, Suzuki and Amaral 2004, Witter, Wouterlood, Naber and Van Haeften 2000b, Johnston and Amaral 2004, Witter 2007, van Strien, Cappaert and Witter 2009, Kondo, Lavenex and Amaral 2009).

Projections from the entorhinal cortex layer 2 reach the granule cells (of which there are 10^6 in the rat) in the dentate gyrus (DG), via the perforant path (pp) (Witter 1993). In the dentate gyrus there is a set inhibitory interneurons that provide recurrent inhibition

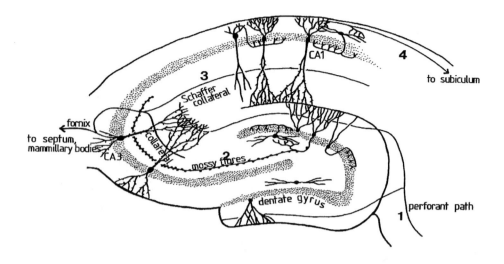

Fig. 24.13 Representation of connections within the hippocampus. Inputs reach the hippocampus through the perforant path (1) which makes synapses with the dendrites of the dentate granule cells and also with the apical dendrites of the CA3 pyramidal cells. The dentate granule cells project via the mossy fibres (2) to the CA3 pyramidal cells. The well-developed recurrent collateral system of the CA3 cells is indicated. The CA3 pyramidal cells project via the Schaffer collaterals (3) to the CA1 pyramidal cells, which in turn have connections (4) to the subiculum.

between the granule cells. (In addition, there are some excitatory interneurons in the hilus that interconnect granule cells.) The granule cells project to CA3 cells via the mossy fibres (mf), which provide a sparse but possibly powerful connection to the 3×10^5 CA3 pyramidal cells in the rat. Each CA3 cell receives approximately 50 mossy fibre inputs, so that the sparseness of this connectivity is thus 0.005%. By contrast, there are many more – probably weaker – direct perforant path inputs also from layer 2 of the entorhinal cortex onto each CA3 cell, in the rat in the order of 4×10^3. The largest number of synapses (about 1.2×10^4 in the rat) on the dendrites of CA3 pyramidal cells is, however, provided by the (recurrent) axon collaterals of CA3 cells themselves (rc) (see Fig. 24.14). It is remarkable that the recurrent collaterals are distributed to other CA3 cells throughout the hippocampus (Amaral and Witter 1989, Amaral and Witter 1995, Amaral, Ishizuka and Claiborne 1990, Ishizuka, Weber and Amaral 1990, Witter 2007), so that effectively the CA3 system provides a single network, with a connectivity of approximately 2% between the different CA3 neurons given that the connections are bilateral. The CA3–CA3 recurrent collateral system is even more extensive in macaques than in rats (Kondo, Lavenex and Amaral 2009). The neurons that comprise CA3, in turn, project to CA1 neurons via the Schaffer collaterals. In addition, projections that terminate in the CA1 region originate in layer 3 of the entorhinal cortex (see Fig. 24.1).

24.3.2 Entorhinal cortex

The entorhinal cortex provides inputs to the hippocampus, as well as receiving the outputs of the hippocampus, so its operation is considered briefly first. The entorhinal cortex performs interesting computations in its own right. In the medial entorhinal cortex, there are grid cells, which have regularly spaced peaks of firing in an environment, so that as a rat runs through an environment, a single neuron increases then decreases its firing a number of times as the rat

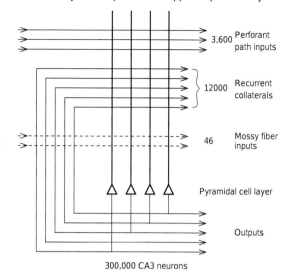

3,600 — Perforant path inputs

12000 — Recurrent collaterals

46 — Mossy fiber inputs

Pyramidal cell layer

Outputs

300,000 CA3 neurons

Fig. 24.14 The numbers of connections onto each CA3 cell from three different sources in the rat. (After Treves and Rolls 1992, and Rolls and Treves 1998.)

traverses the environment (Hafting, Fyhn, Molden, Moser and Moser 2005, Moser, Rowland and Moser 2015). Each grid cell responds to a set of places in a spatial environment, with the places to which a cell responds set out in a regular grid. Different grid cells have different phases (positional offsets) and grid spacings or frequencies (Hafting, Fyhn, Molden, Moser and Moser 2005, Giocomo, Moser and Moser 2011, Moser, Rowland and Moser 2015, Kropff and Treves 2008). The grid cell system appears to provide ring continuous attractors that would be useful not only for spatial path integration (computing position based on self-motion) (Giocomo et al. 2011, Moser, Moser and Roudi 2014a, Zilli 2012), but also for the timing information useful in sequence encoding for non-spatial as well as spatial information, as described in Section 24.3.6.2.

One attractive theory is that different temporal delays in the neural circuitry of the medial entorhinal cortex are related to different temporal adaptation time courses and result in the different sizes of spatial grids that are found in the medial entorhinal cortex (Kropff and Treves 2008). This is essentially a timing hypothesis; and it is an interesting new idea that the timing mechanism that may be inherent in the medial entorhinal cortex might also be a source of the temporal delay information needed to form sequence memories in CA1, using the direct projection from medial entorhinal cortex to CA1 to introduce timing information in the form of neurons that fire at different times in a sequence, which is what is needed for the formation of odor, object and spatial sequence memories in CA1 (see Section 24.3.6.2), and would also be useful in the formation of spatial grid cells in the entorhinal cortex. Consistent with this hypothesis, timing cells are found in the medial entorhinal cortex (Kraus, Brandon, Robinson, Connerney, Hasselmo and Eichenbaum 2015).

The lateral entorhinal cortex may be more involved in representing object information, and provides a route for inputs from the ventral visual system to reach hippocampal circuitry (Hargreaves et al. 2005, Witter 1993, Knierim, Neunuebel and Deshmukh 2014).

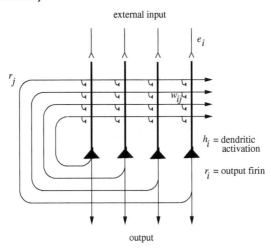

external input

e_i

r_j

w_{ij}

h_i = dendritic activation

r_i = output firin

output

Fig. 24.15 The architecture of an attractor or autoassociation neural network. The architecture of a discrete attractor neural network used for storing discrete patterns is the same as that of a continuous attractor neural network (CANN) used for storing continuous, e.g. spatial, patterns (see further Sections B.3 and B.5).

24.3.3 CA3 as an autoassociation memory

24.3.3.1 Arbitrary associations, and pattern completion in recall

CA3 recurrent collaterals form a single attractor network

Many of the synapses in the hippocampus show associative modification as shown by long-term potentiation, and this synaptic modification appears to be involved in learning (Lynch 2004, Morris 2003, Morris 1989, Morris, Moser, Riedel, Martin, Sandin, Day and O'Carroll 2003, Nakazawa et al. 2003, Nakazawa et al. 2004, Andersen, Morris, Amaral, Bliss and O'Keefe 2007, Wang and Morris 2010, Jackson 2013) (Chapter 9). On the basis of the evidence summarized above, Rolls has proposed that the CA3 stage acts as an autoassociation memory which enables episodic memories to be formed and stored in the CA3 network, and that subsequently the extensive recurrent collateral connectivity allows for the retrieval of a whole representation to be initiated by the activation of some small part of the same representation (the cue) (Rolls 1987, Rolls 1989a, Rolls 1989b, Rolls 1989f, Rolls 1990a, Rolls 1990b). The crucial synaptic modification for this is in the recurrent collateral synapses. (A description of the operation of autoassociative networks is provided in Section B.3 and by Hertz et al. (1991) and Rolls and Treves (1998). The architecture of an autoassociation network is shown in Fig. 24.15.)

The hypothesis is that because the CA3 operates effectively as a single network, it can allow arbitrary associations between inputs originating from very different parts of the cerebral cortex to be formed. These might involve associations between information originating in the temporal visual cortex about the presence of an object, and information originating in the parietal cortex about where it is. I note that although there is some spatial gradient in the CA3 recurrent connections, so that the connectivity is not fully uniform (Ishizuka et al. 1990, Witter 2007), nevertheless the network will still have the properties of a single interconnected autoassociation network allowing associations between arbitrary neurons to be formed, given the presence of many long-range connections which overlap from different CA3 cells. It is very interesting indeed that in primates (macaques), the associational projections from CA3 to CA3 travel extensively along the longitudinal axis, and overall the radial, transverse, and longitudinal gradients of CA3 fiber distribution, clear in the rat, are much more subtle in the nonhuman primate brain (Kondo, Lavenex and Amaral 2009). The implication is that in primates, the CA3 network operates even more as a single network than in rodents.

The autoassociation or attractor memory is different from pattern association memory, in which a visual stimulus might become associated with a taste by associative synaptic modification. Later presentation of the visual stimulus would retrieve the taste representation. However, presentation of the taste would not retrieve the visual representation, and this is an important and fundamental difference between autoassociation and pattern association, as described in detail in Appendix B. Pattern association memories generalize, and auto-association memories complete incomplete representations. Further, competitive networks can generalize, but may not complete perfectly depending on what input is being clamped into the network (Appendix B).

Tests of the attractor hypothesis of CA3 operation based on the effects of CA3 lesions or CA3 NMDA knockouts on pattern completion are described in Section 24.5.1 (Kesner and Rolls 2015). Another type of test of the autoassociation (or attractor) hypothesis for CA3 has been to train rats in different environments, e.g. a square and a circular environment, and then test the prediction of the hypothesis that when presented with an environment am-biguous between these, hippocampal neurons will fall into an attractor state that represents one of the two previously learned environments, but not a mixture of the two environments. Evidence consistent with the hypothesis has been found (Wills, Lever, Cacucci, Burgess and O'Keefe 2005). In a particularly dramatic example, it has been found that within each theta cycle, hippocampal pyramidal neurons may represent one or other of the learned environ-ments (Jezek, Henriksen, Treves, Moser and Moser 2011). This was measured by recording in an ambiguous environment, in which it was found that the CA3 representation 'flickered' between two possible environments. This is an indication, predicted by Rolls and Treves (1998) (page 118), that autoassociative memory recall can take place sufficiently rapidly to be complete within one theta cycle (120 ms), and that theta cycles could provide a mechanism for a fresh retrieval process to occur after a reset caused by the inhibitory part of each theta cycle, so that the memory can be updated rapidly to reflect a continuously changing environment, and not remain too long in an attractor state.

Storage capacity

We have performed quantitative analyses of the storage and retrieval processes in the CA3 network (Treves and Rolls 1991, Treves and Rolls 1992b). We have extended previous formal models of autoassociative memory (Amit 1989) by analyzing a network with graded response units, so as to represent more realistically the continuously variable rates at which neurons fire, and with incomplete connectivity (Treves 1990, Treves and Rolls 1991). We have found that in general the maximum number p_{max} of firing patterns that can be (individually) retrieved is proportional to the number C^{RC} of (associatively) modifiable recurrent collateral synapses per cell, by a factor that increases roughly with the inverse of the population sparseness a^p of the neuronal representation. The sparseness a^p of the representation can be measured, by extending the binary notion of the proportion of neurons that are firing, as

$$a^p = \frac{(\sum_{i=1}^{N} y_i/N)^2}{\sum_{i=1}^{N} y_i^2/N} \tag{24.1}$$

where y_i is the firing rate of the ith neuron in the set of N neurons (Treves and Rolls 1991, Rolls and Treves 1998, Rolls and Deco 2002, Franco, Rolls, Aggelopoulos and Jerez 2007). More precisely, Treves and Rolls (1991) and Rolls, Treves, Foster and Perez-Vicente (1997c) have shown that such a network with graded patterns does operate efficiently as an autoassociative network, and can store (and recall correctly) a number of different patterns p as follows

$$p \approx \frac{C^{RC}}{a^p \ln(\frac{1}{a^p})} k \tag{24.2}$$

where C^{RC} is the number of synapses on the dendrites of each neuron devoted to the recurrent collaterals from other neurons in the network, and k is a factor that depends weakly on the detailed structure of the rate distribution, on the connectivity pattern, etc., but is roughly in the order of 0.2–0.3. For example, for $C^{RC} = 12,000$ and $a^p = 0.02$ (realistic estimates for the rat), p_{max} is calculated to be approximately 36,000. This analysis emphasizes the utility of having a sparse representation in the hippocampus, for this enables many different memories to be stored. Third, in order for most associative networks to store information efficiently, heterosynaptic long term depression (as well as LTP) is required (Fazeli and Collingridge 1996, Rolls and Treves 1990, Treves and Rolls 1991) (see Section B.3). Simulations that are fully consistent with the analytic theory are provided by Simmen, Treves and Rolls (1996b) and Rolls, Treves, Foster and Perez-Vicente (1997c).

We have also indicated how to estimate I, the total amount of information (in bits per synapse) that can be retrieved from the network. I is defined with respect to the information i_p (in bits per cell) contained in each stored firing pattern, by subtracting the amount i_l lost in retrieval and multiplying by p/C^{RC}:

$$I \approx \frac{p}{C^{RC}}(i_p - i_l) \tag{24.3}$$

The maximal value I_{max} of this quantity was found (Treves and Rolls 1991) to be in several interesting cases around 0.2–0.3 bits per synapse, with only a mild dependency on parameters such as the sparseness of coding a^p.

We may then estimate (Treves and Rolls 1992b) how much information has to be stored in each pattern for the network to efficiently exploit its information retrieval capacity I_{max}. The estimate is expressed as a requirement on i_p:

$$i_p > a^p \ln(\frac{1}{a^p}) \tag{24.4}$$

As the information content i_p of each stored pattern depends on the storage process, we see how the retrieval capacity analysis, coupled with the notion that the system is organized so as to be an efficient memory device in a quantitative sense, leads to a constraint on the storage process.

A number of points deserve comment. First, if it is stated that a certain number of memories is the upper limit of what could be stored in a given network, then the question is sometimes asked, what constitutes a memory? The answer is precise. Any one memory is represented by the firing rates of the population of neurons that are stored by the associative synaptic modification, and can be correctly recalled later. The firing rates in the primate hippocampus might be constant for a period of for example 1 s in which the monkey was looking at an object in one position in space; synaptic modification would occur in this time period (cf. the time course of LTP, which is sufficiently rapid for this); and the memory of the event would have been stored. The quantitative analysis shows how many such random patterns of rates of the neuronal population can be stored and later recalled correctly. If the rates were constant for 5 s while the rat was at one place or monkey was looking at an object at one position in space, then the memory would be for the pattern of firing of the neurons in this 5-s period. (The pattern of firing of the population refers to the rate at which each neuron in the population of CA3 neurons is firing.)

Second, the question sometimes arises of whether the CA3 neurons operate as an attractor network. An attractor network is one in which a stable pattern of firing is maintained once it has been started. Autoassociation networks trained with modified Hebb rules can store the number of different memories, each one expressed as a stable attractor, indicated in equation B.15. However, the hippocampal CA3 cells do not necessarily have to operate as a stable attractor: instead, it would be sufficient for the present theory if they can retrieve stored information in response to a partial cue initiating retrieval. The partial cue would remain on during recall, so that the attractor network would be operating in the clamped condition (Rolls and Treves 1998) (see Section B.3). The completion of the partial pattern would then provide more information than entered the hippocampus, and the extra information retrieved would help the next stage to operate. Demonstrations of this by simulations that are fully consistent with the analytic theory are provided by Rolls (1995b), Simmen, Treves and Rolls (1996b), and Rolls, Treves, Foster and Perez-Vicente (1997c).

Third, in order for most associative networks to store information efficiently, hetero-synaptic long term depression (as well as LTP) is required (Rolls and Treves 1990, Treves and Rolls 1991, Rolls 1996b). Without heterosynaptic LTD, there would otherwise always be a correlation between any set of positively firing inputs acting as the input pattern vector to a neuron. LTD effectively enables the average firing of each input axon to be subtracted from its input at any one time, reducing the average correlation between different pattern vectors to be stored to a low value (Rolls 1996b) (see Section B.3).

Fourth, the firing of CA3 cells is relatively sparse, and this helps to decorrelate different population vectors of CA3 cell firing for different memories (Rolls 2013d, Rolls 2013b). (Sparse representations are more likely to be decorrelated with each other (Appendices B and C).) Evidence on the sparseness of the CA3 cell representation in rats includes evidence that CA3 cell ensembles may support the fast acquisition of detailed memories by providing a locally continuous, but globally orthogonal spatial representation, onto which new sensory inputs can rapidly be associated (Leutgeb and Leutgeb 2007). In the macaque hippocampus, in which spatial view cells are found, for the representation of 64 locations around the walls of the room, the mean single cell sparseness a^s was 0.34, and the mean population sparseness a^p was 0.33 (Rolls, Treves, Robertson, Georges-François and Panzeri 1998b, Rolls and Treves 2011). For comparison, the corresponding values for macaque inferior temporal cortex neurons tuned to objects and faces were 0.77 (Franco, Rolls, Aggelopoulos and Jerez 2007, Rolls and Treves 2011); for taste and oral texture neurons in the insular cortex the population sparseness was 0.71; for taste and oral texture neurons in the orbitofrontal cortex was 0.61; and for taste and oral texture neurons in the amygdala was 0.81 (Rolls and Treves 2011). Thus the evidence is that the hippocampal CA3 / pyramidal cell representation is more sparse in macaques than in neocortical areas and the amygdala, and this is consistent with the importance in hippocampal CA3 of using a sparse representation to produce a large memory capacity. Although the value of a for the sparseness of the representation does not seem especially low, it must be remembered that these are graded firing rate representations, and the graded nature appears to increase the value of a (Rolls and Treves 2011). A further analysis of how to interpret these sparseness measures is provided in Section C.3.1.1.

Fifth, given that the memory capacity of the hippocampal CA3 system is limited, it is necessary to have some form of forgetting in this store, or another mechanism, to ensure that its capacity is not exceeded. (Exceeding the capacity can lead to a loss of much of the information retrievable from the network.) Heterosynaptic LTD could help this forgetting, by enabling new memories to overwrite old memories (Rolls 1996b) (see Section 9.3). The limited capacity of the CA3 system does also provide one of the arguments that some transfer of information from the hippocampus to neocortical memory stores may be useful (see Treves and Rolls (1994)). Given its limited capacity, the hippocampus might be a useful store for only a limited period, which might be in the order of days, weeks, or months. This period may well depend on the acquisition rate of new episodic memories. If the animal were in a constant and limited environment, then as new information is not being added to the hippocampus, the representations in the hippocampus would remain stable and persistent. These hypotheses have clear experimental implications, both for recordings from single neurons and for the gradient of retrograde amnesia, both of which might be expected to depend on whether the environment is stable or frequently changing. They show that the conditions under which a gradient of retrograde amnesia might be demonstrable would be when large numbers of new memories are being acquired, not when only a few memories (few in the case of the hippocampus being less than a few hundred) are being learned.

The potential link to the gradient of retrograde amnesia is that the retrograde memories lost in amnesia are those not yet consolidated in longer term storage (in the neocortex). As they are still held in the hippocampus, their number has to be less than the storage capacity of the (presumed) CA3 autoassociative memory. Therefore the time gradient of the amnesia provides not only a measure of a characteristic time for consolidation, but also an upper bound on the rate of storage of new memories in CA3. For example, if one were to take as a measure of the time gradient in the monkey, say, 5 weeks (about 50,000 min) (Squire 1992) and as a reasonable estimate of the capacity of CA3 in the monkey e.g $p = 50,000$, then one would conclude that there is an upper bound on the rate of storage in CA3 of not more than one new memory per minute, on average. (This might be an average over many weeks; the fastest rate might be closer to 1 per s, see Treves and Rolls (1994).) These quantitative considerations are consistent with the concept described in Section 24.2.3 that retrieval of episodic memories from the hippocampus for a considerable time after they have been stored would be useful in helping the neocortex to build a semantic memory (Rolls 1990a, Rolls 1989f, Rolls 1989b, Rolls 1990a, Treves and Rolls 1994, Moscovitch et al. 2005, McClelland et al. 1995).

Recall and Completion

A fundamental property of the autoassociation model of the CA3 recurrent collateral network is that the recall can be symmetric, that is, the whole of the memory can be retrieved from any part. For example, in an object–place autoassociation memory, an object could be recalled from a place retrieval cue, and vice versa. This is not the case with a pattern association network. If for example the CA3 activity represented a place / spatial view, and perforant path inputs with associative synapses to CA3 neurons carried object information (consistent with evidence that the lateral perforant path (LPP) may reflect inputs from the perirhinal cortex connecting via the lateral entorhinal cortex (Hargreaves et al. 2005)), then an object could recall a place, but a place could not recall an object.

Another fundamental property is that the recall can be complete even from a small fragment. Thus, it is a prediction that when an incomplete retrieval cue is given, CA3 may be especially important in the retrieval process. Tests of this prediction are described in Section 24.4. Further evidence for pattern completion has been observed using imaging with voltage-sensitive dye in the CA3 region of a rat hippocampal slice. Following the induction of long-term potentiation from two stimulation sites activated simultaneously, stimulation at either of the two sites produced the whole pattern of activation that could be produced from both stimulation sites before LTP, thus demonstrating pattern completion in CA3 (Jackson 2013).

CA3 as a short term attractor memory

Another fundamental property of the autoassociation model of the CA3 recurrent collateral network is that it can implement a short-term memory by maintaining the firing of neurons using the excitatory recurrent collateral connections. A stable attractor can maintain one memory active in this way for a considerable period, until a new input pushes the attractor to represent a new location or memory (see Section B.3). For example, if one place were being held in a CA3 place short-term memory, then if the rat moved to a new place, the CA3 representation would move to represent the new place, and the short-term memory held in the CA3 network would be lost. It is thus predicted that when the hippocampus is used as a short-term memory with ongoing neuronal activity used to represent the short-term memory, then maintenance of the memory will be very dependent on what happens in the delay period. If the animal is relatively isolated from its environment and does not move around in a well-defined spatial environment (as can be achieved by placing a vertical cylinder / bucket around a rat in the delay period), then the memory may be maintained for a considerable period in the CA3 (and even updated by idiothetic, self-motion, inputs as described below). However, the CA3 short-term memory will be very sensitive to disruption / interference, so that if the rat is allowed to move in the spatial environment during the delay period, then it is predicted that CA3 will not be able to maintain correctly the spatial short-term memory. In the circumstances where a representation must be kept active while the hippocampus (or inferior temporal visual cortex or parietal cortex) must be updating its representation to reflect the new changing perceptual inputs, which must be represented in these brain areas for the ongoing events to have high-level representations, the prefrontal cortex is thought computationally to provide an off-line buffer store, see Section 4.3.1.

The predictions are thus that if there is no task in a delay period (and no need for an off-line buffer store), the hippocampus may be sufficient to perform the short-term memory function (such as remembering a previous location), provided that there is little distraction in the delay period. On the other hand, short-term memory deficits are predicted to be produced by prefrontal cortex lesions if there are intervening stimuli in a delay period in a task.

24.3.3.2 Continuous, spatial, patterns and CA3 representations

The fact that spatial patterns, which imply continuous representations of space, are represented in the hippocampus has led to the application of continuous attractor models to help understand hippocampal function. . This has been necessary, because space is inherently continuous, because the firing of place and spatial view cells is approximately Gaussian as a function of the distance away from the preferred spatial location, because these cells have spatially overlapping fields, and because the theory is that these cells in CA3 are connected by Hebb-modifiable synapses. This specification would inherently lead the system to operate as a continuous attractor network. Such models have been developed with the hippocampus in mind (Samsonovich and McNaughton 1997, Battaglia and Treves 1998b, Rolls, Stringer and Trappenberg 2002, Stringer, Trappenberg, Rolls and De Araujo 2002b, Stringer, Rolls, Trappenberg and De Araujo 2002a, Stringer and Rolls 2002, Stringer, Rolls and Trappenberg 2004, Stringer, Rolls and Trappenberg 2005, Stringer and Rolls 2006) and are described next (see for detailed specification Section B.5).

A class of network that can maintain the firing of its neurons to represent any location along a continuous physical dimension such as spatial position, head direction, etc. is a 'continuous attractor' neural network (CANN). It uses excitatory recurrent collateral connections between the neurons (as are present in CA3) to reflect the distance between the neurons in the state space of the animal (e.g. place or head direction). These networks can maintain the bubble of neural activity constant for long periods wherever it is started to represent the current state (head direction, position, etc) of the animal, and are likely to be involved in many aspects of spatial processing and memory, including spatial vision. Global inhibition is used to keep the number of neurons in a bubble or packet of actively firing neurons relatively constant, and to help to ensure that there is only one activity packet. Continuous attractor networks can be thought of as very similar to autoassociation or discrete attractor networks

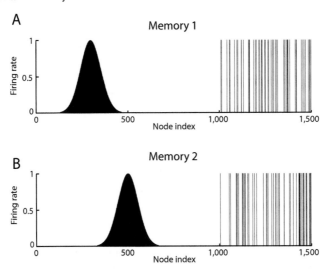

Fig. 24.16 The types of firing patterns stored in continuous attractor networks are illustrated for the patterns present on neurons 1–1,000 for Memory 1 (when the firing is that produced when the spatial state represented is that for location 300), and for Memory 2 (when the firing is that produced when the spatial state represented is that for location 500). The continuous nature of the spatial representation results from the fact that each neuron has a Gaussian firing rate that peaks at its optimal location. This particular mixed network also contains discrete representations that consist of discrete subsets of active binary firing rate neurons in the range 1,001–1,500. The firing of these latter neurons can be thought of as representing the discrete events that occur at the location. Continuous attractor networks by definition contain only continuous representations, but this particular network can store mixed continuous and discrete representations, and is illustrated to show the difference of the firing patterns normally stored in separate continuous attractor and discrete attractor networks. For this particular mixed network, during learning, Memory 1 is stored in the synaptic weights, then Memory 2, etc, and each memory contains a part that is continuously distributed to represent physical space, and a part that represents a discrete event or object. (After Rolls, Stringer and Trappenberg 2002.)

(see Appendix B), and have the same architecture, as illustrated in Fig. 24.15. The main difference is that the patterns stored in a CANN are continuous patterns, with each neuron having broadly tuned firing which decreases with for example a Gaussian function as the distance from the optimal firing location of the cell is varied, and with different neurons having tuning that overlaps throughout the space. Such tuning is illustrated in Fig. 24.16. For comparison, autoassociation networks normally have discrete (separate) patterns (each pattern implemented by the firing of a particular subset of the neurons), with no continuous distribution of the patterns throughout the space (see Fig. 24.16). A consequent difference is that the CANN can maintain its firing at any location in the trained continuous space, whereas a discrete attractor or autoassociation network moves its population of active neurons towards one of the previously learned attractor states, and thus implements the recall of a particular previously learned pattern from an incomplete or noisy (distorted) version of one of the previously learned patterns. The energy landscape of a discrete attractor network (see Section B.3, Hopfield (1982)) has separate energy minima, each one of which corresponds to a learned pattern, whereas the energy landscape of a continuous attractor network is flat, so that the activity packet remains stable with continuous firing wherever it is started in the state space. (The state space refers to the set of possible spatial states of the animal in its environment, e.g. the set of possible places in a room.) Continuous attractor networks and their application to memory systems in the hippocampus are reviewed next with details in Section B.5, as they are very likely to apply to the operation of systems with spatial representations and recurrent connections, such as the CA3 neurons. Continuous attractor networks have been studied by for example Amari (1977), Zhang (1996), and Stringer, Trappenberg, Rolls and De Araujo (2002b).

One key issue in such continuous attractor neural networks is how the synaptic strengths between the neurons in the continuous attractor network could be learned in biological systems, and it has been shown that associative synaptic modification while navigating the environment can implement this, because the spatial representations of nearby locations overlap and therefore become associatively linked (see Section B.5.3 and Stringer, Trappenberg, Rolls and De Araujo (2002b)).

A second key issue in such continuous attractor neural networks is how the bubble of neuronal firing representing one location in the continuous state space should be updated based on non-visual cues to represent a new location in state space (Section B.5.5). This is essentially the problem of path integration: how a system that represents a memory of where the agent is in physical space could be updated based on idiothetic (self-motion) cues such as vestibular cues (which might represent a head velocity signal), or proprioceptive cues (which might update a representation of place based on movements being made in the space, during for example walking in the dark). Examples of classes of neurons that can be updated idiothetically include head direction cells in rats (Muller et al. 1996, Ranck 1985, Taube et al. 1996, Taube et al. 1990a, Bassett and Taube 2005) and primates (Robertson, Rolls, Georges-François and Panzeri 1999), which respond maximally when the animal's head is facing in a preferred direction and can be updated by head rotation in the dark; place cells in rats (O'Keefe 1984, Markus, Qin, Leonard, Skaggs, McNaughton and Barnes 1995, O'Keefe and Dostrovsky 1971, McNaughton et al. 1983, Muller et al. 1991, Jeffery and Hayman 2004, Jeffery et al. 2004) that fire maximally when the animal is in a particular location and can be updated by running in a particular direction in the dark; and spatial view cells in primates that respond when the monkey is looking towards a particular location in space (Georges-François, Rolls and Robertson 1999, Robertson, Rolls and Georges-François 1998, Rolls, Robertson and Georges-François 1997a, Rolls, Treves, Robertson, Georges-François and Panzeri 1998b) and can be updated by eye movements in the dark (Robertson, Rolls and Georges-François 1998).

Path integration is described in Section B.5.5, using as an example the update of a head direction representation by a head velocity signal. For example, a proposal about how the synaptic connections from idiothetic inputs to a continuous attractor network can be learned with some biological plausibility, in that the network can be trained by a self-organizing process, was introduced by Stringer, Trappenberg, Rolls and De Araujo (2002b). The mechanism associates a short-term memory trace of the firing of the neurons in the attractor network reflecting recent movements in the state space (e.g. of places), with an idiothetic velocity of movement input (see Fig. B.27). This has been applied to head direction cells (Stringer, Trappenberg, Rolls and De Araujo 2002b, Stringer and Rolls 2006), rat place cells (Stringer, Trappenberg, Rolls and De Araujo 2002b, Stringer, Rolls, Trappenberg and De Araujo 2002a), and primate spatial view cells (Stringer, Rolls and Trappenberg 2004, Stringer, Rolls and Trappenberg 2005, Rolls and Stringer 2005). These attractor networks provide a basis for understanding cognitive maps, and how they are updated by learning and by self-motion. The implication is that to the extent that path integration of place or spatial view representations is performed within the hippocampus itself, then the CA3 system is the most likely part of the hippocampus to be involved in this, because it has the appropriate recurrent collateral connections. Consistent with this, Whishaw and colleagues (Wallace and Whishaw 2003, Maaswinkel, Jarrard and Whishaw 1999, Whishaw, Hines and Wallace 2001) have shown that path integration is impaired by hippocampal lesions.

Part of the neural mechanism for the updating of place by path integration may be in the entorhinal cortex (McNaughton, Battaglia, Jensen, Moser and Moser 2006) (see Section 24.3.2). Head direction cells are present in the entorhinal cortex, where grid cells (see Section 24.3.4) are also present (Giocomo et al. 2011, Moser, Roudi, Witter, Kentros, Bonhoeffer and Moser 2014b, Moser et al. 2015), and indeed where some cells respond to combinations of grid and head direction (Leutgeb, Leutgeb, Treves, Meyer, Barnes, McNaughton, Moser and Moser 2005, Sargolini, Fyhn, Hafting, McNaughton, Witter, Moser and Moser 2006). Such combination cells in the entorhinal cortex may be part of the mechanism by which the place map is kept in register with the head direction map. The medial entorhinal cortex grid cells are a strong candidate for path integration over distance and direction travelled, for the repeating nature of the grid (see Section 24.3.4 for details and Fig. 24.17) is likely to be a reflection of a continuous attractor cycling once round its ring for every peak observed. The simplification this offers is that the path integration need be learned only once, as it is just an integration over self-movement, and can be used then in any environment. Consistent with this, grid cells are relatively little influenced by the details of the visual cues in the environment in which the rat is tested, apart from the rotation of the whole grid by the head direction cells, which can be set by the environment (Hafting, Fyhn, Molden, Moser and Moser 2005, Fyhn, Molden, Witter, Moser and Moser 2004, Sargolini, Fyhn, Hafting, McNaughton, Witter, Moser and Moser 2006).

If path integration for place is performed by grid cells in the entorhinal cortex (Giocomo et al. 2011, Moser et al. 2015, Kropff and Treves 2008, Zilli 2012, Moser et al. 2014b, Moser et al. 2014a), this leaves open the question of how the distance travelled is incorporated into the spatial or cognitive map of an environment, which will incorporate all the visual landmarks. As these visual cues are not apparently strongly represented in the entorhinal grid cells themselves, the implication is that the distance and direction travelled that is a result of the grid cell path integration needs to be linked into the spatial map of a particular environment, and this is represented as shown by the properties of rat place cells and primate spatial view cells in the hippocampus proper, in CA3 and CA1. The implication is that with an idiothetically defined place computed from the grid cell representation and reflected in the firing of CA3 cells, it then becomes an associative task for the CA3 cells to link associatively that idiothetic representation of place with the visual and other world-based cues, as well as with the objects, odours etc. at those places.

The summary of path integration in the context of grid cells provided by McNaughton et al. (2006) was not self-organizing, that is, the appropriate synaptic strengths for the model to perform the path integration did not become specified as a result of movements in the environment. Self-organizing continuous attractor neural networks that can perform path integration from velocity signals are described in Section B.5.5 and include head direction from head velocity (Stringer, Trappenberg, Rolls and De Araujo 2002b, Stringer and Rolls 2006); place from whole body motion (Stringer, Trappenberg, Rolls and De Araujo 2002b, Stringer, Rolls, Trappenberg and De Araujo 2002a); and spatial view from eye and whole body motion (Stringer, Rolls and Trappenberg 2004, Stringer, Rolls and Trappenberg 2005, Rolls and Stringer 2005). Models such as these provide a basis for understanding the principles of operation of medial entorhinal cortex grid cells (Giocomo, Moser and Moser 2011, Moser, Rowland and Moser 2015, Kropff and Treves 2008, Moser, Roudi, Witter, Kentros, Bonhoeffer and Moser 2014b, Moser, Moser and Roudi 2014a).

A third key issue is how stability in the bubble of activity representing the current location can be maintained without much drift in darkness, when it is operating as a memory system. A solution is described in Section B.5.6 (Stringer et al. 2002b).

A fourth key issue is considered in Section B.5.8 in which networks are described that store both continuous patterns and discrete patterns (see Fig. 24.16). These networks can be used to store for example the location in (continuous, physical) space where an object (an item with a discrete representation) is present (Rolls, Stringer and Trappenberg 2002). *This type of mixed attractor in which discrete representations of objects and continuous representations of space can be combined is important for understanding the operation of the CA3 region of the hippocampus in episodic memory, in which object and place representations must be combined.*

A fifth key issue is the capacity of a continuous attractor network for storing many different environments each with many places. It turns out that the capacity is high, for each environment has its own chart of connected set of places, and the charts for different environments are quite uncorrelated with each other (Battaglia and Treves 1998b) (Section B.5.4). Although the capacity of a continuous attractor network for charts is high, because each chart is relatively uncorrelated with other charts, the capacity is nevertheless more than 100 times lower than for the number of discrete items that can be stored in an attractor network such as CA3 in the hippocampus (Battaglia and Treves 1998b, Treves 2016). The smaller number of charts or maps than discrete items is related to a reduction of approximately 0.3 caused by changing from discrete to continuous, spatial, representations; and to the fact that each chart includes many places (Treves 2016, Papp and Treves 2008). But that might still mean in the order of 200 charts in the rat hippocampus.

Overall, it is argued that the CA3 region is likely to be involved in associating a relatively continuous spatial representation with an object representation as an episodic memory computation, and not with performing a continuous attractor-based idiothetic update of the spatial representation, which latter may be performed by the entorhinal cortex (Rolls 2013d). Consistent with this view, it has been argued that the bumpiness of the CA3 representation of space is more consistent with episodic memory storage, than with spatial path integration using the CA3 system as a continuous attractor network implementing path integration (Cerasti and Treves 2013, Stella, Cerasti and Treves 2013).

24.3.3.3 The dynamics of the recurrent network

The analysis described above of the capacity of a recurrent network such as the CA3 considered steady-state conditions of the firing rates of the neurons. The question arises of how quickly the recurrent network would settle into its final state. With reference to the CA3 network, how long does it take before a pattern of activity, originally evoked in CA3 by afferent inputs, becomes influenced by the activation of recurrent collaterals? In a more general context, recurrent collaterals between the pyramidal cells are an important feature of the connectivity of the cerebral neocortex. How long would it take these collaterals to contribute fully to the activity of cortical cells? If these settling processes took in the order of hundreds of ms, they would be much too slow to contribute usefully to cortical activity, whether in the hippocampus or in the neocortex (Rolls and Deco 2002, Panzeri, Rolls, Battaglia and Lavis 2001, Rolls 1992a, Rolls 2003).

It has been shown that if the neurons are treated not as McCulloch-Pitts neurons that are simply 'updated' at each iteration, or cycle of time steps (and assume the active state if the threshold is exceeded), but instead are analyzed and modelled as 'integrate-and-fire' neurons in real continuous time, then the network can effectively 'relax' into its recall state very rapidly, in one or two time constants of the synapses (Treves 1993, Rolls and Treves 1998, Battaglia and Treves 1998b, Panzeri et al. 2001). This corresponds to perhaps 20 ms in the brain. One factor in this rapid dynamics of autoassociative networks with brain-like 'integrate-and-fire' membrane and synaptic properties is that with some spontaneous activity, some of the neurons in the network are close to threshold already before the recall cue is applied, and hence some of the neurons are very quickly pushed by the recall cue into firing, so that information starts to be exchanged very rapidly (within 1–2 ms of brain time) through the modified synapses by the neurons in the network. The progressive exchange of information starting early on within what would otherwise be thought of as an iteration period (of perhaps 20 ms, corresponding to a neuronal firing rate of 50 spikes/s), is the mechanism accounting for rapid recall in an autoassociative neuronal network made biologically realistic in this way. Further analysis of the fast dynamics of these networks if they are implemented in a biologically plausible way with 'integrate-and-fire' neurons is provided in Section B.6.4, in Appendix A5 of Rolls and Treves (1998), by Treves (1993), by Battaglia and Treves (1998b), and by Panzeri, Rolls, Battaglia and Lavis (2001).

24.3.3.4 Memory for sequences

One of the first extensions of the standard autoassociator paradigm that has been explored in the literature is the capability to store and retrieve not just individual patterns, but whole sequences of patterns. Hopfield (1982) suggested that this could be achieved by adding to the standard connection weights, which associate a pattern with itself, a new, asymmetric component, that associates a pattern with the next one in the sequence. In practice this scheme does not work very well, unless the new component is made to operate on a slower time scale than the purely autoassociative component (Kleinfeld 1986, Sompolinsky and Kanter 1986). With two different time scales, the autoassociative component can stabilize a pattern for a while, before the heteroassociative component moves the network, as it were, into the next pattern. The heteroassociative retrieval cue for the next pattern in the sequence is just the previous pattern in the sequence. A particular type of 'slower' operation occurs if the asymmetric component acts after a delay τ. In this case, the network sweeps through the sequence, staying for a time of order τ in each pattern.

One can see how the necessary ingredient for the storage of sequences is only a minor departure from purely Hebbian learning: in fact, the (symmetric) autoassociative component of the weights can be taken to reflect the Hebbian learning of strictly simultaneous conjunctions of pre- and post-synaptic activity, whereas the (asymmetric) heteroassociative component can be implemented by Hebbian learning of each conjunction of postsynaptic activity with presynaptic activity shifted a time τ in the past. Both components can then be seen as resulting from a generalized Hebbian rule, which increases the weight whenever postsynaptic activity is paired with presynaptic activity occurring within a given time range, which may extend from a few hundred milliseconds in the past up to include strictly simultaneous activity. This is similar to a trace rule (see Chapter 25, and Rolls (2012c)), which itself matches very well the observed conditions for induction of long-term potentiation, and appears entirely plausible. The learning rule necessary for learning sequences, though, is more complex than a simple trace rule in that the time-shifted conjunctions of activity that are encoded in the weights must in retrieval produce activations that are time-shifted as well (otherwise one falls back into the Hopfield (1982) proposal, which does not quite work). The synaptic weights should therefore keep separate 'traces' of what was simultaneous and what was time-shifted during the original experience, and this is not very plausible. Levy and colleagues (Levy, Wu and Baxter 1995, Wu, Baxter and Levy 1996) have investigated these issues further, and the temporal asymmetry that may be present in LTP has been suggested as a mechanism that might provide some of the temporal properties that are necessary for the brain to store and recall sequences (Abbott and Blum 1996, Abbott and Nelson 2000, Minai and Levy 1993, Markram et al. 1998). A problem with this suggestion is that, given that the temporal dynamics of attractor networks are inherently very fast when the networks have continuous dynamics, and that the temporal asymmetry in LTP may be in the order of only milliseconds to a few tens of milliseconds, the recall of the sequences would be very fast, perhaps 10–20 ms per step of the sequence, with every step of a 10-step sequence effectively retrieved and gone in a quick-fire session of 100–200 ms.

Rolls and Stringer (see Rolls and Kesner (2006)) have suggested that the over-rapid replay of a sequence of memories stored in an autoassociation network such as CA3 if it included asymmetric synaptic weights to encode a sequence could be controlled by the physical inputs from the environment. If a sequence of places 1, 2, and 3 had been learned by the use of an asymmetric trace learning rule implemented in the CA3 network, then the firing initiated by place 1 would reflect (by the learned association) a small component of place 2 (which might be used to guide navigation to place 2, and which might be separated out from place 1 better by the competitive network action in CA1), but would not move fully away from representing place 1 until the animal moved away from place 1, because of the clamping effect on the CA3 firing of the external input representing place 1 to the recurrent network. In this way, the physical constraints of movements between the different places in the environment would control the speed of readout from the sequence memory.

The CA1 neurons might make a useful contribution to this sequential recall by acting as a competitive network that would separate out the patterns represented in CA3 for the current and the next place, and this has been tested in the simulation by Rolls and Stringer (see Rolls and Kesner (2006)). What has been found is that if the representation is set to be less sparse in the CA1 than the CA3, and the CA1 neurons operate on the upper part of a sigmoid activation function, then the next step in the sequence can be represented by a population of neurons in the CA1 better than it can in CA3. (This allows the CA3 representation to be kept sparse, which helps to maximize the number of memories that can be stored.) The CA1 neurons representing the next place in the sequence would be the basis for the animal navigating to that next place. There could be heteroassociative firing of the third item in a sequence, but this would be much weaker. This functionality however could help the disambiguation of long sequences with overlapping parts (Jensen and Lisman 1996).

In the network of Rolls and Stringer (see Rolls and Kesner (2006)), CA3 might, by the continuing firing implemented by autoassociation, allow one item to be held in short-term memory until the next item arrives for heteroassociation. This would enable long time gaps within the sequence during training to be bridged.

Lisman and colleagues (Jensen and Lisman 1996, Jensen and Lisman 2005, Lisman and Idiart 1995, Jensen, Idiart and Lisman 1996, Lisman, Talamini and Raffone 2005) have suggested that sequences might be encoded by utilizing the neuronal after-depolarization which follows a spike from a neuron by approximately 200 ms to tend to make neurons fire again after they are activated. Interaction with theta oscillations (5–8 Hz) and gamma oscillations (30–80 Hz) in the model would make a set of neurons coding for one item tend to fire first in a new theta cycle. Neurons encoding a second item in a sequence would tend to fire in the next gamma cycle 20 ms later. Neurons encoding a third item in a sequence would tend to fire in the next gamma cycle 20 ms later. To keep the neurons representing one item synchronized, LTP associated with NMDA receptors with a short time constant of e.g. 12 ms is needed. Every theta cycle, the whole sequence is replayed by the model with 20 ms between items. In one version of their model (Jensen and Lisman 2005) this mechanism is suggested to be implemented in the neocortex. To form associations between successive items in a sequence, LTP associated with NMDA receptors with a long time constant of e.g. 150 ms is needed, and this it is suggested might be implemented in the hippocampus (Jensen and Lisman 2005, Lisman et al. 2005). It is a difficulty of the model that every theta cycle, the whole sequence is replayed by the model with 20 ms between items, for if the next item in a sequence is needed (e.g. the third), then the model just replays at this high speed all the following items in a sequence, so a special decoding system would be needed to know which the next item is. The model is also very susceptible to noise. The model also only works if there is great regularity in the spike trains of the neurons, which should repeat at very regular intervals locked to theta, and this is inconsistent with the almost Poisson nature of the spike trains of single neurons found in most brain areas. Further, no experimental evidence has been found that successive items in a remembered sequence are replayed at gamma frequency every theta cycle.

Another way in which a delay could be inserted in a recurrent collateral path in the brain is by inserting another cortical area in the recurrent path. This could fit in with the cortico-cortical backprojection connections described in Fig. 24.1 with return projections to the entorhinal cortex, which might then send information back into the hippocampus, which would introduce some conduction delay (Panzeri, Rolls, Battaglia and Lavis 2001). Another possibility is that the connections between the deep and the superficial layers of the entorhinal cortex help to close a loop from entorhinal cortex through hippocampal circuitry and back to the entorhinal cortex (van Haeften et al. 2003). Another suggestion is that the CA3 system holds by continuing firing a memory for event 1, which can then be associated with event 2 at the CA3 to CA1 synapses if event 2 activates CA1 neurons by the direct entorhinal input (Levy 1989).

Another proposal is that sequence memory can be implemented by using synaptic adaptation to effectively encode the order of the items in a sequence (Deco and Rolls 2005c) (see Section B.9). Whenever the attractor system is quenched into inactivity, the next member of the sequence emerges out of the spontaneous activity, because the least recently activated member of the sequence has the least synaptic or neuronal adaptation. This could be implemented in recurrent networks such as the CA3 or the prefrontal cortex.

For recency short-term memory, the sequence type of memory would be possible, but difficult because once a sequence has been recalled, a system has to "manipulate" the items, picking out which is later (or earlier) in the sequence. A simpler possibility is to use a short term form of long term potentiation (LTP) which decays, and, then the latest of two items can be picked out because when it is seen again, the neuronal response is larger than to the earlier items (Rolls and Deco 2002, Renart et al. 2001). This type of memory does not explicitly encode sequences, but instead reflects just how recently an item occurred. This could be implemented at any synapses in the system (e.g. in CA3 or CA1), and does not require recurrent collateral connectivity.

Probably the most likely way for the hippocampus to encode sequences is to use the time cells described in Section 24.3.6.2, and associate a particular time in the sequence with for example a place or object. This would enable the temporal order to be recalled correctly. This system is known to act over time periods of the order of up to tens of seconds. This could be implemented either in CA3, or CA1 (which has direct inputs from the medial entorhinal cortex), or both.

24.3.3.5 The dilution of the CA3 recurrent collateral connectivity enhances memory storage capacity and pattern completion

Fig. 24.14 shows that in the rat, there are approximately 300,000 CA3 neurons, but only 12,000 recurrent collateral synapses per neuron. The dilution of the connectivity is thus $12,000 / 300,000 = 0.04$. The connectivity is thus not complete, and complete connectivity in an autoassociation network would make it simple, for the connectivity between the neurons would then be symmetric (i.e. the connection strength from any one neuron to another is matched by a connection of the same strength in the opposite direction), and this guarantees energy minima for the basins of attraction that will be stable, and a memory capacity than can be calculated (Hopfield 1982). We have shown how this attractor type of network can be extended to have similar properties with diluted connectivity, and also with sparse representations with graded firing rates (Treves and Rolls 1991, Treves 1990, Treves 1991b, Rolls and Webb 2012, Webb, Rolls, Deco and Feng 2011).

However, the question has recently been asked about whether there are any advantages to diluted autoassociation or attractor networks compared to fully connected attractor networks

(Rolls 2012a). One biological property that may be a limiting factor is the number of synaptic connections per neuron, which is 12,000 in the CA3-CA3 network just for the recurrent collaterals (see Fig. 24.14). The number may be higher in humans, allowing more memories to be stored in the hippocampus than order 12,000. I note that the storage of large number of memories may be facilitated in humans because the left and right hippocampus appear to be much less connected between the two hemispheres than in the rat, which effectively has a single hippocampus in that the CA3 neurons are connected by commissural connections. In humans, with effectively two separate CA3 networks, one on each side of the brain, the memory storage capacity may be doubled, as the capacity is set by the number of recurrent collaterals per neuron in each attractor network. In humans, the right hippocampus may be devoted to episodic memories with spatial and visual components, whereas the left hippocampus may be devoted to memories with verbal / linguistic components, i.e. in which words may be part of the episode (e.g. who said what to whom and when) (Barkas et al. 2010, Bonelli et al. 2010, Sidhu et al. 2013).

The answer that has been suggested to why the connectivity of the CA3 autoassociation network is diluted (and why neocortical recurrent networks are also diluted), is that this may help to reduce the probability of having two or more synapses between any pair of randomly connected neurons within the network, which it has been shown greatly impairs the number of memories that can be stored in an attractor network, because of the distortion that this produces in the energy landscape (Rolls 2012a) (Section 7.2). In more detail, the hypothesis proposed is that the diluted connectivity allows biological processes that set up synaptic connections between neurons to arrange for there to be only very rarely more than one synaptic connection between any pair of neurons. If probabilistically there were more than one connection between any two neurons, it was shown by simulation of an autoassociation attractor network that such connections would dominate the attractor states into which the network could enter and be stable, thus strongly reducing the memory capacity of the network (the number of memories that can be stored and correctly retrieved), below the normal large capacity for diluted connectivity. Diluted connectivity between neurons in the cortex thus has an important role in allowing high capacity of memory networks in the cortex, and helping to ensure that the critical capacity is not reached at which overloading occurs leading to an impairment in the ability to retrieve any memories from the network (Rolls 2012a). The diluted connectivity is thus seen as an adaptation that simplifies the genetic specification of the wiring of the brain, by enabling just two attributes of the connectivity to be specified (e.g. from a CA3 to another CA3 neuron chosen at random to specify the CA3 to CA3 recurrent collateral connectivity), rather than which particular neuron should connect to which other particular neuron (Rolls 2012a, Rolls and Stringer 2000). Consistent with this hypothesis, there are NMDA receptors with the genetic specification that they are NMDA receptors on neurons of a particular type, CA3 neurons (as shown by the evidence from CA3-specific vs CA1-specific NMDA receptor knockouts) (Nakazawa et al. 2002, Nakazawa et al. 2003, Nakazawa et al. 2004, Rondi-Reig, Libbey, Eichenbaum and Tonegawa 2001). A consequence is that the vector of output neuronal firing in the CA3 region, i.e. the number of CA3 neurons, is quite large (300,000 neurons in the rat). The large number of elements in this vector may have consequences for the noise in the system, as we will see below.

The role of dilution in the connectivity of the CA3 recurrent collateral connectivity includes enabling this large number of separate memories to be recalled from any part of each memory, that is, in pattern completion (Rolls 2012a) (Section 7.2).

The dilution of the CA3-CA3 recurrent collateral connectivity at 0.04 may be greater dilution than that in a local neocortical area, which is in the order of 0.1 (Rolls 2012a) (Chapter 7). This is consistent with the hypothesis that the storage capacity of the CA3 system is at a premium, and so the dilution is kept to a low value (i.e. great dilution), as then there is lower distortion of the basins of attraction and hence the memory capacity is maximized (Rolls 2012a) (Section 7.2).

24.3.3.6 Noise and stability produced by the diluted connectivity and the graded firing rates in the CA3-CA3 attractor network

Many processes in the brain are influenced by the noise or variability of neuronal spike firing (Rolls and Deco 2010, Faisal, Selen and Wolpert 2008, Deco, Rolls, Albantakis and Romo 2013). The action potentials are generated in a way that frequently approximates a Poisson process, in which the spikes for a given mean firing rate occur at times that are essentially random (apart from a small effect of the refractory period), with a coefficient of variation of the interspike interval distribution (CV) near 1.0 (Rolls and Deco 2010). The sources of the noise include quantal transmitter release, and noise in ion channel openings (Faisal, Selen and Wolpert 2008). The membrane potential is often held close to the firing threshold, and then small changes in the inputs and the noise in the neuronal operations cause spikes to be emitted at almost random times for a given mean firing rate. Spiking neuronal networks with balanced inhibition and excitation currents and associatively modified recurrent synaptic connections can be shown to possess a stable attractor state where neuron spiking is approximately Poisson too (Amit and Brunel 1997, Miller and Wang 2006). The noise caused by the variability of individual neuron spiking which then affects other neurons in the network can play an important role in the function of such recurrent attractor networks, by causing for example an otherwise stable network to jump into a decision state (Wang 2002, Deco and Rolls 2006, Rolls and Deco 2010). Attractor networks with this type of spiking-related noise are used in the brain for memory recall, and for decision-making, which in terms of the neural mechanism are effectively the same process. Noise in attractor networks is useful for memory and decision-making, for it makes them non-deterministic, and this contributes to new solutions to problems, and indeed to creativity (Rolls and Deco 2010) (Chapter 5).

To investigate the extent to which the diluted connectivity affects the dynamics of attractor networks in the cerebral cortex (which includes the hippocampus), we simulated an integrate-and-fire attractor network taking decisions between competing inputs with diluted connectivity of 0.25 or 0.1 but the same number of synaptic connections per neuron for the recurrent collateral synapses within an attractor population as for full connectivity (Rolls and Webb 2012). The results indicated that there was less spiking-related noise with the diluted connectivity in that the stability of the network when in the spontaneous state of firing increased, and the accuracy of the correct decisions increased. The decision times were a little slower with diluted than with complete connectivity. Given that the capacity of the network is set by the number of recurrent collateral synaptic connections per neuron, on which there is a biological limit, the findings indicate that the stability of cortical networks, and the accuracy of their correct decisions or memory recall operations, can be increased by utilizing diluted connectivity and correspondingly increasing the number of neurons in the network (which may help to smooth the noise), with little impact on the speed of processing of the cortex. Thus diluted connectivity can decrease cortical spiking-related noise, and thus enhance the reliability of memory recall, which includes completion from a partial recall cue (Rolls and Webb 2012).

Representations in the neocortex and in the hippocampus are often distributed with graded firing rates in the neuronal populations (Rolls and Treves 2011) (Appendix C). The firing rate probability distribution of each neuron to a set of stimuli is often exponential or gamma (Rolls and Treves 2011) (Section C.3.1). These graded firing rate distributed representations are present in the hippocampus, both for place cells in rodents and for spatial view cells in the primate (O'Keefe and Speakman 1987, O'Keefe 1979, McNaughton, Barnes and O'Keefe 1983, Rolls, Robertson and Georges-François 1997a, Rolls, Treves, Robertson,

Georges-François and Panzeri 1998b, Georges-François, Rolls and Robertson 1999, Rolls and Treves 2011). In processes in the brain such as memory recall in the hippocampus or decision-making in the cortex that are influenced by the noise produced by the close to random spike timings of each neuron for a given mean rate, the noise with this graded type of representation may be larger than with the binary firing rate distribution that is usually investigated.

In integrate-and-fire simulations of an attractor decision-making network, we showed that the noise is indeed greater for a given sparseness of the representation for graded, exponential, than for binary firing rate distributions (Webb, Rolls, Deco and Feng 2011). The greater noise was measured by faster escaping times from the spontaneous firing rate state when the decision cues are applied, and this corresponds to faster decision or reaction times. The greater noise was also evident as less stability of the spontaneous firing state before the decision cues are applied. The implication is that spiking-related noise will continue to be a factor that influences processes such as decision-making, signal detection, short-term memory, and memory recall and completion (including in the CA3 network) even with the quite large networks found in the cerebral cortex. In these networks there are several thousand recurrent collateral synapses onto each neuron. The greater noise with graded firing rate distributions has the advantage that it can increase the speed of operation of cortical circuitry (Webb, Rolls, Deco and Feng 2011). The graded firing rates also by operating in a non-linear network effectively increase the sparseness of the representation, and this itself is a pattern separation effect (Webb, Rolls, Deco and Feng 2011).

24.3.3.7 Pattern separation of CA3 cell populations encoding different memories

For the CA3 to operate with high capacity as an autoassociation or attractor memory, the sets of CA3 neurons that represent each event to be stored and later recalled need to be as uncorrelated from each other as possible. Correlations between patterns reduce the memory capacity of an autoassociation network (Marr 1971, Kohonen 1977, Kohonen 1989, Kohonen et al. 1981, Sompolinsky 1987, Rolls and Treves 1998), and because storage capacity is at a premium in an episodic memory system, there are several mechanisms that reduce the correlations between the firing of the population vectors of CA3 neuron firing each one of which represents a different event to be stored in memory. In the theoretical physics approach to the capacity of attractor networks, it is indeed assumed that the different vectors of firing rates to be stored are well separated from each other, by drawing each vector of firing at random, and by assuming very large (infinite) numbers of neurons in each pattern (Hopfield 1982, Rolls and Treves 1998) (Section B.3).

We have proposed that there are several mechanisms that help to achieve this pattern separation, namely the mossy fibre pattern separation effect produced by the small number of connections received by a CA3 neuron from mossy fibers which dominate the CA3 cell firing; the expansion recoding, and the sparse representation provided by the dentate granule cells that form the mossy fiber synapses; and the sparseness of the CA3 cell representation. Neurogenesis of dentate granule cells is a fifth potential contributor to achieving pattern separation of CA3 cell firing. These five mechanisms are described here, and elsewhere (Rolls 2013d, Rolls 2013b, Rolls 2016e). It is remarked that some of this architecture may be special to the hippocampus, and not found in the neocortex, because of the importance of storing and retrieving large numbers of (episodic) memories in the hippocampus. The neocortex in contrast is more concerned with building new representations for which competitive learning is more important, and thus neocortical circuitry does not use a mossy fibre system to produce new random sets of neurons activated.

24.3.3.8 Mossy fibre inputs to the CA3 cells

We hypothesize that the mossy fibre inputs force efficient information storage by virtue of their strong and sparse influence on the CA3 cell firing rates (Rolls 1987, Rolls 1989b, Rolls 1989f, Treves and Rolls 1992b). (The strong effects likely to be mediated by the mossy fibres were also emphasized by McNaughton and Morris (1987) and McNaughton and Nadel (1990).) We hypothesize that the mossy fibre input appears to be particularly appropriate in several ways. First of all, the fact that mossy fibre synapses are large and located very close to the soma makes them relatively powerful in activating the postsynaptic cell. (This should not be taken to imply that a CA3 cell can be fired by a single mossy fibre EPSP.)

Second, the firing activity of dentate granule cells appears to be very sparse (Jung and McNaughton 1993, Leutgeb and Leutgeb 2007) and this, together with the small number of connections on each CA3 cell, produces a sparse signal, which can then be transformed into an even sparser firing activity in CA3 by a threshold effect[19]. The hypothesis is that the mossy fibre sparse connectivity solution performs the appropriate function to enable learning to operate correctly in CA3 (Treves and Rolls 1992b, Cerasti and Treves 2010). The perforant path input would, the quantitative analysis shows, not produce a pattern of firing in CA3 that contains sufficient information for learning (Treves and Rolls 1992b).

Third, non-associative plasticity of mossy fibres (Brown, Ganong, Kairiss, Keenan and Kelso 1989, Brown, Kairiss and Keenan 1990b) might have a useful effect in enhancing the signal-to-noise ratio, in that a consistently firing mossy fibre would produce non-linearly amplified currents in the postsynaptic cell, which would not happen with an occasionally firing fibre (Treves and Rolls 1992b). This plasticity, and also learning in the dentate, would also have the effect that similar fragments of each episode (e.g. the same environmental location) recurring on subsequent occasions would be more likely to activate the same population of CA3 cells, which would have potential advantages in terms of economy of use of the CA3 cells in different memories, and in making some link between different episodic memories with a common feature, such as the same location in space.

Fourth, with only a few, and powerful, active mossy fibre inputs to each CA3 cell, setting a given sparseness of the representation provided by CA3 cells would be simplified, for the EPSPs produced by the mossy fibres would be Poisson distributed with large membrane potential differences for each active mossy fibre. Setting the average firing rate of the dentate granule cells would effectively set the sparseness of the CA3 representation, without great precision being required in the threshold setting of the CA3 cells (Rolls, Treves, Foster and Perez-Vicente 1997c). Part of what is achieved by the mossy fibre input may be setting the sparseness of the CA3 cells correctly, which, as shown above and in Section B.3, is very important in an autoassociative memory store.

Fifth, the non-associative and sparse connectivity properties of the mossy fibre connections to CA3 cells may be appropriate for an episodic memory system that can learn very fast, in one trial. The hypothesis is that the sparse connectivity would help arbitrary relatively uncorrelated sets of CA3 neurons to be activated for even somewhat similar input patterns without the need

---

[19]For example, if only 1 granule cell in 100 were active in the dentate gyrus, and each CA3 cell received a connection from 50 randomly placed granule cells, then the number of active mossy fibre inputs received by CA3 cells would follow a Poisson distribution of average $50/100 = 1/2$, i.e. 60% of the cells would not receive any active input, 30% would receive only one, 7.5% two, little more than 1% would receive three, and so on. (It is easy to show from the properties of the Poisson distribution and our definition of sparseness, that the sparseness of the mossy fibre signal as seen by a CA3 cell would be $x/(1 + x)$, with $x = C^{MF}a_{DG}$, assuming equal strengths for all mossy fibre synapses. C^{MF} is the number of mossy fibre connections to a CA3 neuron, and a_{DG} is the sparseness of the representation in the dentate granule cells.) If three mossy fibre inputs were required to fire a CA3 cell and these were the only inputs available, we see that the activity in CA3 would be roughly as sparse, in the example, as in the dentate gyrus.

for any learning of how best to separate the patterns, which in a self-organizing competitive network would take several repetitions (at least) of the set of patterns. The mossy fibre solution may thus be adaptive in a system that must learn in one trial, and for which the CA3 recurrent collateral learning requires uncorrelated sets of CA3 cells to be allocated for each (one-trial) episodic memory. The hypothesis is that the mossy fibre sparse connectivity solution performs the appropriate function without the mossy fibre system having to learn by repeated presentations how best to separate a set of training patterns (see further Section 24.4).

The argument based on information suggests, then, that an input system with the characteristics of the mossy fibres is essential during learning, in that it may act as a sort of (unsupervised) teacher that effectively strongly influences which CA3 neurons fire based on the pattern of granule cell activity. This establishes an information-rich neuronal representation of the episode in the CA3 network (Treves and Rolls 1992b, Rolls 2013b, Kesner and Rolls 2015, Rolls 2016e). The perforant path input would, the quantitative analysis shows, not produce a pattern of firing in CA3 that contains sufficient information for learning (Treves and Rolls 1992b).

The particular property of the small number of mossy fibre connections onto a CA3 cell, approximately 46 (see Fig. 24.14), is that this has a randomizing effect on the representations set up in CA3, so that they are as different as possible from each other (Rolls 1989b, Rolls 1989e, Treves and Rolls 1992b). (This means for example that place cells in a given environment are well separated to cover the whole space.) The result is that any one event or episode will set up a representation that is very different from other events or episodes, because the set of CA3 neurons activated for each event is random. This is then the optimal situation for the CA3 recurrent collateral effect to operate, for it can then associate together the random set of neurons that are active for a particular event (for example an object in a particular place), and later recall the whole set from any part. It is because the representations in CA3 are unstructured, or random, in this way that large numbers of memories can be stored in the CA3 autoassociation system, and that interference between the different memories is kept as low as possible, in that they are maximally different from each other (Hopfield 1982, Treves and Rolls 1991).

The requirement for a small number of mossy fibre connections onto each CA3 neuron applies not only to discrete (Treves and Rolls 1992b) but also to spatial representations, and some learning in these connections, whether associative or not, can help to select out the small number of mossy fibres that may be active at any one time to select a set of random neurons in the CA3 (Cerasti and Treves 2010). Any learning may help by reducing the accuracy required for a particular number of mossy fibre connections to be specified genetically onto each CA3 neuron. The optimal number of mossy fibres for the best information transfer from dentate granule cells to CA3 cells is in the order of 35–50 (Treves and Rolls 1992b, Cerasti and Treves 2010). The mossy fibres also make connections useful for feedforward inhibition in CA3 (Acsady, Kamondi, Sik, Freund and Buzsaki 1998), which is likely to be useful to help in the sparse representations being formed in CA3.

On the basis of these points, we predict that the mossy fibres may be necessary for new learning in the hippocampus, but may not be necessary for recall of existing memories from the hippocampus. Experimental evidence consistent with this prediction about the role of the mossy fibres in learning has been found in rats with disruption of the dentate granule cells (Lassalle, Bataille and Halley 2000) (see Section 24.4).

I have hypothesized that non-associative plasticity of mossy fibres (Brown et al. 1989, Brown et al. 1990b) might have a useful effect in enhancing the signal-to-noise ratio, in that a consistently firing mossy fibre would produce nonlinearly amplified currents in the postsynaptic cell, which would not happen with an occasionally firing fibre (Treves and Rolls 1992b). This plasticity, and also learning in the dentate, would also have the effect that similar fragments of each episode (e.g. the same environmental location) recurring on subsequent occasions would be more likely to activate the same population of CA3 cells, which would have potential advantages in terms of economy of use of the CA3 cells in different memories, and in making some link between different episodic memories with a common feature, such as the same location in space. Consistent with this, dentate neurons that fire repeatedly are more effective in activating CA3 neurons (Henze, Wittner and Buzsaki 2002).

As acetylcholine does turn down the efficacy of the recurrent collateral synapses between CA3 neurons (Hasselmo et al. 1995, Giocomo and Hasselmo 2007), then cholinergic activation also might help to allow external inputs rather than the internal recurrent collateral inputs to dominate the firing of the CA3 neurons during learning, as the current theory proposes. If cholinergic activation at the same time facilitated LTP in the recurrent collaterals (as it appears to in the neocortex), then cholinergic activation could have a useful double role in facilitating new learning at times of behavioural activation, when presumably it may be particularly relevant to allocate some of the limited memory capacity to new memories.

24.3.3.9 Perforant path inputs to CA3 cells initiate recall in CA3 and contribute to generalization

By calculating the amount of information that would end up being carried by a CA3 firing pattern produced solely by the perforant path input and by the effect of the recurrent connections, we have been able to show (Treves and Rolls 1992b) that an input of the perforant path type, alone, is unable to direct efficient information storage. Such an input is too weak, it turns out, to drive the firing of the cells, as the 'dynamics' of the network is dominated by the randomizing effect of the recurrent collaterals. This is the manifestation, in the CA3 network, of a general problem affecting storage (i.e. learning) in all autoassociative memories. The problem arises when the system is considered to be activated by a set of input axons making synaptic connections that have to compete with the recurrent connections, rather than having the firing rates of the neurons artificially clamped into a prescribed pattern.

An autoassociative memory network needs afferent inputs also in the other mode of operation, i.e. when it retrieves a previously stored pattern of activity. We have shown (Treves and Rolls 1992b) that the requirements on the organization of the afferents are in this case very different, implying the necessity of a second, separate, input system, which we have identified with the perforant path input to CA3. In brief, the argument is based on the notion that the cue available to initiate retrieval might be rather small, i.e. the distribution of activity on the afferent axons might carry a small correlation, $q \ll 1$, with the activity distribution present during learning. In order not to lose this small correlation altogether, but rather transform it into an input current in the CA3 cells that carries a sizable signal – which can then initiate the retrieval of the full pattern by the recurrent collaterals – one needs a large number of associatively modifiable synapses. This is expressed by the following formulas that give the specific signal S produced by sets of associatively modifiable synapses, or by non-associatively modifiable synapses: if C^{AFF} is the number of afferents per cell,

$$S_{\mathrm{ASS}} \approx \frac{\sqrt{C^{\mathrm{AFF}}}}{\sqrt{p}} q \qquad (24.5)$$

$$S_{\mathrm{NONASS}} \approx \frac{1}{\sqrt{C^{\mathrm{AFF}}}} q \qquad (24.6)$$

Associatively modifiable synapses are therefore needed, and are needed in a number C^{AFF} of the same order as the number of concurrently stored patterns p, so that small cues can be effective; whereas non-associatively modifiable synapses – or even more so, non-modifiable ones – produce very small signals, which decrease in size the larger the number of synapses.

In contrast with the storage process, the average strength of these synapses does not now play a crucial role. This suggests that the perforant path system is the one involved in relaying the cues that initiate retrieval.

The concept is that to initiate retrieval, a numerically large input (the perforant path system, see Fig. 24.14) is useful so that even a partial cue is sufficient (see Eqn 17 of Treves and Rolls (1992b)); and that the retrieval cue need not be very strong, as the recurrent collaterals (in CA3) then take over in the retrieval process to produce good recall (Treves and Rolls 1992b). In this scenario, the perforant path to CA3 synapses operate as a pattern associator, the quantitative properties of which are described in Section B.2. If an incomplete recall cue is provided to a pattern association network using distributed input representations, then most of the output pattern will be retrieved, and in this sense pattern association networks generalize between similar retrieval patterns to produce the correct output firing, and this generalization performed at the perforant path synapses to CA3 cells helps in the further completion produced by the recurrent collateral CA3-CA3 autoassociation process.

In contrast, during storage, strong signals, in the order of mV for each synaptic connection, are provided by the mossy fibre inputs to dominate the recurrent collateral activations, so that the new pattern of CA3 cell firing can be stored in the CA3 recurrent collateral connections (Treves and Rolls 1992b).

Before leaving the CA3 cells, it is suggested that separate scaling of the three major classes of excitatory input to the CA3 cells (recurrent collateral, mossy fibre, and perforant path, see Fig. 24.1) could be independently scaled, by virtue of the different classes of inhibitory interneuron which receive their own set of inputs, and end on different parts of the dendrite of the CA3 cells (cf. for CA1 Buhl, Halasy and Somogyi (1994) and Gulyas, Miles, Hajos and Freund (1993)). This possibility is made simpler by having these major classes of input terminate on different segments of the dendrites. Each of these inputs, and the negative feedback produced through inhibitory interneurons when the CA3 cells fire, should for optimal functioning be separately regulated (Rolls 1995b), and the anatomical arrangement of the different types of inhibitory interneuron might be appropriate for achieving this.

24.3.4 Dentate granule cells

The theory is that the dentate granule cell stage of hippocampal processing which precedes the CA3 stage acts in a number of ways to produce during learning the sparse yet efficient (i.e. non-redundant) representation in CA3 neurons that is required for the autoassociation to perform well (Rolls 1989a, Rolls 1989b, Rolls 1989f, Treves and Rolls 1992b, Rolls and Kesner 2006, Rolls 2013b, Kesner and Rolls 2015, Rolls 2016e). An important property for episodic memory is that the dentate by acting in this way would perform pattern separation (or orthogonalization), enabling the hippocampus to store different memories of even similar events, and this prediction has been confirmed as described in Section 24.4. By *pattern separation* I mean that the correlations between different memory patterns represented by a population of neurons become reduced (Rolls 2013b, Rolls 2016e).

24.3.4.1 The dentate cells as a competitive network

The first way in which the dentate granule cell stage helps to produce during learning the sparse yet efficient (i.e. non-redundant) representation in CA3 neurons is that the perforant path – dentate granule cell system with its Hebb-like modifiability is suggested to act as a competitive learning network to remove redundancy from the inputs producing a more orthogonal, sparse, and categorised set of outputs (Rolls 1987, Rolls 1989a, Rolls 1989b, Rolls 1989f, Rolls 1990a, Rolls 1990b). Competitive networks are described in Section B.4.

The non-linearity in the NMDA receptors may help the operation of such a competitive net, for it ensures that only the most active neurons left after the competitive feedback inhibition have synapses that become modified and thus learn to respond to that input (Rolls 1989a) (see Section B.4.9.3). If the synaptic modification produced in the dentate granule cells lasts for a period of more than the duration of learning the episodic memory, then it could reflect the formation of codes for regularly occurring combinations of active inputs that might need to participate in different episodic memories. Because of the non-linearity in the NMDA receptors, the non-linearity of the competitive interactions between the neurons (produced by feedback inhibition and non-linearity in the activation function of the neurons) need not be so great (Rolls 1989a) (see Section B.4). Because of the feedback inhibition, the competitive process may result in a relatively constant number of strongly active dentate neurons relatively independently of the number of active perforant path inputs to the dentate cells. The operation of the dentate granule cell system as a competitive network may also be facilitated by a Hebb rule of the form:

$$\delta w_{ij} = \alpha y_i (x_j - w_{ij}) \tag{24.7}$$

where δw_{ij} is the change of the synaptic weight w_{ij} that results from the simultaneous (or conjunctive) presence of presynaptic firing x_j and postsynaptic firing or activation y_i, and α is a learning rate constant that specifies how much the synapses alter on any one pairing (see Rolls (1989a) and Appendix B). Incorporation of a rule such as this which implies heterosynaptic long-term depression as well as long-term potentiation (Levy and Desmond 1985, Levy, Colbert and Desmond 1990) makes the sum of the synaptic weights on each neuron remain roughly constant during learning (Oja 1982, Rolls and Deco 2002, Rolls and Treves 1998, Rolls 1989a) (see Section B.4).

This functionality could be used to help build hippocampal place cells in rats from the grid cells present in the medial entorhinal cortex (Hafting, Fyhn, Molden, Moser and Moser 2005, Moser, Rowland and Moser 2015). Each grid cell responds to a set of places in a spatial environment, with the places to which a cell responds set out in a regular grid. Different grid cells have different phases (positional offsets) and grid spacings or frequencies (Hafting, Fyhn, Molden, Moser and Moser 2005, Giocomo, Moser and Moser 2011, Moser, Rowland and Moser 2015, Kropff and Treves 2008). We have simulated the dentate granule cells as a system that receives as inputs the activity of a population of grid cells as the animal traverses a spatial environment, and have shown that the competitive net builds dentate-like place cells from such entorhinal grid cell inputs (Rolls, Stringer and Elliot 2006c). This occurs because entorhinal cells with coactive inputs when the animal is in one place become associated together by the action of the competitive net, yet each dentate cell represents primarily one place because the dentate representation is kept sparse, thus helping to implement symmetry-breaking. Simulations to demonstrate this are shown next.

Fig. 24.17a and b show examples of two 2D entorhinal cortex (EC) grid cells (with frequencies of 4 and 7 cycles along the X axis) (Rolls, Stringer and Elliot 2006c). As in the neurophysiological recordings (Hafting et al. 2005), the firing peaks occur at the vertices of a grid of equilateral triangles. Fig. 24.17c and d show the firing rate profiles of two dentate gyrus (DG) cells without training. Multiple peaks in the response profiles are evident, and this shows that the contrast enhancement and the competitive interactions in the DG layer were not sufficient without training and thereby synaptic modification to produce place-like fields in DG.

Fig. 24.18a and b show the firing rate profiles of two DG cells after training with the Hebb rule. This type of relatively small 2D place field was frequently found in the simulations. This result shows that competitive learning was sufficient to produce place-like fields in DG. Fig. 24.18c and d show the firing rate profiles of two DG cells after training with the trace rule,

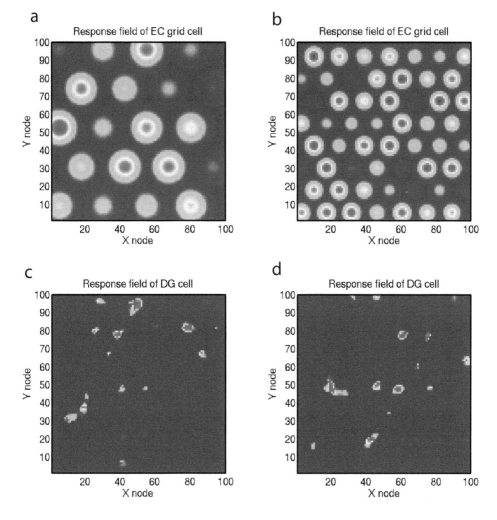

Fig. 24.17 Simulation of competitive learning in the dentate gyrus to produce place cells from the entorhinal cortex grid cell inputs. (a and b) Firing rate profiles of two entorhinal cortex (EC) grid cells with frequencies of 4 and 7 cycles. (c and d) Firing rate profiles of two dentate gyrus (DG) cells with no training. (After Rolls, Stringer and Elliot 2006.)

which is a modified form of equation B.22 on page 752 that has a short-term memory trace of previous neuronal activity as specified in equation 25.3 on page 588. The use of the trace rule is intended to test the concept that by keeping neurons in a modifiable state for a short time after strong activation (using for example the long time constant of NMDA receptors) the inputs from places near to a recently visited place where the activity of the neuron was high will tend to become associated onto the same postsynaptic neuron, thereby enlarging the place field. In the simulations, the place fields tended to be larger with the trace than with the purely associative Hebb rule shown in equation 24.7 (Rolls, Stringer and Elliot 2006c).

Although the DG cells illustrated had one main place field, some remaining grid-like firing was sometimes evident in the place fields with both Hebb and trace rule training, and indeed is evident in Fig. 24.18c and d. These are also characteristics of dentate granule cells recorded under the same conditions as the EC grid cell experiments (E. Moser, personal communica-

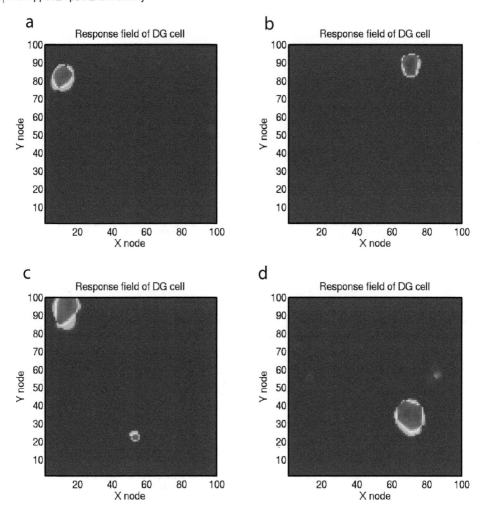

Fig. 24.18 Simulation of competitive learning in the dentate gyrus to produce place cells from the entorhinal cortex grid cell inputs. (a and b) Firing rate profiles of two DG cells after training with the Hebb rule. (c and d) Firing rate profiles of two DG cells after training with the trace rule with $\eta = 0.8$. (After Rolls, Stringer and Elliot 2006.)

tion). These results were found only with training, with untrained networks showing repeating peaks of firing in large parts of the space as illustrated in Fig. 24.17c and d. These results show that competitive learning can account for the mapping of 2D grid cells in the entorhinal cortex to 2D place cells in the dentate gyrus / CA regions (Rolls, Stringer and Elliot 2006c). The role of competitive learning in forming dentate granule cells with single or multiple place fields has been confirmed (Si and Treves 2009).

In primates, there is now evidence that there is a grid-cell like representation in the entorhinal cortex, with neurons having grid-like firing as the monkey moves the eyes across a spatial scene (Killian, Jutras and Buffalo 2012). Similar competitive learning processes may transform these entorhinal cortex 'spatial view grid cells' into hippocampal spatial view cells, and may help with the idiothetic (produced in this case by movements of the eyes) update of spatial view cells (Robertson, Rolls and Georges-François 1998). The presence of spatial view grid cells in the entorhinal cortex of primates (Killian et al. 2012) is of course predicted from the presence of spatial view cells in the primate CA3 and CA1 regions (Rolls and Xiang 2006). Further support of this type of representation of space being viewed 'out there' rather than where one is located as for rat place cells is that neurons in the human entorhinal cortex with spatial view grid-like properties have now been described (Jacobs et al. 2013).

The importance of transforming from a grid cell representation in the entorhinal cortex to a place cell representation (or in primates a spatial view cell representation) in the hippocampus is considered quantitatively next. At least one important function this allows is the formation of memories, formed for example between an object and a place, which is proposed to be a prototypical function of the hippocampus that is fundamental to episodic memory (Rolls 1996c, Rolls and Treves 1998, Rolls and Kesner 2006, Rolls and Xiang 2006). For the formation of such object–place memories in an associative network (in particular, an auto-associative network in the CA3 region of the hippocampus), the place must be made explicit in the representation, and moreover, for high capacity, that is the ability to store and retrieve many memories, the representation must be sparse. By made explicit, I mean that information can be read off easily from the firing rates of the neurons, with different places producing very different sets of neurons firing, so that different representations are relatively orthogonal. The entorhinal cortex representations are not only not sparse (in the simulations of Rolls, Stringer and Elliot (2006c), the sparseness a^p was 0.54), but in addition the typical overlap between the sets of firing of the neuronal populations representing two different places was high (with a mean cosine of the angle or normalized dot product of 0.540). In contrast, the sparseness of the representations formed in the dentate granule cells with Hebb rule training was 0.024, and a typical cosine of the angle was 0.000. (Low values of this measure of sparseness, a^p, defined in equation 24.1, indicate sparse representations. The cosine of the angle between two vectors takes the value 1 if they point in the same direction, and the value 0 if they are orthogonal, as described in Appendix A.) These sparse and orthogonal representations are what is required for high capacity storage of object and place, object and reward, and of in general episodic, memories, and this is the function, I propose, of the mapping from entorhinal cortex to hippocampal cells for which Rolls, Stringer and Elliot (2006c) produced a computational model. This concept maps well onto utility in an environment, for it is the place where we are located or at which we are looking in the world with which we wish to associate objects or rewards, and this is made explicit in the dentate granule cell / hippocampal representation (Rolls 1999c, Rolls et al. 2005c, Rolls and Xiang 2005, Rolls and Kesner 2006, Rolls and Xiang 2006, Kesner and Rolls 2015). In contrast, the entorhinal cortex may represent self-motion space in a way that is suitable for idiothetic path integration in *any* environment.

24.3.4.2 Separation of overlapping inputs

The second way in which the dentate granule cell stage acts to produce during learning the sparse yet efficient (i.e. non-redundant) representation in CA3 neurons is also a result of the competitive learning hypothesized to be implemented by the dentate granule cells (Rolls 1987, Rolls 1989a, Rolls 1989b, Rolls 1989f, Rolls 1990a, Rolls 1990b, Rolls 1994b). It is proposed that this allows overlapping (or very similar) inputs to the hippocampus to be separated, in the following way (see also Rolls (1996c)). Consider three patterns B, W, and BW, where BW is a linear combination of B and W. (To make the example very concrete, we could consider binary patterns where B=10, W=01 and BW=11.) Then the memory system is required to associate B with reward, W with reward, but BW with punishment. Without the hippocampus, rats might have more difficulty in solving such problems, particularly when they are spatial, for the dentate / CA3 system in rodents is characterized by being implicated in spatial memory. However, it is a property of competitive neuronal networks that they can separate such overlapping patterns, as has been shown elsewhere (Rolls and

Treves 1998, Rolls 1989a) (see Section B.4); normalization of synaptic weight vectors is required for this property. It is thus an important part of hippocampal neuronal network architecture that there is a competitive network that precedes the CA3 autoassociation system. Without the dentate gyrus, if a conventional autoassociation network were presented with the mixture BW having learned B and W separately, then the autoassociation network would produce a mixed output state, and would therefore be incapable of storing separate memories for B, W and BW. It is suggested therefore that competition in the dentate gyrus is one of the powerful computational features of the hippocampus, and that could enable it to help solve spatial pattern separation tasks. Consistent with this, rats with dentate gyrus lesions are impaired at a metric spatial pattern separation task (Gilbert, Kesner and Lee 2001, Goodrich-Hunsaker, Hunsaker and Kesner 2005) (see Section 24.4). The recoding of grid cells in the entorhinal cortex (Hafting et al. 2005, Moser et al. 2015) into small place field cells in the dentate granule cells that was modelled by Rolls, Stringer and Elliot (2006c) can also be considered to be a case where overlapping inputs must be recoded so that different spatial components can be treated differently.

It is noted that Sutherland and Rudy's configural learning hypothesis was similar, but was not tested with spatial pattern separation. Instead, when tested with for example tone and light combinations, it was not consistently found that the hippocampus was important (O'Reilly and Rudy 2001, Sutherland and Rudy 1991). I suggest that application of the configural concept, but applied to spatial pattern separation, may capture part of what the dentate gyrus acting as a competitive network could perform, particularly when a large number of such overlapping spatial memories must be stored and retrieved.

24.3.4.3 Low contact probability of dentate with CA3 neurons

The third way in which the dentate gyrus is hypothesized to contribute to the sparse and relatively orthogonal representations in CA3 arises because of the very low contact probability in the mossy fibre–CA3 connections, and has been explained above in Section 24.3.3.8 and by Treves and Rolls (1992b).

24.3.4.4 Dentate mossy fibre synapses to CA3 cells are strong

A fourth way is that as suggested and explained above in Section 24.3.3.8, the dentate granule cell–mossy fibre input to the CA3 cells may be powerful and its use particularly during learning would be efficient in forcing a new pattern of firing onto the CA3 cells during learning.

24.3.4.5 Expansion recoding

Fifth, expansion recoding can decorrelate input patterns, and this can be performed by a stage of competitive learning with a large number of neurons (Appendix B). A mechanism like this appears to be implemented by the dentate granule cells, which are numerous (10^6) in the rat, compared to 300,000 CA3 cells), have associatively modifiable synapses (required for a competitive network), and strong inhibition provided by the inhibitory interneurons. This may not represent expansion of numbers relative to the number of entorhinal cortex cells, but the principle of a large number of dentate granule cells, with competitive learning and strong inhibition through inhibitory interneurons, would produce a decorrelation of signals like that achieved by expansion recoding (Rolls 2013b).

24.3.4.6 Neurogenesis of dentate granule cells

Sixth, adult neurogenesis in the dentate gyrus may perform the computational role of facilitating pattern separation for new patterns, by providing new dentate granule cells with new sets of random connections to CA3 neurons (Rolls 2010b, Aimone et al. 2010, Aimone and Gage 2011, Johnston et al. 2016) (Chapter 13). Consistent with the dentate spatial pattern separation hypothesis reviewed here, in mice with impaired dentate neurogenesis, spatial learning in a delayed non-matching-to-place task in the radial arm maze was impaired for arms that were presented with little separation, but no deficit was observed when the arms were presented farther apart (Clelland, Choi, Romberg, Clemenson, Fragniere, Tyers, Jessberger, Saksida, Barker, Gage and Bussey 2009). Consistently, impaired neurogenesis in the dentate also produced a deficit for small spatial separations in an associative object-in-place task (Clelland et al. 2009).

The dentate granule cell / mossy fiber system may be one of the few places in the mammalian brain where neurogenesis may occur, because here the proposed computational role of the dentate granule cells is just to select new subsets of CA3 neurons for new learning. That is, the dentate granule cells, according to the theory, do not store the information, which instead is stored by the modification that then occurs in the CA3-CA3 recurrent collateral synapses between the selected CA3 neurons. In contrast, in most other brain areas where memories are stored and learning occurs, the learning is implemented by synaptic modification of the neurons in the circuit, and any loss of existing neurons and replacement by new neurons would result in loss of the learning or memory (see Chapter 13).

24.3.5 CA1 cells

24.3.5.1 Associative retrieval at the CA3 to CA1 (Schaffer collateral) synapses

The CA3 cells connect to the CA1 cells by the Schaeffer collateral synapses. The following arguments outline the advantage of this connection being associatively modifiable, and apply independently of the relative extent to which the CA3 or the direct entorhinal cortex inputs to CA1 drive the CA1 cells during the learning phase.

The amount of information about each episode retrievable from CA3 has to be balanced off against the number of episodes that can be held concurrently in storage. The balance is regulated by the sparseness of the coding. Whatever the amount of information per episode in CA3, one may hypothesize that the organization of the structures that follow CA3 (i.e. CA1, the various subicular fields, and the return projections to neocortex shown in Fig. 24.1) should be optimized so as to preserve and use this information content in its entirety. This would prevent further loss of information, after the massive but necessary reduction in information content that has taken place along the sensory pathways and before the autoassociation stage in CA3. We have proposed (Treves and Rolls 1994, Treves 1995, Schultz and Rolls 1999) that the need to preserve the full information content present in the output of an autoassociative memory requires an intermediate recoding stage (CA1) with special characteristics. In fact, a calculation of the information present in the CA1 firing pattern, elicited by a pattern of activity retrieved from CA3, shows that a considerable fraction of the information is lost if the synapses are non-modifiable, and that this loss can be prevented only if the CA3 to CA1 synapses are associatively modifiable. Their modifiability should match the plasticity of the CA3 recurrent collaterals. The additional information that can be retrieved beyond that retrieved by CA3 because the CA3 to CA1 synapses are associatively modifiable is strongly demonstrated by the hippocampal simulation described by Rolls (1995b), and quantitatively analyzed by Schultz and Rolls (1999).

24.3.6 Recoding in CA1 to facilitate retrieval to the neocortex

An additional factor is that if the total amount of information carried by CA3 cells is redistributed over a larger number of CA1 cells, less information needs to be loaded onto each CA1 cell, rendering the code more robust to information loss in the next stages. For example, if each CA3 cell had to code for 2 bits of information, e.g. by firing at one of four equiprobable activity levels, then each CA1 cell (if there were twice as many as there are CA3 cells) could code for just 1 bit, e.g. by firing at one of only two equiprobable levels. Thus the same information content could be maintained in the overall representation while reducing the sensitivity to noise in the firing level of each cell. In fact, there are more CA1 cells (4×10^5) than CA3 cells in (SD) rats. There are even more CA1 cells (4.6×10^6) in humans (and the ratio of CA1 to CA3 cells is greater). The CA1 cells may thus provide the first part of the expansion for the return projections to the enormous numbers of neocortical cells in primates, after the bottleneck of the single network in CA3, the number of neurons in which may be limited because it has to operate as a single network.

Another argument on the operation of the CA1 cells is also considered to be related to the CA3 autoassociation effect. In this, several arbitrary patterns of firing occur together on the CA3 neurons, and become associated together to form an episodic or 'whole scene' memory. It is essential for this operation that several different sparse representations are present conjunctively in order to form the association. Moreover, when completion operates in the CA3 autoassociation system, all the neurons firing in the original conjunction can be brought into activity by only a part of the original set of conjunctive events. For these reasons, a memory in the CA3 cells consists of several different simultaneously active ensembles of activity. To be explicit, the parts A, B, C, D and E of a particular episode would each be represented, roughly speaking, by its own population of CA3 cells, and these five populations would be linked together by autoassociation. It is suggested that the CA1 cells, which receive these groups of simultaneously active ensembles, can detect the conjunctions of firing of the different ensembles that represent the episodic memory, and allocate by competitive learning neurons to represent at least larger parts of each episodic memory (Rolls 1987, Rolls 1989a, Rolls 1989b, Rolls 1989f, Rolls 1990a, Rolls 1990a, Rolls 1990b). In relation to the simple example above, some CA1 neurons might code for ABC, and others for BDE, rather than having to maintain independent representations in CA1 of A, B, C, D, and E. This implies a more efficient representation, in the sense that when eventually after many further stages, neocortical neuronal activity is recalled (as discussed below), each neocortical cell need not be accessed by all the axons carrying each component A, B, C, D and E, but instead by fewer axons carrying larger fragments, such as ABC, and BDE. This process is performed by competitive networks, which self-organize to find categories in the input space, where each category is represented by a set of simultaneously active inputs (Rolls and Deco 2002, Rolls and Treves 1998, Rolls 2000b) (see Section B.4). Concerning the details of operation of the CA1 system, we note that although competitive learning may capture part of how it is able to recode, the competition is probably not global, but instead would operate relatively locally within the domain of the connections of inhibitory neurons. This simple example is intended to show how the coding may become less componential and more conjunctive in CA1 than in CA3, but should not be taken to imply that the representation produced becomes more sparse.

Concerning the details of operation of the CA1 system, we note that although competitive learning may capture part of how it is able to recode, the competition is probably not global, but instead would operate relatively locally within the domain of the connections of inhibitory neurons. This simple example is intended to show how the coding may become less componential and more conjunctive in CA1 than in CA3, but should not be taken to imply that the representation produced becomes more sparse. An example of what might be predicted follows. Given that objects and the places in which they are seen may be associated by learning in the hippocampus, we might predict that relatively more CA3 neurons might be encoding either objects or places (the components to be associated), and relatively more CA1 neurons might be encoding conjunctive combinations of objects and places. Although object, spatial view, and object-spatial view combination neurons have now been shown to be present in the primate hippocampus during a room-based object-spatial view task (Rolls, Xiang and Franco 2005c), the relative proportions of the types of neuronal response in CA3 vs CA1 have not yet been analyzed.

The conjunctive recoding just described applies to any one single item in memory. Successive memory items would be encoded by a temporal sequence of such conjunctively encoded representations. To the extent that CA1 helps to build conjunctively encoded separate memory items, this could help a sequence of such memories (i.e. memory items) to be distinct (i.e. orthogonal to each other). This may be contrasted with successive memory items as encoded in CA3, in which there might be more overlap because each memory would be represented by its components, and some of the components might be common to some of the successive memory items. Episodic memory has been used here to refer to the memory of a single event or item. It can also be used to describe a succession of items that occurred close together in time, usually at one place. CA1 could help the encoding of these successive items that are parts of a multi-item episodic memory by making each item distinct from the next, due to the conjunctive encoding of each item. These arguments suggest that the CA1 could be useful in what is described as temporal pattern separation in Section 24.4.

24.3.6.1 CA1 inputs from CA3 vs direct entorhinal inputs

Another feature of the CA1 network is its double set of afferents, with each of its cells receiving most synapses from the Schaeffer collaterals coming from CA3, but also a proportion (about 1/6 – Amaral et al. (1990)) from direct perforant path projections from the entorhinal cortex. Such projections appear to originate mainly in layer 3 of entorhinal cortex (Witter, Van Hoesen and Amaral 1989b), from a population of cells only partially overlapping with that (mainly in layer 2) giving rise to the perforant path projections to DG and CA3. This suggests that it is useful to include in CA1 not only what it is possible to recall from CA3, but also the detailed information present in the retrieval cue itself (Treves and Rolls 1994).

Another possibility is that the perforant path input provides the strong forcing input to the CA1 neurons during learning, and that the output of the CA3 system is associated with this forced CA1 firing during learning (McClelland et al. 1995). During recall, an incomplete cue could then be completed in CA3, and the CA3 output would then produce firing in CA1 that would correspond to that present during the learning. This suggestion is essentially identical to that of Treves and Rolls (1994) about the backprojection system and recall, except that McClelland et al. (1995) suggest that the output of CA3 is associated at the CA3 to CA1 (Schaeffer collateral) synapses with the signal present during training in CA1, whereas in the theory of Treves and Rolls (1994) the output of the hippocampus consists of CA1 firing which is associated in the entorhinal cortex and earlier cortical stages with the firing present during learning, providing a theory of how the correct recall is implemented at every backprojection stage though the neocortex (see Section 24.3.7).

24.3.6.2 CA1 and sequence memory for objects and odors

Evidence described in Section 24.4 and by Kesner and Rolls (2015) indicates that the CA1, but not (dorsal) CA3, subregion is involved in the memory for sequences of objects and odors. In humans, the hippocampus becomes activated when the temporal order of events is being processed (Lehn, Steffenach, van Strien, Veltman, Witter and Haberg 2009). How might this be implemented? As summarized in Section 24.3.2, with respect to the medial entorhinal cortex grid cells, there are a number of theories of their computational bases (Giocomo et al. 2011, Moser et al. 2014a, Zilli 2012). One attractive theory is that different temporal delays in the neural circuitry of the medial entorhinal cortex related to a temporal adaptation process might result in the different sizes of spatial grids that are found in the medial entorhinal cortex (Kropff and Treves 2008). This is essentially a timing hypothesis; and it is an interesting new idea that the timing mechanism that may be inherent in the medial entorhinal cortex might also be a source of the temporal delay / timing information needed to form sequence memories in CA1, using the direct projection from medial entorhinal cortex to CA1 to introduce timing information in the form of neurons that fire at different temporal times in a sequence, which is what is needed for the formation of odor, object and spatial sequence memories in CA1.

Consistent with this hypothesis, timing cells are found in the medial entorhinal cortex (Kraus, Brandon, Robinson, Connerney, Hasselmo and Eichenbaum 2015)). Object (including odor) information could reach CA1 from the lateral entorhinal cortex. The time and object information might be combined in CA1 by the associative process inherent in a competitive network to form time AND object combination neurons (with time originating from medial and object from lateral entorhinal cortex). Replay of the time information on another trial could allow the same object and time combination neurons to be retrieved by generalization, and thus effectively for the position of the object in the sequence to be retrieved.

This hypothesis is consistent with neurophysiological evidence of MacDonald, Eichenbaum and colleagues (MacDonald, Lepage, Eden and Eichenbaum 2011, Eichenbaum 2014, Howard and Eichenbaum 2015) showing that neurons in the rat hippocampus have firing rates that reflect which temporal part of the task is current. In particular, a sequence of different neurons is activated at successive times during a time delay period. The tasks used included an object-odor paired associate non-spatial task with a 10 s delay period between the visual stimulus and the odor. The new evidence also shows that a large proportion of hippocampal neurons fire in relation to individual events in a sequence being remembered (e.g. a visual object or odor), and some to combinations of the event and the time in the delay period (MacDonald et al. 2011).

These interesting neurophysiological findings indicate that rate encoding is being used to encode time, that is, the firing rates of different neurons are high at different times within a trial, delay period, etc (MacDonald et al. 2011, Kraus et al. 2015).

An alternative hypothesis would implicate the (perhaps ventral part of) CA3 in object-temporal sequence encoding. This hypothesis utilizes the slow transitions from one attractor to another which are a feature that arises at least in some networks in the brain due to the noise-influenced transitions from one state to another, and due to slow dynamics such as those produced by neuronal adaptation (Deco and Rolls 2005c). The process would involve autoassociation between neurons that fire at different times in a trial and neurons that represent the object. These associations would provide the basis for the recall of the object from the time in a trial, or vice versa. The retrieval of object or temporal information from each other would occur in a way that is analogous to that shown for recalling the object from the place, or the place from the object (Rolls, Stringer and Trappenberg 2002) (Fig. 24.16), but substituting the details of the properties of the 'time encoding' neurons (MacDonald et al. 2011) for what was previously the spatial (place) component. In addition, if the time encoding neurons simply cycled through their normal sequence during recall, this would enable the sequence of objects or events associated with each subset of time encoding neurons to be recalled correctly in the order in which they were presented.

24.3.6.3 Subcortical, vs neocortical backprojection, outputs from CA1

The CA1 network is on anatomical grounds thought to be important for the hippocampus to access the neocortex, and the route by which retrieval to the neocortex would occur (see Fig. 24.1). However, as also shown in Fig. 24.1, there are direct subcortical outputs of the hippocampus. One route is that outputs that originate from CA3 project via the fimbria directly to the medial septum (Gaykema et al. 1991) or indirectly via the lateral septum to the medial septum (Swanson and Cowan 1977, Saper, Swanson and Cowan 1979). The medial septum in turn provides cholinergic and GABAergic inputs back to the hippocampus, especially the

CA3 subregion. In addition to projecting back to the hippocampus, the medial septum also projects to subiculum, entorhinal cortex and mammillary bodies. In turn, the mammillary bodies project to the anterior thalamus which projects to the cingulate cortex, and also to the entorhinal cortex via the pre- and parasubiculum. The outputs from CA3 in the fimbria can thus potentially send information to the entorhinal cortex via direct projections from the medial septum, and via the anterior thalamus. A second subcortical output route is provided by the subiculum and CA1 projections via the fimbria/fornix to the lateral septum, and to the mammillary bodies and anterior thalamic nuclei, which in turn project to the cingulate cortex as well as to the prelimbic cortex (rat), which could provide a potential output to behaviour (Vann and Aggleton 2004), and to the entorhinal cortex. A third subcortical output route is that the subiculum and CA1 also project to the nucleus accumbens (Swanson and Cowan 1977). How might the CA1 to entorhinal cortico-cortical backprojection system, and the different subcortical output routes from the hippocampus, implement different functions?

First, we should note that the CA1 to entorhinal cortico-cortical backprojection system is numerically, in terms of the number of connections involved, large, and this as developed below is likely to be crucial in allowing large numbers of memories to be retrieved from the hippocampus back to the neocortex. This recall operation to the neocortex is described below, but we note here that once a memory has been retrieved to the neocortex, the neocortex then provides a way for hippocampal function to influence behaviour. For example, in a place–object recall task, a place cue used to initiate recall of the whole place-object memory in the CA3 would then lead via the backprojections to the object being retrieved (via entorhinal cortex layer 5 and perirhinal cortex) back to the inferior temporal visual cortex (see Fig. 24.1 left), and this object representation could then guide behaviour via the other outputs of the inferior temporal visual cortex (Rolls and Deco 2002). As objects are represented in the neocortex, it is a strong prediction of the theory that place–object recall will involve this CA1 backprojection pathway back to the neocortex. Backprojections from CA1 via entorhinal cortex to parahippocampal gyrus and thus to parietal cortex or other parts of the neocortex with spatial representations could allow spatial representations to be retrieved in for example an object–place recall task, and this provides a possible route to spatial action (though an alternative subcortical route for spatial action is described below). In addition, there are more direct CA1, subicular, and entorhinal connections to the prefrontal cortex (Jay and Witter 1991), which could allow the hippocampus to provide inputs to prefrontal cortex networks involved in short-term memory and in planning actions (Deco and Rolls 2003, Deco and Rolls 2005a).

Second, we note that the part of the second subcortical pathway that projects from CA1 and subiculum to the mammillary bodies and anterior thalamus is also numerically large, with in the order of 10^6 axons in the fornix and mammillo-thalamic tract. This could therefore provide an information-rich output pathway to regions such as the cingulate cortex via which some types of (e.g. spatial) behavioural response could be specified by the hippocampus. In contrast, the CA3 projections via the fimbria to the medial septum described as part of the first subcortical route are much less numerous, with indeed for example only approximately 2000 cholinergic neurons in the rat medial septum (Moore, Lee, Paiva, Walker and Heaton 1998). Given further that this first subcortical route returns mainly to the hippocampus and connected structures, this route may be more involved in regulating hippocampal function (via cholinergic and GABAergic influences), than in directly specifying behaviour (cf. Hasselmo et al. (1995)).

Differential predictions of the functions mediated via the cortical vs subcortical outputs of the hippocampal system are thus as follows. Place-to-object recall should involve the backprojections via CA1 to the neocortex, for that is where the objects are represented. But object to place (or spatial response) tasks might not require CA1 and backprojections to parts of the cortex with spatial representations, in so far as the hippocampal subcortical output needs then only to affect which behavioural (spatial) response is made, which it is conceivable could be produced by fornix/fimbria efferents which indirectly reach structures such as the cingulate cortex. To test the hypothesis of potentially different output systems for different types of behaviour, a new rat paradigm could be run comparing object–place vs place–object cued recall tasks. The prediction is that object–place cued recall might depend on hippocampal outputs to subcortical structures such as the mammillary bodies etc, though cortical outputs might be an alternative route. It would at least be of interest to know whether cued object–place recall is impaired by damaging hippocampal subcortical outputs. The stronger prediction is that place–object cued recall depends on CA1 and recall to the neocortex. It is noted that in either case the hippocampus itself may be useful, because object–place and place–object associations involve places, and cannot be solved by object-reward association memory, which instead depends on the orbitofrontal cortex and amygdala (Rolls 2014a).

24.3.7 Backprojections to the neocortex, episodic memory recall, and consolidation

The need for information to be retrieved from the hippocampus to influence neocortical areas and recall memories from the hippocampus was noted in Section 24.1. David Marr (1971) realised the importance of a theory for this, promised one in his 1971 paper, but never produced a theory. Rolls produced a theory of how the recall of information to the neocortex could be implemented via cortico-cortical backprojections from the hippocampus to the neocortex (Rolls 1989b, Rolls 1989f, Rolls 1990b), and the theory became quantitative (Treves and Rolls 1994), and was tested by simulation (Rolls 1995b). The theory and developments of it are described next.

It is suggested that the modifiable connections from the CA3 neurons to the CA1 neurons allow the whole episode in CA3 to be produced in CA1. This may be assisted as described above by the direct perforant path input to CA1. This might allow details of the input key for the recall process, as well as the possibly less information-rich memory of the whole episode recalled from the CA3 network, to contribute to the firing of CA1 neurons. The CA1 neurons would then activate, via their termination in the deep layers of the entorhinal cortex, at least the pyramidal cells in the deep layers of the entorhinal cortex (see Fig. 24.1). These entorhinal cortex layer 5 neurons would then, by virtue of their backprojections (Lavenex and Amaral 2000, Witter, Naber, van Haeften, Machielsen, Rombouts, Barkhof, Scheltens and Lopes da Silva 2000a) to the parts of cerebral cortex that originally provided the inputs to the hippocampus, terminate in the superficial layers of those neocortical areas, where synapses would be made onto the distal parts of the dendrites of the cortical pyramidal cells (Rolls 1989a, Rolls 1989b, Rolls 1989f, Markov et al. 2014b, Pandya et al. 2015). The areas of cerebral neocortex in which this recall would be produced could include multimodal cortical areas (e.g. the cortex in the superior temporal sulcus which receives inputs from temporal, parietal and occipital cortical areas, and from which it is thought that cortical areas such as 39 and 40 related to language developed); and the orbitofrontal and anterior cingulate cortex to retrieve the reward / affective aspects of an episodic memory (Rolls 2014a, Rolls 2015c); and also areas of unimodal association cortex (e.g. inferior temporal visual cortex).

Thus when an episodic memory is recalled, it is predicted that activity will be reinstated using backprojections in cortical areas that represent the different semantic features of the episodic memory. For example, if a friend on a ski slope was recalled, activity would be recalled in parietal and related areas that represent spatial scenes, in temporal lobe areas that represent faces, and in the orbitofrontal cortex where reward value and emotion are represented. The episodic memory would be represented in CA3. Each of these other cortical areas would have neurons that represent the semantic features of each attribute or component of the episodic memory. The same prediction is made about the effects of words, which similarly from cortical areas where words are represented (Section 8.5) would by backprojections activate many different cortical areas in which the different semantic features are represented. This prediction has been confirmed using fMRI (Huth, de Heer, Griffiths, Theunissen and Gallant 2016). The concept is that the episodic memory would be represented in the hippocampus, the word in language areas (in both using sparse distributed representations), and that each of the components that are active when the episodic memory was formed, or word is normally accessed, using forward inputs from the world, would become activated to represent the relevant semantic parts during the memory recall, or during the presentation of the word.

The backprojections, by recalling previous episodic events, could provide information useful to the neocortex in the building of new representations in the multimodal and unimodal association cortical areas, which by building new long-term representations can be considered as a form of memory consolidation (Rolls 1989a, Rolls 1989b, Rolls 1989f, Rolls 1990a, Rolls 1990b).

The hypothesis of the architecture with which this would be achieved is shown in Fig. 24.1. The feedforward connections from association areas of the cerebral neocortex (solid lines in Fig. 24.1) show major convergence as information is passed to CA3, with the CA3 autoassociation network having the smallest number of neurons at any stage of the processing. The backprojections allow for divergence back to neocortical areas. The way in which I suggest that the backprojection synapses are set up to have the appropriate strengths for recall is as follows (Rolls 1989a, Rolls 1989b, Rolls 1989f, Rolls 2008d, Kesner and Rolls 2015, Rolls 2015b). During the setting up of a new episodic memory, there would be strong feedforward activity progressing towards the hippocampus. During the episode, the CA3 synapses would be modified, and via the CA1 neurons and the subiculum, a pattern of activity would be produced on the backprojecting synapses to the entorhinal cortex. Here the backprojecting synapses from active backprojection axons onto pyramidal cells being activated by the forward inputs to entorhinal cortex would be associatively modified. A similar process would be implemented at (at least some) preceding stages of neocortex, that is in the parahippocampal gyrus/perirhinal cortex stage, and in association cortical areas, as shown in Fig. 24.1.

The concept is that during the learning of an episodic memory, cortical pyramidal cells in at least one of the stages would be driven by forward inputs, but would simultaneously be receiving backprojected activity (indirectly) from the hippocampus which would by pattern association from the backprojecting synapses to the cortical pyramidal cells become associated with whichever cortical cells were being made to fire by the forward inputs. Then later on, during recall, a recall cue from perhaps another part of cortex might reach CA3, where the firing during the original episode would be completed. The resulting backprojecting activity would then, as a result of the pattern association learned previously, bring back the firing in any cortical area that was present during the original episode. Thus retrieval involves reinstating the activity that was present in different cortical areas that was present during the learning of an episode. (The pattern association is also called heteroassociation, to contrast it with autoassociation. The pattern association operates at multiple stages in the backprojection pathway, as made evident in Fig. 24.1). If the recall cue was an object, this might result in recall of the neocortical firing that represented the place in which that object had been seen previously. As noted elsewhere in this Chapter and by McClelland et al. (1995), that recall might be useful to the neocortex to help it build new semantic memories, which might inherently be a slow process and is not part of the theory of recall. It is an interesting possibility that recall might involve several cycles through the recall process. After the information fed back from the first pass with a recall cue from perhaps only one cortical area, information might gradually be retrieved to other cortical areas involved in the original memory, and this would then act as a better retrieval cue for the next pass.

The timing of the backprojecting activity would be sufficiently rapid for this, in that for example inferior temporal cortex (ITC) neurons become activated by visual stimuli with latencies of 90–110 ms and may continue firing for several hundred milliseconds (Rolls 1992a, Rolls and Deco 2002, Rolls 2003); and hippocampal pyramidal cells are activated in visual object-and-place and conditional spatial response tasks with latencies of 120–180 ms (Miyashita, Rolls, Cahusac, Niki and Feigenbaum 1989, Rolls, Miyashita, Cahusac, Kesner, Niki, Feigenbaum and Bach 1989b). Thus, backprojected activity from the hippocampus might be expected to reach association cortical areas such as the inferior temporal visual cortex within 60 – 100 ms of the onset of their firing, and there would be a several hundred millisecond period in which there would be conjunctive feedforward activation present with simultaneous backprojected signals in the association cortex.

During recall, the backprojection connections onto the distal synapses of cortical pyramidal cells would be helped in their efficiency in activating the pyramidal cells by virtue of two factors. The first is that with no forward input to the neocortical pyramidal cells, there would be little shunting of the effects received at the distal dendrites by the more proximal effects on the dendrite normally produced by the forward synapses. Further, without strong forward activation of the pyramidal cells, there would not be very strong feedback and feedforward inhibition via GABA cells, so that there would not be a further major loss of signal due to (shunting) inhibition on the cell body and (subtractive) inhibition on the dendrite. (The converse of this is that when forward inputs are present, as during normal processing of the environment rather than during recall, the forward inputs would, appropriately, dominate the activity of the pyramidal cells, which would be only influenced, not determined, by the backprojecting inputs (Rolls 1989b, Rolls 1989e, Deco and Rolls 2005b).)

The synapses receiving the backprojections would have to be Hebb-modifiable, as suggested by Rolls (1989b) and Rolls (1989e) (Chapter 11). This would solve the de-addressing problem, which is the problem of how the hippocampus is able to bring into activity during recall just those cortical pyramidal cells that were active when the memory was originally being stored. The solution hypothesized (Rolls 1989b, Rolls 1989e) arises because modification occurs during learning of the synapses from active backprojecting neurons from the hippocampal system onto the dendrites of only those neocortical pyramidal cells active at the time of learning. Without this modifiability of cortical backprojections during learning at some cortical stages at least, it is difficult to see how exactly the correct cortical pyramidal cells active during the original learning experience would be activated during recall. Consistent with this hypothesis (Rolls 1989b, Rolls 1989e), there are NMDA receptors present especially in superficial layers of the cerebral cortex (Monaghan and Cotman 1985), implying Hebb-like learning just where the backprojecting axons make synapses with the apical dendrites of cortical pyramidal cells.

The quantitative argument that the backprojecting synapses in at least one stage have to be associatively modifiable parallels that applied to the pattern retrieval performed at the entorhinal to CA3 synapses (Treves and Rolls 1992b) and at the CA3 to CA1 synapses (Schultz and Rolls 1999), and is that the information retrieved would otherwise be very low. The performance of pattern association networks is considered in detail by Rolls and Treves (1990) and Rolls and Treves (1998) (see Appendix B) and other authors (Hertz et al. 1991). An interesting recent development is that dilution in the connectivity of pattern association networks by reducing the probability of two synapses onto a neuron from the same backprojecting input helps to maintain high capacity in the backprojecting system (Rolls 2015b). If there are some instances of two or more synapses onto a neuron from the same input axon in a pattern association network, the capacity of the system degrades because of distortion and interferences between the stored patterns (Rolls 2015b). It is also noted that the somewhat greater anatomical spread of the backprojection than the forward connections between two different stages in the hierarchy shown in Fig. 24.1 would not be a problem, for it would provide every chance for the backprojecting axons to find co-active neurons in an earlier cortical stage that are part of the representation that is relevant to the current memory being formed.

If the backprojection synapses are associatively modifiable, we may consider the duration of the period for which their synaptic modification should persist. What follows from the operation of the system described above is that there would be no point, indeed it would be disadvantageous, if the synaptic modifications lasted for longer than the memory remained in the hippocampal buffer store. What would be optimal would be to arrange for the associative modification of the backprojecting synapses to remain for as long as the memory persists in the hippocampus. This suggests that a similar mechanism for the associative modification within the hippocampus and for that of at least one stage of the backprojecting synapses would be appropriate. It is suggested that the presence of high concentrations of NMDA synapses in the distal parts of the dendrites of neocortical pyramidal cells and within the hippocampus may reflect the similarity of the synaptic modification processes in these two regions (Kirkwood, Dudek, Gold, Aizenman and Bear 1993). It is noted that it would be appropriate to have this similarity of time course (i.e. rapid learning within 1–2 s, and slow decay over perhaps weeks) for at least one stage in the series of backprojecting stages from the CA3 region to the neocortex. Such stages might include the CA1 region, subiculum, entorhinal cortex, and perhaps the parahippocampal gyrus / perirhinal cortex. However from multimodal cortex (e.g. the parahippocampal gyrus) back to earlier cortical stages, it might be desirable for the backprojecting synapses to persist for a long period, so that some types of recall and top-down processing (Rolls and Deco 2002, Rolls 1989b, Rolls 1989e) mediated by the operation of neocortico-neocortical backprojecting synapses could be stable, and might not require modification during the learning of a new episodic memory (see Section 24.5).

An alternative hypothesis to that above is that rapid modifiability of backprojection synapses would be required only at the beginning of the backprojecting stream. Relatively fixed associations from higher to earlier neocortical stages would serve to activate the correct neurons at earlier cortical stages during recall. For example, there might be rapid modifiability from CA3 to CA1 neurons, but relatively fixed connections from there back (McClelland et al. 1995). For such a scheme to work, one would need to produce a theory not only of the formation of semantic memories in the neocortex, but also of how the operations performed according to that theory would lead to recall by setting up appropriately the backprojecting synapses.

We have noted elsewhere that backprojections, which included cortico-cortical backprojections, and backprojections originating from structures such as the hippocampus and amygdala, may have a number of different but compatible functions (Rolls and Deco 2002, Rolls 2005, Rolls 1990a, Rolls 1989b, Rolls 1989e, Rolls 1990b, Rolls 1989a, Rolls 1992a) including implementing top-down attention by biased competition (see Chapters 11 and 6). The particular function with which we have been concerned here is how memories stored in the hippocampus might be recalled in regions of the cerebral neocortex.

24.3.8 Backprojections to the neocortex – quantitative aspects

How many backprojecting fibres does one need to synapse on any given neocortical pyramidal cell, in order to implement the mechanism outlined above? Clearly, if neural network theory were to produce a definite constraint of that sort, quantitative anatomical data could be used for verification or falsification.

Attempts to come up with an estimate of the number of synapses required have sometimes followed the simple line of reasoning presented next, which is shown to be unsatisfactory.

[The type of argument to be described has been applied to analyse the capacity of pattern association and autoassociation memories (see for example Marr (1971) and Willshaw and Buckingham (1990)), and to show that there are limitations in this type of approach, this approach is considered in this paragraph.] Consider first the assumption that hippocampo-cortical connections are monosynaptic and not modifiable, and that all existing synapses have the same efficacy. Consider further the assumption that a hippocampal representation across N cells of an event consists of $a_h N$ cells firing at the same elevated rate (where a_h is the sparseness of the hippocampal representation), while the remaining $(1 - a_h)N$ cells are silent. If each pyramidal cell in the association areas of neocortex receives synapses from an average C^{HBP} hippocampal axons, there will be an average probability $y = C^{HBP}/N$ of finding a synapse from any given hippocampal cell to any given neocortical one. Across neocortical cells, the number A of synapses of hippocampal origin activated by the retrieval of a particular episodic memory will follow a Poisson distribution of average $y a_h N = a_h C^{HBP}$

$$P(A) = (a_h C^{HBP})^A \exp(-a_h C^{HBP})/A! \qquad (24.8)$$

The neocortical cells activated as a result will be those, in the tail of the distribution, receiving at least T active input lines, where T is a given threshold for activation. Requiring that $P(A > T)$ be at least equal to a_{nc}, the fraction of neocortical cells involved in the neocortical representation of the episode, results in a constraint on C^{HBP}. This simple type of calculation can be extended to the case in which hippocampo-cortical projections are taken to be polysynaptic, and mediated by modifiable synapses. In any case, the procedure does not appear to produce a very meaningful constraint for at least three reasons. First, the resulting minimum value of C^{HBP}, being extracted by looking at the tail of an exponential distribution, varies dramatically with any variation in the assumed values of the parameters a_h, a_{nc}, and T. Second, those parameters are ill-defined in principle: a_h and a_{nc} are used having in mind the unrealistic assumption of binary distributions of activity in both the hippocampus and neocortex (although the sparseness of a representation can be defined in general, as shown above, it is the particular definition pertaining to the binary case that is invoked here); while the definition of T is based on unrealistic assumptions about neuronal dynamics (how coincident does one require the various inputs to be in time, in order to generate a single spike, or a train of spikes of given frequency, in the postsynaptic cell?). Third, the calculation assumes that neocortical cells receive no other inputs, excitatory or inhibitory. Relaxing this assumption, to include for example non-specific activation by subcortical afferents, makes the calculation extremely fragile. This argument applies in general to approaches to neurocomputation which base their calculations on what would happen in the tail of an exponential, Poisson, or binomial distribution. Such calculations must be interpreted with great care for the above reasons.

An alternative way to estimate a constraint on C^{HBP}, still based on very simple assumptions, but which is more robust with respect to relaxing those assumptions, is the following.

Consider a polysynaptic sequence of backprojecting stages, from hippocampus to neocortex, as a string of simple (hetero-)associative memories in which, at each stage, the input lines are those coming from the previous stage (closer to the hippocampus) (Treves and Rolls 1994) (Fig. 24.1). Implicit in this framework is the assumption that the synapses at each stage are modifiable and have been indeed modified at the time of first experiencing each episode, according to some Hebbian associative plasticity rule. A plausible requirement for a successful hippocampo-directed recall operation is that the signal generated from the hippocampally retrieved pattern of activity, and carried backwards towards neocortex, remains undegraded when compared to the noise due, at each stage, to the interference effects caused by the concurrent storage of other patterns of activity on the same backprojecting synaptic systems. That requirement is equivalent to that used in deriving the storage capacity of such a series of heteroassociative memories, and it was shown in Treves and Rolls (1991) that the maximum number of independently generated activity patterns that can be retrieved is given, essentially, by the same formula as equation 24.2 above

$$p \approx \frac{C}{a \ln(\frac{1}{a})} k' \tag{24.9}$$

where, however, a is now the sparseness of the representation at any given backprojection stage, and C is the average number of (back-)projections each cell of that stage receives from cells of the previous one[20]. (k' is a similar slowly varying factor to that introduced above.) If p is equal to the number of memories held in the hippocampal memory, it is limited by the retrieval capacity of the CA3 network, p_{max}. Putting together the formula for the latter with that shown here, one concludes that, roughly, the requirement implies that the number of afferents of (indirect) hippocampal origin to a given neocortical stage (C^{HBP}) must be $C^{HBP} = C^{RC} a_{nc}/a_{CA3}$, where C^{RC} is the number of recurrent collaterals to any given cell in CA3, the average sparseness of a neocortical representation is a_{nc}, and a_{CA3} is the sparseness of memory representations in CA3.

The above requirement is very strong: even if representations were to remain as sparse as they are in CA3, which is unlikely, to avoid degrading the signal, C^{HBP} should be as large as C^{RC}, i.e. 12,000 in the rat. Moreover, other sources of noise not considered in the present calculation would add to the severity of the constraint, and partially compensate for the relaxation in the constraint that would result from requiring that only a fraction of the p episodes would involve any given cortical area. If then C^{HBP} has to be of the same order as C^{RC}, one is led to a very definite conclusion: a mechanism of the type envisaged here could not possibly rely on a set of monosynaptic CA3-to-neocortex backprojections. This would imply that, to make a sufficient number of synapses on each of the vast number of neocortical cells, each cell in CA3 has to generate a disproportionate number of synapses (i.e. C^{HBP} times the ratio between the number of neocortical and that of CA3 cells). The required divergence can be kept within reasonable limits only by assuming that the backprojecting system is polysynaptic, provided that the number of cells involved grows gradually at each stage, from CA3 back to neocortical association areas (Treves and Rolls 1994) (cf. Fig. 24.1).

Although backprojections between any two adjacent areas in the cerebral cortex are approximately as numerous as forward projections, and much of the distal parts of the dendrites of cortical pyramidal cells are devoted to backprojections, the actual number of such connections onto each pyramidal cell may be on average only in the order of thousands. Further, not all might reflect backprojection signals originating from the hippocampus, for there are backprojections which might be considered to originate in the amygdala (Amaral et al. 1992) or in multimodal cortical areas (allowing for example for recall of a visual image by an auditory stimulus with which it has been regularly associated). In this situation, one may consider whether the backprojections from any one of these systems would be sufficiently numerous to produce recall. One factor which may help here is that when recall is being produced by the backprojections, it may be assisted by the local recurrent collaterals between nearby ($\approx$1 mm) pyramidal cells which are a feature of neocortical connectivity. These would tend to complete a partial neocortical representation being recalled by the backprojections into a complete recalled pattern. (Note that this completion would be only over the local information present

---

[20]The interesting and important insight here is that multiple iterations through a recurrent autoassociative network can be treated as being formally similar to a whole series of feedforward associative networks, with each new feedforward network equivalent to another iteration in the recurrent network (Treves and Rolls 1991). This formal similarity enables the quantitative analysis developed in general for autoassociation nets (see Treves and Rolls (1991)) to be applied directly to the series of backprojection networks (Treves and Rolls 1994).

within a cortical area about e.g. visual input or spatial input; it provides a local 'clean-up' mechanism, and could not replace the global autoassociation performed effectively over the activity of very many cortical areas which the CA3 could perform by virtue of its widespread recurrent collateral connectivity.) There are two alternative possibilities about how this would operate. First, if the recurrent collaterals showed slow and long-lasting synaptic modification, then they would be useful in completing the whole of long-term (e.g. semantic) memories. Second, if the neocortical recurrent collaterals showed rapid changes in synaptic modifiability with the same time course as that of hippocampal synaptic modification, then they would be useful in filling in parts of the information forming episodic memories which could be made available locally within an area of the cerebral neocortex.

This theory of recall by the backprojections thus provides a quantitative account of why the cerebral cortex has as many backprojection as forward projection connections (Chapter 11).

A recent development has been on understanding the diluted connectivity in the back-projection pathways. The backprojection pathways have diluted connectivity in that each neocortical neuron may receive up to in the order of 10,000 backprojection synapses, yet the total number of neurons at each neocortical stage in the backprojection hierarchy shown in Fig. 24.1 has very many more neurons than this, at least 10 times as many neurons as there are synapses per neuron. This diluted connectivity in the backprojection pathways is likely to reduce the probability of more than one connection onto a receiving neuron in the backprojecting pathways. It has now been shown that if there were more than one connection onto each neuron in a pattern associator, this would reduce the capacity of the system, which in this case is the number of memories that can be recalled from the hippocampus to the neocortex (Rolls 2015b).

These concepts show how the backprojection system to neocortex can be conceptualized in terms of pattern completion, as follows. The information that is present when a memory is formed may be present in different areas of the cerebral cortex, for example of a face in a temporal cortex face area (Rolls 2012c), of a spatial location in a neocortical location area, and of a reward received in the orbitofrontal cortex (Rolls 2014a). To achieve detailed retrieval of the memory, reinstatement of the activity during recall of the neuronal activity during the original memory formation may be needed. This is what the backprojection system described could achieve, and is a form of completion of the information that was represented in the different cortical areas when the memory was formed. In particular, the concept of completion here is that if a recall cue from a visual object area is provided, then the emotional parts of the episodic memory can be recalled in the orbitofrontal cortex, and the spatial parts in parietal cortical areas, with the result that a complete memory is retrieved, with activity recalled into several higher-order cortical areas. Because such a wide set of different neocortical areas must be content-addressed, a multistage feedback system is required, to keep the number of synapses per neuron in the backprojection pathways down to reasonable numbers. (Having CA1 directly address neocortical areas would require each CA1 neuron to have tens of millions of synapses with cortical neurons. That is part of the computational problem solved by the multistage backprojection system shown in Fig. 24.1. Thus, the backprojection system with its series of pattern associators can each be thought of as retrieving the complete pattern of cortical activity in many different higher-order cortical areas that was present during the original formation of the episodic memory.

24.3.9 Simulations of hippocampal operation

In order to test the operation of the whole system for individual parts of which an analytic theory is available (Rolls and Treves 1998, Treves and Rolls 1994, Treves and Rolls 1992b, Schultz and Rolls 1999), Rolls (1995b) simulated a scaled down version of the part of the

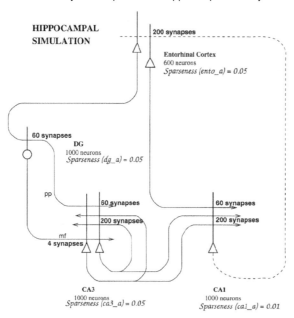

**HIPPOCAMPAL
SIMULATION**

200 synapses

Entorhinal Cortex
600 neurons
Sparseness (ento_a) = 0.05

60 synapses

DG
1000 neurons
Sparseness (dg_a) = 0.05

pp

60 synapses

60 synapses

200 synapses

200 synapses

mf
4 synapses

CA3
1000 neurons
Sparseness (ca3_a) = 0.05

CA1
1000 neurons
Sparseness (ca1_a) = 0.01

Fig. 24.19 Simulation of a scaled down model of the hippocampus and recall to the neocortex. The numbers of synapses and the sparseness values used are shown. (After Rolls 1995b.)

architecture shown in Fig. 24.1 from the entorhinal cortex to the hippocampus and back to the entorhinal cortex. The network included competitive learning in the dentate gyrus to form a sparse representation, using this to drive the CA3 cells by diluted non-associative mossy fibre connectivity during learning, competitive learning in CA1, and pattern association in the retrieval back to entorhinal cortex (Fig. 24.19). The analytic approaches to the storage capacity of the CA3 network, the role of the mossy fibres and of the perforant path, the functions of CA1, and the operation of the backprojections in recall, were all shown to be computationally plausible in the computer simulations. In the simulation, during recall, partial keys are presented to the entorhinal cortex, completion is produced by the CA3 autoassociation network, and recall is produced in the entorhinal cortex of the original learned vector. The network, which has 1,000 neurons at each stage, can recall large numbers, which approach the calculated storage capacity, of different sparse random vectors.

One of the points highlighted by the simulation is that the network operated much better if the CA3 cells operated in binary mode (either firing or not), rather than having continuously graded firing rates (Rolls 1995b). The reason for this is that given that the total amount of information that can be stored in a recurrent network such as the CA3 network is approximately constant independently of how graded the firing rates are in each pattern (Treves 1990), then if much information is used to store the graded firing rates in the firing of CA3 cells, fewer patterns can be stored. The implication of this is that in order to store many memories in the hippocampus, and to be able to recall them at later stages of the system in for example the entorhinal cortex and beyond, it may be advantageous to utilize relatively binary firing rates in the CA3 part at least of the hippocampus.

This finding has been confirmed and clarified by simulation of the CA3 autoassociative system alone, and it has been suggested that the advantage of operation with binary firing rates may be related to the low firing rates characteristic of hippocampal neurons (Rolls, Treves, Foster and Perez-Vicente 1997c).

(a)   (b)

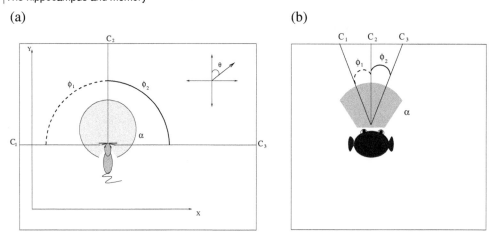

Fig. 24.20 Rat place cells (a) vs primate spatial view cells (b). Sketch of the square containment area in which the agent is situated. The containment area contains a grid of 200 × 200 possible positions x, y which the agent visits during the simulations, and at each position the agent rotates its head direction θ through 360 degrees in increments of 5 degrees. In addition, the containment area has a number of visual cues distributed evenly around the perimeter. As the agent explores its environment, individual hippocampal cells are stimulated by specific visual cues currently within the field of view of the agent α. On the left is shown the case for a rat with a 270° field of view, while on the right is shown the case for a primate with a 30° field of view, where the shaded areas correspond to the sizes of the fields of view of the agents. In the model, the firing rates of the hippocampal cells are dependent on the angles ϕ subtended by the visual cues (e.g. objects such as buildings in fixed positions in space) currently within the field of view of the agent. (After DeAraujo, Rolls and Stringer 2001.)

Another aspect of the theory emphasized by the results of the simulation was the importance of having effectively a single network provided in the hippocampus by the CA3 recurrent collateral network, for only if this operated as a single network (given the constraint of some topography present at earlier stages), could the whole of a memory be completed from any of its parts (Rolls 1995b).

Another aspect of the theory illustrated by these simulations was the information retrieval that can occur at the CA3–CA1 (Schaeffer collateral) synapses if they are associatively modifiable (Schultz and Rolls 1999).

24.3.10 The learning of spatial view and place cell representations from visual inputs

In the ways just described, the dentate granule cells could be particularly important in helping to build and prepare spatial representations for the CA3 network. The actual representation of space in the primate hippocampus includes a representation of spatial view, whereas in the rat hippocampus it is of the place where the rat is. The representation in the rat may be related to the fact that with a much less developed visual system than the primate, the rat's representation of space may be defined more by the olfactory and tactile as well as distant visual cues present, and may thus tend to reflect the place where the rat is. However, the spatial representations in the rat and primate could arise from essentially the same computational process as follows (Rolls 1999c, De Araujo, Rolls and Stringer 2001).

The starting assumption is that in both the rat and the primate, the dentate granule cells (and the CA3 and CA1 pyramidal cells) respond to combinations of the inputs received. In the case of the primate, a combination of visual features in the environment will, because of the fovea providing high spatial resolution over a typical viewing angle of perhaps 10–20 degrees, result in the formation of a spatial view cell, the effective trigger for which will thus be a combination of visual features within a relatively small part of space. In contrast, in the rat, given the very extensive visual field subtended by the rodent retina, which may extend over 180–270 degrees, a combination of visual features formed over such a wide visual angle would effectively define a position in space that is a place (see Fig. 24.20). The actual processes by which the hippocampal formation cells would come to respond to feature combinations could be similar in rats and monkeys, involving for example competitive learning in the dentate granule cells, as utilized in the simulations of De Araujo, Rolls and Stringer (2001) which produced neurons with place-like and spatial view-like tuning. This competitive learning has the required properties, of associating together co-occurrent features, and of ensuring by the competition that each feature combination (representing a place or spatial view) is very different from that of other places or spatial views (pattern separation, and orthogonalizing effect), and is also a sparse representation. Consistent with this, dentate place fields in rats are small (Jung and McNaughton 1993). Although this computation could be performed to some extent also by autoassociation learning in CA3 pyramidal cells, and competitive learning in CA1 pyramidal cells (Rolls and Treves 1998, Treves and Rolls 1994), the combined effect of the dentate competitive learning and the mossy fibre low probability but strong synapses to CA3 cells would enable this part of the hippocampal circuitry to be especially important in separating spatial representations, which as noted above are inherently continuous. The prediction thus is that during the learning of spatial tasks in which the spatial discrimination is difficult (for example because the places are close together), then the dentate system should be especially important. Tests of this prediction are described in Section 24.4.

24.3.11 Linking the inferior temporal visual cortex to spatial view and place cell representations

It is now possible to propose a unifying hypothesis of the relation between the ventral visual system, and hippocampal spatial representations (Rolls, Tromans and Stringer 2008f). Let us consider a computational architecture in which a fifth layer is added to the VisNet architecture described in Chapter 25, as illustrated in Fig. 24.21. In the anterior inferior temporal visual cortex, which corresponds to the fourth layer of VisNet, neurons respond to objects, but several objects close to the fovea (within approximately 10°) can be represented because many object-tuned neurons have asymmetric receptive fields with respect to the fovea (Aggelopoulos and Rolls 2005) (see Section 25.5.10). If the fifth layer of VisNet performs the same operation as previous layers, it will form neurons that respond to combinations of objects in the scene with the positions of the objects relative spatially to each other incorporated into the representation (as described in Section 25.5.5). It was found that training now with whole scenes that consist of a set of objects in a given fixed spatial relation to each other results in neurons in the added, fifth, layer that respond to one of the trained whole scenes, but do not respond if the objects in the scene are rearranged to make a new scene from the same objects (Rolls, Tromans and Stringer 2008f). The formation of these scene-specific representations in the added layer is related to the fact that in the inferior temporal cortex, and, we showed, in the VisNet model, the receptive fields of inferior temporal cortex neurons shrink and become asymmetric when multiple objects are present simultaneously in a natural scene. This also provides a solution to the issue of the representation of multiple objects, and their relative spatial positions, in complex natural scenes (Rolls, Tromans and Stringer 2008f). The implication of the findings is that in a unifying computational approach, competitive network processes operating in areas such as the parahippocampal cortex, the entorhinal cortex, and/or the dentate granule cells could form unique views of scenes by forming a sparse representation of these object or feature

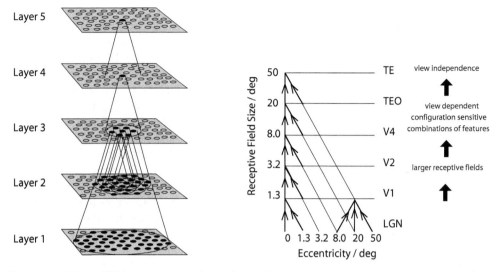

Fig. 24.21 Adding a fifth layer, corresponding to the parahippocampal gyrus / hippocampal system, after the inferior temporal visual cortex (corresponding to layer 4) may lead to the self-organization of spatial view / place cells in layer 5 when whole scenes are presented (see text). Convergence in the visual system is shown in the earlier layers (see Chapter 25). Right – as it occurs in the brain. V1, visual cortex area V1; TEO, posterior inferior temporal cortex; TE, inferior temporal cortex (IT). Left – as implemented in VisNet (layers 1–4). Convergence through the network is designed to provide fourth layer neurons with information from across the entire input retina.

tuned inferior temporal cortex ventral visual stream representations which have some spatial asymmetry into spatial view cells in the hippocampus (Rolls, Tromans and Stringer 2008f) (see further Section 25.5.15).

It is an interesting part of the hypothesis just described that because spatial views and places are defined by the relative spatial positions of fixed landmarks (such a buildings), slow learning of such representations over a number of trials might be useful, so that the neurons come to represent spatial views or places, and do not learn to represent a random collection of moveable objects seen once in conjunction. In this context, an alternative brain region to the dentate gyrus for this next layer of VisNet-like processing might be the parahippocampal areas that receive from the inferior temporal visual cortex. Spatial view cells are present in the parahippocampal areas (Georges-François, Rolls and Robertson 1999, Robertson, Rolls and Georges-François 1998, Rolls, Robertson and Georges-François 1997a, Rolls, Treves, Robertson, Georges-François and Panzeri 1998b, Rolls, Xiang and Franco 2005c). These spatial view representations could be formed in these regions as, effectively, an added layer to VisNet. Moreover, these cortical regions have recurrent collateral connections that could implement a continuous attractor representation. Alternatively, it is possible that these parahippocampal spatial representations reflect the effects of backprojections from the hippocampus to the entorhinal cortex and thus to parahippocampal areas. In either case, it is an interesting and unifying hypothesis that an effect of adding an additional layer to VisNet-like ventral stream visual processing might with training in a natural environment lead to the self-organization, using the same principles as in the ventral visual stream, of spatial view or place representations in parahippocampal or hippocampal areas (Rolls, Tromans and Stringer 2008f).

24.3.12 A scientific theory of the art of memory: scientia artis memoriae

Simonides of Ceos lived to tell the story of how when a banquet hall collapsed in an earthquake, he could identify all the victims by recalling from each place at the table who had been sitting there (Cicero 55 BC). This way of remembering items was developed into what has become known as *ars memoriae* by Roman senators, who presented the steps of a complex legal argument in a speech that might last a whole day by associating each step in the argument with a location in a spatial scene through which their memory could progress from one end to the other during the speech to recall each item in the correct order (Yates 1992).

Why is *ars memoriae* so successful in helping to remember complex series of points or arguments or people or objects?

The theory is proposed that *ars memoriae* in implemented in the CA3 region of the hippocampus where there are spatial view cells in primates that allow a particular view to be associated with a particular object in an event or episodic memory. Given that the CA3 cells with their extensive recurrent collateral system connecting different CA3 cells, and associative synaptic modifiability, form an autoassociation or attractor network, the spatial view cells with their approximately Gaussian view fields become linked in a continuous attractor. In this continuous attractor network, the strength of the connections thus reflects the proximity of two spatial views. As the view space is traversed (for example by self-motion), the views are therefore successively recalled in the correct order, with no view missing. Given that each view has been associated with a different discrete item, the items are recalled in the correct order, with none missing (Rolls, Stringer and Trappenberg 2002). There is some forgetting in the system, so the same view can be used on a later occasion to associate each view in the scene with a new set of items. This is the theory proposed for *ars memoriae* (Rolls 2017g).

24.4 Tests of the theory of hippocampal cortex operation

Empirical evidence that tests the theory comes from a wide range of investigation methods, including the effects of selective lesions to different subregions of the hippocampal system, the activity of single neurons in different subregions of the hippocampal system, pharmacological and genetic manipulations of different parts of the system, the detailed neuroanatomy of the system, and functional neuroimaging studies and clinical neuropsychological studies in humans (Kesner and Rolls 2015).

Because the theory makes predictions about the functions of different parts of the hippocampal system (e.g. dentate, CA3, and CA1), evidence from subregion analysis is very relevant to testing the theory, and such evidence is included in the following assessment, which starts from the dentate gyrus, and works through to CA1. The evidence comes from the effects of selective lesions to different subregions of the hippocampal system, the activity of single neurons in different subregions, together with pharmacological and genetic manipulations of different parts of the system (Kesner and Rolls 2015).

24.4.1 Dentate gyrus (DG) subregion of the hippocampus

Based on the anatomy of the DG, its input and output pathways, and the development of a computational model, it has been suggested (see Section 24.3, Rolls (1989b), and Rolls (1996c)) that the DG can act as a competitive learning network with Hebb-like modifiability to remove redundancy from the inputs producing a more orthogonal, sparse, and categorized set of outputs. One example of this in the theory is that to the extent that some entorhinal cortex neurons represent space as a grid (Moser et al. 2015), the correlations in this type of encoding are removed by competitive learning to produce a representation of place, with each place encoded differently from other places (Rolls, Stringer and Elliot 2006c)[21]. To the extent then that DG acts to produce separate representations of different places, it is predicted that the DG will be especially important when memories must be formed about similar places. To form spatial representations, learned conjunctions of sensory inputs including vestibular, olfactory, visual, auditory, and somatosensory may be involved. DG may help to form orthogonal representations based on all these inputs, and could thus help to form orthogonal non-spatial as well as orthogonal spatial representations for use in CA3. In any case, the model predicts that for spatial information, the DG should play an important role in hippocampal memory functions when the spatial information is very similar, for example, when the places are close together.

24.4.1.1 Spatial pattern separation

To examine the contribution of the DG to spatial pattern separation, Gilbert, Kesner and Lee (2001) tested rats with DG lesions using a paradigm that measured one-trial short term memory for spatial location information as a function of spatial similarity between two spatial locations (Gilbert, Kesner and DeCoteau 1998). Rats were trained to displace an object that was randomly positioned to cover a baited food well in 1 of 15 locations along a row of food wells. Following a short delay, the rats were required to choose between two objects identical to the sample phase object. One object was in the same location as the sample phase object and the second object was in a different location along the row of food wells. A rat was rewarded for displacing the object in the same position as the sample phase object (correct choice) but received no reward for displacing the foil object (incorrect choice). Five spatial separations, from 15 cm to 105 cm, were used to separate the correct object from the foil object on the choice phase. The results showed that rats with DG lesions were significantly impaired at short spatial separations; however, the performance of the DG lesioned rats increased as a function of increased spatial separation between the correct object and the foil on the choice phases. The performance of rats with DG lesions matched controls at the largest spatial separation. The graded nature of the impairment and the significant linear increase in performance as a function of increased separation illustrate the deficit in pattern separation produced by DG lesions (see Fig. 24.22). Based on these results, it can be concluded that lesions of the DG decrease efficiency in spatial pattern separation, which resulted in impairments on trials with increased spatial proximity and hence increased spatial similarity among working memory representations (Gilbert, Kesner and Lee 2001). In the same study it was found that CA1 lesions do not produce a deficit in this task.

Additional evidence comes from a study (Goodrich-Hunsaker, Hunsaker and Kesner 2008) using a modified version of an exploratory paradigm developed by Poucet (1989), in which rats with DG lesions and controls were tested on tasks involving a metric spatial manipulation. In this task, a rat was allowed to explore two different visual objects that were separated by a specific distance on a cheese-board maze. On the initial presentation of the objects, the rat explored each object. However, across subsequent presentations of the objects in their respective locations, the rat habituated and eventually spent less time exploring the objects. Once the rat had habituated to the objects and their locations, the metric spatial distance between the two objects was manipulated so the two objects were either closer together or further apart. The time the rat spent exploring each object was recorded. The results showed

---

[21] Indeed, it was a prediction of the theory that the mapping from entorhinal cortex grid cells to dentate / CA place cells could be achieved by competitive learning, and this prediction was verified by simulation (Rolls, Stringer and Elliot 2006c).

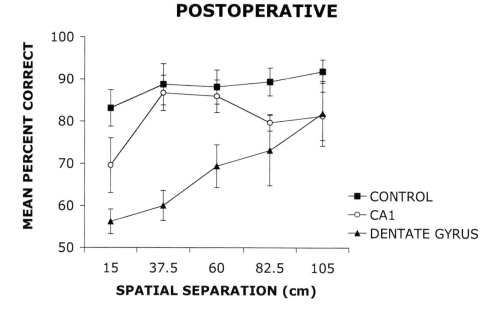

Fig. 24.22 Spatial pattern separation impairment produced by dentate gyrus lesions. Mean percent correct performance as a function of spatial separation of control group, CA1 lesion group, and dentate gyrus lesion group on postoperative trials. A graded impairment was found as a function of the distance between the places only following dentate gyrus lesions. (After Gilbert, Kesner and Lee, 2001.)

that DG lesions impaired detection of the metric distance change in that rats with DG lesions spent significantly less time exploring the two objects that were displaced.

The results of both experiments provide empirical validation of the role of DG in spatial pattern separation, and support the prediction from the computational model presented in this chapter.

There are neurophysiological results that are also consistent, in that McNaughton, Barnes, Meltzer and Sutherland (1989) found that following colchicine-induced lesions of the DG, there is a significant decrease in the reliability of CA3 place-related firing. Therefore, if CA3 cells display less reliability following DG lesions, the cells may not form accurate representations of space due to decreased efficiency in pattern separation. Also consistent are the findings that the place fields of DG cells and specifically dentate granular cells (Jung and McNaughton 1993, Leutgeb, Leutgeb, Moser and Moser 2007) are small and highly reliable, and this may reflect the role of DG in pattern separation.

These studies thus indicate that the DG plays an important role in spatial pattern separation in spatial memory tasks. Building and utilizing separate representations for other classes of stimuli engage other brain areas, including for objects the inferior temporal visual cortex (Rolls 2000a, Rolls 2012c), for reward value the orbitofrontal cortex (Rolls 2014a) and amygdala (Gilbert and Kesner 2002), and for motor responses the caudate nucleus (Kesner and Gilbert 2006).

24.4.1.2 Pattern separation and neurogenesis

Based on the observation that neurogenesis occurs in the DG and that new DG granule cells can be formed across time, it has been proposed that the dDG mediates a spatial pattern separation mechanism as well as generates patterns of episodic memories within remote memory (Aimone and Gage 2011). Thus far, it has been shown in mice that disruption of neurogenesis using low-dose x-irradiation was sufficient to produce a loss of newly born dDG cells. Further testing indicated impairments in spatial learning in a delayed non-matching-to-place task in the radial-arm maze. Specifically, impairment occurred for arms that were presented with little separation, but no deficit was observed when the arms were presented farther apart (Clelland et al. 2009), suggesting a spatial pattern separation deficit. These results are similar to those described for rats in the study by Morris, Churchwell, Kesner and Gilbert (2012). Another study in mice provided evidence that the disruption of neurogenesis using lentivirus expression of a dominant Wnt protein produced a loss of newly born dDG cells; and these mice were tested in an associative object-in-place task with different spatial separations and observed to be impaired as a function of the degree of separation, again suggesting a spatial pattern separation deficit (Clelland et al. 2009). In a more recent study (Kesner, Hui, Sommer, Wright, Barrera and Fanselow 2014) it was shown that DNA methyltransferase 1-c knockout mice are impaired relative to controls in the spatial pattern separation task (Goodrich-Hunsaker, Hunsaker and Kesner 2008). These data suggest that neurogenesis in the dDG may contribute to the operation of spatial pattern separation. Thus, spatial pattern separation may play an important role in the acquisition of new spatial information, and there is a good possibility that the dDG may have been the subregion responsible for the impairments in the various tasks described above.

24.4.1.3 Encoding versus retrieval

The theory postulates that the dentate / mossy fibre system is necessary for setting up the appropriate conditions for optimal storage of new information in the CA3 system (which could be called encoding); and that (especially when an incomplete cue is provided) the retrieval of information from CA3 is optimally cued by the direct entorhinal to CA3 connections.

In strong support of the role of the DG in encoding, a study conducted in mice showed that the mossy fibre projections to CA3 are essential for the encoding of spatial information in a water maze reference memory task, but are not necessary for retrieval (Lassalle, Bataille and Halley 2000).

Another type of evidence comes from a study in which rats had 10 learning trials per day in a Hebb–Williams maze. Lee and Kesner (2004b) found that based on a within-day analysis DG lesions (but not lesions of the perforant path input to CA3) impaired the acquisition of this task, consistent with an encoding or learning impairment. However, when tested using a between-days analysis, retrieval of what had been learned previously was not impaired by DG lesions, but was impaired by lesions of the perforant path input to CA3. This double dissociation is consistent with the hypothesis that the DG is important for optimal encoding, and the entorhinal to CA3 connections for optimal retrieval.

Further evidence for a DG mediation of spatial encoding comes from the observation that rats with DG lesions are impaired in learning the Morris water maze task when the start location varied on each trial (Nanry, Mundy and Tilson 1989, Sutherland, Whishaw and Kolb 1983, Xavier, Oliveira-Filho and Santos 1999). Under these conditions, spatial pattern separation may be at a premium, for different, partially overlapping, subsets of spatial cues are likely to be visible from the different starting locations.

Further evidence consistent with the hypothesis that the DG is important in spatial learning (acquisition) is that rats with DG lesions are impaired on acquisition of spatial contextual fear conditioning (Lee and Kesner 2004a).

However, we note that in the studies described in this section (encoding vs retrieval), during acquisition of these tasks over a number of trials both the storage and the retrieval processes are likely to influence the rate of behavioural learning. Thus, even though in behavioural tasks the deficit can be described as an encoding deficit because it is apparent during initial learning, it is not easy to separate the actual underlying processes by which the DG may be important for setting up the representations for new learning in the CA3 system, and the perforant path to CA3 connections for initiating retrieval especially with a partial cue from CA3.

24.4.2 CA3 subregion of the hippocampus

In the model the CA3 system acts as an autoassociation system. This enables arbitrary (especially spatial in animals and probably language for humans as well) associations to be formed between whatever is represented in the hippocampus, in that for example any place could be associated with any object, and in that the object could be recalled with a spatial recall cue, or the place with an object recall cue. The same system must be capable of fast (one-trial) learning if it is to contribute to forming new episodic memories, though it could also play an important role in the encoding of new information requiring multiple trials. The same system is also predicted to be important in retrieval of hippocampus-dependent information when there is an incomplete retrieval cue. The same system has the property that it can maintain firing activity because of the associative recurrent collateral connections, and for this reason is likely to be important in hippocampus-dependent delay / short-term memory tasks. The CA3 recurrent collateral system could also make a contribution to the learning of sequences if it has some temporal asymmetry in its associative synapses, as described in Section 24.3.3.4. The CA3 system might also contribute to spatial path integration, though this could be performed outside the hippocampus. Tests of these proposed functions are described next.

24.4.2.1 Rapid encoding of new information

The theory proposes that associative learning in the CA3 recurrent collateral connections can occur rapidly, in as little as one trial, so that it can contribute to episodic memory. Evidence for one-trial learning between places and objects or odors, in the hippocampal CA3 is provided in Section 24.4.2.5.

The situation is a little different in a task in which there is one-trial learning of just a spatial location. In one such task, delayed non-matching-to-place (DNMP) for a single spatial location in an 8-arm maze, Lee and Kesner (2002) and Lee and Kesner (2003a) showed that blockade of CA3 NMDA receptors with AP5 or CA3 (neurotoxic) lesions did not impair one-trial learning in a familiar environment, but did impair it in an initial period in a novel environment. The implication is that a place can be remembered over a 10 s delay even without the CA3, but that learning of new spatial environments does require the CA3 subregion. Consistent with this observation is the finding that hippocampal CA3 (dorsal plus ventral) lesions disrupt learning of the standard 8-arm maze where the rats have to learn not to go back to arms previously visited (Handelmann and Olton 1981). Rats tend to use different sequences on every trial, thus requiring new learning on every trial. When learning a new spatial environment, associations may be made between the different visual cues and their spatial locations, and this type of learning is computationally similar to that involved in object–place or odour–place association learning, providing a consistent interpretation of CA3 function.

Nakazawa et al. (2003) also reported similar findings with mice in which the function of CA3 NMDA receptors was disrupted. These mice were impaired in learning a novel platform location in a working memory water maze task, whereas they were normal in finding familiar platform locations.

Evidence that rapid learning is reflected in the responses of CA3 neurons comes from a study in which it was found that CA3 neurons alter their responses rapidly when rats encounter novel configurations of familiar cues for the first time (Lee, Rao and Knierim 2004). Specifically, rats were trained to run clockwise on a ring track whose surface was composed of four different texture cues (local cues). The ring track was positioned in the centre of a curtained area in which various visual landmarks were also available along the curtained walls. To produce a novel cue configuration in the environment, distal landmarks and local cues on the track were rotated in opposite directions (distal landmarks were rotated clockwise and local cues were rotated counterclockwise by equal amounts).

In sum, these results and those in Section 24.4.2.5 strongly suggest that rapid plastic changes in the CA3 network are essential in encoding novel information involving associations between objects and places, odours and places, or between landmark visual cues and spatial locations, and NMDA receptor-mediated plasticity mechanisms appear to play a significant role in the process.

24.4.2.2 The types of information associated in CA3

A prediction is that the CA3 recurrent collateral associative connections enable arbitrary associations to be formed between whatever is represented in the hippocampus, in that, for example, any place could be associated with any object. Arbitrary in this context refers to the possibility of what is provided for by non-localized connectivity within the CA3–CA3 system, in that any one representation in CA3 can potentially become associated with any other representation in CA3. The prediction is neutral with respect to what is actually represented in the CA3 system, which we note from the effects of lesions almost always includes space, at least in animals, but could include language in humans. The object representations might originate in the temporal cortical visual areas, and the spatial representations might utilize some information from the parietal cortex such as vestibular inputs related to self-motion (see Section 24.3). Two important issues need to be clarified. First, is the hippocampus the only neural system that supports all types of stimulus-stimulus associations (Eichenbaum and Cohen 2001)? Based on the observations that the primate orbitofrontal cortex is involved in some types of stimulus–stimulus association, including visual-to-taste and olfactory-to-taste association learning and more generally in rapid and reversible stimulus–reinforcer association learning (Rolls 2004a, Rolls 2014a, Grabenhorst and Rolls 2011), there appear to be other neural systems that can learn stimulus-stimulus associations. Second, what is the nature of the information that can be associated in this arbitrary way in the CA3 recurrent collateral system? The studies described next indicate that at least the non-human hippocampus is involved in associations when one of the components is spatial.

In order to directly test the involvement of the CA3 subregion of the hippocampus in spatial paired-associate learning, rats were trained on a successive discrimination Go/No-Go task to examine object–place and odour–place paired associate learning. Normal rats learn these tasks in approximately 240 trials. CA3 lesions impaired the learning of these object–place and odour–place paired associations (Gilbert and Kesner 2003). Furthermore, lesions of DG or CA1 did not produce a deficit, and the reasons for this are considered later.

In contrast, object–odour paired associate learning was not impaired by CA3 lesions (even with a delay between the object and the odour) (Kesner, Hunsaker and Gilbert 2005). Thus, not all types of paired associate learning require the rat hippocampus. This conclusion is supported by the fact that (large or total) hippocampal lesions do not impair the following types of association including odour–odour (Bunsey and Eichenbaum 1996, Li, Matsumoto and Watanabe 1999), odour–reward (Wood, Agster and Eichenbaum 2004), auditory–visual (Jarrard and Davidson 1990), and object–object associations (Cho and Kesner 1995, Murray, Gaffan and Mishkin 1993). Thus, in rats paired associate learning appears to be impaired by CA3 lesions only when one of the associates is place.

24.4.2.3 Acquisition/encoding associated with multiple trials

The CA3 subregion also appears to be necessary in some tasks that require multiple trials to acquire the task, and a common feature of these tasks is that a new environment must be learned. For example, lesions of the CA3 (but not the CA1) subregion impair the acquisition of object–place and odour–place paired associate learning, a task that requires multiple trials to learn (Gilbert and Kesner 2003). Another multiple trial task in which a learning set is needed that is affected by CA3, but not by CA1, lesions is acquisition of a non-match to sample one-trial spatial location task on an 8-arm maze with 10 s delays (delayed non-match to place, DNMP). In this case the output from CA3 is via the fimbria, in that fimbria lesions that interrupt the CA3 hippocampal output without affecting the input also impair acquisition of this task. Finally, CA3 lesioned rats are impaired in the standard water maze task which requires multiple trials (Brun, Otnass, Molden, Steffenach, Witter, Moser and Moser 2002, Florian and Roullet 2004) (although mice that lacked NMDA receptors in CA3 do not appear to be impaired in learning the water maze (Nakazawa et al. 2002)). These findings are consistent with the hypothesis that when a new environment is learned this may involve multiple associations between different environmental cues, both with each other and with idiothetic (self-motion cues), and that this type of learning depends on new associations formed in the CA3–CA3 connections (see Section 24.3.3.2).

24.4.2.4 Pattern completion in CA3

The CA3 system is predicted to be important in the retrieval of hippocampus-dependent information when there is an incomplete retrieval cue. Support for the pattern completion process in CA3 can be found in lesion studies. In one study (Gold and Kesner 2005), rats were tested on a cheese board with a black curtain with four extramaze cues surrounding the apparatus. (The cheese board is like a dry land water maze with 177 holes on a 119 cm diameter board.) Rats were trained to move a sample phase object covering a food well that could appear in one of five possible spatial locations. During the test phase of the task, following a 30 s delay, the animal needs to find the same food well in order to receive reinforcement with the object now removed. After reaching stable performance in terms of accuracy to find the correct location, rats received lesions in CA3. During post-surgery testing, four extramaze cues were always available during the sample phase. However, during the test phase zero, one, two, or three cues were removed in different combinations. The results indicate that controls performed well on the task regardless of the availability of one, two, three, or all cues, suggesting intact spatial pattern completion. Following the CA3 lesion, however, there was an impairment in accuracy compared to the controls especially when only one or two cues were available, suggesting impairment in spatial pattern completion in CA3 lesioned rats (Gold and Kesner 2005) (see Fig. 24.23). A useful aspect of this task is that the ability to remember a spatial location learned in one presentation can be tested with varying number of available cues, and many times in which the locations vary, to allow for accurate measurement of pattern completion ability when the information stored on the single presentation must be recalled. Based on the observation that the lateral perforant path input into the dentate gyrus mediates visual object information via activation of opioid receptors, it was shown that naloxone (a μ opioid receptor antagonist) injected into the CA3 produced the same pattern completion problem (Kesner and Warthen 2010).

In another study Nakazawa, Quirk, Chitwood, Watanabe, Yeckel, Sun, Kato, Carr, Johnston, Wilson and Tonegawa (2002) trained CA3 NMDA receptor-knockout mice in an analogous task, using the water maze. When the animals were required to perform the task in an environment where some of the familiar cues were removed, they were impaired in performing the task. The result suggests that the NMDA receptor-dependent synaptic plasticity mechanisms in CA3 are critical to perform the pattern completion process in the hippocampus.

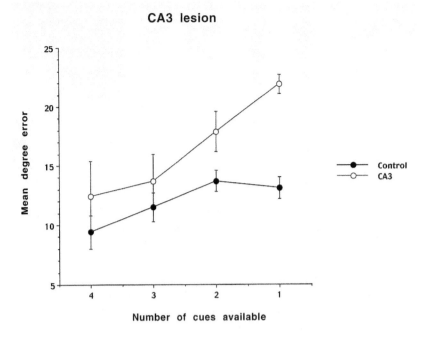

Fig. 24.23 Pattern completion impairment produced by CA3 lesions. The mean (and sem) degree of error in finding the correct place in the cheese-board task when rats were tested with 1, 2, 3 or 4 of the cues available. A graded impairment in the CA3 lesion group as a function of the number of cues available was found. The task was learned in the study phase with the 4 cues present. The performance of the control group is also shown. (After Gold and Kesner 2005.)

24.4.2.5 Recall and arbitrary associations in CA3

A prediction of the theory is that the CA3 recurrent collateral associative connections enable arbitrary associations to be formed between whatever is represented in the hippocampus, in that for example any place could be associated with any object, and in that the object could be recalled with a spatial recall cue, or the place with an object recall cue.

In one test of this, Day, Langston and Morris (2003) trained rats in a study phase to learn in one trial an association between two flavours of food and two spatial locations. During a recall test phase they were presented with a flavour which served as a cue for the selection of the correct location. They found that injections of an NMDA blocker (AP5) or AMPA blocker (CNQX) to the dorsal hippocampus prior to the study phase impaired encoding, but injections of AP5 prior to the test phase did not impair the place recall, whereas injections of CNQX did impair the place recall. The interpretation is that somewhere in the hippocampus NMDA receptors are necessary for forming one-trial odour–place associations, and that recall can be performed without further involvement of NMDA receptors.

In a development related to the Day, Langston and Morris (2003) experiment, Kesner, Hunsaker and Warthen (2008) designed a task in which a visual object could be recalled from a spatial location, and in an important extension, a spatial location could be recalled from an object. In this task, after training to displace objects, rats in the study phase are placed in the start box and when the door in front of the start box is opened the rats are allowed to displace one object in one location, and then after returning to the start box, the door is opened again and the rats are allowed to displace a second object in another location. There are 50 possible objects and 48 locations. In the test phase, the rat is shown one object (first or second randomized) in the start box as a cue, and then, after a 10-s delay, the door is opened and the rats must go to the correct location (choosing and displacing one of two identical neutral objects). The rats receive a reward for selecting the correct location that was associated with the object cue. A spatial location-recall for a visual object task was also used. For the spatial recall for a visual object task, the study phase is the same, but in this case in the test phase when the door is opened the rat is allowed to displace a neutral object in one location (first or second randomized) on the maze as a location cue, return to the start box, and then, after a 10 s delay, the door is opened and the rats must select the correct object (choosing and displacing one of two visual objects). The rats received a reward for selecting the correct visual object that was associated with the location cue. It was found that CA3a,b lesions produce chance performance on both the one-trial object-place recall task, and on the one-trial place-object recall task (Kesner et al. 2008). The potential implications of such results are that indeed the CA3a,b supports arbitrary associations as well as episodic memory based on one-trial learning. A control fixed visual conditional to place task with the same delay was not impaired, showing that it is recall after one-trial (or rapid) learning that is impaired.

Additional support comes from the finding that in a similar one-trial objectâ^place learning followed by recall of the spatial position in which to respond when shown the object, Rolls and Xiang (2005) showed that some primate hippocampal (including CA3) neurons respond to an object cue with the spatial position in which the object had been shown earlier in the trial. Thus, some hippocampal neurons appear to reflect spatial recall given an object recall cue.

Evidence that the CA3 is not necessarily required during recall in a reference memory spatial task, such as the water maze spatial navigation for a single spatial location task, is that CA3 lesioned rats are not impaired during recall of a previously learned water maze task (Brun et al. 2002, Florian and Roullet 2004). However, if completion from an incomplete cue is needed, then CA3 NMDA receptors are necessary (presumably to ensure satisfactory CA3–CA3 learning) even in a reference memory task (Nakazawa, Quirk, Chitwood, Watanabe, Yeckel, Sun, Kato, Carr, Johnston, Wilson and Tonegawa 2002). Thus, the CA3 system appears to be especially needed in rapid, one-trial object–place recall, and when completion from an incomplete cue is needed.

24.4.2.6 Functional analysis of mossy fibre vs direct perforant path input into CA3

The computational model suggests that the dentate granule cell / mossy fibre pathway to CA3 may be important during the learning of new associations in the CA3 network, and that part of the way in which it is important is that it helps by pattern separation to produce relatively sparse and orthogonal representations in CA3 (see Section 24.3.3.8). In contrast, the theory predicts that the direct perforant path input to CA3 is important in initiating retrieval from the CA3 autoassociation network, especially with an incomplete retrieval cue (see Section 24.3.3.9).

Support for this hypothesis comes from the findings of Lee and Kesner (2004b) and Jerman, Kesner and Hunsaker (2006) that lesions of the DG or CA3 (or a crossed lesion) disrupt within-day learning on the Hebb–Williams maze, but that retrieval of information at the start of the following day is not impaired. In contrast, lesions of the perforant path input to CA3 from entorhinal cortex disrupt retrieval (i.e. initial performance on the following day), but not learning within a day (Lee and Kesner 2004b). These findings support the hypothesis that the CA3 processes both inputs from the DG to support pattern separation; and inputs from the entorhinal cortex via the perforant path to support pattern completion.

The perforant path can be divided into a medial and lateral component. It has been suggested that the medial component processes spatial information and that the lateral component processes non-spatial (e.g objects, odours) information (Hargreaves et al. 2005, Witter, Groenewegen, Lopes da Silva and Lohman 1989a). In one study, Ferbinteanu, Holsinger and McDonald (1999) showed that lesions of the medial perforant path (MPP) disrupted water maze learning, whereas lateral perforant path (LPP) lesions had no effect. As predicted by the theory, there is associative LTP between the medial or lateral perforant path and the intrinsic commissural/associational-CA3 synapses) (Martinez, Do, Martinez and Derrick 2002). Either place or object recall cues could thus be introduced by the associative MPP and LPP connections to CA3 cells.

In addition, Martinez et al. (2002) demonstrated associative (cooperative) LTP between the medial and lateral perforant path inputs to the CA3 neurons. This could provide a mechanism for object (LPP) – place (MPP) associative learning, with either the object or the place during recall activating a CA3 neuron. However, this LPP to MPP cooperative LTP onto a CA3 neuron would be a low capacity memory system, in that there are only 3,300 PP inputs to a CA3 cell, compared to 12,000 recurrent collaterals in the CA3–CA3 connections, and these numbers are the leading term in the memory capacity of pattern association and autoassociation memory systems (see Appendix B).

We note that disruption of DG and mossy fibre input into CA3 does not produce a disruption in the acquisition of the object–place paired associate task unless the stimuli are close together, implying that the DG contribution is important particularly when pattern separation is needed (Gilbert and Kesner 2003). The implication is that sufficient input for object–place learning can be introduced into the CA3 system (which is required for this object–place learning) by the perforant path inputs provided that spatial pattern separation is not at a premium (or perhaps the storage of large numbers of different object–place associations is not at a premium, as this is the other condition for which the computational theory indicates that the DG is needed to help produce sparse and orthogonal representations in the CA3 system).

24.4.2.7 Orthogonal representations in CA3

The CA3 subregion may have more distinct representations of different environments than CA1. This may be consistent with the computational point that if CA3 is an autoassociator, the pattern representations in it should be as orthogonal as possible to maximize memory capacity and minimize interference. The actual pattern separation may be performed, the theory holds, as a result of the operation of the dentate granule cells as a competitive net, and the nature of the mossy fibre connections to CA3 cells. Some of the empirical evidence is as follows.

Tanila (1999) showed that CA3 place cells were able to maintain distinct representations of two visually identical environments, and selectively reactivate either one of the representation patterns depending on the experience of the rat. Also, Leutgeb et al. (2004) showed that when rats experienced a completely different environment, CA3 place cells developed orthogonal representations of those different environments by changing their firing rates between the two environments, whereas CA1 place cells maintained similar responses. Evidence on the sparseness of the CA3 cell representation in rats includes evidence that CA3 cell ensembles may support the fast acquisition of detailed memories by providing a locally continuous, but globally orthogonal spatial representation, onto which new sensory inputs can rapidly be associated (Leutgeb and Leutgeb 2007). Further, CA3 cells are very sensitive to the place where the rat is, and so potentially help to provide a basis for object-place memory in which the place is precisely specified (Leutgeb and Leutgeb 2007).

In the computational account, each environment would be a separate chart, and the number of charts that could be stored in CA3 would be high if the representations in each chart are relatively orthogonal to those in other charts (see Section B.5.4 and Battaglia and Treves

(1998b)), and further, the charts could operate independently (Stringer, Rolls and Trappenberg 2004). Any one chart of one spatial environment can be understood as a continuous attractor network, with place cells with Gaussian shaped place fields, which overlap continuously with each other. In different charts (different spatial environments) the neurons may represent very different parts of space, and neurons representing close places in one environment may represent distant places in another environment (chart).

24.4.2.8 Short-term memory, path integration, and CA3

The CA3 recurrent collateral associative connections suggest that the CA3 system can operate as an attractor network which can be useful in some types of working memory. Support for this idea comes from a variety of studies analysed by Kesner and Rolls (2001), including a study where in the short-term (30 s delay) memory for spatial location task to measure spatial pattern separation, it was shown that CA3 lesioned rats were impaired for all pattern separations, consistent with the hypothesis that the rats could not remember the correct spatial location (Gilbert and Kesner 2006). Furthermore, in the short-term memory-based delayed non-matching to sample place task mentioned above (Lee and Kesner 2003b), lesions of CA3 impaired the acquisition of this task with 10 s delays. Also single neuron activity has been recorded in CA3 during the delay period in rats in a spatial position short-term memory task (Hampson, Hedberg and Deadwyler 2000) and in monkeys in an object–place and a location-scene association short-term memory task (Cahusac, Miyashita and Rolls 1989). The finding that neuronal representations in the hippocampus switch abruptly from representing a square vs a circular environment on which rats have been trained when the environment changes incrementally between the two has also been suggested as evidence that the hippocampus implements attractor dynamics (Wills et al. 2005).

In a task in which rats were required to remember multiple places, CA3 and CA1 lesions produced a deficit. In the task, during the study phase rats were presented with four different places within sections that were sequentially visited on a newly devised maze (i.e. Tulum maze). Each place was cued by a unique object that was specifically associated with each location within the section during the study phase. Following a 15-s delay and during the test phase in the absence of the cued object, rats were required to recall and revisit the place within one section of the maze that had been previously visited. Both CA1 and CA3 lesions disrupted accurate relocation of a previously visited place (Lee, Jerman and Kesner 2005). Thus, short-term memory for multiple places in a one-trial multiple place task depends on both CA3 and CA1. As an attractor network can in general hold only one item active in a delay period by maintained firing, this type of multiple item short-term memory is computationally predicted to require synaptic modification to store each item, and both CA3 and CA1 appear to contribute to this.

As described in Section 24.3.3.2, to the extent that path integration of place or spatial view representations is performed within the hippocampus itself, then the CA3 system is the most likely subregion of the hippocampus to be involved in this, because it has the appropriate recurrent collateral connections. Consistent with this, Whishaw and colleagues (Wallace and Whishaw 2003, Maaswinkel et al. 1999, Whishaw et al. 2001) and Save, Guazzelli and Poucet (2001) have shown that path integration is impaired by hippocampal lesions. Path integration of head direction is reflected in the firing of neurons in the presubiculum, and mechanisms outside the hippocampus probably implement path integration for head direction (Sharp 2002). Place cells are also updated idiothetically in the dark (Mizumori, Ragozzino, Cooper and Leutgeb 1999), as are CA3 and CA1 primate spatial view cells (Robertson, Rolls and Georges-François 1998).

24.4.2.9 Cholinergic modulation of CA3 towards storage vs recall including pattern completion

Acetylcholine may act to facilitate memory storage by increasing LTP in the CA3–CA3 recurrent collaterals, and reducing the recurrent collateral synaptic recall of previous interfering memories (Hasselmo et al. 1995, Giocomo and Hasselmo 2007). Consistent with this, scopolamine, but not physostigmine, infusions, into CA3 disrupt encoding. In contrast, physostigmine, but not scopolamine infusions, into CA3 disrupt recall. This was observed during a spatial exploration paradigm (Hunsaker, Rogers and Kesner 2007), during Hebbâ^Williams maze learning (Rogers and Kesner 2003), and during delay fear-conditioning (Rogers and Kesner 2004).These findings are consistent with the hypothesis that acetylcholine in CA3 when low supports recall which involves pattern completion implemented by the recurrent collaterals; and that acetylcholine when high facilitates new learning.

24.4.3 CA1 subregion of the hippocampus

The CA1 subregion of the hippocampus receives inputs from two major sources: the Schaffer collateral inputs from CA3, and the perforant path inputs from the entorhinal cortex (see Fig. 24.1). CA1 outputs are directed towards the subiculum, entorhinal cortex, prefrontal cortex and many other neural regions including the lateral septum, anterior thalamus, and the mammillary bodies (Amaral and Witter 1995). The anatomical and physiological characteristics suggest that the CA1 can operate as a competitive network and have triggered the development of computational models of CA1 in which for example the CA1 system is involved in the recall process by which backprojections to the cerebral neocortex allow neuronal activity during recall to reflect that present during the original learning (see Section 24.3.7). In this terminology, cortico-cortical backprojections originate from deep (layer 5) pyramidal cells, and terminate on the apical dendrites of pyramidal cells in layer 1 of the preceding cortical area (see Fig. 24.1). In contrast, cortico-cortical forward projections originate from superficial (layers 2 and 3) pyramidal cells, and terminate in the deep layers of the hierarchically next cortical area (see Fig. 24.1). Within this context, the projections from the entorhinal cortex to DG, CA3 and CA1 can be thought of as forward projections; and from CA1 to entorhinal cortex, subiculum etc. as the start of the backprojection system (see Fig. 24.1).

The evidence reviewed next of the effects of selective damage to CA1 indicates that it makes a special contribution to temporal aspects of memory including associations over a delay period, and sequence memory. There is also evidence relating it to intermediate memory as well as consolidation (see for more detail Rolls and Kesner (2006)). The effects of damage to the CA1 may not be identical to the effects of damage to CA3, because CA1 has some inputs that bypass CA3 (e.g. the direct entorhinal / perforant path input); and CA3 has some outputs that bypass CA1. In particular, the CA3 output through the fimbria projects directly to the lateral septum and then to the medial septum or directly to the medial septum (Amaral and Witter 1995, Risold and Swanson 1997). The medial septum, in turn, provides cholinergic and GABAergic inputs to the hippocampus (Amaral and Witter 1995).

24.4.3.1 Sequence memory and CA1

In Section 24.3.3.4 I describe how a sequence memory could be implemented in a recurrent collateral network such as CA3 with some temporal asymmetry in the associative synaptic modification learning rule; with an alternative hypothesis that pairing places with time cells might implement spatial sequence learning.

To test this, Hunsaker, Lee and Kesner (2008) trained rats to remember unique sequences of spatial locations along a runway box. Rats with lesions to either CA3 or CA1 had difficulty with this spatial sequence memory task, although CA1 lesioned animals had a much greater deficit. However, when animals were trained on a non-episodic version of the same task, hippocampal lesions had no effect. The results suggest that the CA3 and CA1 contribute to episodic based spatial sequence memory and pattern completion (Hunsaker et al. 2008). Further results are described by Kesner and Rolls (2015).

24.4.3.2 Associations across time, and CA1

There is evidence implicating the hippocampus in mediating associations across time (Rawlins 1985, Kesner 1998), and hypotheses on the mechanisms are described in Sections 24.3.3.4 and 24.3.6.2.

The CA1 subregion of the hippocampus may play a role in influencing the formation of associations whenever a time component (requiring a memory trace) is introduced between any two stimuli that need to be associated. Support for this idea comes from studies based on a classical conditioning paradigm. Lesions of the hippocampus in rabbits or rats disrupt the acquisition of eye-blink trace conditioning. In trace conditioning a short delay intervenes between the conditioned stimulus (CS) and the unconditioned stimulus (UCS). When, however, a UCS and CS overlap in time (delay conditioning), rabbits with hippocampal damage typically perform as well as normals (Moyer, Deyo and Disterhoft 1990, Weiss, Bouwmeester, Power and Disterhoft 1999). Based on a subregional analysis, it has now been shown that ventral, but not dorsal, CA1 lesions impair at least the retention (tested after 48 h) of trace fear conditioning (Rogers, Hunsaker and Kesner 2006). Similar learning deficits in trace fear conditioning (but not in conditioning without a delay between the CS and UCS), were observed for mice that lacked NMDA receptors in both dorsal and ventral CA1 subregions of the hippocampus (Huerta et al. 2000).

Based on a previous finding that the acquisition of an object-odor association is not dependent on the hippocampus, would adding a temporal component to an object-odor association task recruit the hippocampus and more specifically the CA1 region? To test this idea rats were given dorsal CA1, CA3, or control lesions prior to learning an object-delay-odor task. Rats that had dorsal CA1 lesions were unable to make the association and never performed above chance, however CA3 lesioned rats displayed no deficit (Kesner, Hunsaker and Gilbert 2005). The observation of time cells during the trace in the object-trace-odor task supports the idea that the CA1 supports temporal processing of object and odor information. These results support the idea that the CA1 region is at least on the route for forming arbitrary associations across time even when there is no spatial component.

In a subsequent experiment rats with dorsal CA1, CA3 or control lesions were tested on an object-trace-place task (Hunsaker, Thorup, Welch and Kesner 2006). In this case both CA1 and CA3 were impaired suggesting that CA3 contributes to temporal processing of information only when place is one of the components. Thus, it is likely that the dorsal CA3 is a source for timing information which contributes to spatial temporal processing. However, the dorsal CA3 does not appear to contribute timing information for object and odor information. A possible suggestion is that the ventral CA3 supports the processing of odor information which then could contribute timing information to CA1. Another possibility is that object and odor information can reach CA1 from the lateral entorhinal cortex and since timing cells are found in the medial entorhinal time and object as well as time and odor could form time & object, or odor & object combination neurons.

24.4.3.3 Order memory, and CA1

There are also data on order memory, which does not necessarily imply that a particular sequence has been learned, and could be recalled. Estes (1986) summarized data demonstrating that in human memory there are fewer errors for distinguishing items (by specifying the order in which they occurred) that are far apart in a sequence than those that are temporally adjacent.

Other studies have also shown that order judgements improve as the number of items in a sequence between the test items increases (Banks 1978, Chiba, Kesner and Reynolds 1994, Madsen and Kesner 1995). This phenomenon is referred to as a temporal distance effect (sometimes referred to as a temporal pattern separation effect). It is assumed to occur because there is more interference for temporally proximal events than temporally distant events.

Based on these findings, Gilbert, Kesner and Lee (2001) tested memory for the temporal order of items in a one-trial sequence learning paradigm. In the task, each rat was given one daily trial consisting of a sample phase followed by a choice phase. During the sample phase, the animal visited each arm of an 8-arm radial maze once in a randomly predetermined order and was given a reward at the end of each arm. The choice phase began immediately following the presentation of the final arm in the sequence. In the choice phase, two arms were opened simultaneously and the animal was allowed to choose between the two arms. To obtain a food reward, the animal had to enter the arm that occurred earliest in the sequence that it had just followed. Temporal separations of 0, 2, 4, and 6 were randomly selected for each choice phase. These values represented the number of arms in the sample phase that intervened between the two arms that were to be used in the test phase. After reaching criterion rats received CA1 lesions. Following surgery, control rats matched their preoperative performance across all temporal separations. In contrast, rats with CA1 lesions performed at chance across 0, 2, or 4 temporal separations and a little better than chance in the case of a 6 separation. The results suggest that the CA1 subregion is involved in memory for spatial location as a function of temporal separation of spatial locations and that lesions of the CA1 decrease efficiency in temporal pattern separation. CA1 lesioned rats cannot separate events across time, perhaps due to an inability to inhibit interference that may be associated with sequentially occurring events. The increase in temporal interference impairs the rat's ability to remember the order of specific events.

Even though CA1 lesions produced a deficit in temporal pattern separation, some computational models (Levy 1996, Lisman and Idiart 1999, Wallenstein and Hasselmo 1997) and the model described above in Section 24.3 have suggested that the CA3 region is an appropriate part of the hippocampus to form a sequence memory, for example by utilising synaptic associativity in the CA3–CA3 recurrent collaterals that has a temporally asymmetric component (Section 24.3.3.4). It was thus of interest to use the temporal order task to determine whether the CA3 region plays an important role in sequence memory. The CA3 lesions impaired the same sequence learning task described above that was also impaired by lesions of the CA1 region (Gilbert and Kesner 2006). However, this is likely to be the case only for spatial information.

The hippocampus is known to process spatial and temporal information independently (O'Keefe and Nadel 1978, Kesner 1998). In the previous experiments sequence learning and temporal pattern separation were assessed using spatial cues. Therefore, one possibility is that the CA1 and CA3 deficits were due to the processing of spatial rather than non-spatial temporal information. To determine if the hippocampus is critical for processing all domains of temporal information, it is necessary to use a task that does not depend on spatial information. Since the hippocampus does not mediate short-term (delayed match to sample) memory for odours (Dudchenko, Wood and Eichenbaum 2000, Otto and Eichenbaum 1992), it is possible to test whether the hippocampus plays a role in memory for the temporal sequence of odours (i.e. temporal separation effect). Memory for the temporal order for a sequence of odours was assessed in rats based on a varied sequence of five odours, using a similar paradigm described for sequences of spatial locations. Rats with hippocampal lesions were impaired relative to control animals for memory for all temporal distances for the odours, yet the rats were able to discriminate between the odours (Kesner, Gilbert and Barua 2002). In a further subregional analysis, rats with dorsal CA1 lesions show a mild impairment, but rats with ventral CA1 lesions show a severe impairment in memory for the temporal distance for odours (Kesner, Hunsaker and Ziegler 2011). Thus, the CA1 appears to be involved in separating events in time for spatial and non-spatial information, so that one event can be remembered distinct from another event, but the dorsal CA1 might play a more important role than the ventral CA1 for spatial information, and conversely ventral CA1 might play a more important role than the dorsal CA1 for odour information (Kesner and Rolls 2015).

In a more recent experiment using a paradigm described by Hannesson, Howland and Phillips (2004), it was shown that temporal order information for visual objects is impaired only by CA1, but not CA3 lesions (Hoge and Kesner 2007). Thus, with respect to sequence learning or memory for order information one hypothesis that would be consistent with the data is that the CA1 is the critical substrate for sequence learning and temporal order or temporal pattern separation. This would be consistent with CA1 deficits in sequence completion of a spatial task and CA1 deficits in temporal order for spatial locations, odour, and visual objects. Dorsal CA3 contributes to this temporal order or sequence process whenever spatial location is also important. When visual objects are used the CA3 region does not play a role.

In these order tasks the animals are not required to recall the sequence (for example by retracing their steps). Instead, the animals are asked which of two items occurred earlier in the list. To implement this type of memory, some temporally decaying memory trace might provide a model (Marshuetz 2005), and in such a model, temporally adjacent items would have memory traces of more similar strength and would be harder to discriminate than the strengths of the memory traces of more temporally distant items. (We assume that the animal could learn the rule of responding to the stronger or weaker trace.) This temporal decay memory depends on both the CA3 and CA1 (Gilbert, Kesner and Lee 2001).

24.4.3.4 Intermediate memory, and CA1

There is some evidence that the CA1 region is especially involved in an intermediate term memory. For example, rats with CA3, but not CA1 lesions were impaired in the acquisition of a delayed non-match to place task (in an 8-arm maze) with a 10 s delay. Also, when transferred to a new maze in a different room and using a 10 s delay, there was a deficit for CA3, but not CA1 lesioned rats. Deficits for CA1 emerged when rats were transferred to a 5 min delay. Comparable deficits at a 5 min delay were also found for CA3 lesioned rats. It should be noted that similar results as described for the lesion data were obtained following AP5 injections, which impaired transfer to the new environment when made into CA3 but not CA1. At 5 min delays AP5 injections into CA1 produced a sustained deficit in performance, whereas AP5 injections to CA3 did not produce a sustained deficit (Lee and Kesner 2002, Lee and Kesner 2003a).

Although the CA3 system is involved even at short delays in acquiring this spatial short-term memory task, it is not apparently involved in the task once it has been acquired (Lee and Kesner 2003a). The implication is that (after acquisition) the CA3 system is not required to operate as a short-term attractor working memory to hold on line in the 10 s delay the place that has just been shown, although it could contribute to acquisition by operating in this way. The CA3 could also help in acquisition by providing a good spatial representation for the task, and in particular would enable the spatial cues in the environment to be learned. The suggestion therefore is that at the 5 min delay, the system must operate by associative connections from entorhinal cortex to CA1, which enable the sample place in the 8-arm maze to be remembered until the non-match part of the trial 5 min later, and the AP5 impairment after injection into CA1 is consistent with this.

Further evidence for some dissociation of an intermediate-term from a short-term memory in the hippocampus is described by Kesner and Rolls (2015).

How could memories become hippocampal-independent after long time delays? One possibility is that there is active recall to the neocortex which can then use episodic information to build new semantic memories, as described in Section 24.2.3 (Rolls 1989b, Rolls 1989e, Rolls 1990a, Treves and Rolls 1994, McClelland et al. 1995). In this case, the hippocampus could be said to play a role in the consolidation of memories in the neocortex. We note that in another sense a consolidation process is required in the neocortex during the original learning of the episodic memory, in that we hypothesize that associative connections between hippocampal backprojections and neocortical neurons activated by forward inputs are formed (and presumably consolidated) during the original learning, so that later a hippocampal signal can recall neocortical representations. Another possibility is that there are more passive effects of what has been learned in the hippocampus on the neocortex, occurring for example during sleep (Wilson 2002, Wilson and McNaughton 1994). Indeed, given that the hippocampus is massively connected to the neocortex by backprojections as described above, it is very likely that any alteration of the functional connectivity within the hippocampus, for example the coactivity of place neurons representing nearby places in an environment which has just been learned, will have an influence on the correlations found between neocortical neurons also representing the places of events (Wilson 2002, Wilson and McNaughton 1994). In this case, there could be synaptic modification in the neocortex between newly correlated neuronal activity, and this would effectively produce a form of new learning ('consolidation') in the neocortex. Another sense in which consolidation might be used is to refer to the findings considered in this section that there is a time-dependency of the learning in CA1 that seems special, enabling CA1 to contribute to the formation of memories ('consolidation') with long time delays, and also, perhaps with a different time course, to their subsequent forgetting or reconsolidation. The term 'consolidation' could thus be used to refer to many different types of computational process, which should be distinguished.

Many more tests of the theory of hippocampal operation, and a summary of the interactions and dissociations between CA3 and CA1, are provided by Kesner and Rolls (2015). In addition, there is a Special Issue of *Neurobiology of Learning and Memory* devoted to pattern separation and pattern completion in the hippocampal system (Rolls and Kesner 2016).

24.5 Evaluation of the theory of hippocampal cortex operation

24.5.1 Tests of the theory by hippocampal system subregion analyses

24.5.1.1 Dentate gyrus/mossy fibre system

The studies showing that the dentate/mossy fibre system is especially involved in hippocampal learning and is not necessary for retrieval (Lassalle et al. 2000, Lee and Kesner 2004b) provide strong support for the hypothesis that the DG system sets up the appropriate representations in CA3 for efficient learning, by helping to ensure that the different memory patterns to be stored in CA3 are relatively orthogonal; and that a separate system (the perforant path input to the CA3 network in the theory) initiates retrieval from the CA3 recurrent collateral network (Treves and Rolls 1992b). The evidence that spatial metrically close place memory is especially dependent on the dentate gyrus (Gilbert, Kesner and Lee 2001, Goodrich-Hunsaker, Hunsaker and Kesner 2005) is consistent with the hypothesis that the dentate granule cells help CA3 function by setting up relatively uncorrelated representations of spatial information so that the CA3 recurrent collateral system can form distinct memories (Rolls 1989b, Treves and Rolls 1992b, Treves and Rolls 1994) (Section 24.3.4). An important part of the theory is that the dentate / mossy fibre system by helping to set up relatively orthogonal representations in CA3 for different memories helps to increase the number of memories that can be stored in the CA3 system (which is calculated to be tens of thousands) (Treves and Rolls 1994). Consistent with the dentate spatial pattern separation hypothesis, in mice with impaired dentate neurogenesis (Aimone and Gage 2011), spatial learning in a delayed non-matching-to-place task in the radial arm maze was impaired for arms that were presented with little separation, but no deficit was observed when the arms were presented farther apart (Clelland et al. 2009) (Section 24.4.1.2).

It is useful to note that the non-associative mossy fibre to CA3 synapses are appropriate for a system that performs one-trial (episodic) learning, for it helps to set up orthogonal representations in CA3 even on one trial, without learning. (The CA3 to CA3 synapses would then learn that unique pattern of firing that would be produced on that trial by the mossy fibre pattern separation effect on CA3 firing.)

24.5.1.2 CA3 system

The hypothesis that CA3 operates as an autoassociation or attractor network, and can thereby contribute to pattern completion, short-term memory, recall, rapid (one-trial) encoding, and sequence memory, for certain types of information, receives considerable support from the subregion analyses (Section 24.4.2). The evidence that pattern completion when an incomplete retrieval cue is presented depends on the CA3 recurrent collateral system (Gold and Kesner 2005, Nakazawa et al. 2002) is consistent with a major prediction of the theory. The evidence that some types of short-term memory depend on the CA3 system (see Sections 24.4.2.8 and 24.4.3.2) is consistent with the hypothesis that it can operate as an attractor network, which has the property that it can keep a memory active by continuing firing of CA3 neurons. A property of this type of memory is that it can be interrupted by new processing or distraction, and it is of interest that in the CA3-dependent short-term memory tasks, the environment is often partially hidden by a bucket, which would tend to help keep the attractor memory from being interrupted by a distracter. It is predicted that these memory tasks will be susceptible to distracters presented in the delay period. The predictions are thus that if there is no task in a delay period (and no need for an off-line buffer store), the hippocampus may be sufficient to perform the short-term memory function (such as remembering a previous location), provided that there is little distraction in the delay period. On the other hand, if there are intervening stimuli in the delay period, then the prefrontal cortex is predicted computationally to be involved. Thus, lesions of the prefrontal cortex are predicted to impair spatial short-term memory tasks particularly when there are intervening items, or much distraction or processing of other stimuli, in the delay period (see Section 24.3.3.1).

The evidence that primate CA3 and CA1 neurons can combine information about objects and places in an object–place memory task (and that other neurons encode just the places or just the objects (Rolls, Xiang and Franco 2005c)), and about rewards and places in a reward–place memory task (Rolls and Xiang 2005); and that CA3 lesions disrupt object–place and odour–place learning in rats (Gilbert and Kesner 2003) is important evidence that the types of information required for these arbitrary types of association are present in the primate and rat hippocampus, and that continuous and discrete representations can be represented in the same network (Rolls, Stringer and Trappenberg 2002) (see Fig. 24.16). Subregion analyses show that place information requires especially the CA3 subregion, and that some types of temporal information (including odour and spatial location order tasks and trace conditioning) may require a combination of CA3 and CA1. The evidence that some hippocampal pyramidal neurons reflect the place to be recalled in an object–place one-trial recall task (Rolls and Xiang 2006) also indicates that the system is useful in episodic (rapid, one-trial) memories. In these cases, the crucial site for the associativity would be the CA3 recurrent collateral system. Subregion analysis in the rat shows that the CA3 subregion is required for one-trial object–place recall (Kesner et al. 2008). Subregion analyses provide additional evidence that the CA3 system is important in rapid (one trial) encoding, in for example a one-trial non matching to place task (Lee and Kesner 2002, Lee and Kesner 2003a), in the one-trial object–place recall task with many objects and many places just cited, and in one-trial learning for spatial location in the water maze (Nakazawa, Sun, Quirk, Rondi-Reig, Wilson and Tonegawa 2003).

Sequence memory could be implemented by temporally asymmetric synaptic modifiability in the CA3 system, or by associations in CA3 or CA1 with timing inputs from the medial entorhinal cortex (Section 24.3.6.2). The evidence reviewed in Section 24.3.6.2 suggests that CA1 may be especially important with this, consistent with the hypothesis that an association between the time cell inputs to CA1 and other CA1 representations is important.

24.5.1.3 CA1 system

As described in Section 24.4.3, CA1 subregion lesions disrupt memory of the temporal order of spatial or odour items (Section 24.3.6.2. They also impair trace conditioning, and starting in early positions when performing a fixed sequence spatial task that has been learned previously. They also impair learning with long (e.g. 5 min) delays in a non-matching to sample place task (intermediate memory); and in the Hebb–Williams maze impair performance on the first trials on the following day when retrieval is at a premium. All these tasks are also impaired by disruption of direct entorhinal to CA1 connections (Section 24.4.3). This evidence suggests that associations in CA1 with timing inputs from the medial entorhinal cortex are important in these functions (Section 24.3.6.2).

As noted in Section 24.4.3.4 on intermediate memory, a hypothesis that may help to unify these findings is that the time course of synaptic changes in CA1 has a different, and longer, time course to that in for example CA3, which allows it to play a particular role in learning these time delay types of memory task.

Computationally, it is noted that the direct entorhinal to CA1 connections could be used in recall, in a way that is analogous to the entorhinal to CA3 projections. Further, the tasks impaired by CA1 but not by CA3 lesions tend to involve non-spatial information, and temporal delays. However, as elaborated below, a main prediction of the computational theory with respect to CA1 is that CA1, and the CA3 to CA1 connections, will be especially important when recall of information is required back to the neocortex using backprojections, in for example place–object recall, and this has not yet been tested by subregion analysis.

A further analysis of the interactions and dissociations between CA3 and CA1 is provided by Kesner and Rolls (2015).

24.5.2 Comparison with other theories of hippocampal function

The theory described here is quantitative, and supported by both formal analyses and quantitative simulations. Many of the points made, such as on the number of memories that can be stored in autoassociative networks, the utility of sparse representations, and the dynamics of the operation of networks with recurrent connections, are quite general, and will apply to networks in a number of different brain areas. With respect to the hippocampus, the theory specifies the maximum number of memories that could be stored in it, and this has implications for how it could be used biologically. It indicates that if this number is approached, it will be useful to have a mechanism for recalling information from the hippocampus for incorporation into memories elsewhere. With respect to recall, the theory provides a quantitative account for why there are as many backprojections as forward projections in the cerebral cortex. Overall, the theory provides an explanation for how this part of the brain could work, and even if this theory needs to be revised, it is suggested that a fully quantitative computational theory along the lines proposed which is based on the evidence available from a wide range of techniques will be essential before we can say that we understand how a part of the cortex operates. This theory is almost unique in being a *quantitative* theory of hippocampal function.

Hypotheses have been described in this chapter about how a number of different parts of hippocampal and related circuitry might operate. Although these hypotheses are consistent with a theory of how the hippocampus operates, some of these hypotheses could be incorporated into other views or theories. In order to highlight the differences between alternative theories, and in order to lead to constructive analyses that can test them, the theory described above is compared with other theories of hippocampal function in this section. Although the differences between the theories are highlighted in this section, the overall view described here is close in different respects to the views of a number of other investigators (Marr 1971, Brown and Zador 1990, Eichenbaum, Otto and Cohen 1992, McNaughton and Nadel 1990, Squire 1992, McClelland, McNaughton and O'Reilly 1995, Moscovitch, Rosenbaum, Gilboa, Addis, Westmacott, Grady, McAndrews, Levine, Black, Winocur and Nadel 2005, Wang and Morris 2010, Preston and Eichenbaum 2013, Winocur and Moscovitch 2011) and of course priority is not claimed on all the propositions put forward here.

Some theories postulate that the hippocampus performs spatial computation. The theory of O'Keefe and Nadel (1978) that the hippocampus implements a cognitive map, placed great emphasis on spatial function. It supposed that the hippocampus at least holds information about allocentric space in a form which enables rats to find their way in an environment even when novel trajectories are necessary, that is it permits an animal to "go from one place to another independent of particular inputs (cues) or outputs (responses), and to link together conceptually parts of the environment which have never been experienced at the same time". O'Keefe (1990) extended this analysis and produced a computational theory of the hippocampus as a cognitive map, in which the hippocampus performs geometric spatial computations. Key aspects of the theory are that the hippocampus stores the centroid and slope of the distribution of landmarks in an environment, and stores the relationships between the centroid and the individual landmarks. The hippocampus then receives as inputs information about where the rat currently is, and where the rat's target location is, and computes geometrically the body turns and movements necessary to reach the target location. In this sense, the hippocampus is taken to be a spatial computer, which produces an output which is very different from its inputs. This is in contrast to the present theory, in which the hippocampus is a memory device, which is able to recall what was stored in it, using as input a partial cue. The theory of O'Keefe postulates that the hippocampus actually performs a spatial computation. The well-deserved award of a Nobel prize to John O'Keefe and colleagues explicitly recognized the spatial aspects of his approach, referring to an "inner GPS" in the brain. Updates to the theory (Burgess, Jackson, Hartley and O'Keefe 2000, Burgess, Recce and O'Keefe 1994) also make the same postulate, but now the firing of place cells is determined by the distance and approximate bearing to landmarks, and the navigation is performed by increasing the strength of connections from place cells to "goal cells", and then performing gradient-ascent style search for the goal using the network. Rolls' discoveries and theory are thus somewhat different from those of O'Keefe, in that Rolls has shown that spatial view cells may be especially relevant to the operation of the hippocampus in primates including humans; and in that the roles of the hippocampal system in memory are emphasized.

McNaughton, Chen and Markus (1991) have also proposed that the hippocampus is involved in spatial computation. They propose a "compass" solution to the problem of spatial navigation along novel trajectories in known environments, postulating that distances and bearings (i.e. vector quantities) from landmarks are stored, and that computation of a new trajectory involves vector subtraction by the hippocampus. They postulate that a linear associative mapping is performed, using as inputs a "cross-feature" (combination) representation of (head) angular velocity and (its time integral) head direction, to produce as output the future value of the integral (head direction) after some specified time interval. The system can be reset by learned associations between local views of the environment and head direction, so that when later a local view is seen, it can lead to an output from the network which is a (corrected) head direction. They suggest that some of the key signals in the computational system can be identified with the firing of hippocampal cells (e.g. local view cells) and subicular cells (head direction cells). It should be noted that this theory requires a (linear) associative mapping with an output (head direction) different in form from the inputs (head angular velocity over a time period, or local view). This is pattern association (with the conditioned stimulus local view, and the unconditioned stimulus head direction), not autoassociation, and it has been postulated that this pattern association can be performed by the hippocampus (cf. McNaughton and Morris (1987)). This theory is again in contrast to the present theory, in which the hippocampus operates as a memory to store events that occur at the same time, and can recall the whole memory from any part of what was stored. (A pattern associator uses a conditioned stimulus to map an input to a pattern of firing in an output set of neurons which is like that produced in the output neurons by the unconditioned stimulus. A description of pattern associators and autoassociators in a neurobiological context is provided in Appendix B. The present theory is fully consistent with the presence of 'spatial view' cells and whole body motion cells in the primate hippocampus (Rolls and Xiang 2006, Rolls 1999c, Rolls and O'Mara 1993) (or place or local view cells in the rat hippocampus, and head direction cells in the presubiculum), for it is often important to store and later recall where one has been (views of the environment, body turns made, etc), and indeed such (episodic) memories are required for navigation by "dead reckoning" in small environments.

The present theory thus holds that the hippocampus is used for the formation of episodic memories using autoassociation. This function is often necessary for successful spatial computation, but is not itself spatial computation. Instead, we believe that spatial computation is more likely to be performed in the neocortex (utilising information if necessary recalled from the hippocampus). Consistent with this view, hippocampal damage impairs the ability to learn new environments but not to perform spatial computations such as finding one's way to a place in a familiar environment, whereas damage to the parietal cortex and parahippocampal cortex can lead to problems such as topographical and other spatial agnosias, in humans (Grusser and Landis 1991, Kolb and Whishaw 2015). This is consistent with spatial computations normally being performed in the neocortex. (In monkeys, there is evidence for a role of the parietal cortex in allocentric spatial computation. For example, monkeys with parietal cortex lesions are impaired at performing a landmark task, in which the object to be chosen is signified by the proximity to it of a "landmark" (another object) (Ungerleider and Mishkin 1982).)

A theory closely related to the present theory of how the hippocampus operates has been developed by McClelland, McNaughton and O'Reilly (1995). It is very similar to the theory we have developed (Rolls 1987, Rolls 1989a, Rolls 1989b, Rolls 1989f, Treves and Rolls 1992b, Treves and Rolls 1994, Rolls 1996c, Rolls and Treves 1998, Rolls and Kesner 2006, Rolls 2008e, Rolls 2010b, Kesner and Rolls 2015) at the systems level, except that it takes a stronger position on the gradient of retrograde amnesia, emphasises that recall from the hippocampus of episodic information is used to help build semantic representations in the neocortex, and holds that the last set of synapses that are modified rapidly during the learning of each episode are those between the CA3 and the CA1 pyramidal cells (see Fig. 24.1). It also emphasizes the important point that the hippocampal and neocortical memory systems may be quite different, with the hippocampus specialized for the rapid learning of single events or episodes, and the neocortex for the slower learning of semantic representations which may necessarily benefit from the many exemplars needed to shape the semantic representation. In the formulation by McClelland, McNaughton and O'Reilly (1995), the entorhinal cortex connections via the perforant path onto the CA1 cells are non-modifiable (in the short term), and allow a representation of neocortical long-term memories to activate the CA1 cells. The new information learned in an episode by the CA3 system is then linked to existing long-term memories by the CA3 to CA1 rapidly modifiable synapses. All the connections from the CA1 back via the subiculum, entorhinal cortex, parahippocampal cortex etc. to the association neocortex are held to be unmodifiable in the short term, during the formation of an episodic memory. The formal argument that leads us to suggest that the backprojecting synapses are associatively modifiable during the learning of an episodic memory is similar to that which we have used to show that for efficient recall, the synapses which initiate recall in the CA3 system (identified above with the perforant path projection to CA3) must be associatively modifiable if recall is to operate efficiently (Treves and Rolls 1992b). The present theory holds that it is possible that for several stages back into neocortical processing, the backprojecting synapses should be associatively modifiable, with a similar time course to the time it takes to learn a new episodic memory. It may well be that at earlier stages of cortical processing, for example from inferior temporal visual cortex to V4, and from V4 to V2, the backprojections are relatively more fixed, being formed (still associatively) during early developmental plasticity or during the formation of new long-term semantic memory structures. Having such relatively fixed synaptic strengths in these earlier cortical backprojection systems could ensure that whatever is recalled in higher cortical areas, such as objects, will in turn recall relatively fixed and stable representations of parts of objects or features. Given that the functions of backprojections may include many top-down processing operations, including attention (Deco and Rolls 2005a) and priming, it may be useful to ensure that there is consistency in how higher cortical areas affect activity in earlier "front-end" or preprocessing cortical areas. Indeed, the current theory shows that at least one backprojection stage, the hippocampo-cortical connections must be associatively modifiable during the learning of an episodic memory, but does not require the associative backprojection learning to occur at all backprojection stages during the learning of the episodic memory. Crucial stages might include CA1 to subiculum or to entorhinal cortex (see Fig. 24.1), or entorhinal cortex to parahippocampal gyrus and perirhinal cortex. It would be interesting to test this using local inactivation of NMDA receptors at different stages of the backprojection system to determine where this impairs the learning of for example place–object recall, i.e. a task that is likely to utilize the hippocampo-cortical backprojection pathways.

If a model of the hippocampal / neocortical memory system could store only a small number of patterns, it would not be a good model of the real hippocampal / neocortical memory system in the brain. Indeed, this appears to be a major limitation of another model presented by Alvarez and Squire (1994). The model specifies that the hippocampus helps the operation of a neocortical multimodal memory system in which all memories are stored by associative recurrent collaterals between the neocortical neurons. Although the idea worked in the model with twenty neurons and two patterns to be learned (Alvarez and Squire 1994), the whole idea is computationally not feasible, because the number of memories that can be stored in a single autoassociative network of the type described is limited by the number of inputs per neuron from other neurons, not by the number of neurons in the network (Treves and Rolls 1991, Treves and Rolls 1994). This would render the capacity of the whole neocortical multimodal (or amodal) memory store very low (in the order of the number of inputs per neuron from the other neurons, that is in the order of 5,000–10,000) (O'Kane and Treves 1992). This example makes it clear that it is important to take into account analytic and quantitative approaches when investigating memory systems. The current work is an attempt to do this.

The theory of Lisman et al. (2005) of the memory for sequences has been described and evaluated in Section 24.3.3.4. This theory of sequential recall within the hippocampus is inextricably linked to the internal timing within the hippocampus imposed he believes by the theta and gamma oscillations, and this makes it difficult to recall each item in the sequence as it is needed. It is not specified how one would read out the sequence information, given that the items are only 12 ms apart. The Jensen and Lisman (1996) model requires short time constant NMDA channels, and is therefore unlikely to be implemented in the hippocampus. Hasselmo and Eichenbaum (2005) have taken up some of these sequence ideas and incorporated them into their model, which has its origins in the Rolls and Treves model (Rolls 1989b, Treves and Rolls 1992b, Treves and Rolls 1994, Rolls and Treves 1998). The proposal that acetylcholine could be important during encoding by facilitating CA3–CA3 LTP, and should be lower during retrieval (Hasselmo, Schnell and Barkai 1995), is a useful concept.

Another type of sequence memory uses synaptic adaptation to effectively encode the order of the items in a sequence (Deco and Rolls 2005c). This could be implemented in recurrent networks such as the CA3 or the prefrontal cortex.

The aim of this comparison of the present theory with other theories has been to highlight differences between the theories, to assist in the future assessment of the utility and the further development of each of the theories.

Teyler and DiScenna (1986) proposed that the hippocampus stores pointers to neocortical areas, and uses those pointers to recall memories in the cortex. The present theory is different, for it emphasizes that spatial and timing representations are present in the hippocampus, and can be combined with object information to store episodic memories in the hippocampus by forming associations between these components. However, the theory also shows how recall of these episodic memories within the hippocampus can then using backprojections to neocortical areas allow neocortical representations present at the time of the encoding to be retrieved, and then potentially stored in the neocortex in a schema or semantic memory. The actual retrieval process to neocortex is implemented by using associatively modified synapses in at least one stage of the backprojection system, and this provides a deaddressing method for the hippocampus to retrieve a content-addressable neocortical memory. This deaddressing process for recall to the neocortex that is part of the current theory can be likened to the use of a pointer. The present theory (Rolls 2008d, Rolls 2010b, Kesner and Rolls 2015) is I believe unique in that it enables both the hippocampal storage process, and the recall to the neocortex, to be understood quantitatively.

Overall, the approach taken in this chapter shows how it is now possible to understand quantitatively how the circuitry in a major brain system could actually operate to store and later recall memories. The principles of operation of the cortex uncovered illustrate important foundations of many types of cortical computation, including forming new representations (in this case of spatial view or place) as a result of hierarchical processing (Chapter 2), the functions of attractor networks in the brain (Chapter 4), the importance of the sparseness of the representation (Chapter 8), the recall of information from a memory system (Chapter 12 and Appendix B), the functions of backprojections between cortical areas (Chapter 11), and the importance of diluted connectivity in the cerebral cortex (Chapter 7).

24.6 Highlights

1. A description of the operation of the hippocampus in episodic memory is provided, with a theory of how the hippocampus operates, and how the hippocampal memory system illustrates many of the principles of operation of the cerebral cortex.

2. A feature of the hippocampal network is that the CA3 system has a highly developed recurrent collateral system which enables it to associate together as an autoassociation or attractor network places and objects, with the inputs coming from very different neocortical systems. The single attractor network in CA3 enables any object to be

associated with any place in an episodic or event memory, and for the place to be recalled from the object, and the object from the place. This is completion of a memory in an autoassociation network.

3. Although the representation of place in rodents is of the place where the rodent is located, in primates there is a representation of spatial view, enabling humans to remember where they have seen a person or object.

4. The number of memories that can be stored in the hippocampus is determined largely by the number of recurrent collateral connections onto any one neuron (in the order of 12,000 in the rat) and the sparseness of the representation.

5. The dentate gyrus input to CA3 performs pattern separation of the inputs before CA3 to help achieve the relatively uncorrelated representations that help to achieve high memory capacity in CA3. The dentate granule cells perform this operation by competitive learning; by the low probability of contact of the dentate granule cell to CA3 connections (0.00015 in the rat: a granule cell makes 46 mossy fibre connections onto 300,000 CA3 cells); and by neurogenesis with the new dentate granule cells helping to ensure that new memories are distinct from old memories by making new random connections to CA3 neurons. The dentate gyrus is the only part of cortex where new neurons can develop in adulthood, for here the premium is on selecting new random sets of neurons in CA3 for new memories distinct from previous memories, whereas elsewhere in cortex the premium is on building and reusing useful new representations.

6. The CA1 which follows the CA3 prepares for the recall of memories, completed in CA3 by a partial retrieval cue, back to the neocortex. The CA1 has the architecture of a competitive network, useful for recoding the parts of an episodic memory necessarily represented as separate in CA3 for autoassociation in CA3.

7. Backprojections from the hippocampal system to neocortical areas have a large number of synapses onto each neocortical neuron to achieve high recall capacity. This accounts for why there are as many backprojections as forward connections between adjacent cortical areas in a hierarchy. For the recall to work efficiently, the backprojection synapses at one or more stages of the backprojection pathway must be associatively modifiable.

8. Once set up in this way, the backprojection connections can also be used for top-down attentional control (see Chapter 6).

9. In addition to place cells in rodents and spatial view cells in primates, there are also time cells in the hippocampus which increase their firing at different times in a delay period. These neurons may be thought of as similar to place cells, and when associated with objects or places, enable the order in which the objects or places occurred to be recalled. CA1 is especially implicated in this operation.

10. The medial entorhinal cortex contains grid cells in rodents which fire with several peaks of firing as a rodent traverses places. In primates, there are spatial view grid cells, which are likely to provide inputs to the spatial view cells found in the primate hippocampus. Different entorhinal grid cells operate at different spatial scales. The medial entorhinal cortex appears to operate as a continuous attractor network which can perform path integration, updating the spatial representation based on motion signals. The same type of attractor network may implement the time cells also present in the medial entorhinal cortex. The lateral entorhinal cortex provides a route for object-related information from the ventral cortical systems to reach the hippocampus.

25 Invariant visual object recognition learning

25.1 Introduction

The ventral visual system and its role in the major computation of view invariant object recognition has been selected for this Chapter because it illustrates many of the principles of operation of the cerebral cortex and in particular the neocortex, and shows how these principles of operation combine together to produce important functions in vision. While separate principles have been the focus of much of the rest of this book, this Chapter comes close to a theory of how the ventral visual system may operate.

One of the major problems that is solved by the visual system in the cerebral cortex is the building of a representation of visual information that allows object and face recognition to occur relatively independently of size, contrast, spatial frequency, position on the retina, angle of view, etc. The importance of and requirements for these representations are described by Rolls (2014a). These transform-invariant representations of objects, provided by the inferior temporal visual cortex (as described in Section 25.2, Appendix C, and Section 24.2.6), are extremely important for the operation of other systems in the brain, for if there is an invariant representation, it is possible to learn on a single trial about reward/punishment associations of the object, the place where that object is located, and whether the object has been seen recently, and then to correctly generalize to other views etc. of the same object. The way in which these transform-invariant representations of objects are formed is a major issue in understanding the principles of operation of the neocortex, for with this type of operation, we must not only store and retrieve information, but we must solve in addition the major computational problem of how all the different images on the retina (position, size, view etc) of an object can be mapped to the same representation of that object in the cortex. It is the principles of cortical operation that underlie this major computation with which we are concerned in this Chapter.

In the first part of this Chapter, we consider the nature of the representations of objects and faces that are found in the inferior temporal visual cortex. Then later in the Chapter we consider the major computational issue of how these invariant representations of objects and faces may be formed by self-organizing learning in the cerebral cortex.

Before considering invariant object and face recognition in the inferior temporal visual cortex, we note that this builds on the preprocessing performed in V1, the striate visual cortex of primates (described for example in Chapter 2 of Rolls and Deco (2002) and in Kandel et al. (2013)). An outline of the overall architecture of the visual pathways is shown in Fig. 25.1. The model to be described in Section 25.5 also builds on what is known about visual processing in extrastriate visual areas (Rolls and Deco 2002, Kandel et al. 2013)), and the inferior temporal visual cortex (described next in Section 25.2) (Rolls and Deco 2002, Rolls 2007e, Rolls 2008b, Rolls 2012c).

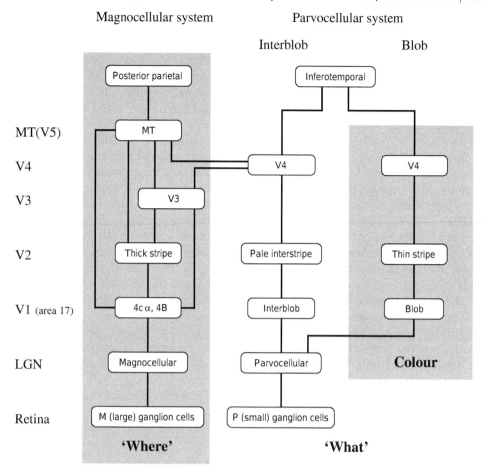

Fig. 25.1 A schematic diagram of the visual pathways from the retina to visual cortical areas. LGN, lateral geniculate nucleus; V1, primary visual cortex; V2, V3, V4, and MT, other visual cortical areas; M, magnocellular; P, parvocellular.

25.2 Invariant representations of faces and objects in the inferior temporal visual cortex

Desirable properties of an object recognition system include position, size and view invariant representations, and distributed representations that allow for generalization (Rolls 2008d). These are properties of the inferior temporal visual cortex, as described next in this Section (25.2).

25.2.1 Processing to the inferior temporal cortex in the primate visual system

A schematic diagram to indicate some aspects of the processing involved in object identification from the primary visual cortex, V1, through V2 and V4 to the posterior inferior temporal cortex (TEO) and the anterior inferior temporal cortex (TE) is shown in Fig. 2.1. The approximate location of these visual cortical areas on the brain of a macaque monkey is shown in Figs. 25.2 and 1.10, which also show that TE has a number of different subdivisions. The different TE areas all contain visually responsive neurons, as do many of the areas within the cortex in the superior temporal sulcus (Baylis, Rolls and Leonard 1987). For the purposes of this summary, these areas will be grouped together as the anterior inferior temporal cortex (IT), except where otherwise stated. A description of some of the specializations within this region, and a hypothesis about why there is specialization, is provided in Section 25.2.10. Some of the information processing that takes place through these pathways that must be addressed by computational models is described in the following subsections. Fuller accounts are provided elsewhere (Rolls 2000a, Rolls and Deco 2002, Rolls 2007e, Rolls 2007c, Rolls 2008b, Tompa and Sary 2010, Rolls 2011d, Rolls 2012c, DiCarlo, Zoccolan and Rust 2012, Tsao 2014).

Many of the studies on neurons in the anterior inferior temporal cortex and cortex in the superior temporal sulcus have been performed with neurons that respond particularly to faces, because such neurons can be found regularly in recordings in this region, and therefore provide a good population for systematic studies (Rolls 2000a, Rolls and Deco 2002, Rolls 2007e, Rolls and Deco 2006, Rolls 2011d, Rolls 2012c). The face-selective neurons described in this book are found mainly between 7 mm and 3 mm posterior to the sphenoid reference, which in a 3–4 kg macaque corresponds to approximately 11 to 15 mm anterior to the interaural plane (Baylis, Rolls and Leonard 1987, Rolls 2007e, Rolls 2007c, Rolls 2011d, Rolls 2012c). The 'middle face patch' of Tsao, Freiwald, Tootell and Livingstone (2006) was at A6, which is probably part of the posterior inferior temporal cortex. In the anterior inferior temporal cortex areas we have investigated, there are separate regions specialized for face identity in areas TEa and TEm on the ventral lip of the superior temporal sulcus and the adjacent gyrus, and for face expression and movement in the cortex deep in the superior temporal sulcus (Hasselmo, Rolls and Baylis 1989a, Baylis, Rolls and Leonard 1987, Rolls 2007e, Rolls 2011d, Rolls 2012c) (see Sections 25.2.9 and 25.2.10).

Examples of some of the objects encoded by anterior inferior temporal cortex neurons as described in this book (for example in this Chapter, in Section 24.2.6, and in Appendix C) are shown in Figs. 24.11 and C.7.

25.2.2 Translation invariance and receptive field size

There is convergence from each small part of a region to the succeeding region (or layer in the hierarchy) in such a way that the receptive field sizes of neurons (for example 1 degree near the fovea in V1) become larger by a factor of approximately 2.5 with each succeeding stage (see Fig. 2.1). [The typical parafoveal receptive field sizes found would not be inconsistent with the calculated approximations of, for example, 8 degrees in V4, 20 degrees in TEO, and 50 degrees in inferior temporal cortex (Boussaoud, Desimone and Ungerleider 1991) (see Fig. 2.1).] Such zones of convergence would overlap continuously with each other (Fig. 2.1). This connectivity provides part of the basis for the fact that many neurons in the temporal cortical visual areas respond to a stimulus relatively independently of where it is in their receptive field, and moreover maintain their stimulus selectivity when the stimulus appears in different parts of the visual field (Gross, Desimone, Albright and Schwartz 1985, Tovee, Rolls and Azzopardi 1994, Rolls, Aggelopoulos and Zheng 2003a). This is called translation or shift invariance. In addition to having topologically appropriate connections, it is necessary for the connections to have the appropriate synaptic weights to perform the mapping of each set of features, or object, to the same set of neurons in IT. How this could be achieved is addressed in the computational neuroscience models described later in this Chapter and elsewhere (Wallis and Rolls 1997, Rolls and Deco 2002, Rolls and Stringer 2006, Stringer, Perry, Rolls and Proske 2006, Rolls 2012c).

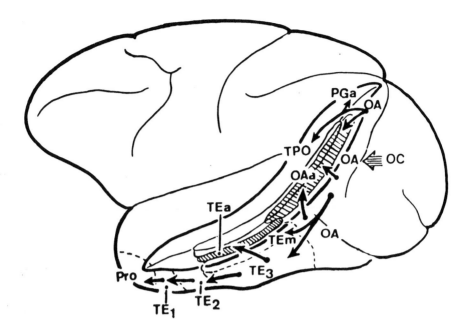

Fig. 25.2 Lateral view of the macaque brain (left hemisphere) showing the different architectonic areas (e.g. TEm, TEa) in and bordering the anterior part of the superior temporal sulcus (STS) of the macaque (see text). The STS has been drawn opened to reveal to reveal the cortical areas inside it, and is circumscribed by a thick line.

25.2.3 Reduced translation invariance in natural scenes, and the selection of a rewarded object

Until recently, research on translation invariance considered the case in which there is only one object in the visual field. What happens in a cluttered, natural, environment? Do all objects that can activate an inferior temporal neuron do so whenever they are anywhere within the large receptive fields of inferior temporal neurons (cf. Sato (1989))? If so, the output of the visual system might be confusing for structures that receive inputs from the temporal cortical visual areas. If one of the objects in the visual field was associated with reward, and another with punishment, would the output of the inferior temporal visual cortex to emotion-related brain systems be an amalgam of both stimuli? If so, how would we be able to choose between the stimuli, and have an emotional response to one but not perhaps the other, and select one for action and not the other (see Fig. 25.3)?

In an investigation of this, it was found that the mean firing rate across all cells to a fixated effective face with a non-effective face in the parafovea (centred 8.5 degrees from the fovea) was 34 spikes/s. On the other hand, the average response to a fixated non-effective face with an effective face in the periphery was 22 spikes/s (Rolls and Tovee 1995a). Thus these cells gave a reliable output about which stimulus is actually present at the fovea, in that their response was larger to a fixated effective face than to a fixated non-effective face, even when there were other parafoveal stimuli effective for the neuron. Thus the neurons provide information biased towards what is present at the fovea, and not equally about what is present anywhere in the visual field. This makes the interface to action simpler, in that what is at the fovea can be interpreted (e.g. by an associative memory) partly independently of the surroundings, and

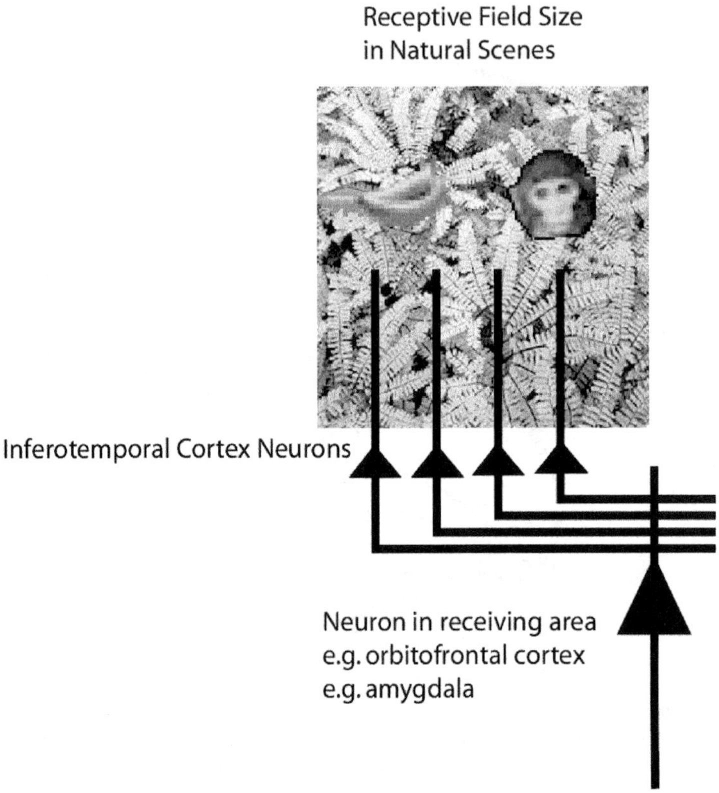

Receptive Field Size
in Natural Scenes

Inferotemporal Cortex Neurons

Neuron in receiving area
e.g. orbitofrontal cortex
e.g. amygdala

Fig. 25.3 Objects shown in a natural scene, in which the task was to search for and touch one of the stimuli. The objects in the task as run were smaller. The diagram shows that if the receptive fields of inferior temporal cortex neurons are large in natural scenes with multiple objects (in this scene, bananas and a face), then any receiving neuron in structures such as the orbitofrontal cortex and amygdala would receive information from many stimuli in the field of view, and would not be able to provide evidence about each of the stimuli separately.

choices and actions can be directed if appropriate to what is at the fovea (cf. Ballard (1993)). These findings are a step towards understanding how the visual system functions in a normal environment (see also Gallant, Connor and Van-Essen (1998), Stringer and Rolls (2000), and Rolls and Deco (2006)).

To investigate further how information is passed from the inferior temporal cortex (IT) to other brain regions to enable stimuli to be selected from natural scenes for action, Rolls, Aggelopoulos and Zheng (2003a) analysed the responses of single and simultaneously recorded IT neurons to stimuli presented in complex natural backgrounds. In one situation, a visual fixation task was performed in which the monkey fixated at different distances from the effective stimulus. In another situation the monkey had to search for two objects on a screen, and a touch of one object was rewarded with juice, and of another object was punished with saline (see Fig. 25.3 for a schematic overview, and Fig. 25.59 on page 643 for the actual display). In both situations neuronal responses to the effective stimuli for the neurons were compared when the objects were presented in the natural scene or on a plain background. It was found that the overall response of the neuron to objects was sometimes somewhat reduced when they were presented in natural scenes, though the selectivity of the neurons remained. However, the main finding was that the magnitudes of the responses of the neurons typically

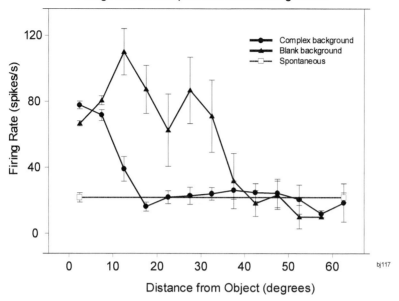

Fig. 25.4 Firing of a temporal cortex cell to an effective stimulus presented either in a blank background or in a natural scene, as a function of the angle in degrees at which the monkey was fixating away from the effective stimulus. The task was to search for and touch the stimulus. (After Rolls, Aggelopoulos and Zheng 2003.)

became much less in the real scene the further the monkey fixated in the scene away from the object (see Figs. 25.4 and 25.60, and Section 25.5.9.1). It is proposed that this reduced translation invariance in natural scenes helps an unambiguous representation of an object which may be the target for action to be passed to the brain regions that receive from the primate inferior temporal visual cortex. It helps with the binding problem, by reducing in natural scenes the effective receptive field of inferior temporal cortex neurons to approximately the size of an object in the scene. The computational utility and basis for this is considered in Section 25.5.9 and by Rolls and Deco (2002), Trappenberg, Rolls and Stringer (2002), Deco and Rolls (2004), Aggelopoulos and Rolls (2005), and Rolls and Deco (2006), and includes an advantage for what is at the fovea because of the large cortical magnification of the fovea, and shunting interactions between representations weighted by how far they are from the fovea.

These findings indicate that the principle of providing strong weight to whatever is close to the fovea is an important principle governing the operation of the inferior temporal visual cortex, and in general of the output of the ventral visual system in natural environments. This principle of operation is very important in interfacing the visual system to action systems, because the effective stimulus in making inferior temporal cortex neurons fire is in natural scenes usually on or close to the fovea. This means that the spatial coordinates of where the object is in the scene do not have to be represented in the inferior temporal visual cortex, nor passed from it to the action selection system, as the latter can assume that the object making IT neurons fire is close to the fovea in natural scenes. Thus the position in visual space being fixated provides part of the interface between sensory representations of objects and their coordinates as targets for actions in the world. The small receptive fields of IT neurons in natural scenes make this possible. After this, local, egocentric, processing implemented in the dorsal visual processing stream using e.g. stereodisparity may be used to guide action towards objects being fixated (Rolls and Deco 2002, Rolls 2008d).

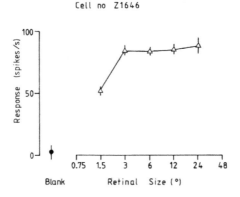

Fig. 25.5 Typical response of an inferior temporal cortex face-selective neuron to faces of different sizes. The size subtended at the retina in degrees is shown. (From Experimental Brain Research, Size and contrast have only small effects on the responses to faces of neurons in the cortex of the superior temporal sulcus of the monkey. 65, 1986, 38–48. E. T. Rolls, G. C. Baylis. Copyright ©1986, Springer-Verlag, with permission of Springer.)

The reduced receptive field size in complex natural scenes also enables emotions to be selective to just what is being fixated, because this is the information that is transmitted by the firing of IT neurons to structures such as the orbitofrontal cortex and amygdala.

Interestingly, although the size of the receptive fields of inferior temporal cortex neurons become reduced in natural scenes so that neurons in IT respond primarily to the object being fixated, there is nevertheless frequently some asymmetry in the receptive fields (see Section 25.5.10 and Fig. 25.64). This provides a partial solution to how multiple objects and their position in a scene can be captured with a single glance (Aggelopoulos and Rolls 2005).

25.2.4 Size and spatial frequency invariance

Some neurons in the inferior temporal visual cortex and cortex in the anterior part of the superior temporal sulcus (IT/STS) respond relatively independently of the size of an effective face stimulus, with a mean size invariance (to a half maximal response) of 12 times (3.5 octaves) (Rolls and Baylis 1986). An example of the responses of an inferior temporal cortex face-selective neuron to faces of different sizes is shown in Fig. 25.5. This is not a property of a simple single-layer network (see Fig. 25.16), nor of neurons in V1, which respond best to small stimuli, with a typical size-invariance of 1.5 octaves. Also, the neurons typically responded to a face when the information in it had been reduced from 3D to a 2D representation in grey on a monitor, with a response that was on average 0.5 of that to a real face.

Another transform over which recognition is relatively invariant is spatial frequency. For example, a face can be identified when it is blurred (when it contains only low spatial frequencies), and when it is high-pass spatial frequency filtered (when it looks like a line drawing). If the face images to which these neurons respond are low-pass filtered in the spatial frequency domain (so that they are blurred), then many of the neurons still respond when the images contain frequencies only up to 8 cycles per face. Similarly, the neurons still respond to high-pass filtered images (with only high spatial frequency edge information) when frequencies down to only 8 cycles per face are included (Rolls, Baylis and Leonard 1985). Face recognition shows similar invariance with respect to spatial frequency (see Rolls, Baylis and Leonard (1985)). Further analysis of these neurons with narrow (octave) bandpass spatial frequency filtered face stimuli shows that the responses of these neurons to an unfiltered face can not be predicted from a linear combination of their responses to the narrow band stimuli (Rolls, Baylis and Hasselmo 1987). This lack of linearity of these neurons, and their responsiveness to a wide range of spatial frequencies (see also their broad critical band masking (Rolls 2008b)), indicate that in at least this part of the primate visual system recognition does not occur using Fourier analysis of the spatial frequency components of images.

The utility of this representation for memory systems in the brain is that the output of the visual system will represent an object invariantly with respect to position on the retina, size, etc, and this simplifies the functionality required of the (multiple) memory systems, which need then simply associate the object representation with reward (orbitofrontal cortex and amygdala), associate it with position in the environment (hippocampus), recognise it as familiar (perirhinal cortex), associate it with a motor response in a habit memory (basal ganglia), etc. (Rolls 2007d, Rolls 2008d, Rolls 2012c, Rolls 2014a). The associations can be relatively simple, involving for example Hebbian associativity (see chapters throughout this book, and Appendix B).

Some neurons in the temporal cortical visual areas actually represent the absolute size of objects such as faces independently of viewing distance (Rolls and Baylis 1986). This could be called neurophysiological size constancy. The utility of this representation by a small population of neurons is that the absolute size of an object is a useful feature to use as an input to neurons that perform object recognition. Faces only come in certain sizes.

25.2.5 Combinations of features in the correct spatial configuration

Many neurons in this processing stream respond to combinations of features (including objects), but not to single features presented alone, and the features must have the correct spatial arrangement. This has been shown, for example, with faces, for which it has been shown by masking out or presenting parts of the face (for example eyes, mouth, or hair) in isolation, or by jumbling the features in faces, that some cells in the cortex in IT/STS respond only if two or more features are present, and are in the correct spatial arrangement (Perrett, Rolls and Caan 1982, Rolls, Tovee, Purcell, Stewart and Azzopardi 1994b). Fig. 25.6 shows examples of four neurons, the top one of which responds only if all the features are present, and the others of which respond not only to the full face, but also to one or more features. Corresponding evidence has been found for non-face cells. For example, Tanaka, Saito, Fukada and Moriya (1990) showed that some posterior inferior temporal cortex neurons might only respond to the combination of an edge and a small circle if they were in the correct spatial relationship to each other.

Evidence consistent with the suggestion that neurons are responding to combinations of a few variables represented at the preceding stage of cortical processing is that some neurons in V2 and V4 respond to end-stopped lines, to tongues flanked by inhibitory subregions, to combinations of lines, or to combinations of colours (see Rolls and Deco (2002), Hegde and Van Essen (2000) and Ito and Komatsu (2004)). Neurons that respond to combinations of features but not to single features indicate that the system is non-linear (Elliffe, Rolls and Stringer 2002).

The fact that some temporal cortex neurons respond only to objects or faces consisting of a set of features only if the whole combination of features is present and they are in the correct spatial arrangement with respect to each other (not jumbled) is part of the evidence that some inferior temporal cortex neurons are tuned to respond to objects or faces.

If a task requires in addition to the normal representation of objects by the inferior temporal cortex, the ability to discriminate between stimuli composed of highly overlapping feature conjunctions in a low-dimensional feature space, then the perirhinal cortex may contribute to this type of discrimination, as described in Section 24.2.6.

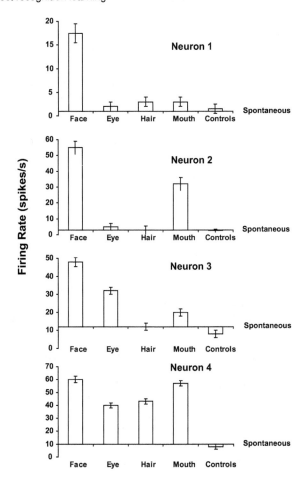

Fig. 25.6 Responses of four temporal cortex neurons to whole faces and to parts of faces. The mean firing rate ± sem are shown. The responses are shown as changes from the spontaneous firing rate of each neuron. Some neurons respond to one or several parts of faces presented alone. Other neurons (of which the top one is an example) respond only to the combination of the parts (and only if they are in the correct spatial configuration with respect to each other as shown by Rolls et al 1994b). The control stimuli were non-face objects. (After Perrett, Rolls and Caan 1982.)

25.2.6 A view-invariant representation

For recognizing and learning about objects (including faces), it is important that an output of the visual system should be not only translation- and size-invariant, but also relatively view-invariant. In an investigation of whether there are such neurons, we found that some temporal cortical neurons reliably responded differently to the faces of two different individuals independently of viewing angle (Hasselmo, Rolls, Baylis and Nalwa 1989b) (see example in Fig. 25.7 upper), although in most cases (16/18 neurons) the response was not perfectly view-independent. Mixed together in the same cortical regions there are neurons with view-dependent responses (Hasselmo, Rolls, Baylis and Nalwa 1989b, Rolls and Tovee 1995b, Chang and Tsao 2017) (see example in Fig. 25.7 lower). Such neurons might respond, for example, to a view of a profile of a monkey but not to a full-face view of the same monkey (Perrett, Smith, Potter, Mistlin, Head, Milner and Jeeves 1985b, Hasselmo, Rolls, Baylis and Nalwa 1989b).

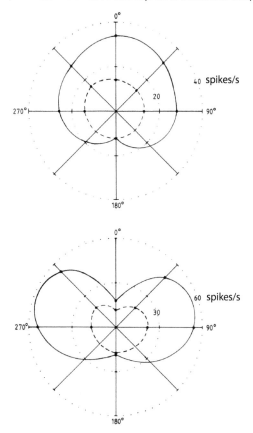

Fig. 25.7 Upper: View invariant response of an inferior temporal cortex neuron to two faces. In the polar response plot, $0°$ is the front of the face, and $90°$ the right profile. The firing rate in spikes/s is shown by the distance from the centre of the plot. The responses to two different faces are shown by the solid and dashed lines. The neuron discriminated between the faces, apart from the view of the back of the head. Lower: View dependent response of an inferior temporal cortex neuron to two faces. The neuron responded best to the left and right profiles. Some neurons respond to only one of the profiles, and this neuron had a small preference for the left profile. (After Hasselmo, Rolls, Baylis and Nalwa 1989b.)

These findings of view-dependent, partially view-independent, and view-independent representations in the same cortical regions are consistent with the hypothesis discussed below that view-independent representations are being built in these regions by associating together the outputs of neurons that have different view-dependent responses to the same individual. These findings also provide evidence that one output of the visual system includes representations of what is being seen, in a view-independent way that would be useful for object recognition and for learning associations about objects; and that another output is a view-based representation that would be useful in social interactions to determine whether another individual is looking at one, and for selecting details of motor responses, for which the orientation of the object with respect to the viewer is required (Rolls and Deco 2002, Rolls 2008d).

Further evidence that some neurons in the temporal cortical visual areas have object-based rather than view-based responses comes from a study of a population of neurons that responds to moving faces (Hasselmo, Rolls, Baylis and Nalwa 1989b). For example, four neurons responded vigorously to a head undergoing ventral flexion, irrespective of whether

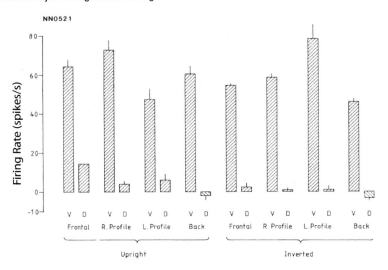

Fig. 25.8 Object-centred encoding: Neuron responding to ventral flexion (V) of the head independently of view even when the head was inverted. Ventral flexion is tilting the head from the full frontal view forwards to face down at 45 degrees. Dorsal flexion is tilting the head until it is looking 45 degrees up. The response is shown as a change in firing rate in spikes/s ± sem from the spontaneous firing rate. (After Hasselmo, Rolls, Baylis and Nalwa 1989b.)

the view of the head was full face, of either profile, or even of the back of the head. These different views could only be specified as equivalent in object-based coordinates. Further, the movement specificity was maintained across inversion, with neurons responding for example to ventral flexion of the head irrespective of whether the head was upright or inverted (see example in Fig. 25.8). In this procedure, retinally encoded or viewer-centered movement vectors are reversed, but the object-based description remains the same.

Also consistent with object-based encoding is the finding of a small number of neurons that respond to images of faces of a given absolute size, irrespective of the retinal image size or distance (Rolls and Baylis 1986).

Neurons with view invariant responses to objects seen naturally by macaques have also been described (Booth and Rolls 1998). The stimuli were presented for 0.5 s on a colour video monitor while the monkey performed a visual fixation task. The stimuli were images of 10 real plastic objects that had been in the monkey's cage for several weeks, to enable him to build view invariant representations of the objects. Control stimuli were views of objects that had never been seen as real objects. The neurons analyzed were in the TE cortex in and close to the ventral lip of the anterior part of the superior temporal sulcus. Many neurons were found that responded to some views of some objects. However, for a smaller number of neurons, the responses occurred only to a subset of the objects (using ensemble encoding), irrespective of the viewing angle. Moreover, the firing of a neuron on any one trial, taken at random and irrespective of the particular view of any one object, provided information about which object had been seen, and this information increased approximately linearly with the number of neurons in the sample (see Fig. 25.9). This is strong quantitative evidence that some neurons in the inferior temporal cortex provide an invariant representation of objects. Moreover, the results of Booth and Rolls (1998) show that the information is available in the firing rates, and has all the desirable properties of distributed representations described above, including exponentially high coding capacity, and rapid speed of readout of the information.

Further evidence consistent with these findings is that some studies have shown that

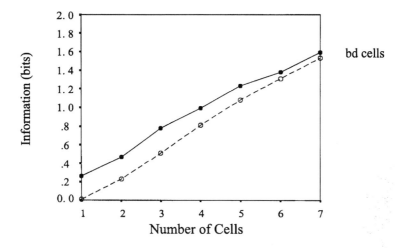

Fig. 25.9 View-independent object encoding: information in a population of different numbers of inferior temporal cortex cells available from a single trial in which one view was shown about which of 10 objects each with 4 different views in the stimulus set had been seen. The solid symbols show the information decoded with a Bayesian probability estimator algorithm, and the open symbols with dot product decoding. (After Booth and Rolls 1998.)

the responses of some visual neurons in the inferior temporal cortex do not depend on the presence or absence of critical features for maximal activation (Perrett, Rolls and Caan 1982, Tanaka 1993, Tanaka 1996). For example, the responses of neuron 4 in Fig. 25.6 responded to several of the features in a face when these features were presented alone (Perrett, Rolls and Caan 1982). In another example, Mikami, Nakamura and Kubota (1994) showed that some TE cells respond to partial views of the same laboratory instrument(s), even when these partial views contain different features. Such functionality is important for object recognition when part of an object is occluded, by for example another object. In a different approach, Logothetis, Pauls, Bulthoff and Poggio (1994) have reported that in monkeys extensively trained (over thousands of trials) to treat different views of computer generated wire-frame 'objects' as the same, a small population of neurons in the inferior temporal cortex did respond to different views of the same wire-frame object (see also Logothetis and Sheinberg (1996)). However, extensive training is not necessary for invariant representations to be formed, and indeed no explicit training in invariant object recognition was given in the experiment by Booth and Rolls (1998), as Rolls' hypothesis (1992a) is that view invariant representations can be learned by associating together the different views of objects as they are moved and inspected naturally in a period that may be in the order of a few seconds. Evidence for this is described in Section 25.2.7.

25.2.7 Learning of new representations in the temporal cortical visual areas

To investigate the idea that visual experience might guide the formation of the responsiveness of neurons so that they provide an economical and ensemble-encoded representation of items actually present in the environment, the responses of inferior temporal cortex face-selective neurons have been analyzed while a set of new faces were shown. Some of the neurons studied in this way altered the relative degree to which they responded to the different members of the

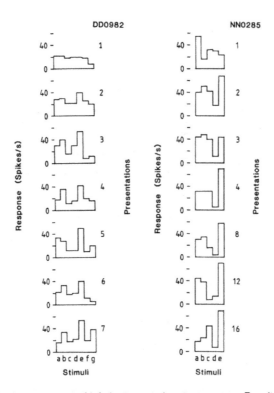

Fig. 25.10 Learning in the responses of inferior temporal cortex neurons. Results for two face-selective neurons are shown. On presentation 1 a set of 7 completely novel faces was shown when recording from neuron DD0982. The firing rates elicited on 1-sec presentations of each stimulus (a-g) are shown. After the first presentation, the set of stimuli was repeated in random sequence for a total of 7 iterations. After the first two presentations, the relative responses of the neuron to the different stimuli settled down to a fairly reliable response profile to the set of stimuli (as shown by statistical analysis). NN0285 – data from a similar experiment with another neuron, and another completely new set of face stimuli. (After Rolls, Baylis, Hasselmo and Nalwa 1989a.)

set of novel faces over the first few (1–2) presentations of the set (Rolls, Baylis, Hasselmo and Nalwa 1989a) (see examples in Fig. 25.10). If in a different experiment a single novel face was introduced when the responses of a neuron to a set of familiar faces were being recorded, the responses to the set of familiar faces were not disrupted, while the responses to the novel face became stable within a few presentations. Alteration of the tuning of individual neurons in this way may result in a good discrimination over the population as a whole of the faces known to the monkey. This evidence is consistent with the categorization being performed by self-organizing competitive neuronal networks, as described in Section B.4. Consistent findings have been described more recently (Freedman 2015).

Further evidence that these neurons can learn new representations very rapidly comes from an experiment in which binarized black and white (two-tone) images of faces that blended with the background were used (see Fig. 25.11, left). These did not activate face-selective neurons. Full grey-scale images of the same photographs were then shown for ten 0.5 s presentations (Fig. 25.11, middle). In a number of cases, if the neuron happened to be responsive to that face, when the binarized version of the same face was shown next (Fig. 25.11, right), the neurons responded to it (Tovee, Rolls and Ramachandran 1996). This is a direct parallel to the same phenomenon that is observed psychophysically, and provides dramatic evidence that

Objects

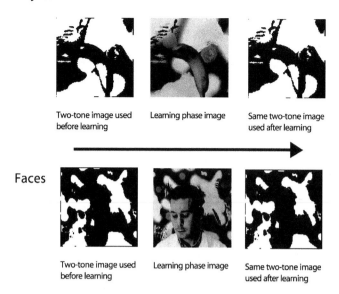

Faces

Fig. 25.11 Images of the type used to investigate rapid learning in the neurophysiological experiments of Tovee, Rolls and Ramachandran (1996) and the PET imaging study of Dolan, Fink, Rolls et al. (1997). When the black and white (two-tone) images at the left are shown, the objects or faces are not generally recognized. After the full grey scale images (middle) have been shown for a few seconds, humans and inferior temporal cortex neurons respond to the faces or objects in the black and white images (right).

these neurons are influenced by only a very few seconds (in this case 5 s) of experience with a visual stimulus. We have shown a neural correlate of this effect using similar stimuli and a similar paradigm in a PET (positron emission tomography) neuroimaging study in humans, with a region showing an effect of the learning found for faces in the right temporal lobe, and for objects in the left temporal lobe (Dolan, Fink, Rolls, Booth, Holmes, Frackowiak and Friston 1997).

Such rapid learning of representations of new objects appears to be a major type of learning in which the temporal cortical areas are involved. Ways in which this learning could occur are considered later in this Chapter. In addition, some of these neurons may be involved in a short term memory for whether a particular familiar visual stimulus (such as a face) has been seen recently. The evidence for this is that some of these neurons respond differently to recently seen stimuli in short term visual memory tasks (Baylis and Rolls 1987, Miller and Desimone 1994, Xiang and Brown 1998). In the inferior temporal visual cortex proper, neurons respond more to novel than to familiar stimuli, but treat the stimuli as novel if more than one other stimuli intervene between the first (novel) and second (familiar) presentations of a particular stimulus (Baylis and Rolls 1987). More ventrally, in what is in or close to the perirhinal cortex, these memory spans may hold for several intervening stimuli in the same task (Xiang and Brown 1998) (see Section 24.2.6). Some neurons in these areas respond more when a sample stimulus reappears in a delayed match to sample task with intervening stimuli (Miller and Desimone 1994), and the basis for this using a short term memory implemented in the prefrontal cortex is described in Section 4.3.1. Neurons in the more ventral (perirhinal) cortical area respond during the delay in a match to sample task with a delay between the sample stimulus and the to-be-matched stimulus (Miyashita 1993, Renart, Parga and Rolls 2000) (see Section 24.2.6).

25.2.8 Distributed encoding

An important question for understanding brain function is whether a particular object (or face) is represented in the brain by the firing of one or a few gnostic (or 'grandmother') cells (Barlow 1972), or whether instead the firing of a group or ensemble of cells each with somewhat different responsiveness provides the representation (see Appendix C for a quantitative analysis of information encoding by inferior temporal cortex and other neurons). Advantages of distributed codes (see Section 8.2, Appendix B, Appendix C, Rolls and Treves (1998), Rolls (2008d), and Rolls and Treves (2011)) include generalization and graceful degradation (fault tolerance), and a potentially very high capacity in the number of stimuli that can be represented (that is exponential growth of capacity with the number of neurons in the representation). If the ensemble encoding is sparse, this provides a good input to an associative memory, for then large numbers of stimuli can be stored (see Appendix B of this book, Chapters 2 and 3 of Rolls and Treves (1998), and Rolls (2008d). We have shown that in the inferior temporal visual cortex and cortex in the anterior part of the superior temporal sulcus (IT/STS), responses of a group of neurons, but not of a single neuron, provide evidence on which face was shown. We showed, for example, that these neurons typically respond with a graded set of firing to different faces, with firing rates from 120 spikes/s to the most effective face, to no response at all to a number of the least effective faces (Baylis, Rolls and Leonard 1985, Rolls and Tovee 1995b, Rolls and Deco 2002). In fact, the firing rate probability distribution of a single neuron to a set of stimuli is approximately exponential (Rolls and Tovee 1995b, Treves, Panzeri, Rolls, Booth and Wakeman 1999, Rolls and Deco 2002, Baddeley, Abbott, Booth, Sengpiel, Freeman, Wakeman and Rolls 1997, Franco, Rolls, Aggelopoulos and Jerez 2007). To provide examples, Fig. 16.15 shows typical firing rate changes of a single neuron on different trials to each of several different faces. This makes it clear that from the firing rate on any one trial, information is available about which stimulus was shown, and that the firing rate is graded, with a different firing rate response of the neuron to each stimulus (see further Section C.3.1).

The distributed nature of the encoding typical for neurons in the inferior temporal visual cortex is illustrated in Fig. 25.12, which shows that temporal cortical neurons typically responded to several members of a set of five faces, with each neuron having a different profile of responses to each face (Baylis, Rolls and Leonard 1985). It would be difficult for most of these single cells to tell which of even five faces, let alone which of hundreds of faces, had been seen. Yet across a population of such neurons, much information about the particular face that has been seen is provided, as shown below.

The single neuron selectivity or sparseness a^s of the activity of inferior temporal cortex neurons was 0.65 over a set of 68 stimuli including 23 faces and 45 non-face natural scenes, and a measure called the response sparseness a_r^s of the representation, in which the spontaneous rate was subtracted from the firing rate to each stimulus so that the responses of the neuron were being assessed, was 0.38 across the same set of stimuli (Rolls and Tovee 1995b). [For the definition of sparseness see Section C.3.1. For binary neurons (firing for example either at a high rate or not at all), the single neuron sparseness is the proportion of stimuli that a single neuron responds to. These definitions, and what is found in cortical neuronal representations, are described further in Sections 8.2, C.3.1 and C.3.2, by Rolls and Treves (2011), and by Franco, Rolls, Aggelopoulos and Jerez (2007).]

It has been possible to apply information theory to show that each neuron conveys on average approximately 0.4 bits of information about which face in a set of 20 faces has been seen (Tovee and Rolls 1995, Tovee, Rolls, Treves and Bellis 1993, Rolls, Treves, Tovee and Panzeri 1997d). If a neuron responded to only one of the faces in the set of 20, then it could convey (if noiseless) 4.6 bits of information about one of the faces (when that face was

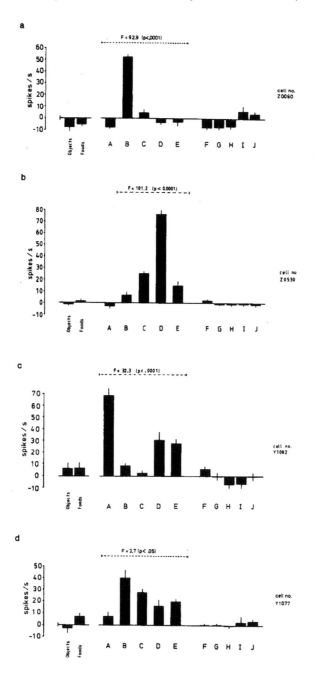

Fig. 25.12 Responses of four different temporal cortex visual neurons to a set of five faces (A–E), and, for comparison, to a wide range of non-face objects and foods. F–J are non-face stimuli. The means and standard errors of the responses computed over 8–10 trials are shown. (Reproduced with permission from Brain Research, 342: 91–102. Selectivity between faces in the responses of a population of neurons in the cortex in the superior temporal sulcus of the monkey. G.C. Baylis, E.T. Rolls, C.M. Leonard. Copyright ©1985 Published by Elsevier B.V.)

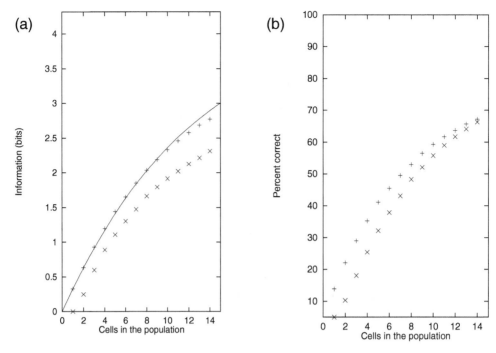

Fig. 25.13 (a) The values for the average information available in the responses of different numbers of these neurons on each trial, about which of a set of 20 face stimuli has been shown. The decoding method was Dot Product (DP, ×) or Probability Estimation (PE, +). The full line indicates the amount of information expected from populations of increasing size, when assuming random correlations within the constraint given by the ceiling (the information in the stimulus set, $I = 4.32$ bits). (b) The percent correct for the corresponding data to those shown in (a). The measurement period was 500 ms. (After Rolls, Treves and Tovee 1997b.)

shown). If, at the other extreme, it responded to half the faces in the set, it would convey 1 bit of information about which face had been seen on any one trial. In fact, the average maximum information about the best stimulus was 1.8 bits of information. This provides good evidence not only that the representation is distributed, but also that it is a sufficiently reliable representation that useful information can be obtained from it (see Section C.3.2).

An important result is that when the information available from a population of neurons about which of 20 faces has been seen is considered, the information increases approximately linearly as the number of cells in the population increases from 1 to 14 (Rolls, Treves and Tovee 1997b, Abbott, Rolls and Tovee 1996) (see Fig. 25.13 and Section C.3.5). Remembering that the information in bits is a logarithmic measure, this shows that the representational capacity of this population of cells increases exponentially (see Fig. 25.14). This is the case both when an optimal, probability estimation, form of decoding of the activity of the neuronal population is used, and also when the neurally plausible dot product type of decoding is used (Fig. 25.13). (The dot product decoding assumes that what reads out the information from the population activity vector is a neuron or a set of neurons that operates just by forming the dot product of the input population vector and its synaptic weight vector – see Rolls, Treves and Tovee (1997b), and Appendix B.) By simulation of further neurons and further stimuli, we have shown that the capacity grows very impressively, approximately as shown in Fig. 25.14 (Abbott, Rolls and Tovee 1996). The result has been replicated with simultaneously recorded neurons (Rolls, Franco, Aggelopoulos and Reece 2003b, Rolls, Aggelopoulos, Franco and

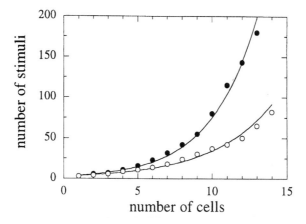

Fig. 25.14 The number of stimuli (in this case from a set of 20 faces) that are encoded in the responses of different numbers of neurons in the temporal lobe visual cortex, based on the results shown in Fig. 25.13. The solid circles are the result of the raw calculation, and the open circles correspond to the cross-validated case. (After Rolls, Treves and Tovee 1997b; and Abbott, Rolls and Tovee 1996.)

Treves 2004) (see further Section C.3.5). This result is exactly what would be hoped for from a distributed representation. This result is not what would be expected for local encoding, for which the number of stimuli that could be encoded would increase linearly with the number of cells. (Even if the grandmother cells were noisy, adding more replicates to increase reliability would not lead to more than a linear increase in the number of stimuli that can be encoded as a function of the number of cells.) Moreover, the encoding in the inferior temporal visual cortex about objects remains based on the spike count from each neuron, and not on the relative time of firing of each neuron or stimulus-dependent synchronization, when analysed with simultaneous single neuron recording (Rolls, Franco, Aggelopoulos and Reece 2003b, Rolls, Aggelopoulos, Franco and Treves 2004, Franco, Rolls, Aggelopoulos and Treves 2004) even in natural scenes when an attentional task is being performed (Aggelopoulos, Franco and Rolls 2005). Further, much of the information is available in short times of e.g. 20 or 50 ms (Tovee and Rolls 1995, Rolls, Franco, Aggelopoulos and Jerez 2006b), so that the receiving neuron does not need to integrate over a long time period to estimate a firing rate.

These findings provide very firm evidence that the encoding built at the end of the visual system is distributed, and that part of the power of this representation is that by receiving inputs from relatively small numbers of such neurons, neurons at the next stage of processing (for example in memory structures such as the hippocampus, amygdala, and orbitofrontal cortex) would obtain information about which of a very great number of objects or faces had been shown (Rolls and Treves 2011).

In this sense, the inferior temporal visual cortex provides a representation of objects and faces, in which information about which object or face is shown is made explicit in the firing of the neurons in such a way that the information can be read off very simply by memory systems such as the orbitofrontal cortex, amygdala, and perirhinal cortex / hippocampal systems. The information can be read off using dot product decoding, that is by using a synaptically weighted sum of inputs from inferior temporal cortex neurons (see further Appendix C and Section 24.2.6). Examples of some of the types of objects and faces that are encoded in this way by anterior inferior temporal cortex neurons are shown in Figs. 24.11 and C.7.

This representational capacity of neuronal populations has fundamental implications for the connectivity of the brain, for it shows that neurons need not have hundreds of thousands or millions of inputs to have available to them information about what is represented in another population of cells, but that instead the real numbers of perhaps 8,000–10,000 synapses per neuron would be adequate for them to receive considerable information from the several different sources between which this set of synapses is allocated.

It may be noted that it is unlikely that there are further processing areas beyond those described where ensemble coding changes into grandmother cell encoding. Anatomically, there does not appear to be a whole further set of visual processing areas present in the brain; and outputs from the temporal lobe visual areas such as those described are taken directly to limbic and related regions such as the amygdala and orbitofrontal cortex, and via the perirhinal and entorhinal cortex to the hippocampus (see Chapter 24, Rolls (2000a), Rolls (2008d), and Rolls (2011d)). Indeed, tracing this pathway onwards, we have found a population of neurons with face-selective responses in the amygdala, and in the majority of these neurons, different responses occur to different faces, with ensemble (not local) coding still being present (Leonard, Rolls, Wilson and Baylis 1985). The amygdala, in turn, projects to another structure that may be important in other behavioural responses to faces, the ventral striatum, and comparable neurons have also been found in the ventral striatum (Williams, Rolls, Leonard and Stern 1993). We have also recorded from face-responding neurons in the part of the orbitofrontal cortex that receives from the IT/STS cortex, and have found that the encoding there is also not local but is distributed (Rolls, Critchley, Browning and Inoue 2006a).

25.2.9 Face expression, gesture, and view represented in a population of neurons in the cortex in the superior temporal sulcus

In addition to the population of neurons that code for face identity, which tend to have object-based representations and are in areas TEa and TEm on the ventral bank of the superior temporal sulcus, there is a separate population in the cortex in the superior temporal sulcus (e.g. area TPO) that conveys information about facial expression (Hasselmo, Rolls and Baylis 1989a) (see e.g. Fig. 25.15). Some of the neurons in this region tend to have view-based representations (so that information is conveyed for example about whether the face is looking at one, or is looking away), and might respond to moving faces, and to facial gesture (Hasselmo, Rolls, Baylis and Nalwa 1989b).

Thus information in cortical areas that project to the amygdala and orbitofrontal cortex is about face identity, and about face expression and gesture (Rolls, Critchley, Browning and Inoue 2006a, Rolls 2011d). Both types of information are important in social and emotional responses to other primates (including humans), which must be based on who the individual is as well as on the face expression or gesture being made (Rolls 2014a). One output from the amygdala for this information is probably via the ventral striatum, for a small population of neurons has been found in the ventral striatum with responses selective for faces (Rolls and Williams 1987a, Williams, Rolls, Leonard and Stern 1993).

25.2.10 Specialized regions in the temporal cortical visual areas

As we have just seen, some neurons respond to face identity, and others to face expression (Hasselmo, Rolls and Baylis 1989a, Rolls 2011d, Rolls 2014a). The neurons responsive to expression were found primarily in the cortex in the superior temporal sulcus, while the neurons responsive to identity were found in the inferior temporal gyrus.

A further way in which some of these neurons in the cortex in the superior temporal sulcus may be involved in social interactions is that some of them respond to gestures, for example

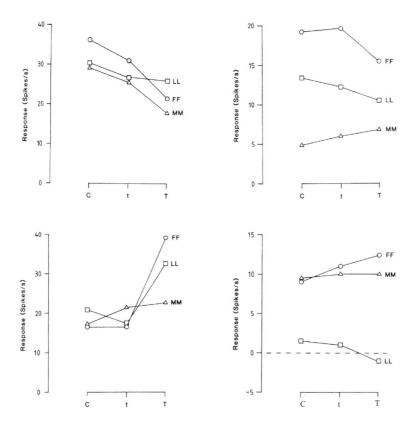

Fig. 25.15 There is a population of neurons in the cortex in the superior temporal sulcus with responses tuned to respond differently to different face expressions. The cells in the two left panels did not discriminate between individuals (faces MM, FF, and MM), but did discriminate between different expressions on the faces of those individuals (C, calm expression; t, mild threat; T, strong threat). In contrast, the cells in the right two panels responded differently to different individuals, and did not discriminate between different expressions. The neurons that discriminated between expressions were found mainly in the cortex in the fundus of the superior temporal sulcus; the neurons that discriminated between identity were in contrast found mainly in the cortex in the lateral part of the ventral lip of the superior temporal sulcus (areas TEa and TEm). (Reproduced with permission from Behavioural Brain Research. 32: 203–218, The role of expression and identity in the face-selective responses of neurons in the temporal visual cortex of the monkey. Michael E. Hasselmo, Edmund T. Rolls, Gordon C. Baylis. Copyright ©1989 Published by Elsevier B.V.)

to a face undergoing ventral flexion (i.e. the head bending at the neck forwards and down) (Perrett, Smith, Potter, Mistlin, Head, Milner and Jeeves 1985b, Hasselmo, Rolls, Baylis and Nalwa 1989b). The interpretation of these neurons as being useful for social interactions is that in some cases these neurons respond not only to ventral head flexion, but also to the eyes lowering and the eyelids closing (Hasselmo, Rolls, Baylis and Nalwa 1989b). These movements (turning the head away, breaking eye contact, and eyelid lowering) often occur together when a monkey is breaking social contact with another, and neurons that respond to these components could be built by associative synaptic modification. It is also important when decoding facial expression to retain some information about the direction of the head relative to the observer, for this is very important in determining whether a threat is being made in your direction. The presence of view–dependent, head and body gesture (Hasselmo, Rolls, Baylis and Nalwa 1989b), and eye gaze (Perrett, Smith, Potter, Mistlin, Head, Milner and Jeeves 1985b), representations in some of these cortical regions where face expression is represented is consistent with this requirement. In contrast, the TE areas (more ventral, mainly in the macaque inferior temporal gyrus), in which neurons tuned to face identity (Hasselmo, Rolls and Baylis 1989a) and with view–independent responses (Hasselmo, Rolls, Baylis and Nalwa 1989b) are more likely to be found, may be more related to an object–based representation of identity. Of course, for appropriate social and emotional responses, both types of subsystem would be important, for it is necessary to know both the direction of a social gesture, and the identity of the individual, in order to make the correct social or emotional response (Rolls 2014a).

Further evidence for specialization of function in the different architectonically defined areas of the temporal cortex (Seltzer and Pandya 1978) (see Fig. 25.2) was found by Baylis, Rolls and Leonard (1987). Areas TPO, PGa and IPa are multimodal, with neurons that respond to visual, auditory and/or somatosensory inputs. The more ventral areas in the inferior temporal gyrus (areas TE3, TE2, TE1, TEa and TEm) are primarily unimodal visual areas. Areas in the cortex in the anterior and dorsal part of the superior temporal sulcus (e.g. TPO, IPa and IPg) have neurons specialized for the analysis of moving visual stimuli. Neurons responsive primarily to faces are found more frequently in areas TPO, TEa and TEm, where they comprise approximately 20% of the visual neurons responsive to stationary stimuli, in contrast to the other temporal cortical areas in which they comprise 4–10%. The stimuli that activate other cells in these TE regions include simple visual patterns such as gratings, and combinations of simple stimulus features (Gross, Desimone, Albright and Schwartz 1985, Baylis, Rolls and Leonard 1987, Tanaka, Saito, Fukada and Moriya 1990). If patches are identified by fMRI, the proportion of neurons tuned to for example faces may be high (Tsao et al. 2006, Freiwald, Tsao and Livingstone 2009, Tsao 2014). Due to the fact that face-selective neurons have a wide distribution (Baylis, Rolls and Leonard 1987), and also occur in different patches (Tsao et al. 2006, Tsao 2014), it might be expected that only large lesions, or lesions that interrupt outputs of these visual areas, would produce readily apparent face-processing deficits.

Another specialization is that areas TEa and TEm, which receive inter alia from the cortex in the intraparietal sulcus, have neurons that are tuned to binocular disparity, so that information derived from stereopsis about the 3D structure of objects is represented in the inferior temporal cortical visual areas (Janssen, Vogels and Orban 1999, Janssen, Vogels and Orban 2000). Interestingly, these neurons respond only when the black and white regions in the two eyes are correlated (that is, when a white patch in one eye corresponds to a white patch in the other eye) (Janssen, Vogels, Liu and Orban 2003). This corresponds to what we see perceptually and consciously. In contrast, in V1, MT and MST, depth-tuned neurons can respond to anticorrelated depth images, that is where white in one eye corresponds to black in the other (Parker, Cumming and Dodd 2000, DeAngelis, Cumming and Newsome 2000, Parker 2007), and this may be suitable for eye movement control, but much less for shape and object discrimination and identification. It may be expected that other depth cues, such as perspective, surface shading, and occlusion, affect the response properties of some neurons in inferior temporal cortex visual areas. Binocular disparity, and information from these other depth cues, may be used to compute the absolute size of objects, which is represented independently of the distance of the object for a small proportion of inferior temporal cortex neurons (Rolls and Baylis 1986). Knowing the absolute size of an object is useful evidence to include in the identification of an object. Although cues from binocular disparity can thus drive some temporal cortex visual neurons, and there is a small proportion of inferior temporal neurons that respond better to real faces and objects than to 2D representations on a monitor, it is found that the majority of TE neurons respond as well or almost as well to 2D images on a video monitor as to real faces or objects (Perrett, Rolls and Caan 1982). Moreover, the tuning of inferior temporal cortex neurons to images of faces or objects on a video monitor is similar to that to real objects (personal observations of E. T. Rolls).

Neuroimaging data, while not being able to address the details of what is encoded in a brain area or of how it is encoded, does provide evidence consistent with the neurophysiology that there are different face processing systems in the human brain, and different processing subsystems for faces, objects, moving objects, and scenes (Spiridon, Fischl and Kanwisher 2006, Grill-Spector and Malach 2004, Epstein and Kanwisher 1998, Morin, Hadj-Bouziane, Stokes, Ungerleider and Bell 2015, Nasr, Liu, Devaney, Yue, Rajimehr, Ungerleider and Tootell 2011, Weiner and Grill-Spector 2015, Deen, Koldewyn, Kanwisher and Saxe 2015). For example, Kanwisher, McDermott and Chun (1997) and Ishai, Ungerleider, Martin, Schouten and Haxby (1999) have shown activation by faces of an area in the fusiform gyrus; Hoffman and Haxby (2000) have shown that distinct areas are activated by eye gaze and face identity; Dolan, Fink, Rolls, Booth, Holmes, Frackowiak and Friston (1997) have shown that a fusiform gyrus area becomes activated after humans learn to identify faces in complex scenes; and the amygdala (Morris, Fritch, Perrett, Rowland, Young, Calder and Dolan 1996) and orbitofrontal cortex (Blair, Morris, Frith, Perrett and Dolan 1999, Rolls 2014a) may become activated particularly by certain face expressions. Consistent with the neurophysiology described below and by Baylis, Rolls and Leonard (1987), human fMRI studies are able to detect small patches of non-face and other types of responsiveness in areas that appeared to respond primarily to one category such as faces (Grill-Spector, Sayres and Ress 2006, Haxby 2006).

Different inferior temporal cortex neurons in macaques not only provide different types of information about different aspects of faces, as just described, but also respond differently to different *categories* of visual stimuli, with for example some neurons conveying information primarily about faces, and others about objects (Rolls and Tovee 1995b, Rolls, Treves, Tovee and Panzeri 1997d, Booth and Rolls 1998, Rolls and Treves 2011, Rolls 2012c) (see Appendix C). In fact, when recording in the inferior temporal cortex, one finds small clustered groups of neurons, with neurons within a cluster responding to somewhat similar attributes of stimuli, and different clusters responding to different categories or types of visual stimuli. For example, within the set of neurons responding to moving faces, there are some neuronal clusters that respond to for example ventral and dorsal flexion of the head on the body; others that respond to axial rotation of the head; others that respond to the sight of mouth movements; and others that respond to the gaze direction in a face being viewed (Hasselmo, Rolls, Baylis and Nalwa 1989b, Perrett, Smith, Potter, Mistlin, Head, Milner and Jeeves 1985b). Within the domain of objects, some neuronal clusters respond based on the shapes of the object (e.g. responding to elongated and not round objects), others respond based on the texture of the object being viewed, and others are sensitive to what the object is doing (Hasselmo, Rolls, Baylis and Nalwa 1989b, Perrett, Smith, Mistlin, Chitty, Head, Potter, Broennimann, Milner and Jeeves 1985a). Within a cluster, each neuron is tuned differently, and may combine at least partly with information predominant in other clusters, so that at least some of the neurons in a cluster can become quite selective between different objects, with the preference for individual objects showing the usual exponential firing rate distribution (see Figs. C.4 and C.6). Using optical imaging, Tanaka and colleagues (Tanaka 1996) have seen comparable evidence for different localized clusters of neuronal activation produced by different types or categories of stimuli.

The principle that Rolls proposes underlies this local clustering is the representation of the high dimensional space of objects and their features into a two-dimensional cortical sheet using the local self-organizing mapping principles described in Section B.4.6. In this situation, these principles produce maps that have many fractures, but nevertheless include local clusters of similar neurons. Exactly this clustering can be produced in the model of visual object recognition, VisNet, described in Section 25.5 by replacing the lateral inhibition filter shown in Fig. 25.18 with a difference of Gaussians filter of the type illustrated in Fig. B.20. Such a difference of Gaussians filter would reflect the effects of short range excitatory (recurrent collateral) connections between cortical pyramidal cells, and longer range inhibition produced through inhibitory interneurons. These local clusters minimize the wiring length between neurons, if the computations involve exchange of information within neurons of a similar general type, as will usually be the case (see Section B.4.6). Minimizing wiring length is crucial in keeping brain size relatively low (see Section B.4.6). Another useful property of such self-organizing maps is that they encourage distributed representations in which semi-continuous similarity functions are mapped. This is potentially useful in building representations that then generalize usefully when new stimuli are shown between representations that are already set up continuously in the map.

Another fundamental contribution and key aspect of neocortical design facilitated by the short-range recurrent excitatory connections is that the attractors that are formed are local. This is fundamentally important, for if the neocortex had long-range connectivity, it would tend to be able to form only as many attractor states as there are connections per neocortical neuron, which would be a severe limit on cortical memory capacity (O'Kane and Treves 1992) (see Sections 1.10.4 and 4.10).

It is consistent with this general conceptual background that Kreiman, Koch and Freid (2000) and others (see Fried, Rutishauser, Cerf and Kreiman (2014) and Rolls (2015d)) have described some neurons in the human temporal lobe that seem to respond to categories of object. This is consistent with the principles just described, although in humans backprojections from language or other cognitive areas concerned for example with tool use might also influence the categories represented in high order cortical areas, as described in Sections B.4.5 and 11, and by Farah, Meyer and McMullen (1996) and Farah (2000).

25.3   Approaches to invariant object recognition

A goal of this book is to describe some of the principles of operation of the cerebral cortex. Some of the principles of operation of the cortical visual systems are described in this section (25.3), and include a hierarchical feed-forward series of competitive networks using convergence from stage to stage; and the use of a modified Hebb synaptic learning rule that incorporates a short-term memory trace of previous neuronal activity to help learn the invariant properties of objects from the temporo-spatial statistics produced by the normal viewing of objects (Rolls 1992a, Wallis and Rolls 1997, Rolls and Milward 2000, Stringer and Rolls 2000, Rolls and Stringer 2001a, Elliffe, Rolls and Stringer 2002, Stringer and Rolls 2002, Rolls and Deco 2002, Deco and Rolls 2004, Rolls and Stringer 2006, Rolls 2008d, Rolls 2012c).

We start by emphasizing that generalization to different positions, sizes, views etc. of an object is not a simple property of one-layer neural networks. Although neural networks do generalize well, the type of generalization they show naturally is to vectors which have a high dot product or correlation with what they have already learned. To make this clear, Fig. 25.16 is a reminder that the activation h_i of each neuron is computed as

$$h_i = \sum_j x_j w_{ij} \qquad (25.1)$$

where the sum is over the C input axons, indexed by j. Now consider translation (or shift) of the input pattern vector by one position. The dot product will now drop to a low level, and the neuron will not respond, even though it is the same pattern, just shifted by one location. This makes the point that special processes are needed to compute invariant representations.

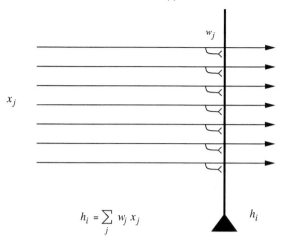

$$h_i = \sum_j w_j x_j$$

Fig. 25.16 A neuron that computes a dot product of the input pattern with its synaptic weight vector generalizes well to other patterns based on their similarity measured in terms of dot product or correlation, but shows no translation (or size, etc.) invariance.

Network approaches to such invariant pattern recognition are described in this chapter, and the nature of the problem to be solved is described further in Section B.4.7 and Fig. B.23 on page 750. Once an invariant representation has been computed by a sensory system, it is in a form that is suitable for presentation to a pattern association or autoassociation neural network (see Appendix B).

A number of different computational approaches that have been taken both in artificial vision systems and as suggestions for how the brain performs invariant object recognition are described in this section (25.3). This places in context the approach that appears to be taken in the brain and that forms the basis for VisNet described in Section 25.5.

25.3.1 Feature spaces

One very simple possibility for performing object classification is based on feature spaces, which amount to lists of the features that are present in a particular object. The features might consist of textures, colours, areas, ratios of length to width, etc. The spatial arrangement of the features is not taken into account. If n different properties are used to characterize an object, each viewed object is represented by a set of n real numbers. It then becomes possible to represent an object by a point R^n in an n-dimensional space (where R is the resolution of the real numbers used). Such schemes have been investigated (Tou and Gonzalez 1974, Gibson 1950, Gibson 1979, Bolles and Cain 1982, Mundy and Zisserman 1992, Selfridge 1959, Mel 1997), but, because the relative positions of the different parts are not implemented in the object recognition scheme, are not sensitive to spatial jumbling of the features. For example, if the features consisted of nose, mouth, and eyes, such a system would respond to faces with jumbled arrangements of the eyes, nose, and mouth, which does not match human vision, nor the responses of macaque inferior temporal cortex neurons, which are sensitive to the spatial arrangement of the features in a face (Rolls, Tovee, Purcell, Stewart and Azzopardi 1994b). Similarly, such an object recognition system might not distinguish a normal car from a car with the back wheels removed and placed on the roof. Such systems do not therefore perform shape recognition (where shape implies something about the spatial arrangement of features within an object, see further Ullman (1996)), and something more is needed, and is implemented in the primate visual system. However, I note that the features that are present in objects, e.g. a furry texture, are useful to incorporate in object recognition systems, and the brain may well use, and the model VisNet in principle can use, evidence from which features are present in an object as part of the evidence for identification of a particular object. I note that the features might consist also of for example the pattern of movement that is characteristic of a particular object (such as a buzzing fly), and might use this as part of the input to final object identification.

The capacity to use shape in invariant object recognition is fundamental to primate vision, but may not be used or fully implemented on the visual systems of some other animals with less developed visual systems. For example, pigeons may correctly identify pictures containing people, a particular person, trees, pigeons etc, but may fail to distinguish a figure from a scrambled version of a figure (Herrnstein 1984, Cerella 1986). Thus their object recognition may be based more on a collection of parts than on a direct comparison of complete figures in which the relative positions of the parts are important. Even if the details of the conclusions reached from this research are revised (Wasserman, Kirkpatrick-Steger and Biederman 1998), it nevertheless does appear that at least some birds may use computationally simpler methods than those needed for invariant shape recognition. For example, it may be that when some birds are trained to discriminate between images in a large set of pictures, they tend to rely on some chance detail of each picture (such as a spot appearing by mistake on the picture), rather than on recognition of the shapes of the object in the picture (Watanabe, Lea and Dittrich 1993).

25.3.2 Structural descriptions and syntactic pattern recognition

A second approach to object recognition is to decompose the object or image into parts, and to then produce a structural description of the relations between the parts. The underlying assumption is that it is easier to capture object invariances at a level where parts have been identified. This is the type of scheme for which Marr and Nishihara (1978) and Marr (1982) opted. The particular scheme (Binford 1981) they adopted consists of generalized cones, series of which can be linked together to form structural descriptions of some, especially animate, stimuli (see Fig. 25.17). Such schemes assume that there is a 3D internal model (structural description) of each object. Perception of the object consists of parsing or segmenting the scene into objects, and then into parts, then producing a structural description of the object, and then testing whether this structural description matches that of any known object stored in the system. Other examples of structural description schemes include those of Winston (1975), Sutherland (1968), and Milner (1974). The relations in the structural description may need to be quite complicated, for example 'connected together', 'inside of', 'larger than' etc.

Perhaps the most developed model of this type is the recognition by components (RBC) model of Biederman (1987), implemented in a computational model by Hummel and Biederman (1992). His small set (less than 50) of primitive parts named 'geons' include simple 3D shapes such as boxes, cylinders and wedges. Objects are described by a syntactically linked list of the relations between each of the geons of which they are composed. Describing a table in this way (as a flat top supported by three or four legs) seems quite economical. The emphasis more recently has been on the non-accidental properties of objects, which are maintained when objects transform (Kim and Biederman 2012). Other schemes use 2D surface patches as their primitives (Dane and Bajcsy 1982, Brady, Ponce, Yuille and Asada 1985, Faugeras and Hebert 1986, Faugeras 1993). When 3D objects are being recognized, the implication is that the structural description is a 3D description. This is in contrast to feature hierarchical systems, in which recognition of a 3D object from any view might be accomplished by storing a set of associated 2D views (see below, Section 25.3.5).

There are a number of difficulties with schemes based on structural descriptions, some

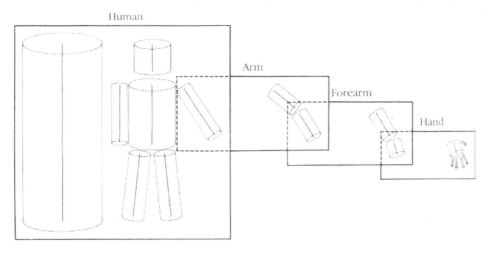

Fig. 25.17 A 3D structural description of an object based on generalized cone parts. Each box corresponds to a 3D model, with its model axis on the left side of the box and the arrangement of its component axes on the right. In addition, some component axes have 3D models associated with them, as indicated by the way the boxes overlap. (After Marr and Nishihara 1978.)

general, and some with particular reference to the potential difficulty of their implementation in the brain. First, it is not always easy to decompose the object into its separate parts, which must be performed before the structural description can be produced. For example, it may be difficult to produce a structural description of a cat curled up asleep from separately identifiable parts. Identification of each of the parts is also frequently very difficult when 3D objects are seen from different viewing angles, as key parts may be invisible, occluded, or highly distorted. This is particularly likely to be difficult in 3D shape perception. It appears that being committed to producing a correct description of the parts before other processes can operate is making too strong a commitment early on in the recognition process.

A second difficulty is that many objects or animals that can be correctly recognized have rather similar structural descriptions. For example, the structural descriptions of many four-legged animals are rather similar. Rather more than a structural description seems necessary to identify many objects and animals.

A third difficulty, which applies especially to biological systems, is the difficulty of implementing the syntax needed to hold the structural description as a 3D model of the object, of producing a syntactic structural description on the fly (in real time, and with potentially great flexibility of the possible arrangement of the parts), and of matching the syntactic description of the object in the image to all the stored representations in order to find a match. An example of a structural description for a limb might be body > thigh > shin > foot > toes. In this description > means 'is linked to', and this link must be between the correct pair of descriptors. If we had just a set of parts, without the syntactic or relational linking, then there would be no way of knowing whether the toes are attached to the foot or to the body. In fact, worse than this, there would be no evidence about what was related to what, just a set of parts. Such syntactical relations are difficult to implement in neuronal networks, because if the representations of all the features or parts just mentioned were active simultaneously, how would the spatial relations between the features also be encoded? (How would it be apparent just from the firing of neurons that the toes were linked to the rest of the foot but not to the body?) It would be extremely difficult to implement this 'on the fly' syntactic binding in a biologically plausible network (though cf. Hummel and Biederman (1992)), and the only suggested mechanism for flexible syntactic binding, temporal synchronization of the firing of different neurons, is not well supported as a quantitatively important mechanisms for information encoding in the ventral visual system, and would have major difficulties in implementing correct, relational, syntactic binding (see Sections 25.5.5.1 and C.3.7).

A fourth difficulty of the structural description approach is that segmentation into objects must occur effectively before object recognition, so that the linked structural description list can be of one object. Given the difficulty of segmenting objects in typical natural cluttered scenes (Ullman 1996), and the compounding problem of overlap of parts of objects by other objects, segmentation as a first necessary stage of object recognition adds another major difficulty for structural description approaches.

A fifth difficulty is that metric information, such as the relative size of the parts that are linked syntactically, needs to be specified in the structural description (Stankiewicz and Hummel 1994), which complicates the parts that have to be syntactically linked.

It is because of these difficulties that even in artificial vision systems implemented on computers, where almost unlimited syntactic binding can easily be implemented, the structural description approach to object recognition has not yet succeeded in producing a scheme which actually works in more than an environment in which the types of objects are limited, and the world is far from the natural world, consisting for example of 2D scenes (Mundy and Zisserman 1992).

Although object recognition in the brain is unlikely to be based on the structural description approach, for the reasons given above, and the fact that the evidence described in this chapter supports a feature hierarchy rather than the structural description implementation in the brain, it is certainly the case that humans can provide verbal, syntactic, descriptions of objects in terms of the relations of their parts, and that this is often a useful type of description. Humans may therefore it is suggested supplement a feature hierarchical object recognition system built into their ventral visual system with the additional ability to use the type of syntax that is necessary for language to provide another level of description of objects. This is of course useful in, for example, engineering applications.

25.3.3 Template matching and the alignment approach

Another approach is template matching, comparing the image on the retina with a stored image or picture of an object. This is conceptually simple, but there are in practice major problems. One major problem is how to align the image on the retina with the stored images, so that all possible images on the retina can be compared with the stored template or templates of each object.

The basic idea of the alignment approach (Ullman 1996) is to compensate for the transformations separating the viewed object and the corresponding stored model, and then compare them. For example, the image and the stored model may be similar, except for a difference in size. Scaling one of them will remove this discrepancy and improve the match between them. For a 2D world, the possible transforms are translation (shift), scaling, and rotation. Given for example an input letter of the alphabet to recognize, the system might, after segmentation (itself a very difficult process if performed independently of (prior to) object recognition), compensate for translation by computing the centre of mass of the object, and shifting the character to a 'canonical location'. Scale might be compensated for by calculating the convex hull (the smallest envelope surrounding the object), and then scaling the image. Of course how the shift and scaling would be accomplished is itself a difficult point – easy to perform on a computer using matrix multiplication as in simple computer graphics, but not the sort of computation that could be performed easily or accurately by any biologically plausible network. Compensating for rotation is even more difficult (Ullman 1996). All this has to happen before the segmented canonical representation of the object is compared to the stored object templates with the same canonical representation. The system of course becomes vastly more complicated when the recognition must be performed of 3D objects seen in a 3D world, for now the particular view of an object after segmentation must be placed into a canonical form, regardless of which view, or how much of any view, may be seen in a natural scene with occluding contours. However, this process is helped, at least in computers that can perform high precision matrix multiplication, by the fact that (for many continuous transforms such as 3D rotation, translation, and scaling) all the possible views of an object transforming in 3D space can be expressed as the linear combination of other views of the same object (see Chapter 5 of Ullman (1996); Koenderink and van Doorn (1991); and Koenderink (1990)).

This alignment approach is the main theme of the book by Ullman (1996), and there are a number of computer implementations (Lowe 1985, Grimson 1990, Huttenlocher and Ullman 1990, Shashua 1995). However, as noted above, it seems unlikely that the brain is able to perform the high precision calculations needed to perform the transforms required to align any view of a 3D object with some canonical template representation. For this reason, and because the approach also relies on segmentation of the object in the scene before the template alignment algorithms can start, and because key features may need to be correctly identified to be used in the alignment (Edelman 1999), this approach is not considered further here.

However, it is certainly the case that humans are very sensitive to precise feature configurations (Ullman, Assif, Fetaya and Harari 2016).

We may note here in passing that some animals with a less computationally developed visual system appear to attempt to solve the alignment problem by actively moving their heads or eyes to see what template fits, rather than starting with an image on the eye and attempting to transform it into canonical coordinates. This 'active vision' approach used for example by some invertebrates has been described by Land (1999) and Land and Collett (1997).

25.3.4 Invertible networks that can reconstruct their inputs

Hinton, Dayan, Frey and Neal (1995) and Hinton and Ghahramani (1997) (see Section B.14) have argued that cortical computation is invertible, so that for example the forward transform of visual information from V1 to higher areas loses no information, and there can be a backward transform from the higher areas to V1. A comparison of the reconstructed representation in V1 with the actual image from the world might in principle be used to correct all the synaptic weights between the two (in both the forward and the reverse directions), in such a way that there are no errors in the transform. This suggested reconstruction scheme would seem to involve non-local synaptic weight correction (though see Section B.14 for a suggested, although still biologically implausible, neural implementation), or other biologically implausible operations. The scheme also does not seem to provide an account for why or how the responses of inferior temporal cortex neurons become the way they are (providing information about which object is seen relatively independently of position on the retina, size, or view). The whole forward transform performed in the brain seems to lose much of the information about the size, position and view of the object, as it is evidence about which object is present invariant of its size, view, etc. that is useful to the stages of processing about objects that follow. Because of these difficulties, and because the backprojections are needed for processes such as recall (see Chapter 11 and Section 24.3.7), this invertible network approach is not considered further in this book.

In the context of recall, if the visual system were to perform a reconstruction in V1 of a visual scene from what is represented in the inferior temporal visual cortex, then it might be supposed that remembered visual scenes might be as information-rich (and subjectively as full of rich detail) as seeing the real thing. This is not the case for most humans, and indeed this point suggests that at least what reaches consciousness from the inferior temporal visual cortex (which is activated during the recall of visual memories) is the identity of the object (as made explicit in the firing rate of the neurons), and not the low-level details of the exact place, size, and view of the object in the recalled scene, even though, according to the reconstruction argument, that information should be present in the inferior temporal visual cortex.

25.3.5 Feature hierarchies and 2D view-based object recognition

The feature hierarchy approach to object recognition is described in Section 2.2.1.

After the immediately following description of early models of a feature hierarchy approach implemented in the Cognitron and Neocognitron, we turn for the remainder of this chapter to analyses of how a feature hierarchy approach to invariant visual object recognition might be implemented in the brain, and how key computational issues could be solved by such a system. The analyses are developed and tested with a model, VisNet, which will shortly be described. Much of the data we have on the operation of the high order visual cortical areas (see Section 25.2) (Rolls 2007e, Rolls 2008d, Rolls 2011d, Rolls 2012c) suggest that they implement a feature hierarchy approach to visual object recognition, as is made evident in the remainder of this chapter.

25.3.5.1 The Cognitron and Neocognitron

An early computational model of a hierarchical feature-based approach to object recognition, joining other early discussions of this approach (Selfridge 1959, Sutherland 1968, Barlow 1972, Milner 1974), was proposed by Fukushima (1975, 1980, 1989, 1989, 1991). His model used two types of cell within each layer to approach the problem of invariant representations. In each layer, a set of 'simple cells', with defined position, orientation, etc. sensitivity for the stimuli to which they responded, was followed by a set of 'complex cells', which generalized a little over position, orientation, etc. This simple cell – complex cell pairing within each layer provided some invariance. When a neuron in the network using competitive learning with its stimulus set, which was typically letters on a 16×16 pixel array, learned that a particular feature combination had occurred, that type of feature analyzer was replicated in a non-local manner throughout the layer, to provide further translation invariance. Invariant representations were thus learned in a different way from VisNet. Up to eight layers were used. The network could learn to differentiate letters, even with some translation, scaling, or distortion. Although internally it is organized and learns very differently to VisNet, it is an independent example of the fact that useful invariant pattern recognition can be performed by multilayer hierarchical networks. A major biological implausibility of the system is that once one neuron within a layer learned, other similar neurons were set up throughout the layer by a non-local process. A second biological limitation was that no learning rule or self-organizing process was specified as to how the complex cells can provide translation invariant representations of simple cell responses – this was simply handwired. Solutions to both these issues are provided by VisNet.

25.4 Hypotheses about the computational mechanisms in the visual cortex for object recognition

The neurophysiological findings described in Section 25.2, and wider considerations on the possible computational properties of the cerebral cortex (Rolls 1989b, Rolls 1989f, Rolls

1992a, Rolls 1994a, Rolls 1995a, Rolls 1997c, Rolls 2000a, Rolls and Treves 1998, Rolls and Deco 2002, Rolls 2008d), lead to the following outline working hypotheses on object recognition by visual cortical mechanisms (see Rolls (1992a) and Rolls (2012c)). The principles underlying the processing of faces and other objects may be similar, but more neurons may become allocated to represent different aspects of faces because of the need to recognize the faces of many different individuals, that is to identify many individuals within the category faces.

Cortical visual processing for object recognition is considered to be organized as a set of hierarchically connected cortical regions consisting at least of V1, V2, V4, posterior inferior temporal cortex (TEO), inferior temporal cortex (e.g. TE3, TEa and TEm), and anterior temporal cortical areas (e.g. TE2 and TE1). (This stream of processing has many connections with a set of cortical areas in the anterior part of the superior temporal sulcus, including area TPO.) There is convergence from each small part of a region to the succeeding region (or layer in the hierarchy) in such a way that the receptive field sizes of neurons (e.g. 1 degree near the fovea in V1) become larger by a factor of approximately 2.5 with each succeeding stage (and the typical parafoveal receptive field sizes found would not be inconsistent with the calculated approximations of e.g. 8 degrees in V4, 20 degrees in TEO, and 50 degrees in the inferior temporal cortex (Boussaoud, Desimone and Ungerleider 1991)) (see Fig. 2.1 on page 41). Such zones of convergence would overlap continuously with each other (see Fig. 2.1). This connectivity would be part of the architecture by which translation invariant representations are computed.

Each layer is considered to act partly as a set of local self-organizing competitive neuronal networks with overlapping inputs. (The region within which competition would be implemented would depend on the spatial properties of inhibitory interneurons, and might operate over distances of 1–2 mm in the cortex.) These competitive nets operate by a single set of forward inputs leading to (typically non-linear, e.g. sigmoid) activation of output neurons; of competition between the output neurons mediated by a set of feedback inhibitory interneurons which receive from many of the principal (in the cortex, pyramidal) cells in the net and project back (via inhibitory interneurons) to many of the principal cells and serve to decrease the firing rates of the less active neurons relative to the rates of the more active neurons; and then of synaptic modification by a modified Hebb rule, such that synapses to strongly activated output neurons from active input axons strengthen, and from inactive input axons weaken (see Section B.4). A biologically plausible form of this learning rule that operates well in such networks is

$$\delta w_{ij} = \alpha y_i (x_j - w_{ij}) \tag{25.2}$$

where α is a learning rate constant, and x_j and w_{ij} are in appropriate units (see Section B.4). Such competitive networks operate to detect correlations between the activity of the input neurons, and to allocate output neurons to respond to each cluster of such correlated inputs. These networks thus act as categorizers. In relation to visual information processing, they would remove redundancy from the input representation, and would develop low entropy representations of the information (cf. Barlow (1985), Barlow, Kaushal and Mitchison (1989)). Such competitive nets are biologically plausible, in that they utilize Hebb-modifiable forward excitatory connections, with competitive inhibition mediated by cortical inhibitory neurons. The competitive scheme I suggest would not result in the formation of 'winner-take-all' or 'grandmother' cells, but would instead result in a small ensemble of active neurons representing each input (Rolls 1989b, Rolls 1989f, Rolls and Treves 1998, Rolls 2012c) (see Section B.4). The scheme has the advantages that the output neurons learn better to distribute themselves between the input patterns (cf. Bennett (1990)), and that the sparse representations formed have utility in maximizing the number of memories that can be stored when, towards the end of the visual system, the visual representation of objects is interfaced to associative memory (Rolls 1989b, Rolls 1989f, Rolls and Treves 1998)[22].

Translation invariance would be computed in such a system by utilizing competitive learning to detect regularities in inputs when real objects are translated in the physical world. The hypothesis is that because objects have continuous properties in space and time in the world, an object at one place on the retina might activate feature analyzers at the next stage of cortical processing, and when the object was translated to a nearby position, because this would occur in a short period (e.g. 0.5 s), the membrane of the postsynaptic neuron would still be in its 'Hebb-modifiable' state (caused for example by calcium entry as a result of the voltage-dependent activation of NMDA receptors), and the presynaptic afferents activated with the object in its new position would thus become strengthened on the still-activated postsynaptic neuron. It is suggested that the short temporal window (e.g. 0.5 s) of Hebb-modifiability helps neurons to learn the statistics of objects moving in the physical world, and at the same time to form different representations of different feature combinations or objects, as these are physically discontinuous and present less regular correlations to the visual system. Földiák (1991) has proposed computing an average activation of the postsynaptic neuron to assist with the same problem. One idea here is that the temporal properties of the biologically implemented learning mechanism are such that it is well suited to detecting the relevant continuities in the world of real objects. Another suggestion is that a memory trace for what has been seen in the last 300 ms appears to be implemented by a mechanism as simple as continued firing of inferior temporal neurons after the stimulus has disappeared, as has been found in masking experiments (Rolls and Tovee 1994, Rolls, Tovee, Purcell, Stewart and Azzopardi 1994b, Rolls, Tovee and Panzeri 1999b, Rolls 2003, Rolls 2008d).

I also suggest that other invariances, for example size, spatial frequency, and rotation invariance, could be learned by a comparable process. (Early processing in V1 which enables different neurons to represent inputs at different spatial scales would allow combinations of the outputs of such neurons to be formed at later stages. Scale invariance would then result from detecting at a later stage which neurons are almost conjunctively active as the size of an object alters.) It is suggested that this process takes place at each stage of the multiple-layer cortical processing hierarchy, so that invariances are learned first over small regions of space, and then over successively larger regions. This limits the size of the connection space within which correlations must be sought.

Increasing complexity of representations could also be built in such a multiple layer hierarchy by similar mechanisms. At each stage or layer the self-organizing competitive nets would result in combinations of inputs becoming the effective stimuli for neurons. In order to avoid the combinatorial explosion, it is proposed, following Feldman (1985), that low-order combinations of inputs would be what is learned by each neuron. (Each input would not be represented by activity in a single input axon, but instead by activity in a set of active input axons.) Evidence consistent with this suggestion that neurons are responding to combinations of a few variables represented at the preceding stage of cortical processing is that some

---

[22] In that each neuron has graded responses centred about an optimal input, the proposal has some of the advantages with respect to hypersurface reconstruction described by Poggio and Girosi (1990a). However, the system I propose learns differently, in that instead of using perhaps non-biologically plausible algorithms to optimally locate the centres of the receptive fields of the neurons, the neurons use graded competition to spread themselves throughout the input space, depending on the statistics of the inputs received, and perhaps with some guidance from backprojections (see below). In addition, the competitive nets I propose use as a distance function the dot product between the input vector to a neuron and its synaptic weight vector, whereas radial basis function networks use a Gaussian measure of distance (see Section B.4.8). Both systems benefit from the finite width of the response region of each neuron which tapers from a maximum, and is important for enabling the system to generalize smoothly from the examples with which it has learned (cf. Poggio and Girosi (1990b), Poggio and Girosi (1990a)), to help the system to respond for example with the correct invariances as described below.

neurons in V1 respond to combinations of bars or edges (see Section 2.5 of Rolls and Deco (2002); Sillito, Grieve, Jones, Cudeiro and Davis (1995); and Shevelev, Novikova, Lazareva, Tikhomirov and Sharaev (1995)); V2 and V4 respond to end-stopped lines, to angles formed by a combination of lines, to acute curvature, to tongues flanked by inhibitory subregions, and to combinations of colours (Hegde and Van Essen 2000, Rolls and Deco 2002, Ito and Komatsu 2004, Carlson, Rasquinha, Zhang and Connor 2011, Kourtzi and Connor 2011); in posterior inferior temporal cortex to stimuli which may require two or more simple features to be present (Tanaka, Saito, Fukada and Moriya 1990); and in the temporal cortical face processing areas to images that require the presence of several features in a face (such as eyes, hair, and mouth) in order to respond (Perrett, Rolls and Caan 1982, Yamane, Kaji and Kawano 1988) (see Chapter 5 of Rolls and Deco (2002), Rolls (2007e), Rolls (2008d), Rolls (2011d), and Fig. 25.6). (Precursor cells to face-responsive neurons might, it is suggested, respond to combinations of the outputs of the neurons in V1 that are activated by faces, and might be found in areas such as V4.)

It is an important part of this suggestion that some local spatial information would be inherent in the features which were being combined. For example, cells might not respond to the combination of an edge and a small circle unless they were in the correct spatial relation to each other. (This is in fact consistent with the data of Tanaka, Saito, Fukada and Moriya (1990), and with our data on face neurons, in that some faces neurons require the face features to be in the correct spatial configuration, and not jumbled, Rolls, Tovee, Purcell, Stewart and Azzopardi (1994b).) The local spatial information in the features being combined would ensure that the representation at the next level would contain some information about the (local) arrangement of features. Further low-order combinations of such neurons at the next stage would include sufficient local spatial information so that an arbitrary spatial arrangement of the same features would not activate the same neuron, and this is the proposed, and limited, solution which this mechanism would provide for the feature binding problem (Elliffe, Rolls and Stringer 2002). By this stage of processing a view-dependent representation of objects suitable for view-dependent processes such as behavioural responses to face expression and gesture would be available.

It is suggested that view-independent representations could be formed by the same type of computation, operating to combine a limited set of views of objects. The plausibility of providing view-independent recognition of objects by combining a set of different views of objects has been proposed by a number of investigators (Koenderink and Van Doorn 1979, Poggio and Edelman 1990, Logothetis, Pauls, Bulthoff and Poggio 1994, Ullman 1996). Consistent with the suggestion that the view-independent representations are formed by combining view-dependent representations in the primate visual system, is the fact that in the temporal cortical areas, neurons with view-independent representations of faces are present in the same cortical areas as neurons with view-dependent representations (from which the view-independent neurons could receive inputs) (Hasselmo, Rolls, Baylis and Nalwa 1989b, Perrett, Smith, Potter, Mistlin, Head, Milner and Jeeves 1985b, Booth and Rolls 1998). This solution to 'object-based' representations is very different from that traditionally proposed for artificial vision systems, in which the coordinates in 3D space of objects are stored in a database, and general-purpose algorithms operate on these to perform transforms such as translation, rotation, and scale change in 3D space (e.g. Marr (1982)). In the present, much more limited but more biologically plausible scheme, the representation would be suitable for recognition of an object, and for linking associative memories to objects, but would be less good for making actions in 3D space to particular parts of, or inside, objects, as the 3D coordinates of each part of the object would not be explicitly available. It is therefore proposed that visual fixation is used to locate in foveal vision part of an object to which movements must be made, and that local disparity and other measurements of depth then provide sufficient information for the motor system to make actions relative to the small part of space in which a local, view-dependent, representation of depth would be provided (cf. Ballard (1990)).

The computational processes proposed above operate by an unsupervised learning mechanism, which utilizes statistical regularities in the physical environment to enable representations to be built. In some cases it may be advantageous to utilize some form of mild teaching input to the visual system, to enable it to learn for example that rather similar visual inputs have very different consequences in the world, so that different representations of them should be built. In other cases, it might be helpful to bring representations together, if they have identical consequences, in order to use storage capacity efficiently. It is proposed elsewhere (Rolls 1989b, Rolls 1989f, Rolls and Treves 1998, Rolls 2008d) (see Chapter 11) that the backprojections from each adjacent cortical region in the hierarchy (and from the amygdala and hippocampus to higher regions of the visual system) play such a role by providing guidance to the competitive networks suggested above to be important in each cortical area. This guidance, and also the capability for recall, are it is suggested implemented by Hebb-modifiable connections from the backprojecting neurons to the principal (pyramidal) neurons of the competitive networks in the preceding stages (Rolls 1989b, Rolls 1989f, Rolls and Treves 1998, Rolls 2008d) (see Section B.4).

The computational processes outlined above use sparse distributed coding with relatively finely tuned neurons with a graded response region centred about an optimal response achieved when the input stimulus matches the synaptic weight vector on a neuron. The distributed nature of the coding but with fine tuning would help to limit the combinatorial explosion, to keep the number of neurons within the biological range. The graded response region would be crucial in enabling the system to generalize correctly to solve for example the invariances. However, such a system would need many neurons, each with considerable learning capacity, to solve visual perception in this way. This is fully consistent with the large number of neurons in the visual system, and with the large number of, probably modifiable, synapses on each neuron (e.g. 10,000). Further, the fact that many neurons are tuned in different ways to faces is consistent with the fact that in such a computational system, many neurons would need to be sensitive (in different ways) to faces, in order to allow recognition of many individual faces when all share a number of common properties.

25.5 The feature hierarchy approach to invariant object recognition: computational issues

The feature hierarchy approach to invariant object recognition was introduced in Sections 2.2.1 and 25.3.5, and advantages and disadvantages of it were discussed. Hypotheses about how object recognition could be implemented in the brain which are consistent with much of the neurophysiology discussed in Section 25.2 and elsewhere (Rolls and Deco 2002, Rolls 2007e, Rolls 2008d, Rolls 2011d, Rolls 2012c) were set out in Section 25.4. These hypotheses effectively incorporate a feature hierarchy system while encompassing much of the neurophysiological evidence. In this Section (25.5), we consider the computational issues that arise in such feature hierarchy systems, and in the brain systems that implement visual object recognition. The issues are considered with the help of a particular model, VisNet, which requires precise specification of the hypotheses, and at the same time enables them to be explored and tested numerically and quantitatively. However, we emphasize that the issues to be covered in Section 25.5 are key and major computational issues for architectures of this feature hierarchical type (Rolls and Stringer 2006), and are very relevant to understanding how invariant object recognition is implemented in the cerebral cortex.

A feature of the research with VisNet is that most of the investigations have been performed to account for particular neurophysiological findings, and the results obtained with VisNet have then been compared with the neuronal responses found in the cerebral cortex, in particular in the inferior temporal visual cortex (Rolls 2012c, Robinson and Rolls 2015).

VisNet is a model of invariant object recognition based on Rolls' (1992a) hypotheses. It is a computer simulation that allows hypotheses to be tested and developed about how multilayer hierarchical networks of the type believed to be implemented in the visual cortical pathways operate (Rolls 2008d, Rolls 2012c). The architecture captures a number of aspects of the architecture of the visual cortical pathways, and is described next. The model of course, as with all models, requires precise specification of what is to be implemented, and at the same time involves specified simplifications of the real architecture, as investigations of the fundamental aspects of the information processing being performed are more tractable in a simplified and at the same time quantitatively specified model. First the architecture of the model is described, and this is followed by descriptions of key issues in such multilayer feature hierarchical models, such as the issue of feature binding, the optimal form of training rule for the whole system to self-organize, the operation of the network in natural environments and when objects are partly occluded, how outputs about individual objects can be read out from the network, and the capacity of the system.

25.5.1 The architecture of VisNet

Fundamental elements of Rolls' (1992a) theory for how cortical networks might implement invariant object recognition are described in Section 25.4. They provide the basis for the design of VisNet, and can be summarized as:

- A series of competitive networks, organized in hierarchical layers, exhibiting mutual inhibition over a short range within each layer. These networks allow combinations of features or inputs occurring in a given spatial arrangement to be learned by neurons, ensuring that higher order spatial properties of the input stimuli are represented in the network.
- A convergent series of connections from a localized population of cells in preceding layers to each cell of the following layer, thus allowing the receptive field size of cells to increase through the visual processing areas or layers.
- A modified Hebb-like learning rule incorporating a temporal trace of each cell's previous activity, which, it is suggested, will enable the neurons to learn transform invariances.

The first two elements of Rolls' theory are used to constrain the general architecture of a network model, VisNet, of the processes just described that is intended to learn invariant representations of objects. The simulation results described in this chapter using VisNet show that invariant representations can be learned by the architecture. It is moreover shown that successful learning depends crucially on the use of the modified Hebb rule. The general architecture simulated in VisNet, and the way in which it allows natural images to be used as stimuli, has been chosen to enable some comparisons of neuronal responses in the network and in the brain to similar stimuli to be made.

25.5.1.1 The trace rule

The learning rule implemented in the VisNet simulations utilizes the spatio-temporal constraints placed upon the behaviour of 'real-world' objects to learn about natural object transformations. By presenting consistent sequences of transforming objects the cells in the network can learn to respond to the same object through all of its naturally transformed states, as described by Földiák (1991), Rolls (1992a), Wallis, Rolls and Földiák (1993), and Wallis and Rolls (1997). The learning rule incorporates a decaying trace of previous cell activity and is henceforth referred to simply as the 'trace' learning rule. The learning paradigm we describe here is intended in principle to enable learning of any of the transforms tolerated by inferior temporal cortex neurons, including position, size, view, lighting, and spatial frequency (Rolls 1992a, Rolls 1994a, Rolls 1995a, Rolls 1997c, Rolls 2000a, Rolls and Deco 2002, Rolls 2007e, Rolls 2007c, Rolls 2008b, Rolls 2008d, Rolls 2012c, Rolls and Webb 2014, Webb and Rolls 2014, Robinson and Rolls 2015).

To clarify the reasoning behind this point, consider the situation in which a single neuron is strongly activated by a stimulus forming part of a real world object. The trace of this neuron's activation will then gradually decay over a time period in the order of 0.5 s. If, during this limited time window, the net is presented with a transformed version of the original stimulus then not only will the initially active afferent synapses modify onto the neuron, but so also will the synapses activated by the transformed version of this stimulus. In this way the cell will learn to respond to either appearance of the original stimulus. Making such associations works in practice because it is very likely that within short time periods different aspects of the same object will be being inspected. The cell will not, however, tend to make spurious links across stimuli that are part of different objects because of the unlikelihood in the real world of one object consistently following another.

Various biological bases for this temporal trace have been advanced[23]:

- The persistent firing of neurons for as long as 100–400 ms observed after presentations of stimuli for 16 ms (Rolls and Tovee 1994) could provide a time window within which to associate subsequent images. Maintained activity may potentially be implemented by recurrent connections between as well as within cortical areas (Rolls and Treves 1998, Rolls and Deco 2002) (see Section 4.3.1)[24].

- The binding period of glutamate in the NMDA channels, which may last for 100 ms or more, may implement a trace rule by producing a narrow time window over which the *average* activity at each presynaptic site affects learning (Rolls 1992a, Rhodes 1992, Földiák 1992, Spruston, Jonas and Sakmann 1995, Hestrin, Sah and Nicoll 1990).

- Chemicals such as nitric oxide may be released during high neural activity and gradually decay in concentration over a short time window during which learning could be enhanced (Földiák 1992, Montague, Gally and Edelman 1991).

The trace update rule used in the baseline simulations of VisNet (Wallis and Rolls 1997) is equivalent to both Földiák's used in the context of translation invariance and to the earlier rule of Sutton and Barto (1981) explored in the context of modelling the temporal properties of classical conditioning, and can be summarized as follows:

$$\delta w_j = \alpha \overline{y}^\tau x_j \tag{25.3}$$

where

---

[23] The precise mechanisms involved may alter the precise form of the trace rule which should be used. Földiák (1992) describes an alternative trace rule which models individual NMDA channels. Equally, a trace implemented by extended cell firing should be reflected in representing the trace as an external firing rate, rather than an internal signal.

[24] The prolonged firing of inferior temporal cortex neurons during memory delay periods of several seconds, and associative links reported to develop between stimuli presented several seconds apart (Miyashita 1988) are on too long a time scale to be immediately relevant to the present theory. In fact, associations between visual events occurring several seconds apart would, under *normal* environmental conditions, be detrimental to the operation of a network of the type described here, because they would probably arise from different objects. In contrast, the system described benefits from associations between visual events which occur close in time (typically within 1 s), as they are likely to be from the same object.

$$\overline{y}^\tau = (1 - \eta)y^\tau + \eta\overline{y}^{\tau-1} \tag{25.4}$$

and

x_j: j^{th} input to the neuron.

$\overline{y}^\tau$:   Trace value of the output of the neuron at time step τ.

w_j: Synaptic weight between j^{th} input and the neuron.

y: Output from the neuron.

α: Learning rate. Annealed between unity and zero.

η: Trace value. The optimal value varies with presentation sequence length.

To bound the growth of each neuron's synaptic weight vector, $\mathbf{w}_i$ for the ith neuron, its length is explicitly normalized (a method similarly employed by von der Malsburg (1973) which is commonly used in competitive networks, see Section B.4). An alternative, more biologically relevant implementation, using a local weight bounding operation which utilizes a form of heterosynaptic long-term depression (see Section 1.5), has in part been explored using a version of the Oja (1982) rule (see Wallis and Rolls (1997)).

25.5.1.2 The network implemented in VisNet

The network itself is designed as a series of hierarchical, convergent, competitive networks, in accordance with the hypotheses advanced above. The actual network consists of a series of four layers, constructed such that the convergence of information from the most disparate parts of the network's input layer can potentially influence firing in a single neuron in the final layer – see Fig. 2.1. This corresponds to the scheme described by many researchers (Van Essen, Anderson and Felleman 1992, Rolls 1992a, for example) as present in the primate visual system – see Fig. 2.1. The forward connections to a cell in one layer are derived from a topologically related and confined region of the preceding layer. The choice of whether a connection between neurons in adjacent layers exists or not is based upon a Gaussian distribution of connection probabilities which roll off radially from the focal point of connections for each neuron. (A minor extra constraint precludes the repeated connection of any pair of cells.) In particular, the forward connections to a cell in one layer come from a small region of the preceding layer defined by the radius in Table 25.1 which will contain approximately 67% of the connections from the preceding layer. Table 25.1 shows the dimensions for VisNetL, the system we have been using since 2008, which is a (16x) larger version of the version of VisNet than used in most of our earlier investigations, which utilized 32x32 neurons per layer.

Table 25.1 VisNet dimensions

	Dimensions	# Connections	Radius
Layer 4	128x128	100	48
Layer 3	128x128	100	36
Layer 2	128x128	100	24
Layer 1	128x128	272	24
Input layer	256x256x32	–	–

Figure 2.1 shows the general convergent network architecture used. Localization and limitation of connectivity in the network is intended to mimic cortical connectivity, partially because of the clear retention of retinal topology through regions of visual cortex. This

architecture also encourages the gradual combination of features from layer to layer which has relevance to the binding problem, as described in Section 25.5.5[25].

25.5.1.3 Competition and lateral inhibition

In order to act as a competitive network some form of mutual inhibition is required within each layer, which should help to ensure that all stimuli presented are evenly represented by the neurons in each layer. This is implemented in VisNet by a form of lateral inhibition. The idea behind the lateral inhibition, apart from this being a property of cortical architecture in the brain, was to prevent too many neurons that received inputs from a similar part of the preceding layer responding to the same activity patterns. The purpose of the lateral inhibition was to ensure that different receiving neurons coded for different inputs. This is important in reducing redundancy (see Section B.4). The lateral inhibition is conceived as operating within a radius that was similar to that of the region within which a neuron received converging inputs from the preceding layer (because activity in one zone of topologically organized processing within a layer should not inhibit processing in another zone in the same layer, concerned perhaps with another part of the image)[26].

The lateral inhibition and contrast enhancement just described is actually implemented in VisNet2 (Rolls and Milward 2000) in two stages, to produce filtering of the type illustrated in Fig. 25.18. This lateral inhibition is implemented by convolving the activation of the neurons in a layer with a spatial filter, I, where δ controls the contrast and σ controls the width, and a and b index the distance away from the centre of the filter

$$
I_{a,b} = \begin{cases} -\delta e^{-\frac{a^2+b^2}{\sigma^2}} & \text{if } a \neq 0 \text{ or } b \neq 0, \\ 1 - \sum\limits_{a \neq 0, b \neq 0} I_{a,b} & \text{if } a = 0 \text{ and } b = 0. \end{cases} \tag{25.5}
$$

This is a filter that leaves the average activity unchanged. A modified version of this filter designed as a difference of Gaussians with the same inhibition but shorter range local excitation is being tested to investigate whether the self-organizing maps that this promotes (Section B.4.6) helps the system to provide some continuity in the representations formed. The concept

---

[25] Modelling topological constraints in connectivity leads to an issue concerning neurons at the edges of the network layers. In principle these neurons may either receive no input from beyond the edge of the preceding layer, or have their connections repeatedly sample neurons at the edge of the previous layer. In practice either solution is liable to introduce artificial weighting on the few active inputs at the edge and hence cause the edge to have unwanted influence over the development of the network as a whole. In the real brain such edge-effects would be naturally smoothed by the transition of the locus of cellular input from the fovea to the lower acuity periphery of the visual field. However, it poses a problem here because we are in effect only simulating the small high acuity foveal portion of the visual field in our simulations. As an alternative to the former solutions Wallis and Rolls (1997) elected to form the connections into a toroid, such that connections wrap back onto the network from opposite sides. This wrapping happens at all four layers of the network, and in the way an image on the 'retina' is mapped to the input filters. This solution has the advantage of making all of the boundaries effectively invisible to the network. Further, this procedure does not itself introduce problems into evaluation of the network for the problems set, as many of the critical comparisons in VisNet involve comparisons between a network with the same architecture trained with the trace rule, or with the Hebb rule, or not trained at all. In practice, it is shown below that only the network trained with the trace rule solves the problem of forming invariant representations.

[26] Although the extent of the lateral inhibition actually investigated by Wallis and Rolls (1997) in VisNet operated over adjacent pixels, the lateral inhibition introduced by Rolls and Milward (2000) in what they named VisNet2 and which has been used in subsequent simulations operates over a larger region, set within a layer to approximately half of the radius of convergence from the preceding layer. Indeed, Rolls and Milward (2000) showed in a problem in which invariant representations over 49 locations were being used with a 17 face test set, that the best performance was with intermediate range lateral inhibition, using the parameters for σ shown in Table 25.3. These values of σ set the lateral inhibition radius within a layer to be approximately half that of the spread of the excitatory connections from the preceding layer.

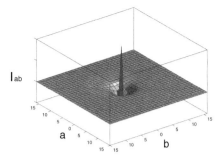

Fig. 25.18 Contrast-enhancing filter, which has the effect of local lateral inhibition. The parameters δ and σ are variables used in equation 25.5 to modify the amount and extent of inhibition respectively.

is that this may help the system to code efficiently for large numbers of untrained stimuli that fall between trained stimuli in similarity space.

The second stage involves contrast enhancement. In VisNet (Wallis and Rolls 1997), this was implemented by raising the neuronal activations to a fixed power and normalizing the resulting firing within a layer to have an average firing rate equal to 1.0. In VisNet2 (Rolls and Milward 2000) and in subsequent simulations a more biologically plausible form of the activation function, a sigmoid, was used:

$$y = f^{\text{sigmoid}}(r) = \frac{1}{1 + e^{-2\beta(r-\alpha)}} \tag{25.6}$$

where r is the activation (or firing rate) of the neuron after the lateral inhibition, y is the firing rate after the contrast enhancement produced by the activation function, and β is the slope or gain and α is the threshold or bias of the activation function. The sigmoid bounds the firing rate between 0 and 1 so global normalization is not required. The slope and threshold are held constant within each layer. The slope is constant throughout training, whereas the threshold is used to control the sparseness of firing rates within each layer. The (population) sparseness of the firing within a layer is defined (Rolls and Treves 1998) as:

$$a = \frac{(\sum_i y_i/n)^2}{\sum_i y_i^2/n} \tag{25.7}$$

where n is the number of neurons in the layer. To set the sparseness to a given value, e.g. 5%, the threshold is set to the value of the 95th percentile point of the activations within the layer. (Unless otherwise stated here, the neurons used the sigmoid activation function as just described.)

In most simulations with VisNet2 and later, the sigmoid activation function was used with parameters (selected after a number of optimization runs) as shown in Table 25.2.

Table 25.2 Sigmoid parameters for the runs with 25 locations by Rolls and Milward (2000)

Layer	1	2	3	4
Percentile	99.2	98	88	91
Slope β	190	40	75	26

In addition, the lateral inhibition parameters normally used in VisNet2 simulations are as shown in Table 25.3[27].

Table 25.3 Lateral inhibition parameters for the 25-location runs

Layer	1	2	3	4
Radius, σ	1.38	2.7	4.0	6.0
Contrast, δ	1.5	1.5	1.6	1.4

25.5.1.4 The input to VisNet

VisNet is provided with a set of input filters which can be applied to an image to produce inputs to the network which correspond to those provided by simple cells in visual cortical area 1 (V1). The purpose of this is to enable within VisNet the more complicated response properties of cells between V1 and the inferior temporal cortex (IT) to be investigated, using as inputs natural stimuli such as those that could be applied to the retina of the real visual system. This is to facilitate comparisons between the activity of neurons in VisNet and those in the real visual system, to the same stimuli. In VisNet no attempt is made to train the response properties of simple cells, but instead we start with a defined series of filters to perform fixed feature extraction to a level equivalent to that of simple cells in V1, as have other researchers in the field (Hummel and Biederman 1992, Buhmann, Lange, von der Malsburg, Vorbrüggen and Würtz 1991, Fukushima 1980), because we wish to simulate the more complicated response properties of cells between V1 and the inferior temporal cortex (IT). In many papers before 2012 we used difference of Gaussian filters, but more recently we have used Gabor filters to simulate V1 similar to those described and used by Deco and Rolls (2004).

The Gabor filters are described next. Following Daugman (1988) the receptive fields of the simple cell-like input neurons are modelled by 2D-Gabor functions. The Gabor receptive fields have five degrees of freedom given essentially by the product of an elliptical Gaussian and a complex plane wave. The first two degrees of freedom are the 2D-locations of the receptive field's centre; the third is the size of the receptive field; the fourth is the orientation of the boundaries separating excitatory and inhibitory regions; and the fifth is the symmetry. This fifth degree of freedom is given in the standard Gabor transform by the real and imaginary part, i.e by the phase of the complex function representing it, whereas in a biological context this can be done by combining pairs of neurons with even and odd receptive fields. This design is supported by the experimental work of Pollen and Ronner (1981), who found simple cells in quadrature-phase pairs. Even more, Daugman (1988) proposed that an ensemble of simple cells is best modelled as a family of 2D-Gabor wavelets sampling the frequency domain in a log-polar manner as a function of eccentricity. Experimental neurophysiological evidence constrains the relation between the free parameters that define a 2D-Gabor receptive field (De Valois and De Valois 1988). There are three constraints fixing the relation between the width, height, orientation, and spatial frequency (Lee 1996). The first constraint posits that the aspect ratio of the elliptical Gaussian envelope is 2:1. The second constraint postulates that the plane wave tends to have its propagating direction along the short axis of the elliptical Gaussian. The third constraint assumes that the half-amplitude bandwidth of the frequency response is about 1 to 1.5 octaves along the optimal orientation. Further, we assume that the mean is zero in order to have an admissible wavelet basis (Lee 1996).

---

[27]Where a power activation function was used in the simulations of Wallis and Rolls (1997), the power for layer 1 was 6, and for the other layers was 2.

In more detail, the Gabor filters are constructed as follows (Deco and Rolls 2004). We consider a pixelized grey-scale image given by a $N \times N$ matrix $\Gamma_{ij}^{\text{orig}}$. The subindices ij denote the spatial position of the pixel. Each pixel value is given a grey level brightness value coded in a scale between 0 (black) and 255 (white). The first step in the preprocessing consists of removing the DC component of the image (i.e. the mean value of the grey-scale intensity of the pixels). (The equivalent in the brain is the low-pass filtering performed by the retinal ganglion cells and lateral geniculate cells. The visual representation in the LGN is essentially a contrast invariant pixel representation of the image, i.e. each neuron encodes the relative brightness value at one location in visual space referred to the mean value of the image brightness.) We denote this contrast-invariant LGN representation by the $N \times N$ matrix Γ_{ij} defined by the equation

$$\Gamma_{ij} = \Gamma_{ij}^{\text{orig}} - \frac{1}{N^2} \sum_{i=1}^{N} \sum_{j=1}^{N} \Gamma_{ij}^{\text{orig}}. \tag{25.8}$$

Feedforward connections to a layer of V1 neurons perform the extraction of simple features like bars at different locations, orientations and sizes. Realistic receptive fields for V1 neurons that extract these simple features can be represented by 2D-Gabor wavelets. Lee (1996) derived a family of discretized 2D-Gabor wavelets that satisfy the wavelet theory and the neurophysiological constraints for simple cells mentioned above. They are given by an expression of the form

$$G_{pqkl}(x, y) = a^{-k} \Psi_{\Theta_l}(a^{-k}(x - 2p), a^{-k}(y - 2q)) \tag{25.9}$$

where

$$\Psi_{\Theta_l} = \Psi(x \cos(l\Theta_0) + y \sin(l\Theta_0), -x \sin(l\Theta_0) + y \cos(l\Theta_0)), \tag{25.10}$$

and the mother wavelet is given by

$$\Psi(x, y) = \frac{1}{\sqrt{2\pi}} e^{-\frac{1}{8}(4x^2 + y^2)} [e^{i\kappa x} - e^{-\frac{\kappa^2}{2}}]. \tag{25.11}$$

In the above equations $\Theta_0 = \pi/L$ denotes the step size of each angular rotation; l the index of rotation corresponding to the preferred orientation $\Theta_l = l\pi/L$; k denotes the octave; and the indices pq the position of the receptive field centre at $c_x = p$ and $c_y = q$. In this form, the receptive fields at all levels cover the spatial domain in the same way, i.e. by always overlapping the receptive fields in the same fashion. In the model we use $a = 2$, $b = 1$ and $\kappa = \pi$ corresponding to a spatial frequency bandwidth of one octave. We used symmetric filters with the angular spacing between the different orientations set to 45 degrees; and with 4 filter frequencies spaced one octave apart starting with 0.5 cycles per pixel, and with the sampling from the spatial frequencies set as shown in Table 25.4. Each individual filter is tuned to spatial frequency (0.0625 to 0.5 cycles / pixel over four octaves); orientation ($0°$ to $135°$ in steps of $45°$); and sign (± 1).

Table 25.4 VisNet layer 1 connectivity. The frequency is in cycles per pixel

Frequency	0.5	0.25	0.125	0.0625
# Connections	201	50	13	8

Of the 272 layer 1 connections, the number to each group is as shown in Table 25.4.

Cells of layer 1 of VisNet receive a topologically consistent, localized, random selection of the filter responses in the input layer, under the constraint that each cell samples every filter

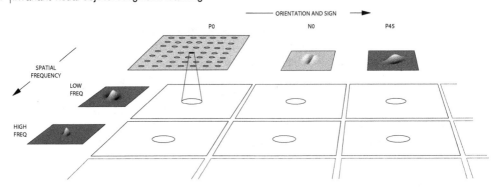

Fig. 25.19 The filter sampling paradigm. Here each square represents the retinal image presented to the network after being filtered by a difference of gaussian filter of the appropriate orientation sign and frequency. The circles represent the consistent retinotopic coordinates used to provide input to a layer 1 cell. The filters double in spatial frequency towards the reader. Left to right the orientation tuning increases from $0°$ in steps of $45°$, with segregated pairs of positive (P) and negative (N) filter responses.

spatial frequency and receives a constant number of inputs. Figure 25.19 shows pictorially the general filter sampling paradigm, and Fig. 25.20 the typical connectivity to a layer 1 cell from the filters of the input layer. The blank squares indicate that no connection exists between the layer 1 cell chosen and the filters of that particular orientation, sign, and spatial frequency.

25.5.1.5 Measures for network performance

A neuron can be said to have learnt an invariant representation if it discriminates one set of stimuli from another set, across all transformations. For example, a neuron's response is translation invariant if its response to one set of stimuli irrespective of presentation is consistently higher than for all other stimuli irrespective of presentation location. Note that we state 'set of stimuli' since neurons in the inferior temporal cortex are not generally selective for a single stimulus but rather a subpopulation of stimuli (Baylis, Rolls and Leonard 1985, Abbott, Rolls and Tovee 1996, Rolls, Treves and Tovee 1997b, Rolls and Treves 1998, Rolls and Deco 2002, Rolls 2007e, Franco, Rolls, Aggelopoulos and Jerez 2007, Rolls and Treves 2011) (see Appendix C). The measure of network performance used in VisNet1 (Wallis and Rolls 1997), the 'Fisher metric' (referred to in some figure labels as the Discrimination Factor), reflects how well a neuron discriminates between stimuli, compared to how well it discriminates between different locations (or more generally the images used rather than the objects, each of which is represented by a set of images, over which invariant stimulus or object representations must be learned). The Fisher measure is very similar to taking the ratio of the two F values in a two-way ANOVA, where one factor is the stimulus shown, and the other factor is the position in which a stimulus is shown. The measure takes a value greater than 1.0 if a neuron has more different responses to the stimuli than to the locations. That is, values greater than 1 indicate invariant representations when this measure is used in the following figures. Further details of how the measure is calculated are given by Wallis and Rolls (1997).

Measures of network performance based on information theory and similar to those used in the analysis of the firing of real neurons in the brain (see Appendix C) were introduced by Rolls and Milward (2000) for VisNet2, and are used in later papers. A single cell information measure was introduced which is the maximum amount of information the cell has about any one stimulus / object independently of which transform (e.g. position on the retina) is shown. Because the competitive algorithm used in VisNet tends to produce local representations (in which single cells become tuned to one stimulus or object), this information measure can

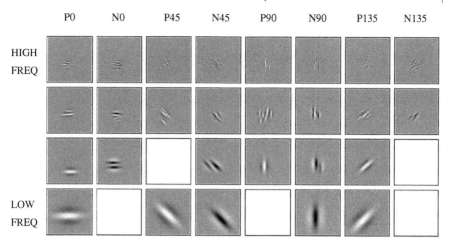

Fig. 25.20 Typical connectivity before training between a single cell in the first layer of the network and the input layer, represented by plotting the receptive fields of every input layer cell connected to the particular layer 1 cell. Separate input layer cells have activity that represents a positive (P) or negative (N) output from the bank of filters which have different orientations in degrees (the columns) and different spatial frequencies (the rows). Here the overall receptive field of the layer 1 cell is centred just below the centre-point of the retina. The connection scheme allows for relatively fewer connections to lower frequency cells than to high frequency cells in order to cover a similar region of the input at each frequency. A blank square indicates that there is no connection to the layer 1 neuron from an input neuron with that particular filter type.

approach $\log_2 N_S$ bits, where N_S is the number of different stimuli. Indeed, it is an advantage of this measure that it has a defined maximal value, which enables how well the network is performing to be quantified. Rolls and Milward (2000) showed that the Fisher and single cell information measures were highly correlated, and given the advantage just noted of the information measure, it was adopted in Rolls and Milward (2000) and subsequent papers. Rolls and Milward (2000) also introduced a multiple cell information measure, which has the advantage that it provides a measure of whether all stimuli are encoded by different neurons in the network. Again, a high value of this measure indicates good performance.

For completeness, we provide further specification of the two information theoretic measures, which are described in detail by Rolls and Milward (2000) (see Appendix C for introduction of the concepts). The measures assess the extent to which either a single cell, or a population of cells, responds to the same stimulus invariantly with respect to its location, yet responds differently to different stimuli. The measures effectively show what one learns about which stimulus was presented from a single presentation of the stimulus at any randomly chosen location. Results for top (4th) layer cells are shown. High information measures thus show that cells fire similarly to the different transforms of a given stimulus (object), and differently to the other stimuli. The single cell stimulus-specific information, $I(s, R)$, is the amount of information the set of responses, R, has about a specific stimulus, s (see Rolls, Treves, Tovee and Panzeri (1997d) and Rolls and Milward (2000)). $I(s, R)$ is given by

$$I(s, R) = \sum_{r \in R} P(r|s) \log_2 \frac{P(r|s)}{P(r)} \qquad (25.12)$$

where r is an individual response from the set of responses R of the neuron. For each cell the performance measure used was the maximum amount of information a cell conveyed about any one stimulus. This (rather than the mutual information, $I(S, R)$ where S is the whole set

Fig. 25.21 The three stimuli used in the first two experiments.

of stimuli s), is appropriate for a competitive network in which the cells tend to become tuned to one stimulus[28].

If all the output cells of VisNet learned to respond to the same stimulus, then the information about the set of stimuli S would be very poor, and would not reach its maximal value of $\log_2$ of the number of stimuli (in bits). The second measure that is used here is the information provided by a set of cells about the stimulus set, using the procedures described by Rolls, Treves and Tovee (1997b) and Rolls and Milward (2000). The multiple cell information is the mutual information between the whole set of stimuli S and of responses R calculated using a decoding procedure in which the stimulus s' that gave rise to the particular firing rate response vector on each trial is estimated. (The decoding step is needed because the high dimensionality of the response space would lead to an inaccurate estimate of the information if the responses were used directly, as described by Rolls, Treves and Tovee (1997b) and Rolls and Treves (1998).) A probability table is then constructed of the real stimuli s and the decoded stimuli s'. From this probability table, the mutual information between the set of actual stimuli S and the decoded estimates S' is calculated as

$$I(S, S') = \sum_{s,s'} P(s, s') \log_2 \frac{P(s, s')}{P(s)P(s')} \tag{25.13}$$

This was calculated for the subset of cells which had as single cells the most information about which stimulus was shown. In particular, in Rolls and Milward (2000) and subsequent papers, the multiple cell information was calculated from the first five cells for each stimulus that had maximal single cell information about that stimulus, that is from a population of 35 cells if there were seven stimuli (each of which might have been shown in for example 9 or 25 positions on the retina).

25.5.2 Initial experiments with VisNet

Having established a network model, Wallis and Rolls (1997) (following a first report by Wallis, Rolls and Földiák (1993)) described four experiments in which the theory of how invariant representations could be formed was tested using a variety of stimuli undergoing a number of natural transformations. In each case the network produced neurons in the final layer whose responses were largely invariant across a transformation and highly discriminating between stimuli or sets of stimuli.

25.5.2.1 'T','L' and '+' as stimuli: learning translation invariance

One of the classical properties of inferior temporal cortex face cells is their invariant response to face stimuli translated across the visual field (Tovee, Rolls and Azzopardi 1994). In this first experiment, the learning of translation invariant representations by VisNet was investigated.

---

[28] $I(s, R)$ has more recently been called the stimulus-specific surprise, see DeWeese and Meister (1999). Its average across stimuli is the mutual information $I(S, R)$.

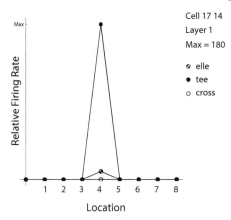

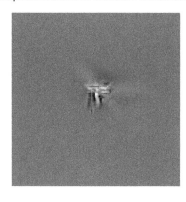

Fig. 25.22 The left graph shows the response of a layer 1 neuron to the three training stimuli for the nine training locations. Alongside this are the results of summating all the filter inputs to the neuron. The discrimination factor for this cell was 1.04.

In order to test the network a set of three stimuli, based upon probable 3D edge cues – consisting of a 'T', 'L' and '+' shape – was constructed[29]. The actual stimuli used are shown in Fig. 25.21. These stimuli were chosen partly because of their significance as form cues, but on a more practical note because they each contain the same fundamental features – namely a horizontal bar conjoined with a vertical bar. In practice this means that the oriented simple cell filters of the input layer cannot distinguish these stimuli on the basis of which features are present. As a consequence of this, the representation of the stimuli received by the network is non-orthogonal and hence considerably more difficult to classify than was the case in earlier experiments involving the trace rule described by Földiák (1991). The expectation is that layer 1 neurons would learn to respond to spatially selective combinations of the basic features thereby helping to distinguish these non-orthogonal stimuli. The trajectory followed by each stimulus consisted of sweeping left to right horizontally across three locations in the top row, and then sweeping back, right to left across the middle row, before returning to the right hand side across the bottom row – tracing out a 'Z' shape path across the retina. Unless stated otherwise this pattern of nine presentation locations was adopted in all image translation experiments described by Wallis and Rolls (1997).

Training was carried out by permutatively presenting all stimuli in each location a total of 800 times. The sequence described above was followed for each stimulus, with the sequence start point and direction of sweep being chosen at random for each of the 800 training trials.

Figures 25.22 and 25.23 show the response after training of a first layer neuron selective for the 'T' stimulus. The weighted sum of all filter inputs reveals the combination of horizontally and vertically tuned filters in identifying the stimulus. In this case many connections to the lower frequency filters have been reduced to zero by the learning process, except at the relevant orientations. This contrasts strongly with the random wiring present before training, as seen previously in Fig. 25.20. It is important that neurons at early stages of feature hierarchy networks respond to combinations of features in defined relative spatial positions, before invariance is built into the system, as this is part of the way that the binding problem is solved, as described in more detail in Section 25.5.5 and by Elliffe, Rolls and Stringer (2002). The feature combination tuning is illustrated by the VisNet layer 1 neuron shown in Figs. 25.22 and 25.23.

[29]Chakravarty (1979) describes the application of these shapes as cues for the 3D interpretation of edge junctions, and Tanaka et al. (1991) have demonstrated the existence of cells responsive to such stimuli in IT.

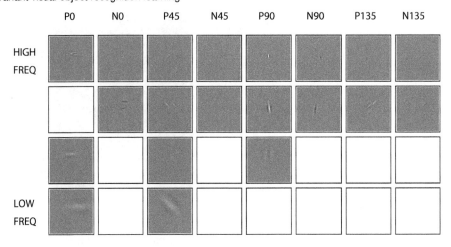

| | P0 | N0 | P45 | N45 | P90 | N90 | P135 | N135 |

Fig. 25.23 The same cell as in the previous figure and the same input reconstruction results but separated into four rows of differing spatial frequency, and eight columns representing the four filter tuning orientations in positive and negative complementary pairs.

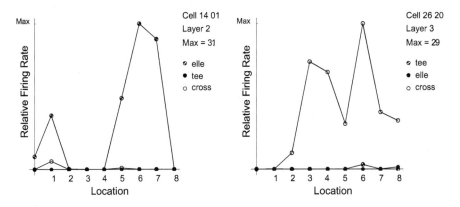

Fig. 25.24 Response profiles for two intermediate layer neurons – discrimination factors 1.34 and 1.64 – in the L, T, and + experiment.

Likewise, Fig. 25.24 depicts two neural responses, but now from the two intermediate layers of the network, taken from the top 30 most highly invariant cells, not merely the top two or three. The gradual increase in the discrimination indicates that the tolerance to shifts of the preferred stimulus gradually builds up through the layers.

The results for layer 4 neurons are illustrated in Fig. 25.25. By this stage translation-invariant, stimulus-identifying, cells have emerged. The response profiles confirm the high level of neural selectivity for a particular stimulus irrespective of location.

Figure 25.26 contrasts the measure of invariance, or discrimination factor, achieved by cells in the four layers, averaged over five separate runs of the network. Translation invariance clearly increases through the layers, with a considerable increase in translation invariance between layers 3 and 4. This sudden increase may well be a result of the geometry of the network, which enables cells in layer 4 to receive inputs from any part of the input layer.

Having established that invariant cells have emerged in the final layer, we now consider the role of the trace rule, by assessing the network tested under two new conditions. Firstly, the performance of the network was measured before learning occurs, that is with its initially

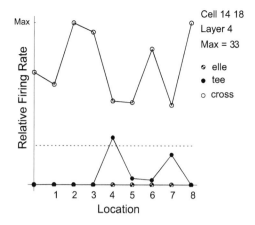
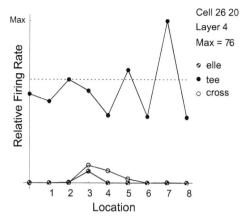

Fig. 25.25 Response profiles for two fourth layer neurons – discrimination factors 4.07 and 3.62 – in the L, T, and + experiment.

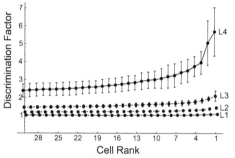
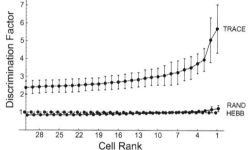

Fig. 25.26 Variation in neural discrimination factors as a measure of performance for the top 30 most highly discriminating cells through the four layers of the network, averaged over five runs of the network in the L, T, and + experiment.

Fig. 25.27 Variation in neural discrimination factors as a measure of performance for the top 30 most highly discriminating cells in the fourth layer for the three training regimes, averaged over five runs of the network.

random connection weights. Secondly, the network was trained with η in the trace rule set to 0, which causes learning to proceed in a traceless, standard Hebbian, fashion. (Hebbian learning is purely associative, as shown for example in equation B.19.)

Figure 25.27 shows the results under the three training conditions. The results show that the trace rule is the decisive factor in establishing the invariant responses in the layer 4 neurons. It is interesting to note that the Hebbian learning results are actually *worse* than those achieved by chance in the untrained net. In general, with Hebbian learning, the most highly discriminating cells barely rate higher than 1. This value of discrimination corresponds to the case in which a cell responds to only one stimulus and in only one location. The poor performance with the Hebb rule comes as a direct consequence of the presentation paradigm being employed. If we consider an image as representing a vector in multidimensional space, a particular image in the top left-hand corner of the input retina will tend to look more like any other image in that same location than the same image presented elsewhere. A simple competitive network using just Hebbian learning will thus tend to categorize images by *where* they are rather than what they are – the exact opposite of what the net was intended to learn. This comparison thus indicates that a small memory trace acting in the standard Hebbian learning

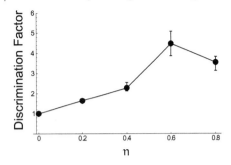

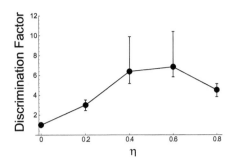

Fig. 25.28 Variation in network performance as a function of the trace rule parameter η in neurons of layers 2 to 4 – over nine locations in the L, T and + experiment.

Fig. 25.29 Variation in network performance as a function of the trace rule parameter η in neurons of layers 2 to 4 – over five presentation locations in the L, T and + experiment.

paradigm can radically alter the normal vector averaging, image classification, performed by a Hebbian-based competitive network.

One question that emerges about the representation in the final layer of the network relates to how evenly the network divides up its resources to represent the learnt stimuli. It is conceivable that one stimulus stands out among the set of stimuli by containing very distinctive features which would make it easier to categorize. This may produce an unrepresentative number of neurons with high discrimination factors which are in fact all responding to the same stimulus. It is important that at least some cells code for (or provide information about) each of the stimuli. As a simple check on this, the preferred stimulus of each cell was found and the associated measure of discrimination added to a total for each stimulus. This measure in practice never varied by more than a factor of 1.3:1 for all stimuli. The multiple cell information measure used in some later figures addresses the same issue, with similar results.

25.5.2.2 'T','L', and '+' as stimuli: Optimal network parameters

The second series of investigations described by Wallis and Rolls (1997) using the 'T','L' and '+' stimuli, centred upon finding optimal parameters for elements of the network, such as the optimal trace time constant η, which controls the relative effect of previous activities on current learning as described above. The network performance was gauged using a single 800 epoch training run of the network with the median discrimination factor (with the upper and lower quartile values) for the top sixteen cells of the fourth layer being displayed at each parameter value.

Figure 25.28 displays the effect of varying the value of η for the nine standard presentation locations. The optimal value of η might conceivably change with the alteration of the number of training locations, and indeed one might predict that it would be smaller if the number of presentation locations was reduced. To confirm this, network performance was also measured for presentation sweeps over only five locations. Figure 25.29 shows the results of this experiment, which confirm the expected shift in the general profile of the curve towards shorter time constant values. Of course, the optimal value of η derived is in effect a compromise between optimal values for the three layers in which the trace operates. Since neurons in each layer have different effective receptive field sizes, one would expect each layer's neurons to be exposed to different portions of the full sweep of a particular stimulus. This would in turn suggest that the optimal value of η will grow through the layers.

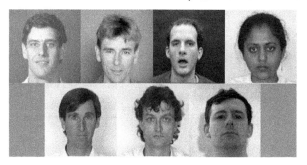

Fig. 25.30 Seven faces used as stimuli in the face translation experiment.

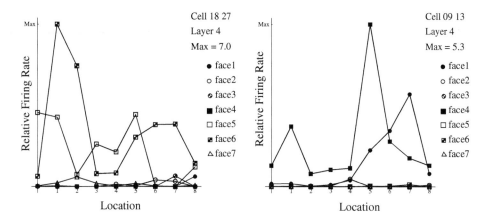

Fig. 25.31 Response profiles for two neurons in the fourth layer – discrimination factors 2.64 and 2.10. The net was trained on 7 faces each in 9 locations.

25.5.2.3 Faces as stimuli: translation invariance

The aim of the next set of experiments described by Wallis and Rolls (1997) was to start to address the issues of how the network operates when invariant representations must be learned for a larger number of stimuli, and whether the network can learn when much more complicated, real biological stimuli (faces) are used. The set of face images used appears in Fig. 25.30. In practice, to equalize luminance the D.C. component of the images was removed. In addition, so as to minimize the effect of cast shadows, an oval Hamming window was applied to the face image which also served to remove any hard edges of the image relative to the plain background upon which they were set.

The results of training in the translation invariance paradigm with 7 faces each in 9 locations are shown in Figs. 25.31, 25.32 and 25.33. The network produces neurons with high discrimination factors, and this only occurs if it is trained with the trace rule. Some layer 4 neurons showed a somewhat distributed representation, as illustrated in the examples of layer 4 neurons shown in Fig. 25.31.

In order to check that there was an invariant representation in layer 4 of VisNet that could be read by a receiving population of neurons, a fifth layer was added to the net which fully sampled the fourth layer cells. This layer was in turn trained in a supervised manner using gradient descent or with a Hebbian associative learning rule. Wallis and Rolls (1997) showed that the object classification performed by the layer 5 network was better if the network had been trained with the trace rule than when it was untrained or was trained with a Hebb rule.

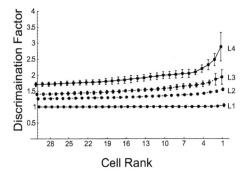

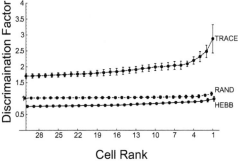

Fig. 25.32 Variation in network performance for the top 30 most highly discriminating cells through the four layers of the network, averaged over five runs of the network. The net was trained on 7 faces each in 9 locations.

Fig. 25.33 Variation in network performance for the top 30 most highly discriminating cells in the fourth layer for the three training regimes, averaged over five runs of the network. The net was trained on 7 faces each in 9 locations.

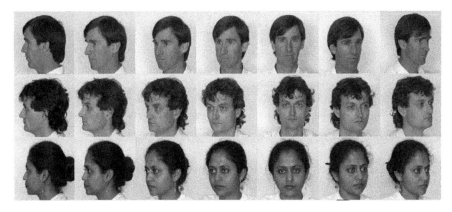

Fig. 25.34 Three faces in seven different views used as stimuli in an experiment by Wallis and Rolls (1997).

25.5.2.4 Faces as stimuli: view invariance

Given that the network had been shown to be able to operate usefully with a more difficult translation invariance problem, we next addressed the question of whether the network can solve other types of transform invariance, as we had intended. The next experiment addressed this question, by training the network on the problem of 3D stimulus rotation, which produces non-isomorphic transforms, to determine whether the network can build a view-invariant categorization of the stimuli (Wallis and Rolls 1997). The trace rule learning paradigm should, in conjunction with the architecture described here, prove capable of learning any of the transforms tolerated by IT neurons, so long as each stimulus is presented in short sequences during which the transformation occurs and can be learned. This experiment continued with the use of faces but now presented them centrally in the retina in a sequence of different views of a face. The images used are shown in Fig. 25.34. The faces were again smoothed at the edges to erase the harsh image boundaries, and the D.C. term was removed. During the 800 epochs of learning, each stimulus was chosen at random, and a sequence of preset views of it was shown, rotating the face either to the left or to the right.

Although the actual number of images being presented is smaller, some 21 views in all, there is good reason to think that this problem may be harder to solve than the previous

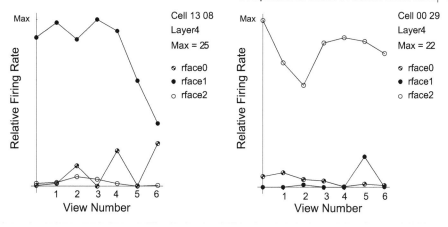

Fig. 25.35 Response profiles for cells in the last two layers of the network – discrimination factors 11.12 and 12.40 – in the experiment with seven different views of each of three faces.

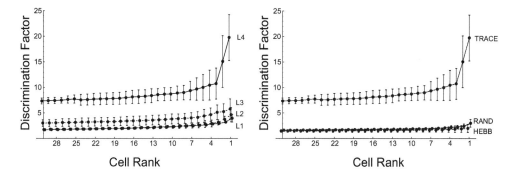

Fig. 25.36 Variation in network performance for the top 30 most highly discriminating cells through the four layers of the network, averaged over five runs of the network in the experiment with seven different views of each of three faces.

Fig. 25.37 Variation in network performance for the top 30 most highly discriminating cells in the fourth layer for the three training regimes, averaged over five runs of the network in the experiment with seven different views of each of three faces.

translation experiments. This is simply due to the fact that all 21 views exactly overlap with one another. The net was indeed able to solve the invariance problem, with examples of invariant layer 4 neuron response profiles appearing in Fig. 25.35.

Figure 25.36 confirms the improvement in invariant stimulus representation found through the layers, and that layer 4 provides a considerable improvement in performance over the previous layers. Figure 25.37 shows the Hebb trained and untrained nets performing equally poorly, whilst the trace trained net shows good invariance across the entire 30 cells selected.

25.5.3 The optimal parameters for the temporal trace used in the learning rule

The trace used in VisNet enables successive features that, based on the natural statistics of the visual input, are likely to be from the same object or feature complex to be associated together. For good performance, the temporal trace needs to be sufficiently long that it covers the period in which features seen by a particular neuron in the hierarchy are likely to come from the same object. On the other hand, the trace should not be so long that it produces associations between features that are parts of different objects, seen when for example the eyes move to another object. One possibility is to reset the trace during saccades between different objects. If explicit trace resetting is not implemented, then the trace should, to optimize the compromise implied by the above, lead to strong associations between temporally close stimuli, and increasingly weaker associations between temporally more distant stimuli. In fact, the trace implemented in VisNet has an exponential decay, and it has been shown that this form is optimal in the situation where the exact duration over which the same object is being viewed varies, and where the natural statistics of the visual input happen also to show a decreasing probability that the same object is being viewed as the time period in question increases (Wallis and Baddeley 1997). Moreover, as is made evident in Figs. 25.29 and 25.28, performance can be enhanced if the duration of the trace does at the same time approximately match the period over which the input stimuli are likely to come from the same object or feature complex. Nevertheless, good performance can be obtained in conditions under which the trace rule allows associations to be formed only between successive items in the visual stream (Rolls and Milward 2000, Rolls and Stringer 2001a).

It is also the case that the optimal value of η in the trace rule is likely to be different for different layers of VisNet, and for cortical processing in the 'what' visual stream. For early layers of the system, small movements of the eyes might lead to different feature combinations providing the input to cells (which at early stages have small receptive fields), and a short duration of the trace would be optimal. However, these small eye movements might be around the same object, and later layers of the architecture would benefit from being able to associate together their inputs over longer times, in order to learn about the larger scale properties that characterize individual objects, including for example different views of objects observed as an object turns or is turned. Thus the suggestion is made that the temporal trace could be effectively longer at later stages (e.g. inferior temporal visual cortex) compared to early stages (e.g. V2 and V4) of processing in the visual system. In addition, as will be shown in Section 25.5.5, it is important to form feature combinations with high spatial precision before invariance learning supported by a temporal trace starts, in order that the feature combinations and not the individual features have invariant representations. This leads to the suggestion that the trace rule should either not operate, or be short, at early stages of cortical visual processing such as V1. This is reflected in the operation of VisNet2, which does not use a temporal trace in layer 1 (Rolls and Milward 2000).

25.5.4 Different forms of the trace learning rule, and their relation to error correction and temporal difference learning

The original trace learning rule used in the simulations of Wallis and Rolls (1997) took the form

$$\delta w_j = \alpha \bar{y}^\tau x_j^\tau \tag{25.14}$$

where the trace $\bar{y}^\tau$ is updated according to

$$\bar{y}^\tau = (1 - \eta)y^\tau + \eta \bar{y}^{\tau-1}. \tag{25.15}$$

The parameter $\eta \in [0, 1]$ controls the relative contributions to the trace $\bar{y}^\tau$ from the instantaneous firing rate y^τ and the trace at the previous time step $\bar{y}^{\tau-1}$, where for $\eta = 0$ we have $\bar{y}^\tau = y^\tau$ and equation 25.14 becomes the standard Hebb rule

$$\delta w_j = \alpha y^\tau x_j^\tau. \tag{25.16}$$

At the start of a series of investigations of different forms of the trace learning rule, Rolls and Milward (2000) demonstrated that VisNet's performance could be greatly enhanced with a modified Hebbian learning rule that incorporated a trace of activity from the preceding time steps, with no contribution from the activity being produced by the stimulus at the current time step. This rule took the form

$$\delta w_j = \alpha \bar{y}^{\tau-1} x_j^\tau. \tag{25.17}$$

The trace shown in equation 25.17 is in the postsynaptic term, and similar effects were found if the trace was in the presynaptic term, or in both the pre- and the postsynaptic terms. The crucial difference from the earlier rule (see equation 25.14) was that the trace should be calculated up to only the preceding timestep, with no contribution to the trace from the firing on the current trial to the current stimulus. How might this be understood?

One way to understand this is to note that the trace rule is trying to set up the synaptic weight on trial τ based on whether the neuron, based on its previous history, is responding to that stimulus (in other transforms, e.g. position). Use of the trace rule at $\tau - 1$ does this, that is it takes into account the firing of the neuron on previous trials, with no contribution from the firing being produced by the stimulus on the current trial. On the other hand, use of the trace at time τ in the update takes into account the current firing of the neuron to the stimulus in that particular position, which is not a good estimate of whether that neuron should be allocated to invariantly represent that stimulus. Effectively, using the trace at time τ introduces a Hebbian element into the update, which tends to build position-encoded analyzers, rather than stimulus-encoded analyzers. (The argument has been phrased for a system learning translation invariance, but applies to the learning of all types of invariance.) A particular advantage of using the trace at $\tau - 1$ is that the trace will then on different occasions (due to the randomness in the location sequences used) reflect previous histories with different sets of positions, enabling the learning of the neuron to be based on evidence from the stimulus present in many different positions. Using a term from the current firing in the trace (i.e. the trace calculated at time τ) results in this desirable effect always having an undesirable element from the current firing of the neuron to the stimulus in its current position.

25.5.4.1 The modified Hebbian trace rule and its relation to error correction

The rule of equation 25.17 corrects the weights using a postsynaptic trace obtained from the previous firing (produced by other transforms of the same stimulus), with no contribution to the trace from the current postsynaptic firing (produced by the current transform of the stimulus). Indeed, insofar as the current firing y^τ is not the same as $\bar{y}^{\tau-1}$, this difference can be thought of as an error. This leads to a conceptualization of using the difference between the current firing and the preceding trace as an error correction term, as noted in the context of modelling the temporal properties of classical conditioning by Sutton and Barto (1981), and developed next in the context of invariance learning (see Rolls and Stringer (2001a)).

First, we re-express the rule of equation 25.17 in an alternative form as follows. Suppose we are at timestep τ and have just calculated a neuronal firing rate y^τ and the corresponding trace $\bar{y}^\tau$ from the trace update equation 25.15. If we assume $\eta \in (0, 1)$, then rearranging equation 25.15 gives

$$\bar{y}^{\tau-1} = \frac{1}{\eta}(\bar{y}^\tau - (1 - \eta)y^\tau), \tag{25.18}$$

and substituting equation 25.18 into equation 25.17 gives

$$\begin{aligned}\delta w_j &= \alpha \frac{1}{\eta}(\bar{y}^\tau - (1 - \eta)y^\tau)x_j^\tau \\ &= \alpha \frac{1-\eta}{\eta}(\frac{1}{1-\eta}\bar{y}^\tau - y^\tau)x_j^\tau \\ &= \hat{\alpha}(\hat{\beta}\bar{y}^\tau - y^\tau)x_j^\tau \end{aligned} \tag{25.19}$$

where $\hat{\alpha} = \alpha\frac{1-\eta}{\eta}$ and $\hat{\beta} = \frac{1}{1-\eta}$. The modified Hebbian trace learning rule (25.17) is thus equivalent to equation 25.19 which is in the general form of an error correction rule (Hertz, Krogh and Palmer 1991). That is, rule (25.19) involves the subtraction of the current firing rate y^τ from a target value, in this case $\hat{\beta}\overline{y}^\tau$.

Although above we have referred to rule (25.17) as a modified Hebbian rule, we note that it is only associative in the sense of associating *previous* cell firing with the current cell inputs. In the next section we continue to explore the error correction paradigm, examining five alternative examples of this sort of learning rule.

25.5.4.2 Five forms of error correction learning rule

Error correction learning rules are derived from gradient descent minimization (Hertz, Krogh and Palmer 1991), and continually compare the current neuronal output to a target value t and adjust the synaptic weights according to the following equation at a particular timestep τ

$$\delta w_j = \alpha(t - y^\tau)x_j^\tau. \tag{25.20}$$

In this usual form of gradient descent by error correction, the target t is fixed. However, in keeping with our aim of encouraging neurons to respond similarly to images that occur close together in time it seems reasonable to set the target at a particular timestep, t^τ, to be some function of cell activity occurring close in time, because encouraging neurons to respond to temporal classes will tend to make them respond to the different variants of a given stimulus (Földiák 1991, Rolls 1992a, Wallis and Rolls 1997). For this reason, Rolls and Stringer (2001a) explored a range of error correction rules where the targets t^τ are based on the trace of neuronal activity calculated according to equation 25.15. We note that although the target is not a fixed value as in standard error correction learning, nevertheless the new learning rules perform gradient descent on each timestep, as elaborated below. Although the target may be varying early on in learning, as learning proceeds the target is expected to become more and more constant, as neurons settle to respond invariantly to particular stimuli. The first set of five error correction rules we discuss are as follows.

$$\delta w_j = \alpha(\beta\overline{y}^{\tau-1} - y^\tau)x_j^\tau, \tag{25.21}$$

$$\delta w_j = \alpha(\beta y^{\tau-1} - y^\tau)x_j^\tau, \tag{25.22}$$

$$\delta w_j = \alpha(\beta\overline{y}^\tau - y^\tau)x_j^\tau, \tag{25.23}$$

$$\delta w_j = \alpha(\beta\overline{y}^{\tau+1} - y^\tau)x_j^\tau, \tag{25.24}$$

$$\delta w_j = \alpha(\beta y^{\tau+1} - y^\tau)x_j^\tau, \tag{25.25}$$

where updates (25.21), (25.22) and (25.23) are performed at timestep τ, and updates (25.24) and (25.25) are performed at timestep $\tau + 1$. (The reason for adopting this convention is that the basic form of the error correction rule (25.20) is kept, with the five different rules simply replacing the term t.) It may be readily seen that equations (25.22) and (25.25) are special cases of equations (25.21) and (25.24) respectively, with $\eta = 0$.

These rules are all similar except for their targets t^τ, which are all functions of a temporally nearby value of cell activity. In particular, rule (25.23) is directly related to rule (25.19), but is more general in that the parameter $\hat{\beta} = \frac{1}{1-\eta}$ is replaced by an unconstrained parameter β. In addition, we also note that rule (25.21) is closely related to a rule developed in Peng, Sha, Gan and Wei (1998) for view invariance learning. The above five error correction rules are biologically plausible in that the targets t^τ are all local cell variables (see Appendix B and Rolls and Treves (1998)). In particular, rule (25.23) uses the trace $\overline{y}^\tau$ from the current time

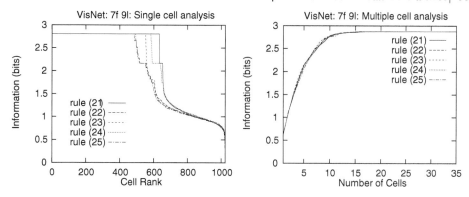

Fig. 25.38 Numerical results with the five error correction rules (25.21), (25.22), (25.23), (25.24), (25.25) (with positive clipping of synaptic weights) trained on 7 faces in 9 locations. On the left are single cell information measures, and on the right are multiple cell information measures. (After Rolls and Stringer 2001a.)

level τ, and rules (25.22) and (25.25) do not need exponential trace values $\overline{y}$, instead relying only on the instantaneous firing rates at the current and immediately preceding timesteps. However, all five error correction rules involve decrementing of synaptic weights according to an error which is calculated by subtracting the current activity from a target.

Numerical results with the error correction rules trained on 7 faces in 9 locations are presented in Fig. 25.38. For all the results shown the synaptic weights were clipped to be positive during the simulation, because it is important to test that decrementing synaptic weights purely within the positive interval $w \in [0, \infty]$ will provide significantly enhanced performance. That is, it is important to show that error correction rules do not necessarily require possibly biologically implausible modifiable negative weights. For each of the rules (25.21), (25.22), (25.23), (25.24), (25.25), the parameter β has been individually optimized to the following respective values: 4.9, 2.2, 2.2, 3.8, 2.2. On the left and right are results with the single and multiple cell information measures, respectively. Comparing Fig. 25.38 with Fig. 25.39 shows that all five error correction rules offer considerably improved performance over both the standard trace rule (25.14) and rule (25.17). From the left-hand side of Fig. 25.38 it can be seen that rule (25.21) performs best, and this is probably due to two reasons. Firstly, rule (25.21) incorporates an exponential trace $\overline{y}^{\tau-1}$ in its target t^{τ}, and we would expect this to help neurons to learn more quickly to respond invariantly to a class of inputs that occur close together in time. Hence, setting $\eta = 0$ as in rule (25.22) results in reduced performance. Secondly, unlike rules (25.23) and (25.24), rule (25.21) does not contain any component of y^{τ} in its target. If we examine rules (25.23), (25.24), we see that their respective targets $\beta\overline{y}^{\tau}$, $\beta\overline{y}^{\tau+1}$ contain significant components of y^{τ}.

25.5.4.3 Relationship to temporal difference learning

Rolls and Stringer (2001a) not only considered the relationship of rule (25.17) to error correction, but also considered how the error correction rules shown in equations (25.21), (25.22), (25.23), (25.24) and (25.25) are related to temporal difference learning (Sutton 1988, Sutton and Barto 1998). Sutton (1988) described temporal difference methods in the context of prediction learning. These methods are a class of incremental learning techniques that can learn to predict final outcomes through comparison of successive predictions from the preceding time steps. This is in contrast to traditional supervised learning, which involves the comparison of predictions only with the final outcome. Consider a series of multistep prediction problems in which for each problem there is a sequence of observation vectors, $\mathbf{x}^1$,

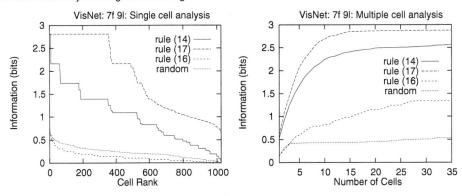

Fig. 25.39 Numerical results with the standard trace rule (25.14), learning rule (25.17), the Hebb rule (25.16), and random weights, trained on 7 faces in 9 locations: single cell information measure (left), multiple cell information measure (right). (After Rolls and Stringer 2001a.)

$\mathbf{x}^2$, ..., $\mathbf{x}^m$, at successive timesteps, followed by a final scalar outcome z. For each sequence of observations temporal difference methods form a sequence of predictions y^1, y^2, ..., y^m, each of which is a prediction of z. These predictions are based on the observation vectors $\mathbf{x}^\tau$ and a vector of modifiable weights $\mathbf{w}$; i.e. the prediction at time step τ is given by $y^\tau(\mathbf{x}^\tau, \mathbf{w})$, and for a linear dependency the prediction is given by $y^\tau = \mathbf{w}^T\mathbf{x}^\tau$. (Note here that $\mathbf{w}^T$ is the transpose of the weight vector $\mathbf{w}$.) The problem of prediction is to calculate the weight vector $\mathbf{w}$ such that the predictions y^τ are good estimates of the outcome z.

The supervised learning approach to the prediction problem is to form pairs of observation vectors $\mathbf{x}^\tau$ and outcome z for all time steps, and compute an update to the weights according to the gradient descent equation

$$\delta\mathbf{w} = \alpha(z - y^\tau)\nabla_\mathbf{w} y^\tau \tag{25.26}$$

where α is a learning rate parameter and $\nabla_\mathbf{w}$ indicates the gradient with respect to the weight vector $\mathbf{w}$. However, this learning procedure requires all calculation to be done at the end of the sequence, once z is known. To remedy this, it is possible to replace method (25.26) with a temporal difference algorithm that is mathematically equivalent but allows the computational workload to be spread out over the entire sequence of observations. Temporal difference methods are a particular approach to updating the weights based on the values of successive predictions, y^τ, $y^{\tau+1}$. Sutton (1988) showed that the following temporal difference algorithm is equivalent to method (25.26)

$$\delta\mathbf{w} = \alpha(y^{\tau+1} - y^\tau)\sum_{k=1}^{\tau}\nabla_\mathbf{w} y^k, \tag{25.27}$$

where $y^{m+1} \equiv z$. However, unlike method (25.26) this can be computed incrementally at each successive time step since each update depends only on $y^{\tau+1}$, y^τ and the sum of $\nabla_\mathbf{w} y^k$ over previous time steps k. The next step taken in Sutton (1988) is to generalize equation (25.27) to the following final form of temporal difference algorithm, known as 'TD(λ)'

$$\delta\mathbf{w} = \alpha(y^{\tau+1} - y^\tau)\sum_{k=1}^{\tau}\lambda^{\tau-k}\nabla_\mathbf{w} y^k \tag{25.28}$$

where $\lambda \in [0, 1]$ is an adjustable parameter that controls the weighting on the vectors $\nabla_\mathbf{w} y^k$. Equation (25.28) represents a much broader class of learning rules than the more usual gradient descent-based rule (25.27), which is in fact the special case TD(1).

A further special case of equation (25.28) is for $\lambda = 0$, i.e. TD(0), as follows

$$\delta\mathbf{w} = \alpha(y^{\tau+1} - y^{\tau})\nabla_{\mathbf{w}}y^{\tau}. \tag{25.29}$$

But for problems where y^{τ} is a linear function of $\mathbf{x}^{\tau}$ and $\mathbf{w}$, we have $\nabla_{\mathbf{w}}y^{\tau} = \mathbf{x}^{\tau}$, and so equation (25.29) becomes

$$\delta\mathbf{w} = \alpha(y^{\tau+1} - y^{\tau})\mathbf{x}^{\tau}. \tag{25.30}$$

If we assume the prediction process is being performed by a neuron with a vector of inputs $\mathbf{x}^{\tau}$, synaptic weight vector $\mathbf{w}$, and output $y^{\tau} = \mathbf{w}^T\mathbf{x}^{\tau}$, then we see that the TD(0) algorithm (25.30) is identical to the error correction rule (25.25) with $\beta = 1$. In understanding this comparison with temporal difference learning, it may be useful to note that the firing at the end of a sequence of the transformed exemplars of a stimulus is effectively the temporal difference target z. This establishes a link to temporal difference learning (described further in Section B.16.3). Further, we note that from learning epoch to learning epoch, the target z for a given neuron will gradually settle down to be more and more fixed as learning proceeds.

We now explore in more detail the relation between the error correction rules described above and temporal difference learning. For each sequence of observations with a single outcome the temporal difference method (25.30), when viewed as an error correction rule, is attempting to adapt the weights such that $y^{\tau+1} = y^{\tau}$ for all successive pairs of time steps – the same general idea underlying the error correction rules (25.21), (25.22), (25.23), (25.24), (25.25). Furthermore, in Sutton and Barto (1998), where temporal difference methods are applied to reinforcement learning, the TD(λ) approach is again further generalized by replacing the target $y^{\tau+1}$ by any weighted average of predictions y from arbitrary future timesteps, e.g. $t^{\tau} = \frac{1}{2}y^{\tau+3} + \frac{1}{2}y^{\tau+7}$, including an exponentially weighted average extending forward in time. So a more general form of the temporal difference algorithm has the form

$$\delta\mathbf{w} = \alpha(t^{\tau} - y^{\tau})\mathbf{x}^{\tau}, \tag{25.31}$$

where here the target t^{τ} is an arbitrary weighted average of the predictions y over future timesteps. Of course, with standard temporal difference methods the target t^{τ} is always an average over *future* timesteps $k = \tau + 1, \tau + 2$, etc. But in the five error correction rules this is only true for the last exemplar (25.25). This is because with the problem of prediction, for example, the ultimate target of the predictions $y^1, ..., y^m$ is a final outcome $y^{m+1} \equiv z$. However, this restriction does not apply to our particular application of neurons trained to respond to temporal classes of inputs within VisNet. Here we only wish to set the firing rates $y^1, ..., y^m$ to the same value, not some final given value z. However, the more general error correction rules clearly have a close relationship to standard temporal difference algorithms. For example, it can be seen that equation (25.22) with $\beta = 1$ is in some sense a temporal mirror image of equation (25.30), particularly if the updates δw_j are added to the weights w_j only at the end of a sequence. That is, rule (25.22) will attempt to set $y^1, ..., y^m$ to an *initial* value $y^0 \equiv z$. This relationship to temporal difference algorithms allowed us to begin to exploit established temporal difference analyses to investigate the convergence properties of the error correction methods (Rolls and Stringer 2001a).

Although the main aim of Rolls and Stringer (2001a) in relating error correction rules to temporal difference learning was to begin to exploit established temporal difference analyses, they observed that the most general form of temporal difference learning, TD(λ), in fact suggests an interesting generalization to the existing error correction learning rules for which we currently have $\lambda = 0$. Assuming $y^{\tau} = \mathbf{w}^T\mathbf{x}^{\tau}$ and $\nabla_{\mathbf{w}}y^{\tau} = \mathbf{x}^{\tau}$, the general equation (25.28) for TD(λ) becomes

$$\delta\mathbf{w} = \alpha(y^{\tau+1} - y^{\tau})\sum_{k=1}^{\tau}\lambda^{\tau-k}\mathbf{x}^k \tag{25.32}$$

where the term $\sum_{k=1}^{\tau} \lambda^{\tau-k} \mathbf{x}^k$ is a weighted sum of the vectors $\mathbf{x}^k$. This suggests generalizing the original five error correction rules (25.21), (25.22), (25.23), (25.24), (25.25) by replacing the term x_j^τ by a weighted sum $\hat{x}_j^\tau = \sum_{k=1}^{\tau} \lambda^{\tau-k} x_j^k$ with $\lambda \in [0, 1]$. In Sutton (1988) $\hat{x}_j^\tau$ is calculated according to

$$\hat{x}_j^\tau = x_j^\tau + \lambda \hat{x}_j^{\tau-1} \tag{25.33}$$

with $\hat{x}_j^0 \equiv 0$. This gives the following five temporal difference-inspired error correction rules

$$\delta w_j = \alpha(\beta \overline{y}^{\tau-1} - y^\tau)\hat{x}_j^\tau, \tag{25.34}$$

$$\delta w_j = \alpha(\beta y^{\tau-1} - y^\tau)\hat{x}_j^\tau, \tag{25.35}$$

$$\delta w_j = \alpha(\beta \overline{y}^\tau - y^\tau)\hat{x}_j^\tau, \tag{25.36}$$

$$\delta w_j = \alpha(\beta \overline{y}^{\tau+1} - y^\tau)\hat{x}_j^\tau, \tag{25.37}$$

$$\delta w_j = \alpha(\beta y^{\tau+1} - y^\tau)\hat{x}_j^\tau, \tag{25.38}$$

where it may be readily seen that equations (25.35) and (25.38) are special cases of equations (25.34) and (25.37) respectively, with $\eta = 0$. As with the trace $\overline{y}^\tau$, the term $\hat{x}_j^\tau$ is reset to zero when a new stimulus is presented. These five rules can be related to the more general TD(λ) algorithm, but continue to be biologically plausible using only local cell variables. Setting $\lambda = 0$ in rules (25.34), (25.35), (25.36), (25.37), (25.38), gives us back the original error correction rules (25.21), (25.22), (25.23), (25.24), (25.25) which may now be related to TD(0).

Numerical results with error correction rules (25.34), (25.35), (25.36), (25.37), (25.38), and $\hat{x}_j^\tau$ calculated according to equation (25.33) with $\lambda = 1$, with positive clipping of weights, trained on 7 faces in 9 locations are presented in Fig. 25.40. For each of the rules (25.34), (25.35), (25.36), (25.37), (25.38), the parameter β has been individually optimized to the following respective values: 1.7, 1.8, 1.5, 1.6, 1.8. On the left and right are results with the single and multiple cell information measures, respectively. Comparing these five temporal difference-inspired rules it can be seen that the best performance is obtained with rule (25.38) where many more cells reach the maximum level of performance possible with respect to the single cell information measure. In fact, this rule offered the best such results. This may well be due to the fact that this rule may be directly compared to the standard TD(1) learning rule, which itself may be related to classical supervised learning for which there are well known optimality results, as discussed further by Rolls and Stringer (2001a).

From the simulations described by Rolls and Stringer (2001a) it appears that the form of optimization described above associated with TD(1) rather than TD(0) leads to better performance within VisNet. Comparing Figs. 25.38 and 25.40 shows that the TD(1)-like rule (25.38) with $\lambda = 1.0$ and $\beta = 1.8$ gives considerably superior results to the TD(0)-like rule (25.25) with $\beta = 2.2$. In fact, the former of these two rules provided the best single cell information results in these studies. We hypothesize that these results are related to the fact that only a finite set of image sequences is presented to VisNet, and so the type of optimization performed by TD(1) for repeated presentations of a finite data set is more appropriate for this problem than the form of optimization performed by TD(0).

25.5.4.4 Evaluation of the different training rules

In terms of biological plausibility, we note the following. First, all the learning rules investigated by Rolls and Stringer (2001a) are local learning rules, and in this sense are biologically plausible (see Appendix B). (The rules are local in that the terms used to modify the synaptic weights are potentially available in the pre- and post-synaptic elements.)

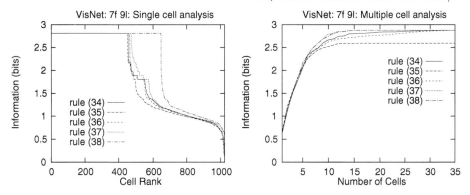

Fig. 25.40 Numerical results with the five temporal difference-inspired error correction rules (25.34), (25.35), (25.36), (25.37), and (25.38), and $\hat{x}_j^\tau$ calculated according to equation (25.33) (with positive clipping of synaptic weights) trained on 7 faces in 9 locations. On the left are single cell information measures, and on the right are multiple cell information measures. (After Rolls and Stringer 2001a.)

Second we note that all the rules do require some evidence of the activity on one or more previous stimulus presentations to be available when the synaptic weights are updated. Some of the rules, e.g. learning rule (25.23), use the trace $\bar{y}^\tau$ from the current time level, while rules (25.22) and (25.25) do not need to use an exponential trace of the neuronal firing rate, but only the instantaneous firing rates y at two successive time steps. It is known that synaptic plasticity does involve a combination of separate processes each with potentially differing time courses (Koch 1999), and these different processes could contribute to trace rule learning. Another mechanism suggested for implementing a trace of previous neuronal activity is the continuing firing for often 300 ms produced by a short (16 ms) presentation of a visual stimulus (Rolls and Tovee 1994) which is suggested to be implemented by local cortical recurrent attractor networks (Rolls and Treves 1998).

Third, we note that in utilizing the trace in the targets t^τ, the error correction (or temporal difference inspired) rules perform a comparison of the instantaneous firing y^τ with a temporally nearby value of the activity, and this comparison involves a subtraction. The subtraction provides an error, which is then used to increase or decrease the synaptic weights. This is a somewhat different operation from long-term depression (LTD) as well as long term potentiation (LTP), which are *associative* changes which depend on the pre- and post-synaptic activity. However, it is interesting to note that an error correction rule which appears to involve a subtraction of current firing from a target might be implemented by a combination of an associative process operating with the trace, and an anti-Hebbian process operating to remove the effects of the current firing. For example, the synaptic updates $\delta w_j = \alpha(t^\tau - y^\tau)x_j^\tau$ can be decomposed into two separate associative processes $\alpha t^\tau x_j^\tau$ and $-\alpha y^\tau x_j^\tau$, that may occur independently. (The target, t^τ, could in this case be just the trace of previous neural activity from the preceding trials, excluding any contribution from the current firing.) Another way to implement an error correction rule using associative synaptic modification would be to force the post-synaptic neuron to respond to the error term. Although this has been postulated to be an effect that could be implemented by the climbing fibre system in the cerebellum (Ito 1989, Ito 1984, Ito 2013) (Chapter 23), there is no similar system known for the neocortex, and it is not clear how this particular implementation of error correction might operate in the neocortex.

In Section 25.5.4.2 we describe five learning rules as error correction rules. We now discuss an interesting difference of these error correction rules from error correction rules as conventionally applied. It is usual to derive the general form of error correction learning rule from gradient descent minimization in the following way (Hertz, Krogh and Palmer 1991). Consider the idealized situation of a single neuron with a number of inputs x_j and output $y = \sum_j w_j x_j$, where w_j are the synaptic weights. We assume that there are a number of input patterns and that for the kth input pattern, $\mathbf{x}^k = [x_1^k, x_2^k, ...]^T$, the output y^k has a target value t^k. Hence an error measure or cost function can be defined as

$$e(\mathbf{w}) = \frac{1}{2}\sum_k (t^k - y^k)^2 = \frac{1}{2}\sum_k (t^k - \sum_j w_j x_j^k)^2. \tag{25.39}$$

This cost function is a function of the input patterns $\mathbf{x}^k$ and the synaptic weight vector $\mathbf{w} = [w_1, w_2, ...]^T$. With a fixed set of input patterns, we can reduce the error measure by employing a gradient descent algorithm to calculate an improved set of synaptic weights. Gradient descent achieves this by moving downhill on the error surface defined in $\mathbf{w}$ space using the update

$$\delta w_j = -\alpha \frac{\partial e}{\partial w_j} = \alpha \sum_k (t^k - y^k)x_j^k. \tag{25.40}$$

If we update the weights after each pattern k, then the update takes the form of an error correction rule

$$\delta w_j = \alpha(t^k - y^k)x_j^k, \tag{25.41}$$

which is also commonly referred to as the delta rule or Widrow–Hoff rule (see Widrow and Hoff (1960) and Widrow and Stearns (1985)). Error correction rules continually compare the neuronal output with its pre-specified target value and adjust the synaptic weights accordingly. In contrast, the way Rolls and Stringer (2001a) introduced of utilizing error correction is to specify the target as the activity trace based on the firing rate at nearby timesteps. Now the actual firing at those nearby time steps is not a pre-determined fixed target, but instead depends on how the network has actually evolved. This effectively means the cost function $e(\mathbf{w})$ that is being minimized changes from timestep to timestep. Nevertheless, the concept of calculating an error, and using the magnitude and direction of the error to update the synaptic weights, is the similarity Rolls and Stringer (2001a) made to gradient descent learning.

To conclude this evaluation, the error correction and temporal difference rules explored by Rolls and Stringer (2001a) provide interesting approaches to help understand invariant pattern recognition learning. Although we do not know whether the full power of these rules is expressed in the brain, we provided suggestions about how they might be implemented. At the same time, we note that the original trace rule used by Földiák (1991), Rolls (1992a), and Wallis and Rolls (1997) is a simple associative rule, is therefore biologically very plausible, and, while not as powerful as many of the other rules introduced by Rolls and Stringer (2001a), can nevertheless solve the same class of problem. Rolls and Stringer (2001a) also emphasized that although they demonstrated how a number of new error correction and temporal difference rules might play a role in the context of view invariant object recognition, they may also operate elsewhere where it is important for neurons to learn to respond similarly to temporal classes of inputs that tend to occur close together in time.

25.5.5 The issue of feature binding, and a solution

In this section we investigate two key issues that arise in hierarchical layered network architectures, such as VisNet, other examples of which have been described and analyzed by Fukushima (1980), Ackley, Hinton and Sejnowski (1985), Rosenblatt (1961), Riesenhuber and Poggio (1999b) and Serre et al. (2007b) (see also Section 25.6). One issue is whether the network can discriminate between stimuli that are composed of the same basic alphabet of features. The second issue is whether such network architectures can find solutions to the spatial binding problem. These issues are addressed next and by Elliffe, Rolls and Stringer (2002).

The first issue investigated is whether a hierarchical layered network architecture of the type exemplified by VisNet can discriminate stimuli that are composed of a limited set of features and where the different stimuli include cases where the feature sets are subsets and supersets of those in the other stimuli. An issue is that if the network has learned representations of both the parts and the wholes, will the network identify that the whole is present when it is shown, and not just that one or more parts is present. (In many investigations with VisNet, complex stimuli (such as faces) were used where each stimulus might contain unique features not present in the other stimuli.) To address this issue, Elliffe, Rolls and Stringer (2002) used stimuli that are composed from a set of four features which are designed so that each feature is spatially separate from the other features, and no unique combination of firing caused for example by overlap of horizontal and vertical filter outputs in the input representation distinguishes any one stimulus from the others. The results described in Section 25.5.5.4 show that VisNet can indeed learn correct invariant representations of stimuli which do consist of feature sets where individual features do not overlap spatially with each other and where the stimuli can be composed of sets of features which are supersets or subsets of those in other stimuli. Fukushima and Miyake (1982) did not address this crucial issue where different stimuli might be composed of subsets or supersets of the same set of features, although they did show that stimuli with partly overlapping features could be discriminated by the Neocognitron.

In Section 25.5.5.5 we address the spatial binding problem in architectures such as VisNet. This computational problem that needs to be addressed in hierarchical networks such as the primate visual system and VisNet is how representations of features can be (e.g. translation) invariant, yet can specify stimuli or objects in which the features must be specified in the correct spatial arrangement. This is the feature binding problem, discussed for example by von der Malsburg (1990a), and arising in the context of hierarchical layered systems (Ackley, Hinton and Sejnowski 1985, Fukushima 1980, Rosenblatt 1961). The issue is whether or not features are bound into the correct combinations in the correct relative spatial positions, or if alternative combinations of known features or the same features in different relative spatial positions would elicit the same responses. All this has to be achieved while at the same time producing position invariant recognition of the whole combination of features, that is, the object. This is a major computational issue that needs to be solved for memory systems in the brain to operate correctly. This can be achieved by what is effectively a learning process that builds into the system a set of neurons in the hierarchical network that enables the recognition process to operate correctly with the appropriate position, size, view etc. invariances.

25.5.5.1 Syntactic binding of separate neuronal ensembles by synchronization

The problem of syntactic binding of neuronal representations, in which some features must be bound together to form one object, and other simultaneously active features must be bound together to represent another object, has been addressed by von der Malsburg (see von der Malsburg (1990a)). He has proposed that this could be performed by temporal synchronization of those neurons that were temporarily part of one representation in a different time slot from other neurons that were temporarily part of another representation. The idea is attractive in allowing arbitrary relinking of features in different combinations. Singer, Engel, Konig, and colleagues (Singer, Gray, Engel, Konig, Artola and Brocher 1990, Engel, Konig, Kreiter, Schillen and Singer 1992, Singer and Gray 1995, Singer 1999, Fries 2015), and others (Abeles 1991) have obtained some evidence that when features must be bound, synchronization of neuronal populations can occur (but see Shadlen and Movshon (1999)), and this has been modelled (Hummel and Biederman 1992) (Section 8.3).

Synchronization to implement syntactic binding has a number of disadvantages and limitations (see also Rolls (2008d), Riesenhuber and Poggio (1999a) and Section 8.3). The greatest computational problem is that synchronization does not by itself define the spatial relations between the features being bound, so is not just as a binding mechanism adequate for shape recognition. For example, temporal binding might enable features 1, 2 and 3, which might define one stimulus to be bound together and kept separate from for example another stimulus consisting of features 2, 3 and 4, but would require a further temporal binding (leading in the end potentially to a combinatorial explosion) to indicate the relative spatial positions of the 1, 2 and 3 in the 123 stimulus, so that it can be discriminated from e.g. 312.

A second problem with the synchronization approach to the spatial binding of features is that, when stimulus-dependent temporal synchronization has been rigourously tested with information theoretic approaches, it has so far been found that most of the information available is in the number of spikes, with rather little, less than 5% of the total information, in stimulus-dependent synchronization (Aggelopoulos, Franco and Rolls 2005, Franco, Rolls, Aggelopoulos and Treves 2004, Rolls, Aggelopoulos, Franco and Treves 2004) (see Section C.3.7). For example, Aggelopoulos, Franco and Rolls (2005) showed that when macaques used object-based attention to search for one of two objects to touch in a complex natural scene, between 99% and 94% of the information was present in the firing rates of inferior temporal cortex neurons, and less that 5% in any stimulus-dependent synchrony that was present between the simultaneously recorded inferior temporal cortex neurons. The implication of these results is that any stimulus-dependent synchrony that is present is not quantitatively important as measured by information theoretic analyses under natural scene conditions when feature binding, segmentation of objects from the background, and attention are required. This has been found for the inferior temporal cortex, a brain region where features are put together to form representations of objects (Rolls and Deco 2002), and where attention has strong effects, at least in scenes with blank backgrounds (Rolls, Aggelopoulos and Zheng 2003a). It would of course also be of interest to test the same hypothesis in earlier visual areas, such as V4, with quantitative, information theoretic, techniques. In connection with rate codes, it should be noted that a rate code implies using the number of spikes that arrive in a given time, and that this time can be very short, as little as 20–50 ms, for very useful amounts of information to be made available from a population of neurons (Tovee, Rolls, Treves and Bellis 1993, Rolls and Tovee 1994, Rolls, Tovee, Purcell, Stewart and Azzopardi 1994b, Tovee and Rolls 1995, Rolls, Tovee and Panzeri 1999b, Rolls 2003, Rolls, Franco, Aggelopoulos and Jerez 2006b) (see Section C.3.4).

In the context of VisNet, and how the real visual system may operate to implement object recognition, the use of synchronization does not appear to match the way in which the visual system is organized. For example, von der Malsburg's argument would indicate that, using only a two-layer network, synchronization could provide the necessary feature linking to perform object recognition with relatively few neurons, because they can be reused again and again, linked differently for different objects. In contrast, the primate uses a considerable part of its cortex, perhaps 50% in monkeys, for visual processing, with therefore what could be in the order of 6×10^8 neurons and 6×10^{12} synapses involved (estimating from the values given in Table 1.1), so that the solution adopted by the real visual system may be one which relies on many neurons with simpler processing than arbitrary syntax implemented by synchronous firing of separate assemblies suggests. On the other hand, a solution such as that investigated by VisNet, which forms low-order combinations of what is represented in previous layers, is very demanding in terms of the number of neurons required, and this matches what is found in the primate visual system. It will be fascinating to see how research on these different approaches to processing in the primate visual system develops. For the development of both approaches, the use of well-defined neuronal network models is proving to be very helpful.

25.5.5.2 Sigma-Pi neurons

Another approach to a binding mechanism is to group spatial features based on local mechanisms that might operate for closely adjacent synapses on a dendrite (in what is a Sigma-Pi type of neuron, see Sections 25.6 and A.2.3) (Finkel and Edelman 1987, Mel, Ruderman and Archie 1998). A problem for such architectures is how to force one particular neuron to respond to the same feature combination invariantly with respect to all the ways in which that feature combination might occur in a scene.

25.5.5.3 Binding of features and their relative spatial position by feature combination neurons

The approach to the spatial binding problem that is proposed for VisNet is that individual neurons at an early stage of processing are set up (by learning) to respond to low order combinations of input features occurring in a given relative spatial arrangement and position on the retina (Rolls 1992a, Rolls 1994a, Rolls 1995a, Wallis and Rolls 1997, Rolls and Treves 1998, Elliffe, Rolls and Stringer 2002, Rolls and Deco 2002, Rolls 2008d) (cf. Feldman (1985)). (By low order combinations of input features we mean combinations of a few input features. By forming neurons that respond to combinations of a few features in the correct spatial arrangement the advantages of the scheme for syntactic binding are obtained, yet without the combinatorial explosion that would result if the feature combination neurons responded to combinations of many input features so producing potentially very specifically tuned neurons which very rarely responded.) Then invariant representations are developed in the next layer from these feature combination neurons which already contain evidence on the local spatial arrangement of features. Finally, in later layers, only one stimulus would be specified by the particular set of low order feature combination neurons present, even though each feature combination neuron would itself be somewhat invariant. The overall design of the scheme is shown in Fig. 2.2. Evidence that many neurons in V1 respond to combinations of spatial features with the correct spatial configuration is now starting to appear (see Section 25.4), and neurons that respond to feature combinations (such as two lines with a defined angle between them, and overall orientation) are found in V2 (Hegde and Van Essen 2000, Ito and Komatsu 2004). The tuning of a VisNet layer 1 neuron to a combination of features in the correct relative spatial position is illustrated in Figs. 25.22 and 25.23.

25.5.5.4 Discrimination between stimuli with super- and sub-set feature combinations

Some investigations with VisNet (Wallis and Rolls 1997) have involved groups of stimuli that might be identified by some unique feature common to all transformations of a particular stimulus. This might allow VisNet to solve the problem of transform invariance by simply learning to respond to a unique feature present in each stimulus. For example, even in the case where VisNet was trained on invariant discrimination of T, L, and +, the representation of the T stimulus at the spatial filter level inputs to VisNet might contain unique patterns of filter outputs where the horizontal and vertical parts of the T join. The unique filter outputs thus formed might distinguish the T from for example the L.

Elliffe, Rolls and Stringer (2002) tested whether VisNet is able to form transform invariant cells with stimuli that are specially composed from a common alphabet of features, with no stimulus containing any firing in the spatial filter inputs to VisNet not present in at least one of the other stimuli. The limited alphabet enables the set of stimuli to consist of feature sets which are subsets or supersets of those in the other stimuli.

Feature-Combination Stimulus Set

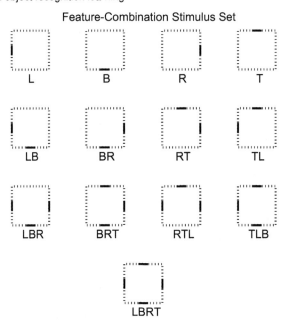

Fig. 25.41 Merged feature objects. All members of the full object set are shown, using a dotted line to represent the central 32×32 square on which the individual features are positioned, with the features themselves shown as dark line segments. Nomenclature is by acronym of the features present. (After Elliffe, Rolls and Stringer 2002.)

For these experiments the common pool of stimulus features chosen was a set of two horizontal and two vertical 8×1 bars, each aligned with the sides of a 32×32 square. The stimuli can be constructed by arbitrary combination of these base level features. We note that effectively the stimulus set consists of four features, a top bar (T), a bottom bar (B), a left bar (L), and a right bar (R). Figure 25.41 shows the complete set used, containing every possible image feature combination. (Note that the two double-feature combinations where the features are parallel to each other are not included, in the interests of retaining symmetry and equal inter-object overlap within each feature-combination level.) Subsequent discussion will group these objects by the number of features each contains: single-; double-; triple-; and quadruple-feature objects correspond to the respective rows of Fig. 25.41. Stimuli are referred to by the list of features they contain; e.g. 'LBR' contains the left, bottom, and right features, while 'TL' contains top and left only. Further details of how the stimuli were prepared are provided by Elliffe, Rolls and Stringer (2002).

To train the network a stimulus was presented in a randomized sequence of nine locations in a square grid across the 128×128 input retina. The central location of the square grid was in the centre of the 'retina', and the eight other locations were offset 8 pixels horizontally and/or vertically from this. Two different learning rules were used, 'Hebbian' (25.16), and 'trace' (25.17), and also an untrained condition with random weights. As in earlier work (Wallis and Rolls 1997, Rolls and Milward 2000) only the trace rule led to any cells with invariant responses, and the results shown here are for networks trained with the trace rule.

The results with VisNet trained on the set of stimuli shown in Fig. 25.41 with the trace rule are as follows. Firstly, it was found that single neurons in the top layer learned to differentiate between the stimuli in that the responses of individual neurons were maximal for one of the stimuli and had no response to any of the other stimuli invariantly with respect

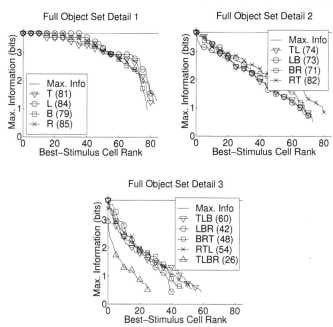

Fig. 25.42 Performance of VisNet2 on the full set of stimuli shown in Fig. 25.41. Separate graphs showing the information available about the stimulus for cells tuned to respond best to each of the stimuli are shown. The number of cells responding best to each of the stimuli is indicated in parentheses. The information values are shown for the different cells ranked according to how much information about that stimulus they encode. Separate graphs are shown for cells tuned to stimuli consisting of single features, pairs of features, and triples of features as well as the quadruple feature stimulus TLBR. (After Elliffe, Rolls and Stringer 2002.)

to location. Secondly, to assess how well every stimulus was encoded for in this way, Fig. 25.42 shows the information available about each of the stimuli consisting of feature singles, feature pairs, feature triples, and the quadruple-feature stimulus 'TLBR'. The single cell information available from the 26–85 cells with best tuning to each of the stimuli is shown. The cells in general conveyed translation invariant information about the stimulus to which they responded, with indeed cells which perfectly discriminated one of the stimuli from all others over every testing position (for all stimuli except 'RTL' and 'TLBR').

The results presented show clearly that the VisNet paradigm can accommodate networks that can perform invariant discrimination of objects which have a subset–superset relationship. The result has important consequences for feature binding and for discriminating stimuli for other stimuli which may be supersets of the first stimulus. For example, a VisNet cell which responds invariantly to feature combination TL can genuinely signal the presence of exactly that combination, and will not necessarily be activated by T alone, or by TLB. The basis for this separation by competitive networks of stimuli that are subsets and supersets of each other is described in Sections B.4 and 7.4, and by Rolls and Treves (1998, Section 4.3.6).

25.5.5.5 Feature binding in a hierarchical network with invariant representations of local feature combinations

In this section we consider the ability of output layer neurons to learn new stimuli if the lower layers are trained solely through exposure to simpler feature combinations from which the new stimuli are composed. A key question we address is how invariant representations of low order feature combinations in the early layers of the visual system are able to uniquely specify the correct spatial arrangement of features in the overall stimulus and contribute to preventing false recognition errors in the output layer.

The problem, and its proposed solution, can be treated as follows. Consider an object 1234 made from the features 1, 2, 3 and 4. The invariant low order feature combinations might represent 12, 23, and 34. Then if neurons at the next layer respond to combinations of the activity of these neurons, the only neurons in the next layer that would respond would be those tuned to 1234, not to for example 3412, which is distinguished from 1234 by the input of a pair neuron responding to 41 rather than to 23. The argument (Rolls 1992a) is that low-order spatial feature combination neurons in the early stage contain sufficient spatial information so that a particular combination of those low-order feature combination neurons specifies a unique object, even if the relative positions of the low-order feature combination neurons are not known, because they are somewhat invariant.

The architecture of VisNet is intended to solve this problem partly by allowing high spatial precision combinations of input features to be formed in layer 1. The actual input features in VisNet are, as described above, the output of oriented spatial-frequency tuned filters, and the combinations of these formed in layer 1 might thus be thought of in a simple way as for example a T or an L or for that matter a Y. Then in layer 2, application of the trace rule might enable neurons to respond to a T with limited spatial invariance (limited to the size of the region of layer 1 from which layer 2 cells receive their input). Then an 'object' such as H might be formed at a higher layer because of a conjunction of two Ts in the same small region.

To show that VisNet can actually solve this problem, Elliffe, Rolls and Stringer (2002) performed the experiments described next. They trained the first two layers of VisNet with feature pair combinations, forming representations of feature pairs with some translation invariance in layer 2. Then they used feature triples as input stimuli, allowed no more learning in layers 1 and 2, and then investigated whether layers 3 and 4 could be trained to produce invariant representations of the triples where the triples could only be distinguished if the local spatial arrangement of the features within the triple had effectively to be encoded in order to distinguish the different triples. For this experiment, they needed stimuli that could be specified in terms of a set of different features (they chose vertical (1), diagonal (2), and horizontal (3) bars) each capable of being shown at a set of different relative spatial positions (designated A, B and C), as shown in Fig. 25.43. The stimuli are thus defined in terms of what features are present and their precise spatial arrangement with respect to each other. The length of the horizontal and vertical feature bars shown in Fig. 25.43 is 8 pixels. To train the network a stimulus (that is a pair or triple feature combination) is presented in a randomized sequence of nine locations in a square grid across the 128×128 input retina. The central location of the square grid is in the centre of the 'retina', and the eight other locations are offset 8 pixels horizontally and/or vertically from this. We refer to the two and three feature stimuli as 'pairs' and 'triples', respectively. Individual stimuli are denoted by three numbers which refer to the individual features present in positions A, B and C, respectively. For example, a stimulus with positions A and C containing a vertical and diagonal bar, respectively, would be referred to as stimulus 102, where the 0 denotes no feature present in position B. In total there are 18 pairs (120, 130, 210, 230, 310, 320, 012, 013, 021, 023, 031, 032, 102, 103, 201, 203, 301, 302) and 6 triples (123, 132, 213, 231, 312, 321). This nomenclature not only defines which features are present within objects, but also the spatial relationships of their component features. Then the computational problem can be illustrated by considering the triple 123. If invariant representations are formed of single features, then there would be no way that neurons higher in the hierarchy could distinguish the object 123 from 213 or any other arrangement of the three features. An approach to this problem (see e.g. Rolls (1992a)) is to form early on in the processing neurons that respond to overlapping combinations of

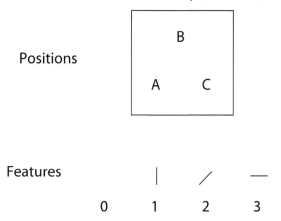

Positions

Features

0   1   2   3

Fig. 25.43 Feature combinations for experiments of Section 25.5.5.5: there are 3 features denoted by 1, 2 and 3 (including a blank space 0) that can be placed in any of 3 positions A, B, and C. Individual stimuli are denoted by three consecutive numbers which refer to the individual features present in positions A, B and C respectively. In the experiments in Section 25.5.5.5, layers 1 and 2 were trained on stimuli consisting of pairs of the features, and layers 3 and 4 were trained on stimuli consisting of triples. Then the network was tested to show whether layer 4 neurons would distinguish between triples, even though the first two layers had only been trained on pairs. In addition, the network was tested to show whether individual cells in layer 4 could distinguish between triples even in locations where the triples were not presented during training. (After Elliffe, Rolls and Stringer 2002.)

Table 25.5 The different training regimes used in VisNet experiments 1–4 of Section 25.5.5.5. In the no training condition the synaptic weights were left in their initial untrained random values.

	Layers 1, 2	Layers 3, 4
Experiment 1	trained on pairs	trained on triples
Experiment 2	no training	no training
Experiment 3	no training	trained on triples
Experiment 4	trained on triples	trained on triples

features in the correct spatial arrangement, and then to develop invariant representations in the next layer from these neurons which already contain evidence on the local spatial arrangement of features. An example might be that with the object 123, the invariant feature pairs would represent 120, 023, and 103. Then if neurons at the next layer correspond to combinations of these neurons, the only next layer neurons that would respond would be those tuned to 123, not to for example 213. The argument is that the low-order spatial feature combination neurons in the early stage contain sufficient spatial information so that a particular combination of those low-order feature combination neurons specifies a unique object, even if the relative positions of the low-order feature combination neurons are not known because these neurons are somewhat translation invariant (cf. also Fukushima (1988)).

The stimuli used in the experiments of Elliffe, Rolls and Stringer (2002) were constructed from pre-processed component features as discussed in Section 25.5.5.4. That is, base stimuli containing a single feature were constructed and filtered, and then the pairs and triples were constructed by merging these pre-processed single feature images. In the first experiment layers 1 and 2 of VisNet were trained with the 18 feature pairs, each stimulus being presented in sequences of 9 locations across the input. This led to the formation of neurons that responded to the feature pairs with some translation invariance in layer 2. Then they trained layers 3 and

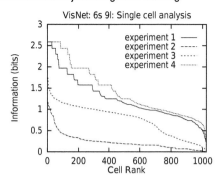 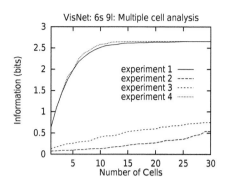

Fig. 25.44 Numerical results for experiments 1–4 as described in Table 25.5, with the trace learning rule (25.17). On the left are single cell information measures, and on the right are multiple cell information measures. (After Elliffe, Rolls and Stringer 2002.)

4 on the 6 feature triples in the same 9 locations, while allowing no more learning in layers 1 and 2, and examined whether the output layer of VisNet had developed transform invariant neurons to the 6 triples. The idea was to test whether layers 3 and 4 could be trained to produce invariant representations of the triples where the triples could only be distinguished if the local spatial arrangement of the features within the triple had effectively to be encoded in order to distinguish the different triples. The results from this experiment were compared and contrasted with results from three other experiments which involved different training regimes for layers 1,2 and layers 3,4. All four experiments are summarized in Table 25.5. Experiment 2 involved no training in layers 1,2 and 3,4, with the synaptic weights left unchanged from their initial random values. These results are included as a baseline performance with which to compare results from the other experiments 1, 3 and 4. The model parameters used in these experiments were as described by Rolls and Milward (2000) and Rolls and Stringer (2001a).

In Fig. 25.44 we present numerical results for the four experiments listed in Table 25.5. On the left are the single cell information measures for all top (4th) layer neurons ranked in order of their invariance to the triples, while on the right are multiple cell information measures. To help to interpret these results we can compute the maximum single cell information measure according to

$$\text{Maximum single cell information } = \log_2(\text{Number of triples}), \qquad (25.42)$$

where the number of triples is 6. This gives a maximum single cell information measure of 2.6 bits for these test cases. First, comparing the results for experiment 1 with the baseline performance of experiment 2 (no training) demonstrates that even with the first two layers trained to form invariant responses to the pairs, and then only layers 3 and 4 trained on feature triples, layer 4 is indeed capable of developing translation invariant neurons that can discriminate effectively between the 6 different feature triples. Indeed, from the single cell information measures it can be seen that a number of cells have reached the maximum level of performance in experiment 1. In addition, the multiple cell information analysis presented in Fig. 25.44 shows that all the stimuli could be discriminated from each other by the firing of a number of cells. Analysis of the response profiles of individual cells showed that a fourth layer cell could respond to one of the triple feature stimuli and have no response to any other of the triple feature stimuli invariantly with respect to location.

A comparison of the results from experiment 1 with those from experiment 3 (see Table 25.5 and Fig. 25.44) reveals that training the first two layers to develop neurons that respond invariantly to the pairs (performed in experiment 1) actually leads to improved invariance of

4th layer neurons to the triples, as compared with when the first two layers are left untrained (experiment 3).

Two conclusions follow from these results (Elliffe, Rolls and Stringer 2002). First, a hierarchical network that seeks to produce invariant representations in the way used by VisNet can solve the feature binding problem. In particular, when feature pairs in layer 2 with some translation invariance are used as the input to later layers, these later layers can nevertheless build invariant representations of objects where all the individual features in the stimulus must occur in the correct spatial position relative to each other. This is possible because the feature combination neurons formed in the first layer (which could be trained just with a Hebb rule) do respond to combinations of input features in the correct spatial configuration, partly because of the limited size of their receptive fields (see e.g. Fig. 25.20).

The second conclusion is that even though early layers can in this case only respond to small feature subsets, these provide, with no further training of layers 1 and 2, an adequate basis for learning to discriminate in layers 3 and 4 stimuli consisting of combinations of larger numbers of features. Indeed, comparing results from experiment 1 with experiment 4 (in which all layers were trained on triples, see Table 25.5) demonstrates that training the lower layer neurons to develop invariant responses to the pairs offers almost as good performance as training all layers on the triples (see Fig. 25.44).

25.5.5.6 Stimulus generalization to new locations

Another important aspect of the architecture of VisNet is that it need not be trained with every stimulus in every possible location. Indeed, part of the hypothesis (Rolls 1992a) is that training early layers (e.g. 1–3) with a wide range of visual stimuli will set up feature analyzers in these early layers which are appropriate later on with no further training of early layers for new objects. For example, presentation of a new object might result in large numbers of low order feature combination neurons in early layers of VisNet being active, but the particular set of feature combination neurons active would be different for the new object. The later layers of the network (in VisNet, layer 4) would then learn this new set of active layer 3 neurons as encoding the new object. However, if the new object was then shown in a new location, the same set of layer 3 neurons would be active because they respond with spatial invariance to feature combinations, and given that the layer 3 to 4 connections had already been set up by the new object, the correct layer 4 neurons would be activated by the new object in its new untrained location, and without any further training.

To test this hypothesis, Elliffe, Rolls and Stringer (2002) repeated the general procedure of experiment 1 of Section 25.5.5.5, training layers 1 and 2 with feature pairs, but then instead trained layers 3 and 4 on the triples in only 7 of the original 9 locations. The crucial test was to determine whether VisNet could form top layer neurons that responded invariantly to the 6 triples when presented over all nine locations, not just the seven locations at which the triples had been presented during training. The results are presented in Fig. 25.45, with single cell information measures on the left and multiple cell information measures on the right. VisNet is still able to develop some fourth layer neurons with perfect invariance, that is which have invariant responses over all nine location, as shown by the single cell information analysis. The response profiles of individual fourth layer cells showed that they can continue to discriminate between the triples even in the two locations where the triples were not presented during training. In addition, the multiple cell analysis shown in Fig. 25.45 demonstrates that a small population of cells was able to discriminate between all of the stimuli irrespective of location, even though for two of the test locations the triples had not been trained at those particular locations during the training of layers 3 and 4.

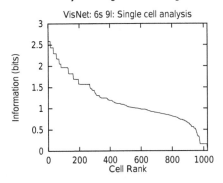
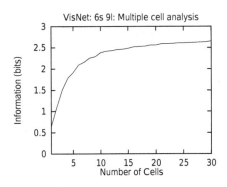

Fig. 25.45 Generalization to new locations: numerical results for a repeat of experiment 1 of Section 25.5.5.5 with the triples presented at only 7 of the original 9 locations during training, and with the trace learning rule (25.17). On the left are single cell information measures, and on the right are multiple cell information measures.

25.5.5.7 Discussion of feature binding in hierarchical layered networks

Elliffe, Rolls and Stringer (2002) thus first showed (see Section 25.5.5.4) that hierarchical feature detecting neural networks can learn to respond differently to stimuli that consist of unique combinations of non-unique input features, and that this extends to stimuli that are direct subsets or supersets of the features present in other stimuli.

Second, Elliffe, Rolls and Stringer (2002) investigated (see Section 25.5.5.5) the hypothesis that hierarchical layered networks can produce identification of unique stimuli even when the feature combination neurons used to define the stimuli are themselves partly translation invariant. The stimulus identification should work correctly because feature combination neurons in which the spatial features are bound together with high spatial precision are formed in the first layer. Then at later layers when neurons with some translation invariance are formed, the neurons nevertheless contain information about the relative spatial position of the original features. There is only then one object which will be consistent with the set of active neurons at earlier layers, which though somewhat translation invariant as combination neurons, reflect in the activity of each neuron information about the original spatial position of the features. We note that the trace rule training used in early layers (1 and 2) in Experiments 1 and 4 would set up partly invariant feature combination neurons, and yet the late layers (3 and 4) were able to produce during training neurons in layer 4 that responded to stimuli that consisted of unique spatial arrangements of lower order feature combinations. Moreover, and very interestingly, Elliffe, Rolls and Stringer (2002) were able to demonstrate that VisNet layer 4 neurons would respond correctly to visual stimuli at untrained locations, provided that the feature subsets had been trained in early layers of the network at all locations, and that the whole stimulus had been trained at some locations in the later layers of the network.

The results described by Elliffe, Rolls and Stringer (2002) thus provide one solution to the feature binding problem. The solution that has been shown to work in the model is that in a multilayer competitive network, feature combination neurons which encode the spatial arrangement of the bound features are formed at intermediate layers of the network. Then neurons at later layers of the network which respond to combinations of active intermediate layer neurons do contain sufficient evidence about the local spatial arrangement of the features to identify stimuli because the local spatial arrangement is encoded by the intermediate layer neurons. The information required to solve the visual feature binding problem thus becomes encoded by self-organization into what become hard-wired properties of the network. In this sense, feature binding is not solved at run time by the necessity to instantaneously set up arbitrary syntactic links between sets of co-active neurons. The computational solution proposed to the superset/subset aspect of the binding problem will apply in principle to other multilayer competitive networks, although the issues considered here have not been explicitly addressed in architectures such as the Neocognitron (Fukushima and Miyake 1982).

Consistent with these hypotheses about how VisNet operates to achieve, by layer 4, position-invariant responses to stimuli defined by combinations of features in the correct spatial arrangement, investigations of the effective stimuli for neurons in intermediate layers of VisNet showed as follows. In layer 1, cells responded to the presence of individual features, or to low order combinations of features (e.g. a pair of features) in the correct spatial arrangement at a small number of nearby locations. In layers 2 and 3, neurons responded to single features or to higher order combinations of features (e.g. stimuli composed of feature triples) in more locations. These findings provide direct evidence that VisNet does operate as described above to solve the feature binding problem.

A further issue with hierarchical multilayer architectures such as VisNet is that false binding errors might occur in the following way (Mozer 1991, Mel and Fiser 2000). Consider the output of one layer in such a network in which there is information only about which pairs are present. How then could a neuron in the next layer discriminate between the whole stimulus (such as the triple 123 in the above experiment) and what could be considered a more distributed stimulus or multiple different stimuli composed of the separated subparts of that stimulus (e.g. the pairs 120, 023, 103 occurring in 3 of the 9 training locations in the above experiment)? The problem here is to distinguish a single object from multiple other objects containing the same component combinations (e.g. pairs). We proposed that part of the solution to this general problem in real visual systems is implemented through lateral inhibition between neurons in individual layers, and that this mechanism, implemented in VisNet, acts to reduce the possibility of false recognition errors in the following two ways.

First, consider the situation in which neurons in layer N have learned to represent low order feature combinations with location invariance, and where a neuron n in layer $N + 1$ has learned to respond to a particular set Ω of these feature combinations. The problem is that neuron n receives the same input from layer N as long as the same set Ω of feature combinations is present, and cannot distinguish between different spatial arrangements of these feature combinations. The question is how can neuron n respond only to a particular favoured spatial arrangement Ψ of the feature combinations contained within the set Ω. We suggest that as the favoured spatial arrangement Ψ is altered by rearranging the spatial relationships of the component feature combinations, the new feature combinations that are formed in new locations will stimulate additional neurons nearby in layer $N + 1$, and these will tend to inhibit the firing of neuron n. Thus, lateral inhibition within a layer will have the effect of making neurons more selective, ensuring neuron n responds only to a single spatial arrangement Ψ from the set of feature combinations Ω, and hence reducing the possibility of false recognition.

The second way in which lateral inhibition may help to reduce binding errors is through limiting the sparseness of neuronal firing rates within layers. In our discussion above the spurious stimuli we suggested that might lead to false recognition of triples were obtained from splitting up the component feature combinations (pairs) so that they occurred in separate training locations. However, this would lead to an increase in the number of features present in the complete stimulus; triples contain 3 features while their spurious counterparts would contain 6 features (resulting from 3 separate pairs). For this simple example, the increase in the number of features is not dramatic, but if we consider, say, stimuli composed of 4 features where the component feature combinations represented by lower layers might be triples, then to form spurious stimuli we need to use 12 features (resulting from 4 triples occurring in separate locations). But if the lower layers also represented all possible pairs then the number of features required in the spurious stimuli would increase further. In fact, as the size of the stimulus increases in terms of the number of features, and as the size of the component feature combinations represented by the lower layers increases, there is a combinatorial explosion in terms of the number of features required as we attempt to construct spurious stimuli to trigger false recognition. And the construction of such spurious stimuli will then be prevented through setting a limit on the sparseness of firing rates within layers, which will in turn set a limit on the number of features that can be represented. Lateral inhibition is likely to contribute in both these ways to the performance of VisNet when the stimuli consist of subsets and supersets of each other, as described in Section 25.5.5.4.

Another way is which the problem of multiple objects is addressed is by limiting the size of the receptive fields of inferior temporal cortex neurons so that neurons in IT respond primarily to the object being fixated, but with nevertheless some asymmetry in the receptive fields (see Section 25.5.10). Multiple objects are then 'seen' by virtue of being added to a visuo-spatial scratchpad, as addressed in Section 25.7.

A related issue that arises in this class of network is whether forming neurons that respond to feature combinations in the way described here leads to a combinatorial explosion in the number of neurons required. The solution to this issue that is proposed is to form only low-order combinations of features at any one stage of the network (Rolls (1992a); cf. Feldman (1985)). Using low-order combinations limits the number of neurons required, yet enables the type of computation that relies on feature combination neurons that is analyzed here to still be performed. The actual number of neurons required depends also on the redundancies present in the statistics of real-world images. Even given these factors, it is likely that a large number of neurons would be required if the ventral visual system performs the computation of invariant representations in the manner captured by the hypotheses implemented in VisNet. Consistent with this, a considerable part of the non-human primate brain is devoted to visual information processing. The fact that large numbers of neurons and a multilayer organization are present in the primate ventral visual system is actually thus consistent with the type of model of visual information processing described here.

25.5.6 Operation in a cluttered environment

In this section we consider how hierarchical layered networks of the type exemplified by VisNet operate in cluttered environments. Although there has been much work involving object recognition in cluttered environments with artificial vision systems, many such systems typically rely on some form of explicit segmentation followed by search and template matching procedure (see Ullman (1996) for a general review). In natural environments, objects may not only appear against cluttered (natural) backgrounds, but also the object may be partially occluded. Biological nervous systems operate in quite a different manner to those artificial vision systems that rely on search and template matching, and the way in which biological systems cope with cluttered environments and partial occlusion is likely to be quite different also.

One of the factors that will influence the performance of the type of architecture considered here, hierarchically organized series of competitive networks, which form one class of approaches to biologically relevant networks for invariant object recognition (Fukushima 1980, Rolls 1992a, Wallis and Rolls 1997, Poggio and Edelman 1990, Rolls and Treves 1998, Rolls 2008d, Rolls 2012c), is how lateral inhibition and competition are managed within a layer. Even if an object is not obscured, the effect of a cluttered background will be to fire additional neurons, which will in turn to some extent compete with and inhibit those neurons that are specifically tuned to respond to the desired object. Moreover, where the clutter is adjacent to part of the object, the feature analysing neurons activated against a blank background

Fig. 25.46 Cluttered backgrounds used in VisNet simulations: backgrounds 1 and 2 are on the left and right respectively.

might be different from those activated against a cluttered background, if there is no explicit segmentation process. We consider these issues next, following investigations of Stringer and Rolls (2000).

25.5.6.1   VisNet simulations with stimuli in cluttered backgrounds

In this section we show that recognition of objects learned previously against a blank background is hardly affected by the presence of a natural cluttered background. We go on to consider what happens when VisNet is set the task of learning new stimuli presented against cluttered backgrounds.

The images used for training and testing VisNet in the simulations described next performed by Stringer and Rolls (2000) were specially constructed. There were 7 face stimuli approximately 64 pixels in height constructed without backgrounds from those shown in Fig. 25.30. In addition there were 3 possible backgrounds: a blank background (greyscale 127, where the range is 0–255), and two cluttered backgrounds as shown in Fig. 25.46 which are 128×128 pixels in size. Each image presented to VisNet's 128×128 input retina was composed of a single face stimulus positioned at one of 9 locations on either a blank or cluttered background. The cluttered background was intended to be like the background against which an object might be viewed in a natural scene. If a background is used in an experiment described here, the same background is always used, and it is always in the same position, with stimuli moved to different positions on it. The 9 stimulus locations are arranged in a square grid across the background, where the grid spacings are 32 pixels horizontally or vertically. Before images were presented to VisNet's input layer they were pre-processed by the standard set of input filters which accord with the general tuning profiles of simple cells in V1 (Hawken and Parker 1987); full details are given in Rolls and Milward (2000). To train the network a sequence of images is presented to VisNet's retina that corresponds to a single stimulus occurring in a randomized sequence of the 9 locations across a background. At each presentation the activation of individual neurons is calculated, then their firing rates are calculated, and then the synaptic weights are updated. After a stimulus has been presented in all the training locations, a new stimulus is chosen at random and the process repeated. The presentation of all the stimuli across all locations constitutes 1 epoch of training. In this manner the network is trained one layer at a time starting with layer 1 and finishing with

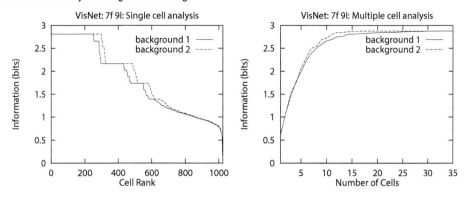

Fig. 25.47 Numerical results for experiment 2, with the 7 faces presented on a blank background during training and a cluttered background during testing. On the left are single cell information measures, and on the right are multiple cell information measures.

layer 4. In the investigations described in this subsection, the numbers of training epochs for layers 1–4 were 50, 100, 100 and 75 respectively.

In this experiment (see Stringer and Rolls (2000), experiment 2), VisNet was trained with the 7 face stimuli presented on a blank background, but tested with the faces presented on each of the 2 cluttered backgrounds. Figure 25.47 shows results for experiment 2, with single and multiple cell information measures on the left and right respectively. It can be seen that a number of cells have reached the maximum possible single cell information measure of 2.8 bits ($\log_2$ of the number of stimuli) for this test case, and that the multiple cell information measures also reach the 2.8 bits indicating perfect performance. Compared to performance when shown against a blank background, there was very little deterioration in performance when testing with the faces presented on either of the two cluttered backgrounds. This is an interesting result to compare with many artificial vision systems that would need to carry out computationally intensive serial searching and template matching procedures in order to achieve such results. In contrast, the VisNet neural network architecture is able to perform such recognition relatively quickly through a simple feedforward computation. Further results from this experiment are presented in Fig. 25.48 where we show the response profiles of a 4th layer neuron to the 7 faces presented on cluttered background 1 during testing. It can be seen that this neuron achieves excellent invariant responses to the 7 faces even with the faces presented on a cluttered background. The response profiles are independent of location but differentiate between the faces in that the responses are maximal for only one of the faces and minimal for all other faces.

This is an interesting and important result, for it shows that after learning, special mechanisms for segmentation and for attention are not needed in order for neurons already tuned by previous learning to the stimuli to be activated correctly in the output layer. Although the experiments described here tested for position invariance, we predict and would expect that the same results would be demonstrable for size and view invariant representations of objects.

In experiments 3 and 4 of Stringer and Rolls (2000), VisNet was trained with the 7 face stimuli presented on either one of the 2 cluttered backgrounds, but tested with the faces presented on a blank background. Results for this experiment showed poor performance. The results of experiments 3 and 4 suggest that in order for a cell to *learn* invariant responses to different transforms of a stimulus when it is presented during training in a cluttered background, some form of segmentation is required in order to separate the figure (i.e. the stimulus or object) from the background. This segmentation might be performed using evidence in the visual scene about different depths, motions, colours, etc. of the object from

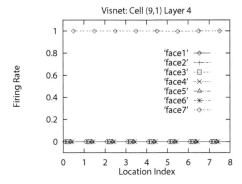

Fig. 25.48 Response profiles of a top layer neuron to the 7 faces from experiment 2 of Stringer and Rolls (2000), with the faces presented against cluttered background 1 during testing.

its background. In the visual system, this might mean combining evidence represented in different cortical areas, and might be performed by cross-connections between cortical areas to enable such evidence to help separate the representations of objects from their backgrounds in the form-representing cortical areas.

Another mechanism that helps the operation of architectures such as VisNet and the primate visual system to learn about new objects in cluttered scenes is that the receptive fields of inferior temporal cortex neurons become much smaller when objects are seen against natural backgrounds (Sections 25.5.9.1 and 25.5.9). This will help greatly to learn about new objects that are being fixated, by reducing responsiveness to other features elsewhere in the scene.

Another mechanism that might help the learning of new objects in a natural scene is attention. An attentional mechanism might highlight the current stimulus being attended to and suppress the effects of background noise, providing a training representation of the object more like that which would be produced when it is presented against a blank background. The mechanisms that could implement such attentional processes are described in Chapter 6. If such attentional mechanisms do contribute to the development of view invariance, then it follows that cells in the temporal cortex may only develop transform invariant responses to objects to which attention is directed.

Part of the reason for the poor performance in experiments 3 and 4 was probably that the stimuli were always presented against the same fixed background (for technical reasons), and thus the neurons learned about the background rather than the stimuli. Part of the difficulty that hierarchical multilayer competitive networks have with learning in cluttered environments may more generally be that without explicit segmentation of the stimulus from its background, at least some of the features that should be formed to encode the stimuli are not formed properly, because the neurons learn to respond to combinations of inputs which come partly from the stimulus, and partly from the background. To investigate this, Stringer and Rolls (2000) performed experiment 5 in which layers 1–3 were pretrained with stimuli to ensure that good feature combination neurons for stimuli were available, and then allowed learning in only layer 4 when stimuli were presented in the cluttered backgrounds. Layer 4 was then trained in the usual way with the 7 faces presented against a cluttered background. The results for this experiment are shown in Fig. 25.49, with single and multiple cell information measures on the left and right respectively. It was found that prior random exposure to the face stimuli led to much improved performance. Indeed, it can be seen that a number of cells have reached the maximum possible single cell information measure of 2.8 bits for this test case, although the

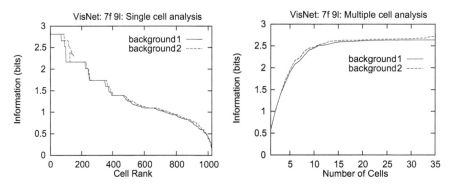

Fig. 25.49 Numerical results for experiment 5 of Stringer and Rolls (2000). In this experiment VisNet is first exposed to a completely random sequence of faces in different positions against a blank background during which layers 1–3 are allowed to learn. This builds general feature detecting neurons in the lower layers that are tuned to the face stimuli, but cannot develop view invariance since there is no temporal structure to the order in which different views of different faces occur. Then layer 4 is trained in the usual way with the 7 faces presented against a cluttered background, where the images are now presented such that different views of the same face occur close together in time. On the left are single cell information measures, and on the right are multiple cell information measures.

multiple cell information measures do not quite reach the 2.8 bits that would indicate perfect performance for the complete face set.

These results demonstrate that the problem of developing position invariant neurons to stimuli occurring against cluttered backgrounds may be ameliorated by the prior existence of stimulus-tuned feature-detecting neurons in the early layers of the visual system, and that these feature-detecting neurons may be set up through previous exposure to the relevant class of objects. When tested in cluttered environments, the background clutter may of course activate some other neurons in the output layer, but at least the neurons that have learned to respond to the trained stimuli are activated. The result of this activity is sufficient for the activity in the output layer to be useful, in the sense that it can be read off correctly by a pattern associator connected to the output layer. Indeed, Stringer and Rolls (2000) tested this by connecting a pattern associator to layer 4 of VisNet. The pattern associator had seven neurons, one for each face, and 1,024 inputs, one from each neuron in layer 4 of VisNet. The pattern associator learned when trained with a simple associative Hebb rule (equation 25.16 on page 604) to activate the correct output neuron whenever one of the faces was shown in any position in the uncluttered environment. This ability was shown to be dependent on invariant neurons for each stimulus in the output layer of VisNet, for the pattern associator could not be taught the task if VisNet had not been previously trained with a trace learning rule to produce invariant representations. Then it was shown that exactly the correct neuron was activated when any of the faces was shown in any position with the cluttered background. This read-off by a pattern associator is exactly what we hypothesize takes place in the brain, in that the inferior temporal visual cortex (where neurons with invariant responses are found) projects to structures such as the orbitofrontal cortex and amygdala, where associations between the invariant visual representations and stimuli such as taste and touch are learned (Rolls and Treves 1998, Rolls 1999a, Rolls 2014a) (see Chapter 15). Thus testing whether the output of an architecture such as VisNet can be used effectively by a pattern associator is a very biologically relevant way to evaluate the performance of this class of architecture.

25.5.6.2 Learning invariant representations of an object with multiple objects in the scene and with cluttered backgrounds

The results of the experiments just described suggest that in order for a neuron to *learn* invariant responses to different transforms of a stimulus when it is presented during training in a cluttered background, some form of segmentation is required in order to separate the figure (i.e. the stimulus or object) from the background. This segmentation might be performed using evidence in the visual scene about different depths, motions, colours, etc. of the object from its background. In the visual system, this might mean combining evidence represented in different cortical areas, and might be performed by cross-connections between cortical areas to enable such evidence to help separate the representations of objects from their backgrounds in the form-representing cortical areas.

A second way in which training a feature hierarchy network in a cluttered natural scene may be facilitated follows from the finding that the receptive fields of inferior temporal cortex neurons shrink from in the order of 70 degrees in diameter when only one object is present in a blank scene to much smaller values of as little as 5–10 degrees close to the fovea in complex natural scenes (Rolls, Aggelopoulos and Zheng 2003a). The proposed mechanism for this is that if there is an object at the fovea, this object, because of the high cortical magnification factor at the fovea, dominates the activity of neurons in the inferior temporal cortex by competitive interactions (Trappenberg, Rolls and Stringer 2002, Deco and Rolls 2004) (see Section 25.5.9). This allows primarily the object at the fovea to be represented in the inferior temporal cortex, and, it is proposed, for learning to be about this object, and not about the other objects in a whole scene.

Third, top-down spatial attention (Deco and Rolls 2004, Deco and Rolls 2005a, Rolls and Deco 2006, Rolls 2008d, Rolls 2008f) (see Chapter 6) could bias the competition towards a region of visual space where the object to be learned is located.

Fourth, if object 1 is presented during training with different objects present on different trials, then the competitive networks that are part of VisNet will learn to represent each object separately, because the features that are part of each object will be much more strongly associated together, than are those features with the other features present in the different objects seen on some trials during training (Stringer and Rolls 2008, Stringer, Rolls and Tromans 2007b). It is a natural property of competitive networks that input features that co-occur very frequently together are allocated output neurons to represent the pattern as a result of the learning. Input features that do not co-occur frequently, may not have output neurons allocated to them. This principle may help feature hierarchy systems to learn representations of individual objects, even when other objects with some of the same features are present in the visual scene, but with different other objects on different trials. With this fundamental and interesting property of competitive networks, it has now become possible for VisNet to self-organize invariant representations of individual objects, even though each object is always presented during training with at least one other object present in the scene (Stringer and Rolls 2008, Stringer, Rolls and Tromans 2007b).

25.5.6.3 VisNet simulations with partially occluded stimuli

In this section we examine the recognition of partially occluded stimuli. Many artificial vision systems that perform object recognition typically search for specific markers in stimuli, and hence their performance may become fragile if key parts of a stimulus are occluded. However, in contrast we demonstrate that the model of invariance learning in the brain discussed here can continue to offer robust performance with this kind of problem, and that the model is able to correctly identify stimuli with considerable flexibility about what part of a stimulus is visible.

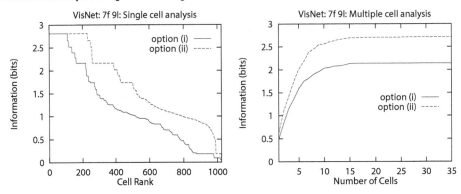

Fig. 25.50 Effects of partial occlusion of a stimulus: numerical results for experiment 6 of Stringer and Rolls, 2000, with the 7 faces presented on a blank background during both training and testing. Training was performed with the whole face. However, during testing there are two options: either (i) the top half of all the faces are occluded, or (ii) the bottom half of all the faces are occluded. On the left are single cell information measures, and on the right are multiple cell information measures.

In these simulations (Stringer and Rolls 2000), training and testing was performed with a blank background to avoid confounding the two separate problems of occlusion and background clutter. In object recognition tasks, artificial vision systems may typically rely on being able to locate a small number of key markers on a stimulus in order to be able to identify it. This approach can become fragile when a number of these markers become obscured. In contrast, biological vision systems may generalize or complete from a partial input as a result of the use of distributed representations in neural networks, and this could lead to greater robustness in situations of partial occlusion.

In this experiment (6 of Stringer and Rolls (2000)), the network was first trained with the 7 face stimuli without occlusion, but during testing there were two options: either (i) the top halves of all the faces were occluded, or (ii) the bottom halves of all the faces were occluded. Since VisNet was tested with either the top or bottom half of the stimuli no stimulus features were common to the two test options. This ensures that if performance is good with both options, the performance cannot be based on the use of a single feature to identify a stimulus. Results for this experiment are shown in Fig. 25.50, with single and multiple cell information measures on the left and right respectively. When compared with the performance without occlusion (Stringer and Rolls 2000), Fig. 25.50 shows that there is only a modest drop in performance in the single cell information measures when the stimuli are partially occluded.

For both options (i) and (ii), even with partially occluded stimuli, a number of cells continue to respond maximally to one preferred stimulus in all locations, while responding minimally to all other stimuli. However, comparing results from options (i) and (ii) shows that the network performance is better when the bottom half of the faces is occluded. This is consistent with psychological results showing that face recognition is performed more easily when the top halves of faces are visible rather than the bottom halves (see Bruce (1988)). The top half of a face will generally contain salient features, e.g. eyes and hair, that are particularly helpful for recognition of the individual, and it is interesting that these simulations appear to further demonstrate this point. Furthermore, the multiple cell information measures confirm that performance is better with the upper half of the face visible (option (ii)) than the lower half (option (i)). When the top halves of the faces are occluded the multiple cell information measure asymptotes to a suboptimal value reflecting the difficulty of discriminating between these more difficult images. Further results from experiment 6 are presented in Fig. 25.51 where we show the response profiles of a 4th layer neuron to the 7 faces, with the bottom half

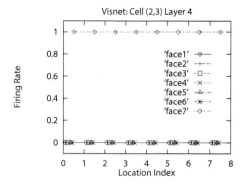

Fig. 25.51 Effects of partial occlusion of a stimulus. Response profiles of a top layer neuron to the 7 faces from experiment 6 of Stringer and Rolls (2000), with the bottom half of all the faces occluded during testing.

of all the faces occluded during testing. It can be seen that this neuron continues to respond invariantly to the 7 faces, responding maximally to one of the faces but minimally for all other faces.

Thus this model of the ventral visual system offers robust performance with this kind of problem, and the model is able to correctly identify stimuli with considerable flexibility about what part of a stimulus is visible, because it is effectively using distributed representations and associative processing.

25.5.7 Learning 3D transforms

In this section we describe investigations of Stringer and Rolls (2002) which show that trace learning can in the VisNet architecture solve the problem of in-depth rotation invariant object recognition by developing representations of the transforms which features undergo when they are on the surfaces of 3D objects. Moreover, it is shown that having learned how features on 3D objects transform as the object is rotated in depth, the network can correctly recognize novel 3D variations within a generic view of an object which is composed of previously learned feature combinations.

Rolls' hypothesis of how object recognition could be implemented in the brain postulates that trace rule learning helps invariant representations to form in two ways (Rolls 1992a, Rolls 1994a, Rolls 1995a, Rolls 2000a, Rolls 2008d, Rolls 2012c). The first process enables associations to be learned between different generic 3D views of an object where there are different qualitative shape descriptors. One example of this would be the front and back views of an object, which might have very different shape descriptors. Another example is provided by considering how the shape descriptors typical of 3D shapes, such as Y vertices, arrow vertices, cusps, and ellipse shapes, alter when most 3D objects are rotated in 3 dimensions. At some point in the 3D rotation, there is a catastrophic rearrangement of the shape descriptors as a new generic view can be seen (Koenderink 1990). An example of a catastrophic change to a new generic view is when a cup being viewed from slightly below is rotated so that one can see inside the cup from slightly above. The bottom surface disappears, the top surface of the cup changes from a cusp to an ellipse, and the inside of the cup with a whole set of new features comes into view. The second process is that within a generic view, as the object is rotated in depth, there will be no catastrophic changes in the qualitative 3D shape descriptors, but instead the quantitative values of the shape descriptors alter. For example, while the cup is being rotated within a generic view seen from somewhat below, the curvature of the cusp forming the top boundary will alter, but the qualitative shape descriptor will remain a cusp. Trace learning could help with both processes. That is, trace learning could help to associate together qualitatively different sets of shape descriptors that occur close together in time, and describe for example the generically different views of a cup. Trace learning could also help with the second process, and learn to associate together the different quantitative values of shape descriptors that typically occur when objects are rotated within a generic view.

We note that there is evidence that some neurons in the inferior temporal cortex may show the two types of 3D invariance. First, Booth and Rolls (1998) showed that some inferior temporal cortex neurons can respond to different generic views of familiar 3D objects. Second, some neurons do generalize across quantitative changes in the values of 3D shape descriptors while faces (Hasselmo, Rolls, Baylis and Nalwa 1989b) and objects (Tanaka 1996, Logothetis, Pauls and Poggio 1995) are rotated within generic views. Indeed, Logothetis, Pauls and Poggio (1995) showed that a few inferior temporal cortex neurons can generalize to novel (untrained) values of the quantitative shape descriptors typical of within-generic view object rotation.

In addition to the qualitative shape descriptor changes that occur catastrophically between different generic views of an object, and the quantitative changes of 3D shape descriptors that occur within a generic view, there is a third type of transform that must be learned for correct invariant recognition of 3D objects as they rotate in depth. This third type of transform is that which occurs to the surface features on a 3D object as it transforms in depth. The main aim here is to consider mechanisms that could enable neurons to learn this third type of transform, that is how to generalize correctly over the changes in the surface markings on 3D objects that are typically encountered as 3D objects rotate within a generic view. Examples of the types of perspectival transforms investigated are shown in Fig. 25.52. Surface markings on the sphere that consist of combinations of three features in different spatial arrangements undergo characteristic transforms as the sphere is rotated from 0 degrees towards –60 degrees and +60 degrees. We investigated whether the class of architecture exemplified by VisNet, and the trace learning rule, can learn about the transforms that surface features of 3D objects typically undergo during 3D rotation in such a way that the network generalizes across the change of the quantitative values of the surface features produced by the rotation, and yet still discriminates between the different objects (in this case spheres). In the cases being considered, each object is identified by surface markings that consist of a different spatial arrangement of the same three features (a horizontal, vertical, and diagonal line, which become arcs on the surface of the object).

We note that it has been suggested that the finding that neurons may offer some degree of 3D rotation invariance after training with a single view (or limited set of views) represents a challenge for existing trace learning models, because these models assume that an initial exposure is required during learning to every transformation of the object to be recognized (Riesenhuber and Poggio 1998). Stringer and Rolls (2002) showed as described here that this is not the case, and that such models can generalize to novel within-generic views of an object provided that the characteristic changes that the features show as objects are rotated have been learned previously for the sets of features when they are present in different objects.

Elliffe, Rolls and Stringer (2002) demonstrated for a 2D system how the existence of translation invariant representations of low order feature combinations in the early layers of the visual system could allow correct stimulus identification in the output layer even when the stimulus was presented in a novel location where the stimulus had not previously occurred during learning. The proposal was that the low-order spatial feature combination neurons in the early stages contain sufficient spatial information so that a particular combination of those low-order feature combination neurons specifies a unique object, even if the relative positions of the low-order feature combination neurons are not known because these neurons are somewhat translation invariant (see Section 25.5.5.5). Stringer and Rolls (2002) extended

Triple 123

Triple 132

Triple 213

Triple 231

Triple 312

Triple 321

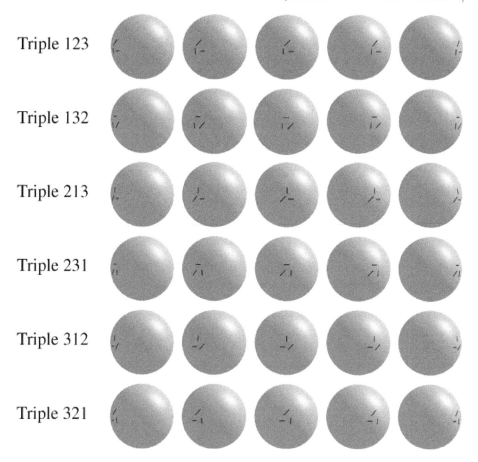

Fig. 25.52 Learning 3D perspectival transforms of features. Representations of the 6 visual stimuli with 3 surface features (triples) presented to VisNet during the simulations described in Section 25.5.7. Each stimulus is a sphere that is uniquely identified by a unique combination of three surface features (a vertical, diagonal and horizontal arc), which occur in 3 relative positions A, B, and C. Each row shows one of the stimuli rotated through the 5 different rotational views in which the stimulus is presented to VisNet. From left to right the rotational views shown are: (i) −60 degrees, (ii) −30 degrees, (iii) 0 degrees (central position), (iv) +30 degrees, and (v) +60 degrees. (After Stringer and Rolls 2002.)

this analysis to feature combinations on 3D objects, and indeed in their simulations described in this section therefore used surface markings for the 3D objects that consisted of triples of features.

The images used for training and testing VisNet were specially constructed for the purpose of demonstrating how the trace learning paradigm might be further developed to give rise to neurons that are able to respond invariantly to novel within-generic view perspectives of an object, obtained by rotations in-depth up to 30 degrees from any perspectives encountered during learning. The stimuli take the form of the surface feature combinations of 3-dimensional rotating spheres, with each image presented to VisNet's retina being a 2-dimensional projection of the surface features of one of the spheres. Each stimulus is uniquely identified by two or three surface features, where the surface features are (1) vertical, (2) diagonal, and (3) horizontal arcs, and where each feature may be centred at three different spatial positions, designated A, B, and C, as shown in Fig. 25.52. The stimuli are thus defined in terms of what features are present and their precise spatial arrangement with respect to each other. We refer to the two and three feature stimuli as 'pairs' and 'triples', respectively. Individual stimuli are denoted by three numbers which refer to the individual features present in positions A, B and C, respectively. For example, a stimulus with positions A and C containing a vertical and diagonal bar, respectively, would be referred to as stimulus 102, where the 0 denotes no feature present in position B. In total there are 18 pairs (120, 130, 210, 230, 310, 320, 012, 013, 021, 023, 031, 032, 102, 103, 201, 203, 301, 302) and 6 triples (123, 132, 213, 231, 312, 321).

To train the network each stimulus was presented to VisNet in a randomized sequence of five orientations with respect to VisNet's input retina, where the different orientations are obtained from successive in-depth rotations of the stimulus through 30 degrees. That is, each stimulus was presented to VisNet's retina from the following rotational views: (i) $-60°$, (ii) $-30°$, (iii) $0°$ (central position with surface features facing directly towards VisNet's retina), (iv) $30°$, (v) $60°$. Figure 25.52 shows representations of the 6 visual stimuli with 3 surface features (triples) presented to VisNet during the simulations. (For the actual simulations described here, the surface features and their deformations were what VisNet was trained and tested with, and the remaining blank surface of each sphere was set to the same greyscale as the background.) Each row shows one of the stimuli rotated through the 5 different rotational views in which the stimulus is presented to VisNet. At each presentation the activation of individual neurons is calculated, then the neuronal firing rates are calculated, and then the synaptic weights are updated. Each time a stimulus has been presented in all the training orientations, a new stimulus is chosen at random and the process repeated. The presentation of all the stimuli through all 5 orientations constitutes 1 epoch of training. In this manner the network was trained one layer at a time starting with layer 1 and finishing with layer 4. In the investigations described here, the numbers of training epochs for layers 1–4 were 50, 100, 100 and 75, respectively.

In experiment 1, VisNet was trained in two stages. In the first stage, the 18 feature pairs were used as input stimuli, with each stimulus being presented to VisNet's retina in sequences of five orientations as described above. However, during this stage, learning was only allowed to take place in layers 1 and 2. This led to the formation of neurons which responded to the feature pairs with some rotation invariance in layer 2. In the second stage, we used the 6 feature triples as stimuli, with learning only allowed in layers 3 and 4. However, during this second training stage, the triples were only presented to VisNet's input retina in the first 4 orientations (i)–(iv). After the two stages of training were completed, Stringer and Rolls (2002) examined whether the output layer of VisNet had formed top layer neurons that responded invariantly to the 6 triples when presented in all 5 orientations, not just the 4 in which the triples had been presented during training. To provide baseline results for comparison, the results from experiment 1 were compared with results from experiment 2 which involved no training in layers 1,2 and 3,4, with the synaptic weights left unchanged from their initial random values.

In Fig. 25.53 numerical results are given for the experiments described. On the left are the single cell information measures for all top (4th) layer neurons ranked in order of their invariance to the triples, while on the right are multiple cell information measures. To help to interpret these results we can compute the maximum single cell information measure according to

$$\text{Maximum single cell information} = \log_2(\text{Number of triples}), \qquad (25.43)$$

where the number of triples is 6. This gives a maximum single cell information measure of 2.6 bits for these test cases. The information results from the experiment demonstrate that even with the triples presented to the network in only four of the five orientations during training, layer 4 is indeed capable of developing rotation invariant neurons that can

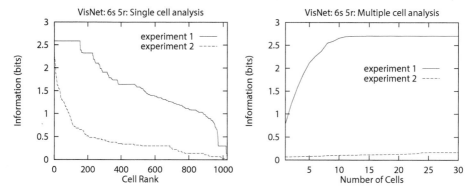

Fig. 25.53 Learning 3D perspectival transforms of features. Numerical results for experiments 1 and 2: on the left are single cell information measures, and on the right are multiple cell information measures.

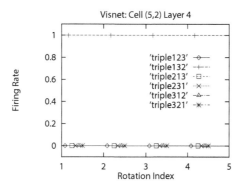

Fig. 25.54 Learning 3D perspectival transforms of features. Numerical results for experiment 1: response profiles of a top layer neuron to the 6 triples in all 5 orientations.

discriminate effectively between the 6 different feature triples in all 5 orientations, that is with correct recognition from all five perspectives. In addition, the multiple cell information for the experiment reaches the maximal level of 2.6 bits, indicating that the network as a whole is capable of perfect discrimination between the 6 triples in any of the 5 orientations. These results may be compared with the very poor baseline performance from the control experiment, where no learning was allowed before testing. Further results from experiment 1 are presented in Fig. 25.54 where we show the response profiles of a top layer neuron to the 6 triples. It can be seen that this neuron has achieved excellent invariant responses to the 6 triples: the response profiles are independent of orientation, but differentiate between triples in that the responses are maximal for triple 132 and minimal for all other triples. In particular, the cell responses are maximal for triple 132 presented in all 5 of the orientations.

Stringer and Rolls (2002) also performed a control experiment to show that the network really had learned invariant representations specific to the kinds of 3D deformations undergone by the surface features as the objects rotated in-depth. In the control experiment the network was trained on 'spheres' with non-deformed surface features; and then as predicted the network failed to operate correctly when it was tested with objects with the features present in the transformed way that they appear on the surface of a real 3D object.

Stringer and Rolls (2002) were thus able to show how trace learning can form neurons that can respond invariantly to novel rotational within-generic view perspectives of an object, obtained by within-generic view 3D rotations up to 30 degrees from any view encountered during learning. They were able to show in addition that this could occur for a novel view of an object which was not an interpolation from previously shown views. This was possible given that the low order feature combination sets from which an object was composed had been learned about in early layers of VisNet previously. The within-generic view transform invariant object recognition described was achieved through the development of true 3-dimensional representations of objects based on 3-dimensional features and feature combinations, which, unlike 2-dimensional feature combinations, are invariant under moderate in-depth rotations of the object. Thus, in a sense, these rotation invariant representations encode a form of 3-dimensional knowledge with which to interpret the visual input from the real world, that is able provide a basis for robust rotation invariant object recognition with novel perspectives. The particular finding in the work described here was that VisNet can learn how the surface features on 3D objects transform as the object is rotated in depth, and can use knowledge of the characteristics of the transforms to perform 3D object recognition. The knowledge embodied in the network is knowledge of the 3D properties of objects, and in this sense assists the recognition of 3D objects seen from different views.

The process investigated by Stringer and Rolls (2002) will only allow invariant object recognition over moderate 3D object rotations, since rotating an object through a large angle may lead to a catastrophic change in the appearance of the object that requires the new qualitative 3D shape descriptors to be associated with those of the former view. In that case, invariant object recognition must rely on the first process referred to at the start of this Section (25.5.7) in order to associate together the different generic views of an object to produce view invariant object identification. For that process, association of a few cardinal or generic views is likely to be sufficient (Koenderink 1990). The process described in this section of learning how surface features transform is likely to make a major contribution to the within-generic view transform invariance of object identification and recognition.

Further investigation of how VisNet can learn to recognise objects despite catastrophic changes in view during 3D object rotation is described in Section 25.6 (Robinson and Rolls 2015).

25.5.8 Capacity of the architecture, and incorporation of a trace rule into a recurrent architecture with object attractors

One issue that has not been considered extensively so far is the capacity of hierarchical feedforward networks of the type exemplified by VisNet that are used for invariant object recognition. One approach to this issue is to note that VisNet operates in the general mode of a competitive network, and that the number of different stimuli that can be categorized by a competitive network is in the order of the number of neurons in the output layer, as described in Section B.4. Given that the successive layers of the real visual system (V1, V2, V4, posterior inferior temporal cortex, anterior inferior temporal cortex) are of the same order of magnitude[30], VisNet is designed to work with the same number of neurons in each successive layer. The hypothesis is that because of redundancies in the visual world, each layer of the system by its convergence and competitive categorization can capture sufficient of the statistics of the visual input at each stage to enable correct specification of the properties of the world that specify objects. For example, V1 does not compute all possible combinations of a few lateral geniculate inputs, but instead represents linear series of geniculate inputs to form edge-like and bar-like feature analyzers, which are the dominant arrangement of pixels found at the small scale in natural visual scenes. Thus the properties of the visual world at this

---

[30]Of course the details are worth understanding further. V1 is for example somewhat larger than earlier layers, but on the other hand serves the dorsal as well as the ventral stream of visual cortical processing.

stage can be captured by a small proportion of the total number of combinations that would be needed if the visual world were random. Similarly, at a later stage of processing, just a subset of all possible combinations of line or edge analyzers would be needed, partly because some combinations are much more frequent in the visual world, and partly because the coding because of convergence means that what is represented is for a larger area of visual space (that is, the receptive fields of the neurons are larger), which also leads to economy and limits what otherwise would be a combinatorial need for feature analyzers at later layers. The hypothesis thus is that the effects of redundancies in the input space of stimuli that result from the statistical properties of natural images (Field 1987), together with the convergent architecture with competitive learning at each stage, produces a system that can perform invariant object recognition for large numbers of objects. Large in this case could be within one or two orders of magnitude of the number of neurons in any one layer of the network (or cortical area in the brain). The extent to which this can be realized can be explored with simulations of the type implemented in VisNet, in which the network can be trained with natural images which therefore reflect fully the natural statistics of the stimuli presented to the real brain.

We should note that a rich variety of information in perceptual space may be represented by subtle differences in the distributed representation provided by the output of the visual system. At the same time, the actual number of different patterns that may be stored in for example a pattern associator connected to the output of the visual system is limited by the number of input connections per neuron from the output neurons of the visual system (see Section B.2). One essential function performed by the ventral visual system is to provide an invariant representation which can be read by a pattern associator in such a way that if the pattern associator learns about one view of the object, then the visual system allows generalization to another view of the same object, because the same output neurons are activated by the different view. In the sense that any view can and must activate the same output neurons of the visual system (the input to the associative network), then we can say the invariance is made explicit in the representation. Making some properties of an input representation explicit in an output representation has a major function of enabling associative networks that use visual inputs in for example recognition, episodic memory, emotion and motivation to generalize correctly, that is invariantly with respect to image transforms that are all consistent with the same object in the world (Rolls and Treves 1998, Rolls 2008d, Rolls 2012c).

Another approach to the issue of the capacity of networks that use trace-learning to associate together different instances (e.g. views) of the same object is to reformulate the issue in the context of autoassociation (attractor) networks, where analytic approaches to the storage capacity of the network are well developed (see Section B.3, Amit (1989), and Rolls and Treves (1998)). This approach to the storage capacity of networks that associate together different instantiations of an object to form invariant representations has been developed by Parga and Rolls (1998) and Elliffe, Rolls, Parga and Renart (2000), and is described next.

In this approach, the storage capacity of a *recurrent* network which performs for example view invariant recognition of objects by associating together different views of the same object which tend to occur close together in time, was studied (Parga and Rolls 1998, Elliffe, Rolls, Parga and Renart 2000). The architecture with which the invariance is computed is a little different to that described earlier. In the model of Rolls ((1992a), (1994a), (1995a), Wallis and Rolls (1997), Rolls and Milward (2000), and Rolls and Stringer (2006)), the postsynaptic memory trace enabled different afferents from the preceding stage to modify onto the same postsynaptic neuron (see Fig. 25.55). In that model there were no recurrent connections between the neurons, although such connections were one way in which it was postulated the memory trace might be implemented, by simply keeping the representation of one view or aspect active until the next view appeared. Then an association would occur between representations that were active close together in time (within e.g. 100–300 ms).

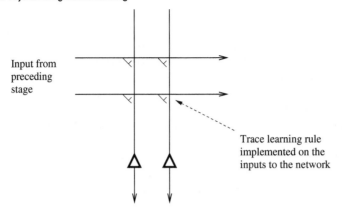

Fig. 25.55 The learning scheme implemented in VisNet. A trace learning rule is implemented in the feedforward inputs to a competitive network.

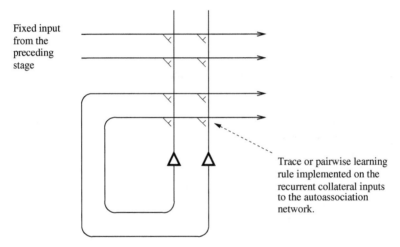

Fig. 25.56 The learning scheme considered by Parga and Rolls (1998) and Elliffe, Rolls, Parga and Renart (2000). There are inputs to the network from the preceding stage via unmodifiable synapses, and a trace or pairwise associative learning rule is implemented in the recurrent collateral synapses of an autoassociative memory to associate together the different exemplars (e.g. views) of the same object.

In the model developed by Parga and Rolls (1998) and Elliffe, Rolls, Parga and Renart (2000), there is a set of inputs with fixed synaptic weights to a network. The network itself is a recurrent network, with a trace rule incorporated in the recurrent collaterals (see Fig. 25.56). When different views of the same object are presented close together in time, the recurrent collaterals learn using the trace rule that the different views are of the same object. After learning, presentation of any of the views will cause the network to settle into an attractor that represents all the views of the object, that is which is a view invariant representation of an object[31].

We envisage a set of neuronal operations which set up a synaptic weight matrix in the

---

[31] In this Section, the different exemplars of an object which need to be associated together are called views, for simplicity, but could at earlier stages of the hierarchy represent for example similar feature combinations (derived from the same object) in different positions in space.

	O_1v_1	O_1v_2	O_1v_3	O_1v_4	O_1v_5	O_2v_1	O_2v_2	O_2v_3	O_2v_4	O_2v_5	. . .
O_1v_1	1	1	1	1	1						
O_1v_2	1	1	1	1	1						
O_1v_3	1	1	1	1	1						
O_1v_4	1	1	1	1	1						
O_1v_5	1	1	1	1	1						
O_2v_1						1	1	1	1	1	
O_2v_2						1	1	1	1	1	
O_2v_3						1	1	1	1	1	
O_2v_4						1	1	1	1	1	
O_2v_5						1	1	1	1	1	
.											
.											
.											

Fig. 25.57 A schematic illustration of the first type of associations contributing to the synaptic matrix considered by Parga and Rolls (1998). Object 1 (O_1) has five views labelled v_1 to v_5, etc. The matrix is formed by associating the pattern presented in the columns with itself, that is with the same pattern presented as rows.

recurrent collaterals by associating together because of their closeness in time the different views of the same object.

In more detail, Parga and Rolls (1998) considered two main approaches. First, one could store in a synaptic weight matrix the s views of an object. This consists of equally associating all the views to each other, including the association of each view with itself. Choosing in Fig. 25.57 an example such that objects are defined in terms of five different views, this might produce (if each view produced firing of one neuron at a rate of 1) a block of 5×5 pairs of views contributing to the synaptic efficacies each with value 1. Object 2 might produce another block of synapses of value 1 further along the diagonal, and symmetric about it. Each object or memory could then be thought of as a single attractor with a distributed representation involving five elements (each element representing a different view). Then the capacity of the system in terms of the number P_o of objects that can be stored is just the number of separate attractors which can be stored in the network. For random fully distributed patterns this is as shown numerically by Hopfield (1982)

$$P_o = 0.14\,C \tag{25.44}$$

where there are C inputs per neuron (and $N = C$ neurons if the network is fully connected). Now the synaptic matrix envisaged here does not consist of random fully distributed binary elements, but instead we will assume has a sparseness $a = s/N$, where s is the number of views stored for each object, from any of which the whole representation of the object must be recognized. In this case, one can show (Gardner 1988, Tsodyks and Feigel'man 1988, Treves and Rolls 1991) that the number of objects that can be stored and correctly retrieved is

$$P_o = \frac{k\,C}{a\,\ln(1/a)} \tag{25.45}$$

where C is the number of synapses on each neuron devoted to the recurrent collaterals from other neurons in the network, and k is a factor that depends weakly on the detailed structure of the rate distribution, on the connectivity pattern, etc., but is approximately in the order of 0.2–0.3. A problem with this proposal is that as the number of views of each object increases to a large number (e.g. >20), the network will fail to retrieve correctly the internal representation

	O_1v_1	O_1v_2	O_1v_3	O_1v_4	O_1v_5	O_2v_1	O_2v_2	O_2v_3	O_2v_4	O_2v_5	. . .
O_1v_1	1	b	b	b	b						
O_1v_2	b	1	b	b	b						
O_1v_3	b	b	1	b	b						
O_1v_4	b	b	b	1	b						
O_1v_5	b	b	b	b	1						
O_2v_1						1	b	b	b	b	
O_2v_2						b	1	b	b	b	
O_2v_3						b	b	1	b	b	
O_2v_4						b	b	b	1	b	
O_2v_5						b	b	b	b	1	
.											
.											
.											

Fig. 25.58 A schematic illustration of the second and main type of associations contributing to the synaptic matrix considered by Parga and Rolls (1998) and Elliffe, Rolls, Parga and Renart (2000). Object 1 (O_1) has five views labelled v_1 to v_5, etc. The association of any one view with itself has strength 1, and of any one with another view of the same object has strength b.

of the object starting from any one view (which is only a fraction $1/s$ of the length of the stored pattern that represents an object).

The second approach, taken by Parga and Rolls (1998) and Elliffe, Rolls, Parga and Renart (2000), is to consider the operation of the network when the associations between pairs of views can be described by a matrix that has the general form shown in Fig. 25.58. Such an association matrix might be produced by different views of an object appearing after a given view with equal probability, and synaptic modification occurring of the view with itself (giving rise to the diagonal term), and of any one view with that which immediately follows it. The same weight matrix might be produced not only by pairwise association of successive views because the association rule allows for associations over the short time scale of e.g. 100–200 ms, but might also be produced if the synaptic trace had an exponentially decaying form over several hundred milliseconds, allowing associations with decaying strength between views separated by one or more intervening views. The existence of a regime, for values of the coupling parameter between pairs of views in a finite interval, such that the presentation of any of the views of one object leads to the same attractor regardless of the particular view chosen as a cue, is one of the issues treated by Parga and Rolls (1998) and Elliffe, Rolls, Parga and Renart (2000). A related problem also dealt with was the capacity of this type of synaptic matrix: how many objects can be stored and retrieved correctly in a view invariant way? Parga and Rolls (1998) and Elliffe, Rolls, Parga and Renart (2000) showed that the number grows linearly with the number of recurrent collateral connections received by each neuron. Some of the groundwork for this approach was laid by the work of Amit and collaborators (Griniasty, Tsodyks and Amit 1993, Amit 1989).

A variant of the second approach is to consider that the remaining entries in the matrix shown in Fig. 25.58 all have a small value. This would be produced by the fact that sometimes a view of one object would be followed by a view of a different object, when for example a large saccade was made, with no explicit resetting of the trace. On average, any one object would follow another rarely, and so the case is considered when all the remaining associations between pairs of views have a low value.

Parga and Rolls (1998) and Elliffe, Rolls, Parga and Renart (2000) were able to show that invariant object recognition is feasible in attractor neural networks in the way described. The system is able to store and retrieve in a view invariant way an extensive number of objects, each defined by a finite set of views. What is implied by extensive is that the number of objects is proportional to the size of the network. The crucial factor that defines this size is the number of connections per neuron. In the case of the fully connected networks considered in this section, the size is thus proportional to the number of neurons. To be particular, the number of objects that can be stored is 0.081 $N/5$, when there are five views of each object. The number of objects is 0.073 $N/11$, when there are eleven views of each object. This is an interesting result in network terms, in that s views each represented by an independent random set of active neurons can, in the network described, be present in the same 'object' attraction basin. It is also an interesting result in neurophysiological terms, in that the number of objects that can be represented in this network scales linearly with the number of recurrent connections per neuron. That is, the number of objects P_o that can be stored is approximately

$$P_o = \frac{k\,C}{s} \tag{25.46}$$

where C is the number of synapses on each neuron devoted to the recurrent collaterals from other neurons in the network, s is the number of views of each object, and k is a factor that is in the region of 0.07–0.09 (Parga and Rolls 1998).

Although the explicit numerical calculation was done for a rather small number of views for each object (up to 11), the basic result, that the network can support this kind of 'object' phase, is expected to hold for any number of views (the only requirement being that it does not increase with the number of neurons). This is of course enough: once an object is defined by a set of views, when the network is presented with a somewhat different stimulus or a noisy version of one of them it will still be in the attraction basin of the object attractor.

Parga and Rolls (1998) thus showed that multiple (e.g. 'view') patterns could be within the basin of attraction of a shared (e.g. 'object') representation, and that the capacity of the system was proportional to the number of synapses per neuron divided by the number of views of each object.

Elliffe, Rolls, Parga and Renart (2000) extended the analysis of Parga and Rolls (1998) by showing that correct retrieval could occur where retrieval 'view' cues were distorted; where there was some association between the views of different objects; and where there was only partial and indeed asymmetric connectivity provided by the associatively modified recurrent collateral connections in the network. The simulations also extended the analysis by showing that the system can work well with sparse patterns, and indeed that the use of sparse patterns increases (as expected) the number of objects that can be stored in the network.

Taken together, the work described by Parga and Rolls (1998) and Elliffe, Rolls, Parga and Renart (2000) introduced the idea that the trace rule used to build invariant representations could be implemented in the recurrent collaterals of a neural network (as well as or as an alternative to its incorporation in the forward connections from one layer to another incorporated in VisNet), and provided a precise analysis of the capacity of the network if it operated in this way. In the brain, it is likely that the recurrent collateral connections between cortical pyramidal cells in visual cortical areas do contribute to building invariant representations, in that if they are associatively modifiable, as seems likely, and because there is continuing firing for typically 100–300 ms after a stimulus has been shown, associations between different exemplars of the same object that occur together close in time would almost necessarily become built into the recurrent synaptic connections between pyramidal cells.

Invariant representation of faces in the context of attractor neural networks has also been discussed by Bartlett and Sejnowski (1997) in terms of a model where different views of faces are presented in a fixed sequence (Griniasty, Tsodyks and Amit 1993). This is not however the general situation; normally any pair of views can be seen consecutively and they will become associated. The model described by Parga and Rolls (1998) treats this more general situation.

I wish to note the different nature of the invariant object recognition problem studied here, and the paired associate learning task studied by Miyashita and colleagues (Miyashita and Chang 1988, Miyashita 1988, Sakai and Miyashita 1991, Fujimichi, Naya, Koyano, Takeda, Takeuchi and Miyashita 2010). In the invariant object recognition case no particular learning protocol is required to produce an activity of the inferior temporal cortex cells responsible for invariant object recognition that is maintained for 300 ms. The learning can occur rapidly, and the learning occurs between stimuli (e.g. different views) which occur with no intervening delay. In the paired associate task, which had the aim of providing a model of semantic memory, the monkeys must learn to associate together two stimuli that are separated in time (by a number of seconds), and this type of learning can take weeks to train. During the delay period the sustained activity is rather low in the experiments, and thus the representation of the first stimulus that remains is weak, and can only poorly be associated with the second stimulus. However, formally the learning mechanism could be treated in the same way as that used by Parga and Rolls (1998) for invariant object recognition. The experimental difference is just that in the paired associate task used by Miyashita et al., it is the weak memory of the first stimulus that is associated with the second stimulus. In contrast, in the invariance learning, it would be the firing activity being produced by the first stimulus (not the weak memory of the first stimulus) that can be associated together. It is possible that the perirhinal cortex makes a useful contribution to invariant object recognition by providing a short-term memory that helps successive views of the same objects to become associated together (Buckley, Booth, Rolls and Gaffan 2001).

The mechanisms described here using an attractor network with a trace associative learning rule would apply most naturally when a small number of representations need to be associated together to represent an object. One example is associating together what is seen when an object is viewed from different perspectives. Another example is scale, with respect to which neurons early in the visual system tolerate scale changes of approximately 1.5 octaves, so that the whole scale range could be covered by associating together a limited number of such representations (see Chapter 5 of Rolls and Deco (2002) and Fig. 2.1). The mechanism would not be so suitable when a large number of different instances would need to be associated together to form an invariant representation of objects, as might be needed for translation invariance. For the latter, the standard model of VisNet with the associative trace learning rule implemented in the feedforward connections (or trained by continuous spatial transformation learning as described in Section 25.5.11) would be more appropriate. However, both types of mechanism, with the trace rule in the feedforward or in recurrent collateral synapses, could contribute (separately or together) to achieve invariant representations. Part of the interest of the attractor approach described in this section is that it allows analytic investigation.

Another approach to training invariance is the purely associative mechanism continuous spatial transformation learning, described in Section 25.5.11. With this training procedure, the capacity is increased with respect to the number of training locations, with for example 169 training locations producing translation invariant representations for two face stimuli (Perry, Rolls and Stringer 2006). It will be of interest in future research to investigate how the VisNet architecture, whether trained with a trace or purely associative rule, scales up with respect to capacity as the number of neurons in the system increases. More distributed representations in the output layer, which may be facilitated by encouraging the system to form a self-organising map, by including short range excitation as well as longer range inhibition between the neurons (see Section 25.5.1), may also help to increase the capacity.

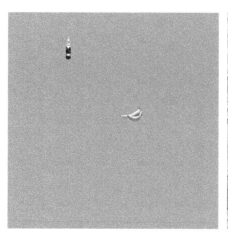

Fig. 25.59 The visual search task. The monkey had to search for and touch an object (in this case a banana) when shown in a complex natural scene, or when shown on a plain background. In each case a second object is present (a bottle) which the monkey must not touch. The stimuli are shown to scale. The screen subtended 70 deg × 55 deg. (After Rolls, Aggelopoulos and Zheng 2003.)

25.5.9 Vision in natural scenes – effects of background versus attention

Object-based attention refers to attention to an object. For example, in a visual search task the object might be specified as what should be searched for, and its location must be found. In spatial attention, a particular location in a scene is pre-cued, and the object at that location may need to be identified. Here we consider some of the neurophysiology of object selection and attention in the context of a feature hierarchy approach to invariant object recognition. The computational mechanisms of attention, including top-down biased competition, are described in Chapter 6.

25.5.9.1 Neurophysiology of object selection in the inferior temporal visual cortex

Much of the neurophysiology, psychophysics, and modelling of attention has been with a small number, typically two, of objects in an otherwise blank scene. In this Section, I consider how attention operates in complex natural scenes, and in particular describe how the inferior temporal visual cortex operates to enable the selection of an object in a complex natural scene (see also Rolls and Deco (2006)). The inferior temporal visual cortex contains distributed and invariant representations of objects and faces (Rolls 2000a, Booth and Rolls 1998, Rolls, Treves and Tovee 1997b, Rolls and Tovee 1995b, Tovee, Rolls and Azzopardi 1994, Hasselmo, Rolls and Baylis 1989a, Rolls and Baylis 1986, Rolls 2007e, Rolls 2007f, Rolls 2007c, Rolls 2008d, Rolls 2011d, Rolls 2012c).

To investigate how attention operates in complex natural scenes, and how information is passed from the inferior temporal cortex (IT) to other brain regions to enable stimuli to be selected from natural scenes for action, Rolls, Aggelopoulos and Zheng (2003a) analyzed the responses of inferior temporal cortex neurons to stimuli presented in complex natural backgrounds. The monkey had to search for two objects on a screen, and a touch of one object was rewarded with juice, and of another object was punished with saline (see Fig. 25.3 for a schematic illustration and Fig. 25.59 for a version of the display with examples of the stimuli shown to scale). Neuronal responses to the effective stimuli for the neurons were compared when the objects were presented in the natural scene or on a plain background. It

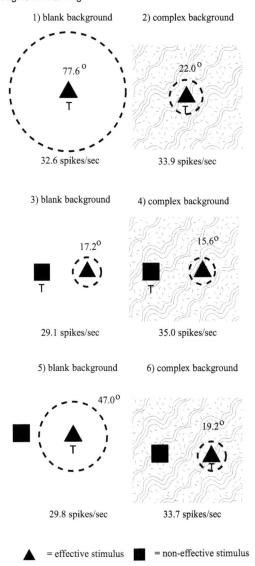

1) blank background   2) complex background

77.6°

T

32.6 spikes/sec   33.9 spikes/sec

22.0°

T

3) blank background   4) complex background

17.2°

T

29.1 spikes/sec   35.0 spikes/sec

15.6°

T

5) blank background   6) complex background

47.0°

T

29.8 spikes/sec   33.7 spikes/sec

19.2°

T

▲ = effective stimulus   ■ = non-effective stimulus

Fig. 25.60 Summary of the receptive field sizes of inferior temporal cortex neurons to a 5 degree effective stimulus presented in either a blank background (blank screen) or in a natural scene (complex background). The stimulus that was a target for action in the different experimental conditions is marked by **T**. When the target stimulus was touched, a reward was obtained. The mean receptive field diameter of the population of neurons analyzed, and the mean firing rate in spikes/s, is shown. The stimuli subtended 5 deg × 3.5 deg at the retina, and occurred on each trial in a random position in the 70 deg × 55 deg screen. The dashed circle is proportional to the receptive field size. Top row: responses with one visual stimulus in a blank (left) or complex (right) background. Middle row: responses with two stimuli, when the effective stimulus was not the target of the visual search. Bottom row: responses with two stimuli, when the effective stimulus was the target of the visual search. (After Rolls, Aggelopoulos, and Zheng 2003.)

was found that the overall response of the neuron to objects was hardly reduced when they were presented in natural scenes, and the selectivity of the neurons remained. An example of a response of one of these neurons is illustrated in Fig. 10.1. However, the main finding was that the magnitudes of the responses of the neurons typically became much less in the real scene the further the monkey fixated in the scene away from the object (see Fig. 25.4). A small receptive field size has also been found in inferior temporal cortex neurons when monkeys have been trained to discriminate closely spaced small visual stimuli (DiCarlo and Maunsell 2003).

It is proposed that this reduced translation invariance in natural scenes helps an unambiguous representation of an object which may be the target for action to be passed to the brain regions that receive from the primate inferior temporal visual cortex. It helps with the binding problem, by reducing in natural scenes the effective receptive field of at least some inferior temporal cortex neurons to approximately the size of an object in the scene.

It is also found that in natural scenes, the effect of object-based attention on the response properties of inferior temporal cortex neurons is relatively small, as illustrated in Fig. 25.60 (Rolls, Aggelopoulos and Zheng 2003a).

25.5.9.2 Attention in natural scenes – a computational account

The results summarized in Fig. 25.60 for 5 degree stimuli show that the receptive fields were large (77.6 degrees) with a single stimulus in a blank background (top left), and were greatly reduced in size (to 22.0 degrees) when presented in a complex natural scene (top right). The results also show that there was little difference in receptive field size or firing rate in the complex background when the effective stimulus was selected for action (bottom right, 19.2 degrees), and when it was not (middle right, 15.6 degrees) (Rolls, Aggelopoulos and Zheng 2003a). (For comparison, the effects of attention against a blank background were much larger, with the receptive field increasing from 17.2 degrees to 47.0 degrees as a result of object-based attention, as shown in Fig. 25.60, left middle and bottom.)

Trappenberg, Rolls and Stringer (2002) have suggested what underlying mechanisms could account for these findings, and simulated a model to test the ideas. The model utilizes an attractor network representing the inferior temporal visual cortex (implemented by the recurrent connections between inferior temporal cortex neurons), and a neural input layer with several retinotopically organized modules representing the visual scene in an earlier visual cortical area such as V4 (see Fig. 25.61). The attractor network aspect of the model produces the property that the receptive fields of IT neurons can be large in blank scenes by enabling a weak input in the periphery of the visual field to act as a retrieval cue for the object attractor. On the other hand, when the object is shown in a complex background, the object closest to the fovea tends to act as the retrieval cue for the attractor, because the fovea is given increased weight in activating the IT module because the magnitude of the input activity from objects at the fovea is greatest due to the higher magnification factor of the fovea incorporated into the model. This results in smaller receptive fields of IT neurons in complex scenes, because the object tends to need to be close to the fovea to trigger the attractor into the state representing that object. (In other words, if the object is far from the fovea, then it will not trigger neurons in IT which represent it, because neurons in IT are preferentially being activated by another object at the fovea.) This may be described as an attractor model in which the competition for which attractor state is retrieved is weighted towards objects at the fovea.

Attentional top-down object-based inputs can bias the competition implemented in this attractor model, but have relatively minor effects (in for example increasing receptive field size) when they are applied in a complex natural scene, as then as usual the stronger forward inputs dominate the states reached. In this network, the recurrent collateral connections may be thought of as implementing constraints between the different inputs present, to help arrive at firing in the network which best meets the constraints. In this scenario, the preferential weighting of objects close to the fovea because of the increased magnification factor at the fovea is a useful principle in enabling the system to provide useful output. The attentional

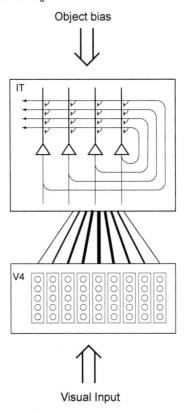

Fig. 25.61 The architecture of the inferior temporal cortex (IT) model of Trappenberg, Rolls and Stringer (2002) operating as an attractor network with inputs from the fovea given preferential weighting by the greater magnification factor of the fovea. The model also has a top-down object-selective bias input. The model was used to analyze how object vision and recognition operate in complex natural scenes.

object biasing effect is much more marked in a blank scene, or a scene with only two objects present at similar distances from the fovea, which are conditions in which attentional effects have frequently been examined. The results of the investigation (Trappenberg, Rolls and Stringer 2002) thus suggest that top-down attention may be a much more limited phenomenon in complex, natural, scenes than in reduced displays with one or two objects present. The results also suggest that the alternative principle, of providing strong weight to whatever is close to the fovea, is an important principle governing the operation of the inferior temporal visual cortex, and in general of the output of the visual system in natural environments. This principle of operation is very important in interfacing the visual system to action systems, because the effective stimulus in making inferior temporal cortex neurons fire is in natural scenes usually on or close to the fovea. This means that the spatial coordinates of where the object is in the scene do not have to be represented in the inferior temporal visual cortex, nor passed from it to the action selection system, as the latter can assume that the object making IT neurons fire is close to the fovea in natural scenes.

There may of course be in addition a mechanism for object selection that takes into account the locus of covert attention when actions are made to locations not being looked at. However, the simulations described in this section suggest that in any case covert attention is likely to be a much less significant influence on visual processing in natural scenes than in reduced scenes with one or two objects present.

Given these points, one might question why inferior temporal cortex neurons can have such large receptive fields, which show translation invariance. At least part of the answer to this may be that inferior temporal cortex neurons must have the capability to be large if they are to deal with large objects. A V1 neuron, with its small receptive field, simply could not receive input from all the features necessary to define an object. On the other hand, inferior temporal cortex neurons may be able to adjust their size to approximately the size of objects, using in part the interactive effects described in Chapter 6, and need the capability for translation invariance because the actual relative positions of the features of an object could be at different relative positions in the scene. For example, a car can be recognized whichever way it is viewed, so that the parts (such as the bonnet or hood) must be identifiable as parts wherever they happen to be in the image, though of course the parts themselves also have to be in the correct relative positions, as allowed for by the hierarchical feature analysis architecture described in this chapter.

Some details of the simulations follow. Each independent module within 'V4' in Fig. 25.61 represents a small part of the visual field and receives input from earlier visual areas represented by an input vector for each possible location which is unique for each object. Each module was 6 degrees in width, matching the size of the objects presented to the network. For the simulations Trappenberg, Rolls and Stringer (2002) chose binary random input vectors representing objects with $N^{V4}a^{V4}$ components set to ones and the remaining $N^{V4}(1 - a^{V4})$ components set to zeros. N^{V4} is the number of nodes in each module and a^{V4} is the sparseness of the representation which was set to be $a^{V4} = 0.2$ in the simulations.

The structure labelled 'IT' represents areas of visual association cortex such as the inferior temporal visual cortex and cortex in the anterior part of the superior temporal sulcus in which neurons provide distributed representations of faces and objects (Booth and Rolls 1998, Rolls 2000a). Nodes in this structure are governed by leaky integrator dynamics (similar to those used in the mean field approach described in Section B.8.1) with time constant τ

$$\tau \frac{dh_i^{IT}(t)}{dt} = -h_i^{IT}(t) + \sum_j (w_{ij}^{IT} - c^{IT})y_j^{IT}(t) + \sum_k w_{ik}^{IT-V4} y_k^{V4}(t) + k^{IT\_BIAS} I_i^{OBJ}.$$

(25.47)

The firing rate y_i^{IT} of the ith node is determined by a sigmoidal function from the activation h_i^{IT} as follows

$$y_i^{IT}(t) = \frac{1}{1 + \exp\left[-2\beta(h_i^{IT}(t) - \alpha)\right]},$$

(25.48)

where the parameters $\beta = 1$ and $\alpha = 1$ represent the gain and the bias, respectively.

The recognition functionality of this structure is modelled as an attractor neural network (ANN) with trained memories indexed by μ representing particular objects. The memories are formed through Hebbian learning on sparse patterns,

$$w_{ij}^{IT} = k^{IT} \sum_\mu (\xi_i^\mu - a^{IT})(\xi_j^\mu - a^{IT}),$$

(25.49)

where k^{IT} (set to 1 in the simulations) is a normalization constant that depends on the learning rate, $a^{IT} = 0.2$ is the sparseness of the training pattern in IT, and ξ_i^μ are the components of the pattern used to train the network. The constant c^{IT} in equation 25.47 represents the strength of the activity-dependent global inhibition simulating the effects of inhibitory interneurons. The external 'top-down' input vector I^{OBJ} produces object-selective inputs, which are used as the attentional drive when a visual search task is simulated. The strength of this object bias is modulated by the value of k^{IT_BIAS} in equation (25.47).

The weights $w_{ij}^{\text{IT}-\text{V4}}$ between the V4 nodes and IT nodes were trained by Hebbian learning of the form

$$w_{ij}^{\text{IT}-\text{V4}} = k^{\text{IT}-\text{V4}}(k) \sum_{\mu} (\xi_i^{\mu} - a^{V4})(\xi_j^{\mu} - a^{IT}). \tag{25.50}$$

to produce object representations in IT based on inputs in V4. The normalizing modulation factor $k^{\text{IT}-\text{V4}}(k)$ allows the gain of inputs to be modulated as a function of their distance from the fovea, and depends on the module k to which the presynaptic node belongs. The model supports translation invariant object recognition of a single object in the visual field if the normalization factor is the same for each module and the model is trained with the objects placed at every possible location in the visual field. The translation invariance of the weight vectors between each 'V4' module and the IT nodes is however explicitly modulated in the model by the module-dependent modulation factor $k^{\text{IT}-\text{V4}}(k)$ as indicated in Fig. 25.61 by the width of the lines connecting V4 with IT. The strength of the foveal V4 module is strongest, and the strength decreases for modules representing increasing eccentricity. The form of this modulation factor was derived from the parameterization of the cortical magnification factors given by Dow, Snyder, Vautin and Bauer (1981)[32].

To study the ability of the model to recognize trained objects at various locations relative to the fovea the system was trained on a set of objects. The network was then tested with distorted versions of the objects, and the 'correlation' between the target object and the final state of the attractor network was taken as a measure of the performance. The correlation was estimated from the normalized dot product between the target object vector that was used during training the IT network, and the state of the IT network after a fixed amount of time sufficient for the network to settle into a stable state. The objects were always presented on backgrounds with some noise (introduced by flipping 2% of the bits in the scene which were not the test stimulus) in order to utilize the properties of the attractor network, and because the input to IT will inevitably be noisy under normal conditions of operation.

In the first simulation only one object was present in the visual scene in a plain (blank) background at different eccentricities from the fovea. As shown in Fig. 25.62a by the line labelled 'blank background', the receptive fields of the neurons were very large. The value of the object bias $k^{\text{IT-BIAS}}$ was set to 0 in these simulations. Good object retrieval (indicated by large correlations) was found even when the object was far from the fovea, indicating large IT receptive fields with a blank background. The reason that any drop is seen in performance as a function of eccentricity is because flipping 2% of the bits outside the object introduces some noise into the recall process. This demonstrates that the attractor dynamics can support translation invariant object recognition even though the translation invariant weight vectors between V4 and IT are explicitly modulated by the modulation factor $k^{\text{IT}-\text{V4}}$ derived from the cortical magnification factor.

In a second simulation individual objects were placed at all possible locations in a natural and cluttered visual scene. The resulting correlations between the target pattern and the asymptotic IT state are shown in Fig. 25.62a with the line labelled 'natural background'. Many objects in the visual scene are now competing for recognition by the attractor network, and the objects around the foveal position are enhanced through the modulation factor derived from the cortical magnification factor. This results in a much smaller size of the receptive field of IT neurons when measured with objects in natural backgrounds.

In addition to this major effect of the background on the size of the receptive field, which parallels and may account for the physiological findings outlined above and in Section

---

[32]This parameterization is based on V1 data. However, it was shown that similar forms of the magnification factor hold also in V4 (Gattass, Sousa and Covey 1985). Similar results to the ones presented here can also be achieved with different forms of the modulation factor such as a shifted Gaussian.

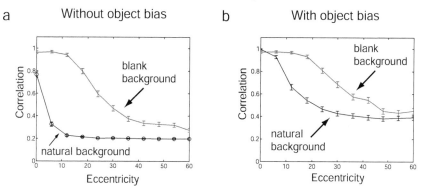

Fig. 25.62 Correlations as measured by the normalized dot product between the object vector used to train IT and the state of the IT network after settling into a stable state with a single object in the visual scene (blank background) or with other trained objects at all possible locations in the visual scene (natural background). There is no object bias included in the results shown in graph a, whereas an object bias is included in the results shown in b with $k^{\text{IT-BIAS}}$=0.7 in the experiments with a natural background and $k^{\text{IT-BIAS}}$=0.1 in the experiments with a blank background. (After Trappenberg, Rolls and Stringer 2002.)

25.5.9.1, there is also a dependence of the size of the receptive fields on the level of object bias provided to the IT network. Examples are shown in Fig. 25.62b where an object bias was used. The object bias biases the IT network towards the expected object with a strength determined by the value of $k^{\text{IT-BIAS}}$, and has the effect of increasing the size of the receptive fields in both blank and natural backgrounds (see Fig. 25.62b compared to a). This models the effect found neurophysiologically (Rolls, Aggelopoulos and Zheng 2003a)[33].

Some of the conclusions are as follows (Trappenberg, Rolls and Stringer 2002). When single objects are shown in a scene with a blank background, the attractor network helps neurons to respond to an object with large eccentricities of this object relative to the fovea of the agent. When the object is presented in a natural scene, other neurons in the inferior temporal cortex become activated by the other effective stimuli present in the visual field, and these forward inputs decrease the response of the network to the target stimulus by a competitive process. The results found fit well with the neurophysiological data, in that IT operates with almost complete translation invariance when there is only one object in the scene, and reduces the receptive field size of its neurons when the object is presented in a cluttered environment. The model described here provides an explanation of the responses of real IT neurons in natural scenes.

In natural scenes, the model is able to account for the neurophysiological data that the IT neuronal responses are larger when the object is close to the fovea, by virtue of fact that objects close to the fovea are weighted by the cortical magnification factor related modulation $k^{\text{IT-V4}}$.

The model accounts for the larger receptive field sizes from the fovea of IT neurons in natural backgrounds if the target is the object being selected compared to when it is not selected (Rolls, Aggelopoulos and Zheng 2003a). The model accounts for this by an effect of top-down bias which simply biases the neurons towards particular objects compensating for their decreasing inputs produced by the decreasing magnification factor modulation with increasing distance from the fovea. Such object-based attention signals could originate in the

---

[33] $k^{\text{IT-BIAS}}$ was set to 0.7 in the experiments with a natural background and to 0.1 with a blank background, reflecting the fact that more attention may be needed to find objects in natural cluttered environments because of the noise present than in blank backgrounds. Equivalently, a given level of attention may have a smaller effect in a natural scene than in a blank background, as found neurophysiologically (Rolls, Aggelopoulos and Zheng 2003a).

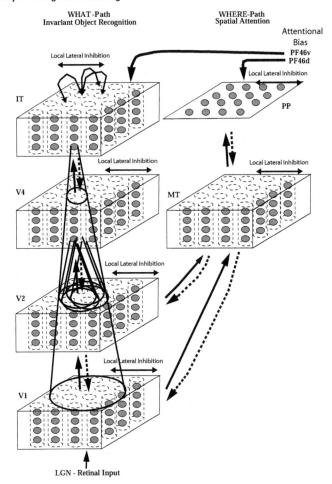

WHAT-Path
Invariant Object Recognition

WHERE-Path
Spatial Attention

Fig. 25.63 Cortical architecture for hierarchical and attention-based visual perception after Deco and Rolls (2004). The system is essentially composed of five modules structured such that they resemble the two known main visual paths of the mammalian visual cortex. Information from the retino-geniculo-striate pathway enters the visual cortex through area V1 in the occipital lobe and proceeds into two processing streams. The occipital-temporal stream leads ventrally through V2–V4 and IT (inferior temporal visual cortex), and is mainly concerned with object recognition. The occipito-parietal stream leads dorsally into PP (posterior parietal complex), and is responsible for maintaining a spatial map of an object's location. The solid lines with arrows between levels show the forward connections, and the dashed lines the top-down backprojections. Short-term memory systems in the prefrontal cortex (PF46) apply top-down attentional bias to the object or spatial processing streams. (After Deco and Rolls 2004.)

prefrontal cortex and could provide the object bias for the inferior temporal visual cortex (Renart, Parga and Rolls 2000) (see Chapter 6).

Important properties of the architecture for obtaining the results just described are the high magnification factor at the fovea and the competition between the effects of different inputs, implemented in the above simulation by the competition inherent in an attractor network.

We have also been able to obtain similar results in a hierarchical feedforward network where each layer operates as a competitive network (Deco and Rolls 2004). This network thus captures many of the properties of our hierarchical model of invariant object recognition (Rolls 1992a, Wallis and Rolls 1997, Rolls and Milward 2000, Stringer and Rolls 2000, Rolls and Stringer 2001a, Elliffe, Rolls and Stringer 2002, Stringer and Rolls 2002, Rolls and Deco 2002, Stringer, Perry, Rolls and Proske 2006, Rolls and Stringer 2007, Rolls and Stringer 2006, Rolls 2008d, Rolls 2012c), but incorporates in addition a foveal magnification factor and top-down projections with a dorsal visual stream so that attentional effects can be studied, as shown in Fig. 25.63.

Deco and Rolls (2004) trained the network shown in Fig. 25.63 with two objects, and used the trace learning rule (Wallis and Rolls 1997, Rolls and Milward 2000) in order to achieve translation invariance. In a first experiment we placed only one object on the retina at different distances from the fovea (i.e. different eccentricities relative to the fovea). This corresponds to the blank background condition. In a second experiment, we also placed the object at different eccentricities relative to the fovea, but on a cluttered natural background. Larger receptive fields were found with the blank as compared to the cluttered natural background.

Deco and Rolls (2004) also studied the influence of object-based attentional top-down bias on the effective size of the receptive field of an inferior temporal cortex neuron for the case of an object in a blank or a cluttered background. To do this, they repeated the two simulations but now considered a non-zero top-down bias coming from prefrontal area 46v and impinging on the inferior temporal cortex neuron specific for the object tested. When no attentional object bias was introduced, a shrinkage of the receptive field size was observed in the complex vs the blank background. When attentional object bias was introduced, the shrinkage of the receptive field due to the complex background was somewhat reduced. This is consistent with the neurophysiological results (Rolls, Aggelopoulos and Zheng 2003a). In the framework of the model (Deco and Rolls 2004), the reduction of the shrinkage of the receptive field is due to the biasing of the competition in the inferior temporal cortex layer in favour of the specific IT neuron tested, so that it shows more translation invariance (i.e. a slightly larger receptive field). The increase of the receptive field size of an IT neuron, although small, produced by the external top-down attentional bias offers a mechanism for facilitation of the search for specific objects in complex natural scenes (see further Chapter 6).

I note that it is possible that a 'spotlight of attention' (Desimone and Duncan 1995) can be moved covertly away from the fovea as described in Chapter 6. However, at least during normal visual search tasks in natural scenes, the neurons are sensitive to the object at which the monkey is looking, that is primarily to the object that is on the fovea, as shown by Rolls, Aggelopoulos and Zheng (2003a) and Aggelopoulos and Rolls (2005), and described in Sections 25.5.9.1 and 25.5.10.

25.5.10   The representation of multiple objects in a scene

When objects have distributed representations, there is a problem of how multiple objects (whether the same or different) can be represented in a scene, because the distributed representations overlap, and it may not be possible to determine whether one has an amalgam of several objects, or a new object (Mozer 1991), or multiple instances of the same object, let alone the relative spatial positions of the objects in a scene. Yet humans can determine the relative spatial locations of objects in a scene even in short presentation times without eye movements (Biederman 1972) (and this has been held to involve some spotlight of attention). Aggelopoulos and Rolls (2005) analyzed this issue by recording from single inferior temporal cortex neurons with five objects simultaneously present in the receptive field. They found that although all the neurons responded to their effective stimulus when it was at the fovea, some could also respond to their effective stimulus when it was in some but not other parafoveal positions 10 degrees from the fovea. An example of such a neuron is shown in Fig. 25.64. The asymmetry is much more evident in a scene with 5 images present (Fig. 25.64A) than when only one image is shown on an otherwise blank screen (Fig. 25.64B). Competition between

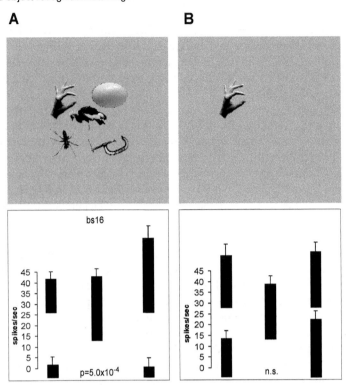

Fig. 25.64 (A). The responses (firing rate with the spontaneous rate subtracted, means ± sem) of an inferior temporal cortex neuron when tested with 5 stimuli simultaneously present in the close (10 deg) configuration with the parafoveal stimuli located 10 degrees from the fovea. (B). The responses of the same neuron when only the effective stimulus was presented in each position. The firing rate for each position is that when the effective stimulus (in this case the hand) for the neuron was in that position. The p value is that from the ANOVA calculated over the four parafoveal positions. (After Aggelopoulos and Rolls 2005.)

different stimuli in the receptive field thus reveals the asymmetry in the receptive field of inferior temporal visual cortex neurons.

The asymmetry provides a way of encoding the position of multiple objects in a scene. Depending on which asymmetric neurons are firing, the population of neurons provides information to the next processing stage not only about which image is present at or close to the fovea, but where it is with respect to the fovea. This information is provided by neurons that have firing rates that reflect the relevant information, and stimulus-dependent synchrony is not necessary. Top-down attentional biasing input could thus, by biasing the appropriate neurons, facilitate bottom-up information about objects without any need to alter the time relations between the firing of different neurons. The exact position of the object with respect to the fovea, and effectively thus its spatial position relative to other objects in the scene, would then be made evident by the subset of asymmetric neurons firing.

This is thus the solution that these experiments indicate is used for the representation of multiple objects in a scene, an issue that has previously been difficult to account for in neural systems with distributed representations (Mozer 1991) and for which 'attention' has been a proposed solution. The learning of invariant representations of objects when multiple objects are present in a scene is considered in Section 25.5.6.2.

25.5.11 Learning invariant representations using spatial continuity: Continuous Spatial Transformation learning

The temporal continuity typical of objects has been used in an associative learning rule with a short-term memory trace to help build invariant object representations in the networks described previously in this chapter. Stringer, Perry, Rolls and Proske (2006) showed that spatial continuity can also provide a basis for helping a system to self-organize invariant representations. They introduced a new learning paradigm 'continuous spatial transformation (CT) learning' which operates by mapping spatially similar input patterns to the same postsynaptic neurons in a competitive learning system. As the inputs move through the space of possible continuous transforms (e.g. translation, rotation, etc.), the active synapses are modified onto the set of postsynaptic neurons. Because other transforms of the same stimulus overlap with previously learned exemplars, a common set of postsynaptic neurons is activated by the new transforms, and learning of the new active inputs onto the same postsynaptic neurons is facilitated.

The concept is illustrated in Fig. 25.65. During the presentation of a visual image at one position on the retina that activates neurons in layer 1, a small winning set of neurons in layer 2 will modify (through associative learning) their afferent connections from layer 1 to respond well to that image in that location. When the same image appears later at nearby locations, so that there is spatial continuity, the same neurons in layer 2 will be activated because some of the active afferents are the same as when the image was in the first position. The key point is that if these afferent connections have been strengthened sufficiently while the image is in the first location, then these connections will be able to continue to activate the same neurons in layer 2 when the image appears in overlapping nearby locations. Thus the same neurons in the output layer have learned to respond to inputs that have similar vector elements in common.

As can be seen in Fig. 25.65, the process can be continued for subsequent shifts, provided that a sufficient proportion of input cells stay active between individual shifts. This whole process is repeated throughout the network, both horizontally as the image moves on the retina, and hierarchically up through the network. Over a series of stages, transform invariant (e.g. location invariant) representations of images are successfully learned, allowing the network to perform invariant object recognition. A similar CT learning process may operate for other kinds of transformation, such as change in view or size.

Stringer, Perry, Rolls and Proske (2006) demonstrated that VisNet can be trained with continuous spatial transformation learning to form view invariant representations. They showed that CT learning requires the training transforms to be relatively close together spatially so that spatial continuity is present in the training set; and that the order of stimulus presentation is not crucial, with even interleaving with other objects possible during training, because it is spatial continuity rather the temporal continuity that drives the self-organizing learning with the purely associative synaptic modification rule.

Perry, Rolls and Stringer (2006) extended these simulations with VisNet of view invariant learning using CT to more complex 3D objects, and using the same training images in human psychophysical investigations, showed that view invariant object learning can occur when spatial but not temporal continuity applies in a training condition in which the images of different objects were interleaved. However, they also found that the human view invariance learning was better if sequential presentation of the images of an object was used, indicating that temporal continuity is an important factor in human invariance learning.

Perry, Rolls and Stringer (2010) extended the use of continuous spatial transformation learning to translation invariance. They showed that translation invariant representations can be learned by continuous spatial transformation learning; that the transforms must be close for this to occur; that the temporal order of presentation of each transformed image during

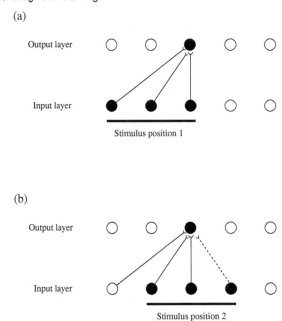

Fig. 25.65 An illustration of how continuous spatial transformation (CT) learning would function in a network with a single layer of forward synaptic connections between an input layer of neurons and an output layer. Initially the forward synaptic weights are set to random values. The top part (a) shows the initial presentation of a stimulus to the network in position 1. Activation from the (shaded) active input cells is transmitted through the initially random forward connections to stimulate the cells in the output layer. The shaded cell in the output layer wins the competition in that layer. The weights from the active input cells to the active output neuron are then strengthened using an associative learning rule. The bottom part (b) shows what happens after the stimulus is shifted by a small amount to a new partially overlapping position 2. As some of the active input cells are the same as those that were active when the stimulus was presented in position 1, the same output cell is driven by these previously strengthened afferents to win the competition again. The rightmost shaded input cell activated by the stimulus in position 2, which was inactive when the stimulus was in position 1, now has its connection to the active output cell strengthened (denoted by the dashed line). Thus the same neuron in the output layer has learned to respond to the two input patterns that have similar vector elements in common. As can be seen, the process can be continued for subsequent shifts, provided that a sufficient proportion of input cells stay active between individual shifts. (After Stringer, Rolls, Perry and Proske 2006.)

training is not crucial for learning to occur; that relatively large numbers of transforms can be learned; and that such continuous spatial transformation learning can be usefully combined with temporal trace training as explored further (Spoerer, Eguchi and Stringer 2016).

25.5.12 Lighting invariance

Object recognition should occur correctly even despite variations of lighting. In an investigation of this, Rolls and Stringer (2006) trained VisNet on a set of 3D objects generated with OpenGL in which the viewing angle and lighting source could be independently varied (see Fig. 25.66). After training with the trace rule on all the 180 views (separated by 1 deg, and rotated about the vertical axis in Fig. 25.66) of each of the four objects under the left lighting condition, we tested whether the network would recognize the objects correctly when they were shown again, but with the source of the lighting moved to the right so that the objects appeared different (see Fig. 25.66). Figure 25.67 shows that the single and multiple

Left Lighting

Right Lighting

Fig. 25.66 Lighting invariance. VisNet was trained on a set of 3D objects (cube, tetrahedron, octahedron and torus) generated with OpenGL in which for training the objects had left lighting, and for testing the objects had right lighting. Just one view of each object is shown in the Figure, but for training and testing 180 views of each object separated by 1 deg were used. (After Rolls and Stringer 2006.)

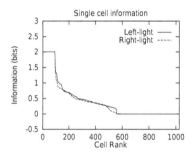

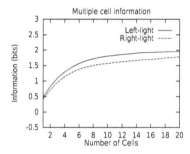

Fig. 25.67 Lighting invariance. The performance of the network after training with 180 views of each object lit from the left, when tested with the lighting again from the left (Left-light), and when tested with the lighting from the right (Right-light). The single cell information measure shows that many single neurons in layer 4 had the maximum amount of information about the objects, 2 bits, which indicates that they responded to all 180 views of one of the objects, and none of the 180 views of the other objects. The multiple cell information shows that the cells were sufficiently different in the objects to which they responded invariantly that all of the objects were perfectly represented when tested with the training images, and very well represented (with nearly 2 bits of information) when tested in the untrained lighting condition. (After Rolls and Stringer 2006.)

cell information measures for the set of objects tested with the light source in the same position as during training (Left-light), and that the measures were almost as good with testing with the light source moved to the right position (Right-light). Thus lighting invariant object recognition was demonstrated (Rolls and Stringer 2006).

Some insight into the good performance with a change of lighting is that some neurons in the inferior temporal visual cortex respond to the outlines of 3D objects (Vogels and Biederman 2002), and these outlines will be relatively consistent across lighting variations. Although the features about the object represented in VisNet will include more than the representations of the outlines, the network may because it uses distributed representations of each object generalize correctly provided that some of the features are similar to those present during training. Under very difficult lighting conditions, it is likely that the performance of the network could be improved by including variations in the lighting during training, so that the trace rule could help to build representations that are explicitly invariant with respect to lighting.

25.5.13 Invariant global motion in the dorsal visual system

A key issue in understanding the cortical mechanisms that underlie motion perception is how we perceive the motion of objects such as a rotating wheel invariantly with respect to position on the retina, and size. For example, we perceive the wheel shown in Fig. 2.3a rotating clockwise independently of its position on the retina. This occurs even though the local motion for the wheels in the different positions may be opposite. How could this invariance of the visual motion perception of objects arise in the visual system? Invariant motion representations are known to be developed in the cortical dorsal visual system.

In a unifying hypothesis with the design of the ventral cortical visual system, Rolls and Stringer (2007) proposed that the dorsal visual system uses a hierarchical feedforward network architecture (V1, V2, MT, MSTd, parietal cortex) with training of the connections with a short-term memory trace associative synaptic modification rule to capture what is invariant at each stage. This is described in Section 2.2.2.

The interesting and quite new principle is that some of the same mechanisms including trace rule learning and hierarchical organisation that are used in the ventral visual system to compute invariant representations of stationary objects may also be used in the dorsal visual system to compute representations of the global motion of a moving object.

This may well be an example of another *principle of cortical operation, the re-use of the same principles of cortical operation for different computations in different cortical areas.*

25.5.14 Deformation-invariant object recognition

When we see a human sitting down, standing up, or walking, we can recognise one of these poses independently of the individual, or we can recognise the individual person, independently of the pose. The same issues arise for deforming objects. For example, if we see a flag deformed by the wind, either blowing out or hanging languidly, we can usually recognise the flag, independently of its deformation; or we can recognise the deformation independently of the identity of the flag.

We hypothesized that these types of recognition can be implemented by the primate visual system using temporo-spatial continuity as objects transform as a learning principle. In particular, we hypothesized that pose or deformation can be learned under conditions in which large numbers of different people are successively seen in the same pose, or objects in the same deformation. We also hypothesized that person-specific representations that are independent of pose, and object-specific representations that are independent of deformation and view, could be built, when individual people or objects are observed successively transforming from one pose or deformation and view to another (Webb and Rolls 2014).

These hypotheses were tested in VisNet. The images to be learned consisted of for example a set of flags with different deformations, as illustrated in The deformation invariant recognition of flags described above was obtained with a set of 4 flags (each with 5 deformations each with two views, as illustrated in Fig. 25.68). Another set of stimuli consisted of three different people shown with three different poses (sitting, standing, or walking), all presented with 12 different views. It was found that depending on the statistics of the visual input, two types of representation could be built by VisNet. Pose-specific or deformation-specific representations could be built that were invariant with respect to individual and view, if the statistics of the inputs included in temporal proximity the same pose or deformation. This might occur for example if one saw a scene with several people sitting down, and the people were successively fixated. Identity-specific representations could be built that were invariant with respect to pose or deformation and view, if the statistics of the inputs included in temporal proximity the same person in different poses, or the same flag in different deformations. This might occur for example if one saw an individual person moving through different poses (sitting, standing up,

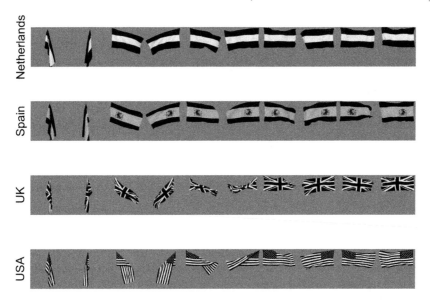

Fig. 25.68 The flag stimuli used to train VisNet. Each flag is shown with different wind forces and rotations. Starting on the left the first pair, both the 0° and 180° views are shown for windspeed 0, and each successive pair is shown for wind force increased by 50 Blender units. (After Webb and Rolls 2014.)

and walking); or if one saw the same flag blowing into different shapes in the wind (Webb and Rolls 2014).

We proposed that this is how pose-specific and pose-invariant, and deformation-specific and deformation-invariant, perceptual representations are built in the brain.

This illustrates an important principle, that information is present in the statistics of the inputs, and can be taken advantage of by the brain to learn different types of representation. This was powerfully illustrated in this investigation in that the functional architecture and stimuli were identical, and it was just the temporal statistics of the inputs that resulted in different types of representation being built.

The utility of these statistics of the input are not taken advantage of in current deep learning models of object recognition (Section B.13).

25.5.15   Learning invariant representations of scenes and places

The primate hippocampal system has neurons that respond to a view of a spatial scene, or when that location in a scene is being looked at in the dark or when it is obscured (Georges-François et al. 1999, Robertson et al. 1998, Rolls et al. 1997a, Rolls et al. 1998b, Rolls and Xiang 2006). The representation is relatively invariant with respect to the position of the macaque in the environment, and of head direction, and eye position. The requirement for these spatial view neurons is that a position in the spatial scene is being looked at (see Chapter 24). (There is an analogous set of place neurons in the rat hippocampus that respond in this case when the rat is in a given position in space, relatively invariantly with respect to head direction (McNaughton et al. 1983, Muller et al. 1991, O'Keefe 1984).) How might these spatial view neurons be set up in primates?

Before addressing this, it is useful to consider the difference between a spatial view or scene representation, and an object representation. An object can be moved to different places in space or in a spatial scene. An example is a motor car that can be moved to different places in space. The object is defined by a combination of features or parts in the correct relative spatial position, but its representation is independent of where it is in space. In contrast, a representation of space has objects in defined relative spatial positions, which cannot be moved relative to one another in space. An example might be Trafalgar Square, in which Nelson's column is in the middle, and the National Gallery and St Martin's in the Fields church are at set relative locations in space, and cannot be moved relative to one another. This draws out the point that there may be some computational similarities between the construction of an objects and of a scene or a representation of space, but there are also important differences in how they are used. In the present context we are interested in how the brain may set up a spatial view representation in which the relative position of the objects in the scene defines the spatial view. That spatial view representation may be relatively invariant with respect to the exact position from which the scene is viewed (though extensions are needed if there are central objects in a space through which one moves).

It is now possible to propose a unifying hypothesis of the relation between the ventral visual system, and primate hippocampal spatial view representations (Rolls 2008d, Rolls, Tromans and Stringer 2008f), which has been introduced in Section 24.3.11. Let us consider a computational architecture in which a fifth layer is added to the VisNet architecture, as illustrated in Fig. 24.21. In the anterior inferior temporal visual cortex, which corresponds to the fourth layer of VisNet, neurons respond to objects, but several objects close to the fovea (within approximately 10°) can be represented because many object-tuned neurons have asymmetric receptive fields with respect to the fovea (Aggelopoulos and Rolls 2005) (see Section 25.5.10). If the fifth layer of VisNet performs the same operation as previous layers, it will form neurons that respond to combinations of objects in the scene with the positions of the objects relative spatially to each other incorporated into the representation (as described in Section 25.5.5). The result will be spatial view neurons in the case of primates when the visual field of the primate has a narrow focus (due to the high resolution fovea), and place cells when as in the rat the visual field is very wide (De Araujo, Rolls and Stringer 2001, Rolls 2008d). The trace learning rule in layer 5 should help the spatial view or place fields that develop to be large and single, because of the temporal continuity that is inherent when the agent moves from one part of the view or place space to another, in the same way as has been shown for the entorhinal grid cell to hippocampal place cell mapping (Rolls, Stringer and Elliot 2006c, Rolls 2008d)).

The hippocampal dentate granule cells form a network expected to be important in this competitive learning of spatial view or place representations based on visual inputs. As the animal navigates through the environment, different spatial view cells would be formed. Because of the overlapping fields of adjacent spatial view neurons, and hence their coactivity as the animal navigates, recurrent collateral associative connections at the next stage of the system, CA3, could form a continuous attractor representation of the environment (Rolls 2008d). We thus have a hypothesis for how the spatial representations are formed as a natural extension of the hierarchically organised competitive networks in the ventral visual system. The expression of such spatial representations in CA3 may be particularly useful for associating those spatial representations with other inputs, such as objects or rewards (Rolls 2008d) (Chapter 24).

We have performed simulations to test this hypothesis with VisNet simulations with conceptually a fifth layer added (Rolls, Tromans and Stringer 2008f). Training now with whole scenes that consist of a set of objects in a given fixed spatial relation to each other results in neurons in the added layer that respond to one of the trained whole scenes, but do not respond if the objects in the scene are rearranged to make a new scene from the same objects. The formation of these scene-specific representations in the added layer is related to the fact that in the inferior temporal cortex (Aggelopoulos and Rolls 2005), and in the VisNet model (Rolls et al. 2008f), the receptive fields of inferior temporal cortex neurons shrink and become asymmetric when multiple objects are present simultaneously in a natural scene. This also provides a solution to the issue of the representation of multiple objects, and their relative spatial positions, in complex natural scenes (Rolls 2008d, Rolls 2012c).

Consistently, in a more artificial network trained by gradient ascent with a goal function that included forming relatively time invariant representations and decorrelating the responses of neurons within each layer of the 5-layer network, place-like cells were formed at the end of the network when the system was trained with a real or simulated robot moving through spatial environments (Wyss, Konig and Verschure 2006), and slowness as an asset in learning spatial representations has also been investigated by others (Wiskott and Sejnowski 2002, Wiskott 2003, Franzius, Sprekeler and Wiskott 2007). It will be interesting to test whether spatial view cells develop in a VisNet fifth layer if trained with foveate views of the environment, or place cells if trained with wide angle views of the environment (cf. De Araujo et al. (2001), and the utility of testing this with a VisNet-like architecture is that it is embodies a biologically plausible implementation based on neuronally plausible competitive learning and a short-term memory trace learning rule.

25.5.16 Finding and recognising objects in natural scenes: complementary computations in the dorsal and ventral visual systems

Here I consider how the brain solves the major computational task of recognising objects in complex natural scenes, still a major challenge for computer vision approaches.

One mechanism that the brain uses to simplify the task of recognising objects in complex natural scenes is that the receptive fields of inferior temporal cortex neurons change from approximately 70° in diameter when tested under classical neurophysiology conditions with a single stimulus on a blank screen to as little as a radius of 8° (for a 5° stimulus) when tested in a complex natural scene (Rolls, Aggelopoulos and Zheng 2003a, Aggelopoulos and Rolls 2005) (with consistent findings described by Sheinberg and Logothetis (2001)) (see Section 25.5.9 and Fig. 25.60). This greatly simplifies the task for the object recognition system, for instead of dealing with the whole scene as in traditional computer vision approaches, the brain processes just a small fixated region of a complex natural scene at any one time, and then the eyes are moved to another part of the screen. During visual search for an object in a complex natural scene, the primate visual system, with its high resolution fovea, therefore keeps moving the eyes until they fall within approximately 8° of the target, and then inferior temporal cortex neurons respond to the target object, and an action can be initiated towards the target, for example to obtain a reward (Rolls et al. 2003a). The inferior temporal cortex neurons then respond to the object being fixated with view, size, and rotation invariance (Rolls 2012c), and also need some translation invariance, for the eyes may not be fixating the centre of the object when the inferior temporal cortex neurons respond (Rolls et al. 2003a).

The questions then arise of how the eyes are guided in a complex natural scene to fixate close to what may be an object; and how close the fixation is to the centre of typical objects for this determines how much translation invariance needs to be built into the ventral visual system. It turns out that the dorsal visual system (Ungerleider and Mishkin 1982, Ungerleider and Haxby 1994) implements bottom-up saliency mechanisms by guiding saccades to salient stimuli, using properties of the stimulus such as high contrast, colour, and visual motion (Miller and Buschman 2013). (Bottom-up refers to inputs reaching the visual system from the retina.) One particular region, the lateral intraparietal cortex (LIP), which is an area in the dorsal visual system, seems to contain saliency maps sensitive to strong sensory inputs (Arcizet et al. 2011). Highly salient, briefly flashed, stimuli capture both behavior and the response of LIP neurons (Goldberg et al. 2006, Bisley and Goldberg 2003, Bisley and Goldberg 2006). Inputs reach

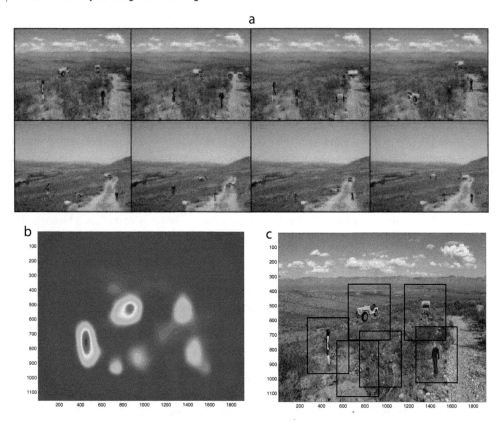

Fig. 25.69 a. Eight of the 12 test scenes. Each scene has 4 objects, each in one of its four views. b. The bottom up saliency map generated by the GBVS code for one of the scenes. The highest levels in the saliency map are red, and the lowest blue. c. Rectangles (384x384 pixels) placed around each peak in the scene for which the bottom-up saliency map is illustrated in (b). (Reproduced from Frontiers in Computational Neuroscience. 8: 85, Finding and recognizing objects in natural scenes: complementary computations in the dorsal and ventral visual systems. Rolls ET and Webb TJ. ©2014 Rolls and Webb.)

LIP via dorsal visual stream areas including area MT, and via V4 in the ventral stream (Miller and Buschman 2013, Soltani and Koch 2010). Although top-down attention using biased competition can facilitate the operation of attentional mechanisms, and is a subject of great interest (Desimone and Duncan 1995, Rolls and Deco 2002, Deco and Rolls 2005a, Miller and Buschman 2013), top-down object-based attention makes only a small contribution to visual search in a natural scene for an object, increasing the receptive field size from a radius of approximately $7.8°$ to approximately $9.6°$ (Rolls, Aggelopoulos and Zheng 2003a), and is not considered further here.

In research described by Rolls and Webb (2014) we investigated computationally how a bottom-up saliency mechanism in the dorsal visual stream reaching for example area LIP could operate in conjunction with invariant object recognition performed by the ventral visual stream reaching the inferior temporal visual cortex to provide for invariant object recognition in natural scenes. The hypothesis is that the dorsal visual stream, in conjunction with structures such as the superior colliculus (Knudsen 2011), uses saliency to guide saccadic eye movements to salient stimuli in large parts of the visual field, and that once a stimulus has been fixated, the ventral visual stream performs invariant object recognition on the region being fixated. The dorsal visual stream in this process knows little about invariant object recognition, so cannot identify objects in natural scenes. Similarly, the ventral visual stream cannot perform the whole process, for it cannot efficiently find possible objects in a large natural scene, because its receptive fields are only approximately $9°$ in radius in complex natural scenes. It is how the dorsal and ventral streams work together to implement invariant object recognition in natural scenes that we investigated. By investigating this computationally, we were able to test whether the dorsal visual stream can find objects with sufficient accuracy to enable the ventral visual stream to perform the invariant object recognition. The issue here is that the ventral visual stream has in practice some translation invariance in natural scenes, but this is limited to approximately $9°$ (Rolls et al. 2003a, Aggelopoulos and Rolls 2005). The computational reason why the ventral visual stream does not compute translation invariant representations over the whole visual field as well as view, size and rotation invariance, is that the computation is too complex. Indeed, it is a problem that has not been fully solved in computer vision systems when they try to perform invariant object recognition over a large natural scene. The brain takes a different approach, of simplifying the problem by fixating on one part of the scene at a time, and solving the somewhat easier problem of invariant representations within a region of approximately $9°$.

For this scenario to operate, the ventral visual stream needs then to implement view-invariant recognition, but to combine it with some translation invariance, as the fixation position produced by bottom up saliency will not be at the centre of an object, and indeed may be considerably displaced from the centre of an object. In the model of invariant visual object recognition that we have developed, VisNet, which models the hierarchy of visual areas in the ventral visual stream by using competitive learning to develop feature conjunctions supplemented by a temporal trace or by spatial continuity or both, all previous investigations have explored either view or translation invariance learning, but not both (Rolls 2012c). Combining translation and view invariance learning is a considerable challenge, for the number of transforms becomes the product of the numbers of each transform type, and it is not known how VisNet (or any other biologically plausible approach to invariant object recognition) will perform with the large number, and with the two types of transform combined. Indeed, an important part of the research was to investigate how well architectures of the VisNet type generalize between both trained locations and trained views. This is important for setting the numbers of different views and translations of each object that must be trained. Rolls and Webb (2014) described an investigation that takes these points into account, and measured the performance of a model of the dorsal visual system with VisNet to test how the system would perform to implement view-invariant object recognition in natural scenes.

The operation of the dorsal visual system in providing a saliency map that would guide the locations to which visual fixations would occur was simulated with a bottom up saliency algorithm that is one of the standard ones that has been developed, which adopts the Itti and Koch (2000) approach to visual saliency, and implements it by graph-based visual saliency (GBVS) algorithms (Harel, Koch and Perona 2006a, Harel, Koch and Perona 2006b). The basis for the saliency map consists of features such as high contrast edges, and the system knows nothing about objects, people, vehicles etc. This system performs well, that is similarly to humans, in many bottom-up saliency tasks. With the scene illustrated in Fig. 25.69a, the saliency map that was produced is illustrated in Fig. 25.69b. The peaks in this saliency map were used as the site of successive 'fixations', at each of which a rectangles (of 384x384 pixels) was placed, and was used as the input image to VisNet as illustrated in Fig. 25.69c. VisNet had been trained on four views spaced $45°$ apart of each of the 4 objects as illustrated in Fig. 25.70.

VisNet was trained on a 25-location grid with size 64×64 with spacing of 16 pixels, and with 4 different views of each object, as it was not expected that the peak of a saliency map would be the centre of an object. Part of the aim was to investigate how much translation

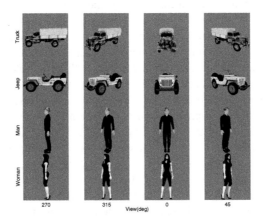

Fig. 25.70 Training images for the identification of objects in a natural scene: 4 views of each of 4 objects were used. Each image was 256x256 pixels. (Reproduced from Frontiers in Computational Neuroscience. 8: 85, Finding and recognizing objects in natural scenes: complementary computations in the dorsal and ventral visual systems. Rolls ET and Webb TJ. ©2014 Rolls and Webb.)

invariance needs to be present in the ventral visual system, in the context of where the dorsal visual system found salient regions in a scene. We found that performance was reasonably good, 90% correct where chance was 25% correct, for which object had been shown. That is, even when the fixation was not on the centre of the object, performance was good. Moreover, the performance was good independently of the view of the person or object, showing that in VisNet both view and position invariance can be trained into the system (Rolls and Webb 2014). Further, the system also generalized reasonably to views between the training views which were 45° apart. Further, this good performance was obtained when inevitably what was extracted as what was close to the fovea included parts of the background scene within the rectangles shown in Fig. 25.69c.

This investigation elucidated show how the brain may solve this major computational problem of recognition of multiple objects seen in different views in complex natural scenes, by moving the eyes to fixate close to objects in a natural scene using bottom-up saliency implemented in the dorsal visual system, and then performing object recognition successively for each of the fixated regions using the ventral visual system modelled to have both translation and view invariance in VisNet. The research described by Rolls and Webb (2014) emphasises that because the eyes do not locate the centre of objects based on saliency, then translation invariance as well as view, size etc invariance needs to be implemented in the ventral visual system. The research showed how a model of invariant object recognition in the ventral visual system, VisNet, can perform the required combination of translation and view invariant recognition, and moreover can generalize between views of objects that are 45° apart during training, and can also generalize to intermediate locations when trained in a coarse training grid with the spacing between trained locations equivalent to 1–3°. Part of the utility of this research is that it helps to identify how much translation invariance needs to be incorporated into the ventral visual system to enable object recognition with successive fixations guided by saliency to different positions in a complex natural scene.

25.6 Further approaches to invariant object recognition

25.6.1 Other types of slow learning

In a more artificial network trained by gradient ascent with a goal function that included forming relatively time invariant representations and decorrelating the responses of neurons within each layer of the 5-layer network, place-like cells were formed at the end of the network when the system was trained with a real or simulated robot moving through spatial environments (Wyss, Konig and Verschure 2006), and slowness as an asset in learning spatial representations has also been investigated by others (Wiskott and Sejnowski 2002, Wiskott 2003, Franzius, Sprekeler and Wiskott 2007). It will be interesting to test whether spatial view cells develop in a VisNet fifth layer if trained with foveate views of the environment, or place cells if trained with wide angle views of the environment (cf. De Araujo et al. (2001), and the utility of testing this with a VisNet-like architecture is that it is embodies a biologically plausible implementation based on neuronally plausible competitive learning and a short-term memory trace learning rule.

25.6.2 HMAX

A related approach to invariant object recognition is HMAX (Riesenhuber and Poggio 1999b, Riesenhuber and Poggio 1999a, Riesenhuber and Poggio 2000, Serre, Oliva and Poggio 2007b, Serre, Wolf, Bileschi, Riesenhuber and Poggio 2007c), which builds on the hypothesis that not just shift invariance (as implemented in the Neocognitron of Fukushima (1980)), but also other invariances such as scale, rotation and even view, could be built into a feature hierarchy system such as VisNet (Rolls 1992a, Wallis, Rolls and Földiák 1993, Wallis and Rolls 1997, Rolls and Milward 2000, Rolls and Stringer 2007, Rolls 2008d, Rolls 2012c). HMAX is a feature hierarchy approach that uses alternate 'simple or S cell' and 'complex or C cell' layers in a way analogous to Fukushima (1980) (see Fig. 25.71). The inspiration for this architecture Riesenhuber and Poggio (1999b) may have come from the simple and complex cells found in V1 by Hubel and Wiesel (1968).

The function of each S cell layer is to represent more complicated features from the inputs, and works by template matching. In particular, when images are presented to HMAX, a set of firings is produced in a patch of S cells, and this patch is propagated laterally so that the same processing is performed throughout the layer. The use of patches of activity produced by an input image to select synaptic weights for the patch while not typical learning, does mean that the synaptic weights in the network are set up to reflect some effects produced by input images. Learning within the model is performed by sampling patches of the C layer output that are produced by a randomly selected image from the training set, and setting the S neurons that receive from the patch to have the weights that would enable the S neuron to be selective for that patch. Then in forward passes, a C layer set of firings is convolved with the S filters that follow the C layer to produce the next S level activity. The convolution is a non-local operator that acts over the whole C layer and effectively assumes that a given type of S neuron has identical connectivity throughout the spatial extent of the layer.

The function of each 'C' cell layer is to provide some translation invariance over the features discovered in the preceding simple cell layer (as in Fukushima (1980)), and operates by performing a MAX function on the inputs. The non-linear MAX function makes a complex cell respond only to whatever is the highest activity input being received, and is part of the process by which invariance is achieved according to this proposal. This C layer process involves 'implicitly scanning over afferents of the same type differing in the parameter of the transformation to which responses should be invariant (for instance, feature size for scale

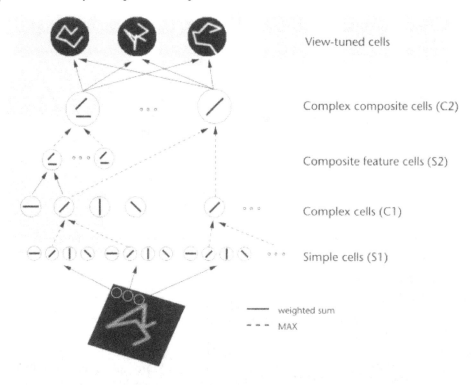

View-tuned cells

Complex composite cells (C2)

Composite feature cells (S2)

Complex cells (C1)

Simple cells (S1)

——— weighted sum
- - - MAX

Fig. 25.71 Sketch of Riesenhuber and Poggio's (1999a,b) HMAX model of invariant object recognition. The model includes layers of 'S' cells which perform template matching (solid lines), and 'C' cells (solid lines) which pool information by a non-linear MAX function to achieve invariance (see text). (After Riesenhuber and Poggio 1999a, b.)

invariance), and then selecting the best-matching afferent' (Riesenhuber and Poggio 1999b). Brain mechanisms by which this computation could be set up are not part of the scheme.

The final complex cell layer (of in some instantiations several S-C pairs of layers) is then typically used as an input to a non-biologically plausible support vector machine or least squares computation to perform classification of the representations of the final layer into object classes. This is a supervised type of training, in which a target is provided from the outside world for each neuron in the last C layer. The inputs to both HMAX and VisNet are Gabor-filtered images intended to approximate V1. One difference is that VisNet is normally trained on images generated by objects as they transform in the world, so that view, translation, size, rotation etc invariant representations of objects can be learned by the network. In contrast, HMAX is typically trained with large databases of pictures of different exemplars of for example hats and beer mugs as in the Caltech databases, which do not provide the basis for invariant representations of objects to be learned, but are aimed at object classification.

Robinson and Rolls (2015) compared the performance of HMAX and VisNet in order to help identify which principles of operation of these two models of the ventral visual system best accounted for the responses of inferior temporal cortex neurons. The aim was to identify which of their different principles of operation might best capture the principles of cortical computation. The outputs of both systems were measured in the same way, with a biologically plausible pattern association network receiving input from the final layer of HMAX or VisNet, and with one neuron for each class to be identified. That is, a non-biologically plausible support vector machine or least squares regression were not used to measure the performance.

A standard HMAX model (Mutch and Lowe 2008, Serre, Wolf, Bileschi, Riesenhuber and Poggio 2007c, Serre, Oliva and Poggio 2007b) was used in the implementation by Mutch and Lowe (2008) which has two S-C layers. We note that an HMAX family model has in the order of 10 million computational units (Serre et al. 2007b), which is at least 100 times the number contained within the current implementation of VisNet (which uses 128x128 neurons in each of 4 layers, i.e. 65,536 neurons).

First, Robinson and Rolls (2015) tested the performance of both nets on an HMAX type of problem, learning to identify classes of object (e.g. hats vs bears) from a set of exemplars, chosen in this case from the CalTech database. The performance of the nets in terms of percentage correct was similar, but a major difference was that it was found that the final layer C neurons of HMAX have a very non-sparse representation (unlike that in the brain or in VisNet) that provides little information in the single neuron responses about the object class. This highlighted an important principle of cortical computation, that the neuronal representation must be in a form in which it can be read, by for example a pattern association network, which is what the cortex and VisNet achieve by producing a sparse distributed representation (Chapter 8 and Fig. 8.2).

Second, when trained with different views of each object, HMAX performed very poorly because it has no mechanism to learn view invariance, i.e. that somewhat different images produced by a single object seen in different views are in fact of the same object. In contrast, VisNet learned this well, because its short-term memory trace learning rule enabled the different views of the same object appearing in short time periods to be associated together. This highlighted another important principle of cortical computation, that advantage is taken of the statistics of inputs from the world to help learning, with in this case temporal and spatial continuity being relevant.

Third, it was shown that VisNet neurons, like many neurons in the inferior temporal visual cortex (Rolls, Tovee, Purcell, Stewart and Azzopardi 1994b, Perrett, Rolls and Caan 1982), do not respond to scrambled images of faces, and thus encode shape information. HMAX neurons responded with similarly high rates to the unscrambled and scrambled faces, indicating that low level features including texture may be relevant to HMAX performance. The VisNet and inferior temporal cortex neurons at the same time encoded the identity of the unscrambled faces. Some other neurons in the inferior temporal visual cortex respond to parts of faces such as eyes or mouth (Perrett, Rolls and Caan 1982, Issa and DiCarlo 2012), and this is consistent with the hypothesis that the inferior temporal visual cortex builds configuration-specific whole face or object representations from their parts, helped by feature combination neurons learned at earlier stages of the ventral visual system hierarchy (Rolls 1992a, Rolls 2012c). HMAX has no learning to enable it to respond to non-linear combinations of features in the correct relative spatial positions (Robinson and Rolls 2015). This highlights another important principal of cortical architecture, that neurons learn to respond to non-linear combinations of their inputs using processes such as competitive learning (Section B.4), and that this is essential for correct operation of most sensory systems.

Fourth, it was shown that VisNet can learn to recognise objects even when the view provided by the object changes catastrophically as it transforms, whereas HMAX has no learning mechanism in its S-C hierarchy that provides for view-invariant learning. The objects used in the experiment are shown in Fig. 25.72. There were two objects, two cups, each with four views, constructed with Blender. VisNet was trained with all views of one object shown in random permuted sequence, then all views of the other object shown in random permuted sequence, to enable VisNet to use its temporal trace learning rule to learn about the different images that occurring together in time were likely to be different views of the same object. VisNet performed 100% correct in this task by forming neurons in its layer 4 that responded either to all views of one cup (labelled 'Bill') and to no views of the other cup (labelled

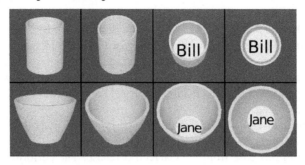

Fig. 25.72 View invariant representations of cups. The two objects, each with four views.

'Jane'), or vice versa. HMAX neurons did not discriminate between the objects, but instead responded more to the images of each object that contained text. This dominance by text is consistent with the fact that HMAX is operating to a considerable extent as a set of image filters, the activity in which included much produced by the text. HMAX has no mechanism within its S-C layers that enables it to learn which input images belong to one object vs another, whereas VisNet can solve this computational problem, by using temporal and spatial continuity present in the way that objects are viewed in a natural environment.

These investigations (Robinson and Rolls 2015) draw out a fundamental difference between VisNet and HMAX. The output layer neurons of VisNet can represent transform-invariant properties of objects, and can form single neurons that respond to the different views of objects even when the images of the different views may be quite different, as is the case for many real-world objects when they transform in the world. Thus basing object recognition on image statistics, and categorisation based on these, is insufficient for transform-invariant object recognition. VisNet can learn to respond to the different transforms of objects using the trace learning rule to capture the properties of objects as they transform in the world. In contrast, HMAX up to the C2 layer sets some of its neurons to respond to exemplars in the set of images, but has no way of knowing which exemplars may be of the same object, and no way therefore to learn about the properties of objects as they transform in the real world, showing catastrophic changes in the image as they transform (Koenderink 1990), exemplified in the example in experiment 4 by the new views as the objects transform from not showing to showing writing in the base of the cup (Fig. 25.72). Moreover, because the C2 neurons reflect mainly the way in which all the Gabor filters respond to image exemplars, the firing of C2 neurons is typically very similar and non-sparse to different images, though if the images have very different statistics in terms of for example text or not, it is these properties that dominate the firing of the C2 neurons.

Further evidence consistent with the approach developed in the investigations of VisNet described in this chapter comes from psychophysical studies. Wallis and Bülthoff (1999) and Perry, Rolls and Stringer (2006) describe psychophysical evidence for learning of view invariant representations by experience, in that the learning can be shown in special circumstances to be affected by the temporal sequence in which different views of objects are seen.

Some other approaches to biologically plausible invariant object recognition are being developed with hierarchies that may be allowed unsupervised learning (Pinto, Doukhan, DiCarlo and Cox 2009, DiCarlo et al. 2012, Yamins, Hong, Cadieu, Solomon, Seibert and DiCarlo 2014). For example, a hierarchical network has been trained with unsupervised learning, and with many transforms of each object to help the system to learn invariant representations in an analogous way to that in which VisNet is trained, but the details of the network architecture are selected by finding parameter values for the specification of the network structure that produce good results on a benchmark classification task (Pinto et al. 2009). However, formally these are convolutional networks, so that the neuronal filters for one local region are replicated over the whole of visual space, which is computationally efficient but biologically implausible. Further, a general linear model is used to decode the firing in the output level of the model to assess performance, so it is not clear whether the firing rate representations of objects in the output layer of the model are very similar to that of the inferior temporal visual cortex. In contrast, with VisNet (Rolls and Milward 2000, Rolls 2012c) the information measurement procedures that we use (Rolls et al. 1997d, Rolls et al. 1997b) are the same as those used to measure the representation that is present in the inferior temporal visual cortex (Rolls and Treves 2011, Tovee et al. 1993, Rolls and Tovee 1995b, Tovee and Rolls 1995, Abbott et al. 1996, Rolls et al. 1997b, Baddeley et al. 1997, Rolls et al. 1997d, Treves et al. 1999, Panzeri, Treves, Schultz and Rolls 1999b, Rolls et al. 2004, Franco et al. 2004, Aggelopoulos et al. 2005, Franco et al. 2007, Rolls et al. 2006b, Rolls 2012c).

In addition, an important property of inferior temporal cortex neurons is that they convey information about the individual object or face, not just about a class such as face vs non-face, or animal vs non-animal (Rolls and Tovee 1995b, Rolls, Treves and Tovee 1997b, Abbott, Rolls and Tovee 1996, Rolls, Treves, Tovee and Panzeri 1997d, Baddeley, Abbott, Booth, Sengpiel, Freeman, Wakeman and Rolls 1997, Rolls 2008d, Rolls and Treves 2011, Rolls 2011d, Rolls 2012c). This key property is essential for recognising a particular person or object, and is frequently not addressed in models of invariant object recognition, which still typically focus on classification into e.g. animal vs non-animal, or classes such as hats and bears from databases such as the CalTech (Serre, Wolf, Bileschi, Riesenhuber and Poggio 2007c, Serre, Oliva and Poggio 2007b, Mutch and Lowe 2008, Serre, Kreiman, Kouh, Cadieu, Knoblich and Poggio 2007a, Yamins, Hong, Cadieu, Solomon, Seibert and DiCarlo 2014). It is clear that VisNet has this key property of representing individual objects, faces, etc., as is illustrated in the investigations of Robinson and Rolls (2015), and previously (Rolls and Milward 2000, Stringer and Rolls 2000, Stringer and Rolls 2000, Stringer and Rolls 2002, Rolls and Webb 2014, Webb and Rolls 2014, Stringer, Perry, Rolls and Proske 2006, Perry, Rolls and Stringer 2006, Perry, Rolls and Stringer 2010, Rolls 2012c). VisNet achieves this by virtue of its competitive learning, in combination with its trace learning rule to learn that different images are of the same object. The investigations of Robinson and Rolls (2015) provide evidence that HMAX categorises together images with similar low level feature properties (such as the presence of text), and does not perform shape recognition relevant to the identification of an individual in which the spatial arrangements of the parts is important. An important point here is that testing whether networks categorise images into different classes of object is insufficient as a measure of whether a network is operating biologically plausibly (Yamins et al. 2014), for there are many more criteria that must be satisfied, including many highlighted here, including correct identification of individuals with all the perceptual invariances, operation using combinations of features in the correct spatial configuration, and producing a biologically plausible sparse distributed representation, all without an external teacher for the output neurons (cf. Pinto, Cox and DiCarlo (2008), Pinto, Doukhan, DiCarlo and Cox (2009), DiCarlo et al. (2012), Cadieu, Hong, Yamins, Pinto, Ardila, Solomon, Majaj and DiCarlo (2014), and Yamins, Hong, Cadieu, Solomon, Seibert and DiCarlo (2014)).

These investigations (Robinson and Rolls 2015) thus highlighted several principles of cortical computation. One is that advantage is taken of the statistics of inputs from the world to help learning, with for example temporal and spatial continuity being relevant. Another is that neurons need to learn to respond to non-linear combinations of their inputs, in the case of vision including their spatial arrangement which is provided by the convergent topology from area to area of the visual cortex. Another principle is that the system must be able to form sparse distributed representations with neurons that encode perceptual and invariance properties, so that the next stage of cortical processing can read the information using dot product decoding as in a pattern associator, autoassociator, or competitive network. None of these properties are provided by HMAX, and all are provided by VisNet, highlighting the importance of these principles of cortical computation.

25.6.3 Sigma-Pi synapses

Another approach to the implementation of invariant representations in the brain is the use of neurons with Sigma-Pi synapses. Sigma-Pi synapses, described in Section A.2.3, effectively allow one input to a synapse to be multiplied or gated by a second input to the synapse. The multiplying input might gate the appropriate set of the other inputs to a synapse to produce the shift or scale change required. For example, the x^c input in equation A.14 could be a signal that varies with the shift required to compute translation invariance, effectively mapping the appropriate set of x_j inputs through to the output neurons depending on the shift required (Mel, Ruderman and Archie 1998, Mel and Fiser 2000, Olshausen, Anderson and Van Essen 1993, Olshausen, Anderson and Van Essen 1995). Local operations on a dendrite could be involved in such a process (Mel, Ruderman and Archie 1998). The explicit neural implementation of the gating mechanism seems implausible, given the need to multiply and thus remap large parts of the retinal input depending on shift and scale modifying connections to a particular set of output neurons. Moreover, the explicit control signal to set the multiplication required in V1 has not been identified. Moreover, if this was the solution used by the brain, the whole problem of shift and scale invariance could in principle be solved in one layer of the system, rather than with the multiple hierarchically organized set of layers actually used in the brain, as shown schematically in Fig. 2.1. The multiple layers actually used in the brain are much more consistent with the type of scheme incorporated in VisNet. Moreover, if a multiplying system of the type hypothesized by Mel, Ruderman and Archie (1998), Olshausen, Anderson and Van Essen (1993) and Olshausen, Anderson and Van Essen (1995) was implemented in a multilayer hierarchy with the shift and scale change emerging gradually, then the multiplying control signal would need to be supplied to every stage of the hierarchy. A further problem with such approaches is how the system is trained in the first place.

25.6.4 Deep learning

Another approach, from the machine learning area, is that of convolution networks (LeCun, Kavukcuoglu and Farabet 2010, LeCun, Bengio and Hinton 2015), which are described in Section B.13. Convolution networks are a biologically-inspired trainable architecture that can learn invariant features. Each stage in a ConvNet is composed of a filter bank, some non-linearities, and feature pooling layers. With multiple stages, a ConvNet can learn multi-level hierarchies of features (LeCun et al. 2010, LeCun et al. 2015). Non-linearities that include rectification and local contrast normalization are important in such systems (Jarrett, Kavukcuoglu, Ranzato and LeCun 2009) (and are of course properties of VisNet).

Convolution networks are very biologically implausible, in that each unit (or 'neuron') receives from only typically a 3 x 3 or 2 x 2 unit patch of the preceding area; by using lateral weight copying within a layer; by having up to 140 or more layers stacked on top of each other in the hierarchy; by using the non-biologically plausible backpropagation of error training algorithm; and by using a teacher for every neuron in the output layer. Interestingly, unlike VisNet, there is no attempt in general to teach the network transform invariance by presenting images with spatial continuity, and no attempt to take advantage of the statistics of the world to help it learn which transforms are probably of the same object by capitalising on temporal continuity. Further, convolution networks cannot explain the sensitivity of humans to precise feature configurations (Ullman et al. 2016).

25.7 Visuo-spatial scratchpad memory, and change blindness

Given the fact that the responses of inferior temporal cortex neurons are quite locked to the stimulus being viewed, it is unlikely that IT provides the representation of the visual world that we think we see, with objects at their correct places in a visual scene. In fact, we do not really see the whole visual scene, as most of it is a memory reflecting what was seen when the eyes were last looking at a particular part of the scene. The evidence for this statement comes from change blindness experiments, which show that humans rather remarkably do not notice if, while they are moving their eyes and cannot respond to changes in the visual scene, a part of the scene changes (O'Regan, Rensink and Clark 1999, Rensink 2000, Rensink 2014). A famous example is that in which a baby was removed from the mother's arms during the subject's eye movement, and the subject failed to notice that the scene was any different. Similarly, unless we are fixating the location that is different in two alternated versions of a visual scene, we are remarkably insensitive to differences in the scene, such as a glass being present on a dining table in one but not another picture of a dining room. Given then that much of the apparent richness of our visual world is actually based on what was seen at previously fixated positions in the scene (with this being what the inferior temporal visual cortex represents), we may ask where this 'visuo-spatial' scratchpad (short-term, iconic, memory) (Rensink 2014) is located in the brain. One possibility is in the right parieto-occipital area, for patients with lesions in the parieto-occipital region have (dorsal) simultanagnosia, in which they can recognize objects, but cannot see more than one object at a time (Farah 2004). Alternatively, the network could be in the left inferior temporo-occipital regions, for patients with (ventral) simultanagnosia cannot recognize but can see multiple objects in a scene (Farah 2004).

The computational basis for this could be a number of separate, that is local, attractors each representing part of the space, and capable of being loaded by inputs from the inferior temporal visual cortex. According to this computational model, the particular attractor network in the visuo-spatial scratchpad memory would be addressed by information based on the position of the eyes, of covert attention, and probably the position of the head. While being so addressed, inputs to the scratch-pad from the inferior temporal cortex neurons would then provide information about the object at the fovea, together with some information about the location of objects in different parafoveal regions (see Section 25.5.10), and this would then enable object information to be associated by synaptic modification with the active neurons in the attractor network representing that location (see Fig. 25.73). Because there are separate spatially local attractors for each location, each of the attractors with its associated object information could be kept active simultaneously, to maintain for short periods information about the relative spatial position of multiple objects in a scene. The attractor for each spatial location would need to represent some information about the object present at that location (as otherwise a binding problem would arise between the multiple objects and the multiple locations), and this may be a reason why the object information in the visuo-spatial scratchpad is not detailed. The suggestion of different attractors for different regions of the scene is consistent with the architecture of the cerebral neocortex, in which the high density of connections including those of inhibitory interneurons is mainly local, within a range of 1–2 mm (see Section 1.10). (If the inhibitory connections were more widespread, so that the inhibition became more global within the visuo-spatial scratchpad memory, the system would

Local attractor networks for a visuo–spatial scratchpad

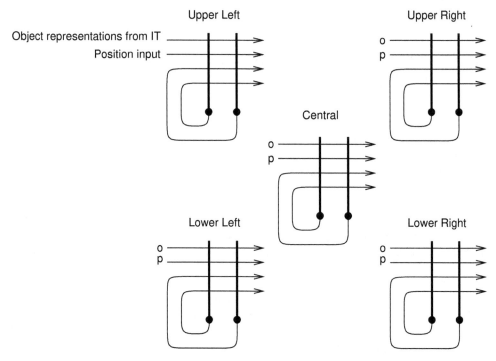

Fig. 25.73 A schematic model of a visuo-spatial scratchpad memory: multiple separate attractor networks could form associations between objects and their place in a scene, and maintain the activity of all the attractors active simultaneously. IT, inferior temporal visual cortex.

be more like a continuous attractor network with multiple activity packets (Stringer, Rolls and Trappenberg 2004) (see Section B.5.4).)

It may only be when the inferior temporal visual cortex is representing objects at or close to the fovea in a complex scene that a great deal of information is present about an object, given the competitive processes described in this chapter combined with the large cortical magnification factor of the fovea. The perceptual system may be built in this way for the fundamental reasons described in this chapter (which enable a hierarchical competitive system to learn invariant representations that require feature binding), which account for why it cannot process a whole scene simultaneously. The visual system can, given the way in which it is built, thereby give an output in a form that is very useful for memory systems, but mainly about one or a few objects close to the fovea.

25.8 Different processes involved in different types of object identification

To conclude this chapter, it is proposed that there are (at least) three different types of process that could be involved in object identification. The first is the simple situation where different objects can be distinguished by different non-overlapping sets of features (see Section 25.3.1). An example might be a banana and an orange, where the list of features of the banana might include yellow, elongated, and smooth surface; and of the orange its orange colour, round shape, and dimpled surface. Such objects could be distinguished just on the basis of a list of the properties, which could be processed appropriately by a competitive network, pattern associator, etc. No special mechanism is needed for view invariance, because the list of properties is very similar from most viewing angles. Object recognition of this type may be common in animals, especially those with visual systems less developed than those of primates. However, this approach does not describe the shape and form of objects, and is insufficient to account for primate vision. Nevertheless, the features present in objects are valuable cues to object identity, and are naturally incorporated into the feature hierarchy approach.

A second type of process might involve the ability to generalize across a small range of views of an object, that is within a generic view, where cues of the first type cannot be used to solve the problem. An example might be generalization across a range of views of a cup when looking into the cup, from just above the near lip until the bottom inside of the cup comes into view. This type of process includes the learning of the transforms of the surface markings on 3D objects which occur when the object is rotated, as described in Section 25.5.7. Such generalization would work because the neurons are tuned as filters to accept a range of variation of the input within parameters such as relative size and orientation of the components of the features. Generalization of this type would not be expected to work when there is a catastrophic change in the features visible, as for example occurs when the cup is rotated so that one can suddenly no longer see inside it, and the outside bottom of the cup comes into view.

The third type of process is one that can deal with the sudden catastrophic change in the features visible when an object is rotated to a completely different view, as in the cup example just given (cf. Koenderink (1990)). Another example, quite extreme to illustrate the point, might be when a card with different images on its two sides is rotated so that one face and then the other is in view. This makes the point that this third type of process may involve arbitrary pairwise association learning, to learn which features and views are different aspects of the same object. Another example occurs when only some parts of an object are visible. For example, a red-handled screwdriver may be recognized either from its round red handle, or from its elongated silver-coloured blade.

The full view-invariant recognition of objects that occurs even when the objects share the same features, such as colour, texture, etc. is an especially computationally demanding task which the primate visual system is able to perform with its highly developed temporal lobe cortical visual areas. The neurophysiological evidence and the neuronal networks described in this chapter provide clear hypotheses about how the primate visual system may perform this task.

25.9 Highlights

1. We have seen in this chapter that the feature hierarchy approach has a number of advantages in performing object recognition over other approaches (see Section 25.3), and that some of the key computational issues that arise in these architectures have solutions (see Sections 25.4 and 25.5). The neurophysiological and computational approach taken here focuses on a feature hierarchy model in which invariant representations can be built by self-organizing learning based on the statistics of the visual input.

2. The model can use temporal continuity in an associative synaptic learning rule with a short-term memory trace, and/or it can use spatial continuity in Continuous Spatial Transformation learning.

3. The model of visual processing in the ventral cortical stream can build representations of objects that are invariant with respect to translation, view, size, and lighting.

4. The model uses a feature combination neuron approach with the relative spatial positions of the objects specified in the feature combination neurons, and this provides a solution to the binding problem.

5. The model has been extended to provide an account of invariant representations in the dorsal visual system of the global motion produced by objects such as looming, rotation, and object-based movement.

6. The model has been extended to incorporate top-down feedback connections to model the control of attention by biased competition in for example spatial and object search tasks (see further Chapter 6).

7. The model has also been extended to account for how the visual system can select single objects in complex visual scenes, how multiple objects can be represented in a scene, and how invariant representations of single objects can be learned even when multiple objects are present in the scene.

8. The model has also been extended to account for how the visual system can select multiple objects in complex visual scenes using a simulation of saliency computations in the dorsal visual system, and then with fixations on the salient parts of the scene perform view-invariant visual object recognition using the simulation of the ventral visual stream, VisNet.

9. It has also been suggested in a unifying proposal that adding a fifth layer to the model and training the system in spatial environments will enable hippocampus-like spatial view neurons or place cells to develop, depending on the size of the field of view (Section 24.3.11).

10. We have thus seen how many of the major computational issues that arise when formulating a theory of object recognition in the ventral visual system (such as feature binding, invariance learning, the recognition of objects when they are in cluttered natural scenes, the representation of multiple objects in a scene, and learning invariant representations of single objects when there are multiple objects in the scene), could be solved in the cortex, with tests of the hypotheses performed by simulations that are consistent with complementary neurophysiological results.

11. The approach described in this chapter is unifying in a number of ways. First, a set of simple organizational principles involving a hierarchy of cortical areas with convergence from stage to stage, and competitive learning using a modified associative learning rule with a short-term memory trace of preceding neuronal activity, provide a basis for understanding much processing in the ventral visual stream, from V1 to the inferior temporal visual cortex. Second, the same principles help to understand some of the processing in the dorsal visual stream by which invariant representations of the global motion of objects may be formed. Third, the same principles continued from the ventral visual stream onwards to the hippocampus help to show how spatial view and place representations may be built from the visual input. Fourth, in all these cases, the learning is possible because the system is able to extract invariant representations because it can utilize the spatio-temporal continuities and statistics in the world that help to define objects, moving objects, and spatial scenes. Fifth, a great simplification and economy in terms of brain design is that the computational principles need not be different in each of the cortical areas in these hierarchical systems, for some of the important properties of the processing in these systems to be performed.

12. The principles of cortical operation that are illustrated include the following:

One is that advantage is taken of the statistics of inputs from the world to help learning, with for example temporal and spatial continuity being relevant.

Another is that neurons need to learn to respond to non-linear combinations of their inputs, in the case of vision including their spatial arrangement which is provided by the convergent topology from area to area of the visual cortex, using principles such as competitive learning.

Another principle is that the system must be able to form sparse distributed representations with neurons that encode perceptual and invariance properties, so that the next stage of cortical processing can read the information using dot product decoding as in a pattern associator, autoassociator, or competitive network.

Another principle is the use of hierarchical cortical computation with convergence from stage to stage, which breaks the computation down into neuronally manageable computations.

Another principle is breaking the computation down into manageable parts, by for example not trying to analyze the whole of a scene simultaneously, but instead using successive fixations to objects in different parts of the scene, and maintaining in short-term memory a limited representation of the whole scene.

13. In conclusion, we have seen in this chapter how a major form of perception, the invariant visual recognition of objects, involves not only the storage and retrieval of information, but also major computations to produce invariant representations. Once these invariant representations have been formed, they are used for many processes including not only recognition memory (see Section 24.2.6), but also associative learning of the rewarding and punishing properties of objects for emotion and motivation (see Chapter 15), the memory for the spatial locations of objects and rewards (see Chapter 24), the building of spatial representations based on visual input (Section 24.3.11), and as an input to short-term memory (Section 4.3.1), attention (Chapter 6), and decision systems (Section 5.6).

26 Synthesis

This Chapter describes some syntheses of points that arise from what is presented in this book.

26.1 Principles of cortical operation, not a single theory

In this book, I have not attempted to produce a single computational theory of how the cortex operates. Instead, I have highlighted many different principles of cortical function, most of which are likely to be building blocks of how our cortex operates. The reason for this approach is that many of the principles may well be correct, and useful in understanding how the cortex operates, but some might turn out not to be useful or correct. The aim of this book is therefore to propose some of the fundamental principles of operation of the cerebral cortex, many or most of which will provide a foundation for understanding the operation of the cortex, rather than to produce a single theory of operation of the cortex, which might be disproved if any one of its elements was found to be weak.

Indeed, there is in a sense no single theory of operation for the cerebral cortex, but instead there are theories of how each cortical area operates to perform its computations using a subset of all the principles described, and performing an operation that must reflect both its inputs, and how they are used in terms of which other brain areas they influence. Examples of these theories of particular brain areas and systems are provided as examples in Chapter 24 on the hippocampus and memory, and Chapter 25 on invariant visual object recognition.

26.2 Levels of explanation, and the mind-brain problem

We can now understand brain processing from the level of ion channels in neurons, through neuronal biophysics, to neuronal firing, through the computations performed by populations of neurons, and how their activity is reflected by functional neuroimaging, to behavioural and cognitive effects (Rolls 2008d, Rolls and Deco 2010, Rolls 2014a). Activity at any one level can be used to understand activity at the next. This raises the philosophical issue of how we should consider causality with these different levels (Rolls 2012d). Does the brain cause effects in the mind, or do events at the mental, mind, level influence brain activity?

What is the relation between the mind and the brain? This is the mind–brain or mind–body problem. Do mental, mind, events cause brain events? Do brain events cause mental effects? What can we learn from the relation between software and hardware in a computer about mind–brain interactions and how causality operates? Neuroscience shows that there is a close relation between mind and matter (captured by the following inverted saying: 'Never matter, no mind').

My view is that the relationship between mental events and neurophysiological events is similar (apart from the problem of consciousness) to the relationship between the program running in a computer and the hardware of the computer. In a sense, the program (the software loaded onto the computer usually written in a high-level language such as C or Matlab) 'causes' the logic gates (TTL, transistor-transistor logic) of the hardware to move to the next state. This hardware state change 'causes' the program to move to its next step or state. However, I think that instead when we are looking at different levels of what is overall the operation of a system, causality can usefully be understood as operating within levels (causing one step of the program to move to the next; or the neurons to move from one state to another), but not between levels (e.g. software to hardware and vice versa). That is, if the events at the different levels of explanation are occurring simultaneously, without a time delay, then my view is that we should not think of causality as operating between levels, but just that what happens at a higher level may be an emergent property of what happens at a lower level. This is the solution I propose to this aspect of the mind–body (or mind–brain) problem.

Following this thinking, when one step of a process at one level of explanation moves to the next step in time, we can speak of causality that would meet the criteria for Granger causality where one time series, including the time series being considered, can be used to predict the next step in time (Section 6.3.3) (Granger 1969, Ge, Feng, Grabenhorst and Rolls 2012, Bressler and Seth 2011). In contrast, when we consider the relationship between processes described at different levels of explanation, such as the relation between a step in the hardware in a computer and a step in the software, then these processes may occur simultaneously, and be inextricably linked with each other, and just be different ways of describing the same process, so that temporal (Granger) causality does not apply to this relation.

Thus my view of the mind–brain issue is that we are considering the process as a mechanism with different levels of explanation. As described above, we can now understand brain processing from the level of ion channels in neurons, through neuronal biophysics, to neuronal firing, through the computations performed by populations of neurons, to behavioural and cognitive effects, and even perhaps to the phenomenal (feeling) aspects of consciousness as described in Chapter 22. The whole processing is now specified from the mechanistic level of neuronal firings, etc. up through the computational level to the cognitive and behavioural level. Sometimes the cognitive effects seem remarkable, for example the recall of a whole memory from a part of it, and we describe this as an 'emergent property', but once understood from the mechanistic level upwards, the functions implemented are elegant and wonderful, but understandable and not magical or poorly understood (Rolls 2008d, Rolls 2012d). The point I make is that however one thinks about causality in such a mechanistic system with interesting 'emergent' computational properties, the system is now well-defined, is no longer mysterious or magical, and we have now from a combination of neuroscience and analyses of the type used in theoretical physics a clear understanding of the properties of neural systems and how cognition emerges from neural mechanisms. There are of course particular problems that remain to be resolved with this approach, such as that of how language is implemented in the brain, but my point is that this mechanistic approach, supported by parsimony, appears to be capable of leading us to a full understanding of brain function, cognition, and behaviour. However, the property of phenomenal consciousness is a big step for an 'emergent property', and that hard problem is therefore considered in more detail in Chapter 22.

Overall, understanding brain activity at these different levels provides a unifying approach to understanding brain function, which is proving to be so powerful that the fundamental operations involved in many aspects of brain function can be understood in principle, though with of course many details still to be discovered. These functions include many aspects of perception including visual face and object recognition, and taste, olfactory and related processing; short-term memory; long-term memory; attention; emotion; and decision-making. Predictions made at one level can be tested at another. Conceptually this is an enormous advance. But it is also of great practical importance, in medicine. For example, we now have new ways of predicting effects of possible pharmacological treatments for brain diseases by a developing understanding of how drugs affect synaptic receptors, which in turn affect neuronal activity, which in turn affect the stability of the whole network of neurons and hence cognitive symptoms such as attention vs distractibility (see Chapter 16).

26.3 Brain computation compared to computation on a digital computer

An important component of the approach taken in this book to understanding the principles of operation of the cerebral cortex is computational neuroscience, because this formalises the operations being considered, and enables the different levels of analysis in neuroscience to be brought together in a multilevel causal account.

However, the types of computation that are performed by cortical brain systems, and the computational style, are very different from the type of computation performed by a digital computer, which performs specified logical / syntactic operations on exact data retrieved from memory, and then stores the exact result back in memory. To highlight some of the principles of brain computation by the cortex described in this book, and to emphasize how the type of computation is very different from that performed by digital computers, the principles of computation by the brain with those of a digital computer are compared next.

Data addressing. An item of data is retrieved from the memory of a digital computer by providing the address of the data in memory, and then the data can be manipulated (moved, compared, added to the data at another address in the computer etc.) using typically a 32-bit or 64-bit binary word of data. Pointers to memory locations are thus used extensively. In contrast, in the cortex, the data are used as the access key (in for example a pattern associator and autoassociator), and the neurons with synaptic weights that match the data respond. Memory in the brain is thus **content-addressable**. In one time constant of the synapses / cell membranes the brain has thus found the correct output. In contrast, in a digital computer a serial search is required, in which the data at every address must be retrieved and compared in turn to the test data to discover if there is a match.

Vector similarity vs logical operations. Cortical computation including that performed by associative memories and competitive networks operates by vector similarity – the dot product of the input and of the synaptic weight vector are compared, and the neurons with the highest dot product will be most activated (Fig. 8.2). Even if an exact match is not found, some output is likely to result. In contrast, in a digital computer, logic operations (such as AND, OR, XOR) and exact mathematical operations (such as addition, subtraction, multiplication, and division) are computed. (There is no bit-wise similarity between the binary representations of 7 (0111) and 8 (1000).) The similarity computations performed by the brain may be very useful, in enabling similarities to be seen and parallels to be drawn, and this may be an interesting aspect of human creativity, realized for example in *Finnegans's Wake* by James Joyce in which thoughts reached by associative thinking abound. However, the lateral thinking must be controlled, to prevent bizarre similarities being found, and this is argued to be related to the symptoms of schizophrenia in Section 16.1.

Fault tolerance. Because exact computations are performed in a digital computer, there is no in-built fault tolerance or graceful degradation. If one bit of a memory has a fault, the whole memory chip must be discarded. In contrast, the brain is naturally fault tolerant, because it uses vector similarity (between its input firing rate vector and synaptic weight vectors) in its calculations, and linked to this, distributed representations. This makes the brain robust developmentally with respect to 'missing synapses', and robust with respect to later losing some synapses or neurons (see Appendix B).

Word length. To enable the vector similarity comparison to have high capacity (for example memory capacity) the 'word length' in the brain is typically long, with between 10,000 and 50,000 synapses onto every neuron being common in cortical areas. (Remember that the leading term in the factor that determines the storage capacity of an associative memory is the number of synapses per neuron – see Sections B.2 and B.3.) In contrast, the word length in typical digital computers at 32 or 64 bits is much shorter, though with the binary and exact encoding used this allows great precision in a digital computer.

Readability of the code. To comment further on the encoding: in the cortex, the code must not be too compact, so that it can be read by neuronally plausible dot product decoding, as shown in Section 8.2. In contrast, the binary encoding used in a digital computer is optimally efficient, with one bit stored and retrievable for each binary memory location. However, the computer binary code cannot be read by neuronally plausible dot-product decoding.

Precision. The precision of the components in a digital computer is that every modifiable memory location must store one bit accurately. In contrast, it is of interest that synapses in the brain need not work with exact precision, with for example typically less that one bit per synapse being usable in associative memories (Treves and Rolls 1991, Rolls and Treves 1998). The precision of the encoding of information in the firing rate of a neuron is likely to be a few bits – perhaps 3 – as judged by the standard deviation and firing rate range of individual cortical neurons (Appendix C (Rolls 2008d)).

The speed of computation. This brings us to the speed of computation. In the brain, considerable information can be read in 20 ms from the firing rate of an individual neuron (e.g. 0.2 bits), leading to estimates of 10–30 bits/s for primate temporal cortex visual neurons (Rolls, Treves and Tovee 1997b, Rolls and Tovee 1994), and 2–3 bits/s for rat hippocampal cells (Skaggs, McNaughton, Gothard and Markus 1993, Rolls 2008d, Rolls and Treves 2011) (Appendix C). Though this is very slow compared to a digital computer, the brain does have the advantage that a single neuron receives spikes from thousands of individual neurons, and computes its output from all of these inputs within a period of approximately 10–20 ms (determined largely by the time constant of the synapses) (Rolls 2008d). Moreover, each neuron, up to at least the order of tens of neurons, conveys independent information, as described in Section 8.2 and Appendix C.

Parallel vs serial processing. Computation in a conventional digital computer is inherently serial, with a single central processing unit that must fetch the data from a memory address, manipulate the word of data, and store it again at a memory address. In contrast, brain computation is parallel in at least three senses.

First, an individual neuron in performing a dot product between its input firing rate vector and its synaptic weight vector does operate in an analog way to sum all the injected currents through the thousands of synapses to calculate the activation h_i, and fire if a threshold is reached, in a time in the order of the synaptic time constant. To implement this on a digital computer would take $2C$ operations (C multiply operations, and C add operations, where C is the number of synapses per neuron – see equation 1.1).

Second, each neuron in a single network (e.g. a small region of the cortex with of the order of hundreds of thousands of neurons) does this dot product computation in parallel, followed by interaction through the GABA inhibitory neurons, which again is fast. (It is in the order of the time constant of the synapses involved, operates in continuous time, and does not have to wait at all until the dot product operation of the pyramidal cells has been completed by all neurons given the spontaneous neuronal activity which allows some neurons to reflect their changed inputs rapidly in when the next spike occurs.) This interaction sets the threshold in associative and competitive networks, and helps to set the sparseness of the representation of the population of neurons.

Third, different brain areas operate in parallel. An example is that the ventral visual stream computes object representations, while simultaneously the dorsal visual stream computes (inter alia) the types of global motion described by Rolls and Stringer (2007), including for example a wheel rotating in the same direction as it traverses the visual field. Another example is that within a hierarchical system in the brain, every stage operates simultaneously, as a pipeline processor, with a good example being V1–V2–V4–IT, which can all operate simultaneously as the data are pipelined through (Chapter 25).

We could refer to the computation that takes place in different modules, that is in networks that are relatively separate in terms of the number of connections between modules relative to those within modules, such as those in the dorsal and ventral visual streams, as being parallel computation. Within a single module or network, such as the CA3 region of the hippocampus, or inferior temporal visual cortex, we could refer to the computation as being *parallel distributed computation*, in that the closely connected neurons in the network all contribute to the result of the computation. For example, with distributed representations in an attractor network, all the neurons interact with each other directly and through the inhibitory interneurons to retrieve and then maintain a stable pattern in short-term memory (Section B.3). In a competitive network involved in pattern categorization, all the neurons interact through the inhibitory interneurons to result in an active population of neurons that represents the best match between the input stimulus and what has been learned previously by the network, with neurons with a poor match being inhibited by neurons with a good match (Section B.4). In a more complicated scenario with closely connected interacting modules, such as the prefrontal cortex and the inferior temporal cortex during top-down attention tasks and more generally forward and backward connections between adjacent cortical areas, we might also use the term parallel distributed computation, as the bottom-up and top-down interactions may be important in how the whole dynamical system of interconnected networks settles (see examples in Sections 4.3.1, 6 and Appendix B).

Stochastic dynamics and probabilistic computation. Digital computers do not have noise to contend with as part of the computation, as they use binary logic levels, and perform exact computation. In contrast, brain computation is inherently noisy, and this gives it a non-exact, probabilistic, character. One of the sources of noise in the brain is the spiking activity of each neuron. Each neuron must transmit information by spikes, for an all-or-none spike carried along an axon ensures that the signal arrives faithfully, and is not subject to the uncertain cable transmission line losses of analog potentials. But once a neuron needs to spike, then it turns out to be important to have spontaneous activity, so that neurons do not all have to charge up from a hyperpolarized baseline whenever a new input is received. The fact that neurons are kept near threshold, with therefore some spontaneous spiking, is inherent to the rapid operation of for example autoassociative retrieval, as described in Section B.3. But keeping the neurons close to threshold, and the spiking activity received from other neurons, results in spontaneous spike trains that are approximately Poisson, that is randomly timed for a given mean firing rate. The result of the interaction of all these randomly timed inputs is that in a network of finite size (i.e. with a limited number of neurons) there will be statistical fluctuations, which influence which memory is recalled, which decision is taken, etc. as described in Section 5.6. Thus brain computation is inherently noisy and probabilistic, and this has many advantages, as described in Section 5.6 and by Rolls and Deco (2010).

Syntax. Digital computers can perform arbitrary syntactical operations on operands, because they use pointers to address each of the different operands required (corresponding even for example to the subject, the verb, and the object of a sentence). In contrast, as data are not accessed in the brain by pointers that can point anywhere, but instead just having neurons firing to represent a data item, a real problem arises in specifying which neurons firing represent for example the subject, the verb, and the object, and distributed representations potentially make this even more difficult. The brain thus inherently finds syntactical operations difficult (as explained in Chapter 17). We do not know how the brain implements the syntax required for language. But we do know that the firing of neurons conveys 'meaning' based on spatial location in the brain. For example, a neuron firing in V1 indicates that a bar or edge matching the filter characteristic of the neuron is present at a particular location in space. Another neuron in V1 encodes another feature at another position in space. A neuron in the inferior temporal visual cortex indicates (with other neurons helping to form a distributed representation) that a particular object or face is present in the visual scene. Perhaps the implementation of the syntax required for language that is implemented in the brain also utilizes the spatial location of the network in the cortex to help specify what syntactical role the representation should perform. This is a suggestion I make, as it is one way that the brain could deal with the implementation of the syntax required for language (Chapter 17).

Modifiable connectivity. The physical architecture (what is connected to what) of a digital computer is fixed. In contrast, the connectivity of the brain alters as a result of experience and learning, and indeed it is alterations in the strength of the synapses (which implement the connectivity) that underlie learning and memory. Indeed, self-organization in for example competitive networks has a strong influence on how the brain is matched to the statistics of the incoming signals from the world, and of the architecture that develops. In a digital computer, every connection must be specified. In contrast, in the brain there are far too few genes (of order 25,000) for the synaptic connections in the brain (of order 10^{15}, given approximately 10^{11} neurons each with in the order of 10^4 synapses) for the genes to specify every connection[34]. The genes must therefore specify some much more general rules, such as that each CA3 neuron should make approximately 12,000 synapses with other CA3 neurons, and receive approximately 48 synapses from dentate granule cells (see Chapter 24). The actual connections made would then be made randomly within these constraints, and then strengthened or lost as a result of self-organization based on for example conjunctive pre- and postsynaptic activity. Some of the rules that may be specified genetically have been suggested on the basis of a comparison of the architecture of different brain areas (Rolls and Stringer 2000) (Chapter 19). Moreover, it has been shown that if these rules are selected by a genetic algorithm based on the fitness of the network that self-organizes and learns based on these rules, then architectures are built that solve different computational problems in one-layer networks, including pattern association learning, autoassociation memory, and competitive learning (Rolls and Stringer 2000) (Chapter 19). The architecture of the brain is thus interestingly adaptive, but guided in the long term by genetic selection of the building rules.

Logic. The learning rules that are implemented in the brain that are most widely accepted are associative, as exemplified by LTP and LTD. This, and the vector similarity operations implemented by neurons, set the stage for processes such as pattern association, autoassociation,

---

[34]For comparison, a computer with 1 Gb of memory has approximately 10^{10} modifiable locations, and if it had a 100 Gb disk that would have approximately 10^{12} modifiable locations.

and competitive learning to occur naturally, but not for logical operations such as XOR and NAND or arithmetic operations. Of course, the non-linearity inherent in the firing threshold of neurons is important in many of the properties of associative memories and competitive learning, as described in Appendix B, and indeed are how some of the non-linearities that can be seen with attention can arise (Deco and Rolls 2005b).

Dynamical interaction between modules. Because the brain has populations of neurons that are simultaneously active (operating in parallel), but are interconnected, many properties arise naturally in dynamical neural systems, including the interactions that give rise to top-down attention (Chapter 6), the effects of mood on memory (Rolls 2008d) etc. Because simultaneous activity of different computational nodes does not occur in digital computers, these dynamical systems properties that arise from interacting subsystems do not occur naturally, though they can be simulated.

The cortex has recurrent excitatory connections within a cortical area, and reciprocal, forward and feedback, connections between adjacent cortical areas in the hierarchy. The excitatory connections enable cortical activity to be maintained over short periods, making short-term memory an inherent property of the cortex. They also provide the autoassociative long-term memory with completion from a partial cue (given associative synaptic modifiability in these connections). However, completion on a digital computer is a difficult and serial process to identify a possible correct partial match. Another comparison is that the short-term memory property of the cortex is part of what makes the cortex a dynamical interacting system, with for example what is in short-term memory in the prefrontal cortex acting to influence memory recall, perception, and even what decision is taken, in other networks, by top-down biased competition (see Chapters 6 and 4). There is a price that the brain pays for this positive feedback inherent in its recurrent cortical circuitry, which is that this circuitry is inherently unstable, and requires strong control by inhibitory interneurons to minimize the risk of epilepsy and other disorders (Chapter 16).

Modular organization. Brain organization is modular, with many relatively independent modules each performing a different function, whereas digital computers typically have a single central processing unit connected to memory. The cortex has many localized modules with dense connectivity within a module, and then connections to a few other modules. The reasons for the modularity of the brain are considered in Section 26.4 and Chapters 2 and 3.

Hierarchical organization. As described in Chapter 2, many cortical systems are organized hierarchically. A major reason for this is that this enables the connectivity to be kept within the limits of which neurons appear capable (up to 50,000 synapses per neuron), yet for global computation (such as the presence of a particular object anywhere in the visual field) to be achieved, as exemplified by VisNet, a model of invariant visual object recognition (see Fig. 2.1 and Chapter 25). Another important reason is that this simplifies the learning that is required at each stage and enables it to be a local operation, in contrast to backpropagation of error networks where similar problems could in principle be solved in a two-layer network (with one hidden layer), but would require training with a non-local learning rule (Appendix C) as well as potentially neurons with very large numbers of connections. Another feature of cortical organization is that the number of areas in any hierarchy is not more than 4 or 5, because each area requires 20–30 ms of computation and transmission time. (This of course contrasts with the hundreds of successive layers being explored with artificial neural networks, as described in Section B.13, which is completely biologically implausible.)

26.4 Understanding how the brain works

I further think that it is very interesting to remark on how tractable the brain is to understand as a computer, relative to what it might have been, or relative to trying to reverse engineer a digital computer. This tractability applies to many brain systems, including those involved in perception, memory, attention, decision-making, and emotion (Rolls 2008d, Rolls 2014a, Rolls 2012d, Rolls and Deco 2010), though not to language, where the issue of how the syntax required for language is implemented in the brain is still an enormous and fascinating mystery (Chapter 17), or to consciousness, which is still a problematic issue (Chapter 22).

The encoding scheme. One reason for the tractability of the brain is its encoding scheme, whereby much information can be read off or decoded from the firing rates of single neurons and of populations of neurons (Chapter 8, Appendix C, and Rolls and Treves (2011)). The deep computational reason for this appears to be that neurons decode the information by dot product decoding (Fig. 8.2 and equation 1.1), and the consequence is that each of the independent inputs to a neuron adds information to what can be categorized by the neuron (Appendix B). The brain would have been much less tractable if binary encoding of the type used in a computer was used, as this is a combinatorial code, and any single bit in the computer word, or any subset of bits, yields little evidence on its own about the particular item being represented.

Neurons have a single output. A second reason for the tractability of the vertebrate brain is that each neuron has a single output, and this enables whatever information the neuron is representing and transmitting to be read (by scientists, and by neurons) from the spikes (action potentials) emitted by the neuron. This is much simpler than many invertebrate neurons, which have multiple output connections to other neurons, each of which could convey a different signal, and each of which may be difficult to record. The reason for spiking activity in mammalian neurons is that the information may need to be transmitted long distances along the axon, and the all-or-none self-propagating action potential is an accurate way to transmit the information without signal loss due to voltage degradation along the axon. But that biological need makes the vertebrate system (relatively) easy for scientists to decode and understand, as each vertebrate neuron does have a single output signal (sent to many other neurons) (Rolls and Treves 1998).

Modular organization. A third reason is that the brain is inherently modular, so that damage to one part can be used to analyze and understand particular functions being implemented, and how each function can be dissociated from other functions. (The methodology used is referred to as double dissociation.) One deep reason for this is that the cortex must operate as a set of local networks, for otherwise the total memory storage capacity of the whole cortex would be that of one of its modules or columns, given that storage capacity is determined by the number of connections C onto each neuron, not by the number of neurons in the whole system (O'Kane and Treves 1992) (Chapters 3 and 4, and Appendix B).

A second deep reason for this is that evolution by natural selection operates most efficiently when functions can be dissociated from each other and genes affect individual functions (implemented in the case of the brain by individual networks in different brain areas, including different cortical areas). Selection for particular characteristics or functionality can then proceed rapidly, for any change of fitness caused by gene changes related to a particular function feeds back rapidly into the genes selected for the next generation.

Another important advantage of modularity is that it helps to minimize the lengths of the connections, the axons, between the computing elements, the neurons, and thus helps to keep brain weight down, an important factor in evolution (Chapter 18). Minimizing connection length also simplifies the way in which brain design is specified by genes, for just connecting locally is a simple genetic specification, and reduces the need for special instructions to find the correct possibly distant neurons in the brain to which to connect (Chapter 19) (Rolls and Stringer 2000). Having all the neurons present while the brain is developing also helps neurons to find the correct populations to which to connect during brain development. In fact, neuron numbers tend to reduce somewhat after birth, and pruning connections, and neurons, is much simpler than trying to specify which new neurons to connect to in a possibly very distant brain region in the adult brain. The only recognized case where some neurons may be added after birth is dentate granule cells of the hippocampal system, which connect only locally to nearby CA3 cells, and may help to ensure that new episodic memories that are stored are different from previous memories by helping to select new subsets of hippocampal CA3 cells (Chapters 13 and 24) (Kesner and Rolls 2015, Rolls 2008d, Clelland et al. 2009, Rolls 2010b).

Conservation of architectures. A fourth reason for the tractability of the brain is that the same principles of operation implemented by similar network architectures are used in different brain areas for different purposes. An example is the use of autoassociation networks for short-term memory, episodic memory, semantic memory, and decision-making, in different brain areas. A very interesting example is that of backprojections in the neocortex, which illustrate how two simple functions, pattern association and autoassociation, can be combined (Chapter 11). This reflects the conservative nature of evolution: if a gene-specified implementation can be adapted for a different function, that is likely to be quicker and less risky than a completely new solution. (An example is the bones of our middle ear, the malleus, incus, and stapes, used to transmit sound to the cochlea, which were originally fish gills.) The consequence is that once we understand how the architecture is being used for one function, much of the theory can be used to understand how the same or a similar structure (network, architecture) is used elsewhere in the brain for another function.

However, we should note that the conservative nature of evolution (in a literal sense: earlier design features are often conserved and remain, even when different and better features evolve) can lead to some added complexity of the system. In this case evolutionarily old systems (such as the amygdalar control of autonomic function, or the basal ganglia system for habits) remain and account for some behaviour, even though newer systems such as goal directed action, and rational planning, provide newer and more powerful routes to action, and enable long-term planning. In this case, to understand behaviour, we must understand that different brain systems, perhaps with different goals, may all be vying for behavioural output, and one brain system may not entirely know what another one is doing (Chapter 22).

Ergodicity. A fifth reason for the tractability of the mammalian brain is that distributed representations are used, which involve large numbers of neurons. The result is that knowing how a subset of the neurons in a given population responds enables one to build a statistical understanding of how the whole population works computationally. Using the terminology of theoretical physics, a principle of weak *ergodicity* applies, by which studying the properties of a single element of the system for a long time (e.g. the firing rate response of a single neuron to a large set of stimuli) can yield information that can also be obtained from many elements for a short time (e.g. the firing rates of each of a large population of neurons to a single stimulus). We have shown that this principle, which holds if the elements (neurons) are statistically independent in what they respond to, does apply in the mammalian cortex (Franco, Rolls, Aggelopoulos and Jerez 2007). A condition for weak ergodicity is that the neurons are independent, and this is what the multiple cell information analyses show (Sections 8.2.2 and C.3.1.2). This statistical property enables a lot to be learned about the whole population if only a subset of the neurons is recorded, which is what is feasible. (Therefore the fact that we cannot record all neurons in the brain simultaneously need not limit our understanding.)

This principle also allows statistical mechanics to make quantitative predictions about for example the storage capacity of networks of neurons in the brain (Hopfield 1982, Amit 1989, Treves and Rolls 1991, Rolls 2008d). The deep reason why brain design (by Darwinian evolution) has resulted in this property is that having independent tuning by neurons to a set of stimuli (produced for example by competitive learning, Section B.4) enables the information that can be represented to increase linearly with the number of neurons (and thus the number of stimuli that can be represented to increase exponentially with the number of neurons, Section 8.2 and Appendix C), and also enables the properties of associative memories to arise that include generalization, completion, and graceful degradation so that the system can operate with diluted connectivity (Chapter 7 and Appendix B).

Gene specification of rewards, and emotion. A sixth reason for the tractability of the brain is that emotion allows genes to specify behaviour in a simple way, by specifying the goals for actions (specific rewards / subjective pleasures etc.), rather than the very complex specification that would be required if actions / movements had to be gene-specified in all but the simplest of cases (Rolls 2014a), as described in Chapter 15.

26.5 Synthesis on principles of operation of the cerebral cortex

In this section I draw together some of the principles of operation of the cerebral cortex, to help the reader understand the progression that has been made in the course of this book. I provide pointers to the chapters where the principles are described, but add to what is in those chapters by new thoughts on how some of the principles relate to each other to provide foundations for cortical operation. I also introduce some new thoughts here that point the way forward, with new ideas and concepts that may help to provide a foundation for further understanding and investigations of how the cerebral cortex operates in health and disease.

26.5.1 Hierarchical organization

(Chapter 2.) The hierarchical organization of the cerebral cortex enables complex computations to be solved, such as invariant visual object recognition, in a 'divide and conquer' strategy that breaks the computation up into multiple steps, in which each step can be performed by neurons limited to in the order of 10,000–20,000 inputs and outputs, in which the connectivity is mainly local, and in which brain size, weight, and the complexity and genetic specification of the wiring is kept tractable. Hierarchical organization also enables different types of representation to be kept independent, for example representations of objects from representations of their reward value. The hierarchical organization of the neocortex also enables outputs to be taken from each cortical stage of processing, from the L5 pyramidal cells (Chapter 18). One example is outputs from L5 neurons of the primary visual cortex, which may be used for convergent eye movements in connection with stereopsis. Another example is the projection from the inferior temporal visual cortex to the striatum (to the tail of the caudate nucleus), which by responding to novel objects (Caan, Perrett and Rolls 1984), may enable behaviour to be directed towards novel (i.e. changed) *objects*.

26.5.2 Localization of function

(Chapter 3.) The advantages of localization of function include many of those described for 'hierarchical organization'.

26.5.3 Recurrent collateral excitatory connections between the pyramidal cells, and the resulting attractor dynamics

(Chapter 4.) The use of excitatory recurrent collateral connections, and the potential that this provides for attractor network operation, is for me the major feature of the cerebral cortex, and a major step in evolution, because this principle of operation allows information to be held on-line, in for example short-term memory, which provides the foundation for planning ahead, and for top-down attention. The same attractor properties are invaluable for long-term memory including episodic and semantic memory, for via the property of *completion*, attractor network operations allow a whole memory to be completed from any of its parts. Attractor networks also provide an excellent solution to decision-making, for they enable the results of the decision to be held on-line, in short-term memory, until an action has been completed. A key principle of operation of the neocortex is that the attractors implemented by the recurrent collateral connections are **local attractor networks**. In a complementary way, the key concept of hippocampal operation is that the CA3 attractor is a **global attractor network** allowing associations between any representations in the hippocampus to be formed, in for example episodic memory (Chapters 24 and 12). These recurrent collateral synaptic connections, and their implications for understanding cortical computation, are often hardly mentioned by those who make 'wiring diagrams' of the circuitry of the cerebral cortex, and a highlight of this book which I hope will stand the test of time is that the excitatory recurrent collateral synaptic connections are the key to understanding cortical operation and computation.

However, we need to know much more than we know now about cortical attractor networks. Associative synaptic plasticity including heterosynaptic long-term depression in these connections is a key component of this concept of cortical operation (Section B.3 and Rolls and Treves (1990)), and we need to know more about this. We also need to know whether pyramidal cells operate as a single component of an attractor network with their recurrent collateral synapses on the apical and basal dendrites, and whether there are any key differences in the operation of the recurrent collateral synapses on the apical and basal dendrites, for example in their origin, time constants, or adaptation properties (see below). We also need to know more about the extent to which the superficial and deep layers of the neocortex can implement somewhat different attractors (see Section 26.5.18).

Another way of understanding how all the constraints from the recurrent collaterals, backprojections, and the inputs to an area of cortex shape the activity on the network is to consider all these as providing constraints on the dynamics of the system. Even if the network does not have sufficient time to fall into a stable attractor state, a succession of states may be visited by the system, and the trajectory of these states may be quite stable (Rabinovich, Huerta and Laurent 2008, Rabinovich, Varona, Tristan and Afraimovich 2014). A possible way to understand the operation of such a system is that of a stable heteroclinic channel. A stable heteroclinic channel is defined by a sequence of successive metastable ('saddle') states. Under the proper conditions, all the trajectories in the neighborhood of these saddle points remain in the channel, ensuring robustness and reproducibility of the dynamical trajectory through the state space (Rabinovich et al. 2008, Rabinovich et al. 2014). This conceptualization might find application in the dynamics of the networks involved in language that are considered in Chapter 17.

26.5.4 The noisy cortex

(Chapter 5.) The 'noise' in the cortex includes the randomness introduced by the almost Poisson nature of the spike timing of neurons for a given mean firing rate. This noise has a great impact especially in attractor networks, because the statistical fluctuations caused by the noise can move these non-linear systems in one direction or another, and once the non-linear positive feedback takes over, the system remains stable in that decision or recall state. The non-linearity of the positive feedback in attractor networks thus amplifies the effects of the noise. This has enormous implications, ranging from probabilistic decision-making, probabilistic memory recall which is an important component of creativity, and unpredictable behaviour. This book, and previous books (Rolls 2008d, Rolls and Deco 2010), emphasize that this noise can be advantageous (Chapter 5); and also that some psychiatric disorders including schizophrenia, obsessive-compulsive disorders, depression, and also normal aging, can be elucidated by considering the **stability** (Chapter 16) (understability or overstability) in attractor networks in particular cortical areas.

26.5.5 Top-down attention

(Chapter 6.) Top-down attention is conceptualized as a gentle interaction effect between attractors, made possible by the feedback as well as feedforward connections between adjacent areas in the cerebral cortical hierarchy. (The effect is described as a gentle interaction effect, because the top-down selective attention effect must not dominate the bottom-up input from the environment as otherwise we might miss important information from the environment; and the attractors must be separate, as described in Chapter 6.) The impact of top-down attention is to guide lower-level networks towards solutions that are beneficial in the light of the current interests of an organism highlighted by what is the subject of top-down attention. It is a way of tuning the cortex to be more efficient, because of top-down guidance. It can also help an organism to focus its cortical processing on the current task, using what is held in the short-term attractor network that guides attention.

26.5.6 Diluted connectivity

(Chapter 7.) Diluted connectivity in the cerebral cortex, in the recurrent collateral connections in particular, is seen as an optimization to reduce the probability that there will be multiple synaptic connections between cortical networks involved in autoassociation attractor networks, which reduce the memory capacity of these association networks. In pattern association networks, such as those in the backprojection pathways, diluted connectivity also helps to reduce the probability of multiple synapses between pairs of neurons, which would reduce the memory capacity. Diluted connectivity in competitive networks can help the categorisation, and can help the stability, of competitive networks (Section 7.4).

26.5.7 Sparse distributed coding using graded firing rates with relatively independent representations by different neurons and a code that can be read by neuronally plausible dot product decoding

(Chapter 8 and Appendix C.) The available evidence shows that sparse distributed encoding using graded firing rates is the main method of encoding used in the (primate including human) cerebral cortex; that it has high representational capacity; and that it can be read by neuronally plausible dot product decoding. This is essentially a place code (though not using local encoding), and this may provide a fundamental advantage if used in syntactical operations in the neocortex (Chapter 17).

26.5.8 Synaptic modification

(Chapter 9.) The associative learning rule implied by long-term potentiation is excellent for autoassociation memories and pattern association memories (though more investigation of whether the computationally useful heterosynaptic long-term depression is also present in the relevant synapses is needed). The associative learning rule implied by long-term potentiation is excellent also for competitive networks (though more investigation of whether the computationally useful heterosynaptic long-term depression that depends on the existing synaptic strength is also present in the relevant synapses is needed). There is some evidence that dopamine may be a third term involved in the modification of some synapses, which may help learning in some brain systems to reflect positive reward prediction errors.

It is an interesting and demanding question about whether associative synaptic modification is sufficient for operation of the cerebral cortex. It may be; or there may be new principles of synaptic modification to be discovered, such as those used in artificial neural networks that use error correction. However, the functional architecture of the neocortex can use backprojections for operations such as recall (Chapters 11 and 24) and top-down attention (Chapter 6), and this may make their use not possible in addition for other types of computation such as error backpropagation. However, one simple modification that does make a large difference beyond pure associativity is some temporal smoothing in the learning system using a trace learning rule, for as shown in Chapter 25, this allows the temporal structure of the sensory input to be used to build transform-invariant representations.

26.5.9 Adaptation and facilitation

(Chapter 10.) Adaptation, whether synaptic or neuronal, can play an important role in cortical computation by for example reducing the stability of cortical attractor networks so that they move on to something else in reasonable time; in sequence memory; and in implementing computations such as sensory-specific satiety. However, adaptation differs between different cortical neuronal types (Markram et al. 2015), and this may have hitherto unpredicted useful consequences. For example, it was suggested in Chapter 18 that L5 cortical neurons might have a shorter time constant, and be more 'bursty', than L2/L3 pyramidal cells, because of the importance of precise dynamics when initiating motor responses. It is now suggested that this might be implemented by having faster synaptic or neuronal adaptation associated with L5 than with L2/L3 pyramidal cells. This could be investigated (see further Section 26.5.18).

Synaptic facilitation can also be useful computationally, for example in maintaining active many separate short-term memory states in an attractor network (Rolls, Dempere-Marco and Deco 2013).

26.5.10 Backprojections

(Chapter 11.) There are as many backprojections as forward projections between each pair of adjacent areas in a cortical hierarchy, and this provides for high capacity of memory recall (Chapter 24). Backprojections are also used for top-down attention (Chapter 6). Backprojections do not simply transfer what is represented at higher cortical areas to earlier cortical areas.

26.5.11   Neurogenesis

(Chapter 13.) The dentate gyrus input to hippocampal CA3 performs pattern separation of the inputs before CA3 to help achieve the relatively uncorrelated representations that help to achieve high memory capacity in CA3. The dentate granule cells perform this operation by competitive learning; by the low probability of contact of the dentate granule cell to CA3 connections (0.00015 in the rat); and by neurogenesis with the new dentate granule cells helping to ensure that new memories are distinct from old memories by making new random connections to CA3 neurons.

The dentate gyrus is the only part of cortex where new neurons can develop in adulthood, for here the premium is on selecting new random sets of neurons in CA3 for new memories distinct from previous memories; whereas elsewhere in cortex the premium is on building useful new representations (which may represent new categories, as in competitive learning), and then reaccessing those representations later. Further, once the neocortex has developed it would be difficult for long-range connections to be guided to the correct destinations far away in the full-size adult brain. For these reasons, neurogenesis is not a property of the cerebral cortex in adulthood, apart from in the specialized hippocampal dentate granule cell system.

26.5.12   Binding and syntax

(Chapter 17.) It is shown that a solution to the binding problem in vision is provided by the formation of feature-combination neurons selective to the spatial arrangement of the features present in the preceding cortical area (Sections 25.5.5, 25.3.2, and 25.4), rather than stimulus-dependent synchrony (Sections 8.3, C.2.4 and C.2.5) or coherence (Section 8.3).

Building on this evidence, it is suggested that similar place encoding may be useful as a solution to the syntax used in language (Chapter 17). A key investigation here is to test in humans (in investigations performed for clinical purposes) whether some single neurons respond to a noun such as 'cat' differently when it is the subject or object of a sentence (Chapter 17). Language might then be implemented as a trajectory guided by stronger forward than backward projections through a state space of local neocortical attractors (Chapter 17 and Rolls and Deco (2015a)), but this is very much a hypothesis still in development.

26.5.13   Evolution of the cerebral cortex

(Chapter 18.) The great development of the orbitofrontal cortex, temporal lobes, and frontal cortical areas in primates including humans is highlighted, and how with increasing emphasis on the cortical processing of stimuli, even systems as apparently evolutionarily old as the taste system become differently connected to emphasize cortical processing in primates.

The fascinating and fundamental problems of how lamination helps cortical operation, and how different types of cortex compare in their operation, are considered in Sections 18.2 and 26.5.18.

The ways in which cortical design offers much beyond the evolutionarily much older basal ganglia are considered in Chapters 20 and 4. Chapter 23 highlights how a very different type of cortex, cerebellar cortex, again has quantitatively well-defined rules of connectivity, which again provide a foundation for understanding how different types of cortex operate.

26.5.14   Genetic specification of cortical design

(Chapter 19.) Comparison of the functional architecture of different cortical areas suggests hypotheses about the relatively few parameters required to specify each cortical area, which specify the classes of neuron and the class-to-class rules of connectivity, yet leave open to development and self-organizing learning the details of cortical cell-to-cell wiring and synaptic strength. These hypotheses are amenable to further test with genetic algorithms, and eventually to better empirical understanding of what is encoded by genes.

26.5.15   The cortical systems for emotion

(Chapter 15.) The organization of cortical systems for emotion in primates including humans illustrates important principles of cortical function.

One principle is that different types of processing are performed independently in different localised and hierarchically organised areas of neocortex, including primarily unimodal processing of visual, taste, and olfactory processing to form first 'what' levels of representation; a second tier of processing with reward value, emotion-related, representations in which the sensory modalities are combined; and a third level of representation in which decisions are taken, and actions are initiated (Fig. 2.4). This design feature allows objects to be remembered, including where they have been seen, independently of their reward value. This segregation of object from value representations is far from complete in rodents, making rodents a poor model of the cortical systems involved in emotion and motivation in humans.

Another key principle is that there are many output systems for emotion-related stimuli, which probably represent new processing systems added in evolution, and which operate to some extent in parallel, though each performs a different computation (Figs. 2.4 and 22.3). These systems include those for brainstem reflexes, for unlearned and learned autonomic responses (including the amygdala, orbitofrontal cortex, and hypothalamus); for stimulus-response habits (the basal ganglia); for goal-directed actions (the orbitofrontal cortex to compute the value of possible goals, and the anterior cingulate cortex for action–goal outcome learning); and for actions computed with explicit reasoning (Fig. 22.3). The system involved in reasoning has a credit assignment problem, as it may not be evident if a plan fails which step in a multi-step plan (which requires syntax) was faulty. With this as the background, it is suggested that the operation of a syntactic monitoring systems that corrects the lower order syntactic processing may be related to the computations involved in consciousness, in what is a higher order syntactic thought (HOST) theory of consciousness (Chapter 22).

Another principle is that there may be conflicts between these processing systems, with the gene-defined goals that are related to emotion and motivation being in the interests of the genes; with the reasoned goals for action being potentially in the interests of the individual and not the genes (Section 22.3); and with noise affecting the decisions made between these two systems (Chapter 5 and Section 22.3).

Another key principle is that value systems represent the value of different stimuli by different neurons in the orbitofrontal cortex using sparse distributed representations, which, together with synaptic adaptation (Section 10.5), account for sensory-specific satiety, a property of all reward systems that is an important evolutionary principle of cortical function (Sections 2.2.3.3 and 15.8).

Another key principle of cortical design by genes is that the goals for action, the rewards and punishers, are a much more flexible way of influencing behaviour than defining behavioural responses, for then any action can be learned to obtain a goal, greatly simplifying the genetic specification of behaviour that is adaptive for the genes.

26.5.16   Memory systems

(Chapters 12 and 24.) Chapter 24 provides a description of the ways in which many principles of operation of the cortex combine together to perform a major brain computation, the storage and recall of episodic memory, with an introduction provided in Chapter 12. The principles of operation that are combined and illustrated include pattern separation, neurogenesis, pattern completion, a single global attractor network (in CA3), backprojections, memory recall, the capacity of attractor networks and pattern association networks, diluted connectivity, and sparse distributed representations.

26.5.17   Visual cortical processing for invariant visual object recognition

(Chapters 25 and 14.) Chapter 25 provides a description and simulation model of the ways in which many principles of operation of the cortex combine together to perform a major brain computation, the transform-invariant visual recognition of objects, with an introduction provided in Chapter 14. The principles of operation that are combined and illustrated include the use of competitive networks to build new representations; the use of diluted connectivity in a competitive network; the convergent combination of inputs received from different parts of a preceding cortical area in a cortical hierarchy; the modification of an associative synaptic learning rule by a short-term memory trace to enable a competitive network to build representations that reflect the temporal and spatial statistics of the inputs thereby helping to build transform-invariant representations; the operation of a cortical stream with natural inputs that helps to reveal computation restricted to a small part of visual space which helps to solve binding problems; the solution of the binding problem by forming feature combination neurons selective to the spatial arrangement of the features present in the preceding cortical area; the role of the diluted connectivity in enabling the location of objects relative to the fovea to be encoded; the role of top-down attention in cluttered visual scenes; the use of sparse distributed representations that can be read by biologically plausible dot product decoding; the independent representation of information by neurons in an ensemble of neurons active to a stimulus; and the complementary contributions of the dorsal visual cortical stream to find and fixate salient regions in a natural scene, and of the ventral visual cortical visual stream to perform invariant object recognition at the fixated locations.

26.5.18   Cortical lamination, operation, and evolution

(Section 18.2, with introduction in Section 1.10.) In Section 18.2.1 new hypotheses are introduced about neocortical computation. These are extended here, partly with the aim of formulating hypotheses for future research (see also Rolls and Mills (2017)).

New concepts in Section 18.2.1 include hypotheses that the superficial and deep layers support somewhat different attractors, with the superficial layers being specialised for feedforward processing involving categorisation into discrete representations using competitive learning to discover new categories, based on the convergence of inputs from different parts of the preceding cortical area. As a result, the superficial layers would implement discrete attractor networks with the recurrent collaterals. In contrast, the deep layers, in particular layer 5 with its outputs to motor systems including via the striatum / basal ganglia may need a different type of attractor, which though initiated by the superficial layers, might be smoother spatially to make movements smooth, by implementing what is more like a continuous attractor network; and might be temporally faster, as the premium is now on what may sometimes be a requirement for fast but brief output to suddenly drive motor systems when necessary. (The superficial layers may benefit from slower integration of information, to help build useful new categories, and to allow representations to be influenced by not only the recurrent collaterals, but also by what may be being backprojected from the next cortical stage in the hierarchy (cf. Fig. 4.5 on page 88).) Consistent with these hypotheses, the deep layers of the cerebral cortex are described as having more 'bursty' firing than the more steady and sustained firing of the superficial layers. How might the deep layers of the neocortex be set up to produce these somewhat different types of representation?

The faster time course of the deeper layers might, I suggest, be implemented by having relatively fast synaptic or neuronal adaptation (Chapter 10, cf. Markram et al. (2015) who show that adaptation varies greatly between different classes of cortical neuron). Consistent with this hypothesis, more adaptation was found in deep cortical layers than in superficial layers of the rat auditory cortex to tones (Szymanski, Garcia-Lazaro and Schnupp 2009). A faster time course in deep layers might also be produced by a relatively higher proportion of AMPA or kainate receptors with their short time constant (5–10 ms) in the deep layers, with relatively more NMDA receptors with their relatively long time constants (100 ms) in superficial layers. Consistent with this hypothesis, NMDA receptors are present in relatively higher densities in superficial layers (L2/L3), and kainate receptors in relatively higher densities in deep layers (L5 and L6) of the human fusiform gyrus, a ventral stream visual area involved in face and object processing (Caspers, Palomero-Gallagher, Caspers, Schleicher, Amunts and Zilles 2015). In addition, the pyramidal cells in the deep layers probably have different time constants, for they tend to be large, related in part to the fact that many of their axons have to travel long distances to reach their motor targets and therefore need fast conduction velocities.

A shallower energy landscape of the deep layers may contribute to less discrete representations than the superficial layers (making the deeper layers more like a continuous attractor), and this could be produced by for example the faster adaptation already mentioned in the deep layers which would tend to produce more continuous spatial scales for the deep layer neurons due to the relation between time and space, and perhaps by different inhibition, which could be related to the fact that some of the inhibitory neurons in the cortex have layer-specific connections (Sections 1.10 and 18.2). These hypotheses are eminently testable.

Further factors in the cortical lamination are that because the targets and the afferent connections of the neurons in each cortical layer are different for the types of computational reason being discussed, it may be efficient when building the cortex to use different molecular recognition mechanisms to define the connections of the classes of neurons in each layer; that lamination may help to simplify the local connectivity rules and therefore the genetic specification by making cortical layer a parameter that can be gene-specified; that cortical lamination may help to reduce the length of the interconnections because the neurons that need to communicate with each other most are close together in the same lamina; and that cortical lamination facilitates the use of separate classes of inhibitory neurons (Section 1.10.3) that can normalize the activity of each class of cortical excitatory neuron separately, which is important as it is suggested that the different excitatory populations perform different computations and therefore require different normalizations (Chapter 19 and Section 18.2).

A related key issue is why the backprojections to the preceding cortical layer in the hierarchy tend to come from the deep layers, especially L5B (Markov et al. 2014b). It is as if the preceding cortical area needs to know what is being output to motor systems. Perhaps the more diffuse anatomical connectivity of the backprojections is quite suitable for memory recall, so that neurons over a wide area in the preceding cortical area that are associated with a higher level representation can be brought into firing during recall. Perhaps also the shallower basins of attraction are more suitable for the memory recall process for the preceding cortical area. There are important issues to be better understood here.

As far as layer 6 is concerned, my working hypothesis can be summarised as follows (for introduction see Sections 1.10 and 18.2.1). The thalamus provides pattern separation for inputs to neocortex, because its lateral inhibition acts as a high-pass spatial filter. This applies especially to primary sensory cortical areas, and is supplemented in these areas by many stellate / granule cells that perform further pattern separation. Via these neurons, as well as by direct thalamo-cortical synapses, inputs are introduced (in primary sensory cortical areas) into the superficial neocortical layers. However, primarily for gain control, it is helpful to have backprojections to the thalamus to provide for negative feedback gain control, and the layer 6 pyramidal cells are specialised for this. Consistent with this gain control role, the layer 6 neurons also have some terminals in layer 4, where they may perform feedforward control of the effects produced through thalamo-cortical synapses. Segregation of layer 6 as a separate layer in neocortex may again be related to simplicity of local connectivity rules, as well as the different molecular recognition needed to specify the connections of layer 6 neurons (Chapter 19).

Another key issue for neocortex is whether the apical and basal dendrites of cortical pyramidal cells operate as a single system, or whether there is some separation of function. Clearly the backprojections ending in layer 1 at the ends of the apical dendrites will have relatively weaker effects than synapses closer to the cell body electrically, especially because the synaptic currents produced by the layer 1 synapses will be electrically shunted by synapses more proximal to the cell body when the more proximal synapses are active. But do the recurrent collaterals on the basal and apical dendrites operate as a uniform system, or is there some specialization? The forward inputs to superficial pyramidal cells are often onto the basal dendrites (Chapter 11), which is suggestive of some specialization. There is much to be learned here, and as always the quantitative data on the connectivity (e.g. the numbers of synapses) are important, and need much more investigation (see also Rolls and Mills (2017)).

Another key issue is how the different cell types in the neocortex (Fig. 18.1), hippocampal cortex (Fig. 18.3), and pyriform cortex (Fig. 18.2) relate to each other. (I have used corresponding colours in these Figures to facilitate the comparison, and this discussion.) Some comparisons follow, to help produce usefully integrated concepts, building on what is presented in Section 18.2.

The dentate granule cells in the hippocampus and the stellate cells in the neocortex both appear to perform pattern separation. A difference is that in the hippocampus, during learning, the dentate granule cells with their low contact probability and non-associative synaptic modification just perform pattern separation, selecting a new subset of CA3 cells for each new episodic memory (Chapter 24). In this situation, a separate input is needed to implement recall, and this is identified as the associatively modifiable direct perforant path input to CA3 neurons (Chapter 24). In contrast, in the neocortex, it is hypothesized that competitive learning takes place on the feedforward inputs to superficial pyramidal cells (with synapses from stellate cells, or from the thalamus, or from the preceding cortical area in non-primary sensory cortical areas). This competitive learning enables new categories to be formed, and this process is facilitated by the feed-forward convergence. Because the feedforward synapses are modified, they can be used to activate the correct set of L2/L3 pyramidal cells the next time that a similar input is presented. In comparison, the pyriform cortex has no granule cells (Fig. 18.2), although the pyriform cortex does receive from, in addition to the olfactory bulb, some areas that may do pre-processing (Section 18.2.2).

The hippocampal cortex CA3 neurons may correspond to those in the superficial layers of the neocortex (L2/L3), and to the pyramidal cells in especially the anterior pyriform cortex. All appear to operate as attractor networks. In the pyriform cortex, this enables representations produced by odours that activate different glomeruli to be learned, and later recognised when that subset of cells is active (Section 18.2.2). In the hippocampal CA3, everything appears to be organised to maximise the storage capacity in a single attractor network for episodic memory, to allow arbitrary associations to be learned (Chapter 24). In the neocortex, the superficial pyramidal cells are hypothesized to do more, because they perform competitive learning to form new categories, and in this situation the recurrent collaterals help to perform constraint satisfaction when stimuli are presented, to help the discrete category most consistent with the input to be selected (Section 18.2.1). In the neocortex, the attractors must be local, both because different categories and information are represented in different parts of the neocortex, but also in order not to limit the total storage capacity of the neocortex.

The L5 neocortical pyramidal cells send outputs to the striatum and other motor structures (and the striatum would have had mainly motor-related functions at earlier stages of evolution). Large cells may be needed in L5, because some of the connection distances may be large. These outputs also appear to be involved in recall, and are the origin of many cortico-cortical backprojections (Section 18.2.1 and Chapters 24 and 11). In a corresponding way, the CA1 cells of the hippocampus have outputs to the striatum, and are involved in recall, in that they are en route back to the neocortex (Chapter 24). In both hippocampal CA1 and the neocortical L5 pyramidal cells, recoding of the parts of a representation (necessarily represented as separate parts in the preceding autoassociation network), is performed by competitive learning to form a more compact retrieval cue, in which the parts need no longer be represented, but now need only be represented by neurons that encode the combination of the parts, as explained in Section 24.3.6. This recoding for recall may be an important similarity of the hippocampal CA1 and neocortical L5 computation. In the posterior pyriform cortex, there may be some similarity, in that these neurons also send outputs to some motor-related structures such as the striatum; and also in a sense can send the whole recoded combination of odour components as an output to the next stage of processing, in contrast to the anterior pyriform cortex where the parts of an odour (what is represented by each receptor type and glomerulus) may be associated together (Section 18.2.2).

Although these thoughts extend beyond present evidence, it is concepts of this computational type that will I believe be important in helping to set up testable hypotheses for future empirical investigations into the principles of operation of the neocortex.

26.6 Highlights

To draw the book together, and to highlight some of the principles of cortical computation elucidated in this book, the following points are made:

1. Principles of cortical computation, and not a whole theory of cortical computation, are described. However, Chapter 18 on the evolution of the cortex and Section 26.5.18 come close to presenting hypotheses about how cortical microcircuitry operates; and as examples of how the principles combine to provide a theory of how cortical systems operate, I outline a theory of the operation of the hippocampal system for episodic memory in Chapter 24, and a theory of the operation of the ventral visual system for transform invariant object recognition in Chapter 25.

2. The mind-brain problem is considered in the light of different levels of explanation, which offer a solution.

3. A comparison of cortical computation and computation in a digital computer highlights many fundamentally different principles of brain computation, and computation in a digital computer.

4. Reasons why understanding cortical computation is making rapid progress are described.

5. The advances in understanding the emotional, rational, and decision-making systems in the brain have implications for understanding in a wide range of areas, including aesthetics, art, ethics, economics, and social interactions, as considered elsewhere (Rolls 2012d, Rolls 2017d).

6. Section 26.5 draws together some of the principles of operation of the cerebral cortex, to help the reader understand the progression that has been made in the course of this book. Pointers are provided to the chapters where the principles are described, and new thoughts are added on how some of the principles relate to each other to provide foundations for cortical operation. New thoughts are also added on cortical lamination and the types of computation performed in different layers, and in different types of cortex, including neocortex, pyriform cortex, and hippocampal cortex. New ideas and concepts are also added that may help to provide a foundation for further understanding and investigations of how the cerebral cortex operates in health and disease.

Appendix 1 Introduction to linear algebra for neural networks

In this Appendix we review some simple elements of linear algebra relevant to understanding neural networks. This will provide a useful basis for a quantitative understanding of how neural networks operate (see Appendix B).

A.1   Vectors

A vector is an ordered set of numbers. An example of a vector is the set of numbers

$$\begin{bmatrix} 7 \\ 4 \end{bmatrix}$$

If we denote the jth element of this vector as w_j, then $w_1 = 7$, and $w_2 = 4$. We can denote the whole vector by $\mathbf{w}$. This notation is very economical. If the vector has 10,000 elements, then we can still refer to it in mathematical operations as $\mathbf{w}$. $\mathbf{w}$ might refer to the vector of 10,000 synaptic weights on the dendrites of a neuron. Another example of a vector is the set of firing rates of the axons that make synapses onto a dendrite, as shown in Fig. 1.3. The firing rate x of each axon forming the input vector can be indexed by j, and is denoted by x_j. The vector would be denoted by $\mathbf{x}$.

Certain mathematical operations can be performed with vectors. We start with the operation which is fundamental to simple models of neural networks, the inner product or dot product of two vectors.

A.1.1   The inner or dot product of two vectors

The operation of computing the activation h of a neuron from the firing rate on its input axons multiplied by the corresponding synaptic weight can be expressed as:

$$h = \sum_j x_j w_j \tag{A.1}$$

where $\sum_j$ indicates that the sum is over the C input axons to each neuron, indexed by j. Denoting the firing rate vector as $\mathbf{x}$ and the synaptic weight vector as $\mathbf{w}$, we can write

$$h = \mathbf{x} \cdot \mathbf{w} \qquad . \tag{A.2}$$

If the weight vector is

$$\mathbf{w} = \begin{bmatrix} 9 \\ 5 \\ 2 \end{bmatrix}$$

and the firing rate input vector is

$$\mathbf{x} = \begin{bmatrix} 3 \\ 6 \\ 7 \end{bmatrix}$$

then we can write

$$\mathbf{x} \cdot \mathbf{w} = (3 \cdot 9) + (6 \cdot 5) + (7 \cdot 2) = 71 \quad . \tag{A.3}$$

Thus in the inner or dot product, we multiply the corresponding terms, and then sum the result. As this is the simple mathematical operation that is used to compute the activation h in the most simplified abstraction of a neuron (see Chapter 1), we see that it is indeed the fundamental operation underlying many types of neural network. We will shortly see that some of the properties of neuronal networks can be understood in terms of the properties of the dot product. We next review a number of basic aspects of vectors and inner products between vectors.

There is a simple geometrical interpretation of vectors, at least in low-dimensional spaces. If we define, for example, x and y axes at right angles to each other in a two-dimensional space, then any two-component vector can be thought of as having a direction and length in that space that can be defined by the values of the two elements of the vector. If the first element is taken to correspond to x and the second to y, then the x axis lies in the direction [1,0] in the space, and the y axis in the direction [0,1], as shown in Fig. A.1. The line to point [1,1] in the space then lies at $45°$ to both axes, as shown in Fig. A.1.

A.1.2   The length of a vector

Consider taking the inner product of a vector

$$\mathbf{w} = \begin{bmatrix} 4 \\ 3 \end{bmatrix}$$

with itself. Then

$$\|\mathbf{w}\| = \sqrt{\mathbf{w} \cdot \mathbf{w}} = \sqrt{4^2 + 3^2} = 5. \tag{A.4}$$

This is the length of the vector. We can represent this operation in the two-dimensional graph shown in Fig. A.1. In this case, the coordinates where vector $\mathbf{w}$ ends in the space are [1,1]. The length of the vector (from [0,0]) to [1,1]) is obtained by Pythagoras' theorem. Pythagoras' theorem states that the length of the vector $\mathbf{w}$ is equal to the square root of the sum of the squares of the two sides. Thus we define the length of the vector $\mathbf{w}$ as

$$\|\mathbf{w}\| = \sqrt{\mathbf{w} \cdot \mathbf{w}} \tag{A.5}$$

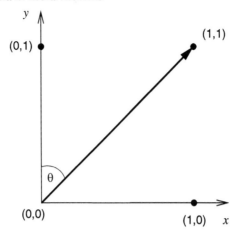

Fig. A.1 Illustration of a vector in a two-dimensional space. The basis for the space is made up of the x axis in the [1,0] direction, and the y axis in the [0,1] direction. (The first element of each vector is then the x value, and the second the y value. The values of x and y for different points, marked by a dot, in the space are shown. The origins of the axes are at point 0,0.) The [1,1] vector projects in the [1,1] (or 45°) direction to the point 1,1, with length 1.414.

In the [1,1] case, this value is $\sqrt{2} = 1.414$.

A.1.3 Normalizing the length of a vector

We can scale a vector in such a way that its length is equal to 1 by dividing it by its length. If we form the dot product of two normalized vectors, its maximum value will be 1, and its minimum value -1.

A.1.4 The angle between two vectors: the normalized dot product

The angle between two vectors $\mathbf{x}$ and $\mathbf{w}$ is defined in terms of the inner product as follows:

$$\cos \theta = \frac{\mathbf{x} \cdot \mathbf{w}}{\|\mathbf{x}\| \|\mathbf{w}\|} \tag{A.6}$$

For example, the angle between two vectors

$$\mathbf{x} = \begin{bmatrix} 0 \\ 1 \end{bmatrix} \quad \text{and} \quad \mathbf{w} = \begin{bmatrix} 1 \\ 1 \end{bmatrix}$$

where the length of vector $\mathbf{x}$ is $\sqrt{0.0 + 1.1} = 1$ and of vector $\mathbf{w}$ is $\sqrt{1.1 + 1.1} = \sqrt{2}$ is

$$\cos \theta = \frac{(0.1) + (1.1)}{1.\sqrt{2}} = 0.707. \tag{A.7}$$

Thus $\theta = \cos^{-1}(0.707) = 45°$.

We can give a simple geometrical interpretation of this as shown in Fig. A.1. However, equation A.6 is much easier to use in a high-dimensional space!

The dot product reflects the similarity between two vectors. Once the length of the vectors is fixed, the higher their dot product, the more similar are the two vectors. By normalizing the dot product, that is by dividing by the lengths of each vector as shown in equation A.6, we obtain a value that varies from -1 to $+1$. This normalized dot product is then just the cosine of the angle between the vectors, and is a very useful measure of the similarity between any two vectors, because it always lies in the range -1 to $+1$. It is closely related to the (Pearson product–moment) correlation coefficient between any two vectors, as we see if we write the equation in terms of its components

$$\cos\theta = \frac{\sum_j x_j w_j}{(\sum_j x_j^2)^{1/2}(\sum_j w_j^2)^{1/2}} \tag{A.8}$$

which is just the formula for the correlation coefficient between two sets of numbers with zero mean (or with the mean value removed by subtracting the mean of the components of each vector from each component of that vector).

Now consider two vectors that have a dot product of zero, that is where $\cos\theta = 0$ or the angle between the vectors is $90°$. Such vectors are described as orthogonal (literally at right angles) to each other. If our two orthogonal vectors were $\mathbf{x}$ and $\mathbf{w}$, then the activation of the neuron, measured by the dot product of these two vectors, would be zero. If our two orthogonal vectors each had a mean of zero, their correlation would also be zero: the two vectors can then be described as unrelated or independent.

If, instead, the two vectors had zero angle between them, that is if $\cos\theta = 1$, then the dot product would be maximal (given the vectors' lengths), the normalized dot product would be 1, and the two vectors would be described as identical to each other apart from their length. Note that in this case their correlation would also be 1, even if the two vectors did not have zero mean components.

For intermediate similarities of the two vectors, the degree of similarity would be expressed by the relative magnitude of the dot product, or by the normalized dot product of the two vectors, which is just the cosine of the angle between them. These measures are closely related to the correlation between two vectors.

Thus we can think of the simple operation performed by neurons as measuring the similarity between their current input vector and their synaptic weight vector. Their activation, h, is this dot product. It is because of this simple operation that neurons can generalize to similar inputs; can still produce useful outputs if some of their inputs or synaptic weights are damaged or missing, that is they can show graceful degradation or fault tolerance; and can be thought of as learning to point their weight vectors towards input patterns, which is very useful in enabling neurons to categorize their inputs in competitive networks (see Section B.4).

A.1.5 The outer product of two vectors

Let us take a row vector having as components the firing rates of a set of output neurons in a pattern associator or competitive network, which we might denote as $\mathbf{y}$, with components y_i and the index i running from 1 to the number N of output neurons. $\mathbf{y}$ is then a shorthand for writing down each component, e.g. [7,2,5,2,...], to indicate that the firing rate of neuron 1 is 7, etc. To avoid confusion, we continue in the following to denote the firing rate of input neuron j as x_j. Now recall (see Chapter 1 and Section B.2) how the synaptic weights are formed in a pattern associator using a Hebb rule as follows:

$$\delta w_{ij} = \alpha y_i x_j \tag{A.9}$$

where δw_{ij} is the change of the synaptic weight w_{ij} which results from the simultaneous (or conjunctive) presence of presynaptic firing x_j and postsynaptic firing or activation y_i, and α is a learning rate constant which specifies how much the synapses alter on any one pairing. In a more compact vector notation, this expression would be

$$\delta \mathbf{w}_i = \alpha \mathbf{y}_i \mathbf{x}' \tag{A.10}$$

where the firing rates on the axons form a column vector with the values, for example, as follows[35]:

$$\mathbf{x}' = \begin{bmatrix} 2 \\ 0 \\ 3 \end{bmatrix}$$
$$....$$

The weights are then updated by a change proportional (the α factor) to the following matrix (Table A.1):

Table A.1 Multiplication of a row vector [7 2 5 ....] by a column vector to form the external or tensor product, representing for example the changes to a matrix of synaptic weights $\mathbf{W}$

	$\begin{bmatrix} 7 & 2 & 5 & ..... \end{bmatrix}$			
$\begin{bmatrix} 2 \\ 0 \\ 3 \end{bmatrix}$	14	4	10	......
	0	0	0	......
	21	6	15	......
.....	.....	.....	.....	.....

This multiplication of the two vectors is called the outer, or tensor, product, and forms a matrix, in this case of (alterations to) synaptic weights. Thus we see that the operation of altering synaptic weights in a network can be thought of as forming a matrix of weight changes, which can then be used to alter the existing matrix of synaptic weights.

A.1.6 Linear and non-linear systems

The operations with which we have been concerned in this Appendix so far are linear operations. We should note that if two matrices operate linearly, we can form their product by matrix multiplication, and then replace the two matrices with the single matrix that is their product. We can thus effectively replace two synaptic matrices in a linear multilayer neural network with one synaptic matrix, the product of the two matrices. For this reason, multilayer neural networks if linear cannot achieve more than can be achieved in a single-layer linear network. It is only in non-linear networks that more can be achieved, in terms of mapping input vectors through the synaptic weight matrices, to produce particular mappings to output vectors. Much of the power of many networks in the brain comes from the fact that they are multilayer non-linear networks (in that the computing elements in each network, the neurons, have non-linear properties such as thresholds, and saturation at high levels of output). Because the matrix by matrix multiplication operations of linear algebra cannot be applied directly to the operation of neural networks in the brain, we turn instead back to other aspects of linear algebra, which can help us to understand which classes of pattern can be successfully learned by different types of neural network.

---

[35] The prime after the $\mathbf{x}$ is used here to remind us that this vector is a column vector, which can be thought of as a transformed row vector, and the prime indicates the transformed vector. We do not use the prime for most of this book in order to keep the notation uncluttered.

A.1.7 Linear combinations of vectors, linear independence, and linear separability

We can multiply a vector by a scalar (a single value, e.g. 2) thus:

$$2 \cdot \begin{bmatrix} 4 \\ 1 \\ 3 \end{bmatrix} = \begin{bmatrix} 8 \\ 2 \\ 6 \end{bmatrix}$$

We can add two vectors thus:

$$\begin{bmatrix} 4 \\ 1 \\ 3 \end{bmatrix} + \begin{bmatrix} 2 \\ 7 \\ 2 \end{bmatrix} = \begin{bmatrix} 6 \\ 8 \\ 5 \end{bmatrix}$$

The sum of the two vectors is an example of a linear combination of vectors, which is in general a weighted sum of several vectors, component by component. Thus, the linear combination of vectors v_1, v_2, ... to form a vector v_s is expressed by the sum

$$v_s = c_1 v_1 + c_2 v_2 + ... \tag{A.11}$$

where c_1 and c_2 are scalars.

By adding vectors in this way, we can produce any vector in the space spanned by a set of vectors as a linear combination of vectors in the set. If in a set of n vectors at least one can be written as a linear combination of the others, then the vectors are described as **linearly dependent**. If in a set of n vectors none can be written as a linear combination of the others, then the vectors are described as **linearly independent**. A linearly independent set of vectors has the properties that any vector in the space spanned by the set can be written in only one way as a linear combination of the set, and the space has dimension $d = n$. In contrast, a vector in a space spanned by a linearly dependent set can be written in an infinite number of equivalent ways, and the dimension d of the space is less than n.

Consider a set of linearly dependent vectors and the d-dimensional space they span. Two subsets of this set are described as **linearly separable** if the vectors of one subset (that is, their endpoints) can be separated from those of the other by a hyperplane, that is a subspace of dimension $d - 1$. *Subsets formed from a set of linearly independent vectors are always linearly separable.* For example, the four vectors:

$$\begin{bmatrix} 0 \\ 0 \end{bmatrix} \quad \begin{bmatrix} 0 \\ 1 \end{bmatrix} \quad \begin{bmatrix} 1 \\ 0 \end{bmatrix} \quad \begin{bmatrix} 1 \\ 1 \end{bmatrix}$$

are linearly dependent, because the fourth can be formed by a linear combination of the second and third (and also because the first, being the null vector, can be formed by multiplying any other vector by zero, a specific linear combination). In fact, $n = 4$ and $d = 2$. If we split this set into subset A including the first and fourth vector, and subset B including the second and third, the two subsets are not linearly separable, because there is no way to draw a line (which is the subspace of dimension $d - 1 = 1$) to separate the two subsets A and B. We will encounter this set of vectors in Appendix B, and this is the geometrical interpretation of why a one-layer, one-output neuron network cannot separate these patterns. Such a network (a simple perceptron) is equivalent to its (single) weight vector, and in turn the weight vector defines a set of parallel $d-1$ dimensional hyperplanes. (Here $d = 2$, so a hyperplane is simply a line, any line perpendicular to the weight vector.) No line can be found that separates the first and fourth vector from the second and third, whatever the weight vector the line is perpendicular to, and hence no perceptron exists that performs the required classification (see Section A.2.1). To separate such patterns, a multilayer network with non-linear neurons is needed (see Appendix B).

Any set of linearly independent vectors comprise the basis of the space they span, and they are called basis vectors. All possible vectors in the space spanned by these vectors can be formed as linear combinations of these vectors. If the vectors of the basis are in addition mutually orthogonal, the basis is an orthogonal basis, and it is, further, an orthonormal basis if the vectors are chosen to be of unit length. Given any space of vectors with a preassigned meaning to each of their components (for example the space of patterns of activation, in which each component is the activation of a particular unit), the most natural, canonical choice for a basis is the set of vectors in which each vector has one component, in turn, with value 1, and all the others with value 0. For example, in the $d = 2$ space considered earlier, the natural choice is to take as basis vectors

$$\begin{bmatrix} 1 \\ 0 \end{bmatrix}$$

and

$$\begin{bmatrix} 0 \\ 1 \end{bmatrix}$$

from which all vectors in the space can be created. This can be seen from Fig. A.1. (A vector in the $[-1,-1]$ direction would have the opposite direction of the vector shown in Fig. A.1.)

If we had three vectors that were all in the same plane in a three-dimensional (x, y, z) space, then the space they spanned would be less than three-dimensional. For example, the three vectors

$$\begin{bmatrix} 1 \\ 0 \\ 0 \end{bmatrix} \qquad \begin{bmatrix} 0 \\ 1 \\ 0 \end{bmatrix} \qquad \begin{bmatrix} -1 \\ -1 \\ 0 \end{bmatrix}$$

all lie in the same z plane, and span only a two-dimensional space. (All points in the space could be shown in the plane of the paper in Fig. A.1.)

A.2   Application to understanding simple neural networks

The operation of simple one-layer networks can be understood in terms of these concepts.

A.2.1 Capability and limitations of single-layer networks: linear separability and capacity

Single-layer perceptrons perform pattern classification, and can be trained by an associative (Hebb) learning rule or by an error-correction (delta) rule (see Appendix B). That is, each neuron classifies the input patterns it receives into classes determined by the teacher. Single-layer perceptrons are thus supervised networks, with a separate teacher for each output neuron. The classification is most clearly understood if the output neurons are binary, or are strongly non-linear, but the network will still try to obtain an optimal mapping with linear or near-linear output neurons.

When each neuron operates as a binary classifier, we can consider how many input patterns p can be classified by each neuron, and the classes of pattern that can be correctly classified. The result is that the maximum number of patterns that can be correctly classified by a neuron with C inputs is

$$p_{\mathrm{max}} = 2C \qquad (A.12)$$

when the inputs have random continuous-valued inputs, but the patterns must be linearly separable (see Hertz et al. (1991)). More generally, a network with a single binary unit can implement a classification between two subspaces of a space of possible input patterns provided that the p actual patterns given as examples of the correct classification are linearly separable.

The linear separability requirement can be made clear by considering a geometric interpretation of the logical AND problem, which is linearly separable, and the XOR (exclusive OR) problem, which is not linearly separable. The truth tables for the AND and XOR functions are shown in Table A.2 (there are two inputs, x_1 and x_2, and one output neuron):

Table A.2 Truth table for AND and XOR functions performed by a single output neuron with two inputs. 1 = active or firing; 0 = inactive.

Inputs		Output	
x_1	x_2	AND	XOR
0	0	0	0
1	0	0	1
0	1	0	1
1	1	1	0

For the AND function, we can plot the mapping required in a 2D graph as shown in Fig. A.2. A line can be drawn to separate the input coordinates for which 0 is required as the output from those for which 1 is required as the output. The problem is thus linearly separable. A neuron with two inputs can set its weights to values which draw this line through this space, and such a one-layer network can thus solve the AND function.

For the XOR function, we can plot the mapping required in a 2D graph as shown in Fig. A.3. No straight line can be drawn to separate the input coordinates for which 0 is required as the output from those for which 1 is required as the output. The problem is thus not linearly separable. For a one-layer network, no set of weights can be found that will perform the XOR, or any other non-linearly separable function.

Although the inability of one-layer networks with binary neurons to solve non-linearly separable problems is a limitation, it is not in practice a major limitation on the processing that can be performed in a neural network for a number of reasons. First, if the inputs can take continuous values, then if the patterns are drawn from a random distribution, the one-layer

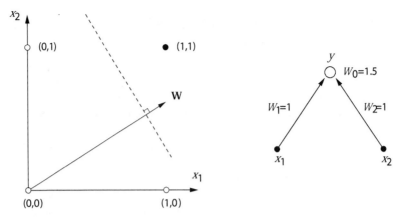

Fig. A.2 Left: the AND function shown in a 2D space. Input values for the two neurons are shown along the two axes of the space. The outputs required are plotted at the coordinates where the inputs intersect, and the values of the output required are shown as an open circle for 0, and a filled circle for 1. The AND function is linearly separable, in that a line can be drawn in the space which separates the coordinates for which 0 output is required from those from which a 1 output is required. **w** shows the direction of the weight vector. Right: a one-layer neural network can set its two weights w_1 and w_2 to values which allow the output neuron to be activated only if both inputs are present. In this diagram, w_0 is used to set a threshold for the neuron, and is connected to an input with value 1. The neuron thus fires only if the threshold of 1.5 is exceeded, which happens only if both inputs to the neuron are 1.

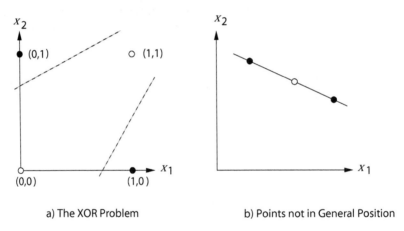

a) The XOR Problem   b) Points not in General Position

Fig. A.3 The XOR function shown in a 2D space. Input values for the two neurons are shown along the two axes of the space. The outputs required are plotted at the coordinates where the inputs intersect, and the values of the output required are shown as an open circle for 0, and a filled circle for 1. The XOR function is not linearly separable, in that a line cannot be drawn in the space to separate the coordinates for those from which a 0 output is required from those from which a 1 output is required. A one-layer neural network cannot set its two weights to values which allow the output neuron to be activated appropriately for the XOR function.

network can map up to $2C$ of them. Second, as described for pattern associators, and for one-layer error-correcting perceptrons (see Appendix B), these networks could be preceded by an expansion recoding network such as a competitive network with more output than input neurons. This effectively provides a two-layer network for solving the problem, and multilayer networks are in general capable of solving arbitrary mapping problems. Ways in which such multilayer networks might be trained are discussed in Chapter 25 and Appendix B.

More generally, a binary output unit provides by its operation a hyperplane (the hyperplane orthogonal to its synaptic weight vector as shown in Fig. A.2) that divides the input space in two. The input space is of dimension C, if C is the number of input axons or connections. A one-layer network with a number n of binary output units is equivalent to n hyperplanes, that could potentially divide the input space into as many as 2^n regions, each corresponding to input patterns leading to a different output. However the number p of *arbitrary* examples of the correct classification (each example consisting of an input pattern and its required correct output) that the network may be able to implement is well below 2^n, and in fact depends on C not on n. This is because for p too large it will be impossible to position the n weight vectors such that all examples of input vectors for which the first output unit is required to be 'on' fall on one side of the hyperplane associated with the first weight vector, all those for which it is required to be 'off' fall on the other side, and simultaneously the same holds with respect to the second output unit (a different dichotomy), the third, and so on. The limit on p, which can be thought of also as the number of independent associations implemented by the network, when this is viewed as a heteroassociator (i.e. pattern associator) with binary outputs, can be calculated with the Gardner method (Gardner 1987, Gardner 1988) and depends on the statistics of the patterns. For input patterns that are also binary, random and with equal probability for each of the two states on every unit, the limit is $p_c = 2C$ (see further Appendix B, and Rolls and Treves (1998) Appendix A3).

A.2.2 Non-linear networks: neurons with non-linear activation functions

These concepts also help one to understand further the limitation of linear systems, and the power of non-linear systems. Consider the dot product operation by which the neuronal activation h is computed:

$$h = \sum_j x_j w_j. \tag{A.13}$$

If the output firing is just a linear function of the activation, any input pattern will produce a non-zero output unless it happens to be exactly orthogonal to the weight vector. For positive-only firing rates and synaptic weights, being orthogonal means taking non-zero values only on non-corresponding components. Since with distributed representations the non-zero components of different input firing vectors will in general be overlapping (i.e. some corresponding components in both firing rate vectors will be on, that is the vectors will overlap), this will result effectively in interference between any two different patterns that for example have to be associated to different outputs. Thus a basic limitation of linear networks is that they can perform pattern association perfectly only if the input patterns $\mathbf{x}$ are orthogonal; and for positive-only patterns that represent actual firing rates only if the different firing rate vectors are non-overlapping. Further, linear networks cannot of course perform any classification, just because they act linearly. (Classification implies producing output states that are clearly defined as being in one class, and not in other classes.) For example, in a linear network, if a pattern is presented that is intermediate between two patterns $\mathbf{v}_1$ and $\mathbf{v}_2$, such as $c_1\mathbf{v}_1 + c_2\mathbf{v}_2$, then the output pattern will be a linear combination of the outputs produced by $\mathbf{v}_1$ and $\mathbf{v}_2$ (e.g. $c_1\mathbf{o}_1 + c_2\mathbf{o}_2$), rather than being classified into $\mathbf{o}_1$ or $\mathbf{o}_2$. In contrast, with non-linear neurons, the patterns need not be orthogonal, only linearly separable, for a one-layer network to be able to correctly classify the patterns (provided that a sufficiently powerful learning rule is used – see Appendix B).

The networks just described, and most of those described in this book, are trained with a local learning rule, in which the pre- and post-synaptic terms needed to alter the synaptic weights are available locally in the synapses, in terms for example of the release of transmitter from the presynaptic terminal, and the depolarization of the postsynaptic neuron. This type of network is considered because this is a biologically plausible constraint (see Section B.12). It is much less biologically plausible to use an algorithm such as multilayer error backpropagation which calculates the correction to the value of a synapse that is needed taking into account the values of the errors of the neurons at later stages of the system and the strengths of all the synapses to these neurons (see Section B.11). This use of a local learning rule is a major difference of the networks described in this book, which is directed at neurally plausible computation, from connectionist networks, which typically assume non-local learning rules and which therefore operate very differently from real neural networks in the brain (see Section B.11 and McLeod, Plunkett and Rolls (1998)).

A.2.3 Non-linear networks: neurons with non-linear activations

Most of the networks described in this book calculate the activation h of each neuron as the linear product of the input firing weighted by the synaptic weight vector (see equation A.13). This corresponds in a real neuron to receiving currents from each of its synapses which sum to produce depolarization of the neuronal cell body and the spike initiation region which is located very close to the cell body. This is a reasonable reflection of what does happen in many neurons, especially those with large dendrites such as pyramidal cells (Koch 1999). This calculation of the activation h by a linear summation not only approximates to what happens in many real neurons, but is also a useful simplification which makes tractable the analysis of many classes of network that utilize such neurons. These analyses provide insight into the operation of networks of neurons, even if the linear summation assumption is not perfectly realized. Having computed the activation linearly, the neurons do of course for essentially all the networks described in this book, then utilize a non-linear activation function, which, as described above, provides the networks with much of their interesting computational power. Given that the activation functions of the neurons are non-linear, some non-linearity in the summation expressed in equation A.13 may in practice be lumped into the non-linearity in the activation function.

However, another class of neuron that is implemented in some networks in the brain utilizes non-linearity in the calculation of the activation h of the neuron, which reflects a local product of two inputs to a neuron. This could arise for example if one synapse makes a presynaptic contact with another synapse which in turn connects to the dendrite, or if two synapses are close together on a thin dendrite. In such situations, the current injected into the neuron could reflect the conjoint firing of the two classes of input (Koch 1999). The dendrite as a whole could then sum all such products into the cell body, leading to the description **Sigma-Pi**. This could be expressed by equation A.14

$$h = \sum_j \sum_k w_{jk} x_j x^c{}_k \tag{A.14}$$

where x_j is the firing rate of input cell j, $x^c{}_k$ is the firing rate of input cell k of class c, and w_{jk} is the connection strength. Such Sigma-Pi neurons were utilized in the model described in Section B.5.5 of how idiothetic inputs could update a continuous attractor network. Another possible application is to learning invariant representations in neural networks. For example, the x^c input in equation A.14 could be a signal that varies with the shift required to compute translation invariance, effectively mapping the appropriate set of x_j inputs through to the output neurons depending on the shift required (Mel, Ruderman and Archie 1998, Mel and Fiser 2000, Olshausen, Anderson and Van Essen 1993, Olshausen, Anderson and Van Essen 1995).

To train such a Sigma-Pi network requires that combinations of the two presynaptic inputs to a neuron be learned onto a neuron, using for example associativity with the post-synaptic term y, as exemplified in equation A.15

$$\delta w_{jk} = \alpha y x_j x^c{}_k.$$ (A.15)

This learning principle is exemplified in the model described in Section B.5.5 of how idiothetic inputs could update a continuous attractor network to perform path integration.

Sigma-Pi networks are clearly very powerful, but require rather specialized anatomical and biophysical arrangements (see Koch (1999)), and hence we do not use them unless they become very necessary in models of neural network operations in the brain. We have shown that in at least some applications such as path integration, it is possible to replace a Sigma-Pi network with a competitive network followed by a pattern association network (Stringer and Rolls 2006).

Appendix 2 Neuronal network models

B.1 Introduction

Formal models of neural networks are needed in order to provide a basis for understanding the processing and memory functions performed by real neuronal networks in the brain. The formal models included in this Appendix all describe fundamental types of network found in different brain regions, and the computations they perform. Each of the types of network described can be thought of as providing one of the fundamental building blocks that the brain uses. Often these building blocks are combined within a brain area to perform a particular computation.

The aim of this Appendix is to describe a set of fundamental networks used by the brain, including the parts of the brain involved in memory, attention, decision-making, and the building of perceptual representations. As each type of network is introduced, we will point briefly to parts of the brain in which each network is found. Understanding these models provides a basis for understanding the theories of how different types of memory functions are performed. The descriptions of these networks are kept relatively concise in this Appendix. More detailed descriptions of some of the quantitative aspects of storage in pattern associators and autoassociators are provided in the Appendices of Rolls and Treves (1998) *Neural Networks and Brain Function*. Another book that provides a clear and quantitative introduction to some of these networks is Hertz, Krogh and Palmer (1991) *Introduction to the Theory of Neural Computation*, and other useful sources include Dayan and Abbott (2001), Gerstner, Kistler, Naud and Paninski (2014) (who focus on neuronal dynamics), Amit (1989) (for attractor networks), Koch (1999) (for a biophysical approach), Wilson (1999)(on spiking networks), and Rolls (2008d).

Some of the background to the operation of the types of neuronal network described here, including a brief review of the evidence on neuronal structure and function, and on synaptic plasticity and the rules by which synaptic strength is modified, much based on studies with long-term potentiation, is provided in Chapter 1.

The network models on which we focus in this Appendix utilize a local learning rule, that is a rule for synaptic modification, in which the signals needed to alter the synaptic strength are present in the pre- and post-synaptic neurons. We focus on these networks because use of a local learning rule is biologically plausible. We discuss the issue of biological plausibility of the networks described, and show how they differ from less biologically plausible networks such as multilayer backpropagation of error networks, in Section B.12.

B.2 Pattern association memory

A fundamental operation of most nervous systems is to learn to associate a first stimulus with a second that occurs at about the same time, and to retrieve the second stimulus when the first is presented. The first stimulus might be the sight of food, and the second stimulus the taste of food. After the association has been learned, the sight of food would enable its taste to be retrieved. In classical conditioning, the taste of food might elicit an unconditioned response of

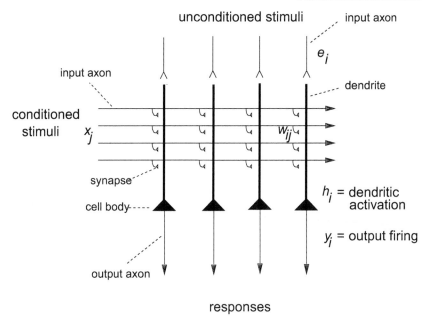

Fig. B.1 A pattern association memory. An unconditioned stimulus has activity or firing rate e_i for the ith neuron, and produces firing y_i of the ith neuron. An unconditioned stimulus may be treated as a vector, across the set of neurons indexed by i, of activity $\mathbf{e}$. The firing rate response can also be thought of as a vector of firing $\mathbf{y}$. The conditioned stimuli have activity or firing rate x_j for the jth axon, which can also be treated as a vector $\mathbf{x}$.

salivation, and if the sight of the food is paired with its taste, then the sight of that food would by learning come to produce salivation. Pattern associators are thus used where the outputs of the visual system interface to learning systems in the orbitofrontal cortex and amygdala that learn associations between the sight of objects and their taste or touch in stimulus–reinforcer association learning (see Chapter 15). Pattern association is also used throughout the cerebral (neo)cortical areas, as it is the architecture that describes the backprojection connections from one cortical area to the preceding cortical area (see Chapters 1, 24 and 6). Pattern association thus contributes to implementing top-down influences in attention, including the effects of attention from higher to lower cortical areas, and thus between the visual object and spatial processing streams (Rolls and Deco 2002) (see Chapter 6); the effects of mood on memory and visual information processing (Rolls and Stringer 2001b); the recall of visual memories; and the operation of short-term memory systems (see Section 4.3.1).

B.2.1 Architecture and operation

The essential elements necessary for pattern association, forming what could be called a prototypical pattern associator network, are shown in Fig. B.1. What we have called the second or unconditioned stimulus pattern is applied through unmodifiable synapses generating an input to each neuron, which, being external with respect to the synaptic matrix we focus on, we can call the external input e_i for the ith neuron. [We can also treat this as a vector, $\mathbf{e}$, as indicated in the legend to Fig. B.1. Vectors and simple operations performed with them are summarized in Appendix A. This unconditioned stimulus is dominant in producing or forcing the firing of the output neurons (y_i for the ith neuron, or the vector $\mathbf{y}$)]. At the same time, the first or conditioned stimulus pattern consisting of the set of firings on the horizontally running input axons in Fig. B.1 (x_j for the jth axon) (or equivalently the vector $\mathbf{x}$) is applied through modifiable synapses w_{ij} to the dendrites of the output neurons. The synapses are modifiable in such a way that if there is presynaptic firing on an input axon x_j paired during learning with postsynaptic activity on neuron i, then the strength or weight w_{ij} between that axon and the dendrite increases. This simple learning rule is often called the Hebb rule, after Donald Hebb who in 1949 formulated the hypothesis that if the firing of one neuron was regularly associated with another, then the strength of the synapse or synapses between the neurons should increase[36]. After learning, presenting the pattern $\mathbf{x}$ on the input axons will activate the dendrite through the strengthened synapses. If the cue or conditioned stimulus pattern is the same as that learned, the postsynaptic neurons will be activated, even in the absence of the external or unconditioned input, as each of the firing axons produces through a strengthened synapse some activation of the postsynaptic element, the dendrite. The total activation h_i of each postsynaptic neuron i is then the sum of such individual activations. In this way, the 'correct' output neurons, that is those activated during learning, can end up being the ones most strongly activated, and the second or unconditioned stimulus can be effectively recalled. The recall is best when only strong activation of the postsynaptic neuron produces firing, that is if there is a threshold for firing, just like real neurons. The advantages of this are evident when many associations are stored in the memory, as will soon be shown.

Next we introduce a more precise description of the above by writing down explicit mathematical rules for the operation of the simple network model of Fig. B.1, which will help us to understand how pattern association memories in general operate. (In this description we introduce simple vector operations, and, for those who are not familiar with these, refer the reader to Appendix A.) We have denoted above a conditioned stimulus input pattern as $\mathbf{x}$. Each of the axons has a firing rate, and if we count or index through the axons using the subscript j, the firing rate of the first axon is x_1, of the second x_2, of the jth x_j, etc. The whole set of axons forms a vector, which is just an ordered (1, 2, 3, etc.) set of elements. The firing rate of each axon x_j is one element of the firing rate vector $\mathbf{x}$. Similarly, using i as the index, we can denote the firing rate of any output neuron as y_i, and the firing rate output vector as $\mathbf{y}$. With this terminology, we can then identify any synapse onto neuron i from neuron j as w_{ij} (see Fig. B.1). In this book, the first index, i, always refers to the receiving neuron (and thus signifies a dendrite), while the second index, j, refers to the sending neuron (and thus signifies a conditioned stimulus input axon in Fig. B.1). We can now specify the learning and retrieval operations as follows:

B.2.1.1 Learning

The firing rate of every output neuron is forced to a value determined by the unconditioned (or external or forcing stimulus) input e_i. In our simple model this means that for any one neuron i,

$$y_i = \mathrm{f}(e_i) \tag{B.1}$$

which indicates that the firing rate is a function of the dendritic activation, taken in this case to reduce essentially to that resulting from the external forcing input (see Fig. B.1). The function f is called the activation function (see Fig. 1.4), and its precise form is irrelevant, at least during this learning phase. For example, the function at its simplest could be taken to be linear, so that the firing rate would be just proportional to the activation.

---

[36] In fact, the terms in which Hebb put the hypothesis were a little different from an association memory, in that he stated that if one neuron regularly comes to elicit firing in another, then the strength of the synapses should increase. He had in mind the building of what he called cell assemblies. In a pattern associator, the conditioned stimulus need not produce before learning any significant activation of the output neurons. The connection strengths must simply increase if there is associated pre- and postsynaptic firing when, in pattern association, most of the postsynaptic firing is being produced by a different input.

The Hebb rule can then be written as follows:

$$\delta w_{ij} = \alpha y_i x_j \tag{B.2}$$

where δw_{ij} is the change of the synaptic weight w_{ij} that results from the simultaneous (or conjunctive) presence of presynaptic firing x_j and postsynaptic firing or activation y_i, and α is a learning rate constant that specifies how much the synapses alter on any one pairing.

The Hebb rule is expressed in this multiplicative form to reflect the idea that both pre-synaptic and postsynaptic activity must be present for the synapses to increase in strength. The multiplicative form also reflects the idea that strong pre- and postsynaptic firing will produce a larger change of synaptic weight than smaller firing rates. It is also assumed for now that before any learning takes place, the synaptic strengths are small in relation to the changes that can be produced during Hebbian learning. We will see that this assumption can be relaxed later when a modified Hebb rule is introduced that can lead to a reduction in synaptic strength under some conditions.

B.2.1.2 Recall

When the conditioned stimulus is present on the input axons, the total activation h_i of a neuron i is the sum of all the activations produced through each strengthened synapse w_{ij} by each active neuron x_j. We can express this as

$$h_i = \sum_{j=1}^{C} x_j w_{ij} \tag{B.3}$$

where $\sum_{j=1}^{C}$ indicates that the sum is over the C input axons (or connections) indexed by j to each neuron.

The multiplicative form here indicates that activation should be produced by an axon only if it is firing, and only if it is connected to the dendrite by a strengthened synapse. It also indicates that the strength of the activation reflects how fast the axon x_j is firing, and how strong the synapse w_{ij} is. The sum of all such activations expresses the idea that summation (of synaptic currents in real neurons) occurs along the length of the dendrite, to produce activation at the cell body, where the activation h_i is converted into firing y_i. This conversion can be expressed as

$$y_i = f(h_i) \tag{B.4}$$

where the function f is again the activation function. The form of the function now becomes more important. Real neurons have thresholds, with firing occurring only if the activation is above the threshold. A threshold linear activation function is shown in Fig. 1.4b on page 7. This has been useful in formal analysis of the properties of neural networks. Neurons also have firing rates that become saturated at a maximum rate, and we could express this as the sigmoid activation function shown in Fig. 1.4c. Yet another simple activation function, used in some models of neural networks, is the binary threshold function (Fig. 1.4d), which indicates that if the activation is below threshold, there is no firing, and that if the activation is above threshold, the neuron fires maximally. Whatever the exact shape of the activation function, some non-linearity is an advantage, for it enables small activations produced by interfering memories to be minimized, and it can enable neurons to perform logical operations, such as to fire or respond only if two or more sets of inputs are present simultaneously.

B.2.2   A simple model

An example of these learning and recall operations is provided in a simple form as follows. The neurons will have simple firing rates, which can be 0 to represent no activity, and 1 to indicate high firing. They are thus binary neurons, which can assume one of two firing rates. If we have a pattern associator with six input axons and four output neurons, we could represent the network before learning, with the same layout as in Fig. B.1, as shown in Fig. B.2:

```
                    U  C  S
                    1  1  0  0
                    ↓  ↓  ↓  ↓
        CS
        1 →         0  0  0  0
        0 →         0  0  0  0
        1 →         0  0  0  0
        0 →         0  0  0  0
        1 →         0  0  0  0
        0 →         0  0  0  0
```

Fig. B.2 Pattern association: before synaptic modification. The unconditioned stimulus (UCS) firing rates are shown as 1 if high and 0 if low as a row vector being applied to force firing of the four output neurons. The six conditioned stimulus (CS) firing rates are shown as a column vector being applied to the vertical dendrites of the output neurons which have initial synaptic weights of 0.

where x or the conditioned stimulus (CS) is 101010, and y or the firing produced by the unconditioned stimulus (UCS) is 1100. (The arrows indicate the flow of signals.) The synaptic weights are initially all 0.

After pairing the CS with the UCS during one learning trial, some of the synaptic weights will be incremented according to equation B.2, so that after learning this pair the synaptic weights will become as shown in Fig. B.3:

```
                    U  C  S
                    1  1  0  0
                    ↓  ↓  ↓  ↓
        CS
        1 →         1  1  0  0
        0 →         0  0  0  0
        1 →         1  1  0  0
        0 →         0  0  0  0
        1 →         1  1  0  0
        0 →         0  0  0  0
```

Fig. B.3 Pattern association: after synaptic modification. The synapses where there is conjunctive pre- and post-synaptic activity have been strengthened to value 1.

We can represent what happens during recall, when, for example, we present the CS that has been learned, as shown in Fig. B.4:

```
CS
1 →    1  1  0  0
0 →    0  0  0  0
1 →    1  1  0  0
0 →    0  0  0  0
1 →    1  1  0  0
0 →    0  0  0  0
       ↓  ↓  ↓  ↓

       3  3  0  0  Activation hᵢ
       1  1  0  0  Firing yᵢ
```

Fig. B.4 Pattern association: recall. The activation h_i of each neuron i is converted with a threshold of 2 to the binary firing rate y_i (1 for high, and 0 for low).

The activation of the four output neurons is 3300, and if we set the threshold of each output neuron to 2, then the output firing is 1100 (where the binary firing rate is 0 if below threshold, and 1 if above). The pattern associator has thus achieved recall of the pattern 1100, which is correct.

We can now illustrate how a number of different associations can be stored in such a pattern associator, and retrieved correctly. Let us associate a new CS pattern 110001 with the UCS 0101 in the same pattern associator. The weights will become as shown next in Fig. B.5 after learning:

```
            U  C  S
            0  1  0  1
            ↓  ↓  ↓  ↓
CS
1 →    1  2  0  1
1 →    0  1  0  1
0 →    1  1  0  0
0 →    0  0  0  0
0 →    1  1  0  0
1 →    0  1  0  1
```

Fig. B.5 Pattern association: synaptic weights after learning a second pattern association.

If we now present the second CS, the retrieval is as shown in Fig. B.6:

```
CS
1 →    1  2  0  1
1 →    0  1  0  1
0 →    1  1  0  0
0 →    0  0  0  0
0 →    1  1  0  0
1 →    0  1  0  1
       ↓  ↓  ↓  ↓

       1  4  0  3  Activation hᵢ
       0  1  0  1  Firing yᵢ
```

Fig. B.6 Pattern association: recall with the second CS.

The binary output firings were again produced with the threshold set to 2. Recall is perfect.

This illustration shows the value of some threshold non-linearity in the activation function of the neurons. In this case, the activations did reflect some small cross-talk or interference from the previous pattern association of CS1 with UCS1, but this was removed by the threshold operation, to clean up the recall firing. The example also shows that when further associations are learned by a pattern associator trained with the Hebb rule, equation B.2, some synapses will reflect increments above a synaptic strength of 1. It is left as an exercise to the reader to verify that recall is still perfect to CS1, the vector 101010. (The activation vector **h** is 3401, and the output firing vector **y** with the same threshold of 2 is 1100, which is perfect recall.)

B.2.3 The vector interpretation

The way in which recall is produced, equation B.3, consists for each output neuron i of multiplying each input firing rate x_j by the corresponding synaptic weight w_{ij} and summing the products to obtain the activation h_i. Now we can consider the firing rates x_j where j varies from 1 to N', the number of axons, to be a vector. (A vector is simply an ordered set of numbers – see Appendix A.) Let us call this vector **x**. Similarly, on a neuron i, the synaptic weights can be treated as a vector, $\mathbf{w}_i$. (The subscript i here indicates that this is the weight vector on the ith neuron.) The operation we have just described to obtain the activation of an output neuron can now be seen to be a simple multiplication operation of two vectors to produce a single output value (called a scalar output). This is the inner product or dot product of two vectors, and can be written

$$h_i = \mathbf{x} \cdot \mathbf{w}_i. \tag{B.5}$$

The inner product of two vectors indicates how similar they are. If two vectors have corresponding elements the same, then the dot product will be maximal. If the two vectors are similar but not identical, then the dot product will be high. If the two vectors are completely different, the dot product will be 0, and the vectors are described as orthogonal. (The term orthogonal means at right angles, and arises from the geometric interpretation of vectors, which is summarized in Appendix A.) Thus the dot product provides a direct measure of how similar two vectors are.

It can now be seen that a fundamental operation many neurons perform is effectively to compute how similar an input pattern vector **x** is to their stored weight vector $\mathbf{w}_i$. The similarity measure they compute, the dot product, is a very good measure of similarity, and indeed, the standard (Pearson product–moment) correlation coefficient used in statistics is the same as a normalized dot product with the mean subtracted from each vector, as shown in Appendix A. (The normalization used in the correlation coefficient results in the coefficient varying always between $+1$ and -1, whereas the actual scalar value of a dot product clearly depends on the length of the vectors from which it is calculated.)

With these concepts, we can now see that during learning, a pattern associator adds to its weight vector a vector $\delta\mathbf{w}_i$ that has the same pattern as the input pattern $\mathbf{x}$, if the postsynaptic neuron i is strongly activated. Indeed, we can express equation B.2 in vector form as

$$\delta\mathbf{w}_i = \alpha y_i \mathbf{x}. \qquad (\text{B.6})$$

We can now see that what is recalled by the neuron depends on the similarity of the recall cue vector $\mathbf{x}_r$ to the originally learned vector $\mathbf{x}$. The fact that during recall the output of each neuron reflects the similarity (as measured by the dot product) of the input pattern $\mathbf{x}_r$ to each of the patterns used originally as $\mathbf{x}$ inputs (conditioned stimuli in Fig. B.1) provides a simple way to appreciate many of the interesting and biologically useful properties of pattern associators, as described next.

B.2.4 Properties

B.2.4.1 Generalization

During recall, pattern associators generalize, and produce appropriate outputs if a recall cue vector $\mathbf{x}_r$ is similar to a vector that has been learned already. This occurs because the recall operation involves computing the dot (inner) product of the input pattern vector $\mathbf{x}_r$ with the synaptic weight vector $\mathbf{w}_i$, so that the firing produced, y_i, reflects the similarity of the current input to the previously learned input pattern $\mathbf{x}$. (Generalization will occur to input cue or conditioned stimulus patterns $\mathbf{x}_r$ that are incomplete versions of an original conditioned stimulus $\mathbf{x}$, although the term completion is usually applied to the autoassociation networks described in Section B.3.)

This is an extremely important property of pattern associators, for input stimuli during recall will rarely be absolutely identical to what has been learned previously, and automatic generalization to similar stimuli is extremely useful, and has great adaptive value in biological systems.

Generalization can be illustrated with the simple binary pattern associator considered above. (Those who have appreciated the vector description just given might wish to skip this illustration.) Instead of the second CS, pattern vector 110001, we will use the similar recall cue 110100, as shown in Fig. B.7:

```
CS
1 →    1  2  0  1
1 →    0  1  0  1
0 →    1  1  0  0
1 →    0  0  0  0
0 →    1  1  0  0
0 →    0  1  0  1
       ↓  ↓  ↓  ↓

       1  3  0  2  Activation $h_i$
       0  1  0  1  Firing $y_i$
```

Fig. B.7 Pattern association: generalization using an input vector similar to the second CS.

It is seen that the output firing rate vector, 0101, is exactly what should be recalled to CS2 (and not to CS1), so correct generalization has occurred. Although this is a small network trained with few examples, the same properties hold for large networks with large numbers of stored patterns, as described more quantitatively in Section B.2.7.1 on capacity below and in Appendix A3 of Rolls and Treves (1998).

B.2.4.2 Graceful degradation or fault tolerance

If the synaptic weight vector w_i (or the weight matrix, which we can call W) has synapses missing (e.g. during development), or loses synapses, then the activation h_i or h is still reasonable, because h_i is the dot product (correlation) of x with w_i. The result, especially after passing through the activation function, can frequently be perfect recall. The same property arises if for example one or some of the conditioned stimulus (CS) input axons are lost or damaged. This is a very important property of associative memories, and is not a property of conventional computer memories, which produce incorrect data if even only 1 storage location (for 1 bit or binary digit of data) of their memory is damaged or cannot be accessed. This property of graceful degradation is of great adaptive value for biological systems.

We can illustrate this with a simple example. If we damage two of the synapses in Fig. B.6 to produce the synaptic matrix shown in Fig. B.8 (where x indicates a damaged synapse which has no effect, but was previously 1), and now present the second CS, the retrieval is as follows:

```
CS
1 →    1  2  0  1
1 →    0  1  0  x
0 →    1  1  0  0
0 →    0  0  0  0
0 →    1  x  0  0
1 →    0  1  0  1
       ↓  ↓  ↓  ↓

       1  4  0  2  Activation $h_i$
       0  1  0  1  Firing $y_i$
```

Fig. B.8 Pattern association: graceful degradation when some synapses are damaged (x).

The binary output firings were again produced with the threshold set to 2. The recalled vector, 0101, is perfect. This illustration again shows the value of some threshold non-linearity in the activation function of the neurons. It is left as an exercise to the reader to verify that recall is still perfect to CS1, the vector 101010. (The output activation vector **h** is 3301, and the output firing vector **y** with the same threshold of 2 is 1100, which is perfect recall.)

B.2.4.3 The importance of distributed representations for pattern associators

A distributed representation is one in which the firing or activity of all the elements in the vector is used to encode a particular stimulus. For example, in a conditioned stimulus vector CS1 that has the value 101010, we need to know the state of all the elements to know which stimulus is being represented. Another stimulus, CS2, is represented by the vector 110001. We can represent many different events or stimuli with such overlapping sets of elements, and because in general any one element cannot be used to identify the stimulus, but instead the information about which stimulus is present is distributed over the population of elements or neurons, this is called a distributed representation (see Section 8.2). If, for binary neurons, half the neurons are in one state (e.g. 0), and the other half are in the other state (e.g. 1), then the representation is described as fully distributed. The CS representations above are thus fully distributed. If only a smaller proportion of the neurons is active to represent a stimulus, as in the vector 100001, then this is a sparse representation. For binary representations, we can quantify the sparseness by the proportion of neurons in the active (1) state.

In contrast, a local representation is one in which all the information that a particular stimulus or event has occurred is provided by the activity of one of the neurons, or elements in the vector. One stimulus might be represented by the vector 100000, another stimulus by the vector 010000, and a third stimulus by the vector 001000. The activity of neuron or element 1 would indicate that stimulus 1 was present, and of neuron 2, that stimulus 2 was present. The representation is local in that if a particular neuron is active, we know that the stimulus represented by that neuron is present. In neurophysiology, if such cells were present, they might be called 'grandmother cells' (cf. Barlow (1972), (1995); see Chapters 1 and 8 and Appendix C), in that one neuron might represent a stimulus in the environment as complex and specific as one's grandmother. Where the activity of a number of cells must be taken into account in order to represent a stimulus (such as an individual taste), then the representation is sometimes described as using ensemble encoding.

The properties just described for associative memories, generalization, and graceful degradation are only implemented if the representation of the CS or **x** vector is distributed. This occurs because the recall operation involves computing the dot (inner) product of the input pattern vector $\mathbf{x}_r$ with the synaptic weight vector $\mathbf{w}_i$. This allows the activation h_i to reflect the similarity of the current input pattern to a previously learned input pattern **x** only if several or many elements of the **x** and $\mathbf{x}_r$ vectors are in the active state to represent a pattern. If local encoding were used, e.g. 100000, then if the first element of the vector (which might be the firing of axon 1, i.e. x_1, or the strength of synapse $i1$, w_{i1}) is lost, the resulting vector is not similar to any other CS vector, and the activation is 0. In the case of local encoding, the important properties of associative memories, generalization and graceful degradation do not thus emerge. Graceful degradation and generalization are dependent on distributed representations, for then the dot product can reflect similarity even when some elements of the vectors involved are altered. If we think of the correlation between Y and X in a graph, then this correlation is affected only a little if a few X, Y pairs of data are lost (see Appendix A).

B.2.5 Prototype extraction, extraction of central tendency, and noise reduction

If a set of similar conditioned stimulus vectors **x** are paired with the same unconditioned stimulus e_i, the weight vector $\mathbf{w}_i$ becomes (or points towards) the sum (or with scaling, the average) of the set of similar vectors **x**. This follows from the operation of the Hebb rule in equation B.2. When tested at recall, the output of the memory is then best to the average input pattern vector denoted $< \mathbf{x} >$. If the average is thought of as a prototype, then even though the prototype vector $< \mathbf{x} >$ itself may never have been seen, the best output of the neuron or network is to the prototype. This produces 'extraction of the prototype' or 'central tendency'. The same phenomenon is a feature of human memory performance (see McClelland and Rumelhart (1986) Chapter 17), and this simple process with distributed representations in a neural network accounts for the psychological phenomenon.

　　If the different exemplars of the vector **x** are thought of as noisy versions of the true input pattern vector $< \mathbf{x} >$ (with incorrect values for some of the elements), then the pattern associator has performed 'noise reduction', in that the output produced by any one of these vectors will represent the output produced by the true, noiseless, average vector $< \mathbf{x} >$.

B.2.6 Speed

Recall is very fast in a real neuronal network, because the conditioned stimulus input firings x_j ($j = 1, C$ axons) can be applied simultaneously to the synapses w_{ij}, and the activation h_i can be accumulated in one or two time constants of the dendrite (e.g. 10–20 ms). Whenever the threshold of the cell is exceeded, it fires. Thus, in effectively one step, which takes the brain no more than 10–20 ms, all the output neurons of the pattern associator can be firing with rates that reflect the input firing of every axon. This is very different from a conventional digital computer, in which computing h_i in equation B.3 would involve C multiplication and addition operations occurring one after another, or $2C$ time steps.

　　The brain performs parallel computation in at least two senses in even a pattern associator. One is that for a single neuron, the separate contributions of the firing rate x_j of each axon j multiplied by the synaptic weight w_{ij} are computed in parallel and added in the same timestep. The second is that this can be performed in parallel for all neurons $i = 1, N$ in the network, where there are N output neurons in the network. It is these types of parallel and time-continuous (see Section B.6) processing that enable these classes of neuronal network in the brain to operate so fast, in effectively so few steps.

　　Learning is also fast ('one-shot') in pattern associators, in that a single pairing of the conditioned stimulus **x** and the unconditioned stimulus (UCS) **e** which produces the unconditioned output firing **y** enables the association to be learned. There is no need to repeat the pairing in order to discover over many trials the appropriate mapping. This is extremely important for biological systems, in which a single co-occurrence of two events may lead to learning that could have life-saving consequences. (For example, the pairing of a visual stimulus with a potentially life-threatening aversive event may enable that event to be avoided in future.) Although repeated pairing with small variations of the vectors is used to obtain the useful properties of prototype extraction, extraction of central tendency, and noise reduction, the essential properties of generalization and graceful degradation are obtained with just one pairing. The actual time scales of the learning in the brain are indicated by studies of associative synaptic modification using long-term potentiation paradigms (LTP, see Section 1.5). Co-occurrence or near simultaneity of the CS and UCS is required for periods of as little as 100 ms, with expression of the synaptic modification being present within typically a few seconds.

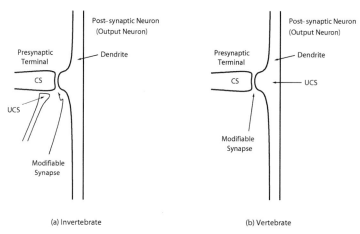

(a) Invertebrate   (b) Vertebrate

Fig. B.9 (b) In vertebrate pattern association learning, the unconditioned stimulus (UCS) may be made available at all the conditioned stimulus (CS) terminals onto the output neuron because the dendrite of the postsynaptic neuron is electrically short, so that the effect of the UCS spreads for long distances along the dendrite. (a) In contrast, in at least some invertebrate association learning systems, the unconditioned stimulus or teaching input makes a synapse onto the presynaptic terminal carrying the conditioned stimulus.

B.2.7 Local learning rule

The simplest learning rule used in pattern association neural networks, a version of the Hebb rule, is, as shown in equation B.2 above,

$$\delta w_{ij} = \alpha y_i x_j.$$

This is a local learning rule in that the information required to specify the change in synaptic weight is available locally at the synapse, as it is dependent only on the presynaptic firing rate x_j available at the synaptic terminal, and the postsynaptic activation or firing y_i available on the dendrite of the neuron receiving the synapse (see Fig. B.9b). This makes the learning rule biologically plausible, in that the information about how to change the synaptic weight does not have to be carried from a distant source, where it is computed, to every synapse. Such a non-local learning rule would not be biologically plausible, in that there are no appropriate connections known in most parts of the brain to bring in the synaptic training or teacher signal to every synapse.

Evidence that a learning rule with the general form of equation B.2 is implemented in at least some parts of the brain comes from studies of long-term potentiation, described in Section 1.5. Long-term potentiation (LTP) has the synaptic specificity defined by equation B.2, in that only synapses from active afferents, not those from inactive afferents, become strengthened. Synaptic specificity is important for a pattern associator, and most other types of neuronal network, to operate correctly. The number of independently modifiable synapses on each neuron is a primary factor in determining how many different memory patterns can be stored in associative memories (see Sections B.2.7.1 and B.3.3.7).

Another useful property of real neurons in relation to equation B.2 is that the postsynaptic term, y_i, is available on much of the dendrite of a cell, because the electrotonic length of the dendrite is short. In addition, active propagation of spiking activity from the cell body along the dendrite may help to provide a uniform postsynaptic term for the learning. Thus if a neuron is strongly activated with a high value for y_i, then any active synapse onto the cell will be capable of being modified. This enables the cell to learn an association between the pattern of activity on all its axons and its postsynaptic activation, which is stored as an addition to its weight vector $\mathbf{w}_i$. Then later on, at recall, the output can be produced as a vector dot product operation between the input pattern vector $\mathbf{x}$ and the weight vector $\mathbf{w}_i$, so that the output of the cell can reflect the correlation between the current input vector and what has previously been learned by the cell.

It is interesting that at least many invertebrate neuronal systems may operate very differently from those described here, as described by Rolls and Treves (1998) (see Fig. B.9a). If there were 5,000 conditioned stimulus inputs to a neuron, the implication is that every one would need to have a presynaptic terminal conveying the same UCS to each presynaptic terminal, which is hardly plausible. The implication is that at least some invertebrate neural systems operate very differently to those in vertebrates and, in such systems, the useful properties that arise from using distributed CS representations such as generalization would not arise in the same simple way as a property of the network.

B.2.7.1 Capacity

The question of the storage capacity of a pattern associator is considered in detail in Appendix A3 of Rolls and Treves (1998). It is pointed out there that, for this type of associative network, the number of memories that it can hold simultaneously in storage has to be analysed together with the retrieval quality of each output representation, and then only for a given quality of the representation provided in the input. This is in contrast to autoassociative nets (Section B.3), in which a critical number of stored memories exists (as a function of various parameters of the network), beyond which attempting to store additional memories results in it becoming impossible to retrieve essentially anything. With a pattern associator, instead, one will always retrieve something, but this something will be very small (in information or correlation terms) if too many associations are simultaneously in storage and/or if too little is provided as input.

The conjoint quality–capacity input analysis can be carried out, for any specific instance of a pattern associator, by using formal mathematical models and established analytical procedures (see e.g. Treves (1995), Rolls and Treves (1998), Treves (1990) and Rolls and Treves (1990)). This, however, has to be done case by case. It is anyway useful to develop some intuition for how a pattern associator operates, by considering what its capacity would be in certain well-defined simplified cases.

Linear associative neuronal networks These networks are made up of units with a linear activation function, which appears to make them unsuitable to represent real neurons with their positive-only firing rates. However, even purely linear units have been considered as provisionally relevant models of real neurons, by assuming that the latter operate sometimes in the linear regime of their transfer function. (This implies a high level of spontaneous activity, and may be closer to conditions observed early on in sensory systems rather than in areas more specifically involved in memory.) As usual, the connections are trained by a Hebb (or similar) associative learning rule. The capacity of these networks can be defined as the total number of associations that can be learned independently of each other, given that the linear nature of these systems prevents anything more than a linear transform of the inputs. This implies that if input pattern C can be written as the weighted sum of input patterns A and B, the output to C will be just the same weighted sum of the outputs to A and B. If there are N' input axons, then there can be only at most N' mutually independent input patterns (i.e. none able to be written as a weighted sum of the others), and therefore the capacity of linear networks, defined above, is just N', or equal to the number of inputs to each neuron. In general, a random set of less than N' vectors (the CS input pattern vectors) will tend to be mutually independent but not mutually orthogonal (at 90 deg to each other) (see Appendix A). If they are not orthogonal (the normal situation), then the dot product of them is not 0, and the output pattern activated by one of the input vectors will be partially activated by other input pattern vectors, in accordance with how similar they are (see equations B.5 and B.6). This amounts to interference, which is therefore the more serious the less orthogonal, on the whole, is the set of input vectors.

Since input patterns are made of elements with positive values, if a simple Hebbian learning rule like the one of equation B.2 is used (in which the input pattern enters directly with no subtraction term), the output resulting from the application of a stored input vector will be the sum of contributions from all other input vectors that have a non-zero dot product with it (see Appendix A), and interference will be disastrous. The only situation in which this would not occur is when different input patterns activate completely different input lines, but this is clearly an uninteresting circumstance for networks operating with distributed representations. A solution to this issue is to use a modified learning rule of the following form:

$$\delta w_{ij} = \alpha y_i (x_j - x) \qquad (B.7)$$

where x is a constant, approximately equal to the average value of x_j. This learning rule includes (in proportion to y_i) increasing the synaptic weight if $(x_j - x) > 0$ (long-term potentiation), and decreasing the synaptic weight if $(x_j - x) < 0$ (heterosynaptic long-term depression). It is useful for x to be roughly the average activity of an input axon x_j across patterns, because then the dot product between the various patterns stored on the weights and the input vector will tend to cancel out with the subtractive term, except for the pattern equal to (or correlated with) the input vector itself. Then up to N' input vectors can still be learned by the network, with only minor interference (provided of course that they are mutually independent, as they will in general tend to be).

Table B.1 Effects of pre- and post-synaptic activity on synaptic modification

	Post-synaptic activation	
	0	high
Presynaptic firing 0	No change	Heterosynaptic LTD
high	Homosynaptic LTD	LTP

This modified learning rule can also be described in terms of a contingency table (Table B.1) showing the synaptic strength modifications produced by different types of learning rule, where LTP indicates an increase in synaptic strength (called long-term potentiation in neurophysiology), and LTD indicates a decrease in synaptic strength (called long-term depression in neurophysiology). Heterosynaptic long-term depression is so-called because it is the decrease in synaptic strength that occurs to a synapse that is other than that through which the postsynaptic cell is being activated. This heterosynaptic long-term depression is the type of change of synaptic strength that is required (in addition to LTP) for effective subtraction of the average presynaptic firing rate, in order, as it were, to make the CS vectors appear more orthogonal to the pattern associator. The rule is sometimes called the Singer–Stent rule, after work by Singer (1987) and Stent (1973), and was discovered in the brain by Levy (Levy 1985, Levy and Desmond 1985) (see also Brown, Kairiss and Keenan (1990b)). Homosynaptic long-term depression is so-called because it is the decrease in synaptic strength that occurs to a synapse which is (the same as that which is) active. For it to occur, the postsynaptic neuron must simultaneously be inactive, or have only low activity. (This rule is sometimes called the BCM rule after the paper of Bienenstock, Cooper and Munro (1982); see Rolls and Deco (2002), Chapter 7).

Associative neuronal networks with non-linear neurons With non-linear neurons, that is with at least a threshold in the activation function so that the output firing y_i is 0 when the activation h_i is below the threshold, the capacity can be measured in terms of the number of different clusters of output pattern vectors that the network produces. This is because the non-linearities now present (one per output neuron) result in some clustering of the outputs produced by all possible (conditioned stimulus) input patterns **x**. Input patterns that are similar to a stored input vector can produce, due to the non-linearities, output patterns even closer to the stored output; and vice versa, sufficiently dissimilar inputs can be assigned to different output clusters thereby increasing their mutual dissimilarity. As with the linear counterpart, in order to remove the correlation that would otherwise occur between the patterns because the elements can take only positive values, it is useful to use a modified Hebb rule of the form shown in equation B.7.

With fully distributed output patterns, the number p of associations that leads to different clusters is of order C, the number of input lines (axons) per output neuron (that is, of order N' for a fully connected network), as shown in Appendix A3 of Rolls and Treves (1998). If sparse patterns are used in the output, or alternatively if the learning rule includes a non-linear postsynaptic factor that is effectively equivalent to using sparse output patterns, the coefficient of proportionality between p and C can be much higher than one, that is, many more patterns can be stored than inputs onto each output neuron (see Appendix A3 of Rolls and Treves (1998)). Indeed, the number of different patterns or prototypes p that can be stored can be derived for example in the case of binary units (Gardner 1988) to be

$$p \approx C/[a_o log(1/a_o)] \tag{B.8}$$

where a_o is the sparseness of the output firing pattern **y** produced by the unconditioned stimulus. p can in this situation be much larger than C (see Appendix A3 of Rolls and Treves (1998), Rolls and Treves (1990) and Treves (1990)). This is an important result for encoding in pattern associators, for it means that provided that the activation functions are non-linear (which is the case with real neurons), there is a very great advantage to using sparse encoding, for then many more than C pattern associations can be stored. Sparse representations may well be present in brain regions involved in associative memory for this reason (see Appendix C).

The non-linearity inherent in the NMDA receptor-based Hebbian plasticity present in the brain may help to make the stored patterns more sparse than the input patterns, and this may be especially beneficial in increasing the storage capacity of associative networks in the brain by allowing participation in the storage of especially those relatively few neurons with high firing rates in the exponential firing rate distributions typical of neurons in sensory systems (see Appendix C).

B.2.7.2 Interference

Interference occurs in linear pattern associators if two vectors are not orthogonal, and is simply dependent on the angle between the originally learned vector and the recall cue or CS vector (see Appendix A), for the activation of the output neuron depends simply on the dot product of the recall vector and the synaptic weight vector (equation B.5). Also in non-linear pattern associators (the interesting case for all practical purposes), interference may occur if two CS patterns are not orthogonal, though the effect can be controlled with sparse encoding of the UCS patterns, effectively by setting high thresholds for the firing of output units. In

Input A	1	0	1
Input B	0	1	1

Required Output 1 | 1 | 0

Fig. B.10 A non-linearly separable mapping.

other words, the CS vectors need not be strictly orthogonal, but if they are too similar, some interference will still be likely to occur.

The fact that interference is a property of neural network pattern associator memories is of interest, for interference is a major property of human memory. Indeed, the fact that interference is a property of human memory and of neural network association memories is entirely consistent with the hypothesis that human memory is stored in associative memories of the type described here, or at least that network associative memories of the type described represent a useful exemplar of the class of parallel distributed storage network used in human memory.

It may also be suggested that one reason that interference is tolerated in biological memory is that it is associated with the ability to generalize between stimuli, which is an invaluable feature of biological network associative memories, in that it allows the memory to cope with stimuli that will almost never be identical on different occasions, and in that it allows useful analogies that have survival value to be made.

B.2.7.3 Expansion recoding

If patterns are too similar to be stored in associative memories, then one solution that the brain seems to use repeatedly is to expand the encoding to a form in which the different stimulus patterns are less correlated, that is, more orthogonal, before they are presented as CS stimuli to a pattern associator. The problem can be highlighted by a non-linearly separable mapping (which captures part of the eXclusive OR (XOR) problem), in which the mapping that is desired is as shown in Fig. B.10. The neuron has two inputs, A and B.

This is a mapping of patterns that is impossible for a one-layer network, because the patterns are not linearly separable[37]. A solution is to remap the two input lines A and B to three input lines 1–3, that is to use expansion recoding, as shown in Fig. B.11. This can be performed by a competitive network (see Section B.4). The synaptic weights on the dendrite of the output neuron could then learn the following values using a simple Hebb rule, equation B.2, and the problem could be solved as in Fig. B.12. The whole network would look like that shown in Fig. B.11.

Competitive networks could help with this type of recoding, and could provide very useful preprocessing for a pattern associator in the brain (Rolls and Treves 1998, Rolls 2008d). It is possible that the lateral nucleus of the amygdala performs this function, for it receives inputs from the temporal cortical visual areas, and may preprocess them before they become the inputs to associative networks at the next stage of amygdala processing (Rolls 2008d, Rolls 2014a). The granule cells of the cerebellum may operate similarly (Chapter 23).

[37] See Appendix A. There is no set of synaptic weights in a one-layer net that could solve the problem shown in Fig. B.10. Two classes of patterns are not linearly separable if no hyperplane can be positioned in their N-dimensional space so as to separate them (see Appendix A). The XOR problem has the additional constraint that $A = 0, B = 0$ must be mapped to Output = 0.

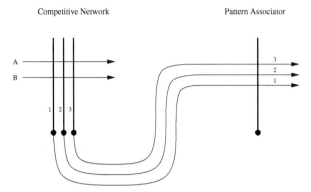

Competitive Network   Pattern Associator

Fig. B.11 Expansion recoding. A competitive network followed by a pattern associator that can enable patterns that are not linearly separable to be learned correctly.

	Synaptic weight
Input 1 (A=1, B=0)	1
Input 2 (A=0, B=1)	1
Input 3 (A=1, B=1)	0

Fig. B.12 Synaptic weights on the dendrite of the output neuron in Fig. B.11.

B.2.8 Implications of different types of coding for storage in pattern associators

Throughout this section, we have made statements about how the properties of pattern associators – such as the number of patterns that can be stored, and whether generalization and graceful degradation occur – depend on the type of encoding of the patterns to be associated. (The types of encoding considered, local, sparse distributed, and fully distributed, are described above.) We draw together these points in Table B.2.

Table B.2 Coding in associative memories*

	Local	Sparse distributed	Fully distributed
Generalization, completion, graceful degradation	No	Yes	Yes
Number of patterns that can be stored	N (large)	of order $C/[a_o \log(1/a_o)]$ (can be larger)	of order C (usually smaller than N)
Amount of information in each pattern (values if binary)	Minimal ($\log(N)$ bits)	Intermediate ($Na_o \log(1/a_o)$ bits)	Large (N bits)

* N refers here to the number of output units, and C to the average number of inputs to each output unit. a_o is the sparseness of output patterns, or roughly the proportion of output units activated by a UCS pattern. Note: logs are to the base 2.

The amount of information that can be stored in each pattern in a pattern associator is considered in Appendix A3 of Rolls and Treves (1998). That Appendix has been made available at http://www.oxcns.org, and contains a quantitative approach to the capacity of pattern association networks that has not been published elsewhere.

In conclusion, the architecture and properties of pattern association networks make them very appropriate for stimulus–reinforcer association learning. Their high capacity enables them to learn the reinforcement associations for very large numbers of different stimuli.

B.3 Autoassociation or attractor memory

Autoassociative memories, or attractor neural networks, store memories, each one of which is represented by a pattern of neural activity. The memories are stored in the recurrent synaptic connections between the neurons of the network, for example in the recurrent collateral connections between cortical pyramidal cells. Autoassociative networks can then recall the appropriate memory from the network when provided with a fragment of one of the memories. This is called completion. Many different memories can be stored in the network and retrieved correctly. A feature of this type of memory is that it is content addressable; that is, the information in the memory can be accessed if just the contents of the memory (or a part of the contents of the memory) are used. This is in contrast to a conventional computer, in which the address of what is to be accessed must be supplied, and used to access the contents of the memory. Content addressability is an important simplifying feature of this type of memory, which makes it suitable for use in biological systems. The issue of content addressability will be amplified below.

An autoassociation memory can be used as a short-term memory, in which iterative processing round the recurrent collateral connection loop keeps a representation active by continuing neuronal firing. The short-term memory reflected in continuing neuronal firing for several hundred milliseconds after a visual stimulus is removed which is present in visual cortical areas such as the inferior temporal visual cortex (see Chapter 25) is probably implemented in this way. This short-term memory is one possible mechanism that contributes to the implementation of the trace memory learning rule which can help to implement invariant object recognition as described in Chapter 25. Autoassociation memories also appear to be used in a short-term memory role in the prefrontal cortex. In particular, the temporal visual cortical areas have connections to the ventrolateral prefrontal cortex which help to implement the short-term memory for visual stimuli (in for example delayed match to sample tasks, and visual search tasks, as described in Section 4.3.1). In an analogous way the parietal cortex has connections to the dorsolateral prefrontal cortex for the short-term memory of spatial responses (see Section 4.3.1). These short-term memories provide a mechanism that enables attention to be maintained through backprojections from prefrontal cortex areas to the temporal and parietal areas that send connections to the prefrontal cortex, as described in Chapter 6. Autoassociation networks implemented by the recurrent collateral synapses between cortical pyramidal cells also provide a mechanism for constraint satisfaction and also noise reduction whereby the firing of neighbouring neurons can be taken into account in enabling the network to settle into a state that reflects all the details of the inputs activating the population of connected neurons, as well as the effects of what has been set up during developmental plasticity as well as later experience. Attractor networks are also effectively implemented by virtue of the forward and backward connections between cortical areas (see Sections 11 and 4.3.1). An autoassociation network with rapid synaptic plasticity can learn each memory in one trial. Because of its 'one-shot' rapid learning, and ability to complete, this type of network is well suited for episodic memory storage, in which each past episode must be stored and recalled later from a fragment, and kept separate from other episodic memories (see Chapter 24).

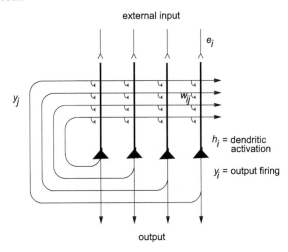

external input

e_i

y_j

w_{ij}

h_i = dendritic activation

y_i = output firing

output

Fig. B.13 The architecture of an autoassociative neural network.

B.3.1 Architecture and operation

The prototypical architecture of an autoassociation memory is shown in Fig. B.13. The external input e_i is applied to each neuron i by unmodifiable synapses. This produces firing y_i of each neuron, or a vector of firing on the output neurons **y**. Each output neuron i is connected by a recurrent collateral connection to the other neurons in the network, via modifiable connection weights w_{ij}. This architecture effectively enables the output firing vector **y** to be associated during learning with itself. Later on, during recall, presentation of part of the external input will force some of the output neurons to fire, but through the recurrent collateral axons and the modified synapses, other neurons in **y** can be brought into activity. This process can be repeated a number of times, and recall of a complete pattern may be perfect. Effectively, a pattern can be recalled or recognized because of associations formed between its parts. This of course requires distributed representations.

Next we introduce a more precise and detailed description of the above, and describe the properties of these networks. Ways to analyze formally the operation of these networks are introduced in Appendix A4 of Rolls and Treves (1998) and by Amit (1989).

B.3.1.1 Learning

The firing of every output neuron i is forced to a value y_i determined by the external input e_i. Then a Hebb-like associative local learning rule is applied to the recurrent synapses in the network:

$$\delta w_{ij} = \alpha y_i y_j. \tag{B.9}$$

It is notable that in a fully connected network, this will result in a symmetric matrix of synaptic weights, that is the strength of the connection from neuron 1 to neuron 2 will be the same as the strength of the connection from neuron 2 to neuron 1 (both implemented via recurrent collateral synapses).

It is a factor that is sometimes overlooked that there must be a mechanism for ensuring that during learning y_i does approximate e_i, and must not be influenced much by activity in the recurrent collateral connections, otherwise the new external pattern **e** will not be stored in the network, but instead something will be stored that is influenced by the previously stored memories. It is thought that in some parts of the brain, such as the hippocampus, there are processes that help the external connections to dominate the firing during learning (see Chapter 24, Treves and Rolls (1992b) and Rolls and Treves (1998)).

B.3.1.2 Recall

During recall, the external input e_i is applied, and produces output firing, operating through the non-linear activation function described below. The firing is fed back by the recurrent collateral axons shown in Fig. B.13 to produce activation of each output neuron through the modified synapses on each output neuron. The activation h_i produced by the recurrent collateral effect on the ith neuron is, in the standard way, the sum of the activations produced in proportion to the firing rate of each axon y_j operating through each modified synapse w_{ij}, that is,

$$h_i = \sum_j y_j w_{ij} \qquad (B.10)$$

where $\sum_j$ indicates that the sum is over the C input axons to each neuron, indexed by j.

The output firing y_i is a function of the activation produced by the recurrent collateral effect (internal recall) and by the external input (e_i):

$$y_i = f(h_i + e_i) \qquad (B.11)$$

The activation function should be non-linear, and may be for example binary threshold, linear threshold, sigmoid, etc. (see Fig. 1.4). The threshold at which the activation function operates is set in part by the effect of the inhibitory neurons in the network (not shown in Fig. B.13). The connectivity is that the pyramidal cells have collateral axons that excite the inhibitory interneurons, which in turn connect back to the population of pyramidal cells to inhibit them by a mixture of shunting (divisive) and subtractive inhibition using GABA (gamma-amino-butyric acid) terminals, as described in Section B.6. There are many fewer inhibitory neurons than excitatory neurons (in the order of 5–10%, see Table 1.1) and of connections to and from inhibitory neurons (see Table 1.1), and partly for this reason the inhibitory neurons are considered to perform generic functions such as threshold setting, rather than to store patterns by modifying their synapses. Similar inhibitory processes are assumed for the other networks described in this Appendix. The non-linear activation function can minimize interference between the pattern being recalled and other patterns stored in the network, and can also be used to ensure that what is a positive feedback system remains stable. The network can be allowed to repeat this recurrent collateral loop a number of times. Each time the loop operates, the output firing becomes more like the originally stored pattern, and this progressive recall is usually complete within 5–15 iterations.

B.3.2 Introduction to the analysis of the operation of autoassociation networks

With complete connectivity in the synaptic matrix, and the use of a Hebb rule, the matrix of synaptic weights formed during learning is symmetric. The learning algorithm is fast, 'one-shot', in that a single presentation of an input pattern is all that is needed to store that pattern.

During recall, a part of one of the originally learned stimuli can be presented as an external input. The resulting firing is allowed to iterate repeatedly round the recurrent collateral system, gradually on each iteration recalling more and more of the originally learned pattern. Completion thus occurs. If a pattern is presented during recall that is similar but not identical to any of the previously learned patterns, then the network settles into a stable recall state in which the firing corresponds to that of the previously learned pattern. The network can thus generalize in its recall to the most similar previously learned pattern. The activation function of the neurons should be non-linear, since a purely linear system would not produce any categorization of the input patterns it receives, and therefore would not be able to effect anything more than a trivial (i.e. linear) form of completion and generalization.

Recall can be thought of in the following way, relating it to what occurs in pattern associators. The external input $\mathbf{e}$ is applied, produces firing $\mathbf{y}$, which is applied as a recall cue on the recurrent collaterals as $\mathbf{y}^T$. (The notation $\mathbf{y}^T$ signifies the transpose of $\mathbf{y}$, which is implemented by the application of the firing of the neurons $\mathbf{y}$ back via the recurrent collateral axons as the next set of inputs to the neurons.) The activity on the recurrent collaterals is then multiplied by the synaptic weight vector stored during learning on each neuron to produce the new activation h_i which reflects the similarity between $\mathbf{y}^T$ and one of the stored patterns. Partial recall has thus occurred as a result of the recurrent collateral effect. The activations h_i after thresholding (which helps to remove interference from other memories stored in the network, or noise in the recall cue) result in firing y_i, or a vector of all neurons $\mathbf{y}$, which is already more like one of the stored patterns than, at the first iteration, the firing resulting from the recall cue alone, $\mathbf{y} = \mathbf{f}(\mathbf{e})$. This process is repeated a number of times to produce progressive recall of one of the stored patterns.

Autoassociation networks operate by effectively storing associations between the elements of a pattern. Each element of the pattern vector to be stored is simply the firing of a neuron. What is stored in an autoassociation memory is a set of pattern vectors. The network operates to recall one of the patterns from a fragment of it. Thus, although this network implements recall or recognition of a pattern, it does so by an association learning mechanism, in which associations between the different parts of each pattern are learned. These memories have sometimes been called autocorrelation memories (Kohonen 1977), because they learn correlations between the activity of neurons in the network, in the sense that each pattern learned is defined by a set of simultaneously active neurons. Effectively each pattern is associated by learning with itself. This learning is implemented by an associative (Hebb-like) learning rule.

The system formally resembles spin glass systems of magnets analyzed quantitatively in statistical mechanics. This has led to the analysis of (recurrent) autoassociative networks as dynamical systems made up of many interacting elements, in which the interactions are such as to produce a large variety of basins of attraction of the dynamics. Each basin of attraction corresponds to one of the originally learned patterns, and once the network is within a basin it keeps iterating until a recall state is reached that is the learned pattern itself or a pattern closely similar to it. (Interference effects may prevent an exact identity between the recall state and a learned pattern.) This type of system is contrasted with other, simpler, systems of magnets (e.g. ferromagnets), in which the interactions are such as to produce only a limited number of related basins, since the magnets tend to be, for example, all aligned with each other. The states reached within each basin of attraction are called attractor states, and the analogy between autoassociator neural networks and physical systems with multiple attractors was drawn by Hopfield (1982) in a very influential paper. He was able to show that the recall state can be thought of as the local minimum in an energy landscape, where the energy would be defined as

$$E = -\frac{1}{2} \sum_{i,j} w_{ij}(y_i - <y>)(y_j - <y>). \tag{B.12}$$

This equation can be understood in the following way. If two neurons are both firing above their mean rate (denoted by $<y>$), and are connected by a weight with a positive value, then the firing of these two neurons is consistent with each other, and they mutually support each other, so that they contribute to the system's tendency to remain stable. If across the whole network such mutual support is generally provided, then no further change will take place, and the system will indeed remain stable. If, on the other hand, either of our pair of neurons was not firing, or if the connecting weight had a negative value, the neurons would not support each other, and indeed the tendency would be for the neurons to try to alter ('flip' in the case of binary units) the state of the other. This would be repeated across the whole network until a situation in which most mutual support, and least 'frustration', was reached. What makes it possible to define an energy function and for these points to hold is that the matrix is symmetric (see Hopfield (1982), Hertz, Krogh and Palmer (1991), Amit (1989)).

Physicists have generally analyzed a system in which the input pattern is presented and then immediately removed, so that the network then 'falls' without further assistance (in what is referred to as the unclamped condition) towards the minimum of its basin of attraction. A more biologically realistic system is one in which the external input is left on contributing to the recall during the fall into the recall state. In this clamped condition, recall is usually faster, and more reliable, so that more memories may be usefully recalled from the network. The approach using methods developed in theoretical physics has led to rapid advances in the understanding of autoassociative networks, and its basic elements are described in Appendix A4 of Rolls and Treves (1998), and by Hertz, Krogh and Palmer (1991) and Amit (1989).

B.3.3 Properties

The internal recall in autoassociation networks involves multiplication of the firing vector of neuronal activity by the vector of synaptic weights on each neuron. This inner product vector multiplication allows the similarity of the firing vector to previously stored firing vectors to be provided by the output (as effectively a correlation), if the patterns learned are distributed. As a result of this type of 'correlation computation' performed if the patterns are distributed, many important properties of these networks arise, including pattern completion (because part of a pattern is correlated with the whole pattern), and graceful degradation (because a damaged synaptic weight vector is still correlated with the original synaptic weight vector). Some of these properties are described next.

B.3.3.1 Completion

Perhaps the most important and useful property of these memories is that they complete an incomplete input vector, allowing recall of a whole memory from a small fraction of it. The memory recalled in response to a fragment is that stored in the memory that is closest in pattern similarity (as measured by the dot product, or correlation). Because the recall is iterative and progressive, the recall can be perfect.

This property and the associative property of pattern associator neural networks are very similar to the properties of human memory. This property may be used when we recall a part of a recent memory of a past episode from a part of that episode. The way in which this could be implemented in the hippocampus is described in Chapter 24.

B.3.3.2 Generalization

The network generalizes in that an input vector similar to one of the stored vectors will lead to recall of the originally stored vector, provided that distributed encoding is used. The principle by which this occurs is similar to that described for a pattern associator.

B.3.3.3 Graceful degradation or fault tolerance

If the synaptic weight vector $\mathbf{w}_i$ on each neuron (or the weight matrix) has synapses missing (e.g. during development), or loses synapses (e.g. with brain damage or ageing), then the activation h_i (or vector of activations $\mathbf{h}$) is still reasonable, because h_i is the dot product (correlation) of $\mathbf{y}^T$ with $\mathbf{w}_i$. The same argument applies if whole input axons are lost. If an output neuron is lost, then the network cannot itself compensate for this, but the next network in the brain is likely to be able to generalize or complete if its input vector has some elements missing, as would be the case if some output neurons of the autoassociation network were damaged.

B.3.3.4 Prototype extraction, extraction of central tendency, and noise reduction

These arise when a set of similar input pattern vectors $\{\mathbf{e}\}$ (which induce firing of the output neurons $\{\mathbf{y}\}$) are learned by the network. The weight vectors $\mathbf{w}_i$ (or strictly $\mathbf{w}_i^T$) become (or point towards) the average $\{< \mathbf{y} >\}$ of that set of similar vectors. This produces 'extraction of the prototype' or 'extraction of the central tendency', and 'noise reduction'. This process can result in better recognition or recall of the prototype than of any of the exemplars, even though the prototype may never itself have been presented. The general principle by which the effect occurs is similar to that by which it occurs in pattern associators. It of course only occurs if each pattern uses a distributed representation.

Related to outputs of the visual system to long-term memory systems (see Chapter 24), there has been intense debate about whether when human memories are stored, a prototype of what is to be remembered is stored, or whether all the instances or the exemplars are each stored separately so that they can be individually recalled (McClelland and Rumelhart (1986), Chapter 17, p. 172). Evidence favouring the prototype view is that if a number of different examples of an object are shown, then humans may report more confidently that they have seen the prototype before than any of the different exemplars, even though the prototype has never been shown (Posner and Keele 1968, Rosch 1975). Evidence favouring the view that exemplars are stored is that in categorization and perceptual identification tasks the responses made are often sensitive to the congruity between particular training stimuli and particular test stimuli (Brooks 1978, Medin and Schaffer 1978, Jacoby 1983a, Jacoby 1983b, Whittlesea 1983). It is of great interest that both types of phenomena can arise naturally out of distributed information storage in a neuronal network such as an autoassociator. This can be illustrated by the storage in an autoassociation memory of sets of stimuli that are all somewhat different examples of the same pattern. These can be generated, for example, by randomly altering each of the input vectors from the input stimulus. After many such randomly altered exemplars have been learned by the network, recall can be tested, and it is found that the network responds best to the original (prototype) input vector, with which it has never been presented. The reason for this is that the autocorrelation components that build up in the synaptic matrix with repeated presentations of the exemplars represent the average correlation between the different elements of the vector, and this is highest for the prototype. This effect also gives the storage some noise immunity, in that variations in the input that are random noise average out, while the signal that is constant builds up with repeated learning.

B.3.3.5 Speed

The recall operation is fast on each neuron on a single iteration, because the pattern $\mathbf{y}^T$ on the axons can be applied simultaneously to the synapses $\mathbf{w}_i$, and the activation h_i can be accumulated in one or two time constants of the dendrite (e.g. 10–20 ms). If a simple implementation of an autoassociation net such as that described by Hopfield (1982) is simulated on a computer, then 5–15 iterations are typically necessary for completion of an incomplete input cue $\mathbf{e}$. This might be taken to correspond to 50–200 ms in the brain, rather too slow for any one local network in the brain to function. However, it has been shown that if the neurons are treated not as McCulloch–Pitts neurons which are simply 'updated' at each iteration, or cycle of timesteps (and assume the active state if the threshold is exceeded), but instead are analyzed and modelled as 'integrate-and-fire' neurons in real continuous time, then the network can effectively 'relax' into its recall state very rapidly, in one or two time constants of the synapses (see Section B.6 and Treves (1993), Battaglia and Treves (1998a) and Appendix A5 of Rolls and Treves (1998)). This corresponds to perhaps 20 ms in the brain.

One factor in this rapid dynamics of autoassociative networks with brain-like 'integrate-and-fire' membrane and synaptic properties is that with some spontaneous activity, some of the neurons in the network are close to threshold already before the recall cue is applied, and hence some of the neurons are very quickly pushed by the recall cue into firing, so that information starts to be exchanged very rapidly (within 1–2 ms of brain time) through the modified synapses by the neurons in the network. The progressive exchange of information starting early on within what would otherwise be thought of as an iteration period (of perhaps 20 ms, corresponding to a neuronal firing rate of 50 spikes/s) is the mechanism accounting for rapid recall in an autoassociative neuronal network made biologically realistic in this way. Further analysis of the fast dynamics of these networks if they are implemented in a biologically plausible way with 'integrate-and-fire' neurons is provided in Section B.6, in Appendix A5 of Rolls and Treves (1998), and by Treves (1993). *The general approach applies to other networks with recurrent connections, not just autoassociators, and the fact that such networks can operate much faster than it would seem from simple models that follow discrete time dynamics is probably a major factor in enabling these networks to provide some of the building blocks of brain function.*

Learning is fast, 'one-shot', in that a single presentation of an input pattern e (producing **y**) enables the association between the activation of the dendrites (the post-synaptic term h_i) and the firing of the recurrent collateral axons $\mathbf{y}^\mathrm{T}$, to be learned. Repeated presentation with small variations of a pattern vector is used to obtain the properties of prototype extraction, extraction of central tendency, and noise reduction, because these arise from the averaging process produced by storing very similar patterns in the network.

B.3.3.6 Local learning rule

The simplest learning used in autoassociation neural networks, a version of the Hebb rule, is (as in equation B.9)

$$\delta w_{ij} = \alpha y_i y_j.$$

The rule is a local learning rule in that the information required to specify the change in synaptic weight is available locally at the synapse, as it is dependent only on the presynaptic firing rate y_j available at the synaptic terminal, and the postsynaptic activation or firing y_i available on the dendrite of the neuron receiving the synapse. This makes the learning rule biologically plausible, in that the information about how to change the synaptic weight does not have to be carried to every synapse from a distant source where it is computed. As with pattern associators, since firing rates are positive quantities, a potentially interfering correlation is induced between different pattern vectors. This can be removed by subtracting the mean of the presynaptic activity from each presynaptic term, using a type of long-term depression. This can be specified as

$$\delta w_{ij} = \alpha y_i (y_j - z) \tag{B.13}$$

where α is a learning rate constant. This learning rule includes (in proportion to y_i) increasing the synaptic weight if $(y_j - z) > 0$ (long-term potentiation), and decreasing the synaptic weight if $(y_j - z) < 0$ (heterosynaptic long-term depression). This procedure works optimally if z is the average activity $< y_j >$ of an axon across patterns.

Evidence that a learning rule with the general form of equation B.9 is implemented in at least some parts of the brain comes from studies of long-term potentiation, described in Section 1.5. One of the important potential functions of heterosynaptic long-term depression is its ability to allow in effect the average of the presynaptic activity to be subtracted from the presynaptic firing rate (see Appendix A3 of Rolls and Treves (1998), and Rolls and Treves (1990)).

Autoassociation networks can be trained with the error-correction or delta learning rule described in Section B.10. Although a delta rule is less biologically plausible than a Hebb-like rule, a delta rule can help to store separately patterns that are very similar (see McClelland and Rumelhart (1988), Hertz, Krogh and Palmer (1991)).

B.3.3.7 Capacity

One measure of storage capacity is to consider how many orthogonal patterns could be stored, as with pattern associators. If the patterns are orthogonal, there will be no interference between them, and the maximum number p of patterns that can be stored will be the same as the number N of output neurons in a fully connected network. Although in practice the patterns that have to be stored will hardly be orthogonal, this is not a purely academic speculation, since it was shown how one can construct a synaptic matrix that effectively orthogonalizes any set of (linearly independent) patterns (Kohonen 1977, Kohonen 1989, Personnaz, Guyon and Dreyfus 1985, Kanter and Sompolinsky 1987). However, this matrix cannot be learned with a local, one-shot learning rule, and therefore its interest for autoassociators in the brain is limited. The more general case of random non-orthogonal patterns, and of Hebbian learning rules, is considered next. However, it is important to reduce the correlations between patterns to be stored in an autoassociation network to not limit the capacity (Marr 1971, Kohonen 1977, Kohonen 1989, Kohonen et al. 1981, Sompolinsky 1987, Rolls and Treves 1998), and in the brain mechanisms to perform pattern separation are frequently present (Rolls 2016e), including granule cells, as shown in many places in this book.

With non-linear neurons used in the network, the capacity can be measured in terms of the number of input patterns $\mathbf{y}$ (produced by the external input $\mathbf{e}$, see Fig. B.13) that can be stored in the network and recalled later whenever the network settles within each stored pattern's basin of attraction. The first quantitative analysis of storage capacity (Amit, Gutfreund and Sompolinsky 1987) considered a fully connected Hopfield (1982) autoassociator model, in which units are binary elements with an equal probability of being 'on' or 'off' in each pattern, and the number C of inputs per unit is the same as the number N of output units. (Actually it is equal to $N - 1$, since a unit is taken not to connect to itself.) Learning is taken to occur by clamping the desired patterns on the network and using a modified Hebb rule, in which the mean of the presynaptic and postsynaptic firings is subtracted from the firing on any one learning trial (this amounts to a covariance learning rule, and is described more fully in Appendix A4 of Rolls and Treves (1998)). With such fully distributed random patterns, the number of patterns that can be learned is (for C large) $p \approx 0.14C = 0.14N$, hence well below what could be achieved with orthogonal patterns or with an 'orthogonalizing' synaptic matrix. Many variations of this 'standard' autoassociator model have been analyzed subsequently.

Treves and Rolls (1991) have extended this analysis to autoassociation networks that are much more biologically relevant in the following ways. First, some or many connections between the recurrent collaterals and the dendrites are missing (this is referred to as diluted connectivity, and results in a non-symmetric synaptic connection matrix in which w_{ij} does not equal w_{ji}, one of the original assumptions made in order to introduce the energy formalism in the Hopfield model). Second, the neurons need not be restricted to binary threshold neurons, but can have a threshold linear activation function (see Fig. 1.4). This enables the neurons to assume real continuously variable firing rates, which are what is found in the brain (Rolls and Tovee 1995b, Treves, Panzeri, Rolls, Booth and Wakeman 1999). Third, the representation need not be fully distributed (with half the neurons 'on', and half 'off'), but instead can have a small proportion of the neurons firing above the spontaneous rate, which is what is found in parts of the brain such as the hippocampus that are involved in memory (see Treves and Rolls (1994), and Chapter 6 of Rolls and Treves (1998)). Such a representation is defined as being sparse, and the sparseness a of the representation can be measured, by extending the binary notion of the proportion of neurons that are firing, as

$$a = \frac{(\sum_{i=1}^{N} y_i/N)^2}{\sum_{i=1}^{N} y_i^2/N} \tag{B.14}$$

where y_i is the firing rate of the ith neuron in the set of N neurons. Treves and Rolls (1991) have shown that such a network does operate efficiently as an autoassociative network, and can store (and recall correctly) a number of different patterns p as follows

$$p \approx \frac{C^{RC}}{a \ln(\frac{1}{a})} k \tag{B.15}$$

where C^{RC} is the number of synapses on the dendrites of each neuron devoted to the recurrent collaterals from other neurons in the network, and k is a factor that depends weakly on the detailed structure of the rate distribution, on the connectivity pattern, etc., but is roughly in the order of 0.2–0.3.

The main factors that determine the maximum number of memories that can be stored in an autoassociative network are thus the number of connections on each neuron devoted to the recurrent collaterals, and the sparseness of the representation. For example, for $C^{RC} = 12,000$ and $a = 0.02$, p is calculated to be approximately $36,000$. This storage capacity can be realized, with little interference between patterns, if the learning rule includes some form of heterosynaptic long-term depression that counterbalances the effects of associative long-term potentiation (Treves and Rolls (1991); see Appendix A4 of Rolls and Treves (1998)). It should be noted that the number of neurons N (which is greater than C^{RC}, the number of recurrent collateral inputs received by any neuron in the network from the other neurons in the network) is not a parameter that influences the number of different memories that can be stored in the network. The implication of this is that increasing the number of neurons (without increasing the number of connections per neuron) does not increase the number of different patterns that can be stored (see Rolls and Treves (1998) Appendix A4), although it may enable simpler encoding of the firing patterns, for example more orthogonal encoding, to be used. This latter point may account in part for why there are generally in the brain more neurons in a recurrent network than there are connections per neuron (see e.g. Chapter 24).

The non-linearity inherent in the NMDA receptor-based Hebbian plasticity present in the brain may help to make the stored patterns more sparse than the input patterns, and this may be especially beneficial in increasing the storage capacity of associative networks in the brain by allowing participation in the storage of especially those relatively few neurons with high firing rates in the exponential firing rate distributions typical of neurons in sensory systems (see Sections B.4.9.3 and C.3.1).

B.3.3.8 Context

The environmental context in which learning occurs can be a very important factor that affects retrieval in humans and other animals. Placing the subject back into the same context in which the original learning occurred can greatly facilitate retrieval.

Context effects arise naturally in association networks if some of the activity in the network reflects the context in which the learning occurs. Retrieval is then better when that context is present, for the activity contributed by the context becomes part of the retrieval cue for the memory, increasing the correlation of the current state with what was stored. (A strategy for retrieval arises simply from this property. The strategy is to keep trying to recall as many fragments of the original memory situation, including the context, as possible, as this will provide a better cue for complete retrieval of the memory than just a single fragment.)

The effects that mood has on memory including visual memory retrieval may be accounted for by backprojections from brain regions such as the amygdala and orbitofrontal cortex in which the current mood, providing a context, is represented, to brain regions involved in memory such as the perirhinal cortex, and in visual representations such as the inferior temporal visual cortex (see Rolls and Stringer (2001b)). The very well-known effects of context in the human memory literature could arise in the simple way just described. An implication of the explanation is that context effects will be especially important at late stages of memory or information processing systems in the brain, for there information from a wide range of modalities will be mixed, and some of that information could reflect the context in which the learning takes place. One part of the brain where such effects may be strong is the hippocampus, which is implicated in the memory of recent episodes, and which receives inputs derived from most of the cortical information processing streams, including those involved in space (see Chapter 24).

B.3.3.9 Mixture states

If an autoassociation memory is trained on pattern vectors $\mathbf{A}$, $\mathbf{B}$, and $\mathbf{A} + \mathbf{B}$ (i.e. $\mathbf{A}$ and $\mathbf{B}$ are both included in the joint vector $\mathbf{A} + \mathbf{B}$; that is if the vectors are not linearly independent), then the autoassociation memory will have difficulty in learning and recalling these three memories as separate, because completion from either $\mathbf{A}$ or $\mathbf{B}$ to $\mathbf{A} + \mathbf{B}$ tends to occur during recall. (The ability to separate such patterns is referred to as configurational learning in the animal learning literature, see e.g. Sutherland and Rudy (1991).) This problem can be minimized by re-representing $\mathbf{A}$, $\mathbf{B}$, and $\mathbf{A} + \mathbf{B}$ in such a way that they are different vectors before they are presented to the autoassociation memory. This can be performed by recoding the input vectors to minimize overlap using, for example, a competitive network, and possibly involving expansion recoding, as described for pattern associators (see Section B.2, Fig. B.11). It is suggested that this is a function of the dentate granule cells in the hippocampus, which precede the CA3 recurrent collateral network (Treves and Rolls 1992b, Treves and Rolls 1994) (see Chapter 24).

B.3.3.10 Memory for sequences

One of the first extensions of the standard autoassociator paradigm that has been explored in the literature is the capability to store and retrieve not just individual patterns, but whole sequences of patterns. Hopfield, in the same 1982 paper, suggested that this could be achieved by adding to the standard connection weights, which associate a pattern with itself, a new, asymmetric component, which associates a pattern with the next one in the sequence. In practice this scheme does not work very well, unless the new component is made to operate on a slower time scale than the purely autoassociative component (Kleinfeld 1986, Sompolinsky and Kanter 1986). With two different time scales, the autoassociative component can stabilize a pattern for a while, before the heteroassociative component moves the network, as it were, into the next pattern. The heteroassociative retrieval cue for the next pattern in the sequence is just the previous pattern in the sequence. A particular type of 'slower' operation occurs if the asymmetric component acts after a delay τ. In this case, the network sweeps through the sequence, staying for a time of order τ in each pattern.

One can see how the necessary ingredient for the storage of sequences is only a minor departure from purely Hebbian learning: in fact, the (symmetric) autoassociative component of the weights can be taken to reflect the Hebbian learning of strictly simultaneous conjunctions of pre- and post-synaptic activity, whereas the (asymmetric) heteroassociative component can be implemented by Hebbian learning of each conjunction of postsynaptic activity with presynaptic activity shifted a time τ in the past. Both components can then be seen as resulting from a generalized Hebbian rule, which increases the weight whenever postsynaptic activity is paired with presynaptic activity occurring within a given time range, which may extend from a few hundred milliseconds in the past up to include strictly simultaneous activity. This is similar to a trace rule (see Chapter 25), which itself matches very well the observed conditions for induction of long-term potentiation, and appears entirely plausible. The learning rule necessary for learning sequences, though, is more complex than a simple trace rule in that the time-shifted conjunctions of activity that are encoded in the weights must in retrieval produce activations that are time-shifted as well (otherwise one falls back into the Hopfield (1982) proposal, which does not quite work). The synaptic weights should therefore keep separate 'traces' of what was simultaneous and what was time-shifted during the original experience, and this is not very plausible.

Levy and colleagues (Levy, Wu and Baxter 1995, Wu, Baxter and Levy 1996) have investigated these issues further, and the temporal asymmetry that may be present in LTP (see Section 1.5) has been suggested as a mechanism that might provide some of the temporal properties that are necessary for the brain to store and recall sequences (Minai and Levy 1993, Abbott and Blum 1996, Markram, Pikus, Gupta and Tsodyks 1998, Abbott and Nelson 2000). A problem with this suggestion is that, given that the temporal dynamics of attractor networks are inherently very fast when the networks have continuous dynamics (see Section B.6), and that the temporal asymmetry in LTP may be in the order of only milliseconds to a few tens of milliseconds (see Section 1.5), the recall of the sequences would be very fast, perhaps 10–20 ms per step of the sequence, with every step of a 10-step sequence effectively retrieved and gone in a quick-fire session of 100–200 ms.

Another way in which a delay could be inserted in a recurrent collateral path in the brain is by inserting another cortical area in the recurrent path. This could fit in with the cortico-cortical backprojection connections described in Chapter 11, which would introduce some conduction delay (see Panzeri, Rolls, Battaglia and Lavis (2001)).

B.3.4 Use of autoassociation networks in the brain

Because of its 'one-shot' rapid learning, and ability to complete, this type of network is well suited for episodic memory storage, in which each episode must be stored and recalled later from a fragment, and kept separate from other episodic memories. It does not take a long time (the 'many epochs' of backpropagation networks) to train this network, because it does not have to 'discover the structure' of a problem. Instead, it stores information in the form in which it is presented to the memory, without altering the representation. An autoassociation network may be used for this function in the CA3 region of the hippocampus (see Chapter 24, and Rolls and Treves (1998) Chapter 6).

An autoassociation memory can also be used as a short-term memory, in which iterative processing round the recurrent collateral loop keeps a representation active until another input cue is received. This may be used to implement many types of short-term memory in the brain (see Section 4.3.1). For example, it may be used in the perirhinal cortex and adjacent temporal lobe cortex to implement short-term visual object memory (Miyashita and Chang 1988, Amit 1995, Hirabayashi, Takeuchi, Tamura and Miyashita 2013); in the dorsolateral prefrontal cortex to implement a short-term memory for spatial responses (Goldman-Rakic 1996); and in the prefrontal cortex to implement a short-term memory for where eye movements should be made in space (see Section 4.3.1 and Rolls (2008d)). Such an autoassociation memory in the temporal lobe visual cortical areas may be used to implement the firing that continues for often 300 ms after a very brief (16 ms) presentation of a visual stimulus (Rolls and Tovee 1994) (see e.g. Fig. C.17), and may be one way in which a short memory trace is implemented to facilitate invariant learning about visual stimuli (see Chapter 25). In all these cases, the short-term memory may be implemented by the recurrent excitatory collaterals that connect nearby pyramidal cells in the cerebral cortex. The connectivity in this system, that is the probability that a neuron synapses on a nearby neuron, may be in the region of 10% (Braitenberg and Schütz 1991, Abeles 1991) (Chapter 7).

The recurrent connections between nearby neocortical pyramidal cells may also be important in defining the response properties of cortical cells, which may be triggered by external inputs (from for example the thalamus or a preceding cortical area), but may be considerably dependent on the synaptic connections received from nearby cortical pyramidal cells.

The cortico-cortical backprojection connectivity described in Chapters 1, 11, and 24 can be interpreted as a system that allows the forward-projecting neurons in one cortical area to be linked autoassociatively with the backprojecting neurons in the next cortical area (see Chapters 11 and 24). This would be implemented by associative synaptic modification in for example the backprojections. This particular architecture may be especially important in constraint satisfaction (as well as recall), that is it may allow the networks in the two cortical areas to settle into a mutually consistent state. This would effectively enable information in higher cortical areas, which would include information from more divergent sources, to influence the response properties of neurons in earlier cortical processing stages. This is an important function in cortical information processing of interacting associative networks.

B.4 Competitive networks, including self-organizing maps

B.4.1 Function

Competitive neural networks learn to categorize input pattern vectors. Each category of inputs activates a different output neuron (or set of output neurons – see below). The categories formed are based on similarities between the input vectors. Similar, that is correlated, input vectors activate the same output neuron. In that the learning is based on similarities in the input space, and there is no external teacher that forces classification, this is an unsupervised network. The term categorization is used to refer to the process of placing vectors into categories based on their similarity. The term classification is used to refer to the process of placing outputs in particular classes as instructed or taught by a teacher. Examples of classifiers are pattern associators, one-layer delta-rule perceptrons, and multilayer perceptrons taught by error backpropagation (see Sections B.2, B.3, B.10 and B.11). In supervised networks there is usually a teacher for each output neuron.

The categorization produced by competitive nets is of great potential importance in perceptual systems including the whole of the visual cortical processing hierarchies, as described in Chapter 25. Each category formed reflects a set or cluster of active inputs x_j that occur together. This cluster of coactive inputs can be thought of as a feature, and the competitive network can be described as building feature analyzers, where a feature can now be defined as a correlated set of inputs. During learning, a competitive network gradually discovers these features in the input space, and the process of finding these features without a teacher is referred to as self-organization.

Another important use of competitive networks is to remove redundancy from the input space, by allocating output neurons to reflect a set of inputs that co-occur.

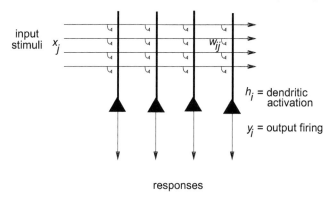

Fig. B.14 The architecture of a competitive network.

Another important aspect of competitive networks is that they separate patterns that are somewhat correlated in the input space, to produce outputs for the different patterns that are less correlated with each other, and may indeed easily be made orthogonal to each other. This has been referred to as orthogonalization.

Another important function of competitive networks is that partly by removing redundancy from the input information space, they can produce sparse output vectors, without losing information. We may refer to this as sparsification.

B.4.2 Architecture and algorithm

B.4.2.1 Architecture

The basic architecture of a competitive network is shown in Fig. B.14. It is a one-layer network with a set of inputs that make modifiable excitatory synapses w_{ij} with the output neurons. The output cells compete with each other (for example by mutual inhibition) in such a way that the most strongly activated neuron or neurons win the competition, and are left firing strongly. The synaptic weights, w_{ij}, are initialized to random values before learning starts. If some of the synapses are missing, that is if there is randomly diluted connectivity, that is not a problem for such networks, and can even help them (see below).

In the brain, the inputs arrive through axons, which make synapses with the dendrites of the output or principal cells of the network. The principal cells are typically pyramidal cells in the cerebral cortex. In the brain, the principal cells are typically excitatory, and mutual inhibition between them is implemented by inhibitory interneurons, which receive excitatory inputs from the principal cells. The inhibitory interneurons then send their axons to make synapses with the pyramidal cells, typically using GABA (gamma-aminobutyric acid) as the inhibitory transmitter.

B.4.2.2 Algorithm

1. Apply an input vector $\mathbf{x}$ and calculate the activation h_i of each neuron

$$h_i = \sum_j x_j w_{ij} \tag{B.16}$$

where the sum is over the C input axons, indexed by j. (It is useful to normalize the length of each input vector $\mathbf{x}$. In the brain, a scaling effect is likely to be achieved both by feedforward inhibition, and by feedback inhibition among the set of input cells (in a preceding network) that give rise to the axons conveying $\mathbf{x}$.)

The output firing y_i^1 is a function of the activation of the neuron

$$y_i^1 = f(h_i). \tag{B.17}$$

The function f can be linear, sigmoid, monotonically increasing, etc. (see Fig. 1.4).

2. Allow competitive interaction between the output neurons by a mechanism such as lateral or mutual inhibition (possibly with self-excitation), to produce a contrast-enhanced version of the firing rate vector

$$y_i = g(y_i^1). \tag{B.18}$$

Function g is typically a non-linear operation, and in its most extreme form may be a winner-take-all function, in which after the competition one neuron may be 'on', and the others 'off'. Algorithms that produce softer competition without a single winner to produce a distributed representation are described in Section B.4.9.4 below.

3. Apply an associative Hebb-like learning rule

$$\delta w_{ij} = \alpha y_i x_j. \tag{B.19}$$

4. Normalize the length of the synaptic weight vector on each dendrite to prevent the same few neurons always winning the competition:

$$\sum_j (w_{ij})^2 = 1. \tag{B.20}$$

(A less efficient alternative is to scale the sum of the weights to a constant, e.g. 1.0.)

5. Repeat steps 1–4 for each different input stimulus **x**, in random sequence, a number of times.

B.4.3 Properties

B.4.3.1 Feature discovery by self-organization

Each neuron in a competitive network becomes activated by a set of consistently coactive, that is correlated, input axons, and gradually learns to respond to that cluster of coactive inputs. We can think of competitive networks as discovering features in the input space, where features can now be defined by a set of consistently coactive inputs. Competitive networks thus show how feature analyzers can be built, with no external teacher. The feature analyzers respond to correlations in the input space, and the learning occurs by self-organization in the competitive network. Competitive networks are therefore well suited to the analysis of sensory inputs. Ways in which they may form fundamental building blocks of sensory systems are described in Chapter 25.

The operation of competitive networks can be visualized with the help of Fig. B.15. The input patterns are represented as dots on the surface of a sphere. (The patterns are on the surface of a sphere because the patterns are normalized to the same length.) The directions of the weight vectors of the three neurons are represented by '×'s. The effect of learning is to move the weight vector of each of the neurons to point towards the centre of one of the clusters of inputs. If the neurons are winner-take-all, the result of the learning is that although there are correlations between the input stimuli, the outputs of the three neurons are orthogonal. In this sense, orthogonalization is performed. At the same time, given that each of the patterns within a cluster produces the same output, the correlations between the patterns within a cluster become higher. In a winner-take-all network, the within-pattern correlation becomes 1, and the patterns within a cluster have been placed within the same category.

(a) 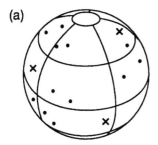   (b)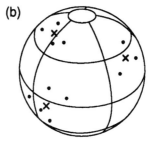

Fig. B.15 Competitive learning. The dots represent the directions of the input vectors, and the '×'s the weights for each of three output neurons. (a) Before learning. (b) After learning. (After Rumelhart and Zipser 1986.)

B.4.3.2 Removal of redundancy

In that competitive networks recode sets of correlated inputs to one or a few output neurons, then redundancy in the input representation is removed. Identifying and removing redundancy in sensory inputs is an important part of processing in sensory systems (cf. Barlow (1989)), in part because a compressed representation is more manageable as an output of sensory systems. The reason for this is that neurons in the receiving systems, for example pattern associators in the orbitofrontal cortex or autoassociation networks in the hippocampus, can then operate with the limited numbers of inputs that each neuron can receive. For example, although the information that a particular face is being viewed is present in the 10^6 fibres in the optic nerve, the information is unusable by associative networks in this form, and is compressed through the visual system until the information about which of many hundreds of faces is present can be represented by less than 100 neurons in the temporal cortical visual areas (Rolls, Treves and Tovee 1997b, Abbott, Rolls and Tovee 1996). (Redundancy can be defined as the difference between the maximum information content of the input data stream (or channel capacity) and its actual content; see Appendix C.)

The recoding of input pattern vectors into a more compressed representation that can be conveyed by a much reduced number of output neurons of a competitive network is referred to in engineering as vector quantization. With a winner-take-all competitive network, each output neuron points to or stands for one of or a cluster of the input vectors, and it is more efficient to transmit the states of the few output neurons than the states of all the input elements. (It is more efficient in the sense that the information transmission rate required, that is the capacity of the channel, can be much smaller.) Vector quantization is of course possible when the input representation contains redundancy.

B.4.3.3 Orthogonalization and categorization

Figure B.15 shows visually how competitive networks reduce the correlation between different clusters of patterns, by allocating them to different output neurons. This is described as orthogonalization. It is a process that is very usefully applied to signals before they are used as inputs to associative networks (pattern associators and autoassociators) trained with Hebbian rules (see Sections B.2 and B.3), because it reduces the interference between patterns stored in these memories. The opposite effect in competitive networks, of bringing closer together very similar input patterns, is referred to as categorization.

These two processes are also illustrated in Fig. B.16, which shows that in a competitive network, very similar input patterns (with correlations higher in this case than approximately 0.8)

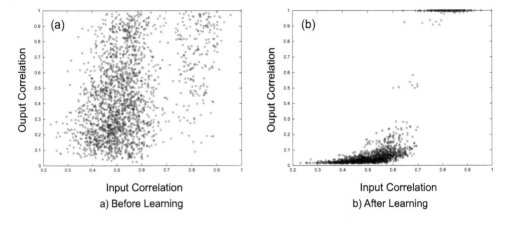

Fig. B.16 Orthogonalization and categorization in a competitive network: (a) before learning; (b) after learning. The correlations between pairs of output vectors (abscissa) are plotted against the correlations of the corresponding pairs of input vectors that generated the output pair, for all possible pairs in the input set. The competitive net learned for 16 cycles. One cycle consisted of presenting the complete input set of stimuli in a renewing random sequence. The correlation measure shown is the cosine of the angle between two vectors (i.e. the normalized dot product). The network used had 64 input axons to each of 8 output neurons. The net was trained with 64 stimuli, made from 8 initial random binary vectors with each bit having a probability of 0.5 of being 1, from each of which 8 noisy exemplars were made by randomly altering 10 % of the 64 elements. Soft competition was used between the output neurons. (A normalized exponential activation function described in Section B.4.9.4 was used to implement the soft competition.) The sparseness a of the input patterns thus averaged 0.5; and the sparseness a of the output firing vector after learning was close to 0.17 (i.e. after learning, primarily one neuron was active for each input pattern; before learning, the average sparseness of the output patterns produced by each of the inputs was 0.39).

produce more similar outputs (close to 1.0), whereas the correlations between pairs of input patterns that are smaller than approximately 0.7 become much smaller in the output representation. (This simulation used soft competition between neurons with graded firing rates.)

Further analyses of the operation of competitive networks, and how diluted connectivity can help the operation of competitive networks, are provided in Section 7.4 (Rolls 2016e).

B.4.3.4 Sparsification

Competitive networks can produce more sparse representations than those that they receive, depending on the degree of competition. With the greatest competition, winner-take-all, only one output neuron remains active, and the representation is at its most sparse. This effect can be understood further using Figs. B.15 and B.16. This sparsification is useful to apply to representations before input patterns are applied to associative networks, because sparse representations allow many different pattern associations or memories to be stored in these networks (see Sections B.2 and B.3).

B.4.3.5 Capacity

In a competitive net with N output neurons and a simple winner-take-all rule for the competition, it is possible to learn up to N output categories, in that each output neuron may be allocated a category. When the competition acts in a less rudimentary way, the number of categories that can be learned becomes a complex function of various factors, including the number of modifiable connections per cell and the degree of dilution, or incompleteness, of the connections. Such a function has not yet been described analytically in general, but an upper bound on it can be deduced for the particular case in which the learning is fast, and can be achieved effectively in one shot, or one presentation of each pattern. In that case, the number of categories that can be learned (by the self-organizing process) will at most be equal to the number of associations that can be formed by the corresponding pattern associators, a process that occurs with the additional help of the driving inputs, which effectively determine the categorization in the pattern associator.

Separate constraints on the capacity result if the output vectors are required to be strictly orthogonal. Then, if the output firing rates can assume only positive values, the maximum number p of categories arises, obviously, in the case when only one output neuron is firing for any stimulus, so that up to N categories are formed. If ensemble encoding of output neurons is used (soft competition), again under the orthogonality requirement, then the number of output categories that can be learned will be reduced according to the degree of ensemble encoding. The p categories in the ensemble-encoded case reflect the fact that the between-cluster correlations in the output space are lower than those in the input space. The advantages of ensemble encoding are that dendrites are more evenly allocated to patterns (see Section B.4.9.5), and that correlations between different input stimuli can be reflected in correlations between the corresponding output vectors, so that later networks in the system can generalize usefully. This latter property is of crucial importance, and is utilized for example when an input pattern is presented that has not been learned by the network. The relative similarity of the input pattern to previously learned patterns is indicated by the relative activation of the members of an ensemble of output neurons. This makes the number of different representations that can be reflected in the output of competitive networks with ensemble encoding much higher than with winner-take-all representations, even though with soft competition all these representations cannot strictly be learned.

B.4.3.6 Separation of non-linearly separable patterns

A competitive network can not only separate (e.g. by activating different output neurons) pattern vectors that overlap in almost all elements, but can also help with the separation of vectors that are not linearly separable. An example is that three patterns **A**, **B**, and **A+B** will lead to three different output neurons being activated (see Fig. B.17). For this to occur, the length of the synaptic weight vectors must be normalized (to for example unit length), so that they lie on the surface of a sphere or hypersphere (see Fig. B.15). (If the weight vectors of each neuron are scaled to the same sum, then the weight vectors do not lie on the surface of a hypersphere, and the ability of the network to separate patterns is reduced.)

The property of pattern separation makes a competitive network placed before an auto-association (or pattern association) network very valuable, for it enables the autoassociator to store the three patterns separately, and to recall **A+B** separately from **A** and **B**. This is referred to as the configuration learning problem in animal learning theory (Sutherland and Rudy 1991). Placing a competitive network before a pattern associator will enable a linearly inseparable problem to be solved. For example, three different output neurons of a two-input competitive network could respond to the patterns 01, 10, and 11, and a pattern associator can learn different outputs for neurons 1–3, which are orthogonal to each other (see Fig. B.11). This is an example of expansion recoding (cf. Marr (1969), who used a different algorithm to obtain the expansion). The sparsification that can be produced by the competitive network can also be advantageous in preparing patterns for presentation to a pattern associator or autoassociator, because the sparsification can increase the number of memories that can be associated or stored.

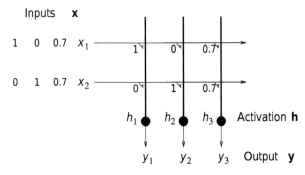

Fig. B.17 Separation of linearly dependent patterns by a competitive network. The network was trained on patterns 10, 01, and 11, applied on the inputs x_1 and x_2. After learning, the network allocated output neuron 1 to pattern 10, neuron 2 to pattern 01, and neuron 3 to pattern 11. The weights in the network produced during the learning are shown. Each input pattern was normalized to unit length, and thus for pattern 11, x_1=0.7 and x_2=0.7, as shown. Because the weight vectors were also normalized to unit length, w_{31}=0.7 and w_{32}=0.7.

B.4.3.7 Stability

These networks are generally stable if the input statistics are stable. If the input statistics keep varying, then the competitive network will keep following the input statistics. If this is a problem, then a critical period in which the input statistics are learned, followed by stabilization, may be useful. This appears to be a solution used in developing sensory systems, which have critical periods beyond which further changes become more difficult. An alternative approach taken by Carpenter and Grossberg in their 'Adaptive Resonance Theory' (Carpenter 1997) is to allow the network to learn only if it does not already have categorizers for a pattern (see Hertz, Krogh and Palmer (1991), p. 228).

Diluted connectivity can help stability, by making neurons tend to find inputs to categorize in only certain parts of the input space, and then making it difficult for the neuron to wander randomly throughout the space later (see further Section 7.4).

B.4.3.8 Frequency of presentation

If some stimuli are presented more frequently than others, then there will be a tendency for the weight vectors to move more rapidly towards frequently presented stimuli, and more neurons may become allocated to the frequently presented stimuli. If winner-take-all competition is used, the result is that the neurons will tend to become allocated during the learning process to the more frequently presented patterns. If soft competition is used, the tendency of neurons to move from patterns that are infrequently or never presented can be reduced by making the competition fairly strong, so that only a few neurons show any learning when each pattern is presented. Provided that the competition is moderately strong (see Section B.4.9.4), the result is that more neurons are allocated to frequently presented patterns, but one or some neurons are allocated to infrequently presented patterns. These points can all be easily demonstrated in simulations.

In an interesting development, it has been shown that if objects consisting of groups of features are presented during training always with another object present, then separate representations of each object can be formed provided that each object is presented many times, but on each occasion is paired with a different object (Stringer and Rolls 2008, Stringer, Rolls and Tromans 2007b). This is related to the fact that in this scenario the frequency of co-occurrence of features within the same object is greater than that of features between different objects (see Section 25.5.6.2).

B.4.3.9 Comparison to principal component analysis (PCA) and cluster analysis

Although competitive networks find clusters of features in the input space, they do not perform hierarchical cluster analysis as typically performed in statistics. In hierarchical cluster analysis, input vectors are joined starting with the most correlated pair, and the level of the joining of vectors is indicated. Competitive nets produce different outputs (i.e. activate different output neurons) for each cluster of vectors (i.e. perform vector quantization), but do not compute the level in the hierarchy, unless the network is redesigned (see Hertz, Krogh and Palmer (1991)).

The feature discovery can also be compared to principal component analysis (PCA). (In PCA, the first principal component of a multidimensional space points in the direction of the vector that accounts for most of the variance, and subsequent principal components account for successively less of the variance, and are mutually orthogonal.) In competitive learning with a winner-take-all algorithm, the outputs are mutually orthogonal, but are not in an ordered series according to the amount of variance accounted for, unless the training algorithm is modified. The modification amounts to allowing each of the neurons in a winner-take-all network to learn one at a time, in sequence. The first neuron learns the first principal component. (Neurons trained with a modified Hebb rule learn to maximize the variance of their outputs – see Hertz, Krogh and Palmer (1991).) The second neuron is then allowed to learn, and because its output is orthogonal to the first, it learns the second principal component. This process is repeated. Details are given by Hertz, Krogh and Palmer (1991), but as this is not a biologically plausible process, it is not considered in detail here.

I note that simple competitive learning is very helpful biologically, because it can separate patterns, but that a full ordered set of principal components as computed by PCA would probably not be very useful in biologically plausible networks. The point here is that biological neuronal networks may operate well if the variance in the input representation is distributed across many input neurons, whereas principal component analysis would tend to result in most of the variance being allocated to a few neurons, and the variance being unevenly distributed across the neurons.

B.4.4 Utility of competitive networks in information processing by the brain

B.4.4.1 Feature analysis and preprocessing

Neurons that respond to correlated combinations of their inputs can be described as feature analyzers. Neurons that act as feature analyzers perform useful preprocessing in many sensory systems (see e.g. Chapter 8 of Rolls and Treves (1998)). The power of competitive networks in multistage hierarchical processing to build combinations of what is found at earlier stages, and thus effectively to build higher-order representations, is also described in Chapter 25 of this book. An interesting development is that competitive networks can learn about individual objects even when multiple objects are presented simultaneously, provided that each object is presented several times more frequently than it is paired with any other individual object (Stringer and Rolls 2008) (see Section 25.5.6.2). This property arises because learning in competitive networks is primarily about forming representations of objects defined by a high correlation of coactive features in the input space (Stringer and Rolls 2008).

B.4.4.2 Removal of redundancy

The removal of redundancy by competition is thought to be a key aspect of how sensory systems, including the ventral cortical visual system, operate. Competitive networks can also be thought of as performing dimension reduction, in that a set of correlated inputs may be responded to as one category or dimension by a competitive network. The concept of redundancy removal can be linked to the point that individual neurons trained with a modified Hebb rule point their weight vector in the direction of the vector that accounts for most of the variance in the input, that is (acting individually) they find the first principal component of the input space (see Section B.4.3.9 and Hertz, Krogh and Palmer (1991)). Although networks with anti-Hebbian synapses between the principal cells (in which the anti-Hebbian learning forces neurons with initially correlated activity to effectively inhibit each other) (Földiák 1991), and networks that perform Independent Component Analysis (Bell and Sejnowski 1995), could in principle remove redundancy more effectively, it is not clear that they are implemented biologically. In contrast, competitive networks are more biologically plausible, and illustrate redundancy reduction. The more general use of an unsupervised competitive preprocessor is discussed below (see Fig. B.24).

B.4.4.3 Orthogonalization

The orthogonalization performed by competitive networks is very useful for preparing signals for presentation to pattern associators and autoassociators, for this re-representation decreases interference between the patterns stored in such networks. Indeed, this can be essential if patterns are overlapping and not linearly independent, e.g. 01, 10, and 11. If three such binary patterns were presented to an autoassociative network, it would not form separate representations of them, because either of the patterns 01 or 10 would result by completion in recall of the 11 pattern. A competitive network allows a separate neuron to be allocated to each of the three patterns, and this set of orthogonal representations can be learned by associative networks (see Fig. B.17).

B.4.4.4 Sparsification

The sparsification performed by competitive networks is very useful for preparing signals for presentation to pattern associators and autoassociators, for this re-representation increases the number of patterns that can be associated or stored in such networks (see Sections B.2 and B.3).

B.4.4.5 Brain systems in which competitive networks may be used for orthogonalization and sparsification

One system is the hippocampus, in which the dentate granule cells are believed to operate as a competitive network in order to prepare signals for presentation to the CA3 autoassociative network (see Chapter 24). In this case, the operation is enhanced by expansion recoding, in that (in the rat) there are approximately three times as many dentate granule cells as there are cells in the preceding stage, the entorhinal cortex. This expansion recoding will itself tend to reduce correlations between patterns (cf. Marr (1970), and Marr (1969)).

Also in the hippocampus, the CA1 neurons are thought to act as a competitive network that recodes the separate representations of each of the parts of an episode that must be separately represented in CA3, into a form more suitable for the recall using pattern association performed by the backprojections from the hippocampus to the cerebral cortex (see Chapter 24 and Rolls and Treves (1998) Chapter 6).

The granule cells of the cerebellum may perform a similar function, but in this case the principle may be that each of the very large number of granule cells receives a very small random subset of inputs, so that the outputs of the granule cells are decorrelated with respect to the inputs (Marr (1969); see Chapter 23).

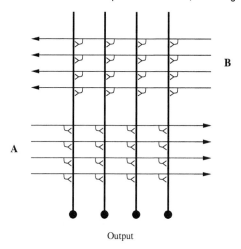

Output

Fig. B.18 Competitive net receiving a normal forward set of inputs **A**, but also another set of inputs **B** that can be used to influence the categories formed in response to **A** inputs. The inputs **B** might be backprojection inputs.

B.4.5 Guidance of competitive learning

Although competitive networks are primarily unsupervised networks, it is possible to influence the categories found by supplying a second input, as follows (Rolls 1989a). Consider a competitive network as shown in Fig. B.18 with the normal set of inputs **A** to be categorized, and with an additional set of inputs **B** from a different source. Both sets of inputs work in the normal way for a competitive network, with random initial weights, competition between the output neurons, and a Hebb-like synaptic modification rule that normalizes the lengths of the synaptic weight vectors onto each neuron. The idea then is to use the **B** inputs to influence the categories formed by the **A** input vectors. The influence of the **B** vectors works best if they are orthogonal to each other. Consider any two **A** vectors. If they occur together with the same **B** vector, then the categories produced by the **A** vectors will be more similar than they would be without the influence of the **B** vectors. The categories will be pulled closer together if soft competition is used, or will be more likely to activate the same neuron if winner-take-all competition is used. Conversely, if any two **A** vectors are paired with two different, preferably orthogonal, **B** vectors, then the categories formed by the **A** vectors will be drawn further apart than they would be without the **B** vectors. The differences in categorization remain present after the learning when just the **A** inputs are used.

This guiding function of one of the inputs is one way in which the consequences of sensory stimuli could be fed back to a sensory system to influence the categories formed when the **A** inputs are presented. This could be one function of backprojections in the cerebral cortex (Rolls 1989c, Rolls 1989a) (Chapter 11). In this case, the **A** inputs of Fig. B.18 would be the forward inputs from a preceding cortical area, and the **B** inputs backprojecting axons from the next cortical area, or from a structure such as the amygdala or hippocampus. If two **A** vectors were both associated with positive reinforcement that was fed back as the same **B** vector from another part of the brain, then the two **A** vectors would be brought closer together in the representational space provided by the output of the neurons. If one of the **A** vectors was associated with positive reinforcement, and the other with negative reinforcement, then the output representations of the two **A** vectors would be further apart. This is one way in which external signals could influence in a mild way the categories formed in sensory systems. Another is that if any **B** vector only occurred for important sensory **A** inputs (as shown by the

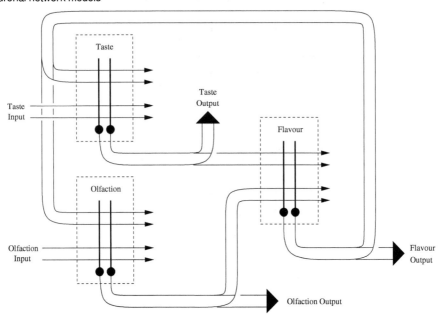

Fig. B.19 A two-layer set of competitive nets in which feedback from layer 2 can influence the categories formed in layer 1. Layer 2 could be a higher cortical visual area with convergence from earlier cortical visual areas (see Chapter 25). In the example, taste and olfactory inputs are received by separate competitive nets in layer 1, and converge into a single competitive net in layer 2. The categories formed in layer 2 (which may be described as representing 'flavour') may be dominated by the relatively orthogonal set of a few tastes that are received by the net. When these layer 2 categories are fed back to layer 1, they may produce in layer 1 categories in, for example, the olfactory network that reflect to some extent the flavour categories of layer 2, and are different from the categories that would otherwise be formed to a large set of rather correlated olfactory inputs. A similar principle may operate in any multilayer hierarchical cortical processing system, such as the ventral visual system, in that the categories that can be formed only at later stages of processing may help earlier stages to form categories relevant to what can be identified at later stages.

immediate consequences of receiving those sensory inputs), then the **A** inputs would simply be more likely to have any representation formed than otherwise, due to strong activation of neurons only when combined **A** and **B** inputs are present.

A similar architecture could be used to provide mild guidance for one sensory system (e.g. olfaction) by another (e.g. taste), as shown in Fig. B.19. (Another example of where this architecture could be used is convergence in the visual system at the next cortical stage of processing, with guiding feedback to influence the categories formed in the different regions of the preceding cortical area, as illustrated in Chapter 11.) The idea is that the taste inputs would be more orthogonal to each other than the olfactory inputs, and that the taste inputs would influence the categories formed in the olfactory input categorizer in layer 1, by feedback from a convergent net in layer 2. The difference from the previous architecture is that we now have a two-layer net, with unimodal or separate networks in layer 1, each feeding forward to a single competitive network in layer 2. The categories formed in layer 2 reflect the co-occurrence of a particular taste with particular odours (which together form flavour in layer 2). Layer 2 then provides feedback connections to both the networks in layer 1. It can be shown in such network that the categories formed in, for example, the olfactory net in layer 1 are influenced by the tastes with which the odours are paired. The feedback signal is built only in layer 2, after there has been convergence between the different modalities. This architecture captures some of the properties of sensory systems, in which there are unimodal processing cortical areas followed by multimodal cortical areas. The multimodal cortical areas can build representations that represent the unimodal inputs that tend to co-occur, and the higher level representations may in turn, by the highly developed cortico-cortical backprojections, be able to influence sensory categorization in earlier cortical processing areas (Rolls 1989a). Another such example might be the effect by which the phonemes heard are influenced by the visual inputs produced by seeing mouth movements (cf. McGurk and MacDonald (1976)). This could be implemented by auditory inputs coming together in the cortex in the superior temporal sulcus onto neurons activated by the sight of the lips moving (recorded during experiments of Baylis, Rolls and Leonard (1987), and Hasselmo, Rolls, Baylis and Nalwa (1989b)), using Hebbian learning with co-active inputs. Backprojections from such multimodal areas to the early auditory cortical areas could then influence the responses of auditory cortex neurons to auditory inputs (see Section 4.10 and Fig. 4.5, and cf. Calvert, Bullmore, Brammer, Campbell, Williams, McGuire, Woodruff, Iversen and David (1997)).

A similar principle may operate in any multilayer hierarchical cortical processing system, such as the ventral visual system, in that the categories that can be formed only at later stages of processing may help earlier stages to form categories relevant to what can be identified at the later stages as a result of the operation of backprojections (Rolls 1989a). The idea that the statistical correlation between the inputs received by neighbouring processing streams can be used to guide unsupervised learning within each stream has also been developed by Becker and Hinton (1992) and others (see Phillips, Kay and Smyth (1995)). The networks considered by these authors self-organize under the influence of collateral connections, such as may be implemented by cortico-cortical connections between parallel processing systems in the brain. They use learning rules that, although somewhat complex, are still local in nature, and tend to optimize specific objective functions. The locality of the learning rule, and the simulations performed so far, raise some hope that, once the operation of these types of networks is better understood, they might achieve similar computational capabilities to backpropagation networks (see Section B.11) while retaining biological plausibility.

B.4.6 Topographic map formation

A simple modification to the competitive networks described so far enables them to develop topological maps. In such maps, the closeness in the map reflects the similarity (correlation) between the features in the inputs. The modification that allows such maps to self-organize is to add short-range excitation and long-range inhibition between the neurons. The function to be implemented has a spatial profile that is described as having a Mexican hat shape (see Fig. B.20). The effect of this connectivity between neurons, which need not be modifiable, is to encourage neurons that are close together to respond to similar features in the input space, and to encourage neurons that are far apart to respond to different features in the input space. When these response tendencies are present during learning, the feature analyzers that are built by modifying the synapses from the input onto the activated neurons tend to be similar if they are close together, and different if far apart. This is illustrated in Figs. B.21 and B.22. Feature maps built in this way were described by von der Malsburg (1973) and Willshaw and von der Malsburg (1976). It should be noted that the learning rule needed is simply the modified Hebb rule described above for competitive networks, and is thus local and biologically plausible. (For computational convenience, the algorithm that Kohonen (Kohonen 1982, Kohonen 1989, Kohonen 1995) has mainly used does not use Mexican hat connectivity between the neurons, but instead arranges that when the weights to a winning neuron are updated, so to a smaller extent are those of its neighbours – see further Hertz, Krogh and Palmer (1991).)

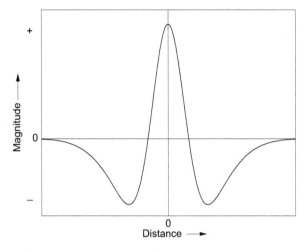

Fig. B.20 Mexican hat lateral spatial interaction profile.

A very common characteristic of connectivity in the brain, found for example throughout the neocortex, consists of short-range excitatory connections between neurons, with inhibition mediated via inhibitory interneurons. The density of the excitatory connectivity even falls gradually as a function of distance from a neuron, extending typically a distance in the order of 1 mm from the neuron (Braitenberg and Schütz 1991), contributing to a spatial function quite like that of a Mexican hat. (Longer-range inhibitory influences would form the negative part of the spatial response profile.) This supports the idea that topological maps, though in some cases probably seeded by chemoaffinity, could develop in the brain with the assistance of the processes just described. It is noted that some cortico-cortical connections even within an area may be longer, skipping past some intermediate neurons, and then making connections after some distance with a further group of neurons. Such longer-range connections are found for example between different columns with similar orientation selectivity in the primary visual cortex. The longer range connections may play a part in stabilizing maps, and again in the exchange of information between neurons performing related computations, in this case about features with the same orientations.

If a low-dimensional space, for example the orientation sensitivity of cortical neurons in the primary visual cortex (which is essentially one-dimensional, the dimension being angle), is mapped to a two-dimensional space such as the surface of the cortex, then the resulting map can have long spatial runs where the value along the dimension (in this case orientation tuning) alters gradually, and continuously. Such self-organization can account for many aspects of the mapping of orientation tuning, and of ocular dominance columns, in V1 (Miller 1994, Harris, Ermentrout and Small 1997). If a high-dimensional information space is mapped to the two-dimensional cortex, then there will be only short runs of groups of neurons with similar feature responsiveness, and then the map must fracture, with a different type of feature mapped for a short distance after the discontinuity. This is exactly what Rolls suggests is the type of topology found in the anterior inferior temporal visual cortex, with the individual groupings representing what can be self-organized by competitive networks combined with a trace rule as described in Section 25.5. Here, visual stimuli are not represented with reference to their position on the retina, because here the neurons are relatively translation invariant. Instead, when recording here, small clumps of neurons with similar responses may be encountered close together, and then one moves into a group of neurons with quite different feature selectivity (personal observations). This topology will arise naturally, given the anatomical

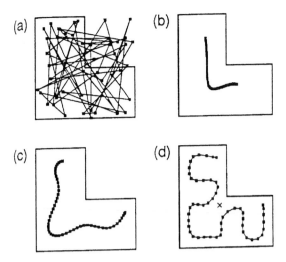

Fig. B.21 Kohonen feature mapping from a two-dimensional L-shaped region to a linear array of 50 units. Each unit has 2 inputs. The input patterns are the X,Y coordinates of points within the L-shape shown. In the diagrams, each point shows the position of a weight vector. Lines connect adjacent units in the 1 D (linear) array of 50 neurons. The weights were initialized to random values within the unit square (a). During feature mapping training, the weights evolved through stages (b) and (c) to (d). By stage (d) the weights have formed so that the positions in the original input space are mapped to a 1 D vector in which adjacent points in the input space activate neighbouring units in the linear array of output units. (Reproduced with permission from Hertz, Krogh and Palmer 1991, Fig. 9.13.)

connectivity of the cortex with its short-range excitatory connections, because there are very many different objects in the world and different types of features that describe objects, with no special continuity between the different combinations of features possible.

Rolls' hypothesis contrasts with the view of Tanaka (1996), who has claimed that the inferior temporal cortex provides an alphabet of visual features arranged in discrete modules. The type of mapping found in higher cortical visual areas as proposed by Rolls implies that topological self-organization is an important way in which maps in the brain are formed, for it seems most unlikely that the locations in the map of different types of object seen in an environment could be specified genetically (Rolls and Stringer 2000). Consistent with this, Tsao, Freiwald, Tootell and Livingstone (2006) described with macaque fMRI 'anterior face patches' at A15 to A22. A15 might correspond to where we have analysed face-selective neurons (it might translate to 3 mm posterior to our sphenoid reference, see Section 25.2), and at this level there are separate regions specialized for face identity in areas TEa and TEm on the ventral lip of the superior temporal sulcus and the adjacent gyrus, and for face expression and movement in the cortex deep in the superior temporal sulcus (Hasselmo, Rolls and Baylis 1989a, Baylis, Rolls and Leonard 1987, Rolls 2007e). The 'middle face patch' of Tsao, Freiwald, Tootell and Livingstone (2006) was at A6, which is probably part of the posterior inferior temporal cortex, and, again consistent with self-organizing map principles, has a high concentration of face-selective neurons within the patch.

The biological utility of developing such topology-preserving feature maps may be that if the computation requires neurons with similar types of response to exchange information more than neurons involved in different computations (which is more than reasonable), then the total length of the connections between the neurons is minimized if the neurons that need

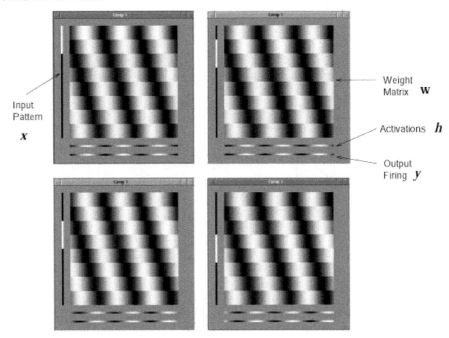

Input
Pattern

x

Weight
Matrix **W**

Activations h

Output
Firing y

Fig. B.22 Example of a one-dimensional topological map that self-organized from inputs in a low-dimensional space. The network has 64 neurons (vertical elements in the diagram) and 64 inputs per neuron (horizontal elements in the diagram). The four different diagrams represent the net tested with different input patterns. The input patterns x are displayed at the left of each diagram, with white representing firing and black not firing for each of the 64 inputs. The central square of each diagram represents the synaptic weights of the neurons, with white representing a strong weight. The row vector below each weight matrix represents the activations of the 64 output neurons, and the bottom row vector the output firing y. The network was trained with a set of 8 binary input patterns, each of which overlapped in 8 of its 16 'on' elements with the next pattern. The diagram shows that as one moves through correlations in the input space (top left to top right to bottom left to bottom right), so the output neurons activated move steadily across the output array of neurons. Closely correlated inputs are represented close together in the output array of neurons. The way in which this occurs can be seen by inspection of the weight matrix. The network architecture was the same as for a competitive net, except that the activations were converted linearly into output firings, and then each neuron excited its neighbours and inhibited neurons further away. This lateral inhibition was implemented for the simulation by a spatial filter operating on the output firings with the following filter weights (cf. Fig. B.20): 5, 5, 5, 5, 5, 5, 5, 5, 5, 10, 10, 10, 10, 10, 10, 10, 10, 5, 5, 5, 5, 5, 5, 5, 5, 5 which operated on the 64-element firing rate vector.

to exchange information are close together (cf. Cowey (1979), Durbin and Mitchison (1990)). Examples of this include the separation of colour constancy processing in V4 from global motion processing in MT as follows (see Chapter 3 of Rolls and Deco (2002)). In V4, to compute colour constancy, an estimate of the illuminating wavelengths can be obtained by summing the outputs of the pyramidal cells in the inhibitory interneurons over several degrees of visual space, and subtracting this from the excitatory central ON colour-tuned region of the receptive field by (subtractive) feedback inhibition. This enables the cells to discount the illuminating wavelength, and thus compute colour constancy. For this computation, no inputs from motion-selective cells (which in the dorsal stream are colour insensitive) are needed. In MT, to compute global motion (e.g. the motion produced by the average flow of local motion elements, exemplified for example by falling snow), the computation can be performed by averaging in the larger (several degrees) receptive fields of MT the local motion inputs received by neurons in earlier cortical areas (V1 and V2) with small receptive fields (see Chapter 3 of Rolls and Deco (2002) and Rolls and Stringer (2007)). For this computation, no input from colour cells is useful. Having separate areas (V4 and MT) for these different computations minimizes the wiring lengths, for having intermingled colour and motion cells in a single cortical area would increase the average connection length between the neurons that need to be connected for the computations being performed. Minimizing the total connection length between neurons in the brain is very important in order to keep the size of the brain relatively small.

Placing close to each other neurons that need to exchange information, or that need to receive information from the same source, or that need to project towards the same destination, may also help to minimize the complexity of the rules required to specify cortical (and indeed brain) connectivity (Rolls and Stringer 2000). For example, in the case of V4 and MT, the connectivity rules can be simpler (e.g. connect to neurons in the vicinity, rather than look for colour-, or motion-marked cells, and connect only to the cells with the correct genetically specified label specifying that the cell is either part of motion or of colour processing). Further, the V4 and MT example shows that how the neurons are connected can be specified quite simply, but of course it needs to be specified to be different for different computations. Specifying a general rule for the classes of neurons in a given area also provides a useful simplification to the genetic rules needed to specify the functional architecture of a given cortical area (Rolls and Stringer 2000). In our V4 and MT example, the genetic rules would need to specify the rules separately for different populations of inhibitory interneurons if the computations performed by V4 and MT were performed with intermixed neurons in a single brain area. Together, these two principles, of minimization of wiring length, and allowing simple genetic specification of wiring rules, may underlie the separation of cortical visual information processing into different (e.g. ventral and dorsal) processing streams. The same two principles operating within each brain processing stream may underlie (taken together with the need for hierarchical processing to enable the computations to be biologically plausible in terms of the number of connections per neuron, and the need for local learning rules, see Section B.12) much of the overall architecture of visual cortical processing, and of information processing and its modular architecture throughout the cortex more generally.

The rules of information exchange just described could also tend to produce more gross topography in cortical regions. For example, neurons that respond to animate objects may have certain visual feature requirements in common, and may need to exchange information about these features. Other neurons that respond to inanimate objects might have somewhat different visual feature requirements for their inputs, and might need to exchange information strongly. (For example, selection of whether an object is a chisel or a screwdriver may require competition by mutual (lateral) inhibition to produce the contrast enhancement necessary to result in unambiguous neuronal responses.) The rules just described would account for neurons with responsiveness to inanimate and animate objects tending to be grouped in separate parts of a cortical map or representation, and thus separately susceptible to brain damage (see e.g. Farah (1990), Farah (2000)).

B.4.7 Invariance learning by competitive networks

In conventional competitive learning, the weight vector of a neuron can be thought of as moving towards the centre of a cluster of similar overlapping input stimuli (Rumelhart and Zipser 1985, Hertz, Krogh and Palmer 1991, Rolls and Treves 1998, Rolls and Deco 2002, Perry, Rolls and Stringer 2006, Rolls 2008d). The weight vector points towards the centre of the set of stimuli in the category. The different training stimuli that are placed into the same

(a) 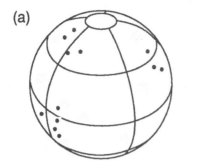   (b)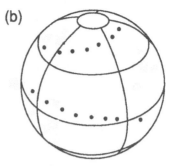

Fig. B.23 (a) Conventional competitive learning. A cluster of overlapping input patterns is categorized as being similar, and this is implemented by a weight vector of an output neuron pointing towards the centre of the cluster. Three clusters are shown, and each cluster might after training have a weight vector pointing towards it. (b) Invariant representation learning. The different transforms of an object may span an elongated region of the space, and the transforms at the far ends of the space may have no overlap (correlation), yet the network must learn to categorize them as similar. The different transforms of two different objects are represented.

category (i.e. activate the same neuron) are typically overlapping in that the pattern vectors are correlated with each other. Figure B.23a illustrates this.

For the formation of invariant representations, there are multiple occurrences of the object at different positions in the space. The object at each position represents a different transform (whether in position, size, view etc.) of the object. The different transforms may be uncorrelated with each other, as would be the case for example with an object translated so far in the space that there would be no active afferents in common between the two transforms. Yet we need these two orthogonal patterns to be mapped to the same output. It may be a very elongated part of the input space that has to be mapped to the same output in invariance learning. These concepts are illustrated in Fig. B.23b.

Objects in the world have temporal and spatial continuity. That is, the statistics of the world are such that we tend to look at one object for some time, during which it may be transforming continuously from one view to another. The temporal continuity property is used in trace rule invariance training, in which a short-term memory in the associative learning rule normally used to train competitive networks is used to help build representations that reflect the continuity in time that characterizes the different transforms of each object, as described in Chapter 25. The transforms of an object also show spatial continuity, and this can also be used in invariance training in what is termed continuous spatial transform learning, described in Section 25.5.11.

In conventional competitive learning the overall weight vector points to the prototypical representation of the object. The only sense in which after normal competitive training (without translations etc) the network generalizes is with respect to the dot product similarity of any input vector compared to the central vector from the training set that the network learns. Continuous spatial transformation learning works by providing a set of training vectors that overlap, and between them cover the whole space over which an invariant transform of the object must be learned. Indeed, it is important for continuous spatial transformation learning that the different exemplars of an object are sufficiently close that the similarity of adjacent training exemplars is sufficient to ensure that the same postsynaptic neuron learns to bridge the continuous space spanned by the whole set of training exemplars of a given object (Stringer,

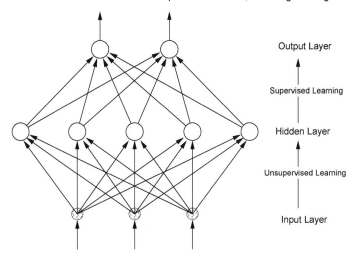

Output Layer

Supervised Learning

Hidden Layer

Unsupervised Learning

Input Layer

Fig. B.24 A hybrid network, in which for example unsupervised learning rapidly builds relatively orthogonal representations based on input differences, and this is followed by a one-layer supervised network (taught for example by the delta rule) that learns to classify the inputs based on the categorizations formed in the hidden/intermediate layer.

Perry, Rolls and Proske 2006, Perry, Rolls and Stringer 2006, Perry, Rolls and Stringer 2006). This will enable the postsynaptic neuron to span a very elongated space of the different transforms of an object, as described in Section 25.5.11.

B.4.8 Radial Basis Function networks

As noted above, a competitive network can act as a useful preprocessor for other networks. In the neural examples above, competitive networks were useful preprocessors for associative networks. Competitive networks are also used as preprocessors in artificial neural networks, for example in hybrid two-layer networks such as that illustrated in Fig. B.24. The competitive network is advantageous in this hybrid scheme, because as an unsupervised network, it can relatively quickly (with a few presentations of each stimulus) discover the main features in the input space, and code for them. This leaves the second layer of the network to act as a supervised network (taught for example by the delta rule, see Section B.10), which learns to map the features found by the first layer into the output required. This learning scheme is very much faster than that of a (two-layer) backpropagation network, which learns very slowly because it takes it a long time to perform the credit assignment to build useful feature analyzers in layer one (the hidden layer) (see Section B.11).

The general scheme shown in Fig. B.24 is used in radial basis function (RBF) neural networks. The main difference from what has been described is that in an RBF network, the hidden neurons do not use a winner-take-all function (as in some competitive networks), but instead use a normalized activation function in which the measure of distance from a weight vector of the neural input is (instead of the dot product $\mathbf{x} \cdot \mathbf{w}_i$ used for most of the networks described in this book), a Gaussian measure of distance:

$$y_i = \frac{\exp[-(\mathbf{x} - \mathbf{w}_i)^2/2\sigma_i^2]}{\sum_k \exp[-(\mathbf{x} - \mathbf{w}_k)^2/2\sigma_k^2]}. \tag{B.21}$$

The effect is that the response y_i of neuron i is a maximum if the input stimulus vector $\mathbf{x}$ is centred at $\mathbf{w}_i$, the weight vector of neuron i (this is the upper term in equation B.21). The magnitude is normalized by dividing by the sum of the activations of all the k neurons in the network. If the input vector $\mathbf{x}$ is not at the centre of the receptive field of the neuron, then the response is decreased according to how far the input vector is from the weight vector $\mathbf{w}_i$ of the neuron, with the weighting decreasing as a Gaussian function with a standard deviation of σ. The idea is like that implemented with soft competition, in that the relative response of different neurons provides an indication of where the input pattern is in relation to the weight vectors of the different neurons. The rapidity with which the response falls off in a Gaussian radial basis function neuron is set by σ_i, which is adjusted so that for any given input pattern vector, a number of RBF neurons are activated. The positions in which the RBF neurons are located (i.e. the directions of their weight vectors, $\mathbf{w}$) are determined usually by unsupervised learning, e.g. the vector quantization that is produced by the normal competitive learning algorithm. The first layer of an RBF network is not different in principle from a network with soft competition, and it is not clear how biologically a Gaussian activation function would be implemented, so the treatment is not developed further here (see Hertz, Krogh and Palmer (1991), Poggio and Girosi (1990a), and Poggio and Girosi (1990b) for further details).

B.4.9 Further details of the algorithms used in competitive networks

B.4.9.1 Normalization of the inputs

Normalization is useful because in step 1 of the training algorithm described in Section B.4.2.2, the neuronal activations, formed by the inner product of the pattern and the normalized weight vector on each neuron, are scaled in such a way that they have a maximum value of 1.0. This helps different input patterns to be equally effective in the learning process. A way in which this normalization could be achieved by a layer of input neurons is given by Grossberg (1976a). In the brain, a number of factors may contribute to normalization of the inputs. One factor is that a set of input axons to a neuron will come from another network in which the firing is controlled by inhibitory feedback, and if the numbers of axons involved is large (hundreds or thousands), then the inputs will be in a reasonable range. Second, there is increasing evidence that the different classes of input to a neuron may activate different types of inhibitory interneuron (e.g. Buhl, Halasy and Somogyi (1994)), which terminate on separate parts of the dendrite, usually close to the site of termination of the corresponding excitatory afferents. This may allow separate feedforward inhibition for the different classes of input. In addition, the feedback inhibitory interneurons also have characteristic termination sites, often on or close to the cell body, where they may be particularly effective in controlling firing of the neuron by shunting (divisive) inhibition, rather than by scaling a class of input (see Section B.6).

B.4.9.2 Normalization of the length of the synaptic weight vector on each dendrite

This is necessary to ensure that one or a few neurons do not always win the competition. (If the weights on one neuron were increased by simple Hebbian learning, and there was no normalization of the weights on the neuron, then it would tend to respond strongly in the future to patterns with some overlap with patterns to which that neuron has previously learned, and gradually that neuron would capture a large number of patterns.) A biologically plausible way to achieve this weight adjustment is to use a modified Hebb rule:

$$\delta w_{ij} = \alpha y_i (x_j - w_{ij}) \tag{B.22}$$

where α is a constant, and x_j and w_{ij} are in appropriate units. In vector notation,

$$\delta \mathbf{w}_i = \alpha y_i (\mathbf{x} - \mathbf{w}_i) \tag{B.23}$$

where $\mathbf{w}_i$ is the synaptic weight vector on neuron i. This implements a Hebb rule that increases synaptic strength according to conjunctive pre- and post-synaptic activity, and also allows the strength of each synapse to decrease in proportion to the firing rate of the postsynaptic neuron (as well as in proportion to the existing synaptic strength). This results in a decrease in synaptic strength for synapses from weakly active presynaptic neurons onto strongly active postsynaptic neurons. Such a modification in synaptic strength is termed heterosynaptic long-term depression in the neurophysiological literature, referring to the fact that the synapses that weaken are other than those that activate the neuron.

This is an important computational use of heterosynaptic long-term depression (LTD). In that the amount of decrease of the synaptic strength depends on how strong the synapse is already, the rule is compatible with what is frequently reported in studies of LTD (see Section 1.5). This rule can maintain the sums of the synaptic weights on each dendrite to be very similar without any need for explicit normalization of the synaptic strengths, and is useful in competitive nets. This rule was used by Willshaw and von der Malsburg (1976). As is made clear with the vector notation above, the modified Hebb rule moves the direction of the weight vector $\mathbf{w}_i$ towards the current input pattern vector $\mathbf{x}$ in proportion to the difference between these two vectors and the firing rate y_i of neuron i.

If explicit weight (vector length) normalization is needed, the appropriate form of the modified Hebb rule is:

$$\delta w_{ij} = \alpha y_i (x_j - y_i w_{ij}). \tag{B.24}$$

This rule, formulated by Oja (1982), makes weight decay proportional to y_i^2, normalizes the synaptic weight vector (see Hertz, Krogh and Palmer (1991)), is still a local learning rule, and is known as the Oja rule.

B.4.9.3 Non-linearity in the learning rule

Non-linearity in the learning rule can assist competition (Rolls 1989b, Rolls 1996c). For example, in the brain, long-term potentiation typically occurs only when strong activation of a neuron has produced sufficient depolarization for the voltage-dependent NMDA receptors to become unblocked, allowing Ca^{2+} to enter the cell (see Section 1.5). This means that synaptic modification occurs only on neurons that are strongly activated, effectively assisting competition to select few winners. The learning rule can be written:

$$\delta w_{ij} = \alpha m_i x_j \tag{B.25}$$

where m_i is a (e.g. threshold) non-linear function of the post-synaptic firing y_i which mimics the operation of the NMDA receptors in learning. (It is noted that in associative networks the same process may result in the stored pattern being more sparse than the input pattern, and that this may be beneficial, especially given the exponential firing rate distribution of neurons, in helping to maximize the number of patterns stored in associative networks (see Sections B.2, B.3, and C.3.1).

B.4.9.4 Competition

In a simulation of a competitive network, a single winner can be selected by searching for the neuron with the maximum activation. If graded competition is required, this can be achieved by an activation function that increases greater than linearly. In some of the networks we have simulated (Rolls 1989b, Rolls 1989a, Wallis and Rolls 1997), raising the activation to a fixed power, typically in the range 2–5, and then rescaling the outputs to a fixed maximum (e.g. 1) is simple to implement. In a real neuronal network, winner-take-all competition can be implemented using mutual (lateral) inhibition between the neurons with non-linear activation functions, and self-excitation of each neuron (see e.g. Grossberg (1976a), Grossberg (1988), Hertz, Krogh and Palmer (1991)).

Another method to implement soft competition in simulations is to use the normalized exponential or 'softmax' activation function for the neurons (Bridle (1990); see Bishop (1995)):

$$y = \exp(h)/\sum_i \exp(h_i) . \tag{B.26}$$

This function specifies that the firing rate of each neuron is an exponential function of the activation, scaled by the whole vector of activations h_i, $i = 1, N$. The exponential function (in increasing supralinearly) implements soft competition, in that after the competition the faster firing neurons are firing relatively much faster than the slower firing neurons. In fact, the strength of the competition can be adjusted by using a 'temperature' T greater than 0 as follows:

$$y = \exp(h/T)/\sum_i \exp(h_i/T). \tag{B.27}$$

Very low temperatures increase the competition, until with $T \to 0$, the competition becomes 'winner-take-all'. At high temperatures, the competition becomes very soft. (When using the function in simulations, it may be advisable to prescale the firing rates to for example the range 0–1, both to prevent machine overflow, and to set the temperature to operate on a constant range of firing rates, as increasing the range of the inputs has an effect similar to decreasing T.)

The softmax function has the property that activations in the range $-\infty$ to $+\infty$ are mapped into the range 0 to 1.0, and the sum of the firing rates is 1.0. This facilitates interpretation of the firing rates under certain conditions as probabilities, for example that the competitive network firing rate of each neuron reflects the probability that the input vector is within the category or cluster signified by that output neuron (see Bishop (1995)).

B.4.9.5 Soft competition

The use of graded (continuous valued) output neurons in a competitive network, and soft competition rather than winner-take-all competition, has the value that the competitive net generalizes more continuously to an input vector that lies between input vectors that it has learned. Also, with soft competition, neurons with only a small amount of activation by any of the patterns being used will nevertheless learn a little, and move gradually towards the patterns that are being presented. The result is that with soft competition, the output neurons all tend to become allocated to one of the input patterns or one of the clusters of input patterns.

B.4.9.6 Untrained neurons

In competitive networks, especially with winner-take-all or finely tuned neurons, it is possible that some neurons remain unallocated to patterns. This may be useful, in case patterns in the unused part of the space occur in future. Alternatively, unallocated neurons can be made to move towards the parts of the space where patterns are occurring by allowing such losers in the competition to learn a little. Another mechanism is to subtract a bias term μ_i from y_i, and to use a 'conscience' mechanism that raises μ_i if a neuron wins frequently, and lowers μ_i if it wins infrequently (Grossberg 1976b, Bienenstock, Cooper and Munro 1982, De Sieno 1988).

B.4.9.7 Large competitive nets: further aspects

If a large neuronal network is considered, with the number of synapses on each neuron in the region of 10,000, as occurs on large pyramidal cells in some parts of the brain, then there is a potential disadvantage in using neurons with synaptic weights that can take on only positive values. This difficulty arises in the following way. Consider a set of positive normalized input firing rates and synaptic weight vectors (in which each element of the vector can take on any value between 0.0 and 1.0). Such vectors of random values will on average be more highly aligned with the direction of the central vector (1,1,1,...,1) than with any other vector. An example can be given for the particular case of vectors evenly distributed on the positive 'quadrant' of a high-dimensional hypersphere: the average overlap (i.e. normalized dot product) between two binary random vectors with half the elements on and thus a sparseness of 0.5 (e.g. a random pattern vector and a random dendritic weight vector) will be approximately 0.5, while the average overlap between a random vector and the central vector will be approximately 0.707. A consequence of this will be that if a neuron begins to learn towards several input pattern vectors it will get drawn towards the average of these input patterns which will be closer to the 1,1,1,...,1 direction than to any one of the patterns. As a dendritic weight vector moves towards the central vector, it will become more closely aligned with more and more input patterns so that it is more rapidly drawn towards the central vector. The end result is that in large nets of this type, many of the dendritic weight vectors will point towards the central vector. This effect is not seen so much in small systems, since the fluctuations in the magnitude of the overlaps are sufficiently large that in most cases a dendritic weight vector will have an input pattern very close to it and thus will not learn towards the centre. In large systems, the fluctuations in the overlaps between random vectors become smaller by a factor of $\frac{1}{\sqrt{N}}$ so that the dendrites will not be particularly close to any of the input patterns.

One solution to this problem is to allow the elements of the synaptic weight vectors to take negative as well as positive values. This could be implemented in the brain by feedforward inhibition. A set of vectors taken with random values will then have a reduced mean correlation between any pair, and the competitive net will be able to categorize them effectively. A system with synaptic weights that can be negative as well as positive is not physiologically plausible, but we can instead imagine a system with weights lying on a hypersphere in the positive quadrant of space but with additional inhibition that results in the cumulative effects of some input lines being effectively negative. This can be achieved in a network by using positive input vectors, positive synaptic weight vectors, and thresholding the output neurons at their mean activation. A large competitive network of this general nature does categorize well, and has been described more fully elsewhere (Bennett 1990). In a large network with inhibitory feedback neurons, and principal cells with thresholds, the network could achieve at least in part an approximation to this type of thresholding useful in large competitive networks.

A second way in which nets with positive-only values of the elements could operate is by making the input vectors sparse and initializing the weight vectors to be sparse, or to have a reduced contact probability. (A measure a of neuronal population sparseness is defined (as before) in equation B.28:

$$a = \frac{(\sum\limits_{i=1}^{N} y_i/N)^2}{\sum\limits_{i=1}^{N} y_i^2/N} \tag{B.28}$$

where y_i is the firing rate of the ith neuron in the set of N neurons.) For relatively small net sizes simulated ($N = 100$) with patterns with a sparseness a of, for example, 0.1 or 0.2, learning onto the average vector can be avoided. However, as the net size increases, the sparseness required does become very low. In large nets, a greatly reduced contact probability between neurons (many synapses kept identically zero) would prevent learning of the average vector, thus allowing categorization to occur (see Section 7.4). Reduced contact probability will, however, prevent complete alignment of synapses with patterns, so that the performance of the network will be affected.

B.5 Continuous attractor networks

B.5.1 Introduction

Single-cell recording studies have shown that some neurons represent the current position along a continuous physical dimension or space even when no inputs are available, for example in darkness (see Chapter 24). Examples include neurons that represent the positions of the eyes (i.e. eye direction with respect to the head), the place where the animal is looking in space, head direction, and the place where the animal is located. In particular, examples of such classes of cells include head direction cells in rats (Ranck 1985, Taube, Muller and Ranck 1990a, Taube, Goodridge, Golob, Dudchenko and Stackman 1996, Muller, Ranck and Taube 1996) and primates (Robertson, Rolls, Georges-François and Panzeri 1999), which respond maximally when the animal's head is facing in a particular preferred direction; place cells in rats (O'Keefe and Dostrovsky 1971, McNaughton, Barnes and O'Keefe 1983, O'Keefe 1984, Muller, Kubie, Bostock, Taube and Quirk 1991, Markus, Qin, Leonard, Skaggs, McNaughton and Barnes 1995) that fire maximally when the animal is in a particular location; and spatial view cells in primates that respond when the monkey is looking towards a particular location in space (Rolls, Robertson and Georges-François 1997a, Georges-François, Rolls and Robertson 1999, Robertson, Rolls and Georges-François 1998). In the parietal cortex there are many spatial representations, in several different coordinate frames (see Chapter 4 of Rolls and Deco (2002) and Andersen, Batista, Snyder, Buneo and Cohen (2000)), and they have some capability to remain active during memory periods when the stimulus is no longer present. Even more than this, the dorsolateral prefrontal cortex networks to which the parietal networks project have the capability to maintain spatial representations active for many seconds or minutes during short-term memory tasks, when the stimulus is no longer present (see Section 4.3.1). In this section, we describe how such networks representing continuous physical space could operate. The locations of such spatial networks in the brain are the parietal areas, the prefrontal areas that implement short-term spatial memory and receive from the parietal cortex (see Section 4.3.1), and the hippocampal system which combines information about objects from the inferior temporal visual cortex with spatial information (see Chapter 24).

A class of network that can maintain the firing of its neurons to represent any location along a continuous physical dimension such as spatial position, head direction, etc. is a 'Continuous Attractor' neural network (CANN). It uses excitatory recurrent collateral connections between the neurons to reflect the distance between the neurons in the state space of the animal (e.g. head direction space). These networks can maintain the bubble of neural activity constant for long periods wherever it is started to represent the current state (head direction, position, etc) of the animal, and are likely to be involved in many aspects of spatial processing and memory, including spatial vision. Global inhibition is used to keep the number of neurons in a bubble or packet of actively firing neurons relatively constant, and to help to ensure that there is only one activity packet. Continuous attractor networks can be thought of as very similar to autoassociation or discrete attractor networks (described in Section B.3), and have the same architecture, as illustrated in Fig. B.25. The main difference is that the patterns stored in a CANN are continuous patterns, with each neuron having broadly tuned firing which decreases with for example a Gaussian function as the distance from the optimal firing location of the cell is varied, and with different neurons having tuning that overlaps throughout the space. Such

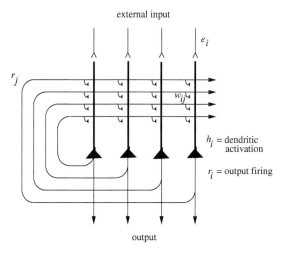

external input

e_i

r_j

w_{ij}

h_i = dendritic
activation

r_i = output firing

output

Fig. B.25 The architecture of a continuous attractor neural network (CANN).

tuning is illustrated in Fig. 24.16. For comparison, the autoassociation networks described in Section B.3 have discrete (separate) patterns (each pattern implemented by the firing of a particular subset of the neurons), with no continuous distribution of the patterns throughout the space (see Fig. 24.16). A consequent difference is that the CANN can maintain its firing at any location in the trained continuous space, whereas a discrete attractor or autoassociation network moves its population of active neurons towards one of the previously learned attractor states, and thus implements the recall of a particular previously learned pattern from an incomplete or noisy (distorted) version of one of the previously learned patterns. The energy landscape of a discrete attractor network (see equation B.12) has separate energy minima, each one of which corresponds to a learned pattern, whereas the energy landscape of a continuous attractor network is flat, so that the activity packet remains stable with continuous firing wherever it is started in the state space. (The state space refers to the set of possible spatial states of the animal in its environment, e.g. the set of possible head directions.)

In Section B.5.2, we first describe the operation and properties of continuous attractor networks, which have been studied by for example Amari (1977), Zhang (1996), and Taylor (1999), and then, following Stringer, Trappenberg, Rolls and De Araujo (2002b), address four key issues about the biological application of continuous attractor network models.

One key issue in such continuous attractor neural networks is how the synaptic strengths between the neurons in the continuous attractor network could be learned in biological systems (Section B.5.3).

A second key issue in such continuous attractor neural networks is how the bubble of neuronal firing representing one location in the continuous state space should be updated based on non-visual cues to represent a new location in state space (Section B.5.5). This is essentially the problem of path integration: how a system that represents a memory of where the agent is in physical space could be updated based on idiothetic (self-motion) cues such as vestibular cues (which might represent a head velocity signal), or proprioceptive cues (which might update a representation of place based on movements being made in the space, during for example walking in the dark).

A third key issue is how stability in the bubble of activity representing the current location can be maintained without much drift in darkness, when it is operating as a memory system (Section B.5.6).

A fourth key issue is considered in Section B.5.8 in which we describe networks that store both continuous patterns and discrete patterns (see Fig. 24.16), which can be used to store for example the location in (continuous, physical) space where an object (a discrete item) is present.

B.5.2   The generic model of a continuous attractor network

The generic model of a continuous attractor is as follows. (The model is described in the context of head direction cells, which represent the head direction of rats (Taube et al. 1996, Muller et al. 1996) and macaques (Robertson, Rolls, Georges-François and Panzeri 1999), and can be reset by visual inputs after gradual drift in darkness.) The model is a recurrent attractor network with global inhibition. It is different from a Hopfield attractor network primarily in that there are no discrete attractors formed by associative learning of discrete patterns. Instead there is a set of neurons that are connected to each other by synaptic weights w_{ij} that are a simple function, for example Gaussian, of the distance between the states of the agent in the physical world (e.g. head directions) represented by the neurons. Neurons that represent similar states (locations in the state space) of the agent in the physical world have strong synaptic connections, which can be set up by an associative learning rule, as described in Section B.5.3. The network updates its firing rates by the following 'leaky-integrator' dynamical equations. The continuously changing activation h_i^{HD} of each head direction cell i is governed by the equation

$$\tau \frac{\mathrm{d}h_i^{\mathrm{HD}}(t)}{\mathrm{d}t} = -h_i^{\mathrm{HD}}(t) + \frac{\phi_0}{C^{\mathrm{HD}}} \sum_j (w_{ij} - w^{\mathrm{inh}}) r_j^{\mathrm{HD}}(t) + I_i^V, \qquad (\mathrm{B}.29)$$

where r_j^{HD} is the firing rate of head direction cell j, w_{ij} is the excitatory (positive) synaptic weight from head direction cell j to cell i, w^{inh} is a global constant describing the effect of inhibitory interneurons, and τ is the time constant of the system[38]. The term $-h_i^{\mathrm{HD}}(t)$ indicates the amount by which the activation decays (in the leaky integrator neuron) at time t. (The network is updated in a typical simulation at much smaller timesteps than the time constant of the system, τ.) The next term in equation B.29 is the input from other neurons in the network r_j^{HD} weighted by the recurrent collateral synaptic connections w_{ij} (scaled by a constant ϕ_0 and C^{HD} which is the number of synaptic connections received by each head direction cell from other head direction cells in the continuous attractor). The term I_i^V represents a visual input to head direction cell i. Each term I_i^V is set to have a Gaussian response profile in most continuous attractor networks, and this sets the firing of the cells in the continuous attractor to have Gaussian response profiles as a function of where the agent is located in the state space (see e.g. Fig. 24.16 on page 496), but the Gaussian assumption is not crucial. (It is known that the firing rates of head direction cells in both rats (Taube, Goodridge, Golob, Dudchenko and Stackman 1996, Muller, Ranck and Taube 1996) and macaques (Robertson, Rolls, Georges-François and Panzeri 1999) is approximately Gaussian.) When the agent is operating without visual input, in memory mode, then the term I_i^V is set to zero. The firing rate r_i^{HD} of cell i is determined from the activation h_i^{HD} and the sigmoid function

$$r_i^{\mathrm{HD}}(t) = \frac{1}{1 + e^{-2\beta(h_i^{\mathrm{HD}}(t) - \alpha)}}, \qquad (\mathrm{B}.30)$$

where α and β are the sigmoid threshold and slope, respectively.

---

[38] Note that for this section, we use r rather than y to refer to the firing rates of the neurons in the network, remembering that, because this is a recurrently connected network (see Fig. B.13), the output from a neuron y_i might be the input x_j to another neuron.

B.5.3 Learning the synaptic strengths between the neurons that implement a continuous attractor network

So far we have said that the neurons in the continuous attractor network are connected to each other by synaptic weights w_{ij} that are a simple function, for example Gaussian, of the distance between the states of the agent in the physical world (e.g. head directions, spatial views etc) represented by the neurons. In many simulations, the weights are set by formula to have weights with these appropriate Gaussian values. However, Stringer, Trappenberg, Rolls and De Araujo (2002b) showed how the appropriate weights could be set up by learning. They started with the fact that since the neurons have broad tuning that may be Gaussian in shape, nearby neurons in the state space will have overlapping spatial fields, and will thus be co-active to a degree that depends on the distance between them. They postulated that therefore the synaptic weights could be set up by associative learning based on the co-activity of the neurons produced by external stimuli as the animal moved in the state space. For example, head direction cells are forced to fire during learning by visual cues in the environment that produce Gaussian firing as a function of head direction from an optimal head direction for each cell. The learning rule is simply that the weights w_{ij} from head direction cell j with firing rate r_j^{HD} to head direction cell i with firing rate r_i^{HD} are updated according to an associative (Hebb) rule

$$\delta w_{ij} = k r_i^{HD} r_j^{HD} \tag{B.31}$$

where δw_{ij} is the change of synaptic weight and k is the learning rate constant. During the learning phase, the firing rate r_i^{HD} of each head direction cell i might be the following Gaussian function of the displacement of the head from the optimal firing direction of the cell

$$r_i^{HD} = e^{-s_{HD}^2/2\sigma_{HD}^2}, \tag{B.32}$$

where s_{HD} is the difference between the actual head direction x (in degrees) of the agent and the optimal head direction x_i for head direction cell i, and σ_{HD} is the standard deviation.

Stringer, Trappenberg, Rolls and De Araujo (2002b) showed that after training at all head directions, the synaptic connections develop strengths that are an almost Gaussian function of the distance between the cells in head direction space, as shown in Fig. B.26 (left). Interestingly if a non-linearity is introduced into the learning rule that mimics the properties of NMDA receptors by allowing the synapses to modify only after strong postsynaptic firing is present, then the synaptic strengths are still close to a Gaussian function of the distance between the connected cells in head direction space (see Fig. B.26, left). They showed that after training, the continuous attractor network can support stable activity packets in the absence of visual inputs (see Fig. B.26, right) provided that global inhibition is used to prevent all the neurons becoming activated. (The exact stability conditions for such networks have been analyzed by Amari (1977)). Thus Stringer, Trappenberg, Rolls and De Araujo (2002b) demonstrated biologically plausible mechanisms for training the synaptic weights in a continuous attractor using a biologically plausible local learning rule.

Stringer, Trappenberg, Rolls and De Araujo (2002b) went on to show that if there was a short term memory trace built into the operation of the learning rule, then this could help to produce smooth weights in the continuous attractor if only incomplete training was available, that is if the weights were trained at only a few locations. The same rule can take advantage in training the synaptic weights of the temporal probability distributions of firing when they happen to reflect spatial proximity. For example, for head direction cells the agent will necessarily move through similar head directions before reaching quite different head directions, and so the temporal proximity with which the cells fire can be used to set up the appropriate synaptic weights. This new proposal for training continuous

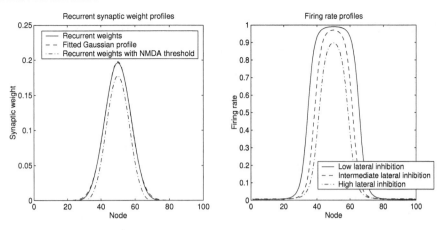

Fig. B.26 Training the weights in a continuous attractor network with an associative rule (equation B.31). Left: the trained recurrent synaptic weights from head direction cell 50 to the other head direction cells in the network arranged in head direction space (solid curve). The dashed line shows a Gaussian curve fitted to the weights shown in the solid curve. The dash-dot curve shows the recurrent synaptic weights trained with rule equation (B.31), but with a non-linearity introduced that mimics the properties of NMDA receptors by allowing the synapses to modify only after strong postsynaptic firing is present. Right: the stable firing rate profiles forming an activity packet in the continuous attractor network during the testing phase when the training (visual) inputs are no longer present. The firing rates are shown after the network has been initially stimulated by visual input to initialize an activity packet, and then allowed to settle to a stable activity profile without visual input. The three graphs show the firing rates for low, intermediate and high values of the lateral inhibition parameter w^{inh}. For both left and right plots, the 100 head direction cells are arranged according to where they fire maximally in the head direction space of the agent when visual cues are available. (After Stringer, Trappenberg, Rolls and de Araujo 2002.)

attractor networks can also help to produce broadly tuned spatial cells even if the driving (e.g. visual) input (I_i^V in equation B.29) during training produces rather narrowly tuned neuronal responses. The learning rule with such temporal properties is a memory trace learning rule that strengthens synaptic connections between neurons, based on the temporal probability distribution of the firing. There are many versions of such rules (Rolls and Milward 2000, Rolls and Stringer 2001a), which are described more fully in Chapter 25, but a simple one that works adequately is

$$\delta w_{ij} = k\bar{r}_i^{HD}\bar{r}_j^{HD} \tag{B.33}$$

where δw_{ij} is the change of synaptic weight, and $\bar{r}^{HD}$ is a local temporal average or trace value of the firing rate of a head direction cell given by

$$\bar{r}^{HD}(t+\delta t) = (1-\eta)r^{HD}(t+\delta t) + \eta\bar{r}^{HD}(t) \tag{B.34}$$

where η is a parameter set in the interval [0,1] which determines the contribution of the current firing and the previous trace. For $\eta = 0$ the trace rule (B.33) becomes the standard Hebb rule (B.31), while for $\eta > 0$ learning rule (B.33) operates to associate together patterns of activity that occur close together in time. The rule might allow temporal associations to influence the synaptic weights that are learned over times in the order of 1 s. The memory trace required for operation of this rule might be no more complicated than the continuing firing that is an inherent property of attractor networks, but it could also be implemented by a number of biophysical mechanisms, discussed in Chapter 25. Finally, we note that some long-term depression (LTD) in the learning rule could help to maintain the weights of different neurons equally potent (see Section B.4.9.2 and equation B.22), and could compensate for irregularity during training in which the agent might be trained much more in some than other locations in the space (see Stringer, Trappenberg, Rolls and De Araujo (2002b)).

B.5.4 The capacity of a continuous attractor network: multiple charts and packets

The capacity of a continuous attractor network can be approached on the following bases. First, as there are no discrete attractor states, but instead a continuous physical space is being represented, some concept of spatial resolution must be brought to bear, that is the number of different positions in the space that can be represented. Second, the number of connections per neuron in the continuous attractor will directly influence the number of different spatial positions (locations in the state space) that can be represented. Third, the sparseness of the representation can be thought of as influencing the number of different spatial locations (in the continuous state space) that can be represented, in a way analogous to that described for discrete attractor networks in equation B.14 (Battaglia and Treves 1998b). That is, if the tuning of the neurons is very broad, then fewer locations in the state space may be represented. Fourth, and very interestingly, if representations of different continuous state spaces, for example maps or charts of different environments, are stored in the same network, there may be little cost of adding extra maps or charts. The reason for this is that the large part of the interference between the different memories stored in such a network arises from the correlations between the different positions in any one map, which are typically relatively high because quite broad tuning of individual cells is common. In contrast, there are in general low correlations between the representations of places in different maps or charts, and therefore many different maps can be simultaneously stored in a continuous attractor network (Battaglia and Treves 1998b).

For a similar reason, it is even possible to have the activity packets that operate in different spaces simultaneously active in a single continuous attractor network of neurons, and to move independently of each other in their respective spaces or charts (Stringer, Rolls and Trappenberg 2004).

B.5.5 Continuous attractor models: path integration

So far, we have considered how spatial representations could be stored in continuous attractor networks, and how the activity can be maintained at any location in the state space in a form of short-term memory when the external (e.g. visual) input is removed. However, many networks with spatial representations in the brain can be updated by internal, self-motion (i.e. idiothetic), cues even when there is no external (e.g. visual) input. Examples are head direction cells in the presubiculum of rats and macaques, place cells in the rat hippocampus, and spatial view cells in the primate hippocampus (see Chapter 24). The major question arises about how such idiothetic inputs could drive the activity packet in a continuous attractor network and, in particular, how such a system could be set up biologically by self-organizing learning.

One approach to simulating the movement of an activity packet produced by idiothetic cues (which is a form of path integration whereby the current location is calculated from recent movements) is to employ a look-up table that stores (taking head direction cells as an example), for every possible head direction and head rotational velocity input generated by the vestibular system, the corresponding new head direction (Samsonovich and McNaughton 1997). Another approach involves modulating the strengths of the recurrent synaptic weights in the continuous attractor on one but not the other side of a currently represented position, so that the stable position of the packet of activity, which requires symmetric connections in different directions from each node, is lost, and the packet moves in the direction of the temporarily increased

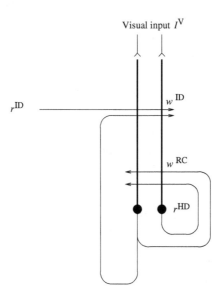

Fig. B.27 General network architecture for a one-dimensional continuous attractor model of head direction cells which can be updated by idiothetic inputs produced by head rotation cell firing r^{ID}. The head direction cell firing is r^{HD}, the continuous attractor synaptic weights are w^{RC}, the idiothetic synaptic weights are w^{ID}, and the external visual input is I^{V}.

weights, although no possible biological implementation was proposed of how the appropriate dynamic synaptic weight changes might be achieved (Zhang 1996). Another mechanism (for head direction cells) (Skaggs, Knierim, Kudrimoti and McNaughton 1995) relies on a set of cells, termed (head) rotation cells, which are co-activated by head direction cells and vestibular cells and drive the activity of the attractor network by anatomically distinct connections for clockwise and counter-clockwise rotation cells, in what is effectively a look-up table. However, no proposal was made about how this could be achieved by a biologically plausible learning process, and this has been the case until recently for most approaches to path integration in continuous attractor networks, which rely heavily on rather artificial pre-set synaptic connectivities.

Stringer, Trappenberg, Rolls and De Araujo (2002b) introduced a proposal with more biological plausibility about how the synaptic connections from idiothetic inputs to a continuous attractor network can be learned by a self-organizing learning process. The essence of the hypothesis is described with Fig. B.27. The continuous attractor synaptic weights w^{RC} are set up under the influence of the external visual inputs I^{V} as described in Section B.5.3. At the same time, the idiothetic synaptic weights w^{ID} (in which the ID refers to the fact that they are in this case produced by idiothetic inputs, produced by cells that fire to represent the velocity of clockwise and anticlockwise head rotation), are set up by associating the change of head direction cell firing that has just occurred (detected by a trace memory mechanism described below) with the current firing of the head rotation cells r^{ID}. For example, when the trace memory mechanism incorporated into the idiothetic synapses w^{ID} detects that the head direction cell firing is at a given location (indicated by the firing r^{HD}) and is moving clockwise (produced by the altering visual inputs I^{V}), and there is simultaneous clockwise head rotation cell firing, the synapses w^{ID} learn the association, so that when that rotation cell firing occurs later without visual input, it takes the current head direction firing in the continuous attractor into account, and moves the location of the head direction attractor in the appropriate direction.

For the learning to operate, the idiothetic synapses onto head direction cell i with firing r_i^{HD} need two inputs: the memory traced term from other head direction cells $\bar{r}_j^{HD}$ (given by equation B.34), and the head rotation cell input with firing r_k^{ID}; and the learning rule can be written

$$\delta w_{ijk}^{ID} = \tilde{k} \, r_i^{HD} \, \bar{r}_j^{HD} \, r_k^{ID}, \tag{B.35}$$

where $\tilde{k}$ is the learning rate associated with this type of synaptic connection. The head rotation cell firing (r_k^{ID}) could be as simple as one set of cells that fire for clockwise head rotation (for which k might be 1), and a second set of cells that fire for anticlockwise head rotation (for which k might be 2).

After learning, the firing of the head direction cells would be updated in the dark (when $I_i^V = 0$) by idiothetic head rotation cell firing r_k^{ID} as follows

$$\tau \frac{dh_i^{HD}(t)}{dt} = -h_i^{HD}(t) + \frac{\phi_0}{C^{HD}} \sum_j (w_{ij} - w^{inh}) r_j^{HD}(t) + I_i^V$$

$$+ \phi_1 \left(\frac{1}{C^{HD \times ID}} \sum_{j,k} w_{ijk}^{ID} r_j^{HD} r_k^{ID} \right). \tag{B.36}$$

Equation B.36 is similar to equation B.29, except for the last term, which introduces the effects of the idiothetic synaptic weights w_{ijk}^{ID}, which effectively specify that the current firing of head direction cell i, r_i^{HD}, must be updated by the previously learned combination of the particular head rotation now occurring indicated by r_k^{ID}, and the current head direction indicated by the firings of the other head direction cells r_j^{HD} indexed through j[39]. This makes it clear that the idiothetic synapses operate using combinations of inputs, in this case of two inputs. Neurons that sum the effects of such local products are termed Sigma-Pi neurons (see Section A.2.3). Although such synapses are more complicated than the two-term synapses used throughout the rest of this book, such three-term synapses appear to be useful to solve the computational problem of updating representations based on idiothetic inputs in the way described. Synapses that operate according to Sigma-Pi rules might be implemented in the brain by a number of mechanisms described by Koch (1999) (Section 21.1.1), Jonas and Kaczmarek (1999), and Stringer, Trappenberg, Rolls and De Araujo (2002b), including having two inputs close together on a thin dendrite, so that local synaptic interactions would be emphasized.

Simulations demonstrating the operation of this self-organizing learning to produce movement of the location being represented in a continuous attractor network were described by Stringer, Trappenberg, Rolls and De Araujo (2002b), and one example of the operation is shown in Fig. B.28. They also showed that, after training with just one value of the head rotation cell firing, the network showed the desirable property of moving the head direction being represented in the continuous attractor by an amount that was proportional to the value of the head rotation cell firing. Stringer, Trappenberg, Rolls and De Araujo (2002b) also describe a related model of the idiothetic cell update of the location represented in a continuous attractor, in which the rotation cell firing directly modulates in a multiplicative way the strength of the recurrent connections in the continuous attractor in such a way that clockwise rotation cells modulate the strength of the synaptic connections in the clockwise direction in the continuous attractor, and vice versa.

---

[39]The term $\phi_1/C^{HD \times ID}$ is a scaling factor that reflects the number $C^{HD \times ID}$ of inputs to these synapses, and enables the overall magnitude of the idiothetic input to each head direction cell to remain approximately the same as the number of idiothetic connections received by each head direction cell is varied.

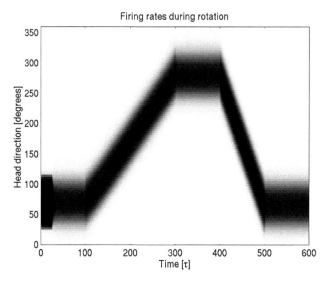

Fig. B.28 Idiothetic update of the location represented in a continuous attractor network. The firing rate of the cells with optima at different head directions (organized according to head direction on the ordinate) is shown by the blackness of the plot, as a function of time. The activity packet was initialized to a head direction of 75 degrees, and the packet was allowed to settle without visual input. For timestep=0 to 100 there was no rotation cell input, and the activity packet in the continuous attractor remained stable at 75 degrees. For timestep=100 to 300 the clockwise rotation cells were active with a firing rate of 0.15 to represent a moderate angular velocity, and the activity packet moved clockwise. For timestep=300 to 400 there was no rotation cell firing, and the activity packet immediately stopped, and remained still. For timestep=400 to 500 the anti-clockwise rotation cells had a high firing rate of 0.3 to represent a high velocity, and the activity packet moved anticlockwise with a greater velocity. For timestep=500 to 600 there was no rotation cell firing, and the activity packet immediately stopped.

It should be emphasized that although the cells are organized in Fig. B.28 according to the spatial position being represented, there is no need for cells in continuous attractors that represent nearby locations in the state space to be close together, as the distance in the state space between any two neurons is represented by the strength of the connection between them, not by where the neurons are physically located. This enables continuous attractor networks to represent spaces with arbitrary topologies, as the topology is represented in the connection strengths (Stringer, Trappenberg, Rolls and De Araujo 2002b, Stringer, Rolls, Trappenberg and De Araujo 2002a, Stringer, Rolls and Trappenberg 2005, Stringer and Rolls 2002). Indeed, it is this that enables many different charts each with its own topology to be represented in a single continuous attractor network (Battaglia and Treves 1998b).

In the network described so far, self-organization occurs, but one set of synapses is Sigma-Pi. We have gone on to show that the Sigma-Pi synapses are not necessary, and can be replaced by a competitive network that learns to respond to combinations of the spatial position and the idiothetic velocity, as illustrated in Fig. B.29 (Stringer and Rolls 2006).

B.5.6 Stabilization of the activity packet within the continuous attractor network when the agent is stationary

With irregular learning conditions (in which identical training with high precision of every node cannot be guaranteed), the recurrent synaptic weights between nodes in the continuous attractor will not be of the perfectly regular and symmetric form normally required in a

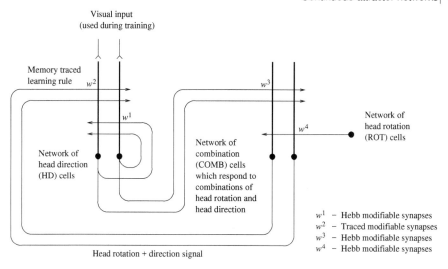

Visual input
(used during training)

Memory traced
learning rule w^2 w^3

w^1 Network of
                                                        head rotation
                                          w^4 (ROT) cells

Network of   Network of
head direction   combination
(HD) cells   (COMB) cells
                                    which respond to
                                    combinations of
                                    head rotation and
                                    head direction w^1 – Hebb modifiable synapses
                                                            w^2 – Traced modifiable synapses
                                                            w^3 – Hebb modifiable synapses
Head rotation + direction signal w^4 – Hebb modifiable synapses

Fig. B.29 Network architecture for a two-layer self-organizing neural network model of the head direction system. The network architecture contains a layer of head direction (HD) cells representing the head direction of the agent, a layer of combination (COMB) cells representing a combination of head direction and rotational velocity, and a layer of rotation (ROT) cells which become active when the agent rotates. There are four types of synaptic connection in the network, which operate as follows. The w^1_{ij} synapses are Hebb-modifiable recurrent connections between head direction cells. These connections help to support stable packets of activity within this continuous attractor layer of head direction cells in the absence of visual input. The combination cells receive inputs from the head direction cells through the Hebb-modifiable w^3_{ij} synapses, and inputs from the rotation cells through the Hebb-modifiable w^4_{ij} synapses. These synaptic inputs encourage combination cells by competitive learning to respond to particular combinations of head direction and rotational velocity. In particular, the combination cells only become significantly active when the agent is rotating. The head direction cells receive inputs from the combination cells through the w^2_{ij} synapses. The $w^2_{i,j}$ synapses are trained using a 'trace' learning rule, which incorporates a temporal trace of recent combination cell activity. This rule introduces an asymmetry into the w^2_{ij} weights, which plays an important role in shifting the activity packet through the layer of head direction cells during path integration in the dark. (After Stringer and Rolls 2006.)

continuous attractor neural network. This can lead to drift of the activity packet within the continuous attractor network of e.g. head direction cells when no visual cues are present, even when the agent is not moving. This drift is a common property of the short-term memories of spatial position implemented in the brain, which emphasizes the computational problems that can arise in continuous attractor networks if the weights between nodes are not balanced in different directions in the space being represented. An approach to stabilizing the activity packet when it should not be drifting in real nervous systems, which does help to minimize the drift that can occur, is now described.

The activity packet may be stabilized within a continuous attractor network, when the agent is stationary and the network is operating in memory mode without an external stabilizing input, by enhancing the firing of those cells that are already firing. In biological systems this might be achieved through mechanisms for short-term synaptic enhancement (Koch 1999). Another way is to take advantage of the non-linearity of the activation function of neurons with NMDA receptors, which only contribute to neuronal firing once the neuron is sufficiently depolarized (Wang 1999). The effect is to enhance the firing of neurons that are already reasonably well activated. The effect has been utilized in a model of a network with recurrent excitatory synapses that can maintain active an arbitrary set of neurons that are initially sufficiently strongly activated by an external stimulus (see Lisman, Fellous and Wang (1998), and, for a discussion on whether these networks could be used to implement short-term memories, see Kesner and Rolls (2001)).

We have incorporated this non-linearity into a model of a head direction continuous attractor network by adjusting the sigmoid threshold α_i (see equation B.30) for each head direction cell i as follows (Stringer, Trappenberg, Rolls and De Araujo 2002b). If the head direction cell firing rate r_i^{HD} is lower than a threshold value, γ, then the sigmoid threshold α_i is set to a relatively high value α^{high}. Otherwise, if the head direction cell firing rate r_i^{HD} is greater than or equal to the threshold value, γ, then the sigmoid threshold α_i is set to a relatively low value α^{low}. It was shown that this procedure has the effect of enhancing the current position of the activity packet within the continuous attractor network, and so prevents the activity packet drifting erratically due to the noise in the recurrent synaptic weights produced for example by irregular learning. An advantage of using the non-linearity in the activation function of a neuron (produced for example by the operation of NMDA receptors) is that this tends to enable packets of activity to be kept active without drift even when the packet is not in one of the energy minima that can result from irregular learning (or from diluted connectivity in the continuous attractor as described below). Thus use of this non-linearity increases the number of locations in the continuous physical state space at which a stable activity packet can be maintained (Stringer, Trappenberg, Rolls and De Araujo 2002b).

The same process might help to stabilize the activity packet against drift caused by the probabilistic spiking of neurons in the network (cf. Chapter 5.6).

B.5.7 Continuous attractor networks in two or more dimensions

Some types of spatial representation used by the brain are of spaces that exist in two or more dimensions. Examples are the two- (or three-) dimensional space representing where one is looking at in a spatial scene. Another is the two- (or three-) dimensional space representing where one is located. It is possible to extend continuous attractor networks to operate in higher dimensional spaces than the one-dimensional spaces considered so far (Taylor 1999, Stringer, Rolls, Trappenberg and De Araujo 2002a). Indeed, it is also possible to extend the analyses of how idiothetic inputs could be used to update two-dimensional state spaces, such as the locations represented by place cells in rats (Stringer, Rolls, Trappenberg and De Araujo 2002a) and the location at which one is looking represented by primate spatial view cells (Stringer, Rolls and Trappenberg 2005, Stringer and Rolls 2002). Interestingly, the number of terms in the synapses implementing idiothetic update do not need to increase beyond three (as in Sigma-Pi synapses) even when higher dimensional state spaces are being considered (Stringer, Rolls, Trappenberg and De Araujo 2002a). Also interestingly, a continuous attractor network can in fact represent the properties of very high dimensional spaces, because the properties of the spaces are captured by the connections between the neurons of the continuous attractor, and these connections are of course, as in the world of discrete attractor networks, capable of representing high dimensional spaces (Stringer, Rolls, Trappenberg and De Araujo 2002a). With these approaches, continuous attractor networks have been developed of the two-dimensional representation of rat hippocampal place cells with idiothetic update by movements in the environment (Stringer, Rolls, Trappenberg and De Araujo 2002a), and of primate hippocampal spatial view cells with idiothetic update by eye and head movements (Stringer, Rolls and Trappenberg 2005, Rolls and Stringer 2005, Stringer and Rolls 2002). Continuous attractor models with some similar properties have also been applied to understanding motor control, for example the generation of a continuous movement in space (Stringer and Rolls 2007, Stringer, Rolls and Taylor 2007a).

B.5.8 Mixed continuous and discrete attractor networks

It has now been shown that attractor networks can store both continuous patterns and discrete patterns, and can thus be used to store for example the location in (continuous, physical) space where an object (a discrete item) is present (see Fig. 24.16 and Rolls, Stringer and Trappenberg (2002)). In this network, when events are stored that have both discrete (object) and continuous (spatial) aspects, then the whole place can be retrieved later by the object, and the object can be retrieved by using the place as a retrieval cue. Such networks are likely to be present in parts of the brain that receive and combine inputs both from systems that contain representations of continuous (physical) space, and from brain systems that contain representations of discrete objects, such as the inferior temporal visual cortex. One such brain system is the hippocampus, which appears to combine and store such representations in a mixed attractor network in the CA3 region, which thus is able to implement episodic memories which typically have a spatial component, for example where an item such as a key is located (see Chapter 24). This network thus shows that in brain regions where the spatial and object processing streams are brought together, then a single network can represent and learn associations between both types of input. Indeed, in brain regions such as the hippocampal system, it is essential that the spatial and object processing streams are brought together in a single network, for it is only when both types of information are in the same network that spatial information can be retrieved from object information, and vice versa, which is a fundamental property of episodic memory (see Chapter 24). It may also be the case that in the prefrontal cortex, attractor networks can store both spatial and discrete (e.g. object-based) types of information in short-term memory (see Section 4.3.1).

B.6 Network dynamics: the integrate-and-fire approach

The concept that attractor (autoassociation) networks can operate very rapidly if implemented with neurons that operate dynamically in continuous time was introduced in Section B.3.3.5. The result described was that the principal factor affecting the speed of retrieval is the time constant of the synapses between the neurons that form the attractor ((Treves 1993, Rolls and Treves 1998, Battaglia and Treves 1998a, Panzeri, Rolls, Battaglia and Lavis 2001). This was shown analytically by Treves (1993), and described by Rolls and Treves (1998) Appendix 5. We now describe in more detail the approaches that produce these results, and the actual results found on the speed of processing.

The networks described so far in this chapter, and analyzed in Appendices 3 and 4 of Rolls and Treves (1998), were described in terms of the steady-state activation of networks of neuron-like units. Those may be referred to as 'static' properties, in the sense that they do not involve the time dimension. In order to address 'dynamical' questions, the time dimension has to be reintroduced into the formal models used, and the adequacy of the models themselves has to be reconsidered in view of the specific properties to be discussed.

Consider for example a real network whose operation has been described by an autoassociative formal model that acquires, with learning, a given attractor structure. How does the state of the network approach, in real time during a retrieval operation, one of those attractors? How long does it take? How does the amount of information that can be read off the network's activity evolve with time? Also, which of the potential steady states is indeed a stable state that can be reached asymptotically by the net? How is the stability of different states modulated by external agents? These are examples of dynamical properties, which to be studied require the use of models endowed with some dynamics.

B.6.1 From discrete to continuous time

Already at the level of simple models in which each unit is described by an input–output relation, one may introduce equally simple 'dynamical' rules, in order both to fully specify the model, and to simulate it on computers. These rules are generally formulated in terms of 'updatings': time is considered to be discrete, a succession of time steps, and at each time step the output of one or more of the units is set, or updated, to the value corresponding to its input variable. The input variable may reflect the outputs of other units in the net as updated at the previous time step or, if delays are considered, the outputs as they were at a prescribed number of time steps in the past. If all units in the net are updated together, the dynamics is referred to as parallel; if instead only one unit is updated at each time step, the dynamics is sequential. (One main difference between the Hopfield (1982) model of an autoassociator and a similar model considered earlier by Little (1974) is that the latter was based on parallel rather than sequential dynamics.) Many intermediate possibilities obviously exist, involving the updating of groups of units at a time. The order in which sequential updatings are performed may for instance be chosen at random at the beginning and then left the same in successive cycles across all units in the net; or it may be chosen anew at each cycle; yet a third alternative is to select at each time step a unit, at random, with the possibility that a particular unit may be selected several times before some of the other ones are ever updated. The updating may also be made probabilistic, with the output being set to its new value only with a certain probability, and otherwise remaining at the current value.

Variants of these dynamical rules have been used for decades in the analysis and computer simulation of physical systems in statistical mechanics (and field theory). They can reproduce in simple but effective ways the stochastic nature of transitions among discrete quantum states, and they have been subsequently considered appropriate also in the simulation of neural network models in which units have outputs that take discrete values, implying that a change from one value to another can only occur in a sudden jump. To some extent, different rules are equivalent, in that they lead, in the evolution of the activity of the net along successive steps and cycles, to the same set of possible steady states. For example, it is easy to realize that when no delays are introduced, states that are stable under parallel updating are also stable under sequential updating. The reverse is not necessarily true, but on the other hand states that are stable when updating one unit at a time are stable irrespective of the updating order. Therefore, static properties, which can be deduced from an analysis of stable states, are to some extent robust against differences in the details of the dynamics assigned to the model. (This is a reason for using these dynamical rules in the study of the thermodynamics of physical systems.) Such rules, however, bear no relation to the actual dynamical processes by which the activity of real neurons evolves in time, and are therefore inadequate for the discussion of dynamical issues in neural networks.

A first step towards realism in the dynamics is the substitution of discrete time with continuous time. This somewhat parallels the substitution of the discrete output variables of the most rudimentary models with continuous variables representing firing rates. Although continuous output variables may evolve also in discrete time, and as far as static properties are concerned differences are minimal, with the move from discrete to continuous outputs the main raison d'etre for a dynamics in terms of sudden updatings ceases to exist, since continuous variables can change continuously in continuous time. A paradox arises immediately, however, if a continuous time dynamics is assigned to firing rates. The paradox is that firing rates, although in principle continuous if computed with a generic time-kernel, tend to vary in jumps as new spikes – essentially discrete events – come to be included in the kernel. To avoid this paradox, a continuous time dynamics can be assigned, instead, to instantaneous continuous variables such as membrane potentials. Hopfield (1984), among others, has introduced a model of an autoassociator in which the output variables represent membrane potentials and evolve continuously in time, and has suggested that under certain conditions the stable states attainable by such a network are essentially the same as for a network of binary units evolving in discrete time. If neurons in the central nervous system communicated with each other via the transmission of graded membrane potentials, as they do in some peripheral and invertebrate neural systems, this model could be an excellent starting point. The fact that, centrally, transmission is primarily via the emission of discrete spikes makes a model based on membrane potentials as output variables inadequate to correctly represent spiking dynamics.

B.6.2 Continuous dynamics with discontinuities

In principle, a solution would be to keep the membrane potential as the basic dynamical variable, evolving in continuous time, and to use as the output variable the spike emission times, as determined by the rapid variation in membrane potential corresponding to each spike. A point-like neuron can generate spikes by altering the membrane potential V according to continuous equations of the Hodgkin–Huxley type:

$$C\frac{dV}{dt} = g_0(V_{\text{rest}} - V) + g_{\text{Na}}mh^3(V_{\text{Na}} - V) + g_{\text{K}}n^4(V_{\text{K}} - V) + I \tag{B.37}$$

$$\tau_m\frac{dm}{dt} = m_\infty(V) - m \tag{B.38}$$

$$\tau_h\frac{dh}{dt} = h_\infty(V) - h \tag{B.39}$$

$$\tau_n\frac{dn}{dt} = n_\infty(V) - n. \tag{B.40}$$

The changes in the membrane potential, driven by the input current I, interact with the opening and closing of intrinsic conductances (here a sodium conductance, whose channels are gated by the 'particles' m and h, and a potassium conductance, whose channels are gated by n; Hodgkin and Huxley (1952)). These equations provide an effective description, phenomenological but broadly based on physical principles, of the conductance changes underlying action potentials, and they are treated in any standard neurobiology text.

From the point of view of formal models of neural networks, this level of description is too complicated to be the basis for an analytic understanding of the operation of networks, and it must be simplified. The most widely used simplification is the so-called integrate-and-fire model (see for example MacGregor (1987) and Brunel and Wang (2001)), which is legitimized by the observation that (sodium) action potentials are typically brief and self-similar events. If, in particular, the only relevant variable associated with the spike is its time of emission (at the soma, or axon hillock), which essentially coincides with the time the potential V reaches a certain threshold level V_{thr}, then the conductance changes underlying the rest of the spike can be omitted from the description, and substituted with the ad hoc prescription that (i) a spike is emitted, with its effect on receiving units and on the unit itself; and (ii) after a brief time corresponding to the duration of the spike plus a refractory period, the membrane potential is reset and resumes its integration of the input current I. After a spike the membrane potential is taken to be reset to a value V_{reset}. This type of simplified dynamics of the membrane potentials is thus in continuous time with added discontinuities: continuous in between spikes, with discontinuities occurring at different times for each neuron in a population, every time a neuron emits a spike.

A leaky integrate-and-fire neuron can be modelled as follows. The model describes the depolarization of the membrane potential V (which typically is dynamically changing as a

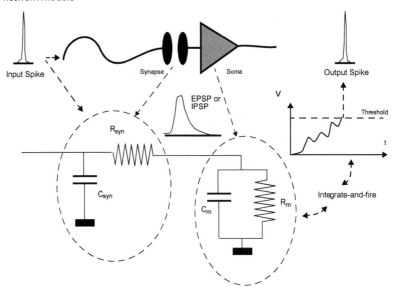

Fig. B.30 Integrate-and-fire neuron. The basic circuit of an integrate-and-fire model consists of the neuron's membrane capacitance C_m in parallel with the membrane's resistance R_m (the reciprocal of the membrane conductance g_m) driven by a synaptic current with a conductance and time constant determined by the synaptic resistance R_{syn} (the reciprocal of the synaptic conductance g_j) and capacitance C_{syn} shown in the Figure. These effects produce excitatory or inhibitory post-synaptic potentials, EPSPs or IPSPs. These potentials are integrated by the cell, and if a threshold V_{thr} is reached a δ-pulse (spike) is fired and transmitted to other neurons, and the membrane potential is reset. (After Deco and Rolls 2003.)

result of the synaptic effects described below between approximately –70 and –50 mV) until threshold V_{thr} (typically –50 mV) is reached when a spike is emitted and the potential is reset to V_{reset} (typically –55 mV). The membrane time constant τ_m is set by the membrane capacitance C_m and the membrane leakage conductance g_m where $\tau_m = C_m/g_m$. V_L denotes the resting potential of the cell, typically –70 mV. Changes in the membrane potential are defined by the following equation

$$C_m \frac{dV(t)}{dt} = g_m\left(V_L - V(t)\right) + \sum_j g_j\left(V_{AMPA} - V(t)\right) + \sum_j g_j\left(V_{GABA} - V(t)\right). \quad \text{(B.41)}$$

The first term on the right of the equation describes how the membrane potential decays back towards the resting potential of the cell depending on how far the cell potential V is from the resting potential V_L, and the membrane leak conductance g_m. The second term on the right represents the excitatory synaptic current that could be injected through AMPA receptors. This is a sum over all AMPA synapses indexed by j on the cell. At each synapse, the current is driven into the cell by the difference between the membrane potential V and the reversal potential V_{AMPA} of the channels opened by AMPA receptors, weighted by the synaptic conductance g_j. This synaptic conductance changes dynamically as a function of the time since a spike reached the synapse, as shown below. Due to their reversal potential V_{AMPA} of typically 0 mV, these currents will tend to depolarize the cell, that is to move the membrane potential towards the firing threshold. The third term on the right represents the inhibitory synaptic current that could be injected through GABA receptors. Due to their reversal potential V_{GABA} of typically –70 mV, these currents will tend to hyperpolarize the cell, that is to move the membrane potential away from the firing threshold. There may be

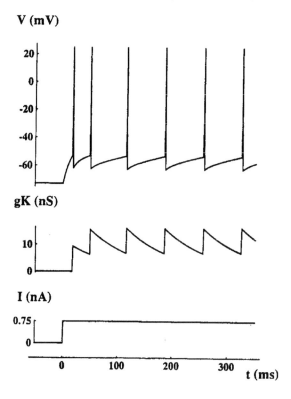

Fig. B.31 Model behaviour of an integrate-and-fire neuron: the membrane potential and adaptation-producing potassium conductance in response to a step of injected current. The spikes were added to the graph by hand, as they do not emerge from the simplified voltage equation. (Reprinted with permission from Network 4, Mean-field analysis of neuronal spike dynamics, Treves, A., 259–284. Copyright ©Taylor and Francis (1993).)

other types of receptor on a cell, for example NMDA receptors, and these operate analogously though with interesting differences, as described in Section B.6.3.

The opening of each synaptic conductance g_j is driven by the arrival of spikes at the presynaptic terminal j, and its closing can often be described as a simple exponential process. A simplified equation for the dynamics of $g_j(t)$ is then

$$\frac{dg_j(t)}{dt} = -\frac{g_j(t)}{\tau} + \Delta g_j \sum_l \delta(t - \Delta t - t_l). \tag{B.42}$$

According to the above equation, each synaptic conductance opens instantaneously by a fixed amount Δg_j a time Δt after the emission of the presynaptic spike at t_l. Δt summarizes delays (axonal, synaptic, and dendritic), and each opening superimposes linearly, without saturating, on previous openings. The value of τ in this equation will be different for AMPA receptors (typically 2–10 ms), NMDA receptors (typically 100 ms), and GABA receptors (typically 10 ms).

In order to model the phenomenon of adaptation in the firing rate, prominent especially with pyramidal cells, it is possible to include a time-varying intrinsic (potassium-like) conductance in the cell membrane (Brown, Gähwiler, Griffith and Halliwell 1990a) (shown as g_K in equations B.43 and B.44). This can be done by specifying that this conductance, which if

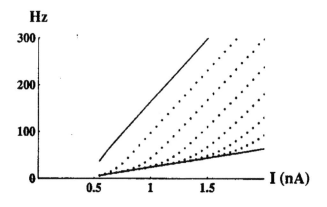

Fig. B.32 Current-to-frequency transduction in a pyramidal cell modelled as an integrate-and-fire neuron. The top solid curve is the firing frequency in the absence of adaptation, $\Delta g_{\mathrm{K}} = 0$. The dotted curves are the instantaneous frequencies computed as the inverse of the ith interspike interval (top to bottom, $i = 1, ..., 6$). The bottom solid curve is the adapted firing curve ($i \to \infty$). With or without adaptation, the input–output transform is close to threshold–linear. (Reprinted with permission from Network 4, Mean-field analysis of neuronal spike dynamics, Treves, A., 259–284. Copyright ©Taylor and Francis (1993).)

open tends to shunt the membrane and thus to prevent firing, opens by a fixed amount with the potential excursion associated with each spike, and then relaxes exponentially to its closed state. In this manner sustained firing driven by a constant input current occurs at lower rates after the first few spikes, in a way similar, if the relevant parameters are set appropriately, to the behaviour observed in vitro of many pyramidal cells (for example, Lanthorn, Storm and Andersen (1984), Mason and Larkman (1990)).

The equations for the dynamics of each neuron with adaptation are then

$$C_{\mathrm{m}}\frac{\mathrm{d}V(t)}{\mathrm{d}t} = g_{\mathrm{m}}\left(V_{\mathrm{L}}-V(t)\right)+\sum_{j} g_j\left(V_{\mathrm{AMPA}}-V(t)\right)+\sum_{j} g_j\left(V_{\mathrm{GABA}}-V(t)\right)+g_{\mathrm{K}}(t)(V_{\mathrm{K}}-V(t))$$

(B.43)

and

$$\frac{\mathrm{d}g_{\mathrm{K}}(t)}{\mathrm{d}t} = -\frac{g_{\mathrm{K}}(t)}{\tau_{\mathrm{K}}} + \sum_{k} \Delta g_{\mathrm{K}}\delta(t - t_k)$$

(B.44)

supplemented by the prescription that when at time $t = t_{k+1}$ the potential reaches the level V_{thr}, a spike is emitted, and hence included also in the sum of equation B.44, and the potential resumes its evolution according to equation B.43 from the reset level V_{reset}. The resulting behaviour is exemplified in Fig. B.31, while Fig. B.32 shows the input–output transform (current to frequency transduction) operated by an integrate-and-fire unit of this type with firing rate adaptation. One should compare this with the transduction operated by real cells, as exemplified for example in Fig. 1.5.

It should be noted that synaptic conductance dynamics is not always included in integrate-and-fire models: sometimes it is substituted with current dynamics, which essentially amounts to neglecting non-linearities due to the appearance of the membrane potential in the driving force for synaptic action (see for example, Amit and Tsodyks (1991), Gerstner (1995)); and sometimes it is simplified altogether by assuming that the membrane potential undergoes small sudden jumps when it receives instantaneous pulses of synaptic current (see the review in Gerstner (1995)). The latter simplification is quite drastic and changes the character of the dynamics markedly; whereas the former can be a reasonable simplification in some circumstances, but it produces serious distortions in the description of inhibitory $GABA_A$ currents, which, having an equilibrium (Cl^-) synaptic potential close to the operating range of the membrane potential, are quite sensitive to the instantaneous value of the membrane potential itself.

B.6.3 An integrate-and-fire implementation

In this subsection the mathematical equations that describe the spiking activity and synapse dynamics in some of the integrate-and-fire simulations performed by Rolls and Deco (Deco and Rolls 2003, Deco, Rolls and Horwitz 2004, Deco and Rolls 2005d, Deco and Rolls 2005c, Deco and Rolls 2005b, Deco and Rolls 2006, Loh, Rolls and Deco 2007a, Rolls, Loh and Deco 2008c, Rolls, Grabenhorst and Deco 2010b, Rolls, Grabenhorst and Deco 2010c, Insabato, Pannunzi, Rolls and Deco 2010, Deco, Rolls and Romo 2010, Rolls and Deco 2011b, Webb, Rolls, Deco and Feng 2011, Rolls and Webb 2012, Rolls, Webb and Deco 2012, Martinez-Garcia, Rolls, Deco and Romo 2011, Rolls, Dempere-Marco and Deco 2013, Rolls and Deco 2015a, Rolls and Deco 2015b, Rolls and Deco 2015b, Rolls and Deco 2016) are set out, in order to show in more detail how an integrate-and-fire simulation is implemented. The equations follow in general the formulation described by Brunel and Wang (2001), though each simulation describes its own architecture to be simulated and neurocomputational questions to be addressed, with additional dynamics introduced where described to implement synaptic facilitation and/or synaptic adaptation (Chapter 10).

Each neuron is described by an integrate-and-fire model. The subthreshold membrane potential $V(t)$ of each neuron evolves according to the following equation:

$$C_m \frac{dV(t)}{dt} = -g_m(V(t) - V_L) - I_{syn}(t) \tag{B.45}$$

where $I_{syn}(t)$ is the total synaptic current flow into the cell, V_L is the resting potential, C_m is the membrane capacitance, and g_m is the membrane conductance. When the membrane potential $V(t)$ reaches the threshold V_{thr} a spike is generated, and the membrane potential is reset to V_{reset}. The neuron is unable to spike during the first τ_{ref} which is the absolute refractory period.

The total synaptic current is given by the sum of glutamatergic excitatory components (NMDA and AMPA) and inhibitory components (GABA). The external excitatory contributions (ext) from outside the network are produced through AMPA receptors ($I_{AMPA,ext}$), while the excitatory recurrent synapses (rec) within the network act through AMPA and NMDA receptors ($I_{AMPA,rec}$ and $I_{NMDA,rec}$). The total synaptic current is therefore given by:

$$I_{syn}(t) = I_{AMPA,ext}(t) + I_{AMPA,rec}(t) + I_{NMDA,rec}(t) + I_{GABA}(t) \tag{B.46}$$

where

$$I_{AMPA,ext}(t) = g_{AMPA,ext}(V(t) - V_E) \sum_{j=1}^{N_{ext}} s_j^{AMPA,ext}(t) \tag{B.47}$$

$$I_{AMPA,rec}(t) = g_{AMPA,rec}(V(t) - V_E) \sum_{j=1}^{N_E} w_j s_j^{AMPA,rec}(t) \tag{B.48}$$

$$I_{NMDA,rec}(t) = \frac{g_{NMDA,rec}(V(t) - V_E)}{(1 + [Mg^{++}] \exp(-0.062V(t))/3.57)} \sum_{j=1}^{N_E} w_j s_j^{NMDA,rec}(t) \tag{B.49}$$

$$I_{\text{GABA}}(t) = g_{\text{GABA}}(V(t) - V_{\text{I}}) \sum_{j=1}^{N_I} s_j^{\text{GABA}}(t) \tag{B.50}$$

In the preceding equations the reversal potential of the excitatory synaptic currents $V_{\text{E}} = 0$ mV and of the inhibitory synaptic currents $V_{\text{I}} = -70$ mV. The different form for the NMDA receptor-activated channels implements the voltage-dependence of NMDA receptors. This voltage-dependency, and the long time constant of the NMDA receptors, are important in the effects produced through NMDA receptors (Brunel and Wang 2001, Wang 1999). The synaptic strengths w_j are specified in the papers by Rolls and Deco cited above, and depend on the architecture being simulated. The fractions of open channels s are given by:

$$\frac{ds_j^{\text{AMPA,ext}}(t)}{dt} = -\frac{s_j^{\text{AMPA,ext}}(t)}{\tau_{\text{AMPA}}} + \sum_k \delta(t - t_j^k) \tag{B.51}$$

$$\frac{ds_j^{\text{AMPA,rec}}(t)}{dt} = -\frac{s_j^{\text{AMPA,rec}}(t)}{\tau_{\text{AMPA}}} + \sum_k \delta(t - t_j^k) \tag{B.52}$$

$$\frac{ds_j^{\text{NMDA,rec}}(t)}{dt} = -\frac{s_j^{\text{NMDA,rec}}(t)}{\tau_{\text{NMDA,decay}}} + \alpha x_j(t)(1 - s_j^{\text{NMDA,rec}}(t)) \tag{B.53}$$

$$\frac{dx_j(t)}{dt} = -\frac{x_j(t)}{\tau_{\text{NMDA,rise}}} + \sum_k \delta(t - t_j^k) \tag{B.54}$$

$$\frac{ds_j^{\text{GABA}}(t)}{dt} = -\frac{s_j^{\text{GABA}}(t)}{\tau_{\text{GABA}}} + \sum_k \delta(t - t_j^k) \tag{B.55}$$

where the sums over k represent a sum over spikes emitted by presynaptic neuron j at time t_j^k. The value of $\alpha = 0.5$ ms^{-1}.

Typical values of the conductances for pyramidal neurons are: $g_{\text{AMPA,ext}}=2.08$, $g_{\text{AMPA,rec}}=0.052$, $g_{\text{NMDA,rec}}=0.164$, and $g_{\text{GABA}}=0.67$ nS; and for interneurons: $g_{\text{AMPA,ext}}=1.62$, $g_{\text{AMPA,rec}}=0.0405$, $g_{\text{NMDA,rec}}=0.129$ and $g_{\text{GABA}}=0.49$ nS.

A full list of the values of the parameters used is given in Section B.8.3 and in each of the papers cited above in this Section B.6.3.

B.6.4 The speed of processing of one-layer attractor networks with integrate-and-fire neurons

Given that the analytic approach to the rapidity of the dynamics of attractor networks with integrate-and-fire dynamics (Treves 1993, Rolls and Treves 1998) applies mainly when the state is close to the attractor basin, it is of interest to check the performance of such networks by simulation when the completion of partial patterns that may be towards the edge of the attractor basin can be tested. Simmen, Rolls and Treves (1996a) and Treves, Rolls and Simmen (1997) made a start with this, and showed that retrieval could indeed be fast, within 1–2 time constants of the synapses. However, they found that they could not load the systems they simulated with many patterns, and the firing rates during the retrieval process tended to be unstable. The cause of this turned out to be that the inhibition they used to maintain the activity level during retrieval was subtractive, and it turns out that divisive (shunting) inhibition is much more effective in such networks, as described by Rolls and Treves (1998) in Appendix 5. Divisive inhibition is likely to be organized by inhibitory inputs that synapse close to the cell body (where the reversal potential is close to that of the channels opened by GABA receptors), in contrast to synapses on dendrites, where the different potentials result in opening of the same channels producing hyperpolarization (that is an effectively subtractive influence with respect to the depolarizing currents induced by excitatory (glutamate-releasing) terminals). Battaglia and Treves (1998a) therefore went on to study networks with neurons where the inhibitory neurons could be made to be divisive by having them synapse close to the cell body in neurons modelled with multiple (ten) dendritic compartments. The excitatory inputs terminated on the compartments more distant from the cell body in the model. They found that with this divisive inhibition, the neuronal firing during retrieval was kept under much better control, and the number of patterns that could be successfully stored and retrieved was much higher. Some details of their simulation follow.

Battaglia and Treves (1998a) simulated a network of 800 excitatory and 200 inhibitory cells in its retrieval of one of a number of memory patterns stored in the synaptic weights representing the excitatory-to-excitatory recurrent connections. The memory patterns were assigned at random, drawing the value of each unit in each of the patterns from a binary distribution with sparseness 0.1, that is a probability of 1 in 10 for the unit to be active in the pattern. No baseline excitatory weight was included, but the modifiable weights were instead constrained to remain positive (by clipping at zero synaptic modifications that would make a synaptic weight negative), and a simple exponential decay of the weight with successive modifications was also applied, to prevent runaway synaptic modifications within a rudimentary model of forgetting. Both excitation on inhibitory units and inhibition were mediated by non-modifiable uniform synaptic weights, with values chosen so as to satisfy stability conditions of the type shown in equations A5.13 of Rolls and Treves (1998). Both inhibitory and excitatory neurons were of the general integrate-and-fire type, but excitatory units had in addition an extended dendritic cable, and they received excitatory inputs only at the more distal end of the cable, and inhibitory inputs spread along the cable. In this way, inhibitory inputs reached the soma of excitatory cells with variable delays, and in any case earlier than synchronous excitatory inputs, and at the same time they could shunt the excitatory inputs, resulting in a largely multiplicative form of inhibition (Abbott 1991). The uniform connectivity was not complete, but rather each type of unit could contact units of the other type with a probability of 0.5, and the same was true for inhibitory-to-inhibitory connections.

After 100 (simulated) ms of activity evoked by external inputs uncorrelated with any of the stored patterns, a cue was provided that consisted of the external input becoming correlated with one of the patterns, at various levels of correlation, for 300 ms. After that, external inputs were removed, but when retrieval operated successfully the activity of the units remained strongly correlated with the memory pattern, or even reached a higher level of correlation if a rather corrupted cue had been used, so that, if during the 300 ms the network had stabilized into a state rather distant from the memory pattern, it got much closer to it once the cue was removed. All correlations were quantified using information measures (see Appendix C)), in terms of mutual information between the firing rate pattern across units and the particular memory pattern being retrieved, or in terms of mutual information between the firing rate of one unit and the set of patterns, or, finally, in terms of mutual information between the decoded firing rates of a subpopulation of 10 excitatory cells and the set of memory patterns. The same algorithms were used to extract information measures as were used for example by Rolls, Treves, Tovee and Panzeri (1997d) with real neuronal data from inferior temporal cortex neurons. The firing rates were measured over sliding windows of 30 ms, after checking that shorter windows produced noisier measures. The effect of using a relatively long window, 30 ms, for measuring rates is an apparent linear early rise in information values with time. Nevertheless, in the real system the activity of these cells is 'read' by other cells receiving inputs from them, and that in turn have their own membrane capacitance-determined characteristic time for integrating input activity, a time broadly in the order of 30 ms. Using

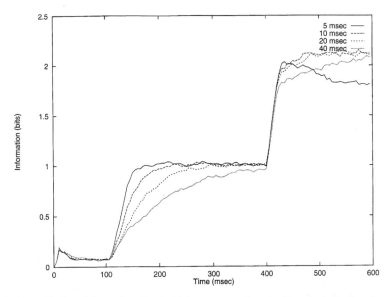

Fig. B.33 Time course of the transinformation about which memory pattern had been selected, as decoded from the firing rates of 10 randomly selected excitatory units. Excitatory conductances closed exponentially with time constants of 5, 10, 20 and 40 ms (curves from top to bottom). A cue of correlation 0.2 with the memory pattern was presented from 100 to 400 ms, uncorrelated external inputs with the same mean strength and sparseness as the cue were applied at earlier times, and no external inputs were applied at later times. (After Battaglia and Treves 1998b.)

such a time window for integrating firing rates did not therefore artificially slow down the read-out process.

It was shown that the time course of different information measures did not depend significantly on the firing rates prevailing during the retrieval state, nor on the resistance–capacitance-determined membrane time constants of the units. Figure B.33 shows that the rise in information after providing the cue at time = 100 ms followed a roughly exponential approach to its steady-state value, which continued until the steady state switched to a new value when the retrieval cue was removed at time = 400 ms. The time constant of the approach to the first steady state was a linear function, as shown in Fig. B.33, of the time constant for excitatory conductances, as predicted by the analysis. (The proportionality factor in the Figure is 2.5, or a collective time constant 2.5 times longer than the synaptic time constant.) The approach to the second steady-state value was more rapid, and the early apparent linear rise prevented the detection of a consistent exponential mode. Therefore, it appears that the cue leads to the basin of attraction of the correct retrieval state by activating transient modes, whose time constant is set by that of excitatory conductances; once the network is in the correct basin, its subsequent reaching the 'very bottom' of the basin after the removal of the cue is not accompanied by any prominent transient mode (see further Battaglia and Treves (1998a)).

Overall, these simulations confirm that recurrent networks, in which excitation is mediated mainly by fast (AMPA, see Section 1.5) channels, can reach asynchronous steady firing states very rapidly, over a few tens of milliseconds, and the rapid approach to steady state is reflected in the relatively rapid rise of information quantities that measure the speed of the operation in functional terms.

An analysis based on integrate-and-fire model units thus indicates that recurrent dynamics can be so fast as to be practically indistinguishable from purely feedforward dynamics, in contradiction to what simple intuitive arguments would suggest. This makes it hazardous to draw conclusions on the underlying circuitry on the basis of the experimentally observed speed with which selective neuronal responses arise, as attempted by Thorpe and Imbert (1989). The results also show that networks that implement feedback processing can settle into a global retrieval state very rapidly, and that rapid processing is not just a feature of feedforward networks.

We return to the intuitive understanding of this rapid processing. The way in which networks with continuous dynamics (such as networks made of real neurons in the brain, and networks modelled with integrate-and-fire neurons) can be conceptualized as settling so fast into their attractor states is that spontaneous activity in the network ensures that some neurons are close to their firing threshold when the retrieval cue is presented, so that the firing of these neurons is influenced within 1–2 ms by the retrieval cue. These neurons then influence other neurons within milliseconds (given the point that some other neurons will be close to threshold) through the modified recurrent collateral synapses that store the information. In this way, the neurons in networks with continuous dynamics can influence each other within a fraction of the synaptic time constant, and retrieval can be very rapid.

B.6.5 The speed of processing of a four-layer hierarchical network with integrate-and-fire attractor dynamics in each layer

Given that the visual system has a whole series of cortical areas organized predominantly hierarchically (e.g. V1 to V2 to V4 to inferior temporal cortex), the issue arises of whether the rapid information processing that can be performed for object recognition is predominantly feedforward, or whether there is sufficient time for feedback processing within each cortical area implemented by the local recurrent collaterals to contribute to the visual information processing being performed. Some of the constraints are as follows.

An analysis of response latencies indicates that there is sufficient time for only 10–20 ms per processing stage in the visual system. In the primate cortical ventral visual system the response latency difference between neurons in layer $4C\beta$ of V1 and inferior temporal cortical cells is approximately 60 ms (Bullier and Nowak 1995, Nowak and Bullier 1997, Schmolesky, Wang, Hanes, Thompson, Leutgeb, Schall and Leventhal 1998). For example, the latency of the responses of neurons in V1 is approximately 30–40 ms (Celebrini, Thorpe, Trotter and Imbert 1993), and in the temporal cortex visual areas approximately 80–110 ms (Baylis, Rolls and Leonard 1987, Sugase, Yamane, Ueno and Kawano 1999). Given that there are 4–6 stages of processing in the ventral visual system from V1 to the anterior inferior temporal cortex, the difference in latencies between each ventral cortical stage is on this basis approximately 10 ms (Rolls 1992a, Oram and Perrett 1994). Information theoretic analyses of the responses of single visual cortical cells in primates reveal that much of the information that can be extracted from neuronal spike trains is often found to be present in periods as short as 20–30 ms (Tovee, Rolls, Treves and Bellis 1993, Tovee and Rolls 1995, Heller, Hertz, Kjaer and Richmond 1995, Rolls, Tovee and Panzeri 1999b). Backward masking experiments indicate that each cortical area needs to fire for only 20–30 ms to pass information to the next stage (Rolls and Tovee 1994, Rolls, Tovee, Purcell, Stewart and Azzopardi 1994b, Kovacs, Vogels and Orban 1995, Rolls, Tovee and Panzeri 1999b, Rolls 2003) (see Section C.3.4). Rapid serial visual presentation of image sequences shows that cells in the temporal visual cortex are still face selective when faces are presented at the rate of 14 ms/image (Keysers, Xiao, Foldiak and Perrett 2001). Finally, event-related potential studies in humans provide strong evidence that the visual system is able to complete some analyses of complex scenes in less than 150 ms (Thorpe, Fize and Marlot 1996).

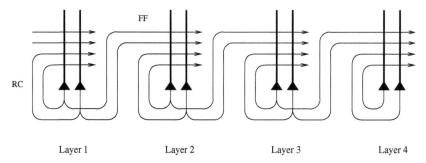

FF

RC

Layer 1   Layer 2   Layer 3   Layer 4

Fig. B.34 The structure of the excitatory connections in the network. There are feedforward (FF) connections between each layer and the next, and excitatory recurrent collaterals (RC) in each layer. Inhibitory connections are also present within each layer, but they are not shown in this Figure. (After Panzeri, Rolls, Battaglia and Lavis 2001.)

To investigate whether feedback processing within each layer could contribute to information processing in such a multilayer system in times as short as 10–20 ms per layer, Panzeri, Rolls, Battaglia and Lavis (2001) simulated a four-layer network with attractor networks in each layer. The network architecture is shown schematically in Fig. B.34. All the neurons realized integrate-and-fire dynamics, and indeed the individual layers and neurons were implemented very similarly to the implementation used by Battaglia and Treves (1998a). In particular, the current flowing from each compartment of the multicompartment neurons to the external medium was expressed as:

$$I(t) = g_{\text{leak}}(V(t) - V^0) + \sum_j g_j(t)(V(t) - V_j),\tag{B.56}$$

where g_{leak} is a constant passive leakage conductance, V^0 the membrane resting potential, $g_j(t)$ the value of the jth synapse conductance at time t, and V_j the reversal potential of the jth synapse. $V(t)$ is the potential in the compartment at time t. The most important parameter in the simulation, the AMPA inactivation time constant, was set to 10 ms. The recurrent collateral (RC) integration time constant of the membrane of excitatory cells was 20 ms long for the simulations presented. The synaptic conductances decayed exponentially in time, obeying the equation

$$\frac{dg_j}{dt} = -\frac{g_j}{\tau_j} + \Delta g_j \sum_k \delta(t - \Delta t - t_k^j),\tag{B.57}$$

where τ_j is the synaptic decay time constant, Δt is a delay term summarizing axonal and synaptic delays, and Δg_j is the amount that the conductance is increased when the presynaptic unit fires a spike. Δg_j thus represents the (unidirectional) coupling strength between the presynaptic and the post-synaptic cell. t_k^j is the time at which the pre-synaptic unit fires its kth spike.

An example of the rapid information processing of the system is shown in Fig. B.35, obtained under conditions in which the local recurrent collaterals can contribute to correct performance because the feedforward (FF) inputs from the previous stage are noisy. (The noise implemented in these simulations was some imperfection in the FF signals produced by some alterations to the FF synaptic weights.) Figure B.35 shows that, when the FF carry an incomplete signal, some information is still transmitted successfully in the 'No RC' condition (in which the recurrent collateral connections in each layer are switched off), and with a relatively short latency. However, the noise term in the FF synaptic strengths makes the retrieval fail more and more layer by layer. When in contrast the recurrent collaterals (RC)

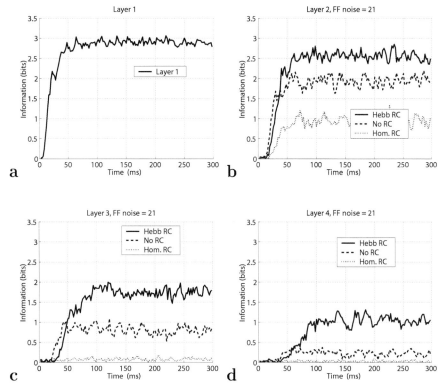

Fig. B.35 The speed of information processing in a 4-layer network with integrate-and-fire neurons. The information time course of the average information carried by the responses of a population of 30 excitatory neurons in each layer. In the simulations considered here, there is noise in the feedforward (FF) synapses. Layer 1 was tested in just one condition. Layers 2–4 are tested in three different conditions: No RC (in which the recurrent collateral synaptic effects do not operate), Hebbian RC (in which the recurrent collaterals have been trained by as associative rule and can help pattern retrieval in each layer), and a control condition named Homogeneous RC (in which the recurrent collaterals could inject current into the neurons, but no useful information was provided by them because they were all set to the same strength). (After Panzeri, Rolls, Battaglia and Lavis 2001.)

are present and operating after Hebbian training, the amount of information retrieved is now much higher, because the RC are able to correct a good part of the erroneous information injected into the neurons by the noisy FF synapses. In Layer 4, 66 ms after cue injection in Layer 1, the information in the Hebbian RC case is 0.2 bits higher than that provided by the FF connections in the 'No RC' condition. This shows that the RC are able to retrieve information in Layer 4 that is not available by any other purely FF mechanism after only roughly 50–55 ms from the time when Layer 1 responds. (This corresponds to 17–18 ms per layer.)

A direct comparison of the latency differences in layers 1–4 of the integrate-and-fire network simulated by Panzeri, Rolls, Battaglia and Lavis (2001) is shown in Fig. B.36. The results are shown for the Hebbian condition illustrated in Fig. B.35, and separate curves are shown for each of the layers 1–4. The Figure shows that, with the time constant of the synapses set to 10 ms, the network can operate with full utilization of and benefit from recurrent processing within each layer in a time which enables the signal to propagate through the 4-layer system with a time course of approximately 17 ms per layer.

The overall results of Panzeri, Rolls, Battaglia and Lavis (2001) were as follows. Through the implementation of continuous dynamics, latency differences were found in information

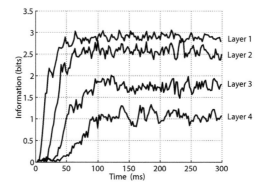

Fig. B.36 The speed of information processing in a 4-layer network with integrate-and-fire neurons. The information time course of the average information carried by the responses of a population of 30 excitatory neurons in each layer. The results are shown for the Hebbian condition illustrated in Fig. B.35, and separate curves are shown for each of the layers 1–4. The Figure shows that, with the time constant of the synapses set to 10 ms, the network can operate with full utilization of and benefit from recurrent processing within each layer in a time in the order of 17 ms per layer. (After Panzeri, Rolls, Battaglia and Lavis 2001.)

retrieval of only 5 ms per layer when local excitation was absent and processing was purely feedforward. However, information latency differences increased significantly when non-associative local excitation (simulating spontaneous firing or unrelated inputs present in the brain) was included. It was also found that local recurrent excitation through associatively modified synapses can contribute significantly to processing in as little as 15 ms per layer, including the feedforward and local feedback processing. Moreover, and in contrast to purely feed-forward processing, the contribution of local recurrent feedback was useful and approximately this rapid even when retrieval was made difficult by noise. These findings provide evidence that cortical information processing can be very fast even when local recurrent circuits are critically involved. The time cost of this recurrent processing is minimal when compared with a feedforward system with spontaneous firing or unrelated inputs already present, and the performance is better than that of a purely feedforward system when noise is present.

It is concluded that local feedback loops within each cortical area can contribute to fast visual processing and cognition.

B.6.6   Spike response model

In this section, we describe another mathematical model that models the activity of single spiking neurons. This model captures the principal effects of real neurons in a realistic way and is simple enough to permit analytical calculations (Gerstner, Ritz and Van Hemmen 1993, Gerstner and Kistler 2002). In contrast to some integrate-and-fire models (Tuckwell 1988), which are essentially given by differential equations, the spike-response model is based on response kernels that describe the integrated effect of spike reception or emission on the membrane potential. In this model, spikes are generated by a threshold process (i.e. the firing time t' is given by the condition that the membrane potential reaches the firing threshold θ; that is, $h(t') = \theta$). Figure B.37 (bottom) shows schematically the spike-generating mechanism.

The membrane potential is given by the integration of the input signal weighted by a kernel defined by the equations

$$h(t') = h^{\text{refr}}(t') + h^{\text{syn}}(t')$$

(B.58)

$$h^{\text{refr}}(t') = \int_0^\infty \eta^{\text{refr}}(z)\delta(t' - z - t'_{\text{last}})dz \tag{B.59}$$

$$h^{\text{syn}}(t') = \sum_j J_j \int_0^\infty \Lambda(z', t' - t'_{\text{last}})s(t' - z')dz' . \tag{B.60}$$

The kernel $\eta^{\text{refr}}(z)$ is the refractory function. If we consider only absolute refractoriness, $\eta^{\text{refr}}(z)$ is given by:

$$\eta^{\text{refr}}(z) = \begin{cases} -\infty & \text{for } 0 < z \leq \tau^{\text{refr}} \\ 0 & \text{for } z \geq \tau^{\text{refr}} \end{cases} \tag{B.61}$$

where τ^{refr} is the absolute refractory time. The time t'_{last} corresponds to the last postsynaptic spike (i.e. the most recent firing of the particular neuron). The second response function is the synaptic kernel $\Lambda(z', t' - t'_{\text{last}})$. It describes the effect of an incoming spike on the membrane potential at the soma of the postsynaptic neuron, and it eventually includes also the dependence on the state of the receiving neuron through the difference $t' - t'_{\text{last}}$ (i.e. through the time that has passed since the last postsynaptic spike). The input spike train yields $s(t' - z') = \sum_i \delta(t' - z' - t_{ij})$, t_{ij} being the ith spike at presynaptic input j. In order to simplify the discussion and without losing generality, let us consider only a single synaptic input, and therefore we can remove the subindex j. In addition, we assume that the synaptic strength J is positive (i.e. excitatory). Integrating equations B.59 and B.60, we obtain

$$h(t') = \eta^{\text{refr}}(t' - t'_{\text{last}}) + J\sum_i \Lambda(t' - t_i, t' - t'_{\text{last}}) \tag{B.62}$$

Synaptic kernels are of the form

$$\Lambda(t' - t_i, t' - t'_{\text{last}}) = \text{H}(t' - t_i)\text{H}(t_i - t'_{\text{last}})\Psi(t' - t_i) \tag{B.63}$$

where $\text{H}(s)$ is the step (or Heaviside) function which vanishes for $s \leq 0$ and takes a value of 1 for $s > 0$. After firing, the membrane potential is reset according to the renewal hypothesis.

This spike-response model is not used in the models described in this book, but is presented to show alternative approaches to modelling the dynamics of network activity.

B.7 Network dynamics: introduction to the mean-field approach

Units whose potential and conductances follow the integrate-and-fire equations in Section B.6 can be assembled together in a model network of any composition and architecture. It is convenient to imagine that units are grouped into classes, such that the parameters quantifying the electrophysiological properties of the units are uniform, or nearly uniform, within each class, while the parameters assigned to synaptic connections are uniform or nearly uniform for all connections from a given presynaptic class to another given postsynaptic class. The parameters that have to be set in a model at this level of description are quite numerous, as listed in Tables B.3 and B.4.

In the limit in which the parameters are constant within each class or pair of classes, a mean-field treatment can be applied to analyze a model network, by summing equations that describe the dynamics of individual units to obtain a more limited number of equations that describe the dynamical behaviour of groups of units (Frolov and Medvedev 1986). The treatment is exact in the further limit in which very many units belong to each class, and is an approximation if each class includes just a few units. Suppose that N_C is the number of

Table B.3 Cellular parameters (chosen according to the class of each unit)

V_{rest}	Resting potential
V_{thr}	Threshold potential
V_{ahp}	Reset potential
V_K	Potassium conductance equilibrium potential
C	Membrane capacitance
τ_K	Potassium conductance time constant
g_0	Leak conductance
Δg_K	Extra potassium conductance following a spike
Δt	Overall transmission delay

Table B.4 Synaptic parameters (chosen according to the classes of presynaptic and postsynaptic units)

V_α	Synapse equilibrium potential
τ_α	Synaptic conductance time constant
Δg_α	Conductance opened by one presynaptic spike
Δt_α	Delay of the connection

classes defined. Summing equations B.43 and B.44 across units of the same class results in N_C functional equations describing the evolution in time of the fraction of cells of a particular class that at a given instant have a given membrane potential. In other words, from a treatment in which the evolution of the variables associated with each unit is followed separately, one moves to a treatment based on density functions, in which the common behaviour of units of the same class is followed together, keeping track solely of the portion of units at any given value of the membrane potential. Summing equation B.42 across connections with the same class of origin and destination results in $N_C \times N_C$ equations describing the dynamics of the overall summed conductance opened on the membrane of a cell of a particular class by all the cells of another given class. A more explicit derivation of mean-field equations is given by Treves (1993) and in Section B.8.

The system of mean-field equations can have many types of asymptotic solutions for long times, including chaotic, periodic, and stationary ones. The stationary solutions are stationary in the sense of the mean fields, but in fact correspond to the units of each class firing tonically at a certain rate. They are of particular interest as the dynamical equivalent of the steady states analyzed by using non-dynamical model networks. In fact, since the neuronal current-to-frequency transfer function resulting from the dynamical equations is rather similar to a threshold linear function (see Fig. B.32), and since each synaptic conductance is constant in time, the stationary solutions are essentially the same as the states described using model networks made up of threshold linear, non-dynamical units. Thus the dynamical formulation reduces to the simpler formulation in terms of steady-state rates when applied to asymptotic stationary solutions; but, among simple rate models, it is equivalent only to those that allow description of the continuous nature of neuronal output, and not to those, for example based on binary units, that do not reproduce this fundamental aspect. The advantages of the dynamical formulation are that (i) it enables one to describe the character and prevalence of other types of asymptotic solutions, and (ii) it enables one to understand how the network reaches, in time, the asymptotic behaviour.

The development of this mean-field approach, and the foundations for its application to models of cortical visual processing and attention, are described in Section B.8.

B.8 Mean-field based neurodynamics

A model of brain functions requires the choice of an appropriate theoretical framework, which permits the investigation and simulation of large-scale biologically realistic neural networks. Starting from the mathematical models of biologically realistic single neurons (i.e. spiking neurons), one can derive models that describe the joint activity of pools of equivalent neurons. This kind of neurodynamical model at the neuronal assembly level is motivated by the experimental observation that cortical neurons of the same type that are near to each other tend to receive similar inputs. As described in the previous section, it is convenient in this simplified approach to neural dynamics to consider all neurons of the same type in a small cortical volume as a computational unit of a neural network. This computational unit is called a neuronal pool or assembly. The mathematical description of the dynamical evolution of neuronal pool activity in multimodular networks, associated with different cortical areas, establishes the roots of the dynamical approach that are used in Chapter 6 and Section 5.6, and in Rolls and Deco (2002) Chapters 9–11. In this Section (B.8), we introduce the mathematical fundamentals utilized for a neurodynamical description of pool activity (see also Section B.8.2). Beginning at the microscopic level and using single spiking neurons to form the pools of a network, we derive the mathematical formulation of the neurodynamics of cell assemblies. Further, we introduce the basic architecture of neuronal pool networks that fulfil the basic mechanisms consistent with the biased competition hypothesis. Each of these networks corresponds to cortical areas that also communicate with each other. We describe therefore the dynamical interaction between different modules or networks, which will be the basis for the implementation of attentional top-down bias.

B.8.1 Population activity

We now introduce thoroughly the concept of a neuronal pool and the differential equations representing the neurodynamics of pool activity.

Starting from individual spiking neurons one can derive a differential equation that describes the dynamical evolution of the averaged activity of a pool of extensively many equivalent neurons. Several areas of the brain contain groups of neurons that are organized in populations of units with (somewhat) similar properties (though in practice the neurons convey independent information, as described in Appendix C). These groups for mean-field modelling purposes are usually called pools of neurons and are constituted by a large and similar population of identical spiking neurons that receive similar external inputs and are mutually coupled by synapses of similar strength. Assemblies of motor neurons (Kandel, Schwartz and Jessell 2000) and the columnar organization in the visual and somatosensory cortex (Hubel and Wiesel 1962, Mountcastle 1957) are examples of these pools. Each single cell in a pool can be described by a spiking model, e.g. the spike response model presented in Section B.6.6. Due to the fact that for large-scale cortical modelling, neuronal pools form a relevant computational unit, we adopt a population code. We take the activity level of each pool of neurons as the relevant dependent variable rather than the spiking activity of individual neurons. We therefore derive a dynamical model for the mean activity of a neural population. In a population of M neurons, the mean activity $A(t)$ is determined by the proportion of active neurons by counting the number of spikes $n_{\text{spikes}}(t, t + \Delta t)$ in a small time interval Δt and dividing by M and by Δt (Gerstner 2000), i.e. formally

$$A(t) = \lim_{\Delta t \to 0} \frac{n_{\text{spikes}}(t, t + \Delta t)}{M \Delta t} . \tag{B.64}$$

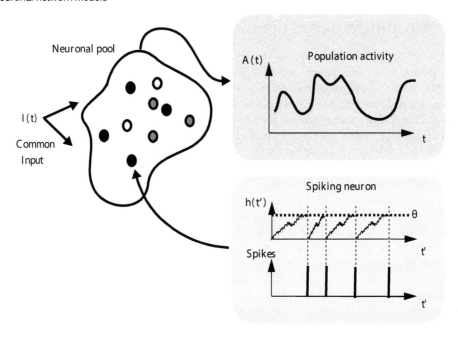

Fig. B.37 Population averaged rate of a neuronal pool of spiking neurons (top) and the action potential generating mechanism of single neurons (bottom). In a neuronal pool, the mean activity $A(t)$ is determined by the proportion of active neurons by counting the number of spikes in a small time interval Δt and dividing by the number of neurons in the pool and by Δt. Spikes are generated by a threshold process. The firing time t' is given by the condition that the membrane potential $h(t')$ reaches the firing threshold θ. The membrane potential $h(t')$ is given by the integration of the input signal weighted by a given kernel (see text for details). (After Rolls and Deco 2002.)

As indicated by Gerstner (2000), and as depicted in Fig. B.37, the concept of pool activity is quite different from the definition of the average firing rate of a single neuron. Contrary to the concept of temporal averaging over many spikes of a single cell, which requires that the input is slowly varying compared with the size of the temporal averaging window, a coding scheme based on pool activity allows rapid adaptation to real-world situations with quickly changing inputs. It is possible to derive dynamical equations for pool activity levels by utilizing the mean-field approximation (Wilson and Cowan 1972, Abbott 1991, Amit and Tsodyks 1991). The mean-field approximation consists of replacing the temporally averaged discharge rate of a cell with an equivalent momentary activity of a neural population (ensemble average) that corresponds to the assumption of ergodicity. According to this approximation, we categorize each cell assembly by means of its activity $A(t)$. A pool of excitatory neurons without external input can be described by the dynamics of the pool activity given by

$$\tau \frac{\partial A(t)}{\partial t} = -A(t) + qF(A(t)) \tag{B.65}$$

where the first term on the right hand side is a decay term and the second term takes into account the excitatory stimulation between the neurons in the pool. In the previous equation, the non-linearity

$$F(x) = \frac{1}{T_r - \tau \log(1 - \frac{1}{\tau x})} \tag{B.66}$$

is the response function (transforming current into discharge rate) for a spiking neuron with deterministic input, membrane time constant τ, and absolute refractory time T_r. Equation B.65

was derived by Gerstner (2000) assuming adiabatic conditions. Gerstner (2000) has shown that the population activity in a homogeneous population of neurons can be described by an integral equation. A systematic reduction of the integral equation to a single differential equation of the form B.65 always supposes that the activity changes only slowly compared with the typical interval length. In other words, the mean-field approach described in the above equations and utilized in parts of Chapters 25 and 6, and Section 5.6 generates a dynamics that neglects fast, transient behaviour. This means that we are assuming that rapid oscillations (and synchronization) do not play a computational role at least for the brain functions that we will consider. Rapid oscillations of neural activity could have a relevant functional role, namely of dynamical cooperation between pools in the same or different brain areas. It is well known in the theory of dynamical systems that the synchronization of oscillators is a cooperative phenomenon. Cooperative mechanisms might complement the competitive mechanisms on which our computational cortical model is based.

An example of the application of the mean field approach, used in a model of decision-making (Deco and Rolls 2006), is provided in Section B.8.2, with the parameters used provided in Section B.8.3. The mean-field analysis described is consistent with the integrate-and-fire spiking simulation described in Section B.6.3, that is, the same parameters used in the mean-field analysis can then be used in the integrate-and-fire simulations. Part of the value of the mean-field analysis is that it provides a way of determining the parameters that will lead to the specified steady state behaviour (in the absence of noise), and these parameters can then be used in a well-defined system for the integrate-and-fire simulations to investigate the full dynamics of the system in the presence of the noise generated by the random spike timings of the neurons.

Ways in which noise can be introduced into the mean-field approach are described by Rolls and Deco (2010).

B.8.2 The mean-field approach used in a model of decision-making

The mean-field approximation used by Deco and Rolls (2006) was derived by Brunel and Wang (2001), assuming that the network of integrate-and-fire neurons is in a stationary state. This mean-field formulation includes synaptic dynamics for AMPA, NMDA and GABA activated ion channels (Brunel and Wang 2001). In this formulation the potential of a neuron is calculated as:

$$\tau_x \frac{dV(t)}{dt} = -V(t) + \mu_x + \sigma_x \sqrt{\tau_x} \eta(t) \tag{B.67}$$

where $V(t)$ is the membrane potential, x labels the populations, τ_x is the effective membrane time constant, μ_x is the mean value the membrane potential would have in the absence of spiking and fluctuations, σ_x measures the magnitude of the fluctuations, and η is a Gaussian process with absolute exponentially decaying correlation function with time constant τ_{AMPA}. The quantities μ_x and σ_x^2 are given by:

$$\mu_x = \frac{(T_{ext}\nu_{ext} + T_{AMPA}n_x^{AMPA} + \rho_1 n_x^{NMDA})V_E + \rho_2 n_x^{NMDA}\langle V \rangle + T_I n_x^{GABA}V_I + V_L}{S_x} \tag{B.68}$$

$$\sigma_x^2 = \frac{g_{AMPA,ext}^2(\langle V \rangle - V_E)^2 N_{ext}\nu_{ext}\tau_{AMPA}^2\tau_x}{g_m^2\tau_m^2}. \tag{B.69}$$

where ν_{ext} Hz is the external incoming spiking rate, $\tau_m = C_m/g_m$ with the values for the excitatory or inhibitory neurons depending on the population considered, and the other quantities are given by:

$$S_x = 1 + T_{\text{ext}}\nu_{\text{ext}} + T_{\text{AMPA}}n_x^{\text{AMPA}} + (\rho_1 + \rho_2)n_x^{\text{NMDA}} + T_{\text{I}}n_x^{\text{GABA}} \tag{B.70}$$

$$\tau_x = \frac{C_m}{g_m S_x} \tag{B.71}$$

$$n_x^{\text{AMPA}} = \sum_{j=1}^{p} r_j w_{jx}^{\text{AMPA}}\nu_j \tag{B.72}$$

$$n_x^{\text{NMDA}} = \sum_{j=1}^{p} r_j w_{jx}^{\text{NMDA}}\psi(\nu_j) \tag{B.73}$$

$$n_x^{\text{GABA}} = \sum_{j=1}^{p} r_j w_{jx}^{\text{GABA}}\nu_j \tag{B.74}$$

$$\psi(\nu) = \frac{\nu\tau_{\text{NMDA}}}{1 + \nu\tau_{\text{NMDA}}}\left(1 + \frac{1}{1 + \nu\tau_{\text{NMDA}}}\sum_{n=1}^{\infty}\frac{(-\alpha\tau_{\text{NMDA,rise}})^n T_n(\nu)}{(n+1)!}\right) \tag{B.75}$$

$$T_n(\nu) = \sum_{k=0}^{n}(-1)^k\binom{n}{k}\frac{\tau_{\text{NMDA,rise}}(1 + \nu\tau_{\text{NMDA}})}{\tau_{\text{NMDA,rise}}(1 + \nu\tau_{\text{NMDA}}) + k\tau_{\text{NMDA,decay}}} \tag{B.76}$$

$$\tau_{\text{NMDA}} = \alpha\tau_{\text{NMDA,rise}}\tau_{\text{NMDA,decay}} \tag{B.77}$$

$$T_{\text{ext}} = \frac{g_{\text{AMPA,ext}}\tau_{\text{AMPA}}}{g_m} \tag{B.78}$$

$$T_{\text{AMPA}} = \frac{g_{\text{AMPA,rec}}N_E\tau_{\text{AMPA}}}{g_m} \tag{B.79}$$

$$\rho_1 = \frac{g_{\text{NMDA}}N_E}{g_m J} \tag{B.80}$$

$$\rho_2 = \beta\frac{g_{\text{NMDA}}N_E(\langle V_x\rangle - V_E)(J-1)}{g_m J^2} \tag{B.81}$$

$$J = 1 + \gamma\exp(-\beta\langle V_x\rangle) \tag{B.82}$$

$$T_{\text{I}} = \frac{g_{\text{GABA}}N_{\text{I}}\tau_{\text{GABA}}}{g_m} \tag{B.83}$$

$$\langle V_x\rangle = \mu_x - (V_{\text{thr}} - V_{\text{reset}})\nu_x\tau_x, \tag{B.84}$$

where p is the number of excitatory populations, r_x is the fraction of neurons in the excitatory population x, w_{jx} the weight of the connections from population x to population j, ν_x is the spiking rate of the population x, $\gamma = [\text{Mg}^{++}]/3.57$, $\beta = 0.062$ and the average membrane potential $\langle V_x\rangle$ has a value between -55 mV and -50 mV.

The spiking rate of a population as a function of the defined quantities is then given by:

$$\nu_x = \phi(\mu_x, \sigma_x), \tag{B.85}$$

where

$$\phi(\mu_x, \sigma_x) = \left(\tau_{\text{rp}} + \tau_x\int_{\beta(\mu_x,\sigma_x)}^{\alpha(\mu_x,\sigma_x)}du\sqrt{\pi}\exp(u^2)[1 + \text{erf}(u)]\right)^{-1} \tag{B.86}$$

$$\alpha(\mu_x, \sigma_x) = \frac{(V_{\text{thr}} - \mu_x)}{\sigma_x}\left(1 + 0.5\frac{\tau_{\text{AMPA}}}{\tau_x}\right) + 1.03\sqrt{\frac{\tau_{\text{AMPA}}}{\tau_x}} - 0.5\frac{\tau_{\text{AMPA}}}{\tau_x} \tag{B.87}$$

$$\beta(\mu_x, \sigma_x) = \frac{(V_{\text{reset}} - \mu_x)}{\sigma_x} \tag{B.88}$$

with $\mathrm{erf}(u)$ the error function and τ_{rp} the refractory period which is considered to be 2 ms for excitatory neurons and 1 ms for inhibitory neurons. To solve the equations defined by (B.85) for all xs we integrate numerically (B.84) and the differential equation B.89, which has fixed point solutions corresponding to equation B.85:

$$\tau_x \frac{d\nu_x}{dt} = -\nu_x + \phi(\mu_x, \sigma_x). \tag{B.89}$$

The values of the parameters used in a mean-field analysis of decision-making are provided in Section B.8.3, with further details of the parameters in Chapter 5.6.

B.8.3 The model parameters used in the mean-field analyses of decision-making

The fixed parameters of the model are shown in Table B.5, and not only provide information about the values of the parameters used in the simulations, but also enable them to be compared to experimentally measured values.

Table B.5 Parameters used in the integrate-and-fire simulations

N_E	800
N_I	200
r	0.1
w_+	2.2
w_I	1.015
N_{ext}	800
ν_{ext}	2.4 kHz
C_m (excitatory)	0.5 nF
C_m (inhibitory)	0.2 nF
g_m (excitatory)	25 nS
g_m (inhibitory)	20 nS
V_L	−70 mV
V_{thr}	−50 mV
V_{reset}	−55 mV
V_E	0 mV
V_I	−70 mV
$g_{AMPA,ext}$ (excitatory)	2.08 nS
$g_{AMPA,rec}$ (excitatory)	0.104 nS
g_{NMDA} (excitatory)	0.327 nS
g_{GABA} (excitatory)	1.25 nS
$g_{AMPA,ext}$ (inhibitory)	1.62 nS
$g_{AMPA,rec}$ (inhibitory)	0.081 nS
g_{NMDA} (inhibitory)	0.258 nS
g_{GABA} (inhibitory)	0.973 nS
$\tau_{NMDA,decay}$	100 ms
$\tau_{NMDA,rise}$	2 ms
τ_{AMPA}	2 ms
τ_{GABA}	10 ms
α	0.5 ms^{-1}

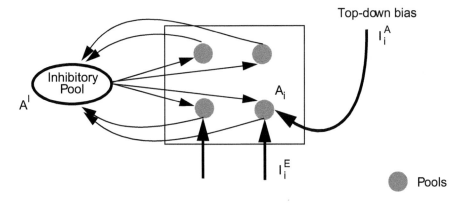

Fig. B.38 Basic computational module for biased competition: a competitive network with external top–down bias. Excitatory pools with activity A_i for the ith pool are connected with a common inhibitory pool with activity A^I in order to implement a competition mechanism. I_i^E is the external sensory input to the cells in pool i, and I_i^A attentional top-down bias, an external input coming from higher modules. The external top-down bias can shift the competition in favour of a specific pool or group of pools. This architecture is similar to that shown in Fig. B.18, but with competition between pools of similar neurons. (After Rolls and Deco 2002.)

B.8.4 A basic computational module based on biased competition

We are interested in the neurodynamics of modules composed of several pools that implement a competitive mechanism[40]. This can be achieved by connecting the pools of a given module with a common inhibitory pool, as is schematically shown in Fig. B.38.

In this way, the more pools of the module that are active, the more active the common inhibitory pool will be, and consequently, the more feedback inhibition will affect the pools in the module, such that only the most excited group of pools will survive the competition. On the other hand, external top-down bias could shift the competition in favour of a specific group of pools. This basic computational module implements therefore the biased competition hypothesis described in Chapter 6. Let us assume that there are m pools in a given module. The system of differential equations describing the dynamics of such a module is given by two differential equations, both of the type of equation B.65. The first differential equation describes the dynamics of the activity level of the excitatory pools (pyramidal neurons) and is mathematically expressed by

$$\tau\frac{\partial A_i(t)}{\partial t} = -A_i(t) + a\mathrm{F}(A_i(t)) - b\mathrm{F}(A^I(t)) +$$
$$I_0 + I_i^E(t) + I_i^A(t) + \nu \quad ; \qquad\qquad \text{for } i = 1, ..., m \qquad (B.90)$$

and the second one describes the dynamics of the activity level of the common inhibitory pool for each feature dimension (stellate neurons)

$$\tau_p\frac{\partial A^I(t)}{\partial t} = -A^I(t) + c\sum_{i=1}^{m}\mathrm{F}(A_i(t)) - d\mathrm{F}(A^I(t)) \qquad\qquad (B.91)$$

where $A_i(t)$ is the activity for pool i, $A^I(t)$ is the activity in the inhibitory pool, I_0 is a diffuse spontaneous background input, $I^E(t)$ is the external sensory input to the cells in pool i, and ν is additive Gaussian noise. The attentional top-down bias $I_i^A(t)$ is defined as an external input coming from higher modules that is not explicitly modelled.

---

[40]These neurodynamics are used in Chapter 6.

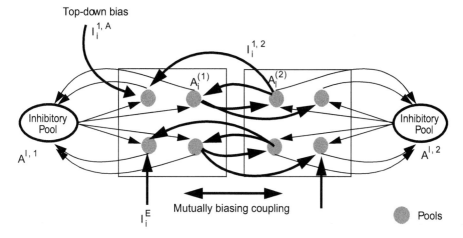

Top-down bias

Inhibitory Pool

Inhibitory Pool

Mutually biasing coupling

Pools

Fig. B.39 Two competitive networks mutually biased through intermodular connections. The activity $A_i^{(1)}$ of the ith excitatory pool in module 1 (on the left) and of the lth excitatory pool in module 2 (on the right) are connected by the mutually biasing coupling $I_i^{1,2}$. The architecture could implement top-down feedback originating from the interaction between brain areas that are explicitly modelled in the system. (Module 2 might be the higher module.) The external top-down bias $I_i^{1,A}$ corresponds to the coupling to pool i of module 1 from brain area A that is not explicitly modelled in the system. (After Rolls and Deco 2002.)

A qualitative description of the main fixed point attractors of the system of differential equations B.90 and B.91 was provided by Usher and Niebur (1996). Basically, we will be interested in the fixed points corresponding to zero activity and larger activation. The parameters will therefore be fixed such that the dynamics evolves to these attractors.

B.8.5 Multimodular neurodynamical architectures

In order to model complex psychophysically and neuropsychologically relevant brain functions such as visual search or object recognition (see e.g. Chapter 6), we must take into account the computational role of individual brain areas and their mutual interaction. The macroscopic phenomenological behaviour will therefore be the result of the mutual interaction of several computational modules.

The dynamical coupling of different basic modules in a multimodular architecture can be described, in our neurodynamical framework, by allowing mutual interaction between pools belonging to different modules. Figure B.39 shows this idea schematically. The system of differential equations describing the global dynamics of such a multimodular system is given by a set of equations of the type of equation B.65.

The excitatory pools belonging to a module obey the following equations:

$$\tau \frac{\partial A_i^{(j)}(t)}{\partial t} = -A_i^{(j)}(t) + aF(A_i^{(j)}(t)) - bF(A^{I,j}(t)) +$$
$$I_i^{j,k} + I_0 + I_i^E(t) + I_i^A(t) + \nu \qquad \text{for } i = 1, ..., m \qquad \text{(B.92)}$$

and the corresponding inhibitory pools evolve according to

$$\tau_p \frac{\partial A^{I,j}(t)}{\partial t} = -A^{I,j}(t) + c \sum_{i=1}^{m} F(A_i^{(j)}(t)) - dF(A^{I,j}(t)). \qquad \text{(B.93)}$$

The mutual coupling $I_i^{j,k}$ between module (j) and (k) is given by

$$I_i^{j,k} = \sum_l W_{il} F(I_l^{(k)}(t)) \tag{B.94}$$

where W_{il} is the synaptic coupling strength between the pool i of module (j) and pool l of module (k). This mutual coupling term can be interpreted as a top-down bias originating from the interaction between brain areas that is explicitly modelled in our system. On the other hand, the external top-down bias I_i^A corresponds to the coupling with brain areas A that are not explicitly modelled in our system.

Additionally, it is interesting to note that the top-down bias in this kind of architecture modulates the response of the pool activity in a multiplicative manner. Responses of neurons in parietal area 7a are modulated by combined eye and head movement, exhibiting a multiplicative gain modulation that modifies the amplitude of the neural responses to retinal input but does not change the preferred retinal location of a cell, nor in general the width of the receptive field (Brotchie, Andersen, Snyder and Goodman 1995). It has also been suggested that multiplicative gain modulation might play a role in translation invariant object representation (Salinas and Abbott 1997). We will use this multiplicative effect for formulating an architecture for attentional gain modulation, which can contribute to correct translation invariant object recognition in ways to be described in Chapter 6.

We show now that multiplicative-like responses can arise from the top-down biased mutual interaction between pools. Another alternative architecture that can also perform product operations on additive synaptic inputs was proposed by Salinas and Abbott (1996). Our basic architecture for showing this multiplicative effect is presented schematically in Fig. B.40a.

Two pools are mutually connected via fixed weights. The first pool or unit receives a bottom-up visual input I_1^V, modelled by the response of a vertically oriented complex cell. The second pool receives a top-down attentional bias $I_2^A = B$. The two pools are mutually coupled with unity weight. The equations describing the activities of the two pools are given by:

$$\tau \frac{\partial A_1(t)}{\partial t} = -A_1(t) + \alpha F(A_1(t)) + I_o + I_1^V + \nu \tag{B.95}$$

$$\tau \frac{\partial A_2(t)}{\partial t} = -A_2(t) + \alpha F(A_2(t)) + I_o + I_2^A + \nu \tag{B.96}$$

where A_1 and A_2 are the activities of pool 1 and pool 2 respectively, $\alpha = 0.95$ is the coefficient of recurrent self-excitation of the pool, ν is the noise input to the pool drawn from a normal distribution $N(\mu = 0, \sigma = 0.02)$, and $I_o = 0.025$ is a direct current biasing input to the pool.

Simulation of the above dynamical equations produces the results shown in Fig. B.40b. The orientation tuning curve of unit (or pool) 1 was modulated by a top-down bias B introduced to unit (pool) 2. The gain modulation was transmitted through the coupling from pool 2 to pool 1 after a few steps of evolution of the dynamical equations. Without the feedback from unit 2, unit 1 exhibits the orientation tuning curve shown as $(B = 0)$. As B increased, the increase in pool 1's response to the vertical bar was significantly greater than the increase in its response to the horizontal bar. Therefore, the attentional gain modulation produced in pool 1 through the mutual coupling was not a simple additive effect, but had a strong multiplicative component. The net effect was that the width of the orientation tuning curve of the cell was roughly preserved under attentional modulation. This was due to the non-linearity in the activation function.

This finding is basically consistent with the effect of attention on the orientation tuning curves of neurons in V4 (McAdams and Maunsell 1999).

(a)

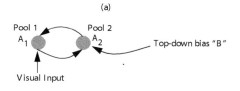

(b)

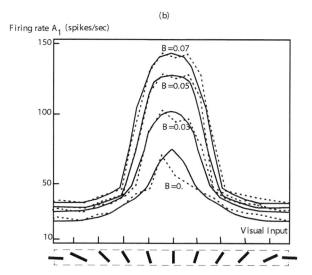

Fig. B.40 (a) The basic building block of the top-down attentional system utilizes non-specific competition between pools within the same module and specific mutual facilitation between pools in different modules. Excitatory neuronal pools within the same module compete with each other through one or more inhibitory neuronal pool(s) I_I with activity A^{I_I}. Excitatory pool 1 (with activity A_1) receives a bottom-up (visual) input I_1^V, and excitatory pool 2 receives a top-down ('attentional') bias input I_2^A. Excitatory neuronal pools in the two different modules can excite each other via mutually biased coupling. (b) The effect of altering the bias input B to pool 2 on the responses or activity of pool 1 to its orientation-tuned visual input (see text). (After Rolls and Deco 2002.)

Summarizing, the neurodynamics of the competitive mechanisms between neuronal pools, and their mutual gain modulation, are the two main ingredients used for proposing a cortical architecture that models attention and different kinds of object search in a visual scene (see Chapter 6 , and Rolls and Deco (2002) Chapters 9–11).

B.9 Sequence memory implemented by adaptation in an attractor network

Sequence memory can be implemented by using synaptic adaptation (Chapter 10) to effectively encode the order of the items in a sequence, as described in Section 10.6 (Deco and Rolls 2005c) and in the model of syntax (Rolls and Deco 2015a) (Chapter 17). Whenever the attractor system is quenched into inactivity, the next member of the sequence emerges out of the spontaneous activity, because the least recently activated member of the sequence has the least synaptic or neuronal adaptation. This mechanism could be implemented in recurrent networks such as the hippocampal CA3, prefrontal cortex, language areas of cortex, etc.

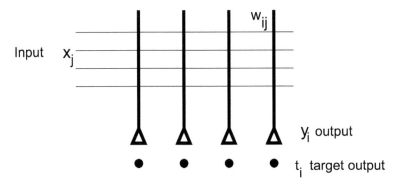

Fig. B.41 One-layer perceptron.

B.10 Perceptrons and one-layer error correction networks

The networks described in the next two Sections (B.10 and B.11) are capable of mapping a set of inputs to a set of required outputs using correction when errors are made. Although some of the networks are very powerful in the types of mapping they can perform, the power is obtained at the cost of learning algorithms that do not use local learning rules. A local learning rule specifies that synaptic strengths should be altered on the basis of information available locally at the synapse, for example the activity of the presynaptic and the post-synaptic neurons. Because the networks described here do not use local learning rules, their biological plausibility remains at present uncertain. One of the aims of future research must be to determine whether comparably difficult problems to those solved by the networks described in Sections B.10 and B.11 can be solved by biologically plausible neuronal networks.

We now describe one-layer networks taught by an error correction algorithm. The term *perceptron* refers strictly to networks with binary threshold activation functions. The outputs might take the values only 1 or 0 for example. The term perceptron arose from networks designed originally to solve perceptual problems (Rosenblatt 1961, Minsky and Papert 1969), and these networks are referred to briefly below. If the output neurons have continuous-valued firing rates, then a more general error-correcting rule called the delta rule is used, and is introduced in this Section (B.10). For such networks, the activation function may be linear, or it may be non-linear but monotonically increasing, without a sharp threshold, as in the sigmoid activation function (see Fig. 1.4).

B.10.1 Architecture and general description

The one-layer error-correcting network has a set of inputs that it is desired to map or classify into a set of outputs (see Fig. B.41). During learning, an input pattern is selected, and produces output firing by activating the output neurons through modifiable synapses, which then fire as a function of their typically non-linear activation function. The output of each neuron is then compared with a target output for that neuron given that input pattern, an error between the actual output and the desired output is determined, and the synaptic weights on that neuron are then adjusted to minimize the error. This process is then repeated for all patterns until the average error across patterns has reached a minimum. A one-layer error-correcting network can thus produce output firing for each pattern in a way that has similarities to a pattern associator. It can perform more powerful mappings than a pattern associator, but requires an error to be computed for each neuron, and for that error to affect the synaptic strength in a way that is not altogether local. A more detailed description follows.

These one-layer networks have a target for each output neuron (for each input pattern). They are thus an example of a supervised network. With the one-layer networks taught with the delta rule or perceptron learning rule described next, there is a separate teacher for each output neuron, as shown in Fig. B.41.

B.10.2   Generic algorithm for a one-layer error correction network

For each input pattern and desired target output:

1. Apply an input pattern to produce input firing **x**, and obtain the activation of each neuron in the standard way by computing the dot product of the input pattern and the synaptic weight vector. The synaptic weight vector can be initially zero, or have random values.

$$h_i = \sum_j x_j w_{ij} \tag{B.97}$$

where $\sum_j$ indicates that the sum is over the C input axons (or connections) indexed by j to each neuron.

2. Apply an activation function to produce the output firing y_i:

$$y_i = f(h_i) . \tag{B.98}$$

This activation function f may be sigmoid, linear, binary threshold, linear threshold, etc. If the activation function is non-linear, this helps to classify the inputs into distinct output patterns, but a linear activation function may be used if an optimal linear mapping is desired (see Adaline and Madaline, below).

3. Calculate the difference for each cell i between the target output t_i and the actual output y_i produced by the input, which is the error Δ_i

$$\Delta_i = t_i - y_i. \tag{B.99}$$

4. Apply the following learning rule, which corrects the (continuously variable) weights according to the error and the input firing x_j

$$\delta w_{ij} = k(t_i - y_i)x_j \tag{B.100}$$

where k is a constant that determines the learning rate. This is often called the delta rule, the Widrow–Hoff rule, or the LMS (least mean squares) rule (see below).

5. Repeat steps 1–4 for all input pattern – output target pairs until the root mean square error becomes zero or reaches a minimum.

In general, networks taught by the delta rule may have linear, binary threshold, or non-linear but monotonically increasing (e.g. sigmoid) activation functions, and may be taught with binary or continuous input patterns (see Rolls and Treves (1998), Chapter 5). The properties of these variations are made clear next.

B.10.3   Capability and limitations of single-layer error-correcting networks

Perceptrons perform pattern classification. That is, each neuron classifies the input patterns it receives into classes determined by the teacher. This is thus an example of a supervised network, with a separate teacher for each output neuron. The classification is most clearly understood if the output neurons are binary, or are strongly non-linear, but the network will still try to obtain an optimal mapping with linear or near-linear output neurons.

When each neuron operates as a binary classifier, we can consider how many input patterns p can be classified by each neuron, and the classes of pattern that can be correctly classified. The result is that the maximum number of patterns that can be correctly classified by a neuron with C inputs is

$$p_{\max} = 2C \qquad (B.101)$$

when the inputs have random continuous-valued inputs, but the patterns must be linearly separable (see Section A.2.1, and Hertz, Krogh and Palmer (1991)). For a one-layer network, no set of weights can be found that will perform the XOR (exclusive OR), or any other non-linearly separable function (see Appendix A).

Although the inability of one-layer networks with binary neurons to solve non-linearly separable problems is a limitation, it is not in practice a major limitation on the processing that can be performed in a neural network for a number of reasons. First, if the inputs can take continuous values, then if the patterns are drawn from a random distribution, the one-layer network can map up to $2C$ of them. Second, as described for pattern associators, the perceptron could be preceded by an expansion recoding network such as a competitive network with more output than input neurons. This effectively provides a two-layer network for solving the problem, and multilayer networks are in general capable of solving arbitrary mapping problems. Ways in which such multilayer networks might be trained are discussed later in this Appendix.

We now return to the issue of the capacity of one-layer perceptrons, that is, how many patterns p can be correctly mapped to correct binary outputs if the input patterns are linearly separable.

B.10.3.1 Output neurons with continuous values, and random patterns

Before treating this case, we note that if the inputs are orthogonal, then just as in the pattern associator, C patterns can be correctly classified, where there are C inputs, x_j, $(j = 1, C)$, per neuron. The argument is the same as for a pattern associator.

We consider next the capacity of a one-layer error-correcting network that learns patterns drawn from a random distribution. For neurons with continuous output values, whether the activation function is linear or not, the capacity (for fully distributed inputs) is set by the criterion that the set of input patterns must be linearly independent (see Hertz, Krogh and Palmer (1991)). (Three patterns are linearly independent if any one cannot be formed by addition (with scaling allowed) of the other two patterns – see Appendix A.) Given that there can be a maximum of C linearly independent patterns in a C-dimensional space (see Appendix A), the capacity of the perceptron with such patterns is C patterns. If we choose p random patterns with continuous values, then they will be linearly independent for $p \leq C$ (except for cases with very low probability when the randomly chosen values may not produce linearly independent patterns). (With random continuous values for the input patterns, it is very unlikely that the addition of any two, with scaling allowed, will produce a third pattern in the set.) Thus with continuous valued input patterns,

$$p_{\max} = C. \qquad (B.102)$$

If the inputs are not linearly independent, networks trained with the delta rule produce a least mean squares (LMS) error (optimal) solution (see below).

B.10.3.2 Output neurons with binary threshold activation functions

Let us consider here strictly defined perceptrons, that is, networks with (binary) threshold output neurons, and taught by the perceptron learning procedure.

Capacity with fully distributed output patterns

The condition here for correct classification is that described in Appendix A for the AND and XOR functions, that the patterns must be linearly separable. If we consider random continuous-valued inputs, then the capacity is

$$p_{\max} = 2C \tag{B.103}$$

(see Cover (1965), Hertz, Krogh and Palmer (1991); this capacity is the case with C large, and the number of output neurons small). The interesting point to note here is that, even with fully distributed inputs, a perceptron is capable of learning more (fully distributed) patterns than there are inputs per neuron. This formula is in general valid for large C, but happens to hold also for the AND function illustrated in Appendix A.2.1.

Sparse encoding of the patterns

If the output patterns y are sparse (but still distributed), then just as with the pattern associator, it is possible to map many more than C patterns to correct outputs. Indeed, the number of different patterns or prototypes p that can be stored is

$$p \approx C/a \tag{B.104}$$

where a is the sparseness of the target pattern t. p can in this situation be much larger than C (cf. Rolls and Treves (1990), and Rolls and Treves (1998) Appendix A3).

Perceptron convergence theorem

It can be proved that such networks will learn the desired mapping in a finite number of steps (Block 1962, Minsky and Papert 1969, Hertz, Krogh and Palmer 1991). (This of course depends on there being such a mapping, the condition for this being that the input patterns are linearly separable.) This is important, for it shows that single-layer networks can be proved to be capable of solving certain classes of problem.

As a matter of history, Minsky and Papert (1969) went on to emphasize the point that no one-layer network can correctly classify non-linearly separable patterns. Although it was clear that multilayer networks can solve such mapping problems, Minsky and Papert were pessimistic that an algorithm for training such a multilayer network would be found. Their emphasis that neural networks might not be able to solve general problems in computation, such as computing the XOR, which is a non-linearly separable mapping, resulted in a decline in research activity in neural networks. In retrospect, this was unfortunate, for humans are rather poor at solving parity problems such as the XOR (Thorpe, O'Regan and Pouget 1989), yet can perform many other useful neural network operations very quickly. Algorithms for training multilayer perceptrons were gradually discovered by a number of different investigators, and became widely known after the publication of the algorithm described by Rumelhart, Hinton and Williams (1986b), and Rumelhart, Hinton and Williams (1986a). Even before this, interest in neural network pattern associators, autoassociators and competitive networks was developing (see Hinton and Anderson (1981), Kohonen (1977), Kohonen (1988)), but the acceptance of the algorithm for training multilayer perceptrons led to a great rise in interest in neural networks, partly for use in connectionist models of cognitive function (McClelland and Rumelhart 1986, McLeod, Plunkett and Rolls 1998), and partly for use in applications (see Bishop (1995)).

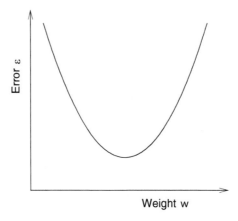

Fig. B.42 The error function ϵ for a neuron in the direction of a particular weight w.

In that perceptrons can correctly classify patterns provided only that they are linearly separable, but pattern associators are more restricted, perceptrons are more powerful learning devices than Hebbian pattern associators.

B.10.3.3 Gradient descent for neurons with continuous-valued outputs

We now consider networks trained by the delta (error correction) rule B.100, and having continuous-valued outputs. The activation function may be linear or non-linear, but provided that it is differentiable (in practice, does not include a sharp threshold), the network can be thought of as gradually decreasing the error on every learning trial, that is as performing some type of gradient descent down a continuous error function. The concept of gradient descent arises from defining an error ϵ for a neuron as

$$\epsilon = \sum_{\mu}(t^{\mu} - y^{\mu})^2 \tag{B.105}$$

where μ indexes the patterns learned by the neuron. The error function for a neuron in the direction of a particular weight would have the form shown in Fig. B.42. The delta rule can be conceptualized as performing gradient descent of this error function, in that for the jth synaptic weight on the neuron

$$\delta w_j = -k\partial\epsilon/\partial w_j \tag{B.106}$$

where $\partial\epsilon/\partial w_j$ is just the slope of the error curve in the direction of w_j in Fig. B.42. This will decrease the weight if the slope is positive and increase the weight if the slope is negative. Given equation B.99, and recalling that $h = \sum_{j} x_j w_j$, equation B.42 becomes

$$\delta w_j = -k\partial/\partial w_j \sum_{\mu}[(t^{\mu} - f(h^{\mu}))^2] \tag{B.107}$$

$$= 2k \sum_{\mu}[(t^{\mu} - y^{\mu})]f'(h)x_j$$

where $f'(h)$ is the derivative of the activation function. Provided that the activation function is monotonically increasing, its derivative will be positive, and the sign of the weight change will only depend on the mean sign of the error. Equation B.108 thus shows one way in which, from a gradient descent conceptualization, equation B.100 can be derived.

With linear output neurons, this gradient descent is proved to reach the correct mapping (see Hertz, Krogh and Palmer (1991)). (As with all single-layer networks with continuous-valued output neurons, a perfect solution is only found if the input patterns are linearly independent. If they are not, an optimal mapping is achieved, in which the sum of the squares of the errors is a minimum.) With non-linear output neurons (for example with a sigmoid activation function), the error surface may have local minima, and is not guaranteed to reach the optimal solution, although typically a near-optimal solution is achieved. Part of the power of this gradient descent conceptualization is that it can be applied to multilayer networks with neurons with non-linear but differentiable activation functions, for example with sigmoid activation functions (see Hertz, Krogh and Palmer (1991)).

B.10.4 Properties

The properties of single-layer networks trained with a delta rule (and of perceptrons) are similar to those of pattern associators trained with a Hebbian rule in many respects (see Section B.2). In particular, the properties of generalization and graceful degradation are similar, provided that (for both types of network) distributed representations are used. The main differences are in the types of pattern that can be separated correctly, the learning speed (in that delta-rule networks can take advantage of many training trials to learn to separate patterns that could not be learned by Hebbian pattern associators), and in that the delta-rule network needs an error term to be supplied for each neuron, whereas an error term does not have to be supplied for a pattern associator, just an unconditioned or forcing stimulus. Given these overall similarities and differences, the properties of one-layer delta-rule networks are considered here briefly.

B.10.4.1 Generalization

During recall, delta-rule one-layer networks with non-linear output neurons produce appropriate outputs if a recall cue vector x_r is similar to a vector that has been learned already. This occurs because the recall operation involves computing the dot (inner) product of the input pattern vector x_r with the synaptic weight vector w_i, so that the firing produced, y_i, reflects the similarity of the current input to the previously learned input pattern x. Distributed representations are needed for this property. If two patterns that a delta-rule network has learned to separate are very similar, then the weights of the network will have been adjusted to force the different outputs to occur correctly. At the same time, this will mean that the way in which the network generalizes in the space between these two vectors will be very sharply defined. (Small changes in the input vector will force it to be classified one way or the other.)

B.10.4.2 Graceful degradation or fault tolerance

One-layer delta-rule networks show graceful degradation provided that the input patterns x are distributed.

B.10.4.3 Prototype extraction, extraction of central tendency, and noise reduction

These occur as for pattern autoassociators.

B.10.4.4 Speed

Recall is very fast in a one-layer pattern associator or perceptron, because it is a feedforward network (with no recurrent or lateral connections). Recall is also fast if the neuron has cell-like properties, because the stimulus input firings x_j ($j = 1, C$ axons) can be applied simultaneously to the synapses w_{ij}, and the activation h_i can be accumulated in one or two time constants of the synapses and dendrite (e.g. 10–20 ms) (see Section B.6.5). Whenever the threshold of the cell is exceeded, it fires. Thus, in effectively one time step, which takes the brain no more than 10–20 ms, all the output neurons of the delta-rule network can be firing with rates that reflect the input firing of every axon.

Learning is as fast ('one-shot') in perceptrons as in pattern associators if the input patterns are orthogonal. If the patterns are not orthogonal, so that the error correction rule has to work in order to separate patterns, then the network may take many trials to achieve the best solution (which will be perfect under the conditions described above).

B.10.4.5 Non-local learning rule

The learning rule is not truly local, as it is in pattern associators, autoassociators, and competitive networks, in that with one-layer delta-rule networks, the information required to change each synaptic weight is not available in the presynaptic terminal (reflecting the presynaptic rate) and the postsynaptic activation. Instead, an error for the neuron must be computed, possibly by another neuron, and then this error must be conveyed back to the postsynaptic neuron to provide the postsynaptic error term, which together with the presynaptic rate determines how much the synapse should change, as in equation B.100,

$$\delta w_{ij} = k(t_i - y_i)x_j$$

where $(t_i - y_i)$ is the error.

A rather special architecture would be required if the brain were to utilize delta-rule error-correcting learning. One such architecture might require each output neuron to be supplied with its own error signal by another neuron. The possibility (Albus 1971) that this is implemented in one part of the brain, the cerebellum, is described in Rolls and Treves (1998) Chapter 9. Another functional architecture would require each neuron to compute its own error by subtracting its current activation by its x inputs from another set of afferents providing the target activation for that neuron. A neurophysiological architecture and mechanism for this is not currently known.

B.10.4.6 Interference

Interference is less of a property of single-layer delta rule networks than of pattern autoassociators and autoassociators, in that delta rule networks can learn to separate patterns even when they are highly correlated. However, if patterns are not linearly independent, then the delta rule will learn a least mean squares solution, and interference can be said to occur.

B.10.4.7 Expansion recoding

As with pattern associators and autoassociators, expansion recoding can separate input patterns into a form that makes them learnable, or that makes learning more rapid with only a few trials needed, by delta rule networks. It has been suggested that this is the role of the granule cells in the cerebellum, which provide for expansion recoding by 1,000:1 of the mossy fibre inputs before they are presented by the parallel fibres to the cerebellar Purkinje cells (Marr 1969, Albus 1971, Rolls and Treves 1998) (Chapter 23).

B.10.4.8 Utility of single-layer error-correcting networks in information processing by the brain

In the cerebellum, each output cell, a Purkinje cell, has its own climbing fibre, that distributes from its inferior olive cell its terminals throughout the dendritic tree of the Purkinje cell. It is this climbing fibre that controls whether learning of the x inputs supplied by the parallel fibres onto the Purkinje cell occurs, and it has been suggested that the function of this architecture is for the climbing fibre to bring the error term to every part of the postsynaptic neuron (Chapter 23). This rather special arrangement with each output cell apparently having its own teacher is probably unique in the brain, and shows the lengths to which the brain might need to go to implement a teacher for each output neuron. The requirement for error-correction learning is to have the neuron forced during a learning phase into a state that reflects its error while presynaptic afferents are still active, and rather special arrangements are needed for this.

B.11 Multilayer perceptrons: backpropagation of error networks

B.11.1 Introduction

So far, we have considered how error can be used to train a one-layer network using a delta rule. Minsky and Papert (1969) emphasized the fact that one-layer networks cannot solve certain classes of input–output mapping problems (as described above). It was clear then that these restrictions would not apply to the problems that can be solved by feedforward multilayer networks, if they could be trained. A multilayer feedforward network has two or more connected layers, in which connections allow activity to be projected forward from one layer to the next, and in which there are no lateral connections within a layer. Such a multilayer network has an output layer (which can be trained with a standard delta rule using an error provided for each output neuron), and one or more hidden layers, in which the neurons do not receive separate error signals from an external teacher. (Because they do not provide the outputs of the network directly, and do not directly receive their own teaching error signal, these layers are described as hidden.) To solve an arbitrary mapping problem (in which the inputs are not linearly separable), a multilayer network could have a set of hidden neurons that would remap the inputs in such a way that the output layer can be provided with a linearly separable problem to solve using training of its weights with the delta rule. The problem was: how could the synaptic weights into the hidden neurons be trained in such a way that they would provide an appropriate representation? Minsky and Papert (1969) were pessimistic that such a solution would be found and, partly because of this, interest in computations in neural networks declined for many years. Although some work in neural networks continued in the following years (e.g. (Marr 1969, Marr 1970, Marr 1971, Willshaw and Longuet-Higgins 1969, Willshaw 1981, Malsburg 1973, Grossberg 1976a, Grossberg 1976b, Arbib 1964, Amari 1982, Amari, Yoshida and Kanatani 1977)), widespread interest in neural networks was revived by the type of approach to associative memory and its relation to human memory taken by the work described in the volume edited by Hinton and Anderson (1981), and by Kohonen (Kohonen 1977, Kohonen 1989). Soon after this, a solution to training a multilayer perceptron using backpropagation of error became widely known (Rumelhart, Hinton and Williams 1986b, Rumelhart, Hinton and Williams 1986a) (although earlier solutions had been found), and very great interest in neural networks and also in neural network approaches to cognitive processing (connectionism) developed (Rumelhart and McClelland 1986, McClelland and Rumelhart 1986, McLeod, Plunkett and Rolls 1998).

B.11.2 Architecture and algorithm

An introduction to the way in which a multilayer network can be trained by backpropagation of error is described next. Then we consider whether such a training algorithm is biologically

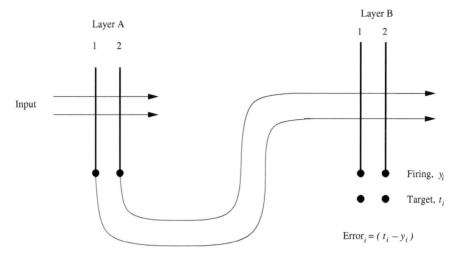

Fig. B.43 A two-layer perceptron. Inputs are applied to layer A through modifiable synapses. The outputs from layer A are applied through modifiable synapses to layer B. Layer B can be trained using a delta rule to produce firing y_i which will approach the target t_i. It is more difficult to modify the weights in layer A, because appropriate error signals must be backpropagated from layer B.

plausible. A more formal account of the training algorithm for multilayer perceptrons (some-times abbreviated MLP) is given by Rumelhart, Hinton and Williams (1986b), Rumelhart, Hinton and Williams (1986a), and Hertz, Krogh and Palmer (1991), and below.

Consider the two-layer network shown in Fig. B.43. Inputs to the hidden neurons in layer A feed forward activity to the output neurons in layer B. The neurons in the network have a sigmoid activation function. One reason for such an activation function is that it is non-linear, and non-linearity is needed to enable multilayer networks to solve difficult (non-linearly sep-arable) problems. (If the neurons were linear, the multilayer network would be equivalent to a one-layer network, which cannot solve such problems.) Neurons B1 and B2 of the output layer, B, are each trained using a delta rule and an error computed for each output neuron from the target output for that neuron when a given input pattern is being applied to the network. Consider now the error that needs to be used to train neuron A1 by a delta rule. This error clearly influences the error of neuron B1 in a way that depends on the magnitude of the synaptic weight from neuron A1 to B1; and on the error of neuron B2 in a way that depends on the magnitude of the synaptic weight from neuron A1 to B2. In other words, the error for neuron A1 depends on:

the weight from A1 to B1 (w_{11}) · error of neuron B1
+ the weight from A1 to B2 (w_{21}) · error of neuron B2.

In this way, the error calculation can be propagated backwards through the network to any neuron in any hidden layer, so that each neuron in the hidden layer can be trained, once its error is computed, by a delta rule (which uses the computed error for the neuron and the presynaptic firing at the synapse to correct each synaptic weight). For this to work, the way in which each neuron is activated and sends a signal forward must be continuous (not binary), so that the extent to which there is an error in, for example, neuron B1 can be related back in a graded way to provide a continuously variable correction signal to previous stages. This is one of the requirements for enabling the network to descend a continuous error surface. The activation function must be non-linear (e.g. sigmoid) for the network to learn more than

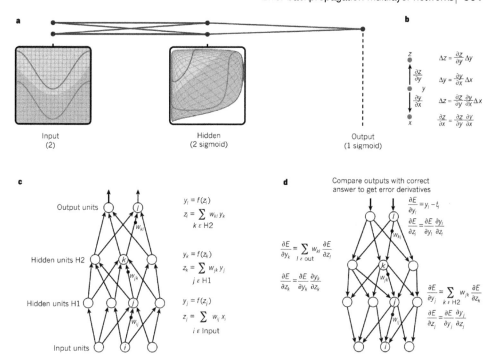

Fig. B.44 Multilayer neural networks and backpropagation (see text). a. How a multilayer neural network (shown by the connected dots) can distort the input space to make the classes of data (examples of which are on the red and blue lines) linearly separable. b. A chain rule of derivatives. c. Equations used for computing the forward pass in a neural net with two hidden layers and one output layer. d. The equations used for computing the backward pass (see text). Note that the indices use different conventions to those used in the remainder of this book. (Reprinted by permission from Macmillan Publishers Ltd: Nature, 521: 436–444, Deep learning, Yann LeCun, Yoshua Bengio, Geoffrey Hinton. Copyright ©2015. a) Reproduced with permission from C. Olah, http://colah.github.io/)

could be learned by a single-layer network. (Remember that a multilayer linear network can always be made equivalent to a single-layer linear network, and that there are some problems that cannot be solved by single-layer networks.) For the way in which the error of each output neuron should be taken into account to be specified in the error correction rule, the position at which the output neuron is operating on its activation function must also be taken into account. For this, the slope of the activation function is needed, and because the slope is needed, the activation function must be differentiable. Although we indicated use of a sigmoid activation function, other activation functions that are non-linear and monotonically increasing (and differentiable) can be used. (For further details, see Rumelhart, Hinton and Williams (1986b), Rumelhart, Hinton and Williams (1986a), and Hertz, Krogh and Palmer (1991)).

More formally, the operation of a backpropagation of error network can be described as follows (Fig. B.44) (LeCun, Bengio and Hinton 2015). Fig. B.44a shows how a multilayer neural network (shown by the connected dots) can distort the input space to make the classes of data (examples of which are on the red and blue lines) linearly separable. Note how a regular grid (shown on the left) in the input space is also transformed (shown in the middle panel) by hidden units. This is an illustrative example with only two input units, two hidden units and one output unit, but the networks may contain hundreds of thousands of units.

Fig. B.44b shows how the chain rule of derivatives tells us how two small effects (that of a small change of x on y, and that of y on z) are composed. A small change Δx in x

gets transformed first into a small change Δx in y by getting multiplied by $\partial y / \partial x$ (that is, the definition of partial derivative). Similarly, the change Δy creates a change Δz in z. Substituting one equation into the other gives the chain rule of derivatives – how Δx gets turned into Δz through multiplication by the product of $\partial y / \partial x$ and $\partial z / \partial x$. It also works when x, y and z are vectors (and the derivatives are Jacobian matrices).

Fig. B.44c shows the equations used for computing the forward pass in a neural net with two hidden layers and one output layer, each constituting a module through which one can backpropagate gradients. At each layer, we first compute the total input z to each unit, which is a weighted sum of the outputs of the units in the layer below. Then a non-linear function $f(.)$ is applied to z to get the output of the unit. For simplicity, bias terms are omitted. The non-linear function used is now typically the rectified linear unit (ReLU) $f(z) = \max(0, z)$ (known in neuroscience as a threshold linear activation function), although sigmoid activation functions have been used.

Fig. B.44d shows the equations used for computing the backward pass. At each hidden layer we compute the error derivative with respect to the output of each unit, which is a weighted sum of the error derivatives with respect to the total inputs to the units in the layer above. We then convert the error derivative with respect to the output into the error derivative with respect to the input by multiplying it by the gradient of $f(z)$. At the output layer, the error derivative with respect to the output of a unit is computed by differentiating the cost function. This gives $y_l - t_l$ if the cost function for unit l is $0.5(y_l - t_l)^2$, where t_l is the target value. Once the $\partial E / \partial z_k$ is known, the error-derivative for the weight w_{jk} on the connection from unit j in the layer below is $y_j \partial E / \partial z_k$.

Once we have this error derivative for a single neuron, we can update the weights into that neuron by performing a gradient descent step in the direction specified by the derivative to minimize the cost function (see LeCun, Bengio and Hinton (2015)).

B.11.3 Properties of multilayer networks trained by error backpropagation

B.11.3.1 Arbitrary mappings

Arbitrary mappings of non-linearly separable patterns can be achieved. For example, such networks can solve the XOR problem, and parity problems in general of which XOR is a special case. (The parity problem is to determine whether the sum of the (binary) bits in a vector is odd or even.) Multilayer feedforward backpropagation of error networks are not guaranteed to converge to the best solution, and may become stuck in local minima in the error surface. However, they generally perform very well.

B.11.3.2 Fast operation

The network operates as a feedforward network, without any recurrent or feedback processing. Thus (once it has learned) the network operates very quickly, with a time proportional to the number of layers.

B.11.3.3 Learning speed

The learning speed can be very slow, taking many thousands of trials. The network learns to gradually approximate the correct input–output mapping required, but the learning is slow because of the credit assignment problem for neurons in the hidden layers. The credit assignment problem refers to the issue of how much to correct the weights of each neuron in the hidden layer. As the example above shows, the error for a hidden neuron could influence the errors of many neurons in the output layers, and the error of each output neuron reflects the error from many hidden neurons. It is thus difficult to assign credit (or blame) on any single trial to any particular hidden neuron, so an error must be estimated, and the net run until the weights of the crucial hidden neurons have become altered sufficiently to allow good performance of the network. Another factor that can slow learning is that if a neuron operates close to a horizontal part of its activation function, then the output of the neuron will depend rather little on its activation, and correspondingly the error computed to backpropagate will depend rather little on the activation of that neuron, so learning will be slow.

More general approaches to this issue suggest that the number of training trials for such a network will (with a suitable training set) be of the same order of magnitude as the number of synapses in the network (see Cortes, Jaeckel, Solla, Vapnik and Denker (1996)).

B.11.3.4 Number of hidden neurons and generalization

Backpropagation networks are generally intended to discover regular mappings between the input and output, that is mappings in which generalization will occur usefully. If there were one hidden neuron for every combination of inputs that had to be mapped to an output, then this would constitute a look-up table, and no generalization between similar inputs (or inputs not yet received) would occur. The best way to ensure that a backpropagation network learns the structure of the problem space is to set the number of neurons in the hidden layers close to the minimum that will allow the mapping to be implemented. This forces the network not to operate as a look-up table. A problem is that there is no general rule about how many hidden neurons are appropriate, given that this depends on the types of mappings required. In practice, these networks are sometimes trained with different numbers of hidden neurons, until the minimum number required to perform the required mapping has been approximated.

B.12 Biologically plausible networks vs backpropagation

Given that the error for a hidden neuron in an error backpropagation network is calculated by propagating backwards information based on the errors of all the output neurons to which a hidden neuron is connected, and all the relevant synaptic weights, and the activations of the output neurons to define the part of the activation function on which they are operating, it is implausible to suppose that the correct information to provide the appropriate error for each hidden neuron is propagated backwards between real neurons. A hidden neuron would have to 'know', or receive information about, the errors of all the neurons to which it is connected, and its synaptic weights to them, and their current activations. If there were more than one hidden layer, this would be even more difficult.

To expand on the difficulties: first, there would have to be a mechanism in the brain for providing an appropriate error signal to each output neuron in the network. With the possible exception of the cerebellum, an architecture where a separate error signal could be provided for each output neurons is difficult to identify in the brain. Second, any retrograde passing of messages across multiple-layer forward-transmitting pathways in the brain that could be used for backpropagation seems highly implausible, not only because of the difficulty of getting the correct signal to be backpropagated, but also because retrograde signals passed by axonal transport in a multilayer net would take days to arrive, long after the end of any feedback given in the environment indicating a particular error. Third, as noted in Chapter 11, the backprojection pathways that are present in the cortex seem suited to perform recall, and this would make it difficult for them also to have the correct strength to carry the correct error signal.

A problem with the backpropagation of error approach in a biological context is thus that in order to achieve their competence, backpropagation networks use what is almost certainly a learning rule that is much more powerful than those that could be implemented biologically, and achieve their excellent performance by performing the mapping though a minimal number of hidden neurons. In contrast, real neuronal networks in the brain probably use much less powerful learning rules, in which errors are not propagated backwards, and at the same time have very large numbers of hidden neurons, without the bottleneck that helps to provide backpropagation networks with their good performance. A consequence of these differences between backpropagation and biologically plausible networks may be that the way in which biological networks solve difficult problems may be rather different from the way in which backpropagation networks find mappings. Thus the solutions found by connectionist systems may not always be excellent guides to how biologically plausible networks may perform on similar problems. Part of the challenge for future work is to discover how more biologically plausible networks than backpropagation networks can solve comparably hard problems, and then to examine the properties of these networks, as a perhaps more accurate guide to brain computation.

As stated above, it is a major challenge for brain research to discover whether there are algorithms that will solve comparably difficult problems to backpropagation, but with a local learning rule. Such algorithms may be expected to require many more hidden neurons than backpropagation networks, in that the brain does not appear to use information bottlenecks to help it solve difficult problems. The issue here is that much of the power of backpropagation algorithms arises because there is a minimal number of hidden neurons to perform the required mapping using a final one-layer delta-rule network. Useful generalization arises in such networks because with a minimal number of hidden neurons, the net sets the representation they provide to enable appropriate generalization. The danger with more hidden neurons is that the network becomes a look-up table, with one hidden neuron for every required output, and generalization when the inputs vary becomes poor. The challenge is to find a more biologically plausible type of network that operates with large numbers of neurons, and yet that still provides useful generalization. An example of such an approach is described in Chapter 25.

B.13 Convolution networks

Although not biologically plausible, a class of artificial network that was inspired by neuroscience is the convolution network (LeCun, Kavukcuoglu and Farabet 2010, LeCun, Bengio and Hinton 2015, Bengio, Goodfellow and Courville 2017). This type of network is described here, to compare to biologically plausible networks. A convolution network is a multilayer network that is trained by backpropagation of error and that has convergence from stage to stage. When used, the term 'deep learning' usually refers to convolution learning, and this is an example of a 'deep neural network'.

In more detail, the architecture of a typical convolution network (ConvNet) is structured as a series of stages (LeCun, Bengio and Hinton 2015). The first few stages are composed of two types of layers: convolutional layers and pooling layers.

Units in a convolutional layer are organized in feature maps, within which each unit is connected to local patches in the feature maps of the previous layer through a set of weights called a filter bank. The result of this local weighted sum is then passed through a non-linearity such as a ReLU (a threshold-linear activation function). All units in a feature map share the same filter bank. To be more specific: the local patch of one layer that connects to a neuron in the next layer may be 3x3 units. Whatever weights are learned by backpropagation are the same for all the units across the layer, and this is achieved artificially, effectively by copying the weights laterally. There may be different feature maps in a layer (e.g. for different colours), and each uses a different filter bank. The reason for this architecture is twofold. First, in array data such as images, local groups of values are often highly correlated, forming distinctive local motifs that are easily detected. Second, the local statistics of images and other signals are invariant to location. In other words, if a motif can appear in one part of the image, it could appear anywhere, hence the idea of units at different locations sharing the same weights and detecting the same pattern in different parts of the array. Mathematically, the filtering operation performed by a feature map is a discrete convolution, hence the name.

Although the role of the convolutional layer is to detect local conjunctions of features from the previous layer, the role of the pooling layer is to merge semantically similar features into one. Because the relative positions of the features forming a motif can vary somewhat, reliably detecting the motif can be done by coarse-graining the position of each feature. A typical pooling unit computes the maximum of a local patch of units in one feature map (or in a few feature maps). To be more specific: a pooling neuron may take the maximum input from a 2 unit x 2 unit patch of units in its input convolution layer. Neighbouring pooling units take input from patches that are shifted by more than one row or column, thereby reducing the dimension of the representation and creating an invariance to small shifts and distortions. The 'stride' may be 2. A convolution net architecture might have these tiny local convolution then pooling of 'max' operator pairs of layers stacked 140 or more on top of one another. Or the architecture might be varied, with several convolution layers before the next pooling layer. At the top of the system, there may be more convolutional and fully-connected pooling layers. To train a ConvNet, stochastic gradient descent is used in conjunction with the gradients computed by backpropagation of error to update the weights to minimize the errors of the output units. The targets for the individual units in the output layer are decided by the experimenter, and might include for one neuron cars, for another neuron bicycles, etc. The whole net is trained with in the order of one million exemplars.

Thus in a convolution network, layers typically alternate, with one layer performing the convolution and filtering followed by a non-linear function such as a threshold linear function (an 'S' layer), and the next layer performing 'pooling' over the neurons in the previous layer (a 'C' layer), to help with transform invariance. The alternation of 'S' and 'C' layers follows the suggestion of Hubel and Wiesel (1962) that complex cells might compute their responses by summing over a set of simple cells. However, there is no evidence for any alternation of S and C layers in most of the primate ventral visual system. The general idea though is that some neurons in the net need to learn to respond to feature combinations, and this is the function of the convolution layer; and the system also needs to do something to cope with transform invariance, and this is done by the max function layers.

The concept of copying weights laterally was used in the Neocognitron (Fukushima 1980) (Section 25.3.5.1), and has the machine advantages that only one small region is learned, and that the 'filters' or neurons are uniformly good (or bad) throughout a layer. The convolution operator in a convolution network has this effect. This lateral copying is of course biologically implausible, and is not used in VisNet (Chapter 25).

With its powerful backpropagation training algorithm, and very large numbers of training trials, a convolution net can learn to activate the correct output neuron in the output layer for a category of object with good accuracy which may be commercially useful (LeCun, Bengio and Hinton 2015). However, unlike VisNet, there is no attempt in general to teach the network transform invariance by presenting images with spatial continuity, and no attempt to take advantage of the statistics of the world to help it learn which transforms are probably of the same object by capitalising on temporal continuity. The network is very biologically implausible, in that each unit (or 'neuron') receives from only typically a 3 x 3 or 2 x 2 unit patch of the preceding area (i.e. the receptive fields are small); by using lateral weight copying within a layer; by having up to 140 or more layers stacked on top of each other in the hierarchy; by using the non biologically plausible backpropagation of error training algorithm; and by using a teacher for every neuron in the output layer.

B.14 Contrastive Hebbian learning: the Boltzmann machine

In a move towards a learning rule that is more local than in backpropagation networks, yet that can solve similar mapping problems in a multilayer architecture, we describe briefly contrastive Hebbian learning. The multilayer architecture has forward connections through the network to the output layer, and a set of matching backprojections from the output layer through each of the hidden layers to the input layer. The forward connection strength between any pair of neurons has the same value as the backward connection strength between the same two neurons, resulting in a symmetric set of forward and backward connection strengths. An input pattern is applied to the multilayer network, and an output is computed using normal feedforward activation processing with neurons with a sigmoid (non-linear and monotonically increasing) activation function. The output firing then via the backprojections is used to create firing of the input neurons. This process is repeated until the firing rates settle down, in an iterative way (which is similar to the settling of the autoassociative nets described in Section B.3). After settling, the correlations between any two neurons are remembered, for this type of unclamped operation, in which the output neurons fire at the rates that the process just described produces. The correlations reflect the normal presynaptic and postsynaptic terms used in the Hebb rule, e.g. $(x_j y_i)^{uc}$, where 'uc' refers to the unclamped condition, and as usual x_j is the firing rate of the input neuron, and y_i is the activity of the receiving neuron. The output neurons are then clamped to their target values, and the iterative process just described is repeated, to produce for every pair of synapses in the network $(x_j y_i)^c$, where the c refers now to the clamped condition. An error correction term for each synapse is then computed from the difference between the remembered correlation of the unclamped and the clamped conditions, to produce a synaptic weight correction term as follows:

$$\delta w_{ij} = k[(x_j y_i)^c - (x_j y_i)^{uc}], \tag{B.108}$$

where k is a learning rate constant. This process is then repeated for each input pattern to output pattern to be learned. The whole process is then repeated many times with all patterns until the output neurons fire similarly in the clamped and unclamped conditions, that is until the errors have become small. Further details are provided by Hinton and Sejnowski (1986). The version described above is the mean field (or deterministic) Boltzmann machine (Peterson and Anderson 1987, Hinton 1989). It is sometimes called the '**wake‑sleep**' algorithm (Hinton et al. 1995), because when the system is unclamped, with no input, it might be likened to a sleeping or dreaming state. More traditionally, a Boltzmann machine updates one randomly chosen neuron at a time, and each neuron fires with a probability that depends on its activation (Ackley, Hinton and Sejnowski 1985, Hinton and Sejnowski 1986). The latter version makes fewer theoretical assumptions, while the former may operate an order of magnitude faster (Hertz, Krogh and Palmer 1991).

In terms of biological plausibility, it certainly is the case that there are backprojections between adjacent cortical areas (see Chapters 1 and 11). Indeed, there are as many back-projections between adjacent cortical areas as there are forward projections. The backward projections seem to be more diffuse than the forward projections, in that they connect to a wider region of the preceding cortical area than the region that sends the forward projections. If the backward and the forward synapses in such an architecture were Hebb-modifiable, then there is a possibility that the backward connections would be symmetric with the forward connections. Indeed, such a connection scheme would be useful to implement top-down recall, as summarized in Chapter 24 and described by Rolls and Treves (1998) in their Chapter 6. What seems less biologically plausible is that after an unclamped phase of operation, the correlations between all pairs of neurons would be remembered, there would then be a clamped phase of operation with each output neuron clamped to the required rate for that particular input pattern, and then the synapses would be corrected by an error correction rule that would require a comparison of the correlations between the neuronal firing of every pair of neurons in the unclamped and clamped conditions.

Although this algorithm has the disadvantages that it is not very biologically plausible, and does not operate as well as standard backpropagation, it has been made use of by O'Reilly and Munakata (2000) in approaches to connectionist modelling in cognitive neuroscience.

B.15   Deep Belief Networks

Another artificial network is a Deep Belief Network which is a probabilistic model composed of many Restricted Boltzmann Machines (RBMs) stacked on top of each other with each layer trained on the output of the previous one (Hinton, Osindero and Teh 2006). A restricted Boltzmann Machine (in contrast to a traditional Boltzmann Machine) has no lateral connections within layers, and the connectivity is relaxed to be undirected. This results in the units in a particular layer being conditionally independent given a set of observed activations in the other. This is a key difference that allows for the efficient learning and inference within these models. Each subsequent layer acts like constraints on the activities of the one below. This is a greedy training process and will probably not yield an optimum representation for the entire hierarchy, primarily because the individual layers are trained in isolation. (Once a layer is trained the weights are frozen.) This inefficiency however can be remedied by a relatively small amount of fine tuning (generative or discriminative) after the stack is complete (Hinton and Salakhutdinov 2006). It should be noted that a Deep Belief Network itself is not a deep Restricted Boltzmann Machine. The lower layers do not themselves define an undirected model, and only the top-most two layers keep their undirected nature. However there has been work on modifications that restore the undirected nature of the lower layers allowing inference within the lower layers to make use of top-down connectivity as well as the usual bottom-up connectivity (Salakhutdinov and Larochelle 2010).

Deep belief networks do have the advantage, like VisNet, that they are unsupervised, and do not need a teacher for every output neuron to help teach the system how to respond. However, scaling such models to full-sized, high-dimensional images remains a difficult problem. To address this a pooling operation has been tested within a deep belief network, and this can help to form transform-invariant representations (Lee, Grosse, Ranganath and Ng 2011). However, their training algorithm remains very biologically implausible.

B.16   Reinforcement learning

In supervised networks, an error signal is provided for each output neuron in the network, and whenever an input to the network is provided, the error signals specify the magnitude and direction of the error in the output produced by each neuron. These error signals are then used to correct the synaptic weights in the network in such a way that the output errors for each input pattern to be learned gradually diminish over trials (see Sections B.10 and B.11). These networks have an architecture that might be similar to that of the pattern associator shown in

Fig. B.1, except that instead of an unconditioned stimulus, there is an error correction signal provided for each output neuron. Such a network trained by an error correcting (or delta) rule is known as a one-layer perceptron. The architecture is not very plausible for most brain regions, in that it is not clear how an individual error signal could be computed for each of thousands of neurons in a network, and fed into each neuron as its error signal and then used in a delta rule synaptic correction (see Section B.10).

The architecture can be generalized to a multilayer feedforward architecture with many layers between the input and output (Rumelhart, Hinton and Williams 1986a), but the learning is very non-local and rather biologically implausible (see Section B.11), in that an error term (magnitude and direction) for each neuron in the network must be computed from the errors and synaptic weights of all subsequent neurons in the network that any neuron influences, usually on a trial-by-trial basis, by a process known as error backpropagation. Thus although computationally powerful, an issue with perceptrons and multilayer perceptrons that makes them generally biologically implausible for many brain regions is that a separate error signal must be supplied for each output neuron, and that with multilayer perceptrons, computed error backpropagation must occur.

When operating in an environment, usually a simple binary or scalar signal representing success or failure of the whole network or organism is received. This is usually action-dependent feedback that provides a single evaluative measure of the success or failure. Evaluative feedback tells the learner whether or not, and possibly by how much, its behaviour has improved; or it provides a measure of the 'goodness' of the behaviour. Evaluative feedback does not directly tell the learner what it should have done, and although it may provide an index of the degree (i.e. magnitude) of success, it does not include directional information telling the learner how to change its behaviour towards a target, as does error-correction learning (see Barto (1995)). Partly for this reason, there has been some interest in networks that can be taught with such a single reinforcement signal. In this Section (B.16), approaches to such networks are described. It is noted that such networks are classified as reinforcement networks in which there is a single teacher, and that these networks attempt to perform an optimal mapping between an input vector and an output neuron or set of neurons. They thus solve the same class of problems as single layer and multilayer perceptrons. They should be distinguished from pattern-association networks in the brain, which might learn associations between previously neutral stimuli and primary reinforcers such as taste (signals which might be interpreted appropriately by a subsequent part of the brain), but do not attempt to produce arbitrary mappings between an input and an output, using a single reinforcement signal.

A class of problems to which such reinforcement networks might be applied are motor-control problems. It was to such a problem that Barto and Sutton (Barto 1985, Sutton and Barto 1981) applied a reinforcement learning algorithm, the associative reward–penalty algorithm described next. The algorithm can in principle be applied to multilayer networks, and the learning is relatively slow. The algorithm is summarized in Section B.16.1 and by Hertz, Krogh and Palmer (1991). More recent developments in reinforcement learning (see Sections B.16.2 and B.16.3) are described by Sutton and Barto (1998) and reviewed by Dayan and Abbott (2001).

B.16.1 Associative reward–penalty algorithm of Barto and Sutton

The terminology of Barto and Sutton is followed here (see Barto (1985)).

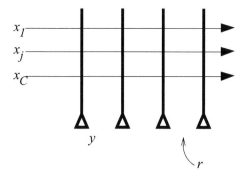

Fig. B.45 A network trained by a single reinforcement input r. The inputs to each neuron are $x_j, j = 1, C$; and y is the output of one of the output neurons.

B.16.1.1 Architecture

The architecture, shown in Fig. B.45, uses a single reinforcement signal, r, = +1 for reward, and −1 for penalty. The inputs x_j take real (continuous) values. The output of a neuron, y, is binary, +1 or −1. The weights on the output neuron are designated w_j.

B.16.1.2 Operation

1. An input vector is applied to the network, and produces activation, h, in the normal way as follows:

$$h = \sum_{j=1}^{C} x_j w_j \qquad (B.109)$$

where $\sum_{j=1}^{C}$ indicates that the sum is over the C input axons (or connections) indexed by j to each neuron.

  2. The output y is calculated from the activation with a noise term η included. The principle of the network is that if the added noise on a particular trial helps performance, then whatever change it leads to should be incorporated into the synaptic weights, in such a way that the next time that input occurs, the performance is improved.

$$y = \begin{cases} +1 & \text{if } h + \eta \geq 0, \\ -1 & \text{else.} \end{cases} \qquad (B.110)$$

where η = the noise added on each trial.

  3. Learning rule. The weights are changed as follows:

$$\delta w_j = \begin{cases} \rho(y - E[y|h])x_j & \text{if } r = +1, \\ \rho\lambda(-y - E[y|h])x_j & \text{if } r = -1. \end{cases} \qquad (B.111)$$

ρ and λ are learning-rate constants. (They are set so that the learning rate is higher when positive reinforcement is received than when negative reinforcement is received.) $E[y|h]$ is the expectation of y given h (usually a sigmoidal function of h with the range ±1). $E[y|h]$ is a (continuously varying) indication of how the neuron usually responds to the current input pattern, i.e. if the actual output y is larger than normally expected, by computing $h = \sum w_j x_j$, because of the noise term, and the reinforcement is +1, increase the weight from x_j; and vice versa. The expectation could be the prediction generated before the noise term is incorporated.

This network combines an associative capacity with its properties of generalization and graceful degradation, with a single 'critic' or error signal for the whole network (Barto 1985). [The term $y - \mathrm{E}[y|h]$ in equation B.111 can be thought of as an error for the output of the neuron: it is the difference between what occurred, and what was expected to occur. The synaptic weight is adjusted according to the sign and magnitude of the error of the postsynaptic firing, multiplied by the presynaptic firing, and depending on the reinforcement r received. The rule is similar to a Hebb synaptic modification rule (equation B.2), except that the postsynaptic term is an error instead of the postsynaptic firing rate, and the learning is modulated by the reinforcement.] The network can solve difficult problems (such as balancing a pole by moving a trolley that supports the pole from side to side, as the pole starts to topple). Although described for single-layer networks, the algorithm can be applied to multilayer networks. The learning rate is very slow, for there is a single reinforcement signal on each trial for the whole network, not a separate error signal for each neuron in the network as is the case in a perceptron trained with an error rule (see Section B.10).

This associative reward–penalty reinforcement-learning algorithm is certainly a move towards biological relevance, in that learning with a single reinforcer can be achieved. That single reinforcer might be broadcast throughout the system by a general projection system. It is not clear yet how a biological system might store the expected output $\mathrm{E}[y|h]$ for comparison with the actual output when noise has been added, and might take into account the sign and magnitude of this difference. Nevertheless, this is an interesting algorithm, which is related to the temporal difference reinforcement learning algorithm described in Section B.16.3.

B.16.2 Reward prediction error or delta rule learning, and classical conditioning

In classical or Pavlovian associative learning, a number of different types of association may be learned (Rolls 2014a, Cardinal, Parkinson, Hall and Everitt 2002). This type of associative learning may be performed by networks with the general architecture and properties of pattern associators (see Section B.2 and Fig. B.1). However, the time course of the acquisition and extinction of these associations can be expressed concisely by a modified type of learning rule in which an error correction term is used (introduced in Section B.16.1), rather than the postsynaptic firing y itself as in equation B.2. Use of this modified, error correction, type of learning also enables some of the properties of classical conditioning to be explained (see Dayan and Abbott (2001) for review), and this type of learning is therefore described briefly here. The rule is known in learning theory as the Rescorla–Wagner rule, after Rescorla and Wagner (1972).

The Rescorla–Wagner rule is a version of error correction or delta-rule learning (see Section B.10), and is based on a simple linear prediction of the expected reward value, denoted by v, associated with a stimulus representation x ($x = 1$ if the stimulus is present, and $x = 0$ if the stimulus is absent). The **expected reward value** v is expressed as the input stimulus variable x multiplied by a weight w

$$v = wx. \tag{B.112}$$

The **reward prediction error** is the difference between the expected reward value v and the actual **reward outcome** r obtained, i.e.

$$\Delta = r - v \tag{B.113}$$

where Δ is the reward prediction error. The value of the weight w is learned by a rule designed to minimize the expected squared error $< (r - v)^2 >$ between the actual reward outcome r and the predicted reward value v. The angle brackets indicate an average over the presentations of the stimulus and reward. The delta rule will perform the required type of learning:

$$\delta w = k(r - v)x \qquad (B.114)$$

where δw is the change of synaptic weight, k is a constant that determines the learning rate, and the term $(r-v)$ is the reward prediction error Δ (equivalent to the error in the postsynaptic firing, rather than the postsynaptic firing y itself as in equation B.2). Application of this rule during conditioning with the stimulus x presented on every trial results in the weight w approaching the asymptotic limit $w = r$ exponentially over trials as the error Δ becomes zero. In extinction, when $r = 0$, the weight (and thus the output of the system) exponentially decays to $w = 0$. This rule thus helps to capture the time course over trials of the acquisition and extinction of conditioning. The rule also helps to account for a number of properties of classical conditioning, including blocking, inhibitory conditioning, and overshadowing (see Dayan and Abbott (2001)).

How this functionality is implemented in the brain is not yet clear. We consider one suggestion (Schultz et al. 1995, Schultz 2004, Schultz 2006, Schultz 2013) after we introduce a further sophistication of reinforcement learning which allows the time course of events within a trial to be taken into account.

B.16.3 Temporal Difference (TD) learning

An important advance in the area of reinforcement learning was the introduction of algorithms that allow for learning to occur when the reinforcement is delayed or received over a number of time steps, and which allow effects within a trial to be taken into account (Sutton and Barto 1998, Sutton and Barto 1990). A solution to these problems is the addition of an adaptive critic that learns through a time difference (TD) algorithm how to predict the future value of the reinforcer. The time difference algorithm takes into account not only the current reinforcement just received, but also a temporally weighted average of errors in predicting future reinforcements. The temporal difference error is the error by which any two temporally adjacent error predictions are inconsistent (see Barto (1995)). The output of the critic is used as an effective reinforcer instead of the instantaneous reinforcement being received (see Sutton and Barto (1998), Sutton and Barto (1990), and Barto (1995)). This is a solution to the temporal credit assignment problem, and enables future rewards to be predicted. Summaries are provided by Doya (1999), Schultz, Dayan and Montague (1997), and Dayan and Abbott (2001).

In reinforcement learning, a learning agent takes an *action* $\mathbf{u}(t)$ in response to the *state* $\mathbf{x}(t)$ of the environment, which results in the change of the state

$$\mathbf{x}(t+1) = F(\mathbf{x}(t), \mathbf{u}(t)), \qquad (B.115)$$

and the delivery of the reinforcement signal, or *reward*

$$r(t+1) = R(\mathbf{x}(t), \mathbf{u}(t)). \qquad (B.116)$$

In the above equations, $\mathbf{x}$ is a vector representation of inputs x_j, and equation B.115 indicates that the next state $\mathbf{x}(t+1)$ at time $(t+1)$ is a function F of the state at the previous time step of the inputs and actions at that time step in a closed system. In equation B.116 the reward at the next time step is determined by a reward function R which uses the current sensory inputs and action taken. The time t may refer to time within a trial.

The goal is to find a *policy* function G which maps sensory inputs **x** to actions

$$\mathbf{u}(t) = G(\mathbf{x}(t)) \tag{B.117}$$

which maximizes the cumulative sum of the rewards based on the sensory inputs.

The current action $\mathbf{u}(t)$ affects all future states and accordingly all future rewards. The maximization is realized by the use of the *value function* V of the states to predict, given the sensory inputs **x**, the cumulative sum (possibly discounted as a function of time) of all future rewards V(**x**) (possibly within a learning trial) as follows:

$$V(\mathbf{x}) = E[r(t+1) + \gamma r(t+2) + \gamma^2 r(t+3) + ...] \tag{B.118}$$

where $r(t)$ is the reward at time t, and $E[\cdot]$ denotes the expected value of the sum of future rewards up to the end of the trial. $0 \leq \gamma \leq 1$ is a discount factor that makes rewards that arrive sooner more important than rewards that arrive later, according to an exponential decay function. (If $\gamma = 1$ there is no discounting.) It is assumed that the presentation of future cues and rewards depends only on the current sensory cues and not the past sensory cues. The right-hand side of equation B.118 is evaluated for the dynamics in equations B.115–B.117 with the initial condition $\mathbf{x}(t) = \mathbf{x}$. The two basic ingredients in reinforcement learning are the estimation (which we term $\hat{V}$) of the value function V, and then the improvement of the policy or action **u** using the value function (Sutton and Barto 1998).

The basic algorithm for learning the value function is to minimize the *temporal difference* (TD) *error* $\Delta(t)$ for time t within a trial, and this is computed by a 'critic' for the estimated value predictions $\hat{V}(\mathbf{x}(t))$ at successive time steps as

$$\Delta(t) = [r(t) + \gamma \hat{V}(\mathbf{x}(t))] - \hat{V}(\mathbf{x}(t-1)) \tag{B.119}$$

where $\hat{V}(\mathbf{x}(t)) - \hat{V}(\mathbf{x}(t-1))$ is the difference in the reward value prediction at two successive time steps, giving rise to the terminology temporal difference learning. If we introduce the term $\hat{v}$ as the estimate of the cumulated reward by the end of the trial, we can define it as a function $\hat{V}$ of the current sensory input $\mathbf{x}(t)$, i.e. $\hat{v} = \hat{V}(\mathbf{x})$, and we can also write equation B.119 as

$$\Delta(t) = r(t) + \gamma \hat{v}(t) - \hat{v}(t-1) \tag{B.120}$$

which draws out the fact that it is differences at successive timesteps in the reward value predictions $\hat{v}$ that are used to calculate Δ.

$\Delta(t)$ is used to improve the estimates $\hat{v}(t)$ by the 'critic', and can also be used (by an 'actor') to learn appropriate actions.

For example, when the value function is represented (in the critic) as

$$\hat{V}(\mathbf{x}(t)) = \sum_{j=1}^{n} w_j^C x_j(t) \tag{B.121}$$

the learning algorithm for the (value) weight w_j^C in the critic is given by

$$\delta w_j^C = k_c \Delta(t) x_j(t-1) \tag{B.122}$$

where δw_j^C is the change of synaptic weight, k_c is a constant that determines the learning rate for the sensory input x_j, and $\Delta(t)$ is the Temporal Difference error at time t. Under certain conditions this learning rule will cause the estimate $\hat{v}$ to converge to the true value (Dayan and Sejnowski 1994).

A simple way of improving the policy of the actor is to take a stochastic action

$$u_i(t) = g(\sum_{j=1}^{n} w_{ij}^A x_j(t) + \mu_i(t)),$$
(B.123)

where g() is a scalar version of the policy function G, w_{ij}^A is a weight in the actor, and $\mu_i(t)$ is a noise term. The TD error $\Delta(t)$ as defined in equation B.119 then signals the unexpected delivery of the reward $r(t)$ or the increase in the state value $\hat{V}(\mathbf{x}(t))$ above expectation, possibly due to the previous choice of action $u_i(t-1)$. The learning algorithm for the action weight w_{ij}^A in the actor is given by

$$\delta w_{ij}^A = k_a \Delta(t)(u_i(t-1) - <u_i>)x_j(t-1),$$
(B.124)

where $<u_i>$ is the average level of the action output, and k_a is a learning rate constant in the actor.

Thus, the TD error $\Delta(t)$, which signals the error in the reward prediction at time t, works as the main teaching signal in both learning the value function (implemented in the critic), and the selection of actions (implemented in the actor). The usefulness of a separate critic is that it enables the TD error to be calculated based on the difference in reward value predictions at two successive time steps as shown in equation B.119.

The algorithm has been applied to modelling the time course of classical conditioning (Sutton and Barto 1990). The algorithm effectively allows the future reinforcement predicted from past history to influence the responses made, and in this sense allows behaviour to be guided not just by immediate reinforcement, but also by 'anticipated' reinforcements. Different types of temporal difference learning are described by Sutton and Barto (1998). An application is to the analysis of decisions when future rewards are discounted with respect to immediate rewards (Dayan and Abbott 2001, Tanaka, Doya, Okada, Ueda, Okamoto and Yamawaki 2004). Another application is to the learning of sequences of actions to take within a trial (Suri and Schultz 1998).

The possibility that dopamine neuron firing may provide an error signal useful in training neuronal systems to predict reward has been discussed in Section 9.6. It has been proposed that the firing of the dopamine neurons can be thought of as an error signal about reward prediction, in that the firing occurs in a task when a reward is given, but then moves forward in time within a trial to the time when a stimulus is presented that can be used to predict when the taste reward will be obtained (Schultz et al. 1995, Schultz 2013) (see Fig. 9.2). The argument is that there is no prediction error when the taste reward is obtained if it has been signalled by a preceding conditioned stimulus, and that is why the dopamine midbrain neurons do not respond at the time of taste reward delivery, but instead, at least during training, to the onset of the conditioned stimulus (Waelti, Dickinson and Schultz 2001). If a different conditioned stimulus is shown that normally predicts that no taste reward will be given, there is no firing of the dopamine neurons to the onset of that conditioned stimulus.

This hypothesis has been built into models of learning in which the error signal is used to train synaptic connections in dopamine pathway recipient regions (such as presumably the striatum and orbitofrontal cortex) (Houk, Adams and Barto 1995, Schultz 2004, Schultz, Dayan and Montague 1997, Waelti, Dickinson and Schultz 2001, Dayan and Abbott 2001, Schultz 2013). Some difficulties with the hypothesis are discussed in Section 9.6 on page 234. The difficulties include the fact that dopamine is released in large quantities by aversive stimuli (see Section 9.6); that error computations for differences between the expected reward and the actual reward received on a trial are computed in the primate orbitofrontal cortex, where expected reward, actual reward, and error neurons are all found, and lesions of which impair the ability to use changes in reward contingencies to reverse behaviour (see Section 15.14); that the tonic, sustained, firing of the dopamine neurons in the delay period of a task with probabilistic rewards may reflect reward uncertainty, and not the expected reward, nor the magnitude of the prediction error (see Section 9.6 and Shizgal and Arvanitogiannis (2003)); and that reinforcement learning is suited to setting up connections that might be required in fixed tasks such as motor habit or sequence learning, for reinforcement learning algorithms seek to set weights correctly in an 'actor', but are not suited to tasks where rules must be altered flexibly, as in rapid one-trial reversal, for which a very different type of mechanism is described in Section 15.15 (Deco and Rolls 2005d, Rolls and Deco 2016).

The temporal difference approach to reinforcement learning has a weakness that although it can be used to predict internal signals during reinforcement learning in some tasks, it does not directly address learning with respect to actions that are not taken. Q-learning can be considered as an extension of TD learning which adds additional terms to take into account signals from actions that are not taken, for example information gained by observation of others (Montague, King-Casas and Cohen 2006).

Overall, reinforcement learning algorithms are certainly a move towards biological relevance, in that learning with a single reinforcer can be achieved in systems that might learn motor habits or fixed sequences. Whether a single prediction error is broadcast throughout a neural system by a general projection system, such as the dopamine pathways in the brain, which distribute to large parts of the striatum and the prefrontal cortex, remains to be clearly established (see further Chapter 15, Section 9.6.2), though using a modulator of synaptic learning may be computationally useful (Fremaux and Gerstner 2015).

B.17 Highlights

1. The operation and properties of biologically plausible pattern association networks, autoassociation networks, and competitive networks are described.
2. The operation of perceptrons, backpropagation networks, and deep learning networks is described.

Appendix 3 Information theory, and neuronal encoding

In order to understand the operation of memory and perceptual systems in the brain, it is necessary to know how information is encoded by neurons and populations of neurons. The concepts and results found are essential for understanding cortical function, are introduced in Chapter 8, and are considered in this Appendix and by Rolls and Treves (2011).

We have seen that one parameter that influences the number of memories that can be stored in an associative memory is the sparseness of the representation, and it is therefore important to be able to quantify the sparseness of the representations.

We have also seen that the properties of an associative memory system depend on whether the representation is distributed or local (grandmother cell like), and it is important to be able to assess this quantitatively for neuronal representations.

It is also necessary to know how the information is encoded in order to understand how memory systems operate. Is the information that must be stored and retrieved present in the firing rates (the number of spikes in a fixed time), or is it present in synchronized firing of subsets of neurons? This has implications for how each stage of processing would need to operate. If the information is present in the firing rates, how much information is available from the spiking activity in a short period, of for example 20 or 50 ms? For each stage of cortical processing to operate quickly (in for example 20 ms), it is necessary for each stage to be able to read the code being provided by the previous cortical area within this order of time. Thus understanding the neural code is fundamental to understanding how each stage of processing works in the brain, and for understanding the speed of processing at each stage.

To treat all these questions quantitatively, we need quantitative ways of measuring sparseness, and also ways of measuring the information available from the spiking activity of single neurons and populations of neurons, and these are the topics addressed in this Appendix, together with some of the main results obtained, which provide answers to these questions.

Because single neurons are the computing elements of the brain and send the results of their processing by spiking activity to other neurons, we can understand brain processing by understanding what is encoded by the neuronal firing at each stage of the brain (e.g. each cortical area), and determining how what is encoded changes from stage to stage. Each neuron responds differently to a set of stimuli (with each neuron tuned differently to the members of the set of stimuli), and it is this that allows different stimuli to be represented. We can only address the richness of the representation therefore by understanding the differences in the responses of different neurons, and the impact that this has on the amount of information that is encoded. These issues can only be adequately and directly addressed at the level of the activity of single neurons and of populations of single neurons, and understanding at this neuronal level (rather than at the level of thousands or millions of neurons as revealed by functional neuroimaging) is essential for understanding brain computation.

Information theory provides the means for quantifying how much neurons communicate to other neurons, and thus provides a quantitative approach to fundamental questions about information processing in the brain. To investigate what in neuronal activity carries information, one must compare the amounts of information carried by different codes, that is different descriptions of the same activity, to provide the answer. To investigate the speed of information transmission, one must define and measure information rates from neuronal responses. To investigate to what extent the information provided by different cells is redundant or instead independent, again one must measure amounts of information in order to provide quantitative evidence. To compare the information carried by the number of spikes, by the timing of the spikes within the response of a single neuron, and by the relative time of firing of different neurons reflecting for example stimulus-dependent neuronal synchronization, information theory again provides a quantitative and well-founded basis for the necessary comparisons. To compare the information carried by a single neuron or a group of neurons with that reflected in the behaviour of the human or animal, one must again use information theory, as it provides a single measure which can be applied to the measurement of the performance of all these different cases. In all these situations, there is no quantitative and well-founded alternative to information theory.

This Appendix briefly introduces the fundamental elements of information theory in Section C.1. A more complete treatment can be found in many books on the subject (e.g. Abramson (1963), Hamming (1990), and Cover and Thomas (1991)), including also Rieke, Warland, de Ruyter van Steveninck and Bialek (1997) which is specifically about information transmitted by neuronal firing. Section C.2 discusses the extraction of information measures from neuronal activity, in particular in experiments with mammals, in which the central issue is how to obtain accurate measures in conditions of limited sampling, that is where the numbers of trials of neuronal data that can be obtained are usually limited by the available recording time. Section C.3 summarizes some of the main results obtained so far on neuronal encoding. The essential terminology is summarized in a Glossary at the end of this Appendix in Section C.4. The approach taken in this Appendix is based on and updated from that provided by Rolls and Treves (1998), Rolls (2008d), and Rolls and Treves (2011).

C.1 Information theory and its use in the analysis of formal models

Although information theory was a surprisingly late starter as a mathematical discipline, having being developed and formalized by C. Shannon (1948), the intuitive notion of information is immediate to us. It is also very easy to understand why we use logarithms in order to quantify this intuitive notion, of how much we know about something, and why the resulting quantity is always defined in relative rather than absolute terms. An introduction to information theory is provided next, with a more formal summary given in Section C.1.3.

C.1.1 The information conveyed by definite statements

Suppose somebody, who did not know, is told that Reading is a town west of London. How much information is he given? Well, that depends. He may have known it was a town in England, but not whether it was east or west of London; in which case the new information amounts to the fact that of two *a priori* (i.e. initial) possibilities (E or W), one holds (W). It is also possible to interpret the statement in the more precise sense, that Reading is west of London, rather than east, north or south, i.e. one out of four possibilities; or else, west rather that north-west, north, etc. Clearly, the larger the number k of *a priori* possibilities, the more one is actually told, and a measure of information must take this into account. Moreover, we would like independent pieces of information to just add together. For example, our person may also be told that Cambridge is, out of l possible directions, north of London. Provided nothing was known on the mutual location of Reading and Cambridge, there are now overall

$k \times l$ *a priori* (initial) possibilities, only one of which remains *a posteriori* (after receiving the information). Given that the number of possibilities for independent events are multiplicative, but that we would like the measure of information to be additive, we use logarithms when we measure information, as logarithms have this property. We thus define the amount I of information gained when we are informed in which of k possible locations Reading is located as

$$I(k) = \log_2 k. \tag{C.1}$$

Then when we combine independent information, for example producing $k \times l$ possibilities from independent events with k and l possibilities respectively, we obtain

$$I(k \times l) = \log_2(k \times l) = \log_2 k + \log_2 l = I(k) + I(l). \tag{C.2}$$

Thus in our example, the information about Cambridge adds up to that about Reading. We choose to take logarithms in base 2 as a mere convention, so that the answer to a yes/no question provides one unit, or bit, of information. Here it is just for the sake of clarity that we used different symbols for the number of possible directions with respect to which Reading and Cambridge are localized; if both locations are specified for example in terms of E, SE, S, SW, W, NW, N, NE, then obviously $k = l = 8, I(k) = I(l) = 3$ bits, and $I(k \times l) = 6$ bits. An important point to note is that the *resolution* with which the direction is specified determines the amount of information provided, and that in this example, as in many situations arising when analysing neuronal codings, the resolution could be made progressively finer, with a corresponding increase in information proportional to the log of the number of possibilities.

C.1.2 The information conveyed by probabilistic statements

The situation becomes slightly less trivial, and closer to what happens among neurons, if information is conveyed in less certain terms. Suppose for example that our friend is told, instead, that Reading has odds of 9 to 1 to be west, rather than east, of London (considering now just two *a priori* possibilities). He is certainly given some information, albeit less than in the previous case. We might put it this way: out of 18 equiprobable *a priori* possibilities (9 west + 9 east), 8 (east) are eliminated, and 10 remain, yielding

$$I = \log_2(18/10) = \log_2(9/5) \tag{C.3}$$

as the amount of information given. It is simpler to write this in terms of probabilities

$$I = \log_2 P^{\text{posterior}}(W)/P^{\text{prior}}(W) = \log_2(9/10)/(1/2) = \log_2(9/5). \tag{C.4}$$

This is of course equivalent to saying that the amount of information given by an uncertain statement is equal to the amount given by the absolute statement

$$I = -\log_2 P^{\text{prior}}(W) \tag{C.5}$$

minus the amount of uncertainty remaining after the statement, $I = -\log_2 P^{\text{posterior}}(W)$. A successive clarification that Reading is indeed west of London carries

$$I' = \log_2((1)/(9/10)) \tag{C.6}$$

bits of information, because 9 out of 10 are now the *a priori* odds, while *a posteriori* there is certainty, $P^{\text{posterior}}(W) = 1$. In total we would seem to have

$$I^{\text{TOTAL}} = I + I' = \log_2(9/5) + \log_2(10/9) = 1 \text{ bit} \tag{C.7}$$

as if the whole information had been provided at one time. This is strange, given that the two pieces of information are clearly not independent, and only independent information should be additive. In fact, we have cheated a little. Before the clarification, there was still one residual possibility (out of 10) that the answer was 'east', and this must be taken into account by writing

$$I = \mathrm{P^{posterior}(W)} \log_2 \frac{\mathrm{P^{posterior}(W)}}{\mathrm{P^{prior}(W)}} + \tag{C.8}$$

$$\mathrm{P^{posterior}(E)} \log_2 \frac{\mathrm{P^{posterior}(E}}{\mathrm{P^{prior}(E)}}$$

as the information contained in the first message. This little detour should serve to emphasize two aspects that are easy to forget when reasoning intuitively about information, and that in this example cancel each other. In general, when uncertainty remains, that is there is more than one possible *a posteriori* state, one has to average information values for each state with the corresponding *a posteriori* probability measure. In the specific example, the sum $I + I'$ totals slightly *more* than 1 bit, and the amount in excess is precisely the information 'wasted' by providing *correlated* messages.

C.1.3 Information sources, information channels, and information measures

In summary, the expression quantifying the information provided by a definite statement that event s, which had an *a priori* probability $\mathrm{P}(s)$, has occurred is

$$I(s) = \log_2(1/\mathrm{P}(s)) = -\log_2 \mathrm{P}(s), \tag{C.9}$$

whereas if the statement is probabilistic, that is several *a posteriori* probabilities remain non-zero, the correct expression involves summing over all possibilities with the corresponding probabilities:

$$I = \sum_s \left[\mathrm{P^{posterior}}(s) \log_2 \frac{\mathrm{P^{posterior}}(s)}{\mathrm{P^{prior}}(s)} \right]. \tag{C.10}$$

When considering a discrete set of mutually exclusive events, it is convenient to use the metaphor of a set of *symbols* comprising an *alphabet* S. The occurrence of each event is then referred to as the emission of the corresponding symbol by an information *source*. The *entropy* of the source, H, is the average amount of information per source symbol, where the average is taken across the alphabet, with the corresponding probabilities

$$H(S) = -\sum_{s \in S} \mathrm{P}(s) \log_2 \mathrm{P}(s). \tag{C.11}$$

An information *channel* receives symbols s from an alphabet S and emits symbols s' from alphabet S'. If the *joint* probability of the channel receiving s and emitting s' is given by the product

$$\mathrm{P}(s, s') = \mathrm{P}(s)\mathrm{P}(s') \tag{C.12}$$

for any pair s, s', then the input and output symbols are *independent* of each other, and the channel transmits zero information. Instead of joint probabilities, this can be expressed with conditional probabilities: the conditional probability of s' given s is written $\mathrm{P}(s'|s)$, and if the two variables are independent, it is just equal to the unconditional probability $\mathrm{P}(s')$. In general, and in particular if the channel does transmit information, the variables are not independent, and one can express their joint probability in two ways in terms of conditional probabilities

$$P(s, s') = P(s'|s)P(s) = P(s|s')P(s'),$$ (C.13)

from which it is clear that

$$P(s'|s) = P(s|s')\frac{P(s')}{P(s)},$$ (C.14)

which is called Bayes' theorem (although when expressed as here in terms of probabilities it is strictly speaking an identity rather than a theorem). The information transmitted by the channel conditional to its having emitted symbol s' (or specific transinformation, $I(s')$) is given by equation C.10, once the unconditional probability $P(s)$ is inserted as the prior, and the conditional probability $P(s|s')$ as the posterior:

$$I(s') = \sum_s P(s|s') \log_2 \frac{P(s|s')}{P(s)}.$$ (C.15)

Symmetrically, one can define the transinformation conditional to the channel having received symbol s

$$I(s) = \sum_{s'} P(s'|s) \log_2 \frac{P(s'|s)}{P(s')}.$$ (C.16)

Finally, the average transinformation, or **mutual information**, can be expressed in fully symmetrical form

$$I = \sum_s P(s) \sum_{s'} P(s'|s) \log_2 \frac{P(s'|s)}{P(s')}$$ (C.17)

$$= \sum_{s,s'} P(s, s') \log_2 \frac{P(s, s')}{P(s)P(s')}.$$

The **mutual information** can also be expressed as the entropy of the source using alphabet S minus the *equivocation* of S with respect to the new alphabet S' used by the channel, written

$$I = H(S) - H(S|S') \equiv H(S) - \sum_{s'} P(s')H(S|s').$$ (C.18)

A channel is characterized, once the alphabets are given, by the set of conditional probabilities for the output symbols, $P(s'|s)$, whereas the unconditional probabilities of the input symbols $P(s)$ depend of course on the source from which the channel receives. Then, the *capacity* of the channel can be defined as the maximal mutual information across all possible sets of input probabilities $P(s)$. Thus, the information transmitted by a channel can range from zero to the lower of two independent upper bounds: the entropy of the source, and the capacity of the channel.

C.1.4 The information carried by a neuronal response and its averages

Considering the processing of information in the brain, we are often interested in the amount of information the response r of a neuron, or of a population of neurons, carries about an event happening in the outside world, for example a stimulus s shown to the animal. Once the inputs and outputs are conceived of as sets of symbols from two alphabets, the neuron(s) may be regarded as an information channel. We may denote with $P(s)$ the *a priori* probability that the particular stimulus s out of a given set was shown, while the conditional probability $P(s|r)$ is the *a posteriori* probability, that is updated by the knowledge of the response r. The response-specific transinformation

$$I(r) = \sum_s P(s|r) \log_2 \frac{P(s|r)}{P(s)} \tag{C.19}$$

takes the extreme values of $I(r) = -\log_2 P(s(r))$ if r unequivocally determines $s(r)$ (that is, $P(s|r)$ equals 1 for that one stimulus and 0 for all others);
and $I(r) = \sum_s P(s) \log_2(P(s)/P(s)) = 0$ if there is no relation between s and r, that is they are independent, so that the response tells us nothing new about the stimulus and thus $P(s|r) = P(s)$.

This is the information conveyed by each particular response. One is usually interested in further averaging this quantity over all possible responses r,

$$< I > = \sum_r P(r) \left[\sum_s P(s|r) \log_2 \frac{P(s|r)}{P(s)} \right]. \tag{C.20}$$

The angular brackets $<>$ are used here to emphasize the averaging operation, in this case over responses. Denoting with $P(s, r)$ the *joint probability* of the pair of events s and r, and using Bayes' theorem, this reduces to the symmetric form (equation C.18) for the **mutual information** $I(S, R)$

$$< I > = \sum_{s,r} P(s, r) \log_2 \frac{P(s, r)}{P(s)P(r)} \tag{C.21}$$

which emphasizes that responses tell us about stimuli just as much as stimuli tell us about responses. This is, of course, a general feature, independent of the two variables being in this instance stimuli and neuronal responses. In fact, what is of interest, besides the mutual information of equations C.20 and C.21, is often the information specifically conveyed about each stimulus,

$$I(s) = \sum_r P(r|s) \log_2 \frac{P(r|s)}{P(r)} \tag{C.22}$$

which is a direct quantification of the variability in the responses elicited by that stimulus, compared to the overall variability. Since $P(r)$ is the probability distribution of responses averaged across stimuli, it is again evident that the stimulus-specific information measure of equation C.22 depends not only on the stimulus s, but also on all other stimuli used. Likewise, the mutual information measure, despite being of an average nature, is dependent on what set of stimuli has been used in the average. This emphasizes again the relative nature of all information measures. More specifically, it underscores the relevance of using, while measuring the information conveyed by a given neuronal population, stimuli that are either representative of real-life stimulus statistics, or of particular interest for the properties of the population being examined[41].

---

[41] The quantity $I(s, R)$, which is what is shown in equation C.22 and where R draws attention to the fact that this quantity is calculated across the full set of responses R, has also been called the stimulus-specific surprise, see DeWeese and Meister (1999). Its average across stimuli is the mutual information $I(S, R)$.

C.1.4.1   A numerical example

To make these notions clearer, we can consider a specific example in which the response of a neuron to the presentation of, say, one of four visual stimuli (A, B, C, D) is recorded for 10 ms, during which the neuron emits either 0, 1, or 2 spikes, but no more. Imagine that the neuron tends to respond more vigorously to visual stimulus B, less to C, even less to A, and never to D, as described by the table of conditional probabilities $P(r|s)$ shown in Table C.1. Then, if different visual stimuli are presented with equal probability,

Table C.1 The conditional probabilities $P(r|s)$ that different neuronal responses (r=0, 1, or 2 spikes) will be produced by each of four stimuli (A–D).

	r=0	r=1	r=2
s=A	0.6	0.4	0.0
s=B	0.0	0.2	0.8
s=C	0.4	0.5	0.1
s=D	1.0	0.0	0.0

the table of joint probabilities $P(s, r)$ will be as shown in Table C.2. From these two tables

Table C.2 Joint probabilities $P(s, r)$ that different neuronal responses (r=0, 1, or 2 spikes) will be produced by each of four equiprobable stimuli (A–D).

	r=0	r=1	r=2
s=A	0.15	0.1	0.0
s=B	0.0	0.05	0.2
s=C	0.1	0.125	0.025
s=D	0.25	0.0	0.0

one can compute various information measures by directly applying the definitions above. Since visual stimuli are presented with equal probability, $P(s) = 1/4$, the entropy of the stimulus set, which corresponds to the maximum amount of information any transmission channel, no matter how efficient, could convey on the identity of the stimuli, is $H_s = -\sum_s [P(s) \log_2 P(s)] = -4[(1/4) \log_2(1/4)] = \log_2 4 = 2$ bits. There is a more stringent upper bound on the mutual information that this cell's responses convey on the stimuli, however, and this second bound is the channel capacity T of the cell. Calculating this quantity involves maximizing the mutual information across prior visual stimulus probabilities, and it is a bit complicated to do, in general. In our particular case the maximum information is obtained when only stimuli B and D are presented, each with probability 0.5. The resulting capacity is $T = 1$ bit. We can easily calculate, in general, the entropy of the responses. This is not an upper bound characterizing the source, like the entropy of the stimuli, nor an upper bound characterizing the channel, like the capacity, but simply a bound on the mutual information for this specific combination of source (with its related visual stimulus probabilities) and channel (with its conditional probabilities). Since only three response levels are possible within the short recording window, and they occur with uneven probability, their entropy is considerably lower than H_s, at $H_r = -\sum_r P(r) \log_2 P(r) = -P(0) \log_2 P(0) - P(1) \log_2 P(1) - P(2) \log_2 P(2) = -0.5 \log_2 0.5 - 0.275 \log_2 0.275 - 0.225 \log_2 0.225 = 1.496$ bits. The actual average information I that the responses transmit about the stimuli, which is a measure of the correlation in the variability of stimuli and responses, does not exceed the absolute variability of either stimuli (as quantified by the first bound) or responses (as quantified by the last bound), nor the capacity of the channel. An explicit calculation using the joint probabilities of the second table in expression C.21 yields $I = 0.733$ bits. This is of course only the average value, averaged both across stimuli and across responses.

The information conveyed by a particular response can be larger. For example, when the cell emits two spikes it indicates with a relatively large probability stimulus B, and this is reflected in the fact that it then transmits, according to expression C.19, $I(r = 2) = 1.497$ bits, more than double the average value.

Similarly, the amount of information conveyed about each individual visual stimulus varies with the stimulus, depending on the extent to which it tends to elicit a differential response. Thus, expression C.22 yields that only $I(s = C) = 0.185$ bits are conveyed on average about stimulus C, which tends to elicit responses with similar statistics to the average statistics across stimuli, and are therefore not easily interpretable. On the other hand, exactly 1 bit of information is conveyed about stimulus D, since this stimulus never elicits any response, and when the neuron emits no spike there is a probability of $1/2$ that the stimulus was stimulus D.

C.1.5 The information conveyed by continuous variables

A general feature, relevant also to the case of neuronal information, is that if, among a *continuum* of *a priori* possibilities, only one, or a discrete number, remains *a posteriori*, the information is strictly infinite. This would be the case if one were told, for example, that Reading is exactly $10'$ west, $1'$ north of London. The *a priori* probability of precisely this set of coordinates among the continuum of possible ones is zero, and then the information diverges to infinity. The problem is only theoretical, because in fact, with continuous distributions, there are always one or several factors that limit the resolution in the *a posteriori* knowledge, rendering the information finite. Moreover, when considering the mutual information in the conjoint probability of occurrence of two sets, e.g. stimuli and responses, it suffices that at least one of the sets is discrete to make matters easy, that is, finite. Nevertheless, the identification and appropriate consideration of these resolution-limiting factors in practical cases may require careful analysis.

C.1.5.1 Example: the information retrieved from an autoassociative memory

One example is the evaluation of the information that can be retrieved from an autoassociative memory. Such a memory stores a number of firing patterns, each one of which can be considered, as in Appendix B, as a vector $\mathbf{r}^\mu$ with components the firing rates $\{r_i^\mu\}$, where the subscript i indexes the neuron (and the superscript μ indexes the pattern). In retrieving pattern μ, the network in fact produces a distinct firing pattern, denoted for example simply as $\mathbf{r}$. The quality of retrieval, or the similarity between $\mathbf{r}^\mu$ and $\mathbf{r}$, can be measured by the average mutual information

$$< I(\mathbf{r}^\mu, \mathbf{r}) > \ = \ \sum_{\mathbf{r}^\mu, \mathbf{r}} P(\mathbf{r}^\mu, \mathbf{r}) \log_2 \frac{P(\mathbf{r}^\mu, \mathbf{r})}{P(\mathbf{r}^\mu)P(\mathbf{r})} \tag{C.23}$$

$$\approx \sum_i \sum_{r_i^\mu, r_i} P(r_i^\mu, r_i) \log_2 \frac{P(r_i^\mu, r_i)}{P(r_i^\mu)P(r_i)}.$$

In this formula the 'approximately equal' sign $\approx$ marks a simplification that is not necessarily a reasonable approximation. If the simplification is valid, it means that in order to extract an information measure, one need not compare whole vectors (the entire firing patterns) with each other, and may instead compare the firing rates of individual cells at storage and retrieval, and sum the resulting single-cell information values. The validity of the simplification is a matter that will be discussed later and that has to be verified, in the end, experimentally, but for the purposes of the present discussion we can focus on the single-cell terms. If either

r_i or r_i^μ has a continuous distribution of values, as it will if it represents not the number of spikes emitted in a fixed window, but more generally the firing rate of neuron i computed by convolving the firing train with a smoothing kernel, then one has to deal with probability densities, which we denote as $p(r)dr$, rather than the usual probabilities $P(r)$. Substituting $p(r)dr$ for $P(r)$ and $p(r^\mu, r)drdr^\mu$ for $P(r^\mu, r)$, one can write for each single-cell contribution (omitting the cell index i)

$$< I(r^\mu, r) >_i = \int dr^\mu dr\; p(r^\mu, r) \log_2 \frac{p(r^\mu, r)}{p(r^\mu)p(r)} \tag{C.24}$$

and we see that the differentials $dr^\mu dr$ cancel out between numerator and denominator inside the logarithm, rendering the quantity well defined and finite. If, however, r^μ were to *exactly* determine r, one would have

$$p(r^\mu, r)dr^\mu dr = p(r^\mu)\delta(r - r(r^\mu))dr^\mu dr = p(r^\mu)dr^\mu \tag{C.25}$$

and, by losing one differential on the way, the mutual information would become infinite. It is therefore important to consider what prevents r^μ from fully determining r in the case at hand – in other words, to consider the sources of noise in the system. In an autoassociative memory storing an extensive number of patterns (see Appendix A4 of Rolls and Treves (1998)), one source of noise always present is the interference effect due to the concurrent storage of all other patterns. Even neglecting other sources of noise, this produces a finite resolution width ρ, which allows one to write an expression of the type $p(r|r^\mu)dr = \exp -(r - r(r^\mu))^2/2\rho^2 dr$ which ensures that the information is finite as long as the resolution ρ is larger than zero.

One further point that should be noted, in connection with estimating the information retrievable from an autoassociative memory, is that the mutual information between the current distribution of firing rates and that of the stored pattern does not coincide with the information *gain* provided by the memory device. Even when firing rates, or spike counts, are all that matter in terms of information carriers, as in the networks considered in this book, one more term should be taken into account in evaluating the information gain. This term, to be subtracted, is the information contained in the external input that elicits the retrieval. This may vary a lot from the retrieval of one particular memory to the next, but of course an efficient memory device is one that is able, when needed, to retrieve much more information than it requires to be present in the inputs, that is, a device that produces a large information gain.

Finally, one should appreciate the conceptual difference between the information a firing pattern carries about another one (that is, about the pattern stored), as considered above, and two different notions: (a) the information produced by the network in selecting the correct memory pattern and (b) the information a firing pattern carries about something in the outside world. Quantity (a), the information intrinsic to selecting the memory pattern, is ill defined when analysing a real system, but is a well-defined and particularly simple notion when considering a formal model. If p patterns are stored with equal strength, and the selection is errorless, this amounts to $\log_2 p$ bits of information, a quantity often, but not always, small compared with the information in the pattern itself. Quantity (b), the information conveyed about some outside correlate, is not defined when considering a formal model that does not include an explicit account of what the firing of each cell represents, but is well defined and measurable from the recorded activity of real cells. It is the quantity considered in the numerical example with the four visual stimuli, and it can be generalized to the information carried by the activity of several cells in a network, and specialized to the case that the network operates as an associative memory. One may note, in this case, that the capacity to retrieve memories with high fidelity, or high information content, is only useful to the extent that the representation to be retrieved carries that amount of information about something relevant – or, in other words, that it is pointless to store and retrieve with great care largely meaningless messages. This type of argument has been used to discuss the role of the mossy fibres in the operation of the CA3 network in the hippocampus (Treves and Rolls 1992b, Rolls and Treves 1998).

C.2 Estimating the information carried by neuronal responses

C.2.1 The limited sampling problem

We now discuss in more detail the application of these general notions to the information transmitted by neurons. Suppose, to be concrete, that an animal has been presented with stimuli drawn from a discrete set, and that the responses of a set of C cells have been recorded following the presentation of each stimulus. We may choose any quantity or set of quantities to characterize the responses; for example let us assume that we consider the firing rate of each cell, r_i, calculated by convolving the spike response with an appropriate smoothing kernel. The response space is then C times the continuous set of all positive real numbers, $(\mathbf{R}/2)^C$. We want to evaluate the average information carried by such responses about which stimulus was shown. In principle, it is straightforward to apply the above formulas, e.g. in the form

$$< I(s, \mathbf{r}) > = \sum_s \mathrm{P}(s) \int \Pi_i dr_i \, p(\mathbf{r}|s) \log_2 \frac{p(\mathbf{r}|s)}{p(\mathbf{r})} \tag{C.26}$$

where it is important to note that $p(\mathbf{r})$ and $p(\mathbf{r}|s)$ are now probability densities defined over the high-dimensional vector space of multi-cell responses. The product sign Π signifies that this whole vector space has to be integrated over, along all its dimensions. $p(\mathbf{r})$ can be calculated as $\sum_s p(\mathbf{r}|s)\mathrm{P}(s)$, and therefore, in principle, all one has to do is to estimate, from the data, the conditional probability densities $p(\mathbf{r}|s)$ – the distributions of responses following each stimulus. In practice, however, in contrast to what happens with formal models, in which there is usually no problem in calculating the exact probability densities, real data come in limited amounts, and thus sample only sparsely the vast response space. This limits the accuracy with which, from the experimental *frequency* of each possible response, we can estimate its *probability*, in turn seriously impairing our ability to estimate $< I >$ correctly. We refer to this as the limited sampling problem. This is a purely technical problem that arises, typically when recording from mammals, because of external constraints on the duration or number of repetitions of a given set of stimulus conditions. With computer simulation experiments, and also with recordings from, for example, insects, sufficient data can usually be obtained that straightforward estimates of information are accurate enough (Strong, Koberle, de Ruyter van Steveninck and Bialek 1998, Golomb, Kleinfeld, Reid, Shapley and Shraiman 1994). The problem is, however, so serious in connection with recordings from monkeys and rats in which limited numbers of trials are usually available for neuronal data, that it is worthwhile to discuss it, in order to appreciate the scope and limits of applying information theory to neuronal processing.

In particular, if the responses are continuous quantities, the probability of observing exactly the same response twice is infinitesimal. In the absence of further manipulation, this would imply that each stimulus generates its own set of unique responses, therefore any response that has actually occurred could be associated unequivocally with one stimulus, and the mutual information would always equal the entropy of the stimulus set. This absurdity shows that in order to estimate probability densities from experimental frequencies, one has to resort to some *regularizing* manipulation, such as smoothing the point-like response values by convolution with suitable kernels, or binning them into a finite number of discrete bins.

C.2.1.1 Smoothing or binning neuronal response data

The issue is how to estimate the underlying probability distributions of neuronal responses to a set of stimuli from only a limited number of trials of data (e.g. 10–30) for each stimulus. Several strategies are possible. One is to discretize the response space into bins, and estimate the probability density as the histogram of the fraction of trials falling into each bin. If the bins are too narrow, almost every response is in a different bin, and the estimated information will be overestimated. Even if the bin width is increased to match the standard deviation of each underlying distribution, the information may still be overestimated. Alternatively, one may try to 'smooth' the data by convolving each response with a Gaussian with a width set to the standard deviation measured for each stimulus. Setting the standard deviation to this value may actually lead to an underestimation of the amount of information available, due to oversmoothing. Another possibility is to make a bold assumption as to what the general shape of the underlying densities should be, for example a Gaussian. This may produce closer estimates. Methods for regularizing the data are discussed further by Rolls and Treves (1998) in their Appendix A2, where a numerical example is given.

C.2.1.2 The effects of limited sampling

The crux of the problem is that, whatever procedure one adopts, limited sampling tends to produce distortions in the estimated probability densities. The resulting mutual information estimates are intrinsically biased. The bias, or average error of the estimate, is upward if the raw data have not been regularized much, and is downward if the regularization procedure chosen has been heavier. The bias can be, if the available trials are few, much larger than the true information values themselves. This is intuitive, as fluctuations due to the finite number of trials available would tend, on average, to either produce or emphasize differences among the distributions corresponding to different stimuli, differences that are preserved if the regularization is 'light', and that are interpreted in the calculation as carrying genuine information. This is illustrated with a quantitative example by Rolls and Treves (1998) in their Appendix A2.

Choosing the right amount of regularization, or the best regularizing procedure, is not possible *a priori*. Hertz, Kjaer, Eskander and Richmond (1992) have proposed the interesting procedure of using an artificial neural network to regularize the raw responses. The network can be trained on part of the data using backpropagation, and then used on the remaining part to produce what is in effect a clever data-driven regularization of the responses. This procedure is, however, rather computer intensive and not very safe, as shown by some self-evident inconsistency in the results (Heller, Hertz, Kjaer and Richmond 1995). Obviously, the best way to deal with the limited sampling problem is to try and use as many trials as possible. The improvement is slow, however, and generating as many trials as would be required for a reasonably unbiased estimate is often, in practice, impossible.

C.2.2 Correction procedures for limited sampling

The above point, that data drawn from a single distribution, when artificially paired, at random, to different stimulus labels, results in 'spurious' amounts of apparent information, suggests a simple way of checking the reliability of estimates produced from real data (Optican, Gawne, Richmond and Joseph 1991). One can disregard the true stimulus associated with each response, and generate a randomly reshuffled pairing of stimuli and responses, which should therefore, being not linked by any underlying relationship, carry no mutual information about each other. Calculating, with some procedure of choice, the spurious information obtained in this way, and comparing with the information value estimated with the same procedure for the real pairing, one can get a feeling for how far the procedure goes into eliminating the apparent information due to limited sampling. Although this spurious information, I_s, is only indicative of the amount of bias affecting the original estimate, a simple heuristic trick (called 'bootstrap'[42]) is to subtract the spurious from the original value, to obtain a somewhat 'corrected' estimate. This procedure can result in quite accurate estimates (see Rolls and Treves (1998), Tovee, Rolls, Treves and Bellis (1993))[43].

A different correction procedure (called 'jack-knife') is based on the assumption that the bias is proportional to $1/N$, where N is the number of responses (data points) used in the estimation. One computes, beside the original estimate $< I_N >$, N auxiliary estimates $< I_{N-1} >_k$, by taking out from the data set response k, where k runs across the data set from 1 to N. The corrected estimate

$$< I > = N < I_N > -(1/N) \sum_k (N-1) < I_{N-1} >_k \qquad \text{(C.27)}$$

is free from bias (to leading order in $1/N$), if the proportionality factor is more or less the same in the original and auxiliary estimates. This procedure is very time-consuming, and it suffers from the same imprecision of any algorithm that tries to determine a quantity as the result of the subtraction of two large and nearly equal terms; in this case the terms have been made large on purpose, by multiplying them by N and $N-1$.

A more fundamental approach (Miller 1955) is to derive an analytical expression for the bias (or, more precisely, for its leading terms in an expansion in $1/N$, the inverse of the sample size). This allows the estimation of the bias from the data itself, and its subsequent subtraction, as discussed in Treves and Panzeri (1995) and Panzeri and Treves (1996). Such a procedure produces satisfactory results, thereby lowering the size of the sample required for a given accuracy in the estimate by about an order of magnitude (Golomb, Hertz, Panzeri, Treves and Richmond 1997). However, it does not, in itself, make possible measures of the information contained in very complex responses with few trials. As a rule of thumb, the number of trials per stimulus required for a reasonable estimate of information, once the subtractive correction is applied, is of the order of the effectively independent (and utilized) bins in which the response space can be partitioned (Panzeri and Treves 1996). This correction procedure is the one that we use standardly (Rolls, Treves, Tovee and Panzeri 1997d, Rolls, Critchley and Treves 1996a, Rolls, Treves, Robertson, Georges-François and Panzeri 1998b, Booth and Rolls 1998, Rolls, Tovee and Panzeri 1999b, Rolls, Franco, Aggelopoulos and Jerez 2006b).

C.2.3 The information from multiple cells: decoding procedures

The bias of information measures grows with the dimensionality of the response space, and for all practical purposes the limit on the number of dimensions that can lead to reasonably accurate direct measures, even when applying a correction procedure, is quite low, two to three. This implies, in particular, that it is not possible to apply equation C.26 to extract the information content in the responses of several cells (more than two to three) recorded

---

[42] In technical usage bootstrap procedures utilize random pairings of responses with stimuli with replacement, while shuffling procedures utilize random pairings of responses with stimuli without replacement.

[43] Subtracting the 'square' of the spurious fraction of information estimated by this bootstrap procedure as used by Optican, Gawne, Richmond and Joseph (1991) is unfounded and does not work correctly (see Rolls and Treves (1998) and Tovee, Rolls, Treves and Bellis (1993)).

simultaneously. One way to address the problem is then to apply some strong form of regularization to the multiple cell responses. Smoothing has already been mentioned as a form of regularization that can be tuned from very soft to very strong, and that preserves the structure of the response space. Binning is another form, which changes the nature of the responses from continuous to discrete, but otherwise preserves their general structure, and which can also be tuned from soft to strong. Other forms of regularization involve much more radical transformations, or changes of variables.

Of particular interest for information estimates is a change of variables that transforms the response space into the stimulus set, by applying an algorithm that derives a predicted stimulus from the response vector, i.e. the firing rates of all the cells, on each trial. Applying such an algorithm is called decoding. Of course, the predicted stimulus is not necessarily the same as the actual one. Therefore the term decoding should not be taken to imply that the algorithm works successfully, each time identifying the actual stimulus. The predicted stimulus is simply a function of the response, as determined by the algorithm considered. Just as with any regularizing transform, it is possible to compute the mutual information between actual stimuli s and predicted stimuli s', instead of the original one between stimuli s and responses r. Since information about (real) stimuli can only be lost and not be created by the transform, the information measured in this way is bound to be lower in value than the real information in the responses. If the decoding algorithm is efficient, it manages to preserve nearly all the information contained in the raw responses, while if it is poor, it loses a large portion of it. If the responses themselves provided all the information about stimuli, and the decoding is optimal, then predicted stimuli coincide with the actual stimuli, and the information extracted equals the entropy of the stimulus set.

The procedure for extracting information values after applying a decoding algorithm is indicated in Fig. C.1 (in which s? is s'). The underlying idea indicated in Fig. C.1 is that if we know the average firing rate of each cell in a population to each stimulus, then on any single trial we can guess (or decode) the stimulus that was present by taking into account the responses of all the cells. The decoded stimulus is s', and the actual stimulus that was shown is s. What we wish to know is how the percentage correct, or better still the information, based on the evidence from any single trial about which stimulus was shown, increases as the number of cells in the population sampled increases. We can expect that the more cells there are in the sample, the more accurate the estimate of the stimulus is likely to be. If the encoding was local, the number of stimuli encoded by a population of neurons would be expected to rise approximately linearly with the number of neurons in the population. In contrast, with distributed encoding, provided that the neuronal responses are sufficiently independent, and are sufficiently reliable (not too noisy), information from the ensemble would be expected to rise linearly with the number of cells in the ensemble, and (as information is a log measure) the number of stimuli encodable by the population of neurons might be expected to rise exponentially as the number of neurons in the sample of the population was increased.

Table C.3 Decoding. s' is the decoded stimulus, i.e. that predicted from the neuronal responses r.

$$s \quad \Rightarrow \quad r \quad \rightarrow \quad s'$$
$$I(s,r)$$
$$\overline{\qquad\qquad I(s,s') \qquad\qquad}$$

The procedure is schematized in Table C.3 where the double arrow indicates the transformation from stimuli to responses operated by the nervous system, while the single arrow indicates the further transformation operated by the decoding procedure. $I(s, s')$ is the mutual

How well can one predict which stimulus was
shown on a single trial from the mean responses
of different neurons to each stimulus?

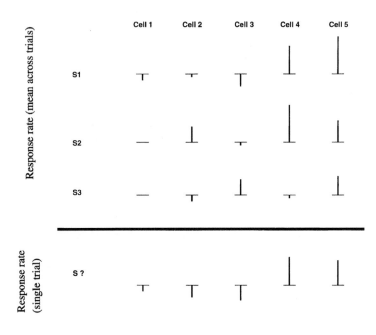

Fig. C.1 This diagram shows the average response for each of several cells (Cell 1, etc.) to each of several stimuli (S1, etc.). The change of firing rate from the spontaneous rate is indicated by the vertical line above or below the horizontal line, which represents the spontaneous rate. We can imagine guessing or predicting from such a table the predicted stimulus S? (i.e. s') that was present on any one trial.

information between the actual stimuli s and the stimuli s' that are predicted to have been shown based on the decoded responses.

A slightly more complex variant of this procedure is a decoding step that extracts from the response on each trial not a single predicted stimulus, but rather probabilities that each of the possible stimuli was the actual one. The joint probabilities of actual and posited stimuli can be averaged across trials, and information computed from the resulting probability matrix $(S \times S)$. Computing information in this way takes into account the relative uncertainty in assigning a predicted stimulus to each trial, an uncertainty that is instead not considered by the previous procedure based solely on the identification of the maximally likely stimulus (Treves 1997). *Maximum likelihood* information values I_{ml} based on a single stimulus tend therefore to be higher than *probability* information values I_p based on the whole set of stimuli, although in very specific situations the reverse could also be true.

The same correction procedures for limited sampling can be applied to information values computed after a decoding step. Values obtained from maximum likelihood decoding, I_{ml}, suffer from limited sampling more than those obtained from probability decoding, I_p, since each trial contributes a whole 'brick' of weight $1/N$ (N being the total number of trials), whereas with probabilities each brick is shared among several slots of the $(S \times S)$ probability matrix. The neural network procedure devised by Hertz, Kjaer, Eskander and Richmond (1992) can in fact be thought of as a decoding procedure based on probabilities, which deals with limited sampling not by applying a correction but rather by strongly regularizing the original responses.

When decoding is used, the rule of thumb becomes that the minimal number of trials per stimulus required for accurate information measures is roughly equal to the size of the stimulus set, if the subtractive correction is applied (Panzeri and Treves 1996). This correction procedure is applied as standard in our multiple cell information analyses that use decoding (Rolls, Treves and Tovee 1997b, Booth and Rolls 1998, Rolls, Treves, Robertson, Georges-François and Panzeri 1998b, Franco, Rolls, Aggelopoulos and Treves 2004, Aggelopoulos, Franco and Rolls 2005, Rolls, Franco, Aggelopoulos and Jerez 2006b).

C.2.3.1 Decoding algorithms

Any transformation from the response space to the stimulus set could be used in decoding, but of particular interest are the transformations that either approach optimality, so as to minimize information loss and hence the effect of decoding, or else are implementable by mechanisms that *could* conceivably be operating in the real system, so as to extract information values that could be extracted by the system itself.

The optimal transformation is in theory well-defined: one should estimate from the data the conditional probabilities $P(r|s)$, and use Bayes' rule to convert them into the conditional probabilities $P(s'|r)$. Having these for any value of r, one could use them to estimate I_p, and, after selecting for each particular real response the stimulus with the highest conditional probability, to estimate I_{ml}. To avoid biasing the estimation of conditional probabilities, the responses used in estimating $P(r|s)$ should not include the particular response for which $P(s'|r)$ is going to be derived (jack-knife cross-validation). In practice, however, the estimation of $P(r|s)$ in usable form involves the fitting of some simple function to the responses. This need for fitting, together with the approximations implied in the estimation of the various quantities, prevents us from defining the really optimal decoding, and leaves us with various algorithms, depending essentially on the fitting function used, which are hopefully close to optimal in some conditions. We have experimented extensively with two such algorithms, that both approximate Bayesian decoding (Rolls, Treves and Tovee 1997b). Both these algorithms fit the response vectors produced over several trials by the cells being recorded to a product of conditional probabilities for the response of each cell given the stimulus. In one case, the single cell conditional probability is assumed to be Gaussian (truncated at zero); in the other it is assumed to be Poisson (with an additional weight at zero). Details of these algorithms are given by Rolls, Treves and Tovee (1997b).

Biologically plausible decoding algorithms are those that limit the algebraic operations used to types that could be easily implemented by neurons, e.g. dot product summations, thresholding and other single-cell non-linearities, and competition and contrast enhancement among the outputs of nearby cells. There is then no need for ever fitting functions or other sophisticated approximations, but of course the degree of arbitrariness in selecting a particular algorithm remains substantial, and a comparison among different choices based on which yields the higher information values may favour one choice in a given situation and another choice with a different data set.

To summarize, the key idea in decoding, in our context of estimating information values, is that it allows substitution of a possibly very high-dimensional response space (which is difficult to sample and regularize) with a reduced object much easier to handle, that is with a discrete set equivalent to the stimulus set. The mutual information between the new set and the stimulus set is then easier to estimate even with limited data, and if the assumptions about population coding, underlying the particular decoding algorithm used, are justified, the value obtained approximates the original target, the mutual information between stimuli and responses. For each response recorded, one can use all the responses except for that one to generate estimates of the average response vectors (the average response for each neuron in the population) to each stimulus. Then one considers how well the selected response vector matches the average response vectors, and uses the degree of matching to estimate, for all stimuli, the probability that they were the actual stimuli. The form of the matching embodies the general notions about population encoding, for example the 'degree of matching' might be simply the dot product between the current vector and the average vector ($\mathbf{r}^{av}$), suitably normalized over all average vectors to generate probabilities

$$P(s'|\mathbf{r}(s)) = \frac{\mathbf{r}(s) \cdot \mathbf{r}^{av}(s')}{\sum_{s''} \mathbf{r}(s) \cdot \mathbf{r}^{av}(s'')} \tag{C.28}$$

where s'' is a dummy variable. (This is called dot product decoding in Fig. 25.13.) One ends up, then, with a table of conjoint probabilities $P(s, s')$, and another table obtained by selecting for each trial the most likely (or predicted) single stimulus s^p, $P(s, s^p)$. Both s' and s^p stand for all possible stimuli, and hence belong to the same set S. These can be used to estimate mutual information values based on probability decoding (I_p) and on maximum likelihood decoding (I_{ml}):

$$< I_p > = \sum_{s \in S} \sum_{s' \in S} P(s, s') \log_2 \frac{P(s, s')}{P(s)P(s')} \tag{C.29}$$

and

$$< I_{ml} > = \sum_{s \in S} \sum_{s^p \in S} P(s, s^p) \log_2 \frac{P(s, s^p)}{P(s)P(s^p)} . \tag{C.30}$$

Examples of the use of these procedures are available (Rolls, Treves and Tovee 1997b, Booth and Rolls 1998, Rolls, Treves, Robertson, Georges-François and Panzeri 1998b, Rolls, Aggelopoulos, Franco and Treves 2004, Franco, Rolls, Aggelopoulos and Treves 2004, Rolls, Franco, Aggelopoulos and Jerez 2006b), and some of the results obtained are described in Section C.3.

C.2.4   Information in the correlations between the spikes of different cells: a decoding approach

Simultaneously recorded neurons sometimes shows cross-correlations in their firing, that is the firing of one is systematically related to the firing of the other cell. One example of this is neuronal response synchronization. The cross-correlation, to be defined below, shows the time difference between the cells at which the systematic relation appears. A significant peak or trough in the cross-correlation function could reveal a synaptic connection from one cell to the other, or a common input to each of the cells, or any of a considerable number of other possibilities. If the synchronization occurred for only some of the stimuli, then the presence of the significant cross-correlation for only those stimuli could provide additional evidence separate from any information in the firing rate of the neurons about which stimulus had been shown. Information theory in principle provides a way of quantitatively assessing the relative contributions from these two types of encoding, by expressing what can be learned from each type of encoding in the same units, bits of information.

Figure C.2 illustrates how synchronization occurring only for some of the stimuli could be used to encode information about which stimulus was presented. In the Figure the spike trains of three neurons are shown after the presentation of two different stimuli on one trial.

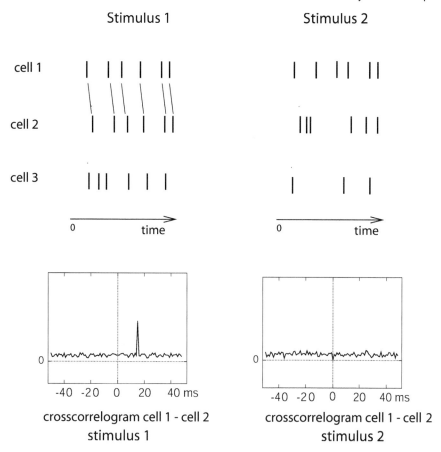

Fig. C.2 Illustration of the information that could be carried by spike trains. The responses of three cells to two different stimuli are shown on one trial. Cell 3 reflects which stimulus was shown in the number of spikes produced, and this can be measured as spike count or rate information. Cells 1 and 2 have no spike count or rate information, because the number of spikes is not different for the two stimuli. Cells 1 and 2 do show some synchronization, reflected in the cross-correlogram, that is stimulus dependent, as the synchronization is present only when stimulus 1 is shown. The contribution of this effect is measured as the stimulus-dependent synchronization information.

As shown by the cross-correlogram in the lower part of the figure, the responses of cell 1 and cell 2 are synchronized when stimulus 1 is presented, as whenever a spike from cell 1 is emitted, another spike from cell 2 is emitted after a short time lag. In contrast, when stimulus 2 is presented, synchronization effects do not appear. Thus, based on a measure of the synchrony between the responses of cells 1 and 2, it is possible to obtain some information about what stimulus has been presented. The contribution of this effect is measured as the stimulus-dependent synchronization information. Cells 1 and 2 have no information about what stimulus was presented from the number of spikes, as the same number is found for both stimuli. Cell 3 carries information in the spike count in the time window (which is also called the firing rate) about what stimulus was presented. (Cell 3 emits 6 spikes for stimulus 1 and 3 spikes for stimulus 2.)

The example shown in Fig. C.2 is for the neuronal responses on a single trial. Given that the neuronal responses are variable from trial to trial, we need a method to quantify the information that is gained from a single trial of spike data in the context of the measured variability in the responses of all of the cells, including how the cells' responses covary in a way which may be partly stimulus-dependent, and may include synchronization effects. The direct approach is to apply the Shannon mutual information measure (Shannon 1948, Cover and Thomas 1991)

$$I(s, \mathbf{r}) = \sum_{s \in S} \sum_{\mathbf{r}} P(s, \mathbf{r}) \log_2 \frac{P(s, \mathbf{r})}{P(s)P(\mathbf{r})}, \tag{C.31}$$

where $P(s, \mathbf{r})$ is a probability table embodying a relationship between the variable s (here, the stimulus) and $\mathbf{r}$ (a vector where each element is the firing rate of one neuron).

However, because the probability table of the relation between the neuronal responses and the stimuli, $P(s, \mathbf{r})$, is so large (given that there may be many stimuli, and that the response space which has to include spike timing is very large), in practice it is difficult to obtain a sufficient number of trials for every stimulus to generate the probability table accurately, at least with data from mammals in which the experiment cannot usually be continued for many hours of recording from a whole population of cells. To circumvent this undersampling problem, Rolls, Treves and Tovee (1997b) developed a decoding procedure (described in Section C.2.3), in which an estimate (or guess) of which stimulus (called s') was shown on a given trial is made from a comparison of the neuronal responses on that trial with the responses made to the whole set of stimuli on other trials. One then obtains a conjoint probability table $P(s, s')$, and then the mutual information based on probability estimation (PE) decoding (I_p) between the estimated stimuli s' and the actual stimuli s that were shown can be measured:

$$< I_p > \ = \sum_{s \in S} \sum_{s' \in S} P(s, s') \log_2 \frac{P(s, s')}{P(s)P(s')} \tag{C.32}$$

$$= \sum_{s \in S} P(s) \sum_{s' \in S} P(s'|s) \log_2 \frac{P(s'|s)}{P(s')}. \tag{C.33}$$

These measurements are in the low dimensional space of the number of stimuli, and therefore the number of trials of data needed for each stimulus is of the order of the number of stimuli, which is feasible in experiments. In practice, it is found that for accurate information estimates with the decoding approach, the number of trials for each stimulus should be at least twice the number of stimuli (Franco, Rolls, Aggelopoulos and Treves 2004).

The nature of the decoding procedure is illustrated in Fig. C.3. The left part of the diagram shows the average firing rate (or equivalently spike count) responses of each of 3 cells (labelled as Rate Cell 1,2,3) to a set of 3 stimuli. The last row (labelled Response single trial) shows the data that might be obtained from a single trial and from which the stimulus that was shown (St. ?) must be estimated or decoded, using the average values across trials shown in the top part of the table, and the probability distribution of these values. The decoding step essentially compares the vector of responses on trial St.? with the average response vectors obtained previously to each stimulus. This decoding can be as simple as measuring the correlation, or dot (inner) product, between the test trial vector of responses and the response vectors to each of the stimuli. This procedure is very neuronally plausible, in that the dot product between an input vector of neuronal activity and the synaptic response vector on a single neuron (which might represent the average incoming activity previously to that stimulus) is the simplest operation that it is conceived that neurons might perform (Rolls and Treves 1998, Rolls and Deco 2002, Rolls 2008d). Other decoding procedures include a Bayesian procedure based on a Gaussian or Poisson assumption of the spike count distributions as described in detail by Rolls, Treves and Tovee (1997b). The Gaussian one is what we have used (Franco, Rolls, Aggelopoulos and Treves 2004, Aggelopoulos, Franco and Rolls 2005), and it is described below.

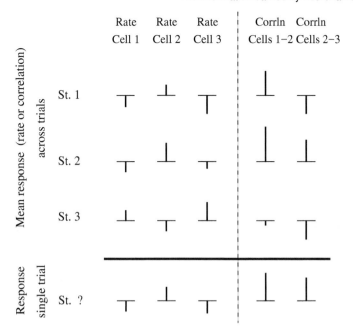

Fig. C.3 The left part of the diagram shows the average firing rate (or equivalently spike count) responses of each of 3 cells (labelled as Rate Cell 1,2,3) to a set of 3 stimuli. The right two columns show a measure of the cross-correlation (averaged across trials) for some pairs of cells (labelled as Corrln Cells 1–2 and 2–3). The last row (labelled Response single trial) shows the data that might be obtained from a single trial and from which the stimulus that was shown (St. ? or s') must be estimated or decoded, using the average values across trials shown in the top part of the table. From the responses on the single trial, the most probable decoded stimulus is stimulus 2, based on the values of both the rates and the cross-correlations. (After Franco, Rolls, Aggelopoulos and Treves 2004.)

The new step taken by Franco, Rolls, Aggelopoulos and Treves (2004) is to introduce into the Table Data(s, r) shown in the upper part of Fig. C.3 new columns, shown on the right of the diagram, containing a measure of the cross-correlation (averaged across trials in the upper part of the table) for some pairs of cells (labelled as Corrln Cells 1–2 and 2–3). The decoding procedure can then take account of any cross-correlations between pairs of cells, and thus measure any contributions to the information from the population of cells that arise from cross-correlations between the neuronal responses. If these cross-correlations are stimulus-dependent, then their positive contribution to the information encoded can be measured. This is the new concept for information measurement from neuronal populations introduced by Franco, Rolls, Aggelopoulos and Treves (2004). We describe next how the cross-correlation information can be introduced into the Table, and then how the information analysis algorithm can be used to measure the contribution of different factors in the neuronal responses to the information that the population encodes.

To test different hypotheses, the decoding can be based on all the columns of the Table (to provide the total information available from both the firing rates and the stimulus-dependent synchronization), on only the columns with the firing rates (to provide the information available from the firing rates), and only on the columns with the cross-correlation values (to provide the information available from the stimulus-dependent cross-correlations). Any information from stimulus-dependent cross-correlations will not necessarily be orthogonal to the rate information, and the procedures allow this to be checked by comparing the total information to that from the sum of the two components. If cross-correlations are present but are not stimulus-dependent, these will not contribute to the information available about which stimulus was shown.

The measure of the synchronization introduced into the Table Data(s, r) on each trial is, for example, the value of the Pearson cross-correlation coefficient calculated for that trial at the appropriate lag for cell pairs that have significant cross-correlations (Franco, Rolls, Aggelopoulos and Treves 2004). This value of this Pearson cross-correlation coefficient for a single trial can be calculated from pairs of spike trains on a single trial by forming for each cell a vector of 0s and 1s, the 1s representing the time of occurrence of spikes with a temporal resolution of 1 ms. Resulting values within the range -1 to 1 are shifted to obtain positive values. An advantage of basing the measure of synchronization on the Pearson cross-correlation coefficient is that it measures the amount of synchronization between a pair of neurons independently of the firing rate of the neurons. The lag at which the cross-correlation measure was computed for every single trial, and whether there was a significant cross-correlation between neuron pairs, can be identified from the location of the peak in the cross-correlogram taken across all trials. The cross-correlogram is calculated by, for every spike that occurred in one neuron, incrementing the bins of a histogram that correspond to the lag times of each of the spikes that occur for the other neuron. The raw cross-correlogram is corrected by subtracting the "shift predictor" cross-correlogram (which is produced by random re-pairings of the trials), to produce the corrected cross-correlogram.

Further details of the decoding procedures are as follows (see Rolls, Treves and Tovee (1997b) and Franco, Rolls, Aggelopoulos and Treves (2004)). The full probability table estimator (PE) algorithm uses a Bayesian approach to extract $P(s'|\mathbf{r})$ for every single trial from an estimate of the probability $P(\mathbf{r}|s')$ of a stimulus–response pair made from all the other trials (as shown in Bayes' rule shown in equation C.34) in a cross-validation procedure described by Rolls et al. (1997b).

$$P(s'|\mathbf{r}) = \frac{P(\mathbf{r}|s')P(s')}{P(\mathbf{r})}. \tag{C.34}$$

where $P(\mathbf{r})$ (the probability of the vector containing the firing rate of each neuron, where each element of the vector is the firing rate of one neuron) is obtained as:

$$P(\mathbf{r}) = \sum_{s'} P(\mathbf{r}|s')P(s'). \tag{C.35}$$

This requires knowledge of the response probabilities $P(\mathbf{r}|s')$ which can be estimated for this purpose from $P(\mathbf{r}, s')$, which is equal to $P(s') \prod_c P(r_c|s')$, where r_c is the firing rate of cell c. We note that $P(r_c|s')$ is derived from the responses of cell c from all of the trials except for the current trial for which the probability estimate is being made. The probabilities $P(r_c|s')$ are fitted with a Gaussian (or Poisson) distribution whose amplitude at r_c gives $P(r_c|s')$. By summing over different test trial responses to the same stimulus s, we can extract the probability that by presenting stimulus s the neuronal response is interpreted as having been elicited by stimulus s',

$$P(s'|s) = \sum_{\mathbf{r} \in \text{test}} P(s'|\mathbf{r})P(\mathbf{r}|s). \tag{C.36}$$

After the decoding procedure, the estimated relative probabilities (normalized to 1) were averaged over all 'test' trials for all stimuli, to generate a (Regularized) table $P^R{}_N(s, s')$ describing the relative probability of each pair of actual stimulus s and posited stimulus s' (computed with N trials). From this probability table the mutual information measure (I_p) was calculated as described above in equation C.33.

We also generate a second (Frequency) table $P^F{}_N(s, s^p)$ from the fraction of times an actual stimulus s elicited a response that led to a predicted (single most likely) stimulus s^p. From this probability Table the mutual information measure based on maximum likelihood decoding (I_{ml}) was calculated with equation C.37:

$$< I_{ml} > = \sum_{s \in S} \sum_{s^p \in S} P(s, s^p) \log_2 \frac{P(s, s^p)}{P(s)P(s^p)} . \tag{C.37}$$

A detailed comparison of maximum likelihood and probability decoding is provided by Rolls, Treves and Tovee (1997b), but we note here that probability estimate decoding is more regularized (see below) and therefore may be safer to use when investigating the effect on the information of the number of cells. For this reason, the results described by Franco, Rolls, Aggelopoulos and Treves (2004) were obtained with probability estimation (PE) decoding. The maximum likelihood decoding does give an immediate measure of the percentage correct.

Another approach to decoding is the dot product (DP) algorithm which computes the normalized dot products between the current firing vector $\mathbf{r}$ on a "test" (i.e. the current) trial and each of the mean firing rate response vectors in the "training" trials for each stimulus s' in the cross-validation procedure. (The normalized dot product is the dot or inner product of two vectors divided by the product of the length of each vector. The length of each vector is the square root of the sum of the squares.) Thus, what is computed are the cosines of the angles of the test vector of cell rates with, in turn for each stimulus, the mean response vector to that stimulus. The highest dot product indicates the most likely stimulus that was presented, and this is taken as the predicted stimulus s^p for the probability table $P(s, s^p)$. (It can also be used to provide percentage correct measures.)

We note that any decoding procedure can be used in conjunction with information estimates both from the full probability table (to produce I_p), and from the most likely estimated stimulus for each trial (to produce I_{ml}).

Because the probability tables from which the information is calculated may be unregularized with a small number of trials, a bias correction procedure to correct for the undersampling is applied, as described in detail by Rolls, Treves and Tovee (1997b) and Panzeri and Treves (1996). In practice, the bias correction that is needed with information estimates using the decoding procedures described by Franco, Rolls, Aggelopoulos and Treves (2004) and by Rolls et al. (1997b) is small, typically less than 10% of the uncorrected estimate of the information, provided that the number of trials for each stimulus is in the order of twice the number of stimuli. We also note that the distortion in the information estimate from the full probability table needs less bias correction than that from the predicted stimulus table (i.e. maximum likelihood) method, as the former is more regularized because every trial makes some contribution through much of the probability table (see Rolls et al. (1997b)). We further note that the bias correction term becomes very small when more than 10 cells are included in the analysis (Rolls et al. 1997b).

Examples of the use of these procedures are available (Franco, Rolls, Aggelopoulos and Treves 2004, Aggelopoulos, Franco and Rolls 2005), and some of the results obtained are described in Section C.3.

C.2.5 Information in the correlations between the spikes of different cells: a second derivative approach

Another information theory-based approach to stimulus-dependent cross-correlation information has been developed as follows by Panzeri, Schultz, Treves and Rolls (1999a) and Rolls, Franco, Aggelopoulos and Reece (2003b). A problem that must be overcome is the fact that with many simultaneously recorded neurons, each emitting perhaps many spikes at different times, the dimensionality of the response space becomes very large, the information tends to be overestimated, and even bias corrections cannot save the situation. The approach described in this Section (C.2.5) limits the problem by taking short time epochs for the information analysis, in which low numbers of spikes, in practice typically 0, 1, or 2, spikes are likely to occur from each neuron.

In a sufficiently short time window, at most two spikes are emitted from a population of neurons. Taking advantage of this, the response probabilities can be calculated in terms of pairwise correlations. These response probabilities are inserted into the Shannon information formula C.38 to obtain expressions quantifying the impact of the pairwise correlations on the information $I(t)$ transmitted in a short time t by groups of spiking neurons:

$$I(t) = \sum_{s \in S} \sum_{\mathbf{r}} P(s, \mathbf{r}) \log_2 \frac{P(s, \mathbf{r})}{P(s)P(\mathbf{r})} \tag{C.38}$$

where $\mathbf{r}$ is the firing rate response vector comprised by the number of spikes emitted by each of the cells in the population in the short time t, and $P(s, \mathbf{r})$ refers to the joint probability distribution of stimuli with their respective neuronal response vectors.

The information depends upon the following two types of correlation:

C.2.5.1 The correlations in the neuronal response variability from the average to each stimulus (sometimes called "noise" correlations) γ:

$\gamma_{ij}(s)$ (for $i \neq j$) is the fraction of coincidences above (or below) that expected from uncorrelated responses, relative to the number of coincidences in the uncorrelated case (which is $\bar{n}_i(s)\bar{n}_j(s)$, the bar denoting the average across trials belonging to stimulus s, where $n_i(s)$ is the number of spikes emitted by cell i to stimulus s on a given trial)

$$\gamma_{ij}(s) = \frac{\overline{n_i(s)n_j(s)}}{(\overline{n_i(s)}\,\overline{n_j(s)})} - 1, \tag{C.39}$$

and is named the 'scaled cross-correlation density'. It can vary from -1 to ∞; negative $\gamma_{ij}(s)$'s indicate anticorrelation, whereas positive $\gamma_{ij}(s)$'s indicate correlation[44]. $\gamma_{ij}(s)$ can be thought of as the amount of trial by trial concurrent firing of the cells i and j, compared to that expected in the uncorrelated case. $\gamma_{ij}(s)$ (for $i \neq j$) is the 'scaled cross-correlation density' (Aertsen, Gerstein, Habib and Palm 1989, Panzeri, Schultz, Treves and Rolls 1999a), and is sometimes called the "noise" correlation (Gawne and Richmond 1993, Shadlen and Newsome 1995, Shadlen and Newsome 1998).

---

[44]$\gamma_{ij}(s)$ is an alternative, which produces a more compact information analysis, to the neuronal cross-correlation based on the Pearson correlation coefficient $\rho_{ij}(s)$ (equation C.40), which normalizes the number of coincidences above independence to the standard deviation of the number of coincidences expected if the cells were independent. The normalization used by the Pearson correlation coefficient has the advantage that it quantifies the strength of correlations between neurons in a rate-independent way. For the information analysis, it is more convenient to use the scaled correlation density $\gamma_{ij}(s)$ than the Pearson correlation coefficient, because of the compactness of the resulting formulation, and because of its scaling properties for small t. $\gamma_{ij}(s)$ remains finite as $t \to 0$, thus by using this measure we can keep the t expansion of the information explicit. Keeping the time-dependence of the resulting information components explicit greatly increases the amount of insight obtained from the series expansion. In contrast, the Pearson noise-correlation measure applied to short timescales approaches zero at short time windows:

$$\rho_{ij}(s) \equiv \frac{\overline{n_i(s)n_j(s)} - \bar{n}_i(s)\bar{n}_j(s)}{\sigma_{n_i(s)}\sigma_{n_j(s)}} \simeq t \qquad \gamma_{ij}(s) = \sqrt{\bar{r}_i(s)\bar{r}_j(s)}, \tag{C.40}$$

where $\sigma_{n_i(s)}$ is the standard deviation of the count of spikes emitted by cell i in response to stimulus s.

C.2.5.2 The correlations in the mean responses of the neurons across the set of stimuli (sometimes called "signal" correlations) ν:

$$\nu_{ij} = \frac{<\overline{n}_i(s)\overline{n}_j(s)>_s}{<\overline{n}_i(s)>_s<\overline{n}_j(s)>_s} - 1 = \frac{<\overline{r}_i(s)\overline{r}_j(s)>_s}{<\overline{r}_i(s)>_s<\overline{r}_j(s)>_s} - 1 \qquad \text{(C.41)}$$

where $\overline{r}_i(s)$ is the mean rate of response of cell i (among C cells in total) to stimulus s over all the trials in which that stimulus was present. ν_{ij} can be thought of as the degree of similarity in the mean response profiles (averaged across trials) of the cells i and j to different stimuli. ν_{ij} is sometimes called the "signal" correlation (Gawne and Richmond 1993, Shadlen and Newsome 1995, Shadlen and Newsome 1998).

C.2.5.3 Information in the cross-correlations in short time periods

In the short timescale limit, the first (I_t) and second (I_{tt}) information derivatives describe the information $I(t)$ available in the short time t

$$I(t) = t\, I_t + \frac{t^2}{2}\, I_{tt} \ . \qquad \text{(C.42)}$$

(The zeroth order, time-independent term is zero, as no information can be transmitted by the neurons in a time window of zero length. Higher order terms are also excluded as they become negligible.)

The instantaneous information rate I_t is[45]

$$I_t = \sum_{i=1}^{C} \left\langle \overline{r}_i(s) \log_2 \frac{\overline{r}_i(s)}{\langle \overline{r}_i(s')\rangle_{s'}} \right\rangle_s . \qquad \text{(C.43)}$$

This formula shows that this information rate (the first time derivative) should not be linked to a high signal to noise ratio, but only reflects the extent to which the mean responses of each cell are distributed across stimuli. It does not reflect anything of the variability of those responses, that is of their noisiness, nor anything of the correlations among the mean responses of different cells.

The effect of (pairwise) correlations between the cells begins to be expressed in the second time derivative of the information. The expression for the instantaneous information 'acceleration' I_{tt} (the second time derivative of the information) breaks up into three terms:

$$
\begin{aligned}
I_{tt} = \ & \frac{1}{\ln 2} \sum_{i=1}^{C}\sum_{j=1}^{C} \langle \overline{r}_i(s)\rangle_s \langle \overline{r}_j(s)\rangle_s \left[\nu_{ij} + (1+\nu_{ij})\ln\left(\frac{1}{1+\nu_{ij}}\right)\right] \\
& + \sum_{i=1}^{C}\sum_{j=1}^{C} \left[\langle \overline{r}_i(s)\overline{r}_j(s)\gamma_{ij}(s)\rangle_s\right] \log_2\left(\frac{1}{1+\nu_{ij}}\right) \\
& + \sum_{i=1}^{C}\sum_{j=1}^{C} \left\langle \overline{r}_i(s)\overline{r}_j(s)(1+\gamma_{ij}(s)) \log_2\left[\frac{(1+\gamma_{ij}(s))\,\langle \overline{r}_i(s')\overline{r}_j(s')\rangle_{s'}}{\langle \overline{r}_i(s')\overline{r}_j(s')(1+\gamma_{ij}(s'))\rangle_{s'}}\right]\right\rangle_s .
\end{aligned} \qquad \text{(C.44)}
$$

The first of these terms is all that survives if there is no noise correlation at all. Thus the *rate component* of the information is given by the sum of I_t (which is always greater than or equal to zero) and of the first term of I_{tt} (which is instead always less than or equal to zero).

---

[45] Note that s' is used in equations C.43 and C.44 just as a dummy variable to stand for s, as there are two summations performed over s.

The second term is non-zero if there is some correlation in the variance to a given stimulus, even if it is independent of which stimulus is present; this term thus represents the contribution of *stimulus-independent noise correlation* to the information.

The third component of I_{tt} represents the contribution of *stimulus-modulated noise correlation*, as it becomes non-zero only for stimulus-dependent noise correlations. These last two terms of I_{tt} together are referred to as the correlational components of the information.

The application of this approach to measuring the information in the relative time of firing of simultaneously recorded cells, together with further details of the method, are described by Panzeri, Treves, Schultz and Rolls (1999b), Rolls, Franco, Aggelopoulos and Reece (2003b), and Rolls, Aggelopoulos, Franco and Treves (2004), and in Section C.3.7.

C.3 Neuronal encoding: results obtained from applying information-theoretic analyses

How is information encoded in cortical areas such as the inferior temporal visual cortex? Can we read the code being used by the cortex? What are the advantages of the encoding scheme used for the neuronal network computations being performed in different areas of the cortex? These are some of the key issues considered in this Section (C.3). Because information is exchanged between the computing elements of the cortex (the neurons) by their spiking activity, which is conveyed by their axon to synapses onto other neurons, the appropriate level of analysis is how single neurons, and populations of single neurons, encode information in their firing. More global measures that reflect the averaged activity of large numbers of neurons (for example, PET (positron emission tomography) and fMRI (functional magnetic resonance imaging), EEG (electroencephalographic recording), and ERPs (event-related potentials)) cannot reveal how the information is represented, or how the computation is being performed.

Although information theory provides the natural mathematical framework for analysing the performance of neuronal systems, its applications in neuroscience have been for many years rather sparse and episodic (e.g. MacKay and McCulloch (1952); Eckhorn and Popel (1974); Eckhorn and Popel (1975); Eckhorn, Grusser, Kroller, Pellnitz and Popel (1976)). One reason for this limited application of information theory has been the great effort that was apparently required, due essentially to the limited sampling problem, in order to obtain accurate results. Another reason has been the hesitation in analysing as a single complex 'black-box' large neuronal systems all the way from some external, easily controllable inputs, up to neuronal activity in some central cortical area of interest, for example including all visual stations from the periphery to the end of the ventral visual stream in the temporal lobe. In fact, two important bodies of work, that have greatly helped revive interest in applications of the theory in recent years, both sidestep these two problems. The problem with analyzing a huge black-box is avoided by considering systems at the sensory periphery; the limited sampling problem is avoided either by working with insects, in which sampling can be extensive (Bialek, Rieke, de Ruyter van Steveninck and Warland 1991, de Ruyter van Steveninck and Laughlin 1996, Rieke, Warland, de Ruyter van Steveninck and Bialek 1997), or by utilizing a formal model instead of real data (Atick and Redlich 1990, Atick 1992). Both approaches have provided insightful quantitative analyses that are in the process of being extended to more central mammalian systems (see e.g. Atick, Griffin and Relich (1996)).

In the treatment provided here, we focus on applications to the mammalian brain, using examples from a whole series of investigations on information representation in visual cortical areas, the original papers on which refer to related publications.

C.3.1 The sparseness of the distributed encoding used by the brain

Some of the types of representation that might be found at the neuronal level are summarized next (cf. Section 8.2). A **local representation** is one in which all the information that a particular stimulus or event occurred is provided by the activity of one of the neurons. This is sometimes called a grandmother cell representation, because in a famous example, a single neuron might be active only if one's grandmother was being seen (see Barlow (1995)). A **fully distributed representation** is one in which all the information that a particular stimulus or event occurred is provided by the activity of the full set of neurons. If the neurons are binary (for example, either active or not), the most distributed encoding is when half the neurons are active for any one stimulus or event. A **sparse distributed representation** is a distributed representation in which a small proportion of the neurons is active at any one time.

C.3.1.1 Single neuron sparseness a^s

Equation C.45 defines a measure of the single neuron sparseness, a^s:

$$a^s = \frac{(\sum\limits_{s=1}^{S} y_s/S)^2}{(\sum\limits_{s=1}^{S} y_s^2)/S} \tag{C.45}$$

where y_s is the mean firing rate of the neuron to stimulus s in the set of S stimuli (Rolls and Treves 1998). For a binary representation, a^s is 0.5 for a fully distributed representation, and $1/S$ if a neuron responds to one of a set of S stimuli. Another measure of sparseness is the kurtosis of the distribution, which is the fourth moment of the distribution. It reflects the length of the tail of the distribution. (An actual distribution of the firing rates of a neuron to a set of 65 stimuli is shown in Fig. C.4. The sparseness a^s for this neuron was 0.69 (see Rolls, Treves, Tovee and Panzeri (1997d).)

It is important to understand and quantify the sparseness of representations in the brain, because many of the useful properties of neuronal networks such as generalization and completion only occur if the representations are not local (see Appendix B), and because the value of the sparseness is an important factor in how many memories can be stored in such neural networks. Relatively sparse representations (low values of a^s) might be expected in memory systems as this will increase the number of different memories that can be stored and retrieved. Less sparse representations might be expected in sensory systems, as this could allow more information to be represented (see Table B.2).

Barlow (1972) proposed a single neuron doctrine for perceptual psychology. He proposed that sensory systems are organized to achieve as complete a representation as possible with the minimum number of active neurons. He suggested that at progressively higher levels of sensory processing, fewer and fewer cells are active, and that each represents a more and more specific happening in the sensory environment. He suggested that 1,000 active neurons (which he called cardinal cells) might represent the whole of a visual scene. An important principle involved in forming such a representation was the reduction of redundancy. The implication of Barlow's (1972) approach was that when an object is being recognized, there are, towards the end of the visual system, a small number of neurons (the cardinal cells) that are so specifically tuned that the activity of these neurons encodes the information that one particular object is being seen. (He thought that an active neuron conveys something of the order of complexity of a word.) The encoding of information in such a system is described as local, in that knowing the activity of just one neuron provides evidence that a particular stimulus (or, more exactly, a given 'trigger feature') is present. Barlow (1972) eschewed 'combinatorial rules of usage of nerve cells', and believed that the subtlety and sensitivity of perception results from the mechanisms determining when a single cell becomes active. In contrast, with distributed or ensemble encoding, the activity of several or many neurons must be known in order to identify which stimulus is present, that is, to read the code. It is the relative firing of the different neurons in the ensemble that provides the information about which object is present.

At the time Barlow (1972) wrote, there was little actual evidence on the activity of neurons in the higher parts of the visual and other sensory systems. There is now considerable evidence, which is now described.

First, it has been shown that the representation of which particular object (face) is present is actually rather distributed. Baylis, Rolls and Leonard (1985) showed this with the responses of temporal cortical neurons that typically responded to several members of a set of five faces, with each neuron having a different profile of responses to each face (see examples in Fig. 25.12 on page 569). It would be difficult for most of these single cells to tell which of even five faces, let alone which of hundreds of faces, had been seen. (At the same time, the neurons discriminated between the faces reliably, as shown by the values of d', taken, in the case of the neurons, to be the number of standard deviations of the neuronal responses that separated the response to the best face in the set from that to the least effective face in the set. The values of d' were typically in the range 1–3.)

Second, the distributed nature of the representation can be further understood by the finding that the firing rate probability distribution of single neurons, when a wide range of natural visual stimuli are being viewed, is approximately exponential, with rather few stimuli producing high firing rates, and increasingly large numbers of stimuli producing lower and lower firing rates, as illustrated in Fig. C.5a (Rolls and Tovee 1995b, Baddeley, Abbott, Booth, Sengpiel, Freeman, Wakeman and Rolls 1997, Treves, Panzeri, Rolls, Booth and Wakeman 1999, Franco, Rolls, Aggelopoulos and Jerez 2007).

For example, the responses of a set of temporal cortical neurons to 23 faces and 42 non-face natural images were measured, and a distributed representation was found (Rolls and Tovee 1995b). The tuning was typically graded, with a range of different firing rates to the set of faces, and very little response to the non-face stimuli (see example in Fig. C.4). The spontaneous firing rate of the neuron in Fig. C.4 was 20 spikes/s, and the histogram bars indicate the change of firing rate from the spontaneous value produced by each stimulus. Stimuli that are faces are marked F, or P if they are in profile. B refers to images of scenes that included either a small face within the scene, sometimes as part of an image that included a whole person, or other body parts, such as hands (H) or legs. The non-face stimuli are unlabelled. The neuron responded best to three of the faces (profile views), had some response to some of the other faces, and had little or no response, and sometimes had a small decrease of firing rate below the spontaneous firing rate, to the non-face stimuli. The sparseness value a^s for this cell across all 68 stimuli was 0.69, and the response sparseness a_r^s (based on the evoked responses minus the spontaneous firing of the neuron) was 0.19. It was found that the sparseness of the representation of the 68 stimuli by each neuron had an average across all neurons of 0.65 (Rolls and Tovee 1995b). This indicates a rather distributed representation. (If neurons had a continuum of firing rates equally distributed between zero and maximum rate, a^s would be 0.75, while if the probability of each response decreased linearly, to reach zero at the maximum rate, a^s would be 0.67).

I comment that these values for a do not seem very sparse. But these values are calculated using the raw firing rates of the neurons, on the basis that these would be what a receiving neuron would receive as its input representation. However, neocortical neurons have a spontaneous firing rate of several spikes/s (with a lower value of 0.75 spikes/s for hippocampal pyramidal cells), and if this spontaneous value is subtracted from the firing rates to yield a 'response sparseness' a_r, this value is considerably lower. For example, the sparseness a of

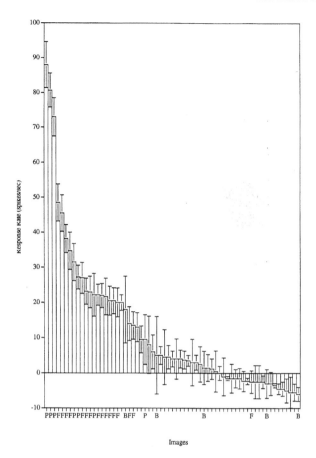

Fig. C.4 Firing rate distribution of a single neuron in the temporal visual cortex to a set of 23 face (F) and 45 non-face images of natural scenes. The firing rate to each of the 68 stimuli is shown. The neuron does not respond to just one of the 68 stimuli. Instead, it responds to a small proportion of stimuli with high rates, to more stimuli with intermediate rates, and to many stimuli with almost no change of firing. This is typical of the distributed representations found in temporal cortical visual areas. (After Rolls and Tovee 1995b.)

inferior temporal cortex responses to a set of 68 stimuli had an average across all neurons that we analyzed in this study of 0.65 (Rolls and Tovee 1995b). If the spontaneous firing rate was subtracted from the firing rate of the neuron to each stimulus, so that the changes of firing rate, i.e., the responses of the neurons, were used in the sparseness calculation, then the 'response sparseness' had a lower value, with a mean of $a_r=0.33$ for the population of neurons, or 0.60 if calculated over the set of faces rather than over all the face and non-face stimuli. Further, the true sparseness of the representation is probably much less than this, for this is calculated only over the neurons that had responses to some of these stimuli. There were many more neurons that had no response to the stimuli. At least 10 times the number of inferior temporal cortex neurons had no responses to this set of 68 stimuli. So the true sparseness would be much lower that this value of 0.33. Further, it is important to remember the relative nature of sparseness measures, which (like the information measures to be discussed below) depend strongly on the stimulus set used. Thus we can reject a cardinal cell representation. As shown below, the readout of information from these cells is actually much better in any case than

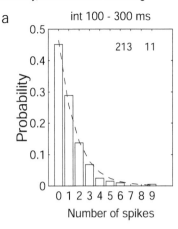 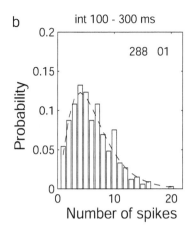

Fig. C.5 Firing rate probability distributions for two neurons in the inferior temporal visual cortex tested with a set of 20 face and non-face stimuli. (a) A neuron with a good fit to an exponential probability distribution (dashed line). (b) A neuron that did not fit an exponential firing rate distribution (but which could be fitted by a gamma distribution, dashed line). The firing rates were measured in an interval 100–300 ms after the onset of the visual stimuli, and similar distributions are obtained in other intervals. (After Franco, Rolls, Aggelopoulos and Jerez 2007.)

would be obtained from a local representation, and this makes it unlikely that there is a further population of neurons with very specific tuning that use local encoding.

These data provide a clear answer to whether these neurons are grandmother cells: they are not, in the sense that each neuron has a graded set of responses to the different members of a set of stimuli, with the prototypical distribution similar to that of the neuron illustrated in Fig. C.4. On the other hand, each neuron does respond very much more to some stimuli than to many others, and in this sense is tuned to some stimuli.

Figure C.5 shows data of the type shown in Fig. C.4 as firing rate probability density functions, that is as the probability that the neuron will be firing with particular rates. These data were from inferior temporal cortex neurons, and show when tested with a set of 20 face and non-face stimuli how fast the neuron will be firing in a period 100–300 ms after the visual stimulus appears (Franco, Rolls, Aggelopoulos and Jerez 2007). Figure C.5a shows an example of a neuron where the data fit an exponential firing rate probability distribution, with many occasions on which the neuron was firing with a very low firing rate, and decreasingly few occasions on which it fired at higher rates. This shows that the neuron can have high firing rates, but only to a few stimuli. Figure C.5b shows an example of a neuron where the data do not fit an exponential firing rate probability distribution, with insufficiently few very low rates. Of the 41 responsive neurons in this data set, 15 had a good fit to an exponential firing rate probability distribution; the other 26 neurons did not fit an exponential but did fit a gamma distribution in the way illustrated in Fig. C.5b. For the neurons with an exponential distribution, the mean firing rate across the stimulus set was 5.7 spikes/s, and for the neurons with a gamma distribution was 21.1 spikes/s (t=4.5, df=25, p < 0.001). It may be that neurons with high mean rates to a stimulus set tend to have few low rates ever, and this accounts for their poor fit to an exponential firing rate probability distribution, which fits when there are many low firing rate values in the distribution as in Fig. C.5a.

The large set of 68 stimuli used by Rolls and Tovee (1995b) was chosen to produce an approximation to a set of stimuli that might be found to natural stimuli in a natural environment, and thus to provide evidence about the firing rate distribution of neurons to natural stimuli. Another approach to the same fundamental question was taken by Baddeley,

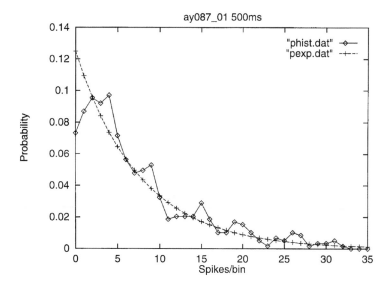

Fig. C.6 The probability of different firing rates measured in short (e.g. 100 ms or 500 ms) time windows of a temporal cortex neuron calculated over a 5 min period in which the macaque watched a video showing natural scenes, including faces. An exponential fit (+) to the data (diamonds) is shown. (After Baddeley, Abbott, Booth, Sengpiel, Freeman, Wakeman and Rolls 1997.)

Abbott, Booth, Sengpiel, Freeman, Wakeman, and Rolls (1997) who measured the firing rates over short periods of individual inferior temporal cortex neurons while monkeys watched continuous videos of natural scenes. They found that the firing rates of the neurons were again approximately exponentially distributed (see Fig. C.6), providing further evidence that this type of representation is characteristic of inferior temporal cortex (and indeed also V1) neurons.

The actual distribution of the firing rates to a wide set of natural stimuli is of interest, because it has a rather stereotypical shape, typically following a graded unimodal distribution with a long tail extending to high rates (see for example Figs. C.5a and C.6). The mode of the distribution is close to the spontaneous firing rate, and sometimes it is at zero firing. If the number of spikes recorded in a fixed time window is taken to be constrained by a fixed maximum rate, one can try to interpret the distribution observed in terms of optimal information transmission (Shannon 1948), by making the additional assumption that the coding is noiseless. An exponential distribution, which maximizes entropy (and hence information transmission for noiseless codes) is the most efficient in terms of energy consumption if its mean takes an optimal value that is a decreasing function of the relative metabolic cost of emitting a spike (Levy and Baxter 1996). This argument would favour sparser coding schemes the more energy expensive neuronal firing is (relative to rest). Although the tail of actual firing rate distributions is often approximately exponential (see for example Figs. C.5a and C.6; Baddeley, Abbott, Booth, Sengpiel, Freeman, Wakeman and Rolls (1997); Rolls, Treves, Tovee and Panzeri (1997d); and Franco, Rolls, Aggelopoulos and Jerez (2007)), the maximum entropy argument cannot apply as such, because noise is present and the noise level varies as a function of the rate, which makes entropy maximization different from information maximization. Moreover, a mode at low but non-zero rate, which is often observed (see e.g. Fig. C.5b), is inconsistent with the energy efficiency hypothesis.

A simpler explanation for the characteristic firing rate distribution arises by appreciating

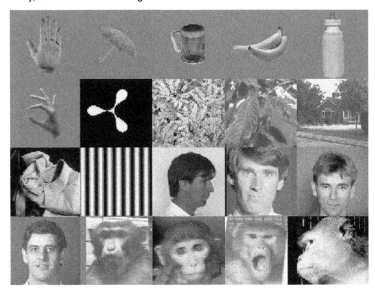

Fig. C.7 The set of 20 stimuli used to investigate the tuning of inferior temporal cortex neurons by Franco, Rolls, Aggelopoulos and Jerez 2007. These objects and faces are typical of those encoded in the ways described here by inferior temporal cortex neurons. The code can be read off simply from the firing rates of the neurons about which object or face was shown, and many of the neurons have invariant responses.

that the value of the activation of a neuron across stimuli, reflecting a multitude of contributing factors, will typically have a Gaussian distribution; and by considering a physiological input–output transform (i.e. activation function), and realistic noise levels. In fact, an input–output transform that is supralinear in a range above threshold results from a fundamentally linear transform and fluctuations in the activation, and produces a variance in the output rate, across repeated trials, that increases with the rate itself, consistent with common observations. At the same time, such a supralinear transform tends to convert the Gaussian tail of the activation distribution into an approximately exponential tail, without implying a fully exponential distribution with the mode at zero. Such basic assumptions yield excellent fits with observed distributions (Treves, Panzeri, Rolls, Booth and Wakeman 1999), which often differ from exponential in that there are too few very low rates observed, and too many low rates (Rolls, Treves, Tovee and Panzeri 1997d, Franco, Rolls, Aggelopoulos and Jerez 2007).

This peak at low but non-zero rates may be related to the low firing rate spontaneous activity that is typical of many cortical neurons. Keeping the neurons close to threshold in this way may maximize the speed with which a network can respond to new inputs (because time is not required to bring the neurons from a strongly hyperpolarized state up to threshold). The advantage of having low spontaneous firing rates may be a further reason why a curve such as an exponential cannot sometimes be exactly fitted to the experimental data.

A conclusion of this analysis was that the firing rate distribution may arise from the threshold non-linearity of neurons combined with short-term variability in the responses of neurons (Treves, Panzeri, Rolls, Booth and Wakeman 1999).

However, given that the firing rate distribution for some neurons is approximately exponential, some properties of this type of representation are worth elucidation. The sparseness of such an exponential distribution of firing rates is 0.5. This has interesting implications, for to the extent that the firing rates are exponentially distributed, this fixes an important parameter of cortical neuronal encoding to be close to 0.5. Indeed, only one parameter specifies the

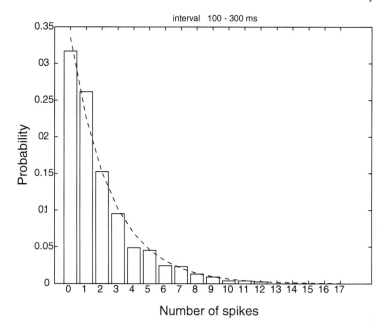

Fig. C.8 An exponential firing rate probability distribution obtained by pooling the firing rates of a population of 41 inferior temporal cortex neurons tested to a set of 20 face and non-face stimuli. The firing rate probability distribution for the 100–300 ms interval following stimulus onset was formed by adding the spike counts from all 41 neurons, and across all stimuli. The fit to the exponential distribution (dashed line) was high. (After Franco, Rolls, Aggelopoulos and Jerez 2007.)

shape of the exponential distribution, and the fact that the exponential distribution is at least a close approximation to the firing rate distribution of some real cortical neurons implies that the sparseness of the cortical representation of stimuli is kept under precise control. The utility of this may be to ensure that any neuron receiving from this representation can perform a dot product operation between its inputs and its synaptic weights that produces similarly distributed outputs; and that the information being represented by a population of cortical neurons is kept high. It is interesting to realize that the representation that is stored in an associative network (see Appendix B) may be more sparse than the 0.5 value for an exponential firing rate distribution, because the non-linearity of learning introduced by the voltage dependence of the NMDA receptors (see Appendix B) effectively means that synaptic modification in, for example, an autoassociative network will occur only for the neurons with relatively high firing rates, i.e. for those that are strongly depolarized.

The single neuron selectivity reflects response distributions of individual neurons across time to different stimuli. As we have seen, part of the interest of measuring the firing rate probability distributions of individual neurons is that one form of the probability distribution, the exponential, maximizes the entropy of the neuronal responses for a given mean firing rate, which could be used to maximize information transmission consistent with keeping the firing rate on average low, in order to minimize metabolic expenditure (Levy and Baxter 1996, Baddeley, Abbott, Booth, Sengpiel, Freeman, Wakeman and Rolls 1997). Franco, Rolls, Aggelopoulos and Jerez (2007) showed that while the firing rates of some single inferior temporal cortex neurons (tested in a visual fixation task to a set of 20 face and non-face stimuli illustrated in Fig. C.7) do fit an exponential distribution, and others with higher spontaneous firing rates do not, as described above, it turns out that there is a very close fit to an exponential distribution of firing rates if all spikes from all the neurons are considered together. This interesting result is shown in Fig. C.8.

One implication of the result shown in Fig. C.8 is that a neuron with inputs from the inferior temporal visual cortex will receive an exponential distribution of firing rates on its afferents, and this is therefore the type of input that needs to be considered in theoretical models of neuronal network function in the brain (see Appendix B). The second implication is that at the level of single neurons, an exponential probability density function is consistent with minimizing energy utilization, and maximizing information transmission, for a given mean firing rate (Levy and Baxter 1996, Baddeley, Abbott, Booth, Sengpiel, Freeman, Wakeman and Rolls 1997).

C.3.1.2 Population sparseness a^{p}

If instead we consider the responses of a population of neurons taken at any one time (to one stimulus), we might also expect a sparse graded distribution, with few neurons firing fast to a particular stimulus. It is important to measure the population sparseness, for this is a key parameter that influences the number of different stimuli that can be stored and retrieved in networks such as those found in the cortex with recurrent collateral connections between the excitatory neurons, which can form autoassociation or attractor networks if the synapses are associatively modifiable (Hopfield 1982, Treves and Rolls 1991, Rolls and Treves 1998, Rolls and Deco 2002, Rolls 2008d) (see Appendix B). Further, in physics, if one can predict the distribution of the responses of the system at any one time (the population level) from the distribution of the responses of a component of the system across time, the system is described as ergodic, and a necessary condition for this is that the components are uncorrelated (Lehky, Sejnowski and Desimone 2005). Considering this in neuronal terms, the average sparseness of a population of neurons over multiple stimulus inputs must equal the average selectivity to the stimuli of the single neurons within the population provided that the responses of the neurons are uncorrelated (Földiák 2003).

The sparseness a^{p} of the population code may be quantified (for any one stimulus) as

$$a^{\mathrm{p}} = \frac{(\sum\limits_{n=1}^{N} y_n/N)^2}{(\sum\limits_{n=1}^{N} y_n^2)/N} \tag{C.46}$$

where y_n is the mean firing rate of neuron n in the set of N neurons.

This measure, a^{p}, of the sparseness of the representation of a stimulus by a population of neurons has a number of advantages. One is that it is the same measure of sparseness that has proved to be useful and tractable in formal analyses of the capacity of associative neural networks and the interference between stimuli that use an approach derived from theoretical physics (Rolls and Treves 1990, Treves 1990, Treves and Rolls 1991, Rolls and Treves 1998) (see Appendix B). We note that high values of a^{p} indicate broad tuning of the population, and that low values of a^{p} indicate sparse population encoding.

Franco, Rolls, Aggelopoulos and Jerez (2007) measured the population sparseness of a set of 29 inferior temporal cortex neurons to a set of 20 stimuli that included faces and objects (see Fig. C.7). Figure C.9a shows, for any one stimulus picked at random, the normalized firing rates of the population of neurons. The rates are ranked with the neuron with the highest rate on the left. For different stimuli, the shape of this distribution is on average the same, though with the neurons in a different order. (The rates of each neuron were normalized to a mean of 10 spikes/s before this graph was made, so that the neurons can be combined in the same graph, and so that the population sparseness has a defined value, as described by Franco,

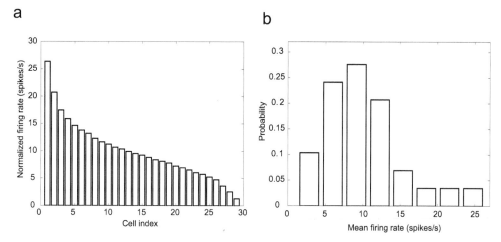

Fig. C.9 Population sparseness. (a) The firing rates of a population of inferior temporal cortex neurons to any one stimulus from a set of 20 face and non-face stimuli. The rates of each neuron were normalized to the same average value of 10 spikes/s, then for each stimulus, the cell firing rates were placed in rank order, and then the mean firing rates of the first ranked cell, second ranked cell, etc. were taken. The graph thus shows how, for any one stimulus picked at random, the expected normalized firing rates of the population of neurons. (b) The population normalized firing rate probability distributions for any one stimulus. This was computed effectively by taking the probability density function of the data shown in (a). (After Franco, Rolls, Aggelopoulos and Jerez 2007.)

Rolls, Aggelopoulos and Jerez (2007).) The population sparseness a^p of this normalized (i.e. scaled) set of firing rates is 0.77.

Figure C.9b shows the probability distribution of the normalized firing rates of the population of (29) neurons to any stimulus from the set. This was calculated by taking the probability distribution of the data shown in Fig. C.9a. This distribution is not exponential because of the normalization of the firing rates of each neuron, but becomes exponential as shown in Fig. C.8 without the normalization step.

A very interesting finding of Franco, Rolls, Aggelopoulos and Jerez (2007) was that when the single cell sparseness a^s and the population sparseness a^p were measured from the same set of neurons in the same experiment, the values were very close, in this case 0.77. (This was found for a range of measurement intervals after stimulus onset, and also for a larger population of 41 neurons.)

The single cell sparseness a^s and the population sparseness a^p can take the same value if the response profiles of the neurons are uncorrelated, that is each neuron is independently tuned to the set of stimuli (Lehky et al. 2005). Franco, Rolls, Aggelopoulos and Jerez (2007) tested whether the response profiles of the neurons to the set of stimuli were uncorrelated in two ways. In a first test, they found that the mean (Pearson) correlation between the response profiles computed over the 406 neuron pairs was low, 0.049 ± 0.013 (sem). In a second test, they computed how the multiple cell information available from these neurons about which stimulus was shown increased as the number of neurons in the sample was increased, and showed that the information increased approximately linearly with the number of neurons in the ensemble. The implication is that the neurons convey independent (non-redundant) information, and this would be expected to occur if the response profiles of the neurons to the stimuli are uncorrelated.

We now consider the concept of ergodicity. The single neuron selectivity, a^s, reflects response distributions of individual neurons across time and therefore stimuli in the world (and has sometimes been termed "lifetime sparseness"). The population sparseness a^p reflects response distributions across all neurons in a population measured simultaneously (to for example one stimulus). The similarity of the average values of a^s and a^p (both 0.77 for inferior temporal cortex neurons (Franco, Rolls, Aggelopoulos and Jerez 2007)) indicates, we believe for the first time experimentally, that the representation (at least in the inferior temporal cortex) is ergodic. The representation is ergodic in the sense of statistical physics, where the average of a single component (in this context a single neuron) across time is compared with the average of an ensemble of components at one time (cf. Masuda and Aihara (2003) and Lehky et al. (2005)). This is described further next.

In comparing the neuronal selectivities a^s and population sparsenesses a^p, we formed a table in which the columns represent different neurons, and the stimuli different rows (Földiák 2003). We are interested in the probability distribution functions (and not just their summary values a^s, and a^p), of the columns (which represent the individual neuron selectivities) and the rows (which represent the population tuning to any one stimulus). We could call the system strongly ergodic (cf. Lehky et al. (2005)) if the selectivity (probability density or distribution function) of each individual neuron is the same as the average population sparseness (probability density function). (Each neuron would be tuned to different stimuli, but have the same shape of the probability density function.) We have seen that this is not the case, in that the firing rate probability distribution functions of different neurons are different, with some fitting an exponential function, and some a gamma function (see Fig. C.5). We can call the system weakly ergodic if individual neurons have different selectivities (i.e. different response probability density functions), but the average selectivity (measured in our case by $<a^s>$) is the same as the average population sparseness (measured by $<a^p>$), where $<ã>$ indicates the ensemble average. We have seen that for inferior temporal cortex neurons the neuron selectivity probability density functions are different (see Fig. C.5), but that their average $<a^s>$ is the same as the average (across stimuli) $<a^p>$ of the population sparseness, 0.77, and thus conclude that the representation in the inferior temporal visual cortex of objects and faces is weakly ergodic (Franco, Rolls, Aggelopoulos and Jerez 2007).

We note that weak ergodicity necessarily occurs if $<a^s>$ and $<a^p>$ are the same and the neurons are uncorrelated, that is each neuron is independently tuned to the set of stimuli (Lehky et al. 2005). The fact that both hold for the inferior temporal cortex neurons studied by Franco, Rolls, Aggelopoulos and Jerez (2007) thus indicates that their responses are uncorrelated, and this is potentially an important conclusion about the encoding of stimuli by these neurons. This conclusion is confirmed by the linear increase in the information with the number of neurons which is the case not only for this set of neurons (Franco, Rolls, Aggelopoulos and Jerez 2007), but also in other data sets for the inferior temporal visual cortex (Rolls, Treves and Tovee 1997b, Booth and Rolls 1998). Both types of evidence thus indicate that the encoding provided by at least small subsets (up to e.g. 20 neurons) of inferior temporal cortex neurons is approximately independent (non-redundant), which is an important principle of cortical encoding.

C.3.1.3 Comparisons of sparseness between areas: the hippocampus, insula, orbitofrontal cortex, and amygdala

In the study of Franco, Rolls, Aggelopoulos and Jerez (2007) on inferior temporal visual cortex neurons, the selectivity of individual cells for the set of stimuli, or single cell sparseness a^s, had a mean value of 0.77. This is close to a previously measured estimate, 0.65, which was obtained with a larger stimulus set of 68 stimuli (Rolls and Tovee 1995b). Thus the single neuron probability density functions in these areas do not produce very sparse representations. Therefore the goal of the computations in the inferior temporal visual cortex may not be to produce sparse representations (as has been proposed for V1 (Field 1994, Olshausen and

Field 1997, Vinje and Gallant 2000, Olshausen and Field 2004)). Instead one of the goals of the computations in the inferior temporal visual cortex may be to compute invariant representations of objects and faces (Rolls 2000a, Rolls and Deco 2002, Rolls 2007e, Rolls and Stringer 2006) (see Chapter 25), and to produce not very sparse distributed representations in order to maximize the information represented (see Table B.2 on page 722). In this context, it is very interesting that the representations of different stimuli provided by a population of inferior temporal cortex neurons are decorrelated, as shown by the finding that the mean (Pearson) correlation between the response profiles to a set of 20 stimuli computed over 406 neuron pairs was low, 0.049 ± 0.013 (sem) (Franco, Rolls, Aggelopoulos and Jerez 2007). The implication is that decorrelation is being achieved in the inferior temporal visual cortex, but not by forming a sparse code. It will be interesting to investigate the mechanisms for this.

In contrast, the representation in some memory systems may be more sparse. For example, in the hippocampus in which spatial view cells are found in macaques, further analysis of data described by Rolls, Treves, Robertson, Georges-François and Panzeri (1998b) shows that for the representation of 64 locations around the walls of the room, the mean single cell sparseness $<a^s>$ was 0.34 ± 0.13 (sd), and the mean population sparseness a^p was 0.33 ± 0.11. The more sparse representation is consistent with the view that the hippocampus is involved in storing memories, and that for this, more sparse representations than in perceptual areas are relevant. These sparseness values are for spatial view neurons, but it is possible that when neurons respond to combinations of spatial view and object (Rolls, Xiang and Franco 2005c), or of spatial view and reward (Rolls and Xiang 2005), the representations are more sparse. It is of interest that the mean firing rate of these spatial view neurons across all spatial views was 1.77 spikes/s (Rolls, Treves, Robertson, Georges-François and Panzeri 1998b). (The mean spontaneous firing rate of the neurons was 0.1 spikes/s, and the average across neurons of the firing rate for the most effective spatial view was 13.2 spikes/s.) It is also notable that weak ergodicity is implied for this brain region too (given the similar values of $<a^s>$ and $<a^p>$), and the underlying basis for this is that the response profiles of the different hippocampal neurons to the spatial views are uncorrelated. Further support for these conclusions is that the information about spatial view increases linearly with the number of hippocampal spatial view neurons (Rolls, Treves, Robertson, Georges-François and Panzeri 1998b), again providing evidence that the response profiles of the different neurons are uncorrelated. The representations in the hippocampus may be more sparse than this, in line with the observation that in rodents, hippocampal neurons may have place fields in only one of several environments.

Further evidence is now available on ergodicity in three further brain areas, the macaque insular primary taste cortex, the orbitofrontal cortex, and the amygdala (Rolls, Critchley, Verhagen and Kadohisa 2010a). In all these brain areas sets of neurons were tested with an identical set of 24 oral taste, temperature, and texture stimuli. (The stimuli were: Taste - 0.1 M NaCl (salt), 1 M glucose (sweet), 0.01 M HCl (sour), 0.001 M quinine HCl (bitter), 0.1 M monosodium glutamate (umami), and water; Temperature - 10°C, 37°C and 42°C; flavour - blackcurrant juice; viscosity - carboxymethyl-cellulose 10 cPoise, 100 cPoise, 1000 cPoise and 10000 cPoise; fatty / oily - single cream, vegetable oil, mineral oil, silicone oil (100 cPoise), coconut oil, and safflower oil; fatty acids - linoleic acid and lauric acid; capsaicin; and gritty texture.) Further analysis of data described by Verhagen, Kadohisa and Rolls (2004) showed that in the primary taste cortex the mean value of a^s across 58 neurons was 0.745 and of a^p (normalized) was 0.708. Further analysis of data described by Rolls, Verhagen and Kadohisa (2003e), Verhagen, Rolls and Kadohisa (2003), Kadohisa, Rolls and Verhagen (2004) and Kadohisa, Rolls and Verhagen (2005a) showed that in the orbitofrontal cortex the mean value of a^s across 30 neurons was 0.625 and of a^p was 0.611. Further analysis of data described by Kadohisa, Rolls and Verhagen (2005b) showed that in the amygdala the mean value of a^s across 38 neurons was 0.811 and of a^p was 0.813. Thus in all these cases, the mean value of a^s is close to that of a^p, and weak ergodicity is implied. The values of a^s and a^p are also relatively high, implying the importance of representing large amounts of information in these brain areas about this set of stimuli by using a very distributed code, and also perhaps about the stimulus set, some members of which may be rather similar to each other.

C.3.2   The information from single neurons

Examples of the responses of single neurons (in this case in the inferior temporal visual cortex) to sets of objects and/or faces (of the type illustrated in Fig. C.7) are shown in Figs. 16.15, 25.12 and C.4. We now consider how much information these types of neuronal response convey about the set of stimuli S, and about each stimulus s in the set. The mutual information $I(S, R)$ that the set of responses R encode about the set of stimuli S is calculated with equation C.21 and corrected for the limited sampling using the analytic bias correction procedure described by Panzeri and Treves (1996) as described in detail by Rolls, Treves, Tovee and Panzeri (1997d). The information $I(s, R)$ about each single stimulus s in the set S, termed the stimulus-specific information (Rolls, Treves, Tovee and Panzeri 1997d) or stimulus-specific surprise (DeWeese and Meister 1999), obtained from the set of the responses R of the single neuron is calculated with equation C.22 and corrected for the limited sampling using the analytic bias correction procedure described by Panzeri and Treves (1996) as described in detail by Rolls, Treves, Tovee and Panzeri (1997d). (The average of $I(s, R)$ across stimuli is the mutual information $I(S, R)$.)

Figure C.10 shows the stimulus-specific information $I(s, R)$ available in the neuronal response about each of 20 face stimuli calculated for the neuron (am242) whose firing rate response profile to the set of 65 stimuli is shown in Fig. C.4. Unless otherwise stated, the information measures given are for the information available on a single trial from the firing rate of the neuron in a 500 ms period starting 100 ms after the onset of the stimuli. It is shown in Fig. C.10 that 2.2, 2.0, and 1.5 bits of information were present about the three face stimuli to which the neuron had the highest firing rate responses. The neuron conveyed some but smaller amounts of information about the remaining face stimuli. The average information $I(S, R)$ about this set (S) of 20 faces for this neuron was 0.55 bits. The average firing rate of this neuron to these 20 face stimuli was 54 spikes/s. It is clear from Fig. C.10 that little information was available from the responses of the neuron to a particular face stimulus if that response was close to the average response of the neuron across all stimuli. At the same time, it is clear from Fig. C.10 that information was present depending on how far the firing rate to a particular stimulus was from the average response of the neuron to the stimuli. Of particular interest, it is evident that information is present from the neuronal response about which face was shown if that neuronal response was below the average response, as well as when the response was greater than the average response.

One intuitive way to understand the data shown in Fig. C.10 is to appreciate that low probability firing rate responses, whether they are greater than or less than the mean response rate, convey much information about which stimulus was seen. This is of course close to the definition of information. Given that the firing rates of neurons are always positive, and follow an asymmetric distribution about their mean, it is clear that deviations above the mean have a different probability to occur than deviations by the same amount below the mean. One may attempt to capture the relative likelihood of different firing rates above and below the mean by computing a z score obtained by dividing the difference between the mean response to each stimulus and the overall mean response by the standard deviation of the response to that stimulus. The greater the number of standard deviations (i.e. the greater the z score) from the mean response value, the greater the information might be expected to be. We therefore show in Fig. C.11 the relation between the z score and $I(s, R)$. (The z score was calculated by

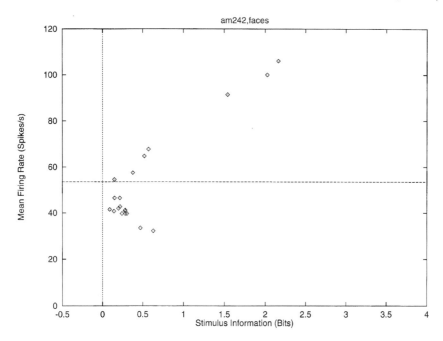

Fig. C.10 The stimulus-specific information $I(s, R)$ available in the response of the same single neuron as in Fig. C.4 about each of the stimuli in the set of 20 face stimuli (abscissa), with the firing rate of the neuron to the corresponding stimulus plotted as a function of this on the ordinate. The horizontal line shows the mean firing rate across all stimuli. (Reproduced from Journal of Computational Neuroscience, 4: 309–333. Information in the neuronal representation of individual stimuli in the primate temporal visual cortex, Rolls, E. T., Treves, A., Tovee, M. and Panzeri, S. Copyright ©1997, Kluwer Academic Publishers, with permission of Springer.)

obtaining the mean and standard deviation of the response of a neuron to a particular stimulus s, and dividing the difference of this response from the mean response to all stimuli by the calculated standard deviation for that stimulus.) This results in a C-shaped curve in Figs. C.10 and C.11, with more information being provided by the cell the further its response to a stimulus is in spikes per second or in z scores either above or below the mean response to all stimuli (which was 54 spikes/s). The specific C-shape is discussed further in Section C.3.4.

The information $I(s, R)$ about each stimulus in the set of 65 stimuli is shown in Fig. C.12 for the same neuron, am242. The 23 face stimuli in the set are indicated by a diamond, and the 42 non-face stimuli by a cross. Using this much larger and more varied stimulus set, which is more representative of stimuli in the real world, a C-shaped function again describes the relation between the information conveyed by the cell about a stimulus and its firing rate to that stimulus. In particular, this neuron reflected information about most, but not all, of the faces in the set, that is those faces that produced a higher firing rate than the overall mean firing rate to all the 65 stimuli, which was 31 spikes/s. In addition, it conveyed information about the majority of the 42 non-face stimuli by responding at a rate below the overall mean response of the neuron to the 65 stimuli. This analysis usefully makes the point that the information available in the neuronal responses about which stimulus was shown is relative to (dependent upon) the nature and range of stimuli in the test set of stimuli.

This evidence makes it clear that a single cortical visual neuron tuned to faces conveys information not just about one face, but about a whole set of faces, with the information

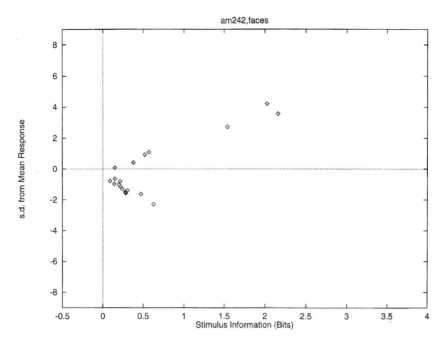

Fig. C.11 The relation for a single cell between the number of standard deviations the response to a stimulus was from the average response to all stimuli (see text, z score) plotted as a function of $I(s, R)$, the information available about the corresponding stimulus, s. (Reproduced from Journal of Computational Neuroscience, 4: 309–333. Information in the neuronal representation of individual stimuli in the primate temporal visual cortex, Rolls, E. T., Treves, A., Tovee, M. and Panzeri, S. Copyright ©1997, Kluwer Academic Publishers, with permission of Springer.)

conveyed on a single trial related to the difference in the firing rate response to a particular stimulus compared to the average response to all stimuli.

The analyses just described for neurons with visual responses are general, in that they apply in a very similar way to olfactory neurons recorded in the macaque orbitofrontal cortex (Rolls, Critchley and Treves 1996a, Rolls, Critchley, Verhagen and Kadohisa 2010a).

The neurons in this sample reflected in their firing rates for the post-stimulus period 100 to 600 ms on average 0.36 bits of mutual information about which of 20 face stimuli was presented (Rolls, Treves, Tovee and Panzeri 1997d). Similar values have been found in other experiments (Tovee, Rolls, Treves and Bellis 1993, Tovee and Rolls 1995, Rolls, Tovee and Panzeri 1999b, Rolls, Franco, Aggelopoulos and Jerez 2006b). The information in short temporal epochs of the neuronal responses is described in Section C.3.4.

C.3.3 The information from single neurons: temporal codes versus rate codes within the spike train of a single neuron

In the third of a series of papers that analyze the response of single neurons in the primate inferior temporal cortex to a set of static visual stimuli, Optican and Richmond (1987) applied information theory in a particularly direct and useful way. To ascertain the relevance of stimulus-locked temporal modulations in the firing of those neurons, they compared the amount of information about the stimuli that could be extracted from just the firing rate, computed over a relatively long interval of 384 ms, with the amount of information that could be extracted from a more complete description of the firing, that included temporal

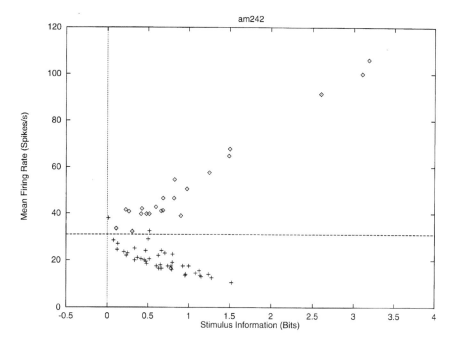

Fig. C.12 The information $I(s, R)$ available in the response of the same neuron about each of the stimuli in the set of 23 face and 42 non-face stimuli (abscissa), with the firing rate of the neuron to the corresponding stimulus plotted as a function of this on the ordinate. The 23 face stimuli in the set are indicated by a diamond, and the 42 non-face stimuli by a cross. The horizontal line shows the mean firing rate across all stimuli. (After Journal of Computational Neuroscience, 4: 309–333. Information in the neuronal representation of individual stimuli in the primate temporal visual cortex, Rolls, E. T., Treves, A., Tovee, M. and Panzeri, S. Copyright ©1997, Kluwer Academic Publishers, with permission of Springer.)

modulation. To derive this latter description (the temporal code within the spike train of a single neuron) they applied principal component analysis (PCA) to the temporal response vectors recorded for each neuron on each trial. The PCA helped to reduce the dimensionality of the neuronal response measurements. A temporal response vector was defined as a vector with as components the firing rates in each of 64 successive 6 ms time bins. The (64×64) covariance matrix was calculated across all trials of a particular neuron, and diagonalized. The first few eigenvectors of the matrix, those with the largest eigenvalues, are the principal components of the response, and the weights of each response vector on these four to five components can be used as a reduced description of the response, which still preserves, unlike the single value giving the mean firing rate along the entire interval, the main features of the temporal modulation within the interval. Thus a four- to five-dimensional temporal code could be contrasted with a one-dimensional rate code, and the comparison made quantitative by measuring the respective values for the mutual information with the stimuli.

Although the initial claim (Optican, Gawne, Richmond and Joseph 1991, Eskandar, Richmond and Optican 1992), that the temporal code carried nearly three times as much information as the rate code, was later found to be an artefact of limited sampling, and more recent analyses tend to minimize the additional information in the temporal description (Tovee, Rolls, Treves and Bellis 1993, Heller, Hertz, Kjaer and Richmond 1995), this type of application has immediately appeared straightforward and important, and it has led to many developments.

By concentrating on the code expressed in the output rather than on the characterization of the neuronal channel itself, this approach is not affected much by the potential complexities of the preceding black box. Limited sampling, on the other hand, is a problem, particularly because it affects much more codes with a larger number of components, for example the four to five components of the PCA temporal description, than the one-dimensional firing rate code. This is made evident in the paper by Heller, Hertz, Kjaer and Richmond (1995), in which the comparison is extended to several more detailed temporal descriptions, including a binary vector description in which the presence or not of a spike in each 1 ms bin of the response constitutes a component of a 320-dimensional vector. Obviously, this binary vector must contain at least all the information present in the reduced descriptions, whereas in the results of Heller, Hertz, Kjaer and Richmond (1995), despite the use of a sophisticated neural network procedure to control limited sampling biases, the binary vector appears to be the code that carries the least information of all. In practice, with the data samples available in the experiments that have been done, and even when using analytic procedures to control limited sampling (Panzeri and Treves 1996), reliable comparison can be made only with up to two- to three-dimensional codes.

Tovee, Rolls, Treves and Bellis (1993) and Tovee and Rolls (1995) obtained further evidence that little information was encoded in the temporal aspects of firing within the spike train of a single neuron in the inferior temporal cortex by taking short epochs of the firing of neurons, lasting 20 ms or 50 ms, in which the opportunity for temporal encoding would be limited (because there were few spikes in these short time intervals). They found that a considerable proportion (30%) of the information available in a long time period of 400 ms utilizing temporal encoding within the spike train was available in time periods as short as 20 ms when only the number of spikes was taken into account.

Overall, the main result of these analyses applied to the responses to static stimuli in the temporal visual cortex of primates is that not much more information (perhaps only up to 10% more) can be extracted from temporal codes than from the firing rate measured over a judiciously chosen interval (Tovee, Rolls, Treves and Bellis 1993, Heller, Hertz, Kjaer and Richmond 1995). Indeed, it turns out that even this small amount of 'temporal information' is related primarily to the onset latency of the neuronal responses to different stimuli, rather than to anything more subtle (Tovee, Rolls, Treves and Bellis 1993). Consistent with this point, in earlier visual areas the additional 'temporally encoded' fraction of information can be larger, due especially to the increased relevance, earlier on, of precisely locked transient responses (Kjaer, Hertz and Richmond 1994, Golomb, Kleinfeld, Reid, Shapley and Shraiman 1994, Heller, Hertz, Kjaer and Richmond 1995). This is because if the responses to some stimuli are more transient and to others more sustained, this will result in more information if the temporal modulation of the response of the neuron is taken into account. However, the relevance of more substantial temporal codes for static visual stimuli remains to be demonstrated. For non-static visual stimuli and for other cortical systems, similar analyses have largely yet to be carried out, although clearly one expects to find much more prominent temporal effects e.g. in the auditory system (Nelken, Prut, Vaadia and Abeles 1994, deCharms and Merzenich 1996), for reasons similar to those just annunciated.

C.3.4 The information from single neurons: the speed of information transfer

It is intuitive that if short periods of firing of single cells are considered, there is less time for temporal modulation effects. The information conveyed about stimuli by the firing rate and that conveyed by more detailed temporal codes become similar in value. When the firing periods analyzed become shorter than roughly the mean interspike interval, even the statistics

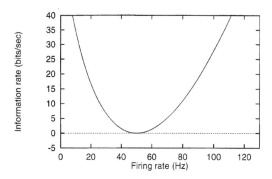

Fig. C.13 Time derivative of the stimulus-specific information as a function of firing rate, for a cell firing at a grand mean rate of 50 Hz. For different grand mean rates, the graph would simply be rescaled.

of firing rate values on individual trials cease to be relevant, and the information content of the firing depends solely on the mean firing rates across all trials with each stimulus. This is expressed mathematically by considering the amount of information provided as a function of the length t of the time window over which firing is analyzed, and taking the limit for $t \to 0$ (Skaggs, McNaughton, Gothard and Markus 1993, Panzeri, Biella, Rolls, Skaggs and Treves 1996). To first order in t, only two responses can occur in a short window of length t: either the emission of an action potential, with probability tr_s, where r_s is the mean firing rate calculated over many trials using the same window and stimulus; or no action potential, with probability $1 - tr_s$. Inserting these conditional probabilities into equation C.22, taking the limit and dividing by t, one obtains for the derivative of the stimulus-specific transinformation

$$dI(s)/dt = r_s \log_2(r_s/ <r>) + (<r> - r_s)/ \ln 2, \qquad \text{(C.47)}$$

where $<r>$ is the grand mean rate across stimuli. This formula thus gives the rate, in bits/s, at which information about a stimulus begins to accumulate when the firing of a cell is recorded. Such an information rate depends only on the mean firing rate to that stimulus and on the grand mean rate across stimuli. As a function of r_s, it follows the U-shaped curve in Fig. C.13. The curve is universal, in the sense that it applies irrespective of the detailed firing statistics of the cell, and it expresses the fact that the emission or not of a spike in a short window conveys information in as much as the mean response to a given stimulus is above or below the overall mean rate. No information is conveyed about those stimuli the mean response to which is the same as the overall mean. In practice, although the curve describes only the universal behaviour of the initial slope of the specific information as a function of time, it approximates well the full stimulus-specific information $I(s, R)$ computed even over rather long periods (Rolls, Critchley and Treves 1996a, Rolls, Treves, Tovee and Panzeri 1997d).

Averaging equation C.47 across stimuli one obtains the time derivative of the mutual information. Further dividing by the overall mean rate yields the adimensional quantity

$$\chi = \sum_s P(s)(r_s/ <r>) \log_2(r_s/ <r>) \qquad \text{(C.48)}$$

which measures, in bits, the mutual information per spike provided by the cell (Bialek, Rieke, de Ruyter van Steveninck and Warland 1991, Skaggs, McNaughton, Gothard and Markus 1993). One can prove that this quantity can range from 0 to $\log_2(1/a)$

$$0 < \chi < \log_2(1/a), \qquad \text{(C.49)}$$

where a is the single neuron sparseness a^s defined in Section C.3.1.1. For mean rates r_s distributed in a nearly binary fashion, χ is close to its upper limit $\log_2(1/a)$, whereas for mean rates that are nearly uniform, or at least unimodally distributed, χ is relatively close to zero (Panzeri, Biella, Rolls, Skaggs and Treves 1996). In practice, whenever a large number of more or less 'ecological' stimuli are considered, mean rates are not distributed in arbitrary ways, but rather tend to follow stereotyped distributions (which for some neurons approximate an exponential distribution of firing rates – see Section C.3.1 (Treves, Panzeri, Rolls, Booth and Wakeman 1999, Baddeley, Abbott, Booth, Sengpiel, Freeman, Wakeman and Rolls 1997, Rolls and Treves 1998, Rolls and Deco 2002, Franco, Rolls, Aggelopoulos and Jerez 2007, Rolls 2008d, Rolls and Treves 2011)), and as a consequence χ and a (or, equivalently, its logarithm) tend to covary (rather than to be independent variables (Skaggs and McNaughton 1992)). Therefore, measuring sparseness is in practice nearly equivalent to measuring information per spike, and the rate of rise in mutual information, $\chi < r >$, is largely determined by the sparseness a and the overall mean firing rate $< r >$.

The important point to note about the single-cell information rate $\chi < r >$ is that, to the extent that different cells express non-redundant codes, as discussed below, the instantaneous *information flow* across a population of C cells can be taken to be simply $C\chi < r >$, and this quantity can easily be measured directly without major limited sampling biases, or else inferred indirectly through measurements of the sparseness a. Values for the information rate $\chi < r >$ that have been published range from 2–3 bits/s for rat hippocampal cells (Skaggs, McNaughton, Gothard and Markus 1993), to 10–30 bits/s for primate temporal cortex visual cells (Rolls, Treves and Tovee 1997b), and could be compared with analogous measurements in the sensory systems of frogs and crickets, in the 100–300 bits/s range (Rieke, Warland and Bialek 1993).

If the first time-derivative of the mutual information measures information flow, successive derivatives characterize, at the single-cell level, different firing modes. This is because whereas the first derivative is universal and depends only on the mean firing rates to each stimulus, the next derivatives depend also on the variability of the firing rate around its mean value, across trials, and take different forms in different firing regimes. Thus they can serve as a measure of discrimination among firing regimes with limited variability, for which, for example, the second derivative is large and positive, and firing regimes with large variability, for which the second derivative is large and negative. Poisson firing, in which in every short period of time there is a fixed probability of emitting a spike irrespective of previous firing, is an example of large variability, and the second derivative of the mutual information can be calculated to be

$$\mathrm{d}^2 I / \mathrm{d}t^2 = [\ln a + (1 - a)] < r >^2 / (a \ln 2), \tag{C.50}$$

where a is the single neuron sparseness a^s defined in Section C.3.1.1. This quantity is always negative. Strictly periodic firing is an example of zero variability, and in fact the second time-derivative of the mutual information becomes infinitely large in this case (although actual information values measured in a short time interval remain of course finite even for exactly periodic firing, because there is still some variability, ± 1, in the number of spikes recorded in the interval). Measures of mutual information from short intervals of firing of temporal cortex visual cells have revealed a degree of variability intermediate between that of periodic and of Poisson regimes (Rolls, Treves, Tovee and Panzeri 1997d). Similar measures can also be used to contrast the effect of the graded nature of neuronal responses, once they are analyzed over a finite period of time, with the information content that would characterize neuronal activity if it reduced to a binary variable (Panzeri, Biella, Rolls, Skaggs and Treves 1996). A binary variable with the same degree of variability would convey information at the same instantaneous rate (the first derivative being universal), but in for example 20–30% reduced amounts when analyzed over times of the order of the interspike interval or longer.

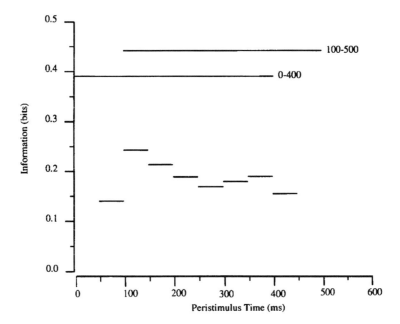

Fig. C.14 The average information I(S,R) available in short temporal epochs (50 ms as compared to 400 ms) of the spike trains of single inferior temporal cortex neurons about which face had been shown. (From Visual Cognition, 2: 35–58, Information encoding in short firing rate epochs by single neurons in the primate temporal visual cortex. Martin J. Tovee & Edmund T. Rolls. Copyright ©1995 Routledge, reprinted by permission of Taylor & Francis Ltd.)

Utilizing these approaches, Tovee, Rolls, Treves and Bellis (1993) and Tovee and Rolls (1995) measured the information available in short epochs of the firing of single neurons, and found that a considerable proportion of the information available in a long time period of 400 ms was available in time periods as short as 20 ms and 50 ms. For example, in periods of 20 ms, 30% of the information present in 400 ms using temporal encoding with the first three principal components was available. Moreover, the exact time when the epoch was taken was not crucial, with the main effect being that rather more information was available if information was measured near the start of the spike train, when the firing rate of the neuron tended to be highest (see Figs. C.14 and C.15). The conclusion was that much information was available when temporal encoding could not be used easily, that is in very short time epochs of 20 or 50 ms.

It is also useful to note from Figs. C.14, C.15 and 16.15 the typical time course of the responses of many temporal cortex visual neurons in the awake behaving primate. Although the firing rate and availability of information is highest in the first 50–100 ms of the neuronal response, the firing is overall well sustained in the 500 ms stimulus presentation period. Cortical neurons in the primate temporal lobe visual system, in the taste cortex (Rolls, Yaxley and Sienkiewicz 1990), and in the olfactory cortex (Rolls, Critchley and Treves 1996a), do not in general have rapidly adapting neuronal responses to sensory stimuli. This may be important for associative learning: the outputs of these sensory systems can be maintained for sufficiently long while the stimuli are present for synaptic modification to occur. Although rapid synaptic adaptation within a spike train is seen in some experiments in brain slices (Markram and Tsodyks 1996, Abbott, Varela, Sen and Nelson 1997), it is not a very marked effect in at least some brain systems in vivo, when they operate in normal physiological conditions with normal levels of acetylcholine, etc.

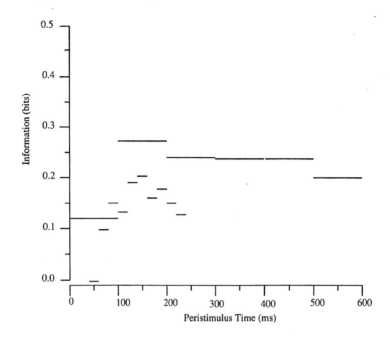

Fig. C.15 The average information I(S,R) available in short temporal epochs (20 ms and 100 ms) of the spike trains of single inferior temporal cortex neurons about which face had been shown. (From Visual Cognition, 2: 35–58, Information encoding in short firing rate epochs by single neurons in the primate temporal visual cortex. Martin J. Tovee & Edmund T. Rolls. Copyright ©1995 Routledge, reprinted by permission of Taylor & Francis Ltd.)

To pursue this issue of the speed of processing and information availability even further, Rolls, Tovee, Purcell, Stewart and Azzopardi (1994b) and Rolls and Tovee (1994) limited the period for which visual cortical neurons could respond by using backward masking. In this paradigm, a short (16 ms) presentation of the test stimulus (a face) was followed after a delay of 0, 20, 40, 60, etc. ms by a masking stimulus (which was a high contrast set of letters) (see Fig. C.16). They showed that the mask did actually interrupt the neuronal response, and that at the shortest interval between the stimulus and the mask (a delay of 0 ms, or a 'Stimulus Onset Asynchrony' of 20 ms), the neurons in the temporal cortical areas fired for approximately 30 ms (see Fig. C.17). Under these conditions, the subjects could identify which of five faces had been shown much better than chance. Interestingly, under these conditions, when the inferior temporal cortex neurons were firing for 30 ms, the subjects felt that they were guessing, and conscious perception was minimal (Rolls, Tovee, Purcell, Stewart and Azzopardi 1994b), the neurons conveyed on average 0.10 bits of information (Rolls, Tovee and Panzeri 1999b). With a stimulus onset asynchrony of 40 ms, when the inferior temporal cortex neurons were firing for 50 ms, not only did the subjects' performance improve, but the stimuli were now perceived clearly, consciously, and the neurons conveyed on average 0.16 bits of information. This has contributed to the view that consciousness has a higher threshold of activity *in a given pathway*, in this case a pathway for face analysis, than does unconscious processing and performance using the same pathway (Rolls 2003, Rolls 2006a).

The issue of how rapidly information can be read from neurons is crucial and fundamental to understanding how rapidly memory systems in the brain could operate in terms of reading

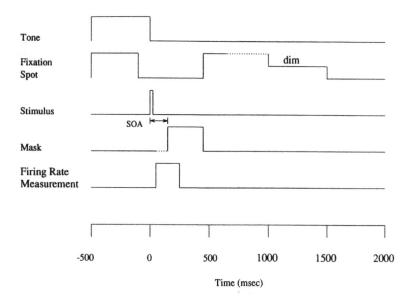

Fig. C.16 Backward masking paradigm. The visual stimulus appeared at time 0 for 16 ms. The time between the start of the visual stimulus and the masking image is the Stimulus Onset Asynchrony (SOA). A visual fixation task was being performed to ensure correct fixation of the stimulus. In the fixation task, the fixation spot appeared in the middle of the screen at time −500 ms, was switched off 100 ms before the test stimulus was shown, and was switched on again at the end of the mask stimulus. Then when the fixation spot dimmed after a random time, fruit juice could be obtained by licking. No eye movements could be performed after the onset of the fixation spot. (After Rolls and Tovee 1994.)

the code from the input neurons to initiate retrieval, whether in a pattern associator or auto-association network (see Appendix B). This is also a crucial issue for understanding how any stage of cortical processing operates, given that each stage includes associative or competitive network processes that require the code to be read before it can pass useful output to the next stage of processing (see Chapter 25; Rolls and Deco (2002); Rolls (2008d); and Panzeri, Rolls, Battaglia and Lavis (2001)). For this reason, we have performed further analyses of the speed of availability of information from neuronal firing, and the neuronal code. A rapid readout of information from any one stage of for example visual processing is important, for the ventral visual system is organized as a hierarchy of cortical areas, and the neuronal response latencies are approximately 100 ms in the inferior temporal visual cortex, and 40–50 ms in the primary visual cortex, allowing only approximately 50–60 ms of processing time for V1–V2–V4–inferior temporal cortex (Baylis, Rolls and Leonard 1987, Nowak and Bullier 1997, Rolls and Deco 2002). There is much evidence that the time required for each stage of processing is relatively short. For example, in addition to the evidence already presented, visual stimuli presented in succession approximately 15 ms apart can be separately identified (Keysers and Perrett 2002); and the reaction time for identifying visual stimuli is relatively short and requires a relatively short cortical processing time (Rolls 2003, Bacon-Mace, Mace, Fabre-Thorpe and Thorpe 2005).

In this context, Delorme and Thorpe (2001) have suggested that just one spike from each neuron is sufficient, and indeed it has been suggested that the order of the first spike in different neurons may be part of the code (Delorme and Thorpe 2001, Thorpe, Delorme and Van Rullen 2001, VanRullen, Guyonneau and Thorpe 2005). (Implicit in the spike order hypothesis is that the first spike is particularly important, for it would be difficult to measure the

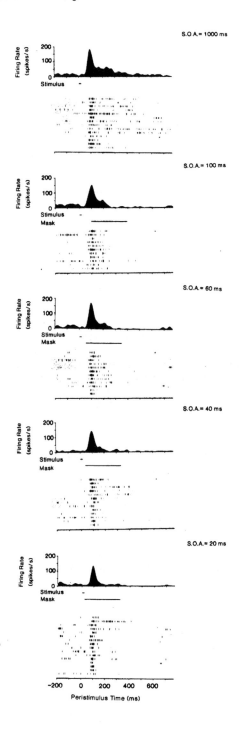

Fig. C.17 Firing of a temporal cortex cell to a 20 ms presentation of a face stimulus when the face was followed with different stimulus onset asynchronies (SOAs) by a masking visual stimulus. At an SOA of 20 ms, when the mask immediately followed the face, the neuron fired for only approximately 30 ms, yet identification above change (by 'guessing') of the face at this SOA by human observers was possible. (After Rolls and Tovee 1994; and Rolls, Tovee, Purcell et al. 1994.)

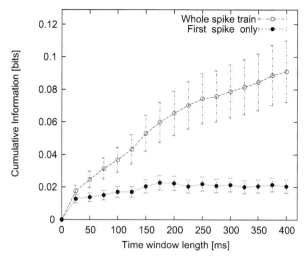

Fig. C.18 Speed of information availability in the inferior temporal visual cortex. Cumulative single cell information from all spikes and from the first spike with the analysis starting at 100 ms after stimulus onset. The mean and sem over 21 neurons are shown. (After Rolls, Franco, Aggelopoulos and Jerez 2006b.)

order for anything other than the first spike.) An alternative view is that the number of spikes in a fixed time window over which a postsynaptic neuron could integrate information is more realistic, and this time might be in the order of 20 ms for a single receiving neuron, or much longer if the receiving neurons are connected by recurrent collateral associative synapses and so can integrate information over time (Deco and Rolls 2006, Rolls and Deco 2002, Panzeri, Rolls, Battaglia and Lavis 2001, Rolls 2008d). Although the number of spikes in a short time window of e.g. 20 ms is likely to be 0, 1, or 2, the information available may be more than that from the first spike alone, and Rolls, Franco, Aggelopoulos and Jerez (2006b) examined this by measuring neuronal activity in the inferior temporal visual cortex, and then applying quantitative information theoretic methods to measure the information transmitted by single spikes, and within short time windows.

The cumulative single cell information about which of the twenty stimuli (Fig. C.7) was shown from all spikes and from the first spike starting at 100 ms after stimulus onset is shown in Fig. C.18. A period of 100 ms is just longer than the shortest response latency of the neurons from which recordings were made, so starting the measure at this time provides the best chance for the single spike measurement to catch a spike that is related to the stimulus. The means and standard errors across the 21 different neurons are shown. The cumulated information from the total number of spikes is larger than that from the first spike, and this is evident and significant within 50 ms of the start of the time epoch. In calculating the information from the first spike, just the first spike in the analysis window starting in this case at 100 ms after stimulus onset was used.

Because any one neuron receiving information from the population being analyzed has multiple inputs, we show in Fig. C.19 the cumulative information that would be available from multiple cells (21) about which of the 20 stimuli was shown, taking both the first spike after the time of stimulus onset (0 ms), and the total number of spikes after 0 ms from each neuron. The cumulative information even from multiple cells is much greater when all the spikes rather than just the first spike are used.

An attractor network might be able to integrate the information arriving over a long time period of several hundred milliseconds (see Chapter 5.6), and might produce the advantage shown in Fig. C.19 for the whole spike train compared to the first spike only. However a

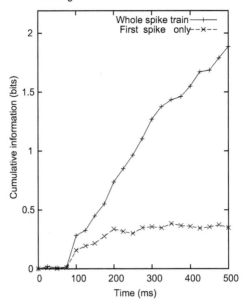

Fig. C.19 Speed of information availability in the inferior temporal visual cortex. Cumulative multiple cell information from all spikes and first spike starting at the time of stimulus onset (0 ms) for the population of 21 neurons about the set of 20 stimuli. (After Rolls, Franco, Aggelopoulos and Jerez 2006b.)

single layer pattern association network might only be able to integrate the information over the time constants of its synapses and cell membrane, which might be in the order of 15–30 ms (Panzeri, Rolls, Battaglia and Lavis 2001, Rolls and Deco 2002) (see Section B.2). In a hierarchical processing system such as the visual cortical areas, there may only be a short time during which each stage may decode the information from the preceding stage, and then pass on information sufficient to support recognition to the next stage (Rolls and Deco 2002) (see Chapter 25). We therefore analyzed the information that would be available in short epochs from multiple inputs to a neuron, and show the multiple cell information for the population of 21 neurons in Fig. C.20 (for 20 ms and 50 epochs). We see in this case that the first spike information, because it is being made available from many different neurons (in this case 21 selective neurons discriminating between the stimuli each with $p<0.001$ in an ANOVA), fares better relative to the information from all the spikes in these short epochs, but is still less than the information from all the spikes (particularly in the 50 ms epoch). In particular, for the epoch starting 100 ms after stimulus onset in Fig. C.21 the information in the 20 ms epoch is 0.37 bits, and from the first spike is 0.24 bits. Correspondingly, for a 50 ms epoch, the values in the epoch starting at 100 ms post stimulus were 0.66 bits for the 50 ms epoch, and 0.40 bits for the first spike. Thus with a population of neurons, having just one spike from each can allow considerable information to be read if only a limited period (of e.g. 20 or 50 ms) is available for the readout, though even in these cases, more information was available if all the spikes in the short window are considered (Fig. C.20).

To show how the information increases with the number of neurons in the ensemble in these short epochs, we show in Fig. C.21 the information from different numbers of neurons for a 20 ms epoch starting at time = 100 ms with respect to stimulus onset, for both the first spike condition and the condition with all the spikes in the 20 ms window. The linear increase in the information in both cases indicates that the neurons provide independent information, which could be because there is no redundancy or synergy, or because these cancel (Rolls, Franco, Aggelopoulos and Reece 2003b, Rolls, Franco, Aggelopoulos and Reece 2003b). It

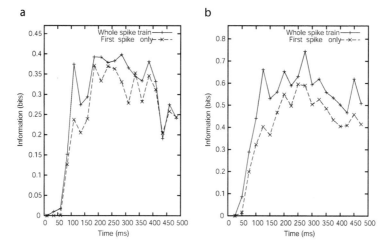

Fig. C.20 Speed of information availability in the inferior temporal visual cortex. (a) Multiple cell information from all spikes and 1 spike in 20 ms time windows taken at different post-stimulus times starting at time 0. (b) Multiple cell information from all spikes and 1 spike in 50 ms time windows taken at different post-stimulus times starting at time 0. (After Rolls, Franco, Aggelopoulos and Jerez 2006b.)

is also clear from Fig. C.21 that even with the population of neurons, and with just a short time epoch of 20 ms, more information is available from the population if all the spikes in 20 ms are considered, and not just the first spike. The 20 ms epoch analyzed for Fig. C.21 is for the post-stimulus time period of 100–120 ms.

To assess whether there is information that is specifically related to the order in which the spikes arrive from the different neurons, Rolls, Franco, Aggelopoulos and Jerez (2006b) computed for every trial the order across the different simultaneously recorded neurons in which the first spike arrived to each stimulus, and used this in the information theoretic analysis. The control condition was to randomly allocate the order values for each trial between the neurons that had any spikes on that trial, thus shuffling or scrambling the order of the spike arrival times in the time window. In both cases, just the first spike in the time window was used in the information analysis. (In both the order and the shuffled control conditions, on some trials some neurons had no spikes, and this itself, in comparison with the fact that some neurons had spiked on that trial, provided some information about which stimulus had been shown. However, by explicitly shuffling in the control condition the order of the spikes for the neurons that had spiked on that trial, comparison of the control with the unshuffled order condition provides a clear measure of whether the order of spike arrival from the different neurons itself carries useful information about which stimulus was shown.) The data set was 36 cells with significantly different ($p<0.05$) responses to the stimulus set where it was possible to record simultaneously from groups of 3 and 4 cells (so that the order on each trial could be measured) in 11 experiments. Taking a 75 ms time window starting 100 ms after stimulus onset, the information with the order of arrival times of the spikes was 0.142 ± 0.02 bits, and in the control (shuffled order) condition was 0.138 ± 0.02 bits (mean across the 11 experiments $\pm$ sem). Thus the information increase by taking into account the order of spike arrival times relative to the control condition was only $(0.142 - 0.138) = 0.004$ bits per experiment (which was not significant). For comparison, the information calculated for the first spike using the same dot product decoding as described above was 0.136 ± 0.03 bits per experiment. Analogous results were obtained for different time windows. Thus taking the spike order into account compared to a control condition in which the spike order was

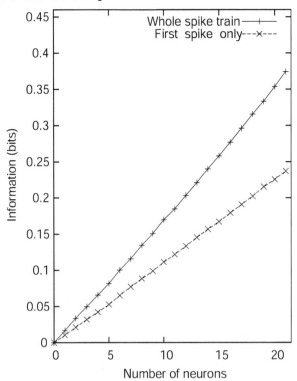

Fig. C.21 Speed of information availability in the inferior temporal visual cortex. Multiple cell information from all spikes and 1 spike in a 20 ms time window starting at 100 ms after stimulus onset as a function of the number of neurons in the ensemble. (After Rolls, Franco, Aggelopoulos and Jerez 2006b.)

scrambled made essentially no difference to the amount of information that was available from the populations of neurons about which stimulus was shown.

The results show that although considerable information is present in the first spike, more information is available under the more biologically realistic assumption that neurons integrate spikes over a short time window (depending on their time constants) of for example 20 ms. The results shown in Fig. C.21 are of considerable interest, for they show that even when one increases the number of neurons in the population, the information available from the number of spikes in a 20 ms time window is larger than the information available from just the first spike. Thus although intuitively one might think that one can compensate by taking a population of neurons rather than just a single neuron when using just the first spike instead of the number of spikes available in a fixed time window, this compensation by increasing neuron numbers is insufficient to make the first spike code as efficient as taking the number of spikes.

Further, in this first empirical test of the hypothesis that there is information that is specifically related to the order in which the spikes arrive from the different neurons, which has been proposed by Thorpe et al (Delorme and Thorpe 2001, Thorpe, Delorme and Van Rullen 2001, VanRullen, Guyonneau and Thorpe 2005), we found that in the inferior temporal visual cortex there was no significant evidence that the order of the spike arrival times from different simultaneously recorded neurons is important. Indeed, the evidence found in the experiments was that the number of spikes in the time window is the important property that is related to the amount of information encoded by the spike trains of simultaneously recorded neurons.

The fact that there was also more information in the number of spikes in a fixed time window than from the first spike only is also evidence that is not consistent with the spike order hypothesis, for the order between neurons can only be easily read from the first spike, and just using information from the first spike would discard extra information available from further spikes even in short time windows.

The encoding of information that uses the number of spikes in a short time window that is supported by the analyses described by Rolls, Franco, Aggelopoulos and Jerez (2006b) deserves further elaboration. It could be thought of as a rate code, in that the number of spikes in a short time window is relevant, but is not a rate code in the rather artificial sense considered by Thorpe et al. (Delorme and Thorpe 2001, Thorpe et al. 2001, VanRullen et al. 2005) in which a rate is estimated from the interspike interval. This is not just artificial, but also begs the question of how, once the rate is calculated from the interspike interval, this decoded rate is passed on to the receiving neurons, or how, if the receiving neurons calculate the interspike interval at every synapse, they utilize it. In contrast, the spike count code in a short time window that is considered here is very biologically plausible, in that each spike would inject current into the post-synaptic neuron, and the neuron would integrate all such currents in a dendrite over a time period set by the synaptic and membrane time constants, which will result in an integration time constant in the order of 15–20 ms. Explicit models of exactly this dynamical processing at the integrate-and-fire neuronal level have been described to define precisely these operations (Deco and Rolls 2003, Deco and Rolls 2005d, Deco, Rolls and Horwitz 2004, Deco and Rolls 2005b, Rolls and Deco 2002, Rolls 2008d). Even though the number of spikes in a short time window of e.g. 20 ms is likely to be 0, 1, or 2, it can be 3 or more for effective stimuli (Rolls, Franco, Aggelopoulos and Jerez 2006b), and this is more efficient than using the first spike.

To add some detail here, a neuron receiving information from a population of inferior temporal cortex neurons of the type described here would have a membrane potential that varied continuously in time reflecting with a time constant in the order of 15–20 ms (resulting from a time constant of order 10 ms for AMPA synapses, 100 ms for NMDA synapses, and 20 ms for the cell membrane) a dot (inner) product over all synapses of each spike count and the synaptic strength. This continuously time varying membrane potential would lead to spikes whenever the results of this integration process produced a depolarization that exceeded the firing threshold. The result is that the spike train of the neuron would reflect continuously with a time constant in the order of 15–20 ms the likelihood that the input spikes it was receiving matched its set of synaptic weights. The spike train would thus indicate in continuous time how closely the stimulus or input matched its most effective stimulus (for a dot product is essentially a correlation). In this sense, no particular starting time is needed for the analysis, and in this respect it is a much better component of a dynamical system than is a decoding that utilizes an order in which the order of the spike arrival times is important and a start time for the analysis must be assumed.

I note that an autoassociation or attractor network implemented by recurrent collateral connections between the neurons can, using its short-term memory, integrate its inputs over much longer periods, for example over 500 ms in a model of how decisions are made (Deco and Rolls 2006) (see Chapter 5.6), and thus if there is time, the extra information available in more than the first spike or even the first few spikes that is evident in Figs. C.18 and C.19 could be used by the brain.

The conclusions from the single cell information analyses are thus that most of the information is encoded in the spike count; that large parts of this information are available in short temporal epochs of e.g. 20 ms or 50 ms; and that any additional information which appears to be temporally encoded is related to the latency of the neuronal response, and reflects sudden changes in the visual stimuli. Therefore a neuron in the next cortical area would obtain considerable information within 20–50 ms by measuring the firing rate of a single neuron. Moreover, if it took a short sample of the firing rate of many neurons in the preceding area, then very much information is made available in a short time, as shown above and in Section C.3.5.

C.3.5 The information from multiple cells: independent information versus redundancy across cells

The rate at which a single cell provides information translates into an instantaneous information flow across a population (with a simple multiplication by the number of cells) only to the extent that different cells provide different (independent) information. To verify whether this condition holds, one cannot extend to multiple cells the simplified formula for the first time-derivative, because it is made simple precisely by the assumption of independence between spikes, and one cannot even measure directly the full information provided by multiple (more than two to three) cells, because of the limited sampling problem discussed above. Therefore one has to analyze the degree of independence (or conversely of redundancy) either directly among pairs – at most triplets – of cells, or indirectly by using decoding procedures to transform population responses. Obviously, the results of the analysis will vary a great deal with the particular neural system considered and the particular set of stimuli, or in general of neuronal correlates, used. For many systems, before undertaking to quantify the analysis in terms of information measures, it takes only a simple qualitative description of the responses to realize that there is a lot of redundancy and very little diversity in the responses. For example, if one selects pain-responsive cells in the somatosensory system and uses painful electrical stimulation of different intensities, most of the recorded cells are likely to convey pretty much the same information, signalling the intensity of the stimulation with the intensity of their single-cell response. Therefore, an analysis of redundancy makes sense only for a neuronal system that functions to represent, and enable discriminations between, a large variety of stimuli, and only when using a set of stimuli representative, in some sense, of that large variety.

Rolls, Treves and Tovee (1997b) measured the information available from a population of inferior temporal cortex neurons using the decoding method described in Section C.2.3, and found that the information increased approximately linearly, as shown in Fig. 25.13 on page 570, and in Fig. C.22 for a 50 ms interval as well as for a 500 ms measuring period. (It is shown below that the increase is limited only by the information ceiling of 4.32 bits necessary to encode the 20 stimuli. If it were not for this approach to the ceiling, the increase would be approximately linear (Rolls, Treves and Tovee 1997b).) To the extent that the information increases linearly with the number of neurons, the neurons convey independent information, and there is no redundancy, at least with numbers of neurons in this range. Although these and some of the other results described in this Appendix are for face-selective neurons in the inferior temporal visual cortex, similar results were obtained for neurons responding to objects in the inferior temporal visual cortex (Booth and Rolls 1998), and for neurons responding to spatial view in the hippocampus (Rolls, Treves, Robertson, Georges-François and Panzeri 1998b).

Although those neurons were not simultaneously recorded, a similar approximately linear increase in the information from *simultaneously* recorded cells as the number of neurons in the sample increased also occurs (Rolls, Franco, Aggelopoulos and Reece 2003b, Rolls, Aggelopoulos, Franco and Treves 2004, Franco, Rolls, Aggelopoulos and Treves 2004, Aggelopoulos, Franco and Rolls 2005, Rolls, Franco, Aggelopoulos and Jerez 2006b). These findings imply little redundancy, and that the number of stimuli that can be encoded increases

(a)
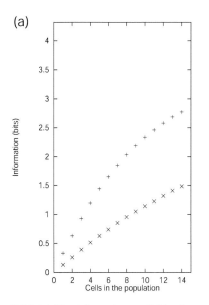

(b)
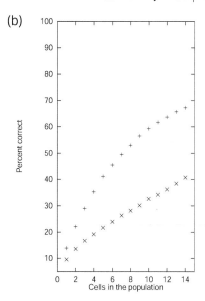

Fig. C.22 (a) The information available about which of 20 faces had been seen that is available from the responses measured by the firing rates in a time period of 500 ms (+) or a shorter time period of 50 ms (x) of different numbers of temporal cortex cells. (b) The corresponding percentage correct from different numbers of cells. (From Experimental Brain Research 114: 149–162. The representational capacity of the distributed encoding of information provided by populations of neurons in the primate temporal visual cortex. Rolls, E. T., Treves, A. and Tovee, M. J. Copyright ©1997, with permission of Springer.)

approximately exponentially with the number of neurons in the population, as illustrated in Figs. 25.14 and C.22.

The issue of redundancy is considered in more detail now. Redundancy can be defined with reference to a multiple channel of capacity $T(C)$ which can be decomposed into C separate channels of capacities $T_i, i = 1, ..., C$:

$$R = 1 - T(C)/\sum_i T_i \qquad (C.51)$$

so that when the C channels are multiplexed with maximal efficiency, $T(C) = \sum_i T_i$ and $R = 0$. What is measured more easily, in practice, is the redundancy defined with reference to a specific source (the set of stimuli with their probabilities). Then in terms of mutual information

$$R' = 1 - I(C)/\sum_i I_i. \qquad (C.52)$$

Gawne and Richmond (1993) measured the redundancy R' among pairs of nearby primate inferior temporal cortex visual neurons, in their response to a set of 32 Walsh patterns. They found values with a mean $< R' > = 0.1$ (and a mean single-cell transinformation of 0.23 bits). Since to discriminate 32 different patterns takes 5 bits of information, in principle one would need at least 22 cells each providing 0.23 bits of strictly orthogonal information to represent the full entropy of the stimulus set. Gawne and Richmond reasoned, however, that, because of the overlap, y, in the information they provided, more cells would be needed than if the redundancy had been zero. They constructed a simple model based on the notion that the overlap, y, in the information provided by any two cells in the population always corresponds to the average redundancy measured for nearby pairs. A redundancy $R' = 0.1$

corresponds to an overlap $y = 0.2$ in the information provided by the two neurons, since, counting the overlapping information only once, two cells would yield 1.8 times the amount transmitted by one cell alone. If a fraction of $1 - y = 0.8$ of the information provided by a cell is novel with respect to that provided by another cell, a fraction $(1 - y)^2$ of the information provided by a third cell will be novel with respect to what was known from the first pair, and so on, yielding an estimate of $I(C) = I(1) \sum_{i=0}^{C-1} (1 - y)^i$ for the total information conveyed by C cells. However such a sum saturates, in the limit of an infinite number of cells, at the level $I(\infty) = I(1)/y$, implying in their case that even with very many cells, no more than $0.23/0.2 = 1.15$ bits could be read off their activity, or less than a quarter of what was available as entropy in the stimulus set! Gawne and Richmond (1993) concluded, therefore, that the average overlap among non-nearby cells must be considerably lower than that measured for cells close to each other.

The model above is simple and attractive, but experimental verification of the actual scaling of redundancy with the number of cells entails collecting the responses of several cells interspersed in a population of interest. Gochin, Colombo, Dorfman, Gerstein and Gross (1994) recorded from up to 58 cells in the primate temporal visual cortex, using sets of two to five visual stimuli, and applied decoding procedures to measure the information content in the population response. The recordings were not simultaneous, but comparison with simultaneous recordings from a smaller number of cells indicated that the effect of recording the individual responses on separate trials was minor. The results were expressed in terms of the *novelty* N in the information provided by C cells, which being defined as the ratio of such information to C times the average single-cell information, can be expressed as

$$N = 1 - R' \tag{C.53}$$

and is thus the complement of the redundancy. An analysis of two different data sets, which included three information measures per data set, indicated a behaviour $N(C) \approx 1/\sqrt{C}$, reminiscent of the improvement in the overall noise-to-signal ratio characterizing C independent processes contributing to the same signal. The analysis neglected however to consider limited sampling effects, and more seriously it neglected to consider saturation effects due to the information content approaching its ceiling, given by the entropy of the stimulus set. Since this ceiling was quite low, for 5 stimuli at $\log_2 5 = 2.32$ bits, relative to the mutual information values measured from the population (an average of 0.26 bits, or 1/9 of the ceiling, was provided by single cells), it is conceivable that the novelty would have taken much larger values if larger stimulus sets had been used.

A simple formula describing the approach to the ceiling, and thus the saturation of information values as they come close to the entropy of the stimulus set, can be derived from a natural extension of the Gawne and Richmond (1993) model. In this extension, the information provided by single cells, measured as a fraction of the ceiling, is taken to coincide with the average overlap among pairs of randomly selected, not necessarily nearby, cells from the population. The actual value measured by Gawne and Richmond would have been, again, $1/22 = 0.045$, below the overlap among nearby cells, $y = 0.2$. The assumption that y, measured across any pair of cells, would have been as low as the fraction of information provided by single cells is equivalent to conceiving of single cells as 'covering' a random portion y of information space, and thus of randomly selected pairs of cells as overlapping in a fraction $(y)^2$ of that space, and so on, as postulated by the Gawne and Richmond (1993) model, for higher numbers of cells. The approach to the ceiling is then described by the formula

$$I(C) \approx H\{1 - \exp[C \ln(1 - y)]\} \tag{C.54}$$

that is, a simple exponential saturation to the ceiling. This simple law indeed describes remarkably well the trend in the data analyzed by Rolls, Treves and Tovee (1997b). Although the model has no reason to be exact, and therefore its agreement with the data should not be expected to be accurate, the crucial point it embodies is that deviations from a purely linear increase in information with the number of cells analyzed are due solely to the ceiling effect. Aside from the ceiling, due to the sampling of an information space of finite entropy, the information contents of different cells' responses are independent of each other. Thus, in the model, the observed redundancy (or indeed the overlap) is purely a consequence of the finite size of the stimulus set. If the population were probed with larger and larger sets of stimuli, or more precisely with sets of increasing entropy, and the amount of information conveyed by single cells were to remain approximately the same, then the fraction of space 'covered' by each cell, again y, would get smaller and smaller, tending to eliminate redundancy for very large stimulus entropies (and a fixed number of cells). The actual data were obtained with limited numbers of stimuli, and therefore cannot probe directly the conditions in which redundancy might reduce to zero. The data are consistent, however, with the hypothesis embodied in the simple model, as shown also by the near exponential approach to lower ceilings found for information values calculated with reduced subsets of the original set of stimuli (Rolls, Treves and Tovee 1997b).

The implication of this set of analyses, some performed towards the end of the ventral visual stream of the monkey, is that the representation of at least some classes of objects in those areas is achieved with minimal redundancy by cells that are allocated each to analyse a different aspect of the visual stimulus. This minimal redundancy is what would be expected of a self-organizing system in which different cells acquired their response selectivities through a random process, with or without local competition among nearby cells (see Section B.4). At the same time, such low redundancy could also very well result in a system that is organized under some strong teaching input, so that the emerging picture is compatible with a simple random process, but could be produced in other ways. The finding that, at least with small numbers of neurons, redundancy may be effectively minimized, is consistent not only with the concept of efficient encoding, but also with the general idea that one of the functions of the early visual system is to progressively minimize redundancy in the representation of visual stimuli (Attneave 1954, Barlow 1961). However, the ventral visual system does much more than produce a non-redundant representation of an image, for it transforms the representation from an image to an invariant representation of objects, as described in Chapter 25. Moreover, what is shown in this section is that the information about objects can be read off from just the spike count of a population of neurons, using decoding as simple as the simplest that could be performed by a receiving neuron, dot product decoding. In this sense, the information about objects is made explicit in the firing rate of the neurons in the inferior temporal cortex, in that it can be read off in this way.

We consider in Section C.3.7 whether there is more to it than this. Does the synchronization of neurons (and it would have to be stimulus-dependent synchronization) add significantly to the information that could be encoded by the number of spikes, as has been suggested by some?

Before this, we consider why encoding by a population of neurons is more powerful than the encoding than is possible by single neurons, adding to previous arguments that a distributed representation is much more computationally useful than a local representation, by allowing properties such as generalization, completion, and graceful degradation in associative neuronal networks (see Appendix B).

C.3.6 Should one neuron be as discriminative as the whole organism, in object encoding systems?

In the analysis of random dot motion with a given level of correlation among the moving dots, single neurons in area MT in the dorsal visual system of the primate can be approximately as sensitive or discriminative as the psychophysical performance of the whole animal (Zohary, Shadlen and Newsome 1994). The arguments and evidence presented here (e.g. in Section C.3.5) suggest that this is not the case for the ventral visual system, concerned with object identification. Why should there be this difference?

Rolls and Treves (1998) suggest that the dimensionality of what is being computed may account for the difference. In the case of visual motion (at least in the study referred to), the problem was effectively one-dimensional, in that the direction of motion of the stimulus along a line in 2D space was extracted from the activity of the neurons. In this low-dimensional stimulus space, the neurons may each perform one of a few similar computations on a particular (local) portion of 2D space, with the side effect that, by averaging over a larger receptive field than in V1, one can extract a signal of a more global nature. Indeed, in the case of more global motion, it is the average of the neuronal activity that can be computed by the larger receptive fields of MT neurons that specifies the average or global direction of motion.

In contrast, in the higher dimensional space of objects, in which there are very many different objects to represent as being different from each other, and in a system that is not concerned with location in visual space but on the contrary tends to be relatively invariant with respect to location, the goal of the representation is to reflect the many aspects of the input information in a way that enables many different objects to be represented, in what is effectively a very high dimensional space. This is achieved by allocating cells, each with an intrinsically limited discriminative power, to sample as thoroughly as possible the many dimensions of the space. Thus the system is geared to use efficiently the parallel computations of all its neurons precisely for tasks such as that of face discrimination, which was used as an experimental probe. Moreover, object representation must be kept higher dimensional, in that it may have to be decoded by dot product decoders in associative memories, in which the input patterns must be in a space that is as high-dimensional as possible (i.e. the activity on different input axons should not be too highly correlated). In this situation, each neuron should act somewhat independently of its neighbours, so that each provides its own separate contribution that adds together with that of the other neurons (in a linear manner, see above and Figs. 25.13, C.22 and 25.14) to provide *in toto* sufficient information to specify which out of perhaps several thousand visual stimuli was seen. The computation involves in this case not an average of neuronal activity (which would be useful for e.g. head direction (Robertson, Rolls, Georges-François and Panzeri 1999)), but instead comparing the dot product of the activity of the population of neurons with a previously learned vector, stored in, for example, associative memories as the weight vector on a receiving neuron or neurons.

Zohary, Shadlen and Newsome (1994) put forward another argument which suggested to them that the brain could hardly benefit from taking into account the activity of more than a very limited number of neurons. The argument was based on their measurement of a small (0.12) correlation between the activity of simultaneously recorded neurons in area MT. They suggested that there would because of this be decreasing signal-to-noise ratio advantages as more neurons were included in the population, and that this would limit the number of neurons that it would be useful to decode to approximately 100. However, a measure of correlations in the activity of different neurons depends entirely on the way the space of neuronal activity is sampled, that is on the task chosen to probe the system. Among face cells in the temporal cortex, for example, much higher correlations would be observed when the task is a simple two-way discrimination between a face and a non-face, than when the task involves finer identification of several different faces. (It is also entirely possible that some face cells could be found that perform as well in a given particular face / non-face discrimination as the whole animal.) Moreover, their argument depends on the type of decoding of the activity of the population that is envisaged (see further Robertson, Rolls, Georges-François and Panzeri (1999)). It implies that the average of the neuronal activity must be estimated accurately. If a set of neurons uses dot product decoding, and then the activity of the decoding population is scaled or normalized by some negative feedback through inhibitory interneurons, then the effect of such correlated firing in the sending population is reduced, for the decoding effectively measures the relative firing of the different neurons in the population to be decoded. This is equivalent to measuring the angle between the current vector formed by the population of neurons firing, and a previously learned vector, stored in synaptic weights. Thus, with for example this biologically plausible decoding, it is not clear whether the correlation Zohary, Shadlen and Newsome (1994) describe would place a severe limit on the ability of the brain to utilize the information available in a population of neurons.

The main conclusion from this and the preceding Section is that the information available from a set or ensemble of temporal cortex visual neurons increases approximately linearly as more neurons are added to the sample. This is powerful evidence that distributed encoding is used by the brain; and the code can be read just by knowing the firing rates in a short time of the population of neurons. The fact that the code can be read off from the firing rates, and by a principle as simple and neuron-like as dot product decoding, provides strong support for the general approach taken in this book to brain function.

It is possible that more information would be available in the relative time of occurrence of the spikes, either within the spike train of a single neuron, or between the spike trains of different neurons, and it is to this that we now turn.

C.3.7   The information from multiple cells: the effects of cross-correlations between cells

Using the second derivative methods described in Section C.2.5 (see Rolls, Franco, Aggelopoulos and Reece (2003b)), the information available from the number of spikes vs that from the cross-correlations between simultaneously recorded cells has been analyzed for a population of neurons in the inferior temporal visual cortex (Rolls, Aggelopoulos, Franco and Treves 2004). The stimuli were a set of 20 objects, faces, and scenes presented while the monkey performed a visual discrimination task. If synchronization was being used to bind the parts of each object into the correct spatial relationship to other parts, this might be expected to be revealed by stimulus-dependent cross-correlations in the firing of simultaneously recorded groups of 2–4 cells using multiple single-neuron microelectrodes.

A typical result from the information analysis described in Section C.2.5 on a set of three simultaneously recorded cells from this experiment is shown in Fig. C.23. This shows that most of the information available in a 100 ms time period was available in the rates, and that there was little contribution to the information from stimulus-dependent ('noise') correlations (which would have shown as positive values if for example there was stimulus-dependent synchronization of the neuronal responses); or from stimulus-independent 'noise' correlation effects, which might if present have reflected common input to the different neurons so that their responses tended to be correlated independently of which stimulus was shown.

The results for the 20 experiments with groups of 2–4 simultaneously recorded inferior temporal cortex neurons are shown in Table C.4. (The total information is the total from equations C.43 and C.44 in a 100 ms time window, and is not expected to be the sum of the contributions shown in Table C.4 because only the information from the cross terms (for $i \neq j$) is shown in the table for the contributions related to the stimulus-dependent contributions and

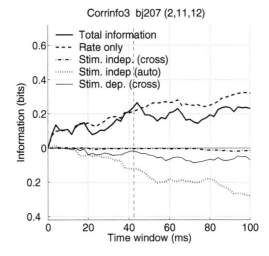

Fig. C.23 A typical result from the information analysis described in Section C.2.5 on a set of 3 simultaneously recorded inferior temporal cortex neurons in an experiment in which 20 complex stimuli effective for IT neurons (objects, faces and scenes) were shown. The graphs show the contributions to the information from the different terms in equations C.43 and C.44 on page 837, as a function of the length of the time window, which started 100 ms after stimulus onset, which is when IT neurons start to respond. The rate information is the sum of the term in equation C.43 and the first term of equation C.44. The contribution of the stimulus-independent noise correlation to the information is the second term of equation C.44, and is separated into components arising from the correlations between cells (the cross component, for $i \neq j$) and from the autocorrelation within a cell (the auto component, for $i = j$). This term is non-zero if there is some correlation in the variance to a given stimulus, even if it is independent of which stimulus is present. The contribution of the stimulus-dependent noise correlation to the information is the third term of equation C.44, and only the cross term is shown (for $i \neq j$), as this is the term of interest. (After Rolls, Aggelopoulos, Franco and Treves 2004.)

Table C.4 The average contributions (in bits) of different components of equations C.43 and C.44 to the information available in a 100 ms time window from 13 sets of simultaneously recorded inferior temporal cortex neurons when shown 20 stimuli effective for the cells.

rate	0.26
stimulus–dependent "noise" correlation-related, cross term	0.04
stimulus–independent "noise" correlation-related, cross term	-0.05
total information	0.31

the stimulus-independent contributions arising from the 'noise' correlations.) The results show that the greatest contribution to the information is that from the rates, that is from the numbers of spikes from each neuron in the time window of 100 ms. The average value of -0.05 for the cross term of the stimulus independent 'noise' correlation-related contribution is consistent with on average a small amount of common input to neurons in the inferior temporal visual cortex. A positive value for the cross term of the stimulus-dependent 'noise' correlation related contribution would be consistent with on average a small amount of stimulus-dependent synchronization, but the actual value found, 0.04 bits, is so small that for 17 of the 20 experiments it is less than that which can arise by chance statistical fluctuations of the time of arrival of the spikes, as shown by MonteCarlo control rearrangements of the same data. Thus on average there was no significant contribution to the information from stimulus-dependent synchronization effects (Rolls, Aggelopoulos, Franco and Treves 2004).

Thus, this data set provides evidence for considerable information available from the number of spikes that each cell produces to different stimuli, and evidence for little impact of common input, or of synchronization, on the amount of information provided by sets of *simultaneously recorded* inferior temporal cortex neurons. Further supporting data for the inferior temporal visual cortex are provided by Rolls, Franco, Aggelopoulos and Reece (2003b). In that parts as well as whole objects are represented in the inferior temporal cortex (Perrett, Rolls and Caan 1982), and in that the parts must be bound together in the correct spatial configuration for the inferior temporal cortex neurons to respond (Rolls, Tovee, Purcell, Stewart and Azzopardi 1994b), we might have expected temporal synchrony, if used to implement feature binding, to have been evident in these experiments.

We have also explored neuronal encoding under natural scene conditions in a task in which top-down attention must be used, a visual search task. We applied the decoding information theoretic method of Section C.2.4 to the responses of neurons in the inferior temporal visual cortex recorded under conditions in which feature binding is likely to be needed, that is when the monkey had to choose to touch one of two simultaneously presented objects, with the stimuli presented in a complex natural background (Aggelopoulos, Franco and Rolls 2005). The investigation is thus directly relevant to whether stimulus-dependent synchrony contributes to encoding under natural conditions, and when an attentional task was being performed. In the attentional task, the monkey had to find one of two objects and to touch it to obtain reward. This is thus an object-based attentional visual search task, where the top-down bias is for the object that has to be found in the scene (Aggelopoulos, Franco and Rolls 2005). The objects could be presented against a complex natural scene background. Neurons in the inferior temporal visual cortex respond in some cases to object features or parts, and in other cases to whole objects provided that the parts are in the correct spatial configuration (Perrett, Rolls and Caan 1982, Desimone, Albright, Gross and Bruce 1984, Rolls, Tovee, Purcell, Stewart and Azzopardi 1994b, Tanaka 1996), and so it is very appropriate to measure whether stimulus-dependent synchrony contributes to information encoding in the inferior temporal visual cortex when two objects are present in the visual field, and when they must be segmented from the background in a natural visual scene, which are the conditions in which it has been postulated that stimulus-dependent synchrony would be useful (Singer 1999, Singer 2000).

Aggelopoulos, Franco and Rolls (2005) found that between 99% and 94% of the information was present in the firing rates of inferior temporal cortex neurons, and less that 5% in any stimulus-dependent synchrony that was present, as illustrated in Fig. C.24. The implication of these results is that any stimulus-dependent synchrony that is present is not quantitatively important as measured by information theoretic analyses under natural scene conditions. This has been found for the inferior temporal visual cortex, a brain region where features are put together to form representations of objects (Rolls 2008d) (Chapter 25), where attention has strong effects, at least in scenes with blank backgrounds (Rolls, Aggelopoulos and Zheng 2003a), and in an object-based attentional search task.

The finding as assessed by information theoretic methods of the importance of firing rates and not stimulus-dependent synchrony is consistent with previous information theoretic approaches (Rolls, Franco, Aggelopoulos and Reece 2003b, Rolls, Aggelopoulos, Franco and Treves 2004, Franco, Rolls, Aggelopoulos and Treves 2004). It would of course also be of interest to test the same hypothesis in earlier visual areas, such as V4, with quantitative, information theoretic, techniques. In connection with rate codes, it should be noted that the findings indicate that the number of spikes that arrive in a given time is what is important for very useful amounts of information to be made available from a population of neurons; and that this time can be very short, as little as 20–50 ms (Tovee and Rolls 1995, Rolls and Tovee 1994, Rolls, Tovee and Panzeri 1999b, Rolls and Deco 2002, Rolls, Tovee, Purcell, Stewart and Azzopardi 1994b, Rolls 2003, Rolls, Franco, Aggelopoulos and Jerez 2006b).

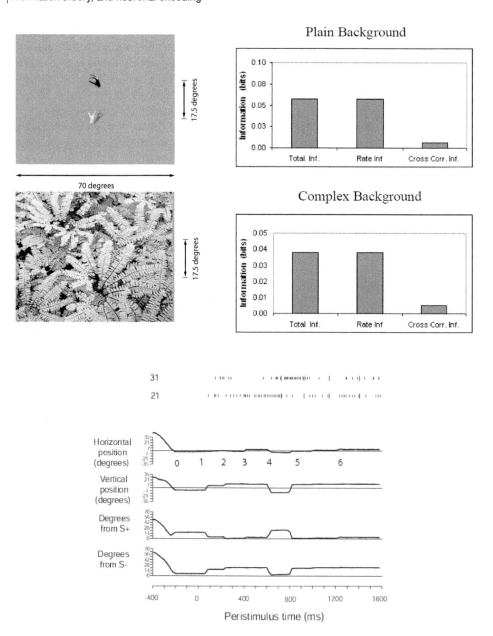

Fig. C.24 Left: the objects against the plain background, and in a natural scene. Right: the information available from the firing rates (Rate Inf) or from stimulus-dependent synchrony (Cross-Corr Inf) from populations of simultaneously recorded inferior temporal cortex neurons about which stimulus had been presented in a complex natural scene. The total information (Total Inf) is that available from both the rate and the stimulus-dependent synchrony, which do not necessarily contribute independently. Bottom: eye position recordings and spiking activity from two neurons on a single trial of the task. (Neuron 31 tended to fire more when the macaque looked at one of the stimuli, S–, and neuron 21 tended to fire more when the macaque looked at the other stimulus, S+. Both stimuli were within the receptive field of the neuron.) (After Aggelopoulos, Franco and Rolls 2005.)

Further, it was shown that there was little redundancy (less than 6%) between the information provided by the spike counts of the simultaneously recorded neurons, making spike counts an efficient population code with a high encoding capacity.

The findings (Aggelopoulos, Franco and Rolls 2005) are consistent with the hypothesis that feature binding is implemented by neurons that respond to features in the correct relative spatial locations (Rolls and Deco 2002, Elliffe, Rolls and Stringer 2002, Rolls 2008d, Rolls 2012c) (Chapter 25), and not by temporal synchrony and attention (Malsburg 1990a, Singer, Gray, Engel, Konig, Artola and Brocher 1990, Abeles 1991, Hummel and Biederman 1992, Singer and Gray 1995, Singer 1999, Singer 2000) (Section 8.3). In any case, the computational point made in Section 25.5.5.1 is that even if stimulus-dependent synchrony was useful for grouping, it would not without much extra machinery be useful for binding the relative spatial positions of features within an object, or for that matter of the positions of objects in a scene which appears to be encoded in a different way (Aggelopoulos and Rolls 2005) (see Section 25.5.10).

So far, we know of no analyses that have shown with information theoretic methods that considerable amounts of information are available about the stimulus from the stimulus-dependent correlations between the responses of neurons in the primate ventral visual system. The use of such methods is needed to test quantitatively the hypothesis that stimulus-dependent synchronization contributes substantially to the encoding of information by neurons.

C.3.8 Conclusions on cortical neuronal encoding

The conclusions emerging from this set of information theoretic analyses, many in cortical areas towards the end of the ventral visual stream of the monkey, and others in the hippocampus for spatial view cells (Rolls, Treves, Robertson, Georges-François and Panzeri 1998b), in the presubiculum for head direction cells (Robertson, Rolls, Georges-François and Panzeri 1999), and in the orbitofrontal cortex and related areas for olfactory and taste cells (Rolls, Critchley and Treves 1996a, Rolls, Critchley, Verhagen and Kadohisa 2010a) for which subsequent analyses have shown a linear increase in information with the number of cells in the population, are as follows (see also Rolls and Treves (2011)).

The representation of at least some classes of objects in those areas is achieved with minimal redundancy by cells that are allocated each to analyze a different aspect of the visual stimulus (Abbott, Rolls and Tovee 1996, Rolls, Treves and Tovee 1997b) (as shown in Sections C.3.5 and C.3.7). This minimal redundancy is what would be expected of a self-organizing system in which different cells acquired their response selectivities through processes that include some randomness in the initial connectivity, and local competition among nearby cells (see Appendix B). Towards the end of the ventral visual stream redundancy may thus be effectively minimized, a finding consistent with the general idea that one of the functions of the early visual system is indeed that of progressively minimizing redundancy in the representation of visual stimuli (Attneave 1954, Barlow 1961). Indeed, the evidence described in Sections C.3.5, C.3.7 and C.3.4 shows that the exponential rise in the number of stimuli that can be decoded when the firing rates of different numbers of neurons are analyzed indicates that the encoding of information using firing rates (in practice the number of spikes emitted by each of a large population of neurons in a short time period) is a very powerful coding scheme used by the cerebral cortex, and that the information carried by different neurons is close to independent provided that the number of stimuli being considered is sufficiently large.

Quantitatively, the encoding of information using firing rates (in practice the number of spikes emitted by each of a large population of neurons in a short time period) is likely to be far more important than temporal encoding, in terms of the number of stimuli that can be encoded. Moreover, the information available from an ensemble of cortical neurons when only the firing rates are read, that is with no temporal encoding within or between neurons, is made available very rapidly (see Figs. C.14 and C.15 and Section C.3.4). Further, the neuronal responses in most ventral or 'what' processing streams of behaving monkeys show sustained firing rate differences to different stimuli (see for example Fig. 16.15 for visual representations, for the olfactory pathways Rolls, Critchley and Treves (1996a), for spatial view cells in the hippocampus Rolls, Treves, Robertson, Georges-François and Panzeri (1998b), and for head direction cells in the presubiculum Robertson, Rolls, Georges-François and Panzeri (1999)), so that it may not usually be necessary to invoke temporal encoding for the information about the stimulus. Further, as indicated in Section C.3.7, information theoretic approaches have enabled the information that is available from the firing rate and from the relative time of firing (synchronization) of inferior temporal cortex neurons to be directly compared with the same metric, and most of the information appears to be encoded in the numbers of spikes emitted by a population of cells in a short time period, rather than by the temporal synchronization of the responses of different neurons when certain stimuli appear (see Section C.3.7 and Aggelopoulos, Franco and Rolls (2005)).

Information theoretic approaches have also enabled different types of readout or decoding that could be performed by the brain of the information available in the responses of cell populations to be compared (Rolls, Treves and Tovee 1997b, Robertson, Rolls, Georges-François and Panzeri 1999). It has been shown for example that the multiple cell representation of information used by the brain in the inferior temporal visual cortex (Rolls, Treves and Tovee 1997b, Aggelopoulos, Franco and Rolls 2005), olfactory cortex (Rolls, Critchley and Treves 1996a), hippocampus (Rolls, Treves, Robertson, Georges-François and Panzeri 1998b), and presubiculum (Robertson, Rolls, Georges-François and Panzeri 1999) can be read fairly efficiently by the neuronally plausible dot product decoding, and that the representation has all the desirable properties of generalization and graceful degradation, as well as exponential coding capacity (see Sections C.3.5 and C.3.7).

Information theoretic approaches have also enabled the information available about different aspects of stimuli to be directly compared. For example, it has been shown that inferior temporal cortex neurons make explicit much more information about what stimulus has been shown rather than where the stimulus is in the visual field (Tovee, Rolls and Azzopardi 1994), and this is part of the evidence that inferior temporal cortex neurons provide translation invariant representations. In a similar way, information theoretic analysis has provided clear evidence that view invariant representations of objects and faces are present in the inferior temporal visual cortex, in that for example much information is available about what object has been shown from any single trial on which any view of any object is presented (Booth and Rolls 1998).

Information theory has also helped to elucidate the way in which the inferior temporal visual cortex provides a representation of objects and faces, in which information about which object or face is shown is made explicit in the firing of the neurons in such a way that the information can be read off very simply by memory systems such as the orbitofrontal cortex, amygdala, and perirhinal cortex / hippocampal systems. The information can be read off using dot product decoding, that is by using a synaptically weighted sum of inputs from inferior temporal cortex neurons (see further Section 24.2.6 and Chapter 25). Moreover, information theory has helped to show that for many neurons considerable invariance in the representations of objects and faces are shown by inferior temporal cortex neurons (e.g. Booth and Rolls (1998)). Examples of some of the types of objects and faces that are encoded in this way are shown in Fig. C.7. Information theory has also helped to show that inferior temporal cortex neurons maintain their object selectivity even when the objects are presented in complex natural backgrounds (Aggelopoulos, Franco and Rolls 2005) (see further Chapter 25 and Section 24.2.6).

Information theory has also enabled the information available in neuronal representations to be compared with that available to the whole animal in its behaviour (Zohary, Shadlen and Newsome 1994) (but see Section C.3.6).

Finally, information theory also provides a metric for directly comparing the information available from neurons in the brain (see Chapter 25 and this Appendix) with that available from single neurons and populations of neurons in simulations of visual information processing (see Chapter 25).

In summary, the evidence from the application of information theoretic and related approaches to how information is encoded in the visual, hippocampal, and olfactory cortical systems described during behaviour leads to the following working hypotheses:

1. Much information is available about the stimulus presented in the number of spikes emitted by single neurons in a fixed time period, the firing rate.

2. Much of this firing rate information is available in short periods, with a considerable proportion available in as little as 20 ms. This rapid availability of information enables the next stage of processing to read the information quickly, and thus for multistage processing to operate rapidly. This time is the order of time over which a receiving neuron might be able to utilize the information, given its synaptic and membrane time constants. In this time, a sending neuron is most likely to emit 0, 1, or 2 spikes.

3. This rapid availability of information is confirmed by population analyses, which indicate that across a population on neurons, much information is available in short time periods.

4. More information is available using this rate code in a short period (of e.g. 20 ms) than from just the first spike.

5. Little information is available by time variations within the spike train of individual neurons for static visual stimuli (in periods of several hundred milliseconds), apart from a small amount of information from the onset latency of the neuronal response. A static stimulus encompasses what might be seen in a single visual fixation, what might be tasted with a stimulus in the mouth, what might be smelled in a single sniff, etc. For a time-varying stimulus, clearly the firing rate will vary as a function of time.

6. Across a population of neurons, the firing rate information provided by each neuron tends to be independent; that is, the information increases approximately linearly with the number of neurons. This applies of course only when there is a large amount of information to be encoded, that is with a large number of stimuli. The outcome is that the number of stimuli that can be encoded rises exponentially in the number of neurons in the ensemble. (For a small stimulus set, the information saturates gradually as the amount of information available from the neuronal population approaches that required to code for the stimulus set.) This applies up to the number of neurons tested and the stimulus set sizes used, but as the number of neurons becomes very large, this is likely to hold less well. An implication of the independence is that the response profiles to a set of stimuli of different neurons are uncorrelated.

7. The information in the firing rate across a population of neurons can be read moderately efficiently by a decoding procedure as simple as a dot product. This is the simplest type of processing that might be performed by a neuron, as it involves taking a dot product of the incoming firing rates with the receiving synaptic weights to obtain the activation (e.g. depolarization) of the neuron. This type of information encoding ensures that the simple emergent properties of associative neuronal networks such as generalization, completion, and graceful

degradation (see Appendix B) can be realized very naturally and simply.

8. There is little additional information to the great deal available in the firing rates from any stimulus-dependent cross-correlations or synchronization that may be present. Stimulus-dependent synchronization might in any case only be useful for grouping different neuronal populations, and would not easily provide a solution to the binding problem in vision. Instead, the binding problem in vision may be solved by the presence of neurons that respond to combinations of features in a given spatial position with respect to each other.

9. There is little information available in the order of the spike arrival times of different neurons for different stimuli that is separate or additional to that provided by a rate code. The presence of spontaneous activity in cortical neurons facilitates rapid neuronal responses, because some neurons are close to threshold at any given time, but this also would make a spike order code difficult to implement.

10. Analysis of the responses of single neurons to measure the sparseness of the representation indicates that the representation is distributed, and not grandmother cell like (or local). Moreover, the nature of the distributed representation, that it can be read by dot product decoding, allows simple emergent properties of associative neuronal networks such as generalization, completion, and graceful degradation (see Appendix B) to be realized very naturally and simply. The evidence becoming available from humans is consistent with this summary (Fried, Rutishauser, Cerf and Kreiman 2014, Rolls 2015d, Rolls 2017a).

11. The representation is not very sparse in the perceptual systems studied (as shown for example by the values of the single cell sparseness a^s), and this may allow much information to be represented. At the same time, the responses of different neurons to a set of stimuli are decorrelated, in the sense that the correlations between the response profiles of different neurons to a set of stimuli are low. Consistent with this, the neurons convey independent information, at least up to reasonable numbers of neurons. The representation may be more sparse in memory systems such as the hippocampus, and this may help to maximize the number of memories that can be stored in associative networks.

12. The nature of the distributed representation can be understood further by the firing rate probability distribution, which has a long tail with low probabilities of high firing rates. The firing rate probability distributions for some neurons fit an exponential distribution, and for others there are too few very low rates for a good fit to the exponential distribution. An implication of an exponential distribution is that this maximizes the entropy of the neuronal responses for a given mean firing rate under some conditions. It is of interest that in the inferior temporal visual cortex, the firing rate probability distribution is very close to exponential if a large number of neurons are included without scaling of the firing rates of each neuron. An implication is that a receiving neuron would see an exponential firing rate probability distribution.

13. The population sparseness a^p, that is the sparseness of the firing of a population of neurons to a given stimulus (or at one time), is the important measure for setting the capacity of associative neuronal networks. In populations of neurons studied in the inferior temporal cortex, hippocampus, and orbitofrontal cortex, it takes the same value as the single cell sparseness a^s, and this is a situation of weak ergodicity that occurs if the response profiles of the different neurons to a set of stimuli are uncorrelated.

Understanding the neuronal code, the subject of this Appendix, is fundamental for understanding how memory and related perceptual systems in the brain operate, as follows:

Understanding the neuronal code helps to clarify what neuronal operations would be useful in memory and in fact in most mammalian brain systems (e.g. dot product decoding, that is taking a sum in a short time of the incoming firing rates weighted by the synaptic weights).

It clarifies how rapidly memory and perceptual systems in the brain could operate, in terms of how long it takes a receiving neuron to read the code.

It helps to confirm how the properties of those memory systems in terms of generalization, completion, and graceful degradation occur, in that the representation is in the correct form for these properties to be realized.

Understanding the neuronal code also provides evidence essential for understanding the storage capacity of memory systems, and the representational capacity of perceptual systems.

Understanding the neuronal code is also important for interpreting functional neuroimaging, for it shows that functional imaging that reflects incoming firing rates and thus currents injected into neurons, and probably not stimulus-dependent synchronization, is likely to lead to useful interpretations of the underlying neuronal activity and processing. Of course, functional neuroimaging cannot address the details of the representation of information in the brain in the way that is essential for understanding how neuronal networks in the brain could operate, for this level of understanding (in terms of all the properties and working hypotheses described above) comes only from an understanding of how single neurons and populations of neurons encode information.

C.4 Information theory terms – a short glossary

1. The **amount of information**, or **surprise**, in the occurrence of an event (or symbol) s_i of probability $P(s_i)$ is

$$I(s_i) = \log_2(1/P(s_i)) = -\log_2 P(s_i). \tag{C.55}$$

(The measure is in bits if logs to the base 2 are used.) This is also the amount of **uncertainty** removed by the occurrence of the event.

2. The average amount of information per source symbol over the whole alphabet (S) of symbols s_i is the **entropy**,

$$H(S) = -\sum_i P(s_i) \log_2 P(s_i) \tag{C.56}$$

(or *a priori* entropy).

3. The probability of the pair of symbols s and s' is denoted $P(s, s')$, and is $P(s)P(s')$ only when the two symbols are **independent**.

4. Bayes theorem (given the output s', what was the input s ?) states that

$$P(s|s') = \frac{P(s'|s)P(s)}{P(s')} \tag{C.57}$$

where $P(s'|s)$ is the **forward** conditional probability (given the input s, what will be the output s' ?), and $P(s|s')$ is the **backward** (or posterior) conditional probability (given the output s', what was the input s ?). The prior probability is $P(s)$.

5. **Mutual information**. Prior to reception of s', the probability of the input symbol s was $P(s)$. This is the *a priori* probability of s. After reception of s', the probability that the input symbol was s becomes $P(s|s')$, the conditional probability that s was sent given that s' was received. This is the *a posteriori* probability of s. The difference between the *a priori* and *a posteriori* uncertainties measures the gain of information due to the reception of s'. Once averaged across the values of both symbols s and s', this is the **mutual information**, or **transinformation**

$$I(S, S') = \sum_{s,s'} P(s, s')\{\log_2[1/P(s)] - \log_2[1/P(s|s')]\} \qquad \text{(C.58)}$$

$$= \sum_{s,s'} P(s, s') \log_2[P(s|s')/P(s)].$$

Alternatively,

$$I(S, S') = H(S) - H(S|S'). \qquad \text{(C.59)}$$

$H(S|S')$ is sometimes called the **equivocation** (of S with respect to S').

C.5 Highlights

1. The encoding of information by neuronal responses is described using Shannon information theory.
2. Information is encoded by sparse distributed place coded representations, with almost independent information conveyed by each neuron, up to reasonable numbers of neurons.
3. Place cell encoding conveys much more information than stimulus-dependent synchronicity ('oscillations', coherence) of firing of groups of neurons in the awake behaving animal.

Appendix 4 Simulation software for neuronal network models

D.1 Introduction

This Appendix of *Cerebral Cortex: Principles of Operation* (Oxford University Press) (Rolls 2016a) describes the Matlab software that has been made available with *Cerebral Cortex: Principles of Operation* (Rolls 2016a) to provide simple demonstrations of the operation of some key neuronal networks related to cortical function. The software is also provided in connection with *Emotion and Decision-Making Explained* (Oxford University Press) (Rolls 2014a) to illustrate the operation of some neuronal networks involved in emotion and decision-making. The aim of providing the software is to enable those who are learning about these networks to understand how they operate and their properties by running simulations and by providing the programs. The code has been kept simple, but can easily be edited to explore the properties of these networks. The code itself contains comments that help to explain how the code works.

The programs are intended to be used with the descriptions of the networks and their properties provided in Appendix B of *Cerebral Cortex: Principles of Operation* (Rolls 2016a), which is available at http://www.oxcns.org. Most of the papers cited in this Appendix can be found at http://www.oxcns.org. Exercises are suggested below that will provide insight into the properties of these networks, and will make the software suitable for class use. The code is available at http://www.oxcns.org. The programs are written in Matlab™ (Mathworks Inc). The programs should work in any version of Matlab, whether it is running under Windows, Linux, or Apple software. The programs will also work under GNU Octave, which is available for free download.

To get started, copy the Matlab program files PatternAssociationDemo.m, AutoAssociationDemo.m, CompetitiveNetDemo.m and SOMdemo.m, and the function files NormVeclen.m and spars.m, into a directory, and in Matlab cd into that directory.

D.2 Autoassociation or attractor networks

The operation and properties of autoassociation (or attractor) neuronal networks are described in Rolls (2016a) Appendix B.3.

D.2.1 Running the simulation

In the command window of Matlab, after you have performed a 'cd' into the directory where the source files are, type 'AutoAssociationDemo' (with every command followed by 'Enter' or 'Return'). The program will run for a bit, and then pause so that you can inspect what has been produced so far. A paused state is indicated by the words 'Paused: Press any key' in the bottom of the Matlab window. Press 'Enter' to move to the next stage, until the program finishes and the command prompt reappears. You can edit the code to comment out some of the 'pause' statements if you do not want them, or to add a 'pause' statement if you would like the program to stop at a particular place so that you can inspect the results. To stop the program and exit from it, use 'Ctrl-C'. When at the command prompt, you can access variables by typing the variable name followed by 'Enter'. (Note: if you set '*display* = 0' near the beginning of the program, there will be no pauses, and you will be able to collect data quickly.)

The fully connected autoassociation network has N=100 neurons with the dendrites that receive the synapses shown as vertical columns in the generated figures, and $nSyn = N$ synapses onto each neuron. At the first pause, you will see a figure showing the 10 random binary training patterns with firing rates of 1 or 0, and with a *Sparseness* (referred to as a in the equations below) of 0.5. As these are binary patterns, the *Sparseness* parameter value is the same as the proportion of high firing rates or 1s. At the second pause you will see distorted versions of these patterns to be used as recall cues later. The number of bits that have been flipped $nFlipBits$=14.

Next you will see a figure showing the synaptic matrix being trained with the 10 patterns. Uncomment the pause in the training loop if you wish to see each pattern being presented sequentially. The synaptic weight matrix *SynMat* (the elements of which are referred to as w_{ij} in the equations below) is initialized to zero, and then each pattern is presented as an external input to set the firing rates of the neurons (the postsynaptic term), which because of the recurrent collaterals become also the presynaptic input. Each time a training pattern is presented, the weight change is calculated by a covariance learning rule

$$\delta w_{ij} = \alpha(y_i - a)(y_j - a) \tag{D.1}$$

where α is a learning rate constant, and y_j is the presynaptic firing rate. This learning rule includes (in proportion to y_i the firing rate of the ith postsynaptic neuron) increasing the synaptic weight if $(y_j - a) > 0$ (long-term potentiation), and decreasing the synaptic weight if $(y_j - a) < 0$ (heterosynaptic long-term depression). As these are random binary patterns with sparseness a (the parameter *Sparseness* in AutoassociationDemo.m), a is the average activity $< y_j >$ of an axon across patterns, and a is also the average activity $< y_i >$ of a neuron across patterns. In the exercises, you can try a form of this learning rule that is more biologically plausible, with only heterosynaptic long-term depression. The change of weight is added to the previous synaptic weight. In the figures generated by the code, the maximum weights are shown as white, and the minimum weights are black. Although both these rules lead to some negative synaptic weights, which are not biologically plausible, this limitation can be overcome, as shown in the exercises. In the display of the synaptic weights, remember that each column represents the synaptic weights on a single neuron. The thin row below the synaptic matrix represents the firing rates of the neurons, and the thin column to the left of the synaptic weight matrix the firing rates of the presynaptic input, which are usually the same during training. Rates of 1 are shown as white, and of 0 as black.

Next, testing with the distorted patterns starts, with the output rates allowed to recirculate through the recurrent collateral axons for 9 recall epochs to produce presynaptic inputs that act through the synaptic weight matrix to produce the firing rate for the next recall epoch. The distorted recall cue is presented only on epoch 1 in what is described as the clamped condition, and after that is removed so that the network has a chance settle into a perfect recall state representing the training pattern without being affected by the distorted recall cue. (In the cortex, this may be facilitated by the greater adaptation of the thalamic inputs than of the recurrent collaterals (Rolls 2016a)). (Thus on epoch 1 the distorted recall cue is shown as the PreSynaptic Input column on the left of the display, and the Activation row and the Firing Rate row below the synaptic weight matrix are the Activations and Rates produced by the recall cue. On later epochs, the Presynaptic Input column shows the output firing Rate from the preceding epoch, recirculated by the recurrent collateral connections.)

We are interested in how perfect the recall is, that is how correlated the recalled firing rate state is with the original training pattern, with this correlation having a maximum value of 1. Every time you press 'Enter' a new recall epoch is performed, and you will see that over one to several recall epochs the recall usually will become perfect, that is the correlation, which is provided for you in the command window, will become 1. However, on some trials, the network does not converge to perfect recall, and this occurs if by chance the distorted recall cue happened to be much closer to one of the other stored patterns. Do note that with randomly chosen training patterns in relatively small networks this will sometimes occur, and that this 'statistical fluctuation' in how close some of the training patterns are to each other, and how close the distorted test patterns are to individual training patterns, is to be expected. These effects smooth out as the network become larger. (Remember that cortical neurons have in the order of 10,000 recurrent collateral inputs to each neuron, so these statistical fluctuations will be largely smoothed out. Do note also that the performance of the network will be different each time it is run, because the random number generator is set with a different seed on each run. (Near the beginning of the program, you could uncomment the command that causes the same seed to be used for the random number generator for each run, to help with further program development that you may try.)

After the last test pattern has been presented, the percentage of the patterns that were correctly recalled is shown, using as a criterion that the correlation of the recalled firing rate with the training pattern $r >= 0.98$. For the reasons just described related to the random generation of the training and distorted test patterns, the percentage correct will vary from run to run, so taking the average of several runs is recommended.

Note that the continuing stable firing after the first few recall epochs models the implementation of short-term memory in the cerebral cortex (Rolls 2008d, Rolls 2016a).

Note that the retrieval of the correct pattern from a distorted version that includes missing 1s models completion in the cerebral cortex, which is important in episodic memory as implemented in the hippocampus in which the whole of the episodic memory can be retrieved from any part (Rolls 2008d, Rolls 2016a, Kesner and Rolls 2015).

Note that the retrieval of the correct pattern from a distorted version that includes missing 1s or has 1s instead of 0s models correct memory retrieval in the cerebral cortex even if there is some distortion of the recall cue (Rolls 2008d, Rolls 2016a). This also enables generalization to similar patterns or stimuli that have been encountered previously, which is highly behaviorally adaptive (Rolls 2008d, Rolls 2016a).

D.2.2 Exercises

1. Measure the percent correct as you increase the number of patterns from 10 to 15 with the Sparseness remaining at 0.5. Plot the result. How close is your result to that found by Hopfield (1982), which was $0.14N$ for a sparseness a of 0.5? You could try increasing N by 10 (to 1000) and multiplying the number of training patterns by 10 and increasing $nFlipBits$ to say 100 to see whether your results, with smaller statistical fluctuations, approximate more closely to the theoretical value.

2. Test the effect of reducing the Sparseness (a) to 0.1, which will increase the number of patterns that can be stored and correctly recalled, i.e. the capacity of the network (Treves and Rolls 1991, Rolls 2016a). (Hint: try a number of patterns in the region of 30.) You could also plot the capacity of the network as a function of the Sparseness (a) with values down to say 0.01 in larger networks, with theoretical results provided by Treves and Rolls (1991) and Rolls (2016a).

3. Test the effect of altering the learning rule, from the covariance rule to the rule with heterosynaptic long term depression (LTD), which is as follows (Rolls 2016a):

$$\delta w_{ij} = \alpha(y_i)(y_j - a). \tag{D.2}$$

(Hint: in the training loop in the program, at about line 102, comment in the heterosynaptic LTD rule.)

4. Test whether the network operates with positive-only synaptic weights. (Hint: there are two lines (close to 131) just after the training loop that if uncommented will add a constant to all the numbers in the synaptic weight matrix to make all the numbers positive (with the minimum weight 0).

5. What happens to the recall if you train more patterns than the critical capacity? (Hint: see Rolls (2016a) section B.16.)

6. Does it in practice make much difference if you allow self-connections, especially in large networks? (Hint: in the program in the training loop if you comment out the line 'if syn $\sim=$ neuron' and its corresponding 'end', self-connections will be allowed.)

D.3 Pattern association networks

The operation and properties of pattern association networks are described in Rolls (2016a) Appendix B.2.

D.3.1 Running the simulation

In the command window of Matlab, after you have performed a 'cd' into the directory where the source files are, type 'PatternAssociationDemo'. The program will run for a bit, and then pause so that you can inspect what has been produced so far. A paused state is indicated by the words 'Paused: Press any key'. Press 'Enter' to move to the next stage, until the program finishes and the command prompt reappears. You can edit the code to comment out some of the 'pause' statements if you do not want them, or to add a 'pause' statement if you would like the program to stop at a particular place so that you can inspect the results. To stop the program and exit from it, use 'Ctrl-C'. When at the command prompt, you can access variables by typing the variable name followed by 'Enter'. (Note: if you set 'display = 0' near the beginning of the program, there will be few pauses, and you will be able to collect data quickly.)

The fully connected pattern association network has N=8 output neurons with the dendrites that receive the synapses shown as vertical columns in the generated figures, and $nSyn$=64 synapses onto each neuron. At the first pause, you will see the 8 random binary conditioned stimulus (CS) training patterns with firing rates of 1 or 0, and with a sparseness of 0.25 (*InputSparseness*). As these are binary patterns, the sparseness is the same as the proportion of high firing rates or 1s. At the second pause you will see distorted versions of these patterns to be used as recall cues later. The number of bits that have been flipped $nFlipBits$=8.

Next you will see the synaptic matrix being trained with the 8 pattern associations between each conditioned stimulus (CS) and its unconditioned stimulus (US), the output firing. In this simulation, the default is for each US pattern to have one bit on, so that the sparseness of the US, *OutputSparseness* is $1/N$. Uncomment the pause in the training loop if you wish to see each pattern being presented sequentially. The synaptic weight matrix *SynMat* (elements of which are referred to as w_{ij} below) is initialized to zero, and then each pattern pair is presented as a CS and US pair, with the CS specifying the presynaptic input, and the US the firing rate in the N neurons present when the CS is presented. Each time a training pattern pair is presented, the weight change is calculated by an associative learning rule that includes heterosynaptic long term depression

$$\delta w_{ij} = \alpha(y_i)(x_j - a). \tag{D.3}$$

where α is a learning rate constant, and x_j is the presynaptic firing rate. This learning rule includes (in proportion to y_i the firing rate of the ith postsynaptic neuron produced by the US) increasing the synaptic weight if $(y_j - a) > 0$ (long-term potentiation), and decreasing the synaptic weight if $(x_j - a) < 0$ (heterosynaptic long-term depression). As the CS patterns are random binary patterns with sparseness a (the parameter *InputSparseness*), a is the average activity $< x_j >$ of an axon across patterns. The reasons for the use of this learning rule are described in Rolls (2016a) section B.3.3.6. In the exercises, you can try a Hebbian associative learning rule that does not include this heterosynaptic long-term depression. The change of weight is added to the previous synaptic weight. In the figures generated by the code, the maximum weights are shown as white, and the minimum weights are black. Although both these rules lead to some negative synaptic weights, which are not biologically plausible, this limitation can be overcome, as shown in the exercises. In the display of the synaptic weights, remember that each column represents the synaptic weights on a single neuron. The top thin row below the synaptic matrix represents the activations of the neurons, the bottom thin row the firing rates of the neurons, and the thin column to the left of the synaptic weight matrix the firing rates of the presynaptic input, i.e. the recall (CS) stimulus. Rates of 1 are shown as white, and of 0 as black.

Next, testing with the distorted CS patterns starts. We are interested in how the pattern association network generalizes with distorted inputs to produce the correct US output that should be produced if there were no distortion. The recall performance is measured by how correlated the recalled firing rate state (the Conditioned Response, CR) is with the output that should be produced to the original training CS pattern, with this correlation having a maximum value of 1. Every time you press 'Enter' a new CS is presented. Do note that with randomly chosen distortions of the CS patterns to produce the recall cue, in relatively small networks the distorted CS will sometimes be close of another CS, and that this 'statistical fluctuation' in how close some of the distorted CSs are to the original CSs is to be expected. These effects smooth out as the network become larger. (Remember that cortical neurons have in the order of 10,000 inputs to each neuron (in addition to the recurrent collaterals, Rolls (2016a)), so these statistical fluctuations will be largely smoothed out. Do note also that the performance of the network will be different each time it is run, because the random number generator is set with a different seed on each run. (Near the beginning of the program, you could uncomment the command that causes the same seed to be used for the random number generator for each run, to help with further program development that you may try.)

After the last distorted testing CS pattern has been presented, the percentage of the patterns that were correctly recalled is shown, using a criterion for the correlation of the recalled firing rate with the training pattern $r >= 0.98$. For the reasons just described related to the random generation of the distorted test patterns, the percentage correct will vary from run to run, so taking the average of several runs is recommended.

Note that the retrieval of the correct output (conditioned response, CR) pattern from a distorted CS recall stimulus models generalization in the cerebral cortex, which is highly behaviorally adaptive (Rolls 2008d, Rolls 2016a).

D.3.2 Exercises

1. Test how well the network generalizes to distorted recall cues, by altering $nFlipBits$ to values between 0 and 14. Remember to take the average of several simulation runs.

  2. Test the effect of altering the learning rule, from the rule with heterosynaptic long term depression (LTD) as in Eqn. D.3, to a simple associative synaptic modification rule which is as follows (Rolls 2016a):

$$\delta w_{ij} = \alpha(y_i)(x_j).\tag{D.4}$$

(Hint: in the training loop in the program, at about line 110, comment out the heterosynaptic LTD rule, and uncomment the associative rule.)
  If you see little difference, can you suggest conditions under which there may be a difference, such as higher loading? (Hint: check Rolls (2016a) section B.2.7.)

  3. Describe how the threshold non-linearity in the activation function helps to remove interference from other training patterns, and from distortions in the recall cue. (Hint: compare the activations to the firing rates of the output neurons.)

  4. In the pattern associator, the conditioned stimulus (CS) retrieves the unconditioned response (UR), that is the effects produced by the unconditioned stimulus (US) with which it has been paired during learning. Can the opposite occur, that is, can the US be used to retrieve the CS? If not, in what type of network is the retrieval symmetrical? (Hint: if you are not sure, read Rolls (2016a) sections B.2 and B.3.)

  5. The network simulated uses local representations for the US, with only one neuron on for each US. If you can program in Matlab, can you write code that would generate distributed representations for the US patterns? Then can you produce new CS patterns with random binary representations and sparsenesses in the range 0.1–0.5, like those used for the autoassociator, so that you can then investigate the storage capacity of pattern association networks. (Hint: reference to Rolls (2016a) section B.2.7 and B.2.8 may be useful.)

D.4 Competitive networks and Self-Organizing Maps

The operation and properties of competitive networks are described in Rolls (2016a) Appendix B.4. This is a self-organizing network, with no teacher. The input patterns are presented in random sequence, and the network learns how to categorise the patterns, placing similar patterns into the same category, and different patterns into different categories. The particular neuron or neurons that represent each category depend on the initial random synaptic weight vectors as well as the learning, so the output neurons are different from simulation run to simulation run. Self-organizing maps are investigated in Exercise 4.

D.4.1 Running the simulation

In the command window of Matlab, after you have performed a 'cd' into the directory where the source files are, type 'CompetitiveNetDemo'. The program will run for a bit, and then pause so that you can inspect what has been produced so far. A paused state is indicated by the words 'Paused: Press any key'. Press 'Enter' to move to the next stage, until the program finishes and the command prompt reappears. You can edit the code to comment out some of the 'pause' statements if you do not want them, or to add a 'pause' statement if you would like the program to stop at a particular place so that you can inspect the results. To stop the program and exit from it, use 'Ctrl-C'. When at the command prompt, you can access variables by typing the variable name followed by 'Enter'. (Note: if you set '*display = 0*' near the beginning of the program, there will be few pauses, and you will be able to collect data quickly.)

The fully connected competitive network has $N=100$ output neurons with the dendrites that receive the synapses shown as vertical columns, and $nSyn=100$ synapses onto each neuron. With the display on (*display = 1*), at the first pause, you will see the 28 binary input patterns for training with firing rates of 1 or 0. Each pattern is 20 elements long, and each pattern is shifted down 3 elements with respect to the previous pattern, with the parameters supplied. (Each pattern thus overlaps in 17 / 20 locations with its predecessor.)

Next you will see the synaptic matrix being trained with the 28 input patterns, which are presented 20 times each (*nepochs* = 20) in random permuted sequence. Uncomment the pause in the training loop if you wish to see each pattern being presented sequentially. The synaptic weight matrix *SynMat* is initialized so that each neuron has random synaptic weights, with the synaptic weight vector on each neuron normalized to have a length of 1. Normalizing the length of the vector during training is important, for it ensures that no one neuron increases its synaptic strengths to very high values and thus wins for every input pattern. (Biological approximations to this are described by Rolls (2016a) in section B.4.9.2, and involve a type of heterosynaptic long-term depression of synaptic strength that is large when the postsynaptic term is high, and the presynaptic firing is low.) (In addition, for convenience, each input pattern is normalized to a length of 1, so that when the dot product is computed with a synaptic weight vector on a neuron, the maximum activation of a neuron will be 1.)

The firing rate of each neuron is computed using a binary threshold activation function (which can be changed in the code to threshold linear). The sparseness of the firing rate representation can be set in the code using *OutputSparseness*, and as supplied is set to 0.01 for the first demonstration, which results in one of the $N=100$ output neurons having a high firing rate. Because the (initially random) synaptic weight vectors of each neuron are different, the activations of each of the neurons will be different.

Learning then occurs, using an associative learning rule with synapses increasing between presynaptic neurons with high rates and postsynaptic neurons with high rates. Think of this as moving the synaptic weight vector of a high firing neuron to point closer to the input pattern that is making the neuron fire. At the same time, the other synapses on the same neuron become weaker due to the synaptic weight normalization, so the synaptic weight vector of the neuron moves away from other, different, patterns. In this way, the neurons self-organize so that some neurons point towards some patterns, and other neurons towards other patterns. You can watch these synaptic weight changes during the learning in the figure generated by the code, in which some of the synapses will become stronger (white) on a neuron and correspond to one or several of the patterns, and at the same time the other synapses on the neuron will become weaker (black). In the display of the synaptic weights, remember that each column represents the synaptic weights on a single neuron. The top thin row below the synaptic matrix represents the activations of the neurons, the bottom thin row the firing rates of the neurons, and the thin column to the left of the synaptic weight matrix the firing rates of the presynaptic input, i.e. the pattern currently being presented. Rates of 1 are shown as white, and of 0 as black. Neurons that remain unallocated, and do not respond to any of the input patterns, appear as vertical columns of random synaptic weights onto a neuron. These neurons remain available to potentially categorise different input patterns.

Next, testing with the different patterns starts. We are interested in whether similar input patterns activate the same output neurons; and different input patterns different output neurons. This is described as categorisation (Rolls 2016a). With the display on (*display* = 1), each time you press 'Enter' the program will step through the patterns in sequence. You will notice in the synaptic matrix that with these initial parameters, only a few neurons have learned. When you step through each testing pattern in sequence, you will observe that one of the neurons responds to several patterns in the sequence, and these patterns are quite closely correlated with each other. Then another neuron becomes active for a further set of close patterns. This is repeated as other patterns are presented, and shows you that similar patterns tend to activate the same output neuron or neurons, and different input patterns tend to activate different output neurons. This illustrates by simulation the categorisation performed by competitive networks.

To illustrate these effects more quantitatively, after the patterns have been presented, you are shown first the correlations between the input patterns, and after that the correlations between the output firing rates produced by each pattern. These results are shown as correlation matrices. You should observe that with the network parameters supplied, the firing rates are grouped into several categories, each corresponding to several input patterns that are highly correlated in the output firing that they produce. The correlation matrix between the output firing rate for the different patterns thus illustrates how a competitive network can categorise similar patterns as similar to each other, and different patterns as different.

D.4.2 Exercises

1. Test the operation of the system with no learning, most easily achieved by setting *nepochs* = 0. This results in the input patterns being mapped to output rates through the random synaptic weight vectors for each neuron. What is your interpretation of what is shown by the correlation matrix between the output firing rates?

(Hint: the randomizing effect of the random synaptic weight vectors on each neuron is to separate out the patterns, with many of the patterns having no correlation with other patterns. In this mode, pattern separation is produced, and not categorisation in that similar inputs are less likely to produce similar outputs, a key property in perception and memory (Rolls 2016a). Pattern separation is important in its own right as a process (Rolls 2016e), and is implemented for example by the random non-associatively modifiable connections of the dentate mossy fibre synapses onto the CA3 neurons (Treves and Rolls 1992b, Rolls 2016a, Rolls 2013b, Kesner and Rolls 2015).)

2. Investigate and describe the effects of altering the similarity between the patterns on the categorization, by for example altering the value of *shift* in the range 2–20.

3. Investigate and describe the effects of making the output firing rate representation (*OutputSparseness*) less sparse. It is suggested that values in the range 0.01–0.1 are investigated.

4. If there is lateral facilitation of nearby neurons, which might be produced by short-range recurrent collaterals in the cortex, then self-organizing maps can be generated, providing a model of topographic maps in the cerebral cortex (Section B.4.6 and Rolls (2016a) Section B.4.6). Investigate this with the modified competitive network 'SOMdemo'. The map is made clear during the testing, in which the patterns are presented in order, and the neurons activated appear in different mapped positions in the firing rate array. Run the program several times to show that the details of the map are different each time, but that the principles are the same, that nearby neurons tend to respond to similar patterns, and that singularities (discontinuities) in the map can occur, just as are found in the primary visual cortex, V1.

The important change of 'SOMdemo' from 'CompetitiveNetDemo' is a local spatial filter for the firing rate array to simulate the effect of facilitation of the firing rates of nearby neurons by the short-range excitatory recurrent collateral connections between nearby cortical pyramidal cells. This is implemented by the *SpatialFilter* kernel the range of which is set by *FilterRange*. Investigate the effects of altering *FilterRange* from its default value of 11 on the topographic maps that are formed. Other changes to produce 'SOMdemo' include modifying the *OutPutSparseness* of the Firing Rate representation to 0.3. Investigate the effect of modifying this on the maps being formed. Another alteration was to alter the activation function from binary threshold to linear threshold to enable graded firing rate representations to be formed.

What are the advantages of the representations formed when self-organizing maps are present? (Hint: Think of interpolation of untrained input patterns close to those already trained; and check Rolls (2016a) Section B.4.6.)

5. Further work. The code supplied may provide a route for readers to develop their own programs. Further examples of Matlab code used to investigate neural systems are available in Anastasio (2010). Matlab itself has a Neural Network Toolbox, with an introduction to its use in modelling simple networks provided in Wallisch, Lusignan, Benayoun, Baker, Dickey and Hatsoupoulos (2009).

D.5 Highlights

1. Appendix D describes Matlab software that has been made available with *Cerebral Cortex: Principles of Operation* at http://www.oxcns.org to provide simple demonstrations of the operation of some key neuronal networks related to cortical function.

2. The software demonstrates the operation of pattern association networks, autoassociation / attractor networks, competitive networks, and self-organizing maps.

References

Abbott, L. F. (1991). Realistic synaptic inputs for model neural networks, *Network* **2**: 245–258.

Abbott, L. F. and Blum, K. I. (1996). Functional significance of long-term potentiation for sequence learning and prediction, *Cerebral Cortex* **6**: 406–416.

Abbott, L. F. and Nelson, S. B. (2000). Synaptic plasticity: taming the beast, *Nature Neuroscience* **3**: 1178–1183.

Abbott, L. F. and Regehr, W. G. (2004). Synaptic computation, *Nature* **431**: 796–803.

Abbott, L. F., Rolls, E. T. and Tovee, M. J. (1996). Representational capacity of face coding in monkeys, *Cerebral Cortex* **6**: 498–505.

Abbott, L. F., Varela, J. A., Sen, K. and Nelson, S. B. (1997). Synaptic depression and cortical gain control, *Science* **275**: 220–224.

Abeles, M. (1991). *Corticonics: Neural Circuits of the Cerebral Cortex*, Cambridge University Press, Cambridge.

Abramson, N. (1963). *Information Theory and Coding*, McGraw-Hill, New York.

Ackley, D. H. (1987). *A Connectionist Machine for Genetic Hill-Climbing*, Kluwer Academic Publishers, Dordrecht.

Ackley, D. H., Hinton, G. E. and Sejnowski, T. J. (1985). A learning algorithm for Boltzmann machines, *Cognitive Science* **9**: 147–169.

Acsady, L., Kamondi, A., Sik, A., Freund, T. and Buzsaki, G. (1998). Gabaergic cells are the major postsynaptic targets of mossy fibers in the rat hippocampus, *J Neurosci* **18**(9): 3386–403.

Adrian, E. D. (1928). *The Basis of Sensations*, Christophers, London.

Aertsen, A. M. H. J., Gerstein, G. L., Habib, M. K. and Palm, G. (1989). Dynamics of neuronal firing correlation: modulation of 'effective connectivity', *Journal of Neurophysiology* **61**: 900–917.

Aggelopoulos, N. C. and Rolls, E. T. (2005). Natural scene perception: inferior temporal cortex neurons encode the positions of different objects in the scene, *European Journal of Neuroscience* **22**: 2903–2916.

Aggelopoulos, N. C., Franco, L. and Rolls, E. T. (2005). Object perception in natural scenes: encoding by inferior temporal cortex simultaneously recorded neurons, *Journal of Neurophysiology* **93**: 1342–1357.

Aggleton, J. P. and Nelson, A. J. D. (2015). Why do lesions in the rodent anterior thalamic nuclei cause such severe spatial deficits?, *Neuroscience & Biobehavioral Reviews* **54**: 108–119.

Aigner, T. G., Mitchell, S. J., Aggleton, J. P., DeLong, M. R., Struble, R. G., Price, D. L., Wenk, G. L., Pettigrew, K. D. and Mishkin, M. (1991). Transient impairment of recognition memory following ibotenic acid lesions of the basal forebrain in macaques, *Experimental Brain Research* **86**: 18–26.

Aimone, J. B. and Gage, F. H. (2011). Modeling new neuron function: a history of using computational neuroscience to study adult neurogenesis, *Eur J Neurosci* **33**(6): 1160–9.

Aimone, J. B., Deng, W. and Gage, F. H. (2010). Adult neurogenesis: integrating theories and separating functions, *Trends Cogn Sci* **14**(7): 325–37.

Akrami, A., Liu, Y., Treves, A. and Jagadeesh, B. (2009). Converging neuronal activity in inferior temporal cortex during the classification of morphed stimuli, *Cerebral Cortex* **19**: 760–776.

Albantakis, L. and Deco, G. (2009). The encoding of alternatives in multiple-choice decision making, *Proceedings of the National Academy of Sciences USA* **106**: 10308–10313.

Alberini, C. M. and Ledoux, J. E. (2013). Memory reconsolidation, *Curr Biol* **23**(17): R746–50.

Albus, J. S. (1971). A theory of cerebellar function, *Mathematical Biosciences* **10**: 25–61.

Aleman, A. and Kahn, R. S. (2005). Strange feelings: do amygdala abnormalities dysregulate the emotional brain in schizophrenia?, *Progress in Neurobiology* **77**(5): 283–298.

Alexander, G. E., Crutcher, M. D. and DeLong, M. R. (1990). Basal ganglia thalamo-cortical circuits: parallel substrates for motor, oculomotor, 'prefrontal' and 'limbic' functions, *Progress in Brain Research* **85**: 119–146.

Alexander, R. D. (1975). The search for a general theory of behavior, *Behavioral Sciences* **20**: 77–100.

Alexander, R. D. (1979). *Darwinism and Human Affairs*, University of Washington Press, Seattle.

Allport, A. (1988). What concept of consciousness?, *in* A. J. Marcel and E. Bisiach (eds), *Consciousness in Contemporary Science*, Oxford University Press, Oxford, pp. 159–182.

Alvarez, P. and Squire, L. R. (1994). Memory consolidation and the medial temporal lobe: a simple network model, *Proceedings of the National Academy of Sciences USA* **91**: 7041–7045.

Amaral, D. G. (1986). Amygdalohippocampal and amygdalocortical projections in the primate brain, *in* R. Schwarcz and Y. Ben-Ari (eds), *Excitatory Amino Acids and Epilepsy*, Plenum Press, New York, pp. 3–18.

Amaral, D. G. (1987). Memory: anatomical organization of candidate brain regions, *in* F. Plum and V. Mountcastle (eds), *Higher Functions of the Brain. Handbook of Physiology, Part I*, American Physiological Society, Washington, DC, pp. 211–294.

Amaral, D. G. (1993). Emerging principles of intrinsic hippocampal organization, *Current Opinion in Neurobiology*

3: 225–229.

Amaral, D. G. and Price, J. L. (1984). Amygdalo-cortical projections in the monkey (Macaca fascicularis), *Journal of Comparative Neurology* **230**: 465–496.

Amaral, D. G. and Witter, M. P. (1989). The three-dimensional organization of the hippocampal formation: a review of anatomical data, *Neuroscience* **31**: 571–591.

Amaral, D. G. and Witter, M. P. (1995). The hippocampal formation, *in* G. Paxinos (ed.), *The Rat Nervous System*, Academic Press, San Diego, pp. 443–493.

Amaral, D. G., Ishizuka, N. and Claiborne, B. (1990). Neurons, numbers and the hippocampal network, *Progress in Brain Research* **83**: 1–11.

Amaral, D. G., Price, J. L., Pitkanen, A. and Carmichael, S. T. (1992). Anatomical organization of the primate amygdaloid complex, *in* J. P. Aggleton (ed.), *The Amygdala*, Wiley-Liss, New York, chapter 1, pp. 1–66.

Amari, S. (1977). Dynamics of pattern formation in lateral-inhibition type neural fields, *Biological Cybernetics* **27**: 77–87.

Amari, S. (1982). Competitive and cooperative aspects in dynamics of neural excitation and self-organization, *in* S. Amari and M. A. Arbib (eds), *Competition and Cooperation in Neural Nets*, Springer, Berlin, chapter 1, pp. 1–28.

Amari, S., Yoshida, K. and Kanatani, K.-I. (1977). A mathematical foundation for statistical neurodynamics, *SIAM Journal of Applied Mathematics* **33**: 95–126.

Amit, D. J. (1989). *Modelling Brain Function*, Cambridge University Press, New York.

Amit, D. J. (1995). The Hebbian paradigm reintegrated: local reverberations as internal representations, *Behavioral and Brain Sciences* **18**: 617–657.

Amit, D. J. and Brunel, N. (1997). Model of global spontaneous activity and local structured activity during delay periods in the cerebral cortex, *Cerebral Cortex* **7**: 237–252.

Amit, D. J. and Tsodyks, M. V. (1991). Quantitative study of attractor neural network retrieving at low spike rates. I. Substrate – spikes, rates and neuronal gain, *Network* **2**: 259–273.

Amit, D. J., Gutfreund, H. and Sompolinsky, H. (1987). Statistical mechanics of neural networks near saturation, *Annals of Physics (New York)* **173**: 30–67.

Anastasio, T. J. (2010). *Tutorial on Neural Systems Modelling*, Sinauer, Sunderland, MA.

Andersen, P., Dingledine, R., Gjerstad, L., Langmoen, I. A. and Laursen, A. M. (1980). Two different responses of hippocampal pyramidal cells to application of gamma-aminobutyric acid, *Journal of Physiology* **307**: 279–296.

Andersen, P., Morris, R., Amaral, D., Bliss, T. and O'Keefe, J. (2007). *The Hippocampus Book*, Oxford University Press, London.

Andersen, R. A. (1995). Coordinate transformations and motor planning in the posterior parietal cortex, *in* M. S. Gazzaniga (ed.), *The Cognitive Neurosciences*, MIT Press, Cambridge, MA, chapter 33, pp. 519–532.

Andersen, R. A., Batista, A. P., Snyder, L. H., Buneo, C. A. and Cohen, Y. E. (2000). Programming to look and reach in the posterior parietal cortex, *in* M. Gazzaniga (ed.), *The New Cognitive Neurosciences*, 2 edn, MIT Press, Cambridge, MA, chapter 36, pp. 515–524.

Anderson, J. R. (1996). ACT: a simple theory of complex cognition, *American Psychologist* **51**: 355–365.

Arbib, M. A. (1964). *Brains, Machines, and Mathematics*, McGraw-Hill, New York (2nd Edn 1987 Springer).

Arcizet, F., Mirpour, K. and Bisley, J. W. (2011). A pure salience response in posterior parietal cortex, *Cerebral Cortex* **21**: 2498–2506.

Armstrong, D. M. and Malcolm, M. (1984). *Consciousness and Causality*, Blackwell, Oxford.

Arnold, P. D., Rosenberg, D. R., Mundo, E., Tharmalingam, S., Kennedy, J. L. and Richter, M. A. (2004). Association of a glutamate (NMDA) subunit receptor gene (GRIN2B) with obsessive-compulsive disorder: a preliminary study, *Psychopharmacology (Berl)* **174**: 530–538.

Arnsten, A. F. and Li, B. M. (2005). Neurobiology of executive functions: catecholamine influences on prefrontal cortical functions, *Biological Psychiatry* **57**: 1377–1384.

Aron, A. R., Robbins, T. W. and Poldrack, R. A. (2014). Inhibition and the right inferior frontal cortex: one decade on, *Trends in Cognitive Sciences* **18**: 177–85.

Artola, A. and Singer, W. (1993). Long term depression: related mechanisms in cerebellum, neocortex and hippocampus, *in* M. Baudry, R. F. Thompson and J. L. Davis (eds), *Synaptic Plasticity: Molecular, Cellular and Functional Aspects*, MIT Press, Cambridge, MA, chapter 7, pp. 129–146.

Atick, J. J. (1992). Could information theory provide an ecological theory of sensory processing?, *Network* **3**: 213–251.

Atick, J. J. and Redlich, A. N. (1990). Towards a theory of early visual processing, *Neural Computation* **2**: 308–320.

Atick, J. J., Griffin, P. A. and Relich, A. N. (1996). The vocabulary of shape: principal shapes for probing perception and neural response, *Network* **7**: 1–5.

Attneave, F. (1954). Some informational aspects of visual perception, *Psychological Review* **61**: 183–193.

Baars, B. J. (1988). *A Cognitive Theory of Consciousness*, Cambridge University Press, New York.

Bacon-Mace, N., Mace, M. J., Fabre-Thorpe, M. and Thorpe, S. J. (2005). The time course of visual processing: backward masking and natural scene categorisation, *Vision Research* **45**: 1459–1469.

Baddeley, R. J., Abbott, L. F., Booth, M. J. A., Sengpiel, F., Freeman, T., Wakeman, E. A. and Rolls, E. T. (1997).

Responses of neurons in primary and inferior temporal visual cortices to natural scenes, *Proceedings of the Royal Society B* **264**: 1775–1783.

Bagal, A. A., Kao, J. P. Y., Tang, C.-M. and Thompson, S. M. (2005). Long-term potentiation of exogenous glutamate responses at single dendritic spines., *Proc Natl Acad Sci U S A* **102**(40): 14434–9.

Balduzzi, D., Vanchinathan, H. and Buhmann, J. (2014). Kickback cuts backprop's red-tape: Biologically plausible credit assignment in neural networks, *arXiv preprint arXiv:1411.6191.*

Ballard, D. H. (1990). Animate vision uses object-centred reference frames, *in* R. Eckmiller (ed.), *Advanced Neural Computers*, North-Holland, Elsevier, Amsterdam, pp. 229–236.

Ballard, D. H. (1993). Subsymbolic modelling of hand-eye co-ordination, *in* D. E. Broadbent (ed.), *The Simulation of Human Intelligence*, Blackwell, Oxford, chapter 3, pp. 71–102.

Balu, D. T. and Coyle, J. T. (2015). The nmda receptor 'glycine modulatory site' in schizophrenia: d-serine, glycine, and beyond, *Curr Opin Pharmacol* **20C**: 109–115.

Banks, S. J., Sziklas, V., Sodums, D. J. and Jones-Gotman, M. (2012). fMRI of verbal and nonverbal memory processes in healthy and epileptogenic medial temporal lobes, *Epilepsy and Behavior* **25**: 42–49.

Banks, W. P. (1978). Encoding and processing of symbolic information in comparative judgements, *in* G. H. Bower (ed.), *The Psychology of Learning and Motivation: Advances in Theory and Research*, Academic Press, pp. 101–159.

Bannon, S., Gonsalvez, C. J., Croft, R. J. and Boyce, P. M. (2002). Response inhibition deficits in obsessive-compulsive disorder, *Psychiatry Research* **110**: 165–174.

Bannon, S., Gonsalvez, C. J., Croft, R. J. and Boyce, P. M. (2006). Executive functions in obsessive-compulsive disorder: state or trait deficits?, *Aust N Z Journal of Psychiatry* **40**: 1031–1038.

Barbas, H. (1993). Organization of cortical afferent input to the orbitofrontal area in the rhesus monkey, *Neuroscience* **56**: 841–864.

Barkas, L. J., Henderson, J. L., Hamilton, D. A., Redhead, E. S. and Gray, W. P. (2010). Selective temporal resections and spatial memory impairment: cue dependent lateralization effects, *Behavioural Brain Research* **208**: 535–544.

Barlow, H. (1995). The neuron doctrine in perception, *in* M. S. Gazzaniga (ed.), *The Cognitive Neurosciences*, MIT Press, Cambridge, MA, chapter 26, pp. 415–435.

Barlow, H. (1997). Single neurons, communal goals, and consciousness, *in* M. Ito, Y. Miyashita and E. T. Rolls (eds), *Cognition, Computation, and Consciousness*, Oxford University Press, Oxford, chapter 7, pp. 121–136.

Barlow, H. B. (1961). Possible principles underlying the transformation of sensory messages, *in* W. Rosenblith (ed.), *Sensory Communication*, MIT Press, Cambridge, MA.

Barlow, H. B. (1972). Single units and sensation: a neuron doctrine for perceptual psychology, *Perception* **1**: 371–394.

Barlow, H. B. (1985). Cerebral cortex as model builder, *in* D. Rose and V. G. Dobson (eds), *Models of the Visual Cortex*, Wiley, Chichester, pp. 37–46.

Barlow, H. B. (1989). Unsupervised learning, *Neural Computation* **1**: 295–311.

Barlow, H. B., Kaushal, T. P. and Mitchison, G. J. (1989). Finding minimum entropy codes, *Neural Computation* **1**: 412–423.

Barnes, C. A. (2003). Long-term potentiation and the ageing brain, *Philosophical Transactions of the Royal Society of London B* **358**: 765–772.

Barnes, D. C., Hofacer, R. D., Zaman, A. R., Rennaker, R. L. and Wilson, D. A. (2008). Olfactory perceptual stability and discrimination, *Nat Neurosci* **11**(12): 1378–80.

Barrett, H. (2012). A hierarchical model of the evolution of human brain specializations, *Proceedings of the National Academy of Sciences* **109**(Supplement 1): 10733–10740.

Bartlett, M. S. and Sejnowski, T. J. (1997). Viewpoint invariant face recognition using independent component analysis and attractor networks, *in* M. Mozer, M. Jordan and T. Petsche (eds), *Advances in Neural Information Processing Systems 9*, MIT Press, Cambridge, MA, pp. 817–823.

Barto, A. G. (1985). Learning by statistical cooperation of self-interested neuron-like computing elements, *Human Neurobiology* **4**: 229–256.

Barto, A. G. (1995). Adaptive critics and the basal ganglia, *in* J. C. Houk, J. L. Davis and D. G. Beiser (eds), *Models of Information Processing in the Basal Ganglia*, MIT Press, Cambridge, MA, chapter 11, pp. 215–232.

Bartol, T. M., Bromer, C., Kinney, J., Chirillo, M. A., Bourne, J. N., Harris, K. M. and Sejnowski, T. J. (2016). Nanoconnectomic upper bound on the variability of synaptic plasticity, *eLife* **4**: e10778.

Bartus, R. T. (2000). On neurodegenerative diseases, models, and treatment strategies: lessons learned and lessons forgotten a generation following the cholinergic hypothesis, *Experimental Neurology* **163**: 495–529.

Bassett, J. and Taube, J. S. (2005). Head direction signal generation: ascending and descending information streams, *in* S. I. Wiener and J. S. Taube (eds), *Head Direction Cells and the Neural Mechanisms of Spatial Orientation*, MIT Press, Cambridge, MA, chapter 5, pp. 83–109.

Battaglia, F. and Treves, A. (1998a). Stable and rapid recurrent processing in realistic autoassociative memories, *Neural Computation* **10**: 431–450.

Battaglia, F. P. and Treves, A. (1998b). Attractor neural networks storing multiple space representations: a model for hippocampal place fields, *Physical Review E* **58**: 7738–7753.

Baxter, M. G. and Murray, E. A. (2001a). Effects of hippocampal lesions on delayed nonmatching-to-sample in monkeys: a reply to Zola and Squire, *Hippocampus* **11**: 201–203.

Baxter, M. G. and Murray, E. A. (2001b). Opposite relationship of hippocampal and rhinal cortex damage to delayed nonmatching-to-sample deficits in monkeys, *Hippocampus* **11**: 61–71.

Baxter, R. D. and Liddle, P. F. (1998). Neuropsychological deficits associated with schizophrenic syndromes, *Schizophrenia Research* **30**: 239–249.

Baylis, G. C. and Rolls, E. T. (1987). Responses of neurons in the inferior temporal cortex in short term and serial recognition memory tasks, *Experimental Brain Research* **65**: 614–622.

Baylis, G. C., Rolls, E. T. and Leonard, C. M. (1985). Selectivity between faces in the responses of a population of neurons in the cortex in the superior temporal sulcus of the monkey, *Brain Research* **342**: 91–102.

Baylis, G. C., Rolls, E. T. and Leonard, C. M. (1987). Functional subdivisions of temporal lobe neocortex, *Journal of Neuroscience* **7**: 330–342.

Baylis, L. L. and Gaffan, D. (1991). Amygdalectomy and ventromedial prefrontal ablation produce similar deficits in food choice and in simple object discrimination learning for an unseen reward, *Experimental Brain Research* **86**: 617–622.

Baylis, L. L. and Rolls, E. T. (1991). Responses of neurons in the primate taste cortex to glutamate, *Physiology and Behavior* **49**: 973–979.

Baylis, L. L., Rolls, E. T. and Baylis, G. C. (1994). Afferent connections of the orbitofrontal cortex taste area of the primate, *Neuroscience* **64**: 801–812.

Bear, M. F. and Singer, W. (1986). Modulation of visual cortical plasticity by acetylcholine and noradrenaline, *Nature* **320**: 172–176.

Beck, A. T. (2008). The evolution of the cognitive model of depression and its neurobiological correlates, *American Journal of Psychiatry* **165**: 969–977.

Beck, D. M. and Kastner, S. (2009). Top-down and bottom-up mechanisms in biasing competition in the human brain, *Vision Research* **49**: 1154–1165.

Becker, S. and Hinton, G. E. (1992). Self-organizing neural network that discovers surfaces in random-dot stereograms, *Nature* **355**: 161–163.

Beckstead, R. M. and Norgren, R. (1979). An autoradiographic examination of the central distribution of the trigeminal, facial, glossopharyngeal, and vagal nerves in the monkey, *Journal of Comparative Neurology* **184**: 455–472.

Beckstead, R. M., Morse, J. R. and Norgren, R. (1980). The nucleus of the solitary tract in the monkey: projections to the thalamus and brainstem nuclei, *Journal of Comparative Neurology* **190**: 259–282.

Bell, A. J. and Sejnowski, T. J. (1995). An information-maximation approach to blind separation and blind deconvolution, *Neural Computation* **7**: 1129–1159.

Benda, J. and Herz, A. V. M. (2003). A universal model for spike-frequency adaptation, *Neural Computation* **15**: 2523–2564.

Bender, W., Albus, M., Moller, H. J. and Tretter, F. (2006). Towards systemic theories in biological psychiatry, *Pharmacopsychiatry* **39 Suppl 1**: S4–S9.

Bengio, Y., Goodfellow, I. J. and Courville, A. (2017). *Deep Learning*, MIT Press, Cambridge, MA.

Bennett, A. (1990). Large competitive networks, *Network* **1**: 449–462.

Berg, F. (1948). A simple objective technique for measuring flexibility in thinking, *Journal of General Psychology* **39**: 15–22.

Berlin, H., Rolls, E. T. and Kischka, U. (2004). Impulsivity, time perception, emotion, and reinforcement sensitivity in patients with orbitofrontal cortex lesions, *Brain* **127**: 1108–1126.

Berlin, H., Rolls, E. T. and Iversen, S. D. (2005). Borderline Personality Disorder, impulsivity, and the orbitofrontal cortex, *American Journal of Psychiatry* **58**: 234–245.

Bermpohl, F., Kahnt, T., Dalanay, U., Hagele, C., Sajonz, B., Wegner, T., Stoy, M., Adli, M., Kruger, S., Wrase, J., Strohle, A., Bauer, M. and Heinz, A. (2010). Altered representation of expected value in the orbitofrontal cortex in mania, *Human Brain Mapping* **31**: 958–969.

Bernard, A., Lubbers, L. S., Tanis, K. Q., Luo, R., Podtelezhnikov, A. A., Finney, E. M., McWhorter, M. M., Serikawa, K., Lemon, T., Morgan, R., Copeland, C., Smith, K., Cullen, V., Davis-Turak, J., Lee, C. K., Sunkin, S. M., Loboda, A. P., Levine, D. M., Stone, D. J., Hawrylycz, M. J., Roberts, C. J., Jones, A. R., Geschwind, D. H. and Lein, E. S. (2012). Transcriptional architecture of the primate neocortex, *Neuron* **73**(6): 1083–99.

Berridge, K. C., Flynn, F. W., Schulkin, J. and Grill, H. J. (1984). Sodium depletion enhances salt palatability in rats, *Behavioral Neuroscience* **98**: 652–660.

Bhattacharyya, S. and Chakraborty, K. (2007). Glutamatergic dysfunction–newer targets for anti-obsessional drugs, *Recent Patents CNS Drug Discovery* **2**: 47–55.

Bi, G.-Q. and Poo, M.-M. (1998). Activity-induced synaptic modifications in hippocampal culture, dependence on spike timing, synaptic strength and cell type, *Journal of Neuroscience* **18**: 10464–10472.

Bi, G.-Q. and Poo, M.-M. (2001). Synaptic modification by correlated activity: Hebb's postulate revisited, *Annual Review of Neuroscience* **24**: 139–166.

Bialek, W., Rieke, F., de Ruyter van Steveninck, R. R. and Warland, D. (1991). Reading a neural code, *Science*

252: 1854–1857.

Biederman, I. (1972). Perceiving real-world scenes, *Science* **177**: 77–80.

Biederman, I. (1987). Recognition-by-components: A theory of human image understanding, *Psychological Review* **94**(2): 115–147.

Bienenstock, E. L., Cooper, L. N. and Munro, P. W. (1982). Theory for the development of neuron selectivity: orientation specificity and binocular interaction in visual cortex, *Journal of Neuroscience* **2**: 32–48.

Bierer, L. M., Haroutunian, V., Gabriel, S., Knott, P. J., Carlin, L. S., Purohit, D. P., Perl, D. P., Schmeidler, J., Kanof, P. and Davis, K. L. (1995). Neurochemical correlates of dementia severity in Alzheimer's disease: relative importance of the cholinergic deficits, *Journal of Neurochemistry* **64**: 749–760.

Binford, T. O. (1981). Inferring surfaces from images, *Artificial Intelligence* **17**: 205–244.

Bishop, C. M. (1995). *Neural Networks for Pattern Recognition*, Clarendon Press, Oxford.

Bisley, J. W. and Goldberg, M. E. (2003). Neuronal activity in the lateral intraparietal area and spatial attention, *Science* **299**: 81–86.

Bisley, J. W. and Goldberg, M. E. (2006). Neural correlates of attention and distractibility in the lateral intraparietal area, *Journal of Neurophysiology* **95**: 1696–1717.

Bisley, J. W. and Goldberg, M. E. (2010). Attention, intention, and priority in the parietal lobe, *Annual Review of Neuroscience* **33**: 1–21.

Blair, R. J., Morris, J. S., Frith, C. D., Perrett, D. I. and Dolan, R. J. (1999). Dissociable neural responses to facial expressions of sadness and anger, *Brain* **122**: 883–893.

Blake, R. and Logothetis, N. K. (2002). Visual competition, *Nature Reviews Neuroscience* **3**: 13–21.

Blakemore, C. and van Sluyters, R. C. (1974). Reversal of the physiological effects of monocular deprivation in kittens: Further evidence for a sensitive period, *Journal of Physiology* **237**: 195–216.

Blakemore, C., Garey, L. J. and Vital-Durand, F. (1978). The physiological effects of monocular deprivation and their reversal in the monkeys visual cortex, *Journal of Physiology* **283**: 223–262.

Bliss, T. V. and Collingridge, G. L. (2013). Expression of NMDA receptor-dependent LTP in the hippocampus: bridging the divide, *Molecular Brain* **6**: 5.

Block, H. D. (1962). The perceptron: a model for brain functioning, *Reviews of Modern Physics* **34**: 123–135.

Block, N. (1995a). On a confusion about a function of consciousness, *Behavioral and Brain Sciences* **18**: 22–47.

Block, N. (1995b). Two neural correlates of consciousness, *Trends in Cognitive Sciences* **9**: 46–52.

Bloomfield, S. (1974). Arithmetical operations performed by nerve cells, *Brain Research* **69**: 115–124.

Bolles, R. C. and Cain, R. A. (1982). Recognizing and locating partially visible objects: The local-feature-focus method, *International Journal of Robotics Research* **1**: 57–82.

Bonelli, S. B., Powell, R. H., Yogarajah, M., Samson, R. S., Symms, M. R., Thompson, P. J., Koepp, M. J. and Duncan, J. S. (2010). Imaging memory in temporal lobe epilepsy: predicting the effects of temporal lobe resection, *Brain* **133**: 1186–1199.

Booth, D. A. (1985). Food-conditioned eating preferences and aversions with interoceptive elements: learned appetites and satieties, *Annals of the New York Academy of Sciences* **443**: 22–37.

Booth, M. C. A. and Rolls, E. T. (1998). View-invariant representations of familiar objects by neurons in the inferior temporal visual cortex, *Cerebral Cortex* **8**: 510–523.

Bornkessel-Schlesewsky, I., Schlesewsky, M., Small, S. L. and Rauschecker, J. P. (2015). Neurobiological roots of language in primate audition: common computational properties, *Trends in Cognitive Sciences* **19**: 142–150.

Borsini, F. and Rolls, E. T. (1984). Role of noradrenaline and serotonin in the basolateral region of the amygdala in food preferences and learned taste aversions in the rat, *Physiology and Behavior* **33**: 37–43.

Boussaoud, D., Desimone, R. and Ungerleider, L. G. (1991). Visual topography of area TEO in the macaque, *Journal of Computational Neurology* **306**: 554–575.

Bovier, A. and Gayrard, V. (1992). Rigorous bounds on the storage capacity of the dilute Hopfield model, *Journal of Statistical Physics* **69**: 597–627.

Brady, M., Ponce, J., Yuille, A. and Asada, H. (1985). Describing surfaces, *A. I. Memo 882, The Artificial Intelligence* **17**: 285–349.

Braitenberg, V. and Schütz, A. (1991). *Anatomy of the Cortex*, Springer-Verlag, Berlin.

Braitenberg, V. and Schütz, A. (1998). *Cortex: Statistics and Geometry of Neuronal Connectivity*, Springer-Verlag, Berlin.

Bressler, S. L. and Seth, A. K. (2011). Wiener-Granger causality: A well established methodology, *Neuroimage* **58**: 323–329.

Bridle, J. S. (1990). Probabilistic interpretation of feedforward classification network outputs, with relationships to statistical pattern recognition, *in* F. Fogelman-Soulie and J. Herault (eds), *Neurocomputing: Algorithms, Architectures and Applications*, Springer-Verlag, New York, pp. 227–236.

Brodmann, K. (1925). *Vergleichende Localisationslehre der Grosshirnrinde. Translated into English by L.J.Garey as Localisation in the Cerebral Cortex 1994 London: Smith-Gordon*, 2 edn, Barth, Leipzig.

Bromberg-Martin, E. S., Matsumoto, M. and Hikosaka, O. (2010a). Dopamine in motivational control: rewarding, aversive, and alerting, *Neuron* **68**: 815–834.

Bromberg-Martin, E. S., Matsumoto, M., Hong, S. and Hikosaka, O. (2010b). A pallidus-habenula-dopamine pathway signals inferred stimulus values, *Journal of Neurophysiology* **104**: 1068–1076.

Brooks, L. R. (1978). Nonanalytic concept formation and memory for instances, *in* E. Rosch and B. B. Lloyd (eds), *Cognition and Categorization*, Erlbaum, Hillsdale, NJ.

Brotchie, P., Andersen, R., Snyder, L. and Goodman, S. (1995). Head position signals used by parietal neurons to encode locations of visual stimuli, *Nature London* **375**: 232–235.

Brown, D. A., Gähwiler, B. H., Griffith, W. H. and Halliwell, J. V. (1990a). Membrane currents in hippocampal neurons, *Progress in Brain Research* **83**: 141–160.

Brown, J. M. (2009). Visual streams and shifting attention, *Prog Brain Res* **176**: 47–63.

Brown, M. and Xiang, J. (1998). Recognition memory: neuronal substrates of the judgement of prior occurrence, *Progress in Neurobiology* **55**: 149–189.

Brown, T. H. and Zador, A. (1990). The hippocampus, *in* G. Shepherd (ed.), *The Synaptic Organization of the Brain*, Oxford University Press, New York, pp. 346–388.

Brown, T. H., Kairiss, E. W. and Keenan, C. L. (1990b). Hebbian synapses: biophysical mechanisms and algorithms, *Annual Review of Neuroscience* **13**: 475–511.

Brown, T. H., Ganong, A. H., Kairiss, E. W., Keenan, C. L. and Kelso, S. R. (eds) (1989). *Long-term Potentiation in Two Synaptic Systems of the Hippocampal Brain Slice*, Academic Press, San Diego.

Bruce, V. (1988). *Recognising Faces*, Erlbaum, Hillsdale, NJ.

Brun, V. H., Otnass, M. K., Molden, S., Steffenach, H. A., Witter, M. P., Moser, M. B. and Moser, E. I. (2002). Place cells and place recognition maintained by direct entorhinal–hippocampal circuitry, *Science* **296**: 2243–2246.

Brunel, N. and Wang, X. J. (2001). Effects of neuromodulation in a cortical network model of object working memory dominated by recurrent inhibition, *Journal of Computational Neuroscience* **11**: 63–85.

Buck, L. and Bargmann, C. I. (2013). Smell and taste: the chemical senses, *in* E. Kandel, J. H. Schwartz, T. H. Jessell, S. A. Siegelbaum and A. J. Hudspeth (eds), *Principles of Neural Science*, 5th edn, McGraw-Hill, New York, chapter 32, pp. 712–742.

Buckley, M. J. and Gaffan, D. (2000). The hippocampus, perirhinal cortex, and memory in the monkey, *in* J. J. Bolhuis (ed.), *Brain, Perception, and Memory: Advances in Cognitive Neuroscience*, Oxford University Press, Oxford, pp. 279–298.

Buckley, M. J. and Gaffan, D. (2006). Perirhinal contributions to object perception, *Trends in Cognitive Sciences* **10**: 100–107.

Buckley, M. J., Booth, M. C. A., Rolls, E. T. and Gaffan, D. (2001). Selective perceptual impairments following perirhinal cortex ablation, *Journal of Neuroscience* **21**: 9824–9836.

Buhl, E. H., Halasy, K. and Somogyi, P. (1994). Diverse sources of hippocampal unitary inhibitory postsynaptic potentials and the number of synaptic release sites, *Nature* **368**: 823–828.

Buhmann, J., Lange, J., von der Malsburg, C., Vorbrüggen, J. C. and Würtz, R. P. (1991). Object recognition in the dynamic link architecture: Parallel implementation of a transputer network, *in* B. Kosko (ed.), *Neural Networks for Signal Processing*, Prentice Hall, Englewood Cliffs, NJ, pp. 121–159.

Bullier, J. and Nowak, L. (1995). Parallel versus serial processing: new vistas on the distributed organization of the visual system, *Current Opinion in Neurobiology* **5**: 497–503.

Bunsey, M. and Eichenbaum, H. (1996). Conservation of hippocampal memory function in rats and humans, *Nature* **379**: 255–257.

Buot, A. and Yelnik, J. (2012). Functional anatomy of the basal ganglia: limbic aspects, *Revue Neurologique (Paris)* **168**: 569–575.

Burgess, N. (2008). Spatial cognition and the brain, *Annals of the New York Academy of Sciences* **1124**: 77–97.

Burgess, N., Recce, M. and O'Keefe, J. (1994). A model of hippocampal function, *Neural Networks* **7**: 1065–1081.

Burgess, N., Jackson, A., Hartley, T. and O'Keefe, J. (2000). Predictions derived from modelling the hippocampal role in navigation, *Biological Cybernetics* **83**: 301–312.

Burgess, N., Maguire, E. A. and O'Keefe, J. (2002). The human hippocampus and spatial and episodic memory, *Neuron* **35**: 625–641.

Burgess, P. W. (2000). Strategy application disorder: the role of the frontal lobes in human multitasking, *Psychological Research* **63**: 279–288.

Burke, S. N. and Barnes, C. A. (2006). Neural plasticity in the ageing brain, *Nature Reviews Neuroscience* **7**: 30–40.

Burton, M. J., Rolls, E. T. and Mora, F. (1976). Effects of hunger on the responses of neurones in the lateral hypothalamus to the sight and taste of food, *Experimental Neurology* **51**: 668–677.

Burwell, R. D., Witter, M. P. and Amaral, D. G. (1995). Perirhinal and postrhinal cortices of the rat: a review of the neuroanatomical literature and comparison with findings from the monkey brain, *Hippocampus* **5**: 390–408.

Bussey, T. J. and Saksida, L. M. (2005). Object memory and perception in the medial temporal lobe: an alternative approach, *Current Opinion in Neurobiology* **15**: 730–737.

Bussey, T. J., Saksida, L. M. and Murray, E. A. (2002). Perirhinal cortex resolves feature ambiguity in complex visual discriminations, *European Journal of Neuroscience* **15**: 365–374.

Bussey, T. J., Saksida, L. M. and Murray, E. A. (2003). Impairments in visual discrimination after perirhinal cortex lesions: testing "declarative" versus "perceptual-mnemonic" views of perirhinal cortex function, *European Journal of Neuroscience* **17**: 649–660.

Bussey, T. J., Saksida, L. M. and Murray, E. A. (2005). The perceptual-mnemonic / feature conjunction model of perirhinal cortex function, *Quarterly Journal of Experimental Psychology* **58B**: 269–282.

Butter, C. M. (1969). Perseveration in extinction and in discrimination reversal tasks following selective prefrontal ablations in Macaca mulatta, *Physiology and Behavior* **4**: 163–171.

Butter, C. M., McDonald, J. A. and Snyder, D. R. (1969). Orality, preference behavior, and reinforcement value of non-food objects in monkeys with orbital frontal lesions, *Science* **164**: 1306–1307.

Buxton, R. B. and Frank, L. R. (1997). A model for the coupling between cerebral blood flow and oxygen metabolism during neural stimulation, *Journal of Cerebral Blood Flow and Metabolism* **17**: 64–72.

Buxton, R. B., Wong, E. C. and Frank, L. R. (1998). Dynamics of blood flow and oxygenation changes during brain activation: the balloon model, *Magnetic Resonance in Medicine* **39**: 855–864.

Buzsáki, G. (2006). *Rhythms of the Brain*, Oxford University Press, Oxford.

Byrne, R. W. and Whiten, A. (1988). *Machiavellian Intelligence: Social Expertise and the Evolution of Intellect in Monkeys, Apes and Humans*, Clarendon Press, Oxford.

Caan, W., Perrett, D. I. and Rolls, E. T. (1984). Responses of striatal neurons in the behaving monkey. 2. Visual processing in the caudal neostriatum, *Brain Research* **290**: 53–65.

Cadieu, C. F., Hong, H., Yamins, D. L. K., Pinto, N., Ardila, D., Solomon, E. A., Majaj, N. J. and DiCarlo, J. J. (2014). Deep neural networks rival the representation of primate IT cortex for core visual object recognition, *PLoS Computational Biology* **10**: e1003963.

Cahusac, P. M. B., Miyashita, Y. and Rolls, E. T. (1989). Responses of hippocampal formation neurons in the monkey related to delayed spatial response and object-place memory tasks, *Behavioural Brain Research* **33**: 229–240.

Cahusac, P. M. B., Rolls, E. T., Miyashita, Y. and Niki, H. (1993). Modification of the responses of hippocampal neurons in the monkey during the learning of a conditional spatial response task, *Hippocampus* **3**: 29–42.

Calvert, G. A., Bullmore, E. T., Brammer, M. J., Campbell, R., Williams, S. C. R., McGuire, P. K., Woodruff, P. W. R., Iversen, S. D. and David, A. S. (1997). Activation of auditory cortex during silent lip-reading, *Science* **276**: 593–596.

Camille, N., Tsuchida, A. and Fellows, L. K. (2011). Double dissociation of stimulus-value and action-value learning in humans with orbitofrontal or anterior cingulate cortex damage, *Journal of Neuroscience* **31**: 15048–15052.

Capuano, B., Crosby, I. T. and Lloyd, E. J. (2002). Schizophrenia: genesis, receptorology and current therapeutics, *Current Medicinal Chemistry* **9**: 521–548.

Cardinal, R. N., Parkinson, J. A., Hall, J. and Everitt, B. J. (2002). Emotion and motivation: the role of the amygdala, ventral striatum, and prefrontal cortex, *Neuroscience and Biobehavioral Reviews* **26**. 321–352.

Carlson, E. T., Rasquinha, R. J., Zhang, K. and Connor, C. E. (2011). A sparse object coding scheme in area v4, *Curr Biol* **21**(4): 288–93.

Carlson, N. R. and Birkett, M. A. (2017). *Physiology of Behavior*, 12th edn, Pearson, Boston.

Carlsson, A. (2006). The neurochemical circuitry of schizophrenia, *Pharmacopsychiatry* **39 Suppl 1**: S10–14.

Carmichael, S. T. and Price, J. L. (1994). Architectonic subdivision of the orbital and medial prefrontal cortex in the macaque monkey, *Journal of Comparative Neurology* **346**: 366–402.

Carmichael, S. T. and Price, J. L. (1995a). Limbic connections of the orbital and medial prefrontal cortex in macaque monkeys, *Journal of Comparative Neurology* **363**: 615–641.

Carmichael, S. T. and Price, J. L. (1995b). Sensory and premotor connections of the orbital and medial prefrontal cortex of macaque monkeys, *Journal of Comparative Neurology* **363**: 642–664.

Carmichael, S. T., Clugnet, M.-C. and Price, J. L. (1994). Central olfactory connections in the macaque monkey, *Journal of Comparative Neurology* **346**: 403–434.

Carpenter, G. A. (1997). Distributed learning, recognition and prediction by ART and ARTMAP neural networks, *Neural Networks* **10**(8): 1473–1494.

Carpenter, R. H. S. and Williams, M. (1995). Neural computation of log likelihood in control of saccadic eye movements, *Nature* **377**: 59–62.

Carruthers, P. (1996). *Language, Thought and Consciousness*, Cambridge University Press, Cambridge.

Carruthers, P. (2000). *Phenomenal Consciousness*, Cambridge University Press, Cambridge.

Carter, C. S., Perlstein, W., Ganguli, R., Brar, J., Mintun, M. and Cohen, J. D. (1998). Functional hypofrontality and working memory dysfunction in schizophrenia, *American Journal of Psychiatry* **155**: 1285–1287.

Caspers, J., Palomero-Gallagher, N., Caspers, S., Schleicher, A., Amunts, K. and Zilles, K. (2015). Receptor architecture of visual areas in the face and word-form recognition region of the posterior fusiform gyrus, *Brain Struct Funct* **220**(1): 205–19.

Cassaday, H. J. and Rawlins, J. N. (1997). The hippocampus, objects, and their contexts, *Behavioral Neuroscience* **111**: 1228–1244.

Castner, S. A., Williams, G. V. and Goldman-Rakic, P. S. (2000). Reversal of antipsychotic-induced working memory deficits by short-term dopamine D1 receptor stimulation, *Science* **287**: 2020–2022.

Cavanna, A. E. and Trimble, M. R. (2006). The precuneus: a review of its functional anatomy and behavioural correlates, *Brain* **129**: 564–583.

Celebrini, S., Thorpe, S., Trotter, Y. and Imbert, M. (1993). Dynamics of orientation coding in area V1 of the awake primate, *Visual Neuroscience* **10**: 811–825.

Cerasti, E. and Treves, A. (2010). How informative are spatial CA3 representations established by the dentate gyrus?, *PLoS Computational Biology* **6**: e1000759.

Cerasti, E. and Treves, A. (2013). The spatial representations acquired in CA3 by self-organizing recurrent connections, *Frontiers in Cellular Neuroscience* **7**: 112.

Cerella, J. (1986). Pigeons and perceptrons, *Pattern Recognition* **19**: 431–438.

Chadwick, M. J., Mullally, S. L. and Maguire, E. A. (2013). The hippocampus extrapolates beyond the view in scenes: an fmri study of boundary extension, *Cortex* **49**: 2067–2079.

Chakrabarty, K., Bhattacharyya, S., Christopher, R. and Khanna, S. (2005). Glutamatergic dysfunction in OCD, *Neuropsychopharmacology* **30**: 1735–1740.

Chakravarty, I. (1979). A generalized line and junction labeling scheme with applications to scene analysis, *IEEE Transactions PAMI* **1**: 202–205.

Chalmers, D. J. (1996). *The Conscious Mind*, Oxford University Press, Oxford.

Chamberlain, S. R., Fineberg, N. A., Blackwell, A. D., Robbins, T. W. and Sahakian, B. J. (2006). Motor inhibition and cognitive flexibility in obsessive-compulsive disorder and trichotillomania, *American Journal of Psychiatry* **163**: 1282–1284.

Chamberlain, S. R., Fineberg, N. A., Menzies, L. A., Blackwell, A. D., Bullmore, E. T., Robbins, T. W. and Sahakian, B. J. (2007). Impaired cognitive flexibility and motor inhibition in unaffected first-degree relatives of patients with obsessive-compulsive disorder, *American Journal of Psychiatry* **164**: 335–338.

Chan, A. M., Dykstra, A. R., Jayaram, V., Leonard, M. K., Travis, K. E., Gygi, B., Baker, J. M., Eskandar, E., Hochberg, L. R., Halgren, E. and Cash, S. S. (2014). Speech-specific tuning of neurons in human superior temporal gyrus, *Cerebral Cortex* **24**: 2679–2693.

Chang, L. and Tsao, D. Y. (2017). The code for facial identity in the primate brain, *Cell* **169**(6): 1013–1028 e14.

Chau, B. K., Sallet, J., Papageorgiou, G. K., Noonan, M. P., Bell, A. H., Walton, M. E. and Rushworth, M. F. (2015). Contrasting roles for orbitofrontal cortex and amygdala in credit assignment and learning in macaques, *Neuron* **87**: 1106–1118.

Chelazzi, L. (1998). Serial attention mechanisms in visual search: a critical look at the evidence, *Psychological Research* **62**: 195–219.

Chelazzi, L., Miller, E., Duncan, J. and Desimone, R. (1993). A neural basis for visual search in inferior temporal cortex, *Nature (London)* **363**: 345–347.

Cheney, D. L. and Seyfarth, R. M. (1990). *How Monkeys See the World*, University of Chicago Press, Chicago.

Cheng, W., Rolls, E. T., Gu, H., Zhang, J. and Feng, J. (2015). Autism: reduced connectivity between cortical areas involved in face expression, theory of mind, and the sense of self, *Brain* **138**: 1382–1398.

Cheng, W., Rolls, E. T., Qiu, J., Liu, W., Tang, Y., Huang, C.-C., Wang, X., Zhang, J., Lin, W., Zheng, L., Pu, J., Tsai, S.-J., Yang, A. C., Lin, C.-P., Wang, F., Xie, P. and Feng, J. (2016). Medial reward and lateral non-reward orbitofrontal cortex circuits change in opposite directions in depression, *Brain* **139**: 3296–3309.

Chiba, A. A., Kesner, R. P. and Reynolds, A. M. (1994). Memory for spatial location as a function of temporal lag in rats: role of hippocampus and medial prefrontal cortex, *Behavioral and Neural Biology* **61**: 123–131.

Cho, Y. H. and Kesner, R. P. (1995). Relational object association learning in rats with hippocampal lesions, *Behavioural Brain Research* **67**: 91–98.

Cho, Y. K., Li, C. S. and Smith, D. V. (2002). Gustatory projections from the nucleus of the solitary tract to the parabrachial nuclei in the hamster, *Chemical Senses* **27**: 81–90.

Chomsky, N. (1965). *Aspects of the Theory of Syntax*, MIT Press, Cambridge, Massachusetts.

Christie, B. R. (1996). Long-term depression in the hippocampus, *Hippocampus* **6**: 1–2.

Chu, J. and Anderson, S. A. (2015). Development of cortical interneurons, *Neuropsychopharmacology* **40**(1): 16–23.

Churchland, A. K., Kiani, R. and Shadlen, M. N. (2008). Decision-making with multiple alternatives, *Nature Neuroscience* **11**: 693–702.

Cicero, M. T. (55 BC). *De Oratore II*, Vol. II, Cicero, Rome.

Clelland, C. D., Choi, M., Romberg, C., Clemenson, G. D., J., Fragniere, A., Tyers, P., Jessberger, S., Saksida, L. M., Barker, R. A., Gage, F. H. and Bussey, T. J. (2009). A functional role for adult hippocampal neurogenesis in spatial pattern separation, *Science* **325**: 210–213.

Colby, C. L., Duhamel, J. R. and Goldberg, M. E. (1993). Ventral intraparietal area of the macaque - anatomic location and visual response properties, *Journal of Neurophysiology* **69**: 902–914.

Collingridge, G. L. and Bliss, T. V. P. (1987). NMDA receptors: their role in long-term potentiation, *Trends in Neurosciences* **10**: 288–293.

Compte, A., Brunel, N., Goldman-Rakic, P. and Wang, X. J. (2000). Synaptic mechanisms and network dynamics underlying spatial working memory in a cortical network model, *Cerebral Cortex* **10**: 910–923.

Connor, C. E., Gallant, J. L., Preddie, D. and Van Essen, D. (1996). Responses in area V4 depend on the spatial relationship between stimulus and attention, *Journal of Neurophysiology* **75**: 1306–1308.

Cooper, J. R., Bloom, F. E. and Roth, R. H. (2003). *The Biochemical Basis of Neuropharmacology*, 8th edn, Oxford University Press, Oxford.

Corchs, S. and Deco, G. (2002). Large-scale neural model for visual attention: integration of experimental single cell and fMRI data, *Cerebral Cortex* **12**: 339–348.

Corr, P. J. and McNaughton, N. (2012). Neuroscience and approach/avoidance personality traits: a two stage (valuation-motivation) approach, *Neuroscience and Biobehavioural Reviews* **36**: 2339–2254.

Corrado, G. S., Sugrue, L. P., Seung, H. S. and Newsome, W. T. (2005). Linear-nonlinear-Poisson models of primate choice dynamics, *Journal of the Experimental Analysis of Behavior* **84**: 581–617.

Cortes, C., Jaeckel, L. D., Solla, S. A., Vapnik, V. and Denker, J. S. (1996). Learning curves: asymptotic values and rates of convergence, *Neural Information Processing Systems* **6**: 327–334.

Cover, T. M. (1965). Geometrical and statistical properties of systems of linear inequalities with applications in pattern recognition, *IEEE Transactions on Electronic Computers* **14**: 326–334.

Cover, T. M. and Thomas, J. A. (1991). *Elements of Information Theory*, Wiley, New York.

Cowell, R. A., Bussey, T. J. and Saksida, L. M. (2006). Why does brain damage impair memory? A connectionist model of object recognition memory in perirhinal cortex, *Journal of Neuroscience* **26**: 12186–12197.

Cowey, A. (1979). Cortical maps and visual perception, *Quarterly Journal of Experimental Psychology* **31**: 1–17.

Coyle, J. T. (2006). Glutamate and schizophrenia: beyond the dopamine hypothesis, *Cellular and Molecular Neurobiology* **26**: 365–384.

Coyle, J. T. (2012). Nmda receptor and schizophrenia: A brief history, *Schizophr Bull* **38**(5): 920–6.

Crane, J. and Milner, B. (2005). What went where? Impaired object-location learning in patients with right hippocampal lesions, *Hippocampus* **15**: 216–231.

Creutzfeldt, O. D. (1995). *Cortex Cerebri. Performance, Structural and Functional Organisation of the Cortex*, Oxford University Press, Oxford.

Crick, F. H. C. (1984). Function of the thalamic reticular complex: the searchlight hypothesis, *Proceedings of the National Academy of Sciences USA* **81**: 4586–4590.

Crick, F. H. C. and Koch, C. (1990). Towards a neurobiological theory of consciousness, *Seminars in the Neurosciences* **2**: 263–275.

Crick, F. H. C. and Mitchison, G. (1995). REM sleep and neural nets, *Behavioural Brain Research* **69**: 147–155.

Critchley, H. D. and Harrison, N. A. (2013). Visceral influences on brain and behavior, *Neuron* **77**: 624–638.

Critchley, H. D. and Rolls, E. T. (1996a). Hunger and satiety modify the responses of olfactory and visual neurons in the primate orbitofrontal cortex, *Journal of Neurophysiology* **75**: 1673–1686.

Critchley, H. D. and Rolls, E. T. (1996b). Olfactory neuronal responses in the primate orbitofrontal cortex: analysis in an olfactory discrimination task, *Journal of Neurophysiology* **75**: 1659–1672.

Critchley, H. D. and Rolls, E. T. (1996c). Responses of primate taste cortex neurons to the astringent tastant tannic acid, *Chemical Senses* **21**: 135–145.

da Costa, N. M. and Martin, K. A. (2010). Whose cortical column would that be?, *Frontiers in Neuroanatomy* **4**: 16.

da Costa, N. M. and Martin, K. A. C. (2013). Sparse reconstruction of brain circuits: or, how to survive without a microscopic connectome, *Neuroimage* **80**: 27–36.

Damasio, A. R. (1994). *Descartes' Error: Emotion, Reason, and the Human Brain*, Grosset/Putnam, New York.

Damasio, A. R. (2003). *Looking for Spinoza*, Heinemann, London.

Dan, Y. and Poo, M.-M. (2004). Spike-timing dependent plasticity of neural circuits, *Neuron* **44**: 23–30.

Dan, Y. and Poo, M.-M. (2006). Spike-timing dependent plasticity: from synapse to perception, *Physiological Reviews* **86**: 1033–1048.

Dane, C. and Bajcsy, R. (1982). An object-centred three-dimensional model builder, *Proceedings of the 6th International Conference on Pattern Recognition*, Munich, pp. 348–350.

Darwin, C. (1859). *The Origin of Species*, John Murray [reprinted (1982) by Penguin Books Ltd], London.

Daugman, J. (1988). Complete discrete 2D-Gabor transforms by neural networks for image analysis and compression, *IEEE Transactions on Acoustic, Speech, and Signal Processing* **36**: 1169–1179.

Davies, M. (2008). Consciousness and explanation, *in* L. Weiskrantz and M. Davies (eds), *Frontiers of Consciousness*, Oxford University Press, Oxford, chapter 1, pp. 1–54.

Davis, M. (1992). The role of the amygdala in conditioned fear, *in* J. P. Aggleton (ed.), *The Amygdala*, Wiley-Liss, New York, chapter 9, pp. 255–306.

Davis, M. (2000). The role of the amygdala in conditioned and unconditioned fear and anxiety, *in* J. P. Aggleton (ed.), *The Amygdala: a Functional Analysis*, 2nd edn, Oxford University Press, Oxford, chapter 6, pp. 213–287.

Davis, M. (2011). NMDA receptors and fear extinction: implications for cognitive behavioral therapy, *Dialogues in Clinical Neuroscience* **13**: 463–474.

Dawkins, M. S. (1986). *Unravelling Animal Behaviour*, 1st edn, Longman, Harlow.

Dawkins, M. S. (1993). *Through Our Eyes Only? The Search for Animal Consciousness*, Freeman, Oxford.

Dawkins, M. S. (1995). *Unravelling Animal Behaviour*, 2nd edn, Longman, Harlow.

Dawkins, R. (1976). *The Selfish Gene*, Oxford University Press, Oxford.

Dawkins, R. (1989). *The Selfish Gene*, 2nd edn, Oxford University Press, Oxford.

Day, M., Langston, R. and Morris, R. G. (2003). Glutamate-receptor-mediated encoding and retrieval of paired-associate learning, *Nature* **424**: 205–209.

Dayan, P. and Abbott, L. F. (2001). *Theoretical Neuroscience*, MIT Press, Cambridge, MA.

Dayan, P. and Sejnowski, T. J. (1994). TD(λ) converges with probability 1, *Machine Learning* **14**: 295–301.

De Araujo, I. E. T. and Rolls, E. T. (2004). Representation in the human brain of food texture and oral fat, *Journal of Neuroscience* **24**: 3086–3093.

De Araujo, I. E. T., Rolls, E. T. and Stringer, S. M. (2001). A view model which accounts for the response properties of hippocampal primate spatial view cells and rat place cells, *Hippocampus* **11**: 699–706.

De Araujo, I. E. T., Kringelbach, M. L., Rolls, E. T. and Hobden, P. (2003a). Representation of umami taste in the human brain, *Journal of Neurophysiology* **90**: 313–319.

De Araujo, I. E. T., Kringelbach, M. L., Rolls, E. T. and McGlone, F. (2003b). Human cortical responses to water in the mouth, and the effects of thirst, *Journal of Neurophysiology* **90**: 1865–1876.

De Araujo, I. E. T., Rolls, E. T., Kringelbach, M. L., McGlone, F. and Phillips, N. (2003c). Taste-olfactory convergence, and the representation of the pleasantness of flavour in the human brain, *European Journal of Neuroscience* **18**: 2059–2068.

De Araujo, I. E. T., Rolls, E. T., Velazco, M. I., Margot, C. and Cayeux, I. (2005). Cognitive modulation of olfactory processing, *Neuron* **46**: 671–679.

De Gelder, B., Vroomen, J., Pourtois, G. and Weiskrantz, L. (1999). Non-conscious recognition of affect in the absence of striate cortex, *NeuroReport* **10**: 3759–3763.

de Lafuente, V., Jazayeri, M. and Shadlen, M. N. (2015). Representation of accumulating evidence for a decision in two parietal areas, *Journal of Neuroscience* **35**: 4306–4318.

de Ruyter van Steveninck, R. R. and Laughlin, S. B. (1996). The rates of information transfer at graded-potential synapses, *Nature* **379**: 642–645.

De Sieno, D. (1988). Adding a conscience to competitive learning, *IEEE International Conference on Neural Networks (San Diego 1988)*, Vol. 1, IEEE, New York, pp. 117–124.

De Valois, R. L. and De Valois, K. K. (1988). *Spatial Vision*, Oxford University Press, New York.

DeAngelis, G. C., Cumming, B. G. and Newsome, W. T. (2000). A new role for cortical area MT: the perception of stereoscopic depth, *in* M. Gazzaniga (ed.), *The New Cognitive Neurosciences, Second Edition*, MIT Press, Cambridge, MA, chapter 21, pp. 305–314.

Debiec, J., LeDoux, J. E. and Nader, K. (2002). Cellular and systems reconsolidation in the hippocampus, *Neuron* **36**: 527–538.

Debiec, J., Doyere, V., Nader, K. and LeDoux, J. E. (2006). Directly reactivated, but not indirectly reactivated, memories undergo reconsolidation in the amygdala, *Proceedings of the National Academy of Sciences USA* **103**: 3428–3433.

deCharms, R. C. and Merzenich, M. M. (1996). Primary cortical representation of sounds by the coordination of action-potential timing, *Nature* **381**: 610–613.

Deco, G. (2001). Biased competition mechanisms for visual attention, *in* S. Wermter, J. Austin and D. Willshaw (eds), *Emergent Neural Computational Architectures Based on Neuroscience*, Springer, Heidelberg, pp. 114–126.

Deco, G. and Lee, T. S. (2002). A unified model of spatial and object attention based on inter-cortical biased competition, *Neurocomputing* **44–46**: 775–781.

Deco, G. and Lee, T. S. (2004). The role of early visual cortex in visual integration: a neural model of recurrent interaction, *European Journal of Neuroscience* **20**: 1089–1100.

Deco, G. and Rolls, E. T. (2002). Object-based visual neglect: a computational hypothesis, *European Journal of Neuroscience* **16**: 1994–2000.

Deco, G. and Rolls, E. T. (2003). Attention and working memory: a dynamical model of neuronal activity in the prefrontal cortex, *European Journal of Neuroscience* **18**: 2374–2390.

Deco, G. and Rolls, E. T. (2004). A neurodynamical cortical model of visual attention and invariant object recognition, *Vision Research* **44**: 621–644.

Deco, G. and Rolls, E. T. (2005a). Attention, short term memory, and action selection: a unifying theory, *Progress in Neurobiology* **76**: 236–256.

Deco, G. and Rolls, E. T. (2005b). Neurodynamics of biased competition and cooperation for attention: a model with spiking neurons, *Journal of Neurophysiology* **94**: 295–313.

Deco, G. and Rolls, E. T. (2005c). Sequential memory: a putative neural and synaptic dynamical mechanism, *Journal of Cognitive Neuroscience* **17**: 294–307.

Deco, G. and Rolls, E. T. (2005d). Synaptic and spiking dynamics underlying reward reversal in the orbitofrontal cortex, *Cerebral Cortex* **15**: 15–30.

Deco, G. and Rolls, E. T. (2006). A neurophysiological model of decision-making and Weber's law, *European Journal of Neuroscience* **24**: 901–916.

Deco, G. and Zihl, J. (2001a). A neurodynamical model of visual attention: Feedback enhancement of spatial resolution in a hierarchical system, *Journal of Computational Neuroscience* **10**: 231–253.

Deco, G. and Zihl, J. (2001b). Top-down selective visual attention: a neurodynamical approach, *Visual Cognition* **8**: 119–140.

Deco, G. and Zihl, J. (2004). A biased competition based neurodynamical model of visual neglect, *Medical Engineering and Physics* **26**: 733–743.

Deco, G., Pollatos, O. and Zihl, J. (2002). The time course of selective visual attention: theory and experiments, *Vision Research* **42**: 2925–2945.

Deco, G., Rolls, E. T. and Horwitz, B. (2004). 'What' and 'where' in visual working memory: a computational neurodynamical perspective for integrating fMRI and single-neuron data, *Journal of Cognitive Neuroscience* **16**: 683–701.

Deco, G., Scarano, L. and Soto-Faraco, S. (2007). Weber's law in decision making: integrating behavioral data in humans with a neurophysiological model, *Journal of Neuroscience* **27**: 11192–11200.

Deco, G., Rolls, E. T. and Romo, R. (2009). Stochastic dynamics as a principle of brain function, *Progress in Neurobiology* **88**: 1–16.

Deco, G., Rolls, E. T. and Romo, R. (2010). Synaptic dynamics and decision-making, *Proceedings of the National Academy of Sciences* **107**: 7545–7549.

Deco, G., Rolls, E. T., Albantakis, L. and Romo, R. (2013). Brain mechanisms for perceptual and reward-related decision-making, *Progress in Neurobiology* **103**: 194–213.

Deen, B., Koldewyn, K., Kanwisher, N. and Saxe, R. (2015). Functional organization of social perception and cognition in the superior temporal sulcus, *Cereb Cortex* **25**(11): 4596–609.

Dehaene, S. and Naccache, L. (2001). Towards a cognitive neuroscience of consciousness: basic evidence and a workspace framework, *Cognition* **79**: 1–37.

Dehaene, S., Changeux, J. P., Naccache, L., Sackur, J. and Sergent, C. (2006). Conscious, preconscious, and subliminal processing: a testable taxonomy, *Trends in Cognitive Sciences* **10**: 204–211.

Dehaene, S., Charles, L., King, J. R. and Marti, S. (2014). Toward a computational theory of conscious processing, *Curr Opin Neurobiol* **25**: 76–84.

Delatour, B. and Witter, M. P. (2002). Projections from the parahippocampal region to the prefrontal cortex in the rat: evidence of multiple pathways, *European Journal of Neuroscience* **15**: 1400–1407.

DeLong, M. and Wichmann, T. (2010). Changing views of basal ganglia circuits and circuit disorders, *Clinical EEG and Neuroscience* **41**: 61–67.

DeLong, M. R., Georgopoulos, A. P., Crutcher, M. D., Mitchell, S. J., Richardson, R. T. and Alexander, G. E. (1984). Functional organization of the basal ganglia: Contributions of single-cell recording studies, *Functions of the Basal Ganglia. CIBA Foundation Symposium*, Pitman, London, pp. 64–78.

Delorme, A. and Thorpe, S. J. (2001). Face identification using one spike per neuron: resistance to image degradations, *Neural Networks* **14**: 795–803.

Delorme, R., Gousse, V., Roy, I., Trandafir, A., Mathieu, F., Mouren-Simeoni, M. C., Betancur, C. and Leboyer, M. (2007). Shared executive dysfunctions in unaffected relatives of patients with autism and obsessive-compulsive disorder, *European Psychiatry* **22**: 32–38.

Deng, W. L., Rolls, E. T., Ji, X., Robbins, T. W., Banaschewski, T., Bokde, A., Bromberg, U., Buechel, C., Desrivieres, S., Conrod, P., Flor, H., Frouin, V., Gallinat, J., Garavan, H., Gowland, P., Heinz, A., Ittermann, B., Martinot, J.-L., Lemaitre, H., Nees, F., Papadopoulos Orfanos, D., Poustka, L., Smolka, M. N., Walter, H., Whelan, R., Schumann, G., Feng, J. and the Imagen consortium (2017). Separate neural systems for behavioral change and for emotional responses to failure during behavioral inhibition, *Human Brain Mapping* p. doi: 10.1002/hbm.23607.

Dennett, D. C. (1987). *The Intentional Stance*, MIT Press, Cambridge, MA.

Dennett, D. C. (1991). *Consciousness Explained*, Penguin, London.

Dere, E., Easton, A., Nadel, L. and Huston, J. P. (eds) (2008). *Handbook of Episodic Memory*, Elsevier, Amsterdam.

Desimone, R. and Duncan, J. (1995). Neural mechanisms of selective visual attention, *Annual Review of Neuroscience* **18**: 193–222.

Desimone, R., Albright, T. D., Gross, C. G. and Bruce, C. (1984). Stimulus-selctive responses of inferior temporal neurons in the macaque, *Journal of Neuroscience* **4**: 2051–2062.

DeWeese, M. R. and Meister, M. (1999). How to measure the information gained from one symbol, *Network* **10**: 325–340.

Di Lorenzo, P. M. (1990). Corticofugal influence on taste responses in the parabrachial pons of the rat, *Brain Research* **530**: 73–84.

Diamond, M. E., Huang, W. and Ebner, F. F. (1994). Laminar comparison of somatosensory cortical plasticity, *Science* **265**: 1885–1888.

DiCarlo, J. J. and Maunsell, J. H. R. (2003). Anterior inferotemporal neurons of monkeys engaged in object recognition can be highly sensitive to object retinal position, *Journal of Neurophysiology* **89**: 3264–3278.

DiCarlo, J. J., Zoccolan, D. and Rust, N. C. (2012). How does the brain solve visual object recognition?, *Neuron* **73**: 415–434.

Dillingham, C. M., Frizzati, A., Nelson, A. J. D. and Vann, S. D. (2015). How do mammillary body inputs contribute to anterior thalamic function?, *Neuroscience & Biobehavioral Reviews* **54**: 108–119.

Ding, M., Chen, Y. and Bressler, S. L. (2006). Granger causality: basic theory and application to neuroscience, in B. Schelter, M. Winterhalder and J. Timmer (eds), *Handbook of Time Series Analysis*, Wiley, Weinheim, chapter 17, pp. 437–460.

Divac, I. (1975). Magnocellular nuclei of the basal forebrain project to neocortex, brain stem, and olfactory bulb. Review of some functional correlates, *Brain Research* **93**: 385–398.

Dolan, R. J., Fink, G. R., Rolls, E. T., Booth, M., Holmes, A., Frackowiak, R. S. J. and Friston, K. J. (1997). How the brain learns to see objects and faces in an impoverished context, *Nature* **389**: 596–599.

Douglas, R. J. and Martin, K. A. (2004). Neuronal circuits of the neocortex, *Annu Rev Neurosci* **27**: 419–51.

Douglas, R. J. and Martin, K. A. (2007). Recurrent neuronal circuits in the neocortex, *Curr Biol* **17**(13): R496–500.

Douglas, R. J. and Martin, K. A. C. (1990). Neocortex, *in* G. M. Shepherd (ed.), *The Synaptic Organization of the Brain*, 3rd edn, Oxford University Press, Oxford, chapter 12, pp. 389–438.

Douglas, R. J., Mahowald, M. A. and Martin, K. A. C. (1996). Microarchitecture of cortical columns, *in* A. Aertsen and V. Braitenberg (eds), *Brain Theory: Biological Basis and Computational Theory of Vision*, Elsevier, Amsterdam.

Douglas, R. J., Markram, H. and Martin, K. A. C. (2004). Neocortex, *in* G. M. Shepherd (ed.), *The Synaptic Organization of the Brain*, 5th edn, Oxford University Press, Oxford, chapter 12, pp. 499–558.

Dow, B. W., Snyder, A. Z., Vautin, R. G. and Bauer, R. (1981). Magnification factor and receptive field size in foveal striate cortex of the monkey, *Experimental Brain Research* **44**: 213–218.

Doya, K. (1999). What are the computations of the cerebellum, the basal ganglia and the cerebral cortex?, *Neural Networks* **12**: 961–974.

Doyere, V., Debiec, J., Monfils, M. H., Schafe, G. E. and LeDoux, J. E. (2007). Synapse-specific reconsolidation of distinct fear memories in the lateral amygdala, *Nature Neuroscience* **10**: 414–416.

Drevets, W. C. (2007). Orbitofrontal cortex function and structure in depression, *Annals of the New York Academy of Sciences* **1121**: 499–527.

Dudai, Y. (2012). The restless engram: consolidations never end, *Annual Review of Neuroscience* **35**: 227–247.

Dudchenko, P. A., Wood, E. R. and Eichenbaum, H. (2000). Neurotoxic hippocampal lesions have no effect on odor span and little effect on odor recognition memory but produce significant impairments on spatial span, recognition, and alternation, *Journal of Neuroscience* **20**: 2964–2977.

Duhamel, J. R., Colby, C. L. and Goldberg, M. E. (1992). The updating of the representation of visual space in parietal cortex by intended eye movements, *Science* **255**: 90–92.

Duncan, J. (1980). The locus of interference in the perception of simultaneous stimuli, *Psychological Review* **87**: 272–300.

Duncan, J. (1996). Cooperating brain systems in selective perception and action, *in* T. Inui and J. L. McClelland (eds), *Attention and Performance XVI*, MIT Press, Cambridge, MA, pp. 549–578.

Duncan, J. and Humphreys, G. (1989). Visual search and stimulus similarity, *Psychological Review* **96**: 433–458.

Durbin, R. and Mitchison, G. (1990). A dimension reduction framework for understanding cortical maps, *Nature* **343**: 644–647.

Durstewitz, D. and Seamans, J. K. (2002). The computational role of dopamine D1 receptors in working memory, *Neural Networks* **15**: 561–572.

Durstewitz, D., Kelc, M. and Gunturkun, O. (1999). A neurocomputational theory of the dopaminergic modulation of working memory functions, *Journal of Neuroscience* **19**: 2807–2822.

Durstewitz, D., Seamans, J. K. and Sejnowski, T. J. (2000a). Dopamine-mediated stabilization of delay-period activity in a network model of prefrontal cortex, *Journal of Neurophysiology* **83**: 1733–1750.

Durstewitz, D., Seamans, J. K. and Sejnowski, T. J. (2000b). Neurocomputational models of working memory, *Nature Neuroscience* **3 Suppl**: 1184–1191.

Easton, A. and Gaffan, D. (2000). Amygdala and the memory of reward: the importance of fibres of passage from the basal forebrain, *in* J. P. Aggleton (ed.), *The Amygdala: a Functional Analysis*, 2nd edn, Oxford University Press, Oxford, chapter 17, pp. 569–586.

Easton, A., Ridley, R. M., Baker, H. F. and Gaffan, D. (2002). Unilateral lesions of the cholinergic basal forebrain and fornix in one hemisphere and inferior temporal cortex in the opposite hemisphere produce severe learning impairments in rhesus monkeys, *Cerebral Cortex* **12**: 729–736.

Eccles, J. C. (1984). The cerebral neocortex: a theory of its operation, *in* E. G. Jones and A. Peters (eds), *Cerebral Cortex: Functional Properties of Cortical Cells*, Vol. 2, Plenum, New York, chapter 1, pp. 1–36.

Eccles, J. C., Ito, M. and Szentagothai, J. (1967). *The Cerebellum as a Neuronal Machine*, Springer-Verlag, New York.

Eckhorn, R. and Popel, B. (1974). Rigorous and extended application of information theory to the afferent visual system of the cat. I. Basic concepts, *Kybernetik* **16**: 191–200.

Eckhorn, R. and Popel, B. (1975). Rigorous and extended application of information theory to the afferent visual system of the cat. II. Experimental results, *Kybernetik* **17**: 7–17.

Eckhorn, R., Grusser, O. J., Kroller, J., Pellnitz, K. and Popel, B. (1976). Efficiency of different neural codes: information transfer calculations for three different neuronal systems, *Biological Cybernetics* **22**: 49–60.

Edelman, S. (1999). *Representation and Recognition in Vision*, MIT Press, Cambridge, MA.

Eichenbaum, H. (1997). Declarative memory: insights from cognitive neurobiology, *Annual Review of Psychology* **48**: 547–572.

Eichenbaum, H. (2014). Time cells in the hippocampus: a new dimension for mapping memories, *Nature Reviews Neuroscience* **15**: 732–744.

Eichenbaum, H. and Cohen, N. J. (2001). *From Conditioning to Conscious Recollection: Memory Systems of the Brain*, Oxford University Press, New York.

Eichenbaum, H., Otto, T. and Cohen, N. J. (1992). The hippocampus - what does it do?, *Behavioural and Neural*

Biology **57**: 2–36.

Ekstrom, A. D., Kahana, M. J., Caplan, J. B., Fields, T. A., Isham, E. A., Newman, E. L. and Fried, I. (2003). Cellular networks underlying human spatial navigation, *Nature* **425**: 184–188.

Elliffe, M. C. M., Rolls, E. T., Parga, N. and Renart, A. (2000). A recurrent model of transformation invariance by association, *Neural Networks* **13**: 225–237.

Elliffe, M. C. M., Rolls, E. T. and Stringer, S. M. (2002). Invariant recognition of feature combinations in the visual system, *Biological Cybernetics* **86**: 59–71.

Elston, G. N., Benavides-Piccione, R., Elston, A., Zietsch, B., Defelipe, J., Manger, P., Casagrande, V. and Kaas, J. H. (2006). Specializations of the granular prefrontal cortex of primates: implications for cognitive processing, *Anatomical Record A Discov Mol Cell Evol Biol* **288**: 26–35.

Engel, A. K., Konig, P., Kreiter, A. K., Schillen, T. B. and Singer, W. (1992). Temporal coding in the visual system: new vistas on integration in the nervous system, *Trends in Neurosciences* **15**: 218–226.

Engel, A. K., Roelfsema, P. R., Fries, P., Brecht, M. and Singer, W. (1997). Role of the temporal domain for response selection and perceptual binding, *Cerebral Cortex* **7**: 571–582.

Enomoto, K., Matsumoto, N., Nakai, S., Satoh, T., Sato, T. K., Ueda, Y., Inokawa, H., Haruno, M. and Kimura, M. (2011). Dopamine neurons learn to encode the long-term value of multiple future rewards, *Proceedings of the National Academy of Sciences U S A* **108**: 15462–15467.

Epstein, J., Stern, E. and Silbersweig, D. (1999). Mesolimbic activity associated with psychosis in schizophrenia. Symptom-specific PET studies, *Annals of the New York Academy of Sciences* **877**: 562–574.

Epstein, R. and Kanwisher, N. (1998). A cortical representation of the local visual environment, *Nature* **392**: 598–601.

Esber, G. R. and Haselgrove, M. (2011). Reconciling the influence of predictiveness and uncertainty on stimulus salience: a model of attention in associative learning, *Proceedings of the Royal Society of London B: Biological Sciences* p. rspb20110836.

Eshel, N. and Roiser, J. P. (2010). Reward and punishment processing in depression, *Biological Psychiatry* **68**: 118–124.

Eskandar, E. N., Richmond, B. J. and Optican, L. M. (1992). Role of inferior temporal neurons in visual memory. I. Temporal encoding of information about visual images, recalled images, and behavioural context, *Journal of Neurophysiology* **68**: 1277–1295.

Espinosa, J. S. and Stryker, M. P. (2012). Development and plasticity of the primary visual cortex, *Neuron* **75**: 230–249.

Estes, W. K. (1986). Memory for temporal information, *in* J. A. Michon and J. L. Jackson (eds), *Time, Mind and Behavior*, Springer-Verlag, New York, pp. 151–168.

Everling, S., Tinsley, C. J., Gaffan, D. and Duncan, J. (2006). Selective representation of task-relevant objects and locations in the monkey prefrontal cortex, *European Journal of Neuroscience* **23**: 2197–2214.

Fahy, F. L., Riches, I. P. and Brown, M. W. (1993). Neuronal activity related to visual recognition memory and the encoding of recency and familiarity information in the primate anterior and medial inferior temporal and rhinal cortex, *Experimental Brain Research* **96**: 457–492.

Faisal, A., Selen, L. and Wolpert, D. (2008). Noise in the nervous system, *Nature Reviews Neuroscience* **9**: 292–303.

Farah, M. J. (1990). *Visual Agnosia*, MIT Press, Cambridge, MA.

Farah, M. J. (2000). *The Cognitive Neuroscience of Vision*, Blackwell, Oxford.

Farah, M. J. (2004). *Visual Agnosia*, 2nd edn, MIT Press, Cambridge, MA.

Farah, M. J., Meyer, M. M. and McMullen, P. A. (1996). The living/nonliving dissociation is not an artifact: giving an *a priori* implausible hypothesis a strong test, *Cognitive Neuropsychology* **13**: 137–154.

Fatt, P. and Katz, B. (1950). Some observations on biological noise, *Nature* **166**: 597–598.

Faugeras (1993). *The Representation, Recognition and Location of 3-D Objects*, MIT Press, Cambridge, MA.

Faugeras, O. D. and Hebert, M. (1986). The representation, recognition and location of 3-D objects, *International Journal of Robotics Research* **5**: 27–52.

Fazeli, M. S. and Collingridge, G. L. (eds) (1996). *Cortical Plasticity: LTP and LTD*, Bios, Oxford.

Feigenbaum, J. D. and Rolls, E. T. (1991). Allocentric and egocentric spatial information processing in the hippocampal formation of the behaving primate, *Psychobiology* **19**: 21–40.

Feldman, D. E. (2009). Synaptic mechanisms for plasticity in neocortex, *Annual Reviews of Neuroscience* **32**: 33–55.

Feldman, D. E. (2012). The spike-timing dependence of plasticity, *Neuron* **75**: 556–571.

Feldman, J. A. (1985). Four frames suffice: a provisional model of vision and space, *Behavioural Brain Sciences* **8**: 265–289.

Felleman, D. J. and Van Essen, D. C. (1991). Distributed hierarchical processing in the primate cerebral cortex, *Cerebral Cortex* **1**: 1–47.

Fellows, L. K. (2007). The role of orbitofrontal cortex in decision making: a component process account, *Annalls of the New York Academy of Sciences* **1121**: 421–430.

Fellows, L. K. (2011). Orbitofrontal contributions to value-based decision making: evidence from humans with frontal lobe damage, *Annals of the New York Academy of Sciences* **1239**: 51–58.

Fellows, L. K. and Farah, M. J. (2003). Ventromedial frontal cortex mediates affective shifting in humans: evidence from a reversal learning paradigm, *Brain* **126**: 1830–1837.

Ferbinteanu, J., Holsinger, R. M. and McDonald, R. J. (1999). Lesions of the medial or lateral perforant path have different effects on hippocampal contributions to place learning and on fear conditioning to context, *Behavioral Brain Research* **101**: 65–84.

Field, D. J. (1987). Relations between the statistics of natural images and the response properties of cortical cells, *Journal of the Optical Society of America, A* **4**: 2379–2394.

Field, D. J. (1994). What is the goal of sensory coding?, *Neural Computation* **6**: 559–601.

Finkel, L. H. and Edelman, G. M. (1987). Population rules for synapses in networks, *in* G. M. Edelman, W. E. Gall and W. M. Cowan (eds), *Synaptic Function*, John Wiley & Sons, New York, pp. 711–757.

Fiorillo, C. D., Tobler, P. N. and Schultz, W. (2003). Discrete coding of reward probability and uncertainty by dopamine neurons, *Science* **299**: 1898–1902.

Florian, C. and Roullet, P. (2004). Hippocampal CA3-region is crucial for acquisition and memory consolidation in Morris water maze task in mice, *Behavioral Brain Research* **154**: 365–374.

Fodor, J. A. (1994). *The Elm and the Expert: Mentalese and its Semantics*, MIT Press, Cambridge, MA.

Földiák, P. (1991). Learning invariance from transformation sequences, *Neural Computation* **3**: 193–199.

Földiák, P. (1992). Models of sensory coding, *Technical Report CUED/F–INFENG/TR 91*, University of Cambridge, Department of Engineering, Cambridge.

Földiák, P. (2003). Sparse coding in the primate cortex, *in* M. A. Arbib (ed.), *Handbook of Brain Theory and Neural Networks*, 2nd edn, MIT Press, Cambridge, MA, pp. 1064–1608.

Forgeard, M. J., Haigh, E. A., Beck, A. T., Davidson, R. J., Henn, F. A., Maier, S. F., Mayberg, H. S. and Seligman, M. E. (2011). Beyond depression: Towards a process-based approach to research, diagnosis, and treatment, *Clinical Psychology (New York)* **18**: 275–299.

Foster, T. C., Castro, C. A. and McNaughton, B. L. (1989). Spatial selectivity of rat hippocampal neurons: dependence on preparedness for movement, *Science* **244**: 1580–1582.

Francis, S., Rolls, E. T., Bowtell, R., McGlone, F., O'Doherty, J., Browning, A., Clare, S. and Smith, E. (1999). The representation of pleasant touch in the brain and its relationship with taste and olfactory areas, *NeuroReport* **10**: 453–459.

Franco, L., Rolls, E. T., Aggelopoulos, N. C. and Treves, A. (2004). The use of decoding to analyze the contribution to the information of the correlations between the firing of simultaneously recorded neurons, *Experimental Brain Research* **155**: 370–384.

Franco, L., Rolls, E. T., Aggelopoulos, N. C. and Jerez, J. M. (2007). Neuronal selectivity, population sparseness, and ergodicity in the inferior temporal visual cortex, *Biological Cybernetics* **96**: 547–560.

Frankland, S. M. and Greene, J. D. (2015). An architecture for encoding sentence meaning in left mid-superior temporal cortex, *Proc Natl Acad Sci U S A* **112**(37): 11732–7.

Franks, K. M., Stevens, C. F. and Sejnowski, T. J. (2003). Independent sources of quantal variability at single glutamatergic synapses, *Journal of Neuroscience* **23**: 3186–3195.

Franzius, M., Sprekeler, H. and Wiskott, L. (2007). Slowness and sparseness lead to place, head-direction, and spatial-view cells, *PLoS Computational Biology* **3**(8): e166.

Freedman, D. J. (2015). Learning-dependent plasticity of visual encoding in inferior temporal cortex, *J Vis* **15**(12): 1420.

Frégnac, Y. (1996). Dynamics of cortical connectivity in visual cortical networks: an overview, *Journal of Physiology, Paris* **90**: 113–139.

Fregnac, Y., Pananceau, M., Rene, A., Huguet, N., Marre, O., Levy, M. and Shulz, D. E. (2010). A re-examination of Hebbian-covariance rules and spike timing-dependent plasticity in cat visual cortex in vivo, *Frontiers in Synaptic Neuroscience* **2**: 147.

Freiwald, W. A., Tsao, D. Y. and Livingstone, M. S. (2009). A face feature space in the macaque temporal lobe, *Nat Neurosci* **12**(9): 1187–96.

Fremaux, N. and Gerstner, W. (2015). Neuromodulated spike-timing-dependent plasticity, and theory of three-factor learning rules, *Frontiers in Neural Circuits* **9**: 85.

Freud, S. (1900). *The Interpretation of Dreams*.

Frey, S. and Petrides, M. (2002). Orbitofrontal cortex and memory formation, *Neuron* **36**: 171–176.

Fried, I., MacDonald, K. A. and Wilson, C. (1997). Single neuron activity in human hippocampus and amygdala during recognition of faces and objects, *Neuron* **18**: 753–765.

Fried, I., Rutishauser, U., Cerf, M. and Kreiman, G. (2014). *Single Neuron Studies of the Human Brain: Probing Cognition*, MIT Press.

Fries, P. (2005). A mechanism for cognitive dynamics: neuronal communication through neuronal coherence, *Trends in Cognitive Sciences* **9**: 474–480.

Fries, P. (2009). Neuronal gamma-band synchronization as a fundamental process in cortical computation, *Annual Reviews of Neuroscience* **32**: 209–224.

Fries, P. (2015). Rhythms for cognition: Communication through coherence, *Neuron* **88**(1): 220–35.

Fries, P., Reynolds, J., Rorie, A. and Desimone, R. (2001). Modulation of oscillatory neuronal synchronization by selective visual attention, *Science* **291**: 1560–1563.

Fries, P., Womelsdorf, T., Oostenveld, R. and Desimone, R. (2008). The effects of visual stimulation and selective visual attention on rhythmic neuronal synchronization in macaque area V4, *Journal of Neuroscience* **28**: 4823–4835.

Friston, K. J., Buechel, C., Fink, G. R., Morris, J., Rolls, E. T. and Dolan, R. J. (1997). Psychophysiological and modulatory interactions in neuroimaging, *Neuroimage* **6**: 218–229.

Frolov, A. A. and Medvedev, A. V. (1986). Substantiation of the "point approximation" for describing the total electrical activity of the brain with use of a simulation model, *Biophysics* **31**: 332–337.

Fuhrmann, G., Markram, H. and Tsodyks, M. (2002a). Spike frequency adaptation and neocortical rhythms, *Journal of Neurophysiology* **88**: 761–770.

Fuhrmann, G., Segev, I., Markram, H. and Tsodyks, M. (2002b). Coding of temporal information by activity-dependent synapses, *Journal of Neurophysiology* **87**: 140–148.

Fujimichi, R., Naya, Y., Koyano, K. W., Takeda, M., Takeuchi, D. and Miyashita, Y. (2010). Unitized representation of paired objects in area 35 of the macaque perirhinal cortex, *European Journal of Neuroscience* **32**: 659–667.

Fukushima, K. (1975). Cognitron: a self-organizing neural network, *Biological Cybernetics* **20**: 121–136.

Fukushima, K. (1980). Neocognitron: a self-organizing neural network model for a mechanism of pattern recognition unaffected by shift in position, *Biological Cybernetics* **36**: 193–202.

Fukushima, K. (1988). Neocognitron: A hierarchical neural network model capable of visual pattern recognition unaffected by shift in position, *Neural Networks* **1**: 119–130.

Fukushima, K. (1989). Analysis of the process of visual pattern recognition by the neocognitron, *Neural Networks* **2**: 413–420.

Fukushima, K. (1991). Neural networks for visual pattern recognition, *IEEE Transactions E* **74**: 179–190.

Fukushima, K. and Miyake, S. (1982). Neocognitron: A new algorithm for pattern recognition tolerant of deformations and shifts in position, *Pattern Recognition* **15**(6): 455–469.

Funahashi, S., Bruce, C. and Goldman-Rakic, P. (1989). Mnemonic coding of visual space in monkey dorsolateral prefrontal cortex, *Journal of Neurophysiology* **61**: 331–349.

Furman, M. and Wang, X.-J. (2008). Similarity effect and optimal control of multiple-choice decision making, *Neuron* **60**: 1153–1168.

Fusi, S. and Abbott, L. F. (2007). Limits on the memory storage capacity of bounded synapses, *Nature Neuroscience* **10**: 485–493.

Fuster, J. M. (1973). Unit activity in prefrontal cortex during delayed-response performance: neuronal correlates of transient memory, *Joural of Neurophysiology* **36**: 61–78.

Fuster, J. M. (1989). *The Prefrontal Cortex*, 2nd edn, Raven Press, New York.

Fuster, J. M. (2000). *Memory Systems in the Brain*, Raven Press, New York.

Fuster, J. M. (2008). *The Prefrontal Cortex*, 4th edn, Academic Press, London.

Fuster, J. M. (2014). The prefrontal cortex makes the brain a preadaptive system, *Proceedings of the IEEE* **102**(4): 417–426.

Fyhn, M., Molden, S., Witter, M. P., Moser, E. I. and Moser, M.-B. (2004). Spatial representation in the entorhinal cortex, *Science* **2004**: 1258–1264.

Gabbott, P. and Rolls, E. T. (2013). Increased neuronal firing in resting and sleep in areas of the macaque medial prefrontal cortex (mPFC) that are part of the default mode network, *European Journal of Neuroscience* **37**: 1737–1746.

Gabbott, P. L., Warner, T. A., Jays, P. R. and Bacon, S. J. (2003). Areal and synaptic interconnectivity of prelimbic (area 32), infralimbic (area 25) and insular cortices in the rat, *Brain Research* **993**: 59–71.

Gaffan, D. (1993). Additive effects of forgetting and fornix transection in the temporal gradient of retrograde amnesia, *Neuropsychologia* **31**: 1055–1066.

Gaffan, D. (1994). Scene-specific memory for objects: a model of episodic memory impairment in monkeys with fornix transection, *Journal of Cognitive Neuroscience* **6**: 305–320.

Gaffan, D. and Harrison, S. (1989a). A comparison of the effects of fornix section and sulcus principalis ablation upon spatial learning by monkeys, *Behavioural Brain Research* **31**: 207–220.

Gaffan, D. and Harrison, S. (1989b). Place memory and scene memory: effects of fornix transection in the monkey, *Experimental Brain Research* **74**: 202–212.

Gaffan, D. and Saunders, R. C. (1985). Running recognition of configural stimuli by fornix transected monkeys, *Quarterly Journal of Experimental Psychology* **37B**: 61–71.

Gaffan, D., Saunders, R. C., Gaffan, E. A., Harrison, S., Shields, C. and Owen, M. J. (1984). Effects of fornix section upon associative memory in monkeys: role of the hippocampus in learned action, *Quarterly Journal of Experimental Psychology* **36B**: 173–221.

Gallant, J. L., Connor, C. E. and Van-Essen, D. C. (1998). Neural activity in areas V1, V2 and V4 during free viewing of natural scenes compared to controlled viewing, *NeuroReport* **9**: 85–90.

Gardner, E. (1987). Maximum storage capacity in neural networks, *Europhysics Letters* **4**: 481–485.

Gardner, E. (1988). The space of interactions in neural network models, *Journal of Physics A* **21**: 257–270.

Gardner-Medwin, A. R. (1976). The recall of events through the learning of associations between their parts, *Proceedings of the Royal Society of London, Series B* **194**: 375–402.

Gattass, R., Sousa, A. P. B. and Covey, E. (1985). Cortical visual areas of the macaque: possible substrates for pattern recognition mechanisms, *Experimental Brain Research, Supplement 11, Pattern Recognition Mechanisms* **54**: 1–20.

Gawne, T. J. and Richmond, B. J. (1993). How independent are the messages carried by adjacent inferior temporal cortical neurons?, *Journal of Neuroscience* **13**: 2758–2771.

Gaykema, R. P., van der Kuil, J., Hersh, L. B. and Luiten, P. G. (1991). Patterns of direct projections from the hippocampus to the medial septum-diagonal band complex: anterograde tracing with phaseolus vulgaris leucoagglutinin combined with immunohistochemistry of choline acetyltransferase, *Neuroscience* **43**: 349–360.

Gazzaniga, M. S. (1988). Brain modularity: towards a philosophy of conscious experience, *in* A. J. Marcel and E. Bisiach (eds), *Consciousness in Contemporary Science*, Oxford University Press, Oxford, chapter 10, pp. 218–238.

Gazzaniga, M. S. (1995). Consciousness and the cerebral hemispheres, *in* M. S. Gazzaniga (ed.), *The Cognitive Neurosciences*, MIT Press, Cambridge, MA, chapter 92, pp. 1392–1400.

Gazzaniga, M. S. and LeDoux, J. (1978). *The Integrated Mind*, Plenum, New York.

Ge, T., Feng, J., Grabenhorst, F. and Rolls, E. T. (2012). Componential Granger causality, and its application to identifying the source and mechanisms of the top-down biased activation that controls attention to affective vs sensory processing, *Neuroimage* **59**: 1846–1858.

Geesaman, B. J. and Andersen, R. A. (1996). The analysis of complex motion patterns by form/cue invariant MSTd neurons, *Journal of Neuroscience* **16**: 4716–4732.

Gennaro, R. J. (2004). *Higher Order Theories of Consciousness*, John Benjamins, Amsterdam.

Georges-François, P., Rolls, E. T. and Robertson, R. G. (1999). Spatial view cells in the primate hippocampus: allocentric view not head direction or eye position or place, *Cerebral Cortex* **9**: 197–212.

Georgopoulos, A. P. (1995). Motor cortex and cognitive processing, *in* M. S. Gazzaniga (ed.), *The Cognitive Neurosciences*, MIT Press, Cambridge, MA, chapter 32, pp. 507–517.

Gerfen, C. R. and Surmeier, D. J. (2011). Modulation of striatal projection systems by dopamine, *Annual Reviews of Neuroscience* **34**: 441–466.

Gerstner, W. (1995). Time structure of the activity in neural network models, *Physical Review E* **51**: 738–758.

Gerstner, W. (2000). Population dynamics of spiking neurons: fast transients, asynchronous states, and locking, *Neural Computation* **12**: 43–89.

Gerstner, W. and Kistler, W. (2002). *Spiking Neuron Models: Single Neurons, Populations and Plasticity*, Cambridge University Press, Cambridge.

Gerstner, W., Ritz, R. and Van Hemmen, L. (1993). A biologically motivated and analytically solvable model of collective oscillations in the cortex, *Biological Cybernetics* **68**: 363–374.

Gerstner, W., Kreiter, A. K., Markram, H. and Herz, A. V. (1997). Neural codes: firing rates and beyond, *Proceedings of the National Academy of Sciences USA* **94**: 12740–12741.

Gerstner, W., Kistler, W., Naud, R. and Paninski, L. (2014). *Neuronal Dynamics – From Single Neurons to Networks and Models of Cognition*, Cambridge University Press, Cambridge.

Gibson, J. J. (1950). *The Perception of the Visual World*, Houghton Mifflin, Boston.

Gibson, J. J. (1979). *The Ecological Approach to Visual Perception*, Houghton Mifflin, Boston.

Gil, Z., Connors, B. W. and Amitai, Y. (1997). Differential regulation of neocortical synapses by neuromodulators and activity, *Neuron* **19**: 679–686.

Gilbert, P. E. and Kesner, R. P. (2002). The amygdala but not the hippocampus is involved in pattern separation based on reward value, *Neurobiology of Learning and Memory* **77**: 338–353.

Gilbert, P. E. and Kesner, R. P. (2003). Localization of function within the dorsal hippocampus: the role of the CA3 subregion in paired-associate learning, *Behavioral Neuroscience* **117**: 1385–1394.

Gilbert, P. E. and Kesner, R. P. (2006). The role of dorsal CA3 hippocampal subregion in spatial working memory and pattern separation, *Behavioural Brain Research* **169**: 142–149.

Gilbert, P. E., Kesner, R. P. and DeCoteau, W. E. (1998). Memory for spatial location: role of the hippocampus in mediating spatial pattern separation, *Journal of Neuroscience* **18**: 804–810.

Gilbert, P. E., Kesner, R. P. and Lee, I. (2001). Dissociating hippocampal subregions: double dissociation between dentate gyrus and CA1, *Hippocampus* **11**: 626–636.

Giocomo, L. M. and Hasselmo, M. E. (2007). Neuromodulation by glutamate and acetylcholine can change circuit dynamics by regulating the relative influence of afferent input and excitatory feedback, *Molecular Neurobiology* **36**: 184–200.

Giocomo, L. M., Moser, M. B. and Moser, E. I. (2011). Computational models of grid cells, *Neuron* **71**: 589–603.

Giugliano, M., La Camera, G., Rauch, A., Luescher, H.-R. and Fusi, S. (2002). Non-monotonic current-to-rate response function in a novel integrate-and-fire model neuron, *in* J. Dorronsoro (ed.), *Proceedings of ICANN 2002, LNCS 2415*, Springer, New York, pp. 141–146.

Giza, B. K. and Scott, T. R. (1983). Blood glucose selectively affects taste-evoked activity in rat nucleus tractus solitarius, *Physiology and Behaviour* **31**: 643–650.

Giza, B. K. and Scott, T. R. (1987a). Blood glucose level affects perceived sweetness intensity in rats, *Physiology and Behaviour* **41**: 459–464.

Giza, B. K. and Scott, T. R. (1987b). Intravenous insulin infusions in rats decrease gustatory-evoked responses to sugars, *American Journal of Physiology* **252**: R994–R1002.

Giza, B. K., Scott, T. R. and Vanderweele, D. A. (1992). Administration of satiety factors and gustatory responsiveness in the nucleus tractus solitarius of the rat, *Brain Research Bulletin* **28**: 637–639.

Giza, B. K., Deems, R. O., Vanderweele, D. A. and Scott, T. R. (1993). Pancreatic glucagon suppresses gustatory responsiveness to glucose, *American Journal of Physiology* **265**: R1231–7.

Glantz, L. A. and Lewis, D. A. (2000). Decreased dendritic spine density on prefrontal cortical pyramidal neurons in schizophrenia, *Arch Gen Psychiatry* **57**(1): 65–73.

Glausier, J. R. and Lewis, D. A. (2013). Dendritic spine pathology in schizophrenia, *Neuroscience* **251**: 90–107.

Gleen, J. F. and Erickson, R. P. (1976). Gastric modulation of gustatory afferent activity, *Physiology and Behaviour* **16**: 561–568.

Glimcher, P. (2003). The neurobiology of visual-saccadic decision making, *Annual Review of Neuroscience* **26**: 133–179.

Glimcher, P. (2004). *Decisions, Uncertainty, and the Brain*, MIT Press, Cambridge, MA.

Glimcher, P. (2005). Indeterminacy in brain and behavior, *Annual Review of Psychology* **56**: 25–56.

Glimcher, P. (2011a). *Foundations of Neuroeconomic Analysis*, Oxford University Press, Oxford.

Glimcher, P. W. (2011b). Understanding dopamine and reinforcement learning: the dopamine reward prediction error hypothesis, *Proceedings of the National Academy of Sciences U S A* **108 Suppl 3**: 15647–15654.

Glimcher, P. W. and Fehr, E. (2013). *Neuroeconomics: Decision-Making and the Brain*, 2nd edn, Academic Press, New York.

Glover, G. H. (1999). Deconvolution of impulse response in event-related BOLD fMRI, *Neuroimage* **9**: 416–429.

Gnadt, J. W. and Andersen, R. A. (1988). Memory related motor planning activity in posterior parietal cortex of macaque, *Experimental Brain Research* **70**: 216–220.

Gochin, P. M., Colombo, M., Dorfman, G. A., Gerstein, G. L. and Gross, C. G. (1994). Neural ensemble encoding in inferior temporal cortex, *Journal of Neurophysiology* **71**: 2325–2337.

Goff, D. C. and Coyle, J. T. (2001). The emerging role of glutamate in the pathophysiology and treatment of schizophrenia, *American Journal of Psychiatry* **158**: 1367–1377.

Gold, A. E. and Kesner, R. P. (2005). The role of the CA3 subregion of the dorsal hippocampus in spatial pattern completion in the rat, *Hippocampus* **15**: 808–814.

Gold, J. I. and Shadlen, M. N. (2007). The neural basis of decision making, *Annual Review of Neuroscience* **30**: 535–574.

Gold, P. W. (2015). The organization of the stress system and its dysregulation in depressive illness, *Molecular Psychiatry* **20**: 32–47.

Goldberg, D. E. (1989). *Genetic Algorithms in Search, Optimization and Machine Learning*, Addison-Wesley Publishing Company, Inc., Boston, MA.

Goldberg, M. E. and Walker, M. F. (2013). The control of gaze, *in* E. R. Kandel, J. H. Schwartz, T. M. Jessell, S. A. Siegelbaum and A. J. Hudspeth (eds), *Principles of Neural Science*, 5th edn, McGraw-Hill, New York, chapter 40, pp. 894–916.

Goldberg, M. E., Bisley, J. W., Powell, K. D. and Gottlieb, J. (2006). Saccades, salience and attention: the role of the lateral intraparietal area in visual behavior, *Progress in Brain Research* **155**: 157–175.

Goldman-Rakic, P. (1994). Working memory dysfunction in schizophrenia, *Journal of Neuropsychology and Clinical Neuroscience* **6**: 348–357.

Goldman-Rakic, P. S. (1996). The prefrontal landscape: implications of functional architecture for understanding human mentation and the central executive, *Philosophical Transactions of the Royal Society B* **351**: 1445–1453.

Goldman-Rakic, P. S. (1999). The physiological approach: functional architecture of working memory and disordered cognition in schizophrenia, *Biological Psychiatry* **46**: 650–661.

Goldman-Rakic, P. S., Castner, S. A., Svensson, T. H., Siever, L. J. and Williams, G. V. (2004). Targeting the dopamine D1 receptor in schizophrenia: insights for cognitive dysfunction, *Psychopharmacology (Berlin)* **174**: 3–16.

Golomb, D., Kleinfeld, D., Reid, R. C., Shapley, R. M. and Shraiman, B. (1994). On temporal codes and the spatiotemporal response of neurons in the lateral geniculate nucleus, *Journal of Neurophysiology* **72**: 2990–3003.

Golomb, D., Hertz, J. A., Panzeri, S., Treves, A. and Richmond, B. J. (1997). How well can we estimate the information carried in neuronal responses from limited samples?, *Neural Computation* **9**: 649–665.

Gonzalez-Burgos, G. and Lewis, D. A. (2012). Nmda receptor hypofunction, parvalbumin-positive neurons, and cortical gamma oscillations in schizophrenia, *Schizophr Bull* **38**(5): 950–7.

Gonzalez-Burgos, G., Hashimoto, T. and Lewis, D. A. (2010). Alterations of cortical gaba neurons and network oscillations in schizophrenia, *Curr Psychiatry Rep* **12**(4): 335–44.

Goodale, M. A. (2004). Perceiving the world and grasping it: dissociations between conscious and unconscious visual processing, *in* M. S. Gazzaniga (ed.), *The Cognitive Neurosciences III*, MIT Press, Cambridge, MA, pp. 1159–1172.

Goodrich-Hunsaker, N. J., Hunsaker, M. R. and Kesner, R. P. (2005). Dissociating the role of the parietal cortex and dorsal hippocampus for spatial information processing, *Behavioral Neuroscience* **119**: 1307–1315.

Goodrich-Hunsaker, N. J., Hunsaker, M. R. and Kesner, R. P. (2008). The interactions and dissociations of the dorsal hippocampus subregions: how the dentate gyrus, ca3, and ca1 process spatial information, *Behav Neurosci* **122**(1): 16–26.

Gothard, K. M., Battaglia, F. P., Erickson, C. A., Spitler, K. M. and Amaral, D. G. (2007). Neural responses to facial expression and face identity in the monkey amygdala, *Journal of Neurophysiology* **97**: 1671–1683.

Gotlib, I. H. and Hammen, C. L. (2009). *Handbook of Depression*, Guilford Press, New York.

Gottfried, J. A. (2010). Central mechanisms of odour object perception, *Nature Reviews Neuroscience* **11**: 628–641.

Gottfried, J. A. (2015). Structural and functional imaging of the human olfactory system, *in* R. L. Doty (ed.), *Handbook of Olfaction and Gustation*, 3rd edn, Wiley Liss, New York, chapter 13, pp. 279–304.

Gottfried, J. A., Winston, J. S. and Dolan, R. J. (2006). Dissociable codes of odor quality and odorant structure in human piriform cortex, *Neuron* **49**: 467–479.

Grabenhorst, F. and Rolls, E. T. (2008). Selective attention to affective value alters how the brain processes taste stimuli, *European Journal of Neuroscience* **27**: 723–729.

Grabenhorst, F. and Rolls, E. T. (2010). Attentional modulation of affective vs sensory processing: functional connectivity and a top down biased activation theory of selective attention, *Journal of Neurophysiology* **104**: 1649–1660.

Grabenhorst, F. and Rolls, E. T. (2011). Value, pleasure, and choice systems in the ventral prefrontal cortex, *Trends in Cognitive Sciences* **15**: 56–67.

Grabenhorst, F., Rolls, E. T., Margot, C., da Silva, M. and Velazco, M. I. (2007). How pleasant and unpleasant stimuli combine in the brain: odor combinations, *Journal of Neuroscience* **27**: 13532–13540.

Grabenhorst, F., Rolls, E. T. and Bilderbeck, A. (2008a). How cognition modulates affective responses to taste and flavor: top-down influences on the orbitofrontal and pregenual cingulate cortices, *Cerebral Cortex* **18**: 1549–1559.

Grabenhorst, F., Rolls, E. T. and Parris, B. A. (2008b). From affective value to decision-making in the prefrontal cortex, *European Journal of Neuroscience* **28**: 1930–1939.

Grabenhorst, F., D'Souza, A., Parris, B. A., Rolls, E. T. and Passingham, R. E. (2010a). A common neural scale for the subjective value of different primary rewards, *Neuroimage* **51**: 1265–1274.

Grabenhorst, F., Rolls, E. T., Parris, B. A. and D'Souza, A. (2010b). How the brain represents the reward value of fat in the mouth, *Cerebral Cortex* **20**: 1082–1091.

Grady, C. L. (2008). Cognitive neuroscience of aging, *Annals of the New York Academy of Sciences* **1124**: 127–144.

Granger, C. W. J. (1969). Investigating causal relations by econometric models and cross-spectral methods, *Econometrica* **37**: 414–438.

Gray, C. M. and Singer, W. (1989). Stimulus-specific neuronal oscillations in orientation columns of cat visual cortex, *Proceedings of National Academy of Sciences USA* **86**: 1698–1702.

Gray, C. M., Konig, P., Engel, A. K. and Singer, W. (1989). Oscillatory responses in cat visual cortex exhibit inter-columnar synchronization which reflects global stimulus properties, *Nature* **338**: 334–337.

Gray, J. A., Young, A. M. J. and Joseph, M. H. (1997). Dopamine's role, *Science* **278**: 1548–1549.

Graybiel, A. M. and Kimura, M. (1995). Adaptive neural networks in the basal ganglia, *in* J. C. Houk, J. L. Davis and D. G. Beiser (eds), *Models of Information Processing in the Basal Ganglia*, MIT Press, Cambridge, MA, chapter 5, pp. 103–116.

Graziano, M. S. A., Andersen, R. A. and Snowden, R. J. (1994). Tuning of MST neurons to spiral motions, *Journal of Neuroscience* **14**: 54–67.

Green, D. and Swets, J. (1966). *Signal Detection Theory and Psychophysics*, Wiley, New York.

Green, M. F. (1996). What are the functional consequences of neurocognitive deficits in schizophrenia?, *American Journal of Psychiatry* **153**: 321–330.

Gregory, R. L. (1970). *The Intelligent Eye*, McGraw-Hill, New York.

Griffin, D. R. (1992). *Animal Minds*, University of Chicago Press, Chicago.

Grill-Spector, K. and Malach, R. (2004). The human visual cortex, *Annual Review of Neuroscience* **27**: 649–677.

Grill-Spector, K., Sayres, R. and Ress, D. (2006). High-resolution imaging reveals highly selective nonface clusters in the fusiform face area, *Nature Neuroscience* **9**: 1177–1185.

Grimson, W. E. L. (1990). *Object Recognition by Computer*, MIT Press, Cambridge, MA.

Griniasty, M., Tsodyks, M. V. and Amit, D. J. (1993). Conversion of temporal correlations between stimuli to spatial correlations between attractors, *Neural Computation* **35**: 1–17.

Gross, C. G. (2002). Genealogy of the "grandmother cell", *Neuroscientist* **8**: 512–518.

Gross, C. G., Rocha-Miranda, C. E. and Bender, D. B. (1972). Visual properties of neurons in inferotemporal cortex of the macaque, *Journal of Neurophysiology* **35**: 96–111.

Gross, C. G., Desimone, R., Albright, T. D. and Schwartz, E. L. (1985). Inferior temporal cortex and pattern recognition, *Experimental Brain Research* **Suppl. 11**: 179–201.

Grossberg, S. (1976a). Adaptive pattern classification and universal recoding i: parallel development and coding of neural feature detectors, *Biological Cybernetics* **23**: 121–134.

Grossberg, S. (1976b). Adaptive pattern classification and universal recoding ii: feedback, expectation, olfaction, illusions, *Biological Cybernetics* **23**: 1187–202.

Grossberg, S. (1988). Non-linear neural networks: principles, mechanisms, and architectures, *Neural Networks*

1: 17–61.

Groves, P. M. (1983). A theory of the functional organization of the neostriatum and the neostriatal control of voluntary movement, *Brain Research Reviews* **5**: 109–132.

Groves, P. M., Garcia-Munoz, M., Linder, J. C., Manley, M. S., Martone, M. E. and Young, S. J. (1995). Elements of the intrinsic organization and information processing in the neostriatum, *in* J. C. Houk, J. L. Davis and D. G. Beiser (eds), *Models of Information Processing in the Basal Ganglia*, MIT Press, Cambridge, MA, chapter 4, pp. 51–96.

Grusser, O.-J. and Landis, T. (1991). *Visual Agnosias*, MacMillan, London.

Guest, S., Grabenhorst, F., Essick, G., Chen, Y., Young, M., McGlone, F., de Araujo, I. and Rolls, E. T. (2007). Human cortical representation of oral temperature, *Physiology and Behavior* **92**: 975–984.

Gulyas, A. I., Miles, R., Hajos, N. and Freund, T. F. (1993). Precision and variability in postsynaptic target selection of inhibitory cells in the hippocampal CA3 region, *European Journal of Neuroscience* **5**: 1729–1751.

Gurney, K., Prescott, T. J. and Redgrave, P. (2001a). A computational model of action selection in the basal ganglia I: A new functional anatomy, *Biological Cybernetics* **84**: 401–410.

Gurney, K., Prescott, T. J. and Redgrave, P. (2001b). A computational model of action selection in the basal ganglia II: Analysis and simulation of behaviour, *Biological Cybernetics* **84**: 411–423.

Gutig, R. and Sompolinsky, H. (2006). The tempotron: a neuron that learns spike timing-based decisions, *Nature Neuroscience* **9**: 420–428.

Haber, S. N. and Knutson, B. (2009). The reward circuit: linking primate anatomy and human imaging, *Neuropsychopharmacology* **35**: 4–26.

Haberly, L. B. (2001). Parallel-distributed processing in olfactory cortex: new insights from morphological and physiological analysis of neuronal circuitry, *Chemical Senses* **26**: 551–576.

Hafner, H., Maurer, K., Loffler, W., an der Heiden, W., Hambrecht, M. and Schultze-Lutter, F. (2003). Modeling the early course of schizophrenia, *Schizophrenia Bulletin* **29**: 325–340.

Hafting, T., Fyhn, M., Molden, S., Moser, M. B. and Moser, E. I. (2005). Microstructure of a spatial map in the entorhinal cortex, *Nature* **436**: 801–806.

Hajnal, A., Takenouchi, K. and Norgren, R. (1999). Effect of intraduodenal lipid on parabrachial gustatory coding in awake rats, *Journal of Neuroscience* **19**: 7182–7190.

Hamani, C., Mayberg, H., Snyder, B., Giacobbe, P., Kennedy, S. and Lozano, A. M. (2009). Deep brain stimulation of the subcallosal cingulate gyrus for depression: anatomical location of active contacts in clinical responders and a suggested guideline for targeting, *Journal of Neurosurgery* **111**: 1209–1215.

Hamani, C., Mayberg, H., Stone, S., Laxton, A., Haber, S. and Lozano, A. M. (2011). The subcallosal cingulate gyrus in the context of major depression, *Biological Psychiatry* **69**: 301–308.

Hamilton, J. P., Chen, M. C. and Gotlib, I. H. (2013). Neural systems approaches to understanding major depressive disorder: an intrinsic functional organization perspective, *Neurobiology of Disease* **52**: 4–11.

Hamilton, W. D. (1964). The genetical evolution of social behaviour, *Journal of Theoretical Biology* **7**: 1–52.

Hamilton, W. D. (1996). *Narrow Roads of Gene Land*, W. H. Freeman, New York.

Hamming, R. W. (1990). *Coding and Information Theory*, 2nd edn, Prentice-Hall, Englewood Cliffs, New Jersey.

Hampshire, A. and Owen, A. M. (2006). Fractionating attentional control using event-related fMRI, *Cerebral Cortex* **16**: 1679–1689.

Hampson, R. E., Hedberg, T. and Deadwyler, S. A. (2000). Differential information processing by hippocampal and subicular neurons, *Annals of the New York Academy of Sciences* **911**: 151–165.

Hampton, A. N. and O'Doherty, J. P. (2007). Decoding the neural substrates of reward-related decision making with functional MRI, *Proceedings of the National Academy of Sciences USA* **104**: 1377–1382.

Hampton, R. R. (2001). Rhesus monkeys know when they can remember, *Proceedings of the National Academy of Sciences of the USA* **98**: 5539–5362.

Hampton, R. R. (2005). Monkey perirhinal cortex is critical for visual memory, but not for visual perception: reexamination of the behavioural evidence from monkeys, *Quarterly Journal of Experimental Psychology* **58B**: 283–299.

Handelmann, G. E. and Olton, D. S. (1981). Spatial memory following damage to hippocampal ca3 pyramidal cells with kainic acid: impairment and recovery with preoperative training, *Brain Research* **217**: 41–58.

Hannesson, D. K., Howland, G. J. and Phillips, A. G. (2004). Interaction between perirhinal and medial prefrontal cortex is required for temporal order but not recognition memory for objects in rats, *Journal of Neuroscience* **24**: 4596–4604.

Harel, J., Koch, C. and Perona, P. (2006a). Graph-based visual saliency, *Advances in Neural Information Processing Systems (NIPS)* **19**: 545–552.

Harel, J., Koch, C. and Perona, P. (2006b). A saliency implementation in MATLAB, p. http://www.klab.caltech.edu/ harel/share/gbvs.php.

Hargreaves, E. L., Rao, G., Lee, I. and Knierim, J. J. (2005). Major dissociation between medial and lateral entorhinal input to dorsal hippocampus, *Science* **308**: 1792–1794.

Harmer, C. J. and Cowen, P. J. (2013). 'it's the way that you look at it'–a cognitive neuropsychological account of ssri action in depression, *Philosophical Transactions of the Royal Society of London B Biological Sciences*

368: 20120407.

Harris, A. E., Ermentrout, G. B. and Small, S. L. (1997). A model of ocular dominance column development by competition for trophic factor, *Proceedings of the National Academy of Sciences USA* **94**: 9944–9949.

Harris, K. D. and Mrsic-Flogel, T. D. (2013). Cortical connectivity and sensory coding, *Nature* **503**: 51–58.

Harris, K. D. and Shepherd, G. M. (2015). The neocortical circuit: themes and variations, *Nature Neuroscience* **18**: 170–181.

Hartley, T., Lever, C., Burgess, N. and O'Keefe, J. (2014). Space in the brain: how the hippocampal formation supports spatial cognition, *Philosophical Transactions of the Royal Society London B Biological Sciences* **369**: 20120510.

Hartston, H. J. and Swerdlow, N. R. (1999). Visuospatial priming and Stroop performance in patients with obsessive compulsive disorder, *Neuropsychology* **13**: 447–457.

Hassabis, D., Chu, C., Rees, G., Weiskopf, N., Molyneux, P. D. and Maguire, E. A. (2009). Decoding neuronal ensembles in the human hippocampus, *Current Biology* **19**: 546–554.

Hasselmo, M. E. and Bower, J. M. (1993). Acetylcholine and memory, *Trends in Neurosciences* **16**: 218–222.

Hasselmo, M. E. and Eichenbaum, H. B. (2005). Hippocampal mechanisms for the context-dependent retrieval of episodes, *Neural Networks* **18**: 1172–1190.

Hasselmo, M. E. and Sarter, M. (2011). Modes and models of forebrain cholinergic neuromodulation of cognition, *Neuropsychopharmacology* **36**: 52–73.

Hasselmo, M. E., Rolls, E. T. and Baylis, G. C. (1989a). The role of expression and identity in the face-selective responses of neurons in the temporal visual cortex of the monkey, *Behavioural Brain Research* **32**: 203–218.

Hasselmo, M. E., Rolls, E. T., Baylis, G. C. and Nalwa, V. (1989b). Object-centered encoding by face-selective neurons in the cortex in the superior temporal sulcus of the monkey, *Experimental Brain Research* **75**: 417–429.

Hasselmo, M. E., Schnell, E. and Barkai, E. (1995). Learning and recall at excitatory recurrent synapses and cholinergic modulation in hippocampal region CA3, *Journal of Neuroscience* **15**: 5249–5262.

Hawken, M. J. and Parker, A. J. (1987). Spatial properties of the monkey striate cortex, *Proceedings of the Royal Society, London B* **231**: 251–288.

Haxby, J. V. (2006). Fine structure in representations of faces and objects, *Nature Neuroscience* **9**: 1084–1086.

Haynes, J. D. and Rees, G. (2005a). Predicting the orientation of invisible stimuli from activity in human primary visual cortex, *Nature Neuroscience* **8**: 686–691.

Haynes, J. D. and Rees, G. (2005b). Predicting the stream of consciousness from activity in human visual cortex, *Current Biology* **15**: 1301–1307.

Haynes, J. D. and Rees, G. (2006). Decoding mental states from brain activity in humans, *Nature Reviews Neuroscience* **7**: 523–534.

Haynes, J. D., Sakai, K., Rees, G., Gilbert, S., Frith, C. and Passingham, R. E. (2007). Reading hidden intentions in the human brain, *Current Biology* **17**: 323–328.

He, C., Chen, F., Li, B. and Hu, Z. (2014). Neurophysiology of HCN channels: from cellular functions to multiple regulations, *Progress in Neurobiology* **112**: 1–23.

Hebb, D. O. (1949). *The Organization of Behavior: a Neuropsychological Theory*, Wiley, New York.

Hedges, J. H., Gartshteyn, Y., Kohn, A., Rust, N. C., Shadlen, M. N., Newsome, W. T. and Movshon, J. A. (2011). Dissociation of neuronal and psychophysical responses to local and global motion, *Current Biology* **21**: 2023–2028.

Hegde, J. and Van Essen, D. C. (2000). Selectivity for complex shapes in primate visual area V2, *Journal of Neuroscience* **20**: RC61.

Heinke, D., Deco, G., Zihl, J. and Humphreys, G. (2002). A computational neuroscience account of visual neglect, *Neurocomputing* **44–46**: 811–816.

Heinrichs, D. W. (2005). Antidepressants and the chaotic brain: implications for the respectful treatment of selves, *Philosophy, Psychiatry, and Psychology* **12**: 215–227.

Heller, J., Hertz, J. A., Kjaer, T. W. and Richmond, B. J. (1995). Information flow and temporal coding in primate pattern vision, *Journal of Computational Neuroscience* **2**: 175–193.

Helmholtz, H. v. (1867). *Handbuch der physiologischen Optik*, Voss, Leipzig.

Hempel, C. M., Hartman, K. H., Wang, X. J., Turrigiano, G. G. and Nelson, S. B. (2000). Multiple forms of short-term plasticity at excitatory synapses in rat medial prefrontal cortex, *Journal of Neurophysiology* **83**: 3031–3041.

Henze, D. A., Wittner, L. and Buzsaki, G. (2002). Single granule cells reliably discharge targets in the hippocampal ca3 network in vivo, *Nat Neurosci* **5**(8): 790–5.

Herrnstein, R. J. (1984). Objects, categories, and discriminative stimuli, *in* H. L. Roitblat, T. G. Bever and H. S. Terrace (eds), *Animal Cognition*, Lawrence Erlbaum and Associates, Hillsdale, NJ, chapter 14, pp. 233–261.

Hertz, J. A., Krogh, A. and Palmer, R. G. (1991). *Introduction to the Theory of Neural Computation*, Addison-Wesley, Wokingham, UK.

Hertz, J. A., Kjaer, T. W., Eskander, E. N. and Richmond, B. J. (1992). Measuring natural neural processing with artificial neural networks, *International Journal of Neural Systems* **3 (Suppl.)**: 91–103.

Hestrin, S., Sah, P. and Nicoll, R. (1990). Mechanisms generating the time course of dual component excitatory synaptic currents recorded in hippocampal slices, *Neuron* **5**: 247–253.

Hevner, R. F. (2016). Evolution of the mammalian dentate gyrus, *J Comp Neurol* **524**(3): 578–94.

Heyes, C. (2008). Beast machines? Questions of animal consciousness, *in* L. Weiskrantz and M. Davies (eds), *Frontiers of Consciousness*, Oxford University Press, Oxford, chapter 9, pp. 259–274.

Hikosaka, K. and Watanabe, M. (2000). Delay activity of orbital and lateral prefrontal neurons of the monkey varying with different rewards, *Cerebral Cortex* **10**: 263–271.

Hill, S. L., Wang, Y., Riachi, I., Schürmann, F. and Markram, H. (2012). Statistical connectivity provides a sufficient foundation for specific functional connectivity in neocortical neural microcircuits, *Proceedings of the National Academy of Sciences* **109**: E2885–E2894.

Hinton, G. E. (1989). Deterministic Boltzmann learning performs steepest descent in weight-space, *Neural Computation* **1**: 143–150.

Hinton, G. E. and Anderson, J. A. (1981). *Parallel Models of Associative Memory*, Erlbaum, Hillsdale, NJ.

Hinton, G. E. and Ghahramani, Z. (1997). Generative models for discovering sparse distributed representations, *Philosophical Transactions of the Royal Society of London, B* **352**: 1177–1190.

Hinton, G. E. and Salakhutdinov, R. R. (2006). Reducing the dimensionality of data with neural networks, *Science* **313**: 504–507.

Hinton, G. E. and Sejnowski, T. J. (1986). Learning and relearning in Boltzmann machines, *in* D. Rumelhart and J. L. McClelland (eds), *Parallel Distributed Processing*, Vol. 1, MIT Press, Cambridge, MA, chapter 7, pp. 282–317.

Hinton, G. E., Dayan, P., Frey, B. J. and Neal, R. M. (1995). The "wake-sleep" algorithm for unsupervised neural networks, *Science* **268**: 1158–1161.

Hinton, G. E., Osindero, S. and Teh, Y.-W. (2006). A fast learning algorithm for deep belief nets, *Neural computation* **18**: 1527–1554.

Hirabayashi, T., Takeuchi, D., Tamura, K. and Miyashita, Y. (2013). Functional microcircuit recruited during retrieval of object association memory in monkey perirhinal cortex, *Neuron* **77**: 192–203.

Hodgkin, A. L. and Huxley, A. F. (1952). A quantitative description of membrane current and its application to conduction and excitation in nerve, *Journal of Physiology* **117**: 500–544.

Hoebel, B. G. (1997). Neuroscience, and appetitive behavior research: 25 years, *Appetite* **29**: 119–133.

Hoebel, B. G., Rada, P., Mark, G. P., Parada, M., Puig de Parada, M., Pothos, E. and Hernandez, L. (1996). Hypothalamic control of accumbens dopamine: a system for feeding reinforcement, *in* G. Bray and D. Ryan (eds), *Molecular and Genetic Aspects of Obesity*, Vol. 5, Louisiana State University Press, Baton Rouge, LA, pp. 263–280.

Hoffman, E. A. and Haxby, J. V. (2000). Distinct representations of eye gaze and identity in the distributed neural system for face perception, *Nature Neuroscience* **3**: 80–84.

Hoge, J. and Kesner, R. P. (2007). Role of CA3 and CA1 subregions of the dorsal hippocampus on temporal processing of objects, *Neurobiology of Learning and Memory* **88**: 225–231.

Holland, J. H. (1975). *Adaptation in Natural and Artificial Systems*, The University of Michigan Press, Michigan, USA.

Holland, P. C. and Gallagher, M. (2004). Amygdala-frontal interactions and reward expectancy, *Current Opinion in Neurobiology* **14**: 148–155.

Hölscher, C. and Rolls, E. T. (2002). Perirhinal cortex neuronal activity is actively related to working memory in the macaque, *Neural Plasticity* **9**: 41–51.

Hölscher, C., Rolls, E. T. and Xiang, J. Z. (2003). Perirhinal cortex neuronal activity related to long term familiarity memory in the macaque, *European Journal of Neuroscience* **18**: 2037–2046.

Hopfield, J. J. (1982). Neural networks and physical systems with emergent collective computational abilities, *Proceedings of the National Academy of Sciences USA* **79**: 2554–2558.

Hopfield, J. J. (1984). Neurons with graded response have collective computational properties like those of two-state neurons, *Proceedings of the National Academy of Sciences USA* **81**: 3088–3092.

Hopfield, J. J. and Herz, A. V. (1995). Rapid local synchronization of action potentials: toward computation with coupled integrate-and-fire neurons, *Proceedings of the National Academy of Sciences USA* **92**: 6655–6662.

Hornak, J., Rolls, E. T. and Wade, D. (1996). Face and voice expression identification in patients with emotional and behavioural changes following ventral frontal lobe damage, *Neuropsychologia* **34**: 247–261.

Hornak, J., Bramham, J., Rolls, E. T., Morris, R. G., O'Doherty, J., Bullock, P. R. and Polkey, C. E. (2003). Changes in emotion after circumscribed surgical lesions of the orbitofrontal and cingulate cortices, *Brain* **126**: 1691–1712.

Hornak, J., O'Doherty, J., Bramham, J., Rolls, E. T., Morris, R. G., Bullock, P. R. and Polkey, C. E. (2004). Reward-related reversal learning after surgical excisions in orbitofrontal and dorsolateral prefrontal cortex in humans, *Journal of Cognitive Neuroscience* **16**: 463–478.

Horne, J. (2006). *Sleepfaring: A Journey Through the Science of Sleep*, Oxford University Press, Oxford.

Horvitz, J. C. (2000). Mesolimbocortical and nigrostriatal dopamine responses to salient non-reward events, *Neuroscience* **96**: 651–656.

Horwitz, B. and Tagamets, M.-A. (1999). Predicting human functional maps with neural net modeling, *Human Brain Mapping* **8**: 137–142.

Horwitz, B., Tagamets, M.-A. and McIntosh, A. R. (1999). Neural modeling, functional brain imaging, and cognition, *Trends in Cognitive Sciences* **3**: 85–122.

Houk, J. C., Adams, J. L. and Barto, A. C. (1995). A model of how the basal ganglia generates and uses neural signals that predict reinforcement, *in* J. C. Houk, J. L. Davies and D. G. Beiser (eds), *Models of Information Processing in the Basal Ganglia*, MIT Press, Cambridge, MA, chapter 13, pp. 249–270.

Howard, M. W. and Eichenbaum, H. (2015). Time and space in the hippocampus, *Brain Research* **1621**: 345–354.

Hubel, D. H. and Wiesel, T. N. (1962). Receptive fields, binocular interaction, and functional architecture in the cat's visual cortex, *Journal of Physiology* **160**: 106–154.

Hubel, D. H. and Wiesel, T. N. (1968). Receptive fields and functional architecture of monkey striate cortex, *Journal of Physiology, London* **195**: 215–243.

Hubel, D. H. and Wiesel, T. N. (1977). Functional architecture of the macaque monkey visual cortex, *Proceedings of the Royal Society, London [B]* **198**: 1–59.

Huerta, P. T., Sun, L. D., Wilson, M. A. and Tonegawa, S. (2000). Formation of temporal memory requires NMDA receptors within CA1 pyramidal neurons, *Neuron* **25**: 473–480.

Hughlings Jackson, J. (1878). *Selected writings of John Hughlings Jackson 2, Edited by J. Taylor, 1932*, Hodder and Staughton, London.

Human Genome Sequencing, C. I. (2004). Finishing the euchromatic sequence of the human genome, *Nature* **431**: 931–945.

Hummel, J. E. and Biederman, I. (1992). Dynamic binding in a neural network for shape recognition, *Psychological Review* **99**: 480–517.

Humphrey, N. K. (1980). Nature's psychologists, *in* B. D. Josephson and V. S. Ramachandran (eds), *Consciousness and the Physical World*, Pergamon, Oxford, pp. 57–80.

Humphrey, N. K. (1986). *The Inner Eye*, Faber, London.

Hunsaker, M. R., Thorup, J. A., Welch, T. and Kesner, R. P. (2006). The role of ca3 and ca1 in the acquisition of an object-trace-place paired-associate task, *Behav Neurosci* **120**(6): 1252–6.

Hunsaker, M. R., Rogers, J. L. and Kesner, R. P. (2007). Behavioral characterization of a transection of dorsal ca3 subcortical efferents: comparison with scopolamine and physostigmine infusions into dorsal ca3, *Neurobiol Learn Mem* **88**(1): 127–36.

Hunsaker, M. R., Lee, B. and Kesner, R. P. (2008). Evaluating the temporal context of episodic memory: the role of ca3 and ca1, *Behav Brain Res* **188**(2): 310–5.

Huth, A. G., de Heer, W. A., Griffiths, T. L., Theunissen, F. E. and Gallant, J. L. (2016). Natural speech reveals the semantic maps that tile human cerebral cortex, *Nature* **532**: 453–458.

Huttenlocher, D. P. and Ullman, S. (1990). Recognizing solid objects by alignment with an image, *International Journal of Computer Vision* **5**: 195–212.

Iadarola, N. D., Niciu, M. J., Richards, E. M., Vande Voort, J. L., Ballard, E. D., Lundin, N. B., Nugent, A. C., Machado-Vieira, R. and Zarate, C. A., J. (2015). Ketamine and other n-methyl-d-aspartate receptor antagonists in the treatment of depression: a perspective review, *Therapeutic Advances in Chronic Disease* **6**: 97–114.

Ingvar, D. H. and Franzen, G. (1974). Abnormalities of cerebral blood flow distribution in patients with chronic schizophrenia, *Acta Psychiatrica Scandinavica* **50**: 425–462.

Insabato, A., Pannunzi, M., Rolls, E. T. and Deco, G. (2010). Confidence-related decision-making, *Journal of Neurophysiology* **104**: 539–547.

Ishai, A., Ungerleider, L. G., Martin, A., Schouten, J. L. and Haxby, J. V. (1999). Distributed representation of objects in the human ventral visual pathway, *Proceedings of the National Academy of Sciences USA* **96**: 9379–9384.

Ishizuka, N., Weber, J. and Amaral, D. G. (1990). Organization of intrahippocampal projections originating from CA3 pyramidal cells in the rat, *Journal of Comparative Neurology* **295**: 580–623.

Ison, M. J., Quian Quiroga, R. and Fried, I. (2015). Rapid encoding of new memories by individual neurons in the human brain, *Neuron* **87**: 220–230.

Issa, E. B. and DiCarlo, J. J. (2012). Precedence of the eye region in neural processing of faces, *The Journal of Neuroscience* **32**: 16666–16682.

Ito, M. (1984). *The Cerebellum and Neural Control*, Raven Press, New York.

Ito, M. (1989). Long-term depression, *Annual Review of Neuroscience* **12**: 85–102.

Ito, M. (1993a). Cerebellar mechanisms of long-term depression, *in* M. Baudry, R. F. Thompson and J. L. Davis (eds), *Synaptic Plasticity: Molecular, Cellular and Functional Aspects*, MIT Press, Cambridge, MA, chapter 6, pp. 117–128.

Ito, M. (1993b). Synaptic plasticity in the cerebellar cortex and its role in motor learning, *Canadian Journal of Neurological Science* **Suppl. 3**: S70–S74.

Ito, M. (2006). Cerebellar circuitry as a neuronal machine, *Progress in Neurobiology* **78**: 272–303.

Ito, M. (2010). Cerebellar cortex, *in* G. M. Shepherd and S. Grillner (eds), *Handbook of Brain Microcircuits*, Oxford University Press, Oxford, chapter 28, pp. 293–300.

Ito, M. (2013). Error detection and representation in the olivo-cerebellar system, *Frontiers in Neural Circuits* **7**: 1.

Ito, M. and Komatsu, H. (2004). Representation of angles embedded within contour stimuli in area V2 of macaque monkeys, *Journal of Neuroscience* **24**: 3313–3324.

Ito, M., Yamaguchi, K., Nagao, S. and Yamazaki, T. (2014). Long-term depression as a model of cerebellar plasticity, *Progress in Brain Research* **210**: 1–30.

Itskov, P. M., Vinnik, E. and Diamond, M. E. (2011). Hippocampal representation of touch-guided behavior in rats: persistent and independent traces of stimulus and reward location, *PLoS One* **6**: e16462.

Itti, L. and Koch, C. (2000). A saliency-based search mechanism for overt and covert shifts of visual attention, *Vision Research* **40**: 1489–1506.

Itti, L. and Koch, C. (2001). Computational modelling of visual attention, *Nature Reviews Neuroscience* **2**: 194–203.

Iversen, S. D. and Mishkin, M. (1970). Perseverative interference in monkey following selective lesions of the inferior prefrontal convexity, *Experimental Brain Research* **11**: 376–386.

Jackendoff, R. (2002). *Foundations of Language*, Oxford University Press, Oxford.

Jackson, B. S. (2004). Including long-range dependence in integrate-and-fire models of the high interspike-interval variability of cortical neurons, *Neural Computation* **16**: 2125–2195.

Jackson, M. B. (2013). Recall of spatial patterns stored in a hippocampal slice by long-term potentiation, *Journal of Neurophysiology* **110**: 2511–2519.

Jacobs, J., Weidemann, C. T., Miller, J. F., Solway, A., Burke, J. F., Wei, X. X., Suthana, N., Sperling, M. R., Sharan, A. D., Fried, I. and Kahana, M. J. (2013). Direct recordings of grid-like neuronal activity in human spatial navigation, *Nature Neuroscience* **16**: 1188–1190.

Jacoby, L. L. (1983a). Perceptual enhancement: persistent effects of an experience, *Journal of Experimental Psychology: Learning, Memory, and Cognition* **9**: 21–38.

Jacoby, L. L. (1983b). Remembering the data: analyzing interaction processes in reading, *Journal of Verbal Learning and Verbal Behavior* **22**: 485–508.

Jahr, C. and Stevens, C. (1990). Voltage dependence of NMDA-activated macroscopic conductances predicted by single-channel kinetics, *Journal of Neuroscience* **10**: 3178–3182.

Jakab, R. L. and Leranth, C. (1995). Septum, in G. Paxinos (ed.), *The Rat Nervous System*, Academic Press, San Diego.

James, W. (1884). What is an emotion?, *Mind* **9**: 188–205.

Janssen, P., Vogels, R. and Orban, G. A. (1999). Macaque inferior temporal neurons are selective for disparity-defined three-dimensional shapes, *Proceedings of the National Academy of Sciences* **96**: 8217–8222.

Janssen, P., Vogels, R. and Orban, G. A. (2000). Selectivity for 3D shape that reveals distinct areas within macaque inferior temporal cortex, *Science* **288**: 2054–2056.

Janssen, P., Vogels, R., Liu, Y. and Orban, G. A. (2003). At least at the level of inferior temporal cortex, the stereo correspondence problem is solved, *Neuron* **37**: 693–701.

Jarrard, L. E. (1993). On the role of the hippocampus in learning and memory in the rat, *Behavioral and Neural Biology* **60**: 9–26.

Jarrard, L. E. and Davidson, T. L. (1990). Acquisition of concurrent conditional discriminations in rats with ibotenate lesions of hippocampus and subiculum, *Psychobiology* **18**: 68–73.

Jarrett, K., Kavukcuoglu, K., Ranzato, M. and LeCun, Y. (2009). What is the best multi-stage architecture for object recognition?, *2009 IEEE 12th International Conference on Computer Vision (ICCV)* pp. 2146–2153.

Jay, T. M. and Witter, M. P. (1991). Distribution of hippocampal CA1 and subicular efferents in the prefrontal cortex of the rat studied by means of anterograde transport of Phaseolus vulgaris–leucoagglutinin, *Journal of Comparative Neurology* **313**: 574–586.

Jeffery, K. J. and Hayman, R. (2004). Plasticity of the hippocampal place cell representation, *Reviews in the Neurosciences* **15**: 309–331.

Jeffery, K. J., Anderson, M. I., Hayman, R. and Chakraborty, S. (2004). A proposed architecture for the neural representation of spatial context, *Neuroscience and Biobehavioral Reviews* **28**: 201–218.

Jensen, O. and Lisman, J. E. (1996). Theta/gamma networks with slow NMDA channels learn sequences and encode episodic memory: role of NMDA channels in recall, *Learning and Memory* **3**: 264–278.

Jensen, O. and Lisman, J. E. (2005). Hippocampal sequence-encoding driven by a cortical multi-item working memory buffer, *Trends in the Neurosciences* **28**: 67–72.

Jensen, O., Idiart, M. A. and Lisman, J. E. (1996). Physiologically realistic formation of autoassociative memory in networks with theta/gamma oscillations: role of fast NMDA channels, *Learning and Memory* **3**: 243–256.

Jerman, T., Kesner, R. P. and Hunsaker, M. R. (2006). Disconnection analysis of CA3 and DG in mediating encoding but not retrieval in a spatial maze learning task, *Learning and Memory* **13**: 458–464.

Jezek, K., Henriksen, E. I., Treves, A., Moser, E. I. and Moser, M.-B. (2011). Theta-paced flickering between place-cell maps in the hippocampus, *Nature* **278**: 246–249.

Jiang, X., Shen, S., Cadwell, C. R., Berens, P., Sinz, F., Ecker, A. S., Patel, S. and Tolias, A. S. (2015). Principles of connectivity among morphologically defined cell types in adult neocortex, *Science* **350**(6264): aac9462.

Johansen-Berg, H. and Lloyd, D. M. (2000). The physiology and psychology of selective attention to touch, *Frontiers in Bioscience* **5**: D894–904.

Johansen-Berg, H., Gutman, D. A., Behrens, T. E., Matthews, P. M., Rushworth, M. F., Katz, E., Lozano, A. M. and Mayberg, H. S. (2008). Anatomical connectivity of the subgenual cingulate region targeted with deep brain stimulation for treatment-resistant depression, *Cerebral Cortex* **18**: 1374–1383.

Johnson-Laird, P. N. (1988). *The Computer and the Mind: An Introduction to Cognitive Science*, Harvard University Press, Cambridge, MA.

Johnston, D. and Amaral, D. (2004). Hippocampus, *in* G. M. Shepherd (ed.), *The Synaptic Organization of the Brain*, 5th edn, Oxford University Press, Oxford, chapter 11, pp. 455–498.

Johnston, S. T., Shtrahman, M., Parylak, S., Goncalves, J. T. and Gage, F. H. (2016). Paradox of pattern separation and adult neurogenesis: A dual role for new neurons balancing memory resolution and robustness, *Neurobiol Learn Mem* **129**: 60–8.

Jonas, E. A. and Kaczmarek, L. K. (1999). The inside story: subcellular mechanisms of neuromodulation, *in* P. S. Katz (ed.), *Beyond Neurotransmission*, Oxford University Press, New York, chapter 3, pp. 83–120.

Jones, B. and Mishkin, M. (1972). Limbic lesions and the problem of stimulus–reinforcement associations, *Experimental Neurology* **36**: 362–377.

Jones, E. G. and Peters, A. (eds) (1984). *Cerebral Cortex, Functional Properties of Cortical Cells*, Vol. 2, Plenum, New York.

Jones, E. G. and Powell, T. P. S. (1970). An anatomical study of converging sensory pathways within the cerebral cortex of the monkey, *Brain* **93**: 793–820.

Jung, M. W. and McNaughton, B. L. (1993). Spatial selectivity of unit activity in the hippocampal granular layer, *Hippocampus* **3**: 165–182.

Kaas, J. H. and Preuss, T. M. (2014). Human brain evolution, *in* L. Squire, D. Berg, S. du Lac, A. Ghosh and N. C. Spitzer (eds), *Fundamental Neuroscience*, 4th edn, Academic Press, London, chapter 42.

Kacelnik, A. and Brito e Abreu, F. (1998). Risky choice and Weber's Law, *Journal of Theoretical Biology* **194**: 289–298.

Kadohisa, M. and Wilson, D. A. (2006). Separate encoding of identity and similarity of complex familiar odors in piriform cortex, *Proceedings of the National Academy of Sciences U S A* **103**: 15206–15211.

Kadohisa, M., Rolls, E. T. and Verhagen, J. V. (2004). Orbitofrontal cortex neuronal representation of temperature and capsaicin in the mouth, *Neuroscience* **127**: 207–221.

Kadohisa, M., Rolls, E. T. and Verhagen, J. V. (2005a). Neuronal representations of stimuli in the mouth: the primate insular taste cortex, orbitofrontal cortex, and amygdala, *Chemical Senses* **30**: 401–419.

Kadohisa, M., Rolls, E. T. and Verhagen, J. V. (2005b). The primate amygdala: neuronal representations of the viscosity, fat texture, grittiness and taste of foods, *Neuroscience* **132**: 33–48.

Kagel, J. H., Battalio, R. C. and Green, L. (1995). *Economic Choice Theory: An Experimental Analysis of Animal Behaviour*, Cambridge University Press, Cambridge.

Kamin, L. J. (1969). Predictability, surprise, attention, and conditioning, *Punishment and Aversive Behavior* pp. 279–296.

Kandel, E. R., Schwartz, J. H. and Jessell, T. H. (eds) (2000). *Principles of Neural Science*, 4th edn, Elsevier, Amsterdam.

Kandel, E. R., Schwartz, J. H., Jessell, T. M., Siegelbaum, S. A. and Hudspeth, A. J. (eds) (2013). *Principles of Neural Science*, 5th edn, McGraw-Hill, New York.

Kanter, I. and Sompolinsky, H. (1987). Associative recall of memories without errors, *Physical Review A* **35**: 380–392.

Kanwisher, N., McDermott, J. and Chun, M. M. (1997). The fusiform face area: a module in human extrastriate cortex specialized for face perception, *Journal of Neuroscience* **17**: 4301–4311.

Karno, M., Golding, J. M., Sorenson, S. B. and Burnam, M. A. (1988). The epidemiology of obsessive-compulsive disorder in five US communities, *Archives of General Psychiatry* **45**: 1094–1099.

Kastner, S., De Weerd, P., Desimone, R. and Ungerleider, L. (1998). Mechanisms of directed attention in the human extrastriate cortex as revealed by functional MRI, *Science* **282**: 108–111.

Kastner, S., Pinsk, M., De Weerd, P., Desimone, R. and Ungerleider, L. (1999). Increased activity in human visual cortex during directed attention in the absence of visual stimulation, *Neuron* **22**: 751–761.

Kelly, K. M., Nadon, N. L., Morrison, J. H., Thibault, O., Barnes, C. A. and Blalock, E. M. (2006). The neurobiology of aging, *Epilepsy Research* **68, Supplement 1**: S5–S20.

Kelly, R. M. and Strick, P. L. (2004). Macro-architecture of basal ganglia loops with the cerebral cortex: use of rabies virus to reveal multisynaptic circuits, *Progress in Brain Research* **143**: 449–459.

Kemp, J. M. and Powell, T. P. S. (1970). The cortico-striate projections in the monkey, *Brain* **93**: 525–546.

Kennard, C. and Swash, M. (eds) (1989). *Hierarchies in Neurology*, Springer, London.

Kepecs, A., Uchida, N., Zariwala, H. A. and Mainen, Z. F. (2008). Neural correlates, computation and behavioural impact of decision confidence, *Nature* **455**: 227–231.

Kesner, R. P. (1998). Neural mediation of memory for time: role of hippocampus and medial prefrontal cortex, *Psychological Bulletin Reviews* **5**: 585–596.

Kesner, R. P. and Gilbert, P. E. (2006). The role of the medial caudate nucleus, but not the hippocampus, in a delayed-matching-to sample task for motor response, *European Journal of Neuroscience* **23**: 1888–1894.

Kesner, R. P. and Rolls, E. T. (2001). Role of long term synaptic modification in short term memory, *Hippocampus* **11**: 240–250.

Kesner, R. P. and Rolls, E. T. (2015). A computational theory of hippocampal function, and tests of the theory: New developments, *Neuroscience and Biobehavioral Reviews* **48**: 92–147.

Kesner, R. P. and Warthen, D. K. (2010). Implications of ca3 nmda and opiate receptors for spatial pattern completion in rats, *Hippocampus* **20**(4): 550–7.

Kesner, R. P., Gilbert, P. E. and Barua, L. A. (2002). The role of the hippocampus in memory for the temporal order of a sequence of odors, *Behavioral Neuroscience* **116**: 286–290.

Kesner, R. P., Lee, I. and Gilbert, P. (2004). A behavioral assessment of hippocampal function based on a subregional analysis, *Reviews in the Neurosciences* **15**: 333–351.

Kesner, R. P., Hunsaker, M. R. and Gilbert, P. E. (2005). The role of CA1 in the acquisition of an object-trace-odor paired associate task, *Behavioral Neuroscience* **119**: 781–786.

Kesner, R. P., Hunsaker, M. R. and Warthen, M. W. (2008). The ca3 subregion of the hippocampus is critical for episodic memory processing by means of relational encoding in rats, *Behavioral Neuroscience* **122**(6): 1217–1225.

Kesner, R. P., Hunsaker, M. R. and Ziegler, W. (2011). The role of the dorsal and ventral hippocampus in olfactory working memory, *Neurobiol Learn Mem* **96**(2): 361–6.

Kesner, R. P., Hui, X., Sommer, T., Wright, C., Barrera, V. R. and Fanselow, M. S. (2014). The role of postnatal neurogenesis in supporting remote memory and spatial metric processing, *Hippocampus* **24**(12): 1663–71.

Keysers, C. and Perrett, D. I. (2002). Visual masking and RSVP reveal neural competition, *Trends in Cognitive Sciences* **6**: 120–125.

Keysers, C., Xiao, D., Foldiak, P. and Perrett, D. (2001). The speed of sight, *Journal of Cognitive Neuroscience* **13**: 90–101.

Khaligh-Razavi, S.-M. and Kriegeskorte, N. (2014). Deep supervised, but not unsupervised, models may explain it cortical representation, *PLoS Computational Biology* **10**(11): e1003915.

Khalil, H. (1996). *Nonlinear Systems*, Prentice Hall, Upper Saddle River, NJ.

Kievit, J. and Kuypers, H. G. J. M. (1975). Subcortical afferents to the frontal lobe in the rhesus monkey studied by means of retrograde horseradish peroxidase transport, *Brain Research* **85**: 261–266.

Killian, N. J., Jutras, M. J. and Buffalo, E. A. (2012). A map of visual space in the primate entorhinal cortex, *Nature* **491**: 761–764.

Kim, J. G. and Biederman, I. (2012). Greater sensitivity to nonaccidental than metric changes in the relations between simple shapes in the lateral occipital cortex, *Neuroimage* **63**(4): 1818–26.

Kircher, T. T. and Thienel, R. (2005). Functional brain imaging of symptoms and cognition in schizophrenia, *Progress in Brain Research* **150**: 299–308.

Kirkwood, A., Dudek, S. M., Gold, J. T., Aizenman, C. D. and Bear, M. F. (1993). Common forms of synaptic plasticity in the hippocampus and neocortex "in vitro", *Science* **260**: 1518–1521.

Kjaer, T. W., Hertz, J. A. and Richmond, B. J. (1994). Decoding cortical neuronal signals: network models, information estimation and spatial tuning, *Journal of Computational Neuroscience* **1**: 109–139.

Kleinfeld, D. (1986). Sequential state generation by model neural networks, *Proceedings of the National Academy of Sciences of the USA* **83**: 9469–9473.

Knierim, J. J. and Neunuebel, J. P. (2016). Tracking the flow of hippocampal computation: Pattern separation, pattern completion, and attractor dynamics, *Neurobiology of Learning and Memory* **129**: 38–49.

Knierim, J. J., Neunuebel, J. P. and Deshmukh, S. S. (2014). Functional correlates of the lateral and medial entorhinal cortex: objects, path integration and local-global reference frames, *Philosophical Transactions of the Royal Society London B Biological Sciences* **369**: 20130369.

Knudsen, E. I. (2011). Control from below: the role of a midbrain network in spatial attention, *European Journal of Neuroscience* **33**: 1961–1972.

Kobayashi, S. (2012). Organization of neural systems for aversive information processing: pain, error, and punishment, *Frontiers in Neuroscience* **6**: 136.

Koch, C. (1999). *Biophysics of Computation*, Oxford University Press, Oxford.

Koch, C. (2004). *The Quest for Consciousness*, Roberts, Englewood, CO.

Koch, K. W. and Fuster, J. M. (1989). Unit activity in monkey parietal cortex related to haptic perception and temporary memory, *Experimental Brain Research* **76**: 292–306.

Koenderink, J. J. (1990). *Solid Shape*, MIT Press, Cambridge, MA.

Koenderink, J. J. and Van Doorn, A. J. (1979). The internal representation of solid shape with respect to vision, *Biological Cybernetics* **32**: 211–217.

Koenderink, J. J. and van Doorn, A. J. (1991). Affine structure from motion, *Journal of the Optical Society of America, A* **8**: 377–385.

Kohonen, T. (1977). *Associative Memory: A System Theoretical Approach*, Springer, New York.

Kohonen, T. (1982). Clustering, taxonomy, and topological maps of patterns, *in* M. Lang (ed.), *Proceedings of the Sixth International Conference on Pattern Recognition*, IEEE Computer Society Press, Silver Spring, MD, pp. 114–125.

Kohonen, T. (1988). *Self-Organization and Associative Memory*, 2nd edn, Springer-Verlag, New York.

Kohonen, T. (1989). *Self-Organization and Associative Memory*, 3rd (1984, 1st edn; 1988, 2nd edn) edn, Springer-Verlag, Berlin.

Kohonen, T. (1995). *Self-Organizing Maps*, Springer-Verlag, Berlin.

Kohonen, T., Oja, E. and Lehtio, P. (1981). Storage and processing of information in distributed memory systems, *in* G. E. Hinton and J. A. Anderson (eds), *Parallel Models of Associative Memory*, Erlbaum, Hillsdale, NJ, chapter 4, pp. 105–143.

Kolb, B. and Whishaw, I. Q. (2015). *Fundamentals of Human Neuropsychology*, 7th edn, Macmillan, New York.

Kondo, H., Lavenex, P. and Amaral, D. G. (2009). Intrinsic connections of the macaque monkey hippocampal formation: II. CA3 connections, *Journal of Comparative Neurology* **515**: 349–377.

Konopaske, G. T., Lange, N., Coyle, J. T. and Benes, F. M. (2014). Prefrontal cortical dendritic spine pathology in schizophrenia and bipolar disorder, *JAMA Psychiatry* **71**(12): 1323–31.

Kosar, E., Grill, H. J. and Norgren, R. (1986). Gustatory cortex in the rat. II. Thalamocortical projections, *Brain Research* **379**: 342–352.

Kosslyn, S. M. (1994). *Image and Brain: The Resolution of the Imagery Debate*, MIT Press, Cambridge, MA.

Kourtzi, Z. and Connor, C. E. (2011). Neural representations for object perception: structure, category, and adaptive coding, *Annu Rev Neurosci* **34**: 45–67.

Kovacs, G., Vogels, R. and Orban, G. A. (1995). Cortical correlate of pattern backward masking, *Proceedings of the National Academy of Sciences USA* **92**: 5587–5591.

Kraus, B. J., Brandon, M. P., Robinson, R. J., n., Connerney, M. A., Hasselmo, M. E. and Eichenbaum, H. (2015). During running in place, grid cells integrate elapsed time and distance run, *Neuron* **88**: 578–589.

Krauzlis, R. J., Lovejoy, L. P. and Zenon, A. (2013). Superior colliculus and visual spatial attention, *Annual Review of Neuroscience* **36**: 165–182.

Krebs, J. R. and Davies, N. B. (1991). *Behavioural Ecology*, 3rd edn, Blackwell, Oxford.

Krebs, J. R. and Kacelnik, A. (1991). Decision making, *in* J. R. Krebs and N. B. Davies (eds), *Behavioural Ecology*, 3rd edn, Blackwell, Oxford, chapter 4, pp. 105–136.

Kreiman, G., Koch, C. and Freid, I. (2000). Category-specific visual responses of single neurons in the human temporal lobe, *Nature Neuroscience* **3**: 946–953.

Kringelbach, M. L. and Rolls, E. T. (2003). Neural correlates of rapid reversal learning in a simple model of human social interaction, *Neuroimage* **20**: 1371–1383.

Kringelbach, M. L. and Rolls, E. T. (2004). The functional neuroanatomy of the human orbitofrontal cortex: evidence from neuroimaging and neuropsychology, *Progress in Neurobiology* **72**: 341–372.

Kringelbach, M. L., O'Doherty, J., Rolls, E. T. and Andrews, C. (2003). Activation of the human orbitofrontal cortex to a liquid food stimulus is correlated with its subjective pleasantness, *Cerebral Cortex* **13**: 1064–1071.

Krizhevsky, A., Sutskever, I. and Hinton, G. E. (2012). Imagenet classification with deep convolutional neural networks, *in* F. Pereira, C. Burges, L. Bottou and K. Weinberger (eds), *Advances in Neural Information Processing Systems 25*, Curran Associates, Inc., pp. 1097–1105.

Kropff, E. and Treves, A. (2005). The storage capacity of Potts models for semantic memory retrieval, *Journal of Statistical Mechanics: Theory and Experiment* **2005**: P08010.

Kropff, E. and Treves, A. (2008). The emergence of grid cells: Intelligent design or just adaptation?, *Hippocampus* **18**: 1256–1269.

Kropff, E., Carmichael, J. E., Moser, M. B. and Moser, E. I. (2015). Speed cells in the medial entorhinal cortex, *Nature* **523**: 419–424.

Kubota, Y. (2014). Untangling gabaergic wiring in the cortical microcircuit, *Current Opinion in Neurobiology* **26**: 7–14.

Lally, N., Nugent, A. C., Luckenbaugh, D. A., Niciu, M. J., Roiser, J. P. and Zarate, C. A., J. (2015). Neural correlates of change in major depressive disorder anhedonia following open-label ketamine, *Journal of Psychopharmacology* **29**: 596–607.

Lamme, V. A. F. (1995). The neurophysiology of figure-ground segregation in primary visual cortex, *Journal of Neuroscience* **15**: 1605–1615.

Land, M. F. (1999). Motion and vision: why animals move their eyes, *Journal of Comparative Physiology A* **185**: 341–352.

Land, M. F. and Collett, T. S. (1997). A survey of active vision in invertebrates, *in* M. V. Srinivasan and S. Venkatesh (eds), *From Living Eyes to Seeing Machines*, Oxford University Press, Oxford, pp. 16–36.

Lange, C. (1885). The emotions, *in* E. Dunlap (ed.), *The Emotions*, 1922 edn, Williams and Wilkins, Baltimore.

Lanthorn, T., Storm, J. and Andersen, P. (1984). Current-to-frequency transduction in CA1 hippocampal pyramidal cells: slow prepotentials dominate the primary range firing, *Experimental Brain Research* **53**: 431–443.

Larochelle, H. and Hinton, G. E. (2010). Learning to combine foveal glimpses with a third-order Boltzmann machine, *Advances in Neural Information Processing Systems (NIPS)* **1**: 1243–1251.

Lassalle, J. M., Bataille, T. and Halley, H. (2000). Reversible inactivation of the hippocampal mossy fiber synapses in mice impairs spatial learning, but neither consolidation nor memory retrieval, in the Morris navigation task, *Neurobiology of Learning and Memory* **73**: 243–257.

Lau, H. and Rosenthal, D. (2011). Empirical support for higher-order theories of conscious awareness, *Trends in Cognitive Sciences* **15**: 365–373.

Lau, H. C., Rogers, R. D. and Passingham, R. E. (2006). On measuring the perceived onsets of spontaneous actions, *Journal of Neuroscience* **26**: 7265–7271.

Lavenex, P. and Amaral, D. G. (2000). Hippocampal-neocortical interaction: a hierarchy of associativity, *Hippocampus* **10**: 420–430.

Lavenex, P., Suzuki, W. A. and Amaral, D. G. (2004). Perirhinal and parahippocampal cortices of the macaque monkey: Intrinsic projections and interconnections, *Journal of Comparative Neurology* **472**: 371–394.

Laxton, A. W., Neimat, J. S., Davis, K. D., Womelsdorf, T., Hutchison, W. D., Dostrovsky, J. O., Hamani, C., Mayberg, H. S. and Lozano, A. M. (2013). Neuronal coding of implicit emotion categories in the subcallosal cortex in patients with depression, *Biological Psychiatry* **74**: 714–719.

Leak, G. K. and Christopher, S. B. (1982). Freudian psychoanalysis and sociobiology: a synthesis, *American Psychologist* **37**: 313–322.

LeCun, Y., Kavukcuoglu, K. and Farabet, C. (2010). Convolutional networks and applications in vision, *2010 IEEE International Symposium on Circuits and Systems* pp. 253–256.

LeCun, Y., Bengio, Y. and Hinton, G. (2015). Deep learning, *Nature* **521**: 436–444.

LeDoux, J. E. (1992). Emotion and the amygdala, *in* J. P. Aggleton (ed.), *The Amygdala*, Wiley-Liss, New York, chapter 12, pp. 339–351.

LeDoux, J. E. (1995). Emotion: clues from the brain, *Annual Review of Psychology* **46**: 209–235.

LeDoux, J. E. (1996). *The Emotional Brain*, Simon and Schuster, New York.

LeDoux, J. E. (2008). Emotional coloration of consciousness: how feelings come about, *in* L. Weiskrantz and M. Davies (eds), *Frontiers of Consciousness*, Oxford University Press, Oxford, pp. 69–130.

LeDoux, J. E. (2011). Rethinking the emotional brain, *Neuron* **73**: 653–676.

Lee, H., Grosse, R., Ranganath, R. and Ng, A. Y. (2011). Unsupervised learning of hierarchical representations with convolutional deep belief networks, *Communications of the ACM* **54**: 95–103.

Lee, I. and Kesner, R. P. (2002). Differential contribution of NMDA receptors in hippocampal subregions to spatial working memory, *Nature Neuroscience* **5**: 162–168.

Lee, I. and Kesner, R. P. (2003a). Differential roles of dorsal hippocampal subregions in spatial working memory with short versus intermediate delay, *Behavioral Neuroscience* **117**: 1044–1053.

Lee, I. and Kesner, R. P. (2003b). Time-dependent relationship between the dorsal hippocampus and the prefrontal cortex in spatial memory, *Journal of Neurosciennce* **23**: 1517–1523.

Lee, I. and Kesner, R. P. (2004a). Differential contributions of dorsal hippocampal subregions to memory acquisition and retrieval in contextual fear-conditioning, *Hippocampus* **14**: 301–310.

Lee, I. and Kesner, R. P. (2004b). Encoding versus retrieval of spatial memory: double dissociation between the dentate gyrus and the perforant path inputs into CA3 in the dorsal hippocampus, *Hippocampus* **14**: 66–76.

Lee, I., Rao, G. and Knierim, J. J. (2004). A double dissociation between hippocampal subfields: differential time course of CA3 and CA1 place cells for processing changed environments, *Neuron* **42**: 803–815.

Lee, I., Jerman, T. S. and Kesner, R. P. (2005). Disruption of delayed memory for a sequence of spatial locations following CA1 or CA3 lesions of the dorsal hippocampus, *Neurobiology of Learning and Memory* **84**: 138–147.

Lee, T. S. (1996). Image representation using 2D Gabor wavelets, *IEEE Transactions on Pattern Analysis and Machine Intelligence* **18,10**: 959–971.

Lee, T. S., Mumford, D., Romero, R. D. and Lamme, V. A. F. (1998). The role of primary visual cortex in higher level vision, *Vision Research* **38**: 2429–2454.

Lee, T. S., Yang, C. Y., Romero, R. D. and Mumford, D. (2002). Neural activity in early visual cortex reflects behavioral experience and higher order perceptual saliency, *Nature Neuroscience* **5**: 589–597.

Lehky, S. R., Sejnowski, T. J. and Desimone, R. (2005). Selectivity and sparseness in the responses of striate complex cells, *Vision Research* **45**: 57–73.

Lehn, H., Steffenach, H. A., van Strien, N. M., Veltman, D. J., Witter, M. P. and Haberg, A. K. (2009). A specific role of the human hippocampus in recall of temporal sequences, *Journal of Neuroscience* **29**: 3475–3484.

Leibowitz, S. F. and Hoebel, B. G. (1998). Behavioral neuroscience and obesity, *in* G. A. Bray, C. Bouchard and P. T. James (eds), *The Handbook of Obesity*, Dekker, New York, pp. 313–358.

Leonard, C. M., Rolls, E. T., Wilson, F. A. W. and Baylis, G. C. (1985). Neurons in the amygdala of the monkey with responses selective for faces, *Behavioural Brain Research* **15**: 159–176.

Leuner, K. and Muller, W. E. (2006). The complexity of the dopaminergic synapses and their modulation by antipsychotics, *Pharmacopsychiatry* **39 Suppl 1**: S15–20.

Leutgeb, J. K., Leutgeb, S., Moser, M. B. and Moser, E. I. (2007). Pattern separation in the dentate gyrus and CA3 of the hippocampus, *Science* **315**: 961–966.

Leutgeb, S. and Leutgeb, J. K. (2007). Pattern separation, pattern completion, and new neuronal codes within a continuous CA3 map, *Learning and Memory* **14**: 745–757.

Leutgeb, S., Leutgeb, J. K., Treves, A., Moser, M. B. and Moser, E. I. (2004). Distinct ensemble codes in hippocampal areas CA3 and CA1, *Science* **305**: 1295–1298.

Leutgeb, S., Leutgeb, J. K., Treves, A., Meyer, R., Barnes, C. A., McNaughton, B. L., Moser, M.-B. and Moser, E. I. (2005). Progressive transformation of hippocampal neuronal representations in "morphed" environments, *Neuron* **48**: 345–358.

LeVay, S., Wiesel, T. N. and Hubel, D. H. (1980). The development of ocular dominance columns in normal and visually deprived monkeys, *Journal of Comparative Neurology* **191**: 1–51.

Levitt, J. B., Lund, J. S. and Yoshioka, T. (1996). Anatomical substrates for early stages in cortical processing of visual information in the macaque monkey, *Behavioural Brain Research* **76**: 5–19.

Levy, W. B. (1985). Associative changes in the synapse: LTP in the hippocampus, *in* W. B. Levy, J. A. Anderson and S. Lehmkuhle (eds), *Synaptic Modification, Neuron Selectivity, and Nervous System Organization*, Erlbaum, Hillsdale, NJ, chapter 1, pp. 5–33.

Levy, W. B. (1989). A computational approach to hippocampal function, *in* R. D. Hawkins and G. H. Bower (eds), *Computational Models of Learning in Simple Neural Systems*, Academic Press, San Diego, pp. 243–305.

Levy, W. B. (1996). A sequence predicting ca3 is a flexible associator that learns and uses context to solve hippocampal-like tasks, *Hippocampus* **6**: 579–590.

Levy, W. B. and Baxter, R. A. (1996). Energy efficient neural codes, *Neural Computation* **8**: 531–543.

Levy, W. B. and Desmond, N. L. (1985). The rules of elemental synaptic plasticity, *in* W. B. Levy, J. A. Anderson and S. Lehmkuhle (eds), *Synaptic Modification, Neuron Selectivity, and Nervous System Organization*, Erlbaum, Hillsdale, NJ, chapter 6, pp. 105–121.

Levy, W. B., Colbert, C. M. and Desmond, N. L. (1990). Elemental adaptive processes of neurons and synapses: a statistical/computational perspective, *in* M. Gluck and D. Rumelhart (eds), *Neuroscience and Connectionist Theory*, Erlbaum, Hillsdale, NJ, chapter 5, pp. 187–235.

Levy, W. B., Wu, X. and Baxter, R. A. (1995). Unification of hippocampal function via computational/encoding considerations, *International Journal of Neural Systems* **6, Suppl.**: 71–80.

Lewis, D. A. (2014). Inhibitory neurons in human cortical circuits: substrate for cognitive dysfunction in schizophrenia, *Curr Opin Neurobiol* **26**: 22–6.

Lewis, M. D. (2005). Bridging emotion theory and neurobiology through dynamic systems modeling, *Behavioral and Brain Sciences* **28**: 169–245.

Li, C. S. and Cho, Y. K. (2006). Efferent projection from the bed nucleus of the stria terminalis suppresses activity of taste-responsive neurons in the hamster parabrachial nuclei, *American Journal of Physiology Regul Integr Comp Physiol* **291**: R914–R926.

Li, C. S., Cho, Y. K. and Smith, D. V. (2002). Taste responses of neurons in the hamster solitary nucleus are modulated by the central nucleus of the amygdala, *Journal of Neurophysiology* **88**: 2979–2992.

Li, H., Matsumoto, K. and Watanabe, H. (1999). Different effects of unilateral and bilateral hippocampal lesions in rats on the performance of radial maze and odor-paired associate tasks, *Brain Research Bulletin* **48**: 113–119.

Libet, B. (2002). The timing of mental events: Libet's experimental findings and their implications, *Consciousness and Cognition* **11**: 291–299; discussion 304–333.

Liddle, P. F. (1987). The symptoms of chronic schizophrenia: a re-examination of the positive-negative dichotomy, *British Journal of Psychiatry* **151**: 145–151.

Lieberman, J. A., Perkins, D., Belger, A., Chakos, M., Jarskog, F., Boteva, K. and Gilmore, J. (2001). The early stages of schizophrenia: speculations on pathogenesis, pathophysiology, and therapeutic approaches, *Biological Psychiatry* **50**: 884–897.

Lisberger, S. G. and Thach, W. T. (2013). The cerebellum, *in* E. Kandel, J. H. Schwartz, T. M. Jessell, S. A. Siegelbaum and A. J. Hudspeth (eds), *Principles of Neural Science*, 5th edn, McGraw-Hill, New York, chapter 24, pp. 960–981.

Lisman, J. E. and Idiart, M. A. (1995). Storage of 7 ± 2 short-term memories in oscillatory subcycles, *Science* **267**: 1512–1515.

Lisman, J. E. and Idiart, M. A. (1999). Relating hippocampal circuitry to function: recall of memory sequences by reciprocal dentate-CA3 interactions, *Neuron* **22**: 233–242.

Lisman, J. E., Fellous, J. M. and Wang, X. J. (1998). A role for NMDA-receptor channels in working memory, *Nature Neuroscience* **1**: 273–275.

Lisman, J. E., Talamini, L. M. and Raffone, A. (2005). Recall of memory sequences by interaction of the dentate and CA3: a revised model of the phase precession, *Neural Networks* **18**: 1191–1201.

Little, W. A. (1974). The existence of persistent states in the brain, *Mathematical Bioscience* **19**: 101–120.

Liu, Y. and Wang, X.-J. (2001). Spike-frequency adaptation of a generalized leaky integrate-and-fire model neuron, *Journal of Computational Neuroscience* **10**: 25–45.

Liu, Y. H. and Wang, X.-J. (2008). A common cortical circuit mechanism for perceptual categorical discrimination and veridical judgment, *PLoS Computational Biology* p. e1000253.

Logothetis, N. K. (2008). What we can do and what we cannot do with fMRI, *Nature* **453**: 869–878.

Logothetis, N. K. and Sheinberg, D. L. (1996). Visual object recognition, *Annual Review of Neuroscience* **19**: 577–621.

Logothetis, N. K., Pauls, J., Bulthoff, H. H. and Poggio, T. (1994). View-dependent object recognition by monkeys, *Current Biology* **4**: 401–414.

Logothetis, N. K., Pauls, J. and Poggio, T. (1995). Shape representation in the inferior temporal cortex of monkeys, *Current Biology* **5**: 552–563.

Logothetis, N. K., Pauls, J., Augath, M., Trinath, T. and Oeltermann, A. (2001). Neurophysiological investigation of the basis of the fMRI signal, *Nature* **412**: 150–157.

Loh, M., Rolls, E. T. and Deco, G. (2007a). A dynamical systems hypothesis of schizophrenia, *PLoS Computational Biology* **3**: e228. doi:10.1371/journal.pcbi.0030228.

Loh, M., Rolls, E. T. and Deco, G. (2007b). Statistical fluctuations in attractor networks related to schizophrenia,

Pharmacopsychiatry **40**: S78–84.

Lowe, D. (1985). *Perceptual Organization and Visual Recognition*, Kluwer, Boston.

Lozano, A. M., Giacobbe, P., Hamani, C., Rizvi, S. J., Kennedy, S. H., Kolivakis, T. T., Debonnel, G., Sadikot, A. F., Lam, R. W., Howard, A. K., Ilcewicz-Klimek, M., Honey, C. R. and Mayberg, H. S. (2012). A multicenter pilot study of subcallosal cingulate area deep brain stimulation for treatment-resistant depression, *Journal of Neurosurgery* **116**: 315–322.

Luce, R. D. (1986). *Response Times: Their Role in Inferring Elementary Mental Organization*, Oxford University Press, New York.

Luck, S. J., Chelazzi, L., Hillyard, S. A. and Desimone, R. (1997). Neural mechanisms of spatial selective attention in areas V1, V2, and V4 of macaque visual cortex, *Journal of Neurophysiology* **77**: 24–42.

Lujan, J. L., Chaturvedi, A., Choi, K. S., Holtzheimer, P. E., Gross, R. E., Mayberg, H. S. and McIntyre, C. C. (2013). Tractography-activation models applied to subcallosal cingulate deep brain stimulation, *Brain Stimulation* **6**: 737–739.

Luk, C. H. and Wallis, J. D. (2009). Dynamic encoding of responses and outcomes by neurons in medial prefrontal cortex, *Journal of Neuroscience* **29**: 7526–7539.

Lund, J. S. (1984). Spiny stellate neurons, *in* A. Peters and E. Jones (eds), *Cerebral Cortex, Vol. 1, Cellular Components of the Cerebral Cortex*, Plenum, New York, chapter 7, pp. 255–308.

Lundy, R. F., J. and Norgren, R. (2004). Activity in the hypothalamus, amygdala, and cortex generates bilateral and convergent modulation of pontine gustatory neurons, *Journal of Neurophysiology* **91**: 1143–1157.

Luo, Q., Ge, T., Grabenhorst, F., Feng, J. and Rolls, E. T. (2013). Attention-dependent modulation of cortical taste circuits revealed by Granger causality with signal-dependent noise, *PLoS Computational Biology* **9**: e1003265.

Luskin, M. B. and Price, J. L. (1983). The topographic organization of associational fibers of the olfactory system in the rat, including centrifugal fibers to the olfactory bulb, *Journal of Comparative Neurology* **216**: 264–291.

Lycan, W. G. (1997). Consciousness as internal monitoring, *in* N. Block, O. Flanagan and G. Guzeldere (eds), *The Nature of Consciousness: Philosophical Debates*, MIT Press, Cambridge, MA, pp. 755–771.

Lynch, G. and Gall, C. M. (2006). AMPAkines and the threefold path to cognitive enhancement, *Trends in Neuroscience* **29**: 554–562.

Lynch, M. A. (2004). Long-term potentiation and memory, *Physiological Reviews* **84**: 87–136.

Ma, Y. (2015). Neuropsychological mechanism underlying antidepressant effect: a systematic meta-analysis, *Molecular Psychiatry* **20**: 311–319.

Maaswinkel, H., Jarrard, L. E. and Whishaw, I. Q. (1999). Hippocampectomized rats are impaired in homing by path integration, *Hippocampus* **9**: 553–561.

MacDonald, C. J., Lepage, K. Q., Eden, U. T. and Eichenbaum, H. (2011). Hippocampal "time cells" bridge the gap in memory for discontiguous events, *Neuron* **71**: 737–749.

MacGregor, R. J. (1987). *Neural and Brain Modelling*, Academic Press, San Diego, CA.

MacKay, D. M. and McCulloch, W. S. (1952). The limiting information capacity of a neuronal link, *Bulletin of Mathematical Biophysics* **14**: 127–135.

Madsen, J. and Kesner, R. P. (1995). The temporal-distance effect in subjects with dementia of the Alzheimer type, *Alzheimer's Disease and Associated Disorders* **9**: 94–100.

Maguire, E. A. (2014). Memory consolidation in humans: new evidence and opportunities, *Experimental Physiology* **99**: 471–486.

Maier, A., Logothetis, N. K. and Leopold, D. A. (2005). Global competition dictates local suppression in pattern rivalry, *Journal of Vision* **5**: 668–677.

Malhotra, A. K., Pinals, D. A., Weingartner, H., Sirocco, K., Missar, C. D., Pickar, D. and Breier, A. (1996). NMDA receptor function and human cognition: the effects of ketamine in healthy volunteers, *Neuropsychopharmacology* **14**: 301–307.

Malkova, L. and Mishkin, M. (2003). One-trial memory for object-place associations after separate lesions of hippocampus and posterior parahippocampal region in the monkey, *Journal of Neuroscience* **23**: 1956–1965.

Malkova, L., Bachevalier, J., Mishkin, M. and Saunders, R. C. (2001). Neurotoxic lesions of perirhinal cortex impair visual recognition memory in rhesus monkeys, *Neuroreport* **12**: 1913–1917.

Malsburg, C. v. d. (1973). Self-organization of orientation-sensitive columns in the striate cortex, *Kybernetik* **14**: 85–100.

Malsburg, C. v. d. (1990a). A neural architecture for the representation of scenes, *in* J. L. McGaugh, N. M. Weinburger and G. Lynch (eds), *Brain Organization and Memory: Cells, Systems and Circuits*, Oxford University Press, Oxford, chapter 18, pp. 356–372.

Malsburg, C. v. d. (1990b). A neural architecture for the representation of scenes, *in* J. L. McGaugh, N. M. Weinberger and G. Lynch (eds), *Brain Organization and Memory: Cells, Systems and Circuits*, Oxford University Press, New York, chapter 19, pp. 356–372.

Mangia, S., Giove, F., Tkac, I., Logothetis, N. K., Henry, P. G., Olman, C. A., Maraviglia, B., Di Salle, F. and Ugurbil, K. (2009). Metabolic and hemodynamic events after changes in neuronal activity: current hypotheses, theoretical predictions and in vivo NMR experimental findings, *Journal of Cerebral Blood Flow and Metabolism* **29**: 441–463.

Manwani, A. and Koch, C. (2001). Detecting and estimating signals over noisy and unreliable synapses: information-theoretic analysis, *Neural Computation* **13**: 1–33.

Markov, N. T. and Kennedy, H. (2013). The importance of being hierarchical, *Current Opinion in Neurobiology* **23**: 187–194.

Markov, N. T., Ercsey-Ravasz, M., Lamy, C., Ribeiro Gomes, A. R., Magrou, L., Misery, P., Giroud, P., Barone, P., Dehay, C., Toroczkai, Z., Knoblauch, K., Van Essen, D. C. and Kennedy, H. (2013a). The role of long-range connections on the specificity of the macaque interareal cortical network, *Proceedings of the National Academy of Sciences USA* **110**: 5187–5192.

Markov, N. T., Ercsey-Ravasz, M., Van Essen, D. C., Knoblauch, K., Toroczkai, Z. and Kennedy, H. (2013b). Cortical high-density counterstream architectures, *Science* **342**(6158): 1238406.

Markov, N. T., Ercsey-Ravasz, M. M., Ribeiro Gomes, A. R., Lamy, C., Magrou, L., Vezoli, J., Misery, P., Falchier, A., Quilodran, R., Gariel, M. A., Sallet, J., Gamanut, R., Huissoud, C., Clavagnier, S., Giroud, P., Sappey-Marinier, D., Barone, P., Dehay, C., Toroczkai, Z., Knoblauch, K., Van Essen, D. C. and Kennedy, H. (2014a). A weighted and directed interareal connectivity matrix for macaque cerebral cortex, *Cerebral Cortex* **24**: 17–36.

Markov, N. T., Vezoli, J., Chameau, P., Falchier, A., Quilodran, R., Huissoud, C., Lamy, C., Misery, P., Giroud, P., Ullman, S., Barone, P., Dehay, C., Knoblauch, K. and Kennedy, H. (2014b). Anatomy of hierarchy: Feedforward and feedback pathways in macaque visual cortex, *Journal of Comparative Neurology* **522**: 225–259.

Markram, H. and Segal, M. (1990). Acetylcholine potentiates responses to N-methyl-D-aspartate in the rat hippocampus, *Neurosci Letters* **113**: 62–65.

Markram, H. and Segal, M. (1992). The inositol 1,4,5-triphosphate pathway mediates cholinergic potentiation of rat hippocampal neuronal responses to NMDA, *Journal of Physiology* **447**: 513–533.

Markram, H. and Tsodyks, M. (1996). Redistribution of synaptic efficacy between neocortical pyramidal neurons, *Nature* **382**: 807–810.

Markram, H., Lübke, J., Frotscher, M. and Sakmann, B. (1997). Regulation of synaptic efficacy by coincidence of postsynaptic APs and EPSPs, *Science* **275**: 213–215.

Markram, H., Pikus, D., Gupta, A. and Tsodyks, M. (1998). Information processing with frequency-dependent synaptic connections, *Neuropharmacology* **37**: 489–500.

Markram, H., Gerstner, W. and Sjöström, P. J. (2012). Spike-timing-dependent plasticity: a comprehensive overview, *Frontiers in Synaptic Neuroscience* **4**: 2.

Markram, H., Muller, E., Ramaswamy, S., Reimann, M. W., Abdellah, M., Sanchez, C. A., Ailamaki, A., Alonso-Nanclares, L., Antille, N., Arsever, S., Kahou, G. A., Berger, T. K., Bilgili, A., Buncic, N., Chalimourda, A., Chindemi, G., Courcol, J. D., Delalondre, F., Delattre, V., Druckmann, S., Dumusc, R., Dynes, J., Eilemann, S., Gal, E., Gevaert, M. E., Ghobril, J. P., Gidon, A., Graham, J. W., Gupta, A., Haenel, V., Hay, E., Heinis, T., Hernando, J. B., Hines, M., Kanari, L., Keller, D., Kenyon, J., Khazen, G., Kim, Y., King, J. G., Kisvarday, Z., Kumbhar, P., Lasserre, S., Le Be, J. V., Magalhaes, B. R., Merchan-Perez, A., Meystre, J., Morrice, B. R., Muller, J., Munoz-Cespedes, A., Muralidhar, S., Muthurasa, K., Nachbaur, D., Newton, T. H., Nolte, M., Ovcharenko, A., Palacios, J., Pastor, L., Perin, R., Ranjan, R., Riachi, I., Rodriguez, J. R., Riquelme, J. L., Rossert, C., Sfyrakis, K., Shi, Y., Shillcock, J. C., Silberberg, G., Silva, R., Tauheed, F., Telefont, M., Toledo-Rodriguez, M., Trankler, T., Van Geit, W., Diaz, J. V., Walker, R., Wang, Y., Zaninetta, S. M., DeFelipe, J., Hill, S. L., Segev, I. and Schurmann, F. (2015). Reconstruction and simulation of neocortical microcircuitry, *Cell* **163**(2): 456–92.

Markus, E. J., Qin, Y. L., Leonard, B., Skaggs, W., McNaughton, B. L. and Barnes, C. A. (1995). Interactions between location and task affect the spatial and directional firing of hippocampal neurons, *Journal of Neuroscience* **15**: 7079–7094.

Marr, D. (1969). A theory of cerebellar cortex, *Journal of Physiology* **202**: 437–470.

Marr, D. (1970). A theory for cerebral cortex, *Proceedings of The Royal Society of London, Series B* **176**: 161–234.

Marr, D. (1971). Simple memory: a theory for archicortex, *Philosophical Transactions of The Royal Society of London, Series B* **262**: 23–81.

Marr, D. (1982). *Vision*, Freeman, San Francisco.

Marr, D. and Nishihara, H. K. (1978). Representation and recognition of the spatial organization of three dimensional structure, *Proceedings of the Royal Society of London B* **200**: 269–294.

Marshuetz, C. (2005). Order information in working memory: an integrative review of evidence from brain and behavior, *Psychological Bulletin* **131**: 323–339.

Marti, D., Deco, G., Del Giudice, P. and Mattia, M. (2006). Reward-biased probabilistic decision-making: mean-field predictions and spiking simulations, *Neurocomputing* **39**: 1175–1178.

Marti, D., Deco, G., Mattia, M., Gigante, G. and Del Giudice, P. (2008). A fluctuation-driven mechanism for slow decision processes in reverberant networks, *PLoS ONE* **3**: e2534. doi:10.1371/journal.pone.0002534.

Martin, K. A. C. (1984). Neuronal circuits in cat striate cortex, *in* E. Jones and A. Peters (eds), *Cerebral Cortex, Vol. 2, Functional Properties of Cortical Cells*, Plenum, New York, chapter 9, pp. 241–284.

Martin, S. J., Grimwood, P. D. and Morris, R. G. (2000). Synaptic plasticity and memory: an evaluation of the hypothesis, *Annual Review of Neuroscience* **23**: 649–711.

Martinez, C. O., Do, V. H., Martinez, J. L. J. and Derrick, B. E. (2002). Associative long-term potentiation (LTP) among extrinsic afferents of the hippocampal CA3 region in vivo, *Brain Research* **940**: 86–94.

Martinez-Garcia, M., Rolls, E. T., Deco, G. and Romo, R. (2011). Neural and computational mechanisms of postponed decisions, *Proceedings of the National Academy of Sciences* **108**: 11626–11631.

Martinez-Trujillo, J. and Treue, S. (2002). Attentional modulation strength in cortical area MT depends on stimulus contrast, *Neuron* **35**: 365–370.

Mascaro, M. and Amit, D. J. (1999). Effective neural response function for collective population states, *Network* **10**: 351–373.

Mason, A. and Larkman, A. (1990). Correlations between morphology and electrophysiology of pyramidal neurones in slices of rat visual cortex. I. Electrophysiology, *Journal of Neuroscience* **10**: 1415–1428.

Masuda, N. and Aihara, K. (2003). Ergodicity of spike trains: when does trial averaging make sense?, *Neural Computation* **15**: 1341–1372.

Matsumoto, M. and Hikosaka, O. (2009). Two types of dopamine neuron distinctly convey positive and negative motivational signals, *Nature* **459**: 837–841.

Matsumoto, M., Matsumoto, K., Abe, H. and Tanaka, K. (2007). Medial prefrontal selectivity signalling prediction errors of action values, *Nature Neuroscience* **10**: 647–656.

Matsumura, N., Nishijo, H., Tamura, R., Eifuku, S., Endo, S. and Ono, T. (1999). Spatial- and task-dependent neuronal responses during real and virtual translocation in the monkey hippocampal formation, *Journal of Neuroscience* **19**: 2318–2393.

Mattia, M. and Del Giudice, P. (2002). Population dynamics of interacting spiking neurons, *Physical Review E* **66**: 051917.

Mattia, M. and Del Giudice, P. (2004). Finite-size dynamics of inhibitory and excitatory interacting spiking neurons, *Physical Review E* **70**: 052903.

Maunsell, J. H. R. (1995). The brain's visual world: representation of visual targets in cerebral cortex, *Science* **270**: 764–769.

Mayberg, H. S. (2003). Positron emission tomography imaging in depression: a neural systems perspective, *Neuroimaging Clinics of North America* **13**: 805–815.

Maynard Smith, J. (1982). *Evolution and the Theory of Games*, Cambridge University Press, Cambridge.

Maynard Smith, J. (1984). Game theory and the evolution of behaviour, *Behavioral and Brain Sciences* **7**: 95–125.

McAdams, C. and Maunsell, J. H. R. (1999). Effects of attention on orientation-tuning functions of single neurons in macaque cortical area V4, *Journal of Neuroscience* **19**: 431–441.

McCabe, C. and Rolls, E. T. (2007). Umami: a delicious flavor formed by convergence of taste and olfactory pathways in the human brain, *European Journal of Neuroscience* **25**: 1855–1864.

McCabe, C., Rolls, E. T., Bilderbeck, A. and McGlone, F. (2008). Cognitive influences on the affective representation of touch and the sight of touch in the human brain, *Social, Cognitive and Affective Neuroscience* **3**: 97–108.

McClelland, J. L. and Rumelhart, D. E. (1986). A distributed model of human learning and memory, *in* J. L. McClelland and D. E. Rumelhart (eds), *Parallel Distributed Processing*, Vol. 2, MIT Press, Cambridge, MA, chapter 17, pp. 170–215.

McClelland, J. L. and Rumelhart, D. E. (1988). *Explorations in Parallel Distributed Processing*, MIT Press, Cambridge, MA.

McClelland, J. L., McNaughton, B. L. and O'Reilly, R. C. (1995). Why there are complementary learning systems in the hippocampus and neocortex: insights from the successes and failures of connectionist models of learning and memory, *Psychological Review* **102**: 419–457.

McClure, S. M., Laibson, D. I., Loewenstein, G. and Cohen, J. D. (2004). Separate neural systems value immediate and delayed monetary rewards, *Science* **306**: 503–507.

McCormick, D. A. and Westbrook, G. L. (2013). Sleep and dreaming, *in* E. Kandel, J. H. Schwartz, T. M. Jessell, S. A. Siegelbaum and A. J. Hudspeth (eds), *Principles of Neural Science*, 5th edn, McGraw-Hill, New York, chapter 51, pp. 1140–1158.

McGaugh, J. L. (2000). Memory - a century of consolidation, *Science* **287**: 248–251.

McGinty, D. and Szymusiak, R. (1988). Neuronal unit activity patterns in behaving animals: brainstem and limbic system, *Annual Review of Psychology* **39**: 135–168.

McGurk, H. and MacDonald, J. (1976). Hearing lips and seeing voices, *Nature* **264**: 746–748.

McLeod, P., Plunkett, K. and Rolls, E. T. (1998). *Introduction to Connectionist Modelling of Cognitive Processes*, Oxford University Press, Oxford.

McNaughton, B. L. (1991). Associative pattern completion in hippocampal circuits: new evidence and new questions, *Brain Research Reviews* **16**: 193–220.

McNaughton, B. L. and Morris, R. G. M. (1987). Hippocampal synaptic enhancement and information storage within a distributed memory system, *Trends in Neuroscience* **10**: 408–415.

McNaughton, B. L. and Nadel, L. (1990). Hebb-Marr networks and the neurobiological representation of action in space, *in* M. A. Gluck and D. E. Rumelhart (eds), *Neuroscience and Connectionist Theory*, Erlbaum, Hillsdale, NJ, pp. 1–64.

McNaughton, B. L., Barnes, C. A. and O'Keefe, J. (1983). The contributions of position, direction, and velocity to single unit activity in the hippocampus of freely-moving rats, *Experimental Brain Research* **52**: 41–49.

McNaughton, B. L., Barnes, C. A., Meltzer, J. and Sutherland, R. J. (1989). Hippocampal granule cells are necessary for normal spatial learning but not for spatially selective pyramidal cell discharge, *Experimental Brain Research* **76**: 485–496.

McNaughton, B. L., Chen, L. L. and Markus, E. J. (1991). "Dead reckoning", landmark learning, and the sense of direction: a neurophysiological and computational hypothesis, *Journal of Cognitive Neuroscience* **3**: 190–202.

McNaughton, B. L., Battaglia, F. P., Jensen, O., Moser, E. I. and Moser, M.-B. (2006). Path integration and the neural basis of the hippocampal map, *Nature Reviews Neuroscience* **7**: 663–676.

Mease, R. A., Krieger, P. and Groh, A. (2014). Cortical control of adaptation and sensory relay mode in the thalamus, *Proceedings of the National Academy of Sciences* **111**: 6798–6803.

Medin, D. L. and Schaffer, M. M. (1978). Context theory of classification learning, *Psychological Review* **85**: 207–238.

Mel, B. W. (1997). SEEMORE: Combining color, shape, and texture histogramming in a neurally-inspired approach to visual object recognition, *Neural Computation* **9**: 777–804.

Mel, B. W. and Fiser, J. (2000). Minimizing binding errors using learned conjunctive features, *Neural Computation* **12**: 731–762.

Mel, B. W., Ruderman, D. L. and Archie, K. A. (1998). Translation-invariant orientation tuning in visual "complex" cells could derive from intradendritic computations, *Journal of Neuroscience* **18**(11): 4325–4334.

Menzies, L., Chamberlain, S. R., Laird, A. R., Thelen, S. M., Sahakian, B. J. and Bullmore, E. T. (2008). Integrating evidence from neuroimaging and neuropsychological studies of obsessive-compulsive disorder: The orbitofronto-striatal model revisited, *Neuroscience and Biobehavioral Reviews* **32**: 525–549.

Mesulam, M.-M. (1990). Human brain cholinergic pathways, *Progress in Brain Research* **84**: 231–241.

Mesulam, M. M. (1998). From sensation to cognition, *Brain* **121**: 1013–1084.

Meunier, M., Bachevalier, J. and Mishkin, M. (1997). Effects of orbital frontal and anterior cingulate lesions on object and spatial memory in rhesus monkeys, *Neuropsychologia* **35**: 999–1015.

Middleton, F. A. and Strick, P. L. (1996a). New concepts about the organization of the basal ganglia, *in* J. A. Obeso (ed.), *Advances in Neurology: The Basal Ganglia and the Surgical Treatment for Parkinson's Disease*, Raven, New York.

Middleton, F. A. and Strick, P. L. (1996b). The temporal lobe is a target of output from the basal ganglia, *Proceedings of the National Academy of Sciences of the USA* **93**: 8683–8687.

Middleton, F. A. and Strick, P. L. (2000). Basal ganglia output and cognition: evidence from anatomical, behavioral, and clinical studies, *Brain and Cognition* **42**: 183–200.

Mikami, A., Nakamura, K. and Kubota, K. (1994). Neuronal responses to photographs in the superior temporal sulcus of the rhesus monkey, *Behavioural Brain Research* **60**: 1–13.

Miller, E. K. and Buschman, T. J. (2013). Cortical circuits for the control of attention, *Current Opinion in Neurobiology* **23**: 216–222.

Miller, E. K. and Desimone, R. (1994). Parallel neuronal mechanisms for short-term memory, *Science* **263**: 520–522.

Miller, E. K., Gochin, P. M. and Gross, C. G. (1993a). Suppression of visual responses of neurons in inferior temporal cortex of the awake macaque by addition of a second stimulus, *Brain Research* **616**: 25–29.

Miller, E. K., Li, L. and Desimone, R. (1993b). Activity of neurons in anterior inferior temporal cortex during a short-term memory task, *Journal of Neuroscience* **13**: 1460–1478.

Miller, E. K., Nieder, A., Freedman, D. J. and Wallis, J. D. (2003). Neural correlates of categories and concepts, *Current Opinion in Neurobiology* **13**: 198–203.

Miller, G. A. (1955). Note on the bias of information estimates, *Information Theory in Psychology; Problems and Methods II-B* pp. 95–100.

Miller, J. F., Neufang, M., Solway, A., Brandt, A., Trippel, M., Mader, I., Hefft, S., Merkow, M., Polyn, S. M., Jacobs, J., Kahana, M. J. and Schulze-Bonhage, A. (2013). Neural activity in human hippocampal formation reveals the spatial context of retrieved memories, *Science* **342**: 1111–1114.

Miller, K. D. (1994). Models of activity-dependent neural development, *Progress in Brain Research* **102**: 303–308.

Miller, K. D. (2016). Canonical computations of cerebral cortex, *Curr Opin Neurobiol* **37**: 75–84.

Miller, P. and Wang, X.-J. (2006). Power-law neuronal fluctuations in a recurrent network model of parametric working memory, *Journal of Neurophysiology* **95**: 1099–1114.

Millikan, R. G. (1984). *Language, Thought, and Other Biological Categories: New Foundation for Realism*, MIT Press, Cambridge, MA.

Milner, A. (2008). Conscious and unconscious visual processing in the human brain, *in* L. Weiskrantz and M. Davies (eds), *Frontiers of Consciousness*, Oxford University Press, Oxford, chapter 5, pp. 169–214.

Milner, A. D. and Goodale, M. A. (1995). *The Visual Brain in Action*, Oxford University Press, Oxford.

Milner, P. (1974). A model for visual shape recognition, *Psychological Review* **81**: 521–535.

Milton, A. L., Lee, J. L., Butler, V. J., Gardner, R. and Everitt, B. J. (2008). Intra-amygdala and systemic antagonism of NMDA receptors prevents the reconsolidation of drug-associated memory and impairs subsequently both novel and previously acquired drug-seeking behaviors, *Journal of Neuroscience* **28**: 8230–8237.

Minai, A. A. and Levy, W. B. (1993). Sequence learning in a single trial, *International Neural Network Society World Congress of Neural Networks* **2**: 505–508.

Minsky, M. L. and Papert, S. A. (1969). *Perceptrons*, expanded 1988 edn, MIT Press, Cambridge, MA.

Miyamoto, S., Duncan, G. E., Marx, C. E. and Lieberman, J. A. (2005). Treatments for schizophrenia: a critical review of pharmacology and mechanisms of action of antipsychotic drugs, *Molecular Psychiatry* **10**: 79–104.

Miyashita, Y. (1988). Neuronal correlate of visual associative long-term memory in the primate temporal cortex, *Nature* **335**: 817–820.

Miyashita, Y. (1993). Inferior temporal cortex: where visual perception meets memory, *Annual Review of Neuroscience* **16**: 245–263.

Miyashita, Y. and Chang, H. S. (1988). Neuronal correlate of pictorial short-term memory in the primate temporal cortex, *Nature* **331**: 68–70.

Miyashita, Y., Rolls, E. T., Cahusac, P. M. B., Niki, H. and Feigenbaum, J. D. (1989). Activity of hippocampal neurons in the monkey related to a conditional spatial response task, *Journal of Neurophysiology* **61**: 669–678.

Miyashita, Y., Okuno, H., Tokuyama, W., Ihara, T. and Nakajima, K. (1996). Feedback signal from medial temporal lobe mediates visual associative mnemonic codes of inferotemporal neurons, *Cognitive Brain Research* **5**: 81–86.

Miyashita, Y., Kameyama, M., Hasegawa, I. and Fukushima, T. (1998). Consolidation of visual associative long-term memory in the temporal cortex of primates, *Neurobiology of Learning and Memory* **1**: 197–211.

Mizumori, S. J. and Tryon, V. L. (2015). Integrative hippocampal and decision-making neurocircuitry during goal-relevant predictions and encoding, *Progress in Brain Research* **219**: 217–242.

Mizumori, S. J., Ragozzino, K. E., Cooper, B. G. and Leutgeb, S. (1999). Hippocampal representational organization and spatial context, *Hippocampus* **9**: 444–451.

Moller, H. J. (2005). Antipsychotic and antidepressive effects of second generation antipsychotics: two different pharmacological mechanisms?, *European Archives of Psychiatry and Clinical Neuroscience* **255**: 190–201.

Molnar, Z., Kaas, J. H., de Carlos, J. A., Hevner, R. F., Lein, E. and Nemec, P. (2014). Evolution and development of the mammalian cerebral cortex, *Brain Behav Evol* **83**(2): 126–39.

Mombaerts, P. (2006). Axonal wiring in the mouse olfactory system, *Annual Review of Cell and Developmental Biology* **22**: 713–737.

Monaghan, D. T. and Cotman, C. W. (1985). Distribution of N-methyl-D-aspartate-sensitive L-[3H]glutamate-binding sites in the rat brain, *Journal of Neuroscience* **5**: 2909–2919.

Mongillo, G., Barak, O. and Tsodyks, M. (2008). Synaptic theory of working memory, *Science* **319**: 1543–1546.

Montagnini, A. and Treves, A. (2003). The evolution of mammalian cortex, from lamination to arealization, *Brain research bulletin* **60**(4): 387–393.

Montague, P. R., Gally, J. A. and Edelman, G. M. (1991). Spatial signalling in the development and function of neural connections, *Cerebral Cortex* **1**: 199–220.

Montague, P. R., King-Casas, B. and Cohen, J. D. (2006). Imaging valuation models in human choice, *Annual Review of Neuroscience* **29**: 417–448.

Moore, D. B., Lee, P., Paiva, M., Walker, D. W. and Heaton, M. B. (1998). Effects of neonatal ethanol exposure on cholinergic neurons of the rat medial septum, *Alcohol* **15**: 219–226.

Mora, F., Rolls, E. T. and Burton, M. J. (1976). Modulation during learning of the responses of neurones in the lateral hypothalamus to the sight of food, *Experimental Neurology* **53**: 508–519.

Mora, F., Avrith, D. B., Phillips, A. G. and Rolls, E. T. (1979). Effects of satiety on self-stimulation of the orbitofrontal cortex in the monkey, *Neuroscience Letters* **13**: 141–145.

Mora, F., Avrith, D. B. and Rolls, E. T. (1980). An electrophysiological and behavioural study of self-stimulation in the orbitofrontal cortex of the rhesus monkey, *Brain Research Bulletin* **5**: 111–115.

Moran, J. and Desimone, R. (1985). Selective attention gates visual processing in the extrastriate cortex, *Science* **229**: 782–784.

Morecraft, R. J. and Tanji, J. (2009). Cingulofrontal interactions and the cingulate motor areas, *in* B. Vogt (ed.), *Cingulate Neurobiology and Disease*, Oxford University Press, Oxford, chapter 5, pp. 113–144.

Morecraft, R. J., Geula, C. and Mesulam, M.-M. (1992). Cytoarchitecture and neural afferents of orbitofrontal cortex in the brain of the monkey, *Journal of Comparative Neurology* **323**: 341–358.

Moreno-Bote, R., Rinzel, J. and Rubin, N. (2007). Noise-induced alternations in an attractor network model of perceptual bistability, *Journal of Neurophysiology* **98**: 1125–1139.

Morin, E. L., Hadj-Bouziane, F., Stokes, M., Ungerleider, L. G. and Bell, A. H. (2015). Hierarchical encoding of social cues in primate inferior temporal cortex, *Cereb Cortex* **25**(9): 3036–45.

Mormann, F., Dubois, J., Kornblith, S., Milosavljevic, M., Cerf, M., Ison, M., Tsuchiya, N., Kraskov, A., Quiroga, R. Q., Adolphs, R., Fried, I. and Koch, C. (2011). A category-specific response to animals in the right human amygdala, *Nature Neuroscience* **14**: 1247–1249.

Morris, A. M., Churchwell, J. C., Kesner, R. P. and Gilbert, P. E. (2012). Selective lesions of the dentate gyrus produce disruptions in place learning for adjacent spatial locations, *Neurobiol Learn Mem* **97**(3): 326–31.

Morris, J. S., Fritch, C. D., Perrett, D. I., Rowland, D., Young, A. W., Calder, A. J. and Dolan, R. J. (1996). A differential neural response in the human amygdala to fearful and happy face expressions, *Nature* **383**: 812–815.

Morris, R. G., Moser, E. I., Riedel, G., Martin, S. J., Sandin, J., Day, M. and O'Carroll, C. (2003). Elements of a neurobiological theory of the hippocampus: the role of activity-dependent synaptic plasticity in memory, *Philosophical Transactions of the Royal Society of London B* **358**: 773–786.

Morris, R. G. M. (1989). Does synaptic plasticity play a role in information storage in the vertebrate brain?, *in* R. G. M. Morris (ed.), *Parallel Distributed Processing: Implications for Psychology and Neurobiology*, Oxford University Press, Oxford, chapter 11, pp. 248–285.

Morris, R. G. M. (2003). Long-term potentiation and memory, *Philosophical Transactions of the Royal Society of London B* **358**: 643–647.

Moscovitch, M., Rosenbaum, R. S., Gilboa, A., Addis, D. R., Westmacott, R., Grady, C., McAndrews, M. P., Levine, B., Black, S., Winocur, G. and Nadel, L. (2005). Functional neuroanatomy of remote episodic, semantic and spatial memory: a unified account based on multiple trace theory, *Journal of Anatomy* **207**: 35–66.

Moser, E. I., Moser, M. B. and Roudi, Y. (2014a). Network mechanisms of grid cells, *Philosophical Transactions of the Royal Society of London B Biological Science* **369**: 20120511.

Moser, E. I., Roudi, Y., Witter, M. P., Kentros, C., Bonhoeffer, T. and Moser, M. B. (2014b). Grid cells and cortical representation, *Nature Reviews Neuroscience* **15**: 466–481.

Moser, M. B., Rowland, D. C. and Moser, E. I. (2015). Place cells, grid cells, and memory, *Cold Spring Harbor Perspectives in Biology* **7**: a021808.

Motter, B. (1994). Neural correlates of attentive selection for colours or luminance in extrastriate area V4, *Journal of Neuroscience* **14**: 2178–2189.

Motter, B. C. (1993). Focal attention produces spatially selective processing in visual cortical areas V1, V2, and V4 in the presence of competing stimuli, *Journal of Neurophysiology* **70**: 909–919.

Mountcastle, V. B. (1957). Modality and topographic properties of single neurons of cat's somatosensory cortex, *Journal of Neurophysiology* **20**: 408–434.

Mountcastle, V. B. (1984). Central nervous mechanisms in mechanoreceptive sensibility, *in* I. Darian-Smith (ed.), *Handbook of Physiology, Section 1: The Nervous System, Vol III, Sensory Processes, Part 2*, American Physiological Society, Bethesda, MD, pp. 789–878.

Movshon, J. A., Adelson, E. H., Gizzi, M. S. and Newsome, W. T. (1985). The analysis of moving visual patterns, *in* C. Chagas, R. Gattass and C. G. Gross (eds), *Pattern Recognition Mechanisms*, Springer-Verlag, New York, pp. 117–151.

Moyer, J. R. J., Deyo, R. A. and Disterhoft, J. F. (1990). Hippocampectomy disrupts trace eye-blink conditioning in rabbits, *Behavioral Neuroscience* **104**: 243–252.

Mozer, M. C. (1991). *The Perception of Multiple Objects: A Connectionist Approach*, MIT Press, Cambridge, MA.

Mueser, K. T. and McGurk, S. R. (2004). Schizophrenia, *Lancet* **363**: 2063–2072.

Muir, J. L., Everitt, B. J. and Robbins, T. W. (1994). AMPA-induced excitotoxic lesions of the basal forebrain: a significant role for the cortical cholinergic system in attentional function, *Journal of Neuroscience* **14**: 2313–2326.

Muller, R. U., Kubie, J. L., Bostock, E. M., Taube, J. S. and Quirk, G. J. (1991). Spatial firing correlates of neurons in the hippocampal formation of freely moving rats, *in* J. Paillard (ed.), *Brain and Space*, Oxford University Press, Oxford, pp. 296–333.

Muller, R. U., Ranck, J. B. and Taube, J. S. (1996). Head direction cells: properties and functional significance, *Current Opinion in Neurobiology* **6**: 196–206.

Mundy, J. and Zisserman, A. (1992). Introduction – towards a new framework for vision, *in* J. Mundy and A. Zisserman (eds), *Geometric Invariance in Computer Vision*, MIT Press, Cambridge, MA, pp. 1–39.

Murray, E. A. and Izquierdo, A. (2007). Orbitofrontal cortex and amygdala contributions to affect and action in primates, *Annals of the New York Academy of Sciences* **1121**: 273–296.

Murray, E. A., Gaffan, D. and Mishkin, M. (1993). Neural substrates of visual stimulus–stimulus association in rhesus monkeys, *Journal of Neuroscience* **13**: 4549–4561.

Murray, E. A., Baxter, M. G. and Gaffan, D. (1998). Monkeys with rhinal cortex damage or neurotoxic hippocampal lesions are impaired on spatial scene learning and object reversals, *Behavioral Neuroscience* **112**: 1291–1303.

Murray, E. A., Moylan, E. J., Saleem, K. S., Basile, B. M. and Turchi, J. (2015). Specialized areas for value updating and goal selection in the primate orbitofrontal cortex, *Elife*.

Mutch, J. and Lowe, D. G. (2008). Object class recognition and localization using sparse features with limited receptive fields, *International Journal of Computer Vision* **80**: 45–57.

Naber, P. A., Lopes da Silva, F. H. and Witter, M. P. (2001). Reciprocal connections between the entorhinal cortex and hippocampal fields CA1 and the subiculum are in register with the projections from CA1 to the subiculum, *Hippocampus* **11**: 99–104.

Nadal, J. P., Toulouse, G., Changeux, J. P. and Dehaene, S. (1986). Networks of formal neurons and memory palimpsests, *Europhysics Letters* **1**: 535–542.

Nagahama, Y., Okada, T., Katsumi, Y., Hayashi, T., Yamauchi, H., Oyanagi, C., Konishi, J., Fukuyama, H. and Shibasaki, H. (2001). Dissociable mechanisms of attentional control within the human prefrontal cortex, *Cerebral Cortex* **11**: 85–92.

Nakazawa, K., Quirk, M. C., Chitwood, R. A., Watanabe, M., Yeckel, M. F., Sun, L. D., Kato, A., Carr, C. A., Johnston, D., Wilson, M. A. and Tonegawa, S. (2002). Requirement for hippocampal CA3 NMDA receptors in associative memory recall, *Science* **297**: 211–218.

Nakazawa, K., Sun, L. D., Quirk, M. C., Rondi-Reig, L., Wilson, M. A. and Tonegawa, S. (2003). Hippocampal CA3 NMDA receptors are crucial for memory acquisition of one-time experience, *Neuron* **38**: 305–315.

Nakazawa, K., McHugh, T. J., Wilson, M. A. and Tonegawa, S. (2004). NMDA receptors, place cells and hippocampal spatial memory, *Nature Reviews Neuroscience* **5**: 361–372.

Nanry, K. P., Mundy, W. R. and Tilson, H. A. (1989). Colchicine-induced alterations of reference memory in rats: role of spatial versus non-spatial task components, *Behavioral Brain Research* **35**: 45–53.

Nasr, S., Liu, N., Devaney, K. J., Yue, X., Rajimehr, R., Ungerleider, L. G. and Tootell, R. B. (2011). Scene-selective cortical regions in human and nonhuman primates, *J Neurosci* **31**(39): 13771–85.

Nelken, I., Prut, Y., Vaadia, E. and Abeles, M. (1994). Population responses to multifrequency sounds in the cat auditory cortex: one- and two-parameter families of sounds, *Hearing Research* **72**: 206–222.

Nesse, R. M. and Lloyd, A. T. (1992). The evolution of psychodynamic mechanisms, *in* J. H. Barkow, L. Cosmides and J. Tooby (eds), *The Adapted Mind*, Oxford University Press, New York, pp. 601–624.

Newcomer, J. W., Farber, N. B., Jevtovic-Todorovic, V., Selke, G., Melson, A. K., Hershey, T., Craft, S. and Olney, J. W. (1999). Ketamine-induced NMDA receptor hypofunction as a model of memory impairment and psychosis, *Neuropsychopharmacology* **20**: 106–118.

Newman, E. L., Gupta, K., Climer, J. R., Monaghan, C. K. and Hasselmo, M. E. (2012). Cholinergic modulation of cognitive processing: insights drawn from computational models, *Frontiers in Behavioural Neuroscience* **6**: 24.

Newsome, W. T., Britten, K. H. and Movshon, J. A. (1989). Neuronal correlates of a perceptual decision, *Nature* **341**: 52–54.

Nicoll, R. A. (1988). The coupling of neurotransmitter receptors to ion channels in the brain, *Science* **241**: 545–551.

Niv, Y., Duff, M. O. and Dayan, P. (2005). Dopamine, uncertainty, and TD learning, *Behavioral and Brain Functions* **1**: 6.

Norgren, R. (1974). Gustatory afferents to ventral forebrain, *Brain Research* **81**: 285–295.

Norgren, R. (1976). Taste pathways to hypothalamus and amygdala, *Journal of Comparative Neurology* **166**: 17–30.

Norgren, R. (1990). Gustatory system, *in* G. Paxinos (ed.), *The Human Nervous System*, Academic Press, San Diego, pp. 845–861.

Norgren, R. and Leonard, C. M. (1971). Taste pathways in rat brainstem, *Science* **173**: 1136–1139.

Norgren, R. and Leonard, C. M. (1973). Ascending central gustatory pathways, *Journal of Comparative Neurology* **150**: 217–238.

Nowak, L. and Bullier, J. (1997). The timing of information transfer in the visual system, *in* K. Rockland, J. Kaas and A. Peters (eds), *Cerebral Cortex: Extrastriate Cortex in Primate*, Plenum, New York, p. 870.

Nugent, A. C., Milham, M. P., Bain, E. E., Mah, L., Cannon, D. M., Marrett, S., Zarate, C. A., Pine, D. S., Price, J. L. and Drevets, W. C. (2006). Cortical abnormalities in bipolar disorder investigated with mri and voxel-based morphometry, *Neuroimage* **30**: 485–497.

Nusslock, R., Young, C. B. and Damme, K. S. (2014). Elevated reward-related neural activation as a unique biological marker of bipolar disorder: assessment and treatment implications, *Behaviour Research and Therapy* **62**: 74–87.

O'Doherty, J., Kringelbach, M. L., Rolls, E. T., Hornak, J. and Andrews, C. (2001). Abstract reward and punishment representations in the human orbitofrontal cortex, *Nature Neuroscience* **4**: 95–102.

Oja, E. (1982). A simplified neuron model as a principal component analyzer, *Journal of Mathematical Biology* **15**: 267–273.

Ojemann, G. A. (2013). Human temporal cortical single neuron activity during language: A review, *Brain Sci* **3**(2): 627–41.

O'Kane, D. and Treves, A. (1992). Why the simplest notion of neocortex as an autoassociative memory would not work, *Network* **3**: 379–384.

O'Keefe, J. (1979). A review of the hippocampal place cells, *Progress in Neurobiology* **13**: 419–439.

O'Keefe, J. (1984). Spatial memory within and without the hippocampal system, *in* W. Seifert (ed.), *Neurobiology of the Hippocampus*, Academic Press, London, pp. 375–403.

O'Keefe, J. (1990). A computational theory of the hippocampal cognitive map, *Progress in Brain Research* **83**: 301–312.

O'Keefe, J. and Dostrovsky, J. (1971). The hippocampus as a spatial map: preliminary evidence from unit activity in the freely moving rat, *Brain Research* **34**: 171–175.

O'Keefe, J. and Nadel, L. (1978). *The Hippocampus as a Cognitive Map*, Clarendon Press, Oxford.

O'Keefe, J. and Speakman, A. (1987). Single unit activity in the rat hippocampus during a spatial memory task, *Experimental Brain Research* **68**(1): 1–27.

Okubo, Y., Suhara, T., Sudo, Y. and Toru, M. (1997a). Possible role of dopamine D1 receptors in schizophrenia, *Molecular Psychiatry* **2**: 291–292.

Okubo, Y., Suhara, T., Suzuki, K., Kobayashi, K., Inoue, O., Terasaki, O., Someya, Y., Sassa, T., Sudo, Y., Matsushima, E., Iyo, M., Tateno, Y. and Toru, M. (1997b). Decreased prefrontal dopamine D1 receptors in schizophrenia revealed by PET, *Nature* **385**: 634–636.

Olshausen, B. A. and Field, D. J. (1997). Sparse coding with an incomplete basis set: a strategy employed by V1, *Vision Research* **37**: 3311–3325.

Olshausen, B. A. and Field, D. J. (2004). Sparse coding of sensory inputs, *Current Opinion in Neurobiology* **14**: 481–487.

Olshausen, B. A., Anderson, C. H. and Van Essen, D. C. (1993). A neurobiological model of visual attention and invariant pattern recognition based on dynamic routing of information, *Journal of Neuroscience* **13**: 4700–4719.

Olshausen, B. A., Anderson, C. H. and Van Essen, D. C. (1995). A multiscale dynamic routing circuit for forming size- and position-invariant object representations, *Journal of Computational Neuroscience* **2**: 45–62.

O'Mara, S. M., Rolls, E. T., Berthoz, A. and Kesner, R. P. (1994). Neurons responding to whole-body motion in the primate hippocampus, *Journal of Neuroscience* **14**: 6511–6523.

O'Neill, M. J. and Dix, S. (2007). AMPA receptor potentiators as cognitive enhancers, *IDrugs* **10**: 185–192.

Ongur, D. and Price, J. L. (2000). The organisation of networks within the orbital and medial prefrontal cortex of rats, monkeys and humans, *Cerebral Cortex* **10**: 206–219.

Ongur, D., Ferry, A. T. and Price, J. L. (2003). Architectonic subdivision of the human orbital and medial prefrontal cortex, *Journal of Comparative Neurology* **460**: 425–449.

Ono, T., Nakamura, K., Nishijo, H. and Eifuku, S. (1993). Monkey hippocampal neurons related to spatial and nonspatial functions, *Journal of Neurophysiology* **70**: 1516–1529.

Optican, L. M. and Richmond, B. J. (1987). Temporal encoding of two-dimensional patterns by single units in primate inferior temporal cortex: III. Information theoretic analysis, *Journal of Neurophysiology* **57**: 162–178.

Optican, L. M., Gawne, T. J., Richmond, B. J. and Joseph, P. J. (1991). Unbiased measures of transmitted information and channel capacity from multivariate neuronal data, *Biological Cybernetics* **65**: 305–310.

Oram, M. W. and Perrett, D. I. (1994). Modeling visual recognition from neurophysiological constraints, *Neural Networks* **7**: 945–972.

O'Regan, J. K., Rensink, R. A. and Clark, J. J. (1999). Change-blindness as a result of 'mudsplashes', *Nature* **398**: 1736–1753.

O'Reilly, J. and Munakata, Y. (2000). *Computational Explorations in Cognitive Neuroscience*, MIT Press, Cambridge, MA.

O'Reilly, R. C. (2006). Biologically based computational models of high-level cognition, *Science* **314**: 91–94.

O'Reilly, R. C. and Rudy, J. W. (2001). Conjunctive representations in learning and memory: principles of cortical and hippocampal function, *Psychological Review* **108**: 311–345.

Otto, T. and Eichenbaum, H. (1992). Neuronal activity in the hippocampus during delayed non-match to sample performance in rats: evidence for hippocampal processing in recognition memory, *Hippocampus* **2**: 323–334.

Owen, M. J., Sawa, A. and Mortensen, P. B. (2016). Schizophrenia, *Lancet*.

Padoa-Schioppa, C. (2011). Neurobiology of economic choice: a good-based model, *Annual Review of Neuroscience* **34**: 333–359.

Padoa-Schioppa, C. and Assad, J. A. (2006). Neurons in the orbitofrontal cortex encode economic value, *Nature* **441**: 223–226.

Palomero-Gallagher, N. and Zilles, K. (2004). Isocortex, *in* G. Paxinos (ed.), *The Rat Nervous System*, Elsevier Academic Press, San Diego, pp. 729–757.

Pandya, D. N., Seltzer, B., Petrides, M. and Cipolloni, P. B. (2015). *Cerebral Cortex: Architecture, Connections, and the Dual Origin Concept*, Oxford University Press, Oxford.

Panksepp, J. (1998). *Affective Neuroscience: The Foundations of Human and Animal Emotions*, Oxford University Press, New York.

Pantelis, C., Barber, F. Z., Barnes, T. R., Nelson, H. E., Owen, A. M. and Robbins, T. W. (1999). Comparison of set-shifting ability in patients with chronic schizophrenia and frontal lobe damage, *Schizophrenia Research* **37**: 251–270.

Panzeri, S. and Treves, A. (1996). Analytical estimates of limited sampling biases in different information measures, *Network* **7**: 87–107.

Panzeri, S., Biella, G., Rolls, E. T., Skaggs, W. E. and Treves, A. (1996). Speed, noise, information and the graded nature of neuronal responses, *Network* **7**: 365–370.

Panzeri, S., Schultz, S. R., Treves, A. and Rolls, E. T. (1999a). Correlations and the encoding of information in the nervous system, *Proceedings of the Royal Society B* **266**: 1001–1012.

Panzeri, S., Treves, A., Schultz, S. and Rolls, E. T. (1999b). On decoding the responses of a population of neurons from short time windows, *Neural Computation* **11**: 1553–1577.

Panzeri, S., Rolls, E. T., Battaglia, F. and Lavis, R. (2001). Speed of feedforward and recurrent processing in multilayer networks of integrate-and-fire neurons, *Network: Computation in Neural Systems* **12**: 423–440.

Papp, G. and Treves, A. (2008). Network analysis of the significance of hippocampal subfields, *in* S. J. Y. Mizumori (ed.), *Hippocampal Place Fields: Relevance to Learning and Memory.*, Oxford University Press, New York, chapter 20, pp. 328–342.

Parga, N. and Rolls, E. T. (1998). Transform invariant recognition by association in a recurrent network, *Neural Computation* **10**: 1507–1525.

Parisi, G. (1986). A memory which forgets, *Journal of Physics A* **19**: L617–L619.

Park, H. and Poo, M.-M. (2013). Neurotrophin regulation of neural circuit development and function, *Nature Reviews Neuroscience* **14**: 7–23.

Parker, A. J. (2007). Binocular depth perception and the cerebral cortex, *Nature Reviews Neuroscience* **8**: 379–391.

Parker, A. J., Cumming, B. G. and Dodd, J. V. (2000). Binocular neurons and the perception of depth, *in* M. Gazzaniga (ed.), *The New Cognitive Neurosciences, Second Edition*, MIT Press, Cambridge, MA, chapter 18, pp. 263–277.

Parkinson, J. K., Murray, E. A. and Mishkin, M. (1988). A selective mnemonic role for the hippocampus in monkeys: memory for the location of objects, *Journal of Neuroscience* **8**: 4059–4167.

Passingham, R. E. P. and Wise, S. P. (2012). *The Neurobiology of the Prefrontal Cortex*, Oxford University Press, Oxford.

Pauls, D. L., Abramovitch, A., Rauch, S. L. and Geller, D. A. (2014). Obsessive-compulsive disorder: an integrative genetic and neurobiological perspective, *Nat Neurosci* **15**(6): 410–24.

Pearce, J. M. and Hall, G. (1980). A model for Pavlovian learning: variations in the effectiveness of conditioned but not of unconditioned stimuli., *Psychological Review* **87**: 532.

Peitgen, H.-O., Jürgens, H. and Saupe, D. (2004). *Chaos and Fractals: New Frontiers of Science*, Springer, New York.

Peled, A. (2004). From plasticity to complexity: a new diagnostic method for psychiatry, *Medical Hypotheses* **63**: 110–114.

Penades, R., Catalan, R., Andres, S., Salamero, M. and Gasto, C. (2005). Executive function and nonverbal memory in obsessive-compulsive disorder, *Psychiatry Research* **133**: 81–90.

Penades, R., Catalan, R., Rubia, K., Andres, S., Salamero, M. and Gasto, C. (2007). Impaired response inhibition in obsessive compulsive disorder, *European Psychiatry* **22**: 404–410.

Peng, H. C., Sha, L. F., Gan, Q. and Wei, Y. (1998). Energy function for learning invariance in multilayer perceptron, *Electronics Letters* **34**(3): 292–294.

Percheron, G., Yelnik, J. and François, C. (1984a). The primate striato-pallido-nigral system: an integrative system for cortical information, *in* J. S. McKenzie, R. E. Kemm and L. N. Wilcox (eds), *The Basal Ganglia: Structure and Function*, Plenum, New York, pp. 87–105.

Percheron, G., Yelnik, J. and François, C. (1984b). A Golgi analysis of the primate globus pallidus. III. Spatial organization of the striato-pallidal complex, *Journal of Comparative Neurology* **227**: 214–227.

Percheron, G., Yelnik, J., François, C., Fenelon, G. and Talbi, B. (1994). Informational neurology of the basal ganglia related system, *Revue Neurologique (Paris)* **150**: 614–626.

Perez Castillo, I. and Skantzos, N. S. (2004). The Little-Hopfield model on a sparse random graph, *Journal of Physics A: Math Gen* **37**: 9087–9099.

Perin, R., Berger, T. K. and Markram, H. (2011). A synaptic organizing principle for cortical neuronal groups, *Proc Natl Acad Sci U S A* **108**: 5419–5424.

Perin, R., Telefont, M. and Markram, H. (2013). Computing the size and number of neuronal clusters in local circuits, *Frontiers in Neuroanatomy* **7**: 1.

Perrett, D. I., Rolls, E. T. and Caan, W. (1982). Visual neurons responsive to faces in the monkey temporal cortex, *Experimental Brain Research* **47**: 329–342.

Perrett, D. I., Smith, P. A. J., Mistlin, A. J., Chitty, A. J., Head, A. S., Potter, D. D., Broennimann, R., Milner, A. D. and Jeeves, M. A. (1985a). Visual analysis of body movements by neurons in the temporal cortex of the macaque monkey: a preliminary report, *Behavioural Brain Research* **16**: 153–170.

Perrett, D. I., Smith, P. A. J., Potter, D. D., Mistlin, A. J., Head, A. S., Milner, D. and Jeeves, M. A. (1985b). Visual cells in temporal cortex sensitive to face view and gaze direction, *Proceedings of the Royal Society of London, Series B* **223**: 293–317.

Perry, C. J. and Fallah, M. (2014). Feature integration and object representations along the dorsal stream visual hierarchy, *Frontiers in Computational Neuroscience*.

Perry, G., Rolls, E. T. and Stringer, S. M. (2006). Spatial vs temporal continuity in view invariant visual object recognition learning, *Vision Research* **46**: 3994–4006.

Perry, G., Rolls, E. T. and Stringer, S. M. (2010). Continuous transformation learning of translation invariant representations, *Experimental Brain Research* **204**: 255–270.

Personnaz, L., Guyon, I. and Dreyfus, G. (1985). Information storage and retrieval in spin-glass-like neural networks, *Journal de Physique Lettres (Paris)* **46**: 359–365.

Pessoa, L. (2009). How do emotion and motivation direct executive control?, *Trends in Cognitive Sciences* **13**: 160–166.

Pessoa, L. and Adolphs, R. (2010). Emotion processing and the amygdala: from a 'low road' to 'many roads' of evaluating biological significance, *Nature Reviews Neuroscience* **11**: 773–783.

Pessoa, L. and Padmala, S. (2005). Quantitative prediction of perceptual decisions during near-threshold fear detection, *Proceedings of the National Academy of Sciences USA* **102**: 5612–5617.

Peters, A. (1984a). Bipolar cells, *in* A. Peters and E. G. Jones (eds), *Cerebral Cortex, Vol. 1, Cellular Components of the Cerebral Cortex*, Plenum, New York, chapter 11, pp. 381–407.

Peters, A. (1984b). Chandelier cells, *in* A. Peters and E. G. Jones (eds), *Cerebral Cortex, Vol. 1, Cellular Components of the Cerebral Cortex*, Plenum, New York, chapter 10, pp. 361–380.

Peters, A. and Jones, E. G. (eds) (1984). *Cerebral Cortex, Vol. 1, Cellular Components of the Cerebral Cortex*, Plenum, New York.

Peters, A. and Regidor, J. (1981). A reassessment of the forms of nonpyramidal neurons in area 17 of the cat visual cortex, *Journal of Comparative Neurology* **203**: 685–716.

Peters, A. and Saint Marie, R. L. (1984). Smooth and sparsely spinous nonpyramidal cells forming local axonal plexuses, *in* A. Peters and E. G. Jones (eds), *Cerebral Cortex, Vol. 1, Cellular Components of the Cerebral Cortex*, New York, Plenum, chapter 13, pp. 419–445.

Peterson, C. and Anderson, J. R. (1987). A mean field theory learning algorithm for neural networks, *Complex Systems* **1**: 995–1015.

Petrides, M. (1985). Deficits on conditional associative-learning tasks after frontal- and temporal-lobe lesions in man, *Neuropsychologia* **23**: 601–614.

Petrides, M. (1996). Specialized systems for the processing of mnemonic information within the primate frontal cortex, *Philosophical Transactions of the Royal Society of London B* **351**: 1455–1462.

Phelps, E. A. and LeDoux, J. E. (2005). Contributions of the amygdala to emotion processing: from animal models to human behavior, *Neuron* **48**: 175–187.

Phillips, W. A., Kay, J. and Smyth, D. (1995). The discovery of structure by multi-stream networks of local processors with contextual guidance, *Network* **6**: 225–246.

Pickens, C. L., Saddoris, M. P., Setlow, B., Gallagher, M., Holland, P. C. and Schoenbaum, G. (2003). Different roles for orbitofrontal cortex and basolateral amygdala in a reinforcer devaluation task, *Journal of Neuroscience* **23**: 11078–11084.

Pinker, S. and Bloom, P. (1992). Natural language and natural selection, *in* J. H. Barkow, L. Cosmides and J. Tooby (eds), *The Adapted Mind*, Oxford University Press, New York, chapter 12, pp. 451–493.

Pinto, N., Cox, D. D. and DiCarlo, J. J. (2008). Why is real-world visual object recognition hard?, *PLoS Computational Biology* **4**: e27.

Pinto, N., Doukhan, D., DiCarlo, J. J. and Cox, D. D. (2009). A high-throughput screening approach to discovering good forms of biologically inspired visual representation, *PLoS Computational Biology* **5**: e1000579.

Pirmoradian, S. and Treves, A. (2013). Encoding words into a Potts attractor network, *in* J. Mayor and P. Gomez (eds), *Computational Models of Cognitive Processes: Proceedings of the 13th Neural Computation and Psychology Workshop (NCPW13)*, World Scientific Press, Singapore, pp. 29–42.

Pitkanen, A., Kelly, J. L. and Amaral, D. G. (2002). Projections from the lateral, basal, and accessory basal nuclei of the amygdala to the entorhinal cortex in the macaque monkey, *Hippocampus* **12**: 186–205.

Pittenger, C. (2015). Glutamatergic agents for ocd and related disorders, *Curr Treat Options Psychiatry* **2**(3): 271–283.

Pittenger, C., Krystal, J. H. and Coric, V. (2006). Glutamate-modulating drugs as novel pharmacotherapeutic agents in the treatment of obsessive-compulsive disorder, *NeuroRx* **3**(1): 69–81.

Pittenger, C., Bloch, M. H. and Williams, K. (2011). Glutamate abnormalities in obsessive compulsive disorder: Neurobiology, pathophysiology, and treatment, *Pharmacology and Therapeutics* **132**: 314–332.

Poggio, T. and Edelman, S. (1990). A network that learns to recognize three-dimensional objects, *Nature* **343**: 263–266.

Poggio, T. and Girosi, F. (1990a). Networks for approximation and learning, *Proceedings of the IEEE* **78**: 1481–1497.

Poggio, T. and Girosi, F. (1990b). Regularization algorithms for learning that are equivalent to multilayer networks, *Science* **247**: 978–982.

Pollen, D. and Ronner, S. (1981). Phase relationship between adjacent simple cells in the visual cortex, *Science* **212**: 1409–1411.

Posner, M. I. and Keele, S. W. (1968). On the genesis of abstract ideas, *Journal of Experimental Psychology* **77**: 353–363.

Poucet, B. (1989). Object exploration, habituation, and response to a spatial change in rats following septal or medial frontal cortical damage, *Behavioral Neuroscience* **103**: 1009–1016.

Powell, T. P. S. (1981). Certain aspects of the intrinsic organisation of the cerebral cortex, *in* O. Pompeiano and C. Ajmone Marsan (eds), *Brain Mechanisms and Perceptual Awareness*, Raven Press, New York, pp. 1–19.

Power, J. M. and Sah, P. (2008). Competition between calcium-activated K+ channels determines cholinergic action on firing properties of basolateral amygdala projection neurons, *Journal of Neuroscience* **28**: 3209–3220.

Preston, A. R. and Eichenbaum, H. (2013). Interplay of hippocampus and prefrontal cortex in memory, *Curr Biol* **23**(17): R764–73.

Preuss, T. M. (1995). Do rats have prefrontal cortex? The Rose-Woolsey-Akert program reconsidered, *Journal of Cognitive Neuroscience* **7**: 1–24.

Price, J. (2006). Connections of orbital cortex, *in* D. H. Zald and S. L. Rauch (eds), *The Orbitofrontal Cortex*, Oxford University Press, Oxford, chapter 3, pp. 39–55.

Price, J. L. and Drevets, W. C. (2012). Neural circuits underlying the pathophysiology of mood disorders, *Trends in Cognitive Science* **16**: 61–71.

Price, J. L., Carmichael, S. T., Carnes, K. M., Clugnet, M.-C. and Kuroda, M. (1991). Olfactory input to the prefrontal cortex, *in* J. L. Davis and H. Eichenbaum (eds), *Olfaction: A Model System for Computational Neuroscience*, MIT Press, Cambridge, MA, pp. 101–120.

Pritchard, T. C., Hamilton, R. B., Morse, J. R. and Norgren, R. (1986). Projections of thalamic gustatory and lingual areas in the monkey, *Journal of Comparative Neurology* **244**: 213–228.

Pritchard, T. C., Hamilton, R. B. and Norgren, R. (1989). Neural coding of gustatory information in the thalamus of macaca mulatta, *Journal of Neurophysiology* **61**: 1–14.

Pryce, C. R., Azzinnari, D., Spinelli, S., Seifritz, E., Tegethoff, M. and Meinlschmidt, G. (2011). Helplessness: a systematic translational review of theory and evidence for its relevance to understanding and treating depression, *Pharmacol Ther* **132**: 242–267.

Quinlan, P. T. and Humphreys, G. W. (1987). Visual search for targets defined by combinations of color, shape, and size: An examination of the task constraints on feature and conjunction searches, *Perception and Psychophysics* **41**: 455–472.

Quiroga, R. Q. (2012). Concept cells: the building blocks of declarative memory functions, *Nature Reviews Neuroscience* **13**: 587–597.

Quiroga, R. Q. (2013). Gnostic cells in the 21st century, *Acta Neurobiol Exp (Warsaw)* **73**: 463–471.

Quiroga, R. Q., Reddy, L., Kreiman, G., Koch, C. and Fried, I. (2005). Invariant visual representation by single neurons in the human brain, *Nature* **453**: 1102–1107.

Quiroga, R. Q., Kreiman, G., Koch, C. and Fried, I. (2008). Sparse but not 'grandmother-cell' coding in the medial temporal lobe, *Trends in Cognitive Sciences* **12**: 87–91.

Rabinovich, M., Huerta, R. and Laurent, G. (2008). Transient dynamics for neural processing, *Science* **321**(5885): 48–50.

Rabinovich, M. I., Varona, P., Tristan, I. and Afraimovich, V. S. (2014). Chunking dynamics: heteroclinics in mind, *Front Comput Neurosci* **8**: 22.

Rachlin, H. (1989). *Judgement, Decision, and Choice: A Cognitive/Behavioural Synthesis*, Freeman, New York.

Rada, P., Mark, G. P. and Hoebel, B. G. (1998). Dopamine in the nucleus accumbens released by hypothalamic stimulation-escape behavior, *Brain Research* **782**: 228–234.

Ranck, Jr., J. B. (1985). Head direction cells in the deep cell layer of dorsolateral presubiculum in freely moving rats, *in* G. Buzsáki and C. H. Vanderwolf (eds), *Electrical Activity of the Archicortex*, Akadémiai Kiadó, Budapest.

Rao, S. C., Rainer, G. and Miller, E. K. (1997). Integration of what and where in the primate prefrontal cortex, *Science* **276**: 821–824.

Ratcliff, R. and Rouder, J. F. (1998). Modeling response times for two-choice decisions, *Psychological Science* **9**: 347–356.

Ratcliff, R., Zandt, T. V. and McKoon, G. (1999). Connectionist and diffusion models of reaction time, *Psychological Reviews* **106**: 261–300.

Rauch, A., La Camera, G., Luescher, H.-R., Senn, W. and Fusi, S. (2003). Neocortical pyramidal cells respond as integrate-and-fire neurons to in vivo-like input currents, *Journal of Neurophysiology* **90**: 1598–1612.

Rauschecker, J. P. (2012). Ventral and dorsal streams in the evolution of speech and language, *Frontiers in Evolutionary Neuroscience* **4**: 7.

Rauschecker, J. P. and Scott, S. K. (2009). Maps and streams in the auditory cortex: nonhuman primates illuminate human speech processing, *Nature Neuroscience* **12**: 718–724.

Rawlins, J. N. P. (1985). Associations across time: the hippocampus as a temporary memory store, *Behavioral Brain Science* **8**: 479–496.

Redgrave, P., Prescott, T. J. and Gurney, K. (1999). Is the short-latency dopamine response too short to signal reward error?, *Trends in Neuroscience* **22**: 146–151.

Reisenzein, R. (1983). The Schachter theory of emotion: two decades later, *Psychological Bulletin* **94**: 239–264.

Renart, A., Parga, N. and Rolls, E. T. (1999a). Associative memory properties of multiple cortical modules, *Network* **10**: 237–255.

Renart, A., Parga, N. and Rolls, E. T. (1999b). Backprojections in the cerebral cortex: implications for memory storage, *Neural Computation* **11**: 1349–1388.

Renart, A., Parga, N. and Rolls, E. T. (2000). A recurrent model of the interaction between the prefrontal cortex and inferior temporal cortex in delay memory tasks, *in* S. Solla, T. Leen and K.-R. Mueller (eds), *Advances in Neural Information Processing Systems*, Vol. 12, MIT Press, Cambridge, MA, pp. 171–177.

Renart, A., Moreno, R., Rocha, J., Parga, N. and Rolls, E. T. (2001). A model of the IT–PF network in object working memory which includes balanced persistent activity and tuned inhibition, *Neurocomputing* **38–40**: 1525–1531.

Rensink, R. A. (2000). Seeing, sensing, and scrutinizing, *Vision Research* **40**: 1469–1487.

Rensink, R. A. (2014). Limits to the usability of iconic memory, *Front Psychol* **5**: 971.

Rescorla, R. A. and Wagner, A. R. (1972). A theory of Pavlovian conditioning: the effectiveness of reinforcement and non-reinforcement, *Classical Conditioning II: Current Research and Theory*, Appleton-Century-Crofts, New York, pp. 64–69.

Reynolds, J. and Desimone, R. (1999). The role of neural mechanisms of attention in solving the binding problem, *Neuron* **24**: 19–29.

Reynolds, J. and Desimone, R. (2003). Interacting roles of attention and visual saliency in V4, *Neuron* **37**: 853–863.

Reynolds, J. H., Chelazzi, L. and Desimone, R. (1999). Competitive mechanisms subserve attention in macaque areas V2 and V4, *Journal of Neuroscience* **19**: 1736–1753.

Reynolds, J. H., Pastemak, T. and Desimone, R. (2000). Attention increases sensitivity of V4 neurons, *Neuron* **26**: 703–714.

Rhodes, P. (1992). The open time of the NMDA channel facilitates the self-organisation of invariant object responses in cortex, *Society for Neuroscience Abstracts* **18**: 740.

Ridley, M. (1993). *The Red Queen: Sex and the Evolution of Human Nature*, Penguin, London.

Rieke, F., Warland, D. and Bialek, W. (1993). Coding efficiency and information rates in sensory neurons, *Europhysics Letters* **22**: 151–156.

Rieke, F., Warland, D., de Ruyter van Steveninck, R. R. and Bialek, W. (1997). *Spikes: Exploring the Neural Code*, MIT Press, Cambridge, MA.

Riesenhuber, M. and Poggio, T. (1998). Just one view: Invariances in inferotemporal cell tuning, *in* M. I. Jordan, M. J. Kearns and S. A. Solla (eds), *Advances in Neural Information Processing Systems*, Vol. 10, MIT Press, Cambridge, MA, pp. 215–221.

Riesenhuber, M. and Poggio, T. (1999a). Are cortical models really bound by the "binding problem"?, *Neuron* **24**: 87–93.

Riesenhuber, M. and Poggio, T. (1999b). Hierarchical models of object recognition in cortex, *Nature Neuroscience* **2**: 1019–1025.

Riesenhuber, M. and Poggio, T. (2000). Models of object recognition, *Nature Neuroscience Supplement* **3**: 1199–1204.

Riley, M. A. and Turvey, M. T. (2001). The self-organizing dynamics of intentions and actions, *American Journal of Psychology* **114**: 160–169.

Risold, P. Y. and Swanson, L. W. (1997). Connections of the rat lateral septal complex, *Brain Research Reviews* **24**: 115–195.

Robertson, R. G., Rolls, E. T. and Georges-François, P. (1998). Spatial view cells in the primate hippocampus: Effects of removal of view details, *Journal of Neurophysiology* **79**: 1145–1156.

Robertson, R. G., Rolls, E. T., Georges-François, P. and Panzeri, S. (1999). Head direction cells in the primate pre-subiculum, *Hippocampus* **9**: 206–219.

Robins, L. N., Helzer, J. E., Weissman, M. M., Orvaschel, H., Gruenberg, E., Burke, J. D. J. and Regier, D. A. (1984). Lifetime prevalence of specific psychiatric disorders in three sites, *Archives of General Psychiatry* **41**: 949–958.

Robinson, L. and Rolls, E. T. (2015). Invariant visual object recognition: biologically plausibile approaches, *Biological Cybernetics* **109**: 505–535.

Roelfsema, P. R., Lamme, V. A. and Spekreijse, H. (1998). Object-based attention in the primary visual cortex of the macaque monkey, *Nature* **395**: 376–381.

Roesch, M. R., Esber, G. R., Li, J., Daw, N. D. and Schoenbaum, G. (2012). Surprise! Neural correlates of Pearce–Hall and Rescorla–Wagner coexist within the brain, *European Journal of Neuroscience* **35**: 1190–1200.

Rogers, J. L. and Kesner, R. P. (2003). Cholinergic modulation of the hippocampus during encoding and retrieval, *Neurobiology of Learning and Memory* **80**: 332–342.

Rogers, J. L. and Kesner, R. P. (2004). Cholinergic modulation of the hippocampus during encoding and retrieval of tone/shock-induced fear conditioning, *Learning and Memory* **11**: 102–107.

Rogers, J. L., Hunsaker, M. R. and Kesner, R. P. (2006). Effects of ventral and dorsal CA1 subregional lesions on trace fear conditioning, *Neurobiology of Learning and Memory* **86**: 72–81.

Rogers, R. D., Andrews, T. C., Grasby, P. M., Brooks, D. J. and Robbins, T. W. (2000). Contrasting cortical and subcortical activations produced by attentional-set shifting and reversal learning in humans, *Journal of Cognitive Neuroscience* **12**: 142–162.

Roitman, J. D. and Shadlen, M. N. (2002). Response of neurons in the lateral intraparietal area during a combined visual discrimination reaction time task, *Journal of Neuroscience* **22**: 9475–9489.

Roland, P. E. and Friberg, L. (1985). Localization of cortical areas activated by thinking, *Journal of Neurophysiology* **53**: 1219–1243.

Rolls, E. T. (1975). *The Brain and Reward*, Pergamon Press, Oxford.

Rolls, E. T. (1981a). Central nervous mechanisms related to feeding and appetite, *British Medical Bulletin* **37**: 131–134.

Rolls, E. T. (1981b). Processing beyond the inferior temporal visual cortex related to feeding, learning, and striatal function, *in* Y. Katsuki, R. Norgren and M. Sato (eds), *Brain Mechanisms of Sensation*, Wiley, New York, chapter 16, pp. 241–269.

Rolls, E. T. (1982). Neuronal mechanisms underlying the formation and disconnection of associations between visual stimuli and reinforcement in primates, *in* C. D. Woody (ed.), *Conditioning: Representation of Involved Neural Functions*, Plenum, New York, pp. 363–373.

Rolls, E. T. (1986a). Neural systems involved in emotion in primates, *in* R. Plutchik and H. Kellerman (eds), *Emotion: Theory, Research, and Experience*, Vol. 3: Biological Foundations of Emotion, Academic Press, New York, chapter 5, pp. 125–143.

Rolls, E. T. (1986b). Neuronal activity related to the control of feeding, *in* R. Ritter, S. Ritter and C. Barnes (eds), *Feeding Behavior: Neural and Humoral Controls*, Academic Press, New York, chapter 6, pp. 163–190.

Rolls, E. T. (1986c). A theory of emotion, and its application to understanding the neural basis of emotion, *in* Y. Oomura (ed.), *Emotions. Neural and Chemical Control*, Japan Scientific Societies Press; and Karger, Tokyo; and Basel, pp. 325–344.

Rolls, E. T. (1987). Information representation, processing and storage in the brain: analysis at the single neuron level, *in* J.-P. Changeux and M. Konishi (eds), *The Neural and Molecular Bases of Learning*, Wiley, Chichester, pp. 503–540.

Rolls, E. T. (1989a). Functions of neuronal networks in the hippocampus and cerebral cortex in memory, *in* R. Cotterill (ed.), *Models of Brain Function*, Cambridge University Press, Cambridge, pp. 15–33.

Rolls, E. T. (1989b). Functions of neuronal networks in the hippocampus and neocortex in memory, *in* J. H. Byrne and W. O. Berry (eds), *Neural Models of Plasticity: Experimental and Theoretical Approaches*, Academic Press, San Diego, CA, chapter 13, pp. 240–265.

Rolls, E. T. (1989c). Information processing and basal ganglia function, *in* C. Kennard and M. Swash (eds), *Hierarchies in Neurology*, Springer-Verlag, London, chapter 15, pp. 123–142.

Rolls, E. T. (1989d). Information processing in the taste system of primates, *Journal of Experimental Biology* **146**: 141–164.

Rolls, E. T. (1989e). Parallel distributed processing in the brain: implications of the functional architecture of neuronal networks in the hippocampus, *in* R. G. M. Morris (ed.), *Parallel Distributed Processing: Implications for Psychology and Neurobiology*, Oxford University Press, Oxford, chapter 12, pp. 286–308.

Rolls, E. T. (1989f). The representation and storage of information in neuronal networks in the primate cerebral cortex and hippocampus, *in* R. Durbin, C. Miall and G. Mitchison (eds), *The Computing Neuron*, Addison-Wesley, Wokingham, England, chapter 8, pp. 125–159.

Rolls, E. T. (1990a). Functions of the primate hippocampus in spatial processing and memory, *in* D. S. Olton and R. P. Kesner (eds), *Neurobiology of Comparative Cognition*, L. Erlbaum, Hillsdale, NJ, chapter 12, pp. 339–362.

Rolls, E. T. (1990b). Theoretical and neurophysiological analysis of the functions of the primate hippocampus in memory, *Cold Spring Harbor Symposia in Quantitative Biology* **55**: 995–1006.

Rolls, E. T. (1990c). A theory of emotion, and its application to understanding the neural basis of emotion, *Cognition and Emotion* **4**: 161–190.

Rolls, E. T. (1992a). Neurophysiological mechanisms underlying face processing within and beyond the temporal cortical visual areas, *Philosophical Transactions of the Royal Society* **335**: 11–21.

Rolls, E. T. (1992b). Neurophysiology and functions of the primate amygdala, *in* J. P. Aggleton (ed.), *The Amygdala*, Wiley-Liss, New York, chapter 5, pp. 143–165.

Rolls, E. T. (1993). The neural control of feeding in primates, *in* D. Booth (ed.), *Neurophysiology of Ingestion*, Pergamon, Oxford, chapter 9, pp. 137–169.

Rolls, E. T. (1994a). Brain mechanisms for invariant visual recognition and learning, *Behavioural Processes* **33**: 113–138.

Rolls, E. T. (1994b). Neurophysiological and neuronal network analysis of how the primate hippocampus functions in memory, *in* J. Delacour (ed.), *The Memory System of the Brain*, World Scientific, London, pp. 713–744.

Rolls, E. T. (1995a). Learning mechanisms in the temporal lobe visual cortex, *Behavioural Brain Research* **66**: 177–185.

Rolls, E. T. (1995b). A model of the operation of the hippocampus and entorhinal cortex in memory, *International Journal of Neural Systems* **6, Supplement**: 51–70.

Rolls, E. T. (1995c). A theory of emotion and consciousness, and its application to understanding the neural basis of emotion, *in* M. S. Gazzaniga (ed.), *The Cognitive Neurosciences*, MIT Press, Cambridge, MA, chapter 72, pp. 1091–1106.

Rolls, E. T. (1996a). The orbitofrontal cortex, *Philosophical Transactions of the Royal Society B* **351**: 1433–1444.

Rolls, E. T. (1996b). Roles of long term potentiation and long term depression in neuronal network operations in the brain, *in* M. S. Fazeli and G. L. Collingridge (eds), *Cortical Plasticity: LTP and LTD*, Bios, Oxford, chapter 11, pp. 223–250.

Rolls, E. T. (1996c). A theory of hippocampal function in memory, *Hippocampus* **6**: 601–620.

Rolls, E. T. (1997a). Brain mechanisms of vision, memory, and consciousness, *in* M. Ito, Y. Miyashita and E. Rolls (eds), *Cognition, Computation, and Consciousness*, Oxford University Press, Oxford, chapter 6, pp. 81–120.

Rolls, E. T. (1997b). Consciousness in neural networks?, *Neural Networks* **10**: 1227–1240.

Rolls, E. T. (1997c). A neurophysiological and computational approach to the functions of the temporal lobe cortical visual areas in invariant object recognition, *in* M. Jenkin and L. Harris (eds), *Computational and Psychophysical Mechanisms of Visual Coding*, Cambridge University Press, Cambridge, chapter 9, pp. 184–220.

Rolls, E. T. (1997d). Taste and olfactory processing in the brain and its relation to the control of eating, *Critical Reviews in Neurobiology* **11**: 263–287.

Rolls, E. T. (1999a). *The Brain and Emotion*, Oxford University Press, Oxford.

Rolls, E. T. (1999b). The functions of the orbitofrontal cortex, *Neurocase* **5**: 301–312.

Rolls, E. T. (1999c). Spatial view cells and the representation of place in the primate hippocampus, *Hippocampus* **9**: 467–480.

Rolls, E. T. (2000a). Functions of the primate temporal lobe cortical visual areas in invariant visual object and face recognition, *Neuron* **27**: 205–218.

Rolls, E. T. (2000b). Memory systems in the brain, *Annual Review of Psychology* **51**: 599–630.

Rolls, E. T. (2000c). Neurophysiology and functions of the primate amygdala, and the neural basis of emotion, *in*

J. P. Aggleton (ed.), *The Amygdala: Second Edition. A Functional Analysis*, Oxford University Press, Oxford, chapter 13, pp. 447–478.

Rolls, E. T. (2000d). The orbitofrontal cortex and reward, *Cerebral Cortex* **10**: 284–294.

Rolls, E. T. (2001). The rules of formation of the olfactory representations found in the orbitofrontal cortex olfactory areas in primates, *Chemical Senses* **26**: 595–604.

Rolls, E. T. (2003). Consciousness absent and present: a neurophysiological exploration, *Progress in Brain Research* **144**: 95–106.

Rolls, E. T. (2004a). The functions of the orbitofrontal cortex, *Brain and Cognition* **55**: 11–29.

Rolls, E. T. (2004b). A higher order syntactic thought (HOST) theory of consciousness, *in* R. J. Gennaro (ed.), *Higher Order Theories of Consciousness*, John Benjamins, Amsterdam, chapter 7, pp. 137–172.

Rolls, E. T. (2005). *Emotion Explained*, Oxford University Press, Oxford.

Rolls, E. T. (2006a). Consciousness absent and present: a neurophysiological exploration of masking, *in* H. Ogmen and B. G. Breitmeyer (eds), *The First Half Second*, MIT Press, Cambridge, MA, chapter 6, pp. 89–108.

Rolls, E. T. (2006b). The neurophysiology and functions of the orbitofrontal cortex, *in* D. H. Zald and S. L. Rauch (eds), *The Orbitofrontal Cortex*, Oxford University Press, Oxford, chapter 5, pp. 95–124.

Rolls, E. T. (2007a). The affective neuroscience of consciousness: higher order syntactic thoughts, dual routes to emotion and action, and consciousness, *in* P. D. Zelazo, M. Moscovitch and E. Thompson (eds), *Cambridge Handbook of Consciousness*, Cambridge University Press, New York, chapter 29, pp. 831–859.

Rolls, E. T. (2007b). A computational neuroscience approach to consciousness, *Neural Networks* **20**: 962–982.

Rolls, E. T. (2007c). Invariant representations of objects in natural scenes in the temporal cortex visual areas, *in* S. Funahashi (ed.), *Representation and Brain*, Springer, Tokyo, chapter 3, pp. 47–102.

Rolls, E. T. (2007d). Memory systems: multiple systems in the brain and their interactions, *in* H. L. Roediger, Y. Dudai and S. M. Fitzpatrick (eds), *Science of Memory: Concepts*, Oxford University Press, New York, chapter 59, pp. 345–351.

Rolls, E. T. (2007e). The representation of information about faces in the temporal and frontal lobes of primates including humans, *Neuropsychologia* **45**: 124–143.

Rolls, E. T. (2007f). Sensory processing in the brain related to the control of food intake, *Proceedings of the Nutrition Society* **66**: 96–112.

Rolls, E. T. (2008a). Emotion, higher order syntactic thoughts, and consciousness, *in* L. Weiskrantz and M. Davies (eds), *Frontiers of Consciousness*, Oxford University Press, Oxford, chapter 4, pp. 131–167.

Rolls, E. T. (2008b). Face representations in different brain areas, and critical band masking, *Journal of Neuropsychology* **2**: 325–360.

Rolls, E. T. (2008c). Functions of the orbitofrontal and pregenual cingulate cortex in taste, olfaction, appetite and emotion, *Acta Physiologica Hungarica* **95**: 131–164.

Rolls, E. T. (2008d). *Memory, Attention, and Decision-Making. A Unifying Computational Neuroscience Approach*, Oxford University Press, Oxford.

Rolls, E. T. (2008e). The primate hippocampus and episodic memory, *in* E. Dere, A. Easton, L. Nadel and J. P. Huston (eds), *Handbook of Episodic Memory*, Elsevier, Amsterdam, chapter 4.2, pp. 417–438.

Rolls, E. T. (2008f). Top-down control of visual perception: attention in natural vision, *Perception* **37**: 333–354.

Rolls, E. T. (2009a). The anterior and midcingulate cortices and reward, *in* B. Vogt (ed.), *Cingulate Neurobiology and Disease*, Oxford University Press, Oxford, chapter 8, pp. 191–206.

Rolls, E. T. (2009b). Functional neuroimaging of umami taste: what makes umami pleasant, *American Journal of Clinical Nutrition* **90**: 803S–814S.

Rolls, E. T. (2009c). The neurophysiology and computational mechanisms of object representation, *in* S. Dickinson, M. Tarr, A. Leonardis and B. Schiele (eds), *Object Categorization: Computer and Human Vision Perspectives*, Cambridge University Press, Cambridge, chapter 14, pp. 257–287.

Rolls, E. T. (2010a). The affective and cognitive processing of touch, oral texture, and temperature in the brain, *Neuroscience and Biobehavioral Reviews* **34**: 237–245.

Rolls, E. T. (2010b). A computational theory of episodic memory formation in the hippocampus, *Behavioural Brain Research* **215**: 180–196.

Rolls, E. T. (2010c). Noise in the brain, decision-making, determinism, free will, and consciousness, *in* E. Perry, D. Collerton, F. LeBeau and H. Ashton (eds), *New Horizons in the Neuroscience of Consciousness*, John Benjamins, Amsterdam, pp. 113–120.

Rolls, E. T. (2011a). Chemosensory learning in the cortex, *Frontiers in Systems Neuroscience* **5**: 78 (1–13).

Rolls, E. T. (2011b). Consciousness, decision-making, and neural computation, *in* V. Cutsuridis, A. Hussain and J. G. Taylor (eds), *Perception-Action Cycle: Models, architecture, and hardware*, Springer, Berlin, chapter 9, pp. 287–333.

Rolls, E. T. (2011c). David Marr's Vision: floreat computational neuroscience, *Brain* **134**: 913–916.

Rolls, E. T. (2011d). Face neurons, *in* A. J. Calder, G. Rhodes, M. H. Johnson and J. V. Haxby (eds), *The Oxford Handbook of Face Perception*, Oxford University Press, Oxford, chapter 4, pp. 51–75.

Rolls, E. T. (2011e). Taste, olfactory, and food texture reward processing in the brain and obesity, *International Journal of Obesity* **35**: 550–561.

Rolls, E. T. (2012a). Advantages of dilution in the connectivity of attractor networks in the brain, *Biologically Inspired Cognitive Architectures* **1**: 44–54.

Rolls, E. T. (2012b). Glutamate, obsessive-compulsive disorder, schizophrenia, and the stability of cortical attractor neuronal networks, *Pharmacology, Biochemistry and Behavior* **100**: 736–751.

Rolls, E. T. (2012c). Invariant visual object and face recognition: neural and computational bases, and a model, VisNet, *Frontiers in Computational Neuroscience* **6**(35): 1–70.

Rolls, E. T. (2012d). *Neuroculture: On the Implications of Brain Science*, Oxford University Press, Oxford.

Rolls, E. T. (2012e). Taste, olfactory, and food texture reward processing in the brain and the control of appetite, *Proceedings of the Nutrition Society* **71**: 488–501.

Rolls, E. T. (2012f). Willed action, free will, and the stochastic neurodynamics of decision-making, *Frontiers in Integrative Neuroscience* **6**: 68.

Rolls, E. T. (2013a). A biased activation theory of the cognitive and attentional modulation of emotion, *Frontiers in Human Neuroscience* **7**: 74.

Rolls, E. T. (2013b). The mechanisms for pattern completion and pattern separation in the hippocampus, *Frontiers in Systems Neuroscience* **7**: 74.

Rolls, E. T. (2013c). On the relation between the mind and the brain: a neuroscience perspective, *Philosophia Scientiae* **17**: 31–70.

Rolls, E. T. (2013d). A quantitative theory of the functions of the hippocampal CA3 network in memory, *Frontiers in Cellular Neuroscience* **7**: 98.

Rolls, E. T. (2013e). What are emotional states, and why do we have them?, *Emotion Review* **5**: 241–247.

Rolls, E. T. (2014a). *Emotion and Decision-Making Explained*, Oxford University Press, Oxford.

Rolls, E. T. (2014b). Emotion and decision-making explained: Précis, *Cortex* **59**: 185–193.

Rolls, E. T. (2014c). Neuroculture: art, aesthetics, and the brain, *Rendiconti Lincei Scienze Fisiche e Naturali* **25**: 291–307.

Rolls, E. T. (2015a). Central neural integration of taste, smell and other sensory modalities, *in* R. L. Doty (ed.), *Handbook of Olfaction and Gustation*, third edn, Wiley, New York, chapter 44, pp. 1027–1048.

Rolls, E. T. (2015b). Diluted connectivity in pattern association networks facilitates the recall of information from the hippocampus to the neocortex, *Progress in Brain Research* **219**: 21–43.

Rolls, E. T. (2015c). Limbic systems for emotion and for memory, but no single limbic system, *Cortex* **62**: 119–157.

Rolls, E. T. (2015d). The neuronal representation of information in the human brain. Review, *Brain* **138**: 3459–3462.

Rolls, E. T. (2015e). Taste, olfactory, and food reward value processing in the brain, *Progress in Neurobiology* **127–128**: 64–90.

Rolls, E. T. (2016a). *Cerebral Cortex: Principles of Operation*, Oxford University Press, Oxford.

Rolls, E. T. (2016b). Functions of the anterior insula in taste, autonomic, and related functions, *Brain and Cognition* **110**: 4–19.

Rolls, E. T. (2016c). Motivation Explained: Ultimate and proximate accounts of hunger and appetite, *Advances in Motivation Science* **3**: 187–249.

Rolls, E. T. (2016d). A non-reward attractor theory of depression, *Neuroscience and Biobehavioral Reviews* **68**: 47–58.

Rolls, E. T. (2016e). Pattern separation, completion, and categorisation in the hippocampus and neocortex, *Neurobiology of Learning and Memory* **129**: 4–28.

Rolls, E. T. (2016f). Reward systems in the brain and nutrition, *Annual Review of Nutrition* **36**: 435–470.

Rolls, E. T. (2017a). Cortical coding, *Language, Cognition and Neuroscience* **32**: 316–329.

Rolls, E. T. (2017b). Evolution of the emotional brain, *in* S. Watanabe, M. A. Hofman and T. Shimiz (eds), *Evolution of Brain, Cognition, and Emotion in Vertebrates*, Springer, Tokyo.

Rolls, E. T. (2017c). Face processing, *in* J. Vonk and T. K. Shackelford (eds), *Encyclopedia of Animal Cognition and Behavior*, Springer Nature.

Rolls, E. T. (2017d). Neurobiological foundations of aesthetics and art, *New Ideas in Psychology* p. doi: 10.1016/j.newideapsych.2017.03.005.

Rolls, E. T. (2017e). The orbitofrontal cortex and emotion in health and disease, including depression, *Neuropsychologia*.

Rolls, E. T. (2017f). The roles of the orbitofrontal cortex via the habenula in non-reward and depression, and in the responses of serotonin and dopamine neurons, *Neuroscience and Biobehavioral Reviews* **75**: 331–334.

Rolls, E. T. (2017g). A scientific theory of *ars memoriae*: spatial view cells in a continuous attractor network with linked items, *Hippocampus* **27**: 570–579.

Rolls, E. T. and Baylis, G. C. (1986). Size and contrast have only small effects on the responses to faces of neurons in the cortex of the superior temporal sulcus of the monkey, *Experimental Brain Research* **65**: 38–48.

Rolls, E. T. and Baylis, L. L. (1994). Gustatory, olfactory and visual convergence within the primate orbitofrontal cortex, *Journal of Neuroscience* **14**: 5437–5452.

Rolls, E. T. and Deco, G. (2002). *Computational Neuroscience of Vision*, Oxford University Press, Oxford.

Rolls, E. T. and Deco, G. (2006). Attention in natural scenes: neurophysiological and computational bases, *Neural Networks* **19**: 1383–1394.

Rolls, E. T. and Deco, G. (2010). *The Noisy Brain: Stochastic Dynamics as a Principle of Brain Function*, Oxford University Press, Oxford.

Rolls, E. T. and Deco, G. (2011a). A computational neuroscience approach to schizophrenia and its onset, *Neuroscience and Biobehavioral Reviews* **35**: 1644–1653.

Rolls, E. T. and Deco, G. (2011b). Prediction of decisions from noise in the brain before the evidence is provided, *Frontiers in Neuroscience* **5**: 33.

Rolls, E. T. and Deco, G. (2015a). Networks for memory, perception, and decision-making, and beyond to how the syntax for language might be implemented in the brain, *Brain Research* **1621**: 316–334.

Rolls, E. T. and Deco, G. (2015b). A stochastic neurodynamics approach to the changes in cognition and memory in aging, *Neurobiology of Learning and Memory* **118**: 150–161.

Rolls, E. T. and Deco, G. (2016). Non-reward neural mechanisms in the orbitofrontal cortex, *Cortex* **83**: 27–38.

Rolls, E. T. and Grabenhorst, F. (2008). The orbitofrontal cortex and beyond: from affect to decision-making, *Progress in Neurobiology* **86**: 216–244.

Rolls, E. T. and Johnstone, S. (1992). Neurophysiological analysis of striatal function, *in* G. Vallar, S. Cappa and C. Wallesch (eds), *Neuropsychological Disorders Associated with Subcortical Lesions*, Oxford University Press, Oxford, chapter 3, pp. 61–97.

Rolls, E. T. and Kesner, R. P. (2006). A theory of hippocampal function, and tests of the theory, *Progress in Neurobiology* **79**: 1–48.

Rolls, E. T. and Kesner, R. P. (2016). Pattern separation and pattern completion in the hippocampal system, *Neurobiology of Learning and Memory* **129**: 1–3.

Rolls, E. T. and McCabe, C. (2007). Enhanced affective brain representations of chocolate in cravers vs non-cravers, *European Journal of Neuroscience* **26**: 1067–1076.

Rolls, E. T. and Mills, W. P. C. (2017). Computations in the deep vs superficial layers of the cerebral cortex.

Rolls, E. T. and Milward, T. (2000). A model of invariant object recognition in the visual system: learning rules, activation functions, lateral inhibition, and information-based performance measures, *Neural Computation* **12**: 2547–2572.

Rolls, E. T. and O'Mara, S. (1993). Neurophysiological and theoretical analysis of how the hippocampus functions in memory, *in* T. Ono, L. Squire, M. Raichle, D. Perrett and M. Fukuda (eds), *Brain Mechanisms of Perception and Memory: From Neuron to Behavior*, Oxford University Press, New York, chapter 17, pp. 276–300.

Rolls, E. T. and O'Mara, S. M. (1995). View-responsive neurons in the primate hippocampal complex, *Hippocampus* **5**: 409–424.

Rolls, E. T. and Rolls, B. J. (1973). Altered food preferences after lesions in the basolateral region of the amygdala in the rat, *Journal of Comparative and Physiological Psychology* **83**: 248–259.

Rolls, E. T. and Rolls, J. H. (1997). Olfactory sensory-specific satiety in humans, *Physiology and Behavior* **61**: 461–473.

Rolls, E. T. and Scott, T. R. (2003). Central taste anatomy and neurophysiology, *in* R. Doty (ed.), *Handbook of Olfaction and Gustation*, 2nd edn, Dekker, New York, chapter 33, pp. 679–705.

Rolls, E. T. and Stringer, S. M. (2000). On the design of neural networks in the brain by genetic evolution, *Progress in Neurobiology* **61**: 557–579.

Rolls, E. T. and Stringer, S. M. (2001a). Invariant object recognition in the visual system with error correction and temporal difference learning, *Network: Computation in Neural Systems* **12**: 111–129.

Rolls, E. T. and Stringer, S. M. (2001b). A model of the interaction between mood and memory, *Network: Computation in Neural Systems* **12**: 89–109.

Rolls, E. T. and Stringer, S. M. (2005). Spatial view cells in the hippocampus, and their idiothetic update based on place and head direction, *Neural Networks* **18**: 1229–1241.

Rolls, E. T. and Stringer, S. M. (2006). Invariant visual object recognition: a model, with lighting invariance, *Journal of Physiology – Paris* **100**: 43–62.

Rolls, E. T. and Stringer, S. M. (2007). Invariant global motion recognition in the dorsal visual system: a unifying theory, *Neural Computation* **19**: 139–169.

Rolls, E. T. and Tovee, M. J. (1994). Processing speed in the cerebral cortex and the neurophysiology of visual masking, *Proceedings of the Royal Society, B* **257**: 9–15.

Rolls, E. T. and Tovee, M. J. (1995a). The responses of single neurons in the temporal visual cortical areas of the macaque when more than one stimulus is present in the visual field, *Experimental Brain Research* **103**: 409–420.

Rolls, E. T. and Tovee, M. J. (1995b). Sparseness of the neuronal representation of stimuli in the primate temporal visual cortex, *Journal of Neurophysiology* **73**: 713–726.

Rolls, E. T. and Treves, A. (1990). The relative advantages of sparse versus distributed encoding for associative neuronal networks in the brain, *Network* **1**: 407–421.

Rolls, E. T. and Treves, A. (1998). *Neural Networks and Brain Function*, Oxford University Press, Oxford.

Rolls, E. T. and Treves, A. (2011). The neuronal encoding of information in the brain, *Progress in Neurobiology* **95**: 448–490.

Rolls, E. T. and Webb, T. J. (2012). Cortical attractor network dynamics with diluted connectivity, *Brain Research* **1434**: 212–225.

Rolls, E. T. and Webb, T. J. (2014). Finding and recognising objects in natural scenes: complementary computations in the dorsal and ventral visual systems, *Frontiers in Computational Neuroscience* **8**: 85.

Rolls, E. T. and Williams, G. V. (1987a). Neuronal activity in the ventral striatum of the primate, *in* M. B. Carpenter and A. Jayamaran (eds), *The Basal Ganglia II – Structure and Function – Current Concepts*, Plenum, New York, pp. 349–356.

Rolls, E. T. and Williams, G. V. (1987b). Sensory and movement-related neuronal activity in different regions of the primate striatum, *in* J. S. Schneider and T. I. Lidsky (eds), *Basal Ganglia and Behavior: Sensory Aspects and Motor Functioning*, Hans Huber, Bern, pp. 37–59.

Rolls, E. T. and Wirth, S. (2017). Spatial representations in the primate pippocampus: evolution and function.

Rolls, E. T. and Xiang, J.-Z. (2005). Reward–spatial view representations and learning in the primate hippocampus, *Journal of Neuroscience* **25**: 6167–6174.

Rolls, E. T. and Xiang, J.-Z. (2006). Spatial view cells in the primate hippocampus, and memory recall, *Reviews in the Neurosciences* **17**: 175–200.

Rolls, E. T., Burton, M. J. and Mora, F. (1976). Hypothalamic neuronal responses associated with the sight of food, *Brain Research* **111**: 53–66.

Rolls, E. T., Judge, S. J. and Sanghera, M. (1977). Activity of neurones in the inferotemporal cortex of the alert monkey, *Brain Research* **130**: 229–238.

Rolls, E. T., Sanghera, M. K. and Roper-Hall, A. (1979). The latency of activation of neurons in the lateral hypothalamus and substantia innominata during feeding in the monkey, *Brain Research* **164**: 121–135.

Rolls, E. T., Burton, M. J. and Mora, F. (1980). Neurophysiological analysis of brain-stimulation reward in the monkey, *Brain Research* **194**: 339–357.

Rolls, E. T., Perrett, D. I., Caan, A. W. and Wilson, F. A. W. (1982). Neuronal responses related to visual recognition, *Brain* **105**: 611–646.

Rolls, E. T., Rolls, B. J. and Rowe, E. A. (1983a). Sensory-specific and motivation-specific satiety for the sight and taste of food and water in man, *Physiology and Behavior* **30**: 185–192.

Rolls, E. T., Thorpe, S. J. and Maddison, S. P. (1983b). Responses of striatal neurons in the behaving monkey. 1. Head of the caudate nucleus, *Behavioural Brain Research* **7**: 179–210.

Rolls, E. T., Thorpe, S. J., Boytim, M., Szabo, I. and Perrett, D. I. (1984). Responses of striatal neurons in the behaving monkey. 3. Effects of iontophoretically applied dopamine on normal responsiveness, *Neuroscience* **12**: 1201–1212.

Rolls, E. T., Baylis, G. C. and Leonard, C. M. (1985). Role of low and high spatial frequencies in the face-selective responses of neurons in the cortex in the superior temporal sulcus, *Vision Research* **25**: 1021–1035.

Rolls, E. T., Baylis, G. C. and Hasselmo, M. E. (1987). The responses of neurons in the cortex in the superior temporal sulcus of the monkey to band-pass spatial frequency filtered faces, *Vision Research* **27**: 311–326.

Rolls, E. T., Scott, T. R., Sienkiewicz, Z. J. and Yaxley, S. (1988). The responsiveness of neurones in the frontal opercular gustatory cortex of the macaque monkey is independent of hunger, *Journal of Physiology* **397**: 1–12.

Rolls, E. T., Baylis, G. C., Hasselmo, M. and Nalwa, V. (1989a). The representation of information in the temporal lobe visual cortical areas of macaque monkeys, *in* J. Kulikowski, C. Dickinson and I. Murray (eds), *Seeing Contour and Colour*, Pergamon, Oxford.

Rolls, E. T., Miyashita, Y., Cahusac, P. M. B., Kesner, R. P., Niki, H., Feigenbaum, J. and Bach, L. (1989b). Hippocampal neurons in the monkey with activity related to the place in which a stimulus is shown, *Journal of Neuroscience* **9**: 1835–1845.

Rolls, E. T., Sienkiewicz, Z. J. and Yaxley, S. (1989c). Hunger modulates the responses to gustatory stimuli of single neurons in the caudolateral orbitofrontal cortex of the macaque monkey, *European Journal of Neuroscience* **1**: 53–60.

Rolls, E. T., Yaxley, S. and Sienkiewicz, Z. J. (1990). Gustatory responses of single neurons in the orbitofrontal cortex of the macaque monkey, *Journal of Neurophysiology* **64**: 1055–1066.

Rolls, E. T., Cahusac, P. M. B., Feigenbaum, J. D. and Miyashita, Y. (1993). Responses of single neurons in the hippocampus of the macaque related to recognition memory, *Experimental Brain Research* **93**: 299–306.

Rolls, E. T., Hornak, J., Wade, D. and McGrath, J. (1994a). Emotion-related learning in patients with social and emotional changes associated with frontal lobe damage, *Journal of Neurology, Neurosurgery and Psychiatry* **57**: 1518–1524.

Rolls, E. T., Tovee, M. J., Purcell, D. G., Stewart, A. L. and Azzopardi, P. (1994b). The responses of neurons in the temporal cortex of primates, and face identification and detection, *Experimental Brain Research* **101**: 474–484.

Rolls, E. T., Critchley, H. D. and Treves, A. (1996a). The representation of olfactory information in the primate orbitofrontal cortex, *Journal of Neurophysiology* **75**: 1982–1996.

Rolls, E. T., Critchley, H. D., Mason, R. and Wakeman, E. A. (1996b). Orbitofrontal cortex neurons: role in olfactory and visual association learning, *Journal of Neurophysiology* **75**: 1970–1981.

Rolls, E. T., Critchley, H. D., Wakeman, E. A. and Mason, R. (1996c). Responses of neurons in the primate taste cortex to the glutamate ion and to inosine $5'$-monophosphate, *Physiology and Behavior* **59**: 991–1000.

Rolls, E. T., Robertson, R. G. and Georges-François, P. (1997a). Spatial view cells in the primate hippocampus, *European Journal of Neuroscience* **9**: 1789–1794.

Rolls, E. T., Treves, A. and Tovee, M. J. (1997b). The representational capacity of the distributed encoding of information provided by populations of neurons in the primate temporal visual cortex, *Experimental Brain Research* 114: 149–162.

Rolls, E. T., Treves, A., Foster, D. and Perez-Vicente, C. (1997c). Simulation studies of the CA3 hippocampal subfield modelled as an attractor neural network, *Neural Networks* 10: 1559–1569.

Rolls, E. T., Treves, A., Tovee, M. and Panzeri, S. (1997d). Information in the neuronal representation of individual stimuli in the primate temporal visual cortex, *Journal of Computational Neuroscience* 4: 309–333.

Rolls, E. T., Critchley, H. D., Browning, A. and Hernadi, I. (1998a). The neurophysiology of taste and olfaction in primates, and umami flavor, *Annals of the New York Academy of Sciences* 855: 426–437.

Rolls, E. T., Treves, A., Robertson, R. G., Georges-François, P. and Panzeri, S. (1998b). Information about spatial view in an ensemble of primate hippocampal cells, *Journal of Neurophysiology* 79: 1797–1813.

Rolls, E. T., Critchley, H. D., Browning, A. S., Hernadi, A. and Lenard, L. (1999a). Responses to the sensory properties of fat of neurons in the primate orbitofrontal cortex, *Journal of Neuroscience* 19: 1532–1540.

Rolls, E. T., Tovee, M. J. and Panzeri, S. (1999b). The neurophysiology of backward visual masking: information analysis, *Journal of Cognitive Neuroscience* 11: 335–346.

Rolls, E. T., Stringer, S. M. and Trappenberg, T. P. (2002). A unified model of spatial and episodic memory, *Proceedings of The Royal Society B* 269: 1087–1093.

Rolls, E. T., Aggelopoulos, N. C. and Zheng, F. (2003a). The receptive fields of inferior temporal cortex neurons in natural scenes, *Journal of Neuroscience* 23: 339–348.

Rolls, E. T., Franco, L., Aggelopoulos, N. C. and Reece, S. (2003b). An information theoretic approach to the contributions of the firing rates and the correlations between the firing of neurons, *Journal of Neurophysiology* 89: 2810–2822.

Rolls, E. T., Kringelbach, M. L. and De Araujo, I. E. T. (2003c). Different representations of pleasant and unpleasant odours in the human brain, *European Journal of Neuroscience* 18: 695–703.

Rolls, E. T., O'Doherty, J., Kringelbach, M. L., Francis, S., Bowtell, R. and McGlone, F. (2003d). Representations of pleasant and painful touch in the human orbitofrontal and cingulate cortices, *Cerebral Cortex* 13: 308–317.

Rolls, E. T., Verhagen, J. V. and Kadohisa, M. (2003e). Representations of the texture of food in the primate orbitofrontal cortex: neurons responding to viscosity, grittiness, and capsaicin, *Journal of Neurophysiology* 90: 3711–3724.

Rolls, E. T., Aggelopoulos, N. C., Franco, L. and Treves, A. (2004). Information encoding in the inferior temporal visual cortex: contributions of the firing rates and the correlations between the firing of neurons, *Biological Cybernetics* 90: 19–32.

Rolls, E. T., Browning, A. S., Inoue, K. and Hernadi, S. (2005a). Novel visual stimuli activate a population of neurons in the primate orbitofrontal cortex, *Neurobiology of Learning and Memory* 84: 111–123.

Rolls, E. T., Franco, L. and Stringer, S. M. (2005b). The perirhinal cortex and long-term familiarity memory, *Quarterly Journal of Experimental Psychology B* 58: 234–245.

Rolls, E. T., Xiang, J.-Z. and Franco, L. (2005c). Object, space and object-space representations in the primate hippocampus, *Journal of Neurophysiology* 94: 833–844.

Rolls, E. T., Critchley, H. D., Browning, A. S. and Inoue, K. (2006a). Face-selective and auditory neurons in the primate orbitofrontal cortex, *Experimental Brain Research* 170: 74–87.

Rolls, E. T., Franco, L., Aggelopoulos, N. C. and Jerez, J. M. (2006b) Information in the first spike, the order of spikes, and the number of spikes provided by neurons in the inferior temporal visual cortex, *Vision Research* 46: 4193–4205.

Rolls, E. T., Stringer, S. M. and Elliot, T. (2006c). Entorhinal cortex grid cells can map to hippocampal place cells by competitive learning, *Network: Computation in Neural Systems* 17: 447–465.

Rolls, E. T., Grabenhorst, F. and Parris, B. (2008a). Warm pleasant feelings in the brain, *Neuroimage* 41: 1504–1513.

Rolls, E. T., Grabenhorst, F., Margot, C., da Silva, M. and Velazco, M. I. (2008b). Selective attention to affective value alters how the brain processes olfactory stimuli, *Journal of Cognitive Neuroscience* 20: 1815–1826.

Rolls, E. T., Loh, M. and Deco, G. (2008c). An attractor hypothesis of obsessive-compulsive disorder, *European Journal of Neuroscience* 28: 782–793.

Rolls, E. T., Loh, M., Deco, G. and Winterer, G. (2008d). Computational models of schizophrenia and dopamine modulation in the prefrontal cortex, *Nature Reviews Neuroscience* 9: 696–709.

Rolls, E. T., McCabe, C. and Redoute, J. (2008e). Expected value, reward outcome, and temporal difference error representations in a probabilistic decision task, *Cerebral Cortex* 18: 652–663.

Rolls, E. T., Tromans, J. M. and Stringer, S. M. (2008f). Spatial scene representations formed by self-organizing learning in a hippocampal extension of the ventral visual system, *European Journal of Neuroscience* 28: 2116–2127.

Rolls, E. T., Grabenhorst, F. and Franco, L. (2009). Prediction of subjective affective state from brain activations, *Journal of Neurophysiology* 101: 1294–1308.

Rolls, E. T., Critchley, H., Verhagen, J. V. and Kadohisa, M. (2010a). The representation of information about taste and odor in the primate orbitofrontal cortex, *Chemosensory Perception* 3: 16–33.

Rolls, E. T., Grabenhorst, F. and Deco, G. (2010b). Choice, difficulty, and confidence in the brain, *Neuroimage*

53: 694–706.

Rolls, E. T., Grabenhorst, F. and Deco, G. (2010c). Decision-making, errors, and confidence in the brain, *Journal of Neurophysiology* **104**: 2359–2374.

Rolls, E. T., Grabenhorst, F. and Parris, B. A. (2010d). Neural systems underlying decisions about affective odors, *Journal of Cognitive Neuroscience* **10**: 1068–1082.

Rolls, E. T., Webb, T. J. and Deco, G. (2012). Communication before coherence, *European Journal of Neuroscience* **36**: 2689–2709.

Rolls, E. T., Dempere-Marco, L. and Deco, G. (2013). Holding multiple items in short term memory: a neural mechanism, *PLoS One* **8**: e61078.

Rolls, E. T., Cheng, W., Gilson, M., Qiu, J., Hu, Z., Li, Y., Huang, C.-C., Yang, A. C., Tsai, S.-J., Zhang, X., Zhuang, K., Lin, C.-P., Deco, G., Xie, P. and Feng, J. (2017a). Effective connectivity in depression, *Biological Psychiatry: Cognitive Neuroscience and Neuroimaging*.

Rolls, E. T., Lu, W., Wan, L., Yan, H., Wang, C., Yang, F., Tan, Y.-L., Li, L., Group, C. S. C., Yu, H., Liddle, P. F., Palaniyappan, L., Zhang, D., Yue, W. and Feng, J. (2017b). Individual differences in schizophrenia, *British Journal of Psychiatry Open*.

Romanski, L. M., Averbeck, B. B. and Diltz, M. (2005). Neural representation of vocalizations in the primate ventrolateral prefrontal cortex, *Journal of Neurophysiology* **93**: 734–747.

Romo, R. and Salinas, E. (2003). Flutter discrimination: Neural codes, perception, memory and decision making, *Nature Reviews Neuroscience* **4**: 203–218.

Romo, R., Hernandez, A. and Zainos, A. (2004). Neuronal correlates of a perceptual decision in ventral premotor cortex, *Neuron* **41**: 165–173.

Rondi-Reig, L., Libbey, M., Eichenbaum, H. and Tonegawa, S. (2001). Ca1-specific n-methyl-d-aspartate receptor knockout mice are deficient in solving a nonspatial transverse patterning task, *Proc Natl Acad Sci U S A* **98**: 3543–8.

Rosch, E. (1975). Cognitive representations of semantic categories, *Journal of Experimental Psychology: General* **104**: 192–233.

Rosenberg, D. R., MacMaster, F. P., Keshavan, M. S., Fitzgerald, K. D., Stewart, C. M. and Moore, G. J. (2000). Decrease in caudate glutamatergic concentrations in pediatric obsessive-compulsive disorder patients taking paroxetine, *Journal of the American Academy of Child and Adolescent Psychiatry* **39**: 1096–1103.

Rosenberg, D. R., MacMillan, S. N. and Moore, G. J. (2001). Brain anatomy and chemistry may predict treatment response in paediatric obsessive–compulsive disorder, *International Journal of Neuropsychopharmacology* **4**: 179–190.

Rosenberg, D. R., Mirza, Y., Russell, A., Tang, J., Smith, J. M., Banerjee, S. P., Bhandari, R., Rose, M., Ivey, J., Boyd, C. and Moore, G. J. (2004). Reduced anterior cingulate glutamatergic concentrations in childhood OCD and major depression versus healthy controls, *Journal of the American Academy of Child and Adolescent Psychiatry* **43**: 1146–1153.

Rosenblatt, F. (1961). *Principles of Neurodynamics: Perceptrons and the Theory of Brain Mechanisms*, Spartan, Washington, DC.

Rosenkilde, C. E., Bauer, R. H. and Fuster, J. M. (1981). Single unit activity in ventral prefrontal cortex in behaving monkeys, *Brain Research* **209**: 375–394.

Rosenthal, D. (1990). A theory of consciousness, *ZIF Report 40/1990. Zentrum für Interdisziplinäre Forschung, Bielefeld*. Reprinted in Block, N., Flanagan, O. and Guzeldere, G. (eds.) (1997) *The Nature of Consciousness: Philosophical Debates*. MIT Press, Cambridge MA, pp. 729–853.

Rosenthal, D. M. (1986). Two concepts of consciousness, *Philosophical Studies* **49**: 329–359.

Rosenthal, D. M. (1993). Thinking that one thinks, *in* M. Davies and G. W. Humphreys (eds), *Consciousness*, Blackwell, Oxford, chapter 10, pp. 197–223.

Rosenthal, D. M. (2004). Varieties of higher order theory, *in* R. J. Gennaro (ed.), *Higher Order Theories of Consciousness*, John Benjamins, Amsterdam, pp. 17–44.

Rosenthal, D. M. (2005). *Consciousness and Mind*, Oxford University Press, Oxford.

Rosenthal, D. M. (2012). Higher-order awareness, misrepresentation, and function, *Philosophical Transactions of the Royal Society B: Biological Sciences* **367**: 1424–1438.

Rossi, A. F., Pessoa, L., Desimone, R. and Ungerleider, L. G. (2009). The prefrontal cortex and the executive control of attention, *Experimental Brain Research* **192**: 489–497.

Rottschy, C., Langner, R., Dogan, I., Reetz, K., Laird, A. R., Schulz, J. B., Fox, P. T. and Eickhoff, S. B. (2012). Modelling neural correlates of working memory: a coordinate-based meta-analysis, *Neuroimage* **60**: 830–846.

Roudi, Y. and Treves, A. (2006). Localized activity profiles and storage capacity of rate-based autoassociative networks, *Physical Review E* **73**: 061904.

Roudi, Y. and Treves, A. (2008). Representing where along with what information in a model of a cortical patch, *PLoS Computational Biology* **4**(3): e1000012.

Roy, R. D., Stefan, M. I. and Rosenmund, C. (2014). Biophysical properties of presynaptic short-term plasticity in hippocampal neurons: insights from electrophysiology, imaging and mechanistic models, *Frontiers in Cellular Neuroscience* **8**: 141.

Rudebeck, P. H. and Murray, E. A. (2011). Dissociable effects of subtotal lesions within the macaque orbital prefrontal cortex on reward-guided behavior, *Journal of Neuroscience* **31**: 10569–10578.

Rudebeck, P. H., Behrens, T. E., Kennerley, S. W., Baxter, M. G., Buckley, M. J., Walton, M. E. and Rushworth, M. F. (2008). Frontal cortex subregions play distinct roles in choices between actions and stimuli, *Journal of Neuroscience* **28**: 13775–13785.

Rumelhart, D. E. and McClelland, J. L. (1986). *Parallel Distributed Processing*, Vol. 1: Foundations, MIT Press, Cambridge, MA.

Rumelhart, D. E. and Zipser, D. (1985). Feature discovery by competitive learning, *Cognitive Science* **9**: 75–112.

Rumelhart, D. E., Hinton, G. E. and Williams, R. J. (1986a). Learning internal representations by error propagation, *in* D. E. Rumelhart, J. L. McClelland and the PDP Research Group (eds), *Parallel Distributed Processing: Explorations in the Microstructure of Cognition*, Vol. 1, MIT Press, Cambridge, MA, chapter 8, pp. 318–362.

Rumelhart, D. E., Hinton, G. E. and Williams, R. J. (1986b). Learning representations by back-propagating errors, *Nature* **323**: 533–536.

Rupniak, N. M. J. and Gaffan, D. (1987). Monkey hippocampus and learning about spatially directed movements, *Journal of Neuroscience* **7**: 2331–2337.

Rushworth, M. F., Noonan, M. P., Boorman, E. D., Walton, M. E. and Behrens, T. E. (2011). Frontal cortex and reward-guided learning and decision-making, *Neuron* **70**: 1054–1069.

Rushworth, M. F., Kolling, N., Sallet, J. and Mars, R. B. (2012). Valuation and decision-making in frontal cortex: one or many serial or parallel systems?, *Current Opinion in Neurobiology* **22**: 946–955.

Rushworth, M. F. S., Walton, M. E., Kennerley, S. W. and Bannerman, D. M. (2004). Action sets and decisions in the medial frontal cortex, *Trends in Cognitive Sciences* **8**: 410–417.

Rushworth, M. F. S., Behrens, T. E., Rudebeck, P. H. and Walton, M. E. (2007a). Contrasting roles for cingulate and orbitofrontal cortex in decisions and social behaviour, *Trends in Cognitive Sciences* **11**: 168–176.

Rushworth, M. F. S., Buckley, M. J., Behrens, T. E., Walton, M. E. and Bannerman, D. M. (2007b). Functional organization of the medial frontal cortex, *Current Opinion in Neurobiology* **17**: 220–227.

Russchen, F. T., Amaral, D. G. and Price, J. L. (1985). The afferent connections of the substantia innominata in the monkey, Macaca fascicularis, *Journal of Comparative Neurology* **242**: 1–27.

Rust, N. C., Mante, V., Simoncelli, E. P. and Movshon, J. A. (2006). How mt cells analyze the motion of visual patterns, *Nat Neurosci* **9**(11): 1421–31.

Sah, P. (1996). Ca^{2+}-activated K^+ currents in neurones: types, physiological roles and modulation, *Trends in Neuroscience* **19**: 150–154.

Sah, P. and Faber, E. S. (2002). Channels underlying neuronal calcium-activated potassium currents, *Progress in Neurobiology* **66**: 345–353.

Saint-Cyr, J. A., Ungerleider, L. G. and Desimone, R. (1990). Organization of visual cortical inputs to the striatum and subsequent outputs to the pallido-nigral complex in the monkey, *Journal of Comparative Neurology* **298**: 129–156.

Sakai, K. and Miyashita, Y. (1991). Neural organisation for the long-term memory of paired associates, *Nature* **354**(6349): 152–155.

Salakhutdinov, R. and Larochelle, H. (2010). Efficient learning of deep boltzmann machines, *International Conference on Artificial Intelligence and Statistics*, pp. 693–700.

Salinas, E. and Abbott, L. F. (1996). A model of multiplicative neural responses in parietal cortex, *Proceedings of the National Academy of Sciences USA* **93**: 11956–11961.

Salinas, E. and Abbott, L. F. (1997). Invariant visual responses from attentional gain fields, *Journal of Neurophysiology* **77**: 3267–3272.

Salinas, E. and Sejnowski, T. J. (2001). Correlated neuronal activity and the flow of neural information, *Nat Rev Neurosci* **2**: 539–550.

Samsonovich, A. and McNaughton, B. (1997). Path integration and cognitive mapping in a continuous attractor neural network model, *Journal of Neuroscience* **17**: 5900–5920.

Sanghera, M. K., Rolls, E. T. and Roper-Hall, A. (1979). Visual responses of neurons in the dorsolateral amygdala of the alert monkey, *Experimental Neurology* **63**: 610–626.

Saper, C. B., Swanson, L. W. and Cowan, W. M. (1979). An autoradiographic study of the efferent connections of the lateral hypothalamic area in the rat, *Journal of Comparative Neurology* **183**: 689–706.

Sargolini, F., Fyhn, M., Hafting, T., McNaughton, B. L., Witter, M. P., Moser, M. B. and Moser, E. I. (2006). Conjunctive representation of position, direction, and velocity in entorhinal cortex, *Science* **312**: 758–762.

Sato, T. (1989). Interactions of visual stimuli in the receptive fields of inferior temporal neurons in macaque, *Experimental Brain Research* **77**: 23–30.

Save, E., Guazzelli, A. and Poucet, B. (2001). Dissociation of the effects of bilateral lesions of the dorsal hippocampus and parietal cortex on path integration in the rat, *Behavavioral Neuroscience* **115**: 1212–1223.

Sawaguchi, T. and Goldman-Rakic, P. S. (1991). D1 dopamine receptors in prefrontal cortex: Involvement in working memory, *Science* **251**: 947–950.

Sawaguchi, T. and Goldman-Rakic, P. S. (1994). The role of D1-dopamine receptor in working memory: local injections of dopamine antagonists into the prefrontal cortex of rhesus monkeys performing an oculomotor delayed-response task, *Journal of Neurophysiology* **71**: 515–528.

Schachter, S. and Singer, J. (1962). Cognitive, social and physiological determinants of emotional state, *Psychological Review* **69**: 378–399.

Scheuerecker, J., Ufer, S., Zipse, M., Frodl, T., Koutsouleris, N., Zetzsche, T., Wiesmann, M., Albrecht, J., Bruckmann, H., Schmitt, G., Moller, H. J. and Meisenzahl, E. M. (2008). Cerebral changes and cognitive dysfunctions in medication-free schizophrenia – An fMRI study, *Journal of Psychiatric Research* **42**: 469–476.

Schiller, D., Monfils, M. H., Raio, C. M., Johnson, D. C., LeDoux, J. E. and Phelps, E. A. (2010). Preventing the return of fear in humans using reconsolidation update mechanisms, *Nature* **463**: 49–53.

Schliebs, R. and Arendt, T. (2006). The significance of the cholinergic system in the brain during aging and in Alzheimer's disease, *Journal of Neural Transmission* **113**: 1625–1644.

Schmolesky, M., Wang, Y., Hanes, D., Thompson, K., Leutgeb, S., Schall, J. and Leventhal, A. (1998). Signal timing across the macaque visual system, *Journal of Neurophysiology* **79**: 3272–3277.

Schnupp, J., Nelken, I. and King, A. (2011). *Auditory Neuroscience: Making Sense of Sound*, MIT Press.

Schoenbaum, G., Roesch, M. R., Stalnaker, T. A. and Takahashi, Y. K. (2009). A new perspective on the role of the orbitofrontal cortex in adaptive behaviour, *Nature Reviews Neuroscience* **10**: 885–892.

Schultz, S. and Rolls, E. T. (1999). Analysis of information transmission in the schaffer collaterals, *Hippocampus* **9**: 582–598.

Schultz, W. (2004). Neural coding of basic reward terms of animal learning theory, game theory, microeconomics and behavioural ecology, *Current Opinion in Neurobiology* **14**: 139–147.

Schultz, W. (2006). Behavioral theories and the neurophysiology of reward, *Annual Review of Psychology* **57**: 87–115.

Schultz, W. (2013). Updating dopamine reward signals, *Current Opinion in Neurobiology* **23**: 229–238.

Schultz, W., Romo, R., Ljunberg, T., Mirenowicz, J., Hollerman, J. R. and Dickinson, A. (1995). Reward-related signals carried by dopamine neurons, *in* J. C. Houk, J. L. Davis and D. G. Beiser (eds), *Models of Information Processing in the Basal Ganglia*, MIT Press, Cambridge, MA, chapter 12, pp. 233–248.

Schultz, W., Dayan, P. and Montague, P. R. (1997). A neural substrate of prediction and reward, *Science* **275**: 1593–1599.

Schwindel, C. D. and McNaughton, B. L. (2011). Hippocampal-cortical interactions and the dynamics of memory trace reactivation, *Progress in Brain Research* **193**: 163–177.

Scott, T. R. and Giza, B. K. (1987). A measure of taste intensity discrimination in the rat through conditioned taste aversions, *Physiology and Behaviour* **41**: 315–320.

Scott, T. R. and Plata-Salaman, C. R. (1999). Taste in the monkey cortex, *Physiology and Behavior* **67**: 489–511.

Scott, T. R. and Small, D. M. (2009). The role of the parabrachial nucleus in taste processing and feeding, *Annals of the New York Academy of Sciences* **1170**: 372–377.

Scott, T. R., Yaxley, S., Sienkiewicz, Z. J. and Rolls, E. T. (1986a). Gustatory responses in the frontal opercular cortex of the alert cynomolgus monkey, *Journal of Neurophysiology* **56**: 876–890.

Scott, T. R., Yaxley, S., Sienkiewicz, Z. J. and Rolls, E. T. (1986b). Taste responses in the nucleus tractus solitarius of the behaving monkey, *Journal of Neurophysiology* **55**: 182–200.

Scott, T. R., Yan, J. and Rolls, E. T. (1995). Brain mechanisms of satiety and taste in macaques, *Neurobiology* **3**: 281–292.

Seamans, J. K. and Yang, C. R. (2004). The principal features and mechanisms of dopamine modulation in the prefrontal cortex, *Progress in Neurobiology* **74**: 1–58.

Seamans, J. K., Gorelova, N., Durstewitz, D. and Yang, C. R. (2001). Bidirectional dopamine modulation of GABAergic inhibition in prefrontal cortical pyramidal neurons, *Journal of Neuroscience* **21**: 3628–3638.

Seeman, P. and Kapur, S. (2000). Schizophrenia: more dopamine, more D2 receptors, *Proceeedings of the National Academy of Sciences USA* **97**: 7673–7675.

Seeman, P., Weinshenker, D., Quirion, R., Srivastava, L. K., Bhardwaj, S. K., Grandy, D. K., Premont, R. T., Sotnikova, T. D., Boksa, P., El-Ghundi, M., O'Dowd, B. F., George, S.-R., Perreault, M. L., Mannisto, P. T., Robinson, S., Palmiter, R. D. and Tallerico, T. (2005). Dopamine supersensitivity correlates with D2 High states, implying many paths to psychosis, *Proceedings of the National Academy of Sciences USA* **102**: 3513–3518.

Seeman, P., Schwarz, J., Chen, J. F., Szechtman, H., Perreault, M., McKnight, G. S., Roder, J. C., Quirion, R., Boksa, P., Srivastava, L. K., Yanai, K., Weinshenker, D. and Sumiyoshi, T. (2006). Psychosis pathways converge via D2 high dopamine receptors, *Synapse* **60**: 319–346.

Seigel, M. and Auerbach, J. M. (1996). Neuromodulators of synaptic strength, *in* M. S. Fazeli and G. L. Collingridge (eds), *Cortical Plasticity*, Bios, Oxford, chapter 7, pp. 137–148.

Seleman, L. D. and Goldman-Rakic, P. S. (1985). Longitudinal topography and interdigitation of corticostriatal projections in the rhesus monkey, *Journal of Neuroscience* **5**: 776–794.

Selfridge, O. G. (1959). Pandemonium: A paradigm for learning, *in* D. Blake and A.Uttley (eds), *The Mechanization of Thought Processes*, H. M. Stationery Office, London, pp. 511–529.

Seltzer, B. and Pandya, D. N. (1978). Afferent cortical connections and architectonics of the superior temporal sulcus and surrounding cortex in the rhesus monkey, *Brain Research* **149**: 1–24.

Senn, W., Markram, H. and Tsodyks, M. (2001). An algorithm for modifying neurotransmitter release probability based on pre- and postsynaptic spike timing, *Neural Computation* **13**: 35–67.

Serre, T., Kreiman, G., Kouh, M., Cadieu, C., Knoblich, U. and Poggio, T. (2007a). A quantitative theory of immediate visual recognition, *Progress in Brain Research* **165**: 33–56.

Serre, T., Oliva, A. and Poggio, T. (2007b). A feedforward architecture accounts for rapid categorization, *Proceedings of the National Academy of Sciences* **104**: 6424–6429.

Serre, T., Wolf, L., Bileschi, S., Riesenhuber, M. and Poggio, T. (2007c). Robust object recognition with cortex-like mechanisms, *IEEE Transactions on Pattern Analysis and Machine Intelligence* **29**: 411–426.

Shackman, A. J., Salomons, T. V., Slagter, H. A., Fox, A. S., Winter, J. J. and Davidson, R. J. (2011). The integration of negative affect, pain and cognitive control in the cingulate cortex, *Nature Reviews Neuroscience* **12**: 154–167.

Shadlen, M. and Newsome, W. (1995). Is there a signal in the noise?, *Current Opinion in Neurobiology* **5**: 248–250.

Shadlen, M. and Newsome, W. (1998). The variable discharge of cortical neurons: implications for connectivity, computation and coding, *Journal of Neuroscience* **18**: 3870–3896.

Shadlen, M. N. and Kiani, R. (2013). Decision making as a window on cognition, *Neuron* **80**: 791–806.

Shadlen, M. N. and Movshon, J. A. (1999). Synchrony unbound: A critical evaluation of the temporal binding hypothesis, *Neuron* **24**: 67–77.

Shallice, T. and Burgess, P. (1996). The domain of supervisory processes and temporal organization of behaviour, *Philosophical Transactions of the Royal Society B*, **351**: 1405–1411.

Shannon, C. E. (1948). A mathematical theory of communication, *AT&T Bell Laboratories Technical Journal* **27**: 379–423.

Sharp, P. E. (1996). Multiple spatial–behavioral correlates for cells in the rat postsubiculum: multiple regression analysis and comparison to other hippocampal areas, *Cerebral Cortex* **6**: 238–259.

Sharp, P. E. (ed.) (2002). *The Neural Basis of Navigation: Evidence From Single Cell Recording*, Kluwer, Boston.

Shashua, A. (1995). Algebraic functions for recognition, *IEEE Transactions on Pattern Analysis and Machine Intelligence* **17**: 779–789.

Sheinberg, D. L. and Logothetis, N. K. (2001). Noticing familiar objects in real world scenes: The role of temporal cortical neurons in natural vision, *Journal of Neuroscience* **21**: 1340–1350.

Shepherd, G. M. (2004). *The Synaptic Organisation of the Brain*, 5th edn, Oxford University Press, Oxford.

Shepherd, G. M. and Grillner, S. (eds) (2010). *Handbook of Brain Microcircuits*, Oxford University Press, Oxford.

Shergill, S. S., Brammer, M. J., Williams, S. C., Murray, R. M. and McGuire, P. K. (2000). Mapping auditory hallucinations in schizophrenia using functional magnetic resonance imaging, *Archives of General Psychiatry* **57**: 1033–1038.

Sherman, S. M. and Guillery, R. W. (2013). *Functional Connections of Cortical Areas: A New View from the Thalamus*, MIT Press, Cambridge, MA.

Shevelev, I. A., Novikova, R. V., Lazareva, N. A., Tikhomirov, A. S. and Sharaev, G. A. (1995). Sensitivity to cross-like figures in cat striate neurons, *Neuroscience* **69**: 51–57.

Shizgal, P. and Arvanitogiannis, A. (2003). Gambling on dopamine, *Science* **299**: 1856–1858.

Si, B. and Treves, A. (2009). The role of competitive learning in the generation of dg fields from ec inputs, *Cogn Neurodyn* **3**(2): 177–87.

Sidhu, M. K., Stretton, J., Winston, G. P., Bonelli, S., Centeno, M., Vollmar, C., Symms, M., Thompson, P. J., Koepp, M. J. and Duncan, J. S. (2013). A functional magnetic resonance imaging study mapping the episodic memory encoding network in temporal lobe epilepsy, *Brain* **136**: 1868–1888.

Sigala, N. and Logothetis, N. K. (2002). Visual categorisation shapes feature selectivity in the primate temporal cortex, *Nature* **415**: 318–320.

Sikström, S. (2007). Computational perspectives on neuromodulation of aging, *Acta Neurochirurgica, Supplement* **97**: 513–518.

Silberberg, G., Wu, C. and Markram, H. (2004). Synaptic dynamics control the timing of neuronal excitation in the activated neocortical microcircuit, *Journal of Physiology* **556**: 19–27.

Silberberg, G., Grillner, S., LeBeau, F. E. N., Maex, R. and Markram, H. (2005). Synaptic pathways in neural microcircuits, *Trends in neurosciences* **28**: 541–551.

Sillito, A. M. (1984). Functional considerations of the operation of GABAergic inhibitory processes in the visual cortex, *in* E. G. Jones and A. Peters (eds), *Cerebral Cortex, Vol. 2, Functional Properties of Cortical Cells*, Plenum, New York, chapter 4, pp. 91–117.

Sillito, A. M., Grieve, K. L., Jones, H. E., Cudeiro, J. and Davis, J. (1995). Visual cortical mechanisms detecting focal orientation discontinuities, *Nature* **378**: 492–496.

Simmen, M. W., Rolls, E. T. and Treves, A. (1996a). On the dynamics of a network of spiking neurons, *in* F. Eekman and J. Bower (eds), *Computations and Neuronal Systems: Proceedings of CNS95*, Kluwer, Boston.

Simmen, M. W., Treves, A. and Rolls, E. T. (1996b). Pattern retrieval in threshold-linear associative nets, *Network* **7**: 109–122.

Singer, W. (1987). Activity-dependent self-organization of synaptic connections as a substrate for learning, *in* J.-P. Changeux and M. Konishi (eds), *The Neural and Molecular Bases of Learning*, Wiley, Chichester, pp. 301–335.

Singer, W. (1995). Development and plasticity of cortical processing architectures, *Science* **270**: 758–764.

Singer, W. (1999). Neuronal synchrony: A versatile code for the definition of relations?, *Neuron* **24**: 49–65.

Singer, W. (2000). Response synchronisation: A universal coding strategy for the definition of relations, *in* M. Gazzaniga (ed.), *The New Cognitive Neurosciences*, 2nd edn, MIT Press, Cambridge, MA, chapter 23, pp. 325–338.

Singer, W. and Gray, C. M. (1995). Visual feature integration and the temporal correlation hypothesis, *Annual Review of Neuroscience* **18**: 555–586.

Singer, W., Gray, C., Engel, A., Konig, P., Artola, A. and Brocher, S. (1990). Formation of cortical cell assemblies, *Cold Spring Harbor Symposium on Quantitative Biology* **55**: 939–952.

Sjöström, P. J., Turrigiano, G. G. and Nelson, S. B. (2001). Rate, timing, and cooperativity jointly determine cortical synaptic plasticity, *Neuron* **32**: 1149–1164.

Skaggs, W. E. and McNaughton, B. L. (1992). Quantification of what it is that hippocampal cell firing encodes, *Society for Neuroscience Abstracts* **18**: 1216.

Skaggs, W. E., McNaughton, B. L., Gothard, K. and Markus, E. (1993). An information theoretic approach to deciphering the hippocampal code, *in* S. Hanson, J. D. Cowan and C. L. Giles (eds), *Advances in Neural Information Processing Systems*, Vol. 5, Morgan Kaufmann, San Mateo, CA, pp. 1030–1037.

Skaggs, W. E., Knierim, J. J., Kudrimoti, H. S. and McNaughton, B. L. (1995). A model of the neural basis of the rat's sense of direction, *in* G. Tesauro, D. S. Touretzky and T. K. Leen (eds), *Advances in Neural Information Processing Systems*, Vol. 7, MIT Press, Cambridge, MA, pp. 173–180.

Sloper, J. J. and Powell, T. P. S. (1979a). An experimental electron microscopic study of afferent connections to the primate motor and somatic sensory cortices, *Philosophical Transactions of the Royal Society of London, Series B* **285**: 199–226.

Sloper, J. J. and Powell, T. P. S. (1979b). A study of the axon initial segment and proximal axon of neurons in the primate motor and somatic sensory cortices, *Philosophical Transactions of the Royal Society of London, Series B* **285**: 173–197.

Small, D. M. (2010). Taste representation in the human insula, *Brain Structre and Function* **214**: 551–561.

Small, D. M. and Scott, T. R. (2009). Symposium overview: What happens to the pontine processing? Repercussions of interspecies differences in pontine taste representation for tasting and feeding, *Annals of the New York Academy of Science* **1170**: 343–346.

Small, D. M., Zald, D. H., Jones-Gotman, M., Zatorre, R. J., Petrides, M. and Evans, A. C. (1999). Human cortical gustatory areas: a review of functional neuroimaing data, *NeuroReport* **8**: 3913–3917.

Smerieri, A., Rolls, E. T. and Feng, J. (2010). Decision time, slow inhibition, and theta rhythm, *Journal of Neuroscience* **30**: 14173–14181.

Smith, M. L. and Milner, B. (1981). The role of the right hippocampus in the recall of spatial location, *Neuropsychologia* **19**: 781–793.

Smith-Swintosky, V. L., Plata-Salaman, C. R. and Scott, T. R. (1991). Gustatory neural coding in the monkey cortex: stimulus quality, *Journal of Neurophysiology* **66**: 1156–1165.

Snyder, A. Z. and Raichle, M. E. (2012). A brief history of the resting state: The washington university perspective, *Neuroimage* **62**: 902–910.

Soltani, A. and Koch, C. (2010). Visual saliency computations: mechanisms, constraints, and the effect of feedback, *Journal of Neuroscience* **30**: 12831–12843.

Soltani, A. and Wang, X.-J. (2006). A biophysically based neural model of matching law behavior: melioration by stochastic synapses, *Journal of Neuroscience* **26**: 3731–3744.

Somogyi, P. (2010). Hippocampus: intrinsic organisation, *in* G. M. Shepherd and S. Grillner (eds), *Handbook of Brain Microcircuits*, Oxford University Press, Oxford, chapter 15, pp. 148–164.

Somogyi, P. and Cowey, A. C. (1984). Double bouquet cells, *in* A. Peters and E. G. Jones (eds), *Cerebral Cortex, Vol. 1, Cellular Components of the Cerebral Cortex*, Plenum, New York, chapter 9, pp. 337–360.

Somogyi, P., Kisvarday, Z. F., Martin, K. A. C. and Whitteridge, D. (1983). Synaptic connections of morphologically identified and physiologically characterized large basket cells in the striate cortex of the cat, *Neuroscience* **10**: 261–294.

Somogyi, P., Tamas, G., Lujan, R. and Buhl, E. H. (1998). Salient features of synaptic organisation in the cerebral cortex, *Brain Research Reviews* **26**: 113–135.

Sompolinsky, H. (1987). The theory of neural networks: the Hebb rule and beyond, *in* L. van Hemmen and I. Morgenstern (eds), *Heidelberg Colloquium on Glassy Dynamics*, Vol. 275, Springer, New York, pp. 485–527.

Sompolinsky, H. and Kanter, I. (1986). Temporal association in asymmetric neural networks, *Physical Review Letters* **57**: 2861–2864.

Song, S., Yao, H. and Treves, A. (2014). A modular latching chain, *Cognitive neurodynamics* **8**: 37–46.

Spiridon, M., Fischl, B. and Kanwisher, N. (2006). Location and spatial profile of category-specific regions in human extrastriate cortex, *Human Brain Mapping* **27**: 77–89.

Spitzer, H., Desimone, R. and Moran, J. (1988). Increased attention enhances both behavioral and neuronal performance, *Science* **240**: 338–340.

Spoerer, C. J., Eguchi, A. and Stringer, S. M. (2016). A computational exploration of complementary learning mechanisms in the primate ventral visual pathway, *Vision Research* **119**: 16–28.

Spruston, N., Jonas, P. and Sakmann, B. (1995). Dendritic glutamate receptor channel in rat hippocampal CA3 and

CA1 pyramidal neurons, *Journal of Physiology* **482**: 325–352.

Squire, L. R. (1992). Memory and the hippocampus: A synthesis from findings with rats, monkeys and humans, *Psychological Review* **99**: 195–231.

Squire, L. R., Stark, C. E. L. and Clark, R. E. (2004). The medial temporal lobe, *Annual Review of Neuroscience* **27**: 279–306.

Stankiewicz, B. and Hummel, J. (1994). Metricat: A representation for basic and subordinate-level classification, *in* G. W. Cottrell (ed.), *Proceedings of the 18th Annual Conference of the Cognitive Science Society*, Erlbaum, San Diego, pp. 254–259.

Stefanacci, L., Suzuki, W. A. and Amaral, D. G. (1996). Organization of connections between the amygdaloid complex and the perirhinal and parahippocampal cortices in macaque monkeys, *Journal of Comparative Neurology* **375**: 552–582.

Stein, D. J., Kogan, C. S., Atmaca, M., Fineberg, N. A., Fontenelle, L. F., Grant, J. E., Matsunaga, H., Reddy, Y. C., Simpson, H. B., Thomsen, P. H., van den Heuvel, O. A., Veale, D., Woods, D. W. and Reed, G. M. (2016). The classification of obsessive-compulsive and related disorders in the icd-11, *J Affect Disord* **190**: 663–74.

Stella, F., Cerasti, E. and Treves, A. (2013). Unveiling the metric structure of internal representations of space, *Frontiers in Neural Circuits* **7**: 81.

Stent, G. S. (1973). A psychological mechanism for Hebb's postulate of learning, *Proceedings of the National Academy of Sciences USA* **70**: 997–1001.

Stephan, K. E. and Roebroeck, A. (2012). A short history of causal modeling of fmri data, *Neuroimage* **62**: 856–863.

Stephan, K. E., Baldeweg, T. and Friston, K. J. (2006). Synaptic plasticity and dysconnection in schizophrenia, *Biological Psychiatry* **59**: 929–939.

Stewart, S. E., Fagerness, J. A., Platko, J., Smoller, J. W., Scharf, J. M., Illmann, C., Jenike, E., Chabane, N., Leboyer, M., Delorme, R., Jenike, M. A. and Pauls, D. L. (2007). Association of the SLC1A1 glutamate transporter gene and obsessive-compulsive disorder, *American Journal of Medical Genetics B: Neuropsychiatric Genetics* **144**: 1027–1033.

Storm-Mathiesen, J., Zimmer, J. and Ottersen, O. P. (1990). Understanding the brain through the hippocampus, *Progress in Brain Research* **83**: 1–.

Strick, P. L., Dum, R. P. and Picard, N. (1995). Macro-organization of the circuits connecting the basal ganglia with the cortical motor areas, *in* J. C. Houk, J. L. Davis and D. G. Beiser (eds), *Models of Information Processing in the Basal Ganglia*, MIT Press, Cambridge, MA, chapter 6, pp. 117–130.

Stringer, S. M. and Rolls, E. T. (2000). Position invariant recognition in the visual system with cluttered environments, *Neural Networks* **13**: 305–315.

Stringer, S. M. and Rolls, E. T. (2002). Invariant object recognition in the visual system with novel views of 3D objects, *Neural Computation* **14**: 2585–2596.

Stringer, S. M. and Rolls, E. T. (2006). Self-organizing path integration using a linked continuous attractor and competitive network: path integration of head direction, *Network: Computation in Neural Systems* **17**: 419–445.

Stringer, S. M. and Rolls, E. T. (2007). Hierarchical dynamical models of motor function, *Neurocomputing* **70**: 975–990.

Stringer, S. M. and Rolls, E. T. (2008). Learning transform invariant object recognition in the visual system with multiple stimuli present during training, *Neural Networks* **21**: 888–903.

Stringer, S. M., Rolls, E. T., Trappenberg, T. P. and De Araujo, I. E. T. (2002a). Self-organizing continuous attractor networks and path integration: Two-dimensional models of place cells, *Network: Computation in Neural Systems* **13**: 429–446.

Stringer, S. M., Trappenberg, T. P., Rolls, E. T. and De Araujo, I. E. T. (2002b). Self-organizing continuous attractor networks and path integration: One-dimensional models of head direction cells, *Network: Computation in Neural Systems* **13**: 217–242.

Stringer, S. M., Rolls, E. T. and Trappenberg, T. P. (2004). Self-organising continuous attractor networks with multiple activity packets, and the representation of space, *Neural Networks* **17**: 5–27.

Stringer, S. M., Rolls, E. T. and Trappenberg, T. P. (2005). Self-organizing continuous attractor network models of hippocampal spatial view cells, *Neurobiology of Learning and Memory* **83**: 79–92.

Stringer, S. M., Perry, G., Rolls, E. T. and Proske, J. H. (2006). Learning invariant object recognition in the visual system with continuous transformations, *Biological Cybernetics* **94**: 128–142.

Stringer, S. M., Rolls, E. T. and Taylor, P. (2007a). Learning movement sequences with a delayed reward signal in a hierarchical model of motor function, *Neural Networks* **29**: 172–181.

Stringer, S. M., Rolls, E. T. and Tromans, J. M. (2007b). Invariant object recognition with trace learning and multiple stimuli present during training, *Network: Computation in Neural Systems* **18**: 161–187.

Strong, S. P., Koberle, R., de Ruyter van Steveninck, R. R. and Bialek, W. (1998). Entropy and information in neural spike trains, *Physical Review Letters* **80**: 197–200.

Sugase, Y., Yamane, S., Ueno, S. and Kawano, K. (1999). Global and fine information coded by single neurons in the temporal visual cortex, *Nature* **400**: 869–873.

Sugrue, L. P., Corrado, G. S. and Newsome, W. T. (2005). Choosing the greater of two goods: neural currencies for valuation and decision making, *Nature Reviews Neuroscience* **6**: 363–375.

Suri, R. E. and Schultz, W. (1998). Learning of sequential movements by neural network model with dopamine-like reinforcement signal, *Experimental Brain Research* **121**: 350–354.

Sutherland, N. S. (1968). Outline of a theory of visual pattern recognition in animal and man, *Proceedings of the Royal Society, B* **171**: 297–317.

Sutherland, R. J. and Rudy, J. W. (1991). Exceptions to the rule of space, *Hippocampus* **1**: 250–252.

Sutherland, R. J., Whishaw, I. Q. and Kolb, B. (1983). A behavioural analysis of spatial localization following electrolytic, kainate- or colchicine-induced damage to the hippocampal formation in the rat, *Behavioral Brain Research* **7**: 133–153.

Sutton, R. S. (1988). Learning to predict by the methods of temporal differences, *Machine Learning* **3**: 9–44.

Sutton, R. S. and Barto, A. G. (1981). Towards a modern theory of adaptive networks: expectation and prediction, *Psychological Review* **88**: 135–170.

Sutton, R. S. and Barto, A. G. (1990). Time-derivative models of Pavlovian reinforcement, *in* M. Gabriel and J. Moore (eds), *Learning and Computational Neuroscience*, MIT Press, Cambridge, MA, pp. 497–537.

Sutton, R. S. and Barto, A. G. (1998). *Reinforcement Learning*, MIT Press, Cambridge, MA.

Suzuki, W. A. and Amaral, D. G. (1994a). Perirhinal and parahippocampal cortices of the macaque monkey – cortical afferents, *Journal of Comparative Neurology* **350**: 497–533.

Suzuki, W. A. and Amaral, D. G. (1994b). Topographic organization of the reciprocal connections between the monkey entorhinal cortex and the perirhinal and parahippocampal cortices, *Journal of Neuroscience* **14**: 1856–1877.

Swann, A. C. (2009). Impulsivity in mania, *Current Psychiatry Reports* **11**: 481–487.

Swanson, L. W. and Cowan, W. M. (1977). An autoradiographic study of the organization of the efferent connections of the hippocampal formation in the rat, *Journal of Comparative Neurology* **172**: 49–84.

Swash, M. (1989). John Hughlings Jackson: a historical introduction, *in* C. Kennard and M. Swash (eds), *Hierarchies in Neurology*, Springer, London, chapter 1, pp. 3–10.

Szabo, M., Almeida, R., Deco, G. and Stetter, M. (2004). Cooperation and biased competition model can explain attentional filtering in the prefrontal cortex, *European Journal of Neuroscience* **19**: 1969–1977.

Szabo, M., Deco, G., Fusi, S., Del Giudice, P., Mattia, M. and Stetter, M. (2006). Learning to attend: Modeling the shaping of selectivity in infero-temporal cortex in a categorization task, *Biological Cybernetics* **94**: 351–365.

Szentagothai, J. (1978). The neuron network model of the cerebral cortex: a functional interpretation, *Proceedings of the Royal Society of London, Series B* **201**: 219–248.

Szymanski, F. D., Garcia-Lazaro, J. A. and Schnupp, J. W. (2009). Current source density profiles of stimulus-specific adaptation in rat auditory cortex, *J Neurophysiol* **102**(3): 1483–90.

Tabuchi, E., Mulder, A. B. and Wiener, S. I. (2003). Reward value invariant place responses and reward site associated activity in hippocampal neurons of behaving rats, *Hippocampus* **13**: 117–132.

Tagamets, M. and Horwitz, B. (1998). Integrating electrophysical and anatomical experimental data to create a large-scale model that simulates a delayed match-to-sample human brain study, *Cerebral Cortex* **8**: 310–320.

Takeuchi, T., Duszkiewicz, A. J. and Morris, R. G. (2014). The synaptic plasticity and memory hypothesis: encoding, storage and persistence, *Philos Trans R Soc Lond B Biol Sci* **369**(1633): 20130288.

Tanaka, K. (1993). Neuronal mechanisms of object recognition, *Science* **262**: 685–688.

Tanaka, K. (1996). Inferotemporal cortex and object vision, *Annual Review of Neuroscience* **19**: 109–139.

Tanaka, K., Saito, C., Fukada, Y. and Moriya, M. (1990). Integration of form, texture, and color information in the inferotemporal cortex of the macaque, *in* E. Iwai and M. Mishkin (eds), *Vision, Memory and the Temporal Lobe*, Elsevier, New York, chapter 10, pp. 101–109.

Tanaka, K., Saito, H., Fukada, Y. and Moriya, M. (1991). Coding visual images of objects in the inferotemporal cortex of the macaque monkey, *Journal of Neurophysiology* **66**: 170–189.

Tanaka, S. C., Doya, K., Okada, G., Ueda, K., Okamoto, Y. and Yamawaki, S. (2004). Prediction of immediate and future rewards differentially recruits cortico-basal ganglia loops, *Nature Neuroscience* **7**: 887–893.

Tanila, H. (1999). Hippocampal place cells can develop distinct representations of two visually identical environments, *Hippocampus* **9**: 235–246.

Taube, J. S., Muller, R. U. and Ranck, Jr., J. B. (1990a). Head-direction cells recorded from the postsubiculum in freely moving rats. I. Description and quantitative analysis, *Journal of Neuroscience* **10**: 420–435.

Taube, J. S., Muller, R. U. and Ranck, Jr., J. B. (1990b). Head-direction cells recorded from the postsubiculum in freely moving rats. II. Effects of environmental manipulations, *Journal of Neuroscience* **10**: 436–447.

Taube, J. S., Goodridge, J. P., Golob, E. G., Dudchenko, P. A. and Stackman, R. W. (1996). Processing the head direction signal: a review and commentary, *Brain Research Bulletin* **40**: 477–486.

Taylor, J. G. (1999). Neural 'bubble' dynamics in two dimensions: foundations, *Biological Cybernetics* **80**: 393–409.

Teyler, T. J. and DiScenna, P. (1986). The hippocampal memory indexing theory, *Behavioral Neuroscience* **100**(2): 147–154.

Thibault, O., Porter, N. M., Chen, K. C., Blalock, E. M., Kaminker, P. G., Clodfelter, G. V., Brewer, L. D. and Landfield, P. W. (1998). Calcium dysregulation in neuronal aging and AlzheimerŠs disease: history and new directions, *Cell Calcium* **24**: 417–433.

Thomson, A. M. and Deuchars, J. (1994). Temporal and spatial properties of local circuits in neocortex, *Trends in Neurosciences* **17**: 119–126.

Thomson, A. M. and Lamy, C. (2007). Functional maps of neocortical local circuitry, *Frontiers in Neuroscience* **1**: 19–42.

Thorpe, S. J. and Imbert, M. (1989). Biological constraints on connectionist models, *in* R. Pfeifer, Z. Schreter and F. Fogelman-Soulie (eds), *Connectionism in Perspective*, Elsevier, Amsterdam, pp. 63–92.

Thorpe, S. J., Maddison, S. and Rolls, E. T. (1979). Single unit activity in the orbitofrontal cortex of the behaving monkey, *Neuroscience Letters* **S3**: S77.

Thorpe, S. J., Rolls, E. T. and Maddison, S. (1983). Neuronal activity in the orbitofrontal cortex of the behaving monkey, *Experimental Brain Research* **49**: 93–115.

Thorpe, S. J., O'Regan, J. K. and Pouget, A. (1989). Humans fail on XOR pattern classification problems, *in* L. Personnaz and G. Dreyfus (eds), *Neural Networks: From Models to Applications*, I.D.S.E.T., Paris, pp. 12–25.

Thorpe, S. J., Fize, D. and Marlot, C. (1996). Speed of processing in the human visual system, *Nature* **381**: 520–522.

Thorpe, S. J., Delorme, A. and Van Rullen, R. (2001). Spike-based strategies for rapid processing, *Neural Networks* **14**: 715–725.

Tompa, T. and Sary, G. (2010). A review on the inferior temporal cortex of the macaque, *Brain Res Rev* **62**(2): 165–82.

Tonegawa, S., Nakazawa, K. and Wilson, M. A. (2003). Genetic neuroscience of mammalian learning and memory, *Philosophical Transactions of the Royal Society of London B Biological Sciences* **358**: 787–795.

Tononi, G. and Cirelli, C. (2014). Sleep and the price of plasticity: from synaptic and cellular homeostasis to memory consolidation and integration, *Neuron* **81**(1): 12–34.

Tou, J. T. and Gonzalez, A. G. (1974). *Pattern Recognition Principles*, Addison-Wesley, Reading, MA.

Tovee, M. J. and Rolls, E. T. (1992). Oscillatory activity is not evident in the primate temporal visual cortex with static stimuli, *Neuroreport* **3**: 369–372.

Tovee, M. J. and Rolls, E. T. (1995). Information encoding in short firing rate epochs by single neurons in the primate temporal visual cortex, *Visual Cognition* **2**: 35–58.

Tovee, M. J., Rolls, E. T., Treves, A. and Bellis, R. P. (1993). Information encoding and the responses of single neurons in the primate temporal visual cortex, *Journal of Neurophysiology* **70**: 640–654.

Tovee, M. J., Rolls, E. T. and Azzopardi, P. (1994). Translation invariance and the responses of neurons in the temporal visual cortical areas of primates, *Journal of Neurophysiology* **72**: 1049–1060.

Tovee, M. J., Rolls, E. T. and Ramachandran, V. S. (1996). Rapid visual learning in neurones of the primate temporal visual cortex, *NeuroReport* **7**: 2757–2760.

Trantham-Davidson, H., Neely, L. C., Lavin, A. and Seamans, J. K. (2004). Mechanisms underlying differential D1 versus D2 dopamine receptor regulation of inhibition in prefrontal cortex, *Journal of Neuroscience* **24**: 10652–10659.

Trappenberg, T. P., Rolls, E. T. and Stringer, S. M. (2002). Effective size of receptive fields of inferior temporal visual cortex neurons in natural scenes, *in* T. G. Dietterich, S. Becker and Z. Gharamani (eds), *Advances in Neural Information Processing Systems*, Vol. 14, MIT Press, Cambridge, MA, pp. 293–300.

Treisman, A. (1982). Perceptual grouping and attention in visual search for features and for objects, *Journal of Experimental Psychology: Human Perception and Performance* **8**: 194–214.

Treisman, A. (1988). Features and objects: The fourteenth Barlett memorial lecture, *The Quarterly Journal of Experimental Psychology* **40A**: 201–237.

Treisman, A. and Gelade, G. (1980). A feature-integration theory of attention, *Cognitive Psychology* **12**: 97–136.

Treisman, A. and Sato, S. (1990). Conjunction search revisited, *Journal of Experimental Psychology: Human Perception and Performance,* **16**: 459–478.

Tremblay, L. and Schultz, W. (1999). Relative reward preference in primate orbitofrontal cortex, *Nature* **398**: 704–708.

Tremblay, L. and Schultz, W. (2000). Modifications of reward expectation-related neuronal activity during learning in primate orbitofrontal cortex, *Journal of Neurophysiology* **83**: 1877–1885.

Treves, A. (1990). Graded-response neurons and information encodings in autoassociative memories, *Physical Review A* **42**: 2418–2430.

Treves, A. (1991a). Are spin-glass effects relevant to understanding realistic auto-associative networks?, *Journal of Physics A* **24**: 2645–2654.

Treves, A. (1991b). Dilution and sparse encoding in threshold-linear nets, *Journal of Physics A: Mathematical and General* **24**: 327–335.

Treves, A. (1993). Mean-field analysis of neuronal spike dynamics, *Network* **4**: 259–284.

Treves, A. (1995). Quantitative estimate of the information relayed by the Schaffer collaterals, *Journal of Computational Neuroscience* **2**: 259–272.

Treves, A. (1997). On the perceptual structure of face space, *Biosystems* **40**: 189–196.

Treves, A. (2003). Computational constraints that may have favoured the lamination of sensory cortex, *Journal of computational neuroscience* **14**: 271–282.

Treves, A. (2005). Frontal latching networks: a possible neural basis for infinite recursion, *Cognitive Neuropsychology* **22**: 276–291.

Treves, A. (2016). The dentate gyrus, defining a new memory of David Marr, *in* L. M. Vaina and R. E. Passingham (eds), *Computational Theories and their Implementation in the Brain: The Legacy of David Marr*, Oxford University Press, Oxford.

Treves, A. and Panzeri, S. (1995). The upward bias in measures of information derived from limited data samples, *Neural Computation* **7**: 399–407.

Treves, A. and Rolls, E. T. (1991). What determines the capacity of autoassociative memories in the brain?, *Network* **2**: 371–397.

Treves, A. and Rolls, E. T. (1992a). Computational analysis of the operation of a real neuronal network in the brain: the role of the hippocampus in memory, *in* I. Aleksander and J. Taylor (eds), *Artificial Neural Networks*, Vol. 2, North-Holland, Amsterdam, pp. 891–898.

Treves, A. and Rolls, E. T. (1992b). Computational constraints suggest the need for two distinct input systems to the hippocampal CA3 network, *Hippocampus* **2**: 189–199.

Treves, A. and Rolls, E. T. (1994). A computational analysis of the role of the hippocampus in memory, *Hippocampus* **4**: 374–391.

Treves, A., Rolls, E. T. and Simmen, M. (1997). Time for retrieval in recurrent associative memories, *Physica D* **107**: 392–400.

Treves, A., Panzeri, S., Rolls, E. T., Booth, M. and Wakeman, E. A. (1999). Firing rate distributions and efficiency of information transmission of inferior temporal cortex neurons to natural visual stimuli, *Neural Computation* **11**: 601–631.

Treves, A., Tashiro, A., Witter, M. P. and Moser, E. I. (2008). What is the mammalian dentate gyrus good for?, *Neuroscience* **154**(4): 1155–72.

Trivers, R. L. (1976). Foreword, *The Selfish Gene by R. Dawkins*, Oxford University Press, Oxford.

Trivers, R. L. (1985). *Social Evolution*, Benjamin, Cummings, CA.

Tsao, D. (2014). The macaque face patch system: A window into object representation, *Cold Spring Harb Symp Quant Biol* **79**: 109–14.

Tsao, D. Y., Freiwald, W. A., Tootell, R. B. and Livingstone, M. S. (2006). A cortical region consisting entirely of face-selective cells, *Science* **311**: 617–618.

Tsodyks, M. V. and Feigel'man, M. V. (1988). The enhanced storage capacity in neural networks with low activity level, *Europhysics Letters* **6**: 101–105.

Tuckwell, H. (1988). *Introduction to Theoretical Neurobiology*, Cambridge University Press, Cambridge.

Turner, B. H. (1981). The cortical sequence and terminal distribution of sensory related afferents to the amygdaloid complex of the rat and monkey, *in* Y. Ben-Ari (ed.), *The Amygdaloid Complex*, Elsevier, Amsterdam, pp. 51–62.

Turova, T. (2012). The emergence of connectivity in neuronal networks: From bootstrap percolation to auto-associative memory, *Brain Research* **1434**: 277–284.

Ullman, S. (1996). *High-Level Vision. Object Recognition and Visual Cognition*, Bradford/MIT Press, Cambridge, MA.

Ullman, S., Assif, L., Fetaya, E. and Harari, D. (2016). Atoms of recognition in human and computer vision, *Proc Natl Acad Sci U S A* **113**(10): 2744–9.

Ungerleider, L. G. (1995). Functional brain imaging studies of cortical mechanisms for memory, *Science* **270**: 769–775.

Ungerleider, L. G. and Haxby, J. V. (1994). 'What' and 'Where' in the human brain, *Current Opinion in Neurobiology* **4**: 157–165.

Ungerleider, L. G. and Mishkin, M. (1982). Two cortical visual systems, *in* D. Ingle, M. A. Goodale and R. J. W. Mansfield (eds), *Analysis of Visual Behaviour*, MIT Press, Cambridge, MA.

Usher, M. and McClelland, J. (2001). On the time course of perceptual choice: the leaky competing accumulator model, *Psychological Reviews* **108**: 550–592.

Usher, M. and Niebur, E. (1996). Modelling the temporal dynamics of IT neurons in visual search: A mechanism for top-down selective attention, *Journal of Cognitive Neuroscience* **8**: 311–327.

van den Heuvel, O. A., Veltman, D. J., Groenewegen, H. J., Cath, D. C., van Balkom, A. J., van Hartskamp, J., Barkhof, F. and van Dyck, R. (2005). Frontal-striatal dysfunction during planning in obsessive-compulsive disorder, *Archives of General Psychiatry* **62**: 301–309.

Van Essen, D., Anderson, C. H. and Felleman, D. J. (1992). Information processing in the primate visual system: an integrated systems perspective, *Science* **255**: 419–423.

van Haeften, T., Baks-te Bulte, L., Goede, P. H., Wouterlood, F. G. and Witter, M. P. (2003). Morphological and numerical analysis of synaptic interactions between neurons in deep and superficial layers of the entorhinal cortex of the rat, *Hippocampus* **13**: 943–952.

Van Hoesen, G. W. (1981). The differential distribution, diversity and sprouting of cortical projections to the amygdala in the rhesus monkey, *in* Y. Ben-Ari (ed.), *The Amygdaloid Complex*, Elsevier, Amsterdam, pp. 77–90.

Van Hoesen, G. W. (1982). The parahippocampal gyrus. New observations regarding its cortical connections in the monkey, *Trends in Neurosciences* **5**: 345–350.

Van Hoesen, G. W., Yeterian, E. H. and Lavizzo-Mourey, R. (1981). Widespread corticostriate projections from temporal cortex of the rhesus monkey, *Journal of Comparative Neurology* **199**: 205–219.

Van Ooyen, A. (2011). Using theoretical models to analyse neural development, *Nature Reviews Neuroscience* **12**: 311–326.

van Strien, N. M., Cappaert, N. L. and Witter, M. P. (2009). The anatomy of memory: an interactive overview of the parahippocampal-hippocampal network, *Nature Reviews Neuroscience* **10**: 272–282.

Vann, S. D. and Aggleton, J. P. (2004). The mammillary bodies: two memory systems in one?, *Nature Reviews Neuroscience* **5**: 35–44.

VanRullen, R., Guyonneau, R. and Thorpe, S. J. (2005). Spike times make sense, *Trends in Neuroscience* **28**: 1–4.

Varela, F., Lachaux, J., Rodriguez, E. and Martinerie, J. (2001). The brainweb: phase synchronization and large-scale integration, *Nature Reviews Neuroscience* **2**: 229–239.

Veale, D. M., Sahakian, B. J., Owen, A. M. and Marks, I. M. (1996). Specific cognitive deficits in tests sensitive to frontal lobe dysfunction in obsessive-compulsive disorder, *Psychological Medicine* **26**: 1261–1269.

Verhagen, J. V., Rolls, E. T. and Kadohisa, M. (2003). Neurons in the primate orbitofrontal cortex respond to fat texture independently of viscosity, *Journal of Neurophysiology* **90**: 1514–1525.

Verhagen, J. V., Kadohisa, M. and Rolls, E. T. (2004). The primate insular taste cortex: neuronal representations of the viscosity, fat texture, grittiness, and the taste of foods in the mouth, *Journal of Neurophysiology* **92**: 1685–1699.

Vigliocco, G., Vinson, D. P., Druks, J., Barber, H. and Cappa, S. F. (2011). Nouns and verbs in the brain: a review of behavioural, electrophysiological, neuropsychological and imaging studies, *Neuroscience and Biobehavioral Reviews* **35**: 407–426.

Vinje, W. E. and Gallant, J. L. (2000). Sparse coding and decorrelation in primary visual cortex during natural vision, *Science* **287**: 1273–1276.

Vogels, R. and Biederman, I. (2002). Effects of illumination intensity and direction on object coding in macaque inferior temporal cortex, *Cerebral Cortex* **12**: 756–766.

Vogt, B. A. (ed.) (2009). *Cingulate Neurobiology and Disease*, Oxford University Press, Oxford.

Von Bonin, G. and Bailey, P. (1947). *The Neocortex of Macaca mulatta*, University of Illinois Press, Urbana.

Von der Malsburg, C. and Bienenstock, E. (1986). Statistical coding and short-term synaptic plasticity: a scheme for knowledge representation in the brain, *in* E. Bienenstock, F. Fogelman-Soulie and G. Weisbach (eds), *Disordered Systems and Biological Organization, NATO ASI Series*, Vol. F20, Springer, Berlin, pp. 247–272.

Waelti, P., Dickinson, A. and Schultz, W. (2001). Dopamine responses comply with basic assumptions of formal learning theory, *Nature* **412**: 43–48.

Walker, M. P. and Stickgold, R. (2006). Sleep, memory, and plasticity, *Annual Review of Psychology* **57**: 139–166.

Wallace, D. G. and Whishaw, I. Q. (2003). NMDA lesions of Ammon's horn and the dentate gyrus disrupt the direct and temporally paced homing displayed by rats exploring a novel environment: evidence for a role of the hippocampus in dead reckoning, *European Journal of Neuroscience* **18**: 513–523.

Wallenstein, G. V. and Hasselmo, M. E. (1997). GABAergic modulation of hippocampal population activity: sequence learning, place field development, and the phase precession effect, *Journal of Neurophysiology* **78**: 393–408.

Wallis, G. and Baddeley, R. (1997). Optimal unsupervised learning in invariant object recognition, *Neural Computation* **9**: 883–894.

Wallis, G. and Bülthoff, H. (1999). Learning to recognize objects, *Trends in Cognitive Sciences* **3**: 22–31.

Wallis, G. and Rolls, E. T. (1997). Invariant face and object recognition in the visual system, *Progress in Neurobiology* **51**: 167–194.

Wallis, G., Rolls, E. T. and Földiák, P. (1993). Learning invariant responses to the natural transformations of objects, *International Joint Conference on Neural Networks* **2**: 1087–1090.

Wallis, J. D. and Miller, E. K. (2003). Neuronal activity in primate dorsolateral and orbital prefrontal cortex during performance of a reward preference task, *European Journal of Neuroscience* **18**: 2069–2081.

Wallis, J. D., Anderson, K. C. and Miller, E. K. (2001). Single neurons in prefrontal cortex encode abstract rules, *Nature* **411**: 953–956.

Wallisch, P., Lusignan, M., Benayoun, M., Baker, T. L., Dickey, A. S. and Hatsoupoulos, N. G. (2009). *MATLAB for Neuroscientists*, Academic Press, Burlington, MA.

Wang, M., Ramos, B. P., Paspalas, C. D., Shu, Y., Simen, A., Duque, A., Vijayraghavan, S., Brennan, A., Dudley, A., Nou, E., Mazer, J. A., McCormick, D. A. and Arnsten, A. F. (2007). Alpha2a-adrenoceptors strengthen working memory networks by inhibiting camp-hcn channel signaling in prefrontal cortex, *Cell* **129**: 397–410.

Wang, S.-H. and Morris, R. G. M. (2010). Hippocampal-neocortical interactions in memory formation, consolidation, and reconsolidation, *Annual Review of Psychology* **61**: 49–79.

Wang, X. J. (1999). Synaptic basis of cortical persistent activity: the importance of NMDA receptors to working memory, *Journal of Neuroscience* **19**: 9587–9603.

Wang, X. J. (2001). Synaptic reverberation underlying mnemonic persistent activity, *Trends in Neurosciences* **24**: 455–463.

Wang, X. J. (2002). Probabilistic decision making by slow reverberation in cortical circuits, *Neuron* **36**: 955–968.

Wang, X.-J. (2008). Decision making in recurrent neuronal circuits, *Neuron* **60**: 215–234.

Wang, X. J. and Kennedy, H. (2016). Brain structure and dynamics across scales: in search of rules, *Curr Opin Neurobiol* **37**: 92–98.

Wang, Y., Markram, H., Goodman, P. H., Berger, T. K., Ma, J. and Goldman-Rakic, P. S. (2006). Heterogeneity in the pyramidal network of the medial prefrontal cortex, *Nature Neuroscience* **9**: 534–542.

Warrington, E. K. and Weiskrantz, L. (1973). An analysis of short-term and long-term memory defects in man, *in* J. A. Deutsch (ed.), *The Physiological Basis of Memory*, Academic Press, New York, chapter 10, pp. 365–395.

Wasserman, E., Kirkpatrick-Steger, A. and Biederman, I. (1998). Effects of geon deletion, scrambling, and movement on picture identification in pigeons, *Journal of Experimental Psychology – Animal Behavior Processes* **24**: 34–46.

Watanabe, S., Lea, S. E. G. and Dittrich, W. H. (1993). What can we learn from experiments on pigeon discrimination?, *in* H. P. Zeigler and H.-J. Bischof (eds), *Vision, Brain, and Behavior in Birds*, MIT Press, Cambridge, MA, pp. 351–376.

Watkins, L. H., Sahakian, B. J., Robertson, M. M., Veale, D. M., Rogers, R. D., Pickard, K. M., Aitken, M. R. and Robbins, T. W. (2005). Executive function in Tourette's syndrome and obsessive-compulsive disorder, *Psychological Medicine* **35**: 571–582.

Webb, T. J. and Rolls, E. T. (2014). Deformation-specific and deformation-invariant visual object recognition: pose vs identity recognition of people and deforming objects, *Frontiers in Computational Neuroscience* **8**: 37.

Webb, T. J., Rolls, E. T., Deco, G. and Feng, J. (2011). Noise in attractor networks in the brain produced by graded firing rate representations, *PLoS One* **6**: e23620.

Weiner, K. S. and Grill-Spector, K. (2015). The evolution of face processing networks, *Trends Cogn Sci* **19**(5): 240–1.

Weiskrantz, L. (1997). *Consciousness Lost and Found*, Oxford University Press, Oxford.

Weiskrantz, L. (1998). *Blindsight*, 2nd edn, Oxford University Press, Oxford.

Weiskrantz, L. (2009). Is blindsight just degraded normal vision?, *Experimental Brain Research* **192**: 413–416.

Weiss, A. P. and Heckers, S. (1999). Neuroimaging of hallucinations: a review of the literature, *Psychiatry Research* **92**: 61–74.

Weiss, C., Bouwmeester, H., Power, J. M. and Disterhoft, J. F. (1999). Hippocampal lesions prevent trace eyeblink conditioning in the freely moving rat, *Behavioural Brain Research* **99**: 123–132.

Weissman, M. M., Bland, R. C., Canino, G. J., Greenwald, S., Hwu, H. G., Lee, C. K., Newman, S. C., Oakley-Browne, M. A., Rubio-Stipec, M., Wickramaratne, P. J. et al. (1994). The cross national epidemiology of obsessive compulsive disorder. The Cross National Collaborative Group, *Journal of Clinical Psychiatry* **55 Suppl**: 5–10.

Welford, A. T. (ed.) (1980). *Reaction Times*, Academic Press, London.

Wessa, M., Kanske, P. and Linke, J. (2014). Bipolar disorder: a neural network perspective on a disorder of emotion and motivation, *Restorative Neurology and Neuroscience* **32**: 51–62.

Whalen, P. J. and Phelps, E. A. (2009). *The Human Amygdala*, Guilford, New York.

Wheeler, E. Z. and Fellows, L. K. (2008). The human ventromedial frontal lobe is critical for learning from negative feedback, *Brain* **131**: 1323–1331.

Whelan, R., Conrod, P. J., Poline, J.-B., Lourdusamy, A., Banaschewski, T., Barker, G. J., Bellgrove, M. A., Büchel, C., Byrne, M., Cummins, T. D. R. et al. (2012). Adolescent impulsivity phenotypes characterized by distinct brain networks, *Nature Neuroscience* **15**: 920–925.

Whishaw, I. Q., Hines, D. J. and Wallace, D. G. (2001). Dead reckoning (path integration) requires the hippocampal formation: evidence from spontaneous exploration and spatial learning tasks in light (allothetic) and dark (idiothetic) tests, *Behavioral Brain Research* **127**: 49–69.

Whitehouse, P., Price, D., Struble, R., Clarke, A., Coyle, J. and Delong, M. (1982). Alzheimer's disease in senile dementia: loss of neurones in the basal forebrain, *Science* **215**: 1237–1239.

Whiten, A. and Byrne, R. W. (1997). *Machiavellian Intelligence II: Extensions and Evaluations*, Cambridge University Press, Cambridge.

Whittlesea, B. W. A. (1983). *Representation and Generalization of Concepts: the Abstractive and Episodic Perspectives Evaluated*, Unpublished doctoral dissertation, MacMaster University.

Widrow, B. and Hoff, M. E. (1960). Adaptive switching circuits, *1960 IRE WESCON Convention Record, Part 4 (Reprinted in Anderson and Rosenfeld, 1988)*, IRE, New York, pp. 96–104.

Widrow, B. and Stearns, S. D. (1985). *Adaptive Signal Processing*, Prentice-Hall, Englewood Cliffs, NJ.

Wiener, N. (1956). The theory of prediction, *in* E. Beckenbach (ed.), *Modern Mathematics for Engineers*, McGraw-Hill, New York, chapter 8, pp. 165–190.

Wiener, S. I. and Taube, J. (eds) (2005). *Head Direction Cells and the Neural Mechanisms Underlying Directional Orientation*, MIT Press, Cambridge, MA.

Williams, G. V., Rolls, E. T., Leonard, C. M. and Stern, C. (1993). Neuronal responses in the ventral striatum of the behaving macaque, *Behavioural Brain Research* **55**: 243–252.

Wills, T. J., Lever, C., Cacucci, F., Burgess, N. and O'Keefe, J. (2005). Attractor dynamics in the hippocampal representation of the local environment, *Science* **308**: 873–876.

Willshaw, D. J. (1981). Holography, associative memory, and inductive generalization, *in* G. E. Hinton and J. A. Anderson (eds), *Parallel Models of Associative Memory*, Erlbaum, Hillsdale, NJ, chapter 3, pp. 83–104.

Willshaw, D. J. and Buckingham, J. T. (1990). An assessment of Marr's theory of the hippocampus as a temporary memory store, *Philosophical Transactions of The Royal Society of London, Series B* **329**: 205–215.

Willshaw, D. J. and Longuet-Higgins, H. C. (1969). The holophone – recent developments, *in* D. Mitchie (ed.),

Machine Intelligence, Vol. 4, Edinburgh University Press, Edinburgh.

Willshaw, D. J. and von der Malsburg, C. (1976). How patterned neural connections can be set up by self-organization, *Proceedings of The Royal Society of London, Series B* **194**: 431–445.

Willshaw, D. J., Dayan, P. and Morris, R. G. (2015). Memory, modelling and marr: a commentary on marr (1971) 'simple memory: a theory of archicortex', *Philos Trans R Soc Lond B Biol Sci*.

Wilson, C. J. (1995). The contribution of cortical neurons to the firing pattern of striatal spiny neurons, *in* J. C. Houk, J. L. Davis and D. G. Beiser (eds), *Models of Information Processing in the Basal Ganglia*, MIT Press, Cambridge, MA, chapter 3, pp. 29–50.

Wilson, D. A. and Sullivan, R. M. (2011). Cortical processing of odor objects, *Neuron* **72**: 506–519.

Wilson, D. A., Xu, W., Sadrian, B., Courtiol, E., Cohen, Y. and Barnes, D. C. (2014). Cortical odor processing in health and disease, *Progress in Brain Research* **208**: 275–305.

Wilson, F. A., O'Scalaidhe, S. P. and Goldman-Rakic, P. S. (1994a). Functional synergism between putative gamma-aminobutyrate-containing neurons and pyramidal neurons in prefrontal cortex, *Proceedings of the National Academy of Sciences U S A* **91**: 4009–4013.

Wilson, F. A. W. and Rolls, E. T. (1990a). Learning and memory are reflected in the responses of reinforcement-related neurons in the primate basal forebrain, *Journal of Neuroscience* **10**: 1254–1267.

Wilson, F. A. W. and Rolls, E. T. (1990b). Neuronal responses related to reinforcement in the primate basal forebrain, *Brain Research* **509**: 213–231.

Wilson, F. A. W. and Rolls, E. T. (1990c). Neuronal responses related to the novelty and familiarity of visual stimuli in the substantia innominata, diagonal band of broca and periventricular region of the primate, *Experimental Brain Research* **80**: 104–120.

Wilson, F. A. W. and Rolls, E. T. (1993). The effects of stimulus novelty and familiarity on neuronal activity in the amygdala of monkeys performing recognition memory tasks, *Experimental Brain Research* **93**: 367–382.

Wilson, F. A. W. and Rolls, E. T. (2005). The primate amygdala and reinforcement: a dissociation between rule-based and associatively-mediated memory revealed in amygdala neuronal activity, *Neuroscience* **133**: 1061–1072.

Wilson, F. A. W., Riches, I. P. and Brown, M. W. (1990). Hippocampus and medial temporal cortex: neuronal responses related to behavioural responses during the performance of memory tasks by primates, *Behavioral Brain Research* **40**: 7–28.

Wilson, F. A. W., O'Sclaidhe, S. P. and Goldman-Rakic, P. S. (1993). Dissociation of object and spatial processing domains in primate prefrontal cortex, *Science* **260**: 1955–1958.

Wilson, F. A. W., O'Scalaidhe, S. P. and Goldman-Rakic, P. (1994b). Functional synergism between putative gamma-aminobutyrate-containing neurons and pyramidal neurons in prefrontal cortex, *Proceedings of the National Academy of Science* **91**: 4009–4013.

Wilson, H. R. (1999). *Spikes, Decisions and Actions: Dynamical Foundations of Neuroscience*, Oxford University Press, Oxford.

Wilson, H. R. and Cowan, J. D. (1972). Excitatory and inhibitory interactions in localised populations of model neurons, *Biophysics Journal* **12**: 1–24.

Wilson, M. A. (2002). Hippocampal memory formation, plasticity, and the role of sleep, *Neurobiology of Learning and Memory* **78**: 565–569.

Wilson, M. A. and McNaughton, B. L. (1994). Reactivation of hippocampal ensemble memories during sleep, *Science* **265**: 603–604.

Winkielman, P. and Berridge, K. C. (2003). What is an unconscious emotion?, *Cognition and Emotion* **17**: 181–211.

Winkielman, P. and Berridge, K. C. (2005). Unconscious affective reactions to masked happy versus angry faces influence consumption behavior and judgments of value, *Personality and Social Psychology Bulletin* **31**: 111–135.

Winocur, G. and Moscovitch, M. (2011). Memory transformation and systems consolidation, *J Int Neuropsychol Soc* **17**(5): 766–80.

Winston, P. H. (1975). Learning structural descriptions from examples, *in* P. H. Winston (ed.), *The Psychology of Computer Vision*, McGraw-Hill, New York, pp. 157–210.

Winterer, G. and Weinberger, D. R. (2004). Genes, dopamine and cortical signal-to-noise ratio in schizophrenia, *Trends in Neurosciences* **27**: 683–690.

Winterer, G., Ziller, M., Dorn, H., Frick, K., Mulert, C., Wuebben, Y., Herrmann, W. M. and Coppola, R. (2000). Schizophrenia: reduced signal-to-noise ratio and impaired phase-locking during information processing, *Clinical Neurophysiology* **111**: 837–849.

Winterer, G., Coppola, R., Goldberg, T. E., Egan, M. F., Jones, D. W., Sanchez, C. E. and Weinberger, D. R. (2004). Prefrontal broadband noise, working memory, and genetic risk for schizophrenia, *American Journal of Psychiatry* **161**: 490–500.

Winterer, G., Musso, F., Beckmann, C., Mattay, V., Egan, M. F., Jones, D. W., Callicott, J. H., Coppola, R. and Weinberger, D. R. (2006). Instability of prefrontal signal processing in schizophrenia, *American Journal of Psychiatry* **163**: 1960–1968.

Wirth, S., Yanike, M., Frank, L. M., Smith, A. C., Brown, E. N. and Suzuki, W. A. (2003). Single neurons in the monkeys hippocampus and learning of new associations, *Science* **300**: 1578–1581.

Wirth, S., Baraduc, P., Plante, A., Pinede, S. and Duhamel, J. R. (2017). Gaze-informed, task-situated representation of space in primate hippocampus during virtual navigation, *PLoS Biology* **15**: e2001045.

Wise, S. P. (2008). Forward frontal fields: phylogeny and fundamental function, *Trends in Neuroscience* **31**: 599–608.

Wiskott, L. (2003). Slow feature analysis: A theoretical analysis of optimal free responses, *Neural Computation* **15**: 2147–2177.

Wiskott, L. and Sejnowski, T. J. (2002). Slow feature analysis: unsupervised learning of invariances, *Neural Computation* **14**: 715–770.

Witter, M. P. (1993). Organization of the entorhinal–hippocampal system: a review of current anatomical data, *Hippocampus* **3**: 33–44.

Witter, M. P. (2007). Intrinsic and extrinsic wiring of ca3: indications for connectional heterogeneity, *Learning and Memory* **14**: 705–713.

Witter, M. P., Groenewegen, H. J., Lopes da Silva, F. H. and Lohman, A. H. M. (1989a). Functional organization of the extrinsic and intrinsic circuitry of the parahippocampal region, *Progress in Neurobiology* **33**: 161–254.

Witter, M. P., Van Hoesen, G. W. and Amaral, D. G. (1989b). Topographical organisation of the entorhinal projection to the dentate gyrus of the monkey, *Journal of Neuroscience* **9**: 216–228.

Witter, M. P., Naber, P. A., van Haeften, T., Machielsen, W. C., Rombouts, S. A., Barkhof, F., Scheltens, P. and Lopes da Silva, F. H. (2000a). Cortico-hippocampal communication by way of parallel parahippocampal-subicular pathways, *Hippocampus* **10**: 398–410.

Witter, M. P., Wouterlood, F. G., Naber, P. A. and Van Haeften, T. (2000b). Anatomical organization of the parahippocampal-hippocampal network, *Annals of the New York Academcy of Sciences* **911**: 1–24.

Wolfe, J. M., Cave, K. R. and Franzel, S. L. (1989). Guided search: An alternative to the feature integration model for visual search, *Journal of Experimental Psychology: Human Perception and Performance* **15**: 419–433.

Wolkin, A., Sanfilipo, M., Wolf, A. P., Angrist, B., Brodie, J. D. and Rotrosen, J. (1992). Negative symptoms and hypofrontality in chronic schizophrenia, *Archives of General Psychiatry* **49**: 959–965.

Womelsdorf, T., Fries, P., Mitra, P. and Desimone, R. (2006). Gamma-band synchronization in visual cortex predicts speed of change detection, *Nature* **439**: 733–736.

Womelsdorf, T., Schoffelen, J. M., Oostenveld, R., Singer, W., Desimone, R., Engel, A. K. and Fries, P. (2007). Modulation of neuronal interactions through neuronal synchronization, *Science* **316**: 1609–1612.

Wong, K. and Wang, X.-J. (2006). A recurrent network mechanism of time integration in perceptual decisions, *Journal of Neuroscience* **26**: 1314–1328.

Wong, K. F., Huk, A. C., Shadlen, M. N. and Wang, X.-J. (2007). Neural circuit dynamics underlying accumulation of time-varying evidence during perceptual decision making, *Frontiers in Computational Neuroscience* **1**: 6.

Wood, E. R., Dudchenko, P. A. and Eichenbaum, H. (1999). The global record of memory in hippocampal neuronal activity, *Nature* **397**: 613–616.

Wood, E. R., Agster, K. M. and Eichenbaum, H. (2004). One-trial odor–reward association: a form of event memory not dependent on hippocampal function, *Behavioral Neuroscience* **118**: 526–539.

Wu, X., Baxter, R. A. and Levy, W. B. (1996). Context codes and the effect of noisy learning on a simplified hippocampal CA3 model, *Biological Cybernetics* **74**: 159–165.

Wyss, R., Konig, P. and Verschure, P. F. (2006). A model of the ventral visual system based on temporal stability and local memory, *PLoS Biology* **4**: e120.

Xavier, G. F., Oliveira-Filho, F. J. and Santos, A. M. (1999). Dentate gyrus-selective colchicine lesion and disruption of performance in spatial tasks: difficulties in "place strategy" because of a lack of flexibility in the use of environmental cues?, *Hippocampus* **9**: 668–681.

Xiang, J. Z. and Brown, M. W. (1998). Differential neuronal encoding of novelty, familiarity and recency in regions of the anterior temporal lobe, *Neuropharmacology* **37**: 657–676.

Xiang, J. Z. and Brown, M. W. (2004). Neuronal responses related to long-term recognition memory processes in prefrontal cortex, *Neuron* **42**: 817–829.

Yamane, S., Kaji, S. and Kawano, K. (1988). What facial features activate face neurons in the inferotemporal cortex of the monkey?, *Experimental Brain Research* **73**: 209–214.

Yamins, D. L., Hong, H., Cadieu, C. F., Solomon, E. A., Seibert, D. and DiCarlo, J. J. (2014). Performance-optimized hierarchical models predict neural responses in higher visual cortex, *Proceedings of the National Academy of Sciences of the U S A* **111**: 8619–8624.

Yates, F. A. (1992). *The Art of Memory*, University of Chicago Press, Chicago, Ill.

Yaxley, S., Rolls, E. T., Sienkiewicz, Z. J. and Scott, T. R. (1985). Satiety does not affect gustatory activity in the nucleus of the solitary tract of the alert monkey, *Brain Research* **347**: 85–93.

Yaxley, S., Rolls, E. T. and Sienkiewicz, Z. J. (1988). The responsiveness of neurones in the insular gustatory cortex of the macaque monkey is independent of hunger, *Physiology and Behavior* **42**: 223–229.

Yaxley, S., Rolls, E. T. and Sienkiewicz, Z. J. (1990). Gustatory responses of single neurons in the insula of the macaque monkey, *Journal of Neurophysiology* **63**: 689–700.

Yelnik, J. (2002). Functional anatomy of the basal ganglia, *Movement Disorders* **17 Suppl 3**: S15–S21.

Yoganarasimha, D., Yu, X. and Knierim, J. J. (2006). Head direction cell representations maintain internal coherence during conflicting proximal and distal cue rotations: comparison with hippocampal place cells, *Journal of*

Neuroscience **26**: 622–631.

Zanos, P., Moaddel, R., Morris, P. J., Georgiou, P., Fischell, J., Elmer, G. I., Alkondon, M., Yuan, P., Pribut, H. J., Singh, N. S., Dossou, K. S., Fang, Y., Huang, X. P., Mayo, C. L., Wainer, I. W., Albuquerque, E. X., Thompson, S. M., Thomas, C. J., Zarate, C. A., J. and Gould, T. D. (2016). Nmdar inhibition-independent antidepressant actions of ketamine metabolites, *Nature* **533**(7604): 481–6.

Zeidman, P. and Maguire, E. A. (2016). Anterior hippocampus: the anatomy of perception, imagination and episodic memory, *Nature Reviews Neuroscience* p. 10.1038/nrn.2015.24.

Zhang, K. (1996). Representation of spatial orientation by the intrinsic dynamics of the head-direction cell ensemble: A theory, *Journal of Neuroscience* **16**: 2112–2126.

Zilli, E. A. (2012). Models of grid cell spatial firing published 2005–2011, *Frontiers in Neural Circuits* **6**: 16.

Zink, C. F., Pagnoni, G., Martin, M. E., Dhamala, M. and Berns, G. S. (2003). Human striatal responses to salient nonrewarding stimuli, *Journal of Neuroscience* **23**: 8092–8097.

Zink, C. F., Pagnoni, G., Martin-Skurski, M. E., Chappelow, J. C. and Berns, G. S. (2004). Human striatal responses to monetary reward depend on saliency, *Neuron* **42**: 509–517.

Zipser, K., Lamme, V. and Schiller, P. (1996). Contextual modulation in primary visual cortex, *Journal of Neuroscience* **16**: 7376–7389.

Zohary, E., Shadlen, M. N. and Newsome, W. T. (1994). Correlated neuronal discharge rate and its implications for psychophysical performance, *Nature* **370**: 140–143.

Zola-Morgan, S., Squire, L. R., Amaral, D. G. and Suzuki, W. A. (1989). Lesions of perirhinal and parahippocampal cortex that spare the amygdala and hippocampal formation produce severe memory impairment, *Journal of Neuroscience* **9**: 4355–4370.

Zola-Morgan, S., Squire, L. R. and Ramus, S. J. (1994). Severity of memory impairment in monkeys as a function of locus and extent of damage within the medial temporal lobe memory system, *Hippocampus* **4**: 483–494.

Zucker, R. S. and Regehr, W. G. (2002). Short-term synaptic plasticity, *Annual Review of Physiology* **64**: 355–405.

Zucker, S. W., Dobbins, A. and Iverson, L. (1989). Two stages of curve detection suggest two styles of visual computation, *Neural Computation* **1**: 68–81.

Index